BIOPROCESS ENGINEERING PRINCIPLES

SECOND EDITION

PAULINE M. DORAN

ELSEVIER

AMSTERDAM • BOSTON • HEIDELBERG • LONDON
NEW YORK • OXFORD • PARIS • SAN DIEGO
SAN FRANCISCO • SINGAPORE • SYDNEY • TOKYO
Academic Press is an imprint of Elsevier

Bioprocess Engineering Principles, 2nd Edition

Pauline Doran

Original English language edition published by Elsevier, Academic Press

ISBN : 978-0-12-220851-5

This edition is authorized by the proprietor for sale in South Korea only.

Reprint ISBN : 978-89-94961-51-4

Elsevier Korea L.L.C

4FL, Chunwoo Bldg, 534 Itaewon-dong,

Yongsan-gu, Seoul 140-861, S. Korea

First Published December 2012

Print in Taiwan

CONTENTS

Preface to the Second Edition

As originally conceived, this book is intended as a text for undergraduate and postgraduate students with little or no engineering background. It seeks to close the gap of knowledge and experience for students trained or being trained in molecular biology, biotechnology, and related disciplines who are interested in how biological discoveries are translated into commercial products and services. Applying biology for technology development is a multidisciplinary challenge requiring an appreciation of the engineering aspects of process analysis, design, and scale-up. Consistent with this overall aim, basic biology is not covered in this book, as a biology background is assumed. Moreover, although most aspects of bioprocess engineering are presented quantitatively, priority has been given to minimising the use of complex mathematics that may be beyond the comfort zone of nonengineering readers. Accordingly, the material has a descriptive focus without a heavy reliance on mathematical detail.

Following publication of the first edition of *Bioprocess Engineering Principles*, I was delighted to find that the book was also being adopted in chemical, biochemical, and environmental engineering programs that offer bioprocess engineering as a curriculum component. For students with several years of engineering training under their belts, the introductory nature and style of the earlier chapters may seem tedious and inappropriate. However, later in the book, topics such as fluid flow and mixing, heat and mass transfer, reaction engineering, and downstream processing are presented in detail as they apply to bioprocessing, thus providing an overview of this specialty stream of traditional chemical engineering.

Because of its focus on underlying scientific and engineering principles rather than on specific biotechnology applications, the material presented in the first edition remains relevant today and continues to provide a sound basis for teaching bioprocess engineering. However, since the first edition was published, there have been several important advances and developments that have significantly broadened the scope and capabilities of bioprocessing. New sections on topics such as sustainable bioprocessing and metabolic engineering are included in this second edition, as these approaches are now integral to engineering design procedures and commercial cell line development.

Expanded coverage of downstream processing operations to include membrane filtration, protein precipitation, crystallisation, and drying is provided. Greater and more in-depth treatment of fluid flow, turbulence, mixing, and impeller design is also available in this edition, reflecting recent advances in our understanding of mixing processes and their importance in determining the performance of cell cultures. More than 100 new illustrations and 150 additional problems and worked examples have been included in this updated edition. A total of over 340 problems now demonstrate how the fundamental principles described in the text are applied in areas such as biofuels, bioplastics, bioremediation, tissue engineering, site-directed mutagenesis, recombinant protein production, and drug development, as well as for traditional microbial fermentation.

I acknowledge with gratitude the feedback and suggestions received from many users of the first edition of *Bioprocess Engineering Principles* over the last 15 years or so. Your input is very welcome and has helped shape the priorities for change and elaboration in the second edition. I would also like to thank Robert Bryson-Richardson and Paulina Mikulic for their special and much appreciated assistance under challenging circumstances in 2011. Bioprocess engineering has an important place in the modern world. I hope that this book will make it easier for students and graduates from diverse backgrounds to appreciate the role of bioprocess engineering in our lives and to contribute to its further progress and development.

Pauline M. Doran
Swinburne University of Technology
Melbourne, Australia

Additional Book Resources

For those who are using this book as a text for their courses, additional teaching resources are available by registering at *www.textbooks.elsevier.com*.

PART 1

INTRODUCTION

CHAPTER

1

Bioprocess Development
An Interdisciplinary Challenge

Bioprocessing is an essential part of many food, chemical, and pharmaceutical industries. Bioprocess operations make use of microbial, animal, and plant cells, and components of cells such as enzymes, to manufacture new products and destroy harmful wastes.

The use of microorganisms to transform biological materials for production of fermented foods has its origins in antiquity. Since then, bioprocesses have been developed for an enormous range of commercial products, from relatively cheap materials such as industrial alcohol and organic solvents, to expensive specialty chemicals such as antibiotics, therapeutic proteins, and vaccines. Industrially useful enzymes and living cells such as bakers' and brewers' yeast are also commercial products of bioprocessing.

Table 1.1 gives examples of bioprocesses employing whole cells. Typical organisms used are also listed. The table is by no means exhaustive; not included are processes for waste water treatment, bioremediation, microbial mineral recovery, and manufacture of traditional foods and beverages such as yoghurt, bread, vinegar, soy sauce, beer, and wine. Industrial processes employing enzymes are also not listed in Table 1.1: these include brewing, baking, confectionery manufacture, clarification of fruit juices, and antibiotic transformation. Large quantities of enzymes are used commercially to convert starch into fermentable sugars, which serve as starting materials for other bioprocesses.

Our ability to harness the capabilities of cells and enzymes is closely related to advances in biochemistry, microbiology, immunology, and cell physiology. Knowledge in these areas has expanded rapidly; tools of modern biotechnology such as recombinant DNA, gene probes, cell fusion, and tissue culture offer new opportunities to develop novel products or improve bioprocessing methods. Visions of sophisticated medicines, cultured human tissues and organs, biochips for new-age computers, environmentally compatible pesticides, and powerful pollution-degrading microbes herald a revolution in the role of biology in industry.

Although new products and processes can be conceived and partially developed in the laboratory, bringing modern biotechnology to industrial fruition requires engineering skills and know-how. Biological systems can be complex and difficult to control;

TABLE 1.1 Examples of Products from Bioprocessing

Product	Typical organism used
BIOMASS	
Agricultural inoculants for nitrogen fixation	*Rhizobium leguminosarum*
Bakers' yeast	*Saccharomyces cerevisiae*
Cheese starter cultures	*Lactococcus* spp.
Inoculants for silage production	*Lactobacillus plantarum*
Single-cell protein	*Candida utilis* or *Pseudomonas methylotrophus*
Yoghurt starter cultures	*Streptococcus thermophilus* and *Lactobacillus bulgaricus*
BULK ORGANICS	
Acetone/butanol	*Clostridium acetobutylicum*
Ethanol (nonbeverage)	*Saccharomyces cerevisiae*
Glycerol	*Saccharomyces cerevisiae*
ORGANIC ACIDS	
Citric acid	*Aspergillus niger*
Gluconic acid	*Aspergillus niger*
Itaconic acid	*Aspergillus itaconicus*
Lactic acid	*Lactobacillus delbrueckii*
AMINO ACIDS	
L-Arginine	*Brevibacterium flavum*
L-Glutamic acid	*Corynebacterium glutamicum*
L-Lysine	*Brevibacterium flavum*
L-Phenylalanine	*Corynebacterium glutamicum*
Others	*Corynebacterium* spp.
NUCLEIC ACID-RELATED COMPOUNDS	
5′-guanosine monophosphate (5′-GMP)	*Bacillus subtilis*
5′-inosine monophosphate (5′-IMP)	*Brevibacterium ammoniagenes*
ENZYMES	
α-Amylase	*Bacillus amyloliquefaciens*
Glucoamylase	*Aspergillus niger*
Glucose isomerase	*Bacillus coagulans*
Pectinases	*Aspergillus niger*
Proteases	*Bacillus* spp.
Rennin	*Mucor miehei* or recombinant yeast

VITAMINS	
Cyanocobalamin (B_{12})	*Propionibacterium shermanii* or *Pseudomonas denitrificans*
Riboflavin (B_2)	*Eremothecium ashbyii*
EXTRACELLULAR POLYSACCHARIDES	
Dextran	*Leuconostoc mesenteroides*
Xanthan gum	*Xanthomonas campestris*
Other	*Polianthes tuberosa* (plant cell culture)
POLY-β-HYDROXYALKANOATE POLYESTERS	
Poly-β-hydroxybutyrate	*Alcaligenes eutrophus*
ANTIBIOTICS	
Cephalosporins	*Cephalosporium acremonium*
Penicillins	*Penicillium chrysogenum*
Aminoglycoside antibiotics (e.g., streptomycin)	*Streptomyces griseus*
Ansamycins (e.g., rifamycin)	*Nocardia mediterranei*
Aromatic antibiotics (e.g., griseofulvin)	*Penicillium griseofulvum*
Macrolide antibiotics (e.g., erythromycin)	*Streptomyces erythreus*
Nucleoside antibiotics (e.g., puromycin)	*Streptomyces alboniger*
Polyene macrolide antibiotics (e.g., candidin)	*Streptomyces viridoflavus*
Polypeptide antibiotics (e.g., gramicidin)	*Bacillus brevis*
Tetracyclines (e.g., 7-chlortetracycline)	*Streptomyces aureofaciens*
ALKALOIDS	
Ergot alkaloids	*Claviceps paspali*
Taxol	*Taxus brevifolia* (plant cell culture)
SAPONINS	
Ginseng saponins	*Panax ginseng* (plant cell culture)
PIGMENTS	
β-Carotene	*Blakeslea trispora*
PLANT GROWTH REGULATORS	
Gibberellins	*Gibberella fujikuroi*
INSECTICIDES	
Bacterial spores	*Bacillus thuringiensis*
Fungal spores	*Hirsutella thompsonii*

(Continued)

TABLE 1.1 Examples of Products from Bioprocessing (Continued)

Product	Typical organism used
MICROBIAL TRANSFORMATIONS	
D-Sorbitol to L-sorbose (in vitamin C production)	*Acetobacter suboxydans*
Steroids	*Rhizopus arrhizus*
VACCINES	
Diphtheria	*Corynebacterium diphtheriae*
Hepatitis B	Surface antigen expressed in recombinant *Saccharomyces cerevisiae*
Mumps	Attenuated viruses grown in chick embryo cell cultures
Pertussis (whooping cough)	*Bordetella pertussis*
Poliomyelitis virus	Attenuated viruses grown in monkey kidney or human diploid cells
Rubella	Attenuated viruses grown in baby hamster kidney cells
Tetanus	*Clostridium tetani*
THERAPEUTIC PROTEINS	
Erythropoietin	Recombinant mammalian cells
Factor VIII	Recombinant mammalian cells
Follicle-stimulating hormone	Recombinant mammalian cells
Granulocyte–macrophage colony-stimulating factor	Recombinant *Escherichia coli*
Growth hormones	Recombinant *Escherichia coli*
Hirudin	Recombinant *Saccharomyces cerevisiae*
Insulin and insulin analogues	Recombinant *Escherichia coli*
Interferons	Recombinant *Escherichia coli*
Interleukins	Recombinant *Escherichia coli*
Platelet-derived growth factor	Recombinant *Saccharomyces cerevisiae*
Tissue plasminogen activator	Recombinant *Escherichia coli* or recombinant mammalian cells
MONOCLONAL ANTIBODIES	
Various, including Fab and Fab_2 fragments	Hybridoma cells
THERAPEUTIC TISSUES AND CELLS	
Cartilage cells	Human (patient) chondrocytes
Skin	Human skin cells

nevertheless, they obey the laws of chemistry and physics and are therefore amenable to engineering analysis. Substantial engineering input is essential in many aspects of bioprocessing, including the design and operation of bioreactors, sterilisers, and equipment for product recovery, the development of systems for process automation and control, and the efficient and safe layout of fermentation factories. The subject of this book, bioprocess engineering, is the study of engineering principles applied to processes involving cell or enzyme catalysts.

1.1 STEPS IN BIOPROCESS DEVELOPMENT: A TYPICAL NEW PRODUCT FROM RECOMBINANT DNA

The interdisciplinary nature of bioprocessing is evident if we look at the stages of development of a complete industrial process. As an example, consider manufacture of a typical recombinant DNA-derived product such as insulin, growth hormone, erythropoietin, or interferon. As shown in Figure 1.1, several steps are required to bring the product into commercial reality; these stages involve different types of scientific expertise.

The first stages of bioprocess development (Steps 1–11) are concerned with genetic manipulation of the host organism; in this case, a gene from animal DNA is cloned into *Escherichia coli*. Genetic engineering is performed in laboratories on a small scale by scientists trained in molecular biology and biochemistry. Tools of the trade include Petri dishes, micropipettes, microcentrifuges, nano- or microgram quantities of restriction enzymes, and electrophoresis gels for DNA and protein fractionation. In terms of bioprocess development, parameters of major importance are the level of expression of the desired product and the stability of the constructed strains.

After cloning, the growth and production characteristics of the recombinant cells must be measured as a function of the culture environment (Step 12). Practical skills in microbiology and kinetic analysis are required; small-scale culture is carried out mostly using shake flasks of 250-ml to 1-litre capacity. Medium composition, pH, temperature, and other environmental conditions allowing optimal growth and productivity are determined. Calculated parameters such as cell growth rate, specific productivity, and product yield are used to describe the performance of the organism.

Once the culture conditions for production are known, scale-up of the process starts. The first stage may be a 1- or 2-litre *bench-top bioreactor* equipped with instruments for measuring and adjusting temperature, pH, dissolved oxygen concentration, stirrer speed, and other process variables (Step 13). Cultures can be more closely monitored in bioreactors than in shake flasks, so better control over the process is possible. Information is collected about the oxygen requirements of the cells, their shear sensitivity, foaming characteristics, and other properties. Limitations imposed by the reactor on the activity of the organism must be identified. For example, if the bioreactor cannot provide dissolved oxygen to an aerobic culture at a sufficiently high rate, the culture will become oxygen-starved. Similarly, in mixing the broth to expose the cells to nutrients in the medium, the stirrer in the reactor may cause cell damage. Whether or not the reactor can provide conditions for optimal activity of the cells is of prime concern. The situation is assessed using measured and calculated parameters such as mass transfer coefficients, mixing time, gas

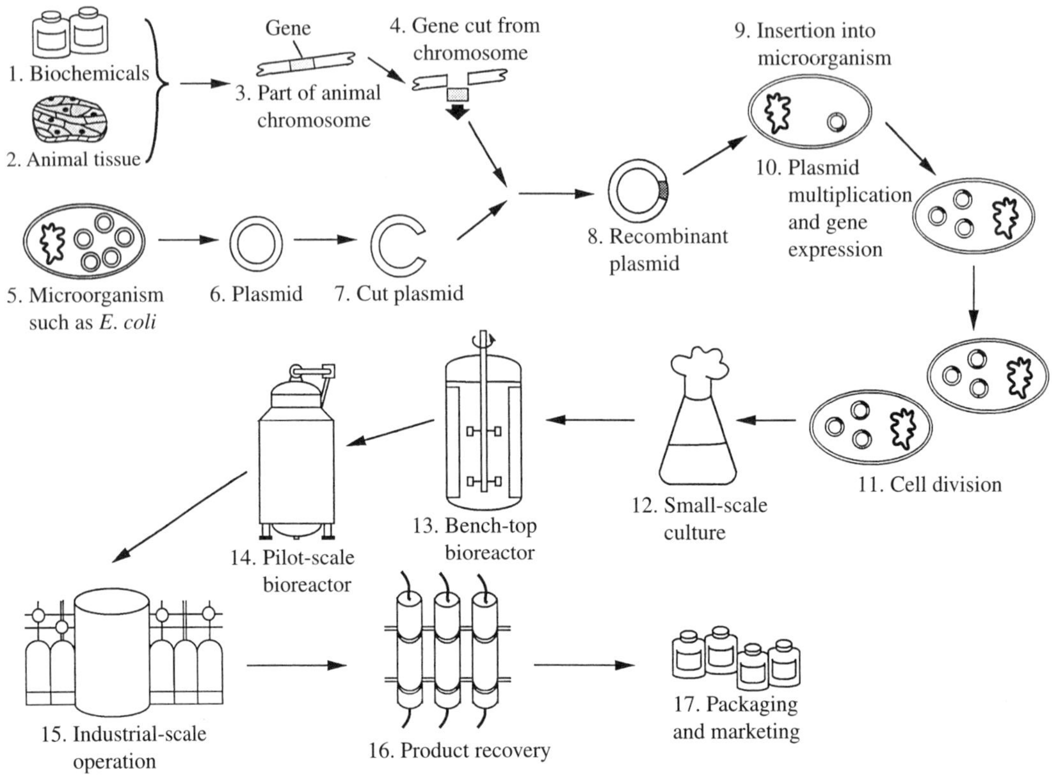

FIGURE 1.1 Steps involved in the development of a new bioprocess for commercial manufacture of a recombinant DNA-derived product.

hold-up, oxygen uptake rate, power number, energy dissipation rate, and many others. It must also be decided whether the culture is best operated as a batch, semi-batch, or continuous process; experimental results for culture performance under various modes of reactor operation may be examined. The viability of the process as a commercial venture is of great interest; information about activity of the cells is used in further calculations to determine economic feasibility.

Following this stage of process development, the system is scaled up again to a *pilot-scale bioreactor* (Step 14). Engineers trained in bioprocessing are normally involved in pilot-scale operations. A vessel of capacity 100 to 1000 litres is built according to specifications determined from the bench-scale prototype. The design is usually similar to that which worked best on the smaller scale. The aim of pilot-scale studies is to examine the response of cells to scale-up. Changing the size of the equipment seems relatively trivial; however, loss or variation of performance often occurs. Even though the reactor geometry, impeller design, method of aeration, and other features may be similar in small and large fermenters, the effect of scale-up on activity of cells can be great. Loss of productivity following scale-up may or may not be recovered; economic projections often need to be reassessed as a result of pilot-scale findings.

If the pilot-scale step is completed successfully, design of the *industrial-scale operation* commences (Step 15). This part of process development is clearly in the territory of bioprocess engineering. As well as the reactor itself, all of the auxiliary service facilities must be designed and tested. These include air supply and sterilisation equipment, steam generator and supply lines, medium preparation and sterilisation facilities, cooling water supply, and process control network. Particular attention is required to ensure that the fermentation can be carried out aseptically. When recombinant cells or pathogenic organisms are involved, design of the process must also reflect containment and safety requirements.

An important part of the total process is *product recovery* (Step 16), also known as *downstream processing*. After leaving the fermenter, raw broth is treated in a series of steps to produce the final product. Product recovery is often difficult and expensive; for some recombinant-DNA-derived products, purification accounts for 80 to 90% of the total processing cost. Actual procedures used for downstream processing depend on the nature of the product and the broth; physical, chemical, or biological methods may be employed. Many operations that are standard in the laboratory become uneconomic or impractical on an industrial scale. Commercial procedures include filtration, centrifugation, and flotation for separation of cells from the liquid; mechanical disruption of the cells if the product is intracellular; solvent extraction; chromatography; membrane filtration; adsorption; crystallisation; and drying. Disposal of effluent after removal of the desired product must also be considered. As with bioreactor design, techniques applied industrially for downstream processing are first developed and tested using small-scale apparatus. Scientists trained in chemistry, biochemistry, chemical engineering, and industrial chemistry play important roles in designing product recovery and purification systems.

After the product has been isolated and brought to sufficient purity, it is packaged and marketed (Step 17). For new biopharmaceuticals such as recombinant proteins and therapeutic agents, medical and clinical trials are required to test the efficacy of the product. Animals are used first, then humans. Only after these trials are carried out and the safety of the product established can it be released for general health care application. Other tests are required for food products. Bioprocess engineers with a detailed knowledge of the production process are often involved in documenting manufacturing procedures for submission to regulatory authorities. Manufacturing standards must be met; this is particularly the case for products derived from genetically modified organisms as a greater number of safety and precautionary measures is required.

As shown in this example, a broad range of disciplines is involved in bioprocessing. Scientists working in this area are constantly confronted with biological, chemical, physical, engineering, and sometimes medical questions.

1.2 A QUANTITATIVE APPROACH

The biological characteristics of cells and enzymes often impose constraints on bioprocessing; knowledge of them is therefore an important prerequisite for rational engineering design. For instance, enzyme thermostability properties must be taken into account when choosing the operating temperature for an enzyme reactor, and the susceptibility of an organism to substrate inhibition will determine whether substrate is fed to the fermenter all at once or intermittently. It is equally true, however, that biologists working in biotechnology must consider the engineering aspects of bioprocessing. Selection or manipulation of organisms should be carried out to achieve the best results in production-scale operations. It would be disappointing, for example, to spend a year or two manipulating an organism to express a foreign gene if the cells in culture produce a highly viscous broth that cannot be adequately mixed or supplied with oxygen in large-scale reactors. Similarly, improving cell permeability to facilitate product excretion has limited utility if the new organism is too fragile to withstand the mechanical forces developed during fermenter operation. Another area requiring cooperation and understanding between engineers and laboratory scientists is medium formation. For example, addition of serum may be beneficial to growth of animal cells, but can significantly reduce product yields during recovery operations and, in large-scale processes, requires special sterilisation and handling procedures.

All areas of bioprocess development—the cell or enzyme used, the culture conditions provided, the fermentation equipment, and the operations used for product recovery—are interdependent. Because improvement in one area can be disadvantageous to another, ideally, bioprocess development should proceed using an integrated approach. In practice, combining the skills of engineers with those of biologists can be difficult owing to the different ways in which biologists and engineers are trained. Biological scientists generally have strong experimental technique and are good at testing qualitative models; however, because calculations and equations are not a prominent feature of the life sciences, biologists are usually less familiar with mathematics. On the other hand, as calculations are important in all areas of equipment design and process analysis, quantitative methods, physics, and mathematical theories play a central role in engineering. There is also a difference in the way biologists and biochemical engineers think about complex processes such as cell and enzyme function. Fascinating as the minutiae of these biological systems may be, in order to build working reactors and other equipment, engineers must take a simplified and pragmatic approach. It is often disappointing for the biology-trained scientist that engineers seem to ignore the wonder, intricacy, and complexity of life to focus only on those aspects that have a significant quantitative effect on the final outcome of the process.

Given the importance of interaction between biology and engineering in bioprocessing, any differences in outlook between engineers and biologists must be overcome. Although it is unrealistic to expect all biotechnologists to undertake full engineering training, there are many advantages in understanding the practical principles of bioprocess engineering if not the full theoretical detail. The principal objective of this book is to teach scientists trained in biology those aspects of engineering science that are relevant to bioprocessing. An adequate background in biology is assumed. At the end of this study, you will have gained a heightened appreciation for bioprocess engineering. You will be able to communicate on a professional level with bioprocess engineers and know how to analyse and critically evaluate new processing proposals. You will be able to carry out routine calculations and checks on processes; in many cases these calculations are not difficult and can be of great value. You will also know what type of expertise a bioprocess engineer can offer and when it is necessary to consult an expert in the field. In the laboratory, your awareness of engineering methods will help avoid common mistakes in data analysis and the design of experimental apparatus.

As our exploitation of biology continues, there is an increasing demand for scientists trained in bioprocess technology who can translate new discoveries into industrial-scale production. As a biotechnologist, you may be expected to work at the interface of biology and engineering science. This textbook on bioprocess engineering is designed to prepare you for that challenge.

CHAPTER

2

Introduction to Engineering Calculations

Calculations used in bioprocess engineering require a systematic approach with well-defined methods and rules. Conventions and definitions that form the backbone of engineering analysis are presented in this chapter. Many of these you will use over and over again as you progress through this text. In laying the foundations for calculations and problem solving, this chapter will be a useful reference that you may need to review from time to time.

The first step in quantitative analysis of systems is to express the system properties using mathematical language. This chapter begins by considering how physical, chemical, and biological processes are characterised mathematically. The nature of physical variables, dimensions, and units is discussed, and formalised procedures for unit conversions are outlined. You will have already encountered many of the concepts used in measurement, such as concentration, density, pressure, temperature, and so on; rules for quantifying these variables are summarised here in preparation for Chapters 4 through 6, where they are first applied to solve processing problems. The occurrence of reactions in biological systems is of particular importance; terminology involved in stoichiometric analysis is considered in this chapter. Finally, as equations representing biological processes often involve terms for the physical and chemical properties of materials, references for handbooks containing this information are provided.

Worked examples and problems are used to illustrate and reinforce the material described in the text. Although the terminology and engineering concepts in these examples may be unfamiliar, solutions to each problem can be obtained using techniques fully explained within this chapter. The context and meaning of many of the equations introduced as problems and examples are explained in more detail in later sections of this book; the emphasis in this chapter is on the use of basic mathematical principles irrespective of the particular application. At the end of the chapter is a checklist so you can be sure you have assimilated all the important points.

2.1 PHYSICAL VARIABLES, DIMENSIONS, AND UNITS

Engineering calculations involve manipulation of numbers. Most of these numbers represent the magnitudes of measurable *physical variables*, such as mass, length, time, velocity, area, viscosity, temperature, density, and so on. Other observable characteristics of nature, such as taste or aroma, cannot at present be described completely using appropriate numbers; we cannot, therefore, include these in calculations.

From all the physical variables in the world, the seven quantities listed in Table 2.1 have been chosen by international agreement as a basis for measurement. Two further supplementary units are used to express angular quantities. These base quantities are called *dimensions*, and it is from these that the dimensions of other physical variables are derived. For example, the dimensions of velocity, which is defined as distance or length travelled per unit time, are LT^{-1}; the dimensions of force, being mass × acceleration, are LMT^{-2}. A list of useful derived dimensional and nondimensional quantities is given in Table 2.2.

Physical variables can be classified into two groups: *substantial variables* and *natural variables*.

2.1.1 Substantial Variables

Examples of substantial variables are mass, length, volume, viscosity, and temperature. Expression of the magnitude of substantial variables requires a precise physical standard against which the measurement is made. These standards are called *units*. You are already familiar with many units: metre, foot, and mile are units of length, for example, and hour and second are units of time. Statements about the magnitude of substantial variables

TABLE 2.1 Base Quantities

Base quantity	Dimensional symbol	Base SI unit	Unit symbol
Length	L	metre	m
Mass	M	kilogram	kg
Time	T	second	s
Electric current	I	ampere	A
Temperature	Θ	kelvin	K
Amount of substance	N	gram-mole	mol or gmol
Luminous intensity	J	candela	cd
Supplementary units			
Plane angle	—	radian	rad
Solid angle	—	steradian	sr

TABLE 2.2 Examples of Derived Dimensional and Dimensionless Quantities

Quantity	Dimensions	Quantity	Dimensions
Acceleration	LT^{-2}	Momentum	LMT^{-1}
Angular velocity	T^{-1}	Osmotic pressure	$L^{-1}MT^{-2}$
Area	L^2	Partition coefficient	1
Atomic weight	1	Period	T
('relative atomic mass')		Power	L^2MT^{-3}
Concentration	$L^{-3}N$	Pressure	$L^{-1}MT^{-2}$
Conductivity	$L^{-3}M^{-1}T^3I^2$	Rotational frequency	T^{-1}
Density	$L^{-3}M$	Shear rate	T^{-1}
Diffusion coefficient	L^2T^{-1}	Shear strain	1
Distribution coefficient	1	Shear stress	$L^{-1}MT^{-2}$
Effectiveness factor	1	Specific death constant	T^{-1}
Efficiency	1	Specific gravity	1
Energy	L^2MT^{-2}	Specific growth rate	T^{-1}
Enthalpy	L^2MT^{-2}	Specific heat capacity	$L^2T^{-2}\Theta^{-1}$
Entropy	$L^2MT^{-2}\Theta^{-1}$	Specific interfacial area	L^{-1}
Equilibrium constant	1	Specific latent heat	L^2T^{-2}
Force	LMT^{-2}	Specific production rate	T^{-1}
Fouling factor	$MT^{-3}\Theta^{-1}$	Specific volume	L^3M^{-1}
Frequency	T^{-1}	Stress	$L^{-1}MT^{-2}$
Friction coefficient	1	Surface tension	MT^{-2}
Gas hold-up	1	Thermal conductivity	$LMT^{-3}\Theta^{-1}$
Half life	T	Thermal resistance	$L^{-2}M^{-1}T^3\Theta$
Heat	L^2MT^{-2}	Torque	L^2MT^{-2}
Heat flux	MT^{-3}	Velocity	LT^{-1}
Heat transfer coefficient	$MT^{-3}\Theta^{-1}$	Viscosity (dynamic)	$L^{-1}MT^{-1}$
Ideal gas constant	$L^2MT^{-2}\Theta^{-1}N^{-1}$	Viscosity (kinematic)	L^2T^{-1}
Illuminance	$L^{-2}J$	Void fraction	1
Maintenance coefficient	T^{-1}	Volume	L^3
Mass flux	$L^{-2}MT^{-1}$	Weight	LMT^{-2}
Mass transfer coefficient	LT^{-1}	Work	L^2MT^{-2}
Molar mass	MN^{-1}	Yield coefficient	1
Molecular weight ('relative molecular mass')	1		

Note: Dimensional symbols are defined in Table 2.1. Dimensionless quantities have dimension 1.

must contain two parts: the number and the unit used for measurement. Clearly, reporting the speed of a moving car as 20 has no meaning unless information about the units, say km h^{-1} or ft s^{-1}, is also included.

As numbers representing substantial variables are multiplied, subtracted, divided, or added, their units must also be combined. The values of two or more substantial variables may be added or subtracted only if their units are the same. For example

$$5.0 \text{ kg} + 2.2 \text{ kg} = 7.2 \text{ kg}$$

On the other hand, the values and units of *any* substantial variables can be combined by multiplication or division; for example

$$\frac{1500 \text{ km}}{12.5 \text{ h}} = 120 \text{ km h}^{-1}$$

The way in which units are carried along during calculations has important consequences. Not only is proper treatment of units essential if the final answer is to have the correct units, units and dimensions can also be used as a guide when deducing how physical variables are related in scientific theories and equations.

2.1.2 Natural Variables

The second group of physical variables are the natural variables. Specification of the magnitude of these variables does not require units or any other standard of measurement. Natural variables are also referred to as *dimensionless variables, dimensionless groups,* or *dimensionless numbers*. The simplest natural variables are ratios of substantial variables. For example, the aspect ratio of a cylinder is its length divided by its diameter; the result is a dimensionless number.

Other natural variables are not as obvious as this, and involve combinations of substantial variables that do not have the same dimensions. Engineers make frequent use of dimensionless numbers for succinct representation of physical phenomena. For example, a common dimensionless group in fluid mechanics is the Reynolds number, *Re*. For flow in a pipe, the Reynolds number is given by the equation:

$$Re = \frac{Du\rho}{\mu} \tag{2.1}$$

where D is the pipe diameter, u is fluid velocity, ρ is fluid density, and μ is fluid viscosity. When the dimensions of these variables are combined according to Eq. (2.1), the dimensions of the numerator exactly cancel those of the denominator. Other dimensionless variables relevant to bioprocess engineering are the Schmidt number, Prandtl number, Sherwood number, Peclet number, Nusselt number, Grashof number, power number, and many others. Definitions and applications of these natural variables are given in later chapters of this book.

In calculations involving rotational phenomena, rotation is described based on the number of radians or revolutions:

$$\text{number of radians} = \frac{\text{length of arc}}{\text{radius}} = \frac{\text{length of arc}}{r} \tag{2.2}$$

$$\text{number of revolutions} = \frac{\text{length of arc}}{\text{circumference}} = \frac{\text{length of arc}}{2\pi r} \tag{2.3}$$

where r is radius. One revolution is equal to 2π radians. Radians and revolutions are nondimensional because the dimensions of length for arc, radius, and circumference in Eqs. (2.2) and (2.3) cancel. Consequently, rotational speed (e.g., number of revolutions per second) and angular velocity (e.g., number of radians per second) have dimensions T^{-1}. Degrees, which are subdivisions of a revolution, are converted into revolutions or radians before application in most engineering calculations. Frequency (e.g., number of vibrations per second) is another variable that has dimensions T^{-1}.

2.1.3 Dimensional Homogeneity in Equations

Rules about dimensions determine how equations are formulated. 'Properly constructed' equations representing general relationships between physical variables must be dimensionally homogeneous. For dimensional homogeneity, the dimensions of terms that are added or subtracted must be the same, and the dimensions of the right side of the equation must be the same as those of the left side. As a simple example, consider the Margules equation for evaluating fluid viscosity from experimental measurements:

$$\mu = \frac{M}{4\pi h\Omega}\left(\frac{1}{R_o^2} - \frac{1}{R_i^2}\right) \tag{2.4}$$

The terms and dimensions in this equation are listed in Table 2.3. Numbers such as 4 have no dimensions; the symbol π represents the number 3.1415926536, which is also dimensionless. As discussed in Section 2.1.2, the number of radians per second represented by Ω has dimensions T^{-1}, so appropriate units would be, for example, s^{-1}. A quick check shows that Eq. (2.4) is dimensionally homogeneous since both sides of the equation have dimensions $L^{-1}MT^{-1}$ and all terms added or subtracted have the same dimensions. Note that when a term such as R_o is raised to a power such as 2, the units and dimensions of R_o must also be raised to that power.

For dimensional homogeneity, the argument of any transcendental function, such as a logarithmic, trigonometric, or exponential function, must be dimensionless. The following examples illustrate this principle.

1. An expression for cell growth is:

$$\ln\frac{x}{x_0} = \mu t \tag{2.5}$$

TABLE 2.3 Terms and Dimensions in Equation (2.4)

Term	Dimensions	SI units
μ (dynamic viscosity)	$L^{-1}MT^{-1}$	pascal second (Pa s)
M (torque)	L^2MT^{-2}	newton metre (N m)
h (cylinder height)	L	metre (m)
Ω (angular velocity)	T^{-1}	radians per second (s^{-1})
R_o (outer radius)	L	metre (m)
R_i (inner radius)	L	metre (m)

where x is cell concentration at time t, x_0 is initial cell concentration, and μ is the specific growth rate. The argument of the logarithm, the ratio of cell concentrations, is dimensionless.

2. The displacement y due to the action of a progressive wave with amplitude A, frequency $\omega/2\pi$ and velocity v is given by the equation:

$$y = A \sin\left[\omega\left(t - \frac{x}{v}\right)\right] \tag{2.6}$$

where t is time and x is distance from the origin. The argument of the sine function, $\omega\left(t - \frac{x}{v}\right)$, is dimensionless.

3. The relationship between α, the mutation rate of *Escherichia coli*, and temperature T, can be described using an Arrhenius-type equation:

$$\alpha = \alpha_0 e^{-E/RT} \tag{2.7}$$

where α_0 is the mutation reaction constant, E is the specific activation energy, and R is the ideal gas constant (see Section 2.5). The dimensions of RT are the same as those of E, so the exponent is as it should be: dimensionless.

The dimensional homogeneity of equations can sometimes be masked by mathematical manipulation. As an example, Eq. (2.5) might be written:

$$\ln x = \ln x_0 + \mu t \tag{2.8}$$

Inspection of this equation shows that rearrangement of the terms to group $\ln x$ and $\ln x_0$ together recovers dimensional homogeneity by providing a dimensionless argument for the logarithm.

Integration and differentiation of terms affect dimensionality. Integration of a function with respect to x increases the dimensions of that function by the dimensions of x. Conversely, differentiation with respect to x results in the dimensions being reduced by the dimensions of x. For example, if C is the concentration of a particular compound expressed as mass per unit volume and x is distance, dC/dx has dimensions $L^{-4}M$, whereas d^2C/dx^2 has dimensions $L^{-5}M$. On the other hand, if μ is the specific growth rate of an organism with dimensions T^{-1} and t is time, then $\int \mu dt$ is dimensionless.

2.1.4 Equations without Dimensional Homogeneity

For repetitive calculations or when an equation is derived from observation rather than from theoretical principles, it is sometimes convenient to present the equation in a nonhomogeneous form. Such equations are called *equations in numerics* or *empirical equations*. In empirical equations, the units associated with each variable must be stated explicitly. An example is Richards' correlation for the dimensionless gas hold-up ε in a stirred fermenter:

$$\left(\frac{P}{V}\right)^{0.4} u^{1/2} = 30\,\varepsilon + 1.33 \tag{2.9}$$

where P is power in units of horsepower, V is ungassed liquid volume in units of ft^3, u is linear gas velocity in units of ft s^{-1}, and ε is fractional gas hold-up, a dimensionless variable. The dimensions of each side of Eq. (2.9) are certainly not the same. For direct application of Eq. (2.9), only those units specified can be used.

2.2 UNITS

Several systems of units for expressing the magnitude of physical variables have been devised through the ages. The metric system of units originated from the National Assembly of France in 1790. In 1960 this system was rationalised, and the SI or Système International d'Unités was adopted as the international standard. Unit names and their abbreviations have been standardised; according to SI convention, unit abbreviations are the same for both singular and plural and are not followed by a period. SI prefixes used to indicate multiples and submultiples of units are listed in Table 2.4. Despite widespread use of SI units, no single system of units has universal application. In particular, engineers in the United States continue to apply British or imperial units. In addition, many physical property data collected before 1960 are published in lists and tables using nonstandard units.

Familiarity with both metric and nonmetric units is necessary. Some units used in engineering, such as the slug (1 slug = 14.5939 kilograms), dram (1 dram = 1.77185 grams), stoke (a unit of kinematic viscosity), poundal (a unit of force), and erg (a unit of energy), are probably not known to you. Although no longer commonly applied, these are legitimate units that may appear in engineering reports and tables of data.

In calculations it is often necessary to convert units. Units are changed using *conversion factors*. Some conversion factors, such as 1 inch = 2.54 cm and 2.20 lb = 1 kg, you probably already know. Tables of common conversion factors are given in Appendix A at the back of this book. Unit conversions are not only necessary to convert imperial units to metric; some physical variables have several metric units in common use. For example, viscosity

TABLE 2.4 SI Prefixes

Factor	Prefix	Symbol	Factor	Prefix	Symbol
10^{-1}	deci*	d	10^{18}	exa	E
10^{-2}	centi*	c	10^{15}	peta	P
10^{-3}	milli	m	10^{12}	tera	T
10^{-6}	micro	μ	10^{9}	giga	G
10^{-9}	nano	n	10^{6}	mega	M
10^{-12}	pico	p	10^{3}	kilo	k
10^{-15}	femto	f	10^{2}	hecto*	h
10^{-18}	atto	a	10^{1}	deka*	da

*Used for areas and volumes.
From J.V. Drazil, 1983, Quantities and Units of Measurement, *Mansell, London.*

may be reported as centipoise or kg h^{-1} m^{-1}; pressure may be given in standard atmospheres, pascals, or millimetres of mercury. Conversion of units seems simple enough; however, difficulties can arise when several variables are being converted in a single equation. Accordingly, an organised mathematical approach is needed.

For each conversion factor, a *unity bracket* can be derived. The value of the unity bracket, as the name suggests, is unity. As an example, the conversion factor:

$$1 \text{ lb} = 453.6 \text{ g} \tag{2.10}$$

can be converted by dividing both sides of the equation by 1 lb to give a unity bracket denoted by vertical bars (| |):

$$1 = \left|\frac{453.6 \text{ g}}{1 \text{ lb}}\right| \tag{2.11}$$

Similarly, division of both sides of Eq. (2.10) by 453.6 g gives another unity bracket:

$$\left|\frac{1 \text{ lb}}{453.6 \text{ g}}\right| = 1 \tag{2.12}$$

To calculate how many pounds are in 200 g, we can multiply 200 g by the unity bracket in Eq. (2.12) or divide 200 g by the unity bracket in Eq. (2.11). This is permissible since the value of both unity brackets is unity, and multiplication or division by 1 does not change the value of 200 g. Using the option of multiplying by Eq. (2.12):

$$200 \text{ g} = 200 \cancel{\text{g}} \cdot \left|\frac{1 \text{ lb}}{453.6 \cancel{\text{g}}}\right| \tag{2.13}$$

On the right side, cancelling the old units leaves the desired unit, lb. Dividing the numbers gives:

$$200\text{ g} = 0.441\text{ lb} \tag{2.14}$$

A more complicated calculation involving a complete equation is given in Example 2.1.

EXAMPLE 2.1 UNIT CONVERSION

Air is pumped through an orifice immersed in liquid. The size of the bubbles leaving the orifice depends on the diameter of the orifice and the properties of the liquid. The equation representing this situation is:

$$\frac{g(\rho_L - \rho_G)D_b^3}{\sigma D_o} = 6$$

where g = gravitational acceleration = 32.174 ft s^{-2}; ρ_L = liquid density = 1 g cm^{-3}; ρ_G = gas density = 0.081 lb ft^{-3}; D_b = bubble diameter; σ = gas–liquid surface tension = 70.8 dyn cm^{-1}; and D_o = orifice diameter = 1 mm.

Calculate the bubble diameter D_b.

Solution

Convert the data to a consistent set of units, for example, g, cm, s. From Appendix A, the conversion factors required are:

- 1 ft = 0.3048 m
- 1 lb = 453.6 g
- 1 dyn cm^{-1} = 1 g s^{-2}

Also:

- 1 m = 100 cm
- 10 mm = 1 cm

Converting units:

$$g = 32.174\frac{\text{ft}}{\text{s}^2} \cdot \left|\frac{0.3048\text{ m}}{1\text{ ft}}\right| \cdot \left|\frac{100\text{ cm}}{1\text{ m}}\right| = 980.7\text{ cm s}^{-2}$$

$$\rho_G = 0.081\frac{\text{lb}}{\text{ft}^3} \cdot \left|\frac{453.6\text{ g}}{1\text{ lb}}\right| \cdot \left|\frac{1\text{ ft}}{0.3048\text{ m}}\right|^3 \cdot \left|\frac{1\text{ m}}{100\text{ cm}}\right|^3 = 1.30 \times 10^{-3}\text{ g cm}^{-3}$$

$$\sigma = 70.8\text{ dyn cm}^{-1} \cdot \left|\frac{1\text{ g s}^{-2}}{1\text{ dyn cm}^{-1}}\right| = 70.8\text{ g s}^{-2}$$

$$D_o = 1\text{ mm} \cdot \left|\frac{1\text{ cm}}{10\text{ mm}}\right| = 0.1\text{ cm}$$

Rearranging the equation to give an expression for D_b^3:

$$D_b^3 = \frac{6\,\sigma D_o}{g\,(\rho_L - \rho_G)}$$

Substituting values gives:

$$D_b^3 = \frac{6\,(70.8\text{ g s}^{-2})\,(0.1\text{ cm})}{980.7\text{ cm s}^{-2}\,(1\text{ g cm}^{-3} - 1.30 \times 10^{-3}\text{ g cm}^{-3})} = 4.34 \times 10^{-2}\text{ cm}^3$$

Taking the cube root:

$$D_b = 0.35\text{ cm}$$

Note that unity brackets are squared or cubed when appropriate, for example, when converting ft^3 to cm^3. This is permissible since the value of the unity bracket is 1, and 1^2 or 1^3 is still 1.

2.3 FORCE AND WEIGHT

According to Newton's law, the force exerted on a body in motion is proportional to its mass multiplied by the acceleration. As listed in Table 2.2, the dimensions of force are LMT^{-2}; the *natural units* of force in the SI system are kg m s^{-2}. Analogously, g cm s^{-2} and lb ft s^{-2} are natural units of force in the metric and British systems, respectively.

Force occurs frequently in engineering calculations, and *derived units* are used more commonly than natural units. In SI, the derived unit for force is the *newton*, abbreviated as N:

$$1\text{ N} = 1\text{ kg m s}^{-2} \tag{2.15}$$

In the British or imperial system, the derived unit for force is the *pound-force*, which is denoted lb_f. One pound-force is defined as (1 lb mass) × (gravitational acceleration at sea level and 45° latitude). In different systems of units, gravitational acceleration g at sea level and 45° latitude is:

$$g = 9.8066\text{ m s}^{-2} \tag{2.16}$$

$$g = 980.66\text{ cm s}^{-2} \tag{2.17}$$

$$g = 32.174\text{ ft s}^{-2} \tag{2.18}$$

Therefore:

$$1\text{ lb}_f = 32.174\text{ lb}_m\text{ ft s}^{-2} \tag{2.19}$$

Note that pound-mass, which is usually represented as lb, has been shown here using the abbreviation lb_m to distinguish it from lb_f. Use of the pound in the imperial system for reporting both mass and force can be a source of confusion and requires care.

To convert force from a defined unit to a natural unit, a special dimensionless unity bracket called g_c is used. The form of g_c depends on the units being converted. From Eqs. (2.15) and (2.19):

$$g_c = 1 = \left|\frac{1\ \text{N}}{1\ \text{kg m s}^{-2}}\right| = \left|\frac{1\ \text{lb}_f}{32.174\ \text{lb}_m\ \text{ft s}^{-2}}\right| \tag{2.20}$$

Application of g_c is illustrated in Example 2.2.

EXAMPLE 2.2 USE OF g_c

Calculate the kinetic energy of 250 lb_m of liquid flowing through a pipe at a speed of 35 ft s^{-1}. Express your answer in units of ft lb_f.

Solution

Kinetic energy is given by the equation:

$$\text{kinetic energy} = E_k = \frac{1}{2} M v^2$$

where M is mass and v is velocity. Using the values given:

$$E_k = \frac{1}{2}(250\ \text{lb}_m)\left(35\ \frac{\text{ft}}{\text{s}}\right)^2 = 1.531 \times 10^5\ \frac{\text{lb}_m\ \text{ft}^2}{\text{s}^2}$$

Multiplying by g_c from Eq. (2.20) gives:

$$E_k = 1.531 \times 10^5\ \frac{\text{lb}_m\ \text{ft}^2}{\text{s}^2} \cdot \left|\frac{1\ \text{lb}_f}{32.174\ \text{lb}_m\ \text{ft s}^{-2}}\right|$$

Calculating and cancelling units gives the answer:

$$E_k = 4760\ \text{ft lb}_f$$

Weight is the force with which a body is attracted by gravity to the centre of the Earth. Therefore, the weight of an object will change depending on its location, whereas its mass will not. Weight changes according to the value of the gravitational acceleration g, which varies by about 0.5% over the Earth's surface. Using Newton's law and depending on the exact value of g, the weight of a mass of 1 kg is about 9.8 newtons; the weight of a mass of 1 lb is about 1 lb_f. Note that although the value of g changes with position on the Earth's surface (or in the universe), the value of g_c within a given system of units does not. g_c is a factor for converting units, not a physical variable.

2.4 MEASUREMENT CONVENTIONS

Familiarity with common physical variables and methods for expressing their magnitude is necessary for engineering analysis of bioprocesses. This section covers some useful definitions and engineering conventions that will be applied throughout the text.

2.4.1 Density

Density is a substantial variable defined as mass per unit volume. Its dimensions are $L^{-3}M$, and the usual symbol is ρ. Units for density are, for example, g cm^{-3}, kg m^{-3}, and lb ft^{-3}. If the density of acetone is 0.792 g cm^{-3}, the mass of 150 cm^3 acetone can be calculated as follows:

$$150\ \text{cm}^3\left(\frac{0.792\ \text{g}}{\text{cm}^3}\right) = 119\ \text{g}$$

Densities of solids and liquids vary slightly with temperature. The density of water at 4°C is 1.0000 g cm^{-3}, or 62.4 lb ft^{-3}. The density of solutions is a function of both concentration and temperature. Gas densities are highly dependent on temperature and pressure.

2.4.2 Specific Gravity

Specific gravity, also known as 'relative density', is a dimensionless variable. It is the ratio of two densities: that of the substance in question and that of a specified reference material. For liquids and solids, the reference material is usually water. For gases, air is commonly used as the reference, but other reference gases may also be specified.

As mentioned previously, liquid densities vary somewhat with temperature. Accordingly, when reporting specific gravity, the temperatures of the substance and its reference material are specified. If the specific gravity of ethanol is given as $0.789^{20°C}_{4°C}$, this means that the specific gravity is 0.789 for ethanol at 20°C referenced against water at 4°C. Since the density of water at 4°C is almost exactly 1.0000 g cm^{-3}, we can say immediately that the density of ethanol at 20°C is 0.789 g cm^{-3}.

2.4.3 Specific Volume

Specific volume is the inverse of density. The dimensions of specific volume are L^3M^{-1}.

2.4.4 Mole

In the SI system, a mole is 'the amount of substance of a system which contains as many elementary entities as there are atoms in 0.012 kg of carbon-12' [1]. This means that a mole in the SI system is about 6.02×10^{23} molecules, and is denoted by the term *gram-mole* or *gmol*. One thousand gmol is called a *kilogram-mole* or *kgmol*. In the American engineering system, the basic mole unit is the *pound-mole* or *lbmol*, which is $6.02 \times 10^{23} \times 453.6$ molecules. The gmol, kgmol, and lbmol therefore represent three different quantities. When molar quantities are specified simply as 'moles', gmol is usually meant.

The number of moles in a given mass of material is calculated as follows:

$$\text{gram-moles} = \frac{\text{mass in grams}}{\text{molar mass in grams}} \tag{2.21}$$

$$\text{lb-moles} = \frac{\text{mass in lb}}{\text{molar mass in lb}} \tag{2.22}$$

Molar mass is the mass of one mole of substance, and has dimensions MN^{-1}. Molar mass is routinely referred to as *molecular weight*, although the molecular weight of a compound is a dimensionless quantity calculated as the sum of the atomic weights of the elements constituting a molecule of that compound. The *atomic weight* of an element is its mass relative to carbon-12 having a mass of exactly 12; atomic weight is also dimensionless. The terms 'molecular weight' and 'atomic weight' are frequently used by engineers and chemists instead of the more correct terms, 'relative molecular mass' and 'relative atomic mass'.

2.4.5 Chemical Composition

Process streams usually consist of mixtures of components or solutions of one or more solutes. The following terms are used to define the composition of mixtures and solutions.

The *mole fraction* of component A in a mixture is defined as:

$$\text{mole fraction A} = \frac{\text{number of moles of A}}{\text{total number of moles}} \tag{2.23}$$

Mole percent is mole fraction × 100. In the absence of chemical reactions and loss of material from the system, the composition of a mixture expressed in mole fraction or mole percent does not vary with temperature.

The *mass fraction* of component A in a mixture is defined as:

$$\text{mass fraction A} = \frac{\text{mass of A}}{\text{total mass}} \tag{2.24}$$

Mass percent is mass fraction × 100; mass fraction and mass percent are also called *weight fraction* and *weight percent*, respectively. Another common expression for composition is weight-for-weight percent (% w/w). Although not so well defined, this is usually considered to be the same as weight percent. For example, a solution of sucrose in water with a concentration of 40% w/w contains 40 g sucrose per 100 g solution, 40 tonnes sucrose per 100 tonnes solution, 40 lb sucrose per 100 lb solution, and so on. In the absence of chemical reactions and loss of material from the system, mass and weight percent do not change with temperature.

Because the composition of liquids and solids is usually reported using mass percent, this can be assumed even if not specified. For example, if an aqueous mixture is reported to contain 5% NaOH and 3% $MgSO_4$, it is conventional to assume that there are 5 g NaOH and 3 g $MgSO_4$ in every 100 g solution. Of course, mole or volume percent may be used for liquid and solid mixtures; however, this should be stated explicitly (e.g., 10 vol%, 50 mol%).

The *volume fraction* of component A in a mixture is:

$$\text{volume fraction A} = \frac{\text{volume of A}}{\text{total volume}} \tag{2.25}$$

Volume percent is volume fraction × 100. Although not as clearly defined as volume percent, volume-for-volume percent (% v/v) is usually interpreted in the same way as volume percent; for example, an aqueous sulphuric acid mixture containing 30 cm^3 acid in 100 cm^3 solution is referred to as a 30% v/v solution. Weight-for-volume percent (% w/v) is also often used; a codeine concentration of 0.15% w/v generally means 0.15 g codeine per 100 ml solution.

Compositions of gases are commonly given in volume percent; if percentage figures are given without specification, volume percent is assumed. According to the *International Critical Tables* [2], the composition of air is 20.99% oxygen, 78.03% nitrogen, 0.94% argon, and 0.03% carbon dioxide; small amounts of hydrogen, helium, neon, krypton, and xenon make up the remaining 0.01%. For most purposes, all inerts are lumped together with nitrogen and the composition of air is taken as approximately 21% oxygen and 79% nitrogen. This means that any sample of air will contain about 21% oxygen *by volume*. At low pressure, gas volume is directly proportional to number of moles; therefore, the composition of air as stated can also be interpreted as 21 *mole%* oxygen. Because temperature changes at low pressure produce the same relative change in the partial volumes of the constituent gases as in the total volume, the volumetric composition of gas mixtures is not altered by variation in temperature. Temperature changes affect the component gases equally, so the overall composition is unchanged.

There are many other choices for expressing the concentration of a component in solutions and mixtures:

1. Moles per unit volume (e.g., gmol l^{-1}, lbmol ft^{-3}).
2. Mass per unit volume (e.g., kg m^{-3}, g l^{-1}, lb ft^{-3}).
3. Parts per million, ppm. This is used for very dilute solutions. Usually, ppm is a mass fraction for solids and liquids and a mole fraction for gases. For example, an aqueous solution of 20 ppm manganese contains 20 g manganese per 10^6 g solution. A sulphur dioxide concentration of 80 ppm in air means 80 gmol SO_2 per 10^6 gmol gas mixture. At low pressures this is equivalent to 80 litres SO_2 per 10^6 litres gas mixture.
4. Molarity, gmol l^{-1}. A molar concentration is abbreviated 1 M.
5. Molality, gmol per 1000 g solvent.
6. Normality, mole equivalents l^{-1}. A normal concentration is abbreviated 1 N and contains one equivalent gram-weight of solute per litre of solution. For an acid or base, an equivalent gram-weight is the weight of solute in grams that will produce or react with one gmol hydrogen ions. Accordingly, a 1 N solution of HCl is the same as a 1 M solution; on the other hand, a 1 N H_2SO_4 or 1 N $Ca(OH)_2$ solution is 0.5 M.
7. Formality, formula gram-weight l^{-1}. If the molecular weight of a solute is not clearly defined, formality may be used to express concentration. A formal solution contains one formula gram-weight of solute per litre of solution. If the formula gram-weight and molecular gram-weight are the same, molarity and formality are the same.

In several industries, concentration is expressed in an indirect way using specific gravity. For a given solute and solvent, the density and specific gravity of solutions are directly dependent on the concentration of solute. Specific gravity is conveniently measured using a hydrometer, which may be calibrated using special scales. The *Baumé scale*, originally developed in France to measure levels of salt in brine, is in common use. One Baumé scale is used for liquids lighter than water; another is used for liquids heavier than water. For liquids heavier than water such as sugar solutions:

$$\text{degrees Baumé (°Bé)} = 145 - \frac{145}{G} \tag{2.26}$$

where G is specific gravity. Unfortunately, the reference temperature for the Baumé and other gravity scales is not standardised worldwide. If the Baumé hydrometer were calibrated at 60°F (15.6°C), G in Eq. (2.26) would be the specific gravity at 60°F relative to water at 60°F; however another common reference temperature is 20°C (68°F). The Baumé scale is used widely in the wine and food industries as a measure of sugar concentration. For example, readings of °Bé from grape juice help determine when grapes should be harvested for wine making. The Baumé scale gives only an approximate indication of sugar levels; there is always some contribution to specific gravity from soluble compounds other than sugar.

Degrees Brix (°Brix), or *degrees Balling*, is another hydrometer scale used extensively in the sugar industry. Brix scales calibrated at 15.6°C and 20°C are in common use. With the 20°C scale, each degree Brix indicates 1 gram of sucrose per 100 g liquid.

2.4.6 Temperature

Temperature is a measure of the thermal energy of a body at thermal equilibrium. As a dimension, it is denoted Θ. Temperature is commonly measured in degrees *Celsius* (centigrade) or *Fahrenheit*. In science, the Celsius scale is most common; 0°C is taken as the ice point of water and 100°C the normal boiling point of water. The Fahrenheit scale is in everyday use in the United States; 32°F represents the ice point and 212°F the normal boiling point of water. Both Fahrenheit and Celsius scales are *relative temperature scales*, meaning that their zero points have been arbitrarily assigned.

Sometimes it is necessary to use *absolute temperatures*. Absolute temperature scales have as their zero point the lowest temperature believed possible. Absolute temperature is used in application of the ideal gas law and many other laws of thermodynamics. A scale for absolute temperature with degree units the same as on the Celsius scale is known as the *Kelvin* scale; the absolute temperature scale using Fahrenheit degree units is the *Rankine* scale. Accordingly, a temperature difference of one degree on the Celsius scale corresponds to a temperature difference of one degree on the Kelvin scale; similarly for the Fahrenheit and Rankine scales. Units on the Kelvin scale used to be termed 'degrees Kelvin' and abbreviated °K. It is modern practice, however, to name the unit simply 'kelvin'; the SI symbol for kelvin is K. Units on the Rankine scale are denoted °R. 0°R = 0 K = −459.67°F = −273.15°C. Comparison of the four temperature scales is shown in Figure 2.1.

Equations for converting temperature units are as follows; T represents the temperature reading:

$$T(\text{K}) = T(°\text{C}) + 273.15 \tag{2.27}$$

$$T(°\text{R}) = T(°\text{F}) + 459.67 \tag{2.28}$$

$$T(°\text{R}) = 1.8\ T(\text{K}) \tag{2.29}$$

$$T(°\text{F}) = 1.8\ T(°\text{C}) + 32 \tag{2.30}$$

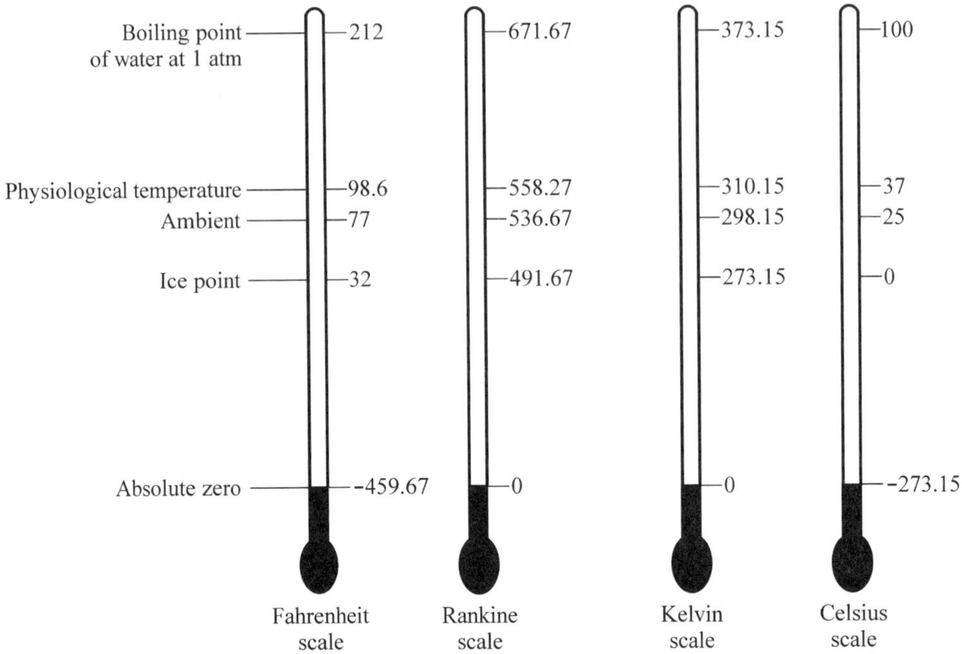

FIGURE 2.1 Comparison of temperature scales.

A temperature difference of one degree on the Kelvin–Celsius scale corresponds to a temperature difference of 1.8 degrees on the Rankine–Fahrenheit scale. This is readily deduced, for example, if we consider the difference between the freezing and boiling points of water, which is 100 degrees on the Kelvin–Celsius scale and $(212 - 32) = 180$ degrees on the Rankine–Fahrenheit scale.

The dimensions of several engineering parameters include temperature. Examples, such as specific heat capacity, heat transfer coefficient, and thermal conductivity, are listed in Table 2.2. These parameters are applied in engineering calculations to evaluate variations in the properties of materials caused by a *change* in temperature. For instance, the specific heat capacity is used to determine the change in enthalpy of a system resulting from a change in its temperature. Therefore, if the specific heat capacity of a certain material is known to be 0.56 kcal kg^{-1} $°C^{-1}$, this is the same as 0.56 kcal kg^{-1} K^{-1}, as any change in temperature measured in units of °C is the same when measured on the Kelvin scale. Similarly, for engineering parameters quantified using the imperial system of units, $°F^{-1}$ can be substituted for $°R^{-1}$ and vice versa.

2.4.7 Pressure

Pressure is defined as force per unit area, and has dimensions $L^{-1}MT^{-2}$. Units of pressure are numerous, including pounds per square inch (psi), millimetres of mercury (mmHg), standard atmospheres (atm), bar, newtons per square metre (N m^{-2}), and many others. The SI pressure unit, N m^{-2}, is called a pascal (Pa). Like temperature, pressure may be expressed using absolute or relative scales.

Absolute pressure is pressure relative to a complete vacuum. Because this reference pressure is independent of location, temperature, and weather, absolute pressure is a precise and invariant quantity. However, absolute pressure is not commonly measured. Most pressure-measuring devices sense the difference in pressure between the sample and the surrounding atmosphere at the time of measurement. These instruments give readings of *relative pressure,* also known as *gauge pressure.* Absolute pressure can be calculated from gauge pressure as follows:

$$\text{absolute pressure} = \text{gauge pressure} + \text{atmospheric pressure} \tag{2.31}$$

As you know from listening to weather reports, atmospheric pressure varies with time and place and is measured using a *barometer*. Atmospheric pressure or *barometric pressure* should not be confused with the standard unit of pressure called the standard atmosphere (atm), defined as 1.013×10^5 N m^{-2}, 14.70 psi, or 760 mmHg at 0°C. Sometimes the units for pressure include information about whether the pressure is absolute or relative. Pounds per square inch is abbreviated *psia* for absolute pressure or *psig* for gauge pressure. *Atma* denotes standard atmospheres of absolute pressure.

Vacuum pressure is another pressure term, used to indicate pressure below barometric pressure. A gauge pressure of −5 psig, or 5 psi below atmospheric, is the same as a vacuum of 5 psi. A perfect vacuum corresponds to an absolute pressure of zero.

2.5 STANDARD CONDITIONS AND IDEAL GASES

A *standard state* of temperature and pressure has been defined and is used when specifying properties of gases, particularly molar volumes. Standard conditions are needed because the volume of a gas depends not only on the quantity present but also on the temperature and pressure. The most widely adopted standard state is 0°C and 1 atm.

Relationships between gas volume, pressure, and temperature were formulated in the eighteenth and nineteenth centuries. These correlations were developed under conditions of temperature and pressure such that the average distance between gas molecules was great enough to counteract the effect of intramolecular forces, and the volume of the molecules themselves could be neglected. A gas under these conditions became known as an *ideal gas*. This term now in common use refers to a gas that obeys certain simple physical laws, such as those of Boyle, Charles, and Dalton. Molar volumes for an ideal gas at standard conditions are:

$$1 \text{ gmol} = 22.4 \text{ litres} \tag{2.32}$$

$$1 \text{ kgmol} = 22.4 \text{ m}^3 \tag{2.33}$$

$$1 \text{ lbmol} = 359 \text{ ft}^3 \tag{2.34}$$

No real gas is an ideal gas at all temperatures and pressures. However, light gases such as hydrogen, oxygen, and air deviate negligibly from ideal behaviour over a wide range of conditions. On the other hand, heavier gases such as sulphur dioxide and hydrocarbons can deviate considerably from ideal, particularly at high pressures. Vapours near the boiling point also deviate markedly from ideal. Nevertheless, for many applications in bioprocess engineering, gases can be considered ideal without much loss of accuracy.

Equations (2.32) through (2.34) can be verified using the *ideal gas law*:

$$pV = nRT \tag{2.35}$$

where p is absolute pressure, V is volume, n is moles, T is absolute temperature, and R is the *ideal gas constant*. Equation (2.35) can be applied using various combinations of units for the physical variables, as long as the correct value and units of R are employed. A list of R values in different systems of units is given in Appendix B.

EXAMPLE 2.3 IDEAL GAS LAW

Gas leaving a fermenter at close to 1 atm pressure and 25°C has the following composition: 78.2% nitrogen, 19.2% oxygen, 2.6% carbon dioxide. Calculate:

(a) The mass composition of the fermenter off-gas
(b) The mass of CO_2 in each cubic metre of gas leaving the fermenter

Solution

Molecular weights:

- Nitrogen = 28
- Oxygen = 32
- Carbon dioxide = 44

(a) Because the gas is at low pressure, the percentages given for composition can be considered mole percentages. Therefore, using the molecular weights, 100 gmol off-gas contains:

$$78.2 \text{ gmol } N_2 \cdot \left| \frac{28 \text{ g } N_2}{1 \text{ gmol } N_2} \right| = 2189.6 \text{ g } N_2$$

$$19.2 \text{ gmol } O_2 \cdot \left| \frac{32 \text{ g } O_2}{1 \text{ gmol } O_2} \right| = 614.4 \text{ g } O_2$$

$$2.6 \text{ gmol } CO_2 \cdot \left| \frac{44 \text{ g } CO_2}{1 \text{ gmol } CO_2} \right| = 114.4 \text{ g } CO_2$$

Therefore, the total mass is (2189.6 +614.4 +114.4) g = 2918.4 g. The mass composition can be calculated as follows:

$$\text{Mass percent } N_2 = \frac{2189.6 \text{ g}}{2918.4 \text{ g}} \times 100 = 75.0\%$$

$$\text{Mass percent } O_2 = \frac{614.4 \text{ g}}{2918.4 \text{ g}} \times 100 = 21.1\%$$

$$\text{Mass percent } CO_2 = \frac{114.4 \text{ g}}{2918.4 \text{ g}} \times 100 = 3.9\%$$

Therefore, the composition of the gas is 75.0 mass% N_2, 21.1 mass% O_2, and 3.9 mass% CO_2.

(b) As the gas composition is given in volume percent, in each cubic metre of gas there must be 0.026 m^3 CO_2. The relationship between moles of gas and volume at 1 atm and 25°C is determined using Eq. (2.35) and an appropriate value of R from Appendix B:

$$(1 \text{ atm})(0.026 \text{ m}^3) = n\left(0.000082057 \frac{\text{m}^3 \text{ atm}}{\text{gmol K}}\right)(298.15 \text{ K})$$

Calculating the moles of CO_2 present:

$$n = 1.06 \text{ gmol}$$

Converting to mass of CO_2:

$$1.06 \text{ gmol} = 1.06 \text{ gmol} \cdot \left|\frac{44 \text{ g}}{1 \text{ gmol}}\right| = 46.8 \text{ g}$$

Therefore, each cubic metre of fermenter off-gas contains 46.8 g CO_2.

2.6 PHYSICAL AND CHEMICAL PROPERTY DATA

Information about the properties of materials is often required in engineering calculations. Because measurement of physical and chemical properties is time-consuming and expensive, handbooks containing this information are a tremendous resource. You may already be familiar with some handbooks of physical and chemical data, including:

- *International Critical Tables* [2]
- *CRC Handbook of Chemistry and Physics* [3]
- *Lange's Handbook of Chemistry* [4]

To these can be added:

- *Perry's Chemical Engineers' Handbook* [5]

and, for information about biological materials:

- *Biochemical Engineering and Biotechnology Handbook* [6]

A selection of physical and chemical property data is included in Appendices C and D.

2.7 STOICHIOMETRY

In chemical or biochemical reactions, atoms and molecules rearrange to form new groups. Mass and molar relationships between the reactants consumed and products formed can be determined using stoichiometric calculations. This information is deduced from correctly written reaction equations and relevant atomic weights.

As an example, consider the principal reaction in alcohol fermentation: conversion of glucose to ethanol and carbon dioxide:

$$C_6H_{12}O_6 \rightarrow 2C_2H_6O + 2CO_2 \tag{2.36}$$

This reaction equation states that one molecule of glucose breaks down to give two molecules of ethanol and two molecules of carbon dioxide. Another way of saying this is that one mole of glucose breaks down to give two moles of ethanol and two moles of carbon dioxide. Applying molecular weights, the equation also shows that reaction of 180 g glucose produces 92 g ethanol and 88 g carbon dioxide.

During chemical or biochemical reactions, the following two quantities are conserved:

1. *Total mass*, so that total mass of reactants = total mass of products
2. *Number of atoms of each element*, so that, for example, the number of C, H, and O atoms in the reactants = the number of C, H, and O atoms, respectively, in the products

Note that there is no corresponding law for conservation of moles: the number of moles of reactants is not necessarily equal to the number of moles of products.

EXAMPLE 2.4 STOICHIOMETRY OF AMINO ACID SYNTHESIS

The overall reaction for microbial conversion of glucose to L-glutamic acid is:

$$\underset{\text{(glucose)}}{C_6H_{12}O_6} + NH_3 + 1.5\,O_2 \rightarrow \underset{\text{(glutamic acid)}}{C_5H_9NO_4} + CO_2 + 3H_2O$$

What mass of oxygen is required to produce 15 g glutamic acid?

Solution

Molecular weights:

- Oxygen = 32
- Glutamic acid = 147

Because stoichiometric equations give relationships between moles, g glutamic acid is first converted to gmol using the unity bracket for molecular weight:

$$15\text{ g glumatic acid} = 15\text{ g glumatic acid} \cdot \left|\frac{1\text{ gmol glutamic acid}}{147\text{ g glutamic acid}}\right| = 0.102\text{ gmol glutamic acid}$$

According to the reaction equation, production of 1 gmol of glutamic acid requires 1.5 gmol O_2. Therefore, production of 0.102 gmol glutamic acid requires $(0.102 \times 1.5) = 0.153$ gmol O_2. This can be expressed as mass of oxygen using the unity bracket for the molecular weight of O_2:

$$0.153 \text{ gmol } O_2 = 0.153 \text{ gmol } O_2 \cdot \left| \frac{32 \text{ g } O_2}{1 \text{ gmol } O_2} \right| = 4.9 \text{ g } O_2$$

Therefore, 4.9 g oxygen is required. More oxygen will be needed if microbial growth also occurs.

By themselves, equations such as Eq. (2.36) suggest that all the reactants are converted into the products specified in the equation, and that the reaction proceeds to completion. This is often not the case for industrial reactions. Because the stoichiometry may not be known precisely, or in order to manipulate the reaction beneficially, reactants are not usually supplied in the exact proportions indicated by the reaction equation. Excess quantities of some reactants may be provided; this excess material is found in the product mixture once the reaction is stopped. In addition, reactants are often consumed in side reactions to make products not described by the principal reaction equation; these side-products also form part of the final reaction mixture. In these circumstances, additional information is needed before the amounts of products formed or reactants consumed can be calculated. Several terms are used to describe partial and branched reactions.

1. The *limiting reactant* or *limiting substrate* is the reactant present in the smallest *stoichiometric* amount. While other reactants may be present in smaller absolute quantities, at the time when the last molecule of the limiting reactant is consumed, residual amounts of all reactants except the limiting reactant will be present in the reaction mixture. As an illustration, for the glutamic acid reaction of Example 2.4, if 100 g glucose, 17 g NH_3, and 48 g O_2 are provided for conversion, glucose will be the limiting reactant even though a greater mass of it is available compared with the other substrates.
2. An *excess reactant* is a reactant present in an amount in excess of that required to combine with all of the limiting reactant. It follows that an excess reactant is one remaining in the reaction mixture once all the limiting reactant is consumed.
   The *percentage excess* is calculated using the amount of excess material relative to the quantity required for complete consumption of the limiting reactant:

$$\% \text{ excess} = \frac{\begin{pmatrix}\text{moles present} - \text{moles required to react} \\ \text{completely with the limiting reactant}\end{pmatrix}}{\begin{pmatrix}\text{moles required to react} \\ \text{completely with the limiting reactant}\end{pmatrix}} \times 100 \tag{2.37}$$

or

$$\% \text{ excess} = \frac{\begin{pmatrix}\text{mass present} - \text{mass required to react} \\ \text{completely with the limiting reactant}\end{pmatrix}}{\begin{pmatrix}\text{mass required to react} \\ \text{completely with the limiting reactant}\end{pmatrix}} \times 100 \tag{2.38}$$

The *required* amount of a reactant is the stoichiometric quantity needed for complete conversion of the limiting reactant. In the preceding glutamic acid example, the required amount of NH_3 for complete conversion of 100 g glucose is 9.4 g; therefore, if 17 g NH_3 are provided, the percent excess NH_3 is 80%. Even if only part of the reaction actually occurs, required and excess quantities are based on the entire amount of the limiting reactant.

Other reaction terms are not as well defined, with multiple definitions in common use:

3. *Conversion* is the fraction or percentage of a reactant converted into products.
4. *Degree of completion* is usually the fraction or percentage of the limiting reactant converted into products.
5. *Selectivity* is the amount of a particular product formed as a fraction of the amount that would have been formed if all the feed material had been converted to that product.
6. *Yield* is the ratio of mass or moles of product formed to the mass or moles of reactant consumed. If more than one product or reactant is involved in the reaction, the particular compounds referred to must be stated, for example, the yield of glutamic acid from glucose was 0.6 g g^{-1}. Because of the complexity of metabolism and the frequent occurrence of side reactions, yield is an important term in bioprocess analysis. Application of the yield concept for cell and enzyme reactions is described in more detail in Chapter 12.

EXAMPLE 2.5 INCOMPLETE REACTION AND YIELD

Depending on culture conditions, glucose can be catabolised by yeast to produce ethanol and carbon dioxide, or can be diverted into other biosynthetic reactions. An inoculum of yeast is added to a solution containing 10 g l^{-1} glucose. After some time, only 1 g l^{-1} glucose remains while the concentration of ethanol is 3.2 g l^{-1}. Determine:

(a) The fractional conversion of glucose to ethanol
(b) The yield of ethanol from glucose

Solution

(a) To find the fractional conversion of glucose to ethanol, we must first determine how much glucose was directed into ethanol biosynthesis. Using a basis of 1 litre, we can calculate the mass of glucose required for synthesis of 3.2 g ethanol. First, g ethanol is converted to gmol using the unity bracket for molecular weight:

$$3.2\text{ g ethanol} = 3.2\text{ g ethanol} \cdot \left|\frac{1\text{ gmol ethanol}}{46\text{ g ethanol}}\right|$$
$$= 0.070\text{ gmol ethanol}$$

According to Eq. (2.36) for ethanol fermentation, production of 1 gmol of ethanol requires 0.5 gmol glucose. Therefore, production of 0.070 gmol ethanol requires

(0.070 × 0.5) = 0.035 gmol glucose. This is converted to g using the molecular weight unity bracket for glucose:

$$0.035 \text{ gmol glucose} = 0.035 \text{ gmol glucose} \cdot \left| \frac{180 \text{ g glucose}}{1 \text{ gmol glucose}} \right|$$
$$= 6.3 \text{ g glucose}$$

Therefore, as 6.3 g glucose was used for ethanol synthesis, based on the total amount of glucose provided per litre (10 g), the fractional conversion of glucose to ethanol was 0.63. Based on the amount of glucose actually consumed per litre (9 g), the fractional conversion to ethanol was 0.70.

(b) Yield of ethanol from glucose is based on the total mass of glucose consumed. Since 9 g glucose was consumed per litre to provide 3.2 g l^{-1} ethanol, the yield of ethanol from glucose was 0.36 g g^{-1}. We can also conclude that, per litre, (9 − 6.3) = 2.7 g glucose was consumed but not used for ethanol synthesis.

2.8 METHODS FOR CHECKING AND ESTIMATING RESULTS

In this chapter, we have considered how to quantify variables and have begun to use different types of equation to solve simple problems. Applying equations to analyse practical situations involves calculations, which are usually performed with the aid of an electronic calculator. Each time you carry out a calculation, are you always happy and confident about the result? How can you tell if you have keyed in the wrong parameter values or made an error pressing the function buttons of your calculator? Because it is relatively easy to make mistakes in calculations, it is a good idea always to review your answers and check whether they are correct.

Professional engineers and scientists develop the habit of validating the outcomes of their mathematical analyses and calculations, preferably using independent means. Several approaches for checking and estimating results are available.

1. Ask yourself whether your answer is reasonable and makes sense. In some cases, judging whether a result is reasonable will depend on your specific technical knowledge and experience of the situation being examined. For example, you may find it difficult at this stage to know whether or not 2×10^{12} is a reasonable value for the Reynolds number in a stirred bioreactor. Nevertheless, you will already be able to judge the answers from other types of calculation. For instance, if you determine using design equations that the maximum cell concentration in a fermenter is 0.002 cells per litre, or that the cooling system provides a working fermentation temperature of 160°C, you should immediately suspect that you have made a mistake.
2. Simplify the calculation and obtain a rough or *order-of-magnitude* estimate of the answer. Instead of using exact numbers, round off the values to integers or powers of 10, and continue rounding off as you progress through the arithmetic. You can verify answers quickly using this method, often without needing a calculator. If the estimated answer is of the same order of magnitude as the result found using exact parameter values, you

can be reasonably sure that the exact result is free from gross error. An order-of-magnitude calculation is illustrated in Example 2.6.

EXAMPLE 2.6 ORDER-OF-MAGNITUDE CALCULATION

The impeller Reynolds number Re_i for fluid flow in a stirred tank is defined as:

$$Re_i = \frac{N_i D_i^2 \rho}{\mu}$$

Re_i was calculated as 1.10×10^5 for the following parameter values: $N_i = 30.6$ rpm, $D_i = 1.15$ m, $\rho = 1015$ kg m^{-3}, and $\mu = 6.23 \times 10^{-3}$ kg m^{-1} s^{-1}. Check this answer using order-of-magnitude estimation.

Solution

The calculation using exact parameter values:

$$Re_i = \frac{(30.6\ \text{min}^{-1}) \cdot \left|\frac{1\ \text{min}}{60\ \text{s}}\right| \cdot (1.15\ \text{m})^2\ (1015\ \text{kg m}^{-3})}{6.23 \times 10^{-3}\ \text{kg m}^{-1}\ \text{s}^{-1}}$$

can be rounded off and approximated as:

$$Re_i = \frac{(30\ \text{min}^{-1}) \cdot \left|\frac{1\ \text{min}}{60\ \text{s}}\right| \cdot (1\ \text{m}^2)\ (1 \times 10^3\ \text{kg m}^{-3})}{6 \times 10^{-3}\ \text{kg m}^{-1}\ \text{s}^{-1}}$$

Combining values and cancelling units gives:

$$Re_i = \frac{0.5 \times 10^6}{6} \approx \frac{0.6 \times 10^6}{6} = 0.1 \times 10^6 = 10^5$$

This calculation is simple enough to be performed without using a calculator. As the rough answer of 10^5 is close to the original result of 1.10×10^5, we can conclude that no gross error was made in the original calculation. Note that units must still be considered and converted using unity brackets in order-of-magnitude calculations.

The following methods for checking calculated results can also be used.

3. Substitute the calculated answer back into the equations for checking. For example, if Re_i (Example 2.6) was determined as 1.10×10^5, this value could be used to back-calculate one of the other parameters in the equation such as ρ:

$$\rho = \frac{Re_i\, \mu}{N_i D_i^2}$$

If the value of ρ obtained in this way is 1015 kg m^{-3}, we would know that our result for Re_i was free of accidental calculator error.

4. If none of the preceding approaches can be readily applied, a final option is to check your answer by repeating the calculation from the beginning. This strategy has the

disadvantage of using a less independent method of checking, so there is a greater chance that you will make the same mistakes in the checking calculation as in the original. It is much better, however, than leaving your answer completely unverified. If possible, you should use a different order of calculator keystrokes in the repeat calculation.

Note that methods 2 through 4 only address the issue of arithmetic or calculation mistakes. Your answer will still be wrong if the equation itself contains an error or if you are applying the wrong equation to solve the problem. You should check for this type of mistake separately. If you get an unreasonable result as described in method 1 but find you have made no calculation error, the equation is likely to be the cause.

SUMMARY OF CHAPTER 2

Having studied the contents of Chapter 2, you should:

- Understand dimensionality and be able to convert units with ease
- Understand the terms *mole, molecular weight, density, specific gravity, temperature,* and *pressure;* know various ways of expressing the *concentration* of solutions and mixtures; and be able to work simple problems involving these concepts
- Be able to apply the ideal gas law
- Understand reaction terms such as *limiting reactant, excess reactant, conversion, degree of completion, selectivity,* and *yield,* and be able to apply stoichiometric principles to reaction problems
- Know where to find physical and chemical property data in the literature
- Be able to perform order-of-magnitude calculations to estimate results and check the answers from calculations

PROBLEMS

2.1 Unit conversion

(a) Convert 1.5×10^{-6} centipoise to kg s^{-1} cm^{-1}.

(b) Convert 0.122 horsepower (British) to British thermal units per minute (Btu min^{-1}).

(c) Convert 10,000 rpm to s^{-1}.

(d) Convert 4335 W m^{-2} $°C^{-1}$ to l atm min^{-1} ft^{-2} K^{-1}.

2.2 Unit conversion

(a) Convert 345 Btu lb^{-1} to kcal g^{-1}.

(b) Convert 670 mmHg ft^3 to metric horsepower h.

(c) Convert 0.554 cal g^{-1} $°C^{-1}$ to kJ kg^{-1} K^{-1}.

(d) Convert 10^3 g l^{-1} to kg m^{-3}.

2.3 Unit conversion

(a) Convert 10^6 μg ml^{-1} to g m^{-3}.

(b) Convert 3.2 centipoise to millipascal seconds (mPa s).

(c) Convert 150 Btu h^{-1} ft^{-2} $(°F\ ft^{-1})^{-1}$ to W m^{-1} K^{-1}.

(d) Convert 66 revolutions per hour to s^{-1}.

2.4 Unit conversion and calculation

The mixing time t_m in a stirred fermenter can be estimated using the following equation:

$$t_m = 5.9\, D_T^{2/3} \left(\frac{\rho V_L}{P}\right)^{1/3} \left(\frac{D_T}{D_i}\right)^{1/3}$$

Evaluate the mixing time in seconds for a vessel of diameter $D_T = 2.3$ m containing liquid volume $V_L = 10{,}000$ litres stirred with an impeller of diameter $D_i = 45$ in. The liquid density $\rho = 65$ lb ft^{-3} and the power dissipated by the impeller $P = 0.70$ metric horsepower.

2.5 Unit conversion and dimensionless numbers

Using Eq. (2.1) for the Reynolds number, calculate *Re* for the two sets of data in the following table.

Parameter	Case 1	Case 2
D	2 mm	1 in.
u	3 cm s^{-1}	1 m s^{-1}
ρ	25 lb ft^{-3}	12.5 kg m^{-3}
μ	10^{-6} cP	0.14×10^{-4} lb$_m$ s^{-1} ft^{-1}

2.6 Property data

Using appropriate handbooks, find values for:

(a) The viscosity of ethanol at 40°C
(b) The diffusivity of oxygen in water at 25°C and 1 atm
(c) The thermal conductivity of Pyrex borosilicate glass at 37°C
(d) The density of acetic acid at 20°C
(e) The specific heat capacity of liquid water at 80°C

Make sure you reference the source of your information, and explain any assumptions you make.

2.7 Dimensionless groups and property data

The rate at which oxygen is transported from gas phase to liquid phase is a very important parameter in fermenter design. A well-known correlation for transfer of gas is:

$$Sh = 0.31\, Gr^{1/3} Sc^{1/3}$$

where *Sh* is the Sherwood number, *Gr* is the Grashof number, and *Sc* is the Schmidt number. These dimensionless numbers are defined as follows:

$$Sh = \frac{k_L D_b}{\mathscr{D}}$$

$$Gr = \frac{D_b^3 \rho_G (\rho_L - \rho_G) g}{\mu_L^2}$$

$$Sc = \frac{\mu_L}{\rho_L \mathscr{D}}$$

where k_L is the mass transfer coefficient, D_b is bubble diameter, $\mathscr{D}$ is the diffusivity of gas in the liquid, ρ_G is the density of the gas, ρ_L is the density of the liquid, μ_L is the viscosity of the liquid, and g is gravitational acceleration.

A gas sparger in a fermenter operated at 28°C and 1 atm produces bubbles of about 2 mm diameter. Calculate the value of the mass transfer coefficient, k_L. Collect property data from, for example, *Perry's Chemical Engineers' Handbook*, and assume that the culture broth has properties similar to those of water. (Do you think this is a reasonable assumption?) Report the literature source for any property data used. State explicitly any other assumptions you make.

2.8 Dimensionless numbers and dimensional homogeneity

The Colburn equation for heat transfer is:

$$\left(\frac{h}{C_pG}\right)\left(\frac{C_p\mu}{k}\right)^{2/3} = \frac{0.023}{\left(\frac{DG}{\mu}\right)^{0.2}}$$

where C_p is heat capacity, Btu lb^{-1} °F^{-1}; μ is viscosity, lb h^{-1} ft^{-1}; k is thermal conductivity, Btu h^{-1} ft^{-2} (°F ft^{-1})$^{-1}$; D is pipe diameter, ft; and G is mass velocity per unit area, lb h^{-1} ft^{-2}. The Colburn equation is dimensionally consistent. What are the units and dimensions of the heat transfer coefficient, h?

2.9 Dimensional homogeneity

The terminal eddy size in a fluid in turbulent flow is given by the Kolmogorov scale:

$$\lambda = \left(\frac{\nu^3}{\varepsilon}\right)^{1/4}$$

where λ is the length scale of the eddies (μm) and ε is the local rate of energy dissipation per unit mass of fluid (W kg^{-1}). Using the principle of dimensional homogeneity, what are the dimensions of the term represented as ν?

2.10 Dimensional homogeneity and g_c

Two students have reported different versions of the dimensionless power number N_P used to relate fluid properties to the power required for stirring:

$$N_P = \frac{Pg}{\rho N_i^3 D_i^5}$$

and

$$N_P = \frac{Pg_c}{\rho N_i^3 D_i^5}$$

where P is power, g is gravitational acceleration, ρ is fluid density, N_i is stirrer speed, D_i is stirrer diameter, and g_c is the force unity bracket. Which equation is correct?

2.11 Mass and weight

The density of water is 62.4 lb$_m$ ft^{-3}. What is the weight of 10 ft^3 of water:

(a) At sea level and 45° latitude?

(b) Somewhere above the Earth's surface where g = 9.76 m s^{-2}?

2.12 Molar units

If a bucket holds 20.0 lb NaOH, how many:

(a) lbmol NaOH

(b) gmol NaOH

(c) kgmol NaOH

does it contain?

2.13 Density and specific gravity

(a) The specific gravity of nitric acid is $1.5129^{20°C}_{4°C}$.

(i) What is its density at 20°C in kg m^{-3}?

(ii) What is its molar specific volume?

(b) The volumetric flow rate of carbon tetrachloride (CCl_4) in a pipe is 50 cm^3 min^{-1}. The density of CCl_4 is 1.6 g cm^{-3}.

(i) What is the mass flow rate of CCl_4?

(ii) What is the molar flow rate of CCl_4?

2.14 Molecular weight

Calculate the average molecular weight of air.

2.15 Mole fraction

A solution contains 30 wt% water, 25 wt% ethanol, 15 wt% methanol, 12 wt% glycerol, 10 wt% acetic acid, and 8 wt% benzaldehyde. What is the mole fraction of each component?

2.16 Solution preparation

You are asked to make up a 6% w/v solution of $MgSO_4 \cdot 7\ H_2O$ in water, and are provided with $MgSO_4 \cdot 7\ H_2O$ crystals, water, a 500-ml beaker, a 250-ml measuring cylinder with 2-ml graduations, a balance that shows weight up to 100 g in increments of 0.1 g, and a stirring rod. Explain how you would prepare the solution in any quantity you wish, explaining carefully each step in the process.

2.17 Moles, molarity, and composition

(a) How many gmol are there in 21.2 kg of isobutyl succinate ($C_{12}H_{22}O_4$)?

(b) If sucrose ($C_{12}H_{22}O_{11}$) crystals flow into a hopper at a rate of 4.5 kg s^{-1}, how many gram-moles of sucrose are transferred in 30 min?

(c) A solution of 75 mM tartaric acid ($C_4H_6O_6$) in water is used as a standard in HPLC analysis. How many grams of $C_4H_6O_6 \cdot H_2O$ are needed to make up 10 ml of solution?

(d) An aqueous solution contains 60 μM salicylaldehyde ($C_7H_6O_2$) and 330 ppm dichloroacetic acid ($C_2H_2Cl_2O_2$). How many grams of each component are present in 250 ml?

2.18 Concentration

A holding tank with capacity 5000 litres initially contains 1500 litres of 25 mM NaCl solution.

(a) What is the final concentration of NaCl if an additional 3000 litres of 25 mM NaCl solution is pumped into the tank?

(b) Instead of (a), if 3000 litres of water are added to the tank, what is the final concentration of NaCl?

(c) Instead of (a) or (b), if an additional 500 litres of 25 mM NaCl solution plus 3000 litres of water are added, what is the final concentration of NaCl:

(i) Expressed as molarity?

(ii) Expressed as % w/v?

(iii) Expressed as g cm^{-3}?

(iv) Expressed as ppm? Justify your answer.

2.19 Gas composition

A gas mixture containing 30% oxygen, 5% carbon dioxide, 2% ammonia, and 63% nitrogen is stored at 35 psia and room temperature (25°C) prior to injection into the headspace of an industrial-scale bioreactor used to culture normal foetal lung fibroblast cells. If the reactor is operated at 37°C and atmospheric pressure, explain how the change in temperature and pressure of the gas as it enters the vessel will affect its composition.

2.20 Specific gravity and composition

Broth harvested from a fermenter is treated for recovery and purification of a pharmaceutical compound. After filtration of the broth and partial evaporation of the aqueous filtrate, a solution containing 38.6% (w/w) pharmaceutical leaves the evaporator at a flow rate of 8.6 litres min^{-1}. The specific gravity of the solution is 1.036. If the molecular weight of the drug is 1421, calculate:

(a) The concentration of pharmaceutical in the solution in units of kg l^{-1}

(b) The flow rate of pharmaceutical in gmol min^{-1}

2.21 Temperature scales

What is −40°F in degrees centigrade? degrees Rankine? kelvin?

2.22 Pressure scales

(a) The pressure gauge on an autoclave reads 15 psi. What is the absolute pressure in the chamber in psi? in atm?

(b) A vacuum gauge reads 3 psi. What is the absolute pressure?

2.23 Gas leak

A steel cylinder containing compressed air is stored in a fermentation laboratory ready to provide aeration gas to a small-scale bioreactor. The capacity of the cylinder is 48 litres, the absolute pressure is 0.35 MPa and the temperature is 22°C. One day in mid-summer when the air conditioning breaks down, the temperature in the laboratory rises to 33°C and the valve at the top of the cylinder is accidentally left open. Estimate the proportion of air that will be lost. What assumptions will you make?

2.24 Gas supply

A small airlift bioreactor is used to culture suspended *Solanum aviculare* (kangaroo apple) plant cells. The reactor contains 1.5 litres of culture broth and is operated at 25°C and atmospheric pressure. The air flow rate under these conditions is 0.8 vvm (1 vvm means 1 volume of gas per volume of liquid per minute). Air is supplied from a 48-litre gas cylinder at 20°C. If, at 4 PM on Friday afternoon, the gauge pressure is 800 psi, is there enough air in the cylinder to operate the reactor over the weekend until 9 AM on Monday morning? At ambient temperature and for pressures up to 100 atm, the ideal gas law can be used to estimate the amount of air present to within 5%.

2.25 Stoichiometry and incomplete reaction

For production of penicillin ($C_{16}H_{18}O_4N_2S$) using *Penicillium* mould, glucose ($C_6H_{12}O_6$) is used as substrate and phenylacetic acid ($C_8H_8O_2$) is added as precursor. The stoichiometry for overall synthesis is:

$$1.67\,C_6H_{12}O_6 + 2\,NH_3 + 0.5\,O_2 + H_2SO_4 + C_8H_8O_2 \rightarrow C_{16}H_{18}O_4N_2S + 2\,CO_2 + 9\,H_2O$$

(a) What is the maximum theoretical yield of penicillin from glucose?
(b) When results from a particular penicillin fermentation were analysed, it was found that 24% of the glucose had been used for growth, 70% for cell maintenance activities (such as membrane transport and macromolecule turnover), and only 6% for penicillin synthesis. Calculate the yield of penicillin from glucose under these conditions.
(c) Batch fermentation under the conditions described in (b) is carried out in a 100-litre tank. Initially, the tank is filled with nutrient medium containing 50 g l^{-1} glucose and 4 g l^{-1} phenylacetic acid. If the reaction is stopped when the glucose concentration is 5.5 g l^{-1}, determine:
  (i) Which is the limiting substrate if NH_3, O_2, and H_2SO_4 are provided in excess
  (ii) The total mass of glucose used for growth
  (iii) The amount of penicillin produced
  (iv) The final concentration of phenylacetic acid

2.26 Stoichiometry, yield, and the ideal gas law

Stoichiometric equations can be used to represent the growth of microorganisms provided a 'molecular formula' for the cells is available. The molecular formula for biomass is obtained by measuring the amounts of C, N, H, O, and other elements in cells. For a particular bacterial strain, the molecular formula was determined to be $C_{4.4}H_{7.3}O_{1.2}N_{0.86}$.
These bacteria are grown under aerobic conditions with hexadecane ($C_{16}H_{34}$) as substrate. The reaction equation describing growth is:

$$C_{16}H_{34} + 16.28\,O_2 + 1.42\,NH_3 \rightarrow 1.65\,C_{4.4}H_{7.3}O_{1.2}N_{0.86} + 8.74\,CO_2 + 13.11\,H_2O$$

(a) Is the stoichiometric equation balanced?
(b) Assuming 100% conversion, what is the yield of cells from hexadecane in g g^{-1}?
(c) Assuming 100% conversion, what is the yield of cells from oxygen in g g^{-1}?
(d) You have been put in charge of a small batch fermenter for growing the bacteria and aim to produce 2.5 kg of cells for inoculation of a pilot-scale reactor.
  (i) What minimum amount of hexadecane substrate must be contained in your culture medium?
  (ii) What must be the minimum concentration of hexadecane in the medium if the fermenter working volume is 3 cubic metres?
  (iii) What minimum volume of air at 20°C and 1 atm pressure must be pumped into the fermenter during growth to produce the required amount of cells?

2.27 Stoichiometry and the ideal gas law

Roots of the *Begonia rex* plant are cultivated in an air-driven bioreactor in medium containing glucose ($C_6H_{12}O_6$) and two nitrogen sources, ammonia (NH_3) and nitrate (HNO_3). The root biomass can be represented stoichiometrically using the formula $CH_{1.63}O_{0.80}N_{0.13}$.
A simplified reaction equation for growth of the roots is:

$$C_6H_{12}O_6 + 3.4\,O_2 + 0.15\,NH_3 + 0.18\,HNO_3 \rightarrow 2.5\,CH_{1.63}O_{0.80}N_{0.13} + 3.5\,CO_2 + 4.3\,H_2O$$

(a) Is the stoichiometric equation balanced?

(b) If the medium contains 30 g l^{-1} glucose, what minimum concentration of nitrate is required to achieve complete conversion of the sugar? Express your answer in units of gmol l^{-1} or M.

(c) If the bioreactor holds 50 litres of medium and there is complete conversion of the glucose, what mass of roots will be generated?

(d) For the conditions described in (c), what minimum volume of air at 20°C and 1 atm pressure must be provided to the bioreactor during growth.

2.28 Stoichiometry, yield, and limiting substrate

Under anoxic conditions, biological denitrification of waste water by activated sludge results in the conversion of nitrate to nitrogen gas. When acetate provides the carbon source, the reaction can be represented as follows:

$$5\,CH_3COOH + 8\,NO_3^- \rightarrow 4\,N_2 + 10\,CO_2 + 6\,H_2O + 8\,OH^-$$

(a) Is the stoichiometric equation balanced?

(b) In the absence of side reactions, what is the yield of nitrogen from acetate in g g^{-1}?

(c) A certain waste water contains 6.0 mM acetic acid and 7 mM $NaNO_3$. If 25% of the acetate and 15% of the nitrate are consumed in other reactions (e.g., for growth of organisms in the sludge), which is the limiting substrate in the denitrification reaction?

(d) For the situation described in (c), what mass of gaseous nitrogen is produced from treatment of 5000 litres of waste water if the reaction is allowed to proceed until the limiting substrate is exhausted?

2.29 Order-of-magnitude calculation

The value of the sigma factor for a tubular-bowl centrifuge is approximated by the equation:

$$\Sigma = \frac{\pi \omega^2 b r^2}{2g}$$

An estimate of Σ corresponding to the following parameter values is required: $\omega = 12{,}000$ rpm, $b = 1.25$ m, $r = 0.37$ m, and $g = 9.8066$ m s^{-2}. Using an order-of-magnitude calculation, determine whether the value of Σ is closer to 100 m^2 or 1000 m^2.

2.30 Order-of-magnitude calculation

Two work-experience students at your start-up company have been asked to evaluate the rate of reaction occurring in a transparent gel particle containing immobilised mouse melanoma cells. The equation for the reaction rate r^*_{As} is:

$$r^*_{As} = \frac{4}{3}\pi R^3 \left(\frac{v_{max} C_{As}}{K_m + C_{As}} \right)$$

where $R = 3.2$ mm, $v_{max} = 0.12$ gmol s^{-1} m^{-3}, $C_{As} = 41$ gmol m^{-3}, and $K_m = 0.8$ gmol m^{-3}. One student reports a reaction rate of 1.6×10^{-8} gmol s^{-1}; the other reports 1.6×10^{-10} gmol s^{-1}. You left your calculator on the bus this morning, but must know quickly which student is correct. Use an order-of-magnitude calculation to identify the right answer.

References

[1] The International System of Units (SI), National Bureau of Standards Special Publication 330, US Government Printing Office, 1977. Adopted by the 14th General Conference on Weights and Measures (1971, Resolution 3).

[2] International Critical Tables (1926–1930), McGraw-Hill; first electronic ed. (2003), International Critical Tables of Numerical Data, Physics, Chemistry and Technology, Knovel/Norwich.

[3] CRC Handbook of Chemistry and Physics, eighty-fourth ed., CRC Press, 2003. (New editions released periodically.)

[4] Lange's Handbook of Chemistry, sixteenth ed., McGraw-Hill, 2005. (New editions released periodically.)

[5] Perry's Chemical Engineers' Handbook, eighth ed., McGraw-Hill, 2008. (New editions released periodically.)

[6] B. Atkinson, F. Mavituna, Biochemical Engineering and Biotechnology Handbook, second ed., Macmillan, 1991.

Suggestions for Further Reading

Units and Dimensions

Cardarelli, F. (1997). *Scientific Unit Conversion*. Springer.

Massey, B. S. (1986). *Measures in Science and Engineering* (Chapters 1–5). Ellis Horwood.

Qasim, S. H. (1977). *SI Units in Engineering and Technology*. Pergamon Press.

Wandmacher, C., & Johnson, A. I. (1995). *Metric Units in Engineering: Going SI*. American Society of Civil Engineers.

Wildi, T. (1991). *Units and Conversion Charts*. Institute of Electrical and Electronics Engineers.

Engineering Variables

Felder, R. M., & Rousseau, R. W. (2005). *Elementary Principles of Chemical Processes* (3rd ed., Chapters 2 and 3). Wiley.

Himmelblau, D. M., & Riggs, J. B. (2004). *Basic Principles and Calculations in Chemical Engineering* (7th ed., Chapters 1–5). Prentice Hall.

CHAPTER

3

Presentation and Analysis of Data

Quantitative information is fundamental to scientific and engineering analysis. Information about bioprocesses, such as the amount of substrate fed into the system, the operating conditions, and properties of the product stream, is obtained by measuring pertinent physical and chemical variables. In industry, data are collected for equipment design, process control, troubleshooting and economic evaluations. In research, experimental data are used to test hypotheses and develop new theories. In either case, quantitative interpretation of data is essential for making rational decisions about the system under investigation. The ability to extract useful and accurate information from data is an important skill for any scientist. Professional presentation and communication of results are also required.

Techniques for data analysis must take into account the existence of error in measurements. Because there is always an element of uncertainty associated with measured data, interpretation calls for a great deal of judgement. This is especially the case when critical decisions in design or operation of processes depend on data evaluation. Although computers and calculators make data processing less tedious, the data analyst must possess enough perception to use these tools effectively.

This chapter discusses sources of error in data and methods of handling errors in calculations. Presentation and analysis of data using graphs and equations and communication of process information using flow sheets are described.

3.1 ERRORS IN DATA AND CALCULATIONS

Measurements are never perfect. Experimentally determined quantities are always somewhat inaccurate due to measurement error; absolutely 'correct' values of physical quantities (time, length, concentration, temperature, etc.) cannot be found. The significance or reliability of conclusions drawn from data must take measurement error into consideration. Estimation of error and principles of error propagation in calculations are important elements of engineering analysis that help prevent misleading representation of data. General principles for estimating and expressing errors are discussed in the following sections.

3.1.1 Significant Figures

Data used in engineering calculations vary considerably in accuracy. Economic projections may estimate market demand for a new biotechnology product to within ±100%; on the other hand, some properties of materials are known to within ±0.0001% or less. The uncertainty associated with quantities should be reflected in the way they are written. The number of figures used to report a measured or calculated variable is an indirect indication of the precision to which that variable is known. It would be absurd, for example, to quote the estimated income from sales of a new product using ten decimal places. Nevertheless, the mistake of quoting too many figures is not uncommon; display of superfluous figures on calculators is very easy but should not be transferred to scientific reports.

A *significant figure* is any digit (i.e., 1–9) used to specify a number. Zero may also be a significant figure when it is not used merely to locate the position of the decimal point. For example, the numbers 6304, 0.004321, 43.55, and 8.063×10^{10} each contain four significant figures. For the number 1200, however, there is no way of knowing whether or not the two zeros are significant figures; a direct statement or an alternative way of expressing the number is needed. For example, 1.2×10^3 has two significant figures, while 1.200×10^3 has four.

A number is rounded to n significant figures using the following rules:

1. If the number in the $(n + 1)$th position is less than 5, discard all figures to the right of the nth place.
2. If the number in the $(n + 1)$th position is greater than 5, discard all figures to the right of the nth place, and increase the nth digit by 1.
3. If the number in the $(n + 1)$th position is exactly 5, discard all figures to the right of the nth place, and increase the nth digit by 1.

For example, when rounding off to four significant figures:

- 1.426348 becomes 1.426
- 1.426748 becomes 1.427
- 1.4265 becomes 1.427

The last rule is not universal but is engineering convention; most electronic calculators and computers round up halves. Generally, rounding off means that the value may be wrong by up to 5 units in the next number-column not reported. Thus, 10.77 kg means that the mass lies somewhere between 10.765 kg and 10.775 kg, whereas 10.7754 kg represents a mass between 10.77535 kg and 10.77545 kg. These rules apply only to quantities based on measured values; some numbers used in calculations refer to precisely known or counted quantities. For example, there is no error associated with the number 1/2 in the equation for kinetic energy:

$$\text{kinetic energy} = E_k = \frac{1}{2} M v^2$$

where M is mass and v is velocity.

It is good practice during calculations to carry along one or two extra significant figures for combination during arithmetic operations; final rounding off should be done only at the end. How many figures should we quote in the final answer? There are several rules of thumb for rounding off after calculations, so rigid adherence to all rules is not always possible. However as a guide, after multiplication or division, the number of significant figures in the result should equal the smallest number of significant figures of any of the quantities involved in the calculation. For example:

$$(6.681 \times 10^{-2})\ (5.4 \times 10^{9}) = 3.608 \times 10^{8} \rightarrow 3.6 \times 10^{8}$$

and

$$\frac{6.16}{0.054677} = 112.6616310 \rightarrow 113$$

For addition and subtraction, look at the position of the last significant figure in each number relative to the decimal point. The position of the last significant figure in the result should be the same as that most to the left. For example:

$$24.335 + 3.90 + 0.00987 = 28.24487 \rightarrow 28.24$$

and

$$121.808 - 112.87634 = 8.93166 \rightarrow 8.932$$

3.1.2 Absolute and Relative Uncertainty

The uncertainty associated with measurements can be stated more explicitly than is possible using the rules of significant figures. For a particular measurement, we should be able to give our best estimate of the parameter value and the interval representing its range of uncertainty. For example, we might be able to say with confidence that the prevailing temperature lies between 23.6°C and 24.4°C. Another way of expressing this result is 24 ± 0.4°C. The value ±0.4°C is known as the *uncertainty*, *error*, *probable error*, or *margin of error* for the measurement, and allows us to judge the quality of the measuring process. Since ±0.4°C represents the actual temperature range by which the reading is uncertain, it is known as the *absolute error*. An alternative expression for 24 ± 0.4°C is 24°C ± 1.7%; in this case the *relative error* is ±1.7%.

Because most values of uncertainty must be estimated rather than measured, there is a rule of thumb that magnitudes of errors should be given with only one or sometimes two significant figures. A flow rate may be expressed as 146 ± 10 gmol h^{-1}, even though this means that two, the 4 and the 6, in the result are uncertain. The number of digits used to express the result should be compatible with the magnitude of its estimated error. For example, in the statement 2.1437 ± 0.12 grams, the estimated uncertainty of 0.12 grams shows that the last two digits in the result are superfluous. Use of more than three significant figures for the result in this case gives a false impression of accuracy.

3.1.3 Propagation of Errors

There are rules for combining errors during mathematical operations. The uncertainty associated with calculated results is determined from the errors associated with the raw data. The simplest type of error to evaluate after combination of measured values is the *maximum possible error*. For addition and subtraction the rule is: *add absolute errors*. The total of the absolute errors becomes the absolute error associated with the final answer. For example, the sum of 1.25 ± 0.13 and 0.973 ± 0.051 is:

$$(1.25 + 0.973) \pm (0.13 + 0.051) = 2.22 \pm 0.2 = 2.22 \pm 8\%$$

Considerable loss of accuracy can occur after subtraction, especially when two large numbers are subtracted to give an answer of small numerical value. Because the absolute error after subtraction of two numbers always increases, the relative error associated with a small-number answer can be very great. For example, consider the difference between two numbers, each with small relative error: $1273 \pm 0.5\%$ and $1268 \pm 0.5\%$. For subtraction, the absolute errors are added:

$$(1273 \pm 6.4) - (1268 \pm 6.3) = (1273 - 1268) \pm (6.4 + 6.3) = 5 \pm 13 = 5 \pm 250\%$$

The maximum possible error in the answer is extremely large compared with the relative errors in the original numbers. For measured values, any small number obtained by subtraction of two large numbers must be examined carefully and with justifiable suspicion. Unless explicit errors are reported, the large uncertainty associated with such results can go unnoticed.

For multiplication and division the rule is: *add relative errors*. The total of the relative errors becomes the relative error associated with the answer. For example, 790 ± 20 divided by 164 ± 1 is the same as $790 \pm 2.5\%$ divided by $164 \pm 0.61\%$:

$$\left(\frac{790}{164}\right) \pm (2.5 + 0.61)\% = 4.8 \pm 3\% = 4.8 \pm 0.1$$

Rules for propagating errors in other types of mathematical expression, such as multiplication by a constant and elevation to a power, can be found in other references (e.g., [1–3]).

Although the concept of maximum possible error is relatively straightforward and can be useful in defining the upper limits of uncertainty, when the values being combined in mathematical operations are independent and their errors randomly distributed, other less extreme estimates of error can be found. Modified rules for propagation of errors are based on the observation that, if two values, A and B, have uncertainties of $\pm a$ and $\pm b$, respectively, it is possible that the error in A is $+a$ while the error in B is $-b$, so the combined error in $A + B$ will be smaller than $a + b$. If the errors are random in nature, there is a 50% chance that an underestimate in A will be accompanied by an overestimate in B. Accordingly, in the example given previously for subtraction of two numbers, it could be argued that the two errors might almost cancel each other if one were +6.4 and the other were −6.3. Although we can never be certain this would occur, uncertainties may still be represented using probable errors rather than maximum possible errors. Rules for estimating probable errors after addition, subtraction, multiplication, division, and other mathematical operations are described in other books dealing with error analysis (e.g., [1, 3]).

So far we have considered the error occurring in single observations. However, as discussed in the following sections, better estimates of errors are obtained by taking repeated measurements. Because this approach is useful only for certain types of measurement error, let us consider the various sources of error in experimental data.

3.1.4 Types of Error

There are two broad classes of measurement error: *systematic* and *random*. A systematic error is one that affects all measurements of the same variable in the same way. If the cause of systematic error is identified, it can be accounted for using a correction factor. For example, errors caused by an imperfectly calibrated analytical balance may be identified using standard weights; measurements with the balance can then be corrected to compensate for the error. Systematic errors easily go undetected; performing the same measurement using different instruments, methods, and observers is required to detect systematic error.

Random or accidental errors are due to unknown causes. Random errors are present in almost all data; they are revealed when repeated measurements of an unchanging quantity give a 'scatter' of different results. As outlined in the next section, scatter from repeated measurements is used in statistical analysis to quantify random error. The term *precision* refers to the *reliability* or *reproducibility* of data, and indicates the extent to which a measurement is free from random error. *Accuracy*, on the other hand, requires both random and systematic errors to be small. Repeated weighings using a poorly calibrated balance can give results that are very precise (because each reading is similar); however, the result would be inaccurate because of the incorrect calibration and systematic error.

During experiments, large, isolated, one-of-a-kind errors can also occur. This type of error is different from the systematic and random errors just mentioned and can be described as a 'blunder'. Accounting for blunders in experimental data requires knowledge of the experimental process and judgement about the likely accuracy of the measurement.

3.1.5 Statistical Analysis

Measurements that contain random errors but are free of systematic errors and blunders can be analysed using statistical procedures. Details are available in standard texts (e.g., [1, 4−6]); only the most basic techniques for statistical treatment will be described here. From readings containing random error, we aim to find the best estimate of the variable measured and to quantify the extent to which random error affects the data.

In the following analysis, errors are assumed to follow a normal or Gaussian distribution. Normally distributed random errors in a single measurement are just as likely to be positive as negative; thus, if an infinite number of repeated measurements were made of the same variable, random error would completely cancel out from the arithmetic mean of these values to give the *population mean* or *true mean*, $\overline{X}$. For less than an infinite number of observations, the arithmetic mean of repeated measurements is still regarded as the best

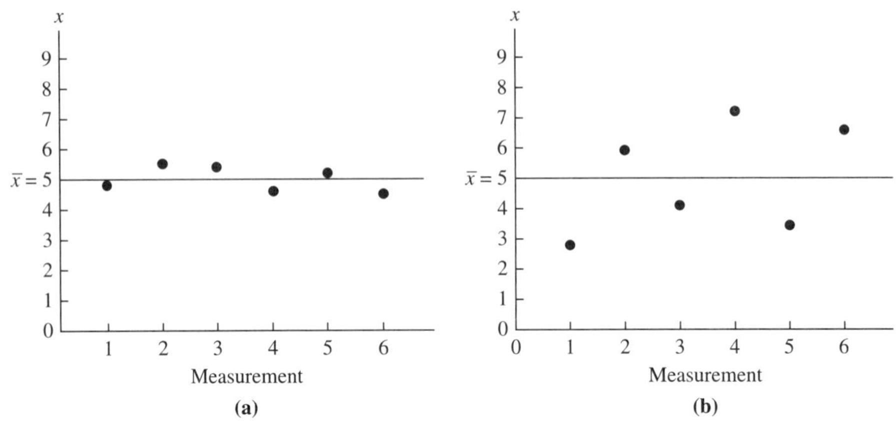

FIGURE 3.1 Two data sets with the same mean but different degrees of scatter.

estimate of $\overline{X}$, provided each measurement is made with equal care and under identical conditions. Taking replicate measurements is therefore standard practice in science; whenever possible, several readings of each datum point should be obtained. For variable x measured n times, the *arithmetic* or *sample mean* is calculated as follows:

$$\bar{x} = \text{mean value of } x = \frac{\sum^{n} x}{n} = \frac{x_1 + x_2 + x_3 + \ldots x_n}{n} \tag{3.1}$$

As indicated, the symbol $\sum^{n}$ represents the sum of n values; $\sum^{n} x$ means the sum of n values of parameter x.

In addition to the sample mean, we are also likely to be interested in how scattered the individual points are about the mean. For example, consider the two sets of data in Figure 3.1, each representing replicate measurements of the same parameter. Although the mean of both sets is the same, the values in Figure 3.1(b) are scattered over a much broader range. The extent of the data scatter or deviation of individual points from the mean reflects the reliability of the measurement technique (or person) used to obtain the data. In this case, we would have more confidence in the data shown in Figure 3.1(a) than in those of Figure 3.1(b).

The deviation of an individual measurement from the mean is known as the *residual*; an example of a residual is $(x_1 - \bar{x})$ where x_1 is one of the measurements in a set of replicates. A simple way of indicating the scatter in data is to report the *maximum error* or maximum residual; this is obtained by finding the datum point furthest from the mean and calculating the difference. For example, the maximum error in the data shown in Figure 3.1(a) is 0.5; the maximum error in Figure 3.1(b) is 2.2. Another indicator of scatter is the *range* of the data, which is the difference between the highest and lowest values in a data set. For the measurements in Figure 3.1(a), the range is 1.0 (from 4.5 to 5.5); in Figure 3.1(b), the range is much larger at 4.4 (from 2.8 to 7.2). Yet, although the maximum error and range tell us about the extreme limits of the measured values, these parameters have the disadvantage of giving no indication about whether only a few or most points are so widely scattered.

The most useful indicator of the amount of scatter in data is the *standard deviation*. For a set of experimental data, the standard deviation σ is calculated from the residuals of all the datum points as follows:

$$\sigma = \sqrt{\frac{\sum^{n}(x-\overline{x})^2}{n-1}} = \sqrt{\frac{(x_1-\overline{x})^2+(x_2-\overline{x})^2+(x_3-\overline{x})^2+\ldots(x_n-\overline{x})^2}{n-1}} \tag{3.2}$$

Equation (3.2) is the definition used by most modern statisticians and manufacturers of electronic calculators; σ as defined in Eq. (3.2) is sometimes called the *sample standard deviation*.

Replicate measurements giving a low value of σ are considered more reliable and generally of better quality than those giving a high value of σ. For data containing random errors, approximately 68% of the measurements can be expected to take values between $\overline{x}-\sigma$ and $\overline{x}+\sigma$. This is illustrated in Figure 3.2, where 13 of 40 datum points lie outside the $\overline{x}\pm\sigma$ range. Also shown in Figure 3.2 is the bell-shaped, normal distribution curve for x. The area enclosed by the curve between $\overline{x}-\sigma$ and $\overline{x}+\sigma$ is 68% of the total area under the curve. As around 95% of the data fall within two standard deviations from the mean, the area enclosed by the normal distribution curve between $\overline{x}-2\sigma$ and $\overline{x}+2\sigma$ accounts for about 95% of the total area.

For less than an infinite number of repeated measurements, if more than one set of replicate measurements is made, it is likely that different sample means will be obtained. For example, if a culture broth is sampled five times for measurement of cell concentration and the mean value from the five measurements is determined, if we take a further five measurements and calculate the mean of the second set of data, a slightly different mean would probably be found. The uncertainty associated with a particular estimate of the

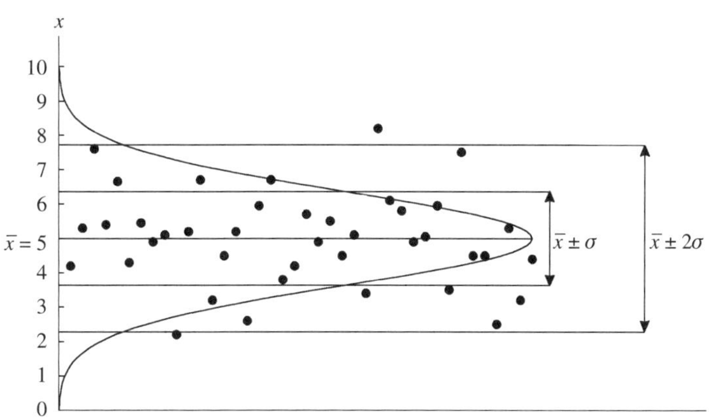

FIGURE 3.2 This image shows the scatter of data within one and two standard deviations from the mean. Each dot represents an individual measurement of x.

mean is expressed using the *standard error* or *standard error of the mean*, σ_m, which can be calculated as:

$$\sigma_m = \frac{\sigma}{\sqrt{n}} \tag{3.3}$$

where n is the number of measurements in the data set and σ is the standard deviation defined in Eq. (3.2). There is a 68% chance that the true population mean $\overline{X}$ falls within the interval $\overline{x} \pm \sigma_m$, and about a 95% chance that it falls within the interval $\overline{x} \pm 2\sigma_m$. The range $\pm 2\sigma_m$ is therefore sometimes called the *95% confidence limit* for the mean. Clearly, the smaller the value of σ_m, the more confidence we can have that the calculated mean $\overline{x}$ is close to the true mean $\overline{X}$.

Therefore, to report the results of repeated measurements, we usually quote the calculated sample mean as the best estimate of the variable, and the standard error as a measure of the confidence we place in the result. In the case of cell concentration, for example, this information could be expressed in the form 1.5 ± 0.3 g l^{-1}, where 1.5 g l^{-1} is the mean cell concentration and 0.3 g l^{-1} is the standard error. However, because error can be represented in a variety of ways, such as as maximum error, standard deviation, standard error, 95% error (i.e., $2\sigma_m$), and so on, the type of error being reported after the $\pm$ sign should always be stated explicitly. The units and dimensions of the mean and standard error are the same as those of x, the variable being measured.

From Eq. (3.3), the magnitude of the standard error σ_m decreases as the number of replicate measurements n increases, thus making the mean more reliable. Information about the sample size should always be provided when reporting the outcome of statistical analysis. In practice, a compromise is usually struck between the conflicting demands of precision and the time and expense of experimentation; sometimes it is impossible to make a large number of replicate measurements. When substantial improvement in the accuracy of the mean and standard error is required, this is generally achieved more effectively by improving the intrinsic accuracy of the measurement rather than by just taking a multitude of repeated readings.

The accuracy of standard deviations and standard errors determined using Eq. (3.2) or Eq. (3.3) is relatively poor. In other words, if many sets of replicate measurements were obtained experimentally, there would be a wide variation in the values of σ and σ_m determined for each set. For typical experiments in biotechnology and bioprocessing where only three or four replicate measurements are made, the error in σ is about 50% [3]. Therefore, when reporting standard deviations and standard errors, only one or perhaps two significant figures are sufficient.

Some data sets contain one or more points that deviate substantially from the others, more than is expected from 'normal' random experimental error. These points known as *outliers* have large residuals and, therefore, strongly influence the results of statistical analysis. It is tempting to explain outliers by thinking of unusual mistakes that could have happened during their measurement, and then eliminating them from the data set. This is a dangerous temptation, however; once the possibility of eliminating data is admitted, it is difficult to know where to stop. It is often inappropriate to eliminate outliers, as they may be legitimate experimental results reflecting the true behaviour of the system. Situations like this call for judgement on the part of the experimenter, based on knowledge of the measurement technique. Guidance can also be obtained from theoretical considerations. As noted before, for a normal distribution of errors in data, the probability of individual readings falling outside the range $\bar{x} \pm 2\sigma$ is about 5%. Further, the probability of points lying outside the range $\bar{x} \pm 3\sigma$ is 0.3%. We could say, therefore, that outliers outside the $\pm 3\sigma$ limit are very likely to be genuine mistakes and candidates for elimination. Confidence in applying this criterion will depend on the accuracy with which we know σ, which depends in turn on the total number of measurements taken. Handling outliers is tricky business; it is worth remembering that some of the greatest discoveries in science have followed from the consideration of outliers and the real reasons for their occurrence.

EXAMPLE 3.1 MEAN, STANDARD DEVIATION, AND STANDARD ERROR

The final concentration of L-lysine produced by a regulatory mutant of *Brevibacterium lactofermentum* is measured 10 times. The results in g l^{-1} are 47.3, 51.9, 52.2, 51.8, 49.2, 51.1, 52.4, 47.1, 49.1, and 46.3. How should the lysine concentration be reported?

Solution

For this sample, $n = 10$. From Eq. (3.1):

$$\bar{x} = \frac{47.3 + 51.9 + 52.2 + 51.8 + 49.2 + 51.1 + 52.4 + 47.1 + 49.1 + 46.3}{10} = 49.84 \text{ g l}^{-1}$$

Substituting this result into Eq. (3.2) gives:

$$\sigma = \sqrt{\frac{\begin{array}{c}(47.3 - 49.84)^2 + (51.9 - 49.84)^2 + (52.2 - 49.84)^2 + (51.8 - 49.84)^2 \\ + (49.2 - 49.84)^2 + (51.1 - 49.84)^2 + (52.4 - 49.84)^2 + (47.1 - 49.84)^2 \\ + (49.1 - 49.84)^2 + (46.3 - 49.84)^2\end{array}}{9}}$$

$$\sigma = \sqrt{\frac{49.24}{9}} = 2.34 \text{ g l}^{-1}$$

Applying Eq. (3.3) for standard error gives:

$$\sigma_m = \frac{2.34 \text{ g l}^{-1}}{\sqrt{10}} = 0.74 \text{ g l}^{-1}$$

Therefore, from 10 repeated measurements, the lysine concentration is 49.8 ± 0.7 g l^{-1}, where $\pm$ indicates standard error.

Methods for combining standard deviations and standard errors in calculations are discussed elsewhere [2, 4, 7, 8]. Remember that standard statistical analysis does not account for systematic error; parameters such as the mean, standard deviation, and standard error are useful only if the measurement error is random. *The effect of systematic error cannot be minimised using standard statistical analysis or by collecting repeated measurements.*

3.2 PRESENTATION OF EXPERIMENTAL DATA

Experimental data are often collected to examine relationships between variables. The role of these variables in the experimental process is clearly defined. *Dependent variables* or *response variables* are uncontrolled during the experiment; dependent variables are measured as they respond to changes in one or more *independent variables* that are controlled or fixed. For example, if we wanted to determine how UV radiation affects the frequency of mutation in a bacterial culture, radiation dose would be the independent variable and number of mutant cells the dependent variable.

There are three general methods for presenting data representing relationships between dependent and independent variables:

- Tables
- Graphs
- Equations

Each has its own strengths and weaknesses. Tables listing data have the highest accuracy, but can easily become too long and the overall result or trend of the data may not be readily discernable. Graphs or plots of data create immediate visual impact because the relationships between the variables are represented directly. Graphs also allow easy interpolation of data, which can be difficult using tables. By convention, independent variables are plotted along the *abscissa* (the x-axis) in graphs, while one or more dependent variables are plotted along the *ordinate* (y-axis). Plots show at a glance the general pattern of the data, and can help identify whether there are anomalous points. It is good practice, therefore, to plot raw experimental data as they are being measured. In addition, graphs can be used directly for quantitative data analysis.

As well as tables and graphs, equations or *mathematical models* can be used to represent phenomena. For example, balanced growth of microorganisms is described using the model:

$$x = x_0\, e^{\mu t} \tag{3.4}$$

where x is the cell concentration at time t, x_0 is the initial cell concentration, and μ is the specific growth rate. Mathematical models can be either mechanistic or empirical. *Mechanistic models* are founded on theoretical assessment of the phenomenon being measured. An example is the Michaelis–Menten equation for enzyme reaction:

$$v = \frac{v_{\max} s}{K_{\rm m} + s} \tag{3.5}$$

where v is the rate of reaction, $v_{\max}$ is the maximum rate of reaction, $K_{\rm m}$ is the Michaelis constant, and s is the substrate concentration. The Michaelis–Menten equation is based on a loose analysis of the reactions supposed to occur during simple enzyme catalysis. On the other hand, *empirical models* are used when no theoretical hypothesis can be postulated. Empirical models may be the only feasible option for correlating data representing complex or poorly understood processes. As an example, the following correlation relates the power required to stir aerated liquids to that required in nonaerated systems:

$$\frac{P_g}{P_0} = 0.10\left(\frac{F_g}{N_i V}\right)^{-0.25}\left(\frac{N_i^2 D_i^4}{g W_i V^{2/3}}\right)^{-0.20} \tag{3.6}$$

In Eq. (3.6) P_g is the power consumption with sparging, P_0 is the power consumption without sparging, F_g is volumetric gas flow rate, N_i is stirrer speed, V is liquid volume, D_i is impeller diameter, g is gravitational acceleration, and W_i is impeller blade width. There is no easy theoretical explanation for this relationship; instead, the equation is based on many observations using different impellers, gas flow rates, and rates of stirring. Equations such as Eq. (3.6) are short, concise means for communicating the results of a large number of experiments. However, they are one step removed from the raw data and can be only an approximate representation of all the information collected.

3.3 DATA ANALYSIS

Once experimental data are collected, what we do with them depends on the information being sought. Data are generally collected for one or more of the following reasons:

- To visualise the general trend of influence of one variable on another
- To test the applicability of a particular model to a process
- To estimate the value of coefficients in mathematical models of the process
- To develop new empirical models

Analysis of data would be simplified enormously if each datum point did not contain error. For example, after an experiment in which the apparent viscosity of a mycelial broth is measured as a function of temperature, if all points on a plot of viscosity versus temperature lay perfectly along a line and there were no scatter, it would be very easy to determine unequivocally the relationship between the variables. In reality, however, procedures for data analysis must be linked closely with statistical mathematics to account for random errors in measurement.

Despite its importance, a detailed treatment of statistical analysis is beyond the scope of this book; there are entire texts devoted to the subject. Rather than presenting methods for data analysis as such, the following sections discuss some of the ideas behind the interpretation of experimental results. Once the general approach is understood, the actual procedures involved can be obtained from the references listed at the end of this chapter.

As we shall see, interpreting experimental data requires a great deal of judgement and sometimes involves difficult decisions. Nowadays, scientists and engineers use computers or calculators equipped with software for data processing. These facilities are very convenient and have removed much of the tedium associated with statistical analysis. There is a danger, however, that software packages can be applied without appreciation of the assumptions inherent to the analysis or its mathematical limitations. Without this knowledge, the user cannot know how valuable or otherwise are the generated results.

As already mentioned in Section 3.1.5, standard statistical methods consider only random error, not systematic error. In practical terms, this means that most procedures for data processing are unsuitable when errors are due to poor instrument calibration, repetition of the same mistakes in measurement, or preconceived ideas about the expected result that may influence the measurement technique. All effort must be made to eliminate these types of error before treating the data. As also noted in Section 3.1.5, the reliability of results from statistical analysis improves as the number of replicate measurements is increased. No amount of sophisticated mathematical or other type of manipulation can make up for sparse, inaccurate data.

3.3.1 Trends

Consider the data plotted in Figure 3.3 representing the consumption of glucose during batch culture of plant cells. If there were serious doubt about the trend of the data, we could present the plot as a scatter of individual points without any lines drawn through them. Sometimes data are simply connected using line segments as shown in Figure 3.3(a); the problem with this representation is that it suggests that the ups and downs of glucose concentration are real. If, as with these data, we are assured that there is a progressive downward trend in sugar concentration despite the occasional apparent increase, we could *smooth* the data by drawing a curve through the points as shown in Figure 3.3(b).

Smoothing moderates the effects of experimental error. By drawing a particular curve we are indicating that, although the scatter of points is considerable, we believe the actual behaviour of the system is smooth and continuous, and that all of the data without experimental error lie on that line. Usually there is great flexibility as to where the smoothing curve is placed, and several questions arise. To which points should the curve pass closest? Should all the datum points be included, or are some points clearly in error or

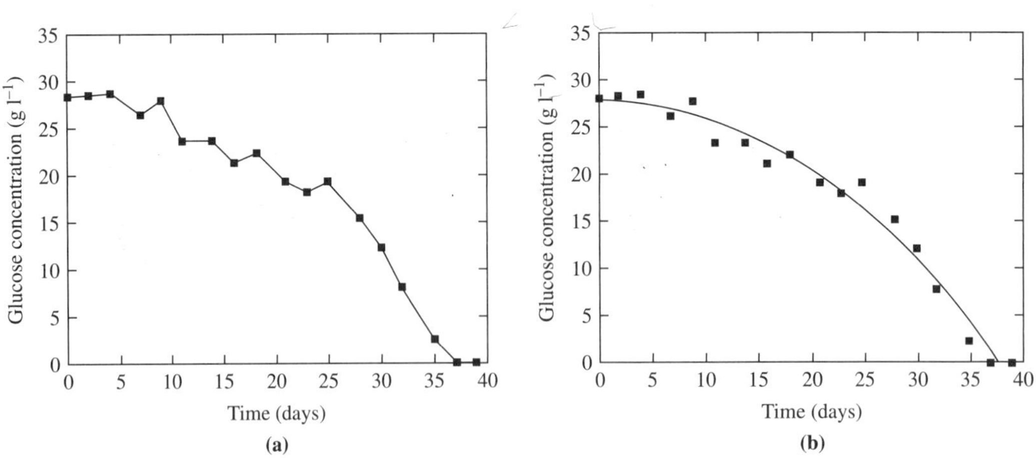

FIGURE 3.3 Glucose concentration during batch culture of plant cells: (a) data connected directly by line segments; (b) data represented by a smooth curve.

expected to lie outside of the general trend? It soon becomes apparent that many equally acceptable curves can be drawn through the data.

Various techniques are available for smoothing. A smooth line can be drawn freehand or with French or flexible curves and other drafting equipment; this is called *hand smoothing*. Procedures for minimising bias during hand smoothing can be applied; some examples are discussed further in Chapter 12. The danger involved in smoothing manually—that we tend to smooth the expected response into the data—is well recognised. Another method is to use a computer software package; this is called *machine smoothing*. Computer routines, by smoothing data according to preprogrammed mathematical or statistical principles, eliminate the subjective element but are still capable of introducing bias into the results. For example, abrupt changes in the trend of data are generally not recognised using these techniques. The advantage of hand smoothing is that judgements about the significance of individual datum points can be taken into account.

Choice of curve is critical if smoothed data are to be applied in subsequent analysis. For example, the data of Figure 3.3 may be used to calculate the *rate* of glucose consumption as a function of time; procedures for this type of analysis are described further in Chapter 12. In rate analysis, different smoothing curves can lead to significantly different results. Because final interpretation of the data depends on decisions made during smoothing, it is important to minimise any errors introduced. One obvious way of doing this is to take as many readings as possible. When smooth curves are drawn through too few points, it is very difficult to justify the smoothing process.

3.3.2 Testing Mathematical Models

Most applications of data analysis involve correlating measured data with existing mathematical models. The model proposes some functional relationship between two or more variables; our primary objective is to compare the properties of the experimental system with those of the model.

As an example, consider Figure 3.4(a), which shows the results from experiments in which rates of heat production and oxygen consumption were measured for two different microbial cultures. Although there is considerable scatter in these data, we could be led to believe that the relationship between rate of heat production and rate of oxygen consumption is linear, as indicated by the straight line in Figure 3.4(a). However, there are an infinite number of ways to represent any set of data; how do we know that this linear relationship is the best? For instance, we might consider whether the data could be fitted by the curve shown in Figure 3.4(b). This nonlinear model seems to follow the data reasonably well; should we conclude that there is a more complex, nonlinear relationship between heat production and oxygen consumption?

Ultimately, we cannot know if a particular relationship holds between variables. This is because we can only test a selection of possible relationships and determine which *of them* fits closest to the data. We can determine which model, linear or nonlinear, is the *better* representation of the data in Figure 3.4, but we can never conclude that the relationship between the variables is *actually* linear or nonlinear. This fundamental limitation of data analysis has important consequences and must be accommodated in our approach. We

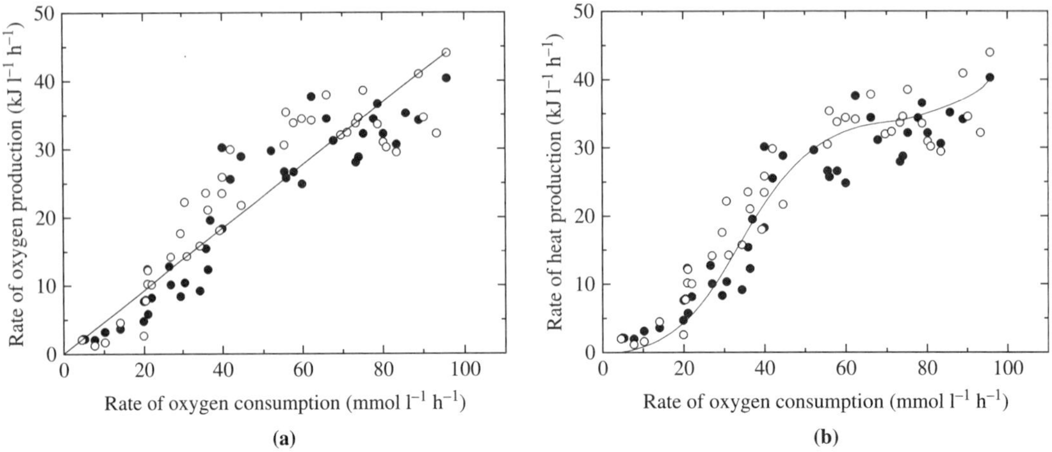

FIGURE 3.4 Experimental data for the rates of heat production and oxygen consumption in two different microbial cultures: (○) *Escherichia coli*, and (•) *Schizosaccharomyces pombe*. (a) The data are fitted with a straight line indicating a linear relationship between the two variables. (b) The same data as in (a) fitted with a nonlinear curve indicating a more complex relationship between the variables.

start off with a hypothesis about how the parameters are related and use data to determine whether this hypothesis is supported. A basic tenet in the philosophy of science is that it is only possible to *disprove* hypotheses by showing that experimental data do not conform to the model. The idea that the primary business of science is to falsify theories, not verify them, was developed last century by the Austrian philosopher, Karl Popper. Popper's philosophical excursions into the meaning of scientific truth make extremely interesting reading (e.g., [9, 10]); his theories have direct application in analysis of measured data. Using experiments, we can never deduce with absolute certainty the physical relationships between variables. The language we use to report the results of data analysis must reflect these limitations; particular models used to correlate data cannot be described as 'correct' or 'true' descriptions of the system but only as 'satisfactory' or 'adequate' for our purposes, keeping in mind the measurement precision.

3.3.3 Goodness of Fit: Least-Squares Analysis

Determining how well data conform to a particular model requires numerical procedures. Generally, these techniques rely on measurement of the deviations or residuals of each datum point from the curve or line representing the model being tested. For example, residuals after correlating cell plasmid content with growth rate using a linear model are shown by the dashed lines in Figure 3.5. A curve or line producing small residuals is required for a good fit of the data.

A popular technique for locating the line or curve that minimises the residuals is *least-squares analysis*. This statistical procedure is based on minimising the *sum of squares of the residuals*. There are several variations of the procedure: Legendre's method minimises the

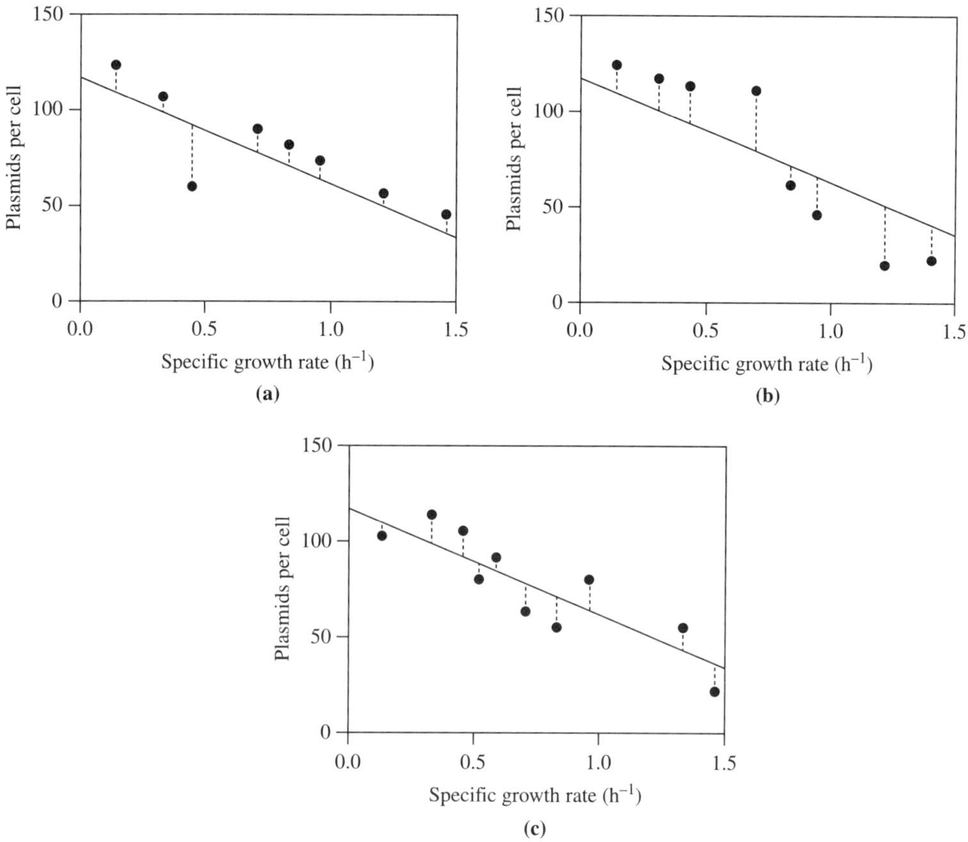

FIGURE 3.5 Residuals in plasmid content after fitting a straight line to experimental data. The residuals are shown as dashed lines.

sum of squares of residuals of the dependent variable; Gauss's and Laplace's methods minimise the sum of squares of weighted residuals where the weighting factors depend on the scatter of replicate datum points. Each method gives different results; it should be remembered that the curve of 'best' fit is ultimately a matter of opinion. For example, by minimising the sum of squares of the residuals, least-squares analysis could produce a curve that does not pass close to particular datum points known beforehand to be more accurate than the rest. Alternatively, we could choose to define the best fit as that which minimises the absolute values of the residuals, or the sum of the residuals raised to the fourth power. The decision to use the sum of squares is an arbitrary one; many alternative approaches are equally valid mathematically.

As well as minimising the residuals, other factors must be taken into account when correlating data. First, the curve used to fit the data should create approximately equal numbers of positive and negative residuals. As shown in Figure 3.5(a), when there are more positive than negative deviations, even though the sum of the residuals is relatively small, the line representing the data cannot be considered a good fit. The fit is also poor when, as shown in Figure 3.5(b), the positive residuals occur mainly at low values of the independent variable while the negative residuals occur at high values. There should be no significant correlation of the residuals with either the dependent or independent variable. The best straight-line fit is shown in Figure 3.5(c); the residuals are relatively small and well distributed in both positive and negative directions, and there is no relationship between the residuals and either variable.

In some data sets, there may be one or more outlier points that deviate substantially from the values predicted by the model. The large residuals associated with these points exert a strong influence on the outcome of regression methods using the sum-of-squares approach. It is usually inappropriate to discard outlier points; they may represent legitimate results and could possibly be explained and fitted using an alternative model not yet considered. The best way to handle outliers is to analyse the data with and without the aberrant values to make sure their elimination does not influence discrimination between models. It must be emphasised that only one point at a time and only very rare datum points, if any, should be eliminated from data sets.

Measuring individual residuals and applying least-squares analysis could be very useful for determining which of the two curves in Figures 3.4(a) and 3.4(b) fits the data more closely. However, as well as mathematical considerations, other factors can influence the choice of model for experimental data. Consider again the data of Figures 3.4(a) and 3.4(b). Unless the fit obtained with the nonlinear model were very much improved compared with the linear model in terms of the magnitude and distribution of the residuals, we might prefer the straight-line correlation because it is simple, and because it conforms with what we know about microbial metabolism and the thermodynamics of respiration. It is difficult to find a credible theoretical justification for representing the relationship with an oscillating curve, so we could be persuaded to reject the nonlinear model even though it fits the data reasonably well. Choosing between models on the basis of supposed mechanism requires a great deal of judgement. Since we cannot know for sure what the relationship is between the two parameters, choosing between models on the basis of supposed mechanism brings in an element of bias. This type of presumptive judgement is the reason that it is so difficult to overturn established scientific theories; even if data are available to support a new hypothesis, there is a tendency to reject it because it does not agree with accepted theory. Nevertheless, if we wanted to fly in the face of convention and argue that an oscillatory relationship between rates of heat evolution and oxygen consumption is more reasonable than a straight-line relationship, we would undoubtedly have to support our claim with more evidence than the data shown in Figure 3.4(b).

3.3.4 Linear and Nonlinear Models

A straight line can be represented by the equation:

$$y = Ax + B \tag{3.7}$$

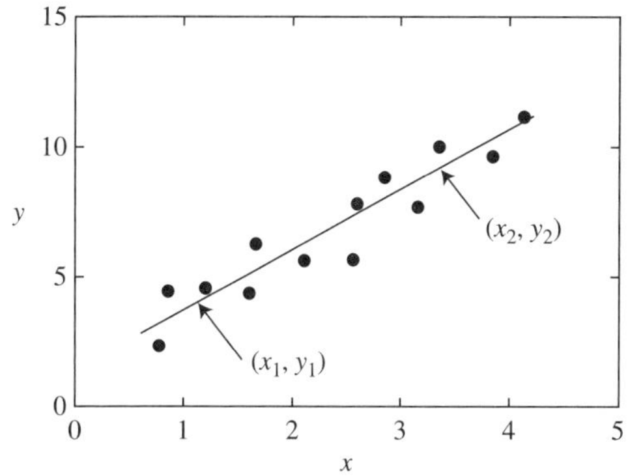

FIGURE 3.6 Straight-line correlation for calculation of model parameters.

where A is the *slope* and B is the *intercept* of the straight line on the ordinate. A and B are also called the *coefficients, parameters,* or *adjustable parameters* of Eq. (3.7). Once a straight line is drawn, A is found by taking any two points (x_1, y_1) and (x_2, y_2) on the line and calculating:

$$A = \frac{y_2 - y_1}{x_2 - x_1} \tag{3.8}$$

As indicated in Figure 3.6, (x_1, y_1) and (x_2, y_2) are points *on the line* through the data; *they are not measured datum points*. Once A is known, B is calculated as:

$$B = y_1 - Ax_1 \qquad \text{or} \qquad B = y_2 - Ax_2 \tag{3.9}$$

Suppose we measure n pairs of values of two variables, x and y, and a plot of the dependent variable y versus the independent variable x suggests a straight-line relationship. In testing correlation of the data with Eq. (3.7), changing the values of A and B will affect how well the model fits the data. Values of A and B giving the best straight line are determined by *linear regression* or *linear least-squares analysis*. This procedure is one of the most frequently used in data analysis; linear regression routines are part of many computer packages and are available on hand-held calculators. Linear regression methods fit data by finding the straight line that minimises the sum of squares of the residuals. Details of the method can be found in statistics texts (e.g., [1, 4, 6, 8, 11]).

Because linear regression is so accessible, it can be applied readily without proper regard for its appropriateness or the assumptions incorporated in its method. Unless the following points are considered before using regression analysis, biased estimates of parameter values will be obtained.

1. Least-squares analysis applies only to data containing random errors.
2. The variables x and y must be independent.
3. Simple linear regression methods are restricted to the special case of all uncertainty being associated with one variable. If the analysis uses a regression of y on x, then y should be the variable involving the largest errors. More complicated techniques are required to deal with errors in x and y simultaneously.

4. Simple linear regression methods assume that each datum point has equal significance. Modified procedures must be used if some points are considered more or less important than others, or if the line must pass through some specified point (e.g., the origin).
5. Each point is assumed to be equally precise, that is, the standard deviation or random error associated with individual readings should be the same for all points. In experiments, the degree of fluctuation in the response variable often changes within the range of interest; for example, measurements may be more or less affected by instrument noise at the high or low end of the scale, or data collected at the beginning of an experiment may have smaller or larger errors compared with those measured at the end. Under these conditions, simple least-squares analysis is flawed.
6. As already mentioned with respect to Figures 3.5(a) and 3.5(b), positive and negative residuals should be approximately evenly distributed, and the residuals should be independent of both x and y variables.

Correlating data with straight lines is a relatively easy form of data analysis. When experimental data deviate markedly from a straight line, correlation using nonlinear models is required. It is usually more difficult to decide which model to test and to obtain parameter values when data do not follow linear relationships. As an example, consider the growth of *Saccharomyces cerevisiae* yeast, which is expected to follow the nonlinear model of Eq. (3.4). We could attempt to check whether measured cell concentration data are consistent with Eq. (3.4) by plotting the values on linear graph paper as shown in Figure 3.7(a). The data appear to exhibit an exponential response typical of simple growth kinetics, but it is not certain that an exponential model is appropriate. It is also difficult to ascertain some of the finer points of the culture behaviour using linear coordinates—for instance, whether the initial points represent a lag phase or whether exponential growth commenced immediately. Furthermore, the value of μ for this culture is not readily discernible from Figure 3.7(a).

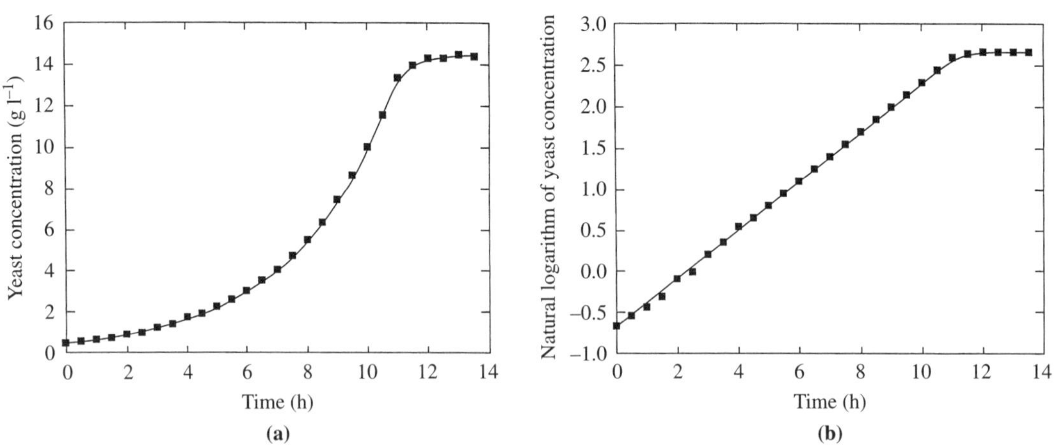

FIGURE 3.7 Growth curve for *Saccharomyces cerevisiae*: (a) data plotted directly on linear graph paper; (b) linearisation of growth data by plotting the logarithms of cell concentration versus time.

TABLE 3.1 Methods for Plotting Data as Straight Lines

$y = Ax^n$	Plot y vs. x on logarithmic coordinates
$y = A + Bx^2$	Plot y vs. x^2 on linear coordinates
$y = A + Bx^n$	First obtain A as the intercept on a plot of y vs. x on linear coordinates, then plot $(y - A)$ vs. x on logarithmic coordinates
$y = B^x$	Plot y vs. x on semi-logarithmic coordinates
$y = A + (B/x)$	Plot y vs. $1/x$ on linear coordinates
$y = \dfrac{1}{Ax + B}$	Plot $1/y$ vs. x on linear coordinates
$y = \dfrac{x}{A + Bx}$	Plot x/y vs. x, or $1/y$ vs. $1/x$, on linear coordinates
$y = 1 + (Ax^2 + B)^{1/2}$	Plot $(y - 1)^2$ vs. x^2 on linear coordinates
$y = A + Bx + Cx^2$	Plot $\dfrac{y - y_n}{x - x_n}$ vs. x on linear coordinates, where (x_n, y_n) are the coordinates of any point on a smooth curve through the experimental points
$y = \dfrac{x}{A + Bx} + C$	Plot $\dfrac{x - x_n}{y - y_n}$ vs. x on linear coordinates, where (x_n, y_n) are the coordinates of any point on a smooth curve through the experimental points

A convenient approach to this problem is to convert the model equation into a linear form. Following the rules for logarithms outlined in Appendix E, taking the natural logarithm of both sides of Eq. (3.4) gives:

$$\ln x = \ln x_0 + \mu t \tag{3.10}$$

Equation (3.10) indicates a linear relationship between $\ln x$ and t, with intercept $\ln x_0$ and slope μ. Accordingly, if Eq. (3.4) is a good model for yeast growth, a plot of the natural logarithm of cell concentration versus time should, during the growth phase, yield a straight line. The results of this linear transformation are shown in Figure 3.7(b). All points before stationary phase appear to lie on a straight line, suggesting the absence of a lag phase. The value of μ is also readily calculated from the slope of the line. Graphical linearisation has the advantage that gross deviations from the model are immediately evident upon visual inspection. Other nonlinear relationships and suggested methods for yielding straight-line plots are given in Table 3.1.

Once data have been transformed to produce straight lines, it is tempting to apply linear least-squares analysis to determine the model parameters. For the data in Figure 3.7(b), we could enter the values of time and the logarithm of cell concentration into a computer or calculator programmed for linear regression. This analysis would give us the straight line through the data that minimises the sum of squares of the residuals. Most users of linear regression choose this technique because they believe it will automatically give them an objective and unbiased analysis of their data. However, *application of linear least-squares analysis to linearised data can result in biased estimates of model parameters*. The reason is

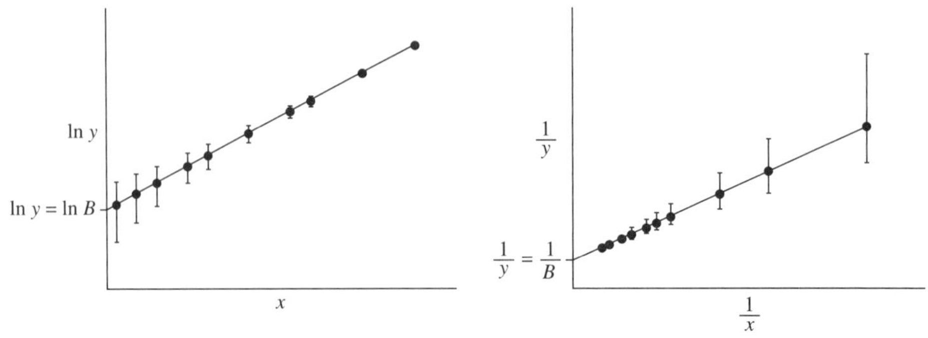

FIGURE 3.8 Transformation of constant errors in y after (a) taking logarithms or (b) inverting the data. Errors in $\ln y$ and $1/y$ vary in magnitude as the value of y changes even though the error in y is constant.

related to the assumption in least-squares analysis that each datum point has equal random error associated with it.

When data are linearised, the error structure is changed so that the distribution of errors becomes distorted. Although the error associated with each raw datum point may be approximately constant, when logarithms are calculated, the transformed errors become dependent on the magnitude of the variable. This effect is illustrated in Figure 3.8(a) where the error bars represent a constant error in y, in this case equal to $B/2$. When logarithms are taken, the resulting error in $\ln y$ is neither constant nor independent of $\ln y$; as shown, the errors in $\ln y$ become larger as $\ln y$ decreases. Similar effects also occur when data are inverted, as in some of the transformations suggested in Table 3.1. As shown in Figure 3.8(b) where the error bars represent a constant error in y of $\pm 0.05B$, small errors in y lead to enormous errors in $1/y$ when y is small; for large values of y the same errors are barely noticeable in $1/y$. When the magnitude of the errors after transformation is dependent on the value of the variable, simple least-squares analysis is compromised.

In such cases, modifications can be made to the analysis. One alternative is to apply *weighted least-squares techniques*. The usual way of doing this is to take replicate measurements of the variable, transform the data, calculate the standard deviations for the transformed variable, and then weight the values by $1/\sigma^2$. Correctly weighted linear regression often gives satisfactory parameter values for nonlinear models; details of the procedures can be found elsewhere [11, 12].

Techniques for *nonlinear regression* usually give better results than weighted linear regression. In nonlinear regression, nonlinear equations such as those in Table 3.1 are fitted directly to the data. However, determining an optimal set of parameters by nonlinear regression can be difficult, and the reliability of the results is harder to interpret. The most common nonlinear methods, such as the Gauss–Newton procedure, available as computer software, are based on gradient, search, or linearisation algorithms and use iterative solution techniques. More information about nonlinear approaches to data analysis is available in other books (e.g., [11]).

In everyday practice, simple linear least-squares methods are applied commonly to linearised data to estimate the parameters of nonlinear models. Linear regression analysis is more readily available on hand-held calculators and in graphics software packages than nonlinear routines, which are generally less easy to use and require more information about the distribution of errors in the data. Nevertheless, you should keep in mind the assumptions associated with linear regression techniques and when they are likely to be violated. A good way to see if linear least-squares analysis has resulted in biased estimates of model parameters is to replot the data and the regression curve on linear coordinates. The residuals revealed on the graph should be relatively small, randomly distributed, and independent of the variables.

3.4 GRAPH PAPER WITH LOGARITHMIC COORDINATES

Two frequently occurring nonlinear functions are the *power law*, $y = Bx^A$, and the *exponential function*, $y = Be^{Ax}$. These relationships are often presented using graph paper with logarithmic coordinates. Before proceeding, it may be necessary for you to review the mathematical rules for logarithms outlined in Appendix E.

3.4.1 Log–Log Plots

When plotted on a linear scale, some data take the form of curves 1, 2, or 3 in Figure 3.9(a), none of which are straight lines. Note that none of these curves intersects either axis except at the origin. If straight-line representation is required, we must transform the data by calculating logarithms; plots of $\log_{10}$ or $\ln y$ versus $\log_{10}$ or $\ln x$ yield straight lines, as shown in Figure 3.9(b). The best straight line through the data can be estimated using a suitable regression analysis, as discussed in Section 3.3.4.

When there are many datum points, calculating the logarithms of x and y can be time-consuming. An alternative is to use a *log–log plot*. The raw data, *not their logarithms*, are plotted directly on log–log graph paper; the resulting graph is as if logarithms were calculated to base e. Graph paper with both axes scaled logarithmically is shown in Figure 3.10; each axis in this example covers two logarithmic cycles. On log–log plots, the origin (0,0) can never be represented; this is because $\ln 0$ (or $\log_{10} 0$) is not defined.

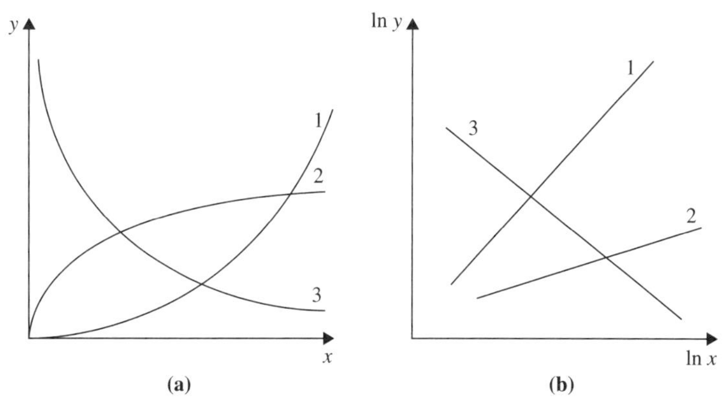

FIGURE 3.9 Equivalent curves on linear and log–log graph paper.

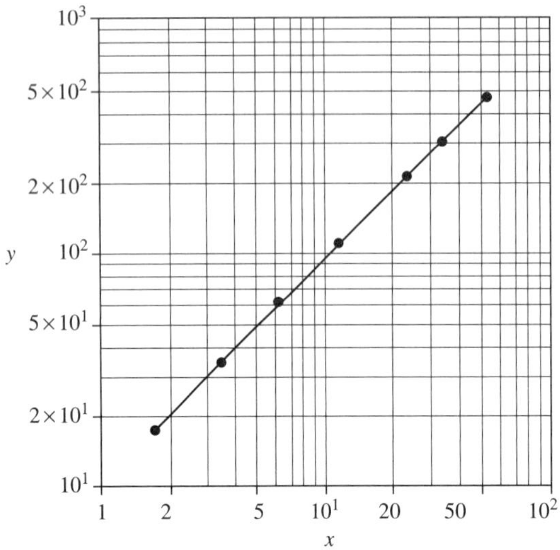

FIGURE 3.10 Log–log plot.

If you are not already familiar with log plots, some practice may be required to get used to the logarithmic scales. Notice that the grid lines on the log-scale axes in Figure 3.10 are not evenly spaced. Within a single log cycle, the grid lines start off wide apart and then become closer and closer. The number at the end of each cycle takes a value 10 times larger than the number at the beginning. For example, on the x-axis of Figure 3.10, 10^1 is 10 times 1 and 10^2 is 10 times 10^1; similarly, on the y-axis, 10^2 is 10 times 10^1 and 10^3 is 10 times 10^2. On the x-axis, 10^1 is midway between 1 and 10^2. This is because $\log_{10}10^1$ (= 1) is midway between $\log_{10}1$ (= 0) and $\log_{10}10^2$ (= 2) or, in terms of natural logs, because $\ln 10^1$ (= 2.3026) is midway between $\ln 1$ (= 0) and $\ln 10^2$ (= 4.6052). Similar relationships can be found between 10^1, 10^2, and 10^3 on the y-axis. The first grid line after 1 is 2, the first grid line after 10 is 20, not 11, the first grid line after 100 is 200, and so on. The distance between 1 and 2 is much greater than the distance between 9 and 10; similarly, the distance between 10 and 20 is much greater than the distance between 90 and 100, and so on. On logarithmic scales, the midpoint between 1 and 10 is about 3.16.

A straight line on log–log graph paper corresponds to the equation:

$$y = Bx^A \tag{3.11}$$

or

$$\ln y = \ln B + A \ln x \tag{3.12}$$

Inspection of Eq. (3.11) shows that, if A is positive, $y = 0$ when $x = 0$. Therefore, a positive value of A corresponds to either curve 1 or curve 2 passing through the origin of Figure 3.9(a). If A is negative, when $x = 0$, y is infinite; therefore, negative A corresponds to curve 3 in Figure 3.9(a), which is asymptotic to both linear axes.

The values of the parameters A and B can be obtained from a straight line on log–log paper as follows. A may be calculated in two ways:

1. A is obtained by reading from the axes the coordinates of two points on the line, (x_1, y_1) and (x_2, y_2), and making the calculation:

$$A = \frac{\ln y_2 - \ln y_1}{\ln x_2 - \ln x_1} = \frac{\ln (y_2/y_1)}{\ln (x_2/x_1)} \tag{3.13}$$

2. Alternatively, if the log–log graph paper is drawn so that the ordinate and abscissa scales are the same—that is, the distance measured with a ruler for a tenfold change in the y variable is the same as for a tenfold change in the x variable—A is the actual slope of the line. A is obtained by taking two points (x_1, y_1) and (x_2, y_2) on the line and measuring the distances between y_2 and y_1 and between x_2 and x_1 with a ruler:

$$A = \frac{\text{distance between } y_2 \text{ and } y_1}{\text{distance between } x_2 \text{ and } x_1} \tag{3.14}$$

Note that all points x_1, y_1, x_2, and y_2 used in these calculations are points *on the line* through the data; *they are not measured datum points.*

Once A is known, B is calculated from Eq. (3.12) as follows:

$$\ln B = \ln y_1 - A \ln x_1 \qquad \text{or} \qquad \ln B = \ln y_2 - A \ln x_2 \tag{3.15}$$

where $B = e^{(\ln B)}$. B can also be determined as the value of y when $x = 1$.

3.4.2 Semi-Log Plots

When plotted on linear-scale graph paper, some data show exponential rise or decay as illustrated in Figure 3.11(a). Curves 1 and 2 can be transformed into straight lines if $\log_{10} y$ or $\ln y$ is plotted against x, as shown in Figure 3.11(b).

An alternative to calculating logarithms is using a *semi-log plot*, also known as a *linear–log plot*. As shown in Figure 3.12, the raw data, *not their logarithms*, are plotted directly on semi-log

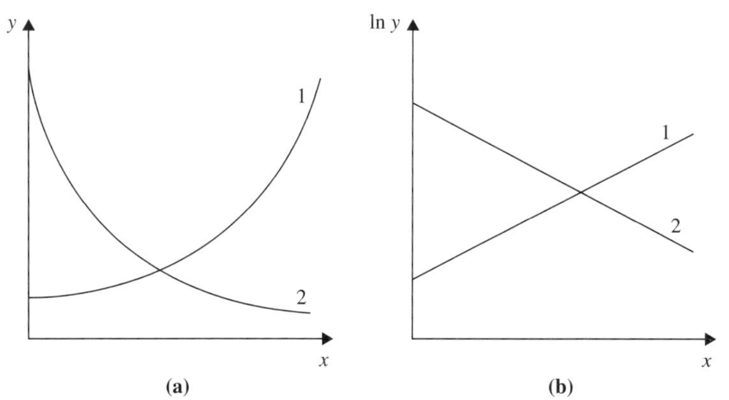

FIGURE 3.11 Graphic representation of equivalent curves on linear and semi-log graph paper.

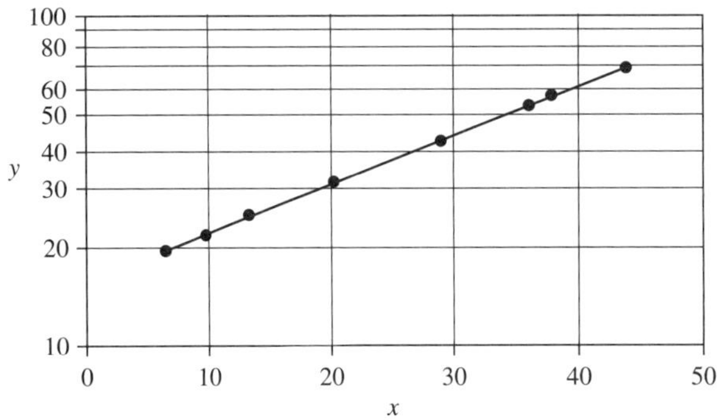

FIGURE 3.12 Semi-log plot.

paper; the resulting graph is as if logarithms of y only were calculated to base e. Zero cannot be represented on the log-scale axis of semi-log plots. In Figure 3.12, values of the dependent variable y were fitted within one logarithmic cycle from 10 to 100; semi-log paper with multiple logarithmic cycles is also available. The features and properties of the log scale used for the y-axis in Figure 3.12 are the same as those described in Section 3.4.1 for log–log plots.

A straight line on semi-log paper corresponds to the equation:

$$y = Be^{Ax} \tag{3.16}$$

or

$$\ln y = \ln B + Ax \tag{3.17}$$

Values of A and B are obtained from the straight line as follows. If two points (x_1, y_1) and (x_2, y_2) are located on the line, A is given by:

$$A = \frac{\ln y_2 - \ln y_1}{x_2 - x_1} = \frac{\ln (y_2/y_1)}{x_2 - x_1} \tag{3.18}$$

B is the value of y at $x = 0$ (i.e., B is the intercept of the line at the ordinate). Alternatively, once A is known, B can be determined as follows:

$$\ln B = \ln y_1 - Ax_1 \quad \text{or} \quad \ln B = \ln y_2 - Ax_2 \tag{3.19}$$

B is calculated as $e^{(\ln B)}$.

EXAMPLE 3.2 CELL GROWTH DATA

Data for cell concentration x versus time t are plotted on semi-log graph paper. Points ($t_1 = 0.5$ h, $x_1 = 3.5$ g l^{-1}) and ($t_2 = 15$ h, $x_2 = 10.6$ g l^{-1}) fall on a straight line passing through the data.

(a) Determine an equation relating x and t.
(b) What is the value of the specific growth rate for this culture?

Solution

(a) A straight line on semi-log graph paper means that x and t can be correlated with the equation $x = Be^{At}$. A and B are calculated using Eqs. (3.18) and (3.19):

$$A = \frac{\ln 10.6 - \ln 3.5}{15 - 0.5} = 0.076$$

and

$$\ln B = \ln 10.6 - (0.076)\,(15) = 1.215$$

or

$$B = 3.37$$

Therefore, the equation for cell concentration as a function of time is:

$$x = 3.37\, e^{0.076t}$$

This result should be checked by, for example, substituting $t_1 = 0.5$:

$$Be^{At_1} = 3.37\, e^{(0.076)(0.5)} = 3.5 = x_1$$

(b) After comparing the empirical equation obtained for x with Eq. (3.4), the specific growth rate μ is 0.076 h^{-1}.

3.5 GENERAL PROCEDURES FOR PLOTTING DATA

Axes on plots must be labelled for the graph to carry any meaning. The units associated with all physical variables should be stated explicitly. If more than one curve is plotted on a single graph, each curve must be identified with a label.

It is good practice to indicate the precision of data on graphs using *error bars*. As an example, consider the data listed in Table 3.2 for monoclonal antibody concentration as a function of medium flow rate during continuous culture of hybridoma cells in a stirred fermenter. The flow rate was measured to within ±0.02 litres per day. Measurement of

TABLE 3.2 Antibody Concentration during Continuous Culture of Hybridoma Cells

Flow rate (l d^{-1})	Antibody concentration (μg ml^{-1})
0.33	75.9
0.40	58.4
0.52	40.5
0.62	28.9
0.78	22.0
1.05	11.5

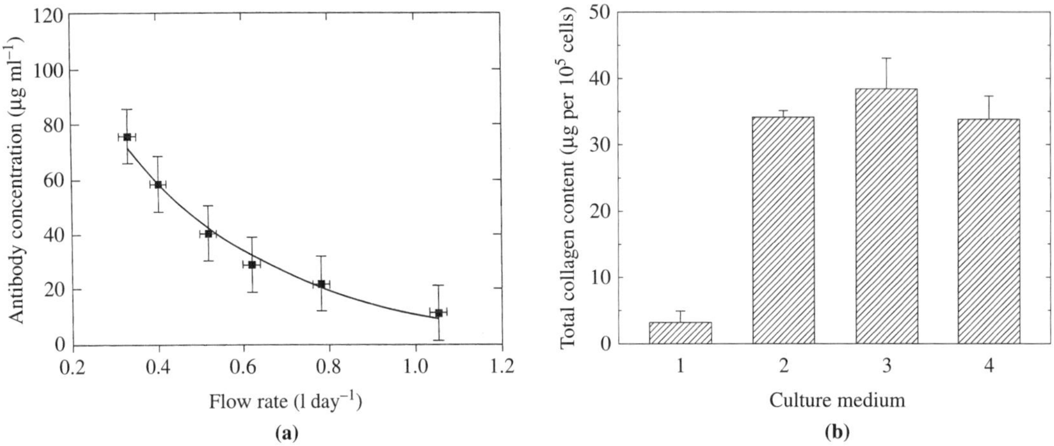

FIGURE 3.13 (a) Error bars indicating the range of values for antibody concentration and medium flow rate measured during continuous culture of hybridoma cells. (b) Error bars indicating the standard error of the mean calculated from triplicate measurements of total collagen content in human tissue-engineered cartilage produced using four different culture media. For each culture medium, the number of replicate measurements $n = 3$.

antibody concentration was more difficult and somewhat imprecise; these values are estimated to involve errors of $\pm 10\ \mu g\ ml^{-1}$. The errors associated with the data are indicated in Figure 3.13(a) using error bars to show the possible range of each variable. When replicate experiments and measurements are performed, error bars can also be used to indicate standard error about the mean. As shown in the bar graph of Figure 3.13(b), data for the total collagen content of tissue-engineered human cartilage produced using four different culture media are plotted with error bars representing the magnitude of the standard errors for each mean value (Section 3.1.5).

3.6 PROCESS FLOW DIAGRAMS

This chapter is concerned with ways of presenting and analysing data. Because of the complexity of large-scale manufacturing processes, communicating information about these systems requires special methods. *Flow diagrams* or *flow sheets* are simplified pictorial representations of processes used to present relevant process information and data. Flow sheets vary in complexity from simple block diagrams to highly complex schematic drawings showing main and auxiliary process equipment such as pipes, valves, pumps, and bypass loops.

Figure 3.14 is a simplified process flow diagram showing the major operations for production of the antibiotic, bacitracin. This qualitative flow sheet indicates the flow of materials, the sequence of process operations, and the principal equipment in use. When flow diagrams are applied in calculations, the operating conditions, masses, and concentrations of material handled by the process are also specified. An example is Figure 3.15, which

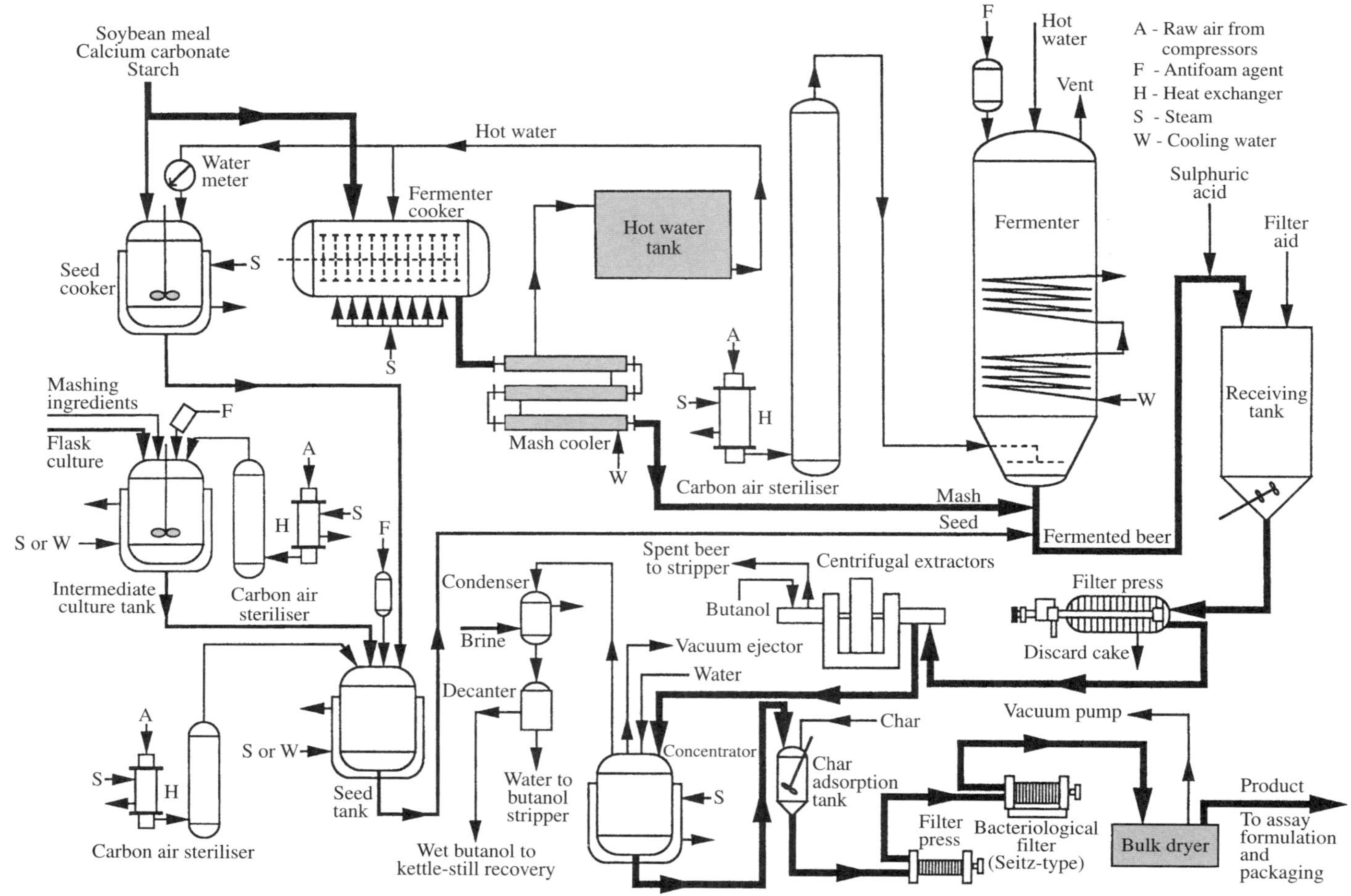

FIGURE 3.14 Process flow sheet showing the major operations for production of bacitracin.
Reprinted with permission from G.C. Inskeep, R.E. Bennett, J.F. Dudley, and M.W. Shepard, 1951, Bacitracin: Product of biochemical engineering, Ind. Eng. Chem. *43, 1488–1498. Copyright 1951, American Chemical Society.*

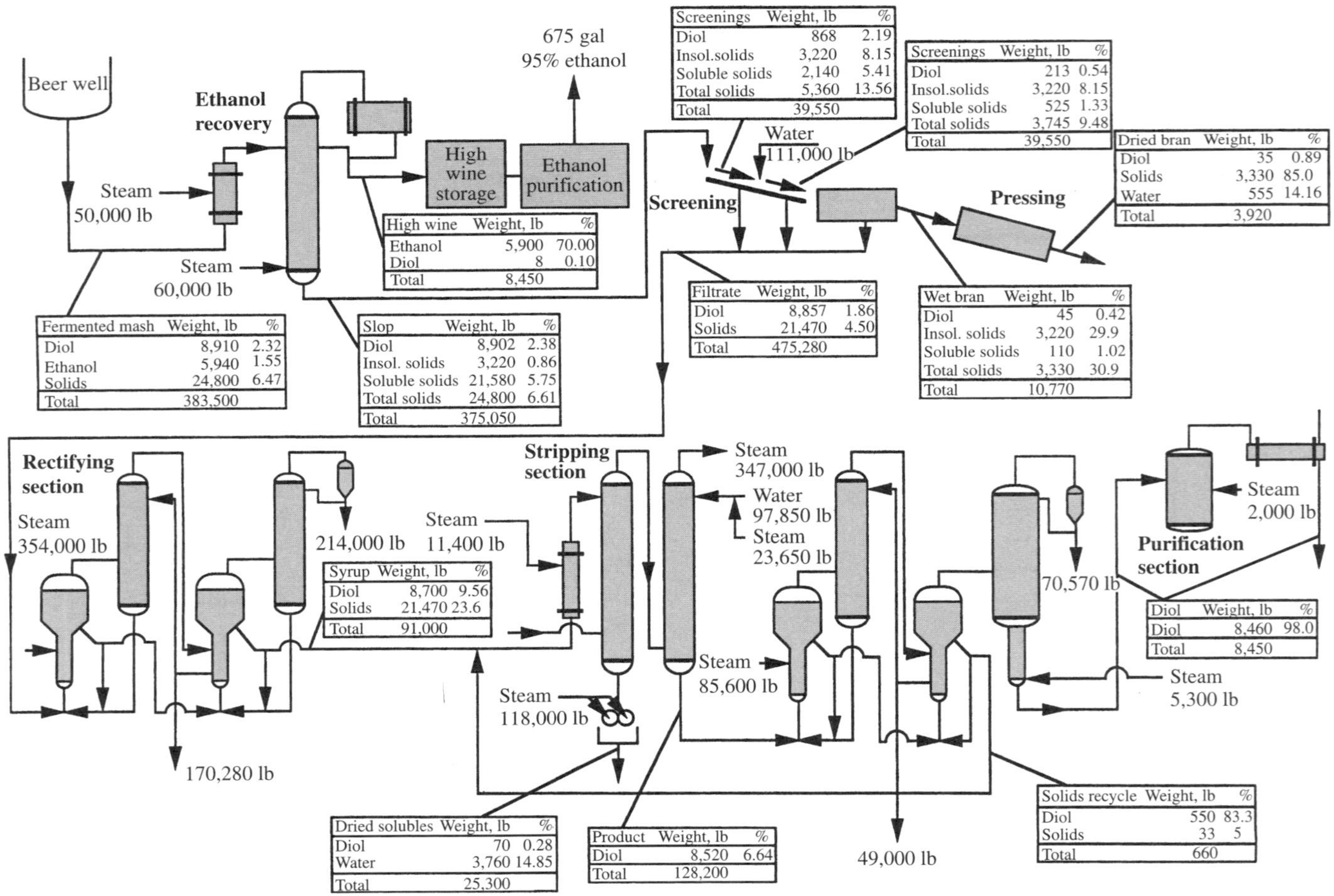

FIGURE 3.15 Quantitative flow sheet for the downstream processing of 2,3-butanediol based on fermentation of 1000 bushels of wheat per day by *Aerobacillus polymyxa*.
From J.A. Wheat, J.D. Leslie, R.V. Tomkins, H.E. Mitton, D.S. Scott, and G.A. Ledingham, 1948, Production and properties of 2,3-butanediol, XXVIII: Pilot plant recovery of levo-*2,3-butanediol from whole wheat mashes fermented by* Aerobacillus polymyxa, Can. J. Res. *26F, 469–496.*

represents the operations used to recover 2,3-butanediol produced by fermentation of whole wheat mash. The quantities and compositions of streams undergoing processes such as distillation, evaporation, screening, and drying are shown to allow calculation of product yields and energy costs.

Detailed engineering flow sheets such as Figure 3.16 are useful for plant construction work and troubleshooting because they show all piping, valves, drains, pumps, and safety equipment. Standard symbols are adopted to convey the information as concisely as possible. Figure 3.16 represents a pilot-scale fermenter with separate vessels for antifoam, acid, and alkali. All air, medium inlet, and harvest lines are shown, as are the steam and condensate drainage lines for *in situ* steam sterilisation of the entire apparatus.

In addition to those illustrated here, other specialised types of flow diagram are used to specify instrumentation for process control networks in large-scale processing plants, and for utilities such as steam, water, fuel, and air supplies. We will not be applying complicated or detailed diagrams such as Figure 3.16 in our analysis of bioprocessing; their use is beyond the scope of this book. However, as simplified versions of flow diagrams are extremely valuable, especially for material and energy balance calculations, we will be applying block-diagram flow sheets in Chapters 4 through 6 for this purpose. You should become familiar with flow diagrams for showing data and other process information.

SUMMARY OF CHAPTER 3

This chapter covers a range of topics related to data presentation and analysis. After studying Chapter 3 you should:

- Understand use of *significant figures*
- Know the types of *error* that affect the accuracy of experimental data, and which errors can be accounted for using statistical techniques
- Be able to report the results of replicate measurements in terms of the *mean* and *standard error*
- Understand the fundamental limitations associated with use of experimental data for testing mathematical models
- Be familiar with *least-squares analysis* and its assumptions
- Be able to analyse linear plots and determine model parameters
- Understand how simple nonlinear models can be linearised to obtain straight-line plots;
- Be able to use *log* and *semi-log graph paper* with ease
- Be familiar with simple *process flow diagrams*

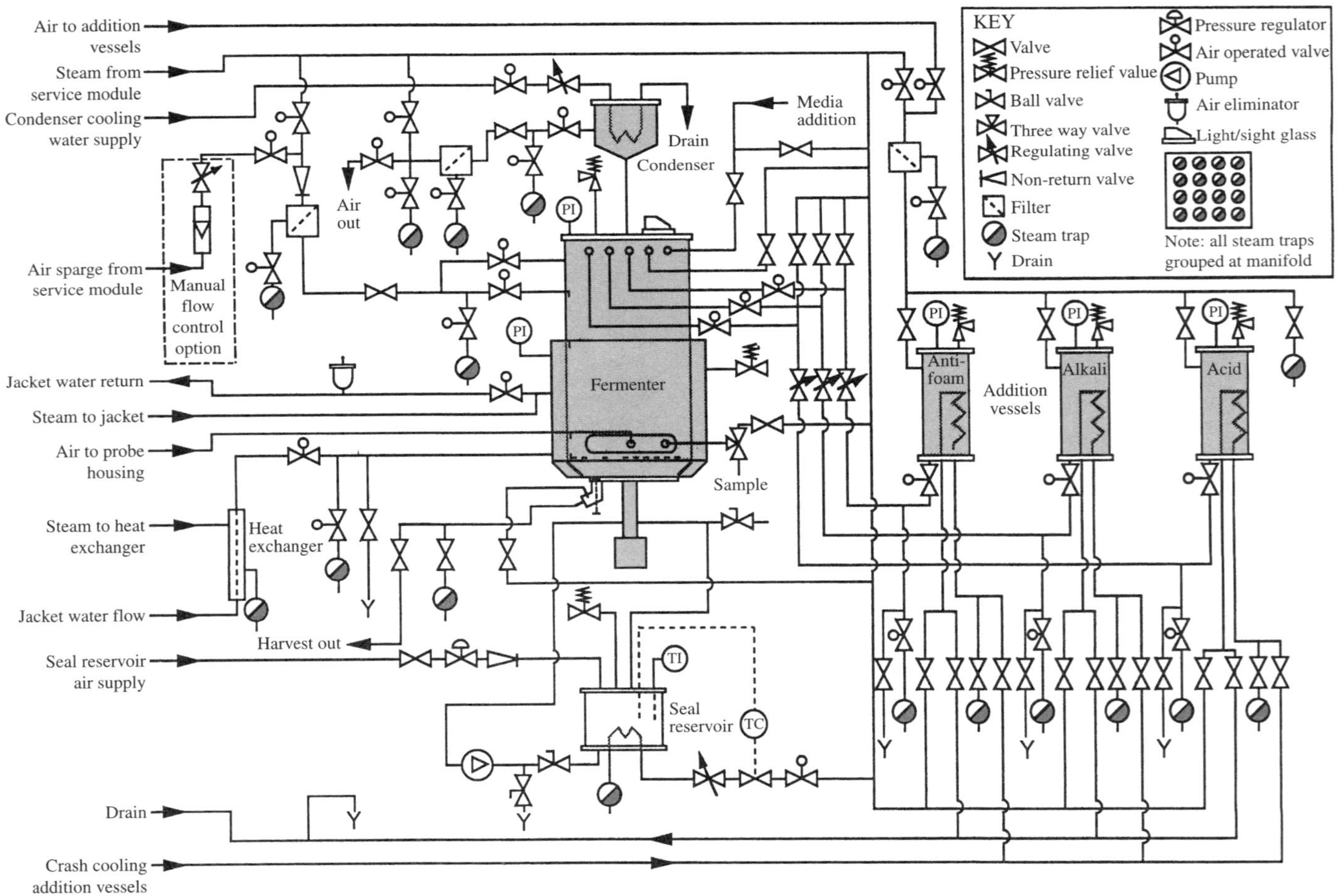

FIGURE 3.16 Detailed equipment diagram for pilot-plant fermentation system.
Reproduced with permission from LH Engineering Ltd., a member of the Inceltech Group of companies. Copyright 1983.

PROBLEMS

3.1 Combination of errors

The oxygen mass transfer coefficient in fermentation vessels is determined from experimental measurements using the formula:

$$k_L a = \frac{OTR}{C^*_{AL} - C_{AL}}$$

where $k_L a$ is the mass transfer coefficient, OTR is the oxygen transfer rate, C^*_{AL} is the solubility of oxygen in the fermentation broth, and C_{AL} is the dissolved oxygen concentration. C^*_{AL} is estimated to be 0.25 mol m^{-3} with an uncertainty of ±4%; C_{AL} is measured as 0.183 mol m^{-3} also with an uncertainty of ±4%. If the OTR is 0.011 mol m^{-3} s^{-1} with an uncertainty of ±5%, what is the maximum possible error associated with $k_L a$?

3.2 Accuracy requirement

For cells growing in liquid medium, the overall biomass yield coefficient Y_{XS} can be estimated as follows:

$$Y_{XS} = \frac{x_f - x_0}{s_0 - s_f}$$

where x_f is the final cell concentration, x_0 is the initial cell concentration, s_0 is the initial substrate concentration, and s_f is the final substrate concentration. For a particular cell culture, $s_0 = 30$ g l^{-1}, $s_f = 0.85$ g l^{-1}, $x_f = 12.7$ g l^{-1}, and $x_0 = 0.66$ g l^{-1}. If the uncertainty associated with the substrate concentration assay is ±0.1 g l^{-1}, how accurate must the measurements for cell concentration be if the maximum possible error in Y_{XS} must be less than 5%?

3.3 Mean and standard error

The pH for maximum activity of β-amylase enzyme is measured six times. The results are 5.15, 5.25, 5.45, 5.20, 5.50, and 5.35.

(a) What is the best estimate of the optimal pH?

(b) How reliable is this value?

(c) If the experiment were stopped after only the first three measurements were taken, what would be the result and its precision?

(d) If an additional six measurements were made with the same results as the previous ones, how would this change the outcome of the experiment?

3.4 Confidence limits for the sample mean

Two years ago, you installed a foam control system on a fermenter used to culture *Hansenula polymorpha* yeast. The control system is based on automatic addition of antifoam agent in response to rising foam levels. Your boss tells you that, since installation, the average consumption of antifoam agent per batch fermentation has been excessive at 2.9 litres. You decide to check her assertion by measuring the volume of antifoam used in each of the next 10 fermentations over a period of three months. The results in litres are 1.4, 1.2, 4.1, 3.3, 2.6, 1.6, 1.4, 3.0, 2.0, and 1.1. Is 2.9 litres a reasonable representation of the volume of antifoam consumed in this fermentation? Explain your answer.

3.5 Measurement accuracy

The accuracy of two steam-sterilisable dissolved oxygen probes is tested. The probes are calibrated, autoclaved, and then used to measure the oxygen level in a solution sparged with gas and known to have a dissolved oxygen tension of 50% air saturation. This procedure is carried out five times for each probe. The results are shown in the following table.

	% air saturation				
Probe 1	51.7	52.6	52.9	49.5	50.2
Probe 2	49.0	48.9	50.1	53.3	53.6

(a) Calculate the sample mean, standard deviation, and standard error for each probe.
(b) Which probe exhibits the greater degree of measurement scatter? Explain your answer.
(c) Which probe is more accurate? Explain your answer.

3.6 Linear and nonlinear models

Determine the equation for y as a function of x using the following information. Reference to coordinate point (x, y) means that x is the abscissa value and y is the ordinate value.

(a) A plot of y versus x on linear graph paper gives a straight line passing through the points (1, 10) and (8, 0.5).
(b) A plot of y versus $x^{1/2}$ on linear graph paper gives a straight line passing through the points (3.2, 14.5) and (8.9, 38.5).
(c) A plot of $1/y$ versus x^2 on linear graph paper gives a straight line passing through the points (5, 6) and (1, 3).
(d) A plot of y versus x on log–log paper gives a straight line passing through (0.5, 25) and (550, 2600).
(e) A plot of y versus x on semi-log paper gives a straight line passing through (1.5, 2.5) and (10, 0.036).

3.7 Calibration curve

Sucrose concentration in a fermentation broth is measured using high-performance liquid chromatography (HPLC). Chromatogram peak areas are measured for five standard sucrose solutions to calibrate the instrument. Measurements are performed in triplicate with results as shown in the following table.

Sucrose concentration (g l^{-1})	Peak area
6.0	55.55, 57.01, 57.95
12.0	110.66, 114.76, 113.05
18.0	168.90, 169.44, 173.55
24.0	233.66, 233.89, 230.67
30.0	300.45, 304.56, 301.11

(a) Determine the mean peak areas for each sucrose concentration, and the standard errors.
(b) Plot the calibration curve relating peak area to sucrose concentration. Plot the standard errors as error bars.

(c) Find an equation for sucrose concentration as a function of peak area.
(d) A sample containing sucrose gives a peak area of 209.86. What is the sucrose concentration?

3.8 Linear-range calibration

An enzyme-linked immunosorbent assay (ELISA) is being developed for measurement of virus concentration in samples of homogenised biomass. Absorbance is measured using a spectrophotometer for a dilution series across a range of virus concentrations. The results are as follows.

Virus concentration (ng ml^{-1})	Absorbance
6.0	2.88
3.0	2.52
1.5	2.22
0.75	2.07
0.38	1.65
0.29	1.35

(a) Within what range of virus concentration is there a linear relationship between virus concentration and absorbance?
(b) Develop an equation for virus concentration as a function of absorbance within the linear range.
(c) Three culture samples give absorbance readings of 2.02, 2.66, and 2.75. What concentrations of virus are present?
(d) In which results from (c) do you have the most confidence? Why?

3.9 Data correlation

The density of a liquid is sometimes expressed as a function of temperature in the following way:

$$\rho = \rho_0 + A\,(T - T_0)$$

where ρ is the density at temperature T, ρ_0 is the density at temperature T_0, and A is a constant. The following density data are obtained as a function of temperature.

Temperature (°C)	Liquid density (g cm^{-3})
0.0	0.665
5.0	0.668
10.0	0.672
15.0	0.673
25.0	0.677
30.0	0.684

Estimate A.

3.10 Linear regression: distribution of residuals

Medium conductivity is sometimes used to monitor cell growth during batch culture. Experiments are carried out to relate the decrease in medium conductivity to the increase in plant cell biomass during culture of *Catharanthus roseus* in an airlift fermenter. The results are tabulated in the following table.

Decrease in medium conductivity (mS cm^{-1})	Increase in biomass concentration (g l^{-1})
0	0
0.12	2.4
0.31	2.0
0.41	2.8
0.82	4.5
1.03	5.1
1.40	5.8
1.91	6.0
2.11	6.2
2.42	6.2
2.44	6.2
2.74	6.6
2.91	6.0
3.53	7.0
4.39	9.8
5.21	14.0
5.24	12.6
5.55	14.6

(a) Plot the points using linear coordinates and obtain an equation for the 'best' straight line through the data using linear least-squares analysis.

(b) Plot the residuals in biomass increase versus decrease in conductivity after comparing the model equation with the actual data. What do you conclude about the goodness of fit for the straight line?

3.11 Nonlinear model: calculation of parameters

The mutation rate of *E. coli* increases with temperature. The following data were obtained by measuring the frequency of mutation of *his*$^-$ cells to produce *his*$^+$ colonies.

Temperature (°C)	Relative mutation frequency, α
15	4.4×10^{-15}
20	2.0×10^{-14}
25	8.6×10^{-14}
30	3.5×10^{-13}
35	1.4×10^{-12}

Mutation frequency is expected to obey an Arrhenius-type equation:

$$\alpha = \alpha_0 e^{-E/RT}$$

where α_0 is the mutation rate parameter, E is activation energy, R is the ideal gas constant, and T is absolute temperature.

(a) Test the model using an appropriate plot with logarithmic coordinates.
(b) What is the activation energy for the mutation reaction?
(c) What is the value of α_0?

3.12 Nonlinear kinetics

A new enzyme has been discovered in bacteria isolated from the Great Southern Ocean near Antarctica. Your research group has been given the task of characterising the enzyme reaction, which converts krilliol to krilloene. Experiments are conducted using a krilliol solution at the optimal reaction temperature of 4°C. Data for krilliol concentration as a function of time are as follows.

Time (min)	Krilliol concentration (g l^{-1})
0	1.50
0.2	1.35
0.5	0.90
1.0	0.75
2.0	0.27

You suspect that the reaction follows first-order kinetics according to the relationship:

$$s = s_0 e^{-kt}$$

where s is the reactant concentration, s_0 is the initial reactant concentration, k is the rate constant, and t is time.

(a) Test the model using an appropriate plot.
(b) Determine the rate constant, k.
(c) How long will it take for the krilliol concentration to reach 0.05 g l^{-1}? How confident are you about this result?

3.13 Nonlinear data correlation

Saccharomyces cerevisiae cells are homogenised to release recombinant hepatitis B surface antigen for vaccine production. During homogenisation, fractional protein release is predicted to follow the equation:

$$1 - S_p = e^{-kt}$$

where S_p is the fractional protein release, k is a constant dependent on the operating pressure of the homogeniser, and t is time. At a fixed pressure, the following data are obtained.

Time (s)	Fractional protein release (−)
20	0.25
30	0.31
60	0.62
90	0.75
120	0.79
150	0.88

(a) Is the predicted model a reasonable one for this system?
(b) What is the value of k?
(c) What time is required to release 50% of the protein?
(d) What time is required to release 95% of the protein? How good is your answer?

3.14 Discriminating between rival models

In bioreactors where the liquid contents are mixed by sparging air into the vessel, the liquid velocity is directly dependent on the gas velocity. The following results were obtained from experiments with 0.15 M NaCl solution.

Gas superficial velocity, u_G (m s^{-1})	Liquid superficial velocity, u_L (m s^{-1})
0.02	0.060
0.03	0.066
0.04	0.071
0.05	0.084
0.06	0.085
0.07	0.086
0.08	0.091
0.09	0.095
0.095	0.095

(a) How well are these data fitted with a linear model? Determine the equation for the 'best' straight line relating gas and liquid velocities.
(b) It has been reported in the literature that fluid velocities in air-driven reactors can be related using the power equation:

$$u_L = \alpha u_G^{v}$$

where α and v are constants. Is this model an appropriate description of the experimental data?
(c) Which equation, linear or nonlinear, is the better model for the reactor system? Explain your answer.

3.15 Nonlinear model: calculation of parameters

When nutrient medium is autoclaved, the number of viable microorganisms decreases with time spent in the autoclave. An experiment is conducted to measure the number of viable cells N in a bottle of glucose solution after various sterilisation times t. Triplicate measurements are taken of the cell number; the mean and standard error of each measurement are listed in the following table.

t (min)	Mean N	Standard error
5	3.6×10^3	0.20×10^3
10	6.3×10^2	1×10^2
15	1.07×10^2	0.4×10^2
20	1.8×10^1	0.5×10^1
30	<1	–

From what is known about thermal death kinetics for microorganisms, it is expected that the relationship between N and t is of the form:

$$N = N_0 \, e^{-k_d t}$$

where k_d is the specific death constant and N_0 is the number of viable cells present before autoclaving begins.

(a) Plot the results in a suitable form to obtain a straight line through the data.
(b) Plot the standard errors on the graph as error bars.
(c) What are the values of k_d and N_0?
(d) What are the units and dimensions of k_d and N_0?

3.16 Nonlinear model

Adsorption of cadmium by brown marine macroalgae is being measured for bioremediation of industrial effluent. Experiments are carried out to measure the concentration of Cd adsorbed by the biomass as a function of the residual Cd concentration in the effluent solution. The equilibrium values of these concentrations are as follows.

Residual liquid-phase Cd concentration (ppm)	Biomass Cd concentration ($\mu g \, g^{-1}$ dry weight)
20	614
50	1360
100	3590
200	8890
500	10,500
1000	15,600
2000	19,700
3500	22,800

The adsorption data are expected to obey a Langmuir model:

$$C^*_{AS} = \frac{C_{ASm}\, K_A\, C^*_A}{1 + K_A C^*_A}$$

where C^*_{AS} is the biomass Cd concentration, C_{ASm} is the maximum biomass Cd concentration corresponding to complete monolayer coverage of all available adsorption sites, C^*_A is the residual Cd concentration in the liquid, and K_A is a constant.

(a) Plot the results in a suitable linearised form using linear coordinates.

(b) What are the values of K_A and C_{ASm}?

(c) What are the units and dimensions of K_A and C_{ASm}?

(d) Five thousand litres of waste water containing 120 ppm Cd are treated using algal biomass in an adsorption column. If the Cd concentration must be reduced to at least 25 ppm by the end of the process, what quantity of algae is required?

References

[1] J.R. Taylor, Introduction to Error Analysis, second ed., University Science Books, 1997.
[2] B.S. Massey, Measures in Science and Engineering, Ellis Horwood, 1986.
[3] W. Lichten, Data and Error Analysis in the Introductory Physics Laboratory, Allyn and Bacon, 1988.
[4] M.L. Samuels, J.A. Witmer, A. Schaffner, Statistics for the Life Sciences, fourth ed., Addison-Wesley, 2011.
[5] A. Graham, Statistics: An Introduction, Hodder and Stoughton, 1995.
[6] T.W. Anderson, J.D. Finn, The New Statistical Analysis of Data, Springer, 1995.
[7] N.C. Barford, Experimental Measurements: Precision, Error and Truth, second ed., Wiley, 1985.
[8] R.E. Walpole, R.H. Myers, S.L. Myers, K.E. Ye, Probability and Statistics for Engineers and Scientists, ninth ed., Prentice Hall, 2011.
[9] K.R. Popper, The Logic of Scientific Discovery, Routledge, 2002.
[10] K.R. Popper, Conjectures and Refutations: The Growth of Scientific Knowledge, fifth ed., Routledge, 2002.
[11] N.R. Draper, H. Smith, Applied Regression Analysis, third ed., Wiley, 1998.
[12] D.C. Baird, Experimentation: An Introduction to Measurement Theory and Experiment Design, third ed., Benjamin Cummings, 1994.

Suggestions for Further Reading

Errors in Measured Data

See references [1] through [3], [7], and [12].

Statistical Analysis of Data

See also references [4], [6] through [8], and [11].

Armitage, P., Berry, G., & Matthews, J. N. S. (2002). *Statistical Methods in Medical Research* (4th ed.). Blackwell Science.

Dowd, J. E., & Riggs, D. S. (1965). A comparison of estimates of Michaelis–Menten kinetic constants from various linear transformations. *J. Biol. Chem.*, *240*, 863–869.

Garfinkel, D., & Fegley, K. A. (1984). Fitting physiological models to data. *Am. J. Physiol.*, *246*, R641–R650.

Mannervik, B. (1982). Regression analysis, experimental error, and statistical criteria in the design and analysis of experiments for discrimination between rival kinetic models. *Methods Enzymology, 87*, 370–390.
Sagnella, G. A. (1985). Model fitting, parameter estimation, linear and nonlinear regression. *Trends Biochem. Sci., 10*, 100–103.
Wardlaw, A. C. (2000). *Practical Statistics for Experimental Biologists* (2nd ed.). Wiley.
Zar, J. H. (2010). *Biostatistical Analysis* (5th ed.). Prentice Hall.

Process Flow Diagrams

Sinnott, R. K. (2005). *Coulson and Richardson's Chemical Engineering, vol. 6: Chemical Engineering Design* (4th ed., Chapters 4 and 5). Elsevier.

PART 2

MATERIAL AND ENERGY BALANCES

CHAPTER

4

Material Balances

One of the simplest concepts in process engineering is the material or mass balance. Because mass is conserved at all times in biological processing, the law of conservation of mass provides the theoretical framework for material balances.

In steady-state material balances, masses entering a process are summed up and compared with the total mass leaving the system; the term *balance* implies that the masses entering and leaving should be equal. Essentially, material balances are accounting procedures: the total mass entering must be accounted for at the end of the process, even if it undergoes heating, mixing, drying, fermentation, or any other operation (except nuclear reaction) within the system. Usually it is not feasible to measure the masses and compositions of all streams entering and leaving a system; unknown quantities can be calculated using mass balance principles. Mass balance problems have a constant theme: given the masses of some input and output streams, calculate the masses of others.

Mass balances provide a very powerful tool in engineering analysis. Many complex situations are simplified by looking at the movement of mass and relating what comes out to what goes in. Questions such as: What is the concentration of carbon dioxide in the fermenter off-gas? What fraction of the substrate consumed is not converted into products? How much reactant is needed to produce x grams of product? How much oxygen must be provided for this fermentation to proceed? can be answered using mass balances. This chapter explains how the law of conservation of mass is applied to atoms, molecular species, and total mass, and sets up formal techniques for solving material balance problems with and without reaction. Aspects of metabolic stoichiometry are also discussed for calculation of nutrient and oxygen requirements during fermentation processes.

4.1 THERMODYNAMIC PRELIMINARIES

Thermodynamics is a fundamental branch of science dealing with the properties of matter. Thermodynamic principles are useful in setting up material balances; some terms borrowed from thermodynamics are defined in the following sections.

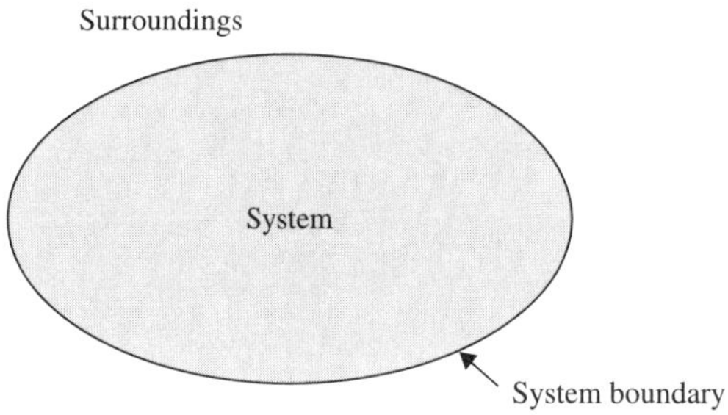

FIGURE 4.1 Thermodynamic system.

4.1.1 System and Process

In thermodynamics, a *system* consists of any matter identified for investigation. As indicated in Figure 4.1, the system is set apart from the *surroundings*, which are the remainder of the *universe*, by a *system boundary*. The system boundary may be real and tangible, such as the walls of a beaker or fermenter, or it may be virtual or notional. If the boundary does not allow mass to pass from system to surroundings and vice versa, the system is a *closed* system with constant mass. Conversely, a system able to exchange mass with its surroundings is an *open* system.

A *process* causes changes in the system or surroundings. Several terms are commonly used to describe processes.

1. A *batch process* operates in a closed system. All materials are added to the system at the start of the process; the system is then closed and the products removed only when the process is complete.
2. A *semi-batch process* allows either input or output of mass.
3. A *fed-batch process* allows input of material to the system but not output.
4. A *continuous process* allows matter to flow in and out of the system. If the rates of mass input and output are equal, continuous processes can be operated indefinitely.

4.1.2 Steady State and Equilibrium

If all properties of a system, such as temperature, pressure, concentration, volume, mass, and so on, do not vary with time, the process is said to be at *steady state*. Thus, if we monitor any variable of a steady-state system, its value will be unchanging with time.

According to this definition of steady state, batch, fed-batch, and semi-batch processes cannot operate under steady-state conditions. The total mass of the system is either increasing or decreasing with time during fed-batch and semi-batch processes; in batch processes, even though the mass may be constant, changes occurring inside the system cause the system properties to vary with time. Such processes are called *transient* or *unsteady-state*. On the other hand, continuous processes may be either steady-state or transient. It is usual to run continuous processes as close to steady state as possible; however, unsteady-state conditions will exist during start-up and for some time after any change in operating conditions.

Steady state is an important and useful concept in engineering analysis. However, it is often confused with another thermodynamic term, *equilibrium*. A system at equilibrium is one in which all opposing forces are exactly counter-balanced so that the properties of the system do not change with time. From experience, we know that systems tend to approach equilibrium conditions when they are isolated from their surroundings. At equilibrium there is no net change in either the system or the universe. Equilibrium implies that there is no net driving force for change; the energy of the system is at a minimum and, in rough terms, the system is 'static', 'unmoving', or 'inert'. For example, when liquid and vapour are in equilibrium in a closed vessel, although there may be constant exchange of molecules between the phases, there is no net change in either the system or the surroundings.

To convert raw materials into useful products there must be an overall change in the universe. Because systems at equilibrium produce no net change, equilibrium is of little value in processing operations. The best strategy is to avoid equilibrium by continuously disturbing the system in such a way that raw material will always be undergoing transformation into the desired product. In continuous processes at steady state, mass is constantly exchanged with the surroundings; this disturbance drives the system away from equilibrium so that a net change in both the system and the universe can occur. Large-scale equilibrium does not often occur in engineering systems; steady states are more common.

4.2 LAW OF CONSERVATION OF MASS

Mass is conserved in ordinary chemical and physical processes. Consider the system of Figure 4.2, in which streams containing glucose enter and leave. The mass of glucose entering the system is M_i kg; the mass of glucose leaving is M_o kg. If M_i and M_o are different, there are four possible explanations:

1. The measurements of M_i and/or M_o are wrong.
2. The system has a leak allowing glucose to enter or escape undetected.
3. Glucose is consumed or generated by chemical reaction within the system.
4. Glucose accumulates within the system.

If we assume that the measurements are correct and there are no leaks, the difference between M_i and M_o must be due to consumption or generation of glucose by reaction and/or accumulation of glucose within the system. A mass balance for the system can be written in a general way to account for these possibilities:

$$\left\{\begin{array}{c}\text{mass in}\\\text{through}\\\text{system}\\\text{boundaries}\end{array}\right\}-\left\{\begin{array}{c}\text{mass out}\\\text{through}\\\text{system}\\\text{boundaries}\end{array}\right\}+\left\{\begin{array}{c}\text{mass}\\\text{generated}\\\text{within}\\\text{system}\end{array}\right\}-\left\{\begin{array}{c}\text{mass}\\\text{consumed}\\\text{within}\\\text{system}\end{array}\right\}=\left\{\begin{array}{c}\text{mass}\\\text{accumulated}\\\text{within}\\\text{system}\end{array}\right\} \quad (4.1)$$

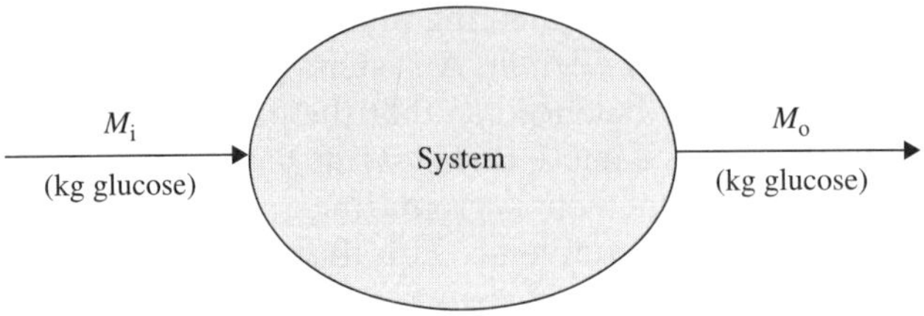

FIGURE 4.2 Flow sheet for a mass balance on glucose.

The accumulation term in the preceding equation can be either positive or negative; negative accumulation represents depletion of preexisting reserves. Equation (4.1) is known as the *general mass balance equation*. The mass referred to in the equation can be total mass, mass of a specific molecular or atomic species, or mass of a particular combination of compounds such as cells or biomass. Use of Eq. (4.1) is illustrated in Example 4.1.

EXAMPLE 4.1 GENERAL MASS BALANCE EQUATION

A continuous process is set up for treatment of waste water. Each day, 10^5 kg cellulose and 10^3 kg bacteria enter in the feed stream, while 10^4 kg cellulose and 1.5×10^4 kg bacteria leave in the effluent. The rate of cellulose digestion by the bacteria is 7×10^4 kg day^{-1}. The rate of bacterial growth is 2×10^4 kg day^{-1}; the rate of cell death by lysis is 5×10^2 kg day^{-1}. Write balances for cellulose and bacteria in the system.

Solution

Cellulose is not generated by the process, only consumed. Using a basis of 1 day, the cellulose balance in kg from Eq. (4.1) is:

$$(10^5 - 10^4 + 0 - 7 \times 10^4) = 2 \times 10^4 = \text{accumulation}$$

Therefore, 2×10^4 kg of cellulose accumulates in the system each day.

Performing the same balance for bacteria:

$$(10^3 - 1.5 \times 10^4 + 2 \times 10^4 - 5 \times 10^2) = 5.5 \times 10^3 = \text{accumulation}$$

Therefore, 5.5×10^3 kg of bacterial cells accumulate in the system each day.

4.2.1 Types of Material Balance

The general mass balance equation (4.1) can be applied with equal ease to two different types of mass balance problem, depending on the data provided. For continuous processes it is usual to collect information about the system referring to a particular instant in time. The amounts of mass entering and leaving are specified using flow rates: for example, molasses enters the system at a rate of 50 lb h^{-1}; at the same instant in time, fermentation broth leaves at a rate of 20 lb h^{-1}. These two quantities can be used directly in Eq. (4.1) as the input and output terms. A mass balance based on rates is called a *differential balance*.

An alternative approach is required for batch and semi-batch processes. Information about these systems is usually collected over a period of time rather than at a particular instant. For example: 100 kg substrate is added to the reactor; after 3 days of incubation, 45 kg product is recovered. Each term of the mass balance equation in this case is a quantity of mass, not a rate. This type of balance is called an *integral balance*.

In this chapter, we will be using differential balances for continuous processes operating at steady state, and integral balances for batch or semi-batch processes between their initial and final states. Calculation procedures for the two types of material balance are very similar.

4.2.2 Simplification of the General Mass Balance Equation

Equation (4.1) can be simplified in certain situations. If a continuous process is at steady state, the accumulation term on the right side of the equation must be zero. This follows from the definition of steady state: because all properties of the system, including its mass, must be unchanging with time, a system at steady state cannot accumulate mass. Under these conditions, Eq. (4.1) becomes:

$$\text{mass in} + \text{mass generated} = \text{mass out} + \text{mass consumed} \tag{4.2}$$

Equation (4.2) is called the *general steady-state mass balance equation*. Equation (4.2) also applies over the entire duration of batch and fed-batch processes; 'mass out' in this case is the total mass harvested from the system so that at the end of the process there is no accumulation.

If reaction does not occur in the system, or if the mass balance is applied to a substance that is neither a reactant nor a product of reaction, the generation and consumption terms in Eqs. (4.1) and (4.2) are zero. Because total mass can be neither created nor destroyed except by nuclear reaction, the generation and consumption terms must also be zero in balances applied to total mass. Similarly, generation and consumption of atomic species such as C, N, O, and so on, cannot occur in normal chemical reaction. Therefore, at steady state, for balances on total mass or atomic species or when reaction does not occur, Eq. (4.2) can be further simplified to:

$$\text{mass in} = \text{mass out} \tag{4.3}$$

Table 4.1 summarises the types of material balance for which direct application of Eq. (4.3) is valid. Because the number of moles does not always balance in systems with reaction, we will carry out all material balances using mass.

4.3 PROCEDURE FOR MATERIAL BALANCE CALCULATIONS

The first step in material balance calculations is to understand the problem. Certain information is available about a process; the task is to calculate unknown quantities. Because it is sometimes difficult to sort through all the details provided, it is best to use standard procedures to translate process information into a form that can be used in calculations.

TABLE 4.1 Application of the Simplified Mass Balance, Eq. (4.3)

Material	At steady state, does mass in = mass out? Without reaction	With reaction
Total mass	Yes	Yes
Total number of moles	Yes	No
Mass of a molecular species	Yes	No
Number of moles of a molecular species	Yes	No
Mass of an atomic species	Yes	Yes
Number of moles of an atomic species	Yes	Yes

Material balances should be carried out in an organised manner; this makes the solution easy to follow, check, or use by others. In this chapter, a formalised series of steps is followed for each mass balance problem. For easier problems these procedures may seem long-winded and unnecessary; however a standard method is helpful when you are first learning mass balance techniques. The same procedures are used in the next chapter as a basis for energy balances.

The following points are essential.

- *Draw a clear process flow diagram showing all relevant information.* A simple box diagram showing all streams entering or leaving the system allows information about a process to be organised and summarised in a convenient way. All given quantitative information should be shown on the diagram. Note that the variables of interest in material balances are masses, mass flow rates, and mass compositions. If information about particular streams is given using volume or molar quantities, mass flow rates and compositions should be calculated before labelling the flow sheet.
- *Select a set of units and state it clearly.* Calculations are easier when all quantities are expressed using consistent units. Units must also be indicated for all variables shown on process diagrams.
- *Select a basis for the calculation and state it clearly.* In approaching mass balance problems, it is helpful to focus on a specific quantity of material entering or leaving the system. For continuous processes at steady state, we usually base the calculation on the amount of material entering or leaving the system within a specified period of time. For batch or semi-batch processes, it is convenient to use either the total amount of material fed to the system or the amount withdrawn at the end. Selection of a basis for calculation makes it easier to visualise the problem; the way this works will become apparent in the worked examples of the next section.
- *State all assumptions applied to the problem.* To solve problems in this and the following chapters, you will need to apply some 'engineering' judgement. Real-life situations are complex, and there will be times when one or more assumptions are required before you can proceed with calculations. To give you experience with this, problems posed in

this text may not give you all the necessary information. The details omitted can be assumed, provided your assumptions are reasonable. Engineers make assumptions all the time; knowing when an assumption is permissible and what constitutes a reasonable assumption is one of the marks of a skilled engineer. When you make assumptions about a problem, it is vitally important that you state them exactly. Other scientists looking through your calculations need to know the conditions under which your results are applicable; they will also want to decide whether your assumptions are acceptable or whether they can be improved.

In this chapter, differential mass balances on continuous processes are performed with the understanding that the system is at steady state; we can assume that mass flow rates and compositions do not change with time and the accumulation term of Eq. (4.1) is zero. If steady state does not prevail in continuous processes, information about the rate of accumulation is required for solution of mass balance problems. This is discussed further in Chapter 6.

Another assumption we must make in mass balance problems is that the system under investigation does not leak. In totalling up all the masses entering and leaving the system, we must be sure that all streams are taken into account. When analysing real systems it is always a good idea to check for leaks before carrying out mass balances.

- *Identify which components of the system, if any, are involved in reaction.* This is necessary for determining which mass balance equation, (4.2) or (4.3), is appropriate. The simpler Eq. (4.3) can be applied to molecular species that are neither reactants nor products of reaction.

EXAMPLE 4.2 SETTING UP A FLOW SHEET

Humid air enriched with oxygen is prepared for a gluconic acid fermentation. The air is prepared in a special humidifying chamber. Liquid water enters the chamber at a rate of 1.5 l h^{-1} at the same time as dry air and 15 gmol min^{-1} of dry oxygen gas. All the water is evaporated. The outflowing gas is found to contain 1% (w/w) water. Draw and label the flow sheet for this process.

Solution

Let us choose units of g and min for this process; the information provided is first converted to mass flow rates in these units. The density of water is taken to be 10^3 g l^{-1}; therefore:

$$1.5\,\mathrm{l\,h^{-1}} = \frac{1.5\,\mathrm{l}}{\mathrm{h}} \times \frac{10^3\,\mathrm{g}}{\mathrm{l}} \cdot \left|\frac{1\,\mathrm{h}}{60\,\mathrm{min}}\right| = 25\,\mathrm{g\,min^{-1}}$$

As the molecular weight of O_2 is 32:

$$15\,\mathrm{gmol\,min^{-1}} = \frac{15\,\mathrm{gmol}}{\mathrm{min}} \cdot \left|\frac{32\,\mathrm{g}}{1\,\mathrm{gmol}}\right| = 480\,\mathrm{g\,min^{-1}}$$

Unknown flow rates are represented with symbols. As shown in Figure 4.3, the flow rate of dry air is denoted D g min^{-1} and the flow rate of humid, oxygen-rich air is H g min^{-1}. The water content in the humid air is shown as 1 mass%.

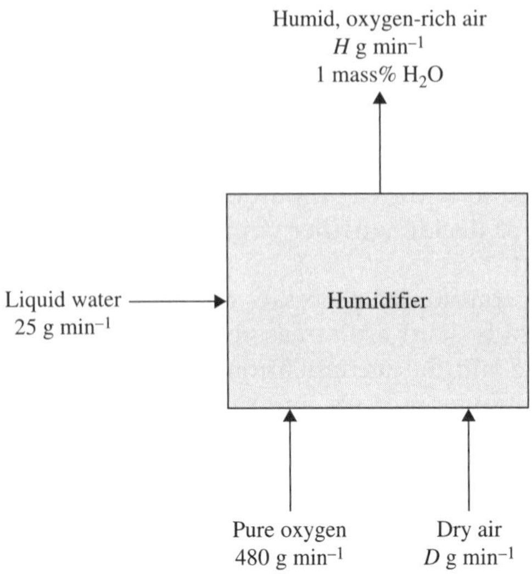

FIGURE 4.3 Flow sheet for oxygen enrichment and humidification of air.

4.4 MATERIAL BALANCE WORKED EXAMPLES

Procedures for setting out mass balance calculations are outlined in this section. Although not the only way to attack these problems, the methods shown will assist your problem-solving efforts by formalising the mathematical approach. Mass balance calculations are divided into four steps: *assemble, analyse, calculate,* and *finalise*. Differential and integral mass balances with and without reaction are illustrated below.

EXAMPLE 4.3 CONTINUOUS FILTRATION

A fermentation slurry containing *Streptomyces kanamyceticus* cells is filtered using a continuous rotary vacuum filter. Slurry is fed to the filter at a rate of 120 kg h^{-1}; 1 kg slurry contains 60 g cell solids. To improve filtration rates, particles of diatomaceous earth filter aid are added at a rate of 10 kg h^{-1}. The concentration of kanamycin in the slurry is 0.05% by weight. Liquid filtrate is collected at a rate of 112 kg h^{-1}; the concentration of kanamycin in the filtrate is 0.045% (w/w). Filter cake containing cells and filter aid is removed continuously from the filter cloth.

(a) What percentage water is the filter cake?

(b) If the concentration of kanamycin dissolved in the liquid within the filter cake is the same as that in the filtrate, how much kanamycin is absorbed per kg filter aid?

Solution

1. Assemble
   - (i) Draw the flow sheet showing all data with units.
     This is shown in Figure 4.4.
   - (ii) Define the system boundary by drawing on the flow sheet.
     The system boundary is shown in Figure 4.4.

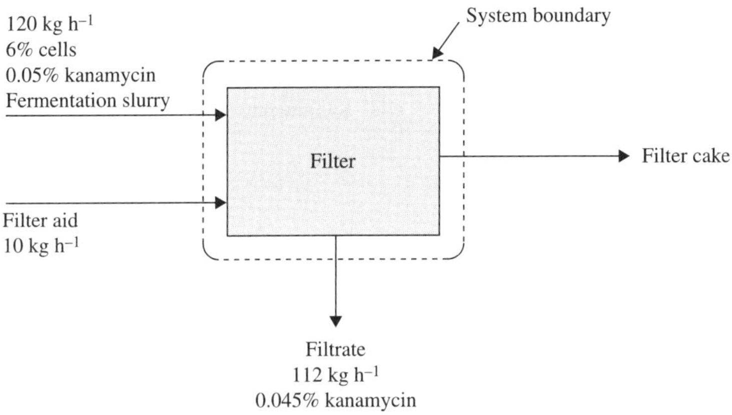

FIGURE 4.4 Flow sheet for continuous filtration.

2. Analyse

(i) State any assumptions.

- process is operating at steady state
- system does not leak
- filtrate contains no solids
- cells do not absorb or release kanamycin during filtration
- filter aid added is dry
- the liquid phase of the slurry, excluding kanamycin, can be considered water

(ii) Collect and state any extra data needed.

No extra data are required.

(iii) Select and state a basis.

The calculation is based on 120 kg slurry entering the filter, or 1 hour.

(iv) List the compounds, if any, that are involved in reaction.

No compounds are involved in reaction.

(v) Write down the appropriate general mass balance equation.

The system is at steady state and no reaction occurs; therefore Eq. (4.3) is appropriate:

$$\text{mass in} = \text{mass out}$$

3. Calculate

(i) Set up a calculation table showing all components of all streams passing across the system boundaries. State the units used for the table. Enter all known quantities.

As shown in Figure 4.4, four streams cross the system boundaries: fermentation slurry, filter aid, filtrate, and filter cake. The components of these streams—cells, kanamycin, filter aid, and water—are represented in Table 4.2. The table is divided into two major sections: In and Out. Masses entering or leaving the system each hour are shown in the table; the units used are kg. Because filtrate and filter cake flow out of the system, there are no entries for these streams on the In side of the table. Conversely, there are no entries for the fermentation slurry and filter aid streams on the Out side of the table. The total mass of each stream is given in the last column on each side of the table. The total amounts of each component flowing in and out of the system are shown in the last row.

TABLE 4.2 Mass Balance Table (kg)

	In					Out				
Stream	Cells	Kanamycin	Filter aid	Water	Total	Cells	Kanamycin	Filter aid	Water	Total
Fermentation slurry	7.2	0.06	0	?	120	–	–	–	–	–
Filter aid	0	0	10	0	10	–	–	–	–	–
Filtrate	–	–	–	–	–	0	0.05	0	?	112
Filter cake	–	–	–	–	–	?	?	?	?	?
Total	?	?	?	?	?	?	?	?	?	?

With all known quantities entered, several masses remain unknown; these quantities are indicated by question marks.

(ii) Calculate unknown quantities; apply the mass balance equation.
To complete Table 4.2, let us consider each row and column separately. In the row representing the fermentation slurry, the total mass of the stream is 120 kg and the masses of each component except water are known. The entry for water can therefore be determined as the difference between 120 kg and the sum of the known components: $(120 - 7.2 - 0.06 - 0)$ kg = 112.74 kg. This mass for water has been entered in Table 4.3. The row for the filter aid stream is already complete in Table 4.2: no cells or kanamycin are present in the diatomaceous earth entering the system; we have also assumed that the filter aid is dry. We can now fill in the final row of the In side of the table; numbers in this row are obtained by adding the values in each column. The total mass of cells input to the system in all streams is 7.2 kg, the total kanamycin entering is 0.06 kg, and so on. The total mass of all components fed into the system is the sum of the last column of the In side: $(120 + 10)$ kg = 130 kg. On the Out side, we can complete the row for filtrate. We have assumed there are no solids such as cells or filter aid in the filtrate; therefore the mass of water in the filtrate is $(112 - 0.05)$ kg = 111.95. As yet, the entire composition and mass of the filter cake remain unknown.

To complete the table, we must consider the mass balance equation relevant to this problem, Eq. (4.3). In the absence of reaction, this equation can be applied to total mass and to the masses of each component of the system.

Total mass balance

$$130 \text{ kg total mass in} = \text{total mass out}$$
$$\therefore \text{Total mass out} = 130 \text{ kg}$$

Cell balance

$$7.2 \text{ kg cells in} = \text{cells out}$$
$$\therefore \text{Cells out} = 7.2 \text{ kg}$$

Kanamycin balance

$$0.06 \text{ kg kanamycin in} = \text{kanamycin out}$$
$$\therefore \text{Kanamycin out} = 0.06 \text{ kg}$$

TABLE 4.3 Completed Mass Balance Table (kg)

	In					Out				
Stream	**Cells**	**Kanamycin**	**Filter aid**	**Water**	**Total**	**Cells**	**Kanamycin**	**Filter aid**	**Water**	**Total**
Fermentation slurry	7.2	0.06	0	112.74	120	–	–	–	–	–
Filter aid	0	0	10	0	10	–	–	–	–	–
Filtrate	–	–	–	–	–	0	0.05	0	111.95	112
Filter cake	–	–	–	–	–	7.2	0.01	10	0.79	18
Total	**7.2**	**0.06**	**10**	**112.74**	**130**	**7.2**	**0.06**	**10**	**112.74**	**130**

Filter aid balance

$$10 \text{ kg filter aid in} = \text{filter aid out}$$
$$\therefore \text{Filter aid out} = 10 \text{ kg}$$

Water balance

$$112.74 \text{ kg water in} = \text{water out}$$
$$\therefore \text{Water out} = 112.74 \text{ kg}$$

These results are entered in the last row of the Out side of Table 4.3. In the absence of reaction, this row is always identical to the final row of the In side. The component masses for the filter cake can now be filled in as the difference between numbers in the final row and the masses of each component in the filtrate. Take time to look over Table 4.3; you should understand how all the numbers shown were obtained.

(iii) Check that your results are reasonable and make sense.
Mass balance calculations must be checked. Make sure that all columns and rows of Table 4.3 add up to the totals shown.

4. Finalise

(i) Answer the specific questions asked in the problem.
The percentage water in the filter cake can be calculated from the results in Table 4.3. Dividing the mass of water in the filter cake by the total mass of this stream, the percentage water is:

$$\frac{0.79 \text{ kg}}{18 \text{ kg}} \times 100 = 4.39\%$$

Kanamycin is dissolved in the water to form the filtrate and the liquid phase retained within the filter cake. If the concentration of kanamycin in the liquid phase is 0.045%, the mass of kanamycin in the filter-cake liquid is:

$$\frac{0.045}{100} \times (0.79 + 0.01) \text{ kg} = 3.6 \times 10^{-4} \text{ kg}$$

However, we know from Table 4.3 that a total of 0.01 kg kanamycin is contained in the filter cake; therefore $(0.01 - 3.6 \times 10^{-4})$ kg = 0.00964 kg kanamycin is so far unaccounted for. Following our assumption that kanamycin is not adsorbed by the cells, 0.00964 kg kanamycin must be retained by the filter aid within the filter cake. As 10 kg filter aid is present, the kanamycin absorbed per kg filter aid is:

$$\frac{0.00964 \text{ kg}}{10 \text{ kg}} = 9.64 \times 10^{-4} \text{ kg kg}^{-1}$$

(ii) State the answers clearly and unambiguously, checking significant figures.

(a) The water content of the filter cake is 4.4%.

(b) The amount of kanamycin absorbed by the filter aid is 9.6×10^{-4} kg kg^{-1}.

Note in Example 4.3 that the complete composition of the fermentation slurry was not provided. Cell and kanamycin concentrations were given; however the slurry most probably contained a variety of other components such as residual carbohydrate, minerals, vitamins, and amino acids, as well as additional fermentation products. These components were ignored in the mass balance; the liquid phase of the slurry was considered to be water only. This assumption is reasonable as the concentration of dissolved substances in fermentation broths is typically very small; water in spent broth usually accounts for more than 90% of the liquid phase.

Note also that the masses of some of the components in Example 4.3 were different by several orders of magnitude; for example, the mass of kanamycin in the filtrate was of the order 10^{-2} kg whereas the total mass of this stream was of the order 10^2 kg. Calculation of the mass of water by difference therefore involved subtracting a very small number from a large one and carrying more significant figures than warranted. This is an unavoidable feature of most mass balances for biological processes, which are characterised by dilute solutions, low product concentrations, and large amounts of water. However, although excess significant figures were carried in the mass balance table, the final answers were reported with due regard to data accuracy.

Example 4.3 illustrates mass balance procedures for a simple steady-state process without reaction. An integral mass balance for a batch system without reaction is outlined in Example 4.4.

EXAMPLE 4.4 BATCH MIXING

Corn-steep liquor contains 2.5% invert sugars and 50% water; the rest can be considered solids. Beet molasses contains 50% sucrose, 1% invert sugars, and 18% water; the remainder is solids. A mixing tank contains 125 kg corn-steep liquor and 45 kg molasses; water is then added to produce a diluted sugar mixture containing 2% (w/w) invert sugars.

(a) How much water is required?
(b) What is the concentration of sucrose in the final mixture?

Solution

1. Assemble
  (i) Flow sheet
  The flow sheet for this batch process is shown in Figure 4.5. Unlike in Figure 4.4 where the streams represented continuously flowing inputs and outputs, the streams in Figure 4.5 represent masses added and removed at the beginning and end of the mixing process, respectively.
  (ii) System boundary
  The system boundary is indicated in Figure 4.5.

2. Analyse
  (i) Assumptions
  – no leaks
  – no inversion of sucrose to reducing sugars, or any other reaction
  (ii) Extra data
  No extra data are required.
  (iii) Basis
  125 kg corn-steep liquor
  (iv) Compounds involved in reaction
  No compounds are involved in reaction.
  (v) Mass balance equation
  The appropriate mass balance equation is Eq. (4.3):

$$\text{mass in} = \text{mass out}$$

3. Calculate
  (i) Calculation table
  Table 4.4 shows all given quantities in kg. Rows and columns on each side of the table have been completed as much as possible from the information provided. Two

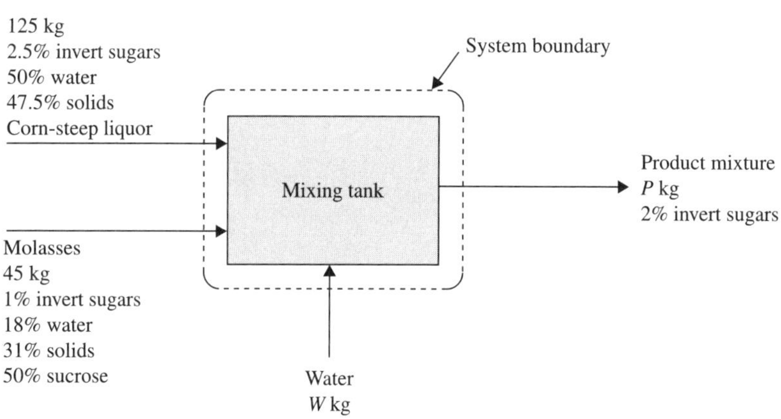

FIGURE 4.5 Graphic of a flow sheet for the batch mixing process.

TABLE 4.4 Mass Balance Table (kg)

	In					Out				
Stream	**Invert sugars**	**Sucrose**	**Solids**	**H_2O**	**Total**	**Invert sugars**	**Sucrose**	**Solids**	**H_2O**	**Total**
Corn-steep liquor	3.125	0	59.375	62.5	125	–	–	–	–	–
Molasses	0.45	22.5	13.95	8.1	45	–	–	–	–	–
Water	0	0	0	W	W	–	–	–	–	–
Product mixture	–	–	–	–	–	$0.02P$	?	?	?	P
Total	**3.575**	**22.5**	**73.325**	**$70.6 + W$**	**$170 + W$**	**$0.02P$**	**?**	**?**	**?**	**P**

unknown quantities are given symbols: the mass of water added is denoted W, and the total mass of product mixture is denoted P.

(ii) Mass balance calculations

Total mass balance

$$(170 + W)\text{ kg total mass in} = P\text{ kg total mass out}$$
$$\therefore 170 + W = P \quad (1)$$

Invert sugars balance

$$3.575\text{ kg invert sugars in} = (0.02P)\text{ kg invert sugars out}$$
$$\therefore 3.575 = 0.02\ P$$
$$P = 178.75\ \text{kg}$$

Using this result in (1):

$$W = 8.75\ \text{kg} \quad (2)$$

Sucrose balance

$$22.5\text{ kg sucrose in} = \text{sucrose out}$$
$$\therefore \text{Sucrose out} = 22.5\text{ kg}$$

Solids balance

$$73.325\text{ kg solids in} = \text{solids out}$$
$$\therefore \text{Solids out} = 73.325\text{ kg}$$

TABLE 4.5 Completed Mass Balance Table (kg)

	In					Out				
Stream	**Invert sugars**	**Sucrose**	**Solids**	**H_2O**	**Total**	**Invert sugars**	**Sucrose**	**Solids**	**H_2O**	**Total**
Corn-steep liquor	3.125	0	59.375	62.5	125	–	–	–	–	–
Molasses	0.45	22.5	13.95	8.1	45	–	–	–	–	–
Water	0	0	0	8.75	8.75	–	–	–	–	–
Product mixture	–	–	–	–	–	3.575	22.5	73.325	79.35	178.75
Total	**3.575**	**22.5**	**73.325**	**79.35**	**178.75**	**3.575**	**22.5**	**73.325**	**79.35**	**178.75**

H_2O balance

$$(70.6 + W)\text{ kg in} = H_2O\text{ out}$$

Using the result from (2):

$$79.35\text{ kg }H_2O\text{ in} = H_2O\text{ out}$$
$$\therefore H_2O\text{ out} = 79.35\text{ kg}$$

These results allow the mass balance table to be completed, as shown in Table 4.5.

(iii) Check the results

All columns and rows of Table 4.5 add up correctly.

4. Finalise

(i) The specific questions

The water required is 8.75 kg. The following is the sucrose concentration in the product mixture:

$$\frac{22.5}{178.75} \times 100 = 12.6\%$$

(ii) Answers

(a) 8.75 kg water is required.

(b) The product mixture contains 13% sucrose.

Material balances on reactive systems are slightly more complicated than Examples 4.3 and 4.4. To solve problems with reaction, stoichiometric relationships must be used in conjunction with mass balance equations. These procedures are illustrated in Examples 4.5 and 4.6.

EXAMPLE 4.5 CONTINUOUS ACETIC ACID FERMENTATION

Acetobacter aceti bacteria convert ethanol to acetic acid under aerobic conditions. A continuous fermentation process for vinegar production is proposed using nongrowing *A. aceti* cells immobilised on the surface of gelatin beads. Air is pumped into the fermenter at a rate of 200 gmol h^{-1}. The production target is 2 kg h^{-1} acetic acid and the maximum acetic acid concentration tolerated by the cells is 12%.

(a) What minimum amount of ethanol is required?
(b) What minimum amount of water must be used to dilute the ethanol to avoid acid inhibition?
(c) What is the composition of the fermenter off-gas?

Solution

1. Assemble

(i) Flow sheet
The flow sheet for this process is shown in Figure 4.6.

(ii) System boundary
The system boundary is shown in Figure 4.6.

(iii) Reaction equation
In the absence of cell growth, maintenance, or other metabolism of substrate, the reaction equation is:

$$\underset{\text{(ethanol)}}{C_2H_5OH} + O_2 \rightarrow \underset{\text{(acetic acid)}}{CH_3COOH} + H_2O$$

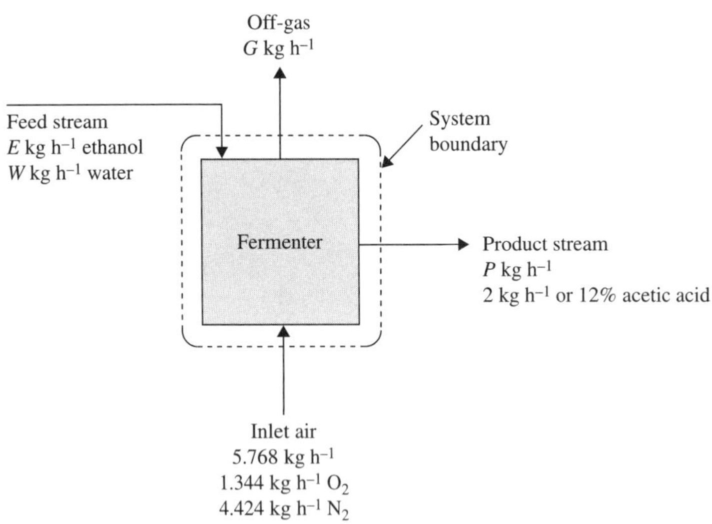

FIGURE 4.6 Flow sheet for continuous acetic acid fermentation.

2. Analyse

(i) Assumptions

- steady state
- no leaks
- inlet air is dry
- gas volume% = mole%
- no evaporation of ethanol, H_2O, or acetic acid
- complete conversion of ethanol
- ethanol is used by the cells for synthesis of acetic acid only; no side-reactions occur
- oxygen transfer is sufficiently rapid to meet the demands of the cells
- solubility of O_2 and N_2 in the liquid phase is negligible
- concentration of acetic acid in the product stream is 12%

(ii) Extra data

Molecular weights:

- Ethanol = 46
- Acetic acid = 60
- $O_2 = 32$
- $N_2 = 28$
- $H_2O = 18$

Composition of air: 21% O_2, 79% N_2 (Section 2.4.5)

(iii) Basis

The calculation is based on 2 kg acetic acid leaving the system, or 1 hour.

(iv) Compounds involved in reaction

The compounds involved in reaction are ethanol, acetic acid, O_2, and H_2O. N_2 is not involved in reaction.

(v) Mass balance equations

For ethanol, acetic acid, O_2, and H_2O, the appropriate mass balance equation is Eq. (4.2):

$$\text{mass in} + \text{mass generated} = \text{mass out} + \text{mass consumed}$$

For total mass and N_2, the appropriate mass balance equation is Eq. (4.3):

$$\text{mass in} = \text{mass out}$$

3. Calculate

(i) Calculation table

The mass balance table listing the data provided is shown as Table 4.6; the units are kg. EtOH denotes ethanol; HAc is acetic acid. If 2 kg acetic acid represents 12 mass% of the product stream, the total mass of the product stream must be $2/0.12 = 16.67$ kg. If we assume complete conversion of ethanol, the only components of the product stream are acetic acid and water; therefore, water must account for 88 mass% of the product stream = 14.67 kg. *E* and *W* denote the unknown quantities of ethanol and water in the feed stream, respectively; *G* represents the total mass of off-gas. The question marks in

TABLE 4.6 Mass Balance Table (kg)

	In						Out					
Stream	**EtOH**	**HAc**	**H_2O**	**O_2**	**N_2**	**Total**	**EtOH**	**HAc**	**H_2O**	**O_2**	**N_2**	**Total**
Feed stream	E	0	W	0	0	$E + W$	–	–	–	–	–	–
Inlet air	0	0	0	1.344	4.424	5.768	–	–	–	–	–	–
Product stream	–	–	–	–	–	–	0	2	14.67	0	0	16.67
Off-gas	–	–	–	–	–	–	0	0	0	?	?	G
Total	E	**0**	W	**1.344**	**4.424**	**5.768 + $E + W$**	**0**	**2**	**14.67**	**?**	**?**	**16.67 + G**

the table show which other quantities must be calculated. In order to represent what is known about the inlet air, some preliminary calculations are needed.

$$O_2 \text{ content} = (0.21)\,(200\text{ gmol}) \cdot \left|\frac{32\text{ g}}{\text{gmol}}\right| = 1344\text{ g} = 1.344\text{ kg}$$

$$N_2 \text{ content} = (0.79)\,(200\text{ gmol}) \cdot \left|\frac{28\text{ g}}{\text{gmol}}\right| = 4424\text{ g} = 4.424\text{ kg}$$

Therefore, the total mass of air in = 5.768 kg. The masses of O_2 and N_2 can now be entered in Table 4.6 as shown.

(ii) Mass balance and stoichiometry calculations

As N_2 is a tie component, its mass balance is straightforward.

N_2 balance

$$4.424\text{ kg } N_2 \text{ in} = N_2 \text{ out}$$
$$\therefore N_2 \text{ out} = 4.424\text{ kg}$$

To deduce the other unknowns, we must use stoichiometric analysis as well as mass balances.

HAc balance

$$0\text{ kg HAc in} + \text{HAc generated} = 2\text{ kg HAc out} + 0\text{ kg HAc consumed}$$
$$\therefore \text{HAc generated} = 2\text{ kg}$$

$$2\text{ kg} = 2\text{ kg} \cdot \left|\frac{1\text{ kgmol}}{60\text{ kg}}\right| = 3.333 \times 10^{-2}\text{ kgmol}$$

From reaction stoichiometry, we know that generation of 3.333×10^{-2} kgmol HAc requires 3.333×10^{-2} kgmol each of EtOH and O_2, and is accompanied by generation of 3.333×10^{-2} kgmol H_2O:

$$3.333 \times 10^{-2}\ \text{kgmol} \cdot \left| \frac{46\ \text{kg}}{1\ \text{kgmol}} \right| = 1.533\ \text{kg EtOH is consumed}$$

$$3.333 \times 10^{-2}\ \text{kgmol} \cdot \left| \frac{32\ \text{kg}}{1\ \text{kgmol}} \right| = 1.067\ \text{kg}\ O_2\ \text{is consumed}$$

$$3.333 \times 10^{-2}\ \text{kgmol} \cdot \left| \frac{18\ \text{kg}}{1\ \text{kgmol}} \right| = 0.600\ \text{kg}\ H_2O\ \text{is generated}$$

We can use this information to complete the mass balances for EtOH, O_2, and H_2O.

EtOH balance

$$\text{EtOH in} + 0\ \text{kg EtOH generated} = 0\ \text{kg EtOH out} + 1.533\ \text{kg EtOH consumed}$$
$$\therefore\ \text{EtOH in} = 1.533\ \text{kg} = E$$

O_2 balance

$$1.344\ \text{kg}\ O_2\ \text{in} + 0\ \text{kg}\ O_2\ \text{generated} = O_2\ \text{out} + 1.067\ \text{kg}\ O_2\ \text{consumed}$$
$$\therefore\ O_2\ \text{out} = 0.277\ \text{kg}$$

Therefore, summing the O_2 and N_2 components of the off gas:

$$G = (0.277 + 4.424)\ \text{kg} = 4.701\ \text{kg}$$

H_2O balance

$$W\ \text{kg}\ H_2O\ \text{in} + 0.600\ \text{kg}\ H_2O\ \text{generated}$$
$$= 14.67\ \text{kg}\ H_2O\ \text{out} + 0\ \text{kg}\ H_2O\ \text{consumed}$$
$$\therefore\ W = 14.07\ \text{kg}$$

These results allow us to complete the mass balance table, as shown in Table 4.7.

(iii) Check the results

All rows and columns of Table 4.7 add up correctly.

4. Finalise

(i) The specific questions

The ethanol required is 1.533 kg. The water required is 14.07 kg. The off-gas contains 0.277 kg O_2 and 4.424 kg N_2. As gas compositions are normally expressed using volume% or mole%, we convert these values to moles:

$$O_2\ \text{content} = 0.277\ \text{kg} \cdot \left| \frac{1\ \text{kgmol}}{32\ \text{kg}} \right| = 8.656 \times 10^{-3}\ \text{kgmol}$$

$$N_2\ \text{content} = 4.424\ \text{kg} \cdot \left| \frac{1\ \text{kgmol}}{28\ \text{kg}} \right| = 0.1580\ \text{kgmol}$$

TABLE 4.7 Completed Mass Balance Table (kg)

	In						Out					
Stream	**EtOH**	**HAc**	**H_2O**	**O_2**	**N_2**	**Total**	**EtOH**	**HAc**	**H_2O**	**O_2**	**N_2**	**Total**
Feed stream	1.533	0	14.07	0	0	15.603	–	–	–	–	–	–
Inlet air	0	0	0	1.344	4.424	5.768	–	–	–	–	–	–
Product stream	–	–	–	–	–	–	0	2	14.67	0	0	16.67
Off-gas	–	–	–	–	–	–	0	0	0	0.277	4.424	4.701
Total	**1.533**	**0**	**14.07**	**1.344**	**4.424**	**21.371**	**0**	**2**	**14.67**	**0.277**	**4.424**	**21.371**

Therefore, the total molar quantity of off-gas is 0.1667 kgmol. The off-gas composition is:

$$\frac{8.656 \times 10^{-3}\ \text{kgmol}}{0.1667\ \text{kgmol}} \times 100 = 5.19\%\ O_2$$

$$\frac{0.1580\ \text{kgmol}}{0.1667\ \text{kgmol}} \times 100 = 94.8\%\ N_2$$

(ii) Answers

Quantities are expressed in kg h^{-1} rather than kg to reflect the continuous nature of the process and the basis used for calculation.

(a) 1.5 kg h^{-1} ethanol is required.

(b) 14 kg h^{-1} water must be used to dilute the ethanol in the feed stream.

(c) The composition of the fermenter off-gas is 5.2% O_2 and 95% N_2.

There are several points to note about the problem and calculation of Example 4.5. First, cell growth and its requirement for substrate were not considered because the cells used in this process were nongrowing. For fermentation with viable cells, growth and other metabolic activity must be taken into account in the mass balance. This requires knowledge of growth stoichiometry, which is considered in Example 4.6 and discussed in more detail in Section 4.6. Use of nongrowing immobilised cells in Example 4.5 meant that the cells were not components of any stream flowing into or out of the process, nor were they generated in reaction. Therefore, cell mass did not have to be included in the calculation.

Example 4.5 illustrates the importance of phase separations. Unreacted oxygen and nitrogen were assumed to leave the system as off-gas rather than as components of the liquid product stream. This assumption is reasonable due to the very poor solubility of oxygen and nitrogen in aqueous liquids: although the product stream most likely contained some dissolved gases, the quantities would be relatively small. This assumption may need to be reviewed for gases with higher solubility, such as ammonia.

In Example 4.5, nitrogen did not react, nor were there more than one stream in and one stream out carrying nitrogen. A material that goes directly from one stream to another is called a *tie component*; the mass balance for a tie component is relatively simple. Tie components are useful because they can provide partial solutions to mass balance problems, making subsequent calculations easier. More than one tie component may be present in a particular process.

One of the listed assumptions in Example 4.5 is rapid oxygen transfer. Because cells use oxygen in dissolved form, oxygen must be transferred into the liquid phase from gas bubbles supplied to the fermenter. The speed of this process depends on the culture conditions and operation of the fermenter as described in more detail in Chapter 10. In mass balance problems we assume that all oxygen required by the stoichiometric equation is immediately available to the cells.

Sometimes it is not possible to solve for unknown quantities in mass balances until near the end of the calculation. In such cases, symbols for various components rather than numerical values must be used in the balance equations. This is illustrated in the integral mass balance of Example 4.6, which analyses the batch culture of growing cells for production of xanthan gum.

EXAMPLE 4.6 XANTHAN GUM PRODUCTION

Xanthan gum is produced using *Xanthomonas campestris* in batch culture. Laboratory experiments have shown that for each gram of glucose utilised by the bacteria, 0.23 g oxygen and 0.01 g ammonia are consumed, while 0.75 g gum, 0.09 g cells, 0.27 g gaseous CO_2, and 0.13 g H_2O are formed. Other components of the system such as phosphate can be neglected. Medium containing glucose and ammonia dissolved in 20,000 litres of water is pumped into a stirred fermenter and inoculated with *X. campestris*. Air is sparged into the fermenter; the total amount of off-gas recovered during the entire batch culture is 1250 kg. Because xanthan gum solutions have high viscosity and are difficult to handle, the final gum concentration should not be allowed to exceed 3.5 wt%.

(a) How much glucose and ammonia are required?
(b) What percentage excess air is provided?

Solution

1. Assemble
   - (i) Flow sheet
     The flow sheet for this process is shown in Figure 4.7.
   - (ii) System boundary
     The system boundary is shown in Figure 4.7.
   - (iii) Reaction equation

$$1\text{ g glucose} + 0.23\text{ g } O_2 + 0.01\text{ g } NH_3 \rightarrow 0.75\text{ g gum} + 0.09\text{ g cells} + 0.27\text{ g } CO_2 + 0.13\text{ g } H_2O$$

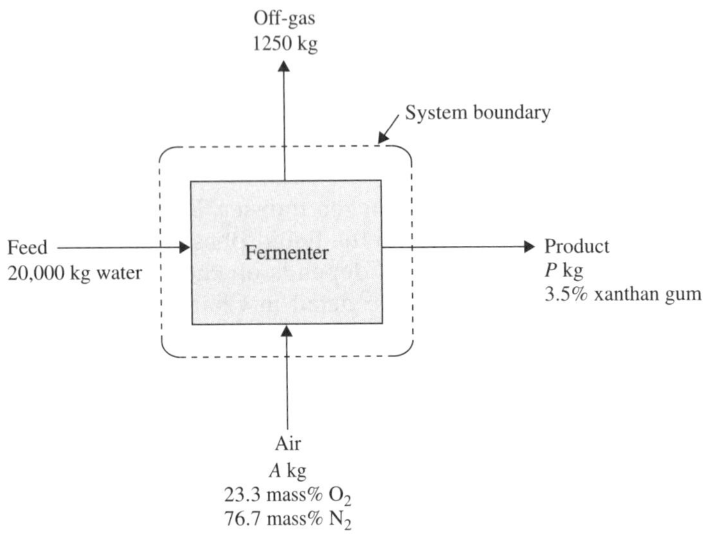

FIGURE 4.7 Graphic of a flow sheet for fermentation of xanthan gum.

2. Analyse

(i) Assumptions

- no leaks
- inlet air and off-gas are dry
- conversion of glucose and NH_3 is 100% complete
- solubility of O_2 and N_2 in the liquid phase is negligible
- CO_2 leaves in the off-gas

(ii) Extra data

Molecular weights:

- $O_2 = 32$
- $N_2 = 28$

Density of water $= 1\ \text{kg l}^{-1}$

Composition of air: 21% O_2, 79% N_2 (Section 2.4.5)

(iii) Basis

1250 kg off-gas

(iv) Compounds involved in reaction

The compounds involved in reaction are glucose, O_2, NH_3, gum, cells, CO_2, and H_2O. N_2 is not involved in reaction.

(v) Mass balance equations

For glucose, O_2, NH_3, gum, cells, CO_2, and H_2O, the appropriate mass balance equation is Eq. (4.2):

$$\text{mass in} + \text{mass generated} = \text{mass out} + \text{mass consumed}$$

For total mass and N_2, the appropriate mass balance equation is Eq. (4.3):

$$\text{mass in} = \text{mass out}$$

3. Calculate

(i) Calculation table

Some preliminary calculations are required to start the mass balance table. First, using $1\ \text{kg l}^{-1}$ as the density of water, 20,000 litres of water is equivalent to 20,000 kg. Let $\underline{A}$ be the unknown mass of air added. Air is composed of 21 mol% O_2 and 79 mol% N_2; we need to determine the composition of air as mass fractions. In 100 gmol air:

$$O_2 \text{ content} = 21 \text{ gmol} \cdot \left| \frac{32 \text{ g}}{1 \text{ gmol}} \right| = 672 \text{ g}$$

$$N_2 \text{ content} = 79 \text{ gmol} \cdot \left| \frac{28 \text{ g}}{1 \text{ gmol}} \right| = 2212 \text{ g}$$

If the total mass of air in 100 gmol is (2212 + 672) = 2884 g, the composition of air is:

$$\frac{672 \text{ g}}{2884 \text{ g}} \times 100 = 23.3 \text{ mass\% } O_2$$

$$\frac{2212 \text{ g}}{2884 \text{ g}} \times 100 = 76.7 \text{ mass\% } N_2$$

Therefore, the mass of O_2 in the inlet air is $0.233A$; the mass of N_2 is $0.767A$. Let F denote the total mass of feed added; let P denote the total mass of product. We will perform the calculation to produce the maximum allowable gum concentration; therefore, the mass of gum in the product is $0.035P$. With the assumption of 100% conversion of glucose and NH_3, these compounds are not present in the product. Quantities known at the beginning of the problem are shown in Table 4.8.

(ii) Mass balance and stoichiometry calculations

Total mass balance

$$(F + A) \text{ kg total mass in} = (1250 + P) \text{ kg total mass out}$$

$$\therefore F + A = 1250 + P \qquad (1)$$

Gum balance

$$0 \text{ kg gum in} + \text{gum generated} = (0.035P) \text{ kg gum out} + 0 \text{ kg gum consumed}$$

$$\therefore \text{Gum generated} = (0.035P) \text{ kg}$$

From reaction stoichiometry, synthesis of $(0.035P)$ kg gum requires:

$$\frac{0.035P}{0.75}(1 \text{ kg}) = (0.0467P) \text{ kg glucose}$$

$$\frac{0.035P}{0.75}(0.23 \text{ kg}) = (0.0107P) \text{ kg } O_2$$

$$\frac{0.035P}{0.75}(0.01 \text{ kg}) = (0.00047P) \text{ kg } NH_3$$

TABLE 4.8 Mass Balance Table (kg)

	In									Out								
Stream	**Glucose**	**O_2**	**N_2**	**CO_2**	**Gum**	**Cells**	**NH_3**	**H_2O**	**Total**	**Glucose**	**O_2**	**N_2**	**CO_2**	**Gum**	**Cells**	**NH_3**	**H_2O**	**Total**
Feed	?	0	0	0	0	0	?	20,000	F	–	–	–	–	–	–	–	–	–
Air	0	0.233A	0.767A	0	0	0	0	0	A	–	–	–	–	–	–	–	–	–
Off-gas	–	–	–	–	–	–	–	–	–	0	?	?	?	0	0	0	0	1250
Product	–	–	–	–	–	–	–	–	–	0	0	0	0	0.035P	?	0	0	P
Total	**?**	**0.233A**	**0.767A**	**0**	**0**	**0**	**?**	**20,000**	**$F+A$**	**0**	**?**	**?**	**?**	**0.035P**	**?**	**0**	**?**	**1250 + P**

and generates:

$$\frac{0.035P}{0.75}(0.09\ \text{kg}) = (0.0042P)\ \text{kg cells}$$

$$\frac{0.035P}{0.75}(0.27\ \text{kg}) = (0.0126P)\ \text{kg}\ CO_2$$

$$\frac{0.035P}{0.75}(0.13\ \text{kg}) = (0.00607P)\ \text{kg}\ H_2O$$

O_2 balance

$$(0.233A)\ \text{kg}\ O_2\ \text{in} + 0\ \text{kg}\ O_2\ \text{generated} = O_2\ \text{out} + (0.0107P)\ \text{kg}\ O_2\ \text{consumed}$$
$$\therefore O_2\ \text{out} = (0.233A - 0.0107P)\ \text{kg} \qquad (2)$$

N_2 balance

N_2 is a tie component.

$$(0.767A)\ \text{kg}\ N_2\ \text{in} = N_2\ \text{out}$$
$$\therefore N_2\ \text{out} = (0.767A)\ \text{kg} \qquad (3)$$

CO_2 balance

$$0\ \text{kg}\ CO_2\ \text{in} + (0.0126P)\ \text{kg}\ CO_2\ \text{generated} = CO_2\ \text{out} + 0\ \text{kg}\ CO_2\ \text{consumed}$$
$$\therefore CO_2\ \text{out} = (0.0126P)\ \text{kg} \qquad (4)$$

The total mass of gas out is 1250 kg. Therefore, adding the amounts of O_2, N_2, and CO_2 out from (2), (3), and (4):

$$1250 = (0.233A - 0.107P) + (0.767A) + (0.0126P)$$
$$1250 = A + 0.0019P$$
$$\therefore A = 1250 - 0.0019P \qquad (5)$$

Glucose balance

$$\text{glucose in} + 0\ \text{kg glucose generated}$$
$$= 0\ \text{kg glucose out} + (0.0467P)\ \text{kg glucose consumed}$$
$$\therefore \text{Glucose in} = (0.0467P)\ \text{kg} \qquad (6)$$

NH_3 balance

$$NH_3\ \text{in} + 0\ \text{kg}\ NH_3\ \text{generated}$$
$$= 0\ \text{kg}\ NH_3\ \text{out} + (0.00047P)\ \text{kg}\ NH_3\ \text{consumed}$$
$$\therefore NH_3\ \text{in} = (0.00047P)\ \text{kg} \qquad (7)$$

We can now calculate the total mass of the feed, F:

$$F = \text{glucose in} + NH_3\ \text{in} + \text{water in}$$

From (6) and (7):

$$F = (0.0467P)\ \text{kg} + (0.00047P)\ \text{kg} + 20{,}000\ \text{kg}$$
$$F = (20{,}000 + 0.04717P)\ \text{kg} \qquad (8)$$

We can now use (8) and (5) in (1):

$$(20{,}000 + 0.04717P) + (1250 - 0.0019P) = 1250 + P$$
$$20{,}000 = 0.95473P$$
$$\therefore P = 20{,}948.3\ \text{kg}$$

Substituting this result in (5) and (8):

$$A = 1210.2\ \text{kg}$$
$$F = 20{,}988.1\ \text{kg}$$

Also:

$$\text{Gum out} = 0.035P = 733.2\ \text{kg}$$

From Table 4.8:

$$O_2\ \text{in} = 282.0\ \text{kg}$$
$$N_2\ \text{in} = 928.2\ \text{kg}$$

Using the results for P, A, and F in (2), (3), (4), (6), and (7):

$$O_2\ \text{out} = 57.8\ \text{kg}$$
$$N_2\ \text{out} = 928.2\ \text{kg}$$
$$CO_2\ \text{out} = 263.9\ \text{kg}$$
$$\text{Glucose in} = 978.3\ \text{kg}$$
$$NH_3\ \text{in} = 9.8\ \text{kg}$$

Cell balance

$$0\ \text{kg cells in} + (0.0042P)\ \text{kg cells generated}$$
$$= \text{cells out} + 0\ \text{kg cells consumed}$$
$$\therefore \text{Cells out} = (0.0042P)\ \text{kg}$$
$$\text{Cells out} = 88.0\ \text{kg}$$

H_2O balance

$$20{,}000\ \text{kg}\ H_2O\ \text{in} + (0.00607P)\ \text{kg}\ H_2O\ \text{generated}$$
$$= H_2O\ \text{out} + 0\ \text{kg}\ H_2O\ \text{consumed}$$
$$\therefore H_2O\ \text{out} = 20{,}000 + (0.00607P)\ \text{kg}$$
$$H_2O\ \text{out} = 20{,}127.2\ \text{kg}$$

These entries are included in Table 4.9.

(iii) Check the results

All the columns and rows of Table 4.9 add up correctly to within round-off error.

4. Finalise

(i) The specific questions

TABLE 4.9 Completed Mass Balance Table (kg)

	In								Out								
Stream	**Glucose**	**O_2**	**N_2**	**CO_2**	**Gum**	**Cells**	**NH_3**	**H_2O**	**Total**	**Glucose**	**O_2**	**N_2**	**CO_2**	**Gum**	**Cells**	**NH_3**	**H_2O**	**Total**
Feed	978.3	0	0	0	0	0	9.8	20,000	20,988.1	–	–	–	–	–	–	–	–	–
Air	0	282.0	928.2	0	0	0	0	0	1210.2	–	–	–	–	–	–	–	–	–
Off-gas	–	–	–	–	–	–	–	–	–	0	57.8	928.2	263.9	0	0	0	0	1250
Product	–	–	–	–	–	–	–	–	–	0	0	0	0	733.2	88.0	0	20,127.2	20,948.3
Total	**978.3**	**282.0**	**928.2**	**0**	**0**	**0**	**9.8**	**20,000**	**22,198.3**	**0**	**57.8**	**928.2**	**263.9**	**733.2**	**88.0**	**0**	**20,127.2**	**22,198.3**

From the completed mass balance table, 978.3 kg glucose and 9.8 kg NH_3 are required. Calculation of the percentage of excess air is based on oxygen, because oxygen is the reacting component of air. The percentage excess can be calculated using Eq. (2.38) in units of kg:

$$\% \text{ excess air} = \frac{\begin{pmatrix}\text{kg } O_2 \text{ present} - \text{kg } O_2 \text{ required to react} \\ \text{completely with the limiting substrate}\end{pmatrix}}{\begin{pmatrix}\text{kg } O_2 \text{ required to react completely} \\ \text{with the limiting substrate}\end{pmatrix}} \times 100$$

In this problem, both glucose and ammonia are limiting substrates. From stoichiometry and the mass balance table, the mass of oxygen required to react completely with 978.3 kg glucose and 9.8 kg NH_3 is:

$$\frac{978.3 \text{ kg}}{1 \text{ kg}} (0.23 \text{ kg}) = 225.0 \text{ kg } O_2$$

The mass provided is 282.0 kg; therefore:

$$\% \text{ excess air} = \frac{282.0 - 225.0}{225.0} \times 100 = 25.3\%$$

(ii) Answers

(a) 980 kg glucose and 9.8 kg NH_3 are required.

(b) 25% excess air is provided.

4.5 MATERIAL BALANCES WITH RECYCLE, BYPASS, AND PURGE STREAMS

So far, we have performed mass balances on simple single-unit processes. However, steady-state systems incorporating recycle, bypass, and purge streams are common in bioprocess industries; flow sheets illustrating these modes of operation are shown in Figure 4.8. Material balance calculations for such systems can be somewhat more involved than those in Examples 4.3 through 4.6; several balances are required before all mass flows can be determined.

As an example, consider the system of Figure 4.9. Because cells are the catalysts in fermentation processes, it is often advantageous to recycle them from spent fermentation broth. Cell recycle requires a separation device, such as a centrifuge or gravity settling tank, to provide a concentrated recycle stream under aseptic conditions. The flow sheet for cell recycle is shown in Figure 4.10; as indicated, at least four different system boundaries can be defined. System I represents the overall recycle process; only the fresh feed and final product streams cross this system boundary. In addition, separate material balances can be performed over each process unit: the mixer, the fermenter, and the settler. Other system boundaries could also be defined; for example, we could group the mixer and

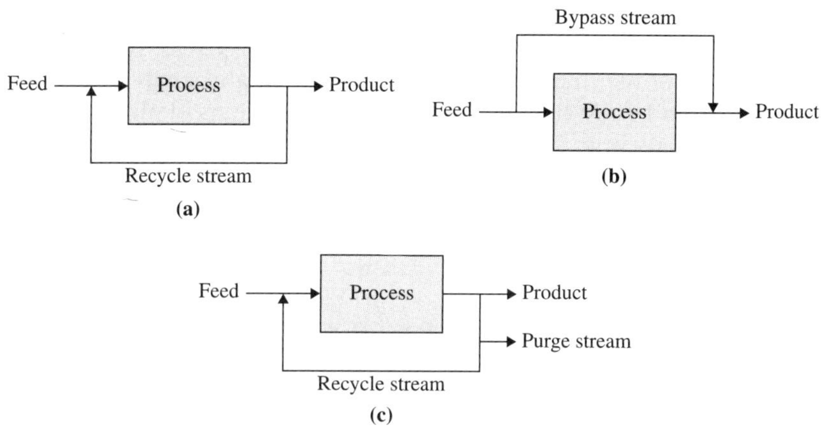

FIGURE 4.8 Flow sheet for processes with (a) recycle, (b) bypass, and (c) purge streams.

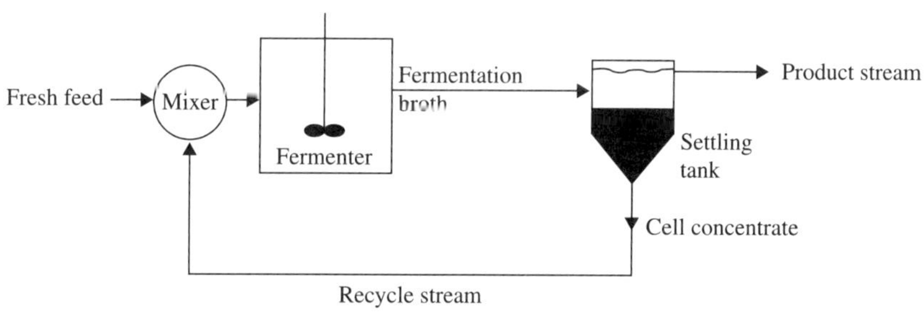

FIGURE 4.9 Fermenter with cell recycle.

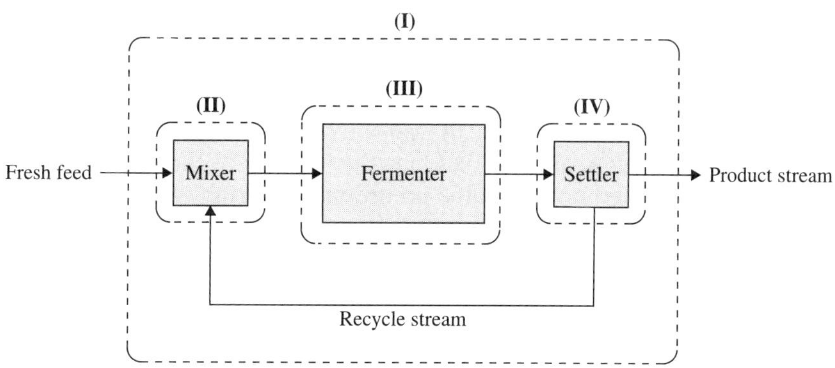

FIGURE 4.10 System boundaries for cell recycle system.

fermenter, or settler and fermenter, together. Material balances with recycle involve carrying out individual mass balance calculations for each designated system. Depending on which quantities are known and what information is sought, analysis of more than one system may be required before the flow rates and compositions of all streams are known.

Mass balances with recycle, bypass, or purge streams usually involve longer calculations than for simple processes, but are not more difficult conceptually. We will not treat these types of process further. Examples of mass balance procedures for multi-unit processes can be found in standard chemical engineering texts (e.g., [1–3]).

4.6 STOICHIOMETRY OF CELL GROWTH AND PRODUCT FORMATION

So far in this chapter, the law of conservation of mass has been used to determine unknown quantities entering or leaving bioprocesses. For mass balances with reaction such as in Examples 4.5 and 4.6, the stoichiometry of conversion must be known before the mass balance can be solved. When cell growth occurs, cells are a product of reaction and must be represented in the reaction equation. A widely used term for cells in fermentation processes is *biomass*. In this section we discuss how reaction equations for biomass growth and product synthesis are formulated. Metabolic stoichiometry has many applications in bioprocessing: as well as in mass and energy balances, it can be used to compare theoretical and actual product yields, check the consistency of experimental fermentation data, and formulate nutrient media.

4.6.1 Growth Stoichiometry and Elemental Balances

Despite its complexity and the thousands of intracellular reactions involved, cell growth obeys the law of conservation of matter. All atoms of carbon, hydrogen, oxygen, nitrogen, and other elements consumed during growth are incorporated into new cells or excreted as products. Confining our attention to those compounds taken up or produced in significant quantity, if the only extracellular products formed are CO_2 and H_2O, we can write the following general equation for aerobic cell growth:

$$C_wH_xO_yN_z + a\,O_2 + b\,H_gO_hN_i \longrightarrow c\,CH_\alpha O_\beta N_\delta + d\,CO_2 + e\,H_2O \tag{4.4}$$

In Eq. (4.4):

- $C_wH_xO_yN_z$ is the chemical formula for the carbon source or substrate (e.g., for glucose $C_6H_{12}O_6$, $w = 6$, $x = 12$, $y = 6$, and $z = 0$). Once the identity of the substrate is known, $C_wH_xO_yN_z$ is fully specified and contains no unknown variables.
- $H_gO_hN_i$ is the chemical formula for the nitrogen source (e.g., for ammonia NH_3, $g = 3$, $h = 0$, and $i = 1$). Once the identity of the nitrogen source is known, $H_gO_hN_i$ contains no unknown variables.
- $CH_\alpha O_\beta N_\delta$ is the chemical 'formula' for dry cells. The formula is a reflection of the dry biomass composition and is based on one C atom: α, β, and δ are the numbers of H, O, and N atoms, respectively, present in the biomass per C atom. There is no fundamental

TABLE 4.10 Elemental Composition of *Escherichia coli* Bacteria

Element	% Dry weight
C	50
O	20
N	14
H	8
P	3
S	1
K	1
Na	1
Ca	0.5
Mg	0.5
Cl	0.5
Fe	0.2
All others	0.3

From R.Y. Stanier, J.L. Ingraham, M.L. Wheelis, and P.R. Painter, 1986, The Microbial World, *5th ed., Prentice Hall, Upper Saddle River, NJ.*

objection to having a molecular formula for cells even if it is not widely applied in biology. As shown in Table 4.10, microorganisms such as *Escherichia coli* contain a wide range of elements; however 90 to 95% of the biomass can be accounted for by four major elements: C, H, O, and N. Compositions of several microbial species in terms of these four elements are listed in Table 4.11. Bacteria tend to have slightly higher nitrogen contents (11–14%) than fungi (6.3–9.0%) [4]. For a particular species, cell composition also depends on the substrate utilised and the culture conditions applied; hence the different entries in Table 4.11 for the same organism. However, the results are remarkably similar for different cells and conditions; $CH_{1.8}O_{0.5}N_{0.2}$ can be used as a general formula for cell biomass when composition analysis is not available. The average 'molecular weight' of cells based on C, H, O, and N content is therefore 24.6, although 5 to 10% residual ash is often added to account for those elements not included in the formula.

- a, b, c, d, and e are stoichiometric coefficients. Because Eq. (4.4) is written using a basis of one mole of substrate, a moles of O_2 are consumed and d moles of CO_2 are formed, for example, per mole of substrate reacted. The total amount of biomass formed during growth is accounted for by the stoichiometric coefficient c.

As illustrated in Figure 4.11, Eq. (4.4) represents a macroscopic view of metabolism; it ignores the detailed structure of the system and considers only those components that have net interchange with the environment. Eq. (4.4) does not include a multitude of compounds such as ATP and NADH that are integral to metabolism and undergo exchange cycles in cells, but are not subject to net exchange with the environment. Components such as

TABLE 4.11 Elemental Composition and Degree of Reduction for Selected Organisms

Organism	Elemental formula	Degree of reduction γ (relative to NH_3)
Bacteria		
Aerobacter aerogenes	$CH_{1.83}O_{0.55}N_{0.25}$	3.98
Escherichia coli	$CH_{1.77}O_{0.49}N_{0.24}$	4.07
Klebsiella aerogenes	$CH_{1.75}O_{0.43}N_{0.22}$	4.23
Klebsiella aerogenes	$CH_{1.73}O_{0.43}N_{0.24}$	4.15
Klebsiella aerogenes	$CH_{1.75}O_{0.47}N_{0.17}$	4.30
Klebsiella aerogenes	$CH_{1.73}O_{0.43}N_{0.24}$	4.15
Paracoccus denitrificans	$CH_{1.81}O_{0.51}N_{0.20}$	4.19
Paracoccus denitrificans	$CH_{1.51}O_{0.46}N_{0.19}$	3.96
Pseudomonas $C_{12}B$	$CH_{2.00}O_{0.52}N_{0.23}$	4.27
Fungi		
Candida utilis	$CH_{1.83}O_{0.54}N_{0.10}$	4.45
Candida utilis	$CH_{1.87}O_{0.56}N_{0.20}$	4.15
Candida utilis	$CH_{1.83}O_{0.46}N_{0.19}$	4.34
Candida utilis	$CH_{1.87}O_{0.56}N_{0.20}$	4.15
Saccharomyces cerevisiae	$CH_{1.64}O_{0.52}N_{0.16}$	4.12
Saccharomyces cerevisiae	$CH_{1.83}O_{0.56}N_{0.17}$	4.20
Saccharomyces cerevisiae	$CH_{1.81}O_{0.51}N_{0.17}$	4.28
Average	$CH_{1.79}O_{0.50}N_{0.20}$	4.19 (standard deviation = 3%)

From J.A. Roels, 1980, Application of macroscopic principles to microbial metabolism, Biotechnol. Bioeng. *22, 2457–2514.*

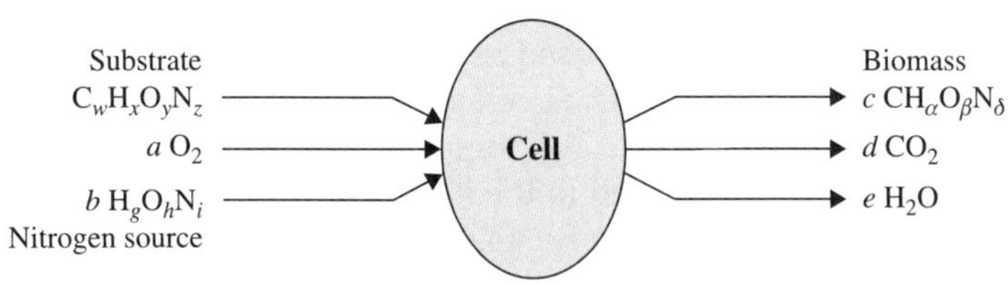

FIGURE 4.11 Conversion of substrate, oxygen, and nitrogen for cell growth.

vitamins and minerals taken up during metabolism could be included; however, since these materials are generally consumed in small quantities, we assume here that their contribution to the stoichiometry and energetics of reaction can be neglected. Other substrates and products can be added easily if appropriate. Despite its simplicity, the macroscopic approach provides a powerful tool for thermodynamic analysis.

Equation (4.4) is not complete unless the stoichiometric coefficients a, b, c, d, and e are known. Once a formula for biomass is obtained, these coefficients can be evaluated using normal procedures for balancing equations, that is, elemental balances and solution of simultaneous equations.

$$\text{C balance: } w = c + d \tag{4.5}$$

$$\text{H balance: } x + bg = c\alpha + 2e \tag{4.6}$$

$$\text{O balance: } y + 2a + bh = c\beta + 2d + e \tag{4.7}$$

$$\text{N balance: } z + bi = c\delta \tag{4.8}$$

Notice that we have five unknown coefficients (a, b, c, d, and e) but only four balance equations. This means that additional information is required before the equations can be solved. Usually this information is obtained from experiments. A useful measurable parameter is the *respiratory quotient*, RQ:

$$RQ = \frac{\text{moles } CO_2 \text{ produced}}{\text{moles } O_2 \text{ consumed}} = \frac{d}{a} \tag{4.9}$$

When an experimental value of RQ is available, Eqs. (4.5) through (4.9) can be solved to determine the stoichiometric coefficients. The results, however, are sensitive to small errors in RQ, which must be measured very accurately. When Eq. (4.4) is completed, the quantities of substrate, nitrogen, and oxygen required for production of biomass can be determined directly.

EXAMPLE 4.7 STOICHIOMETRIC COEFFICIENTS FOR CELL GROWTH

Production of single-cell protein from hexadecane is described by the following reaction equation:

$$C_{16}H_{34} + a\,O_2 + b\,NH_3 \longrightarrow c\,CH_{1.66}O_{0.27}N_{0.20} + d\,CO_2 + e\,H_2O$$

where $CH_{1.66}O_{0.27}N_{0.20}$ represents the biomass. If $RQ = 0.43$, determine the stoichiometric coefficients.

Solution

$$\text{C balance: } 16 = c + d \tag{1}$$

$$\text{H balance: } 34 + 3b = 1.66c + 2e \tag{2}$$

$$\text{O balance: } 2a = 0.27c + 2d + e \tag{3}$$

$$\text{N balance: } b = 0.20c \tag{4}$$

$$RQ\text{: } 0.43 = d/a \tag{5}$$

We must solve this set of simultaneous equations. Solution can be achieved in many different ways; usually it is a good idea to express each variable as a function of only one other variable.

b is already written simply as a function of c in (4); let us try expressing the other variables solely in terms of c. From (1):

$$d = 16 - c \qquad (6)$$

From (5):

$$a = \frac{d}{0.43} = 2.326d \qquad (7)$$

Combining (6) and (7) gives an expression for a in terms of c only:

$$a = 2.326\,(16 - c)$$
$$a = 37.22 - 2.326c \qquad (8)$$

Substituting (4) into (2) gives:

$$34 + 3\,(0.20c) = 1.66c + 2e$$
$$34 = 1.06c + 2e$$
$$e = 17 - 0.53c \qquad (9)$$

Substituting (8), (6), and (9) into (3) gives:

$$2\,(37.22 - 2.326c) = 0.27c + 2\,(16 - c) + (17 - 0.53c)$$
$$25.44 = 2.39c$$
$$c = 10.64$$

Using this result for c in (8), (4), (6), and (9) gives:

$$a = 12.48$$
$$b = 2.13$$
$$d = 5.37$$
$$e = 11.36$$

Check that these coefficient values satisfy Eqs. (1) through (5).

The complete reaction equation is:

$$C_{16}H_{34} + 12.5\ O_2 + 2.13\ NH_3 \longrightarrow 10.6\ CH_{1.66}O_{0.27}N_{0.20} + 5.37\ CO_2 + 11.4\ H_2O$$

Although elemental balances are useful, the presence of water in Eq. (4.4) causes some problems in practical application. Because water is usually present in great excess and changes in water concentration are inconvenient to measure or experimentally verify, H and O balances can present difficulties. Instead, a useful principle is conservation of reducing power or available electrons, which can be applied to determine quantitative relationships between substrates and products. An electron balance shows how available electrons from the substrate are distributed during reaction.

4.6.2 Electron Balances

Available electrons refers to the number of electrons available for transfer to oxygen on combustion of a substance to CO_2, H_2O, and nitrogen-containing compounds. The number of available electrons found in organic material is calculated from the valence of the various elements: 4 for C, 1 for H, −2 for O, 5 for P, and 6 for S. The number of available electrons for N depends on the reference state: −3 if ammonia is the reference, 0 for molecular nitrogen N_2, and 5 for nitrate. The reference state for cell growth is usually chosen to be the same as the nitrogen source in the medium. In the following discussion it will be assumed for convenience that ammonia is used as the nitrogen source; this can be changed easily if other nitrogen sources are employed.

Degree of reduction, γ, is defined as the number of equivalents of available electrons in that quantity of material containing 1 g atom carbon. For substrate $C_wH_xO_yN_z$:

$$\text{Number of available electrons} = 4w + x - 2y - 3z \tag{4.10}$$

Therefore, the degree of reduction of the substrate, γ_S, is:

$$\gamma_S = \frac{4w + x - 2y - 3z}{w} \tag{4.11}$$

Degrees of reduction relative to NH_3 and N_2 for several biological materials are given in Table C.2 in Appendix C. The number of available electrons and the degree of reduction of CO_2, H_2O, and NH_3 are zero. This means that the stoichiometric coefficients for these compounds do not appear in the electron balance, thus simplifying balance calculations.

Available electrons are conserved during metabolism. In a balanced growth equation, the number of available electrons is conserved by virtue of the fact that the amounts of each chemical element are conserved. Applying this principle to Eq. (4.4) with ammonia as the nitrogen source, and recognising that CO_2, H_2O, and NH_3 have zero available electrons, the available electron balance is:

$$\begin{aligned}&\text{Number of available electrons in the substrate}\\&\quad + \text{number of available electrons in } O_2\\&= \text{number of available electrons in the biomass}\end{aligned} \tag{4.12}$$

This relationship can be written as:

$$w\gamma_S - 4a = c\gamma_B \tag{4.13}$$

where γ_S and γ_B are the degrees of reduction of substrate and biomass, respectively. Note that the available-electron balance is not independent of the complete set of elemental balances: if the stoichiometric equation is balanced in terms of each element including H and O, the electron balance is implicitly satisfied.

4.6.3 Biomass Yield

Typically, Eq. (4.13) is used with carbon and nitrogen balances, Eqs. (4.5) and (4.8), and a measured value of *RQ* for evaluation of stoichiometric coefficients. However, as one electron balance, two elemental balances, and one measured quantity are inadequate information for solution of five unknown coefficients, another experimental quantity is required. During cell growth there is, as a general approximation, a linear relationship between the amount of biomass produced and the amount of substrate consumed. This relationship is expressed quantitatively using the *biomass yield*, Y_{XS}:

$$Y_{XS} = \frac{\text{g cells produced}}{\text{g substrate consumed}} \tag{4.14}$$

A large number of factors influence biomass yield, including medium composition, nature of the carbon and nitrogen sources, pH, and temperature. Biomass yield is greater in aerobic than in anaerobic cultures; choice of electron acceptor (e.g., O_2, nitrate, or sulphate) can also have a significant effect.

When Y_{XS} is constant throughout growth, its experimentally determined value can be used to evaluate the stoichiometric coefficient c in Eq. (4.4). Equation (4.14) expressed in terms of the stoichiometric relationship of Eq. (4.4) is:

$$Y_{XS} = \frac{c\ (\text{MW cells})}{\text{MW substrate}} \tag{4.15}$$

where MW is molecular weight and 'MW cells' means the biomass formula-weight plus any residual ash. However, before applying measured values of Y_{XS} and Eq. (4.15) to evaluate c, we must be sure that the experimental culture system is well represented by the stoichiometric equation. For example, we must be sure that substrate is not used in other types of reaction not represented by the reaction equation. One complication with real cultures is that some fraction of substrate consumed is always used for *maintenance activities* such as maintenance of membrane potential and internal pH, turnover of cellular components, and cell motility. These metabolic functions require substrate but do not necessarily produce cell biomass, CO_2, and H_2O in the way described by Eq. (4.4). Maintenance requirements and the difference between observed and true yields are discussed further in Chapter 12. For the time being, we will assume that the available values for biomass yield reflect consumption of substrate only in the reaction represented by the stoichiometric equation.

4.6.4 Product Stoichiometry

In many fermentations, extracellular products are formed during growth in addition to biomass. When this occurs, the stoichiometric equation can be modified to reflect product synthesis. Consider the formation of an extracellular product $C_jH_kO_lN_m$ during growth. Equation (4.4) is extended to include product synthesis as follows:

$$\begin{aligned} &C_wH_xO_yN_z + a\,O_2 + b\,H_gO_hN_i \\ &\quad \longrightarrow c\,CH_\alpha O_\beta N_\delta + d\,CO_2 + e\,H_2O + f\,C_jH_kO_lN_m \end{aligned} \tag{4.16}$$

where f is the stoichiometric coefficient for product. Product synthesis introduces one extra unknown stoichiometric coefficient to the equation; thus, an additional relationship between the reactants and products is required. This is usually provided as another experimentally determined yield coefficient, the *product yield from substrate*, Y_{PS}:

$$Y_{PS} = \frac{\text{g product formed}}{\text{g substrate consumed}} = \frac{f\,(\text{MW product})}{\text{MW substrate}} \tag{4.17}$$

As mentioned above with regard to biomass yields, we must be sure that the experimental system used to measure Y_{PS} conforms to Eq. (4.16). Equation (4.16) does not hold if product formation is not directly linked with growth; accordingly, it cannot be applied for secondary metabolite production such as penicillin fermentation, or for biotransformations such as steroid hydroxylation that involve only a small number of enzymes in cells. In these cases, independent reaction equations must be used to describe growth and product synthesis.

4.6.5 Theoretical Oxygen Demand

As oxygen is often the limiting substrate in aerobic fermentations, oxygen demand is an important parameter in bioprocessing. Oxygen demand is represented by the stoichiometric coefficient a in Eqs. (4.4) and (4.16). The requirement for oxygen is related directly to the electrons available for transfer to oxygen; oxygen demand can therefore be derived from an appropriate electron balance. When product synthesis occurs as represented by Eq. (4.16), the electron balance is:

$$\begin{aligned}&\text{Number of available electrons in the substrate}\\&\quad+\text{ number of available electrons in } O_2\\&=\text{number of available electrons in the biomass}\\&\quad+\text{ number of available electrons in the product}\end{aligned} \tag{4.18}$$

This relationship can be expressed as:

$$w\gamma_S - 4a = c\gamma_B + f\,j\gamma_P \tag{4.19}$$

where γ_P is the degree of reduction of the product. Rearranging gives:

$$a = \frac{1}{4}\,(w\gamma_S - c\gamma_B - f\,j\gamma_P) \tag{4.20}$$

Equation (4.20) is a very useful equation. It means that if we know which organism (γ_B), substrate (w and γ_S), and product (j and γ_P) are involved in cell culture, and the yields of biomass (c) and product (f), we can quickly calculate the oxygen demand. Of course we could also determine a by solving for all the stoichiometric coefficients of Eq. (4.16) as described in Section 4.6.1. However, Eq. (4.20) allows more rapid evaluation, and does not require that the quantities of NH_3, CO_2, and H_2O involved in the reaction be known.

4.6.6 Maximum Possible Yield

From Eq. (4.19) the fractional allocation of available electrons in the substrate can be written as:

$$1 = \frac{4a}{w\gamma_S} + \frac{c\gamma_B}{w\gamma_S} + \frac{fj\gamma_P}{w\gamma_S} \tag{4.21}$$

In Eq. (4.21), the first term on the right side is the fraction of available electrons transferred from the substrate to oxygen, the second term is the fraction of available electrons transferred to the biomass, and the third term is the fraction of available electrons transferred to the product. This relationship can be used to obtain upper bounds for the yields of biomass and product from substrate.

Let us define ζ_B as the fraction of available electrons in the substrate transferred to biomass:

$$\zeta_B = \frac{c\gamma_B}{w\gamma_S} \tag{4.22}$$

In the absence of product formation, if all available electrons were used for biomass synthesis, ζ_B would equal unity. Under these conditions, the maximum value of the stoichiometric coefficient c is:

$$c_{max} = \frac{w\gamma_S}{\gamma_B} \tag{4.23}$$

c_{max} can be converted to a biomass yield with mass units using Eq. (4.15). Therefore, even if we do not know the stoichiometry of growth, we can quickly calculate an upper limit for biomass yield from the molecular formulae for the substrate and product. If the composition of the cells is unknown, γ_B can be taken as 4.2 corresponding to the average biomass formula $CH_{1.8}O_{0.5}N_{0.2}$.

Maximum biomass yields for several substrates are listed in Table 4.12; the maximum biomass yield can be expressed in terms of mass ($Y_{XS,max}$), or as the number of C atoms in the biomass per substrate C-atom consumed (c_{max}/w). These quantities are sometimes known as *thermodynamic maximum biomass yields*. Table 4.12 shows that substrates with high energy content, indicated by high γ_S values, give high maximum biomass yields.

Likewise, the maximum possible product yield in the absence of biomass synthesis can be determined from Eq. (4.21):

$$f_{max} = \frac{w\gamma_S}{j\gamma_P} \tag{4.24}$$

Equation (4.24) allows us to quickly calculate an upper limit for the product yield from the molecular formulae for the substrate and product.

TABLE 4.12 Thermodynamic Maximum Biomass Yields

			Thermodynamic maximum yield corresponding to $\zeta_B = 1$	
Substrate	**Formula**	γ_S	**Carbon yield (c_{max}/w)**	**Mass yield $Y_{XS,max}$**
ALKANES				
Methane	CH_4	8.0	1.9	2.9
Hexane (*n*)	C_6H_{14}	6.3	1.5	2.6
Hexadecane (*n*)	$C_{16}H_{34}$	6.1	1.5	2.5
ALCOHOLS				
Methanol	CH_4O	6.0	1.4	1.1
Ethanol	C_2H_6O	6.0	1.4	1.5
Ethylene glycol	$C_2H_6O_2$	5.0	1.2	0.9
Glycerol	$C_3H_8O_3$	4.7	1.1	0.9
CARBOHYDRATES				
Formaldehyde	CH_2O	4.0	0.95	0.8
Glucose	$C_6H_{12}O_6$	4.0	0.95	0.8
Sucrose	$C_{12}H_{22}O_{11}$	4.0	0.95	0.8
Starch	$(C_6H_{10}O_5)_x$	4.0	0.95	0.9
ORGANIC ACIDS				
Formic acid	CH_2O_2	2.0	0.5	0.3
Acetic acid	$C_2H_4O_2$	4.0	0.95	0.8
Propionic acid	$C_3H_6O_2$	4.7	1.1	1.1
Lactic acid	$C_3H_6O_3$	4.0	0.95	0.8
Fumaric acid	$C_4H_4O_4$	3.0	0.7	0.6
Oxalic acid	$C_2H_2O_4$	1.0	0.24	0.1

From L.E. Erickson, I.G. Minkevich, and V.K. Eroshin, 1978, Application of mass and energy balance regularities in fermentation, Biotechnol. Bioeng. *20, 1595–1621.*

EXAMPLE 4.8 PRODUCT YIELD AND OXYGEN DEMAND

The chemical reaction equation for respiration of glucose is:

$$C_6H_{12}O_6 + 6\,O_2 \longrightarrow 6\,CO_2 + 6\,H_2O$$

Candida utilis cells convert glucose to CO_2 and H_2O during growth. The cell composition is $CH_{1.84}O_{0.55}N_{0.2}$ plus 5% ash. The yield of biomass from substrate is $0.5\ g\ g^{-1}$. Ammonia is used as the nitrogen source.

(a) What is the oxygen demand with growth compared to that without?

(b) *C. utilis* is also able to grow using ethanol as substrate, producing cells of the same composition as above. On a mass basis, how does the maximum possible biomass yield from ethanol compare with the maximum possible yield from glucose?

Solution

Molecular weights:

- Glucose = 180
- Ethanol = 46

MW biomass is (25.44 + ash). Since ash accounts for 5% of the total weight, 95% of the total MW = 25.44. Therefore, MW biomass = 25.44/0.95 = 26.78. From Table C.2, γ_S for glucose is 4.00; γ_S for ethanol is 6.00. $\gamma_B = (4 \times 1 + 1 \times 1.84 - 2 \times 0.55 - 3 \times 0.2) = 4.14$. For glucose $w = 6$; for ethanol $w = 2$.

(a) $Y_{XS} = 0.5 \text{ g g}^{-1}$. Converting this mass yield to a molar yield:

$$Y_{XS} = \frac{0.5 \text{ g biomass}}{\text{g glucose}} \cdot \left| \frac{180 \text{ g biomass}}{1 \text{ gmol glucose}} \right| \cdot \left| \frac{1 \text{ gmol biomass}}{26.78 \text{ g biomass}} \right|$$

$$Y_{XS} = 3.36 \frac{\text{gmol biomass}}{\text{gmol glucose}} = c$$

Oxygen demand is given by Eq. (4.20). In the absence of product formation:

$$a = \frac{1}{4}[6(4.00) - 3.36(4.14)] = 2.52$$

Therefore, the oxygen demand for glucose respiration with growth is 2.5 gmol O_2 per gmol glucose consumed. By comparison with the chemical reaction equation for respiration, this is only about 42% of that required in the absence of growth.

(b) The maximum possible biomass yield is given by Eq. (4.23). Using the data above, for glucose:

$$c_{max} = \frac{6(4.00)}{4.14} = 5.80$$

Converting this to a mass basis:

$$Y_{XS,max} = \frac{5.80 \text{ gmol biomass}}{\text{gmol glucose}} \cdot \left| \frac{1 \text{ gmol glucose}}{180 \text{ g glucose}} \right| \cdot \left| \frac{26.78 \text{ g biomass}}{1 \text{ gmol biomass}} \right|$$

$$Y_{XS,max} = 0.86 \frac{\text{g biomass}}{\text{g glucose}}$$

For ethanol, from Eq. (4.23):

$$c_{max} = \frac{2(6.00)}{4.14} = 2.90$$

and

$$Y_{XS,max} = \frac{2.90 \text{ gmol biomass}}{\text{gmol ethanol}} \cdot \left| \frac{1 \text{ gmol ethanol}}{46 \text{ g ethanol}} \right| \cdot \left| \frac{26.78 \text{ g biomass}}{1 \text{ gmol biomass}} \right|$$

$$Y_{XS,max} = 1.69 \frac{\text{g biomass}}{\text{g ethanol}}$$

Therefore, on a mass basis, the maximum possible amount of biomass produced per gram of ethanol consumed is roughly twice that per gram of glucose consumed. This result is consistent with the data for $Y_{XS,max}$ listed in Table 4.12.

Example 4.8 illustrates two important points. First, the chemical reaction equation for conversion of substrate without growth is a poor approximation of overall stoichiometry when cell growth occurs. When estimating yields and oxygen requirements for any process involving cell growth, the full stoichiometric equation including biomass should be used. Second, the chemical nature or oxidation state of the substrate has a major influence on biomass and product yields through the number of available electrons.

SUMMARY OF CHAPTER 4

At the end of Chapter 4 you should:

- Understand the terms *system*, *surroundings*, *boundary*, and *process* in thermodynamics
- Be able to identify *open* and *closed systems*, and *batch*, *semi-batch*, *fed-batch*, and *continuous processes*
- Understand the difference between *steady state* and *equilibrium*
- Be able to write appropriate equations for conservation of mass for processes with and without reaction
- Be able to solve simple mass balance problems with and without reaction
- Be able to apply stoichiometric principles for macroscopic analysis of cell growth and product formation

PROBLEMS

4.1 Cell concentration using membranes

A battery of cylindrical hollow-fibre membranes is operated at steady state to concentrate a bacterial suspension harvested from a fermenter. Fermentation broth is pumped at a rate of 350 kg min^{-1} through a stack of hollow-fibre membranes as shown in Figure 4P1.1. The broth contains 1% bacteria; the rest may be considered water. Buffer solution enters the annular space around the membrane tubes at a rate of 80 kg min^{-1}; because broth in the membrane tubes is under pressure, water is forced across the membrane into the buffer.

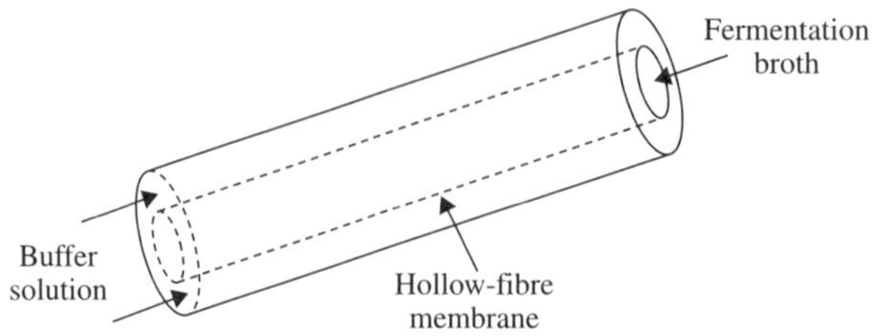

FIGURE 4P1.1 Hollow-fibre membrane for concentration of cells.

Cells in the broth are too large to pass through the membrane and pass out of the tubes as a concentrate.

The aim of the membrane system is to produce a cell suspension containing 6% biomass.

(a) What is the flow rate from the annular space?

(b) What is the flow rate of cell suspension from the membrane tubes?

Assume that the cells are not active, that is, they do not grow. Assume further that the membrane does not allow any molecules other than water to pass from annulus to inner cylinder, or vice versa.

4.2 Membrane reactor

A battery of cylindrical membranes similar to that shown in Figure 4P1.1 is used for an extractive bioconversion. Extractive bioconversion means that fermentation and extraction of product occur at the same time.

Yeast cells are immobilised within the membrane walls. A 10% glucose in water solution is passed through the annular space at a rate of 40 kg h^{-1}. An organic solvent, such as 2-ethyl-1,3-hexanediol, enters the inner tube at a rate of 40 kg h^{-1}.

Because the membrane is constructed of a polymer that repels organic solvents, the hexanediol cannot penetrate the membrane and the yeast is relatively unaffected by its toxicity. On the other hand, because glucose and water are virtually insoluble in 2-ethyl-1,3-hexanediol, these compounds do not enter the inner tube to an appreciable extent. Once immobilised in the membrane, the yeast cells cannot reproduce but convert glucose to ethanol according to the equation:

$$C_6H_{12}O_6 \longrightarrow 2\,C_2H_6O + 2\,CO_2$$

Ethanol is soluble in 2-ethyl-1,3-hexanediol; it diffuses into the inner tube and is carried out of the system. CO_2 gas exits from the membrane unit through an escape valve. In the aqueous stream leaving the annular space, the concentration of unconverted glucose is 0.2% and the concentration of ethanol is 0.5%. If the system operates at steady state:

(a) What is the concentration of ethanol in the hexanediol stream leaving the reactor?

(b) What is the mass flow rate of CO_2?

4.3 Ethanol distillation

Liquid from a brewery fermenter can be considered to contain 10% ethanol and 90% water. This fermentation product is pumped at a rate of 50,000 kg h^{-1} to a distillation column on the factory site. Under current operating conditions, a distillate of 45% ethanol and 55% water is produced from the top of the column at a rate of one-tenth that of the feed.

(a) What is the composition of the waste 'bottoms' from the still?

(b) What is the rate of alcohol loss in the bottoms?

4.4 Raspberry coulis manufacture

A food company produces raspberry coulis as a topping for its best-selling line, double-strength chocolate mousse pie. Fresh raspberries comprising 5% seeds, 20% pulp solids, and 75% water are homogenised and placed in a stainless steel vat. Sugar is added to give a raspberry:sugar mass ratio of 3.5:1. The mixture is blended, strained to remove the seeds, then heated to reduce the water content to 35%. Half a tonne of coulis is produced every day.

(a) What mass of raspberries is required per week?

(b) How much sugar is required per week?

(c) What is the sugar content of the coulis?

4.5 Polyethylene glycol–salt mixture

Aqueous two-phase extraction is used to purify a recombinant HIV–β-galactosidase fusion peptide produced in *Escherichia coli*. For optimum separation, 450 kg of a mixture of 19.7% w/w polyethylene glycol (PEG) and 17.7% w/w potassium phosphate salt in water is needed. Left over from previous pilot-plant trials is 100 kg of a mixture of 20% w/w PEG in water, and 150 kg of a mixture of 20% w/w PEG and 25% w/w salt in water. Also on hand is 200 kg of an aqueous stock solution of 50% w/w PEG, 200 kg of an aqueous stock solution of 40% w/w salt, and an unlimited supply of extra water.

If all of both leftover mixtures must be used, how much of each stock solution and additional water is required?

4.6 Tetracycline crystallisation

Tetracycline produced in *Streptomyces aureus* fermentations is purified by crystallisation. One hundred kg of a supersaturated solution containing 7.7 wt% tetracycline is cooled in a batch fluidised-bed crystalliser. Seed crystals of tetracycline are added at a concentration of 40 ppm to promote crystal growth. At the end of the crystallisation process, the remaining solution contains 2.8% tetracycline.

(a) What is the mass of the residual tetracycline solution?

(b) What mass of tetracycline crystals is produced?

4.7 Flow rate calculation

A solution of 5% NaCl in water is flowing in a stainless steel pipe. To estimate the flow rate, one of your colleagues starts pumping a 30% NaCl tracer solution into the pipe at a rate of 80 ml s^{-1} while you measure the concentration of NaCl downstream after the solutions are well mixed. If the downstream NaCl concentration is 9.5%, use mass balance principles to determine the flow rate of the initial 5% NaCl solution.

4.8 Azeotropic distillation

Absolute or 100% ethanol is produced from a mixture of 95% ethanol and 5% water using the Keyes distillation process. A third component, benzene, is added to lower the volatility of the alcohol. Under these conditions, the overhead product is a constant-boiling mixture of 18.5% ethanol, 7.4% H_2O, and 74.1% benzene. The process is outlined in Figure 4P8.1.

Use the following data to calculate the volume of benzene that should be fed to the still in order to produce 250 litres of absolute ethanol: ρ (100% alcohol) = 0.785 g cm^{-3}; ρ (benzene) = 0.872 g cm^{-3}.

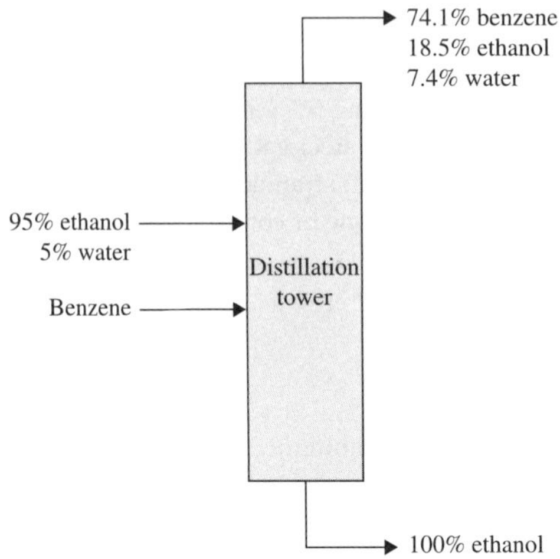

FIGURE 4P8.1 Flow sheet for Keyes distillation process.

4.9 Microparticles for drug release

Poly(DL-lactic-co-glycolic acid)/poly(ethylene glycol) blends are being studied for manufacture of biodegradable microparticles to be used for controlled parenteral delivery of IgG antibody. The first step in preparing the microparticles is to make up a mixture containing 0.2% antibody, 18% polymer blend, 9% water, and 72.8% dichloromethane. Available for this procedure is 1 kg of a blend of 1% polyethylene glycol (PEG) in poly(DL-lactic-co-glycolic acid) (PLGA), 4 kg of a solution of 15% PEG in dichloromethane, and a stock of pure dichloromethane. All of the PEG/PLGA blend and all of the PEG/dichloromethane solution are used.

(a) What quantity of pure dichloromethane is needed?

(b) Antibody is added as an aqueous solution. What quantity and composition of antibody solution are required?

(c) How much final mixture is produced?

(d) What is the concentration of PEG in the final mixture?

4.10 Removal of glucose from dried egg

The enzyme glucose oxidase is used commercially to remove glucose from dehydrated egg to improve colour, flavour, and shelf life. The reaction is:

$$\underset{\text{(glucose)}}{C_6H_{12}O_6} + O_2 + H_2O \longrightarrow \underset{\text{(gluconic acid)}}{C_6H_{12}O_7} + H_2O_2$$

A continuous-flow reactor is set up using immobilised enzyme beads that are retained inside the vessel. Dehydrated egg slurry containing 2% glucose, 20% water, and the remainder unreactive egg solids is available at a rate of 3000 kg h^{-1}. Air is pumped through the reactor contents so that 18 kg oxygen are delivered per hour. The desired glucose level in the dehydrated egg product leaving the enzyme reactor is 0.2%. Determine:

(a) Which is the limiting substrate

(b) The percentage excess substrate/s
(c) The composition of the reactor off-gas
(d) The composition of the final egg product

4.11 Culture of plant roots

Plant roots can be used to produce valuable chemicals in vitro. A batch culture of *Atropa belladonna* roots is established in an air-driven reactor as shown in Figure 4P11.1. The culture is maintained at 25°C. Because roots cannot be removed during operation of the reactor, it is proposed to monitor growth using mass balances.

Nutrient medium containing 3% glucose and 1.75% NH_3 is prepared for the culture; the remainder of the medium can be considered water. Air at 25°C and 1 atm pressure is sparged into the reactor at a rate of 22 $cm^3\ min^{-1}$. During a 10-day culture period, 1425 g of nutrient medium and 47 litres of O_2 are provided, while 15 litres of CO_2 are collected in the off-gas. After 10 days, 1110 g of liquid containing 0.063% glucose and 1.7% dissolved NH_3 is drained from the vessel. The ratio of fresh weight to dry weight for roots is known to be 14:1.

(a) What dry mass of roots is produced in 10 days?
(b) Write the reaction equation for growth, indicating the approximate chemical formula for the roots, $CH_\alpha O_\beta N_\delta$.
(c) What is the limiting substrate?
(d) What is the yield of roots from glucose?

4.12 Production of 1,3-propanediol

A process for microbial synthesis of 1,3-propanediol is being developed for the manufacture of 'green' polyester fabric from renewable resources. Under anaerobic conditions, a selected strain of *Klebsiella pneumoniae* converts glycerol ($C_3H_8O_3$) to 1,3-propanediol ($C_3H_8O_2$) and acetic acid ($C_2H_4O_2$), with minimal formation of other

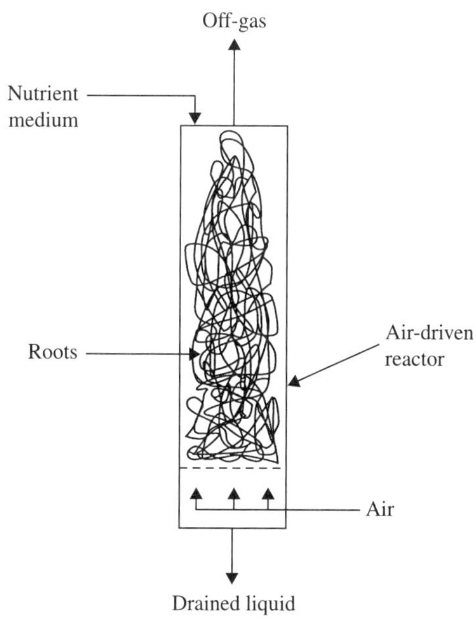

FIGURE 4P11.1 Reactor for culture of plant roots.

fermentation products such as butyric acid, ethanol, and H_2 gas. The fermentation and cell growth equation can be written as:

$$68\ C_3H_8O_3 + 3\ NH_3 \longrightarrow 3\ C_4H_7O_2N + 49\ C_3H_8O_2 + 15\ C_2H_4O_2 + 15\ CO_2 + 40\ H_2O$$

where $C_4H_7O_2N$ represents the biomass.

A continuous fermenter is set up for 1,3-propanediol production at 37°C and atmospheric pressure. Anaerobic conditions are maintained by sparging the broth with nitrogen gas at a flow rate of 1000 litres min^{-1}. The feed rate of medium into the fermenter is 1000 kg h^{-1}; the medium contains ammonia and 14% w/w glycerol. The yield of 1,3-propanediol is strongly affected by glycerol concentration, which must be kept above a certain level to suppress the formation of undesired by-products. Accordingly, the fermentation process is designed so that the product stream contains an unreacted glycerol concentration of 3% w/w.

(a) What is the volumetric flow rate and composition of the off-gas?
(b) What minimum concentration of NH_3 is needed in the feed stream?
(c) If the NH_3 concentration from (b) is used, what is the concentration of 1,3-propanediol in the product stream?

4.13 Cell culture using whey

Waste dairy whey is used to grow *Kluyveromyces fragilis* yeast in a continuous culture system operated at 30°C and 1 atm pressure. Medium containing 4% w/w lactose ($C_{12}H_{22}O_{11}$) and 0.15% w/w NH_3 flows into a specially designed aerated bioreactor at a rate of 200 kg h^{-1}. The reactor is compartmentalised to facilitate gravity settling of the yeast; a suspension containing concentrated cells is drawn off continuously from the bottom of the reactor at a rate of 40 kg h^{-1}, while an aqueous waste stream containing 0.5 kg cells per 100 kg leaves from the top. All of the lactose provided is utilised by the culture. The biomass yield from lactose is known from preliminary experiments to be 0.25 g g^{-1}. The composition of the *K. fragilis* biomass is determined by elemental analysis to be $CH_{1.63}O_{0.54}N_{0.16}$ + 7.5% ash.

(a) What is the *RQ* for this culture?
(b) It is proposed to supply the reactor with air at a maximum flow rate of 305 litres min^{-1}. Will this provide sufficient oxygen? If not, what minimum level of oxygen enrichment of the air is needed if the total gas flow rate is 305 litres min^{-1}?
(c) What is the concentration of residual NH_3 in the aqueous waste stream?
(d) What is the concentration of cells in the cell concentrate?

4.14 Oxygen requirement for growth on glycerol

Klebsiella aerogenes is produced from glycerol in aerobic culture with ammonia as nitrogen source. The biomass contains 8% ash, 0.40 g biomass is produced for each g of glycerol consumed, and no major metabolic products are formed. What is the oxygen requirement for this culture in mass terms?

4.15 Product yield in anaerobic digestion

Anaerobic digestion of volatile acids by methane bacteria is represented by the equation:

$$\underset{\text{(acetic acid)}}{CH_3COOH} + NH_3 \longrightarrow \text{biomass} + CO_2 + H_2O + \underset{\text{(methane)}}{CH_4}$$

The composition of methane bacteria is approximated by the empirical formula $CH_{1.4}O_{0.40}N_{0.20}$. For each kg of acetic acid consumed, 0.67 kg of CO_2 is evolved. How does the yield of methane under these conditions compare with the maximum possible yield?

4.16 Production of PHB

Poly-3-hydroxybutyrate (PHB) is a biodegradable thermoplastic accumulated intracellularly by many microorganisms under unfavourable growth conditions. *Azotobacter chroococcum* is being investigated for commercial PHB production using cheap soluble starch as the raw material and ammonia as the nitrogen source. Synthesis of PHB is observed to be growth-associated with maximum production occurring when the culture is provided with limited oxygen. During steady-state continuous culture of *A. chroococcum*, the concentration of PHB in the cells is 44% w/w and the respiratory coefficient is 1.3. From elemental analysis, *A. chroococcum* biomass without PHB can be represented as $CH_2O_{0.5}N_{0.25}$. The monomeric unit for starch is $C_6H_{10}O_5$; $C_4H_6O_2$ is the monomeric unit for PHB.

(a) Develop an empirical reaction equation for PHB production and cell growth. PHB can be considered a separate product of the culture even though it is not excreted from the biomass.

(b) What is the yield of PHB-containing cells from starch in units of $g\ g^{-1}$?

(c) If downstream recovery of PHB from the cells involves losses of about 35%, how many kg of starch are needed for production and recovery of 25 kg PHB?

4.17 Oxygen consumption by suspended plant cells

Plant cells are cultured in a bioreactor using sucrose ($C_{12}H_{22}O_{11}$) as the carbon source and ammonia (NH_3) as the nitrogen source. The vessel is sparged with air. Biomass is the major product formed; however, because the cells are subject to lysis, significant levels of excreted by-product with the same molecular composition as the biomass are also produced. Elemental analysis of the plant cells gives a molecular formula of $CH_{1.63}O_{0.80}N_{0.13}$ with negligible ash. Yield measurements show that 0.32 g of intact cells is produced per g of sugar consumed, while 0.2 g of by-product is formed per g of intact biomass. If 10 kg sugar is consumed per hour, at what rate must oxygen be provided to the reactor in units of $gmol\ min^{-1}$?

4.18 Substrate requirements for continuous culture

The acetic acid bacterium *Acetobacter pasteurianus* is cultured under aerobic conditions in a continuous fermenter using ethanol (C_2H_6O) as the substrate and ammonia as the nitrogen source. Under these conditions, acetate accumulation is completely suppressed so that biomass is the only major product. The rate of ethanol consumption is $150\ g\ h^{-1}$ and the rate of biomass production is $45\ g\ h^{-1}$.

(a) What is the rate of oxygen consumption during the culture?

(b) Ammonia is fed to the culture at a rate of $20\ g\ h^{-1}$. At what rate does unreacted ammonia leave the reactor?

4.19 Oxygen and sulphur requirements for bacterial culture

Filamentous *Saccharopolyspora erythraea* bacteria are grown in medium containing glucose as the carbon source, ammonia as the nitrogen source, and sulphate as the sulphur source. No major products other than biomass are produced during cell growth. Sulphur is included easily in the stoichiometric equation if we use sulphuric acid (H_2SO_4) to represent the sulphur source and define the valence of S as +6. The molecular formula for *S. erythraea* biomass is found to be $CH_{1.73}O_{0.52}N_{0.17}S_{0.0032}$ and the biomass yield from glucose is 0.29 g g^{-1}.

(a) What are the oxygen requirements for this culture?

(b) If the medium contains 20 g l^{-1} glucose and the culture proceeds until all the glucose is consumed, what minimum concentration of sulphate must also be included in the medium in units of gmol l^{-1}?

4.20 Stoichiometry of single-cell protein synthesis

(a) *Cellulomonas* bacteria used as single-cell protein for human or animal food are produced from glucose under anaerobic conditions. All carbon in the substrate is converted into biomass; ammonia is used as the nitrogen source. The molecular formula for the biomass is $CH_{1.56}O_{0.54}N_{0.16}$; the cells also contain 5% ash. How does the yield of biomass from substrate in mass and molar terms compare with the maximum possible biomass yield?

(b) Another system for manufacture of single-cell protein is *Methylophilus methylotrophus*. This organism is produced aerobically from methanol with ammonia as nitrogen source. The molecular formula for the biomass is $CH_{1.68}O_{0.36}N_{0.22}$; these cells contain 6% ash.

(i) How does the maximum yield of biomass compare with that found in (a)? What is the main reason for the difference?

(ii) If the actual yield of biomass from methanol is 42% of the thermodynamic maximum, what is the oxygen demand?

4.21 Ethanol production by yeast and bacteria

Both *Saccharomyces cerevisiae* yeast and *Zymomonas mobilis* bacteria produce ethanol from glucose under anaerobic conditions without external electron acceptors. The biomass yield from glucose is 0.11 g g^{-1} for yeast and 0.05 g g^{-1} for *Z. mobilis*. In both cases the nitrogen source is NH_3. Both cell compositions are represented by the formula $CH_{1.8}O_{0.5}N_{0.2}$.

(a) What is the yield of ethanol from glucose in both cases?

(b) How do the yields calculated in (a) compare with the thermodynamic maximum?

4.22 Detecting unknown products

Yeast cells growing in continuous culture produce 0.37 g biomass per g glucose consumed; about 0.88 g O_2 is consumed per g cells formed. The nitrogen source is ammonia, and the biomass composition is $CH_{1.79}O_{0.56}N_{0.17}$. Are other products also synthesised?

4.23 Medium formulation

Pseudomonas 5401 is to be used for production of single-cell protein for animal feed. The substrate is fuel oil. The composition of *Pseudomonas* 5401 is $CH_{1.83}O_{0.55}N_{0.25}$. If the final cell concentration is 25 g l^{-1}, what minimum concentration of $(NH_4)_2SO_4$ must be provided in the medium if $(NH_4)_2SO_4$ is the sole nitrogen source?

4.24 Oxygen demand for production of recombinant protein

Recombinant protein is produced by a genetically engineered strain of *Escherichia coli* during cell growth. The recombinant protein can be considered a product of cell culture even though it is not secreted from the cells; it is synthesised in addition to normal *E. coli* biomass. Ammonia is used as the nitrogen source for aerobic respiration of glucose.

The recombinant protein has an overall formula of $CH_{1.55}O_{0.31}N_{0.25}$. The yield of biomass (excluding recombinant protein) from glucose is measured as 0.48 g g^{-1}; the yield of recombinant protein from glucose is about 20% of that for cells.

(a) How much ammonia is required?

(b) What is the oxygen demand?

(c) If the biomass yield remains at 0.48 g g^{-1}, how much different are the ammonia and oxygen requirements for wild-type *E. coli* that is unable to synthesise recombinant protein?

4.25 Effect of growth on oxygen demand

The chemical reaction equation for conversion of ethanol (C_2H_6O) to acetic acid ($C_2H_4O_2$) is:

$$C_2H_6O + O_2 \longrightarrow C_2H_4O_2 + H_2O$$

Acetic acid is produced from ethanol during growth of *Acetobacter aceti*, which has the composition $CH_{1.8}O_{0.5}N_{0.2}$. The biomass yield from substrate is 0.14 g g^{-1}; the product yield from substrate is 0.92 g g^{-1}. Ammonia is used as the nitrogen source. How does growth in this culture affect the oxygen demand for acetic acid production?

4.26 Aerobic sugar metabolism

Candida stellata is a yeast frequently found in wine fermentations. Its sugar metabolism is being studied under aerobic conditions. In continuous culture with fructose ($C_6H_{12}O_6$) as the carbon source and ammonium phosphate as the nitrogen source, the yield of biomass from fructose is 0.025 g g^{-1}, the yield of ethanol (C_2H_6O) from fructose is 0.21 g g^{-1}, and the yield of glycerol ($C_3H_8O_3$) from fructose is 0.07 g g^{-1}.

(a) If the rate of sugar consumption is 190 g h^{-1}, what is the rate of oxygen consumption in g h^{-1}?

(b) What is the *RQ* for this culture?

4.27 Stoichiometry of animal cell growth

Analysis of the stoichiometry of animal cell growth can be complicated because of the large number of macronutrients involved (about 30 amino acids and vitamins plus other organic components and inorganic salts), and because the stoichiometry is sensitive to nutrient concentrations. Nevertheless, glucose ($C_6H_{12}O_6$) and glutamine ($C_5H_{10}O_3N_2$) can be considered the main carbon sources for animal cell growth; glutamine is also the primary nitrogen source. The major metabolic by-products are lactic acid ($C_3H_6O_3$) and ammonia (NH_3).

(a) A simplified stoichiometric equation for growth of hybridoma cells is:

$$C_6H_{12}O_6 + p\, C_5H_{10}O_3N_2 + q\, O_2 + r\, CO_2$$
$$\longrightarrow s\, CH_{1.82}O_{0.84}N_{0.25} + t\, C_3H_6O_3 + u\, NH_3 + v\, CO_2 + w\, H_2O$$

where $CH_{1.82}O_{0.84}N_{0.25}$ represents the biomass and $p, q, r, s, t, u, v,$ and w are stoichiometric coefficients. In a test culture, for every g of glucose consumed, 0.42 g glutamine was taken up and 0.90 g lactic acid and 0.26 g cells were produced.

(i) What is the net carbon dioxide production per 100 g glucose?

(ii) What is the oxygen demand?

(iii) A typical animal cell culture medium contains 11 mM glucose and 2 mM glutamine. Which of these substrates is present in excess?

(iv) To prevent toxic effects on the cells, the concentrations of lactic acid and ammonia must remain below 1 g l^{-1} and 0.07 g l^{-1}, respectively. What maximum concentrations of glucose and glutamine should be provided in the medium used for batch culture of hybridoma cells?

(b) The anabolic component of the stoichiometric equation in (a) can be represented using the equation:

$$C_6H_{12}O_6 + p\, C_5H_{10}O_3N_2 + r\, CO_2 \longrightarrow s\, CH_{1.82}O_{0.84}N_{0.25}$$

(i) What proportion of the carbon in the biomass is derived from glucose, and what proportion is derived from glutamine?

(ii) Considering both equations from (a) and (b), estimate the proportions of the carbon in glucose and glutamine, respectively, that are used for biomass production during culture of hybridoma cells.

4.28 pH as an indicator of yeast growth

pH may be used as an online indicator of biomass concentration in some culture systems. The main contributor to total proton levels in aerobic yeast culture is nitrogen assimilation; when ammonia is the nitrogen source, there is an approximate 1:1 molar ratio between the rate of H^+ production and the rate of N uptake. Under conditions that minimise the formation of other acidic, basic, or nitrogen-containing products except biomass, the stoichiometric equation for yeast growth on glucose can be written as:

$$C_6H_{12}O_6 + a\, O_2 + b\, NH_3 \rightarrow c\, CH_\alpha O_\beta N_\delta + d\, CO_2 + e\, H_2O + f\, H^+$$

where $CH_\alpha O_\beta N_\delta$ is the biomass. If the molecular formula for yeast biomass is $CH_{1.66}O_{0.50}N_{0.15}$, show why the rate of proton evolution measured using a pH probe will be 0.15 times the rate of cell growth expressed on a molar basis.

References

[1] R.M. Felder, R.W. Rousseau, Elementary Principles of Chemical Processes, third ed., Wiley, 2005 (Chapter 4).

[2] D.M. Himmelblau, J.B. Riggs, Basic Principles and Calculations in Chemical Engineering, seventh ed., Prentice Hall, 2004 (Chapters 6–12).

[3] R.K. Sinnott, Coulson and Richardson's Chemical Engineering, volume 6: Chemical Engineering Design, fourth ed., Elsevier, 2005 (Chapter 2).

[4] J.-L. Cordier, B.M. Butsch, B. Birou, U. von Stockar, The relationship between elemental composition and heat of combustion of microbial biomass, Appl. Microbiol. Biotechnol. 25 (1987) 305–312.

Suggestions for Further Reading

Process Mass Balances

See references [1] through [3].

Metabolic Stoichiometry

See also reference [4].

Atkinson, B., & Mavituna, F. (1991). *Biochemical engineering and biotechnology handbook* (2nd ed., Chapter 4). Macmillan.

Erickson, L. E., Minkevich, I. G., & Eroshin, V. K. (1978). Application of mass and energy balance regularities in fermentation. *Biotechnol. Bioeng.*, *20*, 1595–1621.

Roels, J. A. (1983). *Energetics and kinetics in biotechnology* (Chapter 3). Elsevier Biomedical Press.

van Gulik, W. M., ten Hoopen, H. J. G., & Heijnen, J. J. (1992). Kinetics and stoichiometry of growth of plant cell cultures of *Catharanthus roseus* and *Nicotiana tabacum* in batch and continuous fermentors. *Biotechnol. Bioeng.*, *40*, 863–874.

Zeng, A.-P., Hu, W.-S., & Deckwer, W.-D. (1998). Variation of stoichiometric ratios and their correlation for monitoring and control of animal cell cultures. *Biotechnol. Prog.*, *14*, 434–441.

CHAPTER

5

Energy Balances

Unlike many chemical processes, bioprocesses are not particularly energy intensive. Fermenters and enzyme reactors are operated at temperatures and pressures close to ambient; energy input for downstream processing is minimised to avoid damaging heat-labile products. Nevertheless, energy effects are important because biological catalysts are very sensitive to heat and changes in temperature. In large-scale processes, heat released during reaction can cause cell death or denaturation of enzymes if it is not removed quickly. For rational design of temperature-control facilities, energy flows in the system must be determined using energy balances. Energy effects are also important in other areas of bioprocessing such as steam sterilisation.

The law of conservation of energy means that an energy accounting system can be set up to determine the amount of steam or cooling water required to maintain optimum process temperatures. In this chapter, after the necessary thermodynamic concepts are explained, an energy conservation equation applicable to biological processes is derived. The calculation techniques outlined in Chapter 4 are then extended for solution of simple energy balance problems.

5.1 BASIC ENERGY CONCEPTS

Energy takes three forms:

- Kinetic energy, E_k
- Potential energy, E_p
- Internal energy, U

Kinetic energy is the energy possessed by a moving system because of its velocity. *Potential energy* is due to the position of the system in a gravitational or electromagnetic field, or due to the conformation of the system relative to an equilibrium position (e.g., compression of a spring). *Internal energy* is the sum of all molecular, atomic, and subatomic energies of matter. Internal energy cannot be measured directly or known in absolute terms; we can only quantify change in internal energy.

Energy is transferred as either heat or work. *Heat* is energy that flows across system boundaries because of a temperature difference between the system and surroundings. *Work* is energy transferred as a result of any driving force other than temperature difference. There are two types of work: *shaft work* W_s, which is work done by a moving part within the system (e.g., an impeller mixing a fermentation broth), and *flow work* W_f, the energy required to push matter into the system. In a flow-through process, fluid at the inlet has work done on it by fluid just outside of the system, while fluid at the outlet does work on the fluid in front to push the flow along. Flow work is given by the expression:

$$W_f = pV \tag{5.1}$$

where p is pressure and V is volume. (Convince yourself that pV has the same dimensions as work and energy.)

5.1.1 Units

The SI unit for energy is the *joule* (J): 1 J = 1 newton metre (N m). Another unit is the *calorie* (cal), which is defined as the heat required to raise the temperature of 1 g of pure water by 1°C at 1 atm pressure. The quantity of heat according to this definition depends somewhat on the temperature of the water; because there has been no universal agreement on a reference temperature, there are several slightly different calorie units in use. The *international table calorie* (cal_{IT}) is fixed at 4.1868 J exactly. In imperial units, the British thermal unit (Btu) is common; this is defined as the amount of energy required to raise the temperature of 1 lb of water by 1°F at 1 atm pressure. As with the calorie, a reference temperature is required for this definition; 60°F is common although other temperatures are sometimes used.

5.1.2 Intensive and Extensive Properties

Properties of matter fall into two categories: those whose magnitude depends on the quantity of matter present and those whose magnitude does not. Temperature, density, and mole fraction are examples of properties that are independent of the size of the system; these quantities are called *intensive variables*. On the other hand, mass, volume, and energy are *extensive variables*. The values of extensive properties change if the size of the system is altered or if material is added or removed.

Extensive variables can be converted to *specific* quantities by dividing by the mass; for example, specific volume is volume divided by mass. Because specific properties are independent of the mass of the system, they are also intensive variables. In this chapter, for extensive properties denoted by an upper-case symbol, the specific property is given in lower-case notation. Therefore if U is internal energy, u denotes specific internal energy with units of, for example, kJ g^{-1}. Although, strictly speaking, the term *specific* refers to the quantity per unit mass, we will use the same lower-case symbols for molar quantities (with units, e.g., kJ $gmol^{-1}$).

5.1.3 Enthalpy

Enthalpy is a property used frequently in energy balance calculations. It is defined as the combination of two energy terms:

$$H = U + pV \tag{5.2}$$

where H is enthalpy, U is internal energy, p is pressure, and V is volume. Specific enthalpy h is therefore:

$$h = u + pv \tag{5.3}$$

where u is specific internal energy and v is specific volume. Since internal energy cannot be measured or known in absolute terms, neither can enthalpy.

5.2 GENERAL ENERGY BALANCE EQUATIONS

The principle underlying all energy balance calculations is the law of conservation of energy, which states that energy can be neither created nor destroyed. Although this law does not apply to nuclear reactions, conservation of energy remains a valid principle for bioprocesses because nuclear rearrangements are not involved. In the following sections, we will derive the equations used for solution of energy balance problems.

The law of conservation of energy can be written as:

$$\left\{\begin{matrix}\text{energy in through}\\ \text{system boundaries}\end{matrix}\right\} - \left\{\begin{matrix}\text{energy out through}\\ \text{system boundaries}\end{matrix}\right\} = \left\{\begin{matrix}\text{energy accumulated}\\ \text{within the system}\end{matrix}\right\} \tag{5.4}$$

For practical application of this equation, consider the system depicted in Figure 5.1. Mass M_i enters the system while mass M_o leaves. Both masses have energy associated with them in the form of internal, kinetic, and potential energies; flow work is also being done. Energy leaves the system as heat Q; shaft work W_s is performed on the system by the surroundings. We will assume that the system is homogeneous without charge or surface energy effects.

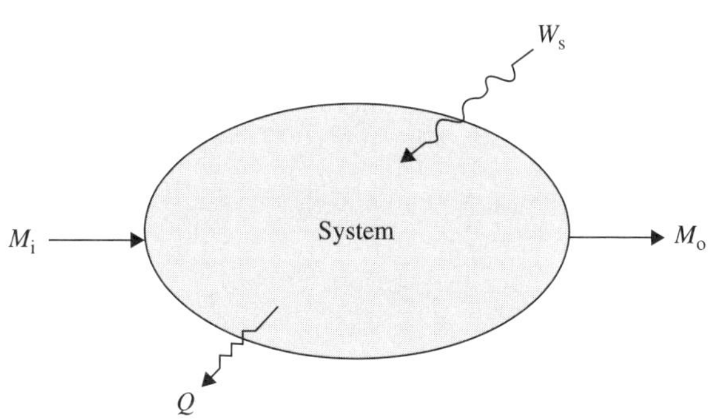

FIGURE 5.1 Flow system for energy balance calculations.

To apply Eq. (5.4), we must identify the forms of energy involved in each term of the expression. If we group together the extensive properties and express them as specific variables multiplied by mass, Eq. (5.4) can be written:

$$M_i\,(u + e_k + e_p + pv)_i - M_o\,(u + e_k + e_p + pv)_o - Q + W_s = \Delta E \tag{5.5}$$

where subscripts i and o refer to inlet and outlet conditions, respectively, and ΔE represents the total change or accumulation of energy in the system. u is specific internal energy, e_k is specific kinetic energy, e_p is specific potential energy, p is pressure, and v is specific volume. All energies associated with the masses crossing the system boundary are added together; the energy transfer terms Q and W_s are considered separately. Shaft work appears explicitly in Eq. (5.5) as W_s; flow work done by the inlet and outlet streams is represented as pv multiplied by mass.

The energy flows represented by Q and W_s can be directed either into or out of the system; appropriate signs must be used to indicate the direction of flow. Because it is usual in bioprocesses that shaft work be done on the system by external sources, in this text we will adopt the convention that work is positive when energy flows *from* the surroundings *to* the system as shown in Figure 5.1. Conversely, work will be considered negative when the system supplies work energy to the surroundings. On the other hand, we will regard heat as positive when the surroundings receive energy from the system—that is, when the temperature of the system is higher than that of the surroundings. Therefore, when W_s and Q are positive quantities, W_s makes a positive contribution to the energy content of the system while Q causes a reduction. These effects are accounted for in Eq. (5.5) by the signs preceding Q and W_s. The opposite sign convention is sometimes used in thermodynamics texts; however, the choice of sign convention is arbitrary if used consistently.

Equation (5.5) refers to a process with only one input and one output stream. A more general equation is Eq. (5.6), which can be used for any number of separate material flows:

$$\sum_{\substack{\text{input}\\\text{streams}}} M\,(u + e_k + e_p + pv) - \sum_{\substack{\text{output}\\\text{streams}}} M\,(u + e_k + e_p + pv) - Q + W_s = \Delta E \tag{5.6}$$

The symbol $\sum$ means summation; the internal, kinetic, potential, and flow work energies associated with all output streams are added together and subtracted from the sum for all input streams. Equation (5.6) is a basic form of the *first law of thermodynamics*, a simple mathematical expression of the law of conservation of energy. The equation can be shortened by substituting enthalpy h for $u + pv$ as defined by Eq. (5.3):

$$\sum_{\substack{\text{input}\\\text{streams}}} M\,(h + e_k + e_p) - \sum_{\substack{\text{output}\\\text{streams}}} M\,(h + e_k + e_p) - Q + W_s = \Delta E \tag{5.7}$$

5.2.1 Special Cases

Equation (5.7) can be simplified considerably if the following assumptions are made:

- Kinetic energy is negligible
- Potential energy is negligible

These assumptions are acceptable for bioprocesses, in which high-velocity motion and large changes in height or electromagnetic field do not generally occur. Thus, the energy balance equation becomes:

$$\sum_{\substack{\text{input}\\\text{streams}}} (Mh) - \sum_{\substack{\text{output}\\\text{streams}}} (Mh) - Q + W_s = \Delta E \tag{5.8}$$

Equation (5.8) can be simplified further in the following special cases:

- *Steady-state flow process.* At steady state, all properties of the system are invariant. Therefore, there can be no accumulation or change in the energy of the system: $\Delta E = 0$. The steady-state energy balance equation is:

$$\sum_{\substack{\text{input}\\\text{streams}}} (Mh) - \sum_{\substack{\text{output}\\\text{streams}}} (Mh) - Q + W_s = 0 \tag{5.9}$$

  Equation (5.9) can also be applied over the entire duration of batch and fed-batch processes if there is no energy accumulation; 'output streams' in this case refers to the harvesting of all mass in the system at the end of the process. Equation (5.9) is used frequently in bioprocess energy balances.
- *Adiabatic process.* A process in which no heat is transferred to or from the system is termed *adiabatic*; if the system has an *adiabatic wall*, it cannot release or receive heat to or from the surroundings. Under these conditions $Q = 0$ and Eq. (5.8) becomes:

$$\sum_{\substack{\text{input}\\\text{streams}}} (Mh) - \sum_{\substack{\text{output}\\\text{streams}}} (Mh) + W_s = \Delta E \tag{5.10}$$

Equations (5.8), (5.9), and (5.10) are energy balance equations that allow us to predict, for example, how much heat must be removed from a fermenter to maintain optimum conditions, or the effect of evaporation on cooling requirements. To apply the equations we must know the specific enthalpy h of flow streams entering or leaving the system. Methods for calculating enthalpy are outlined in the following sections.

5.3 ENTHALPY CALCULATION PROCEDURES

Irrespective of how enthalpy changes occur, certain conventions are used in enthalpy calculations.

5.3.1 Reference States

Specific enthalpy h appears explicitly in energy balance equations. What values of h do we use in these equations if enthalpy cannot be measured or known in absolute terms? Because energy balances are actually concerned with the *difference* in enthalpy between incoming and outgoing streams, we can overcome any difficulties by working always in terms of enthalpy change. In many energy balance problems, changes in enthalpy are evaluated relative to reference states that must be defined at the beginning of the calculation.

Because H cannot be known absolutely, it is convenient to assign $H = 0$ to some reference state. For example, when 1 gmol carbon dioxide is heated at 1 atm pressure from 0°C to 25°C, the change in enthalpy of the gas can be calculated (using methods explained later) as $\Delta H = 0.91$ kJ. If we assign $H = 0$ for CO_2 gas at 0°C, H at 25°C can be considered to be 0.91 kJ. This result does not mean that the absolute value of enthalpy at 25°C is 0.91 kJ; we can say only that the enthalpy at 25°C is 0.91 kJ relative to the enthalpy at 0°C.

We will use various reference states in energy balance calculations to determine enthalpy change. Suppose, for example, that we want to calculate the change in enthalpy as a system moves from State 1 to State 2. If the enthalpies of States 1 and 2 are known relative to the same reference condition H_{ref}, ΔH is calculated as follows:

$$\begin{array}{ccc} \text{State 1} & \xrightarrow{\Delta H} & \text{State 2} \\ \text{Enthalpy} = H_1 - H_{ref} & & \text{Enthalpy} = H_2 - H_{ref} \end{array}$$

$$\Delta H = (H_2 - H_{ref}) - (H_1 - H_{ref}) = H_2 - H_1$$

ΔH is therefore independent of the reference state because H_{ref} cancels out in the calculation.

5.3.2 State Properties

The values of some variables depend only on the state of the system and not on how that state was reached. These variables are called *state properties* or *functions of state*; examples include temperature, pressure, density, and composition. On the other hand, work is a *path function* because the amount of work done depends on the way in which the final state of the system is obtained from previous states.

Enthalpy is a state function. This property of enthalpy is very handy in energy balance calculations. It means that the change in enthalpy of a system can be calculated by taking a series of hypothetical steps or *process path* leading from the initial state and eventually reaching the final state. Change in enthalpy is calculated for each step; the total enthalpy change for the process is then equal to the sum of changes in the hypothetical path. This is true even though the process path used for calculation is not the same as that actually undergone by the system.

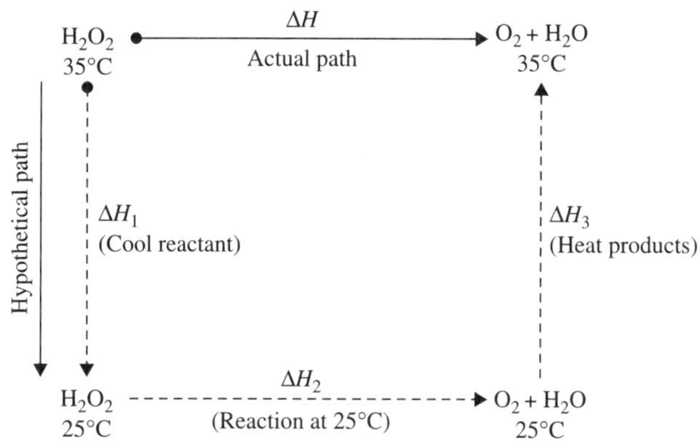

FIGURE 5.2 Hypothetical process path for calculation of enthalpy change.

As an example, consider the enthalpy change for the process shown in Figure 5.2, in which hydrogen peroxide is converted to oxygen and water by catalase enzyme. The enthalpy change for the direct process at 35°C can be calculated using an alternative pathway in which hydrogen peroxide is first cooled to 25°C, oxygen and water are formed by reaction at 25°C, and the products are heated to 35°C. Because the initial and final states for both actual and hypothetical paths are the same, the total enthalpy change is also identical:

$$\Delta H = \Delta H_1 + \Delta H_2 + \Delta H_3 \tag{5.11}$$

The reason for using hypothetical rather than actual pathways to calculate enthalpy change will become apparent later in the chapter.

5.4 ENTHALPY CHANGE IN NONREACTIVE PROCESSES

Change in enthalpy can occur as a result of:

1. Temperature change
2. Change of phase
3. Mixing or solution
4. Reaction

In the remainder of this section we will consider enthalpy changes associated with (1), (2), and (3). We will then consider how the results are used in energy balance calculations. Processes involving reaction will be discussed in Sections 5.8 through 5.11.

5.4.1 Change in Temperature

Heat transferred to raise or lower the temperature of a material is called *sensible heat*; change in the enthalpy of a system due to variation in temperature is called *sensible heat change*. In energy balance calculations, sensible heat change is determined using a property of matter called the *heat capacity at constant pressure*, or just *heat capacity*. We will use the symbol C_p for heat capacity; units for C_p are, for example, J $gmol^{-1}$ K^{-1}, cal g^{-1} $°C^{-1}$, and Btu lb^{-1} $°F^{-1}$. The term *specific heat capacity* or *specific heat* is sometimes used when heat capacity is expressed on a per-unit-mass basis. Values for heat capacity must be known before enthalpy changes from heating or cooling can be determined.

Tables C.3 through C.6 in Appendix C list C_p values for several organic and inorganic compounds. Additional C_p data and information about estimating heat capacities can be found in references such as *Perry's Chemical Engineers' Handbook* [1], *CRC Handbook of Chemistry and Physics* [2], and *International Critical Tables* [3].

There are several methods for calculating enthalpy change using C_p values. When C_p is approximately constant, the change in enthalpy of a substance at constant pressure due to a change in temperature ΔT is:

$$\Delta H = MC_p\Delta T = MC_p(T_2 - T_1) \tag{5.12}$$

where M is either mass or moles of the substance depending on the dimensions of C_p, T_1 is the initial temperature, and T_2 is the final temperature. The corresponding change in specific enthalpy is:

$$\Delta h = C_p\Delta T = C_p(T_2 - T_1) \tag{5.13}$$

EXAMPLE 5.1 SENSIBLE HEAT CHANGE WITH CONSTANT C_p

What is the enthalpy of 150 g formic acid at 70°C and 1 atm relative to 25°C and 1 atm?

Solution

From Table C.5 in Appendix C, C_p for formic acid in the temperature range of interest is 0.524 cal g^{-1} $°C^{-1}$. Substituting into (Eq. 5.12):

$$\Delta H = (150 \text{ g})\,(0.524 \text{ cal g}^{-1}\,°\text{C}^{-1})\,(70 - 25)°\text{C}$$

$$\Delta H = 3537.0 \text{ cal}$$

or

$$\Delta H = 3.54 \text{ kcal}$$

Relative to $H = 0$ at 25°C, the enthalpy of formic acid at 70°C is 3.54 kcal.

Heat capacities for most substances vary with temperature. This means that when we calculate the enthalpy change due to a change in temperature ΔT, the value of C_p itself varies over the range of ΔT. Heat capacities are often tabulated as polynomial functions of temperature, such as:

$$C_p = a + bT + cT^2 + dT^3 \tag{5.14}$$

Coefficients a, b, c, and d for a number of substances are given in Table C.3 in Appendix C.

Sometimes we can assume that the heat capacity is constant; this will give results for sensible heat change that approximate the true value. Because the temperature range of interest in bioprocessing is often relatively small, assuming constant heat capacity for some materials does not introduce large errors. C_p data may not be available at all temperatures; heat capacities such as those listed in Tables C.3 through C.6 are applicable only at a specified temperature or temperature range. As an example, in Table C.5 the heat capacity for liquid acetone between 24.2°C and 49.4°C is given as 0.538 cal g^{-1} °C^{-1}, even though this value will vary within the temperature range.

One method for calculating sensible heat change when C_p varies with temperature involves use of the *mean heat capacity*, C_{pm}. Table C.4 in Appendix C lists mean heat capacities for several common gases. These values are based on changes in enthalpy relative to a single reference temperature, $T_{ref} = 0$°C. To determine the change in enthalpy for a change in temperature from T_1 to T_2, read the values of C_{pm} at T_1 and T_2 and calculate:

$$\Delta H = M\left[(C_{pm})_{T_2}(T_2 - T_{ref}) - (C_{pm})_{T_1}(T_1 - T_{ref})\right] \tag{5.15}$$

5.4.2 Change of Phase

Phase changes, such as vaporisation and melting, are accompanied by relatively large changes in internal energy and enthalpy as bonds between molecules are broken and reformed. Heat transferred to or from a system causing change of phase at constant temperature and pressure is known as *latent heat*. Types of latent heat are:

- *Latent heat of vaporisation* (Δh_v): the heat required to vaporise a liquid
- *Latent heat of fusion* (Δh_f): the heat required to melt a solid
- *Latent heat of sublimation* (Δh_s): the heat required to directly vaporise a solid

Condensation of gas to liquid requires removal rather than addition of heat; the latent heat evolved in condensation is $-\Delta h_v$. Similarly, the latent heat evolved in freezing or solidification of liquid to solid is $-\Delta h_f$.

Latent heat is a property of substances and, like heat capacity, varies with temperature. Tabulated values of latent heats usually apply to substances at their normal boiling, melting, or sublimation point at 1 atm, and are called *standard heats of phase change*. Table C.7 in Appendix C lists latent heats for selected compounds; more values may be found in *Perry's Chemical Engineers' Handbook* [1] and *CRC Handbook of Chemistry and Physics* [2].

The change in enthalpy resulting from phase change is calculated directly from the latent heat. For example, the increase in enthalpy due to evaporation of liquid mass M at constant temperature is:

$$\Delta H = M\Delta h_v \tag{5.16}$$

EXAMPLE 5.2 ENTHALPY OF CONDENSATION

Fifty grams of benzaldehyde vapour is condensed at 179°C. What is the enthalpy of the liquid relative to the vapour?

Solution

From Table C.7 in Appendix C, the molecular weight of benzaldehyde is 106.12, the normal boiling point is 179.0°C, and the standard heat of vaporisation is 38.40 kJ $gmol^{-1}$. For condensation, the latent heat is −38.40 kJ $gmol^{-1}$. The enthalpy change is:

$$\Delta H = 50\,\text{g} \cdot \left|\frac{1\,\text{gmol}}{106.12\,\text{g}}\right| \cdot (-38.40\,\text{kJ gmol}^{-1}) = -18.09\,\text{kJ}$$

Therefore, the enthalpy of 50 g benzaldehyde liquid relative to the vapour at 179°C is −18.1 kJ. As heat is released during condensation, the enthalpy of the liquid is lower than the enthalpy of the vapour.

Phase changes often occur at temperatures other than the normal boiling, melting, or sublimation point; for example, water can evaporate at temperatures higher or lower than 100°C. How can we determine ΔH when the latent heat at the actual temperature of the phase change is not listed in property tables? This problem is overcome by using a hypothetical process path as described in Section 5.3.2. Suppose a liquid is vaporised isothermally at 30°C, but tabulated values for the standard heat of vaporisation refer to 60°C. As shown in Figure 5.3, we can consider a process whereby the liquid is heated from 30°C to 60°C, vaporised at 60°C, and the vapour cooled to 30°C. The total enthalpy change for this hypothetical process is the same as if vaporisation occurred directly at 30°C. ΔH_1 and ΔH_3 are sensible heat changes, which can be calculated using heat capacity values and the methods described in Section 5.4.1. ΔH_2 is the latent heat at standard conditions calculated using Δh_v data available from tables. Because enthalpy is a state property, ΔH for the actual path is the same as $\Delta H_1 + \Delta H_2 + \Delta H_3$.

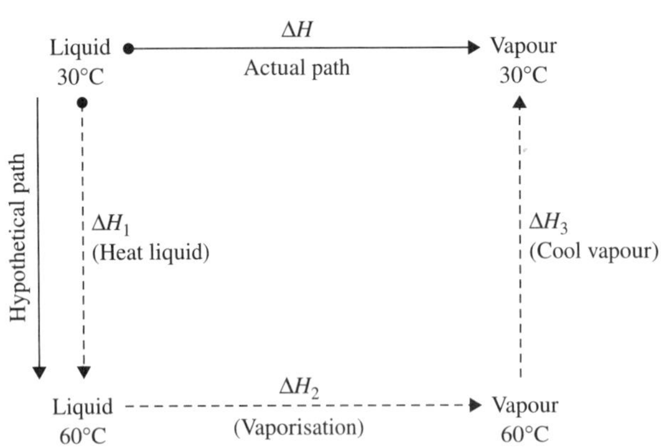

FIGURE 5.3 Process path for calculating latent heat change at a temperature other than the normal boiling point.

5.4.3 Mixing and Solution

So far we have considered enthalpy changes for pure compounds. For an *ideal solution* or *ideal mixture* of several compounds, the thermodynamic properties of the mixture are a simple sum of contributions from the individual components. However, when compounds are mixed or dissolved, bonds between molecules in the solvent and solute are broken and reformed. In *real solutions*, a net absorption or release of energy accompanies these processes, resulting in changes in the internal energy and enthalpy of the mixture. Dilution of sulphuric acid with water is a good example; in this case, energy is released.

For real solutions, there is an additional energy term to consider in evaluating enthalpy: the *integral heat of mixing* or *integral heat of solution*, Δh_m. The integral heat of solution is defined as the change in enthalpy that occurs as one mole of solute is dissolved at constant temperature in a given quantity of solvent. The enthalpy of a nonideal mixture of two compounds A and B is:

$$H_{\text{mixture}} = H_A + H_B + \Delta H_m \tag{5.17}$$

where H_A is the enthalpy of compound A, H_B is the enthalpy of compound B, and ΔH_m is the heat of mixing.

Heat of mixing is a property of the solution components and is dependent on the temperature and concentration of the mixture. As a solution becomes more and more dilute, an asymptotic value of Δh_m is reached. This value is called the *integral heat of solution at infinite dilution*. When water is the primary component of a solution, Δh_m at infinite dilution can be used to calculate the enthalpy of the mixture. Heats of solution and heats of solutions at infinite dilution for selected aqueous systems are listed in *Perry's Chemical Engineers' Handbook* [1], *CRC Handbook of Chemistry and Physics* [2], and *Biochemical Engineering and Biotechnology Handbook* [4].

EXAMPLE 5.3 HEAT OF SOLUTION

Malonic acid and water are available separately at 25°C. If 15 g malonic acid is dissolved in 5 kg water, how much heat must be added for the solution to remain at 25°C? What is the solution enthalpy relative to the components?

Solution

The molecular weight of malonic acid is 104. Because the solution is very dilute (< 0.3% w/w), we can use the integral heat of solution at infinite dilution. From handbooks, Δh_m at room temperature is 4.493 kcal gmol^{-1}. This value is positive; therefore the mixture enthalpy is greater than that of the components and heat is absorbed during solution. The heat required for the solution to remain at 25°C is:

$$\Delta H = 15\,\text{g} \cdot \left| \frac{1\,\text{gmol}}{104\,\text{g}} \right| \cdot (4.493\,\text{kcal gmol}^{-1}) = 0.648\,\text{kcal}$$

Relative to $H = 0$ for water and malonic acid at 25°C, the enthalpy of the solution at 25°C is 0.65 kcal.

In biological systems, significant changes in enthalpy due to heats of mixing do not often occur. Most solutions in fermentation and enzyme processes are dilute aqueous mixtures; in energy balance calculations these solutions are usually considered ideal without much loss of accuracy.

5.5 STEAM TABLES

Steam tables have been used for many years by engineers designing industrial processes and power stations. These tables list the thermodynamic properties of water, including specific volume, internal energy, and enthalpy. As we are concerned here mainly with enthalpies, a list of enthalpy values for steam and water under various conditions has been extracted from the steam tables and given in Appendix D. All enthalpy values must have a reference point; in the steam tables of Appendix D, $H = 0$ for liquid water at the triple point, 0.01°C and 0.6112 kPa pressure. (The triple point is an invariant condition of pressure and temperature at which ice, liquid water, and water vapour are in equilibrium with each other.) Steam tables from other sources may have different reference states. Steam tables eliminate the need for sensible-heat and latent-heat calculations for water and steam, and can be used directly in energy balance calculations.

Tables D.1 and D.2 in Appendix D list enthalpy values for liquid water and *saturated steam*. When liquid and vapour are in equilibrium with each other, they are saturated; a gas saturated with water contains all the water it can hold at the prevailing temperature and pressure. For a pure substance such as water, once the temperature is specified, saturation occurs at only one pressure. For example, from Table D.2 saturated steam at 188°C has a pressure of 1200 kPa. Also from the table, the enthalpy of this steam relative to the triple point of water is 2782.7 kJ kg^{-1}; liquid water in equilibrium with the steam has an enthalpy of 798.4 kJ kg^{-1}. The latent heat of vaporisation under these conditions is the difference between the liquid and vapour enthalpies; as indicated in the middle enthalpy column, Δh_v is 1984.3 kJ kg^{-1}. Table D.1 lists enthalpies of saturated water and steam by temperature; Table D.2 lists these enthalpies by pressure.

It is usual when using steam tables to ignore the effect of pressure on the enthalpy of liquid water. For example, the enthalpy of water at 40°C and 1 atm (101.3 kPa) is found by looking up the enthalpy of saturated water at 40°C in Table D.1, and assuming the value is independent of pressure. This assumption is valid at low pressure, that is, less than about 50 atm. The enthalpy of liquid water at 40°C and 1 atm is therefore 167.5 kJ kg^{-1}.

Enthalpy values for *superheated steam* are given in Table D.3. If the temperature of saturated vapour is increased (or the pressure decreased at constant temperature), the vapour is said to be superheated. A superheated vapour cannot condense until it is returned to saturation conditions. The difference between the temperature of a superheated gas and its saturation temperature is called the *degrees of superheat* of the gas. In Table D.3, enthalpy is listed as a function of temperature at 15 different pressures from 10 kPa to 50,000 kPa; for example, superheated steam at 1000 kPa pressure and 250°C has an enthalpy of 2943 kJ kg^{-1} relative to the triple point. Table D.3 also lists properties at saturation conditions; at 1000 kPa the saturation temperature is 179.9°C, the enthalpy of liquid water under these conditions is 762.6 kJ kg^{-1}, and the enthalpy of saturated vapour is 2776.2 kJ kg^{-1}.

Thus, the degrees of superheat for superheated steam at 1000 kPa and 250°C can be calculated as (250 − 179.9) = 70.1°C. Water under pressure remains liquid even at relatively high temperatures. Enthalpy values for liquid water up to 350°C are found in the upper region of Table D.3 above the line extending to the critical pressure.

5.6 PROCEDURE FOR ENERGY BALANCE CALCULATIONS WITHOUT REACTION

The methods described in Section 5.4 for evaluating enthalpy can be used to solve energy balance problems for systems in which reactions do not occur. Many of the points described in Section 4.3 for material balances also apply when setting out an energy balance.

1. A properly drawn and labelled *flow diagram* is essential to identify all inlet and outlet streams and their compositions. For energy balances, the temperatures, pressures, and phases of the material should also be indicated if appropriate.
2. The *units* selected for the energy balance should be stated; these units are also used when labelling the flow diagram.
3. As in mass balance problems, a *basis* for the calculation is chosen and stated clearly.
4. The *reference state* for $H = 0$ is determined. In the absence of reaction, reference states for each molecular species in the system can be assigned arbitrarily.
5. State all *assumptions* used to solve the problem. Assumptions such as absence of leaks and steady-state operation for continuous processes are generally applicable.

Following on from (5), other assumptions commonly made for energy balances include:

- The system is homogeneous or well mixed. Under these conditions, product streams including gases leave the system at the system temperature.
- Heats of mixing are often neglected for mixtures containing compounds of similar molecular structure. Gas mixtures are always considered ideal.
- Sometimes shaft work can be neglected even though the system is stirred by mechanical means. This assumption may not apply when vigorous agitation is used or when the liquid being stirred is very viscous. When shaft work is not negligible, you will need to know how much mechanical energy is input through the stirrer.
- Evaporation in liquid systems may be considered negligible if the components are not particularly volatile or if the operating temperature is relatively low.
- Heat losses from the system to the surroundings are often ignored; this assumption is generally valid for large insulated vessels when the operating temperature is close to ambient.

5.7 ENERGY BALANCE WORKED EXAMPLES WITHOUT REACTION

As illustrated in the following examples, the format described in Chapter 4 for material balances can be used as a foundation for energy balance calculations.

EXAMPLE 5.4 CONTINUOUS WATER HEATER

Water at 25°C enters an open heating tank at a rate of 10 kg h^{-1}. Liquid water leaves the tank at 88°C at a rate of 9 kg h^{-1}; 1 kg h^{-1} water vapour is lost from the system through evaporation. At steady state, what is the rate of heat input to the system?

Solution

1. Assemble
   (i) Select units for the problem.
   kg, h, kJ, °C
   (ii) Draw the flow sheet showing all data and units.
   The flow sheet is shown in Figure 5.4.
   (iii) Define the system boundary by drawing on the flow sheet.
   The system boundary is indicated in Figure 5.4.
2. Analyse
   (i) State any assumptions.
   - process is operating at steady state
   - system does not leak
   - system is homogeneous
   - evaporation occurs at 88°C
   - vapour is saturated
   - shaft work is negligible
   - no heat losses

   (ii) Select and state a basis.
   The calculation is based on 10 kg water entering the system, or 1 hour.
   (iii) Select a reference state.
   The reference state for water is the same as that used in the steam tables: 0.01°C and 0.6112 kPa.
   (iv) Collect any extra data needed.
   h (liquid water at 88°C) = 368.5 kJ kg^{-1} (Table D.1)
   h (saturated steam at 88°C) = 2656.9 kJ kg^{-1} (Table D.1)
   h (liquid water at 25°C) = 104.8 kJ kg^{-1} (Table D.1)

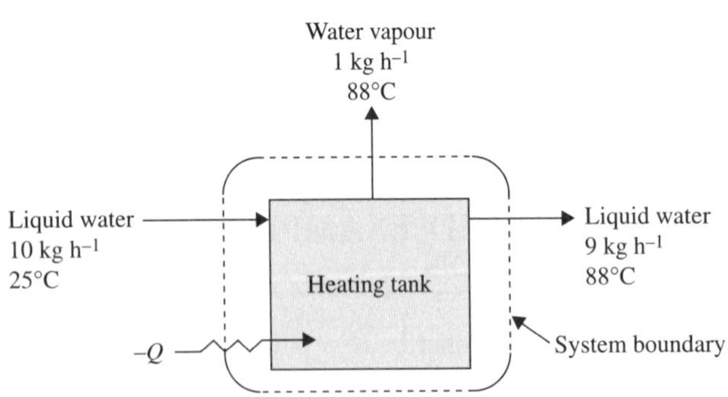

FIGURE 5.4 Flow sheet for continuous water heater.

(v) Determine which compounds are involved in reaction.
No reaction occurs.

(vi) Write down the appropriate mass balance equation.
The mass balance is already complete.

(vii) Write down the appropriate energy balance equation.
At steady state, Eq. (5.9) applies:

$$\sum_{\substack{\text{input}\\\text{streams}}} (Mh) - \sum_{\substack{\text{output}\\\text{streams}}} (Mh) - Q + W_s = 0$$

3. Calculate
Identify the terms in the energy balance equation.
For this problem $W_s = 0$. The energy balance equation becomes:

$$(Mh)_{\text{liq in}} - (Mh)_{\text{liq out}} - (Mh)_{\text{vap out}} - Q = 0$$

Substituting the information available:

$$(10\text{ kg})\,(104.8\text{ kJ kg}^{-1}) - (9\text{ kg})\,(368.5\text{ kJ kg}^{-1}) - (1\text{ kg})\,(2656.9\text{ kJ kg}^{-1}) - Q = 0$$

$$Q = -4925.4\text{ kJ}$$

Q has a negative value. Thus, according to the sign convention outlined in Section 5.2, heat must be supplied to the system from the surroundings.

4. Finalise
Answer the specific questions asked in the problem; check the number of significant figures; state the answers clearly.
The rate of heat input is 4.93×10^3 kJ h^{-1}.

EXAMPLE 5.5 COOLING IN DOWNSTREAM PROCESSING

In downstream processing of gluconic acid, concentrated fermentation broth containing 20% (w/w) gluconic acid is cooled prior to crystallisation. The concentrated broth leaves an evaporator at a rate of 2000 kg h^{-1} and must be cooled from 90°C to 6°C. Cooling is achieved by heat exchange with 2700 kg h^{-1} water initially at 2°C. If the final temperature of the cooling water is 50°C, what is the rate of heat loss from the gluconic acid solution to the surroundings? Assume the heat capacity of gluconic acid is 0.35 cal g^{-1} °C^{-1}.

Solution

1. Assemble

(i) Units
kg, h, kJ, °C

(ii) Flow sheet
The flow sheet is shown in Figure 5.5.

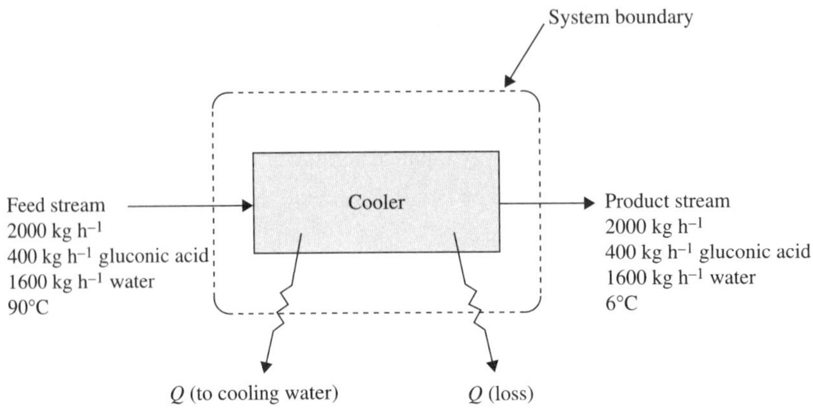

FIGURE 5.5 Flow sheet for cooling gluconic acid solution.

(iii) System boundary

The system boundary indicated in Figure 5.5 separates the gluconic acid solution from the cooling water.

2. Analyse

(i) Assumptions

- steady state
- no leaks
- other components of the fermentation broth can be considered water
- no shaft work

(ii) Basis

2000 kg feed, or 1 hour

(iii) Reference state

$H = 0$ for gluconic acid at 90°C

$H = 0$ for water at its triple point

(iv) Extra data

The heat capacity of gluconic acid is 0.35 cal g^{-1} °C^{-1}; we will assume this C_p remains constant over the temperature range of interest. Converting units:

$$C_p \text{ (gluconic acid)} = \frac{0.35\text{ cal}}{\text{g °C}} \cdot \left|\frac{4.187\text{ J}}{1\text{ cal}}\right| \cdot \left|\frac{1\text{ kJ}}{1000\text{ J}}\right| \cdot \left|\frac{1000\text{ g}}{1\text{ kg}}\right| = 1.47\text{ kJ kg}^{-1}\text{ °C}^{-1}$$

h (liquid water at 90°C) = 376.9 kJ kg^{-1} (Table D.1)

h (liquid water at 6°C) = 25.2 kJ kg^{-1} (Table D.1)

h (liquid water at 2°C) = 8.4 kJ kg^{-1} (Table D.1)

h (liquid water at 50°C) = 209.3 kJ kg^{-1} (Table D.1)

(v) Compounds involved in reaction

No reaction occurs.

(vi) Mass balance equation
The mass balance equation for total mass, gluconic acid, and water is:

$$\text{mass in} = \text{mass out}$$

The mass flow rates are as shown in Figure 5.5.

(vii) Energy balance equation

$$\sum_{\substack{\text{input}\\\text{streams}}} (Mh) - \sum_{\substack{\text{output}\\\text{streams}}} (Mh) - Q + W_s = 0$$

3. Calculate
$W_s = 0$. There are two heat flows out of the system: one to the cooling water (Q) and one representing loss to the surroundings (Q_{loss}). With symbols W = water and G = gluconic acid, the energy balance equation is:

$$(Mh)_{\text{W in}} + (Mh)_{\text{G in}} - (Mh)_{\text{W out}} - (Mh)_{\text{G out}} - Q_{loss} - Q = 0$$

$$(Mh)_{\text{W in}} = (1600\ \text{kg})\ (376.9\ \text{kJ kg}^{-1}) = 6.03 \times 10^5\ \text{kJ}$$

$$(Mh)_{\text{G in}} = 0\ \text{(reference state)}$$

$$(Mh)_{\text{W out}} = (1600\ \text{kg})\ (1.47\ \text{kJ kg}^{-1}) = 4.03 \times 10^4\ \text{kJ}$$

$(Mh)_{\text{G out}}$ at 6°C is calculated as a sensible heat change from 90°C using Eq. (5.12):

$$(Mh)_{\text{G out}} = MC_p(T_2 - T_1) = (400\ \text{kg})\ (1.47\ \text{kJ kg}^{-1}\ {}^\circ\text{C}^{-1})\ (6 - 90)^\circ\text{C}$$

$$\therefore\ (Mh)_{\text{G out}} = -4.94 \times 10^4\ \text{kJ}$$

The heat removed to the cooling water, Q, is equal to the enthalpy change of the cooling water between 2°C and 50°C:

$$Q = (2700\ \text{kg})\ (209.3 - 8.4)\ \text{kJ kg}^{-1} = 5.42 \times 10^5\ \text{kJ}$$

These results can now be substituted into the energy balance equation:

$$(6.03 \times 10^5\ \text{kJ}) + (0\ \text{kJ}) - (4.03 \times 10^4\ \text{kJ}) - (-4.94 \times 10^4\ \text{kJ}) - Q_{loss} - 5.42 \times 10^5\ \text{kJ} = 0$$

$$\therefore\ Q_{loss} = 7.01 \times 10^4\ \text{kJ}$$

4. Finalise
The rate of heat loss to the surroundings is 7.0×10^4 kJ h^{-1}.

It is important to recognise that the final answers to energy balance problems do not depend on the choice of reference states for the components. Although values of h depend on the reference states, as discussed in Section 5.3.1 this dependence disappears when the energy balance equation is applied and the difference between the input and output enthalpies is determined. To prove this point, any of the examples in this chapter can be repeated using different reference conditions to obtain the same final answers.

5.8 ENTHALPY CHANGE DUE TO REACTION

Reactions in bioprocesses occur as a result of enzyme activity and cell metabolism. During reaction, relatively large changes in internal energy and enthalpy occur as bonds between atoms are rearranged. *Heat of reaction* ΔH_{rxn} is the energy released or absorbed during reaction, and is equal to the difference in enthalpy of reactants and products:

$$\Delta H_{rxn} = \sum_{\text{products}} Mh - \sum_{\text{reactants}} Mh \tag{5.18}$$

or

$$\Delta H_{rxn} = \sum_{\text{products}} nh - \sum_{\text{reactants}} nh \tag{5.19}$$

where $\sum$ denotes the sum, M is mass, n is number of moles, and h is specific enthalpy expressed on either a per-mass or per-mole basis. Note that M and n represent the mass and moles actually involved in the reaction, not the total amount present in the system. In an *exothermic reaction* the energy required to hold the atoms of product together is less than for the reactants; surplus energy is released as heat and ΔH_{rxn} is negative in value. On the other hand, energy is absorbed during *endothermic reactions*, the enthalpy of the products is greater than the reactants, and ΔH_{rxn} is positive.

The specific heat of reaction Δh_{rxn} is a property of matter. The value of Δh_{rxn} depends on the reactants and products involved in the reaction and the temperature and pressure. Because any given molecule can participate in a large number of reactions, it is not feasible to tabulate all possible Δh_{rxn} values. Instead, Δh_{rxn} can be calculated from the heats of combustion of the individual components involved in the reaction.

5.8.1 Heat of Combustion

Heat of combustion Δh_c is defined as the heat evolved during reaction of a substance with oxygen to yield certain oxidation products such as CO_2 gas, H_2O liquid, and N_2 gas. The *standard heat of combustion* Δh_c° is the specific enthalpy change associated with this reaction at standard conditions, usually 25°C and 1 atm pressure. By convention, Δh_c° is zero for the products of oxidation (i.e., CO_2 gas, H_2O liquid, N_2 gas, etc.); standard heats of combustion for other compounds are always negative. Table C.8 in Appendix C lists selected values; heats of combustion for other materials can be found in *Perry's Chemical Engineers' Handbook* [1] and *CRC Handbook of Chemistry and Physics* [2]. As an example, the standard heat of combustion for citric acid is given in Table C.8 as -1962.0 kJ gmol^{-1}; this refers to the heat evolved at 25°C and 1 atm in the following reaction:

$$C_6H_8O_7(s) + 4.5\ O_2(g) \longrightarrow 6\ CO_2(g) + 4\ H_2O(l)$$

Standard heats of combustion are used to calculate the *standard heat of reaction* ΔH°_{rxn} for reactions involving combustible reactants and combustion products:

$$\Delta H^{\circ}_{rxn} = \sum_{\text{reactants}} n\Delta h^{\circ}_{c} - \sum_{\text{products}} n\Delta h^{\circ}_{c} \tag{5.20}$$

where n is the moles of reactant or product involved in the reaction, and Δh°_{c} is the standard heat of combustion per mole. The standard heat of reaction is the difference between the heats of combustion of reactants and products.

EXAMPLE 5.6 CALCULATION OF HEAT OF REACTION FROM HEATS OF COMBUSTION

Fumaric acid ($C_4H_4O_4$) is produced from malic acid ($C_4H_6O_5$) using the enzyme fumarase. Calculate the standard heat of reaction for the following enzyme transformation:

$$C_4H_6O_5 \longrightarrow C_4H_4O_4 + H_2O$$

Solution

$\Delta h^{\circ}_{c} = 0$ for liquid water. From Eq. (5.20):

$$\Delta H^{\circ}_{rxn} = (n\Delta h^{\circ}_{c})_{\text{malic acid}} - (n\Delta h^{\circ}_{c})_{\text{fumaric acid}}$$

Table C.8 in Appendix C lists the standard heats of combustion for these compounds:

$$(\Delta h^{\circ}_{c})_{\text{malic acid}} = -1328.8 \text{ kJ gmol}^{-1}$$
$$(\Delta h^{\circ}_{c})_{\text{fumaric acid}} = -1334.0 \text{ kJ gmol}^{-1}$$

Therefore, using a basis of 1 gmol of malic acid converted:

$$\Delta H^{\circ}_{rxn} = 1 \text{ gmol} (-1328.8 \text{ kJ gmol}^{-1}) - 1 \text{ gmol} (-1334.0 \text{ kJ gmol}^{-1})$$
$$\Delta H^{\circ}_{rxn} = 5.2 \text{ kJ}$$

As ΔH°_{rxn} is positive, the reaction is endothermic and heat is absorbed.

5.8.2 Heat of Reaction at Nonstandard Conditions

Example 5.6 shows how to calculate the heat of reaction at standard conditions. However, most reactions do not occur at 25°C and the standard heat of reaction calculated using Eq. (5.20) may not be the same as the actual heat of reaction at the reaction temperature.

Consider the following reaction between compounds A, B, C, and D occurring at temperature T:

$$A + B \longrightarrow C + D$$

The standard heat of reaction at 25°C is known from tabulated heat of combustion data. ΔH_{rxn} at temperature T can be calculated using the alternative reaction pathway outlined in Figure 5.6, in which reaction occurs at 25°C and the reactants and products are heated

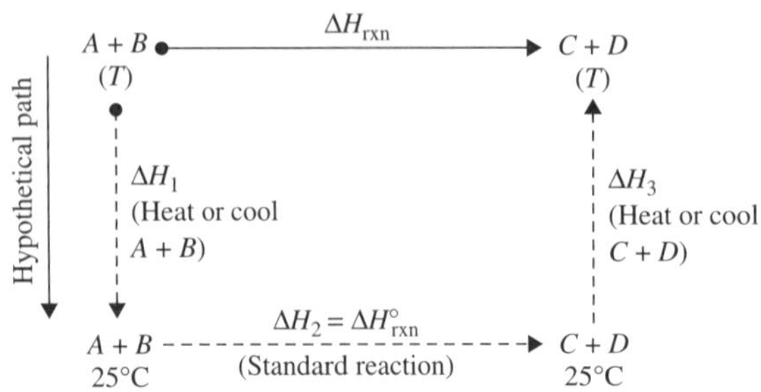

FIGURE 5.6 Hypothetical process path for calculating the heat of the reaction at nonstandard temperature.

or cooled between 25°C and T before and after the reaction. Because the initial and final states for the actual and hypothetical paths are the same, the total enthalpy change is also identical. Therefore:

$$\Delta H_{rxn}(\text{at } T) = \Delta H_1 + \Delta H^\circ_{rxn} + \Delta H_3 \tag{5.21}$$

where ΔH_1 and ΔH_3 are changes in sensible heat and ΔH°_{rxn} is the standard heat of reaction at 25°C. ΔH_1 and ΔH_3 are evaluated using heat capacities and the methods described in Section 5.4.1.

Depending on the magnitude of ΔH°_{rxn} and the extent to which T deviates from 25°C, ΔH_{rxn} may not be much different from ΔH°_{rxn}. For example, consider the reaction for respiration of glucose:

$$C_6H_{12}O_6 + 6\,O_2 \longrightarrow 6\,CO_2 + 6\,H_2O$$

ΔH°_{rxn} for this conversion is -2805.0 kJ; if the reaction occurs at 37°C instead of 25°C, ΔH_{rxn} is -2801.7 kJ. Contributions from sensible heat amount to only 3.3 kJ, which is insignificant compared with the total magnitude of ΔH°_{rxn} and can be ignored without much loss of accuracy. With reference to Figure 5.6, $\Delta H_1 = -4.8$ kJ for cooling 1 gmol glucose and 6 gmol oxygen from 37°C to 25°C; $\Delta H_3 = 8.1$ kJ for heating the products back to 37°C. Having opposite signs, ΔH_1 and ΔH_3 act to cancel each other. This situation is typical of most reactions in bioprocessing where the actual temperature of reaction is not sufficiently different from 25°C to warrant concern about sensible heat changes. When the heat of reaction is substantial compared with other types of enthalpy change, ΔH_{rxn} can be assumed equal to ΔH°_{rxn} irrespective of reaction temperature.

A major exception to this general rule are single-enzyme conversions. Because many single-enzyme reactions involve only small molecular rearrangements, heats of reaction are relatively small. For instance, per mole of substrate, the fumarase reaction of Example 5.6 involves a standard enthalpy change of only 5.2 kJ; other examples are 8.7 kJ gmol^{-1} for the glucose isomerase reaction, -26.2 kJ gmol^{-1} for hydrolysis of sucrose, and -29.4 kJ per gmol glucose for hydrolysis of starch. For conversions such as these, sensible energy changes of 5 to 10 kJ are clearly significant and should not be ignored. Furthermore, calculated standard heats of reaction for enzyme transformations are often imprecise. Being the difference between two or more relatively large heat of combustion values, the small ΔH°_{rxn} for these conversions can carry considerable uncertainty depending on the accuracy of the heat of combustion data. When coupled with usual assumptions, such as constant C_p and Δh_m within the temperature and concentration range of interest, this uncertainty means that estimates of heating and cooling requirements for enzyme reactors are sometimes quite rough.

5.9 HEAT OF REACTION FOR PROCESSES WITH BIOMASS PRODUCTION

Biochemical reactions in cells do not occur in isolation but are linked in a complex array of metabolic transformations. Catabolic and anabolic reactions take place at the same time, so that energy released in one reaction is used in other energy-requiring processes. Cells use chemical energy quite efficiently; however some is inevitably released as heat. How can we estimate the heat of reaction associated with cell metabolism and growth?

5.9.1 Thermodynamics of Cell Growth

As described in Section 4.6.1, a macroscopic view of cell growth is represented by the equation:

$$C_wH_xO_yN_z + a\,O_2 + b\,H_gO_hN_i \longrightarrow c\,CH_\alpha O_\beta N_\delta + d\,CO_2 + e\,H_2O \tag{4.4}$$

where a, b, c, d, and e are stoichiometric coefficients, $C_wH_xO_yN_z$ is the substrate, $H_gO_hN_i$ is the nitrogen source, and $CH_\alpha O_\beta N_\delta$ is dry biomass. Once the stoichiometric coefficients or yields are determined, Eq. (4.4) can be used as the reaction equation in energy balance calculations. We need, however, to determine the heat of reaction for this conversion.

Heats of reaction for cell growth can be estimated using stoichiometry and the concept of available electrons (Section 4.6.2). It has been found empirically that the energy content of organic compounds is related to their degree of reduction as follows:

$$\Delta h^{\circ}_{c} = -q\,\gamma\,x_C \tag{5.22}$$

where Δh°_{c} is the molar heat of combustion at standard conditions, q is the heat evolved per mole of available electrons transferred to oxygen during combustion, γ is the degree of reduction of the compound relative to N_2, and x_C is the number of carbon atoms in the molecular formula. The coefficient q relating Δh°_{c} and γ is relatively constant for a large number of compounds. Patel and Erickson [5] assigned a value of 111 kJ gmol^{-1} to q; in another analysis, Roels [6] determined a value of 115 kJ gmol^{-1}. The correlation found by Roels is based on analysis of several chemical and biochemical compounds including biomass; the results are shown in Figure 5.7.

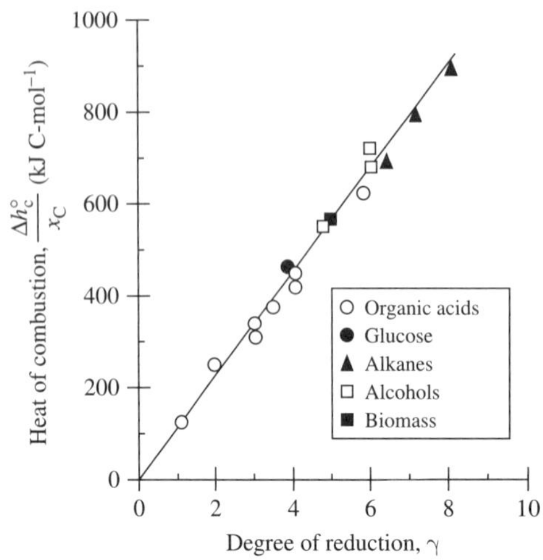

FIGURE 5.7 Relationship between degree of reduction and heat of combustion for various organic compounds. *From J.A. Roels, 1987, Thermodynamics of growth. In: J. Bu'Lock and B. Kristiansen, Eds.,* Basic Biotechnology, *Academic Press, London.*

5.9.2 Heat of Reaction with Oxygen as Electron Acceptor

The direct proportionality between heat of combustion and degree of reduction indicated in Eq. (5.22) and Figure 5.7 has important implications for determining the heat of reaction for aerobic cultures. As the degree of reduction of a substance is related directly to the amount of oxygen required for its complete combustion, when compounds for which Eq. (5.22) applies are involved in reaction with oxygen, the heat produced by the reaction is directly proportional to the amount of oxygen consumed. Because molecular oxygen O_2 accepts four electrons, if one mole of O_2 is consumed during respiration, four moles of electrons must be transferred. Accepting the value of 115 kJ of energy released per gmol of electrons transferred, the amount of energy released from consumption of one gmol O_2 is therefore (4 × 115) kJ, or 460 kJ. The overall result:

$$\Delta H_{rxn} \text{ for fully aerobic metabolism} \simeq -460 \text{ kJ gmol}^{-1} \text{ O}_2 \text{ consumed} \tag{5.23}$$

is verified by the experimental data of Cooney et al. [7] shown in Figure 5.8. Equation (5.23) is quite accurate for fully aerobic cultures under a wide range of conditions, including with product formation. Thus, once the amount of oxygen taken up during aerobic cell culture is known, the heat of reaction can be evaluated immediately.

5.9.3 Heat of Reaction with Oxygen Not the Principal Electron Acceptor

If a fermentation uses electron acceptors other than oxygen, for example in anaerobic culture, the simple relationship for heat of reaction derived in Section 5.9.2 does not apply. Heats of combustion must be used to estimate the heat of reaction for anaerobic conversions. Consider the following reaction equation for anaerobic growth with product formation:

$$\begin{aligned} &C_wH_xO_yN_z + b\,H_gO_hN_i \\ &\longrightarrow c\,CH_\alpha O_\beta N_\delta + d\,CO_2 + e\,H_2O + f\,C_jH_kO_lN_m \end{aligned} \tag{5.24}$$

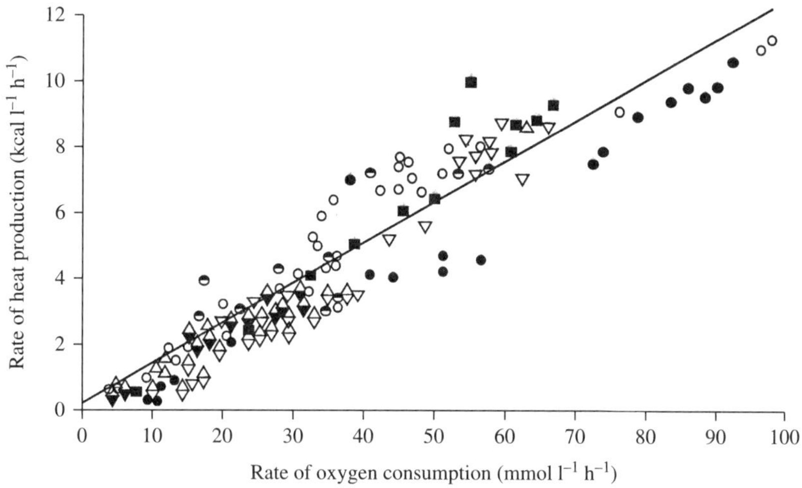

FIGURE 5.8 Correlation between rate of heat evolution and rate of oxygen consumption for a variety of microbial fermentations. (○) *Escherichia coli*, glucose medium; (◆) *Candida intermedia*, glucose medium; (△) *C. intermedia*, molasses medium; (▽) *Bacillus subtilis*, glucose medium; (■) *B. subtilis*, molasses medium; (◓) *B. subtilis*, soybean meal medium; (◇) *Aspergillus niger*, glucose medium; (●) *Asp. niger*, molasses medium.
From C.L. Cooney, D.I.C. Wang, and R.I. Mateles, Measurement of heat evolution and correlation with oxygen consumption during microbial growth, Biotechnol. Bioeng. *11, 269–281; Copyright © 1968. Reprinted by permission of John Wiley.*

where $C_jH_kO_lN_m$ is an extracellular product and f is its stoichiometric coefficient. With ammonia as nitrogen source and heats of combustion of H_2O and CO_2 zero, from Eq. (5.20) the equation for the standard heat of reaction is:

$$\Delta H^{\circ}_{\text{rxn}} = (n\Delta h^{\circ}_{\text{c}})_{\text{substrate}} + (n\Delta h^{\circ}_{\text{c}})_{\text{NH}_3} - (n\Delta h^{\circ}_{\text{c}})_{\text{biomass}} - (n\Delta h^{\circ}_{\text{c}})_{\text{product}} \tag{5.25}$$

where n is the number of moles and $\Delta h^{\circ}_{\text{c}}$ is the standard molar heat of combustion. Heats of combustion for substrate, NH_3, and product can be found in tables, but what is the heat of combustion of biomass?

As shown in Table 4.11, the elemental composition of biomass does not vary a great deal. If we assume an average biomass molecular formula of $CH_{1.8}O_{0.5}N_{0.2}$, the reaction equation for combustion of cells to CO_2, H_2O, and N_2 is:

$$CH_{1.8}O_{0.5}N_{0.2} + 1.2\,O_2 \longrightarrow CO_2 + 0.9\,H_2O + 0.1\,N_2$$

From Table C.2 in Appendix C, the degree of reduction of biomass relative to N_2 is 4.80. Assuming an average of 5% ash associated with the biomass, the cell molecular weight is 25.9. The heat of combustion can be obtained by applying Eq. (5.22) with $q = 115$ kJ gmol^{-1}:

$$(\Delta h^{\circ}_{\text{c}})_{\text{biomass}} = (-115 \text{ kJ gmol}^{-1})\,(4.80)\,(1) \cdot \left|\frac{1 \text{ gmol}}{25.9 \text{ g}}\right|$$

$$(\Delta h^{\circ}_{\text{c}})_{\text{biomass}} = -21.3 \text{ kJ g}^{-1} \tag{5.26}$$

TABLE 5.1 Heats of Combustion of Bacteria and Yeast

Organism	Substrate	Δh_c (kJ g^{-1})
BACTERIA		
Escherichia coli	glucose	-23.04 ± 0.06
	glycerol	-22.83 ± 0.07
Enterobacter cloacae	glucose	-23.22 ± 0.14
	glycerol	-23.39 ± 0.12
Methylophilus methylotrophus	methanol	-23.82 ± 0.06
Bacillus thuringiensis	glucose	-22.08 ± 0.03
YEAST		
Candida lipolytica	glucose	-21.34 ± 0.16
Candida boidinii	glucose	-20.14 ± 0.18
	ethanol	-20.40 ± 0.14
	methanol	-21.52 ± 0.09
Kluyveromyces fragilis	lactose	-21.54 ± 0.07
	galactose	-21.78 ± 0.10
	glucose	-21.66 ± 0.19
	glucose*	-21.07 ± 0.07
		-21.30 ± 0.10
		-20.66 ± 0.26
		-21.22 ± 0.14

**Chemostat rather than batch culture: dilution rates were 0.036 h^{-1}, 0.061 h^{-1}, 0.158 h^{-1}, and 0.227 h^{-1}, respectively.*

From J.-L. Cordier, B.M. Butsch, B. Birou, and U. von Stockar, 1987, The relationship between elemental composition and heat of combustion of microbial biomass, Appl. Microbiol. Biotechnol. *25, 305–312.*

Actual heats of combustion measured by Cordier et al. [8] for a range of microorganisms and culture conditions are listed in Table 5.1. The differences in Δh_c values for bacteria and yeast reflect their slightly different elemental compositions. When the composition of a particular organism is unknown, the heat of combustion can be estimated as shown in the following equations:

$$\Delta h_c \text{ for bacteria} \simeq -23.2 \text{ kJ g}^{-1} \tag{5.27}$$

$$\Delta h_c \text{ for yeast} \simeq -21.2 \text{ kJ g}^{-1} \tag{5.28}$$

These experimentally determined values compare well with that calculated in Eq. (5.26). Once the heat of combustion of the biomass is known, it can be used with the heats of combustion of the substrates and other products to estimate the heat of reaction for anaerobic cultures.

5.9.4 Mixed Aerobic–Anaerobic Metabolism

When energy metabolism in an aerobic culture is completely oxidative, a simple relationship applies between the heat of reaction and the amount of oxygen consumed, as indicated by Eq. (5.23). However, some cultures engage in mixed oxido-reductive metabolism, resulting in a reaction stoichiometry that combines features of both aerobic and anaerobic processing. This can occur, for example, in yeast cultures, under conditions where the maximum respiratory capacity of the cells is insufficient to completely oxidise all the assimilated glucose, so that excess substrate overflows into fermentative pathways for production of ethanol. Animal cell cultures also exhibit mixed metabolism because of the simultaneous activity of several metabolic pathways, including oxidative phosphorylation and glycolysis. Although animal cells require oxygen, lactic acid is produced in anaerobic metabolism, most likely for production of biosynthetic precursors.

When the energy metabolism in a culture is mixed, Eq. (5.23) can significantly underestimate the heat of reaction. This is because additional metabolic activity not linked to oxygen consumption is taking place and releasing additional heat. Therefore, use of the short-cut method of Section 5.9.2 for estimating ΔH_{rxn} should be restricted to fully aerobic systems. Heats of reaction for cultures undergoing oxido-reductive metabolism can be calculated instead using the procedures described in Section 5.9.3 based on heats of combustion.

5.9.5 Magnitude of the Heat of Reaction in Different Cell Cultures

The relationship of Eq. (5.22) indicates that the nature of the substrate, in particular its degree of reduction, plays a role in determining the heat of reaction generated by cell cultures. Substrates with a high degree of reduction γ_S (see Table C.2 in Appendix C) contain large amounts of energy that, in reaction, cannot all be transferred to the biomass and other products, and must therefore be dissipated as heat. This explains why the heat generated by cell growth on highly reduced substrates such as methane is much greater than in carbohydrate medium. For example, when methane ($\gamma_S = 8.0$) is used as the substrate for aerobic cell culture, the amount of heat released per g of biomass produced has been found to be about four times that when less reduced substrates such as glucose ($\gamma_S = 4.0$) are used. Similarly, the heat released using ethanol or methanol ($\gamma_S = 6.0$) as substrate is two to three times that for glucose [6]. Another factor affecting heat release during culture is the intrinsic thermodynamic efficiency of the organism in transferring the energy contained in the substrate to the biomass and products. Different organisms growing on the same substrate can produce different heats of reaction.

Because substrates are not oxidised as far in anaerobic fermentations as in aerobic processes, heats of reaction for anaerobic cultures tend to be considerably lower than for respiratory systems. For example, aerobic growth of yeast on glucose generates about 2000 kJ of heat per gmol of glucose consumed; in contrast, growth of yeast on glucose under anaerobic conditions with production of ethanol generates only about 100 kJ $gmol^{-1}$ glucose [9]. As a consequence, to maintain a constant culture temperature, the cooling requirements for aerobic cultures are typically significantly greater than for anaerobic systems.

5.10 ENERGY BALANCE EQUATION FOR CELL CULTURE

In fermentations, the heat of reaction so dominates the energy balance that small enthalpy effects due to sensible heat changes and heats of mixing can generally be ignored. In this section we incorporate these observations into a simplified energy balance equation for cell processes.

Consider Eq. (5.9) applied to a continuous fermenter. What are the major factors responsible for the enthalpy difference between the input and output streams in fermentations? Because cell culture media are usually dilute aqueous solutions with behaviour close to ideal, even though the composition of the broth may vary as substrates are consumed and products formed, changes in the heats of mixing of these solutes are generally negligible. Similarly, even though there may be a temperature difference between the input and output streams, the overall change in enthalpy due to sensible heat is also small. Usually, the heat of reaction, latent heats of phase change, and shaft work are the only energy effects worth considering in fermentation energy balances. Evaporation is the most likely phase change in fermenter operation; if evaporation is controlled then latent heat effects can also be ignored.

Metabolic reactions typically generate 5 to 20 kJ of heat per second per cubic metre of fermentation broth for growth on carbohydrate, and up to 60 kJ s^{-1} m^{-3} for growth on hydrocarbon substrates. By way of comparison, in aerobic cultures sparged with dry air, evaporation of the fermentation broth removes only about 0.5 kJ s^{-1} m^{-3} as latent heat. The energy input due to shaft work varies between 0.5 and 5 kJ s^{-1} m^{-3} in large vessels and 10 to 20 kJ s^{-1} m^{-3} in small vessels; the heat effects of stirring are therefore more important in small-scale than in large-scale processes. Sensible heats and heats of mixing are generally several orders of magnitude smaller.

Therefore, for cell cultures, we can simplify energy balance calculations by substituting expressions for the heat of reaction and latent heat of vaporisation for the first two terms of Eq. (5.9). From the definition of Eq. (5.18), ΔH_{rxn} is the difference between the product and reactant enthalpies. As the products are contained in the output flow and the reactants in the input, ΔH_{rxn} is approximately equal to the difference in enthalpy between the input and output streams. If evaporation is significant, the enthalpy of the vapour leaving the system will be $M_v \Delta h_v$ greater than the enthalpy of the liquid entering the system or formed by reaction, where M_v is the mass of liquid evaporated and Δh_v is the latent heat of vaporisation.

Taking these factors into account, the modified steady-state energy balance equation for cell culture is as follows:

$$-\Delta H_{rxn} - M_v \Delta h_v - Q + W_s = 0 \tag{5.29}$$

ΔH_{rxn} has a negative sign in Eq. (5.29) because ΔH_{rxn} is equal to [enthalpy of products − enthalpy of reactants], whereas the energy balance equation refers to [enthalpy of input streams − enthalpy of output streams]. As sensible heat effects are considered negligible, the difference between ΔH°_{rxn} and ΔH_{rxn} at the reaction temperature can be ignored. Equation (5.29) applies even if some proportion of the reactants remains unconverted or if there are tie components in the system that do not react. At steady state, any material added to the system that does not participate in reaction must leave in the output stream. Therefore, ignoring enthalpy effects due to change in temperature or solution and unless the material volatilises, the enthalpy of unreacted material in the output stream must be equal to its inlet enthalpy. Thus, unreacted material and tie components make no contribution to the energy balance.

It must be emphasised that Eq. (5.29) is greatly simplified and, as discussed in Section 5.8.2, may not be applicable to single-enzyme conversions. It is, however, a very useful equation for cell culture processes. Because the heat of reaction in aerobic systems is generally higher than for anaerobic metabolism (Section 5.9.5), the principal assumption of Eq. (5.29) that heats of mixing and sensible energy changes are negligible compared with ΔH_{rxn} has greater validity in aerobic cultures.

5.11 CELL CULTURE ENERGY BALANCE WORKED EXAMPLES

For processes involving cell growth and metabolism, the enthalpy change accompanying reaction is relatively large. Energy balances for aerobic and anaerobic cultures can therefore be carried out using the modified energy balance equation (5.29). Because this equation contains no enthalpy terms, it is not necessary to define reference states. Application of Eq. (5.29) to anaerobic fermentation is illustrated in Example 5.7.

EXAMPLE 5.7 CONTINUOUS ETHANOL FERMENTATION

Saccharomyces cerevisiae is grown anaerobically in continuous culture at 30°C. Glucose is used as the carbon source; ammonia is the nitrogen source. A mixture of glycerol and ethanol is produced. At steady state, the net mass flows to and from the reactor are as follows.

glucose in	36.0 kg h^{-1}
NH_3 in	0.40 kg h^{-1}
cells out	2.81 kg h^{-1}
glycerol out	7.94 kg h^{-1}
ethanol out	11.9 kg h^{-1}
CO_2 out	13.6 kg h^{-1}
H_2O out	0.15 kg h^{-1}

Estimate the cooling requirements.

Solution

1. Assemble

(i) Units

kg, kJ, h, °C

(ii) Flow sheet

The flow sheet for this process is shown in Figure 5.9.

(iii) System boundary

The system boundary is shown in Figure 5.9.

2. Analyse

(i) Assumptions:

- steady state
- no leaks
- system is homogeneous
- heat of combustion for yeast is -21.2 kJ g^{-1}
- ideal solutions
- negligible sensible heat change
- no shaft work
- no evaporation

(ii) Basis

36.0 kg glucose, or 1 hour

(iii) Extra data

- MW glucose = 180
- MW NH_3 = 17
- MW glycerol = 92
- MW ethanol = 46

Heats of combustion:

$(\Delta h_c^\circ)_{glucose} = -2805.0$ kJ $gmol^{-1}$ (Table C.8)

$(\Delta h_c^\circ)_{NH_3} = -382.6$ kJ $gmol^{-1}$ (Table C.8)

$(\Delta h_c^\circ)_{glycerol} = -1655.4$ kJ $gmol^{-1}$ (Table C.8)

$(\Delta h_c^\circ)_{ethanol} = -1366.8$ kJ $gmol^{-1}$ (Table C.8)

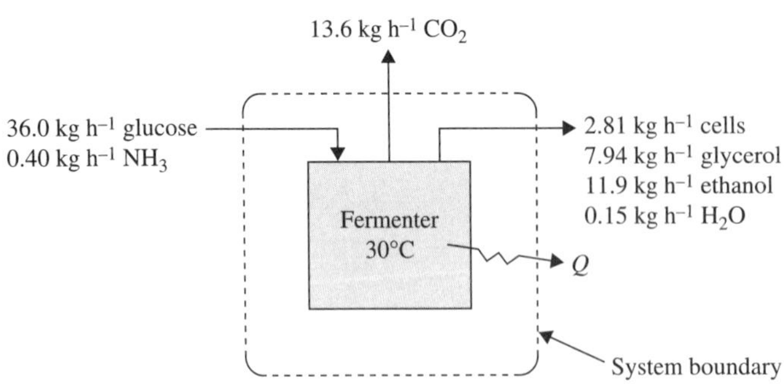

FIGURE 5.9 Flow sheet for anaerobic yeast fermentation.

(iv) Reaction

$$\text{glucose} + NH_3 \longrightarrow \text{biomass} + \text{glycerol} + \text{ethanol} + CO_2 + H_2O$$

All components are involved in reaction.

(v) Mass balance equation

The mass balance is already complete: the total mass of components in equals the total mass of components out.

(vi) Energy balance equation

For cell metabolism, the modified steady-state energy balance equation is Eq. (5.29):

$$-\Delta H_{rxn} - M_v \Delta h_v - Q + W_s = 0$$

3. Calculate

$W_s = 0$; $M_v = 0$. Therefore, the energy balance equation is reduced to:

$$-\Delta H_{rxn} - Q = 0$$

Evaluate the heat of reaction using Eq. (5.20). As the heat of combustion of H_2O and CO_2 is zero, the heat of reaction is:

$$\Delta H_{rxn} = (n\Delta h_c^\circ)_G + (n\Delta h_c^\circ)_A - (n\Delta h_c^\circ)_B - (n\Delta h_c^\circ)_{Gly} - (n\Delta h_c^\circ)_E$$

where G = glucose, A = ammonia, B = cells, Gly = glycerol, and E = ethanol. Because, in this problem, we are given the masses of reactants and products involved in the reaction, we can apply the equation for ΔH_{rxn} in mass terms:

$$\Delta H_{rxn} = (M\Delta h_c^\circ)_G + (M\Delta h_c^\circ)_A - (M\Delta h_c^\circ)_B - (M\Delta h_c^\circ)_{Gly} - (M\Delta h_c^\circ)_E$$

where Δh_c° is expressed per unit mass. Converting the Δh_c° data to kJ kg^{-1}:

$$(\Delta h_c^\circ)_G = -2805.0 \frac{\text{kJ}}{\text{gmol}} \cdot \left|\frac{1\ \text{gmol}}{180\ \text{g}}\right| \cdot \left|\frac{1000\ \text{g}}{1\ \text{kg}}\right| = -1.558 \times 10^4\ \text{kJ kg}^{-1}$$

$$(\Delta h_c^\circ)_A = -382.6 \frac{\text{kJ}}{\text{gmol}} \cdot \left|\frac{1\ \text{gmol}}{17\ \text{g}}\right| \cdot \left|\frac{1000\ \text{g}}{1\ \text{kg}}\right| = -2.251 \times 10^4\ \text{kJ kg}^{-1}$$

$$(\Delta h_c^\circ)_B = -21.2 \frac{\text{kJ}}{\text{g}} \cdot \left|\frac{1000\ \text{g}}{1\ \text{kg}}\right| = -2.120 \times 10^4\ \text{kJ kg}^{-1}$$

$$(\Delta h_c^\circ)_{Gly} = -1655.4 \frac{\text{kJ}}{\text{gmol}} \cdot \left|\frac{1\ \text{gmol}}{92\ \text{g}}\right| \cdot \left|\frac{1000\ \text{g}}{1\ \text{kg}}\right| = -1.799 \times 10^4\ \text{kJ kg}^{-1}$$

$$(\Delta h_c^\circ)_E = -1366.8 \frac{\text{kJ}}{\text{gmol}} \cdot \left|\frac{1\ \text{gmol}}{46\ \text{g}}\right| \cdot \left|\frac{1000\ \text{g}}{1\ \text{kg}}\right| = -2.971 \times 10^4\ \text{kJ kg}^{-1}$$

Therefore:

$$\begin{aligned}\Delta H_{rxn} = {} & (36.0\ \text{kg})(-1.558 \times 10^4\ \text{kJ kg}^{-1}) + (0.4\ \text{kg})(-2.251 \times 10^4\ \text{kJ kg}^{-1}) \\ & - (2.81\ \text{kg})(-2.120 \times 10^4\ \text{kJ kg}^{-1}) - (7.94\ \text{kg})(-1.799 \times 10^4\ \text{kJ kg}^{-1}) \\ & - (11.9\ \text{kg})(-2.971 \times 10^4\ \text{kJ kg}^{-1})\end{aligned}$$

$$\Delta H_{rxn} = -1.392 \times 10^4\ \text{kJ}$$

Substituting this result into the energy balance equation:

$$Q = 1.392 \times 10^4 \text{ kJ}$$

Q is positive, indicating that heat must be removed from the system.

4. Finalise

   1.4×10^4 kJ heat must be removed from the fermenter per hour.

In Example 5.7, the water used as the solvent for components of the nutrient medium was ignored. This water was effectively a tie component, moving through the system unchanged and not contributing to the energy balance. In this problem, the cooling requirements could be determined directly from the heat of reaction.

For aerobic cultures, we can relate the heat of reaction to oxygen consumption, providing a short-cut method for determining ΔH_{rxn}. Heats of combustion are not required in these calculations. Also, as long as the amount of oxygen consumed is known, the mass balance for the problem need not be completed. The procedure for energy balance problems involving aerobic fermentation is illustrated in Example 5.8.

EXAMPLE 5.8 CITRIC ACID PRODUCTION

Citric acid is manufactured using submerged culture of *Aspergillus niger* in a batch reactor operated at 30°C. Over a period of 2 days, 2500 kg glucose and 860 kg oxygen are consumed to produce 1500 kg citric acid, 500 kg biomass, and other products. Ammonia is used as the nitrogen source. Power input to the system by mechanical agitation of the broth is about 15 kW; approximately 100 kg of water is evaporated over the culture period. Estimate the cooling requirements.

Solution

1. Assemble
   - **(i)** Units

     kg, kJ, h, °C
   - **(ii)** Flow sheet

     The flow sheet is shown in Figure 5.10.
   - **(iii)** System boundary

     The system boundary is shown in Figure 5.10.
2. Analyse
   - **(i)** Assumptions
     - system is homogeneous
     - no leaks
     - ideal solutions
     - negligible sensible heat
     - heat of reaction at 30°C is -460 kJ gmol^{-1} O_2 consumed
   - **(ii)** Basis

     1500 kg citric acid produced, or 2 days

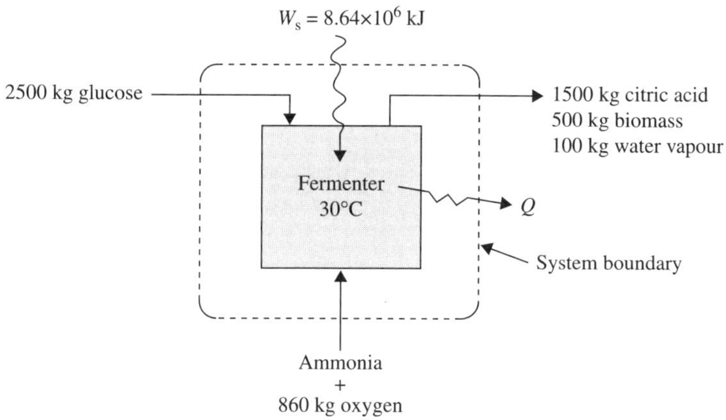

FIGURE 5.10 Flow sheet for microbial production of citric acid.

(iii) Extra data
Δh_v water at 30°C = 2430.7 kJ kg^{-1} (Table D.1)

(iv) Reaction

$$\text{glucose} + O_2 + NH_3 \longrightarrow \text{biomass} + CO_2 + H_2O + \text{citric acid}$$

All components are involved in reaction.

(v) Mass balance
The mass balance need not be completed, as the sensible energies associated with the inlet and outlet streams are assumed to be negligible.

(vi) Energy balance
The aim of the integral energy balance for batch culture is to calculate the amount of heat that must be removed to produce zero accumulation of energy in the system. Equation (5.29) is appropriate:

$$-\Delta H_{rxn} - M_v \Delta h_v - Q + W_s = 0$$

where each term refers to the 2-day culture period.

3. Calculate
ΔH_{rxn} is related to the amount of oxygen consumed:

$$\Delta H_{rxn} = (-460 \text{ kJ gmol}^{-1})\,(860 \text{ kg}) \cdot \left|\frac{1000 \text{ g}}{1 \text{ kg}}\right| \cdot \left|\frac{1 \text{ gmol}}{32 \text{ g}}\right|$$

$$\Delta H_{rxn} = -1.24 \times 10^7 \text{ kJ}$$

Heat lost through evaporation is:

$$M_v \Delta h_v = (100 \text{ kg})\,(2430.7 \text{ kJ kg}^{-1}) = 2.43 \times 10^5 \text{ kJ}$$

Power input by mechanical agitation is 15 kW or 15 kJ s^{-1}. Over a period of 2 days:

$$W_s = (15 \text{ kJ s}^{-1})\,(2 \text{ days}) \cdot \left|\frac{3600 \text{ s}}{1 \text{ h}}\right| \cdot \left|\frac{24 \text{ h}}{1 \text{ day}}\right| = 2.59 \times 10^6 \text{ kJ}$$

These results can now be substituted into the energy balance equation:

$$-(-1.24 \times 10^7 \text{ kJ}) - (2.43 \times 10^5 \text{ kJ}) - Q + (2.59 \times 10^6 \text{ kJ}) = 0$$

$$Q = 1.47 \times 10^7 \text{ kJ}$$

Q is positive, indicating that heat must be removed from the system. Note the relative magnitudes of the energy contributions from heat of reaction, shaft work, and evaporation; the effects of evaporation can often be ignored.

4. Finalise
1.5×10^7 kJ heat must be removed from the fermenter per 1500 kg of citric acid produced.

SUMMARY OF CHAPTER 5

At the end of Chapter 5 you should:

- Know which forms of energy are common in bioprocesses
- Know the *general energy balance* in words and as a mathematical equation, and the simplifications that can be made for bioprocesses
- Be familiar with *heat capacity* tables and be able to calculate *sensible heat changes*
- Be able to calculate *latent heat changes*
- Understand *heats of mixing* for nonideal solutions
- Be able to use *steam tables*
- Be able to determine *standard heats of reaction* from *heats of combustion*
- Know how to determine heats of reaction for aerobic and anaerobic cell cultures
- Be able to carry out energy balance calculations for biological systems with and without reaction

PROBLEMS

5.1 Sensible energy change

Calculate the enthalpy change associated with the following processes:

(a) *m*-Cresol is heated from 25°C to 100°C

(b) Ethylene glycol is cooled from 20°C to 10°C

(c) Succinic acid is heated from 15°C to 120°C

(d) Air is cooled from 150°C to 65°C

5.2 Heat of vaporisation

Nitrogen is sometimes bubbled into fermenters to maintain anaerobic conditions. It does not react, and leaves in the fermenter off-gas. However, it can strip water from the fermenter so that water vapour also leaves in the off-gas. In a continuous fermenter operated at 33°C, 20 g h^{-1} water is evaporated in this way. How much heat must be put into the system to compensate for evaporative cooling and so maintain the temperature at a constant value?

5.3 Steam tables

Use the steam tables to find:

(a) The heat of vaporisation of water at 85°C

(b) The enthalpy of liquid water at 35°C relative to $H = 0$ at 10°C

(c) The enthalpy of saturated water vapour at 40°C relative to $H = 0$ at the triple point

(d) The enthalpy of superheated steam at 2.5 atm absolute pressure and 275°C relative to $H = 0$ at the triple point

5.4 Preheating nutrient medium

Steam is used to heat nutrient medium in a continuous-flow process. Saturated steam at 150°C enters a coil on the outside of the heating vessel and is completely condensed. Liquid medium enters the vessel at 15°C and leaves at 44°C. Heat losses from the jacket to the surroundings are estimated as 0.22 kW. If the flow rate of medium is 3250 kg h^{-1} and its heat capacity is 0.9 cal g^{-1} $°C^{-1}$, how much steam is required?

5.5 Designer coffee mug

An award-winning manufacturer of designer homewares is developing a high-tech coffee mug based on thermodynamic principles. The aim is to keep the contents of the mug at a constant, optimum drinking temperature. When hot coffee is poured into the mug, the heat melts a reservoir of beeswax in an insulated jacket around the outside of the mug. Until all the beeswax is subsequently solidified, the temperature of the coffee will be maintained at 74°C, the melting point of beeswax. Other relevant properties of beeswax are: $\Delta h_f = 190$ J g^{-1}; C_p solid = 1.6 J g^{-1} $°C^{-1}$.

Freshly brewed coffee is poured into the mug at an average temperature of 92°C. If the mug holds 250 mL and is at an initial temperature of 25°C, what is the maximum amount of beeswax that can be used per mug if all the beeswax must be melted by the hot coffee? Assume that the other components of the coffee mug have negligible heat capacity and there are no other heat losses from the coffee.

5.6 Medium preparation

Nutrient medium in an insulated fermentation tank is diluted and brought to sterilisation temperature by injection with clean steam. Ten thousand kg of liquid medium at 130°C is required at the end of the process. Initially, the medium concentrate in the tank is at 40°C; the steam used is saturated at 220°C.

(a) How much steam is required?

(b) How much medium concentrate is required?

5.7 Enzyme conversion

An immobilised enzyme process is used in an ice-cream factory to hydrolyse lactose ($C_{12}H_{22}O_{11}$) to glucose ($C_6H_{12}O_6$) and galactose ($C_6H_{12}O_6$):

$$C_{12}H_{22}O_{11} + H_2O \longrightarrow C_6H_{12}O_6 + C_6H_{12}O_6$$

Gel beads containing β-galactosidase are packed into a column reactor; 2500 kg of lactose enters the reactor per day as a 10% solution in water at 25°C. The reactor operates at steady state and 32°C; all of the lactose is converted. Because the heat of reaction for enzyme conversions is not as great as for cell culture, sensible heat changes and heats of mixing cannot be ignored.

	Δh_c° (kJ gmol^{-1})	C_p (cal g^{-1} °C^{-1})	Δh_m (kcal gmol^{-1})
Lactose	−5652.5	0.30	3.7
Water	–	1.0	–
Glucose	−2805.0	0.30	5.6
Galactose	−2805.7	0.30	5.6

(a) What is the standard heat of reaction for this enzyme conversion?
(b) Estimate the heating or cooling requirements for this process. State explicitly whether heating or cooling is needed.

5.8 Production of glutamic acid

Immobilised cells of a genetically improved strain of *Brevibacterium lactofermentum* are used to convert glucose to glutamic acid for production of MSG (monosodium glutamate). The immobilised cells are unable to grow, but metabolise glucose according to the equation:

$$C_6H_{12}O_6 + NH_3 + 1.5\,O_2 \longrightarrow C_5H_9O_4N + CO_2 + 3\,H_2O$$

A feed stream of 4% glucose in water enters a 25,000-litre reactor at 25°C at a flow rate of 2000 kg h^{-1}. A gaseous mixture of 12% NH_3 in air is sparged into the reactor at 1 atm and 15°C at a flow rate of 4 vvm (1 vvm means 1 vessel volume per minute). The product stream from the reactor contains residual sugar at a concentration of 0.5%.

(a) Estimate the cooling requirements.
(b) How important is cooling in this fermentation? For example, assuming the reaction rate remains constant irrespective of temperature, if cooling were not provided and the reactor operated adiabatically, what would be the temperature? (In fact, the rate of conversion will decline rapidly at high temperatures due to cell death and enzyme deactivation.)

5.9 Bacterial production of alginate

Azotobacter vinelandii is investigated for production of alginate from sucrose. In a continuous fermenter at 28°C with ammonia as nitrogen source, the yield of alginate was found to be 4 g g^{-1} oxygen consumed. It is planned to produce alginate at a rate of 5 kg h^{-1}. Since the viscosity of alginate in aqueous solution is considerable, energy input due to mixing the broth cannot be neglected. The fermenter is equipped with a flat-blade disc turbine; at a satisfactory mixing speed and air flow rate, the power requirements are estimated at 1.5 kW. Calculate the cooling requirements.

5.10 Acid fermentation

Propionibacterium species are tested for commercial-scale production of propionic acid. Propionic and other acids are synthesised in anaerobic culture using sucrose as the substrate and ammonia as the nitrogen source. Overall yields from sucrose are as follows:

propionic acid	40% (w/w)
acetic acid	20% (w/w)
butyric acid	5% (w/w)
lactic acid	3.4% (w/w)
biomass	12% (w/w)

Bacteria are inoculated into a vessel containing sucrose and ammonia; a total of 30 kg sucrose is consumed over a period of 10 days. What are the cooling requirements?

5.11 Ethanol fermentation

A crude fermenter is set up in a shed in the backyard of a suburban house. Under anaerobic conditions with ammonia as the nitrogen source, about 0.45 g ethanol are formed per g glucose consumed. At steady state, the ethanol production rate averages 0.4 kg h^{-1}.

The owner of this enterprise decides to reduce her electricity bill by using the heat released during the fermentation to warm water as an adjunct to the household hot water system. Cold water at 10°C is fed into a jacket surrounding the fermenter at a rate of 2.5 litres h^{-1}. To what temperature is the water heated? Heat losses from the system are negligible. Use a biomass composition of $CH_{1.75}O_{0.58}N_{0.18}$ plus 8% ash.

5.12 Production of bakers' yeast

Bakers' yeast is produced in a 50,000-litre fermenter under aerobic conditions. The carbon substrate is sucrose; ammonia is provided as the nitrogen source. The average biomass composition is $CH_{1.83}O_{0.55}N_{0.17}$ with 5% ash. Under conditions supporting efficient growth, biomass is the only major product and the biomass yield from sucrose is 0.5 g g^{-1}. If the specific growth rate is 0.45 h^{-1}, estimate the rate of heat removal required to maintain constant temperature in the fermenter when the yeast concentration is 10 g l^{-1}.

5.13 Culture kinetic parameters from thermal properties

A 250 litre airlift bioreactor is used for continuous aerobic culture of *Rhizobium etli* bacteria. Succinic acid ($C_4H_6O_4$) is used as the carbon source; ammonium chloride is used as the nitrogen source. The products of the culture are biomass, carbon dioxide, and water only. The bioreactor is equipped with an external jacket through which water flows at a rate of 100 kg h^{-1}. The inlet water temperature is 20°C. The outside of the jacket is covered with insulation to prevent heat losses. At steady state, the cell concentration is 4.5 g l^{-1}, the rate of consumption of succinic acid is 395 g h^{-1}, and the outlet water temperature is 27.5°C.

(a) What is the enthalpy change of the cooling water in the jacket? Use C_p water = 4.2 kJ kg^{-1} $°C^{-1}$.

(b) Assuming that all the heat generated by the culture is absorbed by the cooling water, estimate the rate of oxygen uptake by the cells. What other assumptions are involved in your answer?

(c) Determine the biomass growth rate in units of g h^{-1}.

(d) What is the specific growth rate of the cells in units of h^{-1}?

5.14 Production of snake antivenin

To satisfy strong market demand, a pharmaceutical company has developed a hybridoma cell line that synthesises a monoclonal antibody capable of neutralising the venom of the Australian death adder. In culture, the hybridoma cells exhibit mixed oxido-reductive energy metabolism.

(a) The heat generated by the cells during growth is measured using a flow microcalorimeter attached to a small bioreactor. The relationship between the amount of heat generated and the amount of oxygen consumed is found to be 680 kJ $gmol^{-1}$, which is considerably higher than the 460 kJ $gmol^{-1}$ expected for fully oxidative metabolism. These data are used to estimate the heat of reaction associated with the anaerobic components of hybridoma cell activity. It is hypothesised that the level of production of lactic acid is a direct indicator of anaerobic metabolism, leading to the equation:

$$\Delta H_{rxn}\ (\text{oxido-reductive}) = \Delta H_{rxn}\ (\text{fully oxidative}) + \Delta h_{rxn}\ (\text{anaerobic}) \times \text{gmol lactic acid produced}$$

If the molar ratio of lactic acid production to oxygen consumption measured in the bioreactor is 5.5, estimate Δh_{rxn} for the anaerobic metabolic pathways.

(b) For commercial antivenin production, the hybridoma culture is scaled up to a 500-litre stirred bioreactor. Under these conditions, the specific rate of oxygen uptake is 0.3 mmol $(10^9 \text{ cells})^{-1}\ h^{-1}$, the specific rate of lactic acid production is 1.0 mmol $(10^9 \text{ cells})^{-1}\ h^{-1}$, and the maximum cell density is 7.5×10^6 cells ml^{-1}.

(i) Estimate the maximum rate of cooling required to maintain the culture temperature.

(ii) A new medium is developed along stoichiometric principles to bring the levels of nutrients provided to the cells closer to actual requirements. When the new medium is used, lactic acid production decreases significantly to only 0.05 mmol $(10^9 \text{ cells})^{-1}\ h^{-1}$. If the oxygen demand stays roughly the same, what will be the new cooling requirements?

(iii) The heat generated by the culture is absorbed by cooling water flowing in a coil inside the bioreactor. Water enters the coil at 20°C and leaves at 29°C. What maximum flow rate of cooling water is required for cultures growing in the new medium? Use C_p water = 4.19 kJ $kg^{-1}\ °C^{-1}$.

5.15 Ginseng production

Suspension cultures of *Panax ginseng* plant cells are used to produce ginseng biomass for the health tonic market. The cells are grown in batch culture in a 2500-litre stirred bioreactor operated at 25°C and 1 atm pressure. Over a period of 12 days, 5.5×10^6 litres of air at 25°C and 1 atm are pumped into the reactor; off-gas containing 5.1×10^5 litres of oxygen leaves the vessel over the same period. About 135 kg of water is lost by evaporation during the culture. Estimate the cooling requirements.

5.16 Evaporative cooling

An engineer decides to design a special airlift reactor for growing *Torula utilis* cells. The reactor relies on evaporative cooling only. Water is evaporated from the broth by sparging with dry air; water vapour leaving the vessel is then condensed in an external condenser and returned to the culture. On average, 0.5 kg water is evaporated per day. If the rate of oxygen uptake by the cells is 140 mmol h^{-1} and the culture temperature must be maintained at 30°C, does evaporation provide adequate cooling for this system?

5.17 Penicillin process

Penicillium chrysogenum is used to produce penicillin in a 90,000-litre fermenter. The volumetric rate of oxygen uptake by the cells ranges from 0.45 to 0.85 mmol $l^{-1}\ min^{-1}$ depending on time during the culture. Power input by stirring is 2.9 W l^{-1}. Estimate the cooling requirements.

5.18 Culture of methylotrophic yeast

(a) Cells of *Pichia pastoris* yeast are cultured in medium containing glycerol ($C_3H_8O_3$) as the carbon source and ammonium hydroxide as the nitrogen source. Under aerobic conditions, the yield of biomass from substrate is 0.57 g g^{-1}. Biomass, CO_2, and H_2O are the only major products of this culture. Estimate the cooling requirements per g of biomass produced.

(b) When all the glycerol has been consumed, methanol (CH_4O) is added to the culture. There is enough ammonium hydroxide remaining in the medium to support growth of the biomass with methanol as substrate. The composition of the cells is essentially unchanged with a biomass yield from substrate of 0.44 g g^{-1}. In what way do the cooling demands change per g of biomass produced?

5.19 Algal culture for carotenoid synthesis

A newly discovered, carotenoid-producing microalga isolated from an artesian bore in the Simpson Desert is cultured under aerobic conditions in a 100-litre bubble column reactor. The medium contains glucose as the primary carbon source and ammonia as the nitrogen source. The elemental formula for the biomass is $CH_{1.8}O_{0.6}N_{0.2}$; because carotenoids are present in the cells at levels of only a few μg per g biomass, they do not appreciably alter the biomass formula. The yield of biomass (including carotenoid) from substrate is 0.45 g g^{-1}; biomass, carbon dioxide, and water can be considered the only products formed. The reactor is operated continuously so that the rate of glucose consumption is 77 g h^{-1}. Estimate the cooling requirements. What assumptions are involved in your answer?

5.20 Checking the consistency of measured culture data

The data in the following table are measured during aerobic culture of *Corynebacterium glutamicum* on medium containing molasses, corn extract, and other nutrients for production of lysine.

	Time period of fermentation (h)	
Measured variable	**0–12**	**12–36**
Rate of uptake of reducing sugars (g l^{-1} h^{-1})	0.42	2.0
Rate of biomass production (g l^{-1} h^{-1})	0.29	0.21
Rate of lysine production (g l^{-1} h^{-1})	0.20	0.66
Rate of oxygen uptake (g l^{-1} h^{-1})	0.40	0.75
Rate of heat evolution (kJ l^{-1} h^{-1})	2.5	12.1

(a) Use mass and energy balance principles to check whether the data for the 12–36-h culture period are consistent (e.g., to within ±10%).

(b) It is suspected that amino acids present in the nutrient medium may initially provide an additional source of substrate in this culture. Are the data for the 0–12-h period consistent with this theory?

(c) At the beginning of the culture, accurate measurement of oxygen uptake is difficult so that heat evolution is considered a more reliable indicator of oxygen consumption than direct oxygen measurements. If the amino acids provided in the nutrient medium can be considered to have the same average properties as glutamine, estimate the mass rate of amino acid uptake during the first 12 h of the culture.

References

[1] Perry's Chemical Engineers' Handbook, eighth ed., McGraw-Hill, 2008. (New editions released periodically.)
[2] CRC Handbook of Chemistry and Physics, eighty-fourth ed., CRC Press, Boca Raton, 2003. (New editions released periodically.)
[3] International Critical Tables (1926–1930), McGraw-Hill; 1st electronic ed. (2003), International Critical Tables of Numerical Data, Physics, Chemistry and Technology, Knovel/Norwich.
[4] B. Atkinson, F. Mavituna, Biochemical Engineering and Biotechnology Handbook, second ed., Macmillan, 1991.
[5] S.A. Patel, L.E. Erickson, Estimation of heats of combustion of biomass from elemental analysis using available electron concepts, Biotechnol. Bioeng. 23 (1981) 2051–2067.
[6] J.A. Roels, Energetics and Kinetics in Biotechnology, Elsevier Biomedical Press, 1983 (Chapter 3).
[7] C.L. Cooney, D.I.C. Wang, R.I. Mateles, Measurement of heat evolution and correlation with oxygen consumption during microbial growth, Biotechnol. Bioeng. 11 (1968) 269–281.
[8] J.-L. Cordier, B.M. Butsch, B. Birou, U. von Stockar, The relationship between elemental composition and heat of combustion of microbial biomass, Appl. Microbiol. Biotechnol. 25 (1987) 305–312.
[9] E.H. Battley, Energetics of Microbial Growth, John Wiley, 1987.

Suggestions for Further Reading

Process Energy Balances

Felder, R. M., & Rousseau, R. W. (2005). *Elementary Principles of Chemical Processes* (3rd ed., Chapters 7–9). Wiley.
Himmelblau, D. M., & Riggs, J. B. (2004). *Basic Principles and Calculations in Chemical Engineering* (7th ed., Chapters 21–26 and 28). Prentice Hall.
Sinnott, R. K. (2005). *Coulson and Richardson's Chemical Engineering, volume 6: Chemical Engineering Design* (4th ed., Chapter 3). Elsevier.

Metabolic Thermodynamics

See also references 5 through 9.

Guan, Y., Evans, P. M., & Kemp, R. B. (1998). Specific heat flow rate: an on-line monitor and potential control variable of specific metabolic rate in animal cell culture that combines microcalorimetry with dielectric spectroscopy. *Biotechnol. Bioeng., 58*, 464–477.
Marison, I., & von Stockar, U. (1987). A calorimetric investigation of the aerobic cultivation of *Kluyveromyces fragilis* on various substrates. *Enzyme Microbiol. Technol., 9*, 33–43.
von Stockar, U., & Marison, I. W. (1989). The use of calorimetry in biotechnology. *Adv. Biochem. Eng./Biotechnol., 40*, 93–136.

CHAPTER

6

Unsteady-State Material and Energy Balances

An unsteady-state or transient process is one that causes system properties to vary with time. Batch and semi-batch systems are inherently transient; continuous systems are unsteady during start-up and shut-down. Changing from one set of process conditions to another also creates an unsteady state, as does any fluctuation in input or control variables.

The principles of mass and energy balances developed in Chapters 4 and 5 can be applied to unsteady-state processes. Balance equations are used to determine the rate of change of system parameters; solution of these equations generally requires application of calculus. Questions such as: What is the concentration of product in the reactor as a function of time? and How long will it take to reach a particular temperature after the flow of steam is started? can be answered using unsteady-state mass and energy balances. In this chapter we will consider some simple unsteady-state problems.

6.1 UNSTEADY-STATE MATERIAL BALANCE EQUATIONS

When the mass of a system is not constant, we generally need to know how the mass varies as a function of time. To evaluate the *rate of change* of mass in the system, let us first return to the general mass balance equation introduced in Chapter 4:

$$\left\{\begin{array}{c}\text{mass in}\\\text{through}\\\text{system}\\\text{boundaries}\end{array}\right\}-\left\{\begin{array}{c}\text{mass out}\\\text{through}\\\text{system}\\\text{boundaries}\end{array}\right\}+\left\{\begin{array}{c}\text{mass}\\\text{generated}\\\text{within}\\\text{system}\end{array}\right\}-\left\{\begin{array}{c}\text{mass}\\\text{consumed}\\\text{within}\\\text{system}\end{array}\right\}=\left\{\begin{array}{c}\text{mass}\\\text{accumulated}\\\text{within}\\\text{system}\end{array}\right\} \quad (4.1)$$

Consider the flow system of Figure 6.1 in which reactions are taking place. Species A is involved in the process; M is the mass of A in the system. Using the 'hat' symbol ˆ to denote rate, let $\hat{M}_i$ be the mass flow rate of A entering the system, and $\hat{M}_o$ the mass flow rate of

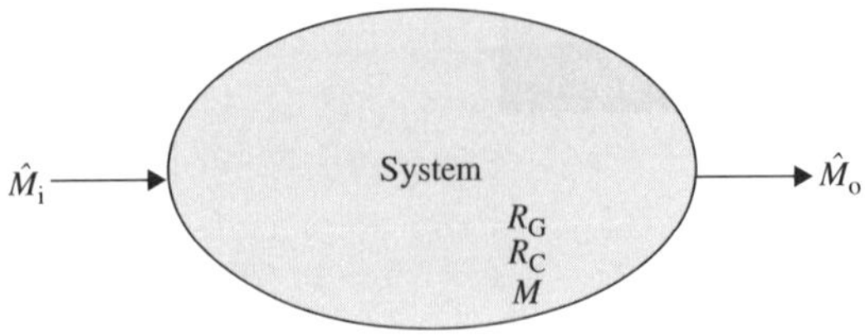

FIGURE 6.1 Flow system for an unsteady-state mass balance.

A leaving. R_G is the mass rate of generation of species A by chemical reaction; R_C is the mass rate of consumption of A by reaction. The dimensions of $\hat{M}_i$, $\hat{M}_o$, R_G, and R_C are MT^{-1} and the units are, for example, g s^{-1}, kg h^{-1}, lb min^{-1}.

All of the variables, $\hat{M}_i$, $\hat{M}_o$, R_G, and R_C, may vary with time. However, let us focus on an infinitesimally small interval of time Δt between times t and $t + \Delta t$. Even though the system variables may be changing, if Δt is sufficiently small we can treat the flow rates $\hat{M}$ and rates of reaction R as if they were constant during this period. Under these circumstances, the terms of the general mass balance equation (4.1) may be written as follows.

- *Mass in*. During period Δt, the mass of species A transported into the system is $\hat{M}_i\,\Delta t$. Note that the dimensions of $\hat{M}_i\,\Delta t$ are M and the units are, for example, g, kg, lb.
- *Mass out*. Similarly, the mass of species A transported out during time Δt is $\hat{M}_o\,\Delta t$.
- *Generation*. The mass of A generated during Δt is $R_G\,\Delta t$.
- *Consumption*. The mass of A consumed during Δt is $R_C\Delta t$.
- *Accumulation*. Let ΔM be the mass of A accumulated in the system during Δt. ΔM may be either positive (accumulation) or negative (depletion).

Entering these terms into the general mass balance equation (4.1) with the accumulation term on the left side gives:

$$\Delta M = \hat{M}_i\Delta t - \hat{M}_o\Delta t + R_G\Delta t - R_C\Delta t \tag{6.1}$$

We can divide both sides of Eq. (6.1) by Δt to give:

$$\frac{\Delta M}{\Delta t} = \hat{M}_i - \hat{M}_o + R_G - R_C \tag{6.2}$$

Equation (6.2) applies when Δt is infinitesimally small. If we take the limit as Δt approaches zero, that is, as t and $t + \Delta t$ become virtually the same, Eq. (6.2) represents the system at an instant rather than over an interval of time. Mathematical techniques for handling this type of situation are embodied in the rules of calculus. In calculus, the *derivative* of y with respect to x, dy/dx, is defined as:

$$\frac{dy}{dx} = \lim_{\Delta x \to 0} \frac{\Delta y}{\Delta x} \tag{6.3}$$

where $\lim_{\Delta x \to 0}$ represents the limit as Δx approaches zero. As Eq. (6.2) is valid for $\Delta t \to 0$, we can write it as:

$$\frac{dM}{dt} = \lim_{\Delta t \to 0} \frac{\Delta M}{\Delta t} = \hat{M}_i - \hat{M}_o + R_G - R_C \tag{6.4}$$

The derivative dM/dt represents the rate of change of mass with time measured at a particular instant. We have thus derived a differential equation for the rate of change of M as a function of the system variables, $\hat{M}_i$, $\hat{M}_o$, R_G, and R_C:

$$\frac{dM}{dt} = \hat{M}_i - \hat{M}_o + R_G - R_C \tag{6.5}$$

At steady state there can be no change in the mass of the system, so the rate of change dM/dt must be zero. Therefore, at steady state, Eq. (6.5) reduces to a form of the familiar steady-state mass balance equation from Chapter 4:

$$\text{mass in} + \text{mass generated} = \text{mass out} + \text{mass consumed} \tag{4.2}$$

Unsteady-state mass balance calculations begin with derivation of a differential equation to describe the process. Equation (6.5) was developed on a mass basis and contains parameters such as mass flow rate $\hat{M}$ and mass rate of reaction R. Another common form of the unsteady-state mass balance is based on volume. The reason for this variation is that reaction rates are usually expressed on a per-volume basis. For example, the rate of a first-order reaction is expressed in terms of the concentration of reactant:

$$r_C = k_1 C_A$$

where r_C is the *volumetric rate of consumption of A by reaction* (with units of, e.g., g cm^{-3} s^{-1}), k_1 is the first-order reaction rate constant, and C_A is the concentration of reactant A. This and other reaction rate equations are described in more detail in Chapter 12. When rate expressions are used in mass and energy balance problems, the relationship between mass and volume must enter the analysis. This is illustrated in Example 6.1.

EXAMPLE 6.1 UNSTEADY-STATE MATERIAL BALANCE FOR A CONTINUOUS STIRRED TANK REACTOR

A continuous stirred tank reactor (CSTR) is operated as shown in Figure 6.2. The volume of liquid in the tank is V. Feed enters with volumetric flow rate F_i; product leaves with volumetric flow rate F_o. The concentration of reactant A in the feed is C_{Ai}; the concentration of A in the exit stream is C_{Ao}. The density of the feed stream is ρ_i; the density of the product stream is ρ_o. The tank is well mixed. The concentration of A in the tank is C_A and the density of liquid in the tank is ρ. In the reactor, compound A undergoes reaction and is transformed into compound B. The volumetric rate of consumption of A by reaction is given by the expression $r_C = k_1 C_A$.

Using unsteady-state balances, derive differential equations for:

(a) Total mass
(b) The mass of component A

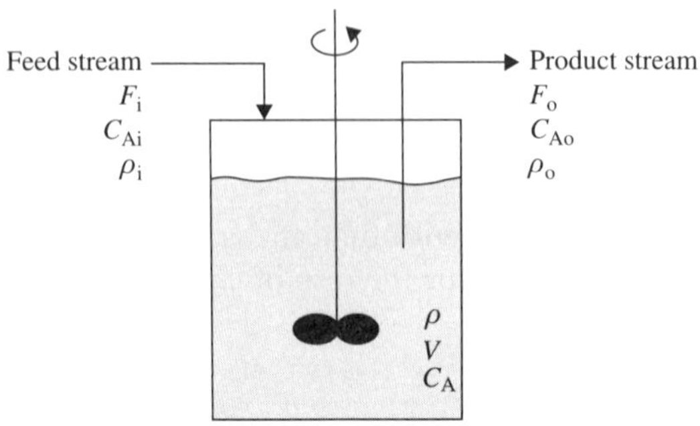

FIGURE 6.2 Continuous stirred tank reactor.

Solution

The general unsteady-state mass balance equation is Eq. (6.5):

$$\frac{dM}{dt} = \hat{M}_i - \hat{M}_o + R_G - R_C$$

(a) For the balance on total mass, R_G and R_C are zero; total mass cannot be generated or consumed by chemical reaction. From the definition of density (Section 2.4.1), total mass can be expressed as the product of volume and density. Similarly, mass flow rate can be expressed as the product of volumetric flow rate and density.

$$\text{Total mass in the tank: } M = \rho V; \text{ therefore } \frac{dM}{dt} = \frac{d(\rho V)}{dt}$$
$$\text{Mass flow rate in: } \hat{M}_i = F_i \rho_i$$
$$\text{Mass flow rate out: } \hat{M}_o = F_o \rho_o$$

Substituting these terms into Eq. (6.5):

$$\frac{d(\rho V)}{dt} = F_i \rho_i - F_o \rho_o \tag{6.6}$$

Equation (6.6) is a differential equation representing an unsteady-state mass balance on total mass.

(b) Compound A is not generated in the reaction; therefore $R_G = 0$. The other terms of Eq. (6.5) can be expressed as follows:

$$\text{Mass of A in the tank: } M = VC_A; \text{ therefore } \frac{dM}{dt} = \frac{d(VC_A)}{dt}$$
$$\text{Mass flow rate of A in: } \hat{M}_i = F_i C_{Ai}$$
$$\text{Mass flow rate of A out: } \hat{M}_o = F_o C_{Ao}$$
$$\text{Rate of consumption of A: } R_C = k_1 C_A V$$

Substituting into Eq. (6.5):

$$\frac{d(VC_A)}{dt} = F_i C_{Ai} - F_o C_{Ao} - k_1 C_A V \tag{6.7}$$

Equation (6.7) is a differential equation representing an unsteady-state mass balance on A.

6.2 UNSTEADY-STATE ENERGY BALANCE EQUATIONS

In Chapter 5, the law of conservation of energy was represented by the equation:

$$\begin{Bmatrix} \text{energy in through} \\ \text{system boundaries} \end{Bmatrix} - \begin{Bmatrix} \text{energy out through} \\ \text{system boundaries} \end{Bmatrix} = \begin{Bmatrix} \text{energy accumulated} \\ \text{within the system} \end{Bmatrix} \tag{5.4}$$

Consider the system shown in Figure 6.3. E is the total energy in the system, $\hat{W}_s$ is the rate at which shaft work is done on the system, $\hat{Q}$ is the rate of heat removal from the system, and $\hat{M}_i$ and $\hat{M}_o$ are mass flow rates to and from the system. All these parameters may vary with time.

Ignoring kinetic and potential energies as discussed in Section 5.2.1, the energy balance equation can be applied over an infinitesimally small interval of time Δt, during which we can treat $\hat{W}_s$, $\hat{Q}$, $\hat{M}_i$, and $\hat{M}_o$ as if they were constant. Under these conditions, the terms of the general energy balance equation are:

- *Input*. During time interval Δt, the amount of energy entering the system is $\hat{M}_i h_i \Delta t + \hat{W}_s \Delta t$, where h_i is the specific enthalpy of the incoming flow stream.
- *Output*. Similarly, the amount of energy leaving the system is $\hat{M}_o h_o \Delta t + \hat{Q} \Delta t$.
- *Accumulation*. Let ΔE be the energy accumulated in the system during time Δt. ΔE may be either positive (accumulation) or negative (depletion). In the absence of kinetic and potential energies, E represents the enthalpy of the system.

Entering these terms into Eq. (5.4) with the accumulation term first:

$$\Delta E = \hat{M}_i h_i \Delta t - \hat{M}_o h_o \Delta t - \hat{Q} \Delta t + \hat{W}_s \Delta t \tag{6.8}$$

We can divide both sides of Eq. (6.8) by Δt:

$$\frac{\Delta E}{\Delta t} = \hat{M}_i h_i - \hat{M}_o h_o - \hat{Q} + \hat{W}_s \tag{6.9}$$

Equation (6.9) is valid for small Δt. The equation for the rate of change of energy at a particular instant in time is determined by taking the limit of Eq. (6.9) as Δt approaches zero:

$$\frac{dE}{dt} = \lim_{\Delta t \to 0} \frac{\Delta E}{\Delta t} = \hat{M}_i h_i - \hat{M}_o h_o - \hat{Q} + \hat{W}_s \tag{6.10}$$

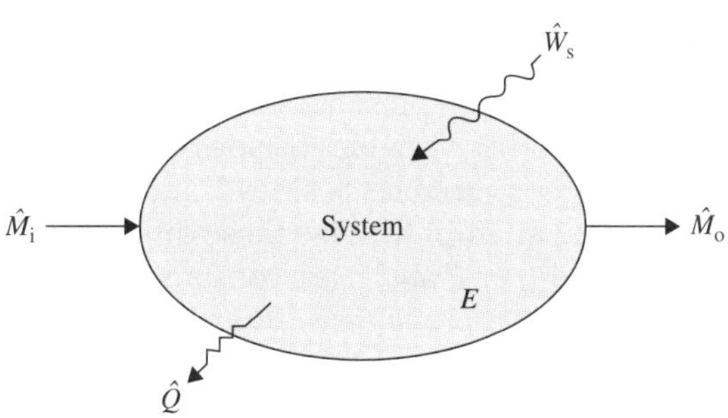

FIGURE 6.3 Flow system for an unsteady-state energy balance.

Equation (6.10) is the unsteady-state energy balance equation for a system with only one inlet and one outlet stream. If there are several such streams, all mass flow rates and enthalpies must be added together:

$$\frac{dE}{dt} = \sum_{\substack{\text{input}\\ \text{streams}}} (\hat{M}h) - \sum_{\substack{\text{output}\\ \text{streams}}} (\hat{M}h) - \hat{Q} + \hat{W}_s \tag{6.11}$$

Equation (6.11) can be simplified for cell culture processes using the same arguments as those presented in Section 5.10. If $\Delta\hat{H}_{rxn}$ is the rate at which heat is absorbed or liberated by reaction, and $\hat{M}_v$ is the mass flow rate of evaporated liquid leaving the system, for cell cultures in which sensible heat changes and heats of mixing can be ignored, the following unsteady-state energy balance equation applies:

$$\frac{dE}{dt} = -\Delta\hat{H}_{rxn} - \hat{M}_v\Delta h_v - \hat{Q} + \hat{W}_s \tag{6.12}$$

For exothermic reactions $\Delta\hat{H}_{rxn}$ is negative; for endothermic reactions $\Delta\hat{H}_{rxn}$ is positive.

6.3 SOLVING DIFFERENTIAL EQUATIONS

As shown in Sections 6.1 and 6.2, unsteady-state mass and energy balances are represented using differential equations. Once the differential equation for a particular system has been found, the equation must be solved to obtain an expression for mass M or energy E as a function of time. Differential equations are solved by integration. Some simple rules for differentiation and integration are outlined in Appendix E. Of course, there are many more rules of calculus than those included in Appendix E; however those shown are sufficient for handling the unsteady-state problems in this chapter. Further details can be found in any elementary calculus textbook, or in mathematics books written especially for biological scientists (e.g., [1–7]).

Before we proceed with solution techniques for unsteady-state mass and energy balances, there are several general points to consider.

1. *A differential equation can be solved directly only if it contains no more than two variables.* For mass and energy balance problems, the differential equation must have the form:

$$\frac{dM}{dt} = f(M,t) \quad \text{or} \quad \frac{dE}{dt} = f(E,t)$$

   where $f(M,t)$ represents some function of M and t, and $f(E,t)$ represents some function of E and t. The function may contain constants, but no other variables besides M and t should appear in the expression for dM/dt, and no other variables besides E and t should appear in the expression for dE/dt. Before you attempt to solve these differential equations, check first that all parameters except M and t, or E and t, are constants.
2. *Solution of differential equations requires knowledge of boundary conditions.* Boundary conditions contain extra information about the system. The number of boundary

conditions required depends on the *order* of the differential equation, which is equal to the order of the highest differential coefficient in the equation. For example, if the equation contains a second derivative (e.g., d^2x/dt^2), the equation is second order. All equations developed in this chapter have been first order because they involve only first-order derivatives of the form dx/dt. One boundary condition is required to solve a first-order differential equation; two boundary conditions are required for a second-order differential equation, and so on. Boundary conditions that apply at the beginning of the process when $t = 0$ are called *initial conditions*.

3. *Not all differential equations can be solved algebraically*, even if the equation contains only two variables and the boundary conditions are available. Solution of some differential equations requires application of numerical techniques, preferably using a computer. In this chapter we will be concerned mostly with simple equations that can be solved using elementary calculus.

The easiest way of solving differential equations is to *separate variables* so that each variable appears on only one side of the equation. For example, consider the simple differential equation:

$$\frac{dx}{dt} = a\,(b - x) \tag{6.13}$$

where a and b are constants. First we must check that the equation contains only two variables x and t, and that all other parameters in the equation are constants. Once this is verified, the equation is separated so that x and t each appear on only one side of the equation. In the case of Eq. (6.13), this is done by dividing each side of the equation by $(b - x)$, and multiplying each side by dt:

$$\frac{dx}{(b - x)} = a\,dt \tag{6.14}$$

The equation is now ready for integration:

$$\int \frac{dx}{(b - x)} = \int a\,dt \tag{6.15}$$

Using integration rules (E.28) and (E.24) from Appendix E:

$$-\ln (b - x) = at + K \tag{6.16}$$

Note that the constants of integration from both sides of the equation have been condensed into one constant K; this is valid because a constant $\pm$ a constant = a constant.

6.4 SOLVING UNSTEADY-STATE MASS BALANCES

Solution of unsteady-state mass balances is sometimes difficult unless certain simplifications are made. Because the aim here is to illustrate the application of unsteady-state balances without becoming too involved in integral calculus, the problems presented in this section will be relatively simple. For the majority of problems in this chapter, analytical solution is possible.

The following restrictions are common in unsteady-state mass balance problems.

- The system is *well mixed* so that properties of the system do not vary with position. If properties within the system are the same at all points, this includes the point from which any product stream is drawn. Accordingly, when the system is well mixed, properties of the outlet stream are the same as those within the system.
- Expressions for reaction rate involve the concentration of only one reactive species. The mass balance equation for this species can then be derived and solved; if other chemical species appear in the kinetic expression, this introduces extra variables into the differential equation making solution more complex.

The following example illustrates solution of an unsteady-state mass balance without reaction.

EXAMPLE 6.2 DILUTION OF SALT SOLUTION

To make 100 litres of solution, 1.5 kg salt is dissolved in water. Pure water is pumped into a tank containing this solution at a rate of 5 l min^{-1}; salt solution overflows at the same rate. The tank is well mixed. How much salt is in the tank at the end of 15 min? Because the salt solution is dilute, assume that its density is constant and equal to that of water.

Solution

1. Flow sheet and system boundary
   These are shown in Figure 6.4.
2. Define variables
   C_A = concentration of salt in the tank; V = volume of solution in the tank; ρ = density of salt solution and water.
3. Assumptions
   – no leaks
   – tank is well mixed
   – density of the salt solution is the same as that of water

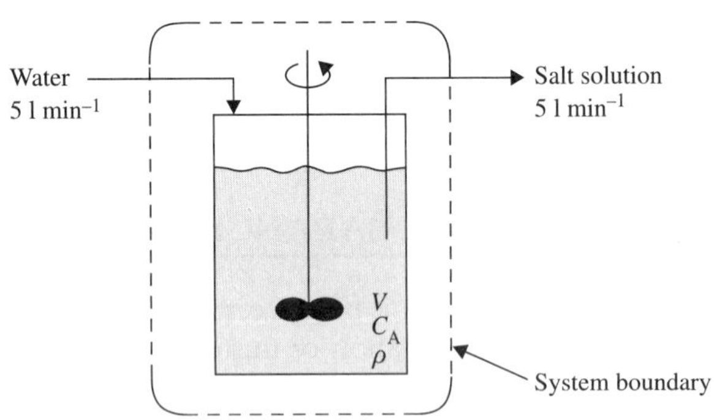

FIGURE 6.4 Well-mixed tank for dilution of salt solution.

4. Boundary conditions

At the beginning of the process, the salt concentration is 1.5 kg in 100 litres, or 0.015 kg l^{-1}. If we call this initial salt concentration C_{A0}, the initial condition is:

$$\text{at } t = 0 \qquad C_A = C_{A0} = 0.015 \text{ kg l}^{-1} \tag{1}$$

We also know that the initial volume of liquid in the tank is 100 l. Therefore, another initial condition is:

$$\text{at } t = 0 \qquad V = V_0 = 100 \text{ l} \tag{2}$$

5. Total mass balance

The unsteady-state balance equation for total mass was derived in Example 6.1 as Eq. (6.6):

$$\frac{d(\rho V)}{dt} = F_i \rho_i - F_o \rho_o$$

In this problem we are told that the volumetric flow rates of the inlet and outlet streams are equal; therefore $F_i = F_o$. In addition, the density of the system is constant so that $\rho_i = \rho_o = \rho$. Under these conditions, the terms on the right side of Eq. (6.6) cancel to zero. On the left side, because ρ is constant it can be taken outside of the differential. Therefore:

$$\rho \frac{dV}{dt} - 0$$

or

$$\frac{dV}{dt} = 0$$

If the derivative of V with respect to t is zero, V must be a constant:

$$V = K$$

where K is the constant of integration. This result means that the volume of the tank is constant and independent of time. Initial condition (2) tells us that $V = 100$ l at $t = 0$; therefore V must equal 100 l at all times. Consequently, the constant of integration K is equal to 100 l, and the volume of liquid in the tank does not vary from 100 l.

6. Mass balance for salt

An unsteady-state mass balance equation for component A such as salt was derived in Example 6.1 as Eq. (6.7):

$$\frac{d(VC_A)}{dt} = F_i C_{Ai} - F_o C_{Ao} - k_1 C_A V$$

In the present problem there is no reaction, so k_1 is zero. Also, $F_i = F_o = F = 5 \text{ l min}^{-1}$. Because the tank is well mixed, the concentration of salt in the outlet stream is equal to that inside the tank, that is, $C_{Ao} = C_A$. In addition, since the inlet stream does not contain salt, $C_{Ai} = 0$. From the balance on total mass, we know that V is constant and therefore can be placed outside of the differential. Taking these factors into consideration, Eq. (6.7) becomes:

$$V\frac{dC_A}{dt} = -FC_A$$

This differential equation contains only two variables C_A and t; F and V are constants. The variables are easy to separate by dividing both sides by VC_A and multiplying by dt:

$$\frac{dC_A}{C_A} = \frac{-F}{V}dt$$

The equation is now ready to integrate:

$$\int\frac{dC_A}{C_A} = \int\frac{-F}{V}dt$$

Using integration rules (E.27) and (E.24) from Appendix E and combining the constants of integration:

$$\ln C_A = \frac{-F}{V}t + K \qquad (3)$$

We have yet to determine the value of K. From initial condition (1), at $t = 0$, $C_A = C_{A0}$. Substituting this information into (3):

$$\ln C_{A0} = K$$

We have thus determined K. Substituting this value for K back into (3):

$$\ln C_A = \frac{-F}{V}t + \ln C_{A0} \qquad (4)$$

This is the solution to the mass balance; it gives an expression for the concentration of salt in the tank as a function of time. Notice that if we had forgotten to add the constant of integration, the answer would not contain the term $\ln C_{A0}$. The equation would then say that at $t = 0$, $\ln C_A = 0$; that is, $C_A = 1$. We know this is not true; instead, at $t = 0$, $C_A = 0.015\ \text{kg l}^{-1}$, so the result without the boundary condition is incorrect. It is important to apply boundary conditions every time you integrate.

The solution equation is usually rearranged to give an exponential expression. This is achieved by subtracting $\ln C_{A0}$ from both sides of (4):

$$\ln C_A - \ln C_{A0} = \frac{-F}{V}t$$

and noting from Eq. (E.9) in Appendix E that $(\ln C_A - \ln C_{A0})$ is the same as $\ln C_A/C_{A0}$:

$$\ln\frac{C_A}{C_{A0}} = \frac{-F}{V}t$$

Taking the antilogarithm of both sides:

$$\frac{C_A}{C_{A0}} = e^{\frac{-F}{V}t}$$

or

$$C_A = C_{A0}\, e^{\frac{-F}{V}t}$$

We can check that this is the correct solution by taking the derivative of both sides with respect to t and making sure that the original differential equation is recovered.

For $F = 5\ \mathrm{l\ min^{-1}}$, $V = 100\ \mathrm{l}$, and $C_{A0} = 0.015\ \mathrm{g\ l^{-1}}$, at $t = 15$ min:

$$C_A = (0.015\ \mathrm{kg\ l^{-1}})\, e^{\left(\frac{-5\ \mathrm{l\ min^{-1}}}{100\ \mathrm{l}}\right)(15\ \mathrm{min})} = 7.09 \times 10^{-3}\ \mathrm{kg\ l^{-1}}$$

The salt concentration after 15 min is $7.09 \times 10^{-3}\ \mathrm{kg\ l^{-1}}$. Therefore:

$$\text{mass of salt} = C_A V = (7.09 \times 10^{-3}\ \mathrm{kg\ l^{-1}})\,(100\ \mathrm{l}) = 0.71\ \mathrm{kg}$$

7. Finalise

After 15 min, the mass of salt in the tank is 0.71 kg.

In Example 6.2 the density of the system was assumed constant. This simplified the mathematics of the problem so that ρ could be taken outside the differential and cancelled from the total mass balance. The assumption of constant density is justified for dilute solutions because the density does not differ greatly from that of the solvent. The result of the total mass balance makes intuitive sense: for a tank with equal flow rates in and out and constant density, the volume of liquid inside the tank should remain constant.

The effect of reaction on the unsteady-state mass balance is illustrated in Example 6.3.

EXAMPLE 6.3 FLOW REACTOR

Rework Example 6.2 to include reaction. Assume that a reaction in the tank consumes salt at a rate given by the first-order equation:

$$r = k_1 C_A$$

where k_1 is the first-order reaction constant and C_A is the concentration of salt in the tank. Derive an expression for C_A as a function of time. If $k_1 = 0.02\ \mathrm{min^{-1}}$, how long does it take for the concentration of salt to fall to a value 1/20 the initial level?

Solution

The flow sheet, boundary conditions, and assumptions for this problem are the same as in Example 6.2. The total mass balance is also the same; total mass in the system is unaffected by reaction.

1. Mass balance for salt
   The unsteady-state mass balance equation for salt with first-order reaction is Eq. (6.7):

$$\frac{d(VC_A)}{dt} = F_i C_{Ai} - F_o C_{Ao} - k_1 C_A V$$

In this problem $F_i = F_o = F$, $C_{Ai} = 0$, and V is constant. Because the tank is well mixed, $C_{Ao} = C_A$. Therefore, Eq. (6.7) becomes:

$$V\frac{dC_A}{dt} = -FC_A - k_1 C_A V$$

This equation contains only two variables, C_A and t; F, V, and k_1 are constants. Separate variables by dividing both sides by VC_A and multiplying by dt:

$$\frac{dC_A}{C_A} = \left(\frac{-F}{V} - k_1\right)dt$$

Integrating both sides gives:

$$\ln C_A = \left(\frac{-F}{V} - k_1\right)t + K$$

where K is the constant of integration. K is determined from initial condition (1) in Example 6.2: at $t = 0$, $C_A = C_{A0}$. Substituting these values gives:

$$\ln C_{A0} = K$$

Substituting this value for K back into the answer:

$$\ln C_A = \left(\frac{-F}{V} - k_1\right)t + \ln C_{A0}$$

or

$$\ln\frac{C_A}{C_{A0}} = \left(\frac{-F}{V} - k_1\right)t$$

For $F = 5\ \mathrm{l\ min^{-1}}$, $V = 100\ \mathrm{l}$, $k_1 = 0.02\ \mathrm{min^{-1}}$, and $C_A/C_{A0} = 1/20$, this equation becomes:

$$\ln\left(\frac{1}{20}\right) = \left(\frac{-5\ \mathrm{l\ min^{-1}}}{100\ \mathrm{l}} - 0.02\ \mathrm{min^{-1}}\right)t$$

or

$$-3.00 = (0.07\ \mathrm{min^{-1}})\ t$$

Solving for t:

$$t = 42.8\ \mathrm{min}$$

2. Finalise
   The concentration of salt in the tank reaches 1/20 its initial level after 43 min.

6.5 SOLVING UNSTEADY-STATE ENERGY BALANCES

Solution of unsteady-state energy balance problems can be mathematically quite complex. In this chapter, only problems with the following characteristics will be treated for ease of mathematical handling.

- The system has at most one input and one output stream; furthermore, these streams have the same mass flow rate. Under these conditions, the total mass of the system is constant.
- The system is well mixed with uniform temperature and composition. Properties of the outlet stream are therefore the same as within the system.
- No chemical reactions or phase changes occur.
- Mixtures and solutions are ideal.
- The heat capacities of the system contents and inlet and outlet streams are independent of composition and temperature, and therefore invariant with time.
- Internal energy U and enthalpy H are independent of pressure.

The principles and equations for unsteady-state energy balances are entirely valid when these conditions are not met; the only difference is that solution of the differential equations is greatly simplified for systems with the above characteristics. The procedure for solution of unsteady-state energy balances is illustrated in Example 6.4.

EXAMPLE 6.4 SOLVENT HEATER

An electric heating coil is immersed in a stirred tank. Solvent at 15°C with heat capacity 2.1 kJ kg^{-1} °C^{-1} is fed into the tank at a rate of 15 kg h^{-1}. Heated solvent is discharged at the same flow rate. The tank is filled initially with 125 kg of cold solvent at 10°C. The rate of heating by the electric coil is 800 W. Calculate the time required for the temperature of the solvent to reach 60°C.

Solution

1. Flow sheet and system boundary
   These are shown in Figure 6.5.
2. Define variables
   If the tank is well mixed, the temperature of the outlet stream is the same as inside the tank. Let T be the temperature in the tank; T_i is the temperature of the incoming stream, M is the mass of solvent in the tank, and $\hat{M}$ is the mass flow rate of solvent to and from the tank.
3. Assumptions
   - no leaks
   - tank is well mixed
   - negligible shaft work
   - heat capacity is independent of temperature
   - no evaporation
   - heat losses to the environment are negligible

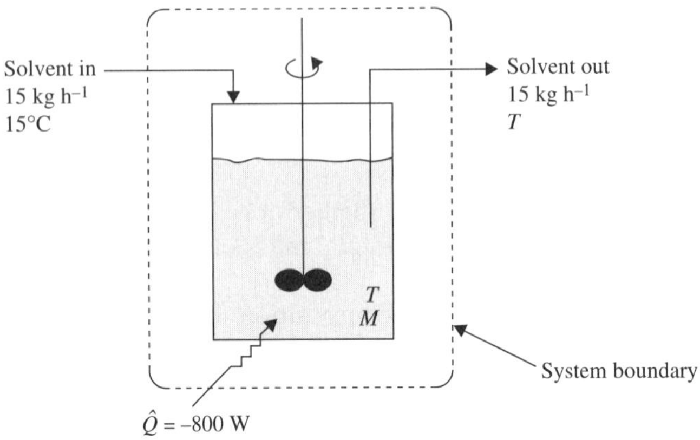

FIGURE 6.5 Continuous process for heating solvent.

4. Reference state
   $H = 0$ for solvent at 10°C: $T_{ref} = 10°C$.
5. Boundary conditions
   The initial condition is:

$$\text{at } t = 0 \quad T = T_0 = 10°C \tag{1}$$

6. Mass balance
   Since the mass flow rates of solvent to and from the tank are equal, the mass M of solvent inside the tank is constant and equal to the initial value, 125 kg. The mass balance is therefore complete.
7. Energy balance
   The unsteady-state energy balance equation for systems with two flow streams is given by Eq. (6.10):

$$\frac{dE}{dt} = \hat{M}_i h_i - \hat{M}_o h_o - \hat{Q} + \hat{W}_s$$

In the absence of phase change, reaction, and heats of mixing, the enthalpies of the input and output streams can be expressed using sensible heats only. With the enthalpy of the solvent defined as zero at T_{ref}, enthalpy h is calculated from the difference between the system temperature and T_{ref}:

$$\text{Flow input: } \hat{M}_i h_i = \hat{M}_i C_p (T_i - T_{ref}) \tag{2}$$

$$\text{Flow output: } \hat{M}_o h_o = \hat{M}_o C_p (T - T_{ref}) \tag{3}$$

Similarly, any change in the energy content of the system E must be reflected as a change in sensible heat and temperature relative to T_{ref}:

$$\text{Accumulation: } \frac{dE}{dt} = \frac{d}{dt}(MC_p \Delta T) = \frac{d}{dt}(MC_p[T - T_{ref}])$$

As M, C_p, and T_{ref} are constants, this equation becomes:

$$\frac{dE}{dt} = MC_p \frac{dT}{dt} \tag{4}$$

We have assumed that shaft work is negligible; therefore:

$$\hat{W}_s = 0 \tag{5}$$

Substituting expressions (2) through (5) into the energy balance equation (6.10) gives:

$$MC_p \frac{dT}{dt} = \hat{M}_i C_p (T_i - T_{ref}) - \hat{M}_o C_p (T - T_{ref}) - \hat{Q}$$

$M = 125$ kg, $C_p = 2.1$ kJ kg^{-1} °C^{-1}, $\hat{M}_i = \hat{M}_o = 15$ kg h^{-1}, $T_i = 15$°C, and $T_{ref} = 10$°C. Converting the data for $\hat{Q}$ into consistent units:

$$\hat{Q} = -800 \text{ W} \cdot \left| \frac{1 \text{ J s}^{-1}}{1 \text{ W}} \right| \cdot \left| \frac{1 \text{ kJ}}{1000 \text{ J}} \right| \cdot \left| \frac{3600 \text{ s}}{1 \text{ h}} \right|$$

$$\hat{Q} = -2.88 \times 10^3 \text{ kJ h}^{-1}$$

$\hat{Q}$ is negative because heat flows into the system. Substituting these values into the energy balance equation:

$$125 \text{ kg } (2.1 \text{ kJ kg}^{-1} {}^\circ\text{C}^{-1}) \frac{dT}{dt} = 15 \text{ kg h}^{-1} (2.1 \text{ kJ kg}^{-1} {}^\circ\text{C}^{-1}) (15^\circ\text{C} - 10^\circ\text{C})$$
$$- 15 \text{ kg h}^{-1} (2.1 \text{ kJ kg}^{-1} {}^\circ\text{C}^{-1}) (T - 10^\circ\text{C}) - (-2.88 \times 10^3 \text{ kJ h}^{-1})$$

Grouping terms gives a differential equation for the rate of temperature change:

$$\frac{dT}{dt} = 12.77 - 0.12T$$

where T has units of °C and t has units of h. Separating variables and integrating:

$$\int \frac{dT}{12.77 - 0.12T} = \int dt$$

Using integration rule (E.28) from Appendix E:

$$\frac{-1}{0.12} \ln(12.77 - 0.12\,T) = t + K$$

The initial condition is given by Eq. (1): at $t = 0$, $T = 10$°C. Therefore, $K = -20.40$ and the solution is:

$$20.40 - \frac{1}{0.12} \ln(12.77 - 0.12T) = t$$

From this equation, at $T = 60$°C, $t = 6.09$ h.

8. Finalise

It takes 6.1 h for the temperature to reach 60°C.

SUMMARY OF CHAPTER 6

At the end of Chapter 6 you should:

- Know what types of process require unsteady-state analysis
- Be able to derive appropriate *differential equations* for unsteady-state mass and energy balances
- Understand the need for *boundary conditions* to solve differential equations representing actual processes
- Be able to solve simple unsteady-state mass and energy balances to obtain equations for system parameters as a function of time

PROBLEMS

6.1 Dilution of sewage

In a sewage treatment plant, a large concrete tank initially contains 440,000 litres of liquid and 10,000 kg of fine suspended solids. To flush this material out of the tank, water is pumped into the vessel at a rate of 40,000 $l\ h^{-1}$. Liquid containing solids leaves at the same rate. Estimate the concentration of suspended solids in the tank at the end of 5 h. State your assumptions.

6.2 Production of fish protein concentrate

Whole gutted fish are dried to make a protein paste. In a batch drier, the rate at which water is removed from the fish is roughly proportional to the moisture content. If a batch of gutted fish loses half its initial moisture content in the first 20 min, how long will the drier take to remove 95% of the water?

6.3 Contamination of vegetable oil

Vegetable oil is used in a food processing factory to prepare instant breadcrumbs. A stirred tank is used to hold the oil. During operation of the breadcrumb process, oil is pumped from the tank at a rate of 4.8 $l\ h^{-1}$. At 8 PM during the night shift, the tank is mistakenly connected to a drum of cod liver oil, which is then pumped into the tank. The volume of vegetable oil in the tank at 8 PM is 60 l.

(a) If the flow rate of cod liver oil into the tank is 7.5 $l\ h^{-1}$ and the tank has a maximum capacity of 100 l, will the tank overflow before the factory manager arrives at 9 AM? Assume that the density of both oils is the same.

(b) If cod liver oil is pumped into the tank at a rate of 4.8 $l\ h^{-1}$ instead of 7.5 $l\ h^{-1}$, what is the composition of oil in the tank at midnight?

6.4 Drainage from mine tailings

A shallow lagoon is fed with run-off from a mine tailings dump at a rate of 1.5×10^9 l day^{-1}. There is no ground seepage from the lagoon. The rate of surface evaporation is such that the volume of water in the lagoon remains roughly constant at 6×10^{11} l. One day, arsenic starts to leach from the mine tailings to give a concentration in the run-off stream of 100 ppm. What is the concentration of arsenic in the lagoon after 2 weeks of leaching?

6.5 Diafiltration

Diafiltration is carried out to exchange the buffers used to suspend a globular protein. Initially, the protein is contained in 3000 litres of solution containing 45 $g\ l^{-1}$ salt. The aim

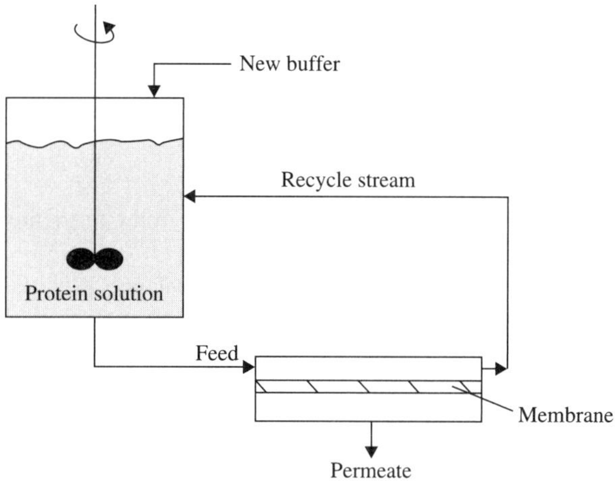

FIGURE 6P5.1 Diafiltration process for buffer exchange.

is to remove 99.9% of the salt and suspend the protein in new, salt-depleted buffer. As indicated in Figure 6P5.1, a microporous ultrafiltration membrane is applied to separate the salt from the protein. The membrane retains the protein but allows passage of all smaller molecules such as salt and water into the permeate stream. The permeate flow rate is 750 l h^{-1}. The total volume of protein solution is kept constant by addition of new buffer.

(a) Estimate the time required to complete this process.
(b) What volume of new buffer is needed?
(c) What mass of salt is present in the new solution?

6.6 Radioactive decay

A radioactive isotope decays at a rate proportional to the amount of isotope present. If the concentration of isotope is C (mg l^{-1}), its rate of decay is:

$$r_C = k_1 C$$

(a) A solution of radioactive isotope is prepared at concentration C_0. Show that the half-life of the isotope, that is, the time required for the isotope concentration to reach half of its original value, is equal to $\ln 2/k_1$.
(b) A solution of the isotope ^{32}P is used to radioactively label DNA for hybridisation studies. The half-life of ^{32}P is 14.3 days. According to institutional safety requirements, the solution cannot be discarded until the activity is 1% of its present value. How long will this take?

6.7 Batch growth of bacteria

During the exponential phase of batch culture, the growth rate of a culture is proportional to the concentration of cells present. When *Streptococcus lactis* bacteria are cultured in milk, the concentration of cells doubles every 45 min. If this rate of growth is maintained for 12 h, what is the final concentration of cells relative to the inoculum level?

6.8 Fed-batch fermentation

A feed stream containing glucose enters a fed-batch fermenter at a constant flow rate. The initial volume of liquid in the fermenter is V_0. Cells in the fermenter consume glucose at a rate given by:

$$r_S = k_1 s$$

where k_1 is the rate constant (h^{-1}) and s is the concentration of glucose in the fermenter ($g\ l^{-1}$).

(a) Assuming constant density, derive an equation for the total mass balance. What is the expression relating volume and time?

(b) Derive a differential equation for the rate of change of substrate concentration.

6.9 Continuous fermentation

A well-mixed fermenter contains cells initially at concentration x_0. A sterile feed enters the fermenter with volumetric flow rate F; fermentation broth leaves at the same rate. The concentration of substrate in the feed is s_i. The equation for the rate of cell growth is:

$$r_X = k_1 x$$

and the equation for the rate of substrate consumption is:

$$r_S = k_2 x$$

where k_1 and k_2 are rate constants with dimensions T^{-1}, r_X and r_S have dimensions $L^{-3}MT^{-1}$, and x is the concentration of cells in the fermenter.

(a) Derive a differential equation for the unsteady-state mass balance of cells.

(b) From this equation, what must be the relationship between F, k_1, and the volume of liquid in the fermenter V at steady state?

(c) Solve the differential equation to obtain an expression for cell concentration in the fermenter as a function of time.

(d) Use the following data to calculate how long it takes for the cell concentration in the fermenter to reach $4.0\ g\ l^{-1}$:

$$F = 2200\ l\ h^{-1}$$
$$V = 10{,}000\ l$$
$$x_0 = 0.5\ g\ l^{-1}$$
$$k_1 = 0.33\ h^{-1}$$

(e) Set up a differential equation for the mass balance of substrate. Substitute the result for x from (c) to obtain a differential equation in which the only variables are substrate concentration and time. (Do you think you would be able to solve this equation algebraically?)

(f) At steady state, what must be the relationship between s and x?

6.10 Plug-flow reactor

When fluid flows through a pipe or channel with sufficiently large Reynolds number, it approximates *plug flow*. Plug flow means that there is no variation of axial velocity over the flow cross-section. When reaction occurs in a plug-flow tubular reactor (PFTR), as reactant is consumed its concentration changes down the length of the tube.

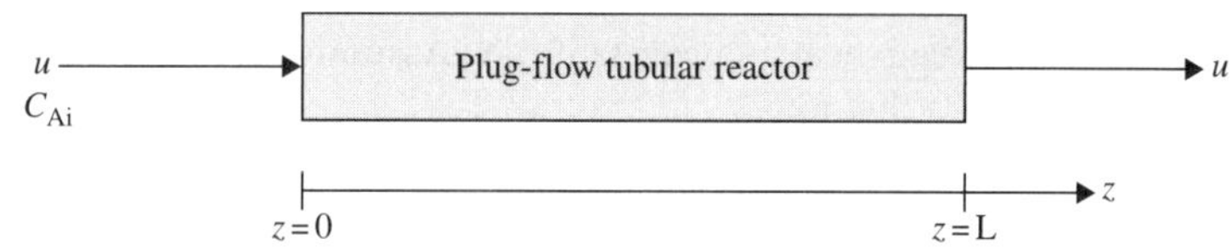

FIGURE 6P10.1 Plug-flow tubular reactor.

(a) Derive a differential equation for the change in reactant concentration with distance at steady state.
(b) What are the boundary conditions?
(c) If the reaction is first order, solve the differential equation to determine an expression for concentration as a function of distance from the front of the tube.
(d) How does this expression compare with that for a well-mixed batch reactor?
Hint: Referring to Figure 6P10.1, consider the accumulation of reactant within a section of the reactor between z and $z + \Delta z$. In this case, the volumetric flow rate of liquid in and out of the section is Au. Use the following symbols:
– A is the reactor cross-sectional area
– u is the fluid velocity
– z is distance along the tube from the entrance
– L is the total length of the reactor
– C_{Ai} is the concentration of reactant in the feed stream
– C_A is the concentration of reactant in the reactor; C_A is a function of z
– r_C is the volumetric rate of consumption of reactant

6.11 Sequential batch reactors

Pseudomonas aeruginosa is used to make antimicrobial rhamnolipids. The bacteria are cultured in a stirred 8000-litre batch fermenter. To prepare the inoculum for the 8000-l fermenter, a smaller seed fermenter of capacity 150 litres is used. The seed fermenter is inoculated with 3 l of broth containing 4.5 g l^{-1} cells. The equation for the rate of cell growth in both fermenters is:

$$r_X = kx$$

where k is the rate constant and x is the concentration of cells. For growth of *P. aeruginosa* in the seed fermenter, the value of k is 0.40 h^{-1}. However, a different medium is used in the production vessel to maximise rhamnolipid synthesis, giving a value of k in the large fermenter of 0.28 h^{-1}. The seed fermenter is shut down when the concentration of bacteria reaches 6.5 g l^{-1}. The broth is then pumped into the 8000-l vessel, and the large-scale fermentation is carried out for 16 h.
(a) What is the duration of the seed fermentation?
(b) What is the final cell concentration in the production vessel?

6.12 Boiling water

A beaker containing 2 litres of water at 18°C is placed on a laboratory hot plate. The water begins to boil in 11 min.
(a) Neglecting evaporation, write the energy balance for the process.

(b) The hot plate delivers heat at a constant rate. Assuming that the heat capacity of water is constant, what is that rate?

6.13 Heating glycerol solution

An adiabatic stirred tank is used to heat 100 kg of a 45% glycerol solution in water. An electrical coil delivers 2.5 kW of power to the tank; 88% of the energy delivered by the coil goes into heating the vessel contents. The glycerol solution is initially at 15°C.

(a) Write a differential equation for the energy balance.

(b) Integrate the equation to obtain an expression for temperature as a function of time.

(c) Assuming glycerol and water form an ideal solution, how long will the solution take to reach 90°C?

6.14 Heating molasses

Diluted molasses is heated in a well-mixed steel tank by saturated steam at 40 psia condensing in a jacket on the outside of the tank. The outer walls of the jacket are insulated. Molasses solution at 20°C enters the tank at a rate of 1020 kg h^{-1} and 1020 kg h^{-1} of heated molasses solution leaves. The rate of heat transfer from the steam through the jacket and to the molasses is given by the equation:

$$\hat{Q} = UA\,(T_{\text{steam}} - T_{\text{molasses}})$$

where $\hat{Q}$ is the rate of heat transfer, U is the overall heat transfer coefficient, A is the surface area for heat transfer, and T is temperature. For this system the value of U is 190 kcal $m^{-2}\,h^{-1}\,°C^{-1}$; C_p for the molasses solution is 0.85 kcal $kg^{-1}\,°C^{-1}$. The initial mass of molasses solution in the tank is 5000 kg; the initial temperature is 20°C. The surface area for heat transfer between the steam and the tank is 1.5 m^2.

(a) Derive a differential equation for the rate of change of temperature in the tank.

(b) Solve the differential equation to obtain an equation relating temperature and time.

(c) Plot the temperature of molasses leaving the tank as a function of time.

(d) What is the maximum temperature that can be achieved in the tank?

(e) Estimate the time required for this system to reach steady state.

(f) How long does it take for the outlet molasses temperature to rise from 20°C to 40°C?

6.15 Preheating culture medium

A glass fermenter used for culture of hybridoma cells contains nutrient medium at 15°C. The fermenter is wrapped in an electrical heating mantle that delivers heat at a rate of 450 W. Before inoculation, the medium and vessel must be at 36°C. The medium is well mixed during heating. Use the following information to determine the time required for medium preheating.

Glass fermenter vessel: mass = 12.75 kg; C_p = 0.20 cal $g^{-1}\,°C^{-1}$

Nutrient medium: mass = 7.50 kg; C_p = 0.92 cal $g^{-1}\,°C^{-1}$

6.16 Water heater

A tank contains 1000 kg of water at 24°C. It is planned to heat this water using saturated steam at 130°C in a coil inside the tank. The rate of heat transfer from the steam is given by the equation:

$$\hat{Q} = UA(T_{\text{steam}} - T_{\text{water}})$$

where $\hat{Q}$ is the rate of heat transfer, U is the overall heat transfer coefficient, A is the surface area for heat transfer, and T is temperature. The heat transfer area provided by the coil is 0.3 m^2; the heat transfer coefficient is 220 kcal m^{-2} h^{-1} °C^{-1}. Liquid condensate leaves the coil saturated.

(a) The tank has a surface area of 0.9 m^2 exposed to the ambient air. The tank exchanges heat through this exposed surface at a rate given by an equation similar to that above. For heat transfer to or from the surrounding air, the heat transfer coefficient is 25 kcal m^{-2} h^{-1} °C^{-1}. If the air temperature is 20°C, calculate the time required to heat the water to 80°C.

(b) What time is saved if the tank is insulated?

Assume the heat capacity of water is constant, and neglect the heat capacity of the tank walls.

6.17 Thermal mixing

Hot water at 90°C enters a mixing tank at a flow rate of 500 kg h^{-1}. At the same time, cool water at 18°C also flows into the tank. At steady state, the temperature of the discharge stream is 30°C and the mass of water in the tank is 1100 kg. Suddenly, the temperature of the cool water being added increases to 40°C. How long does it take the discharge stream to reach 40°C?

6.18 Laboratory heating

A one-room mobile laboratory used for field work is equipped with an electric furnace. The maximum power provided by the furnace is 2400 W. One winter morning when the furnace is switched off, the temperature in the laboratory falls to 5°C, which is equal to the outside air temperature. The furnace is turned on to its maximum setting and, after 30 min, the temperature in the laboratory reaches 20°C. The rate of heat loss from the room through the walls, ceiling, and window panes depends on the outside temperature according to the equation:

$$\hat{Q}_{loss} = k(T - T_o)$$

where $\hat{Q}_{loss}$ is the rate of heat loss, k is a constant, T is the inside temperature, and T_o is the outside temperature. If $k = 420$ kJ °C^{-1} h^{-1}, estimate the 'heat capacity' of the laboratory, that is, the energy required to raise the temperature of the room by 1°C.

References

[1] J.C. Arya, R.W. Lardner, Mathematics for the Biological Sciences, Prentice Hall, 1979.
[2] A. Cornish-Bowden, Basic Mathematics for Biochemists, second ed., Oxford University Press, 1999.
[3] C. Neuhauser, Calculus for Biology and Medicine, third ed., Prentice Hall, 2009.
[4] L.D. Hoffmann, G.L. Bradley, Calculus for Business, Economics, and the Social and Life Sciences, ninth ed., McGraw-Hill, 2007.
[5] E.F. Haeussler, R.S. Paul, Introductory Mathematical Analysis, thirteeth ed., Prentice Hall, 2011.
[6] E. Passow, Schaum's Outline: Understanding Calculus Concepts, McGraw-Hill, 1996.
[7] A. Mizrahi, M. Sullivan, Mathematics: An Applied Approach, seventh ed., Wiley, 2000.

Suggestions for Further Reading

Felder, R. M., & Rousseau, R. W. (2005). *Elementary Principles of Chemical Processes* (3rd ed., Chapter 11). Wiley.

PART 3

PHYSICAL PROCESSES

CHAPTER

7

Fluid Flow

Fluid mechanics is an important area of engineering science concerned with the nature and properties of fluids in motion and at rest. Fluids play a central role in bioprocesses since most of the required physical, chemical, and biological transformations take place in a fluid phase. Because the behaviour of fluids depends to a large extent on their physical characteristics, knowledge of fluid properties and techniques for their measurement is crucial. Fluids in bioprocessing often contain suspended solids, consist of more than one phase, and have non-Newtonian properties; all of these features complicate the analysis of flow behaviour and present many challenges in bioprocess design. In bioreactors, fluid properties play a key role in determining the effectiveness of mixing, gas dispersion, mass transfer, and heat transfer. Together, these processes can exert a significant influence on system productivity and the success of equipment scale-up.

This chapter provides a theoretical basis for Chapters 8, 9, and 10, which consider important physical processes that are coupled to fluid flow. Fluid mechanics accounts for a substantial fraction of the chemical engineering literature; accordingly, complete treatment of the subject is beyond the scope of this book. Here, we content ourselves with study of those aspects of flow behaviour particularly relevant to fermentation fluids and the operation of bioprocessing equipment. Further information can be found in the references at the end of the chapter.

7.1 CLASSIFICATION OF FLUIDS

A fluid is a substance that undergoes continuous deformation when subjected to a shearing force. Shearing forces act tangentially to the surfaces over which they are applied. For example, a simple shear force exerted on a stack of thin parallel plates will cause the plates to slide over each other, as in a pack of cards. Shear can also occur in other geometries; the effect of shear forces in planar and rotational systems is illustrated in Figure 7.1. Shearing in these examples causes *deformation*, which is a change in the relative positions of parts of a body. A shear force must be applied to produce fluid flow.

According to the preceding definition, fluids can be either gases or liquids. Two physical properties, viscosity and density, are used to classify fluids. If the density of a fluid

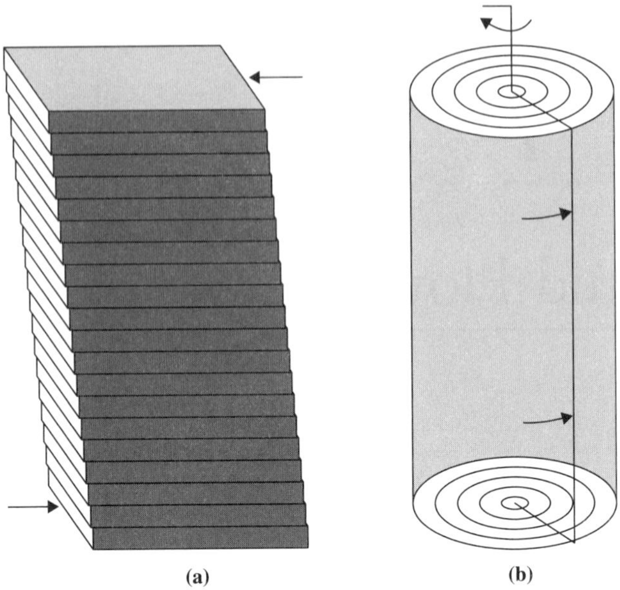

FIGURE 7.1 Laminar deformation due to (a) planar shear, and (b) rotational shear. *From J. R. van Wazer, J.W. Lyons, K. Y. Kim, and R. E. Colwell, 1963,* Viscosity and Flow Measurement, *John Wiley, New York.*

changes with pressure, the fluid is *compressible*. Gases are generally classed as compressible fluids. The density of liquids is practically independent of pressure; liquids are *incompressible* fluids. Sometimes the distinction between compressible and incompressible fluids is not well defined; for example, a gas may be treated as incompressible if the variation in pressure and temperature is small.

Fluids are also classified on the basis of viscosity. Viscosity is the property of fluids responsible for internal friction during flow. An *ideal* or *perfect* fluid is a hypothetical liquid or gas that is incompressible and has zero viscosity. The term *inviscid* applies to fluids with zero viscosity. All *real* fluids have finite viscosity and are therefore called *viscid* or *viscous* fluids.

Fluids can be classified further as *Newtonian* or *non-Newtonian*. This distinction is explained in detail in Sections 7.3 and 7.5.

7.2 FLUIDS IN MOTION

Bioprocesses involve fluids in motion in vessels, pipes, and other equipment. Some general characteristics of fluid flow are described in the following sections.

7.2.1 Streamlines

When a fluid flows through a pipe or over a solid object, the velocity of the fluid varies depending on position. One way of representing variation in velocity is *streamlines*, which follow the flow path. Constant velocity is shown by equidistant spacing of parallel streamlines as shown in Figure 7.2(a). The velocity profile for slow-moving fluid flowing over a

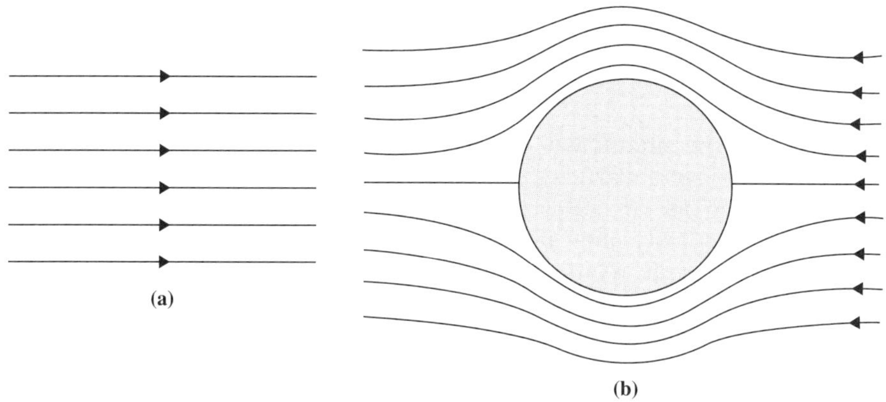

FIGURE 7.2 Streamlines for (a) constant fluid velocity, and (b) steady flow over a submerged object.

submerged object is shown in Figure 7.2(b); reduced spacing between the streamlines indicates that the velocity at the top and bottom of the object is greater than at the front and back.

Streamlines show only the net effect of fluid motion. Although streamlines suggest smooth continuous flow, the individual particles or parcels of fluid may actually be moving in an erratic fashion. The slower the flow, the more closely the streamlines represent actual motion. Slow fluid flow is therefore called *streamline* or *laminar* flow. In contrast, in fast motion, particles of fluid frequently cross and recross the streamlines. This is called *turbulent* flow, which is characterised by the formation of local regions of fluid rotation called *eddies*. The features of turbulent flow are considered in more detail in Section 7.9.

7.2.2 Shear Stress

Shear stresses develop in fluids when different parts of the fluid move relative to each other. Consider a flow stream with adjacent layers of fluid moving steadily at different velocities. Within the fluid there is continuous molecular movement, so that an interchange of molecules occurs between the layers. This interchange influences the velocity of flow in the layers. For example, molecules moving from a slow layer will reduce the speed of an adjacent faster layer, while those coming from a fast layer will have an accelerating effect on a slower layer. This phenomenon is called *viscous drag* and gives rise to *viscous forces* in fluids as neighbouring fluid layers transfer momentum and affect each other's velocity. Such forces are often referred to as *viscous shear stresses*, where shear stress is defined as force divided by the area of interaction between the fluid layers. Shear stresses in fluids are therefore induced by fluid flow and are a consequence of velocity differences within the fluid. The overall effect of viscous shear stress is to reduce the differences in velocity between adjacent fluid layers or streamlines. Shear stress can also be thought of as a source of resistance to changes in fluid motion.

In gases, molecular interchange is the main cause of shear stress. In liquids, there are additional and substantial cohesive or attractive forces between the molecules that resist flow and deformation. Both molecular interchange and cohesion contribute to viscous shear stresses in liquids.

In turbulent flow, the situation is much more complex than that described for steady flow of adjacent fluid layers. Particles of fluid in turbulent flow undergo chaotic and irregular patterns of motion and are subject to large and abrupt changes in velocity. Slow-moving fluid can jump quickly into fast streams of motion; similarly, fast-moving fluid particles frequently arrive in relatively slow-moving regions of flow. The overall effect of this behaviour is a huge increase in the rate of momentum exchange within the flow stream and, consequently, much greater effective shear stresses in the fluid. The extra shear stress contributions due to the strong velocity fluctuations in turbulent flow are known as *Reynolds stresses*. Reynolds stresses are discussed in more detail in Section 7.9.2 (Reynolds Stresses subsection).

7.2.3 Reynolds Number

The transition from laminar to turbulent flow depends not only on the velocity of the fluid, but also on its viscosity and density and the geometry of the flow system. A parameter used to characterise fluid flow is the *Reynolds number*. For full flow in pipes with circular cross-section, the Reynolds number Re is defined as:

$$Re = \frac{Du\rho}{\mu} \tag{7.1}$$

where D is the pipe diameter, u is the average linear velocity of the fluid, ρ is the fluid density, and μ is the fluid viscosity. For stirred vessels there is another definition of Reynolds number:

$$Re_{\mathrm{i}} = \frac{N_{\mathrm{i}} D_{\mathrm{i}}^2 \rho}{\mu} \tag{7.2}$$

where Re_{i} is the *impeller Reynolds number*, N_{i} is the stirrer speed, D_{i} is the impeller diameter, ρ is the fluid density, and μ is the fluid viscosity. The Reynolds number is a dimensionless variable; the units and dimensions of the parameters on the right side of Eqs. (7.1) and (7.2) cancel completely.

The Reynolds number is named after Osborne Reynolds, who, in 1883, published a classical series of papers on the nature of flow in pipes. One of the most significant outcomes of Reynolds's experiments is that there is a *critical Reynolds number* marking the upper boundary for laminar flow in pipes. In smooth pipes, laminar flow is encountered at Reynolds numbers less than 2100. Under normal conditions, flow is turbulent at Re above about 4000. Between 2100 and 4000 is the *transition region* where flow changes between laminar and turbulent; flow characteristics in this region also depend on conditions at the entrance of the pipe and other variables. Flow in stirred tanks may be laminar or turbulent as a function of the impeller Reynolds number. The value of Re_{i} marking the transition depends on the geometry of the impeller and tank; however, for several commonly used mixing systems, flow is laminar at $Re_i < 10$ and turbulent at $Re_i > 10^4$. Thus, unlike for pipe flow, in many stirred tanks there is a relatively large flow-regime transition region of $10 \leq Re_i \leq 10^4$.

Conceptually, *the Reynolds number represents the ratio of inertial forces to viscous forces* in the fluid. It is this ratio that determines whether flow remains laminar or becomes turbulent. If the motion of a particle of fluid flowing in a stream is disturbed towards a new direction of flow, there are two counteracting forces that determine whether the fluid will change its motion or not. On one hand, the inertia of the particle, which is related to its mass and velocity, will tend to carry it in the new direction. However, the particle is also subject to viscous forces due to the presence of the surrounding fluid. As described in Section 7.2.2, viscous forces tend to dampen changes in motion and make the fluid particle conform to the movement of neighbouring regions of the stream. This is illustrated in Figure 7.3(a), which shows streamline flow at very low Reynolds number past a triangular obstacle. Because of the dominance of viscous forces in this flow, any changes to the fluid motion caused by the object are damped and the streamlines perturbed by the obstacle readily reform downstream. As a result, any tendency towards chaotic or turbulent flow is resisted and laminar flow is preserved. However, with increasing Reynolds number, inertial forces become more dominant and viscous forces correspondingly less important, so that significant flow disturbances are likely to occur. For example, as shown in Figure 7.3 (b), a cylindrical object placed across a uniform steady flow can generate considerable downstream instability and turbulence if the Reynolds number is high enough. Generation of turbulence is evident from the presence of rotational flow or eddies behind the cylinder.

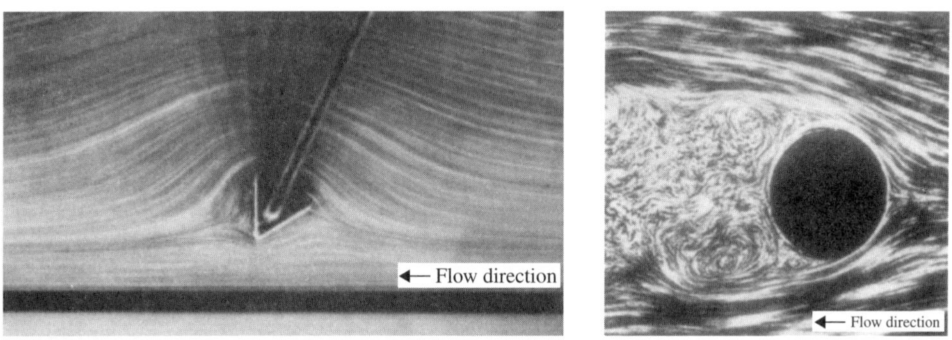

(a) (b)

FIGURE 7.3 Effect of perturbation on fluid flow. (a) Streamline flow past a triangular object at low Reynolds number. The streamlines are preserved downstream of the obstacle.
From M. Lesieur, 1997, Turbulence in Fluids, *Kluwer Academic, Dordrecht; with kind permission of Kluwer Academic and M. Lesieur.*
(b) Flow disturbance due to a cylindrical obstacle placed across a steady flow stream at higher Reynolds number. Turbulent eddies are generated behind the cylinder.
From J. O. Hinze, 1975, Turbulence, *2nd ed., McGraw-Hill, New York; reproduced with permission of The McGraw-Hill Companies.*

7.2.4 Hydrodynamic Boundary Layers

In most practical applications, fluid flow occurs in the presence of a stationary solid surface, such as the walls of a pipe or tank. That part of the fluid where flow is affected by the solid is called the *boundary layer*. As an example, consider flow of fluid parallel to the flat plate shown in Figure 7.4. Contact between the moving fluid and the plate causes the formation of a boundary layer beginning at the leading edge and developing on both the top and bottom of the plate. Figure 7.4 shows only the upper stream; fluid motion below the plate will be a mirror image of that above.

As indicated by the arrows in Figure 7.4(a), the bulk fluid velocity in front of the plate is uniform and of magnitude u_B. The extent of the boundary layer is indicated by the broken line. Above the boundary layer, fluid motion is the same as if the plate were not there. The boundary layer grows in thickness from the leading edge until it develops its full size. The final thickness of the boundary layer depends on the Reynolds number for the bulk flow.

When fluid flows over a stationary object, a thin film of fluid in contact with the surface adheres to it to prevent slippage over the surface. The fluid velocity at the surface of the plate in Figure 7.4 is therefore zero. When part of a flowing fluid has been brought to rest, the flow of adjacent fluid layers will be slowed down by the action of viscous drag as described in Section 7.2.2. This phenomenon is illustrated in Figure 7.4(b). The velocity of fluid within the boundary layer, u, is represented by arrows; u is zero at the surface of the plate. Viscous drag forces are transmitted upwards through the fluid from the stationary layer at the surface. The fluid layer just above the surface moves at a slow but finite velocity; layers further above move at increasing velocity as the drag forces associated with the stationary layer decrease. At the edge of the boundary layer, the fluid is unaffected by the

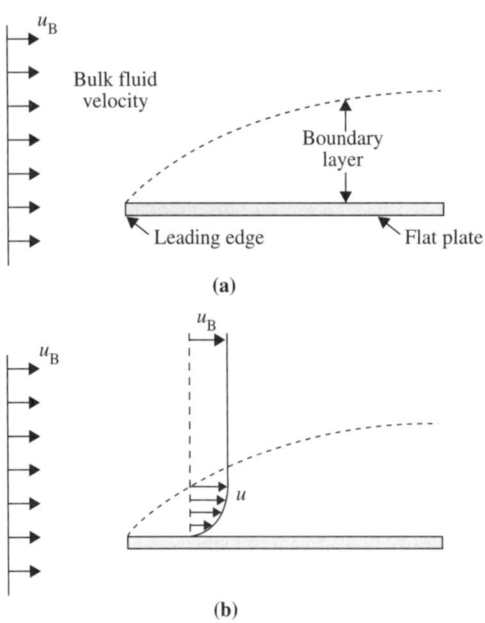

FIGURE 7.4 Fluid boundary layer for flow over a flat plate. (a) The boundary layer forms at the leading edge. (b) Compared with velocity u_B in the bulk fluid, the velocity u in the boundary layer is zero at the plate surface but increases with distance from the plate to reach u_B at the outer limit of the boundary layer.

presence of the plate and the velocity is close to that of the bulk flow, u_B. The magnitude of u at various points in the boundary layer is indicated in Figure 7.4(b) by the length of the arrows in the direction of flow. The line connecting the heads of the velocity arrows shows the *velocity profile* in the fluid. A *velocity gradient*, that is, a change in velocity with distance from the plate, is thus established in a direction perpendicular to the direction of flow. The velocity gradient forms as the drag forces resulting from retardation of fluid flow at the surface are transmitted through the fluid.

Formation of boundary layers is important not only in determining the characteristics of fluid flow, but also for transfer of heat and mass between phases. These topics are discussed further in Chapters 9 and 10.

7.2.5 Boundary Layer Separation

What happens when contact is broken between a fluid and a solid immersed in the flow path? As an example, consider a flat plate aligned perpendicular to the direction of fluid flow, as shown in Figure 7.5. Fluid impinges on the surface of the plate and forms a boundary layer as it flows either up or down the object. When the fluid reaches the top or bottom of the plate, its momentum prevents it from making the sharp turn around the edge. As a result, fluid separates from the plate and proceeds outwards into the bulk fluid. This phenomenon is called *boundary layer separation*. Directly behind the plate is a zone or *wake* of highly decelerating fluid in which large eddies or vortices are formed. Eddies in the wake are kept in rotational motion by the force of the bordering currents.

The effects of boundary layer separation appear in Figure 7.3(b) as well as in Figure 7.5. Boundary layer separation takes place whenever an abrupt change in either the magnitude or direction of the fluid velocity is too great for the fluid to keep to a solid surface. It occurs in sudden contractions, expansions, or bends in a flow channel, or when an object is placed across the flow path. Considerable energy is associated with the wake; this energy is taken from the bulk flow. Formation of wakes should be minimised if large pressure losses in the fluid are to be avoided. However, for some purposes such as promotion of mixing and heat transfer, boundary layer separation may be desirable.

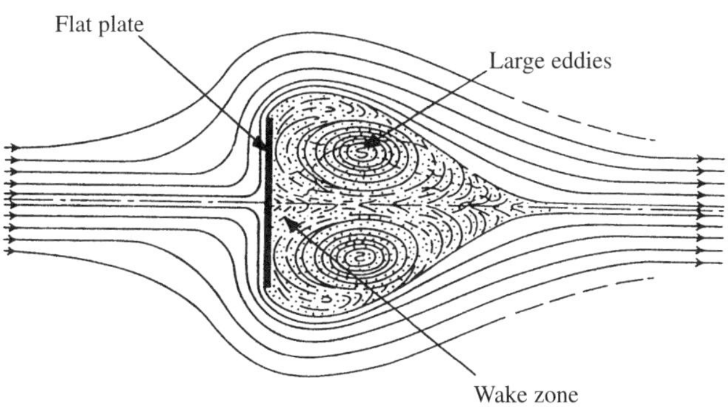

FIGURE 7.5 Flow around a flat plate aligned perpendicular to the direction of flow.
From W.L. McCabe, J.C. Smith, and P. Harriott, 2001, Unit Operations of Chemical Engineering, *6th ed., McGraw-Hill, New York.*

7.3 VISCOSITY

Viscosity is the most important property affecting the flow behaviour of a fluid; it is related to the fluid's resistance to motion. Viscosity has a marked effect on pumping, mixing, mass transfer, heat transfer, and aeration of fluids; these in turn exert a major influence on bioprocess design and economics. The viscosity of fermentation fluids is affected by the presence of cells, substrates, products, and gas.

Viscosity is an important aspect of *rheology*, the science of deformation and flow. Viscosity is the parameter used to relate the shear stress to the velocity gradient in fluids under laminar flow conditions. This relationship can be explained by considering the development of laminar flow between parallel plates, as shown in Figure 7.6. The plates are a relatively short distance apart and, initially, the fluid between them is stationary. The lower plate is then moved steadily to the right with shear force F, while the upper plate remains fixed.

A thin film of fluid adheres to the surface of each plate. Therefore as the lower plate moves, fluid moves with it, while at the surface of the stationary plate the fluid velocity is zero. Due to viscous drag, fluid just above the moving plate is set in motion, but with reduced speed compared with fluid at the surface of the plate. Layers further above also move; however, as we get closer to the top plate, the fluid is affected by viscous drag from the stationary film attached to the upper plate surface. As a consequence, the fluid velocity between the plates decreases from that of the moving plate at $y = 0$ to zero at $y = D$. The velocity at different levels between the plates is indicated in Figure 7.6 by the arrows marked v. Laminar flow due to a moving surface as shown in Figure 7.6 is called *Couette flow*.

When steady Couette flow is attained in simple fluids, the velocity profile is as indicated in Figure 7.6; the slope of the line connecting all the velocity arrows is constant and proportional to the shear force F responsible for motion of the plate. The slope of the line connecting the velocity arrows is the velocity gradient, dv/dy. When the magnitude of the velocity gradient is directly proportional to F, we can write:

$$\frac{dv}{dy} \propto F \tag{7.3}$$

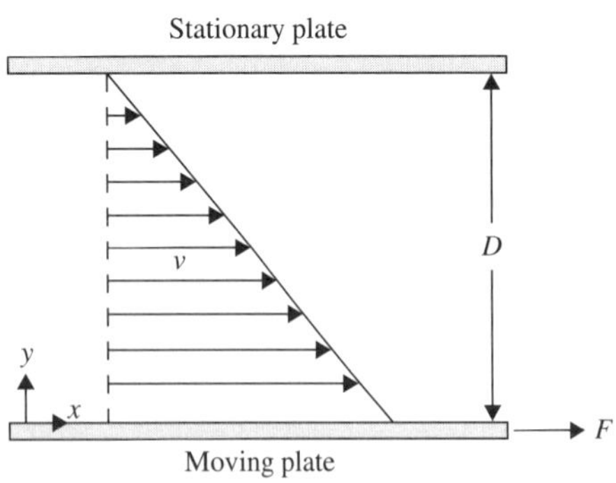

FIGURE 7.6 Velocity profile for Couette flow between parallel plates.

If we define τ as the shear stress at the surface of the lower plate, τ is equal to the shear force F divided by the plate area A:

$$\tau = \frac{F}{A} \tag{7.4}$$

Therefore, it follows from Eq. (7.3) that:

$$\tau \propto \frac{dv}{dy} \tag{7.5}$$

This proportionality is represented by the equation:

$$\tau = -\mu \frac{dv}{dy} \tag{7.6}$$

where μ is the proportionality constant. Equation (7.6) is called *Newton's law of viscosity*, and μ is the viscosity. Newton's law of viscosity is applicable only under laminar flow conditions. The minus sign is necessary in Eq. (7.6) because the velocity gradient is always negative if the direction of F, and therefore τ, is considered positive as illustrated in Figure 7.6. $-dv/dy$ is called the *shear rate* and is often denoted by the symbol $\dot{\gamma}$. Equation (7.6) can therefore also be written as:

$$\tau = \mu\dot{\gamma} \tag{7.7}$$

Viscosity as defined in Eqs. (7.6) and (7.7) is sometimes called *dynamic viscosity*. Because τ has dimensions $L^{-1}MT^{-2}$ and $\dot{\gamma}$ has dimensions T^{-1}, μ must therefore have dimensions $L^{-1}MT^{-1}$. The SI unit of viscosity is the pascal second (Pa s), which is equal to $1\ N\ s\ m^{-2}$ or $1\ kg\ m^{-1}\ s^{-1}$. Other units include centipoise, cP. Direct conversion factors for viscosity units are given in Table A.9 in Appendix A. Of particular interest is the viscosity of water:

$$\text{Viscosity of water at } 20°\text{C} \simeq 1\ \text{cP} = 1\ \text{mPa s} = 10^{-3}\ \text{Pa s} = 10^{-3}\ \text{kg m}^{-1}\ \text{s}^{-1} \tag{7.8}$$

A modified form of viscosity is the *kinematic viscosity*, which is usually given the Greek symbol ν:

$$\nu = \frac{\mu}{\rho} \tag{7.9}$$

where ρ is the fluid density.

Fluids that obey Eq. (7.6) with constant μ are known as *Newtonian fluids*. The *flow curve* or *rheogram* for a Newtonian fluid is shown in Figure 7.7; a plot of τ versus $\dot{\gamma}$ gives a straight line with slope equal to μ. The viscosity of Newtonian fluids remains constant irrespective of changes in shear stress (force applied) or shear rate (velocity gradient). This does not imply that the viscosity is invariant; viscosity depends on many parameters such as temperature, pressure, and fluid composition. However, for a given set of these conditions, the viscosity of Newtonian fluids is independent of the shear stress and shear rate. On the other hand, the ratio between shear stress and shear rate is not constant for

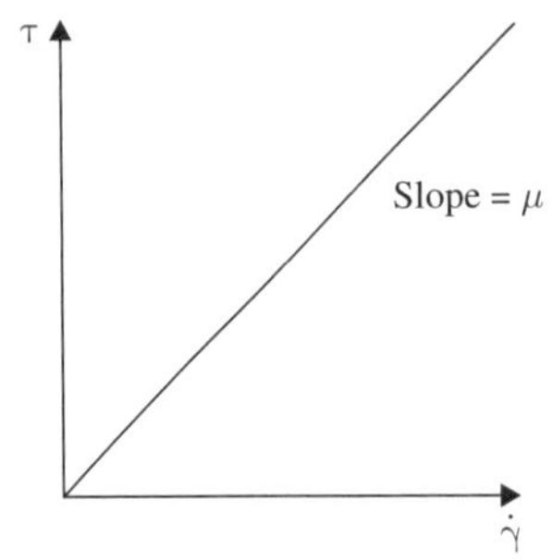

FIGURE 7.7 Flow curve for a Newtonian fluid.

non-Newtonian fluids, but depends on the shear force exerted on the fluid. Accordingly, μ in Eq. (7.6) is not a constant for non-Newtonian fluids, and the velocity profile during Couette flow is not as simple as that shown in Figure 7.6.

7.4 MOMENTUM TRANSFER

Viscous drag forces responsible for the velocity gradient in Figure 7.6 are the instrument of *momentum transfer* in fluids. At $y = 0$ the fluid acquires momentum in the x-direction due to motion of the lower plate. This fluid imparts some of its momentum to the adjacent layer of fluid above the plate, causing it also to move in the x-direction. Momentum in the x-direction is thus transmitted through the fluid in the y-direction.

Momentum transfer in fluids is represented by Eq. (7.6). To interpret this equation in terms of momentum transfer, the shear stress τ is considered to be the *flux* of x-momentum in the y-direction. To specify the directions of both the flux and the momentum, the shear stress in the system of Figure 7.6 could also be given the symbol τ_{yx}. The validity of interpreting shear stress as momentum flux can be verified by checking the dimensions of these parameters. As momentum is given by the expression Mv where M is mass and v is velocity, momentum has dimensions LMT^{-1}. *Flux means rate per unit area*; therefore momentum flux has dimensions $\text{L}^{-1}\text{MT}^{-2}$, which are also the dimensions of τ.

With τ representing momentum flux, according to Eq. (7.6), the flux of momentum is directly proportional to the velocity gradient dv/dy. The negative sign in Eq. (7.6) means that momentum is transferred from regions of high velocity to regions of low velocity—that is, in a direction opposite to the direction of increasing velocity. The magnitude of the velocity gradient dv/dy determines the magnitude of the momentum flux; dv/dy thus acts as the 'driving force' for momentum transfer.

Interpretation of fluid flow as momentum transfer perpendicular to the direction of flow may seem peculiar at first. However, this concept is useful for analysing shear stresses in turbulent flow (Section 7.9.2), and for understanding the parallels between momentum transfer, heat transfer, and mass transfer in terms of their mechanisms and equations. The analogy between these physical processes is discussed further in Chapters 9 and 10.

7.5 NON-NEWTONIAN FLUIDS

Most slurries, suspensions, and dispersions are non-Newtonian, as are homogeneous solutions of long-chain polymers and other large molecules. Many fermentation processes involve materials that exhibit non-Newtonian behaviour, such as starches, extracellular polysaccharides, and culture broths containing suspended cells. Examples of non-Newtonian fluids are listed in Table 7.1.

Classification of non-Newtonian fluids depends on the relationship between shear stress and shear rate in the fluid. Types of non-Newtonian fluid commonly encountered in bioprocessing include *pseudoplastic*, *Bingham plastic*, and *Casson plastic*. The flow curves for these materials are shown in Figure 7.8. In each case, in contrast to Newtonian fluids, the flow curves are not straight lines because the ratio between the shear stress and shear rate is not constant. Nevertheless, an *apparent viscosity*, μ_a, can be defined for non-Newtonian fluids; μ_a is the ratio of shear stress and shear rate at a particular value of $\dot{\gamma}$:

$$\mu_a = \frac{\tau}{\dot{\gamma}} \tag{7.10}$$

In Figure 7.8, μ_a is the slope of the dashed lines in the rheograms for pseudoplastic, Bingham plastic, and Casson plastic fluids; for Newtonian fluids, $\mu_a = \mu$. The apparent viscosity of a non-Newtonian fluid is not a physical property in the same way that Newtonian viscosity is: apparent viscosity depends on the shear force exerted on the fluid. It is therefore meaningless to specify the apparent viscosity of a non-Newtonian fluid without also reporting the shear stress or shear rate at which it is measured.

TABLE 7.1 Examples of Newtonian and Non-Newtonian Fluids

Fluid type	Examples
Newtonian	All gases, water, dispersions of gas in water, low-molecular-weight liquids, aqueous solutions of low-molecular-weight compounds
Non-Newtonian	
Pseudoplastic	Rubber solutions, adhesives, polymer solutions, some greases, starch suspensions, cellulose acetate, mayonnaise, some soap and detergent slurries, some paper pulps, paints, wallpaper paste, biological fluids
Bingham plastic	Some plastic melts, margarine, cooking fats, some greases, toothpaste, some soap and detergent slurries, some paper pulps
Casson plastic	Blood, tomato sauce, orange juice, melted chocolate, printing ink

Adapted from B. Atkinson and F. Mavituna, 1991, Biochemical Engineering and Biotechnology Handbook, *2nd ed., Macmillan, Basingstoke.*

Fluid	Flow curve	Equation	Apparent viscosity μ_a
Newtonian	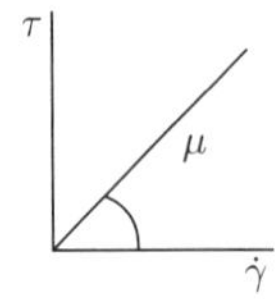	$\tau = \mu\dot{\gamma}$	Constant $\mu_a = \mu$
Pseudoplastic (power law)	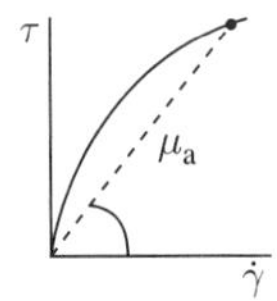	$\tau = K\dot{\gamma}^n$ $n<1$	Decreases with increasing shear rate $\mu_a = K\dot{\gamma}^{n-1}$
Bingham plastic	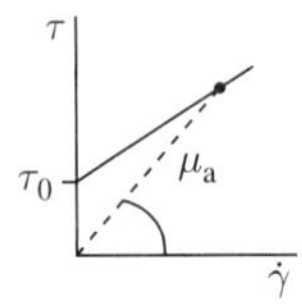	$\tau = \tau_0 + K_p\dot{\gamma}$	Decreases with increasing shear rate when yield stress τ_0 is exceeded $\mu_a = \frac{\tau_0}{\dot{\gamma}} + K_p$
Casson plastic	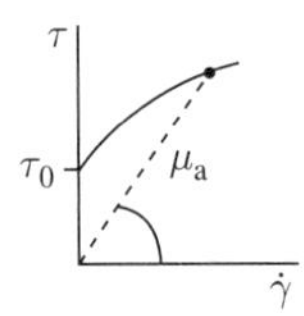	$\tau^{1/2} = \tau_0^{1/2} + K_p\dot{\gamma}^{1/2}$	Decreases with increasing shear rate when yield stress τ_0 is exceeded $\mu_a = \left[\left(\frac{\tau_0}{\dot{\gamma}}\right)^{1/2} + K_p\right]^2$

FIGURE 7.8 Classification of fluids according to their rheological behaviour. *From B. Atkinson and F. Mavituna, 1991,* Biochemical Engineering and Biotechnology Handbook, *2nd ed., Macmillan, Basingstoke.*

7.5.1 Two-Parameter Models

Pseudoplastic fluids obey the *Ostwald–de Waele* or *power law*:

$$\tau = K\dot{\gamma}^n \tag{7.11}$$

where τ is the shear stress, K is the *consistency index*, $\dot{\gamma}$ is the shear rate, and n is the *flow behaviour index*. The parameters K and n characterise the rheology of power-law fluids. The flow behaviour index n is dimensionless; the dimensions of K, $L^{-1}MT^{n-2}$, depend on n. As indicated in Figure 7.8, for pseudoplastic fluids $n<1$. When $n>1$, the fluid is dilatant rather than pseudoplastic. Dilatant is a recognised category of non-Newtonian fluid behaviour; however, dilatant fluids do not occur often in bioprocessing. If $n=1$ the fluid is Newtonian; in this case, Eqs. (7.7) and (7.11) are equivalent with $K=\mu$.

From Eqs. (7.10) and (7.11), the apparent viscosity for power-law fluids is:

$$\mu_a = \frac{\tau}{\dot{\gamma}} = K\dot{\gamma}^{n-1} \tag{7.12}$$

Because $n<1$ for pseudoplastic fluids, Eq. (7.12) indicates that the apparent viscosity decreases with increasing shear rate. Pseudoplastic fluids are therefore also known as *shear-thinning* fluids.

Also included in Figure 7.8 are flow curves for plastic flow. Plastic fluids do not produce motion until some finite force or *yield stress* has been applied; this minimum stress is required to break down to some extent the internal 'structure' of the fluid before any movement can occur. For *Bingham plastic* fluids:

$$\tau = \tau_0 + K_p \dot{\gamma} \tag{7.13}$$

where τ_0 is the yield stress. Once the yield stress is exceeded and flow initiated, Bingham plastics behave like Newtonian fluids; as indicated in Figure 7.8, the rheogram is a straight line for $\tau > \tau_0$ and a constant ratio K_p exists between change in shear stress and change in shear rate. Another common plastic behaviour is described by the *Casson* equation:

$$\tau^{1/2} = \tau_0^{1/2} + K_p \dot{\gamma}^{1/2} \tag{7.14}$$

Once the yield stress τ_0 is exceeded, the behaviour of Casson fluids is pseudoplastic. Several other equations describing non-Newtonian flow have also been developed [1].

7.5.2 Time-Dependent Viscosity

When a shear force is exerted on some fluids, the apparent viscosity either increases or decreases with the duration of the force. If the apparent viscosity increases with time, the fluid is said to be *rheopectic*; however, rheopectic fluids are relatively rare. If the apparent viscosity decreases with time, the fluid is *thixotropic*. Thixotropic behaviour is not uncommon in cultures containing fungal mycelia or extracellular microbial polysaccharides, and appears to be related to reversible 'structure' effects associated with the orientation of cells and macromolecules in the fluid. The rheological properties of thixotropic fluids vary during application of shear forces because it takes time for equilibrium to be established between structure breakdown and redevelopment.

7.5.3 Viscoelasticity

Viscoelastic fluids, such as some polymer solutions, exhibit an elastic response to changes in shear stress. When shear forces are removed from a moving viscoelastic fluid, the direction of flow may be reversed due to elastic forces developed during flow. Most viscoelastic fluids are also pseudoplastic and may exhibit other rheological characteristics such as yield stress. Mathematical analysis of viscoelasticity is therefore quite complex.

7.6 VISCOSITY MEASUREMENT

Many different instruments or *viscometers* are available for measurement of rheological properties. Space does not permit a detailed discussion of viscosity measurement in this text; further information can be found elsewhere [2–7].

The objective of any viscosity measurement system is to create a controlled flow situation where easily measured parameters can be related to the shear stress τ and shear rate $\dot{\gamma}$. Usually the fluid is set in rotational motion and the parameters measured are the torque M and the angular velocity Ω. These quantities are used to calculate τ and $\dot{\gamma}$ using

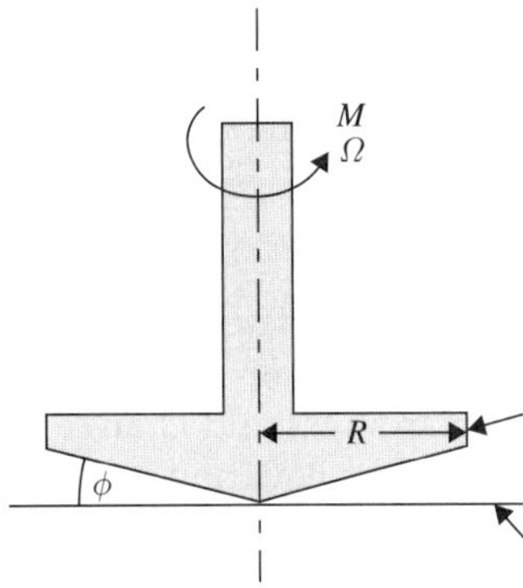

FIGURE 7.9 Cone-and-plate viscometer.

approximate formulae that depend on the geometry of the apparatus; equations for particular viscometers can be found in other texts [2−6]. Once obtained, τ and $\dot{\gamma}$ are applied for evaluation of the viscosity of Newtonian fluids, or viscosity parameters such as K, n, and τ_0 for non-Newtonian fluids. Most modern viscometers use microprocessors to provide automatic readout of parameters such as shear stress, shear rate, and apparent viscosity.

Three types of viscometer commonly used in bioprocessing applications are cone-and-plate, coaxial cylinder, and impeller.

7.6.1 Cone-and-Plate Viscometer

The cone-and-plate viscometer consists of a flat horizontal plate and an inverted cone, the apex of which is in near contact with the plate as shown in Figure 7.9. The angle ϕ between the plate and cone is very small, usually less than 3°, and the fluid to be measured is located in this small gap. Large cone angles are not used in routine work for a variety of reasons, the most important being that analysis of the results for non-Newtonian fluids would be complex or impossible. The cone is rotated in the fluid and the angular velocity Ω and torque M are measured. It is assumed that the fluid undergoes steady laminar flow in concentric circles about the axis of rotation of the cone. This assumption is not always valid; however for ϕ less than about 3°, the error is small. Temperature can be controlled by circulating water from a constant temperature bath beneath the plate; this is effective provided the speed of rotation is not too high. Limitations of the cone-and-plate method for measurement of flow properties, including corrections for edge and temperature effects and turbulence, are discussed elsewhere [2−4].

7.6.2 Coaxial Cylinder Viscometer

The coaxial cylinder viscometer is a popular rotational device for measuring rheological properties. As shown in Figure 7.10, the instrument is designed to shear fluid located in the annulus between two concentric cylinders, one of which is held stationary while the other rotates. A cylindrical bob of radius R_i is suspended in sample fluid held in a stationary cylindrical cup of radius R_o. Liquid covers the bob to a height h from the bottom of

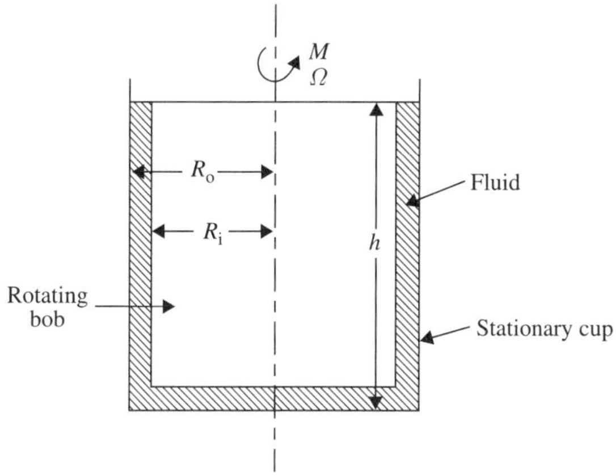

FIGURE 7.10 Coaxial cylinder viscometer.

the outer cup. As the inner cylinder rotates and the fluid undergoes steady laminar flow, the angular velocity Ω and torque M are measured. In some designs the outer cylinder rather than the inner bob rotates; in either case the motion is relative with angular velocity Ω.

Coaxial cylinder viscometers are used with Newtonian and non-Newtonian fluids. When the flow is non-Newtonian, the relationship between shear rate, rotational speed, and geometric factors is not simple and the calculations can be somewhat complicated. Limitations of the coaxial cylinder method, including corrections for end effects, slippage, temperature variation, and turbulence, are discussed elsewhere [2–5].

7.6.3 Impeller Viscometer

Because of the difficulties (discussed in Section 7.6.4) associated with standard rotational viscometers, modified apparatus employing turbine and other impellers have been developed for the study of fermentation fluid rheology [7–9]. Instead of the rotating inner cylinder of Figure 7.10, a small impeller on a stirring shaft is used to shear the fluid sample. As the impeller rotates slowly in the fluid, accurate measurements of torque M and rotational speed N_i are made. The average shear rate in a stirred fluid is proportional to the stirrer speed:

$$\dot{\gamma} = kN_i \tag{7.15}$$

where k is a constant that depends on the geometry of the impeller. For a turbine impeller under laminar flow conditions, the following relationship applies [8]:

$$\tau = \frac{2\pi Mk}{64 D_i^3} \tag{7.16}$$

where D_i is the impeller diameter. Equation (7.15) is experimentally derived; for turbine impellers k is approximately 10. Before carrying out viscosity measurements, the exact value of k for a particular apparatus can be evaluated using liquid with a known viscosity–shear rate relationship.

Because Eq. (7.16) is valid only for laminar flow, viscosity measurements using the impeller method must be carried out under laminar flow conditions. Accordingly, if a turbine impeller is used, Re_i cannot be greater than about 10 as outlined in Section 7.2.3. From Eq. (7.2), Re_i is directly proportional to N_i, which, from Eq. (7.15), determines the value of $\dot{\gamma}$. Therefore, for Re_i restricted to low values because of the necessity for laminar flow, the range of shear rates that may be investigated is very limited. This range can be extended if anchor or helical agitators are used instead of conventional turbines (refer to Figure 8.10 for illustrations of these impellers) as laminar flow is maintained at higher Re_i. The value of k in Eq. (7.15) is also greater for anchor and helical agitators so that higher shear rates can be tested. As Eq. (7.16) is valid only for turbine impellers, the relationship between τ and M must be modified if alternative impellers are used. Application of anchor and helical impellers for viscosity measurement is described in more detail in the literature [10, 11].

Because the flow patterns in stirred fluids are relatively complex, analysis of data from impeller viscometers is not absolutely rigorous from a theoretical point of view. However, the procedure is based on well-proven and widely accepted empirical correlations and is considered the most reliable technique for mycelial broths. As discussed below, the method eliminates many of the operating problems associated with conventional viscometers for the study of fermentation fluids.

7.6.4 Use of Viscometers with Fermentation Broths

Measurement of rheological properties is difficult when the fluid contains suspended solids such as cells. The viscosity of fermentation broths often appears time-dependent due to artefacts associated with the measuring device. With viscometers such as the cone-and-plate and coaxial cylinder, the following problems can arise:

1. The suspension is effectively centrifuged in the viscometer, so that a region with lower cell density is formed near the rotating surface.
2. Solids settle out of suspension during measurement.
3. Cell clumps of about the same size as the gap in the coaxial cylinder viscometer, or about the same size as the cone angle in the cone-and-plate device, interfere with accurate measurement.
4. The measurement depends somewhat on the orientation of cells and cell clumps in the flow field.
5. Some types of cell begin to flocculate or deflocculate when the shear field is applied.
6. Cells can be destroyed during measurement.

The first problem is particularly troublesome because it is hard to detect and can give viscosity results that are too small by a factor of up to 100. For suspensions containing solids, the impeller method offers significant advantages compared with other measurement procedures. Stirring by the impeller prevents sedimentation, promotes uniform distribution of solids throughout the fluid, and reduces time-dependent changes in suspension composition. The impeller method has proved very useful for rheological measurements of microbial suspensions [7].

7.7 RHEOLOGICAL PROPERTIES OF FERMENTATION BROTHS

Rheological data have been reported for a range of fermentation fluids. This information has been obtained using various viscometers and measurement techniques; however, operating problems such as particle settling and broth centrifugation have been ignored in many cases. The rheology of dilute broths and cultures of yeast and non-chain-forming bacteria is usually Newtonian; animal cell suspensions with or without serum also remain Newtonian with viscosity close to that of water. On the other hand, most mycelial and plant cell suspensions are modelled as pseudoplastic fluids or, if there is a yield stress, as Bingham or Casson plastics. The rheological properties of selected microbial and plant cell suspensions are listed in Table 7.2. In most cases, the results are valid over only a limited

TABLE 7.2 Rheological Properties of Microbial and Plant Cell Suspensions

Culture	Shear rate (s^{-1})	Viscometer	Comments	Reference
Saccharomyces cerevisiae (pressed cake diluted with water)	2–100	rotating spindle	Newtonian below 10% solids ($\mu \approx 1-5$ cP); pseudoplastic above 10% solids	[12]
Aspergillus niger (washed cells in buffer)	0–21.6	rotating spindle (guard removed)	pseudoplastic	[13]
Penicillium chrysogenum (whole broth)	1–15	turbine impeller	Casson plastic	[8]
Penicillium chrysogenum (whole broth)	not given	coaxial cylinder	Bingham plastic	[14]
Penicillium chrysogenum (whole broth)	not given	coaxial cylinder	pseudoplastic; *K* and *n* vary with CO_2 content of inlet gas	[15]
Endomyces sp. (whole broth)	not given	coaxial cylinder	pseudoplastic; *K* and *n* vary during batch culture	[16]
Streptomyces noursei (whole broth)	4–28	rotating spindle (guard removed)	Newtonian in batch culture; viscosity 40 cP after 96 h	[17]
Streptomyces aureofaciens (whole broth)	2–58	rotating spindle/coaxial cylinder	initially Bingham plastic due to high starch concentration in the medium; becomes Newtonian as starch is broken down; increasingly pseudoplastic as mycelium concentration increases	[18]

(*Continued*)

TABLE 7.2 Rheological Properties of Microbial and Plant Cell Suspensions (Continued)

Culture	Shear rate (s^{-1})	Viscometer	Comments	Reference
Aureobasidium pullulans (whole broth)	10.2–1020	coaxial cylinder	Newtonian at beginning of culture; increasingly pseudoplastic as concentration of product (exopolysaccharide) increases	[19]
Xanthomonas campestris	0.0035–100	cone-and-plate	pseudoplastic; K increases continually; n levels off when xanthan concentration reaches 0.5%; cell mass (max 0.6%) has relatively little effect on viscosity	[20]
Cellulomonas uda (whole broth)	0.8–100	anchor impeller	shredded newspaper used as substrate; broth pseudoplastic with constant n until end of cellulose degradation; Newtonian thereafter	[11]
Nicotiana tabacum (whole broth)	not given	rotating spindle	pseudoplastic	[21]
Datura stramonium (whole broth)	0–1000	rotating spindle/ parallel plate	pseudoplastic and viscoelastic, with yield stress	[22]
Perilla frutescens (whole broth)	7.2–72	coaxial cylinder	Bingham plastic	[23]

range of shear conditions dictated largely by the choice of viscometer. If a fermentation produces extracellular polymers such as in microbial production of pullulan and xanthan, the rheological characteristics of the broth depend strongly on the properties and concentration of these materials.

7.8 FACTORS AFFECTING BROTH VISCOSITY

The rheology of fermentation broths often changes throughout batch culture. For broths obeying the power law, the flow behaviour index n and consistency index K can vary substantially depending on culture time. As an example, Figure 7.11 shows changes in n and K during batch culture of *Endomyces*. The culture starts off Newtonian ($n = 1$) but quickly becomes pseudoplastic ($n < 1$). K rises steadily throughout most of the batch period; this gives a direct indication of the increase in apparent viscosity since, as indicated in Eq. (7.12), μ_a is directly proportional to K.

Changes in the rheology of fermentation broths are caused by variation in one or more of the following properties:

- Cell concentration
- Cell morphology, including size, shape, mass, and vacuolation

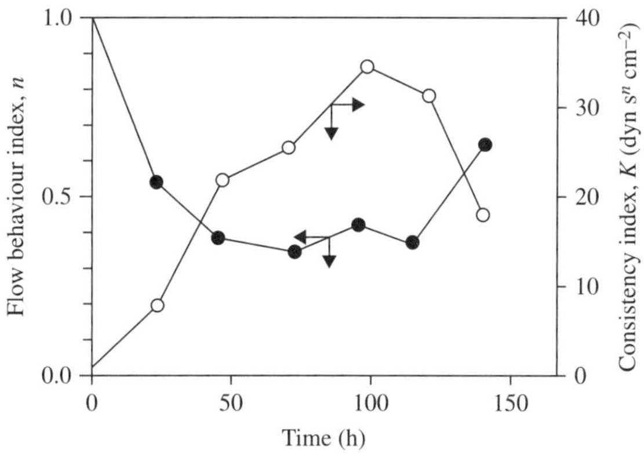

FIGURE 7.11 Variation of the rheological parameters in *Endomyces* fermentation.
From H. Taguchi and S. Miyamoto, 1966, Power requirement in non-Newtonian fermentation broth. Biotechnol. Bioeng. *8, 43–54.*

- Flexibility and deformability of cells
- Osmotic pressure of the suspending fluid
- Concentration of polymeric substrate
- Concentration of polymeric product
- Rate of shear

Some of these parameters are considered in the following sections.

7.8.1 Cell Concentration

The viscosity of a suspension of spheres in Newtonian liquid can be predicted using the *Vand equation*:

$$\mu = \mu_L(1 + 2.5\psi + 7.25\psi^2) \tag{7.17}$$

where μ_L is the viscosity of the suspending liquid and ψ is the volume fraction of solids. Equation (7.17) has been found to hold for yeast and spore suspensions at concentrations up to 14 vol% solids [24]. Many other cell suspensions do not obey Eq. (7.17); cell concentration can have a much stronger influence on rheological properties than is predicted by the Vand equation. As an example, Figure 7.12 shows how cell concentration affects the apparent viscosity of various pseudoplastic plant cell suspensions. In this case, a doubling in cell concentration causes the apparent viscosity to increase by a factor of up to 90. Similar results have been found for mould pellets in liquid culture [25]. When the viscosity is so strongly dependent on cell concentration, a steep drop in viscosity can be achieved by diluting the broth with water or medium. Periodic removal of part of the culture and refilling with fresh medium can thus be used to reduce the viscosity and improve fluid flow in viscous fermentations.

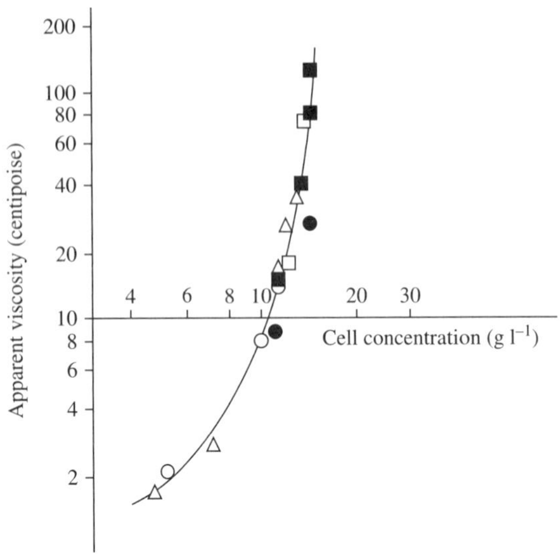

FIGURE 7.12 Relationship between the apparent viscosity and the cell concentration for plant cell suspensions forming aggregates of various size. (○) *Cudrania tricuspidata* 44–149 μm; (●) *C. tricuspidata* 149–297 μm; (□) *Vinca rosea* 44–149 μm; (■) *V. rosea* 149–297 μm; (△) *Nicotiana tabacum* 150–800 μm.
From H. Tanaka, 1982, Oxygen transfer in broths of plant cells at high density. Biotechnol. Bioeng. *24, 425–442.*

7.8.2 Cell Morphology

Small mono-dispersed cells such as bacteria and yeast do not significantly affect the flow properties of fermentation broths. However, the morphological characteristics of other cell types, particularly filamentous microorganisms and plant cells, can exert a profound influence on broth rheology. Filamentous fungi and actinomycetes produce a variety of morphologies depending on the culture conditions. Individual cell filaments may be freely dispersed as shown in Figure 7.13(a), they may form loose clumps of branched and intertwined hyphae as in Figure 7.13(b), or they may develop highly compact cell aggregates and 'hairy' pellets, such as that shown in Figure 7.13(c). These different morphologies have different rheological effects. Interactions between individual filaments and the formation of hyphal networks or loose clumps produce 'structure' in the broth, resulting in high viscosity, pseudoplasticity, and yield stress behaviour. In contrast, broths containing pelleted cells tends to be more Newtonian, depending on how readily the pellets are deformed during flow. The extent of branching of hyphal cells can also affect rheology; cells with a high branching frequency are generally less flexible than nonbranching cells and produce higher viscosities. Factors influencing the morphology of filamentous organisms and their tendency to clump include pH, growth rate, medium composition and ionic strength, dissolved oxygen tension, and agitation intensity.

Sample rheological data for pseudoplastic mycelial broths are shown in Figure 7.14. Suspensions of pelleted mycelia are more closely Newtonian in behaviour than filamentous cells; as illustrated in Figure 7.14(a), the flow behaviour index n for the pellets is closer to unity. As indicated in Figure 7.14(b), the consistency index, and therefore the apparent viscosity, can differ by several orders of magnitude depending on cell morphology.

A more detailed quantitative understanding of the effect of cell morphology on broth viscosity can be obtained using image analysis to characterise the size, shape, and other properties of filamentous cells and cell clumps. Equations have been developed to allow estimation of rheological parameters from morphological properties such as the size, roughness, and compactness of cell clumps and the length and branchedness of dispersed mycelia [7, 26, 27].

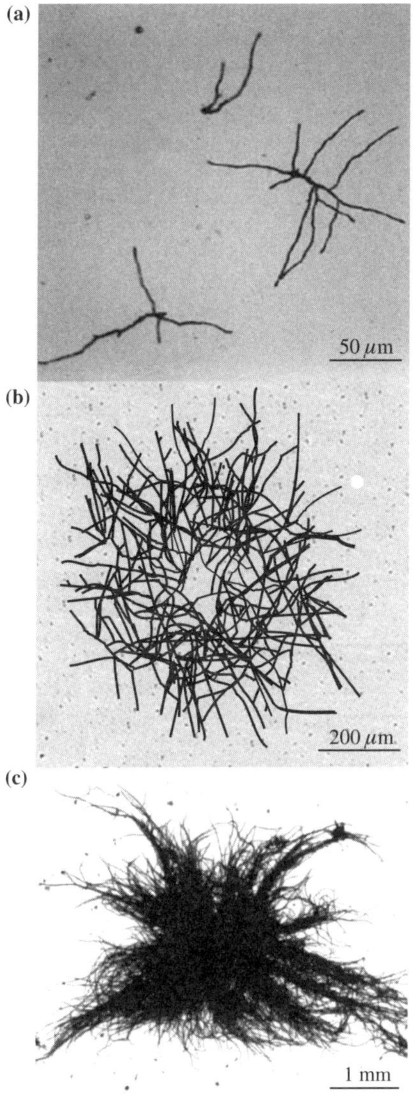

FIGURE 7.13 Different morphologies of filamentous microorganisms in submerged fermentation. (a) Freely dispersed fungal filaments; (b) loose clump of filamentous cells; and (c) hairy pellet with compact core. The bars show approximate scales.
Images provided courtesy of C.R. Thomas and the Image Analysis Group, School of Chemical Engineering, University of Birmingham, U.K.

The viscosity of cell culture broths during fermentation is usually measured by removing a sample from the fermenter and applying conventional viscometers such as those described in Section 7.6. However, because aggregation of filamentous organisms is sensitive to culture conditions, there is an increasing emphasis on *in-situ* and online viscosity measurements to avoid errors associated with morphology change during the time delay between sampling and measurement. Online viscosity readings can be obtained by recirculating broth from the fermenter through an external pipeline or tube viscometer [28], or through an external mixing cell equipped with an impeller viscometer. Alternatively, in some cases the fermenter itself may be used as an impeller viscometer for *in-situ* rheology

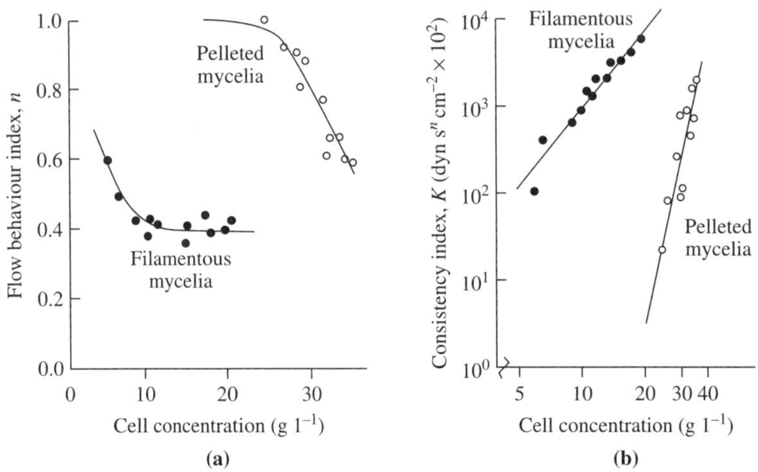

FIGURE 7.14 The effect of morphology on the rheology of mycelial broths.
From J.H. Kim, J.M. Lebeault, and M. Reuss, 1983, Comparative study on rheological properties of mycelial broth in filamentous and pelleted forms. Eur. J. Appl. Microbiol. Biotechnol. *18, 11–16.*

measurements as outlined in Section 7.6.3, provided the impeller speed is reduced to ensure laminar flow.

7.8.3 Osmotic Pressure

The osmotic pressure of the culture medium affects cell turgor pressure. This in turn affects the hyphal flexibility of filamentous cells: increased osmotic pressure gives a lower turgor pressure, making the hyphae more flexible. Improved hyphal flexibility reduces broth viscosity and can also have a marked effect on yield stress.

7.8.4 Product and Substrate Concentrations

When the product of fermentation is a polymer, continued excretion of the product in batch culture raises the broth viscosity. For example, during production of exopolysaccharide by *Aureobasidium pullulans*, the apparent viscosity measured at a shear rate of $1\ s^{-1}$ can reach as high as 24,000 cP [19]. Cell concentration usually has a negligible effect on the overall viscosity in these fermentations: the rheological properties of the fluid are dominated by the dissolved polymer. Other products having a similar effect on culture rheology include dextran, alginate, and xanthan gum.

In contrast, when the fermentation medium contains a polymeric substrate such as starch, the apparent viscosity will decrease as the fermentation progresses and the polymer is broken down. There could also be a progressive change from non-Newtonian to Newtonian behaviour. In mycelial fermentations this change is usually short-lived; as the cells grow and develop a structured filamentous network, the broth becomes increasingly pseudoplastic and viscous even though the polymeric substrate is being consumed.

7.9 TURBULENCE

Turbulence is a feature of virtually all flows of practical engineering interest. Turbulence is essential for effective mixing, mass transfer, and heat transfer in fluids; therefore, achieving turbulence in bioreactors and other mixing equipment is of critical concern. Knowledge of the features and properties of turbulent flow is becoming increasingly important to achieve better design and operation of bioprocessing systems. Yet, despite its significance, our understanding of turbulence is far from complete. Turbulent flow is by nature unpredictable and unsteady, and presents a range of intractable problems for engineering analysis and modelling.

In this section, we consider some basic features of turbulence and methods for its characterisation. In bioprocess engineering, although turbulent flow occurs in a wide range of situations such as in pipes, pumps, reactors, heat exchangers, and downstream processing equipment, we are interested in understanding turbulence mainly with reference to fermenter design, scale-up, and operation. Therefore, most of the fundamental aspects of turbulence covered here are presented with this application in view. In Chapter 8, we consider in more detail the implications of turbulent flow for mixing, power requirements, and shear effects in bioreactors.

7.9.1 Nature of Turbulent Flow

As discussed in Section 7.2.3, generation of turbulent flow depends on the Reynolds number. As the Reynolds number increases, inertial forces dominate viscous forces in the fluid, thus overcoming the tendency of viscous effects to dampen flow instabilities. Turbulence can be regarded as highly disordered fluid motion resulting from the growth of instabilities in an initially laminar flow field.

Mean and Fluctuating Velocities

Turbulent flows are very complex. If we consider a flow stream consisting of individual parcels or particles of fluid, turbulent flow is characterised by random, erratic, and unsteady movements of the fluid particles in different directions. This is illustrated in Figure 7.15, which depicts a turbulent fluid with overall flow direction from right to left. Superimposed on the primary bulk flow are secondary chaotic movements of the fluid

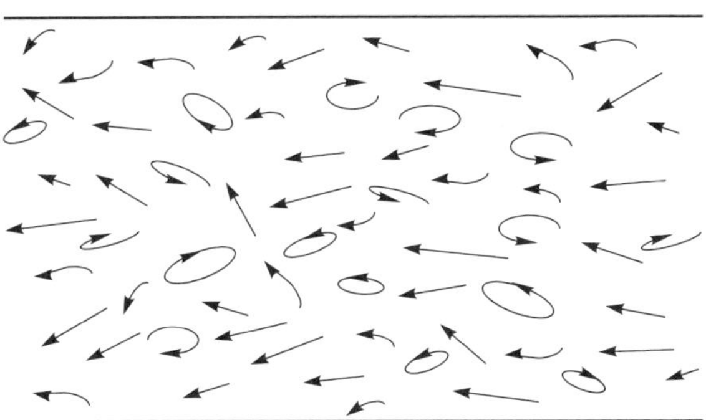

FIGURE 7.15 Turbulent flow stream with primary flow from right to left. Irregular secondary motion is superimposed on the primary flow.

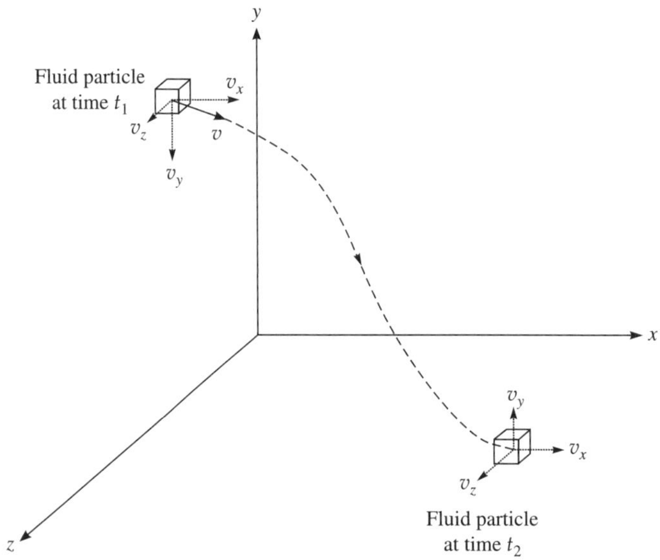

FIGURE 7.16 Movement of a fluid particle in a flow field. The overall velocity at time t_1 is v; the velocity components are v_x, v_y, and v_z.

particles, causing flow deviations and generating vortices of varying shape, size, speed, and rotational direction. As a result, local velocities in the system fluctuate with time and vary substantially with position. Overall, the irregular secondary motion in turbulent flow has profound consequences, affecting mixing efficiency, boundary layer properties, energy losses, and shear intensity.

An important property of turbulent flow is its velocity and the way in which local velocities fluctuate with time. As shown in Figure 7.16, at any instant in time, the velocity v of an infinitesimally small fluid particle in a three-dimensional flow can be represented by velocity components v_x, v_y, and v_z in each of the three directions x, y, and z, respectively. The velocity components are either positive or negative depending on their orientation in the coordinate system. The speed and overall direction of movement of the particle depend on the relative magnitudes of the three velocity components, which vary with time and position in the flow stream.

Let us assume that turbulent flow occurs in a pipe or bioreactor under steady conditions. If we take a series of measurements of v_x, v_y, and v_z at a fixed point in the system as a function of time, the results will look something like Figure 7.17. The magnitudes of the velocity components fluctuate continuously in an apparently random fashion; these fluctuations reflect the erratic and unsteady nature of turbulence. If we consider the instantaneous velocity in turbulent flow to be a stochastic or random variable, we can treat the data using statistical methods.

According to this approach, instantaneous velocity components are comprised of two parts: a *time-averaged value* reflecting the overall or gross characteristics of the flow, and a *fluctuating element* representing the irregular, secondary motion. Therefore we can write:

$$v_x = \overline{v_x} + v'_x \qquad v_y = \overline{v_y} + v'_y \qquad v_z = \overline{v_z} + v'_z \tag{7.18}$$

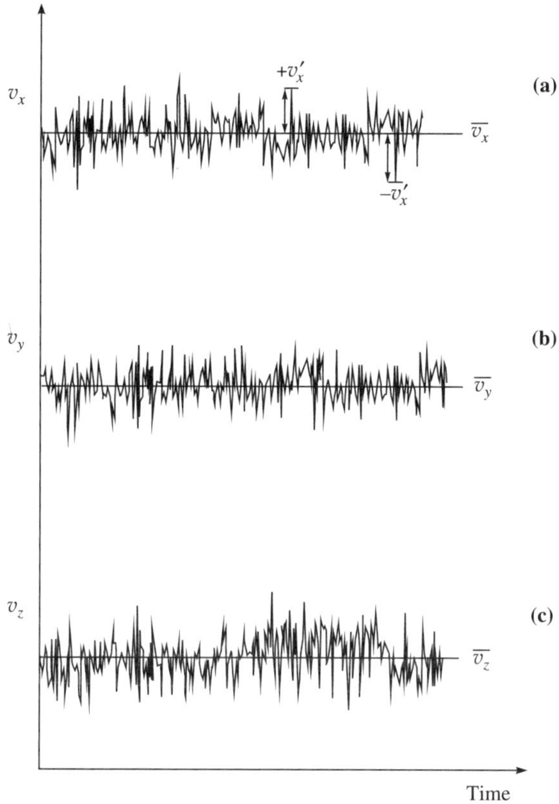

FIGURE 7.17 Time-course measurements of mean and fluctuating velocity components during steady turbulent flow: (a) v_x; (b) v_y; (c) v_z.

where the overbar $^{-}$ indicates the time-averaged or mean value, and the 'prime' symbol$'$ indicates the fluctuating element. For a collection of n values of v_x, v_y, and v_z measured at a fixed position under identical conditions, the time-averaged value can be calculated as the ensemble mean:

$$\overline{v_x} = \frac{1}{n}\sum^{n} v_x \qquad \overline{v_y} = \frac{1}{n}\sum^{n} v_y \qquad \overline{v_z} = \frac{1}{n}\sum^{n} v_z \tag{7.19}$$

where n is a large number and $\sum^{n} v$ is the sum of n values of v. The time-averaged values $\overline{v_x}$, $\overline{v_y}$, and $\overline{v_z}$ for the example of Figure 7.17 are indicated on the diagram; the fluctuating elements v'_x, v'_y, and v'_z take positive or negative values as the velocity components deviate from their respective means. It follows from the definition of the mean that:

$$\overline{v'_x} = \overline{v'_y} = \overline{v'_z} = 0 \tag{7.20}$$

Therefore, over time and for a sufficiently large number of measurements, the sums of the positive and negative fluctuating elements of the velocity components tend to zero.

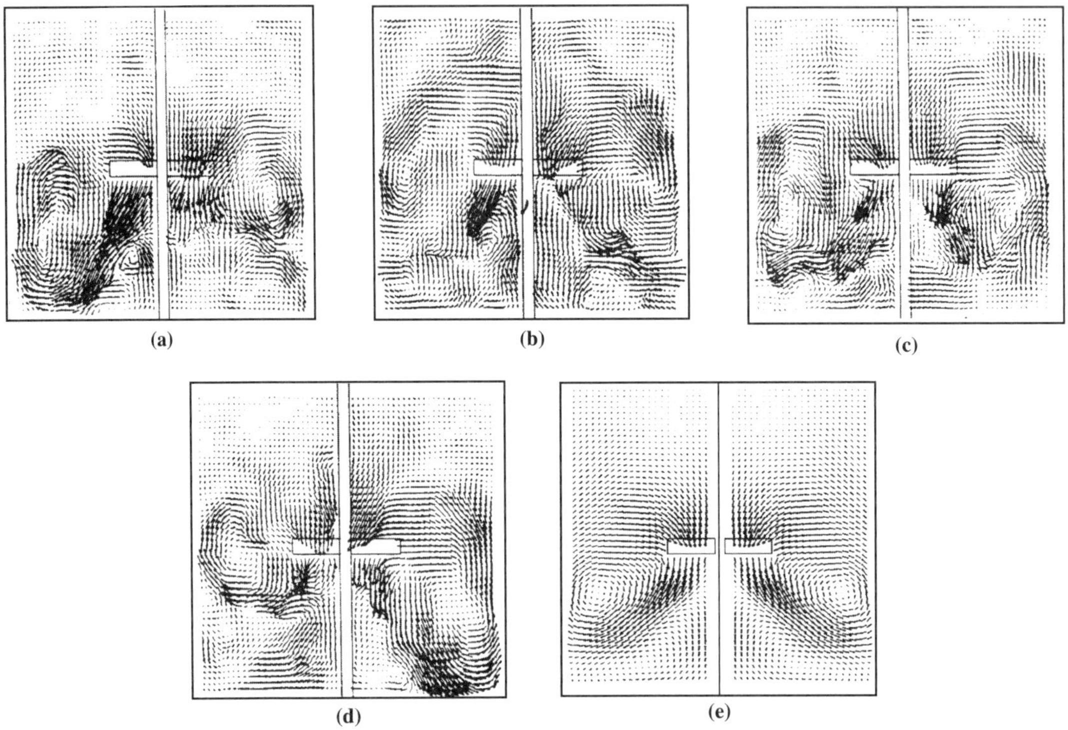

FIGURE 7.18 (a)–(d) Four instantaneous velocity fields in a stirred tank under turbulent flow conditions. A pitched-blade turbine impeller was located centrally in a cylindrical tank containing water. The velocity measurements were made using particle image velocimetry. (e) Time-averaged velocity field determined from 1024 instantaneous measurements such as those shown in (a) to (d) obtained over a period of about 20 min.
From A. Bakker, K.J. Myers, R.W. Ward, and C.K. Lee, 1996, The laminar and turbulent flow pattern of a pitched blade turbine. Trans. IChemE *74A, 485–491.*

A practical example of the velocity fluctuations occurring during turbulent flow is shown in Figure 7.18. Figures 7.18(a) through 7.18(e) are *velocity vector plots* for a stirred tank containing water. In these plots, the length and direction of the arrows located at regular grid points indicate the magnitude and direction of the fluid velocity at that location. The impeller creating the flow is located centrally in the tank. Figures 7.18(a) through 7.18(d) show instantaneous velocity fields in the tank at different times. Although the stirrer is operated in a steady, unvarying manner, the flow field changes significantly and continuously with time. Local regions of high velocity and rotational flow continually develop and subsequently disappear. Figure 7.18(e) shows the time-averaged velocity field, which was determined by averaging the velocities from 1024 instantaneous measurements including those shown in Figures 7.18(a) through 7.18(d). Over time, as indicated by Eq. (7.20), the irregular turbulent velocity fluctuations cancel each other out, leaving the mean flow components only. The mean flow pattern in Figure 7.18(e) suggests that the stirrer generates smooth, regular flow in the tank. In reality, efficient mixing in stirred vessels relies on the action of unpredictable, chaotic turbulent flow structures such as those revealed in Figures 7.18(a) to (d).

The extent to which turbulent velocity in the x, y, or z direction deviates from the mean flow velocity is usually reported as the *root mean square* of the fluctuating velocity component, commonly known as *rms fluctuating velocity*. For n values of v'_x, v'_y, and v'_z measured at a fixed position under identical conditions:

$$\text{rms fluctuating velocity in the } x \text{ direction} = v'_{x,\text{rms}} = \sqrt{\overline{v'^2_x}} = \sqrt{\frac{1}{n}\sum^{n} v'^2_x} \tag{7.21}$$

$$\text{rms fluctuating velocity in the } y \text{ direction} = v'_{y,\text{rms}} = \sqrt{\overline{v'^2_y}} = \sqrt{\frac{1}{n}\sum^{n} v'^2_y} \tag{7.22}$$

$$\text{rms fluctuating velocity in the } z \text{ direction} = v'_{z,\text{rms}} = \sqrt{\overline{v'^2_z}} = \sqrt{\frac{1}{n}\sum^{n} v'^2_z} \tag{7.23}$$

Squaring all the positive and negative values of the fluctuating velocity components before they are added together ensures that $v'_{x,\text{rms}}$, $v'_{y,\text{rms}}$, and $v'_{z,\text{rms}}$ are positive quantities. For example, although $\overline{v'_x} = 0$ in Eq. (7.20) because the positive and negative values of v'_x cancel each other out, $\overline{v'^2_x} \neq 0$ because v'^2_x is always positive. From Eq. (7.18), v'^2_x is equal to $(v_x - \overline{v_x})^2$; therefore the rms velocity is analogous to the standard deviation (Section 3.1.5) as it represents the scatter of velocity values about the mean. Like time-averaged velocity, rms fluctuating velocity is a statistical property of turbulence.

Turbulent flow is not experimentally reproducible in its fine detail; a different flow occurs each time the same experiment is performed. For example, if velocity time-courses such as those shown in Figure 7.17 are measured in a particular flow apparatus, the exact same pattern of fluctuations would not be measured again if the experiment were repeated. The source of this lack of reproducibility is the extreme sensitivity of the flow to small changes in the initial and boundary conditions that cannot be controlled with perfect precision. This apparently random element in turbulent flow behaviour is the basis of the statistical representation of turbulence as outlined in the preceding equations. Nevertheless, *the statistical properties of turbulent flow are assumed to be reproducible* even though the instantaneous properties are not; this is the fundamental reason why a statistical approach to turbulence has been adopted. Therefore, if flow is established under nominally identical conditions, the same mean component velocities and rms fluctuating velocities can be expected to occur.

Turbulent flow can become *steady* or *stationary* if external conditions and other properties of the system are maintained constant. In this sense, 'stationary' means that *the statistical properties of the flow are independent of time*, even though the instantaneous properties continue to fluctuate. If turbulence is steady, the mean properties of the flow field are also considered to be independent of the initial conditions applied to generate the flow. We rely on the concept of *stationarity* during turbulence measurements, as accurate and repeatable time-averaging requires that the mean properties of the flow remain steady during the time needed to take a large number of replicate measurements. For example, the mean velocity field shown in Figure 7.18(e) could be determined by averaging a large number of instantaneous measurements such as those in Figures 7.18(a) to (d) only if the statistical properties of the flow field did not change during the measurement period. In stirred tanks, flow is always unsteady to some degree because of the periodic contribution to velocity from passage of the individual impeller blades. However, by taking a large number of repeated measurements over an appropriate period of time, nonstationary elements can be incorporated into the time-averaging process to yield reproducible and accurate mean and rms velocities.

Eddies and Scales of Turbulence

Turbulent flows develop spinning or swirling fluid structures called *eddies*. Turbulent flow is therefore *rotational* and has *vorticity*. Eddies are capable of stretching, coalescing, and dividing; such eddying motion is a characteristic feature of turbulence. Very steep gradients of velocity are associated with eddies; in other words, in eddies, fluid velocities undergo great changes in magnitude and direction within relatively short distances.

Eddies with a wide range of sizes appear in turbulent flow. The size of the largest eddies is limited by the boundaries of the flow system; for example, in a bioreactor, the diameter of the largest eddies is of the same order of magnitude but somewhat smaller than the diameter of the reactor vessel. At high Reynolds number, turbulence at the large eddy scales is unaffected by the viscosity of the fluid. Large eddies are unstable in turbulent flow and give rise to smaller eddies, which in turn produce even smaller eddies and so on, until the smallest-scale eddies are generated. In low-viscosity fluids such as water, the size of the smallest eddies in turbulent flow is roughly 30 to 100 μm. Therefore, a stirred bioreactor with diameter measured in metres will contain turbulent eddies varying in size over an enormous range of 4 to 5 orders of magnitude—that is, from between 10^{-5} and 10^{-4} m to about 1 m. These different scales of turbulence coexist and are superimposed on the mean flow, with the smaller eddies residing inside the larger ones. The range of eddy sizes becomes wider as the Reynolds number increases because the smallest eddies reduce in size with increasing fluid velocity or decreasing viscosity. The process by which small eddies are produced from large eddies occurs continuously during turbulent flow.

Large eddies are responsible for most of the momentum transport in turbulent flow. They also contain most of the turbulence kinetic energy. On the other hand, fluid motion at the smaller scales is associated with large velocity gradients and correspondingly large shear stresses. In a stirred bioreactor, the large eddies derive their energy from the bulk flow generated by the stirrer. Loss of kinetic energy to heat is insignificant in large eddies; instead, energy is handed down from the large to the small eddies in a series of steps called the *energy cascade*. Vortex stretching in the direction normal to the plane of vortex rotation is considered to be the principal mechanism by which energy is transferred between turbulence scales. In the smallest eddies, the energy received is dissipated as heat through the effects of viscosity and fluid friction, thus destroying the eddy structure and velocity gradients.

The rate of viscous energy dissipation by the smallest eddies depends on the fluid viscosity and the magnitude of the velocity gradient. Therefore, even for flow at high Reynolds number where, in general, inertial forces dominate viscous forces (Section 7.2.3), viscosity plays an important role at the smallest-eddy scale. Because of the high shear stresses developed at the smallest scales of turbulence, rates of energy dissipation are significantly higher in turbulent flow than in laminar flow. Maintaining turbulent flow thus requires a continuous supply of energy to make up for viscous dissipation.

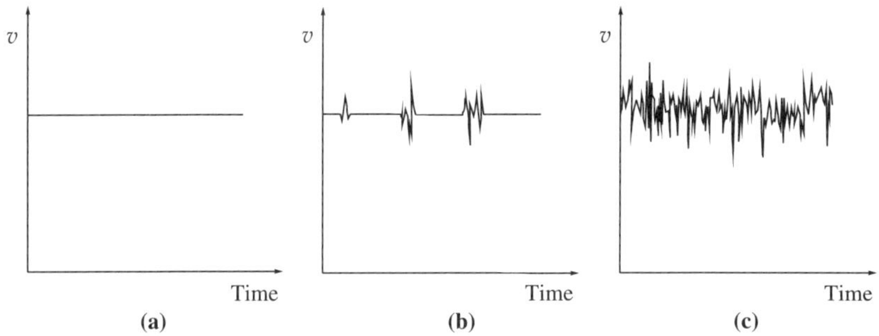

FIGURE 7.19 Time-course measurements of instantaneous velocity in three flow regimes: (a) laminar flow at low *Re*, (b) transition flow at intermediate *Re*, and (c) turbulent flow at high *Re*.

Transition to Turbulence

In flows that are originally laminar, turbulence arises from the development of instabilities at high Reynolds numbers. However, the mechanism of the transition to turbulence is not very well defined or understood. Experiments have shown that the transition is commonly initiated by primary instabilities that produce isolated, two-dimensional, secondary vortex motions. These vortices grow, merge, and become unstable, thus generating intense, localised, three-dimensional disturbances or turbulent 'spots' that arise at random times and positions. Measurements of instantaneous velocity during transition, such as those shown in Figure 7.19, reveal intermittent bursts of local turbulence fluctuations. If these spots grow and merge, an entire field of three dimensional turbulence is formed.

7.9.2 Turbulence Properties

Various concepts and mathematical functions are used to characterise turbulence. Some of these are outlined in the following sections.

Turbulence Intensity

The intensity of turbulence depends on the magnitude of the fluctuating components of the velocity. Because this property is incorporated into the definition of rms fluctuating velocity in Eqs. (7.21) through (7.23), the rms velocities are direct indicators of turbulence intensity. Turbulence intensity I can also be reported as a dimensionless quantity relative to some characteristic or reference velocity; for example:

$$I = \frac{\sqrt{\frac{1}{3}\left(\overline{v_x'^2} + \overline{v_y'^2} + \overline{v_z'^2}\right)}}{\overline{v}} \tag{7.24}$$

where $\overline{v}$ is the uniform time-averaged velocity in the mean direction of flow. Alternatively, in stirred systems, turbulence intensity is often reported relative to the tip speed of the impeller, v_{tip}:

$$I = \frac{\sqrt{\frac{1}{3}\left(\overline{v_x'^2} + \overline{v_y'^2} + \overline{v_z'^2}\right)}}{v_{\text{tip}}} \tag{7.25}$$

v_{tip} is calculated from the impeller size and speed:

$$v_{tip} = \pi N_i D_i \tag{7.26}$$

where N_i is the speed of impeller rotation and D_i is the impeller diameter.

Typically, turbulence intensities range from 0.01 to 0.2 of the average flow velocity. However, in stirred tanks, the turbulence intensity can be as high as 1 close to the impeller blades, dropping rapidly to a maximum of about 0.1 in the rest of the vessel.

Turbulence Kinetic Energy

The turbulence kinetic energy per unit mass of fluid, which is usually given the symbol k, is related to the fluctuating velocity components as follows:

$$k = \frac{1}{2}\left(\overline{v_x'^2} + \overline{v_y'^2} + \overline{v_z'^2}\right) \tag{7.27}$$

Because the fluctuating velocities are measurable quantities, k can be determined experimentally. Time-averages of all three fluctuating velocity components are required.

Turbulence kinetic energy is one of the most important parameters used to characterise flow fields in stirred bioreactors. If the distribution of k throughout the bioreactor vessel is known, the zones of most effective turbulent mixing can be identified.

Reynolds Stresses

In Section 7.3, the shear stress τ for laminar flow of a Newtonian fluid was given by Newton's law of viscosity, Eq. (7.6), in terms of the viscosity μ and the velocity gradient or shear rate dv/dy. As outlined in Section 7.4, shear stress can also be conceptualised as momentum flux. In laminar flow, shear stresses exist in fluids because of molecular interchange between adjacent fluid layers and cohesive forces between liquid molecules (Section 7.2.2). These same effects occur in turbulent flow, but the presence of eddy currents greatly increases the exchange or flux of momentum in all directions. Accordingly, much higher shear stresses are generated for a given velocity gradient in turbulent flow than in laminar flow. This can be represented by the equation:

$$\tau = \tau_{lam} + \tau' \tag{7.28}$$

where τ is the total shear stress in the turbulent flow field, τ_{lam} is the contribution from the mechanisms of momentum transfer in laminar flow, and τ' is the turbulence shear stress due to the fluctuating velocities and eddy motion in turbulent flow. The relative magnitudes of the laminar and turbulent stress contributions vary with position in the flow field; however, in the bulk fluid away from walls and boundary layers:

$$\tau' \gg \tau_{lam} \tag{7.29}$$

In three-dimensional flow, momentum transfer occurs in many different directions. Therefore, there are many different stresses in the fluid depending on the direction of the momentum and the direction in which it is transferred. The directional nature of these stresses can be specified using subscripts on the symbol for stress, τ. For example, τ_{yx} denotes flux in the y-direction of momentum in the x-direction; τ_{xx} is flux in the

x-direction of momentum in the x-direction. τ_{xx}, τ_{yy}, and τ_{zz} are called *normal stresses;* τ_{xy}, τ_{xz}, τ_{yx}, τ_{yz}, τ_{zx}, and τ_{zy} are *shear stresses*. All nine components of the *stress tensor* τ are required to fully describe the stresses in a three-dimensional flow field:

$$\tau = \begin{bmatrix} \tau_{xx} & \tau_{yx} & \tau_{zx} \\ \tau_{xy} & \tau_{yy} & \tau_{zy} \\ \tau_{xz} & \tau_{yz} & \tau_{zz} \end{bmatrix} \tag{7.30}$$

The turbulent stress tensor τ' also has nine components depending on the direction of the momentum and its transfer during turbulent velocity fluctuations. The components of τ' can be expressed in terms of the fluid density ρ and the cross-products of the fluctuating velocity components. For example:

$$\tau'_{xx} = -\rho v'^2_x \qquad \tau'_{yx} = -\rho v'_x v'_y \qquad \tau'_{zx} = -\rho v'_x v'_z \tag{7.31}$$

and so on, for all nine components. Time-averaging these products gives rise to the turbulent *Reynolds stresses*:

$$\begin{array}{lll} \overline{\tau'_{xx}} = -\rho\overline{v'^2_x} & \overline{\tau'_{yx}} = -\rho\overline{v'_x v'_y} & \overline{\tau'_{zx}} = -\rho\overline{v'_x v'_z} \\ \overline{\tau'_{xy}} = -\rho\overline{v'_y v'_x} & \overline{\tau'_{yy}} = -\rho\overline{v'^2_y} & \overline{\tau'_{zy}} = -\rho\overline{v'_y v'_z} \\ \overline{\tau'_{xz}} = -\rho\overline{v'_z v'_x} & \overline{\tau'_{yz}} = -\rho\overline{v'_z v'_y} & \overline{\tau'_{zz}} = -\rho\overline{v'^2_z} \end{array} \tag{7.32}$$

The Reynolds stresses are often written using shorthand notation as $-\rho\overline{v'_i v'_j}$, where the directions indicated by i and j may be the same or different. Conceptually, the Reynolds stresses represent the average rate of momentum transfer per unit area—that is, the momentum flux—due to turbulent velocity fluctuations at a single point in space. Note that the Reynolds stress tensor given in Eq. (7.32) has only six unique components, as $\overline{\tau'_{xy}} = \overline{\tau'_{yx}}$, $\overline{\tau'_{xz}} = \overline{\tau'_{zx}}$, and $\overline{\tau'_{yz}} = \overline{\tau'_{zy}}$.

Homogeneous Turbulence

In homogeneous turbulence, the time-averaged properties of the flow are uniform and independent of position. For example, whereas $\overline{v_x}$, $\overline{v_y}$, *and* $\overline{v_z}$ may differ from each other, each must be constant throughout the system. The same applies to $v'_{x,\text{rms}}$, $v'_{y,\text{rms}}$, and $v'_{z,\text{rms}}$, and to the time-averaged gradients of the fluctuating velocity components, for example, $\overline{\left(\frac{\partial v'_x}{\partial y}\right)^2}$, $\overline{\left(\frac{\partial v'_y}{\partial z}\right)^2}$, and $\overline{\left(\frac{\partial v'_z}{\partial x}\right)^2}$.

Although such a state of motion is not realised readily in experiments, homogeneous turbulence has been given much attention because it greatly simplifies the theoretical treatment of turbulent flow. The assumption of homogeneous turbulence can be justified to a certain extent over small distances somewhat greater than the size of the smallest eddies: at this scale, the mean flow properties are essentially independent of position. However, if turbulence is assumed to be spatially homogeneous it cannot, strictly speaking, also be assumed stationary. From energy balance considerations, a homogeneous turbulent flow field must at the same time be a decaying turbulent flow field; that is, its properties will be changing with time. Fortunately, the rate of decay of the mean flow properties is relatively slow at the smaller scales of turbulence, so that this condition of nonstationarity is not a serious problem in experimental studies that rely on averaging many replicate measurements over time.

Isotropic Turbulence

Analysis of turbulence is greatly facilitated if the turbulence can be assumed isotropic as well as homogeneous. In isotropic turbulence, the statistical properties of the flow are independent of direction and the axes of reference; in other words, the flow in three dimensions has no directional preference. For example, at each point within isotropic turbulence:

$$\begin{gathered}\overline{v_x'^2} = \overline{v_y'^2} = \overline{v_z'^2}\\ v'_{x,\mathrm{rms}} = v'_{y,\mathrm{rms}} = v'_{z,\mathrm{rms}}\\ \overline{\left(\frac{\partial v'_x}{\partial x}\right)^2} = \overline{\left(\frac{\partial v'_y}{\partial y}\right)^2} = \overline{\left(\frac{\partial v'_z}{\partial z}\right)^2}\\ \overline{\left(\frac{\partial v'_x}{\partial y}\right)^2} = \overline{\left(\frac{\partial v'_y}{\partial z}\right)^2} = \overline{\left(\frac{\partial v'_z}{\partial x}\right)^2}\quad \text{etc.}\end{gathered} \tag{7.33}$$

because each of the equated mean properties is measured in the same way for a different selection of axes and velocity components.

The implications of isotropy in turbulence modelling and measurement are profound. True isotropy implies that there can be no mean velocity gradients. In addition, all time-averaged fluctuating cross-velocity terms are zero:

$$\overline{v'_x v'_y} = \overline{v'_x v'_z} = \overline{v'_y v'_z} = 0 \tag{7.34}$$

Equation (7.34) results from the random nature of the fluctuations that would give $v'_x v'_y$, $v'_x v'_z$, and $v'_y v'_z$ just as many positive as negative values in isotropic turbulence, so the averages are zero. An important consequence of Eq. (7.34) is that the components of the Reynolds stress tensor of Eq. (7.32) are zero except for the diagonal elements, $-\rho\overline{v_x'^2}$, $-\rho\overline{v_y'^2}$, and $-\rho\overline{v_z'^2}$, which, from Eq. (7.33), must be identical. Such simplification is typical for isotropic turbulence; the entire turbulence spectrum can be defined using only one velocity component and a minimum number of other quantities and relations, reflecting the lack of directionality of the flow. For example, Eq. (7.27) for turbulence kinetic energy per unit mass can be simplified to:

$$k = \frac{3}{2}\overline{v'^2} \tag{7.35}$$

where $\overline{v'^2}$ represents any one of the three fluctuating velocity components, $\overline{v_x'^2}$, $\overline{v_y'^2}$, or $\overline{v_z'^2}$.

Isotropic turbulence does not exist in many practical applications of turbulent fluids. The orientation of the bulk fluid motion prevents the conditions of isotropy being satisfied in shear flows where there are mean velocity gradients; the large eddies in turbulent flow are therefore anisotropic because they have an overall flow direction. However, the directional elements of the main flow are progressively lost through the energy cascade process (Section 7.9.1, Eddies and Scales of Turbulence subsection) so that, at the smallest scales of turbulence where the eddies do not interact directly with the mean motion, *local isotropy* is often assumed to exist. Analysis of isotropic turbulence therefore has practical application in representing viscous dissipation processes.

Kolmogorov's universal equilibrium theory of local isotropic turbulence provides some insights into the structure and properties of turbulence at the smallest eddy scales. Kolmogorov's theory relates to flows at high Reynolds number in which a broad spectrum of eddy sizes is generated. Under these conditions, there is a substantial size separation between the large eddies that contain most of the turbulence kinetic energy and the small eddies that dissipate it. The smallest, energy-dissipating eddies are considered to have a *universal structure*, or local isotropy. 'Universal' in this context means that the eddies are statistically independent of external conditions such as the boundaries of the system, any special features of the mean flow in which the eddies are embedded, and the larger scales of turbulence. Therefore, at the smallest scales, any sense of overall flow direction is lost. Instead, the properties of eddies in this range depend only on the rate of energy supply from the larger eddies and the rate of energy dissipation by viscosity.

The assumption of *statistical equilibrium* at the small turbulence scales, meaning that the rate of energy supply is equal to the rate of energy dissipation, is one of the premises of Kolmogorov's theory of isotropic turbulence. If energy losses through the energy cascade can be neglected so that the energy supplied to the smallest eddies is practically equal to the total energy dissipated in the system, an expression can be obtained for the characteristic size or length scale of eddies in the dissipative regime:

$$\lambda = \left(\frac{\nu^3}{\varepsilon}\right)^{1/4} \tag{7.36}$$

where λ is the characteristic dimension of the smallest eddies, ν is the kinematic viscosity of the fluid (Section 7.3), and ε is the local rate of dissipation of turbulence kinetic energy per unit mass of fluid. λ is often called the *Kolmogorov scale*. Equation (7.36) provides an estimate of the size of the smallest eddies to order-of-magnitude accuracy, and indicates that the smallest scales of turbulence reduce in size with decreasing fluid viscosity and increasing energy dissipation rate.

An extension of Kolmogorov's theory of isotropic turbulence relates to flows at very high Reynolds numbers. Under these conditions, it is possible that some eddies develop universal structure but are not involved significantly in turbulence energy dissipation. Thus, not only are these eddies independent of external constraints and the mean flow, they are also relatively unaffected by viscosity. This range of eddy sizes is called the *inertial subrange*, as transfer of energy by inertial mechanisms or vortex stretching is its dominant feature. The inertial subrange occurs at scales between the large energetic eddies and the smallest dissipative structures. Although little energy dissipation takes place, the features of turbulence in this range depend entirely on the rate of energy dissipation ε.

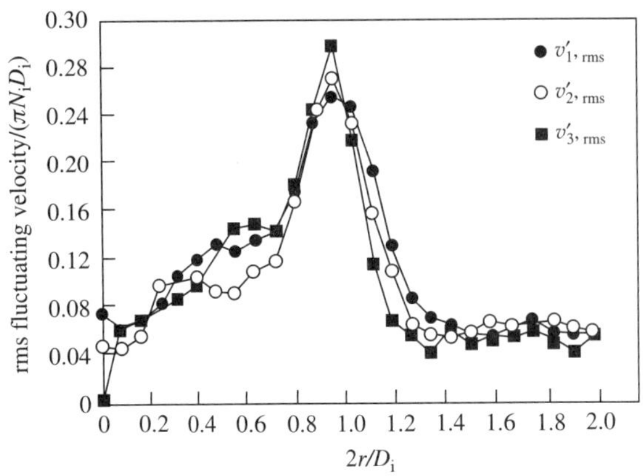

FIGURE 7.20 The magnitude of rms fluctuating velocity components close to the lower edge of a pitched-blade turbine impeller operated at 400 rpm. r is the radial distance from the centre of the impeller, D_i is the impeller diameter, and N_i is the stirrer speed. The rms velocities are plotted as dimensionless parameters relative to the impeller tip speed, $\pi N_i D_i$.
Adapted from S.M. Kresta and P.E. Wood, 1993, The flow field produced by a pitched blade turbine: characterization of the turbulence and estimation of the dissipation rate. Chem. Eng. Sci. *48, 1761–1774.*

This is because the energy that is eventually dissipated must be transferred across the inertial subrange, from the larger to the smaller eddies.

The assumption of local isotropy is often used for analysis of turbulence in stirred bioreactors. Isotropy is usually assumed to apply near the impeller where turbulence is most intense. Because the intensity of turbulence decays very quickly as fluid moves away from the impeller zone, it is unlikely that the criteria for local isotropy can be met in other regions of the vessel. Even so, the assumption of local isotropy near the impeller must be considered an engineering approximation only. Several features of the impeller zone, such as the development of trailing vortices (see Section 8.4.1) and the periodicity of flow from passage of the impeller blades, have the potential to violate the conditions of isotropy.

Mathematical analysis of isotropic turbulence provides a range of tests that can be applied to check whether flow conditions in a particular system are consistent with isotropy. One of these is to demonstrate that all of the rms fluctuating velocity components are equal, thus satisfying one of the criteria for isotropy in Eq. (7.33). The experimental results in Figure 7.20 illustrate that this condition can be shown to apply close to the impeller. While not proving the existence of isotropy, tests such as this have lent confidence to the use of isotropy approximations for analysis of turbulence within restricted regions of stirred vessels [29, 30]. In other work, however, significant deviations from isotropy have been found in the discharge streams of impellers near the impeller blades [31, 32].

Rate of Dissipation of Turbulence Kinetic Energy

A central concept in turbulence modelling and analysis is the rate of dissipation of turbulence kinetic energy per unit mass, ε. As indicated in Eq. (7.36), ε is related closely to the characteristic size of the smallest eddies. As such, it affects a range of processes that are of critical importance in bioreactors and downstream processing equipment. These include mixing, gas bubble dispersion, droplet break-up, coagulation, cell flocculation, and cell damage. Together with the turbulence kinetic energy per unit mass k (see Section

7.9.2, Turbulence Kinetic Energy subsection), the rate of turbulence energy dissipation and its distribution throughout the vessel are among the most important parameters for characterising turbulence in stirred fermenters.

A major difficulty with the rate of dissipation of turbulence kinetic energy is that, unlike for k, direct measurement is not experimentally feasible. Accordingly, many different methods for estimating ε by indirect means have been proposed. From dimensional analysis and with the assumption of local isotropy, one of the most commonly used equations is:

$$\varepsilon = A\frac{v'^3}{L} \tag{7.37}$$

where A is a dimensionless constant, v' is a characteristic fluctuating velocity, and L is a characteristic length scale. However, even though this relationship has been derived, the exact meaning of the parameters in Eq. (7.37) is open to interpretation. Because the larger eddies contain most of the turbulence kinetic energy, L is generally assumed to reflect the properties of the larger turbulence scales in the region of flow being investigated. For example, in several studies of the impeller zone of stirred tanks, L has been given a constant value equal to either one-half the impeller blade width or one-tenth the impeller diameter. v' is usually taken to be the rms fluctuating velocity. Because, as indicated in Eq. (7.33), all three rms fluctuating velocity components are equal under isotropic conditions, either $v'_{x,\mathrm{rms}}$, $v'_{y,\mathrm{rms}}$, or $v'_{z,\mathrm{rms}}$ can be used. The value of A is widely accepted to be about 1 for isotropic turbulence but also depends on the choice of characteristic length L [29].

Equation (7.37) may be expressed in terms of the turbulence kinetic energy k instead of v':

$$\varepsilon = B\frac{k^{3/2}}{L} \tag{7.38}$$

where the constant $B = 0.54A$ and k is given by Eq. (7.35) for isotropic turbulence.

Many other methods and equations have been used to estimate ε in stirred systems. Further details can be found in the literature [29]. Although these relationships, including Eqs. (7.37) and (7.38), have been applied with some success to analyse the impeller region of stirred tanks, because local isotropy applies at most only near the impeller, other approaches are required to determine the rate of turbulence energy dissipation in the remainder of the vessel. At present, the most reliable method is by subtraction—that is, the rate of energy dissipation in the bulk fluid in the remainder of the tank is equal to the total rate of energy input by the stirrer minus the rate of energy dissipation in the impeller region. Using this approach, the proportion of the total energy input that is dissipated in the impeller stream can be obtained, thereby characterising the performance of different impellers or operating procedures.

7.9.3 Turbulence Measurement

Several parameters for characterising and quantifying turbulence are described in Sections 7.9.1 and 7.9.2. Evaluation of these parameters depends on the availability of experimental data for the velocity of turbulent flow. A wide range of turbulence measurement techniques has been developed. Of these, laser-based methods such as laser Doppler velocimetry and particle image velocimetry have been used most frequently to measure turbulent flow in systems relevant to bioprocessing.

As described in Section 7.9.1, turbulence is characterised by unsteady flow patterns that change very quickly and eddies that vary greatly in size. Therefore, to obtain accurate readings of turbulent velocity, the measuring device must be capable of fast data acquisition and must be able to resolve velocity components within a relatively small volume of fluid. Because many replicate measurements are required for meaningful statistical analysis and spatial definition of the flow field, measuring systems should also have constant long-term response with negligible drift.

Instantaneous velocities can be measured directly. Derived parameters such as mean, fluctuating and rms fluctuating velocities, turbulence kinetic energy, Reynolds stresses, and the rate of turbulence energy dissipation require further calculations as outlined in Sections 7.9.1 and 7.9.2.

Laser Doppler Velocimetry

Laser Doppler velocimetry (LDV), which is also sometimes called laser Doppler anemometry (LDA), has been applied extensively to characterise liquid flow, mixing, and energy dissipation in a variety of flow situations, including stirred bioreactors. The high spatial resolution of LDV combined with its ability to follow rapid fluctuations in velocity make it an important tool for the study of turbulence.

A laser Doppler velocimeter measures local, instantaneous fluid velocities by detecting the frequency of light scattered by small particles suspended in the fluid as they pass through a fringe or interference pattern. The main components of a typical LDV system operated in forward scatter mode are shown in Figure 7.21. The instrument comprises a laser light source, optical arrangements to transmit and collect light, a photodetector, and equipment for signal processing. Light from the laser is split by a beam splitter into two beams of equal intensity that cross to form a fringe pattern in the local region of the fluid where velocity measurements are required. If one of the beams is frequency-shifted, the LDV system can measure flow reversals. Scattered light from the measurement volume is collected and the optical signal converted to an electronic signal for processing.

According to the Doppler principle, the frequency of the scattered light will be shifted by an amount directly proportional to the flow velocity. Mean and fluctuating velocity components can be determined online from the frequency record of the photodetector and

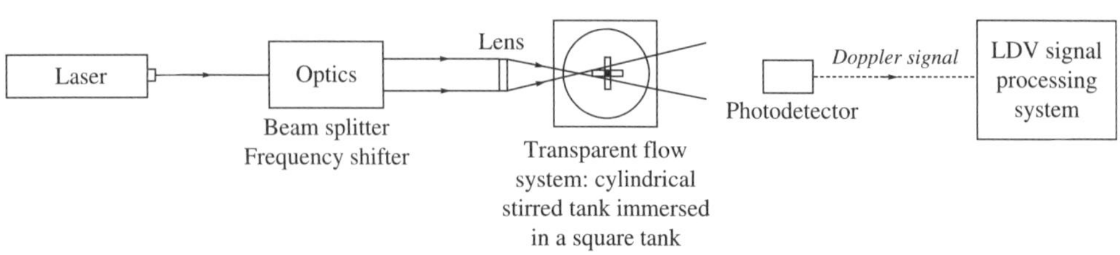

FIGURE 7.21 Apparatus for laser Doppler velocimetry.

the detected Doppler shifts. LDV techniques allow accurate resolution of local velocities within fluid volumes as small as 10^{-6} to 10^{-4} mm^3.

A major advantage associated with LDV is that it is noninvasive; for example, sensor probes do not have to be immersed in the fluid. Consequently, LDV measurements do not cause flow disturbance. In addition, LDV offers high accuracy of measurement without calibration. Light-scattering particles are required in the fluid; when velocities are being measured in water, if appropriate particulate matter is not already present, small (0.1–10 μm) particles of silicone carbide, plastic, polystyrene, or metal-coated material may be added. To allow penetration of light signals into the fluid, the flow system must be constructed from transparent material such as glass, Perspex, or acrylic plastic. When flow fields in cylindrical bioreactors are to be measured, the curvature of the vessel walls and diffraction effects can make accurate positioning of the laser beam difficult. To minimise this problem, the cylindrical vessel is usually immersed in a square tank also made of transparent material and filled with the same fluid as that in the reactor.

Standard LDV systems provide flow information at a single point. To obtain data at different positions in the flow field, either the entire flow system or the laser and optical equipment must be moved around so that the split beams from the laser can intersect at a range of locations within the fluid. This is achieved by mounting the apparatus on a platform equipped with a multidimensional traversing mechanism, which may be operated manually or under computer control. Single-channel LDV measures a single directional component of the fluid velocity; to determine the other two components, the measurements must be repeated after adjusting the orientation of the laser beams. Alternatively, dual-channel LDV measures two velocity components simultaneously; in these systems, the laser light is separated into two beams of different colour—for example, blue and green—before each beam is split. Dual photodetectors are then used to distinguish between the scattered blue and green channel signals. To obtain the third velocity component, a separate measurement is required at the same position in the fluid with a different orientation of the laser beams. Between 1000 and 10,000 measurements can be taken using LDV within just a few seconds to provide statistically reproducible, time-averaged velocity data.

LDV measurements are subject to several types of error. Care is required to accurately determine the position of the probe volume by aligning the laser beams relative to a fixed reference point within the flow field. Errors can also arise because of the finite time taken by the particles to cross the measurement volume, and because of velocity variations within that volume. Typically, errors in mean velocities measured by LDV are 2 to 5%; errors in turbulent fluctuating velocities are in the region of 5 to 10%.

Particle Image Velocimetry

Like laser Doppler velocimetry, particle image velocimetry (PIV) is a laser-based velocity measurement technique. However, whereas LDV measures velocities at one location at a time, PIV can give a global view of instantaneous velocities in two or three dimensions. The majority of PIV systems in use provide two-dimensional velocity data in a planar domain at discrete instants in time.

The basic requirements for a PIV measurement system are shown in Figure 7.22. The components include a laser light source, an optics module for forming a sheet of light to illuminate a plane in the flow field, recording hardware consisting of either a CCD

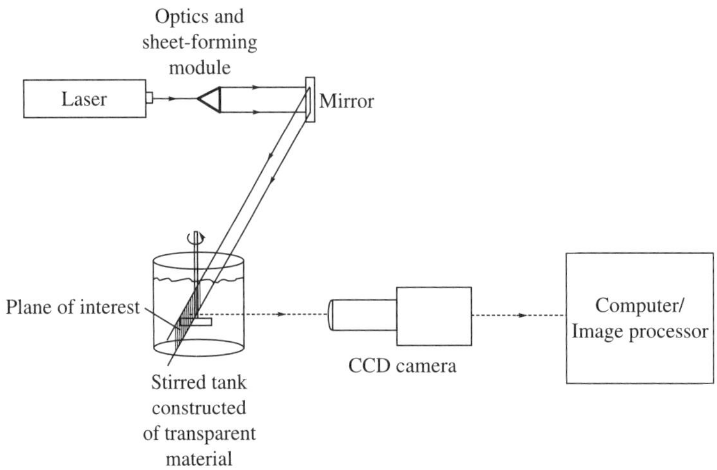

FIGURE 7.22 Basic requirements for a particle image velocimetry measurement system.

(charge-coupled device) or other type of camera, and image-processing equipment and software. Fluid velocities are measured by illuminating particles in the fluid using two successive pulses of light only a fraction of a second apart, recording the positions of the particles using the camera, and then analysing and cross-correlating the two images to determine the particle displacements over the time interval. Because an entire flow plane is imaged, velocities at hundreds or thousands of locations in the fluid are thus obtained rapidly.

As described in the previous section for laser Doppler velocimetry, the flow system must be constructed from transparent material to allow penetration of light signals. Cylindrical vessels such as bioreactors are usually placed inside square tanks containing the same fluid to minimise refractive index variations and wall curvature effects. The fluid is seeded with neutrally buoyant particles; for example, fluorescent polymer spheres of diameter around 80 μm may be used for measurements in water.

Typical PIV results for instantaneous velocities in a stirred tank are shown in Figure 7.18(a) through (d); time-averaged velocities in the plane of illumination are given in Figure 7.18(e). Although most PIV systems provide two-dimensional flow field characterisation, methods are available for simultaneous measurement of all three velocity components in planar domains. These include the use of overlapping or partially overlapping parallel light sheets, and stereoscopic PIV employing two cameras for depth perception. Cinematic holographic PIV allows three-dimensional velocity vector measurement within three- rather than two-dimensional domains; however, these systems are expensive and complex and not often employed. Compared with laser Doppler velocimetry, PIV provides data for overall flow fields in one step rather than point by point. However, sampling is less frequent using PIV so that temporal resolution is lower; very fast velocity fluctuations may therefore be missed during PIV measurements. Nevertheless, the advantages associated with instantaneous spatial resolution often outweigh this drawback, making PIV a valuable tool for analysis of turbulent flow.

7.9.4 Computational Fluid Dynamics

Mapping an entire turbulent velocity field is not an easy experimental task. An alternative is to use computer-based simulation to determine the features of turbulent flow. This approach known as *computational fluid dynamics* (CFD) relies on the availability of high-speed computers and accurate numerical algorithms to solve difficult mathematical equations. In many cases, because of the complexity of turbulence, computational fluid dynamics does not completely eliminate the need for experimental data to characterise flow. At the present time, most CFD applications involve a mixture of theory, experiment, and computer programming.

Fluid Transport Equations

The physical behaviour of any fluid is governed by three fundamental principles: conservation of mass, Newton's second law (force = mass × acceleration), and conservation of energy. These principles can be expressed using mathematical equations, usually in the form of partial differential equations. For isothermal systems in which temperature differences have a negligible effect on flow behaviour, the transport equations needed to solve flow problems are the *equation of continuity* representing the law of conservation of mass, and the *equations of motion* representing the conservation of linear momentum according to Newton's second law. This group of equations is also known as the *Navier–Stokes equations*.

In bioprocess engineering, we are interested in fluid dynamics mainly as it relates to bioreactor design and operation. Accordingly, we will focus on the transport equations for flow of incompressible fluids—that is, liquids with constant density. For simplicity, we will also confine our attention to liquids with Newtonian rheology. Because most bioreactors are cylindrical vessels, the transport equations are most appropriately expressed using cylindrical spatial coordinates. The coordinate system for cylindrical geometry is shown in Figure 7.23; the position of point P is specified using coordinates in the r, θ, and z directions.

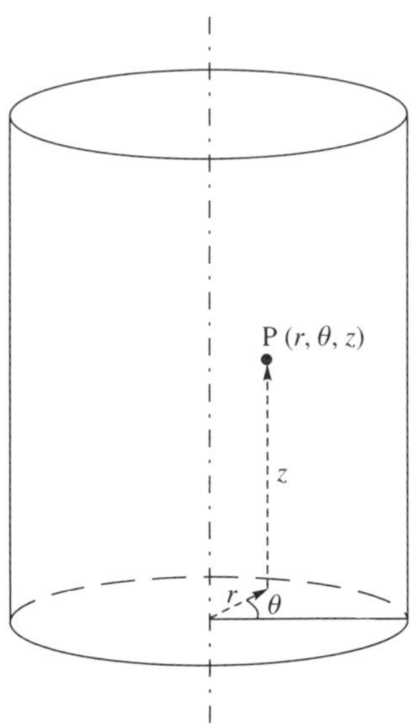

FIGURE 7.23 Cylindrical coordinate system.

TABLE 7.3 Transport Equations in Cylindrical Coordinates for an Incompressible Newtonian Fluid

Equation of continuity
$$\frac{\partial v_r}{\partial r}+\frac{v_r}{r}+\frac{1}{r}\frac{\partial v_\theta}{\partial \theta}+\frac{\partial v_z}{\partial z}=0$$
Navier–Stokes equations
r-component
$$\rho\left(\frac{\partial v_r}{\partial t}+v_r\frac{\partial v_r}{\partial r}+\frac{v_\theta}{r}\frac{\partial v_r}{\partial \theta}-\frac{v_\theta^2}{r}+v_z\frac{\partial v_r}{\partial z}\right)=\frac{-\partial p}{\partial r}+\mu\left[\frac{\partial^2 v_r}{\partial r^2}+\frac{1}{r}\frac{\partial v_r}{\partial r}-\frac{v_r}{r^2}+\frac{1}{r^2}\frac{\partial^2 v_r}{\partial \theta^2}-\frac{2}{r^2}\frac{\partial v_\theta}{\partial \theta}+\frac{\partial^2 v_r}{\partial z^2}\right]+\rho g_r$$
θ-component
$$\rho\left(\frac{\partial v_\theta}{\partial t}+v_r\frac{\partial v_\theta}{\partial r}+\frac{v_\theta}{r}\frac{\partial v_\theta}{\partial \theta}+\frac{v_r v_\theta}{r}+v_z\frac{\partial v_\theta}{\partial z}\right)=\frac{-1}{r}\frac{\partial p}{\partial \theta}+\mu\left[\frac{\partial^2 v_\theta}{\partial r^2}+\frac{1}{r}\frac{\partial v_\theta}{\partial r}-\frac{v_\theta}{r^2}+\frac{1}{r^2}\frac{\partial^2 v_\theta}{\partial \theta^2}+\frac{2}{r^2}\frac{\partial v_r}{\partial \theta}+\frac{\partial^2 v_\theta}{\partial z^2}\right]+\rho g_\theta$$
z-component
$$\rho\left(\frac{\partial v_z}{\partial t}+v_r\frac{\partial v_z}{\partial r}+\frac{v_\theta}{r}\frac{\partial v_z}{\partial \theta}+v_z\frac{\partial v_z}{\partial z}\right)=\frac{-\partial p}{\partial z}+\mu\left[\frac{\partial^2 v_z}{\partial r^2}+\frac{1}{r}\frac{\partial v_z}{\partial r}+\frac{1}{r^2}\frac{\partial^2 v_z}{\partial \theta^2}+\frac{\partial^2 v_z}{\partial z^2}\right]+\rho g_z$$

Transport equations describing the instantaneous motion of Newtonian liquids in three-dimensional cylindrical geometry are listed in Table 7.3. In these equations, v_r, v_θ, and v_z are the directional components of the fluid velocity, t is time, ρ is fluid density, p is fluid pressure, μ is fluid viscosity, and g_r, g_θ, and g_z are the components of gravitational acceleration in the r, θ, and z directions, respectively. $\partial/\partial t$ denotes the partial differential with respect to time; $\partial/\partial r$, $\partial/\partial \theta$, and $\partial/\partial z$ denote the partial differentials with respect to r, θ, and z, respectively. The partial differential $\partial v_r/\partial t$, for example, means the differential of v_r with respect to time, keeping r, θ, and z constant. $\partial^2/\partial \theta^2$ and $\partial^2/\partial z^2$ denote second partial differentials with respect to θ and z, respectively.

General Solution Procedures

The four fluid transport equations in Table 7.3 contain four unknowns, v_r, v_θ, v_z, and p. In principle, therefore, if we specify a time t and a position (r, θ, z) in the fluid, we could solve these equations to determine a unique solution for the prevailing velocity and pressure. Repetition of the solution procedure at different locations would then provide results for the velocity and pressure distributions in the entire flow field. In practice, direct solution of the governing equations of fluid mechanics is rarely carried out. This is because, irrespective of the geometry of the flow system, the equations are mathematically extremely complex. Sets of coupled, nonlinear partial differential equations are not solved readily. Indeed, obtaining solutions to fluid transport equations such as those in Table 7.3 represents one of the most challenging problems in modern science and engineering. As a consequence, a variety of simplifications, approximations, and mathematical shortcuts must be used to obtain solutions that apply more or less reliably under different flow conditions.

Analytical solutions to even the simplest turbulent flows do not exist. Therefore, a calculated description of turbulent flow where the velocity and pressure are known as functions of position and time can only be obtained using numerical simulation techniques and computer programming. To apply these methods, the flow system is first divided into a number of computational cells known as the *grid* or *mesh*. This process identifies discrete points or *nodes* in the fluid for which numerical solution of the transport equations will be found. Similarly, time is divided into discrete intervals for solution at particular times. Therefore, instead of applying the transport equations to obtain solutions over all space and time, the first approximation for numerical analysis is to focus only on particular spatial and temporal locations.

As a second approximation, the partial derivatives in the transport equations are replaced by simpler but less exact algebraic equations that can be applied between two or more of the solution nodes. The process by which differential equations are approximated by a set of algebraic equations at a set of discrete locations in space and time is called *discretisation*. Some of the more common discretisation techniques include *finite difference*, *finite volume*, *finite element*, and *spectral* methods.

An example of a grid and solution domain for CFD analysis of a stirred tank is shown in Figure 7.24. The accuracy of a CFD solution is governed in part by the number of cells in the grid; in general, the greater the number of cells in a given volume, the better is the solution accuracy. On the other hand, the computer time and hardware required, and

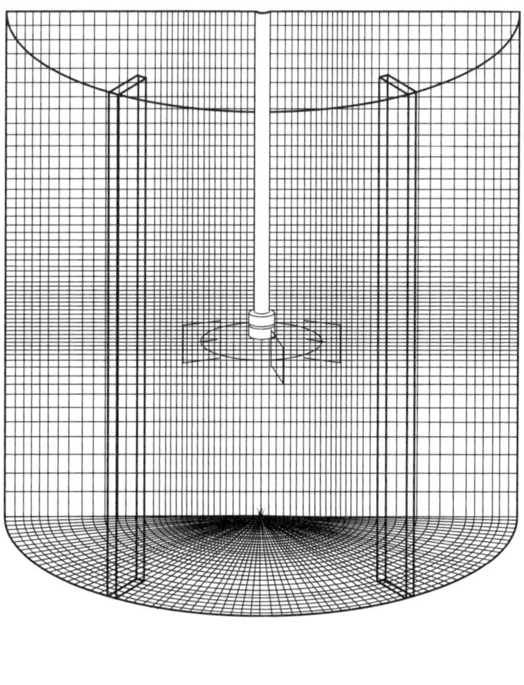

FIGURE 7.24 Grid used to spatially discretise a stirred tank for CFD analysis.
Adapted from V.V. Ranade and V.R. Deshpande, 1999, Gas–liquid flow in stirred reactors: trailing vortices and gas accumulation behind impeller blades. Chem. Eng. Sci. *54, 2305–2315.*

therefore the cost of the solution, increase with the fineness of the grid. Nonuniform meshes are often used; closely spaced nodes are needed in zones where velocity variations are large, while a coarser grid is adequate for regions where there is relatively little change. This is illustrated in Figure 7.24; the mesh is tighter in the impeller discharge region where the greatest velocity gradients are expected. The cells of a grid may be of irregular shape to reflect the geometry of the flow system. Body-fitted coordinates can be used to fit the contours of the volume being analysed; alternatively, a nonstructured grid pattern incorporating hexahedral, tetrahedral, prismatical, and pyramidal cells can be applied if appropriate.

An important problem when discretising flow in a stirred vessel is that the shape and boundaries of the flow domain change with time as a result of the movement of the impeller blades. Incorporating these effects presents a range of computational difficulties. One approach for solving this problem is to use a *deforming* or *moving grid* that changes shape with time. Another approach is the *sliding mesh technique* involving two grids, one of which rotates with the impeller while the other representing the bulk of the tank remains stationary.

Once a grid has been created, another major concern is the application of appropriate *boundary conditions*. As outlined in Section 6.3, the solution of any differential equation representing a real-life process requires one or more boundary conditions that reflect the properties of the particular system being investigated. For unsteady or transient flows, *initial conditions* may also be needed. Because the fluid transport equations are the same for a given fluid and flow geometry irrespective of operating conditions, the choice of boundary conditions dictates the particular solution obtained. Therefore, it is important to select boundary conditions that give an appropriate numerical representation of the actual flow situation.

No-slip and *no-penetration* boundary conditions are often applied at the stationary solid surfaces in the flow field: the no-slip condition assumes that tangential fluid velocity components are zero at the surface; the no-penetration condition assumes zero velocity normal to the surface. When flow is symmetrical about a plane, *symmetry conditions* can be applied; these assume there is no flow through the plane of symmetry. In stirred tanks, the boundary conditions used to specify the impeller have a pronounced influence on the final solution. One option is to treat the region swept out by the impeller as a 'black box'. The transport equations are not solved within this region; however, the values of all the variables must be specified on all surfaces of the box. Alternatively, solution of the equations within the impeller volume may be included in the overall numerical scheme, with the presence of the impeller accounted for using conditions at the impeller discharge boundary. Selection of the impeller boundary conditions usually relies heavily on experimental data such as those obtained using laser Doppler velocimetry, particle image velocimetry, or some other turbulence measurement technique. Thus, empirical information plays a critical role in many applications of computational fluid dynamics to stirred tanks. With more advanced CFD approaches such as the sliding mesh and multiple reference frame techniques, flow around the impeller blades is modelled in detail, thus minimising the need for experimental boundary conditions.

After selection of an appropriate numerical solution method, computer programming is used to solve the equations. The solution technique may employ *direct* or *iterative* procedures. Iterative methods are common: these involve providing an initial guess or estimate of the answer, applying the discretised equations to calculate a new result, then determining the *error* or *residual* by subtracting the initial and calculated values. After a large number of repeated calculations, the residual becomes smaller and smaller as the results from successive iterations tend towards some fixed solution. A final answer or *convergence* is obtained when the error becomes less than some predetermined tolerance limit.

Direct Numerical Simulation

Direct numerical simulation (DNS) means complete three-dimensional and time-dependent solution of the Navier–Stokes equations to obtain results for the instantaneous fluid velocity as a function of position and time. Although this is a desirable goal, unfortunately it is not possible for the vast majority of practical flow situations because of the inherent complexity of the transport equations.

Direct numerical simulation has been applied to a limited number of geometrically simple flow situations, such as flow in a channel or over a flat plate. Elaborate computer programming is required and the amount of computation involved is nontrivial. As in any CFD application, issues related to numerical accuracy, solution stability, convergence, specification of boundary conditions, and optimal use of available computing resources are of concern. For a detailed picture of a turbulent flow field, the equations must be solved using grids with spatial domains fine enough to resolve most of the individual eddies, which may be as small as 30 to 100 μm in low-viscosity fluids such as water. Appropriately small time steps must also be employed to capture the unsteady and fluctuating nature of turbulent flow. Because the scale of the smallest eddies becomes smaller and the fluctuation frequency becomes greater as the fluid Reynolds number increases, the computing requirements for flow resolution at high Reynolds number are enormous. In general, therefore, direct numerical solution is a possibility only for flows with low to moderate levels of turbulence.

At relatively low Reynolds numbers, typical DNS calculations involve resolution of between 5 million and 20 million nodes in the flow field and require 250 to 400 hours of expensive supercomputer time. Application of such techniques to an irregular geometric domain such as the inside of a stirred tank and for highly turbulent flows such as those generated by standard fermenter mixing equipment is not feasible at present because of the enormous computer times and data storage facilities required. Currently, then, rather than providing a means for characterising complex, practical flow systems, DNS is viewed mainly as a research tool for verifying experimental results and for analysing ideal flows that cannot be created easily in the laboratory.

Large Eddy Simulation

As direct numerical simulation of all scales of turbulent flow is limited by available computer power, large eddy simulation (LES) has been developed as a less costly and more approximate approach. In large eddy simulation, the flow equations are represented numerically but the results for only the large eddies are computed. Thus, an approximation of the real flow is obtained in which scales below a certain size are missing. To compensate for this deficiency, the smaller eddies and the energy dissipation process are modelled using additional subgrid terms inserted into the equations of motion.

The premise underlying LES is that small-scale turbulence is weaker and therefore less critical than the larger eddies, which carry most of the energy and turbulent stresses and interact most strongly with the mean flow. Small-scale turbulence is also nearly isotropic and thus more amenable to separate modelling. However, despite the simplifications incorporated into large eddy simulation, the computing requirements for LES analysis of most real-life flow situations remain very large.

Reynolds Averaging of the Transport Equations

As indicated in the two previous sections, obtaining complete solutions to the fluid transport equations is extremely difficult because of their mathematical complexity and corresponding demand for computer time and capacity. Fortunately, however, practical application of the equations can be achieved by recognising that, for most engineering purposes, it is not necessary to resolve the fine details of turbulent fluctuations. In many situations, information about the *mean flow* only is required, as the effects of each and every eddy are not of primary interest. By narrowing our focus and concentrating only on the mean turbulence variables, we can significantly reduce the number of calculation nodes necessary to capture the pertinent features of the flow.

Instead of using grids with cell dimensions of 10 to 100 μm for resolution of the smallest eddies, typical meshes used to analyse the mean flow in stirred tanks have cells around 0.5 cm in size. As a result, the number of computation nodes is reduced to a manageable level, typically 20,000 to 100,000. Similarly, because the fastest events in turbulence take place very quickly, resolution of the instantaneous properties of turbulent flow requires discretised time intervals of about 100 μs. In contrast, as changes in the mean flow occur more slowly, any time-dependence can be resolved using much larger time steps. For many applications such as stirred tank operation, it is reasonable to assume that the mean flow is stationary so that time-dependence can be neglected completely and the equations solved only as a function of position.

To implement this solution approach for mean turbulence properties, we must first rewrite the continuity and Navier–Stokes equations using time-averaged variables rather than instantaneous quantities. By analogy with Eq. (7.18), which separates instantaneous velocity components into time-averaged and fluctuating elements, for cylindrical geometry and including pressure in the time-averaging scheme:

$$v_r = \overline{v_r} + v_r' \qquad v_\theta = \overline{v_\theta} + v_\theta' \qquad v_z = \overline{v_z} + v_z' \qquad p = \overline{p} + p' \tag{7.39}$$

where v_r, v_θ, and v_z are the three directional components of the fluid velocity and p is the fluid pressure. Substituting the terms of Eq. (7.39) into the transport equations of Table 7.3 and taking averages produces a set of equations satisfied by the mean turbulent flow. The equations thus transformed are listed in Table 7.4 [33]. Transformation of the fluid transport equations using statistical averaging was first carried out by Osborne Reynolds. The process is often called *Reynolds averaging* and the equations in Table 7.4 are known as the *Reynolds equations.*

TABLE 7.4 Time-Averaged Transport Equations in Cylindrical Coordinates for an Incompressible Newtonian Fluid

Time-averaged equation of continuity

$$\frac{\partial \overline{v}_r}{\partial r} + \frac{\overline{v}_r}{r} + \frac{1}{r}\frac{\partial \overline{v}_\theta}{\partial \theta} + \frac{\partial \overline{v}_z}{\partial z} = 0$$

Time-averaged Navier–Stokes equations

r-component

$$\rho\left(\frac{\partial \overline{v}_r}{\partial t} + \overline{v}_r\frac{\partial \overline{v}_r}{\partial r} + \frac{\overline{v}_\theta}{r}\frac{\partial \overline{v}_r}{\partial \theta} - \frac{\overline{v}_\theta^{\,2}}{r} + \overline{v}_z\frac{\partial \overline{v}_r}{\partial z}\right) = \frac{-\partial \overline{p}}{\partial r} + \mu\left[\frac{\partial^2 \overline{v}_r}{\partial r^2} + \frac{1}{r}\frac{\partial \overline{v}_r}{\partial r} + \frac{1}{r^2}\frac{\partial^2 \overline{v}_r}{\partial \theta^2} + \frac{\partial^2 \overline{v}_r}{\partial z^2} - \frac{\overline{v}_r}{r^2} - \frac{2}{r^2}\frac{\partial \overline{v}_\theta}{\partial \theta}\right]$$
$$- \rho\frac{\partial \overline{v_r'^2}}{\partial r} - \frac{\rho}{r}\overline{v_r'^2} - \frac{\rho}{r}\frac{\partial(\overline{v_r'v_\theta'})}{\partial \theta} - \rho\frac{\partial(\overline{v_r'v_z'})}{\partial z} + \frac{\rho}{r}\overline{v_\theta'^2} + \rho\overline{g_r}$$

θ-component

$$\rho\left(\frac{\partial \overline{v}_\theta}{\partial t} + \overline{v}_r\frac{\partial \overline{v}_\theta}{\partial r} + \frac{\overline{v}_\theta}{r}\frac{\partial \overline{v}_\theta}{\partial \theta} + \overline{v}_z\frac{\partial \overline{v}_\theta}{\partial z} + \frac{\overline{v}_r\overline{v}_\theta}{r}\right) = \frac{-1}{r}\frac{\partial \overline{p}}{\partial \theta} + \mu\left[\frac{\partial^2 \overline{v}_\theta}{\partial r^2} + \frac{1}{r}\frac{\partial \overline{v}_\theta}{\partial r} + \frac{1}{r^2}\frac{\partial^2 \overline{v}_\theta}{\partial \theta^2} + \frac{\partial^2 \overline{v}_\theta}{\partial z^2} - \frac{\overline{v}_\theta}{r^2} + \frac{2}{r^2}\frac{\partial \overline{v}_r}{\partial \theta}\right]$$
$$- \frac{\rho}{r}\frac{\partial \overline{v_\theta'^2}}{\partial \theta} - \rho\frac{\partial(\overline{v_\theta'v_r'})}{\partial r} - \rho\frac{\partial(\overline{v_\theta'v_z'})}{\partial z} - \frac{2\rho}{r}\overline{v_\theta'v_r'} + \rho\overline{g_\theta}$$

z-component

$$\rho\left(\frac{\partial \overline{v}_z}{\partial t} + \overline{v}_r\frac{\partial \overline{v}_z}{\partial r} + \frac{\overline{v}_\theta}{r}\frac{\partial \overline{v}_z}{\partial \theta} + \overline{v}_z\frac{\partial \overline{v}_z}{\partial z}\right) = \frac{-\partial \overline{p}}{\partial z} + \mu\left[\frac{\partial^2 \overline{v}_z}{\partial r^2} + \frac{1}{r}\frac{\partial \overline{v}_z}{\partial r} + \frac{1}{r^2}\frac{\partial^2 \overline{v}_z}{\partial \theta^2} + \frac{\partial^2 \overline{v}_z}{\partial z^2}\right]$$
$$- \rho\frac{\partial \overline{v_z'^2}}{\partial z} - \rho\frac{\partial(\overline{v_r'v_z'})}{\partial r} - \frac{\rho}{r}\overline{v_r'v_z'} - \frac{\rho}{r}\frac{\partial(\overline{v_\theta'v_z'})}{\partial \theta} + \rho\overline{g_z}$$

When the equations in Tables 7.3 and 7.4 are compared, it is clear that the Reynolds-averaged Navier–Stokes equations (Table 7.4) are longer and contain more unknowns than the original equations written in terms of the instantaneous variables (Table 7.3). From the four unknown instantaneous quantities v_r, v_θ, v_z, and p in Table 7.3, in Table 7.4 there are ten time-averaged values—namely, $\overline{v}_r$, $\overline{v}_\theta$, $\overline{v}_z$, and $\overline{p}$ and the distinct cross-velocity terms $\overline{v_r'^2}$, $\overline{v_\theta'^2}$, $\overline{v_z'^2}$, $\overline{v_r'v_\theta'}$, $\overline{v_r'v_z'}$, and $\overline{v_\theta'v_z'}$. Therefore, the overall effect of the transformation procedure aimed at narrowing the scope of the analysis is that there are more unknowns than the total number of governing equations. This problem in fluid dynamics is referred to as the *closure problem*. It raises fundamental and intractable difficulties for solution of the transport equations over and above the usual complexities of numerical simulation and computation.

Note that the additional cross-velocity terms in the time-averaged equations of Table 7.4 correspond to the Reynolds stresses of Eq. (7.32), which represent the average momentum fluxes due to fluctuations in the turbulent velocity. The closure problem indicates that, as the mean flow is coupled to the turbulent fluctuations via these terms, in general, the mean flow properties cannot be calculated as if they were independent of the turbulence, that is, without knowing the Reynolds stresses. This undesirable situation is a consequence of the inadequacy of the mean flow as a representation of the full range of turbulent—flow behaviour. Estimating the Reynolds stresses, and thus gaining insight into the influence of turbulence on the mean flow, is one of the most important concepts in turbulence studies.

For CFD, closure of the time-averaged transport equations is necessary for solution of the mean flow properties. Further information is required to redefine the problem and establish a closed or solvable mathematical statement. The relationships used to achieve closure must replace certain vital features of turbulence that were eliminated from the equations by averaging. The task of identifying these relationships has inspired many attempts to model the local, time-averaged effects of turbulent fluctuations on the mean flow. *Turbulence modelling* is aimed at developing computational procedures of sufficient accuracy and generality to allow prediction of the Reynolds stresses and, therefore, estimation of the time-averaged properties of turbulent flow.

Classical Turbulence Models

Classical turbulence models are based on the time-averaged Navier–Stokes equations. As outlined in the preceding section, an important component of these equations is the mean momentum flux or Reynolds stress, which represents the effect of turbulence fluctuations on the mean flow. The purpose of classical turbulence modelling is to provide realistic numerical values for the Reynolds stresses by relating them to the mean velocity field in a physically consistent way, preferably without introducing significant additional mathematical complexity.

Many different turbulence models have been developed. A feature of several commonly used models is the concept of *eddy viscosity*. According to this concept, the Reynolds stresses and cross-velocity terms can be related to the mean velocity gradients using equations analogous to the relationship between shear stress and shear rate in laminar flow (Section 7.3). For example, by analogy with Newton's law of viscosity in Eq. (7.6), an expression for the Reynolds stress component $\overline{\tau'_{xy}}$ from Eq. (7.32) is:

$$\overline{\tau'_{xy}} = -\rho\overline{v'_y v'_x} = -\eta\left(\frac{\partial\overline{v_x}}{\partial y} + \frac{\partial\overline{v_y}}{\partial x}\right) \tag{7.40}$$

where η is the eddy viscosity. Equations similar to Eq. (7.40) can be written for the other Reynolds stress components in either the (x, y, z) or (r, θ, z) coordinate system.

Unlike the molecular viscosity μ in Eq. (7.6), η is not a property of the fluid. Instead, the value of η depends on the flow conditions, the position within the flow field, and time. Under fully turbulent conditions, $\eta \gg \mu$ and the contribution of molecular viscosity to the fluid stresses is negligible. Although the eddy viscosity concept expressed in Eq. (7.40) is applied routinely in CFD, it represents one of the crudest approaches to the closure problem and has significant theoretical defects. Nevertheless, in practice, it has proven to be a reasonably successful and convenient mathematical tool for simulation of a wide range of turbulent flows.

For analysis of flow in stirred tanks, the most commonly used turbulence model is the *$k-\varepsilon$ model*, where k represents the turbulence kinetic energy defined in Section 7.9.2 (Turbulence Kinetic Energy subsection) and ε is the rate of dissipation of turbulence kinetic energy defined in Section 7.9.2 (Rate of Dissipation of Turbulence Kinetic Energy subsection). The $k-\varepsilon$ model belongs to the class of *two-equation turbulence models*, in which separate transport equations are solved for two turbulence quantities, in this case k and ε. The transport equations represent the main processes causing changes in these variables—namely, turbulence production, transport, and dissipation. Because the model introduces additional partial differential equations, appropriate boundary conditions for k and ε must be applied; these boundary conditions are usually derived from experimental measurements of the flow field. The $k-\varepsilon$ model also incorporates the eddy viscosity concept to estimate the Reynolds stresses, and employs a supplementary equation relating the eddy viscosity to k and ε:

$$\eta = \rho C \frac{k^2}{\varepsilon} \tag{7.41}$$

where η is the eddy viscosity and C is one of five $k-\varepsilon$ model constants.

The standard $k-\varepsilon$ model assumes equilibrium between the different turbulence scales, implying that turbulence kinetic energy generated at the large-eddy scales is dissipated immediately by the small eddies. This is a restrictive assumption for stirred vessels, as eddies with a wide range of sizes occur between turbulence production and dissipation. In addition, the standard $k-\varepsilon$ model uses a single time scale to characterise turbulence dynamics, even though turbulence fluctuations occur with a spectrum of time scales. Because of these and other assumptions involved in the eddy viscosity hypothesis and in the transport equations for k and ε, the $k-\varepsilon$ model is known to perform less than satisfactorily in a number of flow situations, including where there is strong streamline curvature or vortex generation. The standard $k-\varepsilon$ model also cannot be integrated to a solid boundary without modification to account for turbulence damping in that region. However, despite these limitations, the $k-\varepsilon$ model is one of the most widely used and validated turbulence models. It has been applied successfully to many industrial flow situations and is incorporated into most commercially available CFD software packages. Various modified forms of the $k-\varepsilon$ model have also been developed to increase the accuracy and versatility of flow simulation.

Typical mean velocity flow fields obtained experimentally using particle image velocimetry, and from CFD analysis using a modified $k-\varepsilon$ model, are shown in Figure 7.25. There is a general similarity between the computed and experimental results, in both qualitative and quantitative terms. Yet, when CFD and experimental velocity profiles for stirred vessels are compared in detail, the computed predictions are often inadequate, especially in regions away from the impeller. This can occur irrespective of the mathematical model and numerical methods applied and the techniques used to simulate the impeller boundary conditions. Prediction of turbulence quantities such as the rate of dissipation of turbulence kinetic energy is also often poor, with different CFD methods yielding a wide range of values for the same system. At the present time, therefore, CFD results need

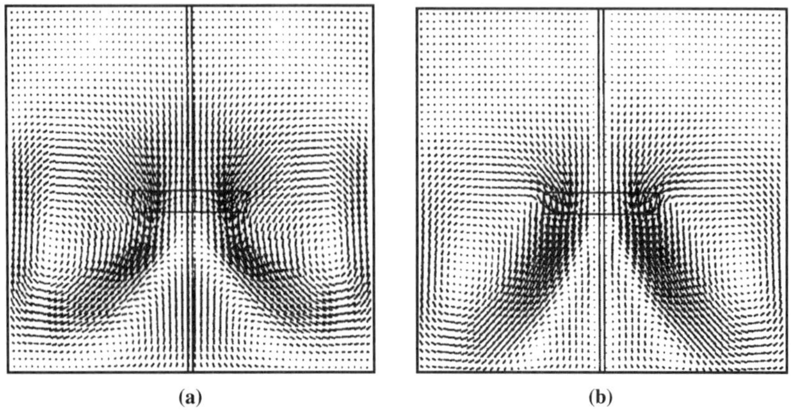

FIGURE 7.25 Mean velocity fields in a stirred tank agitated with a centrally located pitched-blade turbine impeller. (a) Experimental results measured using particle image velocimetry; (b) results from CFD analysis using a modified $k-\varepsilon$ model.

From J. Sheng, H. Meng, and R.O. Fox, 1998, Validation of CFD simulations of a stirred tank using particle image velocimetry data. Can. J. Chem. Eng. *76, 611–625.*

to be validated by comparison with experimental data before they can be considered reliable. Nevertheless, the technology for CFD is still improving. CFD methods suitable for application to more complex systems, such as non-Newtonian fluids, multiple-phase fluids, and reactive processes, are currently being developed.

SUMMARY OF CHAPTER 7

This chapter covers a wide range of topics in rheology and fluid dynamics. At the end of Chapter 7 you should:

- Know the difference between *laminar* and *turbulent* flow
- Understand how shear stresses develop in laminar and turbulent flow
- Be familiar with the concept of the *Reynolds number*
- Be able to describe how fluid *boundary layers* develop in terms of *viscous drag*
- Be able to define *viscosity* in terms of *Newton's law*
- Know what *Newtonian* and *non-Newtonian* fluids are, and the difference between viscosity for Newtonian fluids and *apparent viscosity* for non-Newtonian fluids
- Be familiar with the methods and instruments used to characterise the *rheology* of fermentation broths
- Know which factors affect broth viscosity
- Understand the basis for the statistical approach to analysis of turbulent flow
- Be familiar with terms and parameters used to characterise turbulent flow, such as *stationarity, turbulence intensity, turbulence kinetic energy, rate of dissipation of turbulence kinetic energy, homogeneous turbulence*, and *isotropic turbulence*

- Be able to describe the roles of the different *scales of turbulence* and the *energy cascade* between large and small eddies
- Know the essential features of *Kolmogorov's theory of isotropic turbulence*
- Understand the physical meaning of the *Reynolds stresses* and their importance in *classical turbulence modelling*
- Be familiar with the principles of *laser Doppler velocimetry* and *particle image velocimetry*
- Understand the scope, broad theoretical background, and general methods of application of *computational fluid dynamics* for analysis of fluid flow in bioprocessing

PROBLEMS

7.1 Conditions for turbulence

(a) When the flow rate of water in a garden hose is measured using the bucket-and-stopwatch method, it takes 75 seconds to fill a 5-litre bucket. If the inner diameter of the hose is 15 mm and the density and viscosity of the water are approximately 1000 kg m^{-3} and 1.0 cP, respectively, is flow in the hose laminar or turbulent?

(b) A 5-cm-diameter wooden spoon is used to stir chicken consommé in a saucepan using a regular circular motion at a speed of 1.5 revolutions per second. At the temperature used to warm the broth, the liquid density is 970 kg m^{-3} and the viscosity is 0.45 cP. In this unbaffled system, turbulent flow occurs at impeller Reynolds numbers above about 10^5. Is flow in the saucepan likely to be turbulent?

7.2 Rheology of fermentation broth

The fungus *Aureobasidium pullulans* is used to produce an extracellular polysaccharide by fermentation of sucrose. After a 120-hour fermentation, measurements of shear stress and shear rate are made using a rotating cylinder viscometer.

Shear stress (dyn cm^{-2})	Shear rate (s^{-1})
44.1	10.2
235.3	170
357.1	340
457.1	510
636.8	1020

(a) Plot the rheogram for this fluid.

(b) Determine the appropriate non-Newtonian parameters.

(c) Estimate the apparent viscosity at shear rates of:

(i) 15 s^{-1}

(ii) 200 s^{-1}

7.3 Rheology of yeast suspensions

Apparent viscosities for pseudoplastic cell suspensions at varying cell concentrations are measured using a coaxial cylinder rotary viscometer. The results are shown in the following table.

Cell concentration (%)	Shear rate (s^{-1})	Apparent viscosity (cP)
1.5	10	1.5
	100	1.5
3	10	2.0
	100	2.0
6	20	2.5
	45	2.4
10.5	10	4.7
	20	4.0
	50	4.1
	100	3.8
12	1.8	40
	4.0	30
	7.0	22
	20	15
	40	12
18	1.8	140
	7.0	85
	20	62
	40	55
21	1.8	710
	4.0	630
	7.0	480
	40	330
	70	290

Show on an appropriate plot how K and n vary with cell concentration.

7.4 Impeller viscometer

The rheology of a *Penicillium chrysogenum* broth is examined using an impeller viscometer. The density of the cell suspension is approximately 1000 kg m^{-3}. Samples of broth are stirred under laminar conditions using a Rushton turbine of diameter 4 cm in a glass beaker of diameter 15 cm. The average shear rate generated by the impeller is greater than the stirrer speed by a factor of about 10.2. When the stirrer shaft is attached to a device for measuring torque and rotational speed, the following results are recorded.

Stirrer speed (s^{-1})	Torque (N m)
0.185	3.57×10^{-6}
0.163	3.45×10^{-6}
0.126	3.31×10^{-6}
0.111	3.20×10^{-6}

Can the rheology be described using a power-law model? If so, evaluate K and n.

7.5 Vand equation

A correlation between viscosity and cell concentration is required for suspension cultures of the hemp plant, *Cannabis sativa*. Morphologically, the cells are uniform and close to spherical in shape, and the culture exhibits Newtonian rheology at moderate cell densities. The viscosity of the whole broth including cells is measured using a rotating cylinder viscometer. Samples of broth are then centrifuged at 3000 rpm for 3 min in graduated centrifuge tubes for estimation of the cell volume fraction.

Broth viscosity (cP)	Total volume of broth in centrifuge tube (ml)	Volume of liquid decanted from centrifuge tube (ml)
1.0	10.0	9.6
1.0	11.6	11.0
1.1	11.2	10.4
1.2	9.2	8.3
1.4	10.4	9.2
1.6	10.9	9.3
2.3	11.0	8.8

For all the samples tested, the viscosity of the decanted liquid is 0.9 cP. Can the Vand equation be used to predict the viscosity of this plant cell suspension?

7.6 Viscosity and cell concentration

The filamentous bacterium *Streptomyces levoris* produces pseudoplastic fermentation broths. The rheological properties of the broth are measured at different cell concentrations using a turbine impeller viscometer.

Cell concentration ($g\ l^{-1}$ dry weight)	Shear rate (s^{-1})	Shear stress (Pa)
6.0	5.0	1.0
	10	1.3
	20	1.8
	32	2.3
11.1	5.0	3.0
	10	4.3
	20	5.5
	30	6.7
15.5	5.0	9.2
	10	13.0
	20	17.4
	30	20.5
18.7	5.0	12.8
	10	16.9
	25	23.5

It is proposed to correlate the consistency index K and flow behaviour index n with cell concentration using expressions of the form:

$$K \text{ or } n = Ax^B$$

where A and B are constants and x is the cell concentration.

(a) Is this form of equation appropriate for K and n?

(b) If not, why not? If so, evaluate A and B to determine the equations for K and n as a function of x.

(c) Estimate the apparent viscosity of *S. levoris* broth if the cell concentration is 12.3 g l^{-1} and the shear rate is 8.5 s^{-1}.

7.7 Scale for turbulence dissipation

(a) The rate of turbulence energy dissipation in a stirred fermenter is 0.011 W. This dissipation is assumed to occur within a volume of approximately 200 cm^3. The density of liquid in the vessel is 1000 kg m^{-3} and the viscosity is 2.3×10^{-3} N s m^{-2}. Estimate the scale of the smallest eddies.

(b) Dextran is added to the medium at a concentration of 9.5% w/v. As a result, the liquid viscosity is increased to 1.5×10^{-2} N s m^{-2} while the change in density is negligible. How does this affect the scale of the smallest eddies?

7.8 Size of dissipating eddies

A fermenter contains 5 l of sterile culture medium. With the stirrer operating and the vessel fully insulated, it takes 1 hour for the temperature of the medium to rise by 1°C. The properties of the medium are similar to those of water.

(a) What is the power output of the stirrer?

(b) Estimate the scale of the dissipating eddies. What assumptions are involved in your answer?

References

[1] A.H.P. Skelland, Non-Newtonian Flow and Heat Transfer, John Wiley, 1967.

[2] C.W. Macosko, Rheology: Principles, Measurements, and Applications, VCH, 1994.

[3] J. Ferguson, Z. Kembłowski, Applied Fluid Rheology, Elsevier Applied Science, 1991.

[4] R.W. Whorlow, Rheological Techniques, Ellis Horwood, 1980.

[5] J.R. van Wazer, J.W. Lyons, K.Y. Kim, R.E. Colwell, Viscosity and Flow Measurement, John Wiley, 1963.

[6] H.A. Barnes, J.F. Hutton, K. Walters, An Introduction to Rheology, Elsevier, 1989.

[7] B. Metz, N.W.F. Kossen, J.C. van Suijdam, The rheology of mould suspensions, Adv. Biochem. Eng. 11 (1979) 103–156.

[8] J.A. Roels, J. van den Berg, R.M. Voncken, The rheology of mycelial broths, Biotechnol. Bioeng. 16 (1974) 181–208.

[9] J.J.T.M. Bongenaar, N.W.F. Kossen, B. Metz, F.W. Meijboom, A method for characterizing the rheological properties of viscous fermentation broths, Biotechnol. Bioeng. 15 (1973) 201–206.

[10] J.H. Kim, J.M. Lebeault, M. Reuss, Comparative study on rheological properties of mycelial broth in filamentous and pelleted forms, Eur. J. Appl. Microbiol. Biotechnol. 18 (1983) 11–16.

[11] P. Rapp, H. Reng, D.-C. Hempel, F. Wagner, Cellulose degradation and monitoring of viscosity decrease in cultures of *Cellulomonas uda* grown on printed newspaper, Biotechnol. Bioeng. 26 (1984) 1167–1175.

[12] T.P. Labuza, D. Barrera Santos, R.N. Roop, Engineering factors in single-cell protein production. I. Fluid properties and concentration of yeast by evaporation, Biotechnol. Bioeng. 12 (1970) 123–134.

[13] T. Berkman-Dik, M. Özilgen, T.F. Bozoğlu, Salt, EDTA, and pH effects on rheological behavior of mold suspensions, Enzyme Microbiol. Technol. 14 (1992) 944–948.

[14] F.H. Deindoerfer, E.L. Gaden, Effects of liquid physical properties on oxygen transfer in penicillin fermentation, Appl. Microbiol. 3 (1955) 253–257.
[15] L.-K. Ju, C.S. Ho, J.F. Shanahan, Effects of carbon dioxide on the rheological behavior and oxygen transfer in submerged penicillin fermentations, Biotechnol. Bioeng. 38 (1991) 1223–1232.
[16] H. Taguchi, S. Miyamoto, Power requirement in non-Newtonian fermentation broth, Biotechnol. Bioeng. 8 (1966) 43–54.
[17] F.H. Deindoerfer, J.M. West, Rheological examination of some fermentation broths, J. Biochem. Microbiol. Technol. Eng. 2 (1960) 165–175.
[18] C.M. Tuffile, F. Pinho, Determination of oxygen-transfer coefficients in viscous streptomycete fermentations, Biotechnol. Bioeng. 12 (1970) 849–871.
[19] A. LeDuy, A.A. Marsan, B. Coupal, A study of the rheological properties of a non-Newtonian fermentation broth, Biotechnol. Bioeng. 16 (1974) 61–76.
[20] M. Charles, Technical aspects of the rheological properties of microbial cultures, Adv. Biochem. Eng. 8 (1978) 1–62.
[21] A. Kato, S. Kawazoe, Y. Soh, Viscosity of the broth of tobacco cells in suspension culture, J. Ferment. Technol. 56 (1978) 224–228.
[22] R. Ballica, D.D.Y. Ryu, R.L. Powell, D. Owen, Rheological properties of plant cell suspensions, Biotechnol. Prog. 8 (1992) 413–420.
[23] J.-J. Zhong, T. Seki, S.-I. Kinoshita, T. Yoshida, Rheological characteristics of cell suspension and cell culture of *Perilla frutescens*, Biotechnol. Bioeng. 40 (1992) 1256–1262.
[24] F.H. Deindoerfer, J.M. West, Rheological properties of fermentation broths, Adv. Appl. Microbiol. 2 (1960) 265–273.
[25] J. Laine, R. Kuoppamäki, Development of the design of large-scale fermentors, Ind. Eng. Chem. Process Des. Dev. 18 (1979) 501–506.
[26] K.G. Tucker, C.R. Thomas, Effect of biomass concentration and morphology on the rheological parameters of *Penicillium chrysogenum* fermentation broths, Trans. IChemE 71C (1993) 111–117.
[27] G.L. Riley, K.G. Tucker, G.C. Paul, C.R. Thomas, Effect of biomass concentration and mycelial morphology on fermentation broth rheology, Biotechnol. Bioeng. 60 (2000) 160–172.
[28] U. Björkman, Properties and principles of mycelial flow: a tube rheometer system for fermentation fluids, Biotechnol. Bioeng. 29 (1987) 101–113.
[29] S.M. Kresta, P.E. Wood, The flow field produced by a pitched blade turbine: characterization of the turbulence and estimation of the dissipation rate, Chem. Eng. Sci. 48 (1993) 1761–1774.
[30] N.J. Fentiman, K.C. Lee, G.R. Paul, M. Yianneskis, On the trailing vortices around hydrofoil impeller blades, Trans. IChemE 77A (1999) 731–740.
[31] K.C. Lee, M. Yianneskis, Turbulence properties of the impeller stream of a Rushton turbine, AIChE J. 44 (1998) 13–24.
[32] M. Schäfer, M. Yianneskis, P. Wächter, F. Durst, Trailing vortices around a 45° pitched-blade impeller, AIChE J. 44 (1998) 1233–1246.
[33] J.O. Hinze, Turbulence, second ed., McGraw-Hill, 1975 (Chapter 1).

Suggestions for Further Reading

Introductory Fluid Mechanics

Vardy, A. (1990). *Fluid Principles*. McGraw-Hill.
White, F. M. (2011). *Fluid Mechanics* (7th ed.). McGraw-Hill.

Viscosity and Viscosity Measurement

See also references 1 through 9, 20, and 28.

Allen, D. G., & Robinson, C. W. (1990). Measurement of rheological properties of filamentous fermentation broths. *Chem. Eng. Sci., 45*, 37–48.
Atkinson, B., & Mavituna, F. (1991). *Biochemical Engineering and Biotechnology Handbook* (2nd ed., Chapter 11). Macmillan.
Olsvik, E., & Kristiansen, B. (1994). Rheology of filamentous fermentations. *Biotech. Adv., 12*, 1–39.

Turbulence

See also reference 33.

Kresta, S. (1998). Turbulence in stirred tanks: anisotropic, approximate, and applied. *Can. J. Chem. Eng., 76*, 563–576.

Lesieur, M. (1997). *Turbulence in Fluids* (3rd ed.). Kluwer Academic.

Mathieu, J., & Scott, J. (2000). *Introduction to Turbulent Flow*. Cambridge University Press.

Pope, S. B. (2000). *Turbulent Flows*. Cambridge University Press.

Wu, H., & Patterson, G. K. (1989). Laser-Doppler measurements of turbulent-flow parameters in a stirred mixer. *Chem. Eng. Sci., 44*, 2207–2221.

Turbulence Measurement Techniques

Adrian, R. J. (1991). Particle-imaging techniques for experimental fluid mechanics. *Annu. Rev. Fluid Mech., 23*, 261–304.

Coupland, J. M. (2000). Laser Doppler and pulsed laser velocimetry in fluid mechanics. *Topics Appl. Phys., 77*, 373–412.

Drain, L. E. (1980). *Laser Doppler Technique*. John Wiley.

Prasad, A. K. (2000). Particle image velocimetry. *Curr. Sci., 79*, 51–60.

Computational Fluid Dynamics

Jenne, M., & Reuss, M. (1999). A critical assessment on the use of $k-\varepsilon$ turbulence models for simulation of the turbulent liquid flow induced by a Rushton-turbine in baffled stirred-tank reactors. *Chem. Eng. Sci., 54*, 3921–3941.

Revstedt, J., Fuchs, L., & Trägårdh, C. (1998). Large eddy simulations of the turbulent flow in a stirred reactor. *Chem. Eng. Sci., 53*, 4041–4053.

Sahu, A. K., Kumar, P., Patwardhan, A. W., & Joshi, J. B. (1999). CFD modelling and mixing in stirred tanks. *Chem. Eng. Sci., 54*, 2285–2293.

Shaw, C. T. (1992). *Using Computational Fluid Dynamics*. Prentice Hall.

Speziale, C. G. (1991). Analytical methods for the development of Reynolds-stress closures in turbulence. *Annu. Rev. Fluid Mech., 23*, 107–157.

Versteeg, H. K., & Malalasekera, W. (2007). *Introduction to Computational Fluid Dynamics: The Finite Volume Method* (2nd ed.). Prentice Hall.

Wilcox, D. C. (1998). *Turbulence Modeling for CFD* (2nd ed.). DCW Industries.

CHAPTER

8

Mixing

The physical operation of mixing can determine the success of bioprocesses. In fermentations, single- and multiple-phase mixing occurs in fluids with a range of rheologies. Mixing controls the access of cells to dissolved nutrients and oxygen, and plays a critical role in controlling the culture temperature. The equipment used for mixing has a significant effect on agitation efficiency, power requirements, and operating costs. A consequence of mixing operations is the development of hydrodynamic forces in the fluid. These forces are responsible for important processes in fermenters such as bubble break-up and dispersion; however, cell damage can also occur and must be avoided. Problems with mixing are a major cause of productivity loss after commercial scale up of bioprocesses.

This chapter draws on material introduced in Chapter 7 about fluid properties and flow behaviour. In turn, as mixing underpins effective heat and mass transfer in bioprocesses, this chapter provides the foundations for detailed treatment of these subjects in Chapters 9 and 10.

8.1 FUNCTIONS OF MIXING

Mixing is a physical operation that reduces nonuniformities in fluid by eliminating gradients of concentration, temperature, and other properties. Mixing is accomplished by interchanging material between different locations to produce a mingling of components. If a system is perfectly mixed, there is a random, homogeneous distribution of system properties. Mixing is used in bioprocesses to:

- Blend soluble components of liquid media such as sugars
- Disperse gases such as air through liquids in the form of small bubbles
- Maintain suspension of solid particles such as cells and cell aggregates
- Where necessary, disperse immiscible liquids to form an emulsion or suspension of fine droplets
- Promote heat transfer to or from liquids

Mixing is one of the most important operations in bioprocessing. To create an optimal environment for cell culture, bioreactors must provide the cells with access to all substrates, including oxygen in aerobic cultures. It is not enough to just fill the fermenter with nutrient-rich medium; unless the culture is mixed, zones of nutrient depletion will develop as the cells rapidly consume the materials they need within their local environment. This problem is heightened if mixing does not maintain a uniform suspension of biomass; substrate concentrations can quickly drop to zero within layers of settled cells. We rely on good mixing to distribute any material added during the fermentation, such as fresh medium to feed the cells or concentrated acid or alkali to control the culture pH. If these materials are not mixed rapidly throughout the reactor, their concentration can build up to toxic levels near the feed point with deleterious consequences for the cells in that region. Another important function of mixing is heat transfer. Bioreactors must be capable of transferring heat to or from the broth rapidly enough so that the desired temperature is maintained. Cooling water is used to take up excess heat from fermentations; the rate of heat transfer from the broth to the cooling water depends on mixing conditions.

Mixing can be achieved in many different ways. In this chapter we will concentrate on the most common mixing technique in bioprocessing: mechanical agitation using an impeller.

8.2 MIXING EQUIPMENT

Mixing is carried out most often in cylindrical stirred tanks, such as that shown in Figure 8.1. *Baffles*, which are vertical strips of metal mounted against the wall of the tank, are installed to reduce gross vortexing and swirling of the liquid. Mixing is achieved using an *impeller* mounted on a centrally located *stirrer shaft*. The stirrer shaft is driven rapidly by the *stirrer motor*; the effect of the rotating impeller is to pump the liquid and create a regular flow pattern. Liquid is forced away from the impeller, circulates through the vessel, and periodically returns to the impeller region. In gassed stirred tanks such as bioreactors used for aerobic culture, gas is introduced into the vessel by means of a *sparger* located beneath the impeller.

The equipment chosen for mixing operations exerts a significant influence on the outcome of the process. Aspects of this equipment are outlined in the following sections. The operating characteristics of different impellers are described in detail in Section 8.4.

8.2.1 Vessel Geometry and Liquid Height

The shape of the base of stirred tanks affects the efficiency of mixing. Several base shapes are shown in Figure 8.2. If possible, the base should be rounded at the edges rather than flat; this eliminates sharp corners and pockets into which fluid currents may not penetrate, and discourages the formation of stagnant zones. The energy required to suspend solids in stirred tanks is sensitive to the shape of the vessel base: depending on the type of impeller and the flow pattern generated, the modified geometries shown in Figure 8.2(b) through (e) can be used to enhance particle suspension compared with the flat-bottom tank of Figure 8.2(a). In contrast, sloping sides or a conical base such as that shown in

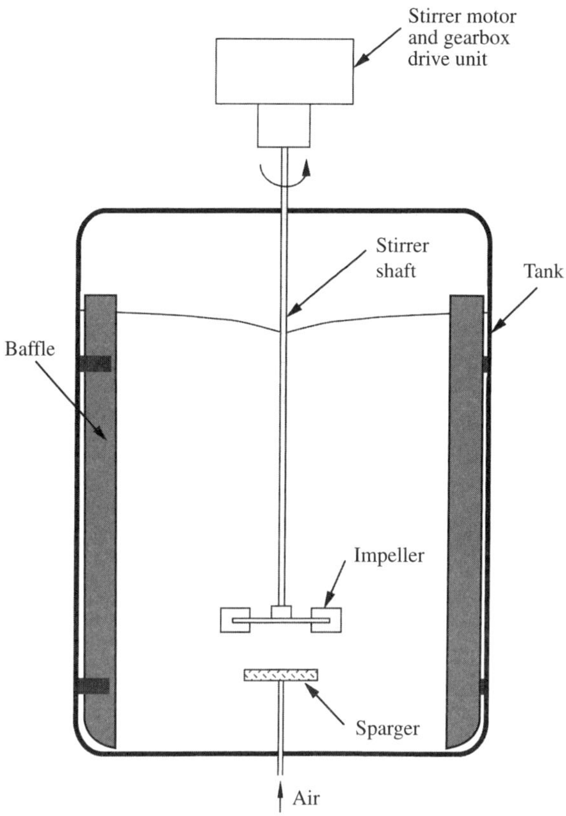

FIGURE 8.1 Typical configuration of a stirred tank.

Figure 8.2(f) promotes settling of solids and should be avoided if solids suspension is required.

Other geometric specifications for stirred tanks are shown in Figure 8.3. For efficient mixing with a single impeller of diameter D_i equal to a 1/4 to 1/2 the tank diameter D_T, the height of liquid in the tank H_L should be no more than 1.0 to 1.25 D_T. Because the intensity of mixing decreases quickly as fluid moves away from the impeller zone, large volumes of liquid in the upper parts of the vessel distant from the impeller are difficult to mix and should be avoided.

Another aspect of vessel geometry influencing mixing efficiency is the clearance C_i between the impeller and the lowest point of the tank floor (Figure 8.3). This clearance affects solids suspension, gas bubble dispersion, and hydrodynamic stability. In most practical stirring operations, C_i is within the range 1/6 to 1/2 the tank diameter.

8.2.2 Baffles

Baffles are standard equipment in stirred tanks. They assist mixing and create turbulence in the fluid by breaking up the circular flow generated by rotation of the impeller.

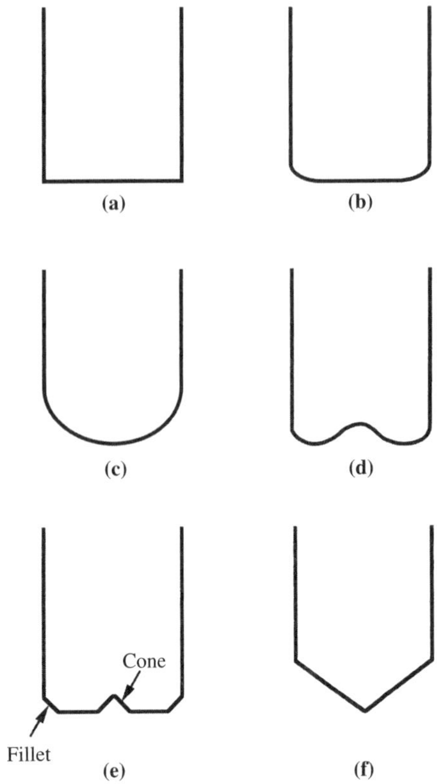

FIGURE 8.2 Different profiles for the base of stirred vessels: (a) flat; (b) dished; (c) round; (d) contoured; (e) cone-and-fillet; (f) conical.

Baffles are attached to the inside vertical walls of the tank by means of welded brackets. Four equally spaced baffles are usually sufficient to prevent liquid swirling and vortex formation. The optimum baffle width W_{BF} depends on the impeller design and fluid viscosity, but is of the order 1/10 to 1/12 the tank diameter. For clean, low-viscosity liquids, baffles are attached perpendicular to the wall as illustrated in Figure 8.4(a). Alternatively, as shown in Figures 8.4(b) and (c), baffles may be mounted away from the wall with clearance $C_{BF} \approx 1/50\ D_T$, or set at an angle. These arrangements prevent the development of stagnant zones and sedimentation along the inner edge of the baffle during mixing of viscous fluids or fluids containing suspended cells or particles.

8.2.3 Sparger

There exists a large variety of sparger designs. These include simple open pipes, perforated tubes, porous diffusers, and complex two-phase injector devices. *Point spargers*, such as open pipe spargers, release bubbles at only one location in the vessel. Other sparger designs such as *ring spargers* have multiple gas outlets so that bubbles are released simultaneously from various locations. Bubbles leaving the sparger usually fall within a relatively narrow size range depending on the sparger type. However, as the bubbles rise

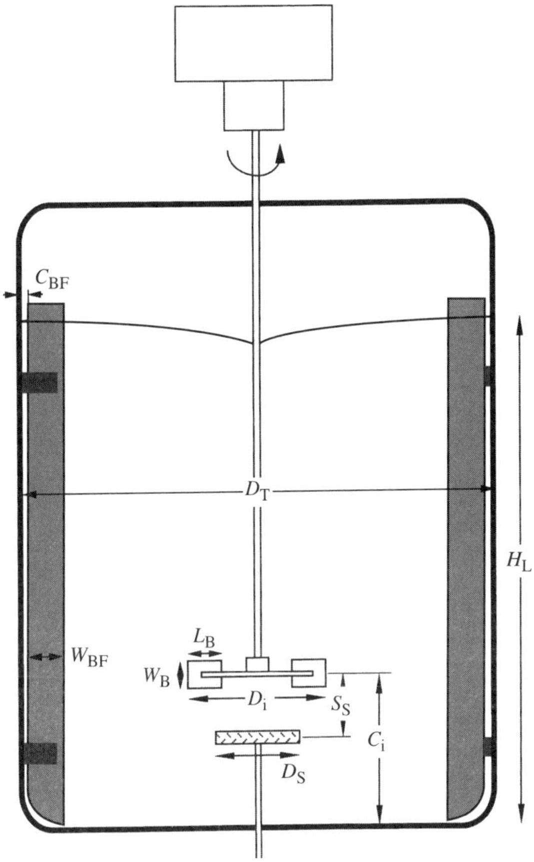

C_{BF} = baffle clearance
C_i = impeller clearance
D_i = impeller diameter
D_S = sparger diameter
D_T = tank diameter
H_L = liquid height
L_B = blade length
S_S = sparger separation
W_B = blade width
W_{BF} = baffle width

FIGURE 8.3 Some geometric specifications for a stirred tank.

from the sparger into the impeller zone, they are subjected to very high shear forces from operation of the stirrer that cause bubble break-up. The resulting small bubbles are flung out by the impeller into the bulk liquid for dispersion throughout the vessel. Although the type of sparger used has a relatively minor influence on the mixing process in most stirred tanks, the diameter D_S of large ring spargers and the separation S_S between the sparger and impeller (Figure 8.3) can have an important influence on the efficiency of gas dispersion.

8.2.4 Stirrer Shaft

The primary function of the stirrer shaft is to transmit *torque* from the stirrer motor to the impeller. Torque is the tendency of a force to cause an object to rotate. The magnitude

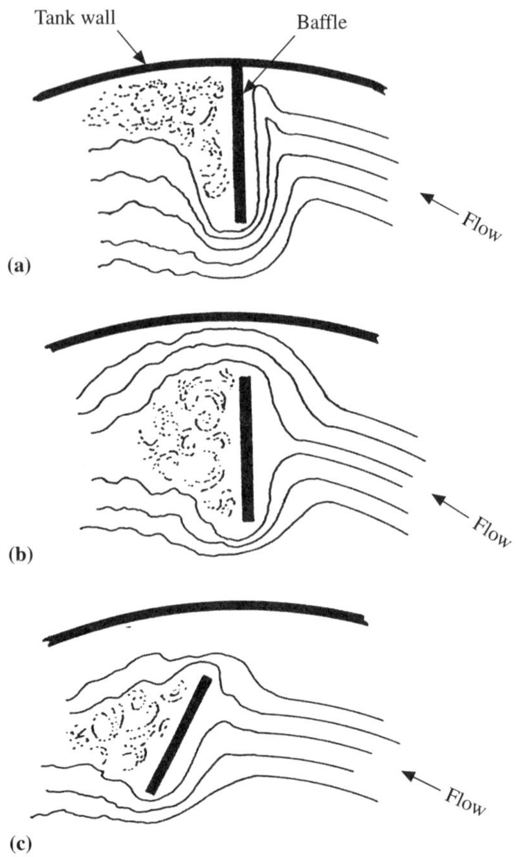

FIGURE 8.4 Baffle arrangements: (a) baffles attached to the wall for low-viscosity liquids; (b) baffles set away from the wall for moderate-viscosity liquids; (c) baffles set away from the wall and at an angle for high-viscosity liquids.
From F.A. Holland and F.S. Chapman, 1966, Liquid Mixing and Processing in Stirred Tanks, *Reinhold, New York.*

of the torque around the shaft axis is related to the power required for operation of the impeller. The stirrer shaft also performs other mechanical functions: it resists the bending forces created by the impeller, it limits any lateral deflections, and it supports the impeller weight. These functions must all be achieved without excessive vibration.

In typical mixing operations, the impeller is attached to a vertical stirrer shaft that passes from the motor through the top of the vessel. However, when headplate access is at a premium because of other devices and instruments located on top of the tank, or if a shorter shaft is required to alleviate mechanical stresses (e.g., when mixing viscous fluids), the stirrer shaft may be designed to enter through the base of the vessel. The vessel configuration for a bottom-entering stirrer is shown in Figure 8.5. The main disadvantage of bottom-entering stirrers is the increased risk of fluid leaks due to failure or wear of the seals between the rotating stirrer shaft and the vessel floor.

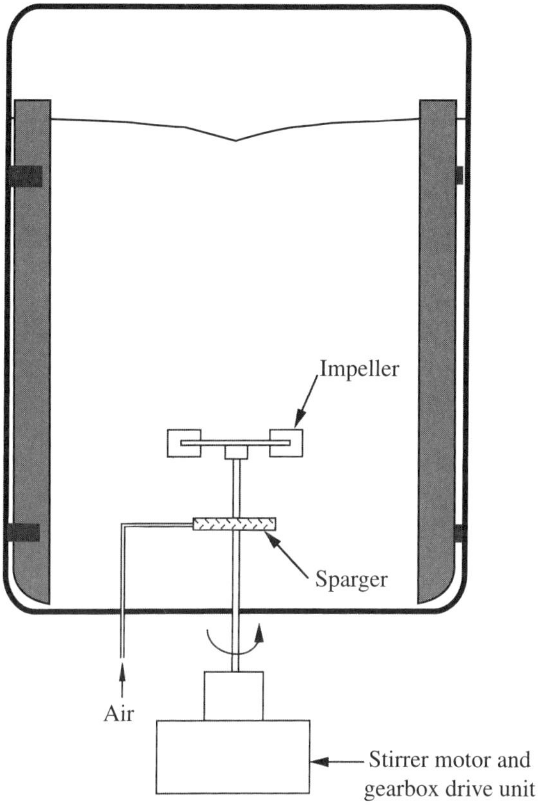

FIGURE 8.5 Stirred tank configuration for a bottom-entry stirrer.

8.3 FLOW PATTERNS IN STIRRED TANKS

The liquid flow pattern established in stirred tanks depends on the impeller design, the size and geometry of the vessel, and the properties of the fluid. There are three directional elements to the flow: *rotational flow* (i.e., around the stirrer shaft), *radial flow* (i.e., from the central axis out to the sides of the tank and back again), and *axial flow* (i.e., up and down the height of the vessel). An effective mixer will cause motion in all three directions; however, radial and axial flows generated at the impeller are primarily responsible for bulk mixing. Impellers are broadly classified as *radial-flow impellers* or *axial-flow impellers* depending on the direction of the flow leaving the impeller; some impellers have both radial- and axial-flow characteristics. When gas is sparged into the tank, two-phase flow patterns are created. The characteristics of this flow depend mainly on the stirrer speed, the gas flow rate, and the fluid properties.

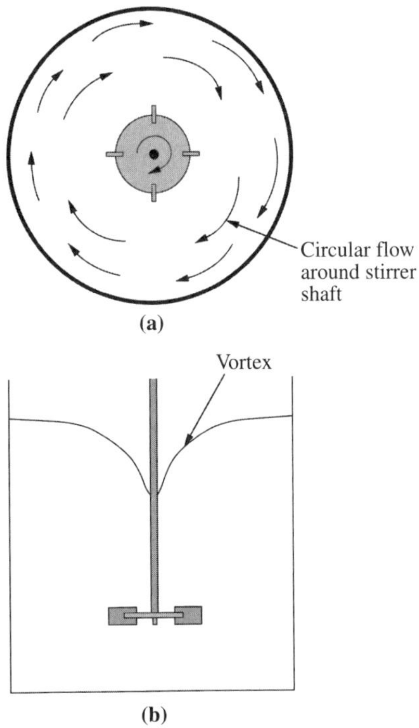

FIGURE 8.6 (a) Circular flow in an unbaffled stirred tank viewed from above. (b) Vortex formation during circular flow.

8.3.1 Rotational Flow

Because stirring is rotational in action, all impellers generate circular flow of liquid around the stirrer shaft as illustrated in Figure 8.6(a). However, simple rotational flow is disadvantageous for mixing and should be avoided as much as possible. In circular flow, liquid moves in a streamline fashion and there is little top-to-bottom mixing between different heights in the vessel. Circular flow also leads to vortex development as shown in Figure 8.6(b). At high impeller speeds, the vortex may reach down to the impeller so that gas from the surrounding atmosphere is drawn into the liquid. This is undesirable as it produces very high mechanical stresses in the stirrer shaft, bearings, and seals. Attenuating circular flow has a high priority in design of mixing systems. It is usually minimised by installing baffles to interrupt the rotational flow pattern and create turbulence in the fluid.

8.3.2 Radial Flow

Radial or horizontal flow is generated by impellers with blades aligned parallel to the stirrer shaft. A typical liquid circulation pattern set up by a high-speed radial-flow impeller is shown in Figure 8.7. Liquid is driven radially from the impeller towards the walls of the tank where it divides into two streams, one flowing up to the top of the tank and the other flowing down to the bottom. These streams eventually reach the central axis of the tank and are drawn back to the impeller. Radial-flow impellers also set up circular flow that must be reduced by baffles.

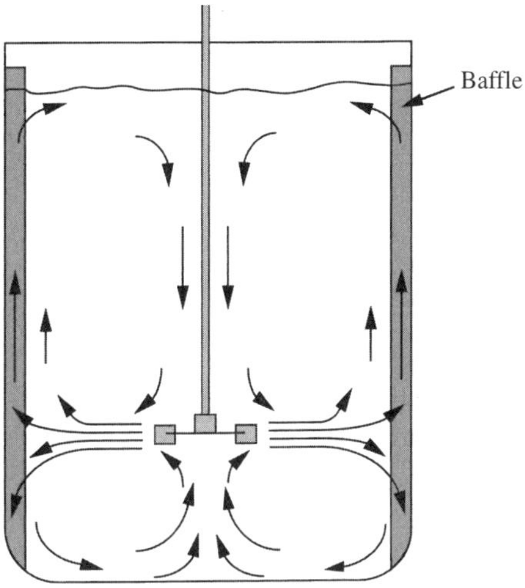

FIGURE 8.7 Flow pattern produced by a radial-flow impeller in a baffled tank.

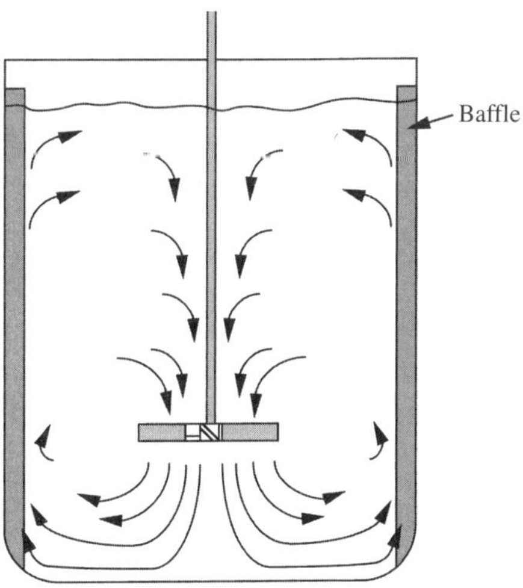

FIGURE 8.8 Flow pattern produced by an axial-flow impeller in a baffled tank.

8.3.3 Axial Flow

Axial flow is necessary for top-to-bottom mixing in stirred tanks. Axial flow is generated by impellers with inclined or pitched blades that make an angle of less than 90° with the plane of rotation. Axial flow is particularly useful when strong vertical currents are required. For example, if the fluid contains solids, a strong downward flow of liquid leaving the impeller will discourage settling at the bottom of the tank. The type of flow pattern set up by a typical axial-flow impeller is illustrated in Figure 8.8. Fluid leaving the impeller is driven at a downward angle until it is deflected from the floor of the vessel. It then spreads out over the floor and flows up along the walls before being drawn back to the impeller. As rotational flow is also generated by axial-flow impellers, baffles are required.

Axial-flow impellers are most commonly operated to generate downward flow of fluid leaving the impeller blades, as illustrated in Figure 8.8. However, if the direction of impeller rotation is reversed, axial-flow impellers may be applied for upward pumping, but this mode of operation is not often used.

8.3.4 Gas Flow Patterns

Sparging stirred tanks with gas creates two-phase flow patterns of bubbles in liquid. Different bubble distributions develop depending on the relative rates of gas flow and stirring. At high gassing rates or low stirrer speeds, the impeller is surrounded by gas and is unable to pump effectively, indicating that the gas-handling capacity of the impeller has been exceeded. This condition is called *impeller flooding*. Flooding should be avoided because an impeller blanketed by gas no longer contacts the liquid properly, resulting in poor mixing and gas dispersion. As the stirrer speed is increased or the gas flow rate reduced, the impeller blades start to process the gas and bubbles are dispersed towards the walls of the tank. This condition is known as *impeller loading*. At even higher stirrer speeds or lower gas flow rates, *complete gas dispersion* is achieved below as well as above the impeller. Complete gas dispersion with homogeneous distribution of gas to all parts of the vessel is the desirable operating condition.

As the stirrer speed is raised above that required to prevent impeller flooding, *gas recirculation* occurs. As well as rising upwards from the sparger to the liquid surface, gas bubbles are increasingly recirculated around the tank via the impeller. The extent of gas recirculation depends on the type of impeller and the stirrer speed, but can become very high. Under these conditions, the total flow rate of gas entering the impeller region at any given time is significantly greater than that supplied directly from the sparger.

Different impellers vary considerably in their ability to handle high gas flow rates without flooding. A parameter giving a rough indication of the tendency of impellers to flood is the impeller *solidity ratio*. The solidity ratio is equal to the projected blade area divided by the area of the circle swept out by the impeller during rotation. This is illustrated in Figure 8.9 for an axial-flow impeller with a relatively low solidity ratio. Impellers with low

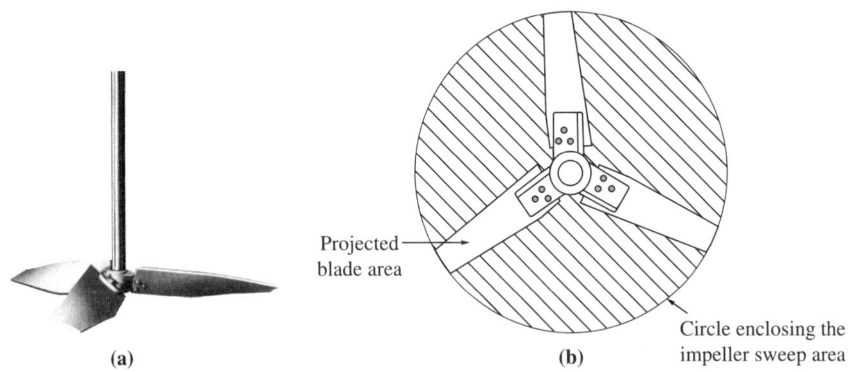

FIGURE 8.9 Illustration of the solidity ratio for an axial-flow impeller.
(a) Low-solidity-ratio Lightnin A310 hydrofoil impeller. *Photograph provided courtesy of Lightnin Mixers, Australia.*
(b) Diagram showing the projected blade area and impeller sweep area. *Adapted from J.Y. Oldshue, 1989, Fluid mixing in 1989.* Chem. Eng. Prog. *85(5), 33–42.*

solidity ratio tend to flood at lower gas velocities than impellers with high solidity ratio operated at the same stirrer speed. Impellers with solidity ratios greater than 90% have been developed for improved gas handling.

8.4 IMPELLERS

Many different types of impeller are available for mixing applications. A small selection is illustrated in Figure 8.10. The choice of impeller depends on several factors, including the liquid viscosity, the need for turbulent shear flows (e.g., for bubble break-up and gas dispersion), and whether strong liquid currents are required. The recommended viscosity ranges for a number of common impellers are indicated in Figure 8.11. Impellers can also be classified broadly depending on whether they produce high levels of turbulence, or whether they have a strong pumping capacity for generation of large-scale flow currents. Both functions are required for good mixing but they usually do not work together. The characteristics of several impellers in terms of their turbulence- and flow-generating properties are indicated in Figure 8.12.

Typically, mixing in fermenters is carried out using *turbines* or *propellers*. These impellers are described in detail in Sections 8.4.1 through 8.4.4. Turbines and propellers are *remote-clearance impellers*; this means they have diameters of 1/4 to 2/3 the tank diameter, thus allowing considerable clearance between the rotating impeller and the vessel walls. They are operated at relatively high speeds to generate impeller tip velocities of the order of 3 m s^{-1}. As indicated in Figure 8.11, turbines and propellers are recommended for mixing liquids with viscosities between 1 and about 10^4 centipoise, which includes most fermentation broths. From Figure 8.12, compared with the broad variety of other stirring devices available, turbines and propellers generate moderate-to-high levels of turbulence while retaining significant pumping capacity.

When remote-clearance agitators are applied in low-viscosity fluids, a turbulent region of high shear and rapid mixing is generated near the impeller. This high-shear region is responsible for bubble break-up in sparged systems. Because the mixing process should

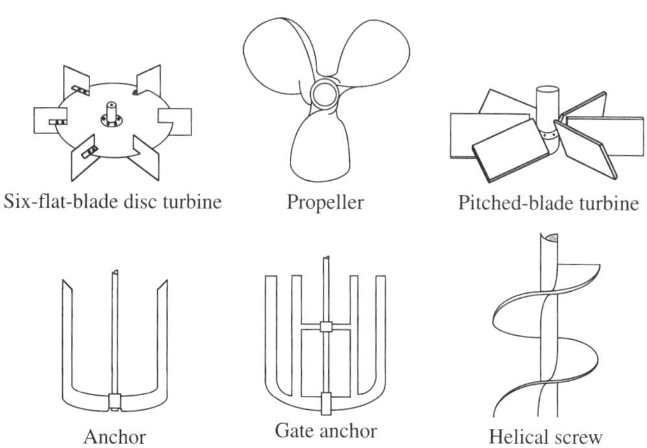

FIGURE 8.10 Selected impeller designs.

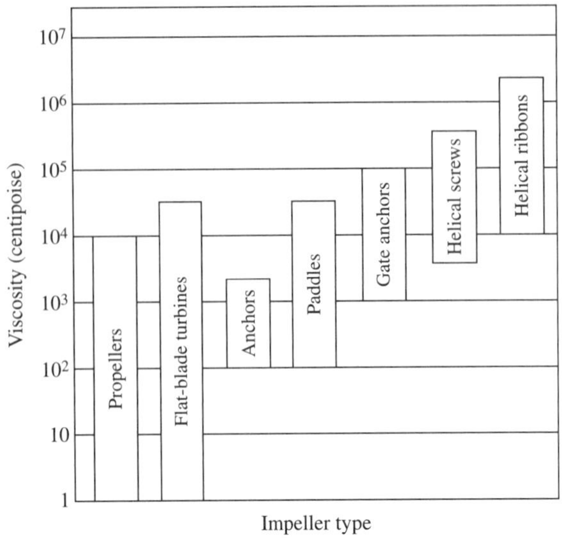

FIGURE 8.11 Viscosity ranges for different impellers.
From F.A. Holland and F.S. Chapman, 1966, Liquid Mixing and Processing in Stirred Tanks, *Reinhold, New York.*

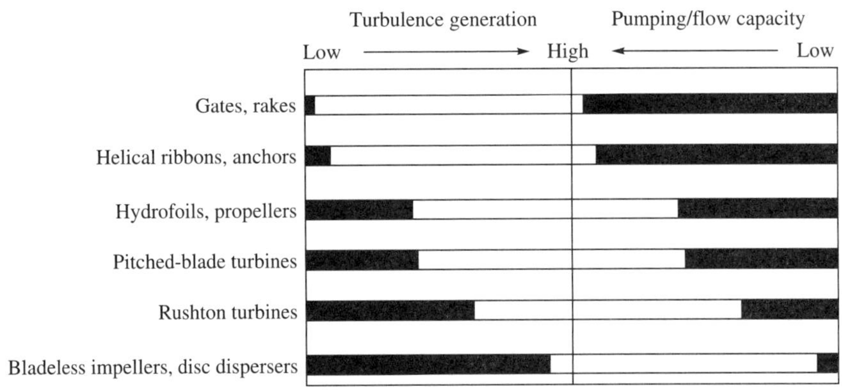

FIGURE 8.12 Characteristics of different impellers for generation of turbulence and liquid pumping.

involve fluid from all parts of the vessel, the impeller must also generate circulatory currents with sufficient velocity to carry material from the impeller to the furthermost regions of the tank and back again. In viscous fluids, it is often impossible for mechanical or economic reasons to rotate the impeller fast enough to generate turbulence; instead, impellers for viscous mixing are designed to provide maximum bulk movement or turnover of material. As indicated in Figures 8.11 and 8.12, impellers with high flow capacity suitable for mixing high-viscosity fluids include *gate, anchor*, and *helical stirrers*. Examples of these impellers are shown in Figure 8.10. All are large agitators installed with small wall clearance (around 1%–5% of the tank diameter) and operated at low stirrer speeds (5–20 rpm) to generate bulk fluid currents. For viscous fluids or when high shear rates must be avoided, slow-speed, low-turbulence, high-flow impellers are preferred to high-speed, high-turbulence, low-flow impellers. Although gate, anchor, and helical stirrers are not often applied for mixing in bioreactors because they do not disperse air bubbles adequately for oxygen supply to the cells, they may be used for other applications in bioprocessing such as blending viscous slurries, pastes, or gums.

Several impellers used in industrial fermentations are described in the following sections.

8.4.1 Rushton Turbine

The most frequently used impeller in the fermentation industry is the six-flat-blade disc-mounted turbine shown in Figures 8.10 and 8.13. This impeller is also known as the *Rushton turbine*. The Rushton turbine has been the impeller of choice for fermentations since the 1950s, largely because it has been well studied and characterised, and because it is very effective for gas dispersion. Although Rushton turbines of diameter one-third the tank diameter have long been used as standard hardware for aerobic fermentations, in recent years it has been recognised that larger impellers of size up to one-half the tank diameter provide considerable benefits for improved mixing and gas distribution.

Without Gassing

A typical mean velocity vector plot for a Rushton turbine is shown in Figure 8.14. In velocity vector plots, the length and direction of the arrows indicate the magnitude and direction of the velocities at discrete locations in the fluid. The velocities in Figure 8.14 were measured using laser Doppler velocimetry—see Section 7.9.3 (Laser Doppler Velocimetry subsection). Figure 8.14 represents the mean liquid flow pattern; as described in Section 7.9.1 (Mean and Fluctuating Velocities subsection), turbulent flow in agitated tanks is characterised by highly chaotic secondary fluid motion and fluctuating velocity fields that disappear when instantaneous flow velocities are averaged over time.

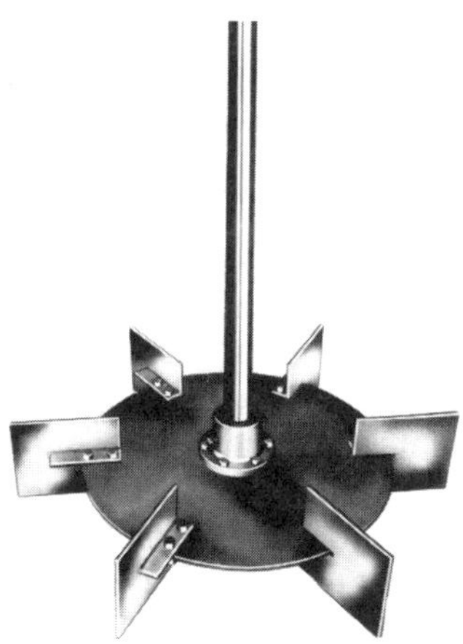

FIGURE 8.13 Rushton turbine.

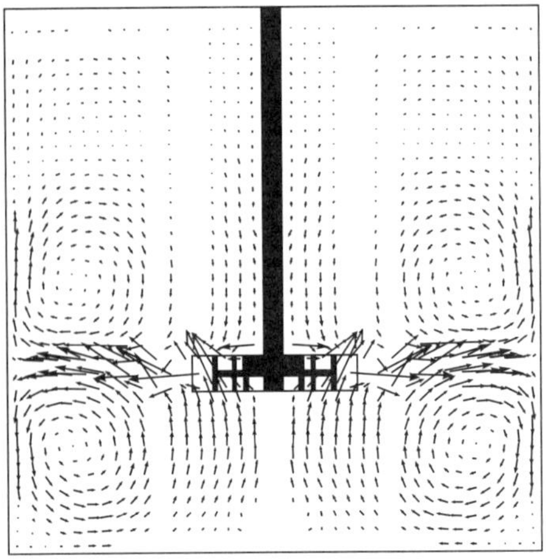

FIGURE 8.14 Typical mean velocity vector plot for a Rushton turbine. The velocities were measured using laser Doppler velocimetry.
Adapted from M. Schäfer, M. Höfken, and F. Durst, 1997, Detailed LDV measurements for visualization of the flow field within a stirred-tank reactor equipped with a Rushton turbine. Trans. IChemE *75A, 729–736.*

As indicated in Figure 8.14, the Rushton turbine is a radial-flow impeller. It generates a jet of high-speed flow directed radially outwards from the impeller; this jet entrains the surrounding fluid to form the impeller discharge stream. The discharge stream slows down as it approaches the tank wall and splits into two sections to create upper and lower circulatory flows. These circulatory currents traverse the remainder of the tank before returning directly to the impeller or becoming entrained again in the impeller outflow. The bulk flow in the vessel therefore comprises two large ring vortices, one above and one below the impeller. For impeller off-bottom clearances of less than one-half the liquid height, liquid velocities in the lower ring vortex are somewhat stronger than those in the upper circulatory stream, which traverses a greater distance into the upper reaches of the vessel during each circuit. Under these conditions, the return axial flow from beneath the impeller can be sufficiently strong to affect the radial discharge pattern, with the result that the outflow issuing from the impeller blades can be inclined slightly upwards instead of purely horizontal, as illustrated in Figure 8.14.

Compared with other types of turbine, Rushton impellers have a low pumping or circulatory capacity per unit power consumed. This is due mainly to a relatively high power requirement, as described in Section 8.5.1. The pumping capacity of impellers is discussed further in Section 8.7.

Operation of the Rushton turbine is characterised by the formation of *two high-speed roll and trailing vortices* in the liquid behind the horizontal edges of each flat blade of the impeller, as shown in Figure 8.15. These vortices play a critical role in determining the performance of the impeller. Most of the mixing in stirred vessels takes place near the vortices issuing from the impeller blades. As discussed further in the next section, the trailing vortex system is responsible for gas dispersion in aerated vessels. It is also the most important flow mechanism for turbulence generation. Steep velocity gradients are associated with trailing vortices; however, as the vortices move out from the blades they lose their identities and break down, thus providing a major source of turbulence in the fluid. The formation of trailing vortices affects the distribution of power dissipation in stirred tanks, which has consequences for phenomena such as cell damage in bioreactors. Fluid entrained by the vortices comprises much of the radial discharge stream generated by Rushton turbines.

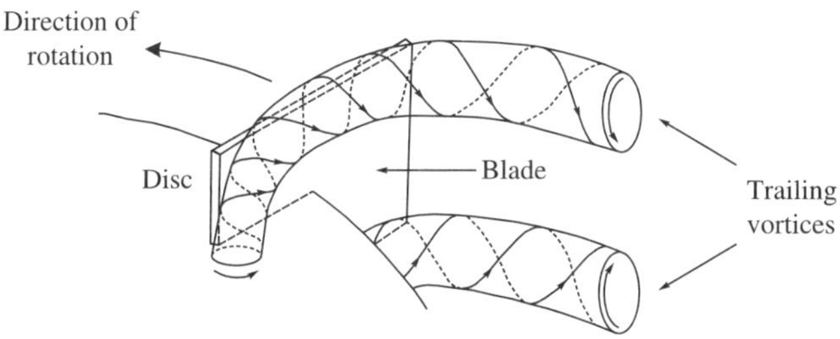

FIGURE 8.15 Roll and trailing vortices generated behind the blade of a Rushton turbine.
From K. van't Riet and J.M. Smith, 1975, The trailing vortex system produced by Rushton turbine agitators. Chem. Eng. Sci. *30, 1093–1105.*

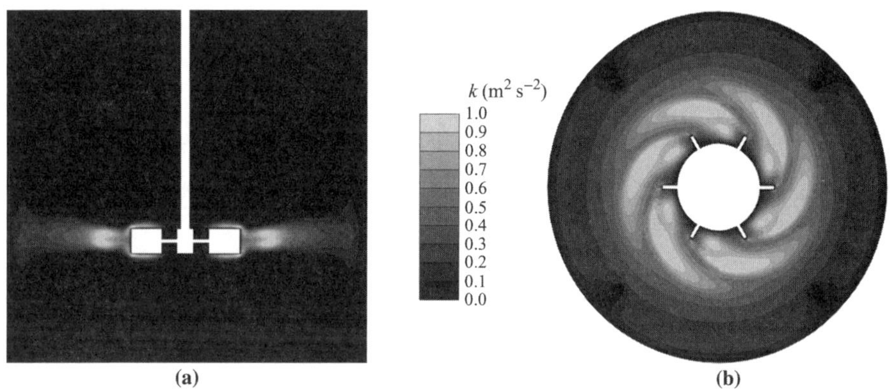

FIGURE 8.16 Distribution of k, the turbulence kinetic energy per unit mass of fluid, in a tank stirred by a Rushton turbine. The data were determined using computational fluid dynamics (CFD) modelling. (a) Distribution of k in the vertical plane; (b) distribution of k in the horizontal plane at impeller height, with the impeller rotating clockwise.
Adapted from K. Ng, N.J. Fentiman, K.C. Lee, and M. Yianneskis, 1998, Assessment of sliding mesh CFD predictions and LDA measurements of the flow in a tank stirred by a Rushton impeller. Trans. IChemE *76A, 737–747. Images provided courtesy of M. Yianneskis, King's College, London.*

Even if the impeller Reynolds number Re_i (Section 7.2.3) is high, indicating that flow is in the turbulent regime, the turbulence intensity in stirred vessels is far from uniform or randomly distributed. Figure 8.16 shows the distribution of turbulence kinetic energy per unit mass of fluid, k (Section 7.9.2, Turbulence Kinetic Energy subsection), in the vertical and horizontal planes of a tank stirred with a Rushton turbine. As illustrated in Figure 8.16(a), the highest values of k occur in the outwardly flowing liquid jet leaving the impeller blades: in this zone, k is at least an order of magnitude greater than in the remainder of the vessel. Figure 8.16(b) shows the turbulence kinetic energy distribution in the horizontal plane of the tank at the height of the impeller. The six plumes of elevated kinetic energy stretching out from the blades identify the trailing vortices generated by the impeller and indicate the extent of their radial spread. The vortices and associated turbulence dominate the horizontal plane of fluid near the impeller; however, these high levels of turbulence kinetic energy are not transmitted very far above or below the impeller, as indicated in Figure 8.16(a).

The distribution of the rate of dissipation of turbulence kinetic energy per unit mass, ε (Section 7.9.2, Rate of Dissipation of Turbulence Kinetic Energy subsection), in a tank

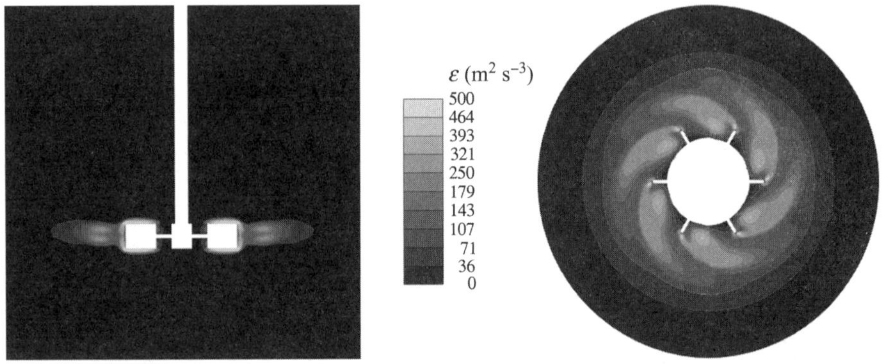

(a) (b)

FIGURE 8.17 Distribution of ε, the rate of dissipation of turbulence kinetic energy per unit mass, in a tank stirred by a Rushton turbine. The data were determined using computational fluid dynamics (CFD) modelling. (a) Distribution of ε in the vertical plane; (b) distribution of ε in the horizontal plane at impeller height, with the impeller rotating clockwise.
Adapted from K. Ng and M. Yianneskis, 2000, Observations on the distribution of energy dissipation in stirred vessels. Trans. IChemE *78A, 334–341. Images provided courtesy of M. Yianneskis, King's College, London.*

stirred by a Rushton turbine is shown in Figure 8.17. The results are qualitatively similar to those for k in Figure 8.16. The values of ε are highest near the blades and in the region dominated by the trailing vortices; in most of the rest of the vessel, ε is 1 to 2 orders of magnitude lower. The intense turbulence generated by the trailing vortices is contained within a relatively small region and dissipates quickly away from the impeller.

With Gassing

Rushton turbines are very effective for gas dispersion. To some extent, this can be attributed to the way the rotating disc on the turbine captures gas released below the impeller and channels it into the regions of high turbulence near the blades. Rushton turbines are often chosen for their gas-handling capacity, as they can be operated with relatively high gas flow rates without impeller flooding.

Typical gas flow patterns generated by a Rushton turbine in a low-viscosity fluid are shown in Figure 8.18. At high gassing rates or low stirrer speeds, the impeller is blanketed by gas, indicating impeller flooding. Under these conditions, as shown in Figure 8.18(a), the flow pattern is dominated by buoyant gas–liquid flow up the middle of the vessel. At higher stirrer speeds or lower gas flow rates, the impeller is loaded as gas is captured behind the impeller blades and dispersed towards the vessel walls, as indicated in Figure 8.18(b). Further increase in stirrer speed or reduction of the gas flow rate produces complete gas dispersion, as illustrated in Figure 8.18(c).

Photographs demonstrating the transition from impeller flooding to complete gas dispersion in a transparent tank stirred by a Rushton turbine are shown in Figure 8.19. In these experiments, the volumetric gas flow rate was held constant as the stirrer speed was increased from 100 rpm to 400 rpm. At the two lowest stirrer speeds in Figure 8.19(a)

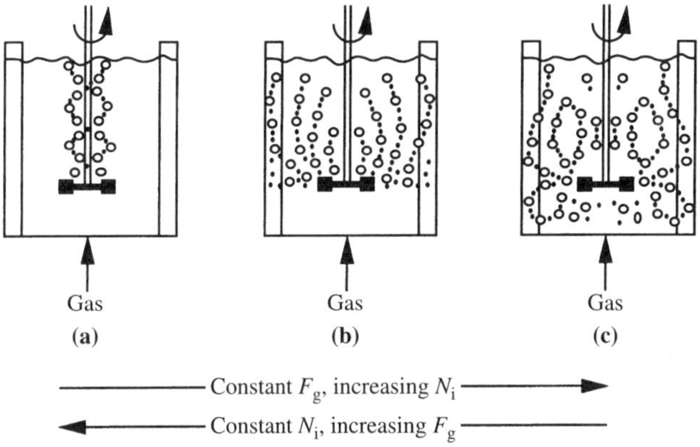

FIGURE 8.18 Patterns of gas distribution in an aerated tank stirred with a Rushton turbine as a function of the impeller speed N_i and gas flow rate F_g. (a) Impeller flooding; (b) impeller loading; (c) complete gas dispersion.
Adapted from A.W. Nienow, M. Konno, and W. Bujalski, 1986, Studies on three-phase mixing: a review and recent results. Chem. Eng. Res. Des. *64*, 35–42.

and (b), the impeller remains flooded as the stirrer is surrounded by gas and there is little outward dispersion of bubbles towards the vessel wall. Liquid motion is weak as blanketing by the gas prevents the impeller from pumping effectively. As indicated in Figure 8.19(c), at higher stirrer speeds the impeller becomes loaded and gas is distributed throughout the upper part of the vessel above the stirrer. Further increases in stirrer speed allow complete dispersion of bubbles below as well as above the impeller, as shown in Figure 8.19(d).

Correlations have been developed for predicting the operating conditions under which impeller loading and complete gas dispersion are achieved using Rushton turbines. These relationships are expressed using two dimensionless variables, the *gas flow number* Fl_g:

$$Fl_g = \frac{F_g}{N_i D_i^3} \tag{8.1}$$

and the *Froude number Fr*:

$$Fr = \frac{N_i^2 D_i}{g} \tag{8.2}$$

where F_g is the volumetric gas flow rate, N_i is stirrer speed, D_i is impeller diameter, and g is gravitational acceleration. Conceptually, Fl_g is the ratio of the gas flow rate to the pumping capacity of the impeller; Fr is the ratio of inertial to gravitational or buoyancy forces. Conditions at the flooding–loading transition for Rushton turbines are represented by the following equation [1]:

$$Fl_g = 30\left(\frac{D_i}{D_T}\right)^{3.5} Fr \quad \text{at the flooding–loading transition} \tag{8.3}$$

The conditions for complete dispersion of gas by Rushton turbines are represented by another equation [1]:

$$Fl_g = 0.2\left(\frac{D_i}{D_T}\right)^{0.5} Fr^{0.5} \quad \text{for complete gas dispersion} \tag{8.4}$$

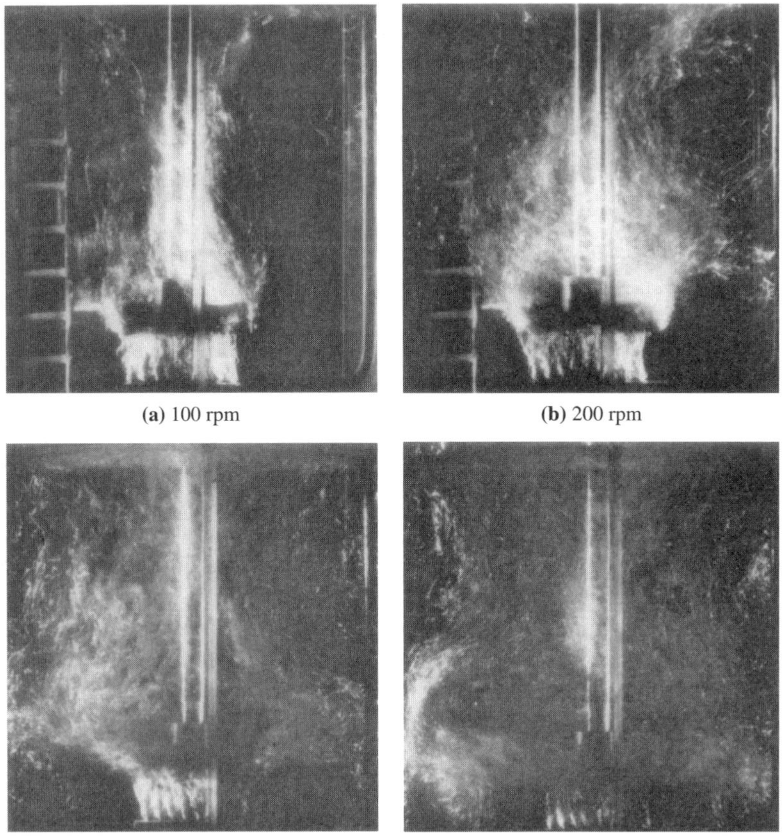

(a) 100 rpm (b) 200 rpm

(c) 300 rpm (d) 400 rpm

FIGURE 8.19 Gas–liquid flow patterns in a tank stirred by a Rushton turbine. The sparger positioned below the impeller is a horizontal steel tube with eight holes on its upper surface.
From K.L. Man, 1985, A study of local heat-transfer coefficients in mechanically agitated gas–liquid vessels. In: Mixing, Proc. 5th Eur. Conf. on Mixing, *Würzburg, Germany, pp. 221–231, BHRA: The Fluid Engineering Centre, Cranfield, U.K.*

Equations (8.3) and (8.4) apply to low-viscosity fluids and were determined using a variety of point and ring (diameter $< D_i$) spargers. The flooding–loading transition is not affected by the impeller off-bottom clearance; however complete gas dispersion is, and Eq. (8.4) was developed for an impeller clearance of a one-quarter the liquid height.

Equations (8.1) through (8.4) demonstrate the strong dependence of gas-handling capacity on the impeller diameter. The volumetric gas flow rate F_g for loading is proportional to $D_i^{7.5}$; F_g for complete gas dispersion is proportional to D_i^4. This means that for a 10% increase in D_i, the gas flow rate that can be handled without flooding increases by about 100%, while the gas flow rate for complete dispersion increases by about 50%. The dependence on stirrer speed is not as strong: a 10% increase in N_i increases F_g for flooding and complete gas dispersion by about 30% and 20%, respectively.

EXAMPLE 8.1 GAS HANDLING WITH A RUSHTON TURBINE

A fermenter of diameter and liquid height 1.4 m is fitted with a Rushton impeller of diameter 0.5 m and off-bottom clearance 0.35 m operated at 75 rpm. The fermentation broth is sparged with air at a volumetric flow rate of 0.28 $m^3\ min^{-1}$. Half-way through the culture some bearings in the stirrer drive begin to fail and the stirrer speed must be reduced to a maximum of 45 rpm for the remainder of the process.

(a) Under normal operating conditions, is the gas completely dispersed?
(b) After the stirrer speed is reduced, is the impeller flooded or loaded?

Solution

(a)

$$N_i = 75\ \text{min}^{-1} \cdot \left|\frac{1\ \text{min}}{60\ \text{s}}\right| = 1.25\ \text{s}^{-1}$$

From Eq. (2.16), $g = 9.81\ \text{m s}^{-2}$. Under normal operating conditions, from Eq. (8.2):

$$Fr = \frac{(1.25\ \text{s}^{-1})^2\ 0.5\ \text{m}}{9.81\ \text{m s}^{-2}} = 0.0796$$

Applying Eq. (8.4) for complete gas dispersion:

$$Fl_g = 0.2\left(\frac{0.5\ \text{m}}{1.4\ \text{m}}\right)^{0.5} (0.0796)^{0.5} = 0.0337$$

Therefore, from Eq. (8.1):

$$F_g = Fl_g N_i D_i^3 = 0.0337(1.25\ \text{s}^{-1})\ (0.5\ \text{m})^3 = 0.00527\ \text{m}^3\ \text{s}^{-1} = 0.32\ \text{m}^3\ \text{min}^{-1}$$

As the air flow rate that can be completely dispersed by the impeller (0.32 $m^3\ min^{-1}$) is greater than the operating flow rate (0.28 $m^3\ min^{-1}$), we can conclude that the air provided is completely dispersed under normal operating conditions.

(b) After the stirrer speed reduction:

$$N_i = 45\ \text{min}^{-1} \cdot \left|\frac{1\ \text{min}}{60\ \text{s}}\right| = 0.75\ \text{s}^{-1}$$

From Eq. (8.2):

$$Fr = \frac{(0.75\ \text{s}^{-1})^2\ 0.5\ \text{m}}{9.81\ \text{m s}^{-2}} = 0.0287$$

Applying Eq. (8.3) for the flooding–loading transition:

$$Fl_g = 30\left(\frac{0.5\ \text{m}}{1.4\ \text{m}}\right)^{3.5} 0.0287 = 0.0234$$

Therefore, from Eq. (8.1):

$$F_g = Fl_g N_i D_i^3 = 0.0234\ (0.75\ \text{s}^{-1})\ (0.5\ \text{m})^3 = 0.00219\ \text{m}^3\ \text{s}^{-1} = 0.13\ \text{m}^3\ \text{min}^{-1}$$

At the reduced stirrer speed, the maximum air flow rate that can be handled without impeller flooding is 0.13 $m^3\ min^{-1}$. As the operating air flow rate (0.28 $m^3\ min^{-1}$) is greater than this, the impeller is flooded.

Gas dispersion by Rushton turbines is related directly to the trailing vortices that develop behind the impeller blades (refer to Figure 8.15). Rolling or rotation of each vortex results in centrifugal acceleration of the liquid and a reduction in pressure along the vortex axis. When the liquid is sparged with gas, the gas accumulates readily in these low-pressure regions, producing *ventilated cavities* behind the blades. A photograph of ventilated cavities behind the blades of a Rushton turbine is shown in Figure 8.20. Bubble dispersion occurs primarily at the outer tails of the cavities, where small bubbles are shed to balance the rate of gas flow into the cavities from under the disc.

The effectiveness of gas dispersion in stirred vessels is controlled by the size and structure of the ventilated cavities behind the impeller blades. The types of cavity formed with Rushton turbines in low-viscosity fluids have been well documented and are illustrated in Figure 8.21. With increasing gas flow rate F_g at constant stirrer speed N_i, vortex cavities at each of the blades give rise to clinging cavities, after which there is a transition to a very stable '3–3' structure characterised by large cavities behind three of the blades and smaller cavities behind the remaining alternate three blades. If the gas flow rate is increased beyond the gas-handling capacity of the impeller, six equal-sized ragged cavities are formed; these cavities are unstable and oscillate violently. The formation of ragged cavities occurs at the point represented in Figure 8.18 as the transition from Figure 8.18(b) to (a): at these high gas flow rates, the impeller ceases to function effectively and is flooded. No particular change in cavity structure has been associated with the transition shown in Figure 8.18 from loading (b) to complete gas dispersion (c). Cavity formation in fluids with viscosity greater than about 20 centipoise is more complex than that represented in Figure 8.21; cavities of different shape and greater stability are produced. Cavities in

FIGURE 8.20 Gas cavities that are formed behind the blades of a 7.6-cm Rushton turbine in water. The impeller was rotated counterclockwise; the camera was located below the impeller and ring sparger.
From W. Bruijn, K. van't Riet, and J.M. Smith, 1974, Power consumption with aerated Rushton turbines. Trans. IChemE *52, 88–104.*

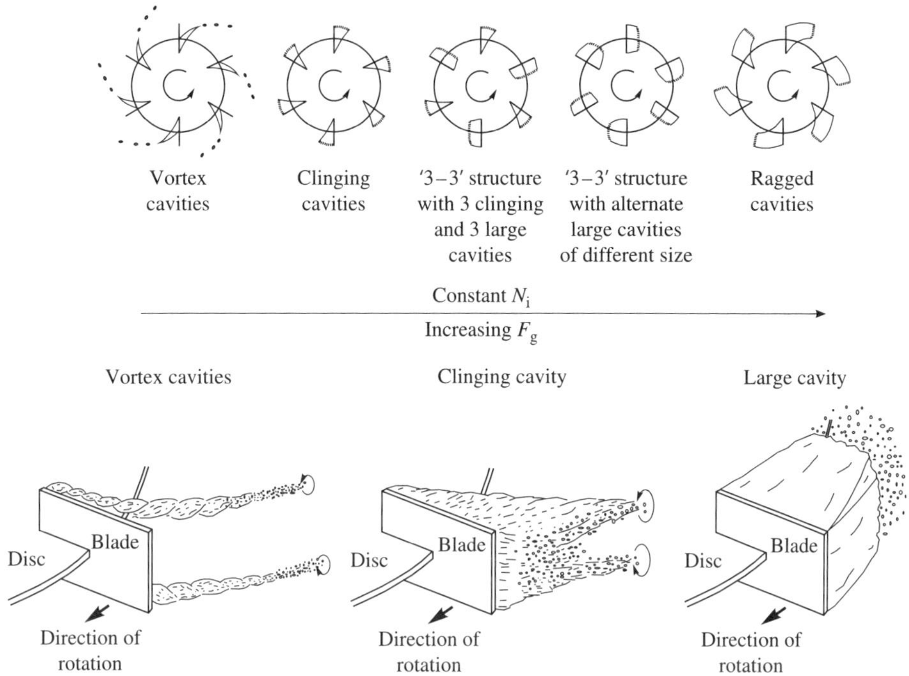

FIGURE 8.21 Changes in ventilated cavity structure for a Rushton turbine at constant stirrer speed N_i and increasing gas flow rate F_g.
Adapted from J.M. Smith and M.M.C.G. Warmoeskerken, 1985, The dispersion of gases in liquids with turbines. In: Mixing, Proc. 5th Eur. Conf. on Mixing, *Würzburg, Germany, pp. 115–126, BHRA: The Fluid Engineering Centre, Cranfield, U.K.*

viscous fluids may be so stable that they persist behind the impeller blades for several hours even after the gas supply is stopped [2].

Solids Suspension

Rushton turbines are effective for solids suspension, including in three-phase (solid–liquid–gas) systems. Suspension of solids is generally improved by reducing the impeller off-bottom clearance, but this can cause flow instabilities when the system is aerated. For three-phase mixing, an impeller clearance of a 1/4 the tank diameter has been recommended for Rushton turbines, as this allows effective solids suspension, gas dispersion under the impeller, and adequate agitation in the upper parts of the vessel [3].

8.4.2 Propellers

A typical three-blade marine-type propeller is illustrated in Figure 8.22. The slope of the individual blades varies continuously from the outer tip to the inner hub. The *pitch* of a propeller is a measure of the angle of the propeller blades. It refers to the properties of the propeller as a segment of a screw: pitch is the advance per revolution, or the distance that

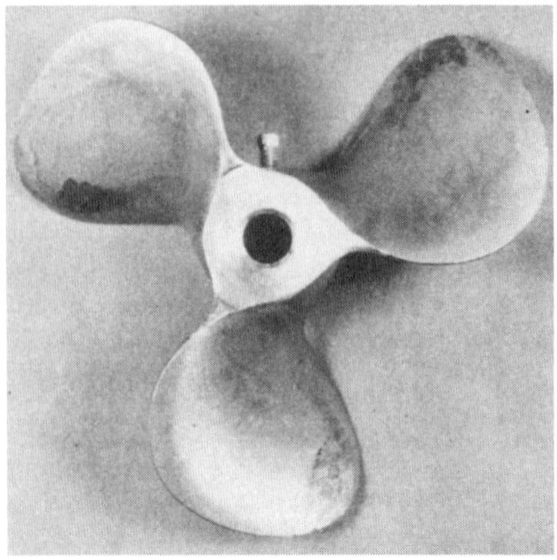

FIGURE 8.22 Propeller.

liquid is displaced along the impeller axis during one full turn. Propellers with square pitch, that is, pitch equal to the impeller diameter, are often used.

Propellers are axial-flow impellers. They may be operated for either downward or upward pumping of the fluid; downward pumping is more common. Propellers have high flow capacity and produce mean flow patterns similar to that shown in Figure 8.8. They are used with low-to-medium viscosity fluids and are usually installed with diameter around one-third the tank diameter. With gassing, propellers operated at high speed can generate flow and torque instabilities. However, propellers are very effective for suspending solids, outperforming Rushton turbines in that respect.

8.4.3 Pitched-Blade Turbines

Pitched-blade turbines have flat inclined blades, as shown in Figure 8.10. Although commonly referred to as axial-flow impellers, pitched-blade turbines generate discharge streams with significant radial as well as axial velocity components. Pitched-blade turbines produce strong liquid flows and have a much higher pumping efficiency than Rushton turbines, making them very effective for blending applications. With gassing, ventilated cavities form behind the blades as a result of underpressure in the trailing vortices in much the same way as for Rushton turbines. However, for pitched-blade turbines, there is only one vortex per blade.

Pitched-blade turbines can be operated in either downward or upward pumping modes, downward being the more common.

Downward Pumping

Figure 8.8 shows the liquid flow pattern typical of downward-pumping pitched-blade turbines. Performance of these impellers is sensitive to several aspects of tank geometry, such as the impeller off-bottom clearance and the sparger size and location. This sensitivity is greater than is found normally with Rushton turbines. Depending on the impeller off-bottom clearance, secondary circulation loops may develop in the lower regions of the tank in addition to the primary flow pattern shown in Figure 8.8. Because the strength of these secondary currents determines the angle of fluid discharge from the impeller, which in turn determines whether the primary circulation currents reach the vessel floor, the flows set up by downward-pumping turbines can be considerably more complex than that shown in Figure 8.8 [4]. Unlike with Rushton turbines, the fluid currents generated by pitched-blade turbines are not compartmentalised into upper and lower circulatory loops. However, because the flow velocity becomes progressively weaker away from the impeller, the primary circulation currents generated by downward-pumping turbines may not reach the upper parts of the tank, even when the liquid height does not exceed the tank diameter [5].

The distribution of turbulence kinetic energy per unit mass of fluid, k (Section 7.9.2, Turbulence Kinetic Energy subsection), in a tank stirred with a downward-pumping pitched-blade turbine is shown in Figure 8.23. Turbulence kinetic energy is not distributed uniformly throughout the tank. The highest values of k are concentrated near the impeller where the turbulence is most intense. As the discharge streams move downwards away from the impeller blades, the turbulence kinetic energy decreases accordingly. Levels of turbulence kinetic energy in the remainder of the tank away from the impeller are up to an order of magnitude lower than the maximum values measured.

With gassing, downward-pumping turbines are prone to flooding, especially if the impeller diameter and solidity ratio are small. The hydrodynamic changes that occur as the stirrer speed is increased at constant gas flow rate are similar to but more complex than those represented in Figure 8.18 for a Rushton turbine; for example, asymmetrical flow patterns may be generated. At low stirrer speeds when the impeller is flooded, liquid circulation is weak and the flow is dominated by gas bubbling up the stirrer shaft. With increasing stirrer speed, streaming ventilated cavities form behind the impeller blades and bubbles breaking away from the cavities are dispersed downwards; however, the large flow of bubbles up the centre of the vessel remains the primary flow pattern. Further increases in speed allow the pumping action of the impeller to dominate, so that gas is vigorously dispersed throughout the vessel. However, this condition is unstable; the flow

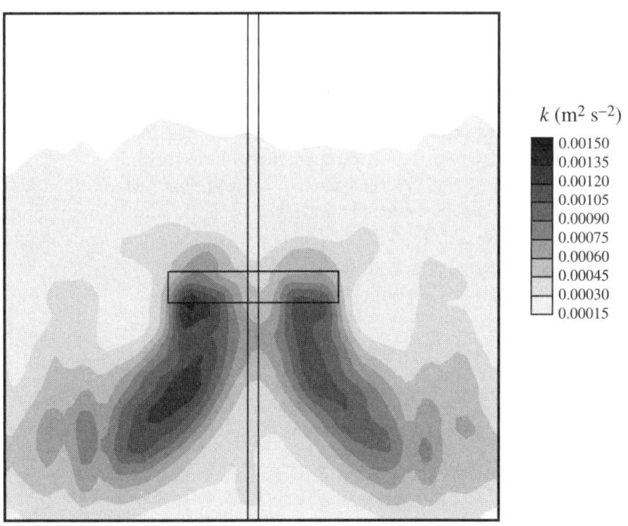

FIGURE 8.23 Distribution of k, the turbulence kinetic energy per unit mass of fluid, in a tank stirred by a downward-pumping pitched-blade turbine located near the centre of the tank. The data were measured using particle image velocimetry. *From J. Sheng, H. Meng, and R.O. Fox, 1998, Validation of CFD simulations of a stirred tank using particle image velocimetry data.* Can. J. Chem. Eng. *76, 611–625. Image provided courtesy of the authors.*

pattern periodically reverts to a nondispersed state and large flow oscillations and torque and power instabilities can occur. Eventually, at high enough stirrer speeds, large gas-filled cavities are formed behind the impeller blades, the instabilities disappear, and the gas remains fully dispersed. Because instability and flow oscillations prior to stable gas dispersion can lead to mechanical problems including vessel vibration, operation in this regime is not recommended. Thus, in contrast with Rushton impellers, for downward-pumping pitched-blade turbines there is no practical range of operating conditions between flooding and complete gas dispersion. As neither flow instability nor incomplete gas dispersion is desirable, the impeller should always be operated at speeds high enough to fully distribute the gas.

The instabilities associated with downward-pumping turbines are generally thought to occur because of the inherent opposition of flow directions generated by the impeller and by the gas; that is, fluid driven downward by the impeller is opposed by bubble flow rising up from the sparger. As shown in Figure 8.24, two flow regimes have been identified for downward-pumping impellers. *Direct loading*, which occurs at low stirrer speeds or high gas flow rates, is characterised by gas entering the impeller region directly from the sparger. In contrast, during *indirect loading* at high stirrer speeds or low gas flows, gas approaching the impeller is swept away by the downward thrust of liquid from the stirrer and only enters the impeller region by recirculation. Flow instabilities are associated with the transition from direct to indirect loading, and coincide with the formation of large gas-filled cavities behind the impeller blades. Problems with instability are greater for impellers with four blades rather than six, and when small impeller-to-tank diameter ratios are used.

Sparger geometry and position have significant effects on the performance of downward-pumping impellers. The use of point spargers increases the likelihood of flow instabilities with gassing. Although pitched-blade impellers are more prone to flooding than Rushton turbines, their gas-handling ability can be improved by replacing point spargers with ring spargers of diameter about 0.8 times the impeller diameter, and optimising the separation between the sparger and impeller [6, 7].

At low gassing rates, suspension of solids is achieved by downward-pumping agitators with very high energy efficiency. However, severe loss of suspension capacity can occur

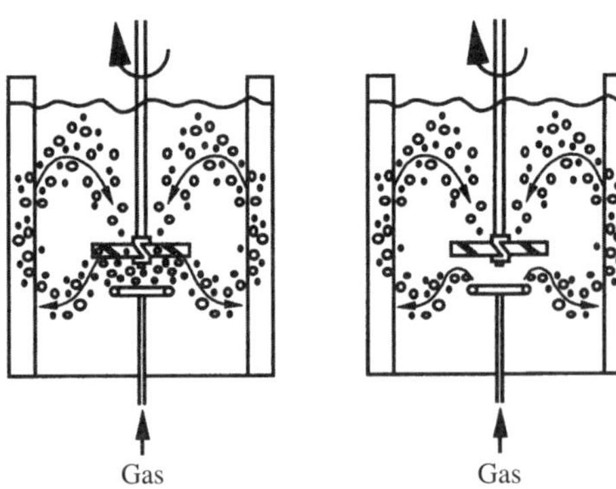

(a) Direct loading (b) Indirect loading

FIGURE 8.24 Direct and indirect gas loading regimes for a downward-pumping pitched-blade turbine.
From M.M.C.G. Warmoeskerken, J. Speur, and J.M. Smith, 1984, Gas–liquid dispersion with pitched blade turbines. Chem. Eng. Commun. *25, 11–29.*

with gassing, especially when there are flow instabilities associated with the direct–indirect loading transition.

Upward Pumping

Many of the problems associated with downward-pumping pitched-blade impellers under gassed conditions can be avoided by reversing the direction of rotation so that the impeller pumps upwards. The upward flow generated is then cocurrent with that of the sparged gas, and the resulting flow patterns are inherently more stable than with downward pumping. Yet, the use of turbine impellers in upward-pumping mode is much less common than downward-pumping operation.

The gas–liquid flow patterns developed by upward-pumping turbines are very different from those produced during downward flow operation. At low stirrer speeds, there is negligible gas dispersion, the impeller is flooded, and there are no gas cavities behind the blades. As the impeller speed is increased, there is a change from flooding to loading as more and more gas is dispersed towards the vessel walls. With loading, streaming vortex gas cavities are formed behind the impeller blades. Clinging cavities develop with further increases in stirrer speed and gas bubbles start to be dispersed below the impeller. At even higher speeds, large clinging cavities are formed and bubbles shed from the tails of these cavities are dispersed throughout the vessel. No significant flow instabilities occur with upward-pumping turbines. Compared with downward-pumping impellers, complete gas dispersion is achieved at lower stirrer speeds and with less energy consumption, and more gas can be handled before flooding occurs.

Because upward-pumping impellers generate relatively small velocities beneath the impeller, the energy required for solids suspension is significantly greater than for downward-pumping impellers. However, an advantage in aerated systems is that both the agitation speed and power required for solids suspension are almost independent of the gassing rate [8].

8.4.4 Alternative Impeller Designs

So far, we have considered the characteristics of several traditional impellers that have been used in the chemical and bioprocessing industries for many decades. More recently, a variety of new agitator configurations has been developed commercially. These modern impellers have a range of technical features aimed at improving mixing in stirred tanks.

Curved-Blade Disc Turbines

Curved-blade disc turbines such as that shown in Figure 8.25 generate primarily radial flow, similar to the Rushton turbine. However, changing the shape of the blades has a significant effect on the impeller power requirements and gas-handling characteristics. Rotation with the concave side forward greatly discourages the development of trailing vortices behind the blades; therefore, with sparging, no large ventilated cavities form on the convex surfaces. A major advantage is that the impeller is more difficult to flood, being able to handle gas flow rates several times higher than those that cause flooding of Rushton turbines [9].

FIGURE 8.25 Scaba 6SRGT six-curved-blade disc turbine.
Photograph courtesy of Scaba AB, Sweden.

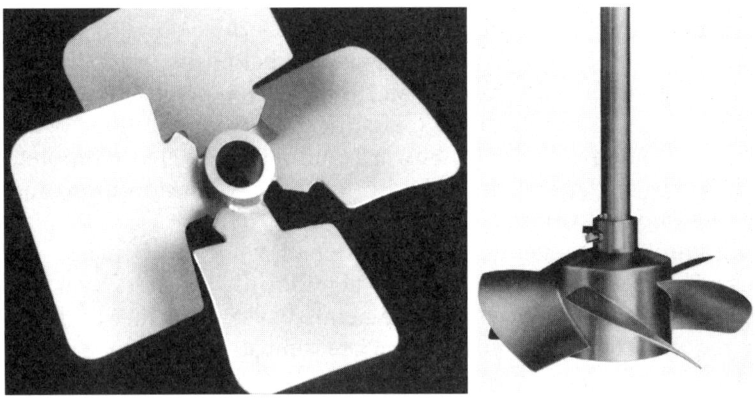

(a) (b)

FIGURE 8.26 (a) Lightnin A315 hydrofoil impeller.
Photograph courtesy of Lightnin Mixers, Australia.
(b) Prochem Maxflo T hydrofoil impeller.
Photograph courtesy of Chemineer Inc., Dayton, OH.

Hydrofoil Impellers

Two different hydrofoil impellers are shown in Figure 8.26. The blade angle and width are varied along the length of hydrofoil blades, and the leading edges are rounded like an airplane wing to reduce form drag and generate a positive lift. The shape of hydrofoil impellers allows for effective pumping and bulk mixing with strong axial velocities and low power consumption. Most hydrofoils are operated for downward pumping, but upward flow is also possible. A typical mean velocity vector plot for a downward-pumping hydrofoil impeller is shown in Figure 8.27. Downward-flowing currents with very strong axial velocity components leave the impeller. Fluid in these currents sweeps the vessel floor in an outward radial direction, then moves up alongside the vessel walls before returning back to the impeller. Liquid velocities in regions of the tank above the main circulation loops are considerably lower than below the impeller.

With aeration, downward-pumping hydrofoil impellers exhibit many of the hydrodynamic properties of downward-pumping pitched-blade turbines. They remain more prone to flooding than Rushton turbines, even when the impeller solidity ratio is large. The types

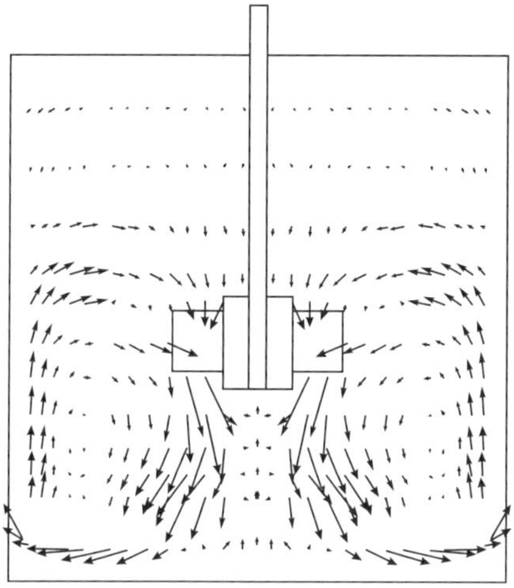

FIGURE 8.27 Mean velocity vector plot for a Prochem Maxflo T hydrofoil impeller. The velocities were measured using laser Doppler velocimetry.
Adapted from Z. Jaworski, A.W. Nienow, and K.N. Dyster, 1996, An LDA study of the turbulent flow field in a baffled vessel agitated by an axial, down-pumping hydrofoil impeller. Can. J. Chem. Eng. *74, 3–15.*

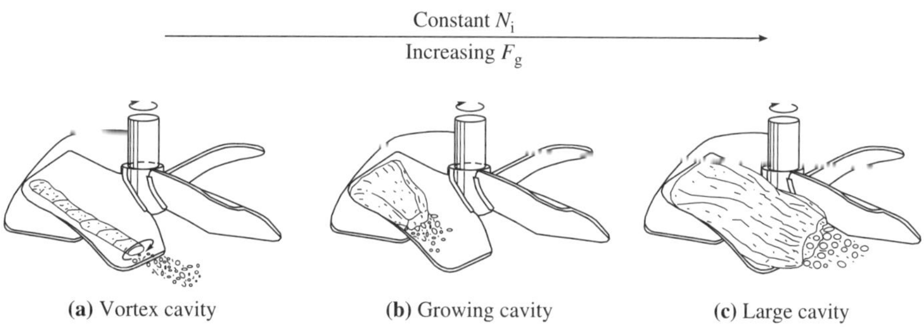

FIGURE 8.28 Changes in ventilated cavity structure for a Lightnin A315 hydrofoil impeller at constant stirrer speed N_i and increasing gas flow rate F_g.
Adapted from A. Bakker and H.E.A. van den Akker, 1994, Gas–liquid contacting with axial flow impellers. Trans. IChemE *72A, 573–582.*

of ventilated cavity formed behind the blades of a downward-pumping hydrofoil impeller in low-viscosity fluid are illustrated in Figure 8.28. The cavities in Figure 8.28(a) and (b) occur during indirect loading (Section 8.4.3, Downward Pumping subsection); at relatively low gas flow rates, only recirculating gas reaches the impeller to form vortex or growing cavities. With increasing gas flow, the growing cavities can exhibit various types of instability. At higher gassing rates when the impeller is no longer capable of deflecting the rising gas flow, direct loading occurs as gas enters the impeller region directly from the sparger to produce large cavities as shown in Figure 8.28(c). Further increases in gas flow lead to impeller flooding.

The flow instabilities generated in gassed systems by downward-pumping hydrofoil impellers can be eliminated using wide-blade, upward-pumping hydrofoils. These impellers also have high gas-handling capacity and a limited tendency to flood at low stirrer speeds.

8.5 STIRRER POWER REQUIREMENTS

Electrical power is used to drive impellers in stirred vessels. The average power consumption per unit volume for industrial bioreactors ranges from 10 kW m^{-3} for small vessels (approx. 0.1 m^3) to 1 to 2 kW m^{-3} for large vessels (approx. 100 m^3). Friction in the stirrer motor gearbox and seals reduces the energy transmitted to the fluid; therefore, the electrical power consumed by stirrer motors is always greater than the mixing power by an amount depending on the efficiency of the drive. Energy costs for operation of stirrers in bioreactors are a major ongoing financial commitment and an important consideration in process economics.

The power required to achieve a given stirrer speed depends on the magnitude of the frictional forces and form drag that resist rotation of the impeller. Friction and form drag give rise to torque on the stirrer shaft; experimentally, the power input for stirring can be determined from measurements of the induced torque M:

$$P = 2\pi N_i M \tag{8.5}$$

where P is power and N_i is the stirrer speed.

General guidelines for estimating the power requirements in stirred tanks are outlined in the following sections.

8.5.1 Ungassed Newtonian Fluids

The power required to mix nonaerated fluids depends on the stirrer speed, the impeller shape and size, the tank geometry, and the density and viscosity of the fluid. The relationship between these variables is usually expressed in terms of dimensionless numbers such as the impeller Reynolds number Re_i:

$$Re_i = \frac{N_i D_i^2 \rho}{\mu} \tag{7.2}$$

and the power number N_P:

$$N_P = \frac{P}{\rho N_i^3 D_i^5} \tag{8.6}$$

In Eqs. (7.2) and (8.6), N_i is stirrer speed, D_i is impeller diameter, ρ is fluid density, μ is fluid viscosity, and P is power. The power number N_P can be considered analogous to a drag coefficient for the stirrer system.

The relationship between Re_i and N_P has been determined experimentally for a range of impeller and tank configurations. The results for five impeller designs—Rushton turbine, downward-pumping pitched-blade turbine, marine propeller, anchor, and helical ribbon—are shown in Figures 8.29 and 8.30. Once the value of N_P is known, the power is calculated from Eq. (8.6) as:

$$P = N_P \rho N_i^3 D_i^5 \tag{8.7}$$

For a given impeller, the general relationship between power number and Reynolds number depends on the flow regime in the tank. The following three flow regimes can be identified in Figures 8.29 and 8.30.

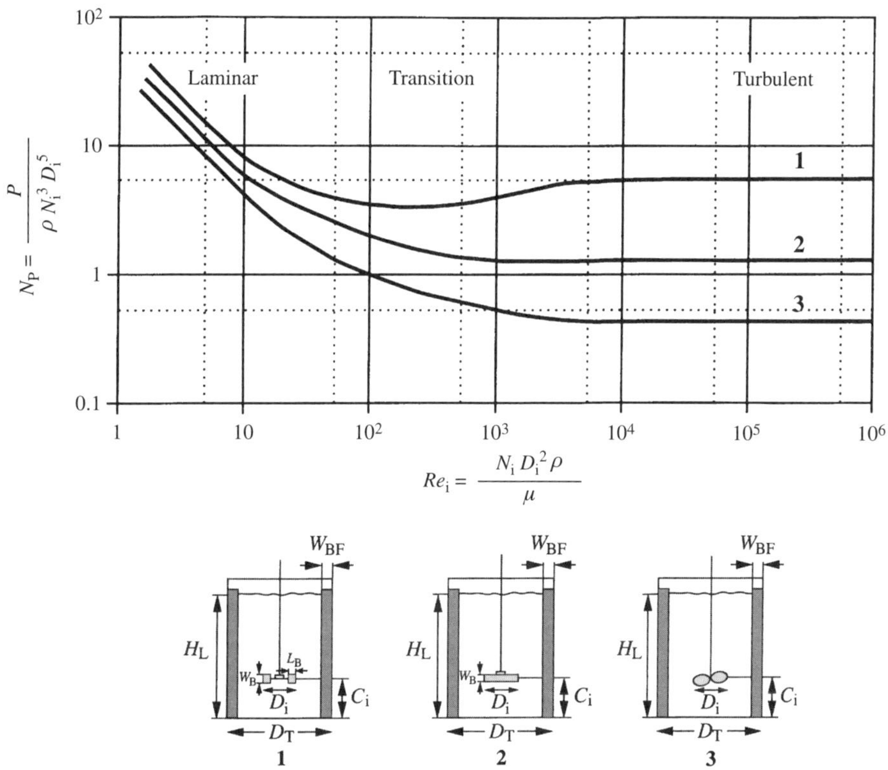

Impeller	D_i/D_T	H_L/D_T	C_i/D_T	Baffles: W_{BF}/D_T	Baffles: Number
1. Rushton turbine $W_B/D_i = 0.2$, $L_B/D_i = 0.25$	0.33	1	0.33	0.1	4
2. Pitched-blade turbine $W_B/D_i = 0.125$, 6 blades, 45°, downward pumping	0.33	1	0.33	0.1	4
3. Marine propeller 3 blades, pitch = D_i	0.33	1	0.33	0.1	4

FIGURE 8.29 Correlations between the Reynolds number and power number for Rushton turbines, downward-pumping pitched-blade turbines, and marine propellers in fluids without gassing.
Data from J.H. Rushton, E.W. Costich, and H.J. Everett, 1950, Power characteristics of mixing impellers. Parts I and II. Chem. Eng. Prog. *46, 395–404, 467–476; and R.L. Bates, P.L. Fondy, and R.R. Corpstein, 1963, An examination of some geometric parameters of impeller power.* Ind. Eng. Chem. Process Des. Dev. *2, 310–314.*

1. *Laminar regime.* The laminar regime corresponds to $Re_i < 10$ for many impellers, including turbines and propellers. For stirrers with small wall-clearance such as anchor and helical ribbon mixers, laminar flow persists until $Re_i = 100$ or greater. In the laminar regime:

$$N_P \propto \frac{1}{Re_i} \qquad \text{or} \qquad P = k_1 \mu N_i^2 D_i^3 \tag{8.8}$$

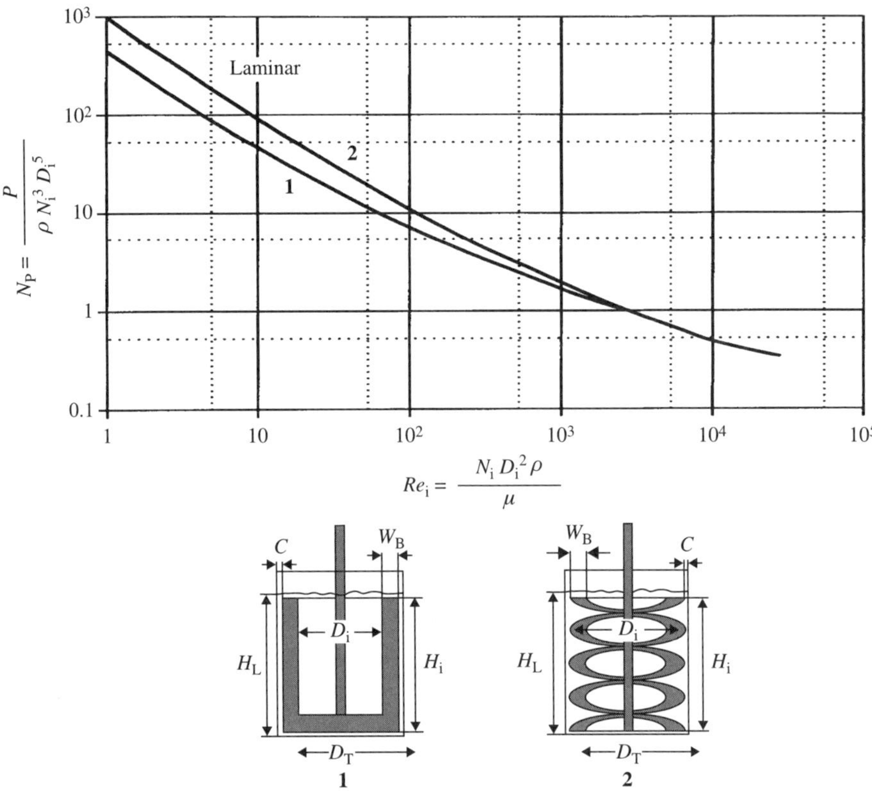

Impeller	D_T/D_i	C/D_i	H_i/D_i	W_B/D_i
1. Anchor	1.02	0.01	1	0.1
2. Herical ribbon	1.02	0.01	1	0.1

FIGURE 8.30 Correlations between the Reynolds number and power number for anchor and helical ribbon impellers in fluids without gassing.
From M. Zlokarnik and H. Judat, 1988, Stirring. In: W. Gerhartz, Ed., Ullmann's Encyclopedia of Industrial Chemistry, *5th ed., vol. B2, pp. 25-1–25-33, VCH, Weinheim, Germany.*

where k_1 is a proportionality constant. Values of k_1 for the impellers represented in Figures 8.29 and 8.30 are listed in Table 8.1 [10, 11]. The power required for laminar flow is independent of the density of the fluid but directly proportional to the fluid viscosity.

2. *Turbulent regime.* The power number is independent of the Reynolds number in turbulent flow. Therefore:

$$P = N'_P \rho N_i^3 D_i^5 \tag{8.9}$$

where N'_P is the constant value of the power number in the turbulent regime. Approximate values of N'_P for the impellers in Figures 8.29 and 8.30 are listed in

TABLE 8.1 Values of the Constants in Eqs. (8.8) and (8.9) for the Stirred Tank Geometries Defined in Figures 8.29 and 8.30

Impeller type	k_1 ($Re_i = 1$)	N'_P ($Re_i = 10^5$)
Rushton turbine	70	5.0
Pitched-blade turbine	50	1.3
Marine propeller	40	0.35
Anchor	420	0.35
Helical ribbon	1000	0.35

Table 8.1 [10, 11]. N'_P for Rushton turbines is significantly higher than for most other impellers, indicating that the Rushton turbine has strong form drag, generates high levels of torque, and transmits more power at the same operating speed than other designs. Values of N'_P for a selection of other impellers and system geometries are listed in Table 8.2. Depending on the number of blades and solidity ratio, hydrofoil impellers have relatively low N'_P; this reflects their aerodynamic blade design, which effectively minimises form drag.

As indicated in Eq. (8.9), the power required for turbulent flow is independent of the viscosity of the fluid but proportional to the density. The turbulent regime is fully developed at $Re_i > 10^4$ for most small impellers in baffled vessels. For the same impellers in vessels without baffles, the power curves are somewhat different from those shown in Figure 8.29. Without baffles, turbulence may not be fully developed until $Re_i > 10^5$; the value of N'_P is also reduced to as little as 10 to 50% of that with baffles [18, 19].

3. *Transition regime.* Between laminar and turbulent flow lies the transition region. Although the viscosity of fermentation broths is not usually sufficient for stirrers in industrial bioreactors to operate in the laminar regime, many broths become sufficiently viscous to give Reynolds numbers in the transition region. Both density and viscosity affect the power requirements in this regime. There is usually a gradual transition from laminar to fully developed turbulent flow in stirred tanks; the Reynolds-number range for transition depends on the system geometry.

Equations (8.8) and (8.9) express the strong dependence of power consumption on stirrer diameter and, to a lesser extent, stirrer speed. Small changes in impeller size have a large effect on power requirements, as would be expected from dependency on impeller diameter raised to the third or fifth power. In the turbulent regime, a 10% increase in impeller diameter increases the power requirements by more than 60%; a 10% increase in stirrer speed raises the power required by over 30%.

Friction and form drag, and therefore the power required for stirring, are sensitive to the detailed geometry of the impeller and configuration of the tank. The curves of Figures 8.29 and 8.30 and the values of N'_P in Tables 8.1 and 8.2 refer to the particular geometries specified and are subject to change if the number, size, or position of the baffles, the number, length, width, pitch, or angle of the impeller blades, the liquid height,

TABLE 8.2 Values of the Turbulent Ungassed Power Number N_P' for a Selection of Impellers and System Geometries

Impeller	System geometry	N_P'	Reference
Rushton turbine	Flat-bottom tank	5.9	[12, 13]
$D_i/D_T = 0.50$	$H_L/D_T = 1.0$		
$W_B/D_i = 0.20$	Number of baffles = 4		
$L_B/D_i = 0.25$	$W_{BF}/D_T = 0.1$		
	$C_i/D_T = 0.25$		
Pitched-blade turbine	Flat-bottom tank	1.8	[14]
Downward pumping	$H_L/D_T = 1.0$		
6 blades, 45°	Number of baffles = 4		
$D_i/D_T = 0.40$	$W_{BF}/D_T = 0.1$		
	$C_i/D_T = 0.25$		
Pitched-blade turbine	Flat-bottom tank	1.6	[13]
Downward pumping	$H_L/D_T = 1.0$		
6 blades, 45°	Number of baffles = 4		
$D_i/D_T = 0.50$	$W_{BF}/D_T = 0.1$		
$W_B/D_i = 0.20$	$C_i/D_T = 0.25$		
Pitched-blade turbine	Flat-bottom tank	1.6	[13, 15]
Upward pumping	$H_L/D_T = 1.0$		
6 blades, 45°	Number of baffles = 4		
$D_i/D_T = 0.5$	$W_{BF}/D_T = 0.1$		
$W_B/D_i = 0.20$	$C_i/D_T = 0.25$		
Curved-blade disc turbine (Scaba 6SRGT)	Flat-bottom tank	1.5	[9]
6 blades	$H_L/D_T = 1.0$		
$D_i/D_T = 0.33$	Number of baffles = 4		
$W_B/D_i = 0.15$	$W_{BF}/D_T = 0.1$		
$L_B/D_i = 0.28$	$C_i/D_T = 0.25$		
Hydrofoil (Lightnin A315)	Flat-bottom tank	0.84	[16]
Downward pumping	$H_L/D_T = 1.0$		
4 blades	Number of baffles = 4		
$D_i/D_T = 0.40$	$W_{BF}/D_T = 0.1$		
	$C_i/D_T = 0.25$		

Hydrofoil (Prochem Maxflo T) Downward pumping 6 blades $D_i/D_T = 0.35$	Flat-bottom tank $H_L/D_T = 1.0$ Number of baffles = 4 $W_{BF}/D_T = 0.1$ $C_i/D_T = 0.45$	1.6	[17]

the impeller clearance from the bottom of the tank, and so on, are altered. For a Rushton turbine under fully turbulent conditions ($Re_i > 10^4$), N'_P lies between about 1 and 7 depending on these parameters [19]. For axial-flow impellers, blade angle has a major influence on power requirements. For example, N'_P for a pitched-blade turbine with six blades set at an angle of 60° to the horizontal is more than fivefold that for the same impeller with blade angle 30° [20]. Impeller pitch has a similarly significant effect on the power number for propellers [19, 21]. Experimental studies have shown that blade thickness and vessel scale can also affect N'_P [13, 22].

EXAMPLE 8.2 CALCULATION OF POWER REQUIREMENTS

A fermentation broth with viscosity 10^{-2} Pa s and density 1000 kg m^{-3} is agitated in a 50-m^3 baffled tank using a marine propeller 1.3 m in diameter. The tank geometry is as specified in Figure 8.29. Calculate the power required for a stirrer speed of 4 s^{-1}.

Solution

From Eq. (7.2):

$$Re_i = \frac{4\ \text{s}^{-1}\ (1.3\ \text{m})^2\ 1000\ \text{kg m}^{-3}}{10^{-2}\ \text{kg m}^{-1}\ \text{s}^{-1}} = 6.76 \times 10^5$$

From Figure 8.29, flow at this Re_i is fully turbulent. From Table 8.1, N'_P is 0.35. Therefore, from Eq. (8.9):

$$P = (0.35)\ 1000\ \text{kg m}^{-3}\ (4\ \text{s}^{-1})^3\ (1.3\ \text{m})^5 = 8.3 \times 10^4\ \text{kg m}^2\ \text{s}^{-3}$$

From Table A.8 in Appendix A, 1 kg m^2 s^{-3} = 1 W. Therefore:

$$P = 83\ \text{kW}$$

8.5.2 Ungassed Non-Newtonian Fluids

Estimation of the power requirements for non-Newtonian fluids is more difficult. It is often impossible with highly viscous fluids to achieve fully developed turbulence; under these conditions N_P is always dependent on Re_i and we cannot use the constant N'_P value in power calculations. In addition, because the viscosity of non-Newtonian liquids varies with shear conditions, the impeller Reynolds number used to correlate power requirements must be redefined. Some power correlations have been developed using an impeller Reynolds number based on the apparent viscosity μ_a (Section 7.5):

$$Re_i = \frac{N_i D_i^2 \rho}{\mu_a} \tag{8.10}$$

Therefore, from Eq. (7.12) for power-law fluids:

$$Re_i = \frac{N_i D_i^2 \rho}{K \dot{\gamma}^{n-1}} \tag{8.11}$$

where n is the flow behaviour index and K is the consistency index. A problem with application of Eq. (8.11) is the evaluation of $\dot{\gamma}$. For stirred tanks, an approximate relationship is often used:

$$\dot{\gamma} = k N_i \tag{7.15}$$

where the value of the constant k depends on the geometry of the impeller. The relationship of Eq. (7.15) is discussed further in Section 8.15; however, for turbine impellers k is about 10. Substituting Eq. (7.15) into Eq. (8.11) gives an appropriate Reynolds number for pseudoplastic fluids:

$$Re_i = \frac{N_i^{2-n} D_i^2 \rho}{K\, k^{n-1}} \tag{8.12}$$

The relationship between the Reynolds number Re_i and the power number N_P for a Rushton turbine in a baffled tank containing pseudoplastic fluid is shown in Figure 8.31. The upper line was measured using Newtonian fluids for which Re_i is defined by Eq. (7.2); this line corresponds to part of the curve already shown in Figure 8.29. The lower line gives the Re_i–N_P relationship for pseudoplastic fluids with Re_i defined by Eq. (8.12). The laminar region extends to higher Reynolds numbers in pseudoplastic fluids than in Newtonian systems. At Re_i below 10 and above 200, the results for Newtonian and

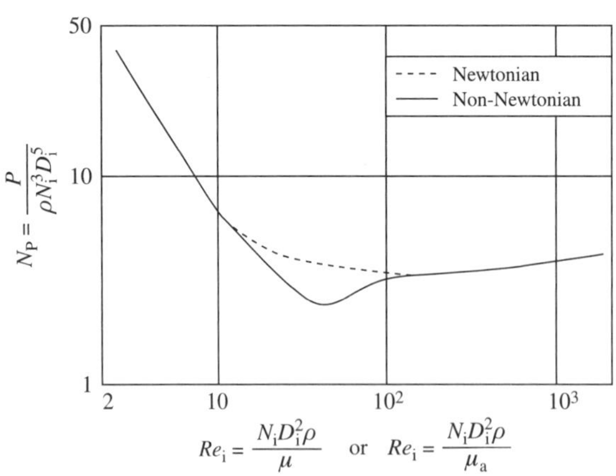

FIGURE 8.31 The correlation between the Reynolds number and power number for a Rushton turbine in ungassed non-Newtonian fluid in a baffled tank.
From A.B. Metzner, R.H. Feehs, H. Lopez Ramos, R.E. Otto, and J.D. Tuthill, 1961, Agitation of viscous Newtonian and non-Newtonian fluids. AIChE J. *7,* 3–9.

non-Newtonian fluids are essentially the same. In the intermediate range, pseudoplastic liquids require less power than Newtonian fluids to achieve the same Reynolds number.

There are several practical difficulties with application of Figure 8.31 to bioreactors. As discussed further in Section 8.15, flow patterns in pseudoplastic and Newtonian fluids differ significantly. Even if there is high turbulence near the impeller in pseudoplastic systems, the bulk liquid may be moving very slowly and consuming relatively little power. Another problem is that, as illustrated in Figure 7.11, the non-Newtonian parameters K and n, and therefore μ_a, can vary substantially during the course of fermentation.

8.5.3 Gassed Fluids

Liquids into which gas is sparged have reduced power requirements for stirring. The presence of gas bubbles decreases the density of the fluid; however, the influence of density as expressed in Eq. (8.9) does not explain adequately all the power characteristics of gas–liquid systems. The main reason that gassing affects power consumption is that bubbles have a profound impact on the hydrodynamic behaviour of fluid around the impeller. As described for Rushton turbines and other impellers in Section 8.4, gas-filled cavities develop behind the stirrer blades in aerated liquids. These cavities decrease the drag forces generated at the impeller and significantly reduce the resistance to impeller rotation, thus causing a substantial drop in the power required to operate the stirrer compared with nonaerated conditions.

The relationship between the power drop with gassing and operating conditions such as the gas flow rate and stirrer speed is often represented using graphs such as those shown in Figure 8.32. In these graphs, P_g is the power required with gassing and P_0 is the power required without gassing; P_0 is the same as the power evaluated using the methods described in Sections 8.5.1 and 8.5.2. The operating conditions are represented by the dimensionless gas flow number Fl_g defined in Eq. (8.1). Figure 8.32 shows reductions in power as a function of the gas flow number for three different impellers in low-viscosity fluid. Each curve represents experimental data for a given impeller operated at constant stirrer speed (i.e., for each curve, the value of Fl_g changes only because of changes in the gas flow rate).

As indicated in Figure 8.32(a), power reductions of up to 70% can occur with Rushton turbines in low-viscosity fluids, although most practical operating conditions give a 40 to 50% loss of power. The rheological properties of the fluid exert a considerable influence: power losses of more than 80% occur with Rushton turbines in viscous or non-Newtonian fluids sparged with gas. The fall in power with increasing gas flow rate is due to the increased size and changes in structure of the ventilated cavities behind the blades of Rushton turbines as illustrated in Figure 8.21. The power demand at low gas flows (low Fl_g) corresponding to the formation of vortex cavities differs little from that without gassing. With the formation of six clinging cavities at higher gas flow rates, the power requirement is reduced by at most 10% compared with ungassed conditions. Development of the '3–3' cavity structure is associated with a significant reduction in power of up to 60% of the ungassed value. If the gas supply is extended beyond the gas-handling capacity

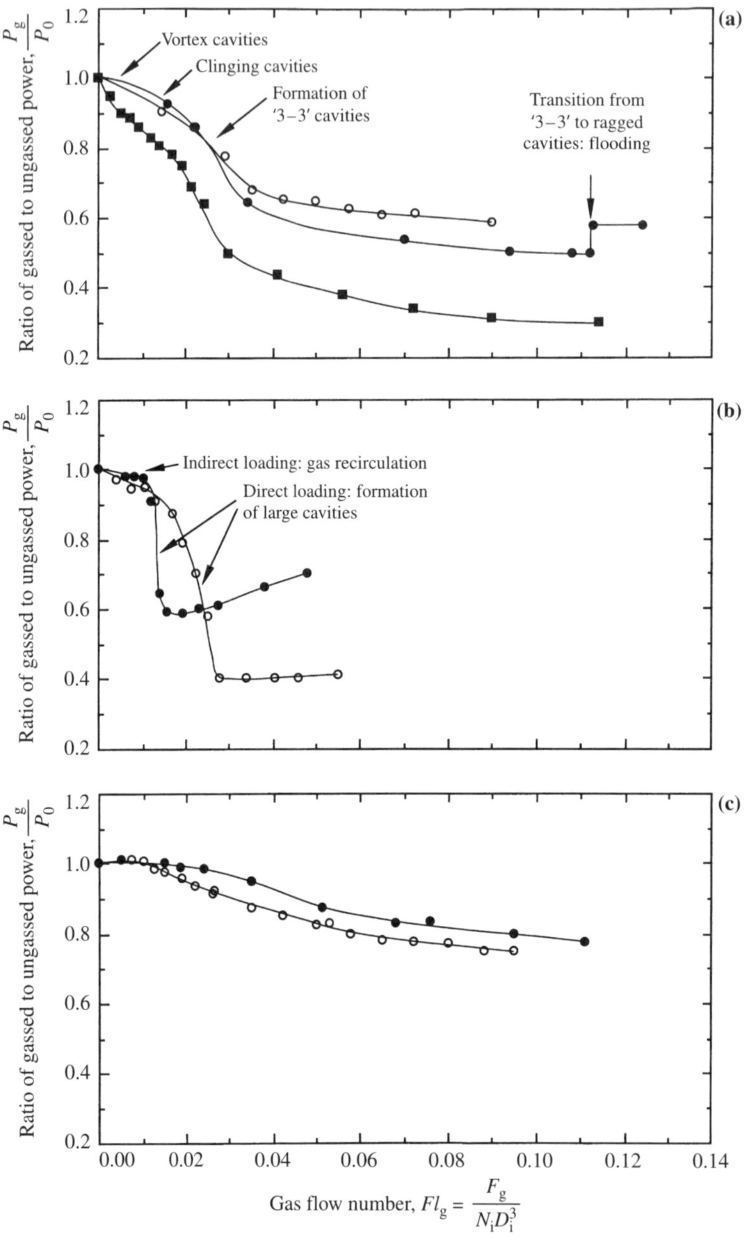

FIGURE 8.32 Variation in power consumption with gassing P_g relative to ungassed power consumption P_0 at constant stirrer speed. Fl_g is dimensionless gas flow number; F_g is volumetric gas flow rate; N_i is stirrer speed; D_i is impeller diameter.
(a) Rushton turbine: (○) $N_i = 120$ rpm; (●) $N_i = 150$ rpm; (■) $N_i = 360$ rpm.
(b) Downward-pumping pitched-blade turbine, six blades are angled 45°: (●) $N_i = 180$ rpm; (○) $N_i = 300$ rpm.
(c) Upward-pumping pitched-blade turbine, six blades are angled 45°: (●) $N_i = 190$ rpm; (○) $N_i = 236$ rpm.

(a) Data from M.M.C.G. Warmoeskerken, J.M. Smith, and M. Konno, 1985, On the flooding/loading transition and the complete dispersal condition in aerated vessels agitated by a Rushton-turbine. In: Mixing, Proc. 5th Eur. Conf. on Mixing, *pp. 143–154, The Fluid Engineering Centre; and M.M.C.G. Warmoeskerken, J. Feijen, and J.M. Smith, 1981, Hydrodynamics and power consumption in stirred gas–liquid dispersions. In: Fluid Mixing,* IChemE Symp. Ser. *64, J1–J14.*
(b) Data from M.M.C.G. Warmoeskerken, J. Speur, and J.M. Smith, 1984, Gas–liquid dispersion with pitched blade turbines. Chem. Eng. Commun. *25, 11–29.*
(c) Data from W. Bujalski, M. Konno, and A.W. Nienow, 1988, Scale-up of 45° pitch blade agitators for gas dispersion and solid suspension. In: Mixing, Proc. 6th Eur. Conf. on Mixing, *pp. 389–398, The Fluid Engineering Centre.*

of the impeller, the formation of six ragged cavities coincides with impeller flooding and a rise in power consumption as indicated in Figure 8.32(a).

Changes in power consumption with gassing for a downward-pumping pitched-blade turbine are shown in Figure 8.32(b). At low gas flow rates corresponding to the indirect loading regime (Section 8.4.3, Downward Pumping subsection), gas flow to the impeller relies on gas recirculation. Under these conditions, the loss of power is relatively small at ≤10%. Depending on the stirrer speed, at higher gas flows there may be a transition to direct loading; this coincides with a sharp decrease in power consumption as large ventilated cavities form behind the impeller blades. The power drawn by downward-pumping pitched-blade turbines can be reduced to as little as 30 to 40% of the power requirements without gassing.

Whereas the power consumption by ungassed upward-pumping pitched-blade turbines is similar to that for downward-pumping turbines of the same geometry (Table 8.2), as indicated in Figure 8.32(c) there is a much smaller reduction in power with aeration for upward-pumping impellers. Power consumption is reduced by a maximum of only about 20% during upward-pumping operation, even though ventilated cavities form behind the impeller blades.

Power characteristics with aeration have also been measured for the impellers described in Section 8.4.4. When the curved-blade disc turbine shown in Figure 8.25 is rotated clockwise with the concave sides of the blades forward, the curvature of the blades ensures that no large ventilated cavities can form on the convex surfaces. As a consequence, in low-viscosity fluids the power consumption with gassing remains close to that without gassing until impeller flooding occurs. In non-Newtonian or viscous fluids, power losses may be greater at up to about 20% [9]. For the hydrofoil impellers shown in Figure 8.26 operated for downward pumping, depending on the stirrer speed, abrupt reductions in power can accompany the transition from indirect to direct loading as large cavities form behind the blades. However, this drop in power is usually less that with Rushton turbines under similar conditions [7]. In contrast, for upward-pumping hydrofoils, there is virtually no reduction in power draw with aeration over a wide range of gas flow rates [23].

As indicated in Figure 8.32, there is a strong correlation between the power consumption in gassed liquids and changes in the structure of the ventilated cavities behind the impeller blades. However, although data such as those shown in Figure 8.32 have been measured for different impellers and illustrate how aeration affects power consumption, these graphs cannot be used to predict impeller power requirements with gassing. This limitation is related to the inadequacy of the gas flow number Fl_g to fully define the hydrodynamic conditions affecting the development of ventilated cavities. Whereas the value of Fl_g reflects the flow rate of gas from the sparger, the amount of gas actually entering the impeller region depends also on gas recirculation within the fluid. As shown in Figure 8.32, the size of the cavities and therefore the power draw vary with stirrer speed at constant Fl_g. This indicates that the results for P_g/P_0 are sensitive to aspects of the mixing system not represented by Fl_g. The exact extent to which power requirements are reduced by sparging is a complex function of the stirrer speed, air flow rate, vessel size, fluid properties, and the geometry of the impeller and tank including the impeller off-bottom clearance and sparger size. Because all of the changes in hydrodynamic behaviour due to gassing are not completely understood, prediction of the power requirements in aerated systems remains difficult and cannot yet be achieved with accuracy.

Although the power requirements in aerated systems depend strongly on the size of the ventilated cavities behind the impeller blades, differences in the power draw with gassing between coalescing and noncoalescing liquids (Section 10.6.2) are small, both for Rushton turbines and for axial-flow impellers such as hydrofoils [2, 24].

Reduction in stirrer power consumption with gassing may seem a desirable feature because of the potential for energy and cost savings during operation of the impeller. However, when all the relevant factors are considered, stirrers with power requirements that are relatively insensitive to gassing are preferred. In the design of fermentation equipment, the stirrer motor is usually sized to allow operation under nonaerated conditions. This is necessary to prevent motor burn-out if there is a failure of air supply during operation of the fermenter: the stirrer motor must be large enough to provide sufficient power during any abrupt change from gassed to ungassed conditions. In addition, medium in fermenters is often mixed without aeration during the heating and cooling cycles of *in situ* batch sterilisation (Section 14.6.1). Therefore, the decrease in impeller power consumption with gassing represents an under-utilisation of the capacity of the stirrer motor. As outlined in Chapter 10 (Section 10.9), the rate of oxygen transfer from gas to liquid in aerated systems depends on the power input to the fluid; therefore, any reduction in power diminishes the effectiveness of mass transfer in the system with potential deleterious consequences for culture performance. Power losses may also reduce the ability of the stirrer to maintain complete suspension of solids. For example, the sudden reductions in power shown in Figure 8.32(a) and (b) can result in severe loss of suspension capacity, with the result that cells begin to settle out on the vessel floor. An additional problem with stirrers that lose power with aeration is our inability to predict the exact extent of the power loss and the conditions under which it will occur. This creates some degree of uncertainty in the operation of gassed stirrer systems. All of these factors have promoted interest in the development of impellers such as the curved-blade disc turbine and upward-pumping hydrofoils, for which there is minimal reduction in power draw with gassing.

8.6 POWER INPUT BY GASSING

In addition to mechanical stirring, gas sparging itself contributes to the total power input to bioreactors during operation under aerated conditions. The power input from sparging, P_v, can be calculated using the equation:

$$P_v = F_g \rho g H_L \tag{8.13}$$

where F_g is the volumetric flow rate of gas at the temperature and average pressure of the liquid in the tank, ρ is the liquid density, g is gravitational acceleration, and H_L is the liquid height. For aerated vessels stirred with an impeller, P_v is usually only a small fraction of the total power input and is often neglected. However, if high gas flow rates are used at low stirrer speeds, for example in reactors that rely mainly on gas sparging for mixing with the stirrer playing a relatively minor role, the contribution of P_v to the total power input can be more substantial.

8.7 IMPELLER PUMPING CAPACITY

Fluid is pumped by the blades of rotating impellers. The volumetric flow rate of fluid leaving the blades varies with operating parameters such as the stirrer speed and size of the impeller, but is also a characteristic of the impeller type or design. The effectiveness of impellers for pumping fluid is represented by a dimensionless number called the *flow number*:

$$Fl = \frac{Q}{N_i D_i^3} \tag{8.14}$$

where Fl is the flow number, Q is the volumetric flow rate of fluid leaving the impeller blades, N_i is the stirrer speed, and D_i is the impeller diameter. The flow number is a measure of the ability of the impeller to generate strong circulatory flows, such as those necessary for blending operations and solids suspension.

The value of Q during impeller operation can be obtained by measuring local fluid velocities near the impeller blades using techniques such as laser Doppler velocimetry (Section 7.9.3). Typical fluid discharge velocity profiles for radial- and axial-flow turbines are shown in Figure 8.33. For radial-flow impellers, the mean discharge velocity is maximum at the centre line of the blade and decays above and below the centre line to form a bell-shaped curve, as illustrated in Figure 8.33(a). As well as fluid leaving the edge of the blade directly, from which Q is evaluated, there is also significant additional flow of entrained fluid above and below the blade that is swept along by the direct discharge stream. The total discharge flow from the impeller region including the entrained fluid can be several times greater than Q. The discharge velocity profile for an axial-flow turbine is shown in Figure 8.33(b). In this case, Q is the volumetric flow rate of fluid leaving directly from the lower edges of the blades, excluding entrained flow from the surrounding region.

Values of the flow number Fl for several impellers operating in the turbulent regime in low-viscosity fluids without gassing are listed in Table 8.3. Fl is dependent on the vessel and blade geometry; however this dependence and the variation of Fl between impeller types is not as great as for the turbulent power number N'_P.

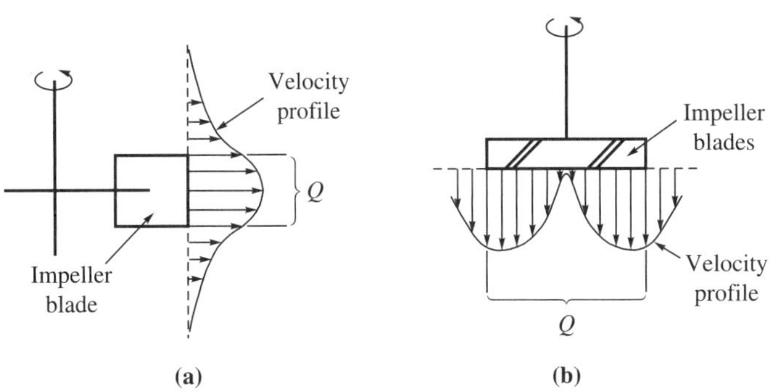

FIGURE 8.33 Graphic of discharge velocity profiles from the blades of: (a) a radial-flow impeller and (b) an axial-flow impeller.

TABLE 8.3 Values of the Turbulent Ungassed Flow Number *Fl* for a Selection of Different Impellers in Baffled Tanks Containing Low-Viscosity Fluid

Impeller	*Fl*	Reference
Rushton turbine	0.78	[25]
Propeller 3 blades	0.73	[26]
Pitched-blade turbine Downward pumping 4 blades, 45°	0.75	[4]
Pitched-blade turbine Downward pumping 6 blades, 45°	0.81	[27]
Hydrofoil (Lightnin A315) Downward pumping 4 blades	0.74	[27]
Hydrofoil (Prochem Maxflo T) Downward pumping 6 blades	0.82	[17]

The effectiveness of different impellers for generating flow can be compared relative to their power requirements. Combination of Eqs. (8.14) and (8.9) yields an expression for the impeller discharge flow rate per unit power consumed:

$$\frac{Q}{P} = \frac{Fl}{N'_P \rho N_i^2 D_i^2} \tag{8.15}$$

Therefore, if two different impellers of the same size are operated in the same fluid at the same stirrer speed, their pumping efficiencies can be compared using the ratio:

$$\frac{\left(\frac{Q}{P}\right)_{\text{impeller 1}}}{\left(\frac{Q}{P}\right)_{\text{impeller 2}}} = \frac{\left(\frac{Fl}{N'_P}\right)_{\text{impeller 1}}}{\left(\frac{Fl}{N'_P}\right)_{\text{impeller 2}}} \tag{8.16}$$

We can compare the pumping efficiencies of Rushton and pitched-blade turbines using Eq. (8.16) and values of N'_P and Fl from Tables 8.2 and 8.3. For a Rushton turbine with $D_i/D_T = 0.50$, $N'_P = 5.9$ and $Fl = 0.78$. For a six-blade downward-pumping pitched-blade turbine of the same size, $N'_P = 1.6$ and $Fl = 0.81$. Therefore:

$$\frac{\left(\frac{Q}{P}\right)_{\text{pitched-blade}}}{\left(\frac{Q}{P}\right)_{\text{Rushton}}} = \frac{\left(\frac{0.81}{1.6}\right)}{\left(\frac{0.78}{5.9}\right)} = 3.8 \tag{8.17}$$

This result indicates that pitched-blade turbines produce almost four times the flow for the same power input as Rushton turbines. The analysis provides an explanation for why Rushton turbines are considered to have relatively low pumping efficiency (Section 8.4.1, Without Gassing subsection), while pitched-blade turbines are recognised for their high pumping capacity and effectiveness for blending operations (Section 8.4.3). A comparison of Rushton turbines with hydrofoil impellers yields similar results. The analysis applies only for ungassed liquids; the effect of gassing on liquid pumping rates and power consumption varies considerably between different impellers.

8.8 SUSPENSION OF SOLIDS

Bioreactors used for cell culture contain biomass solids of varying size. Microorganisms such as single-cell bacteria and yeast are small and finely divided; other cells such as mycelia and plant cells form macroscopic aggregates or clumps depending on the culture conditions. Because cells contain a high percentage of water, the difference in density between the solid phase and the suspending liquid is generally very small. However, in some bioprocesses, cells are immobilised on or in solid matrices of varying material density. These systems include anchorage-dependent animal cells cultured on microcarrier beads, and bacteria attached to sand grains for waste water treatment.

One of the functions of mixing in stirred fermenters is to maintain the cells in suspension. Accumulation of biomass at the bottom of the vessel is highly undesirable, as cells within settled layers have poor access to nutrients and oxygen and can become starved of these components. It is important, therefore, to know what operating conditions are required to completely suspend solids in stirred tanks.

8.8.1 Without Gassing

A common criterion used to define complete suspension of solids is that no particle should remain motionless on the bottom of the vessel for more than 1 to 2 seconds. Applying this criterion, the Zwietering equation [28]:

$$N_{JS} = \frac{S\,\nu_L^{0.1} D_p^{0.2} \left[g(\rho_p - \rho_L)/\rho_L\right]^{0.45} X^{0.13}}{D_i^{0.85}} \tag{8.18}$$

is generally accepted as the best correlation for N_{JS}, the stirrer speed required for just complete suspension of solids in the absence of gassing. In Eq. (8.18), S is a dimensionless parameter dependent on the impeller and tank geometry, ν_L is the liquid kinematic viscosity (Section 7.3), D_p is the diameter of the solid particles, g is gravitational acceleration, ρ_p is the particle density, ρ_L is the liquid density, X is the weight percentage of particles in the suspension, and D_i is the impeller diameter.

Zwietering's equation has been subjected to extensive testing over many years using a wide range of system properties. The exponents in Eq. (8.18) are independent of the tank size, impeller type, impeller-to-tank diameter ratio, and impeller off-bottom clearance;

TABLE 8.4 Values of the Geometric Parameter S in Eq. (8.18) for Flat-Bottom Tanks

Impeller	D_i/D_T	C_i/D_T	S	Reference
Rushton	0.25	0.25	12	[12]
	0.33	0.17	5.8	[12]
	0.33	0.25	6.7	[12]
	0.33	0.50	8.0	[12]
	0.50	0.25	4.25	[12]
	0.50	0.17	3.9	[29]
Propeller	0.33	0.25	6.6	[29]
Pitched-blade turbine	0.33	0.20	5.7	[30]
Downward pumping, 4 blades, 45°	0.33	0.25	6.2	[30]
	0.33	0.33	6.8	[30]
	0.33	0.50	11.5	[30]
	0.50	0.25	5.8	[12]
Pitched-blade turbine Downward pumping, 6 blades, 45°	0.50	0.25	5.7	[29]
Pitched-blade turbine Upward pumping, 6 blades, 45°	0.50	0.25	6.9	[31]

these geometric factors are reflected in the value of S. Table 8.4 lists some values of S for different impeller geometries; these data were obtained using flat-bottomed cylindrical vessels with four baffles of width 1/10 the tank diameter and liquid height equal to the tank diameter.

N_{JS} decreases significantly as the size of the impeller increases, not only because of the direct effect of D_i in Eq. (8.18) but also because the D_i/D_T ratio changes the value of S. For a fixed impeller off-bottom clearance, a general relationship is:

$$S \propto \left(\frac{D_T}{D_i}\right)^{\alpha} \tag{8.19}$$

where α is approximately 1.5 for Rushton turbines and 0.82 for propellers [28]. For many impellers, S is sensitive to the impeller off-bottom clearance ratio C_i/D_T. As shown in Table 8.4 for Rushton and pitched-blade turbines, S at constant D_i/D_T decreases as the impeller clearance is reduced, so that lower stirrer speeds are required for complete suspension. The shape of the base of the vessel (Figure 8.2) also influences the efficiency of solids suspension, with dished, contoured, and cone-and-fillet bases offering advantages in some cases for reducing S. The extent of this effect depends, however, on the type of flow pattern generated by the impeller [14, 29, 32]. The presence of obstructions near the vessel floor, such as large sparger pipes or the bearing and seal housing for bottom-entry stirrers, can significantly impede solids suspension. Under these conditions, application of Eq. (8.18) will result in substantial underestimation of N_{JS}.

Even if a stirrer is operated at speeds equal to or above N_{JS} to obtain complete particle suspension, this does not guarantee that the suspension is homogeneous throughout the tank. Although, in general, speeds considerably higher than N_{JS} are required to achieve uniform particle concentration, for small particles such as dispersed cells with density similar to that of the suspending liquid, a reasonable degree of homogeneity can be expected at N_{JS}. In some respects, the criterion of complete solids suspension is a severe one, as suspension of the final few particles can require a disproportionately large increase in power—for example, as much as 100% [6]. In some systems, an adequate level of solids suspension may be achieved at stirrer speeds lower than N_{JS}.

EXAMPLE 8.3 SOLIDS SUSPENSION

Clump-forming fungal cells are cultured in a flat-bottomed 10-m^3 fermenter of diameter 2.4 m equipped with a Rushton turbine of diameter 1.2 m operated at 50 rpm. The impeller off-bottom clearance is 0.6 m. The density of the fermentation medium is 1000 kg m^{-3} and the viscosity is 0.055 Pa s. The density and diameter of the cell clumps are 1035 kg m^{-3} and 600 μm, respectively. The concentration of cells in the fermenter reaches 40% w/w. Are the cells suspended under these conditions?

Solution

$D_i/D_T = 0.50$ and $C_i/D_T = 0.25$. Therefore, from Table 8.4, $S = 4.25$. From Eq. (2.16), $g = 9.81$ m s^{-2}. From Table A.9 in Appendix A, 1 Pa s = 1 kg m^{-1} s^{-1}. Therefore, using Eq. (7.9) to calculate the kinematic viscosity:

$$\nu_L = \frac{0.055 \text{ kg m}^{-1}\text{ s}^{-1}}{1000 \text{ kg m}^{-3}} = 5.5 \times 10^{-5}\text{ m}^2\text{ s}^{-1}$$

Substituting values into Eq. (8.18) gives:

$$N_{JS} = \frac{4.25\,(5.5 \times 10^{-5}\,\text{m}^2\,\text{s}^{-1})^{0.1}\,(600 \times 10^{-6}\,\text{m})^{0.2}\left(9.81\text{ m s}^{-2}\dfrac{(1035 - 1000)\text{kg m}^{-3}}{1000\text{ kg m}^{-3}}\right)^{0.45} 40^{0.13}}{(1.2\text{ m})^{0.85}}$$

$$N_{JS} = 0.31\text{ s}^{-1} = 18.6\text{ rpm}$$

The operating stirrer speed of 50 rpm is well above the stirrer speed required for solids suspension. The cells are therefore completely suspended.

As illustrated in Example 8.3, complete suspension of cells and small cell clumps is generally achieved at low to moderate stirrer speeds. This is due mainly to the small size and almost neutral density (i.e., density close to that of the suspending liquid) of the cells.

8.8.2 With Gassing

The formation of ventilated cavities behind impeller blades reduces the power draw and liquid pumping capacity of the impeller, with the result that higher stirrer speeds are often required for solids suspension in aerated systems. If the fall in power with gassing is severe—for example, as shown in Figure 8.32(b) for downward-pumping pitched-blade turbines—there may be a corresponding sudden loss of solids suspension capacity.

Equations relating N_{JSg}, the stirrer speed required for just complete suspension of solids in the presence of gassing, to N_{JS} have been developed for various impellers [3, 31]. For Rushton turbines of two different sizes with an impeller off-bottom clearance of one-quarter the tank diameter [3]:

$$N_{JSg} = N_{JS} + 2.4\, F_{gv} \qquad \text{for } D_i/D_T = 0.33 \tag{8.20}$$

$$N_{JSg} = N_{JS} + 0.94\, F_{gv} \qquad \text{for } D_i/D_T = 0.50 \tag{8.21}$$

where F_{gv} is the gas flow rate in units of vvm or volume of gas per volume of liquid per minute, and N_{JS} and N_{JSg} both have units of s^{-1}. The above equations apply when the impeller is not flooded; however, they may not apply for suspension of low-density particles with $(\rho_P - \rho_L)$ less than about 50 kg m^{-3} [33], or at very low gas flow rates [34]. In both these cases, gassing has been shown to aid rather than hinder suspension. Solids suspension in aerated systems is sensitive to several aspects of tank geometry such as the impeller off-bottom clearance and the impeller–sparger separation [3, 14].

8.9 MECHANISMS OF MIXING

In this section we consider the mechanisms controlling the rate of mixing in stirred tanks containing a single liquid phase. As illustrated schematically in Figures 8.7 and 8.8, large liquid circulation loops develop in stirred vessels. For mixing to be effective, the velocity of fluid leaving the impeller must be sufficient to carry material into the most remote regions of the tank; fluid circulated by the impeller must also sweep the entire vessel in a reasonable time. In addition, turbulence must be developed in the fluid as mixing is certain to be poor unless flow is turbulent. All these factors are important in mixing, which can be described as a combination of three physical processes:

- Distribution
- Dispersion
- Diffusion

Distribution is sometimes called *macromixing*; diffusion is also called *micromixing*. Dispersion can be classified as either micro- or macromixing depending on the scale of fluid motion.

The pattern of bulk fluid flow in a baffled vessel stirred by a centrally located radial-flow impeller is shown in detail in Figure 8.34. Due to the periodic pumping action of the impeller, the contents of the vessel are recirculated through the mixing zone in a very regular manner. Near the impeller there is a zone of intense turbulence where fluid currents converge and exchange material. However, as fluid moves away from the impeller, flow

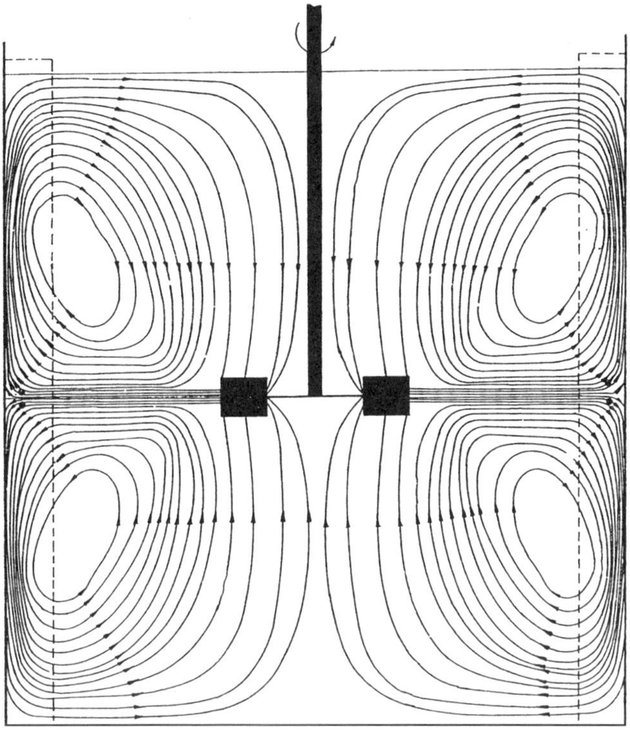

FIGURE 8.34 Flow pattern developed by a centrally located radial-flow impeller. *From R.M. Voncken, J.W. Rotte, and A.Th. ten Houten, 1965, Circulation model for continuous-flow, turbine-stirred, baffled tanks. In:* Mixing—Theory Related to Practice, *Proc. Symp. 10, AIChE–IChemE Joint Meeting, London.*

becomes progressively slower and less turbulent. In large tanks, streamline or laminar flow may develop in these local regions. Under these conditions, because fluid elements move mostly parallel to each other in streamline flow (Section 7.2.1), mixing is not very effective away from the impeller zone.

Let us consider what happens when a small amount of liquid dye is dropped onto the top of the fluid in Figure 8.34. First, the dye is swept by circulating currents down to the impeller. At the impeller there is vigorous and turbulent motion of fluid; the dye is mechanically dispersed into smaller volumes and distributed between the large circulation loops. These smaller parcels of dye are then carried around the tank, dispersing all the while into those parts of the system not yet containing dye. Returning again to the impeller, the dye aliquots are broken up into even smaller volumes for further distribution. After a time, dye is homogeneously distributed throughout the tank and achieves a uniform concentration.

The process whereby dye is transported to all regions of the vessel by bulk circulation currents is called *distribution*. Distribution is an important process in mixing but can be relatively slow. In large tanks, the size of the circulation paths is large and the time taken to traverse them is long; this, together with the regularity of fluid pumping at the impeller, inhibits rapid mixing. Accordingly, *distribution is often the slowest step in the mixing process*. If the stirrer speed is sufficiently high, superimposed on the distribution process is turbulence. In turbulent flow, the fluid no longer travels along streamlines but moves erratically in the form of cross-currents; this enhances the mixing process at scales much smaller than the scale of bulk circulation. As described in Section 7.9.1 (Eddies and Scales of Turbulence subsection), the kinetic energy of turbulent fluid is directed into regions of rotational flow called *eddies*; masses of eddies of various size coexist in turbulent flow. Large eddies are continuously formed from the bulk flow generated by the stirrer; these break down into small eddies that produce even smaller eddies. Eddies, like spinning tops, possess kinetic energy that is transferred to eddies of decreasing size. When the eddies become very small they can no longer sustain rotational motion and their kinetic energy is dissipated as heat. At steady state in a mixed tank, most of the energy from the stirrer is dissipated through the eddies as heat; energy lost in other processes (e.g., fluid collision with the tank walls) is generally negligible.

The process of breaking up the bulk flow into smaller and smaller eddies is called *dispersion*. Dispersion facilitates rapid transfer of material throughout the vessel. The degree of homogeneity possible as a result of dispersion is limited by the size of the smallest eddies that may be formed in a particular fluid. This size is given approximately as the *Kolmogorov scale of mixing* or *scale of turbulence*, λ, defined in Eq. (7.36) as:

$$\lambda = \left(\frac{\nu^3}{\varepsilon}\right)^{1/4} \tag{7.36}$$

where λ is the characteristic dimension of the smallest eddies, ν is the kinematic viscosity of the fluid (Section 7.3), and ε is the local rate of turbulence energy dissipation per unit mass of fluid. At steady state, the average rate of energy dissipation by turbulence over the entire tank is equal to the power input to the fluid by the impeller; this power input is the same as that estimated using the methods of Section 8.5. According to Eq. (7.36), the greater the power input to the fluid, the smaller are the eddies. λ is also dependent on viscosity: at a given power input, smaller eddies are produced in low-viscosity fluids. For low-viscosity liquids such as water, λ is usually in the range 30 to 100 μm. For such fluids, this is the smallest scale of mixing achievable by dispersion.

Within eddies, flow of fluid is rotational and occurs in streamlines. Because streamline flow does not facilitate mixing, to achieve mixing on a scale smaller than the Kolmogorov scale, we must rely on *diffusion*. Molecular diffusion is generally regarded as a slow process; however, over small distances it can be accomplished quite rapidly. Within eddies of diameter 30 to 100 μm, homogeneity is achieved in about 1 s for low-viscosity fluids. Consequently, if the power input to a stirred vessel produces eddies of this dimension, mixing on a molecular scale is accomplished virtually simultaneously.

8.10 ASSESSING MIXING EFFECTIVENESS

As explained in the last section, to achieve rapid mixing in a stirred tank, the agitator must provide good bulk circulation or macromixing. Micromixing at or near the molecular scale is also important, but occurs relatively quickly compared with macromixing. Mixing effectiveness is therefore usually a reflection of the rate of bulk flow.

Mixing time is a useful parameter for assessing the overall speed of mixing in stirred vessels. The mixing time t_m is the time required to achieve a given degree of homogeneity starting from the completely segregated state. It can be measured by injecting a tracer into the vessel and following its concentration at a fixed point in the tank. Tracers in common use include acids, bases, and concentrated salt solutions; corresponding detectors are pH probes and conductivity cells. Mixing time can also be determined by measuring the temperature response after addition of a small quantity of heated liquid.

Let us assume that a small pulse of tracer is added to fluid in a stirred tank already containing tracer material at concentration C_i. When flow in the system is circulatory, the tracer concentration measured at some fixed point in the tank can be expected to follow a pattern similar to that shown in Figure 8.35. Before mixing is complete, a relatively high concentration will be detected every time the bulk flow brings tracer to the measurement point. The peaks in concentration will be separated by a period approximately equal to the average time required for fluid to traverse one bulk circulation loop. In stirred vessels this period is called the *circulation time*, t_c. After several circulations the desired degree of homogeneity is reached.

Definition of the mixing time t_m depends on the degree of homogeneity required. Usually, mixing time is defined as the time after which the concentration of tracer differs from the final concentration C_f by less than 10% of the total concentration difference $(C_f - C_i)$. However, there is no single, universally applied definition of mixing time; sometimes deviations greater or less than 10% are specified. Nevertheless, at t_m the tracer concentration is relatively steady and the fluid composition approaches uniformity. Industrial-scale stirred vessels with working volumes between 1 and 100 m^3 have mixing times between about 30 and 120 s, depending on conditions. For rapid and effective mixing, t_m should be as small as possible.

Intuitively, we can predict that the mixing time in stirred tanks will depend on variables such as the size of the tank and impeller, the fluid properties, and the stirrer speed. The relationship between mixing time and several of these variables has been determined experimentally for different impellers: results for a Rushton turbine in a baffled tank are shown in Figure 8.36. The dimensionless product $N_i t_m$, which is also known as the *homogenisation number* or *dimensionless mixing time*, is plotted as a function of the impeller

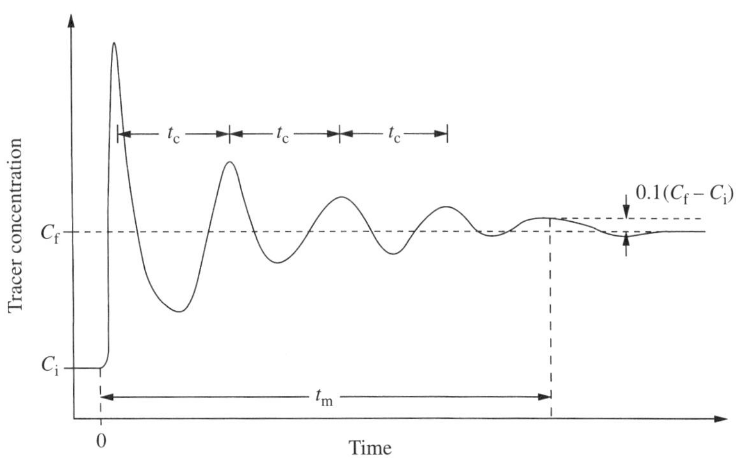

FIGURE 8.35 Concentration response after dye is injected into a stirred tank.

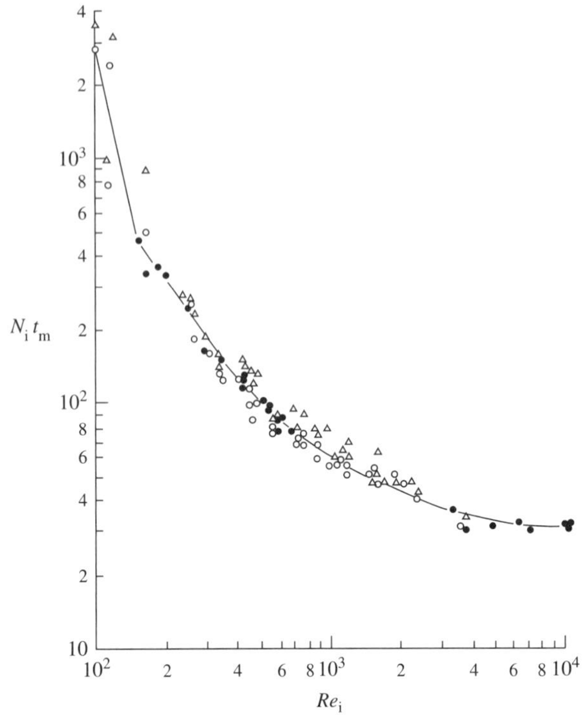

FIGURE 8.36 Variation of mixing time with Reynolds number for a Rushton turbine in a baffled tank. The impeller is located one-third the tank diameter off the floor of the vessel, the impeller diameter is one-third the tank diameter, the liquid height is equal to the tank diameter, and the tank has four baffles of width one-tenth the tank diameter. Several measurement techniques and tank sizes were used: (●) thermal method, vessel diameter 1.8 m; (○) thermal method, vessel diameter 0.24 m; (Δ) decolouration method, vessel diameter 0.24 m.
Reprinted from C.J. Hoogendoorn and A.P. den Hartog, Model studies on mixers in the viscous flow region, Chem. Eng. Sci. *22, 1689–1699. Copyright 1967, with permission from Pergamon Press Ltd, Oxford.*

Reynolds number Re_i; t_m is the mixing time based on a 10% deviation from the total change in conditions, and N_i is the rotational speed of the stirrer. Conceptually, $N_i\ t_m$ represents the number of stirrer rotations required to homogenise the liquid after addition of a small pulse of tracer. At relatively low Reynolds numbers in the laminar–transition regime, $N_i t_m$ increases significantly with decreasing Re_i. However, as the Reynolds number is increased, $N_i t_m$ approaches a constant value that persists into the turbulent regime at Re_i above about 5×10^3. The relationship between $N_i t_m$ and Re_i for most other impellers is qualitatively similar to that shown in Figure 8.36 [10]; in practice, therefore, we can assume that $N_i t_m$ reaches a constant value in turbulent flow. With $N_i t_m$ constant, mixing time reduces in direct proportion to increase in stirrer speed.

An equation has been developed for estimating the mixing time in stirred vessels under turbulent flow conditions. This expression for t_m can be applied irrespective of the type of impeller used [35, 36]:

$$t_m = 5.9\ D_T^{2/3} \left(\frac{\rho V_L}{P}\right)^{1/3} \left(\frac{D_T}{D_i}\right)^{1/3} \tag{8.22}$$

where t_m is the mixing time, D_T is the tank diameter, ρ is the liquid density, V_L is the liquid volume, P is the power input, and D_i is the impeller diameter. Equation (8.22) applies to baffled vessels stirred with a single impeller and with liquid height equal to the tank diameter. The relationship has been verified using a range of different impellers with

$0.2 \le D_i/D_T \le 0.7$ in vessels of diameter up to 2.7 m. The equation is also valid under aerated conditions provided the impeller disperses the gas effectively (i.e., is not flooded) and P is the power drawn with gassing [36]. As Eq. (8.22) does not depend on the type of impeller, we can deduce that all impellers are equally energy efficient with respect to mixing time. Equation (8.22) indicates that, for a tank of fixed diameter and liquid volume, mixing time is reduced if we use a large impeller and a high power input.

For a cylindrical tank with liquid height equal to the tank diameter, the geometric formula for the volume of a cylinder is:

$$V_L = \frac{\pi}{4} D_T^3 \tag{8.23}$$

Also, as Eq. (8.22) applies under turbulent conditions, we can express P in terms of the turbulent power number N'_P using Eq. (8.9). Substituting Eqs. (8.9) and (8.23) into Eq. (8.22) gives:

$$t_m = \frac{5.4}{N_i} \left(\frac{1}{N'_P}\right)^{1/3} \left(\frac{D_T}{D_i}\right)^2 \tag{8.24}$$

Equation (8.24) indicates that mixing time reduces in direct proportion to stirrer speed. This is the same as saying that $N_i t_m$ for a given impeller and tank geometry is constant for turbulent flow, as discussed earlier with reference to Figure 8.36. At constant N_i, t_m is directly proportional to $(D_T/D_i)^2$, showing that mixing times can be reduced significantly using impellers with large D_i/D_T ratio. However, because of the strong influence of impeller diameter on power requirements (Section 8.5.1), increasing D_i also raises the power consumption, so there will be a cost associated with using this strategy to improve mixing.

EXAMPLE 8.4 ESTIMATION OF MIXING TIME

A baffled fermenter with tank diameter and liquid height equal to 1.2 m is stirred using a six-blade downward-pumping Prochem Maxflo T hydrofoil impeller. The impeller diameter is 0.42 m and the stirrer speed is 1.5 s^{-1}. The viscosity of the fermentation broth is 10^{-2} Pa s and the density is 1000 kg m^{-3}. Estimate the mixing time under nonaerated conditions.

Solution

From Eq. (7.2):

$$Re_i = \frac{1.5\ s^{-1}\ (0.42\ m)^2\ 1000\ kg\ m^{-3}}{10^{-2}\ kg\ m^{-1}\ s^{-1}} = 2.6 \times 10^4$$

Flow is turbulent for remote-clearance impellers at this Reynolds number; therefore Eq. (8.24) can be used to calculate t_m. From Table 8.2, N'_P for this hydrofoil impeller is equal to 1.6. Therefore:

$$t_m = \frac{5.4}{1.5\ s^{-1}} \left(\frac{1}{1.6}\right)^{1/3} \left(\frac{1.2\ m}{0.42\ m}\right)^2 = 25.1\ s$$

The mixing time is 25 s.

Because the parameters affecting mixing efficiency also affect stirrer power requirements, it is not always possible to achieve small mixing times without consuming enormous amounts of energy, especially in large vessels. Relationships between power requirements, mixing time, and tank size are explored further in the following section.

8.11 SCALE-UP OF MIXING SYSTEMS

Design of industrial-scale bioprocesses is usually based on the performance of small-scale prototypes. It is always better to know whether a particular process will work properly before it is constructed in full size: determining optimum operating conditions at production scale is expensive and time-consuming. Ideally, scale-up should be carried out so that conditions in the large vessel are as close as possible to those producing good results at the smaller scale. As mixing is of critical importance in bioreactors, it would seem desirable to keep the mixing time constant on scale-up. Unfortunately, as explained in this section, the relationship between mixing time and power consumption makes this rarely possible in practice.

Suppose a cylindrical 1-m^3 pilot-scale stirred tank is scaled up to 100 m^3. Let us consider the power required to maintain the same mixing time in the large and small vessels. From Eq. (8.22), equating t_m values gives:

$$5.9\,D_{T1}^{2/3}\left(\frac{\rho_1 V_{L1}}{P_1}\right)^{1/3}\left(\frac{D_{T1}}{D_{i1}}\right)^{1/3} = 5.9\,D_{T2}^{2/3}\left(\frac{\rho_2 V_{L2}}{P_2}\right)^{1/3}\left(\frac{D_{T2}}{D_{i2}}\right)^{1/3} \tag{8.25}$$

where subscript 1 refers to the small-scale system and subscript 2 refers to the large-scale system. If the tanks are geometrically similar, the ratio D_T/D_i is the same at both scales. Similarly, the fluid density will be the same before and after scale-up, and we can also cancel the constant multiplier from both sides. Therefore, Eq. (8.25) reduces to:

$$D_{T1}^{2/3}\left(\frac{V_{L1}}{P_1}\right)^{1/3} = D_{T2}^{2/3}\left(\frac{V_{L2}}{P_2}\right)^{1/3} \tag{8.26}$$

Cubing both sides of Eq. (8.26) and rearranging gives:

$$P_2 = P_1\left(\frac{D_{T2}}{D_{T1}}\right)^2 \frac{V_{L2}}{V_{L1}} \tag{8.27}$$

The geometric relationship between V_L and D_T is given by Eq. (8.23) for cylindrical tanks with liquid height equal to the tank diameter. Solving Eq. (8.23) for D_T gives:

$$D_T = \left(\frac{4V_L}{\pi}\right)^{1/3} \tag{8.28}$$

Substituting this expression into Eq. (8.27) for both scales gives:

$$P_2 = P_1\left(\frac{V_{L2}}{V_{L1}}\right)^{5/3} \tag{8.29}$$

In our example of scale-up from 1 m^3 to 100 m^3, $V_{L2} = 100\ V_{L1}$. Therefore, the result from Eq. (8.29) is that $P_2 = \sim 2000\ P_1$; that is, the power required to achieve equal mixing time in the 100-m^3 tank is $\sim$2000 times greater than in the 1-m^3 vessel. This represents an extremely large increase in power, much greater than is economically or technically feasible with most equipment used for stirring. This example illustrates why the criterion of constant mixing time can hardly ever be applied for scale-up. Because the implications for power consumption are impractical, it is inevitable that mixing times increase with scale.

Reduced culture performance and productivity often accompany scale-up of bioreactors as a result of lower mixing efficiencies and consequent alteration of the physical environment. One way to improve the design procedure is to use *scale-down methods*. The general idea behind scale-down is that small-scale experiments to determine operating parameters are carried out under conditions that can actually be realised, physically and economically, at the production scale. For example, if we decide that power input to a large-scale vessel cannot exceed a certain limit, we can calculate the corresponding mixing time in the larger vessel, then use an appropriate power input to a small-scale reactor to simulate the mixing conditions that are in the large-scale system. Using this approach, as long as the flow regime (e.g., turbulent flow) is the same in the small- and large-scale fermenters, there is a better chance that the results achieved in the small-scale unit will be reproducible in the larger system.

8.12 IMPROVING MIXING IN FERMENTERS

Because of the limitations outlined in the previous section, longer mixing times are often unavoidable when stirred vessels are scaled up in size. In these circumstances, it is not possible to reduce mixing times sufficiently by simply raising the power input to the stirrer. In this section we consider methods for improving mixing in stirred tanks that do not involve consumption of significantly greater amounts of energy.

8.12.1 Impeller and Vessel Geometry

Mixing can sometimes be improved by changing the system's physical configuration.

- Baffles should be installed; this is routine for stirred fermenters and produces greater turbulence.
- For efficient mixing, the impeller should be mounted below the geometric centre of the vessel. For example, mixing by radial impellers such as Rushton turbines is facilitated when the circulation currents below the impeller are smaller than those above (as shown in Figure 8.7), as this makes the upper and lower circulation loops asynchronous. Under these conditions, fluid particles leaving the impeller at the same instant but entering different circulation paths take different periods of time to return and exchange material. The rate of distribution throughout the vessel is increased when the same fluid particles from different circulation loops do not meet each other every time they return to the impeller region.

- For two- and three-phase mixing (i.e., in gas–liquid and gas–liquid–solid systems), good mixing includes achieving complete gas dispersion and solids suspension. These stirrer functions are sensitive to various aspects of tank geometry, including the impeller off-bottom clearance, type of sparger, clearance between the sparger and the impeller, and base profile of the tank. Optimisation of these features of the system can yield considerable improvements in mixing effectiveness without necessarily requiring large amounts of extra power.
- Although installation of multiple impellers on the same stirrer shaft may seem a solution to problems of poor mixing, as discussed further in Section 8.13, the power required increases substantially when extra impellers are fitted. Furthermore, depending on the impeller design and the separation allowed between the impellers, mixing efficiency can actually be lower with multiple impellers than in single-impeller systems.

8.12.2 Feed Points

Severe mixing problems can occur in industrial-scale fermenters when material is fed into the vessel during operation. Concentrated acid or alkali and antifoam agents are often pumped automatically into the broth for pH and foam control; fermenters operated with continuous flow or in fed-batch mode also have fresh medium and nutrients added during the culture. If mixing and bulk distribution are slow, very high local concentrations of added material develop near the feed point. This problem has been observed in many types of culture but is particularly acute during large-scale production of single-cell protein from methanol. Because high levels of methanol are toxic to cells, biomass yields decrease significantly when mixing of feed material into the broth is slow. Such problems can be alleviated by installing multiple injection points to aid the distribution of substrate throughout the vessel. It is much less expensive to do this than to increase the stirrer speed and power input.

Location of the feed point (or feed points) is also important. In most commercial operations, material is fed into bioreactors using a single inlet delivering to the top surface of the liquid. However, mixing can be improved substantially by feeding directly into the impeller zone. This ensures rapid distribution and dispersion as convective currents and turbulence are strongest in this region. In many respects, use of surface-entry feeding represents the worst choice of feed-point location; as illustrated in Figures 8.14 and 8.27, fluid velocities in the upper reaches of stirred vessels can be very weak. Under these conditions, flow may be virtually stagnant in the regions where feeding is taking place, resulting in very poor rates of blending.

8.13 MULTIPLE IMPELLERS

Typical bioreactors used for aerobic fermentations do not conform to the standard configuration for stirred tanks illustrated in Figure 8.1, where the liquid height is approximately equal to the tank diameter. Instead, aerobic cultures are carried out in tall vessels with liquid heights 2 to 5 times the tank diameter, an aspect ratio of 3:1 being common. The reasons for using this geometry are that relatively high hydrostatic pressures are produced in tall vessels filled with liquid, thus increasing the solubility of oxygen, while rising air bubbles have longer contact time with the liquid, thus improving oxygen transfer from the gas phase.

Mixing in tall fermenters is carried out using more than one impeller mounted on the stirrer shaft. Each impeller generates its own circulation currents, but interaction between the fluid streams from different impellers can produce very complex flow patterns. An important parameter affecting the performance of multiple impellers is the spacing between them.

8.13.1 Multiple Rushton Turbines without Gassing

Because Rushton turbines have been the standard impeller in the bioprocessing industry for many years, the agitation system in many fermentation vessels consists of multiple Rushton turbines. In low-viscosity fluids, if Rushton turbines are spaced adequately apart, they each produce a radial discharge stream and generate independent large-scale circulation loops as illustrated in Figure 8.37. In this case, a vessel equipped with three Rushton turbines is mixed as if three separate stirred tanks were stacked one on top of the other. Under these conditions without gassing, the power required by multiple impellers can be estimated using the following equation:

$$(P)_n = n\,(P)_1 \tag{8.30}$$

where $(P)_n$ is the power required by n impellers and $(P)_1$ is the power required by a single impeller. The minimum spacing between the impellers for the flow pattern of Figure 8.37

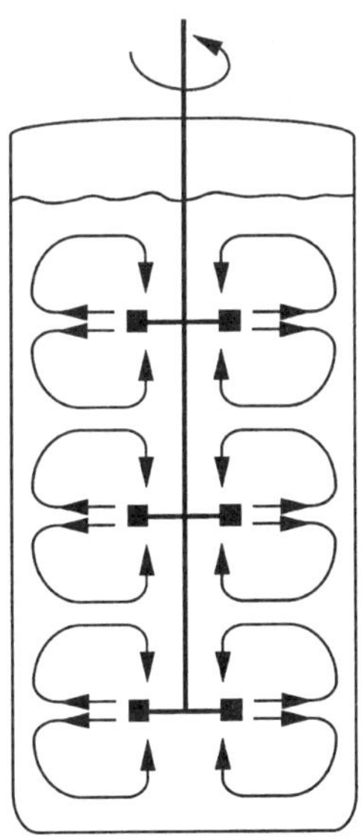

FIGURE 8.37 Independent circulation loops generated by multiple Rushton turbines in a tall fermenter when the spacing between the impellers is relatively large.

to occur, and for Eq. (8.30) to be valid, is not well defined, being reported variously as one to three impeller diameters or one tank diameter [37]. Equation (8.30) indicates that, at this spacing, each impeller draws the same power as if it were operating alone.

An important drawback associated with operation of multiple Rushton turbines is compartmentation of the fluid. As indicated in Figure 8.37, the radial flow pattern generated by Rushton turbines creates separate circulation currents above and below each impeller, providing little opportunity for interaction between the fluid streams emanating from different impellers. The resulting segregation creates a strong barrier to axial flow and top-to-bottom mixing of the contents of the tank over the entire height of the vessel. As a consequence, at a fixed stirrer speed, the overall rate of mixing is lower with multiple Rushton turbines than in a standard single-impeller system. In other words, installation of two Rushton impellers in a vessel with liquid height twice the tank diameter does not achieve the same mixing time as one Rushton impeller in liquid with height equal to the tank diameter. This result reflects the slow speed of fluid exchange between the separate circulation loops induced by multiple Rushton turbines compared with the rate of convective flow developed by a single impeller. In multiple impeller systems, *the rate of exchange flow between the fluid compartments generated by each impeller determines the rate of overall mixing in the vessel*. Therefore, improving the exchange between compartments has a high priority in mixing operations using multiple impellers.

When material is added to fermenters with multiple Rushton turbines, the location of the feed point has a significant effect on mixing time. This is illustrated in Figure 8.38, in which mixing time is plotted as a function of the height of the injection point in a vessel stirred with dual Rushton turbines. In this experiment, the clearance between the impellers was two times the impeller diameter or one tank diameter. The mixing time was lowest when the tracer was injected at the height where the circulation loops from the upper and lower impellers came together. Using this strategy, the barriers to mixing between the circulation loops were minimised as tracer became available to both impellers relatively quickly. The mixing time under these conditions was less than half that when material

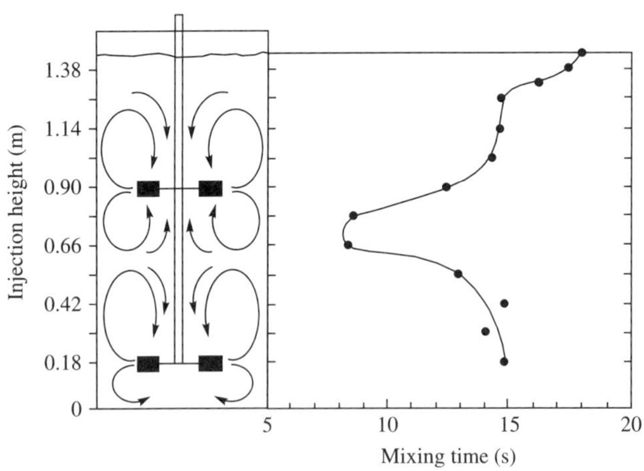

FIGURE 8.38 Variation in mixing time with height of the tracer injection point for dual Rushton turbines with clearance between the impellers equal to the tank diameter.
From D.G. Cronin, A.W. Nienow, and G.W. Moody, 1994, An experimental study of mixing in a proto-fermenter agitated by dual Rushton turbines. Trans. IChemE *72C, 35–40.*

was fed to the top surface of the liquid. If more than two impellers are present, similar advantage could be obtained by injecting feed material separately into each impeller zone.

Impeller spacings other than those giving the flow conditions illustrated in Figure 8.37 have also been tested [38]. If the impellers are moved further apart, the chances of material being exchanged between the circulation loops is reduced further, resulting in poorer overall mixing. On the other hand, if the impellers are moved closer together, the circulation loops from each impeller impinge on each other, allowing better fluid interchange. However, depending on the liquid height and the position of the impellers within the tank, placing impellers close together may leave large volumes of liquid in the upper or lower regions of the vessel less well mixed. Therefore, although inter-impeller mixing is improved, overall mixing is reduced. Additional impellers could be installed to overcome this problem but this would require a substantial increase in the total power input to the system.

8.13.2 Other Impeller Combinations without Gassing

Various combinations of impellers, including Rushton turbines, curved-blade disc turbines, propellers, and pitched-blade and hydrofoil impellers in both downward-pumping and upward-pumping modes, have been tested for multiple-impeller mixing. Combining radial- and axial-flow impellers, or combining different axial-flow turbines, significantly reduces fluid compartmentation and enhances mixing compared with multiple Rushton turbines. However, the flow patterns generated can be complex and show strong sensitivity to impeller geometry, diameter, and clearance.

A typical mean velocity vector plot for a dual-impeller system comprising a Rushton turbine in the lower position and a downward-pumping pitched-blade turbine above is shown in Figure 8.39. The upper impeller generates strong downward flow that enters the lower impeller axially from the top. The lower Rushton turbine generates an outward

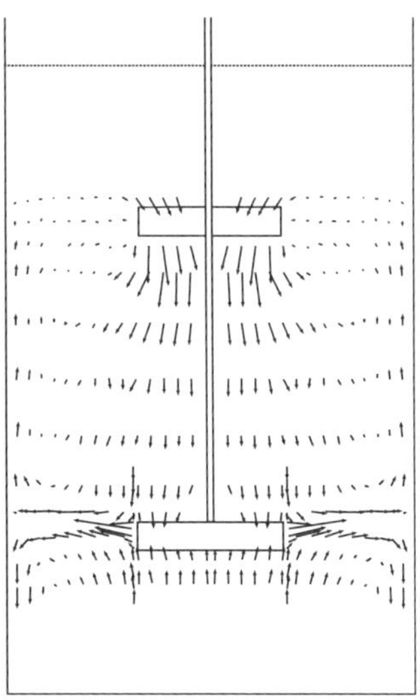

FIGURE 8.39 Mean velocity vector plot for a dual-impeller combination of a Rushton turbine in the lower position and a downward-pumping pitched-blade turbine above. The velocities were measured using laser Doppler velocimetry.

From V.P. Mishra and J.B. Joshi, 1994, Flow generated by a disc turbine. Part IV: Multiple impellers. Trans. IChemE *72A, 657–668.*

radial jet, which divides into two streams near the wall. The fluid flowing downwards at the wall forms a large-scale circulation loop below the Rushton impeller, while the fluid flowing upwards joins the major circulation current generated by the pitched-blade turbine. In this way, in contrast to the independent circulation loops developed by multiple Rushton turbines illustrated in Figure 8.37, the radial-flow–axial-flow combination in Figure 8.39 allows a greater degree of interaction between the fluid currents originating from each impeller. Only two major fluid compartments are generated, in contrast to the four separate circulation loops (i.e., one above and one below each impeller) that would be created with dual Rushton turbines. Generally, the smaller the number of separate large-scale flow loops in the vessel, the better is the mixing.

As discussed in Section 8.10, for single impellers under turbulent flow conditions, the mixing time t_m for a given power input is independent of impeller type. Therefore, Eqs. (8.22) and (8.24) can be applied irrespective of the impeller design. In contrast, in multiple impeller systems, the mixing time depends strongly on the type (or types) of impeller used. This is because the extent of interaction between the flow currents generated by each impeller, which has a significant effect on overall mixing, varies with impeller design.

For operation without gassing and with individual impellers spaced at least one impeller diameter apart, Eq. (8.30) provides a reasonable basis for estimating the power required by combinations of radial- and axial-flow impellers or multiple axial-flow turbines [39, 40]. The total power for the multiple impeller system is approximately equal to the sum of the individual power requirements for each impeller.

8.13.3 Multiple-Phase Operation

When gassing is required in vessels with multiple impellers, Rushton or curved-blade disc turbines are often used in the lowest position closest to the sparger because of their superior ability to handle gas compared with axial-flow turbines. Combining hydrofoil or pitched-blade impellers above a Rushton turbine is very effective in aerated systems: the lower impeller breaks up the gas flow while the upper, high-flow impeller/s distribute the dispersed gas throughout the tank.

With gassing, the power relationship for multiple impellers is not as simple as that in Eq. (8.30). The main reason is that the power required by the individual impellers may be affected by the presence of gas, but each impeller is not affected to the same extent. At the lowest impeller, the formation of ventilated cavities at the impeller blades reduces the power drawn as described in Section 8.5.3. However, as the quantity of gas passing through the upper impellers is much smaller than that handled by the lowest impeller, the reduction in power at the upper impellers is less significant. The effect of gassing on the power required by the upper impellers depends on the distance of the impeller from the sparger and the level of gas recirculation in the system. Although the lower impeller can be flooded if the gas flow rate is too high relative to the stirrer speed, the upper impellers are rarely flooded and continue to pump liquid effectively.

Solids suspension in multiple impeller systems has been studied comparatively little. In general, irrespective of the liquid height, there appears to be no advantage associated with using more than one impeller to suspend solids, with respect to either the stirrer speed or power required. At the same time, however, the use of multiple impellers may improve the uniformity of particle concentration throughout the tank [3, 41].

8.14 RETROFITTING

It is sometimes necessary or desirable to change the type of impeller used in existing fermentation vessels. For many years, Rushton turbines with diameter one-third the tank diameter were used almost exclusively for bioprocessing applications, and most fermentation facilities were constructed for operation with this impeller geometry. However, with increasing recognition of the importance of mixing in bioprocesses, and as modern impellers with improved features are being developed, the idea of replacing Rushton turbines with new impellers in retrofitting operations is now being pursued.

An important requirement for retrofitting is that neither the agitator drive nor the drive assembly be modified. This is because the expense associated with replacing the shaft, seals, bearings, gearbox, and possibly also the motor itself adds considerably to the overall cost. Therefore, ideally, retrofitting is carried out so that the power draw and torque in the new stirrer system are the same as for the old stirrer. From Eq. (8.5), maintaining the same power and torque implies that the operating stirrer speed N_i will also be the same.

For operation in the turbulent regime, we can calculate the size of the new impeller using Eq. (8.9). If the new and old impellers consume the same power, from Eq. (8.9):

$$(N'_P \rho N_i^3 D_i^5)_{new} = (N'_P \rho N_i^3 D_i^5)_{old} \tag{8.31}$$

As the liquid density and stirrer speed will be the same:

$$(D_i)^5_{new} = \frac{(N'_P)_{old}}{(N'_P)_{new}} (D_i)^5_{old} \tag{8.32}$$

or

$$(D_i)_{new} = \left(\frac{(N'_P)_{old}}{(N'_P)_{new}}\right)^{1/5} (D_i)_{old} \tag{8.33}$$

Because Rushton turbines have a relatively high turbulent power number N'_P, if a Rushton turbine is replaced with a low-N'_P impeller such as a hydrofoil or curved-blade disc turbine, Eq. (8.33) indicates that the diameter of the new impeller will be larger than the old. This has implications especially for improved gas handling and dispersion, as discussed in Section 8.4.1 (With Gassing subsection). Additional benefits may also apply: for example, if a curved-blade disc turbine is chosen as the replacement impeller, the loss of power with aeration will be much smaller compared with a Rushton turbine, thus allowing improved rates of oxygen transfer. In multiple impeller systems, replacing upper Rushton turbines with larger-diameter upward- or downward-pumping axial-flow turbines significantly reduces fluid compartmentation and enhances bulk mixing.

While there are advantages associated with retrofitting, there are also potential difficulties. Because of the different directions of fluid discharge from radial- and axial-flow impellers, different stresses are exerted by these types of impeller on the stirrer shaft, gear drive, and tank. The vessel and stirrer assembly must be able to withstand any change in external load after retrofitting. Mechanical as well as hydrodynamic instabilities, including increased vessel vibration, may occur in gassed systems after retrofitting of downward-pumping axial-flow impellers [42]; these problems are largely absent with Rushton turbines. Other factors may also affect retrofitting of larger-diameter impellers: for example, the presence of cooling coils and other fittings inside the vessel could limit the extent to which the impeller size can be increased, or the stirrer assembly may not be capable of supporting the increased impeller weight.

8.15 EFFECT OF RHEOLOGICAL PROPERTIES ON MIXING

As discussed in Sections 7.7 and 7.8, many fermentation broths have high viscosity or exhibit non-Newtonian flow behaviour. These properties have a profound influence on mixing, making it more difficult to achieve small mixing times and homogeneous broth composition. The principal deleterious effects of high fluid viscosity and non-Newtonian rheology are reduced turbulence and the formation of stagnant zones in the vessel.

For effective mixing, flow must be turbulent. As described in Section 8.9, turbulence is responsible for dispersing material at the scale of the smallest eddies. The existence of turbulence is indicated by the value of the impeller Reynolds number Re_i. Turbulence is damped at Re_i below about 10^4 for remote-clearance impellers; as a consequence, mixing times increase significantly as shown in Figure 8.36. Re_i as defined in Eq. (7.2) is inversely proportional to viscosity. Accordingly, nonturbulent flow and poor mixing are likely to occur during agitation of highly viscous fluids. Increasing the power input is an obvious solution; however, raising the power sufficiently to achieve turbulence is often impractical.

Most non-Newtonian fluids in bioprocessing are pseudoplastic. Because the apparent viscosity of these fluids depends on the shear rate, their rheological behaviour varies with the shear conditions in the fermenter. Metzner and Otto [43] proposed that the average shear rate $\dot{\gamma}_{av}$ in a stirred vessel is a linear function of the stirrer speed N_i:

$$\dot{\gamma}_{av} = k N_i \tag{8.34}$$

where k is a constant dependent on the type of impeller. Experimentally determined values of k are listed in Table 8.5. The validity of Eq. (8.34) was established in studies by Metzner et al. [44]. However, like other properties such as fluid velocity and turbulence kinetic energy, the shear rate in stirred vessels is far from uniform, being strongly dependent on distance from the impeller. Figure 8.40 shows estimated values of the shear rate in a pseudoplastic fluid as a function of radial distance from the tip of a Rushton turbine. The maximum shear rate close to the impeller is much higher than the average calculated using Eq. (8.34).

Pseudoplastic fluids are shear-thinning; that is, their apparent viscosity decreases with increasing shear. Accordingly, in stirred vessels, pseudoplastic fluids have relatively low

TABLE 8.5 Observed Values of k in Eq. (8.34)

Impeller type	k
Rushton turbine	10–13
Propeller	10
Anchor	20–25
Helical ribbon	30

From S. Nagata, 1975, Mixing: Principles and Applications, *Kodansha, Tokyo.*

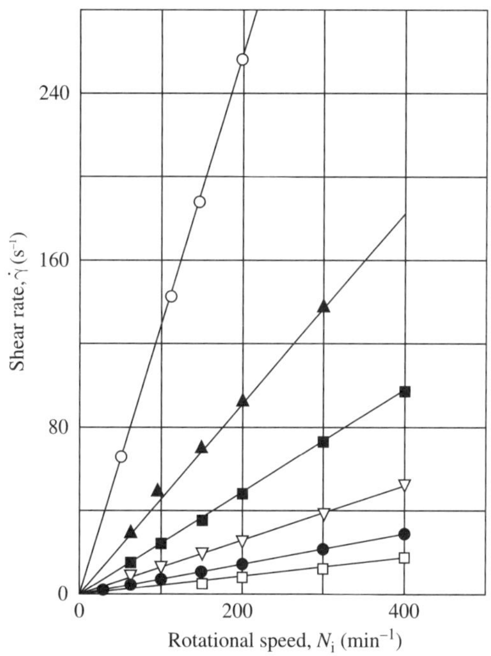

FIGURE 8.40 Shear rates in a pseudoplastic fluid as a function of the stirrer speed and radial distance from the impeller: (○) impeller tip; (▲) 0.10 in.; (■) 0.20 in.; (∇) 0.34 in.; (●) 0.50 in.; (□) 1.00 in.
From A.B. Metzner and J.S. Taylor, 1960, Flow patterns in agitated vessels. AIChE J. *6, 109–114.*

apparent viscosity in the high-shear zone near the impeller, and relatively high apparent viscosity away from the impeller. As a result, flow patterns similar to that shown in Figure 8.41 can develop. Highly shear-thinning fluids with flow behaviour index n (Section 7.5.1) less than 0.2 to 0.3 form *caverns* when subjected to agitation. Caverns are circulating pools of fluid surrounding the impeller; outside the caverns, the bulk liquid scarcely moves at all. Caverns also develop in liquids that exhibit a yield stress (Section 7.5.1); in this case, fluid remains stagnant in regions away from the impeller where the yield stress is not exceeded. Mixing inside caverns is intense, but there is very little exchange of material between the cavern and the rest of the tank. The effect of stirrer speed on cavern size is shown in Figure 8.42. Although the cavern expands as a proportion of the tank volume with increasing stirrer speed, dead zones persist in the peripheral

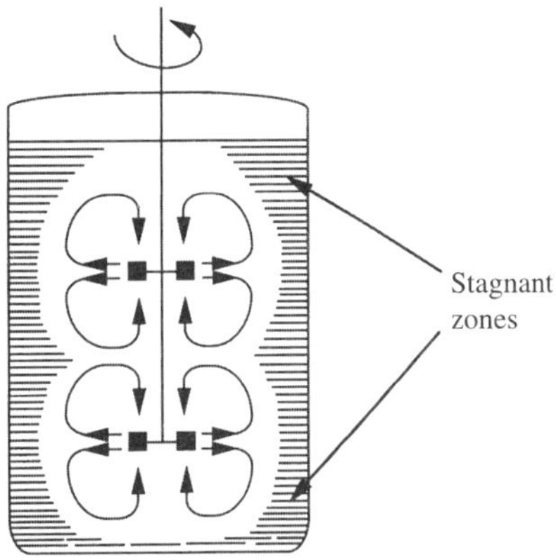

FIGURE 8.41 Mixing pattern for pseudoplastic fluid in a stirred fermenter.

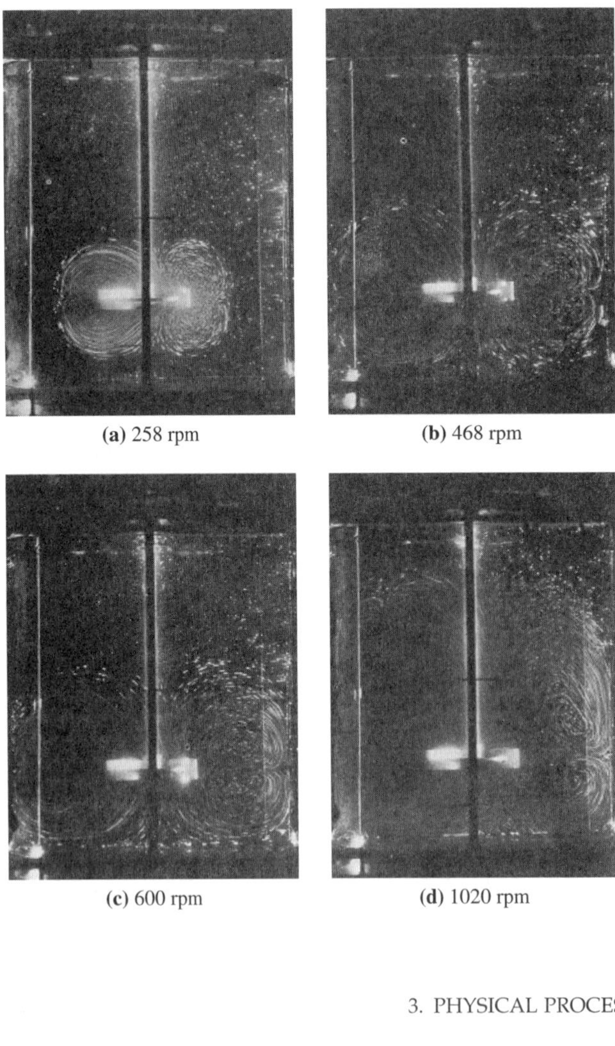

(a) 258 rpm (b) 468 rpm (c) 600 rpm (d) 1020 rpm

FIGURE 8.42 Effect of stirrer speed on cavern size in non-Newtonian, shear-thinning fluid with a yield stress.
From J. Solomon, T.P Elson, A.W. Nienow, and G.C. Pace, 1981, Cavern sizes in agitated fluids with a yield stress. Chem. Eng. Commun. *11, 143–164. Reproduced with permission from Taylor & Francis.*

regions of the vessel even with vigorous agitation and high power input. Therefore, in fermenters containing highly non-Newtonian broths (e.g., for production of biological gums such as xanthan) stagnant zones are very likely to develop in regions of the vessel away from the impeller caverns.

The effects of local fluid thinning in pseudoplastic fluids can be countered by modifying the geometry of the system and/or the impeller design. The use of multiple impellers, even when the liquid height is no greater than the tank diameter, improves the mixing of pseudoplastic fluids significantly. The combination of a Rushton turbine in the lower position and downward-pumping pitched-blade turbines above is particularly effective. Because the size of the caverns increases with impeller diameter, large impellers with diameters greater than one-half the tank diameter are also recommended. Different impeller designs that sweep the entire volume of the vessel may also be beneficial. As discussed in Section 8.4, the most common impellers used for viscous mixing are gate, anchor, and helical stirrers mounted with small clearance between the impeller and tank wall. However, application of these impellers in fermenters is only possible when the culture oxygen demand is low. Although small-clearance impellers operating at relatively slow speed give superior bulk mixing in viscous fluids, high-shear systems with high-speed, remote-clearance impellers are preferable for breaking up gas bubbles and promoting oxygen transfer to the liquid.

8.16 ROLE OF SHEAR IN STIRRED FERMENTERS

As outlined in Section 7.2.2, the development of shear stresses in fluids is related to the presence of velocity gradients. In turbulent flow, unsteady velocity components associated with eddies give rise to additional fluctuating velocity gradients that are superimposed on the mean flow. As a consequence, turbulent shear stresses are much greater than those developed in laminar flow (Section 7.9.2, Reynolds Stresses subsection). Turbulent shear stress also varies considerably with time and position in the fluid.

Mixing in bioreactors used for aerobic cell culture must provide the shear conditions necessary to disperse gas bubbles and, if appropriate, break up liquid droplets or cell flocs. The dispersion of gas bubbles by impellers involves a balance between opposing forces. The interaction of bubbles with turbulent velocity fluctuations in the fluid causes the bubbles to stretch and deform. Bubble break-up occurs if the induced stresses exceed the stabilising forces due to surface tension at the gas–liquid interface, which tend to restore the bubble to its spherical shape. For droplets, the droplet viscosity is an additional stabilising influence opposing break-up. In the case of solid material such as cells or aggregates, break-up occurs if the stresses induced by the turbulent velocity gradients are greater than the mechanical strength of the particles.

Whereas bubble break-up is required in fermenters to facilitate oxygen transfer, disruption of individual cells is undesirable. As indicated in Table 8.6, different cell types have different susceptibilities to damage in the bioreactor environment. This susceptibility is usually referred to as *shear sensitivity*, although the damage observed may not arise necessarily from the effects of fluid velocity gradients. In this context, the term 'shear' is used

TABLE 8.6 Cell Shear Sensitivity

Cell type	Size	Shear sensitivity
Microbial cells	1–10 μm	Low
Microbial pellets/clumps	Up to 1 cm	Moderate
Plant cells	100 μm	Moderate/high
Plant cell aggregates	Up to 1–2 cm	High
Animal cells	20 μm	High
Animal cells on microcarriers	80–200 μm	Very high

imprecisely to mean any mechanism dependent on the hydrodynamic conditions in the vessel that results in cell damage.

Because of their relatively small size, bacteria and yeast are considered generally not to be shear sensitive. The main effect of agitation on filamentous fungi and actinomycetes is a change in morphology between pelleted and dispersed forms; the mean size of cell clumps also varies with hydrodynamic conditions. Insect, mammalian, and plant cells are considered to be particularly sensitive to hydrodynamic forces. Bioreactors used for culture of these cells must take this sensitivity into account while still providing adequate mixing and oxygen transfer. At the present time, the effects of hydrodynamic forces on cells are not understood completely. Cell disruption is an obvious outcome; however more subtle *sublytic effects* such as slower growth or product synthesis, denaturation of extracellular proteins, and thickening of the cell walls may also occur. In some cases, cellular metabolism is stimulated by exposure to hydrodynamic forces. Because there are many gaps in our knowledge of how cells are affected, it is not possible in most cases to predict with confidence what bioreactor operating conditions will or will not damage shear-sensitive organisms. Some of the findings from research into various culture systems, and some of the mechanisms postulated to explain cell damage in bioreactors, are outlined next.

8.16.1 Studies with Animal Cell Cultures

Studies of cell damage in fermenters have been carried out mostly with animal cells. As animal cells lack a protective cell wall, cell injury and loss are significant problems in large-scale culture. Some types of animal cell such as insect, hybridoma, and blood cells grow readily when freely suspended in liquid medium. However, many transformed animal cell lines are *anchorage dependent*: this means that the cells must be attached to a solid surface for survival. In bioreactors, the surface area required for cell attachment is often provided by *microcarrier beads*, which are plastic or polymer beads of diameter 80 to 200 μm. As shown in Figure 8.43, the animal cells cover the surface of the beads, which are then suspended in nutrient medium. Many benefits are associated with the use of microcarriers; however, a disadvantage is that the cells cannot change position or rotate

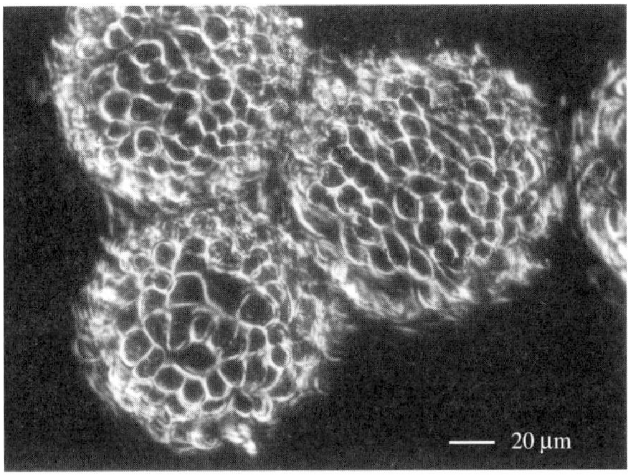

FIGURE 8.43 Chinese hamster ovary (CHO) cells attached to microcarrier beads. *Photograph courtesy of J. Crowley.*

easily in response to hydrodynamic forces in the fluid. This makes animal cells on microcarriers particularly susceptible to shear damage.

Several mechanisms have been postulated to cause cell damage in bioreactors:

- Interaction between cells and turbulent eddies
- Collisions between cells or between microcarrier beads, collision of cells with the impeller, and collision of cells with stationary surfaces in the vessel
- Interaction between cells and shear forces in the boundary layers and wakes near solid objects in the reactor, especially the impeller
- Interaction between cells and the mechanical forces associated with formation of bubbles at the sparger and bubble rise through the liquid
- Interaction between cells and bursting bubbles at the liquid surface

Investigation of these effects has so far resulted in only a partial understanding of animal cell damage in bioreactors. Industrial practice in this area is based largely on assumptions and 'rules of thumb' rather than theoretical or scientific principles. For cells attached to microcarrier beads, bead–bead interactions and interactions between the microcarriers and small turbulent eddies are considered most likely to damage cells. However, it has also been established that if the vessel is sparged with air, severe damage of suspended cells can occur during gas bubble burst at the surface of the liquid.

Interaction between Microcarriers and Turbulent Eddies

Interactions between microcarriers and eddies in turbulent flow have the potential to cause mechanical damage to cells. The intensity of the forces associated with these interactions is dependent on the size of the eddies relative to the microcarrier particles. If the particles are small, they tend to be captured or entrained in the eddies as shown in Figure 8.44(a). As fluid motion within eddies is laminar, if the density of the microcarriers is about the same as the suspending fluid, there is little relative motion of the particles. Accordingly, the velocity difference between the fluid streamlines and the microcarriers is small, except for brief periods of acceleration when the bead enters a new eddy. On

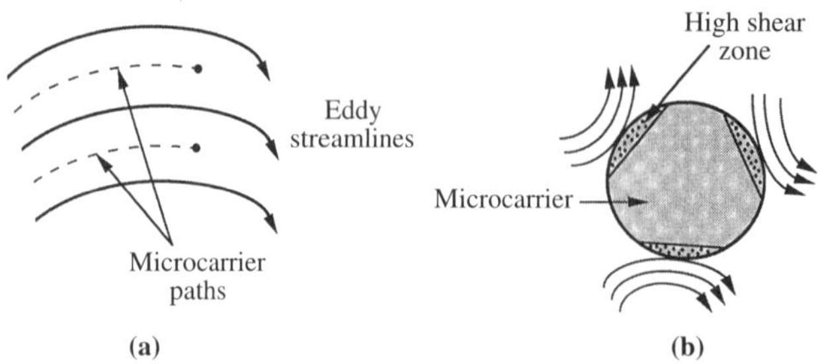

FIGURE 8.44 Eddy–microcarrier interactions. (a) Microcarriers are captured in large eddies and move within the streamline flow. (b) When several small eddies with opposing rotation interact with the microcarrier simultaneously, high levels of shear develop on the bead surface.
From R.S. Cherry and E.T. Papoutsakis, 1986, Hydrodynamic effects on cells in agitated tissue culture reactors. Bioprocess Eng. *1, 29–41.*

average, therefore, if the particles are smaller than the eddies, the shear effects of eddy–cell interactions are minimal.

If the stirrer speed is increased and the average eddy size reduced, interactions between eddies and microcarriers can occur in two possible ways. A single eddy that cannot fully engulf the particle may act on part of its surface and cause the particle to rotate in the fluid; this will result in a relatively low level of shear at the surface of the bead. However, much higher shear stresses develop when several eddies with opposing rotation interact with the particle and dissipate their energies on its surface simultaneously, as illustrated in Figure 8.44(b).

Experimental data for cell damage on microcarriers has been correlated by comparing the microcarrier diameter with the eddy size represented by the Kolmogorov scale as defined in Eq. (7.36):

$$\lambda = \left(\frac{\nu^3}{\varepsilon}\right)^{1/4} \tag{7.36}$$

In this equation, λ is the characteristic size of eddies in the dissipative range of the turbulence spectrum, ν is the kinematic viscosity of the fluid, and ε is the local rate of dissipation of turbulence kinetic energy per unit mass of fluid. For ε calculated as the average rate of energy dissipation over the entire tank volume, it has been found that cells on microcarriers suffer detrimental effects when the Kolmogorov scale drops below 2/3 to 1/2 the diameter of the microcarrier beads [45, 46]. Under these conditions, excessive agitation is considered to generate eddies of small enough size but sufficient energy to cause damage to the cells. The recommendation, therefore, is to operate the bioreactor so that the Kolmogorov scale remains greater than the microcarrier diameter.

As indicated in Eq. (7.36), if the viscosity of the fluid is increased, the size of the smallest eddies also increases. Raising the fluid viscosity should, therefore, reduce cell damage in bioreactors. This effect has been demonstrated by adding thickening agents to animal cell culture medium: Moderate increases in viscosity led to significant reductions in turbulent cell death [46].

Although this approach to predicting turbulent cell damage has been found to apply reasonably well in small-scale vessels, its broader application at larger scales raises several difficulties. Conceptually, the method is based on comparing the size of the microcarriers to the size of the dissipative eddies in the fluid. However, determining the size of the dissipative eddies can be problematical. As discussed in Section 7.9.2 (Isotropic Turbulence subsection), Kolmogorov's theories about scales of turbulence apply only to isotropic turbulence, which does not exist within the entire volume of bioreactors and may not exist even in the region of most intense turbulence near the impeller. Moreover, Kolmogorov's equation gives only an order-of-magnitude estimate of the size of eddies in the dissipative range. On a practical level, the value of λ estimated using Eq. (7.36) is dependent directly on the volume used to evaluate the rate of turbulence energy dissipation per unit mass of fluid, ε. One approach is to assume that the power input by the impeller is dissipated uniformly over the entire tank volume. In this case:

$$\varepsilon = \frac{P}{\rho V_L} \tag{8.35}$$

where P is the power input, ρ is the fluid density, and V_L is the volume of fluid in the vessel. Yet, this is clearly an inaccurate method for estimating ε: as illustrated in Figure 8.17, the rate of turbulence kinetic energy dissipation is far from uniformly distributed in stirred tanks, being much greater near the impeller than in the remainder of the fluid. Therefore, use of the total liquid volume to evaluate ε is questionable and becomes increasingly so at large scales where rates of energy dissipation in most parts of the tank are very low. Application of Eq. (8.35) gives relatively high values of λ for a given power input, thus potentially underestimating the damaging effects of eddies. In contrast, if we assume that the power is dissipated mostly in the region of intense turbulence around the impeller, an alternative expression for ε is [47]:

$$\varepsilon = \frac{P}{\rho D_i^3} \tag{8.36}$$

where D_i is the impeller diameter and D_i^3 represents the approximate volume near the impeller where most of the energy dissipation occurs. Compared with Eq. (8.35), the calculated value of λ is smaller if Eq. (8.36) is used to estimate ε, so that damaging effects will appear more likely. In practice, it is very difficult to know the precise distribution of energy dissipation rates under particular operating conditions, and this affects accurate estimation of ε and thus the size of the dissipative eddies.

Cell Damage from Bursting Bubbles

Because of the much smaller size of individual animal cells ($\sim$20 μm) compared with microcarriers (80–200 μm), individual animal cells in suspension are relatively insensitive to the hydrodynamic forces generated by turbulence unless the agitation intensity is extremely high. However, when liquid containing shear-sensitive cells is sparged with gas, damaging mechanisms other than those associated with turbulence come into play. From research carried out so far, damage of suspended animal cells in bioreactors appears to be caused mainly by bubbles bursting as they disengage from the liquid at the surface of the culture. If the same cells are grown in the absence of air sparging, and the stirrer is prevented from entraining gas from the atmosphere, very high agitation rates can be tolerated without significant cell damage [48, 49].

Figure 8.45 shows the sequence of events associated with rupture of a bubble at a gas–liquid surface. The rising bubble just below the surface pushes liquid at the interface into a hemispherical film cap of thickness 1 to 10 μm. The cap drains under gravity until a critical thickness (typically <0.1 μm) is reached, then ruptures at the thinnest point at the apex. The hole formed expands rapidly in size as the liquid drains away and flows underneath the bubble cavity. When this flow from all sides meets at the bottom of the cavity, if the bubbles are small enough, the resulting fluid pressure creates an upward jet that rises above the surface before disintegrating into droplets and disappearing back into the liquid.

Very high levels of energy dissipation are associated with bubble rupture. The liquid draining back from the film cap reaches velocities of 1 to 50 m s^{-1}, generating shear forces several orders of magnitude greater than known tolerance levels for animal cells. To make matters worse, in typical animal cell cultures, cells attach to rising bubbles and are retained in the thin film of the liquid cap [50], thus being present at the most damaging location as the bubbles burst. After bubble rupture, the cells are subjected to high liquid velocities within the draining film and, as a consequence, suffer severe damage. Measurements with insect cells have shown that, for each 3.5-mm bubble ruptured at the culture

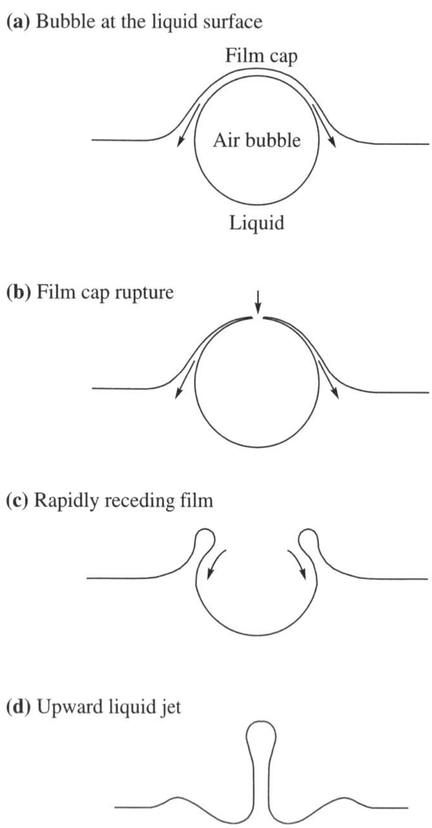

FIGURE 8.45 Sequence of events when a bubble bursts at a gas–liquid interface.

surface, about 10^3 cells are killed from a suspension initially containing approximately 10^6 cells ml^{-1} [51]. At this rate, most of the cells in an average bioreactor would be killed in only a few hours.

The severe damaging effects of bursting bubbles on suspended animal cells can be attenuated by adding protective agents to the medium. Additives such as Pluronic F-68 lower the gas–liquid interfacial tension significantly and prevent the attachment of cells to bubbles, thus removing cells from the damaging zone at the time of bubble rupture. Stable foam layers on the surface of animal cell cultures have also been found to offer some protection from bursting bubbles [52].

8.16.2 Other Studies and Approaches

A common concept in studies of cell damage is that the potential for damage or the extent to which it occurs is related to the rate of dissipation of turbulence kinetic energy in the fluid. For dissipation of a given amount of power, cell damage is likely to depend on whether the power is dissipated within a small volume, in which case the damaging effects will probably be severe, or whether it is dissipated more uniformly within a larger volume at a lower turbulence intensity. The way in which turbulence kinetic energy and rates of energy dissipation are distributed in stirred vessels depends largely on the type of impeller used. Therefore, if the *distribution* of energy dissipation in the tank is important in addition to the overall *rate* of energy dissipated (i.e., the power), we can predict that different degrees of cell damage will occur using different impellers, even if each impeller is dissipating the same amount of power.

The turbulence measurement techniques described in Section 7.9.3 can be used to characterise the performance of impellers in terms of turbulence generation and the distribution of energy dissipation rates. To date, however, mainly because of the practical difficulties associated with evaluating rates of energy dissipation as outlined in Section 7.9.2 (Rate of Dissipation of Turbulence Kinetic Energy subsection) we do not have an unambiguous picture of how different impellers utilise and distribute the power supplied to them. Nevertheless, some success in predicting cell damage has been achieved using empirical correlations derived from our broader understanding of impeller function. In work with fungal hyphae in stirred fermenters, damage measured as a reduction in total hyphal length or cell clump size depended not only on the power input, but also on the size of the impeller, the number of trailing vortices generated per impeller blade, and the frequency with which different impellers circulated fluid through the impeller region [53]. By taking all these factors into account, the extent of damage could be predicted irrespective of the type of impeller used. In these experiments, Rushton turbines were found to be less damaging than pitched-blade turbines, and cell damage increased as the ratio of impeller diameter to tank diameter was reduced. An implication from this work is that the potential for cell damage decreases with scale-up of mixing systems. Although successful with fungal cultures, the applicability of this approach to other cell types remains to be tested.

It is possible that the response of cells to hydrodynamic conditions depends not only on the intensity of the damaging forces generated by turbulent flow, but also on their duration. The cumulative amount of energy dissipated on cells has been used to correlate the damaging effects of turbulence in fungal and plant cell cultures [54–57]. The cumulative energy dissipated on the cells per unit volume can be calculated as:

$$E = \frac{1}{V}\int P\phi \, dt \tag{8.37}$$

where E is the cumulative energy dissipation per unit volume, V is the culture volume, P is the power input, ϕ is the fraction of the culture volume occupied by the cells, and t is time. Under steady-state conditions, for example, during continuous fermentations, Eq. (8.37) reduces to:

$$E = \frac{P\phi\tau}{V} \tag{8.38}$$

where τ is the average residence time in the vessel. In experiments with plant cells, several properties have been measured as indictors of sublytic damage, including cell viability, growth, membrane integrity, cell chain length, protein release into the medium, and cell aggregate size. Depending on the species, cell damage was found at cumulative energy dissipation levels of 10^5 to 10^9 J m^{-3}. The implication of this result is that no damage will occur if cumulative energy dissipation is limited to less than these threshold values.

Another view of the damaging effects of hydrodynamic conditions on cells is related to the fine-scale intermittency of turbulence and the fluctuations in instantaneous velocity, energy dissipation rate, and other properties that are characteristic of turbulent flow. Rather than the average properties of turbulent flow causing cell damage, as is assumed in most other approaches to this problem, it is possible that rare, rapid, and very violent events in turbulence are responsible for rupturing cells and aggregates. Although it has been shown theoretically that strong bursts of energy dissipation have the potential to control cell damage in bioreactors [58], practical application to cell cultures has not yet been demonstrated.

SUMMARY OF CHAPTER 8

This chapter covers various topics related to mixing and cell damage in bioreactors. At the end of Chapter 8 you should:

- Be familiar with the broad range of equipment used for mixing in stirred vessels, including different impeller designs, their operating characteristics, and their suitability for particular mixing applications
- Be able to describe *rotational-*, *radial-*, and *axial-flow* patterns in stirred tanks and the types of impeller that induce them
- Be able to describe the different two-phase flow patterns in stirred vessels with gassing, using terms such as *impeller flooding*, *impeller loading*, *complete gas dispersion*, and *gas recirculation*
- Understand the role of *trailing vortices* and *ventilated cavities* in impeller operation
- Know how impeller size, stirrer speed, tank geometry, liquid properties, and gas sparging affect power consumption in stirred vessels

- Be able to determine the stirrer operating conditions for complete solids suspension
- Understand the mechanisms of mixing
- Know what is meant by *mixing time* and how it is measured
- Be able to describe the effects of scale-up on mixing, and options for improving mixing without the input of extra power
- Understand the factors affecting the performance of multiple-impeller systems
- Know the problems associated with mixing highly pseudoplastic or yield-stress fluids
- Understand how cells can be damaged in stirred and aerated fermenters

PROBLEMS

8.1. Impeller loading and gas dispersion

An aerated bioreactor of diameter 1.2 m is used for batch culture of *Brevibacterium flavum*. The reactor is stirred using a Rushton turbine of diameter 0.4 m. At the beginning of the culture when the cell concentration and culture oxygen requirements are relatively small, the stirrer operating speed of 110 rpm is just sufficient to completely disperse the air bubbles. Towards the end of the culture when the cell density is high, it is decided to double the volumetric flow rate of air to supply more oxygen.

(a) Will the impeller be flooded or loaded under these conditions?

(b) What stirrer speed is needed to achieve complete gas dispersion at the new aeration rate?

8.2. Hydrodynamic conditions for animal cell culture

Mouse hybridoma cells are grown in suspension culture in a stirred tank of diameter 1 m and liquid volume 0.8 m^3. The stirrer speed is 1.5 rps. The vessel has four 10% baffles and is sparged with air at a flow rate of 0.3 vvm (volume of gas per volume of liquid per minute). Pluronic F-68 is added to the medium to protect the cells against the effects of bursting bubbles. The density of the medium is 1000 kg m^{-3} and the viscosity is 1.4 cP. The impeller is a Rushton turbine of diameter 0.4 m.

(a) Is the air dispersed effectively in this system?

(b) Is the flow turbulent?

(c) What proportion of the power input is from sparging?

8.3. Gas dispersion and power requirements

A stirred, baffled fermenter is used to culture *Streptomyces cinnamonensis* for production of monensin. The tank diameter and liquid height are both 1.1 m. The broth density is 1000 kg m^{-3} and the viscosity is 15 cP. The fermenter is mixed using a Rushton turbine of diameter one-half the tank diameter. The air flow rate is 0.66 vvm (volume of gas per volume of liquid per minute).

(a) What stirrer speed is required for complete gas dispersion?

(b) What are the power requirements for complete gas dispersion? Assume the power draw with gassing is 50% of that without gassing.

8.4. Electrical power required for mixing

Laboratory-scale fermenters are usually mixed using small stirrers with electric motors rated between 100 and 500 W. One such motor is used to drive a 7-cm Rushton turbine in a small reactor containing fluid with the properties of water. The stirrer speed is 900 rpm. Estimate the power requirements for this process. How do you explain the difference between the amount of electrical power consumed by the motor and the power required by the stirrer?

8.5. Effect of viscosity on power requirements

A cylindrical bioreactor of diameter 3 m has four baffles. A Rushton turbine mounted in the reactor has a diameter of one-third the tank diameter. The liquid height is equal to the tank diameter and the density of the fluid is approximately 1 g cm^{-3}. The reactor is used to culture an anaerobic organism that does not require gas sparging. The broth can be assumed Newtonian. As the cells grow, the viscosity of the broth increases.

(a) The stirrer is operated at a constant speed of 90 rpm. Compare the power requirements when the viscosity is:

(i) Approximately that of water

(ii) 100 times greater than water

(iii) 2×10^5 times greater than water

(b) The viscosity reaches a value 1000 times greater than water.

(i) What stirrer speed is required to achieve turbulence?

(ii) Estimate the power required to achieve turbulence.

(iii) What power per unit volume is required for turbulence? Is it reasonable to expect to be able to provide this amount of power? Why or why not?

8.6. Power and scale-up

A pilot-scale fermenter of diameter and liquid height 0.5 m is fitted with four baffles of width one-tenth the tank diameter. Stirring is provided using a Scaba 6SRGT curved-blade disc turbine with diameter one-third the tank diameter. The density of the culture broth is 1000 kg m^{-3} and the viscosity is 5 cP. Optimum culture conditions are provided in the pilot-scale fermenter when the stirrer speed is 185 rpm. Following completion of the pilot studies, a larger production-scale fermenter is constructed. The large fermenter has a capacity of 6 m^3, is geometrically similar to the pilot-scale vessel, and is also equipped with a Scaba 6SRGT impeller of diameter one-third the tank diameter.

(a) What is the power consumption in the pilot-scale fermenter?

(b) If the production-scale fermenter is operated so that the power consumed per unit volume is the same as in the pilot-scale vessel, what is the power requirement after scale-up?

(c) For the conditions in (b), what is the stirrer speed after scale-up?

(d) If, instead of (b) and (c), the impeller tip speed ($= \pi N_i D_i$) is kept the same in the pilot- and production-scale fermenters, what is the stirrer speed after scale-up?

(e) For the conditions in (d), what power is required after scale-up?

8.7. Particle suspension and gas dispersion

Bacteria immobilised on particles of gravel are being studied for bioremediation of polychlorinated biphenyls (PCBs). The cells are cultured in an aqueous solution of density 1000 kg m^{-3} and viscosity 0.8 mPa s in a 400-litre bioreactor of diameter 0.8 m. The reactor is stirred with a Rushton turbine of diameter one-third the tank diameter positioned with an impeller off-bottom clearance of a one-quarter the tank diameter. The average particle diameter is 250 μm, the density of gravel is 1.9 g cm^{-3}, and the particle concentration is 15% by weight. The bioreactor is supplied with air at a flow rate of 0.5 vvm (volume of gas per volume of liquid per minute). If the stirrer speed is set so that the solids are just completely suspended in the solution, will the gas be completely dispersed?

8.8. Impeller flooding and power requirements

You are asked to purchase a new Rushton turbine and drive assembly for a fermenter of diameter 2 m equipped with four 10% baffles and with liquid height equal to the tank diameter. The fermenter is used to culture organisms that require an air flow rate of 1.5 vvm (volume of gas per volume of liquid per minute). The broth density is close to 1.0 g cm^{-3} and the viscosity is 0.9 mPa s. Your colleagues have suggested that you design the system for an impeller of diameter one-third the tank diameter, as this is standard company practice. However, after inspecting the vessel internals, you find that an impeller diameter of one-half the tank diameter is also a possibility.

(a) What stirrer speed is required to prevent impeller flooding with $D_i = 1/3\ D_T$?

(b) What stirrer speed is required to prevent impeller flooding with $D_i = 1/2\ D_T$?

Your supervisor is worried about the operating costs associated with the new impeller, and suggests you use the smaller impeller with $D_i = 1/3\ D_T$ to reduce the factory power bill.

(c) Under the operating conditions determined in (a) and (b) to avoid impeller flooding, which impeller consumes the least power? Assume that the % drop in power with gassing is the same for both impellers. Which of the two impellers would you recommend?

8.9. Stirrer effectiveness with sparging

A baffled cylindrical tank of diameter and liquid height 1.15 m is stirred using a four-blade pitched-blade turbine of diameter 0.36 m operated at 200 rpm. The vessel is sparged with air at a volumetric flow rate of 0.036 $m^3\ s^{-1}$. Under these conditions, the turbulent power number for the impeller with gassing is about 1.0. The liquid in the tank has a density of 1 g cm^{-3} and viscosity 1 cP.

(a) Is the impeller likely to be flooded or loaded? What assumptions are involved in your answer?

(b) What is the rate of energy input by the impeller?

(c) What is the rate of energy input by gassing?

(d) In your opinion, is this stirring system effective for mixing and gas dispersion?

8.10. Cell suspension and power requirements

A fermentation broth contains 40 wt% cells of average dimension 10 μm and density 1.04 g cm^{-3}. A marine propeller of diameter 30 cm is used for mixing. The density and viscosity of the medium are approximately the same as water. The fermentation is carried out without gas sparging.

(a) Estimate the stirrer speed required to just completely suspend the cells.

(b) What power is required for cell suspension?

You plan to improve this fermentation process by using a new cell strain immobilised in porous plastic beads of diameter 2 mm and density 1.75 g cm^{-3}. The particle concentration required for comparable rates of fermentation is 10% by weight.

(c) How does changing over to the immobilised cell system affect the stirrer speed and power required for particle suspension?

8.11. Particle suspension and scale-up

Bacteria attached to particles of clinker are being tested for treatment of industrial waste. In the laboratory, the process is carried out under anaerobic conditions in a stirred bioreactor with liquid height equal to the tank diameter. The system is then scaled up to a geometrically similar vessel of volume 90 times that of the laboratory reactor. The suspending fluid and the particle size, density, and concentration remain the same as in the smaller vessel. The type of impeller is also unchanged, as is the impeller-to-tank diameter ratio.

(a) How does the stirrer speed required for suspension of the particles change after scale-up?

(b) Assuming operation in the turbulent regime for both vessels, what effect does scale-up have on:

  (i) The power required for particle suspension?

  (ii) The power per unit volume required for particle suspension?

8.12. Impeller diameter, mixing, and power requirements

Solids suspension and gas dispersion can both be achieved at lower stirrer speeds if the impeller diameter is increased. However, the power requirements for stirring increase with impeller size. In this exercise, the energy efficiencies of small and large impellers are compared for solids suspension and gas dispersion.

(a) For ungassed Rushton turbines operating in the turbulent regime, how does the power required for complete solids suspension vary with impeller diameter if all other properties of the system remain unchanged?

(b) Using the result from (a), compare the power requirements for complete solids suspension using Rushton turbines of diameters one-third and one-half the tank diameter.

(c) All else being equal, how does the power required for complete gas dispersion by Rushton turbines vary with impeller diameter under turbulent flow conditions?

(d) Compare the power requirements for complete gas dispersion using Rushton turbines of diameters one-third and one-half the tank diameter.

8.13. Efficiency of different impellers for solids suspension

Compared with a Rushton turbine of diameter one-half the tank diameter, what are the power requirements for solids suspension by a downward-pumping, six-blade pitched-blade turbine of the same size?

8.14. Power and mixing time with aeration

A cylindrical stirred bioreactor of diameter and liquid height 2 m is equipped with a Rushton turbine of diameter one-third the tank diameter. The bioreactor contains Newtonian culture broth with the same density as water and viscosity 4 cP.

(a) If the specific power consumption must not exceed 1.5 kW m^{-3}, determine the maximum allowable stirrer speed.

(b) What is the mixing time at the stirrer speed determined in (a)?

(c) The tank is now aerated. In the presence of gas bubbles, the approximate relationship between the ungassed turbulent power number $(N'_P)_0$ and the gassed turbulent power number $(N'_P)_g$ is: $(N'_P)_g = 0.5\,(N'_P)_0$. What maximum stirrer speed is now possible in the sparged reactor?

(d) What is the mixing time with aeration at the stirrer speed determined in (c)?

8.15. Scale-up of mixing system

To ensure turbulent conditions during agitation with a turbine impeller, the Reynolds number must be at least 10^4.

(a) A laboratory fermenter of diameter and liquid height 15 cm is equipped with a 5-cm-diameter Rushton turbine operated at 800 rpm. If the density of the broth is close to that of water, what is the upper limit for the broth viscosity if turbulence is to be achieved?

(b) Estimate the mixing time in the laboratory fermenter.

(c) The fermenter is scaled up so the tank and impeller are 15 times the diameter of the laboratory equipment. The stirrer in the large vessel is operated at the same impeller tip speed ($= \pi N_i D_i$) as in the laboratory apparatus. How does scale-up affect the maximum viscosity allowable for maintenance of turbulent conditions?

(d) What effect does scale-up have on the mixing time?

8.16. Effect on mixing of scale-up at constant power per unit volume

A baffled pilot-scale fermenter with liquid height equal to the vessel diameter is scaled up to a geometrically similar production vessel. The working volume of the production fermenter is 50 times greater than that at the pilot scale. If the agitation system is scaled up using the basis of constant power per unit volume, what effect will scale-up have on the mixing time?

8.17. Alternative impellers

Escherichia coli cells are cultured in an industrial-scale fermenter for production of supercoiled plasmid DNA used in gene therapy.

(a) A fermenter of diameter 2.3 m and working volume 10 m^3 is equipped with a Rushton turbine of diameter one-third the tank diameter. The impeller is operated at 60 rpm and the vessel is sparged with air. The density and viscosity of the fermentation fluid are close to those of water, that is, 1000 kg m^{-3} and 1 cP, respectively. The power with gassing is about 60% of the ungassed power.

(i) Calculate the power draw.

(ii) Estimate the mixing time.

(b) To satisfy the burgeoning demand for gene therapy vectors and DNA vaccines, the fermentation factory is being expanded. Two new 10-m^3 fermenters are being designed and constructed. It is decided to investigate the use of different impellers in an effort to reduce the power required but still achieve the same mixing time as that obtained with the Rushton turbine described in (a).

(i) Compared with the Rushton turbine of diameter one-third the tank diameter, what power savings can be made using a Rushton turbine of diameter one-half the tank diameter?

(ii) Compared with the Rushton turbine of diameter one-third the tank diameter, what power savings can be made using a Lightnin A315 hydrofoil impeller of diameter 0.4 times the tank diameter?

(iii) The power with gassing is about 50% of the ungassed power for both the larger Rushton turbine and the A315 hydrofoil. What stirrer speeds are required with these impellers to achieve the same mixing time as that determined in (a) (ii)?

8.18. Retrofitting

A 1.8-m-diameter cylindrical fermenter with four baffles of width one-tenth the tank diameter and working volume 4.6 m^3 is used to produce amylase. The density of the culture broth is 1000 kg m^{-3} and the viscosity is 20 cP. The vessel is equipped with a 0.6-m Rushton turbine. Under these conditions, the motor delivers a maximum stirrer speed of 150 rpm. A sales representative from the local impeller manufacturing company has suggested that better culture performance might be achieved by replacing the Rushton turbine with a more modern impeller design. On offer at discounted prices are curved-blade disc turbines and a downward-pumping hydrofoil with characteristics similar to those of the Lightnin A315.

(a) Estimate the impeller diameters appropriate for retrofitting of the alternative agitators. What assumptions are involved in your answer?

(b) Compare the mixing times expected after retrofitting each of the alternative impellers with the mixing time delivered by the existing Rushton impeller. Considering mixing time only, do you think that the cost of retrofitting a new impeller is justified? Why or why not?

(c) Which of the three impellers would you recommend for use in future fermentations:

(i) Under nonaerated conditions?

(ii) With aeration?

Explain your decision in each case.

8.19. Retrofitting multiple impellers

An industrial fermentation vessel with diameter 1.9 m and aspect ratio 3:1 is used for production of leucine by aerobic cultures of *Serratia marcescens*. At present, the fermenter is fitted with three Rushton turbines of diameter one-third the tank diameter. The impellers are spaced far enough apart so there is no significant interaction between their flow currents. The stirrer motor is rated for a maximum stirrer speed of 1.2 rps. It is proposed to carry out an impeller retrofitting operation to improve the performance of the culture. Although the culture is aerobic, the new stirring system will be designed for operation under nonaerated conditions as a safety precaution against accidental blockage of the air supply. It is decided to replace the two upper Rushton turbines with two identical downward-pumping hydrofoil impellers. The turbulent power number for the hydrofoil impellers is around 0.9. If the bottom Rushton impeller is replaced by a curved-blade disc turbine of diameter one-third the tank diameter, what size hydrofoil impellers are required? What assumptions are involved in your answer?

8.20. Impeller viscometer

An impeller viscometer is being developed to measure the rheological properties of pseudoplastic fermentation broths. The broth density is 1000 kg m^{-3}. Typical rheological parameters for the broth are flow behaviour index $n = 0.2$ and consistency index $K = 0.05$ N s^n m^{-2}. As outlined in Section 7.6.3, impeller viscometers must be operated under laminar flow conditions.

(a) If a Rushton turbine of diameter 4 cm is used at speeds between 2.5 and 10 rpm, is the flow laminar?

(b) What shear rate range does operation with the Rushton turbine provide?

(c) Approximately what range of shear stresses is induced in the Rushton turbine viscometer?

(d) If the Rushton turbine is replaced by a close-clearance helical ribbon impeller of diameter 7.5 cm, what maximum stirrer speed can be used under laminar flow conditions?

(e) If the minimum practical stirrer speed with the helical ribbon impeller is 2.5 rpm, what ranges of shear rates and shear stresses can be investigated using this viscometer? What advantages, if any, does the helical ribbon viscometer offer over the Rushton turbine viscometer?

8.21. Turbulent shear damage

Microcarrier beads 120 μm in diameter are used to culture recombinant CHO cells for production of growth hormone. It is proposed to use a 20-cm Rushton turbine to mix the culture in a 200-litre bioreactor. Oxygen and carbon dioxide are supplied by gas flow through the reactor headspace. The microcarrier suspension has a density of approximately 1010 kg m^{-3} and a viscosity of 10^{-3} Pa s.

(a) Assuming that the power input by the stirrer is dissipated uniformly in the vessel, estimate the maximum allowable stirrer speed that avoids turbulent shear damage of the cells.

(b) How is your estimate affected if the stirrer power is dissipated close to the impeller, within a volume equal to the impeller diameter cubed?

8.22. Avoiding cell damage

Suspended plant cell cultures derived from lemon trees are being used to produce citrus oil for the cosmetic industry. The cells are known to be sensitive to agitation conditions: cell damage occurs and oil production is detrimentally affected if the cumulative energy dissipation level exceeds 10^5 J m^{-3}. The cells are grown in a bioreactor with continuous feeding of nutrient medium and withdrawal of culture broth. At the operating flow rate, the average residence time in the bioreactor is 2.9 days. The diameter of the vessel is 0.73 m, the liquid height is equal to the tank diameter, four 10% baffles are fitted, and stirring is carried out using a 25-cm curved-blade disc turbine with six blades. The vessel is aerated but the effect of gassing on the impeller power draw is negligible. At steady state, the cell concentration is 0.24 v/v, the broth density is 1 g cm^{-3}, and the viscosity is 3.3 mPa s. What is the maximum stirrer speed that can be used without damaging the cells?

References

[1] A.W. Nienow, M.M.C.G. Warmoeskerken, J.M. Smith, M. Konno, On the flooding/loading transition and the complete dispersal condition in aerated vessels agitated by a Rushton-turbine, in: Mixing, Proc. 5th Eur. Conf. on Mixing, Würzburg, Germany, 1985, pp. 143–154; BHRA: The Fluid Engineering Centre, Cranfield, U.K.

[2] W. Bruijn, K. van't Riet, J.M. Smith, Power consumption with aerated Rushton turbines, Trans. IChemE 52 (1974) 88–104.

[3] C.M. Chapman, A.W. Nienow, M. Cooke, J.C. Middleton, Particle–gas–liquid mixing in stirred vessels. Part III: Three phase mixing, Chem. Eng. Res. Des. 61 (1983) 167–181.

[4] S.M. Kresta, P.E. Wood, The mean flow field produced by a 45° pitched blade turbine: changes in the circulation pattern due to off bottom clearance, Can. J. Chem. Eng. 71 (1993) 42–53.

[5] K.J. Bittorf, S.M. Kresta, Active volume of mean circulation for stirred tanks agitated with axial impellers, Chem. Eng. Sci. 55 (2000) 1325–1335.

[6] J.J. Frijlink, M. Kolijn, J.M. Smith, Suspension of solids with aerated pitched blade turbines, in: Fluid Mixing II, IChemE Symp. Ser. 89, U.K. Institution of Chemical Engineers, 1984, pp. 49–58.
[7] A.W. Nienow, Gas dispersion performance in fermenter operation, Chem. Eng. Prog. 86 (2) (1990) 61–71.
[8] A.W. Nienow, M. Konno, W. Bujalski, Studies on three-phase mixing: a review and recent results, Chem. Eng. Res. Des. 64 (1986) 35–42.
[9] F. Saito, A.W. Nienow, S. Chatwin, I.P.T. Moore, Power, gas dispersion and homogenisation characteristics of Scaba SRGT and Rushton turbine impellers, J. Chem. Eng. Japan 25 (1992) 281–287.
[10] M. Zlokarnik, H. Judat, Stirring, in: W. Gerhartz (Ed.), Ullmann's Encyclopedia of Industrial Chemistry, fifth ed., vol. B2, VCH, 1988, pp. 25-1–25-33.
[11] R.L. Bates, P.L. Fondy, R.R. Corpstein, An examination of some geometric parameters of impeller power, Ind. Eng. Chem. Process Des. Dev. 2 (1963) 310–314.
[12] C.M. Chapman, A.W. Nienow, M. Cooke, J.C. Middleton, Particle–gas–liquid mixing in stirred vessels. Part I: Particle–liquid mixing, Chem. Eng. Res. Des. 61 (1983) 71–81.
[13] W. Bujalski, A.W. Nienow, S. Chatwin, M. Cooke, The dependency on scale and material thickness of power numbers of different impeller types, in: Proc. Int. Conf. on Mechanical Agitation, Toulouse, France, ENS Association pour la Promotion du Genie Chimique, 1986, pp. 1-37–1-46 .
[14] J.J. Frijlink, A. Bakker, J.M. Smith, Suspension of solid particles with gassed impellers, Chem. Eng. Sci. 45 (1990) 1703–1718.
[15] S. Ibrahim, A.W. Nienow, Power curves and flow patterns for a range of impellers in Newtonian fluids: $40 < Re < 5 \times 10^5$, Trans. IChemE 73A (1995) 485–491.
[16] C.M. McFarlane, X.-M. Zhao, A.W. Nienow, Studies of high solidity ratio hydrofoil impellers for aerated bioreactors. 2. Air–water studies, Biotechnol. Prog. 11 (1995) 608–618.
[17] Z. Jaworski, A.W. Nienow, K.N. Dyster, An LDA study of the turbulent flow field in a baffled vessel agitated by an axial, down-pumping hydrofoil impeller, Can. J. Chem. Eng. 74 (1996) 3–15.
[18] J.H. Rushton, E.W. Costich, H.J. Everett, Power characteristics of mixing impellers, Part I. Chem. Eng. Prog. 46 (1950) 395–404.
[19] J.H. Rushton, E.W. Costich, H.J. Everett, Power characteristics of mixing impellers. Part II, Chem. Eng. Prog. 46 (1950) 467–476.
[20] M.M.C.G. Warmoeskerken, J. Speur, J.M. Smith, Gas–liquid dispersion with pitched blade turbines, Chem. Eng. Commun. 25 (1984) 11–29.
[21] R.L. Bates, P.L. Fondy, J.G. Fenic, Impeller characteristics and power, in: V.W. Uhl, J.B. Gray (Eds.), Mixing: Theory and Practice, vol. 1, Academic Press, 1966, pp. 111–178.
[22] K. Rutherford, S.M.S. Mahmoudi, K.C. Lee, M. Yianneskis, The influence of Rushton impeller blade and disk thickness on the mixing characteristics of stirred vessels, Trans. IChemE 74A (1996) 369–378.
[23] A.W. Nienow, Gas–liquid mixing studies: a comparison of Rushton turbines with some modern impellers, Trans. IChemE 74A (1996) 417–423.
[24] C.M. McFarlane, A.W. Nienow, Studies of high solidity ratio hydrofoil impellers for aerated bioreactors. 3. Fluids of enhanced viscosity and exhibiting coalescence repression, Biotechnol. Prog. 12 (1996) 1–8.
[25] K.N. Dyster, E. Koutsakos, Z. Jaworski, A.W. Nienow, An LDA study of the radial discharge velocities generated by a Rushton turbine: Newtonian fluids, $Re \geq 5$, Trans. IChemE 71A (1993) 11–23.
[26] V.P. Mishra, K.N. Dyster, Z. Jaworski, A.W. Nienow, J. McKemmie, A study of an up- and a down-pumping wide blade hydrofoil impeller: Part I. LDA measurements, Can. J. Chem. Eng. 76 (1998) 577–588.
[27] A. Bakker, H.E.A. van den Akker, A computational model for the gas–liquid flow in stirred reactors, Trans. IChemE 72A (1994) 594–606.
[28] Th.N. Zwietering, Suspending of solid particles in liquid by agitators, Chem. Eng. Sci. 8 (1958) 244–253.
[29] A.W. Nienow, The suspension of solid particles, in: N. Harnby, M.F. Edwards, A.W. Nienow (Eds.), Mixing in the Process Industries, second ed., Butterworth-Heinemann, 1992, pp. 364–393.
[30] C.W. Wong, J.P. Wang, S.T. Huang, Investigations of fluid dynamics in mechanically stirred aerated slurry reactors, Can. J. Chem. Eng. 65 (1987) 412–419.
[31] W. Bujalski, M. Konno, A.W. Nienow, Scale-up of 45° pitch blade agitators for gas dispersion and solid suspension, in: Mixing, Proc. 6th Eur. Conf. on Mixing, Pavia, Italy, 1988, pp. 389–398, BHRA The Fluid Engineering Centre, Cranfield, U.K.

[32] S. Ibrahim, A.W. Nienow, Particle suspension in the turbulent regime: the effect of impeller type and impeller/vessel configuration, Trans. IChemE 74A (1996) 679–688.
[33] C.M. Chapman, A.W. Nienow, J.C. Middleton, Particle suspension in a gas sparged Rushton-turbine agitated vessel, Trans. IChemE 59 (1981) 134–137.
[34] M.M.C.G. Warmoeskerken, M.C. van Houwelingen, J.J. Frijlink, J.M. Smith, Role of cavity formation in stirred gas–liquid–solid reactors, Chem. Eng. Res. Des. 62 (1984) 197–200.
[35] S. Ruszkowski, A rational method for measuring blending performance, and comparison of different impeller types, in: Mixing, Proc. 8th Eur. Conf. on Mixing, Cambridge, U.K., IChemE Symp. Ser. 136, Institution of Chemical Engineers, 1994, pp. 283–291.
[36] A.W. Nienow, On impeller circulation and mixing effectiveness in the turbulent flow regime, Chem. Eng. Sci. 52 (1997) 2557–2565.
[37] V. Hudcova, V. Machon, A.W. Nienow, Gas–liquid dispersion with dual Rushton turbine impellers, Biotechnol. Bioeng. 34 (1989) 617–628.
[38] D.G. Cronin, A.W. Nienow, G.W. Moody, An experimental study of mixing in a proto-fermenter agitated by dual Rushton turbines, Trans. IChemE 72C (1994) 35–40.
[39] R. Kuboi, A.W. Nienow, The power drawn by dual impeller systems under gassed and ungassed conditions, in: Mixing, Proc. 4th Eur. Conf. on Mixing, Noordwijkerhout, The Netherlands, 1982, pp. 247–261, BHRA Fluid Engineering, Cranfield, U.K.
[40] D. Hari-Prajitno, V.P. Mishra, K. Takenaka, W. Bujalski, A.W. Nienow, J. McKemmie, Gas–liquid mixing studies with multiple up- and down-pumping hydrofoil impellers: power characteristics and mixing time, Can. J. Chem. Eng. 76 (1998) 1056–1068.
[41] A.T.C. Mak, S. Ruszkowski, Solids suspension with multiple impellers. Proc. 1993 IChemE Research Event, Birmingham, U.K., 1993, pp. 756–758, Institution of Chemical Engineers, Rugby, U.K.
[42] A.W. Nienow, G. Hunt, B.C. Buckland, A fluid dynamic study of the retrofitting of large agitated bioreactors: turbulent flow, Biotechnol. Bioeng. 44 (1994) 1177–1185.
[43] A.B. Metzner, R.E. Otto, Agitation of non-Newtonian fluids, AIChE J. 3 (1957) 3–10.
[44] A.B. Metzner, R.H. Feehs, H. Lopez Ramos, R.E. Otto, J.D. Tuthill, Agitation of viscous Newtonian and non-Newtonian fluids, AIChE J. 7 (1961) 3–9.
[45] M.S. Croughan, J.-F. Hamel, D.I.C. Wang, Hydrodynamic effects on animal cells grown in microcarrier cultures, Biotechnol. Bioeng. 29 (1987) 130–141.
[46] M.S. Croughan, E.S. Sayre, D.I.C. Wang, Viscous reduction of turbulent damage in animal cell culture, Biotechnol. Bioeng. 33 (1989) 862–872.
[47] R.S. Cherry, E.T. Papoutsakis, Physical mechanisms of cell damage in microcarrier cell culture bioreactors, Biotechnol. Bioeng. 32 (1988) 1001–1014.
[48] K.T. Kunas, E.T. Papoutsakis, Damage mechanisms of suspended animal cells in agitated bioreactors with and without bubble entrainment, Biotechnol. Bioeng. 36 (1990) 476–483.
[49] S.K.W. Oh, A.W. Nienow, M. Al-Rubeai, A.N. Emery, The effects of agitation intensity with and without continuous sparging on the growth and antibody production of hybridoma cells, J. Biotechnol. 12 (1989) 45–62.
[50] J.J. Chalmers, F. Bavarian, Microscopic visualization of insect cell–bubble interactions. II. The bubble film and bubble rupture, Biotechnol. Prog. 7 (1991) 151–158.
[51] K. Trinh, M. Garcia-Briones, F. Hink, J.J. Chalmers, Quantification of damage to suspended insect cells as a result of bubble rupture, Biotechnol. Bioeng. 43 (1994) 37–45.
[52] J.D. Michaels, J.E. Nowak, A.K. Mallik, K. Koczo, D.T. Wasan, E.T. Papoutsakis, Interfacial properties of cell culture media with cell-protecting additives, Biotechnol. Bioeng. 47 (1995) 420–430.
[53] P. Jüsten, G.C. Paul, A.W. Nienow, C.R. Thomas, Dependence of mycelial morphology on impeller type and agitation intensity, Biotechnol. Bioeng. 52 (1996) 672–684.
[54] M. Reuß, Influence of mechanical stress on the growth of *Rhizopus nigricans* in stirred bioreactors, Chem. Eng. Technol. 11 (1988) 178–187.
[55] E.H. Dunlop, P.K. Namdev, M.Z. Rosenberg, Effect of fluid shear forces on plant cell suspensions, Chem. Eng. Sci. 49 (1994) 2263–2276.
[56] R. Wongsamuth, P.M. Doran, The filtration properties of *Atropa belladonna* plant cell suspensions; effects of hydrodynamic shear and elevated carbon dioxide levels on culture and filtration parameters, J. Chem. Tech. Biotechnol. 69 (1997) 15–26.

[57] P.F. MacLoughlin, D.M. Malone, J.T. Murtagh, P.M. Kieran, The effects of turbulent jet flows on plant cell suspension cultures, Biotechnol. Bioeng. 58 (1998) 595–604.
[58] J. Bałdyga, J.R. Bourne, Interpretation of turbulent mixing using fractals and multifractals, Chem. Eng. Sci. 50 (1995) 381–400.

Suggestions for Further Reading

Mixing Equipment

Oldshue, J. Y. (1983). *Fluid Mixing Technology*. McGraw-Hill.

Power Requirements and Hydrodynamics in Stirred Vessels

See also references 7 and 10.

Harnby, N., Edwards, M. F., & Nienow, A. W. (Eds.), (1992). *Mixing in the Process Industries* (2nd ed.). Butterworth-Heinemann.
McFarlane, C. M., & Nienow, A. W. (1995, 1996). Studies of high solidity ratio hydrofoil impellers for aerated bioreactors: Parts 1–4. *Biotechnol. Prog.*, *11*, 601–607, 608–618; *12*, 1–8, 9–15.
Nienow, A. W. (1998). Hydrodynamics of stirred bioreactors. *Appl. Mech. Rev.*, *51*, 3–32.
Tatterson, G. B. (1991). *Fluid Mixing and Gas Dispersion in Agitated Tanks*. McGraw-Hill.
Tatterson, G. B. (1994). *Scaleup and Design of Industrial Mixing Processes*. McGraw-Hill.

Multiple Impeller Systems

Gogate, P. R., Beenackers, A.A.C.M., & Pandit, A. B. (2000). Multiple-impeller systems with a special emphasis on bioreactors: a critical review. *Biochem. Eng. J.*, *6*, 109–144.

Cell Damage in Fermenters

Chalmers, J. (2000). Animal cell culture, effects of agitation and aeration on cell adaptation. In R. E. Spier (Ed.), *Encyclopedia of Cell Technology* (vol. 1, pp. 41–51). John Wiley.
Namdev, P. K., & Dunlop, E. H. (1995). Shear sensitivity of plant cells in suspensions. *Appl. Biochem. Biotechnol.*, *54*, 109–131.
Papoutsakis, E. T. (1991). Fluid-mechanical damage of animal cells in bioreactors. *Trends in Biotechnol*, *9*, 427–437.
Thomas, C. R., & Zhang, Z. (1998). The effect of hydrodynamics on biological materials. In E. Galindo & O. T. Ramírez (Eds.), *Advances in Bioprocess Engineering II* (pp. 137–170). Kluwer Academic.
Wu, J. (1995). Mechanisms of animal cell damage associated with gas bubbles and cell protection by medium additives. *J. Biotechnol.*, *43*, 81–94.

CHAPTER

9

Heat Transfer

In this chapter we are concerned with the process of heat flow between hot and cold systems. The rate at which heat is transferred depends directly on two variables: the temperature difference between the hot and cold bodies, and the surface area available for heat exchange. The heat transfer rate is also influenced by many other factors, such as the geometry and physical properties of the system and, if fluid is present, the flow conditions. Fluids are often heated or cooled in bioprocessing. Typical examples are the removal of heat during fermenter operation using cooling water, and the heating of raw medium to sterilisation temperature by steam.

As shown in Chapters 5 and 6, energy balances allow us to determine the heating and cooling requirements of fermenters and enzyme reactors. Once the rate of heat transfer for a particular purpose is known, the surface area and other conditions needed to achieve this rate can be calculated using design equations. Estimating the heat transfer area is a central objective in design, as this parameter determines the size and cost of heat exchange equipment. In this chapter, the principles governing heat transfer are outlined for application in bioprocess design.

First, let us look at the types of equipment used for industrial heat exchange.

9.1 HEAT TRANSFER EQUIPMENT

In bioprocessing, heat exchange occurs most frequently between fluids. Equipment is provided to allow the transfer of heat while preventing the fluids from physically contacting each other. In most heat exchangers, heat is transferred through a solid metal wall that separates the fluid streams. Sufficient surface area is provided so that the desired rate of heat transfer can be achieved. Heat transfer is facilitated by agitation and turbulent flow of the fluids.

9.1.1 Bioreactors

Two applications of heat transfer are common in bioreactor operation. The first is in situ batch sterilisation of liquid medium. In this process, the fermenter vessel containing

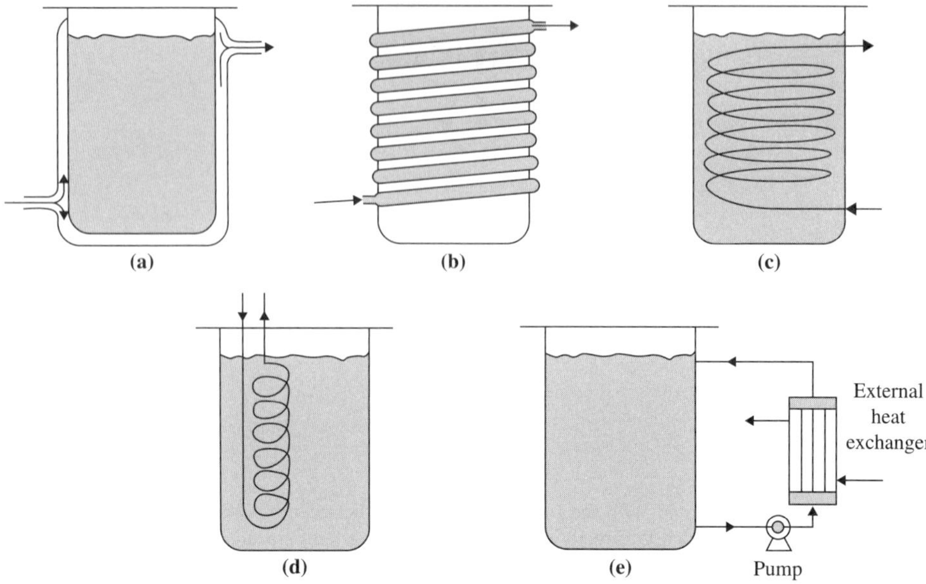

FIGURE 9.1 Heat transfer configurations for bioreactors: (a) jacketed vessel; (b) external coil; (c) internal helical coil; (d) internal baffle-type coil; (e) external heat exchanger.

medium is heated using steam and held at the sterilisation temperature for a period of time. Cooling water is then used to bring the temperature back to normal operating conditions. Sterilisation is discussed in more detail in Chapter 14. The other application of heat transfer is for temperature control during fermenter operation. Most fermentations take place within the range 30 to 37°C; tight control of the temperature to within about 1°C is required. Because the metabolic activity of cells generates a substantial amount of heat, this heat must be removed to avoid temperature increases. In large-scale operations, cooling water is used to achieve temperature control. In small-scale fermenters, the heat exchange requirements are different because the ratio of surface area to volume is much greater and heat losses through the walls of the vessel are more significant. As a result, laboratory-scale fermenters often require heating rather than cooling. Enzyme reactors may also require heating to maintain optimum temperature.

The equipment used for heat exchange in bioreactors usually takes one of the forms illustrated in Figure 9.1. The fermenter may have an external jacket (Figure 9.1(a)) or coil (Figure 9.1(b)) through which steam or cooling water is circulated. Alternatively, helical (Figure 9.1(c)) or baffle-type (Figure 9.1(d)) coils may be located internally. Another method is to pump liquid from the reactor through a separate heat exchange unit as shown in Figure 9.1(e).

The surface area available for heat transfer is lower in the external jacket and coil designs of Figure 9.1(a) and (b) than when internal coils are completely submerged in the reactor contents. External jackets provide sufficient heat transfer area for laboratory and other small-scale bioreactors; however they are generally inadequate for large-scale fermentations. Internal coils (Figure 9.1(c) and (d)) are used frequently in production vessels; the coil

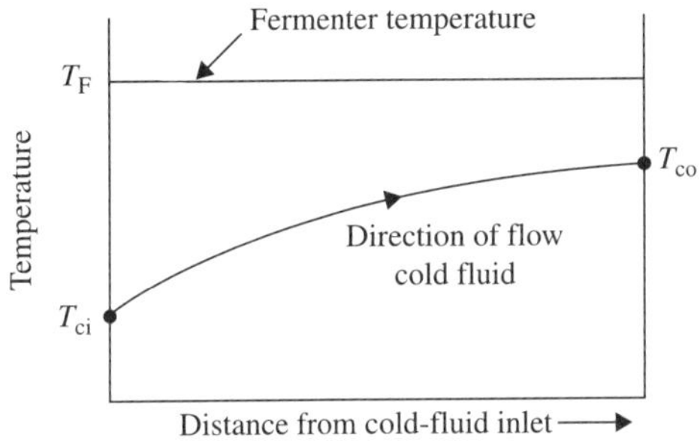

FIGURE 9.2 Temperature changes for control of fermentation temperature using cooling water.

can be operated with high cooling water velocities and the entire tube surface is exposed to the reactor contents providing a relatively large heat transfer area. There are some disadvantages, however, with internal structures, as they interfere with mixing in the vessel and make cleaning of the reactor difficult. Another problem is film growth of cells on the internal heat transfer surfaces. The coil must be able to withstand the thermal and mechanical stresses generated inside the fermenter during sterilisation and agitation; the possibility of nonsterile coolant leaking from fractured metal joints in the cooling coil significantly increases the risk of culture contamination. Because of these problems with internal coils, use of an external jacket is preferable for cell cultures with relatively low cooling requirements. On the other hand, fermentations with high heat loads may require an internal cooling coil together with an external jacket to achieve the necessary rate of cooling. The external heat exchange unit shown in Figure 9.1(e) is independent of the reactor, easy to scale up, and can provide greater heat transfer capacity than any of the other configurations. However, conditions of sterility must be met, the cells must be able to withstand the shear forces imposed during pumping, and, in aerobic fermentations, the residence time in the heat exchanger must be small enough to ensure that the medium does not become depleted of oxygen.

When internal coils such as those in Figure 9.1(c) and (d) are used to carry cooling water for removal of heat from a fermenter, the variation of water temperature with distance through the coil is as shown in Figure 9.2. The temperature of the cooling water rises as it flows through the tube and takes up heat from the fermentation broth. The water temperature increases steadily from the inlet temperature T_{ci} to the outlet temperature T_{co}. On the other hand, if the contents of the fermenter are well mixed, temperature gradients in the bulk fluid are negligible and the fermenter temperature is uniform at T_F.

9.1.2 General Equipment for Heat Transfer

Many types of general-purpose equipment are used industrially for heat exchange operations. The simplest form of heat transfer equipment is the double-pipe heat exchanger. For larger capacities, more elaborate shell-and-tube units containing hundreds of square metres of heat exchange area are required. These devices are described in the following sections.

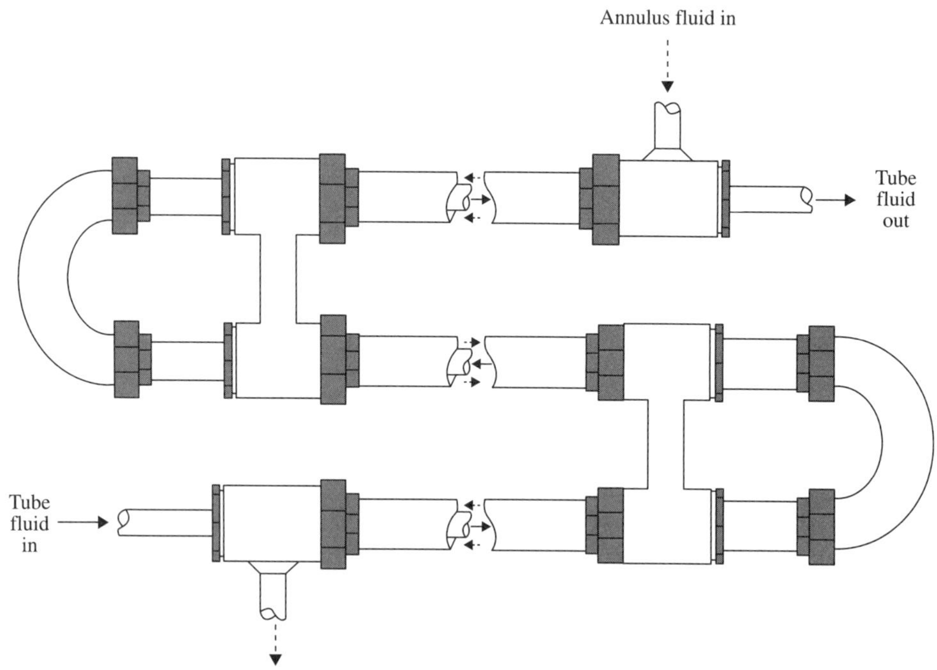

FIGURE 9.3 Double-pipe heat exchanger.
From A.S. Foust, L.A. Wenzel, C.W. Clump, L. Maus, and L.B. Andersen, 1980, Principles of Unit Operations, *2nd ed., John Wiley, New York.*

Double-Pipe Heat Exchanger

A *double-pipe heat exchanger* consists of two metal pipes, one inside the other as shown in Figure 9.3. One fluid flows through the inner tube while the other fluid flows in the annular space between the pipe walls. When one of the fluids is hotter than the other, heat flows from it through the wall of the inner tube into the other fluid. As a result, the hot fluid becomes cooler and the cold fluid becomes warmer.

Double-pipe heat exchangers can be operated with countercurrent or cocurrent flow of fluid. If, as indicated in Figure 9.3, the two fluids enter at opposite ends of the device and pass in opposite directions through the pipes, the flow is *countercurrent*. Cold fluid entering the equipment meets hot fluid just as it is leaving, that is, cold fluid at its lowest temperature is placed in thermal contact with hot fluid also at its lowest temperature. The changes in temperature of the two fluids as they flow countercurrently through the pipes are shown in Figure 9.4. The four terminal temperatures are as follows: T_{hi} is the inlet temperature of the hot fluid, T_{ho} is the outlet temperature of the hot fluid, T_{ci} is the inlet temperature of the cold fluid, and T_{co} is the outlet temperature of the cold fluid leaving the system. A sign of efficient operation is T_{co} close to T_{hi}, or T_{ho} close to T_{ci}.

The alternative to countercurrent flow is *cocurrent* or *parallel* flow. In this mode of operation, both fluids enter their respective tubes at the same end of the exchanger and flow in

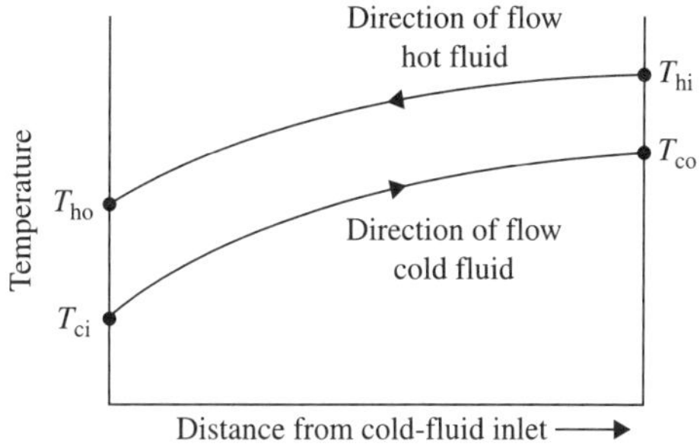

FIGURE 9.4 Temperature changes for countercurrent flow in a double-pipe heat exchanger.

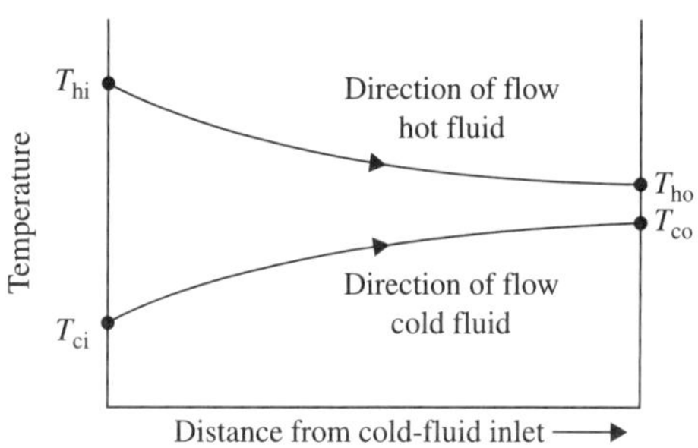

FIGURE 9.5 Temperature changes for cocurrent flow in a double-pipe heat exchanger.

the same direction to the other end. The temperature curves for cocurrent flow are given in Figure 9.5. It is not possible using cocurrent flow to bring the exit temperature of one fluid close to the entrance temperature of the other; instead, the exit temperatures of both streams lie between the two entrance temperatures. Cocurrent flow is not as effective as countercurrent operation because less heat can be transferred; consequently, cocurrent flow is applied less frequently.

Double-pipe heat exchangers can be extended to several passes arranged in a vertical stack, as illustrated in Figure 9.3. However, when large surface areas are needed to achieve the desired rate of heat transfer, the weight of the outer pipe becomes so great that an alternative design, the shell-and-tube heat exchanger, is a better and more economical choice. As a rule of thumb, if the heat transfer area between the fluids must be more than 10 to 15 m^2, a shell-and-tube exchanger is required.

Shell-and-Tube Heat Exchangers

Shell-and-tube heat exchangers are used for heating and cooling of all types of fluid. They have the advantage of containing very large surface areas in a relatively small volume. The simplest form, called a *single-pass shell-and-tube heat exchanger*, is illustrated in Figure 9.6.

Consider the device of Figure 9.6 for exchange of heat from one fluid to another. The heat transfer system is divided into two sections: a *tube bundle* containing pipes through which one fluid flows, and a *shell* or cavity in which the other fluid flows. Hot or cold fluid may be put into either the tubes or the shell. In a single-pass exchanger, the shell and tube fluids pass down the length of the equipment only once. The fluid that is to travel in the tubes enters at the inlet header. The header is divided from the rest of the apparatus by a *tube sheet*. A typical tube sheet is shown in Figure 9.7; this represents an end-view of the heat exchanger from inside the header. Open tubes are fitted into the tube sheet so that fluid in the header, which is prevented from entering the main cavity of the exchanger by the tube sheet, must pass into the tubes. The tube-side fluid leaves the exchanger through another header at the outlet. Shell-side fluid enters the internal cavity of the exchanger and flows around the outside of the tubes in a direction that is largely counter-current to the tube fluid. Heat is exchanged across the tube walls from the hot fluid to the cold fluid.

As shown in Figure 9.6, *baffles* are installed in the shell to divert the shell fluid so that it flows mainly across rather than parallel to the tubes, and to decrease the cross-sectional area for flow. Both these effects promote turbulence in the shell fluid, which improves the rate of heat transfer. As well as directing the flow of shell fluid, baffles support the tube bundle and prevent the tubes from sagging.

The length of the tubes in a single-pass heat exchanger determines the surface area that is available for heat transfer and, therefore, the rate at which heat can be exchanged.

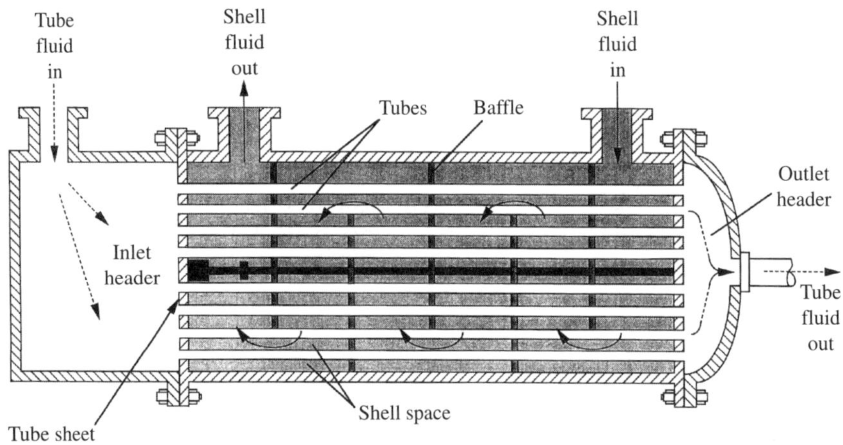

FIGURE 9.6 Single-pass shell-and-tube heat exchanger.
From A.S. Foust, L.A. Wenzel, C.W. Clump, L. Maus, and L.B. Andersen, 1980, Principles of Unit Operations, *2nd ed., John Wiley, New York.*

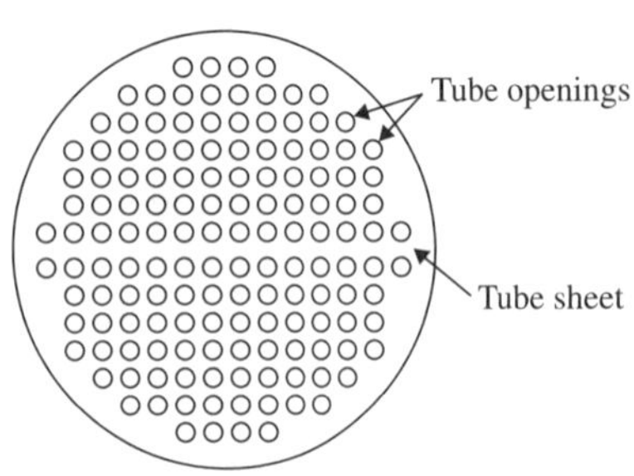

FIGURE 9.7 Tube sheet for a shell-and-tube heat exchanger.

However, there are practical and economic limits to the length of single-pass tubes; if greater heat transfer capacity is required, *multiple-pass heat exchangers* are employed. Heat exchangers containing more that one tube pass are used routinely; Figure 9.8 shows the structure of a heat exchanger with one shell pass and a double tube pass. In this device, the header for the tube fluid is divided into two sections. Fluid entering the header is channelled into the lower half of the tubes and flows to the other end of the exchanger. The header at the other end diverts the fluid into the upper tubes; the tube-side fluid therefore leaves the exchanger at the same end as it entered. On the shell side, the configuration is the same as for the single-pass structure of Figure 9.6; fluid enters one end of the shell and flows around several baffles to the other end.

In a double tube-pass shell-and-tube heat exchanger, flow of the tube and shell fluids is mainly countercurrent for one tube pass and mainly cocurrent for the other. However, because of the action of the baffles, crossflow of shell-side fluid normal to the tubes also occurs. Temperature curves for the exchanger depend on the geometry of the system; this is illustrated in Figure 9.9 for hot fluid flowing in the shell. In the temperature profile of Figure 9.9(b), a *temperature cross* occurs at some point in the exchanger where the temperature of the cold fluid equals the temperature of the hot fluid. This situation should be avoided because, after the cross, the cold fluid is cooled rather than heated. The solution is an increased number of shell passes or, more practically, provision of another heat exchanger in series with the first.

Heat exchangers with multiple shell passes can also be used. However, in comparison with multiple tube-pass equipment, multiple-shell exchangers are complex in construction and are normally applied only in very large installations.

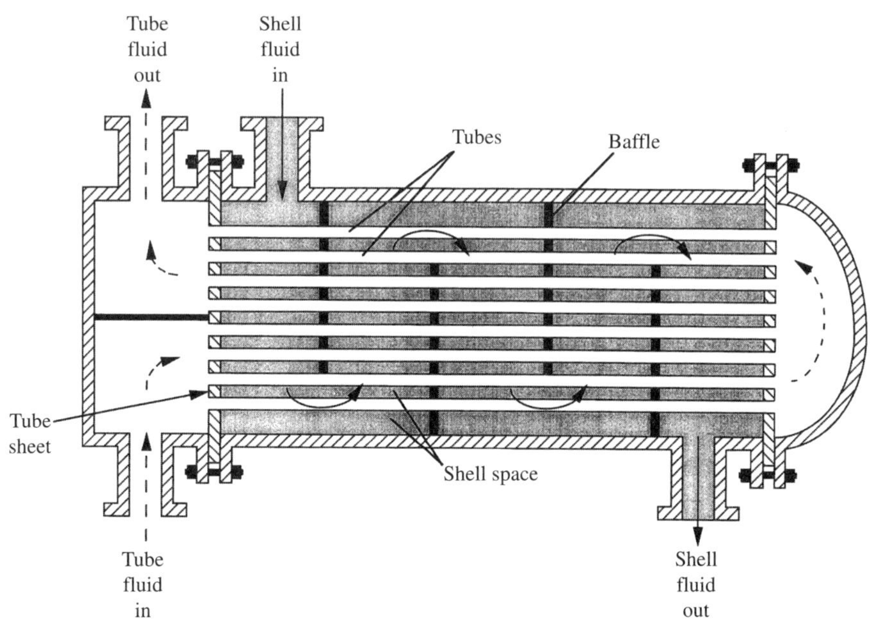

FIGURE 9.8 Double tube-pass heat exchanger.

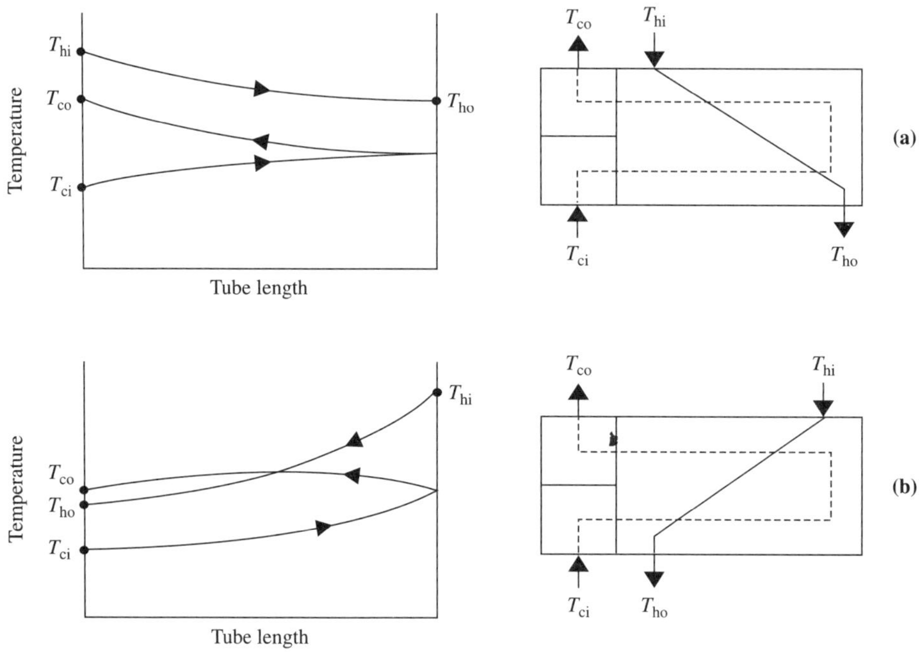

FIGURE 9.9 Temperature changes for a double tube-pass heat exchanger.

9.2 MECHANISMS OF HEAT TRANSFER

Heat transfer occurs by one or more of the following three mechanisms.

- *Conduction*. Heat conduction occurs by transfer of vibrational energy between molecules, or movement of free electrons. Conduction is particularly important in metals and occurs without observable movement of matter.
- *Convection*. Convection requires movement on a macroscopic scale; it is therefore confined to gases and liquids. *Natural convection* occurs when temperature gradients in the system generate localised density differences that result in flow currents. In *forced convection*, flow currents are set in motion by an external agent such as a stirrer or pump and are independent of density gradients. Higher rates of heat transfer are possible with forced convection compared with natural convection.
- *Radiation*. Energy is radiated from all materials in the form of waves. When this radiation is absorbed by matter, it appears as heat. Because radiation is important at much higher temperatures than those normally encountered in biological processing, it will not be considered further.

9.3 CONDUCTION

In most heat transfer equipment, heat is exchanged between fluids separated by a solid wall. Heat transfer through the wall occurs by conduction. In this section we consider equations describing the rate of conduction as a function of operating variables.

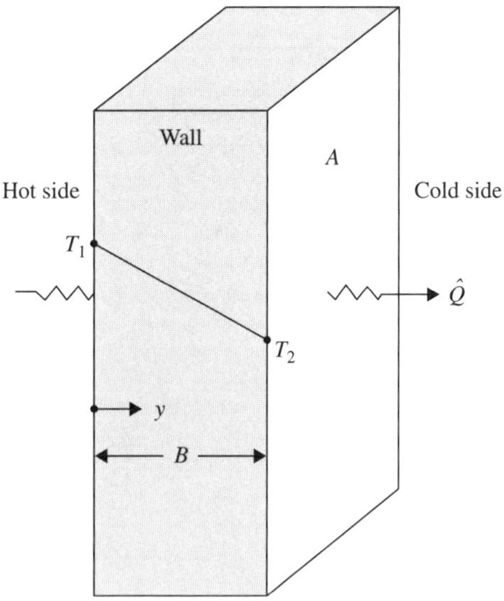

FIGURE 9.10 Heat conduction through a flat wall.

Conduction of heat through a homogeneous solid wall is depicted in Figure 9.10. The wall has thickness B; on one side of the wall the temperature is T_1 and on the other side the temperature is T_2. The area of wall exposed to each temperature is A. The rate of heat conduction through the wall is given by *Fourier's law*.

$$\hat{Q} = -kA\frac{dT}{dy} \tag{9.1}$$

where $\hat{Q}$ is the rate of heat transfer, k is the *thermal conductivity* of the wall, A is the surface area perpendicular to the direction of heat flow, T is temperature, and y is distance measured normal to A. dT/dy is the *temperature gradient*, or change of temperature with distance through the wall. The negative sign in Eq. (9.1) indicates that heat always flows from hot to cold irrespective of whether dT/dy is positive or negative. To illustrate this, when the temperature gradient is negative relative to coordinate y (as shown in Figure 9.10), heat flows in the positive y-direction; conversely, if the gradient were positive (i.e., $T_1 < T_2$), heat would flow in the negative y-direction.

Fourier's law can also be expressed in terms of the *heat flux*, $\hat{q}$. Heat flux is defined as the rate of heat transfer per unit area normal to the direction of heat flow. Therefore, from Eq. (9.1):

$$\hat{q} = -k\frac{dT}{dy} \tag{9.2}$$

The rate of heat transfer $\hat{Q}$ has the same dimensions and units as power. The SI unit for $\hat{Q}$ is the watt (W); in imperial units $\hat{Q}$ is measured in Btu h^{-1}. Corresponding units of $\hat{q}$ are W m^{-2} and Btu h^{-1} ft^{-2}.

Thermal conductivity is a transport property of materials; values can be found in handbooks. The dimensions of k are $LMT^{-3}\Theta^{-1}$; units include W m^{-1} K^{-1} and Btu h^{-1} ft^{-1} $°F^{-1}$. The magnitude of k in Eqs. (9.1) and (9.2) reflects the ease with which heat is conducted;

TABLE 9.1 Thermal Conductivities

Material	Temperature (°C)	k (W m^{-1} °C^{-1})	k (Btu h^{-1} ft^{-1} °F^{-1})
Solids: Metals			
Aluminium	300	230	133
Bronze	–	189	109
Copper	100	377	218
Iron (cast)	53	48	27.6
Iron (wrought)	18	61	35
Lead	100	33	19
Stainless steel	20	16	9.2
Steel (1% C)	18	45	26
Solids: Nonmetals			
Asbestos	0	0.16	0.09
	100	0.19	0.11
	200	0.21	0.12
Bricks (building)	20	0.69	0.40
Cork	30	0.043	0.025
Cotton wool	30	0.050	0.029
Glass	30	1.09	0.63
Glass wool	–	0.041	0.024
Rubber (hard)	0	0.15	0.087
Liquids			
Acetic acid (50%)	20	0.35	0.20
Ethanol (80%)	20	0.24	0.137
Glycerol (40%)	20	0.45	0.26
Water	30	0.62	0.356
	60	0.66	0.381
Gases			
Air	0	0.024	0.014
	100	0.031	0.018
Carbon dioxide	0	0.015	0.0085
Nitrogen	0	0.024	0.0138
Oxygen	0	0.024	0.0141
Water vapour	100	0.025	0.0145

Note: To convert from W m^{-1} °C^{-1} to Btu h^{-1} ft^{-1} °F^{-1}, multiply by 0.578.
From J.M. Coulson, J.F. Richardson, J.R. Backhurst, and J.H. Harker, 1999, Coulson and Richardson's Chemical Engineering, *vol. 1, sixth ed., Butterworth-Heinemann, Oxford.*

the higher the value of k, the faster is the heat transfer. Table 9.1 lists thermal conductivities for some common materials; metals generally have higher thermal conductivities than other substances. Solids with low k values are used as *insulators* to minimise the rate of heat transfer, for example, from steam pipes or in buildings. Thermal conductivity varies somewhat with temperature; however for small ranges of temperature, k can be considered constant.

9.3.1 Analogy between Heat and Momentum Transfer

An analogy exists between the equations for heat and momentum transfer. Newton's law of viscosity given by Eq. (7.6):

$$\tau = -\mu \frac{dv}{dy} \tag{7.6}$$

has the same mathematical form as Eq. (9.2). In heat transfer, the temperature gradient dT/dy is the driving force for heat flow; in momentum transfer the driving force is the velocity gradient dv/dy. Both heat flux and momentum flux are directly proportional to the driving force, and the proportionality constant, μ or k, is a physical property of the material. As we shall see in Chapter 10, this analogy between heat and momentum transfer can be extended to include mass transfer.

9.3.2 Steady-State Conduction

Consider again the conduction of heat through the wall shown in Figure 9.10. At steady state there can be neither accumulation nor depletion of heat within the wall; this means that the rate of heat flow $\hat{Q}$ must be the same at each point in the wall. If k is largely independent of temperature and A is also constant, the only variables in Eq. (9.1) are the temperature T and distance y. We can therefore integrate Eq. (9.1) to obtain an expression for the rate of conduction as a function of the temperature difference across the wall.

Separating variables in Eq. (9.1) gives:

$$\hat{Q}\, dy = -k A\, dT \tag{9.3}$$

Both sides of Eq. (9.3) can be integrated after taking the constants $\hat{Q}$, k, and A outside of the integrals:

$$\hat{Q} \int dy = -kA \int dT \tag{9.4}$$

Therefore, from the rules of integration given in Appendix E:

$$\hat{Q}\, y = -k A T + K \tag{9.5}$$

where K is the integration constant. K is evaluated by applying a single boundary condition; in this case we can use the boundary condition $T = T_1$ at $y = 0$, as indicated in Figure 9.10. Substituting this information into Eq. (9.5) gives:

$$K = kAT_1 \tag{9.6}$$

Using Eq. (9.6) to eliminate K from Eq. (9.5):

$$\hat{Q}\,y = -kA(T - T_1) \tag{9.7}$$

or

$$\hat{Q} = \frac{kA}{y}(T_1 - T) \tag{9.8}$$

Because $\hat{Q}$ at steady state is the same at all points in the wall, Eq. (9.8) holds for all values of y including at $y = B$ where $T = T_2$ (refer to Figure 9.10). Substituting these values into Eq. (9.8) gives the expression:

$$\hat{Q} = \frac{kA}{B}(T_1 - T_2) \tag{9.9}$$

or

$$\hat{Q} = \frac{kA}{B}\Delta T \tag{9.10}$$

Equation (9.10) allows us to calculate $\hat{Q}$ using the total temperature drop across the wall, ΔT.

Equation (9.10) can also be written in the form:

$$\hat{Q} = \frac{\Delta T}{R_w} \tag{9.11}$$

where R_w is the *thermal resistance* to heat transfer offered by the wall:

$$R_w = \frac{B}{kA} \tag{9.12}$$

In heat transfer, the ΔT responsible for flow of heat is known as the *temperature-difference driving force*. Equation (9.11) is an example of the *general rate principle,* which equates the rate of a process to the ratio of the driving force and the resistance. Equation (9.12) can be interpreted as follows: the wall would pose more of a resistance to heat transfer if its thickness B were increased; on the other hand, the resistance is reduced if the surface area A is increased, or if the material in the wall were replaced with a substance of higher thermal conductivity k.

9.3.3 Combining Thermal Resistances in Series

When a system contains several different heat transfer resistances in series, *the overall resistance is equal to the sum of the individual resistances*. For example, if the wall shown in Figure 9.10 were constructed of several layers of different material, each layer would represent a separate resistance to heat transfer. Consider the three-layer system illustrated in Figure 9.11 with surface area A, layer thicknesses B_1, B_2, and B_3, thermal conductivities k_1, k_2, and k_3, and temperature drops across the layers of ΔT_1, ΔT_2, and ΔT_3. If the layers are in perfect thermal contact so there is no temperature drop across the interfaces, the temperature change across the entire structure is:

$$\Delta T = \Delta T_1 + \Delta T_2 + \Delta T_3 \tag{9.13}$$

The rate of heat conduction in this system is given by Eq. (9.11), with the overall resistance R_w equal to the sum of the individual resistances:

$$\hat{Q} = \frac{\Delta T}{R_w} = \frac{\Delta T}{(R_1 + R_2 + R_3)} \tag{9.14}$$

where R_1, R_2, and R_3 are the thermal resistances of the individual layers:

$$R_1 = \frac{B_1}{k_1 A} \quad R_2 = \frac{B_2}{k_2 A} \quad R_3 = \frac{B_3}{k_3 A} \tag{9.15}$$

Equation (9.14) represents the important principle of *additivity of resistances*. We shall use this principle later for analysis of convective heat transfer in pipes and stirred vessels.

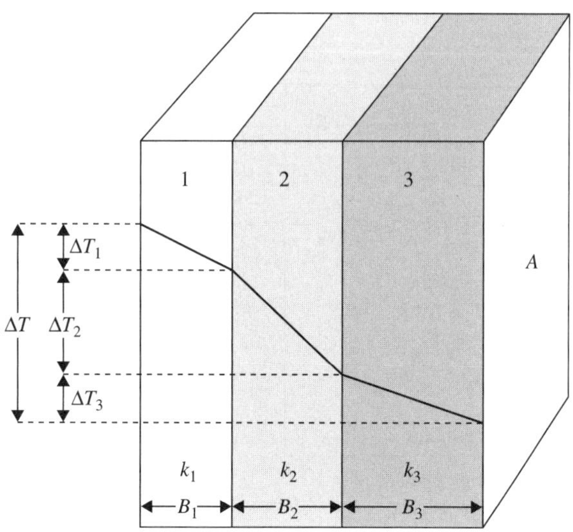

FIGURE 9.11 Heat conduction through three resistances in series.

9.4 HEAT TRANSFER BETWEEN FLUIDS

Convection and conduction both play important roles in heat transfer in fluids. In agitated, single-phase systems, convective heat transfer in the bulk fluid is linked directly to mixing and turbulence and is generally quite rapid. However, in the heat exchange equipment described in Section 9.1, additional resistances to heat transfer are encountered.

9.4.1 Thermal Boundary Layers

Figure 9.12 depicts the heat transfer situation at any point on the pipe wall of a heat exchanger. Figure 9.12(a) identifies a segment of pipe wall separating the hot and cold fluids; Figure 9.12(b) shows the magnified detail of fluid properties at the wall. Hot and cold fluids flow on either side of the wall; we will assume that both fluids are in turbulent flow. The bulk temperature of the hot fluid away from the wall is T_h; T_c is the bulk temperature of the cold fluid. T_{hw} and T_{cw} are the respective temperatures of the hot and cold fluids at the wall.

As explained in Section 7.2.4, when fluid contacts a solid, a fluid boundary layer develops at the surface as a result of viscous drag. Therefore, the hot and cold fluids represented in Figure 9.12 consist of a turbulent core that accounts for the bulk of the fluid, and a thin sublayer or film near the wall where the velocity is relatively low. In the turbulent part of the fluid, rapidly moving eddies transfer heat quickly so that any temperature gradients in the bulk fluid can be neglected. The film of liquid at the wall is

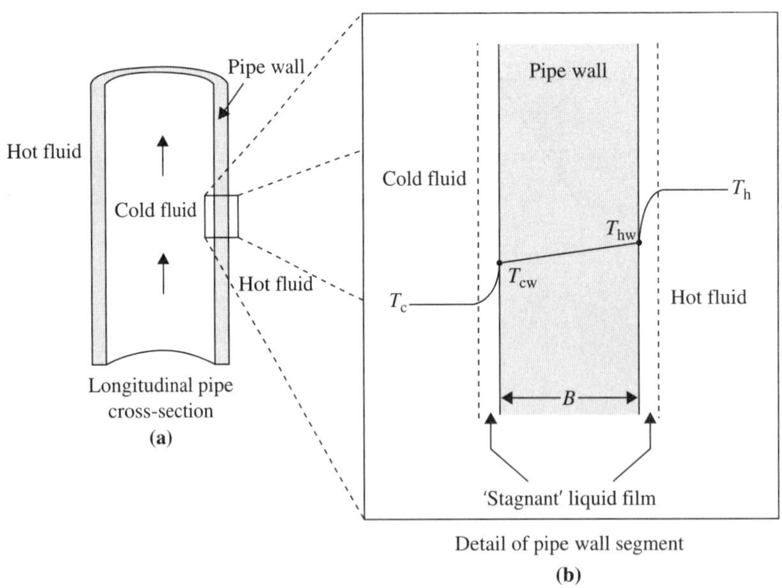

FIGURE 9.12 Graphic representation of the heat transfer between fluids that are separated by a pipe wall: (a) the longitudinal pipe cross-section identifying a segment of the pipe wall; (b) the magnified detail of the pipe wall segment showing boundary layers and temperature gradients at the wall.

called the *thermal boundary layer* or *stagnant film*, although the fluid in it is not actually stationary. This viscous sublayer has an important effect on the rate of heat transfer. Most of the resistance to heat transfer to or from the fluid is contained in the film; the reason for this is that heat flow through the film must occur mainly by conduction rather than convection because of the reduced velocity of the fluid. The width of the film indicated by the broken lines in Figure 9.12(b) is the approximate distance from the wall at which the temperature reaches the bulk fluid temperature, either T_h or T_c. The thickness of the thermal boundary layer in most heat transfer situations is less than the hydrodynamic boundary layer described in Section 7.2.4. In other words, as we move away from the wall, the temperature normally reaches that of the bulk fluid before the velocity reaches that of the bulk flow stream.

9.4.2 Individual Heat Transfer Coefficients

Heat exchanged between the fluids in Figure 9.12 encounters three major resistances in series: the hot-fluid film resistance at the wall, resistance due to the wall itself, and the cold-fluid film resistance. Equations for the rate of conduction through the wall were developed in Section 9.3.2. The rate of heat transfer through each thermal boundary layer in the fluid is given by an equation analogous to Eq. (9.10) for steady-state conduction:

$$\hat{Q} = h\,A\,\Delta T \tag{9.16}$$

where h is the *individual heat transfer coefficient*, A is the area for heat transfer normal to the direction of heat flow, and ΔT is the temperature difference between the wall and the bulk stream. $\Delta T = T_h - T_{hw}$ for the hot-fluid film; $\Delta T = T_{cw} - T_c$ for the cold-fluid film.

Unlike Eq. (9.10) for conduction, Eq. (9.16) does not contain a separate term for the thickness of the boundary layer; this thickness is difficult to measure and depends strongly on the prevailing flow conditions. Instead, the effect of film thickness is included in the value of h so that, unlike thermal conductivity, h is not a transport property of materials and its value cannot be found in handbooks. The heat transfer coefficient h is an empirical parameter incorporating the effects of system geometry, flow conditions, and fluid properties. Because it involves fluid flow, convective heat transfer is a more complex process than conduction. Consequently, there is little theoretical basis for calculation of h; h must be determined experimentally or evaluated using published correlations derived from experimental data. Suitable correlations for heat transfer coefficients are presented in Section 9.5.1. The SI units for h are W m^{-2} K^{-1}; in the imperial system h is expressed as Btu h^{-1} ft^{-2} $°F^{-1}$. Magnitudes of h vary greatly; some typical values are listed in Table 9.2.

According to the general rate principle, the rate of heat transfer $\hat{Q}$ in each fluid boundary layer can be written as the ratio of the temperature-difference driving force and the resistance, analogous to Eq. (9.11). Therefore, from Eq. (9.16), the two resistances to heat transfer on either side of the pipe wall are:

$$R_h = \frac{1}{h_h A} \tag{9.17}$$

TABLE 9.2 Individual Heat Transfer Coefficients

	Range of values of h	
Process	**(W m^{-2} °C^{-1})**	**(Btu h^{-1} ft^{-2} °F^{-1})**
Forced convection		
Heating or cooling air	10–500	2–100
Heating or cooling water	100–20,000	20–4000
Heating or cooling oil	60–2000	10–400
Boiling water flowing		
In a tube	5000–100,000	880–17,600
In a tank	2500–35,000	440–6200
Condensing steam, 1 atm		
On vertical surfaces	4000–11,300	700–2000
Outside horizontal tubes	9500–25,000	1700–4400
Condensing organic vapour	1100–2200	200–400
Superheating steam	30–110	5–20

Note: To convert from W m^{-2} °C^{-1} to Btu h^{-1} ft^{-2} °F^{-1}, multiply by 0.176.
Data from L.C. Thomas, 1992, Heat Transfer, *Prentice Hall, Upper Saddle River, NJ.; J.P. Holman, 1997,* Heat Transfer, *8th ed., McGraw-Hill, New York; and W.H. McAdams, 1954,* Heat Transmission, *3rd ed., McGraw-Hill, New York.*

and

$$R_c = \frac{1}{h_c A} \tag{9.18}$$

where R_h is the resistance to heat transfer in the hot fluid, R_c is the resistance to heat transfer in the cold fluid, h_h is the individual heat transfer coefficient for the hot fluid, h_c is the individual heat transfer coefficient for the cold fluid, and A is the surface area for heat transfer.

9.4.3 Overall Heat Transfer Coefficient

Application of Eq. (9.16) to calculate the rate of heat transfer in each boundary layer requires knowledge of ΔT for each fluid. This is usually difficult because we do not know T_{hw} and T_{cw}; it is much easier and more accurate to measure the bulk temperatures of fluids rather than wall temperatures. This problem is overcome by introducing the *overall heat transfer coefficient*, U, for the total heat transfer process through both fluid boundary layers and the wall. U is defined by the equation:

$$\hat{Q} = UA\,\Delta T \tag{9.19}$$

where ΔT is the overall temperature difference between the bulk hot and cold fluids. The units of U are the same as for h (e.g., W m^{-2} K^{-1} or Btu h^{-1} ft^{-2} °F^{-1}). Equation (9.19) written in terms of the ratio of the driving force ΔT and the resistance yields an expression for the total resistance to heat transfer, R_T:

$$R_T = \frac{1}{UA} \tag{9.20}$$

In Section 9.3.3 it was noted that when thermal resistances occur in series, the total resistance is the sum of the individual resistances. Applying this now to the situation of heat exchange between fluids, R_T is equal to the sum of R_h, R_w, and R_c:

$$R_T = R_h + R_w + R_c \tag{9.21}$$

Combining Eqs. (9.12), (9.17), (9.18), (9.20), and (9.21) gives:

$$\frac{1}{UA} = \frac{1}{h_h A} + \frac{B}{kA} + \frac{1}{h_c A} \tag{9.22}$$

In Eq. (9.22), the surface area A appears in each term. When fluids are separated by a flat wall, the surface area for heat transfer through each boundary layer and the wall is the same, so that A can be cancelled from the equation. However, a minor complication arises for cylindrical geometry such as pipes. Let us assume that hot fluid is flowing inside a pipe while cold fluid flows outside, as shown in Figure 9.13. Because the inside diameter of the pipe is smaller than the outside diameter, the surface areas for heat transfer between the fluid and the pipe wall are different for the two fluids. The surface area of the wall of a cylinder is equal to the circumference multiplied by the length:

$$A = 2\pi R L \tag{9.23}$$

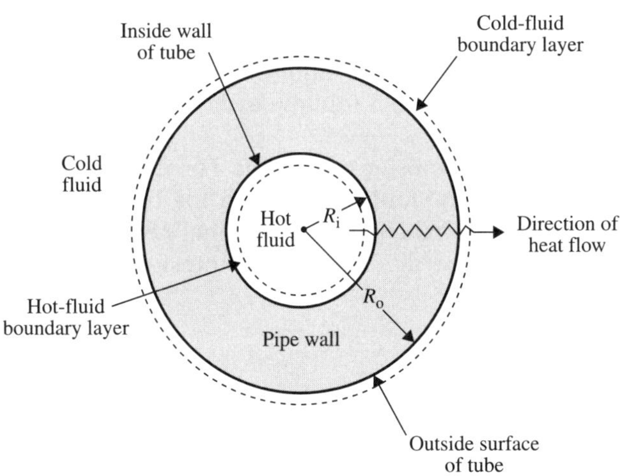

FIGURE 9.13 Effect of pipe wall thickness on the surface area for heat transfer.

where R is the radius of the cylinder and L is its length. Therefore, the heat transfer area at the hot-fluid boundary layer inside the tube is $A_i = 2\pi R_i L$; the heat transfer area at the cold-fluid boundary layer outside the tube is $A_o = 2\pi R_o L$. The surface area available for conduction through the wall varies between A_i and A_o.

The variation of heat transfer area in cylindrical systems depends on the thickness of the pipe wall. For thin walls, the variation will be relatively small because R_i is similar to R_o. In engineering design, variations in surface area are incorporated into the equations for heat transfer. However, for the sake of simplicity, in our analysis we will ignore any differences in surface area; we will assume, in effect, that the pipes are thin-walled. Accordingly, for cylindrical as well as for flat geometry, we can cancel A from Eq. (9.22) and write a simplified equation for U:

$$\frac{1}{U} = \frac{1}{h_h} + \frac{B}{k} + \frac{1}{h_c} \tag{9.24}$$

The overall heat transfer coefficient characterises the physical properties of the system and the operating conditions used for heat transfer. A small value of U for a particular process means that the system has only a limited ability to transfer heat. U can be increased by manipulating operating variables such as the fluid velocity in heat exchangers or the stirrer speed in bioreactors; these changes affect the value of h_c or h_h. The value of U is independent of A. To achieve a particular rate of heat transfer in an exchanger with small U, the heat transfer area must be relatively large; however increasing A raises the cost of the equipment. If U is large, the heat exchanger is well designed and operating under conditions that enhance heat transfer.

9.4.4 Fouling Factors

Heat transfer equipment in service does not remain clean. Dirt and scale are deposited on one or both sides of the pipes, providing additional resistance to heat flow and reducing the overall heat transfer coefficient. The resistances to heat transfer when fouling affects both sides of the heat transfer surface are represented in Figure 9.14. Five resistances are present in series: the thermal boundary layer or liquid film on the hot-fluid side, a fouling layer on the hot-fluid side, the pipe wall, a fouling layer on the cold-fluid side, and the cold-fluid thermal boundary layer.

Each fouling layer has associated with it a heat transfer coefficient. For dirt and scale, the coefficient is called a *fouling factor*. Let h_{fh} be the fouling factor on the hot-fluid side and h_{fc} be the fouling factor on the cold-fluid side. When these additional resistances are present, they must be included in the expression for the overall heat transfer coefficient, U. Equation (9.24) becomes:

$$\frac{1}{U} = \frac{1}{h_{fh}} + \frac{1}{h_h} + \frac{B}{k} + \frac{1}{h_c} + \frac{1}{h_{fc}} \tag{9.25}$$

Adding fouling factors in Eq. (9.25) increases $1/U$, thus decreasing the value of U.

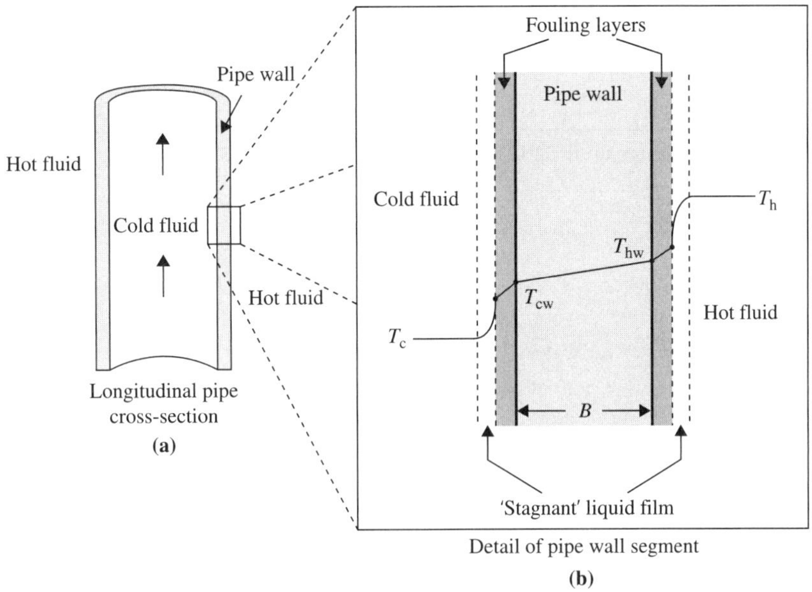

FIGURE 9.14 Heat transfer between fluids separated by a pipe wall with fouling deposits on both surfaces: (a) longitudinal pipe cross-section identifying a segment of the pipe wall; (b) magnified detail of the pipe wall segment showing boundary layers, fouling layers, and temperature gradients at the wall.

Accurate estimation of fouling factors is very difficult. The chemical nature of the deposit and its thermal conductivity depend on the fluid in the tube and the temperature; fouling thickness can also vary between cleanings. Typical values of fouling factors for various fluids are listed in Table 9.3.

9.5 DESIGN EQUATIONS FOR HEAT TRANSFER SYSTEMS

The basic equation for design of heat exchangers is Eq. (9.19). If U, ΔT, and $\hat{Q}$ are known, this equation allows us to calculate A. Specification of A is a major objective of heat exchanger design; the surface area dictates the configuration and size of the equipment and its cost. In the following sections, we will consider procedures for determining U, ΔT, and $\hat{Q}$ for use in Eq. (9.19).

9.5.1 Calculation of Heat Transfer Coefficients

As described in Sections 9.4.3 and 9.4.4, U can be determined as a combination of the individual film heat transfer coefficients, the properties of the separating wall, and, if applicable, any fouling factors. The values of the individual heat transfer coefficients h_h and h_c depend on the thickness of the fluid boundary layers, which, in turn, depends on the flow velocity and fluid properties such as viscosity and thermal conductivity.

TABLE 9.3 Fouling Factors for Scale Deposits

Source of deposit	Fouling factor ($W\ m^{-2}\ °C^{-1}$)	Fouling factor ($Btu\ h^{-1}\ ft^{-2}\ °F^{-1}$)
Water (temperatures up to 52°C, velocities over 1 m s^{-1})		
River water	2800	500
City or well water	5700	1000
Hard water	1900	330
Brackish water	5700	1000
Untreated cooling tower water	1900	330
Seawater	11,400	2000
Steam		
Good quality, oil free	11,400	2000
Liquids		
Industrial organic	5700	1000
Caustic solutions	2800	500
Vegetable oil	1900	330
Fuel oil	1100	200
Gases		
Compressed air	2800	500
Solvent vapour	5700	1000

Note: To convert from W m^{-2} °C^{-1} to Btu h^{-1} ft^{-2} °F^{-1}, multiply by 0.176.
Data from A.C. Mueller, 1985, Process heat exchangers. In: Handbook of Heat Transfer Applications, *2nd ed., W.M. Rohsenow, J.P. Hartnett, and E.N. Ganic (Eds.), pp. 4-78–4-173, McGraw-Hill, New York.*

Increasing the level of turbulence and decreasing the viscosity will reduce the thickness of the liquid film and hence increase the heat transfer coefficient.

Individual heat transfer coefficients for flow in pipes and stirred vessels can be evaluated using empirical correlations expressed in terms of dimensionless numbers. The general form of correlations for heat transfer coefficients is:

$$Nu = f\left(Re \text{ or } Re_i, Pr, Gr, \frac{D}{L}, \frac{\mu_b}{\mu_w}\right) \tag{9.26}$$

where *f* means 'some function of', and:

$$Nu = \text{Nusselt number} = \frac{hD}{k_{fb}} \tag{9.27}$$

$$Re = \text{Reynolds number for pipe flow} = \frac{D\,u\,\rho}{\mu_b} \tag{9.28}$$

$$Re_i = \text{impeller Reynolds number} = \frac{N_i\,D_i^2\,\rho}{\mu_b} \tag{9.29}$$

$$Pr = \text{Prandtl number} = \frac{C_p\,\mu_b}{k_{fb}} \tag{9.30}$$

and

$$Gr = \text{Grashof number for heat transfer} = \frac{D^3\,g\,\rho^2\,\beta\Delta T}{\mu_b^2} \tag{9.31}$$

The parameters in Eqs. (9.26) through (9.31) are as follows: h is the individual heat transfer coefficient, D is the pipe or tank diameter, k_{fb} is the thermal conductivity of the bulk fluid, u is the linear velocity of fluid in the pipe, ρ is the average density of the fluid, μ_b is the viscosity of the bulk fluid, N_i is the rotational speed of the impeller, D_i is the impeller diameter, C_p is the average heat capacity of the fluid, g is gravitational acceleration, β is the coefficient of thermal expansion of the fluid, ΔT is the variation of fluid temperature in the system, L is the pipe length, and μ_w is the viscosity of the fluid at the wall.

The Nusselt number, which contains the heat transfer coefficient h, represents the ratio of the rates of convective and conductive heat transfer. The Prandtl number represents the ratio of molecular momentum and thermal diffusivities; Pr contains physical constants which, for Newtonian fluids, are independent of flow conditions. The Grashof number represents the ratio of buoyancy to viscous forces and appears in correlations only when the fluid is not well mixed. Under these conditions, the fluid density is no longer uniform and natural convection becomes an important heat transfer mechanism. In most bioprocessing applications, heat transfer occurs between fluids in turbulent flow in pipes and stirred vessels; therefore, forced convection dominates natural convection and the Grashof number is not important. The form of the correlation used to evaluate Nu and therefore h depends on the configuration of the heat transfer equipment, the flow conditions, and other factors.

A wide variety of heat transfer situations is met in practice and there are many correlations available to biochemical engineers designing heat exchange equipment. The most common heat transfer applications are as follows:

- Heat flow to or from fluids inside tubes, without phase change
- Heat flow to or from fluids outside tubes, without phase change
- Heat flow from condensing fluids
- Heat flow to boiling liquids

Different equations are needed to evaluate h_h and h_c depending on the flow geometry of the hot and cold fluids. Examples of correlations for heat transfer coefficients relevant to bioprocessing are given in the following sections. Other correlations can be found in the references listed at the end of this chapter.

Flow in Tubes without Phase Change

There are several widely accepted correlations for forced convection in tubes. The heat transfer coefficient for fluid flowing inside a tube can be calculated using the following equation [1]:

$$Nu = 0.023\, Re^{0.8}\, Pr^{0.4} \tag{9.32}$$

Equation (9.32) is valid for either heating or cooling of fluid in the tube, and applies for $Re > 2100$ (turbulent flow) and for fluids with viscosity no greater than 2 mPa s. All of the physical properties used to calculate Nu, Re, and Pr are determined at the mean bulk temperature of the fluid, $(T_i + T_o)/2$, where T_i is the inlet temperature and T_o is the outlet temperature. For application of Eq. (9.32), D in Nu and Re refers to the internal diameter of the tube.

If the tube is wound into a helical shape, for example, to make a cooling coil for a bioreactor, the heat transfer coefficient is somewhat greater than that given by Eq. (9.32) [1]. However, as the effect is relatively small, for simplicity we will consider Eq. (9.32) appropriate for coiled as well as straight tubes.

Application of Eq. (9.32) to evaluate the tube-side heat transfer coefficient in a heat exchanger is illustrated in Example 9.1.

EXAMPLE 9.1 TUBE-SIDE HEAT TRANSFER COEFFICIENT

A single-pass shell-and-tube heat exchanger is used to heat a dilute salt solution for large-scale protein chromatography. Salt solution passes through 42 parallel tubes inside the heat exchanger at a total volumetric rate of 25.5 $m^3\,h^{-1}$; the internal diameter of the tubes is 1.5 cm and the tube length is 4 m. The viscosity of the salt solution is 10^{-3} kg $m^{-1}\,s^{-1}$, the density is 1010 kg m^{-3}, the average heat capacity is 4 kJ $kg^{-1}\,°C^{-1}$, and the thermal conductivity is 0.64 W $m^{-1}\,°C^{-1}$. Calculate the tube-side heat transfer coefficient.

Solution

First we must evaluate Re and Pr. All parameter values for calculation of these dimensionless groups are known except u, the linear fluid velocity. u is obtained by dividing the volumetric flow rate of the fluid by the total cross-sectional area for flow.

$$\begin{aligned}\text{Total cross-sectional area for flow} &= \text{number of tubes} \times \text{cross-sectional area of each tube}\\ &= 42(\pi R^2)\\ &= 42\pi\left(\frac{1.5\times10^{-2}\text{ m}}{2}\right)^2\\ &= 7.42\times10^{-3}\text{ m}^2\end{aligned}$$

Therefore:

$$u = \frac{\text{volumetric flow rate}}{\text{total cross-sectional area for flow}} = \frac{25.5\text{ m}^3\text{ h}^{-1}}{7.42\times10^{-3}\text{ m}^2}\cdot\left|\frac{1\text{ h}}{3600\text{ s}}\right| = 0.95\text{ m s}^{-1}$$

From Eq. (9.28):

$$Re = \frac{1.5 \times 10^{-2}\ \text{m}\ (0.95\ \text{m s}^{-1})\ (1010\ \text{kg m}^{-3})}{10^{-3}\ \text{kg m}^{-1}\ \text{s}^{-1}} = 1.44 \times 10^{4}$$

From Eq. (9.30):

$$Pr = \frac{4 \times 10^{3}\ \text{J kg}^{-1}\ {}^{\circ}\text{C}^{-1}\ (10^{-3}\ \text{kg m}^{-1}\ \text{s}^{-1})}{0.64\ \text{J s}^{-1}\ \text{m}^{-1}\ {}^{\circ}\text{C}^{-1}} = 6.25$$

As $Re > 2100$ and the viscosity is less than 2 mPa s, the conditions for application of Eq. (9.32) are satisfied. Therefore:

$$Nu = 0.023\ (1.44 \times 10^{4})^{0.8}\ (6.25)^{0.4} = 101.6$$

Calculating h for this value of Nu from Eq. (9.27):

$$h = \frac{Nu\ k_{fb}}{D} = \frac{101.6\ (0.64\ \text{W m}^{-1}\ {}^{\circ}\text{C}^{-1})}{1.5 \times 10^{-2}\ \text{m}} = 4335\ \text{W m}^{-2}\ {}^{\circ}\text{C}^{-1}$$

The heat transfer coefficient is 4.3 kW m^{-2} °C^{-1}.

For very viscous liquids, because of the temperature variation across the thermal boundary layer, there may be a significant difference between the viscosity of the fluid in bulk flow and the viscosity of the fluid adjacent to the wall. A modified form of Eq. (9.32) includes a viscosity correction term [1, 2]:

$$Nu = 0.027\ Re^{0.8}\ Pr^{0.33} \left(\frac{\mu_b}{\mu_w}\right)^{0.14} \tag{9.33}$$

where μ_b is the viscosity of the bulk fluid and μ_w is the viscosity of the fluid at the wall.

When flow in pipes is laminar rather than turbulent, Eqs. (9.32) and (9.33) do not apply. Under these conditions, fluid buoyancy and natural convection play a more important heat transfer role than in turbulent flow. Heat transfer correlations for laminar flow in tubes can be found in other texts (e.g., [3, 4]).

Flow across a Single Tube without Phase Change

A correlation for the average heat transfer coefficient for flow of gas or liquid across a cylindrical tube is [4]:

$$Nu = C\ Re^{n}\ Pr^{0.33} \tag{9.34}$$

where C and n are constants depending on the Reynolds number as listed in Table 9.4. Re is evaluated from Eq. (9.28) with D equal to the outside diameter of the tube and using the linear velocity and properties of the fluid flowing across the tube. For flow outside tubes, D in Nu is also taken to be the outside tube diameter. The physical properties are calculated at the mean film temperature of the fluid, which is the average of the mean bulk fluid temperature and the temperature at the surface of the tube.

TABLE 9.4 Values of the Constants C and n in Eq. (9.34)

Re	*C*	*n*
0.4–4	0.989	0.330
4–40	0.911	0.385
40–4000	0.683	0.466
$4000–4\times10^4$	0.193	0.618
$4\times10^4–4\times10^5$	0.0266	0.805

From J.P. Holman, 1997, Heat Transfer, *8th ed., McGraw-Hill, New York.*

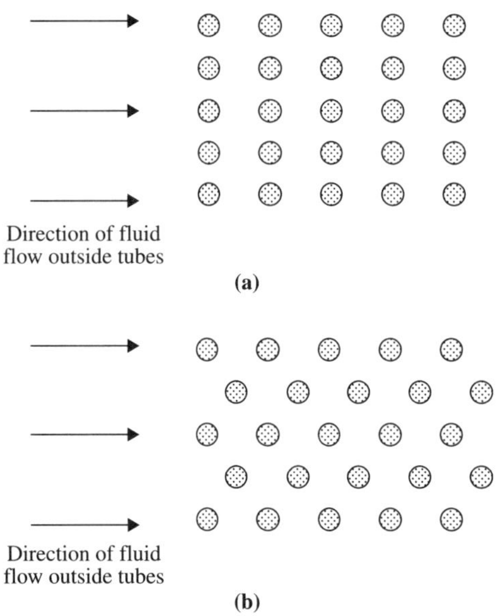

FIGURE 9.15 Configuration of tubes in shell-and-tube heat exchangers: (a) tubes 'in line'; (b) 'staggered' tubes.

Flow at Right Angles to a Bank of Tubes without Phase Change

In the shell section of shell-and-tube heat exchangers, fluid flows around the outside of the tubes. The degree of turbulence, and therefore the heat transfer coefficient for the shell-side fluid, depend in part on the geometric arrangement of the tubes and baffles. Tubes in the exchanger may be arranged 'in line' as shown in Figure 9.15(a), or 'staggered' as shown in Figure 9.15(b). The level of turbulence is considerably lower for tubes in line than for staggered tubes. Because the area for flow across a bank of staggered tubes is continually changing, this is a difficult system to analyse.

The following correlation has been proposed for flow of fluid at right angles to a bank of tubes more than 10 rows deep [3]:

$$Nu = F\,Re_{\max}^{m}\,Pr^{0.34} \tag{9.35}$$

TABLE 9.5 Values of the Constants F and m in Eq. (9.35)

	'In line' tube array		'Staggered' tube array	
Re_{max}	F	m	F	m
10–300	0.742	0.431	1.309	0.360
$300–2 \times 10^5$	0.211	0.651	0.273	0.635
$2 \times 10^5–2 \times 10^6$	0.116	0.700	0.124	0.700

From G.F. Hewitt, G.L. Shires, and T.R. Bott, 1994, Process Heat Transfer, *CRC Press, Boca Raton, FL.*

In Eq. (9.35), F and m are constants dependent on the geometry of the tube bank and the value of Re_{max}, as indicated in Table 9.5. Re_{max} is the Reynolds number evaluated using Eq. (9.28) with D equal to the outside tube diameter and u equal to the maximum fluid velocity based on the minimum free area available for fluid flow. Methods for estimating the maximum fluid velocity for different tube-bank geometries are available in the literature [3]. Re_{max} and Pr are based on fluid properties at the mean bulk temperature of the fluid flowing across the tubes; the mean bulk temperature is equal to $(T_i + T_o)/2$, where T_i and T_o are the inlet and outlet temperatures, respectively. For flow outside tubes, D in Nu is taken as the outside tube diameter. Corrections to Eq. (9.35) for banks of fewer than 10 rows of tubes and for fluids with bulk-to-wall property variations are available [3].

Stirred Liquids

The heat transfer coefficient in stirred vessels depends on the degree of agitation and the properties of the fluid. When heat is transferred to or from a helical coil in the vessel, h for the tank side of the coil can be determined using the following equation [1, 5]:

$$Nu = 0.9\, Re_i^{0.62}\, Pr^{0.33} \left(\frac{\mu_b}{\mu_w}\right)^{0.14} \tag{9.36}$$

where Re_i is given by Eq. (9.29), and D in Nu refers to the inside diameter of the tank. For low-viscosity fluids such as water, the viscosity at the wall μ_w can be assumed equal to the bulk viscosity μ_b. For viscous fermentation broths, because the broth temperature at the wall of the cooling coil is lower than in the bulk liquid, μ_w may be greater than μ_b so that Nu is reduced somewhat. Equation (9.36) can be applied with all types of impeller and has an accuracy to within about 20% in ungassed systems. A range of alternative correlations for helical coils in stirred vessels is available in the literature [3].

When heat is transferred to or from a jacket rather than a coil, the correlation is slightly modified [5]:

$$Nu = 0.36\, Re_i^{0.67} Pr^{0.33} \left(\frac{\mu_b}{\mu_w}\right)^{0.14} \tag{9.37}$$

EXAMPLE 9.2 HEAT TRANSFER COEFFICIENT FOR A STIRRED VESSEL

A stirred fermenter of diameter 5 m contains an internal helical coil for heat transfer. The fermenter is mixed using a turbine impeller 1.8 m in diameter operated at 60 rpm. The fermentation broth has the following properties: $\mu_b = 5 \times 10^{-3}$ Pa s, $\rho = 1000$ kg m^{-3}, $C_p = 4.2$ kJ kg^{-1} °C^{-1}, $k_{fb} = 0.70$ W m^{-1} °C^{-1}. Neglecting viscosity changes at the wall, calculate the heat transfer coefficient for the liquid film on the outside of the coil.

Solution

From Eqs. (9.29) and (9.30):

$$Re_i = \frac{60\ \text{min}^{-1} \cdot \left|\frac{1\ \text{min}}{60\ \text{s}}\right| \cdot (1.8\ \text{m})^2 (1000\ \text{kg m}^{-3})}{5 \times 10^{-3}\ \text{kg m}^{-1}\ \text{s}^{-1}} = 6.48 \times 10^5$$

$$Pr = \frac{4.2 \times 10^3\ \text{J kg}^{-1}\ {}^\circ\text{C}^{-1} (5 \times 10^{-3}\ \text{kg m}^{-1}\ \text{s}^{-1})}{0.70\ \text{J s}^{-1}\ \text{m}^{-1}\ {}^\circ\text{C}^{-1}} = 30$$

To calculate Nu, these values are substituted into Eq. (9.36) with $\mu_b = \mu_w$:

$$Nu = 0.9\ (6.48 \times 10^5)^{0.62}\ (30)^{0.33} = 1.11 \times 10^4$$

Calculating h for this value of Nu from Eq. (9.27):

$$h = \frac{Nu\ k_{fb}}{D} = \frac{1.11 \times 10^4\ (0.70\ \text{W m}^{-1}\ {}^\circ\text{C}^{-1})}{5\ \text{m}} = 1554\ \text{W m}^{-2}\ {}^\circ\text{C}^{-1}$$

The heat transfer coefficient is 1.55 kW m^{-2} °C^{-1}.

9.5.2 Logarithmic-Mean Temperature Difference

Application of the heat exchanger design equation, Eq. (9.19), requires knowledge of the temperature-difference driving force for heat transfer, ΔT. ΔT is the difference between the bulk temperatures of the hot and cold fluids. However, as we have seen in Figures 9.2, 9.4, 9.5, and 9.9, bulk fluid temperatures vary with position in heat exchangers, so that the temperature difference between the hot and cold fluids changes between one end of the equipment and the other. Thus, the driving force for heat transfer varies from point to point in the system. For application of Eq. (9.19), this difficulty is overcome using an average ΔT.

For single-pass heat exchangers in which flow is either cocurrent or countercurrent, the *logarithmic-mean temperature difference* is used. In this case, ΔT is given by the equation:

$$\Delta T = \frac{\Delta T_2 - \Delta T_1}{\ln (\Delta T_2 / \Delta T_1)} \tag{9.38}$$

where subscripts 1 and 2 denote the ends of the equipment. ΔT_1 and ΔT_2 are the temperature differences between the hot and cold fluids at the ends of the exchanger calculated using the values for T_{hi}, T_{ho}, T_{ci}, and T_{co}. For convenience and to eliminate negative numbers and their logarithms, subscripts 1 and 2 can refer to either end of the exchanger. Equation (9.38) has been derived using the following assumptions:

- The overall heat transfer coefficient U is constant.
- The specific heat capacities of the hot and cold fluids are constant or, alternatively, any phase change in the fluids (e.g., boiling or condensation) occurs isothermally.
- Heat losses from the system are negligible.
- The system is at steady state in either countercurrent or cocurrent flow.

The most questionable of these assumptions is that of constant U, since this coefficient varies with the temperature of the fluids and may be affected significantly by phase change. However, if the change in U with temperature is gradual, or if temperature differences in the system are moderate, this assumption is not seriously in error. Other details of the derivation of Eq. (9.38) can be found in other texts [3, 4, 6].

EXAMPLE 9.3 LOG-MEAN TEMPERATURE DIFFERENCE

A liquid stream is cooled from 70°C to 32°C in a double-pipe heat exchanger. Fluid flowing countercurrently with this stream is heated from 20°C to 44°C. Calculate the log-mean temperature difference.

Solution

The heat exchanger configuration is shown in Figure 9.16. At one end of the equipment, $\Delta T_1 = (32 - 20)°C = 12°C$. At the other end, $\Delta T_2 = (70 - 44)°C = 26°C$. From Eq. (9.38):

$$\Delta T = \frac{(26 - 12)°C}{\ln (26/12)} = 18.1°C$$

The log-mean temperature difference is 18°C.

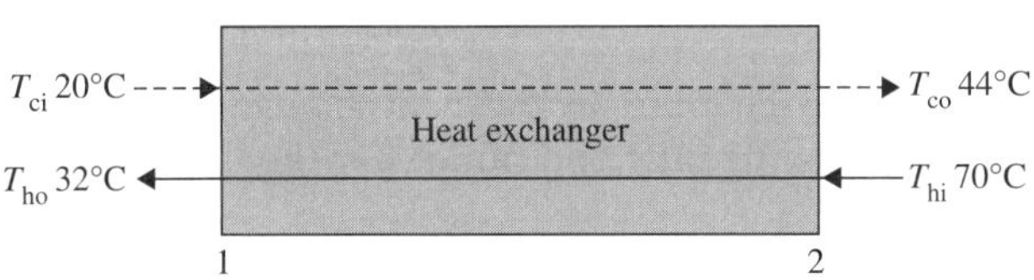

FIGURE 9.16 Flow configuration for a heat exchanger.

The logarithmic-mean temperature difference as defined in Eq. (9.38) is also applicable when one fluid in the heat exchange system remains at a constant temperature. For the case of a fermenter at constant temperature T_F cooled by water in a cooling coil (Figure 9.2), Eq. (9.38) can be simplified to:

$$\Delta T = \frac{T_{co} - T_{ci}}{\ln\left(\frac{T_F - T_{ci}}{T_F - T_{co}}\right)} \quad (9.39)$$

where T_{ci} is the inlet temperature of the cooling water and T_{co} is the outlet temperature of the cooling water.

In multiple-pass shell-and-tube heat exchangers, for example, in the double tube-pass unit illustrated in Figure 9.8, flow is neither purely countercurrent nor purely cocurrent. Flow patterns in multiple-pass equipment can be complex, with cocurrent, countercurrent, and cross flow all present. Under these conditions, the log-mean temperature difference does not accurately represent the average temperature difference in the system. For shell-and-tube heat exchangers with more than a single tube pass or a single shell pass, the log-mean temperature difference must be used with an appropriate correction factor to account for the geometry of the exchanger. Correction factors for a range of equipment configurations have been determined and are available in other references [3, 4, 6].

The log-mean temperature difference is not applicable to heat exchangers in which one or both fluids in the system change phase. Design procedures for heat transfer operations involving boiling liquids or condensing steam can be found elsewhere [3].

9.5.3 Energy Balance

In heat exchanger design, energy balances are applied to determine $\hat{Q}$ and all inlet and outlet temperatures required to specify ΔT. These energy balances are based on the general equations for flow systems derived in Chapters 5 and 6.

Let us first consider the equations for double-pipe or shell-and-tube heat exchangers. From Eq. (6.10), under steady-state conditions ($\mathrm{d}E/\mathrm{d}t = 0$) and as no shaft work is performed during pipe flow ($\hat{W}_s = 0$), the energy balance equation is:

$$\hat{M}_i h_i - \hat{M}_o h_o - \hat{Q} = 0 \tag{9.40}$$

where $\hat{M}_i$ is the mass flow rate in, $\hat{M}_o$ is the mass flow rate out, h_i is the specific enthalpy of the incoming stream, h_o is the specific enthalpy of the outgoing stream, and $\hat{Q}$ is the rate of heat transfer to or from the system. Unfortunately, the conventional symbols for individual heat transfer coefficient and specific enthalpy are the same: h. In this section, h in Eqs. (9.40) through (9.45) and (9.48) denotes specific enthalpy; otherwise in this chapter, h represents the individual heat transfer coefficient.

Equation (9.40) can be applied separately to each fluid in the heat exchanger. As the mass flow rate of each fluid does not vary between its inlet and outlet, for the hot fluid:

$$\hat{M}_h (h_{hi} - h_{ho}) - \hat{Q}_h = 0 \tag{9.41}$$

or

$$\hat{M}_h (h_{hi} - h_{ho}) = \hat{Q}_h \tag{9.42}$$

where subscript h denotes the hot fluid and $\hat{Q}_h$ is the rate of heat transfer from that fluid. Equations similar to Eqs. (9.41) and (9.42) can be derived for the cold fluid:

$$\hat{M}_c (h_{ci} - h_{co}) + \hat{Q}_c = 0 \tag{9.43}$$

or

$$\hat{M}_c (h_{co} - h_{ci}) = \hat{Q}_c \tag{9.44}$$

where subscript c refers to the cold fluid. $\hat{Q}_c$ is the rate of heat flow into the cold fluid; therefore $\hat{Q}_c$ is added rather than subtracted in Eq. (9.43) according to the sign convention for energy flows explained in Section 5.2.

When there are no heat losses from the exchanger, all heat removed from the hot stream must be taken up by the cold stream. Thus, we can equate $\hat{Q}$ terms in Eqs. (9.42) and (9.44): $\hat{Q}_h = \hat{Q}_c = \hat{Q}$. Therefore:

$$\hat{M}_h(h_{hi} - h_{ho}) = \hat{M}_c(h_{co} - h_{ci}) = \hat{Q} \tag{9.45}$$

If neither fluid changes phase (e.g., by boiling or condensation) so that only sensible heat is exchanged, the enthalpy differences in Eq. (9.45) can be expressed in terms of the heat capacity C_p and the temperature change for each fluid (Section 5.4.1). If we assume C_p is constant over the temperature range in the exchanger, Eq. (9.45) becomes:

$$\hat{M}_h C_{ph}(T_{hi} - T_{ho}) = \hat{M}_c C_{pc}(T_{co} - T_{ci}) = \hat{Q} \tag{9.46}$$

where C_{ph} is the heat capacity of the hot fluid, C_{pc} is the heat capacity of the cold fluid, T_{hi} is the inlet temperature of the hot fluid, T_{ho} is the outlet temperature of the hot fluid, T_{ci} is the inlet temperature of the cold fluid, and T_{co} is the outlet temperature of the cold fluid.

In heat exchanger design, Eq. (9.46) is used to determine $\hat{Q}$ and the inlet and outlet conditions of the fluid streams in the absence of phase change. This is illustrated in Example 9.4.

EXAMPLE 9.4 HEAT EXCHANGER

Hot, freshly sterilised nutrient medium is cooled in a double-pipe heat exchanger before being used in a fermentation. Medium leaving the steriliser at 100°C enters the exchanger at a flow rate of 10 $m^3\ h^{-1}$; the desired outlet temperature is 30°C. Heat from the medium is used to raise the temperature of 25 $m^3\ h^{-1}$ of cooling water entering the exchanger at 15°C. The system operates at steady state. Assume that the nutrient medium has the properties of water.

(a) What rate of heat transfer is required?
(b) Calculate the final temperature of the cooling water as it leaves the heat exchanger.

Solution

The density of water and medium is 1000 kg m^{-3}. Therefore:

$$\hat{M}_h = 10\ \text{m}^3\ \text{h}^{-1} \cdot \left|\frac{1\ \text{h}}{3600\ \text{s}}\right| \cdot (1000\ \text{kg m}^{-3}) = 2.78\ \text{kg s}^{-1}$$

$$\hat{M}_c = 25\ \text{m}^3\ \text{h}^{-1} \cdot \left|\frac{1\ \text{h}}{3600\ \text{s}}\right| \cdot (1000\ \text{kg m}^{-3}) = 6.94\ \text{kg s}^{-1}$$

From Table C.3 in Appendix C, the heat capacity of water can be taken as 75.4 J $gmol^{-1}\ °C^{-1}$ for most of the temperature range of interest. Therefore:

$$C_{ph} = C_{pc} = 75.4\ \text{J gmol}^{-1}\ °\text{C}^{-1} \cdot \left|\frac{1\ \text{gmol}}{18\ \text{g}}\right| \cdot \left|\frac{1000\ \text{g}}{1\ \text{kg}}\right| = 4.19 \times 10^3\ \text{J kg}^{-1}\ °\text{C}^{-1}$$

(a) From Eq. (9.46) for the hot fluid:

$$\hat{Q} = 2.78 \text{ kg s}^{-1} \, (4.19 \times 10^3 \text{ J kg}^{-1} \,°\text{C}^{-1}) \, (100 - 30)°\text{C}$$

$$\hat{Q} = 8.15 \times 10^5 \text{ J s}^{-1} = 815 \text{ kW}$$

The rate of heat transfer required is 815 kJ s^{-1}.

(b) For the cold fluid, from Eq. (9.46):

$$T_{\text{co}} = T_{\text{ci}} + \frac{\hat{Q}}{\hat{M}_{\text{c}} C_{pc}}$$

$$T_{\text{co}} = 15°\text{C} + \frac{8.15 \times 10^5 \text{ J s}^{-1}}{6.94 \text{ kg s}^{-1}(4.19 \times 10^3 \text{ J kg}^{-1} \,°\text{C}^{-1})} = 43.0°\text{C}$$

The exit cooling water temperature is 43°C.

Equation (9.46) can also be applied to the situation of heat removal from a bioreactor for the purpose of temperature control. In this case, the temperature of the hot fluid (i.e., the fermentation broth) is uniform throughout the system so that the left side of Eq. (9.46) is zero. If energy is absorbed by the cold fluid as sensible heat, the energy balance equation becomes:

$$\hat{M}_{\text{c}} C_{pc} (T_{\text{co}} - T_{\text{ci}}) = \hat{Q} \tag{9.47}$$

To use Eq. (9.47) for bioreactor design we must know $\hat{Q}$. $\hat{Q}$ is found by considering all significant heat sources and sinks in the system. An expression involving $\hat{Q}$ for fermentation systems was presented in Chapter 6 based on relationships derived in Chapter 5:

$$\frac{dE}{dt} = -\Delta\hat{H}_{\text{rxn}} - \hat{M}_{\text{v}} \Delta h_{\text{v}} - \hat{Q} + \hat{W}_{\text{s}} \tag{6.12}$$

where $\Delta\hat{H}_{\text{rxn}}$ is the rate of heat absorption or evolution due to metabolic reaction, $\hat{M}_{\text{v}}$ is the mass flow rate of evaporated liquid leaving the system, Δh_{v} is the latent heat of vaporisation, and $\hat{W}_{\text{s}}$ is the rate of shaft work done on the system. For exothermic reactions $\Delta\hat{H}_{\text{rxn}}$ is negative; for endothermic reactions $\Delta\hat{H}_{\text{rxn}}$ is positive. In most fermentation systems, the only source of shaft work is the stirrer; therefore $\hat{W}_{\text{s}}$ is the power dissipated by the impeller. Methods for estimating the power required for stirrer operation are described in Section 8.5. Equation (6.12) represents a considerable simplification of the energy balance. It is applicable to systems in which the heat of reaction dominates the energy balance so that contributions from sensible heat and heats of solution can be ignored. In large fermenters, metabolic activity is the dominant source of heat; the energy input by stirring and heat losses from evaporation may also be worth considering in some cases. The other heat sources and sinks are relatively minor and can generally be neglected.

At steady state $dE/dt = 0$ and Eq. (6.12) becomes:

$$\hat{Q} = -\Delta\hat{H}_{rxn} - \hat{M}_v \Delta h_v + \hat{W}_s \tag{9.48}$$

Application of Eq. (9.48) to determine $\hat{Q}$ is illustrated in Examples 5.7 and 5.8. Once $\hat{Q}$ has been estimated, Eq. (9.47) is used to evaluate unknown operating conditions as shown in Example 9.5.

EXAMPLE 9.5 COOLING COIL

A 150-m^3 bioreactor is operated at 35°C to produce fungal biomass from glucose. The rate of oxygen uptake by the culture is 1.5 kg m^{-3} h^{-1}; the agitator dissipates energy at a rate of 1 kW m^{-3}. Cooling water available from a nearby river at 10°C is passed through an internal coil in the fermentation tank at a rate of 60 m^3 h^{-1}. If the system operates at steady state, what is the exit temperature of the cooling water?

Solution

The rate of heat generation by aerobic cultures can be calculated directly from the oxygen demand. As described in Section 5.9.2, approximately 460 kJ of heat is released for each gmol of oxygen consumed. Therefore, from Eq. (5.23) the metabolic heat load is:

$$\Delta\hat{H}_{rxn} = \frac{-460\text{ kJ}}{\text{gmol}} \cdot \left|\frac{1\text{ gmol}}{32\text{ g}}\right| \cdot \left|\frac{1000\text{ g}}{1\text{ kg}}\right| \cdot (1.5\text{ kg m}^{-3}\text{ h}^{-1}) \cdot \left|\frac{1\text{ h}}{3600\text{ s}}\right| \cdot 150\text{ m}^3 = -898\text{ kJ s}^{-1}$$

$$\Delta\hat{H}_{rxn} = -898\text{ kW}$$

$\Delta\hat{H}_{rxn}$ is negative because fermentation is exothermic. The rate of heat dissipation by the agitator is:

$$1\text{ kW m}^{-3}\,(150\text{ m}^3) = 150\text{ kW}$$

We can now calculate $\hat{Q}$ from Eq. (9.48), assuming there is negligible evaporation from the bioreactor:

$$\hat{Q} = -\Delta\hat{H}_{rxn} - \hat{M}_v\,\Delta h_v + \hat{W}_s = (898 - 0 + 150)\text{ kW} = 1048\text{ kW}$$

The density of the cooling water is 1000 kg m^{-3}; therefore:

$$\hat{M}_c = 60\text{ m}^3\text{ h}^{-1} \cdot \left|\frac{1\text{ h}}{3600\text{ s}}\right| \cdot (1000\text{ kg m}^{-3}) = 16.7\text{ kg s}^{-1}$$

From Table C.3 in Appendix C, the heat capacity of water is 75.4 J $gmol^{-1}$ $°C^{-1}$. Therefore:

$$C_{pc} = 75.4\text{ J gmol}^{-1}\,°\text{C}^{-1} \cdot \left|\frac{1\text{ gmol}}{18\text{ g}}\right| \cdot \left|\frac{1000\text{ g}}{1\text{ kg}}\right| = 4.19 \times 10^3\text{ J kg}^{-1}\,°\text{C}^{-1}$$

We can now apply Eq. (9.47) after rearranging and solving for T_{co}:

$$T_{co} = T_{ci} + \frac{\hat{Q}}{\hat{M}_c C_{pc}}$$

$$T_{co} = 10°C + \frac{1048 \times 10^3 \text{ J s}^{-1}}{16.7 \text{ kg s}^{-1}(4.19 \times 10^3 \text{ J kg}^{-1} \text{ °C}^{-1})} = 25.0°C$$

The outlet water temperature is 25°C.

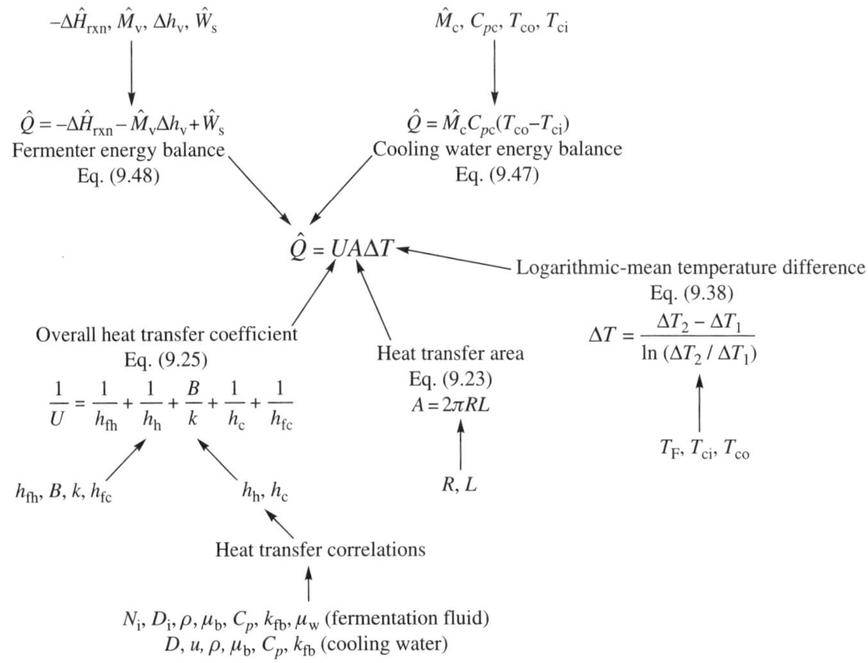

FIGURE 9.17 Summary of relationships and equations for design of a fermenter cooling coil.

9.6 APPLICATION OF THE DESIGN EQUATIONS

The equations in Sections 9.4 and 9.5 provide the essential elements for design of heat transfer systems. Figure 9.17 summarises the relationships involved. Equation (9.19) is used as the design equation; $\hat{Q}$ and the inlet and outlet temperatures are available from energy balance calculations as described in Section 9.5.3. The overall heat transfer coefficient is evaluated from correlations such as those given in Section 9.5.1; additional terms are included if the heat transfer surfaces are fouled. The temperature-difference driving force is estimated from the fermentation and cooling water temperatures. With these parameters at hand, the required heat transfer area can be determined.

Because metabolic rates, and therefore rates of heat production, vary during fermentation, the rate at which heat is removed from the bioreactor must also be varied to maintain constant temperature. Heat transfer design is based on the maximum heat load for the system. When the rate of heat generation drops, operating conditions can be changed to reduce the rate of heat removal. The simplest way to achieve this is to decrease the cooling water flow rate.

The procedures outlined in this chapter represent the simplest and most direct approach to heat transfer design. If several independent variables remain unfixed prior to the design calculations, many different design outcomes are possible. When variables such as the type of fluid, the mass flow rates, and the terminal temperatures are unspecified, they can be manipulated to produce an *optimum design*, for example, the design yielding the lowest total cost per year of operation. Computer packages that optimise heat exchanger design are available commercially.

9.6.1 Heat Exchanger Design for Fermentation Systems

Calculating the equipment requirements for heating or cooling of fermenters can sometimes be simplified by considering the relative importance of each heat transfer resistance.

- For large fermentation vessels containing cooling coils, the fluid velocity in the vessel is generally much slower than in the coils; accordingly, the tube-side thermal boundary layer is relatively thin and most of the heat transfer resistance is located on the fermenter side. Especially when there is no fouling in the tubes, the heat transfer coefficient for the cooling water can often be omitted when calculating *U*.
- Likewise, the pipe wall resistance can sometimes be ignored as conduction of heat through metal is generally very rapid. An exception is stainless steel, which is used widely in the fermentation industry because it does not corrode in mild acid environments, withstands repeated exposure to clean steam, and does not have a toxic effect on cells. The low thermal conductivity of this material means that wall resistance may be important unless the pipe wall is very thin.

The correlations used to estimate fermenter-side heat transfer coefficients, such as Eqs. (9.36) and (9.37), were not developed for fermentation systems and must not be considered to give exact values. Most of the available correlations were developed using small-scale equipment; little information is available for industrial-size reactors.

As described in Section 8.13, aerobic fermentations are typically carried out in tall vessels using multiple impellers; however, there have been relatively few studies of heat transfer using this vessel configuration. For helical cooling coils, geometric parameters such as the vertical separation between individual coils and the space between the coil and tank wall have been found to affect heat transfer, particularly for viscous fluids. Yet these factors are not included in most commonly used heat transfer correlations.

Equations (9.36) and (9.37) were developed for ungassed systems but are applied routinely for aerobic fermentations. Gassing alters the value of fermenter-side heat transfer coefficients, but the magnitude of the effect and whether gassing causes an increase or decrease in heat transfer cannot yet be predicted. The rate of heat transfer in gas–liquid systems appears to depend on the distribution of bubbles in the vessel [7]; for example, accumulation of bubbles on or around the heat transfer surface is deleterious. Taking all of the above factors into account, it is apparent that correlations such as Eqs. (9.36) and (9.37) can provide only a starting point or rough estimate for evaluation of heat transfer coefficients in fermenters.

When estimating heat transfer coefficients for non-Newtonian broths, the apparent viscosity (Section 7.5) can be substituted for μ_b in the correlation equations and dimensionless groups. However, this substitution is not straightforward when rheological parameters such as the flow behaviour index n, the consistency index K, and the yield stress τ_0 change during the culture. The apparent viscosity also depends on the shear rate in the fermenter, which varies greatly throughout the vessel. These factors make evaluation of heat transfer coefficients for non-Newtonian systems difficult.

Application of the heat exchanger design equations to specify a fermenter cooling system is illustrated in Example 9.6.

EXAMPLE 9.6 COOLING-COIL LENGTH IN FERMENTER DESIGN

A fermenter used for antibiotic production must be kept at 35°C. After considering the oxygen demand of the organism and the heat dissipation from the stirrer, the maximum heat transfer rate required is estimated as 550 kW. Cooling water is available at 10°C; the exit temperature of the cooling water is calculated using an energy balance as 25°C. The heat transfer coefficient for the fermentation broth is estimated from Eq. (9.36) as 2150 W m^{-2} °C^{-1}. The heat transfer coefficient for the cooling water is calculated as 14 kW m^{-2} °C^{-1}. It is proposed to install a helical cooling coil inside the fermenter; the outer diameter of the pipe is 8 cm, the pipe wall thickness is 5 mm, and the thermal conductivity of the steel is 60 W m^{-1} °C^{-1}. An average internal fouling factor of 8500 W m^{-2} °C^{-1} is expected; the fermenter-side surface of the coil is kept relatively clean. What length of cooling coil is required?

Solution

$\hat{Q} = 550 \times 10^3$ W. As the temperature in the fermenter is constant, ΔT is calculated from Eq. (9.39):

$$\Delta T = \frac{(25-10)°\text{C}}{\ln\left(\dfrac{35-10}{35-25}\right)} = 16.4°\text{C}$$

U is calculated using Eq. (9.25) after omitting h_{fh}, as there is no fouling layer on the hot side of the coil:

$$\frac{1}{U} = \left(\frac{1}{2150 \text{ W m}^{-2}\,°\text{C}^{-1}} + \frac{5\times10^{-3}\text{ m}}{60 \text{ W m}^{-1}\,°\text{C}^{-1}} + \frac{1}{14\times10^{3}\text{ W m}^{-2}\,°\text{C}^{-1}} + \frac{1}{8500 \text{ W m}^{-2}\,°\text{C}^{-1}}\right)$$
$$= (4.65\times10^{-4} + 8.33\times10^{-5} + 7.14\times10^{-5} + 1.18\times10^{-4})\text{ m}^2\,°\text{C W}^{-1}$$
$$= 7.38\times10^{-4}\text{ m}^2\,°\text{C W}^{-1}$$

$$U = 1355 \text{ W m}^{-2}\,°\text{C}^{-1}$$

Note the relative magnitudes of the four contributions to U: the cooling water film coefficient and the wall resistance make comparatively minor contributions and can often be neglected in design calculations.

We can now apply Eq. (9.19) to evaluate the required surface area A:

$$A = \frac{\hat{Q}}{U\,\Delta T} = \frac{550 \times 10^3\ \text{W}}{1355\ \text{W m}^{-2}\ {}^\circ\text{C}^{-1}\ (16.4^\circ\text{C})} = 24.75\ \text{m}^2$$

Equation (9.23) for the area of a cylinder can be used to evaluate the pipe length, L. As we have information for both the outer pipe diameter (= 8 cm) and the inner pipe diameter (= outer pipe diameter − 2 × wall thickness = 8 cm − 2 × 5 mm = 7 cm), we can use an average pipe radius to determine L:

$$L = \frac{A}{2\pi R} = \frac{24.75\ \text{m}^2}{2\pi\left(\dfrac{0.5\,(8 \times 10^{-2} + 7 \times 10^{-2})}{2}\right)\text{m}} = 105.0\ \text{m}$$

The length of coil required is 105 m. The cost of such a length of pipe is a significant factor in the overall cost of the fermenter.

Mixing and heat transfer are not independent functions in bioreactors. The impeller size and stirrer speed affect the value of the heat transfer coefficient in the fermentation fluid; increased turbulence in the reactor decreases the thickness of the thermal boundary layer and facilitates rapid heat transfer. However, stirring also generates heat that must be removed from the reactor to maintain constant temperature. Although this energy contribution may be small in low-viscosity fluids because of the dominance of the heat of reaction (Section 5.10), heat removal can be a severe problem in bioreactors containing highly viscous fluids. Turbulent flow and high heat transfer coefficients are difficult to achieve in viscous liquids without enormous power input, which itself generates an extra heat load. The effect of gas sparging on the heat load is usually neglected for stirred bioreactors. The energy associated with sparging can be evaluated as described in Section 8.6; however, the contribution to overall cooling requirements is minor.

Internal cooling coils typically add 15 to 25% to the cost of fermentation vessels. The length of coil that can be used is limited by the size of the tank. Because the temperature of fermentations must be maintained within a very narrow range, if the culture generates a large heat load, providing sufficient cooling can be a challenge. As discussed further in Section 9.6.2, heat transfer can be the limiting factor in fermentations. This is of particular concern for high-density cultures growing on carbon substrates with a high degree of reduction (Section 5.9.5), as large amounts of energy are released. Elevated broth viscosities compound the problem by dampening turbulence and preventing good mixing, thus reducing the rate of heat transfer. Because biological reactions take place at near-ambient temperatures, the temperature of the cooling water is always close to that of the fermentation; therefore, a relatively small driving force for heat transfer or ΔT value is inevitable. Improving the situation by refrigerating the cooling water involves a substantial increase in capital and operating costs and is generally not economically feasible. It may be possible to raise the cooling water flow rate, but this is also limited by equipment size and cost considerations. Although using an external heat exchanger such as that illustrated in Figure 9.1(e) is an option for increasing the heat transfer capacity of fermenters, as described in Section 9.1.1 this introduces extra contamination risks and also adds significantly to the overall cost of the equipment. In some cases, the only feasible heat transfer solution for fermentations with high heat loads is to slow down the rate of fermentation, thus slowing down the rate of heat generation. This might be achieved, for example, by reducing the cell density, decreasing the concentration of rate-limiting nutrients in the medium, or using a different carbon source.

9.6.2 Relationship between Heat Transfer and Cell Concentration

The design equation, Eq. (9.19), and the steady-state energy balance equation, Eq. (9.48), allow us to derive some important relationships for fermenter operation. Because cell metabolism is usually the largest source of heat in fermenters, the capacity of the system for heat removal can be linked directly to the maximum cell concentration in the reactor. Assuming that the heat dissipated from the stirrer and the cooling effects of evaporation are negligible compared with the heat of reaction, Eq. (9.48) becomes:

$$\hat{Q} = -\Delta \hat{H}_{rxn} \tag{9.49}$$

In aerobic fermentations, the heat of reaction is related to the rate of oxygen consumption by the cells. As outlined in Section 5.9.2, approximately 460 kJ of heat is released for each gmol of oxygen consumed. Therefore, if Q_O is the rate of oxygen uptake per unit volume in the fermenter:

$$\Delta \hat{H}_{rxn} = (-460 \text{ kJ gmol}^{-1})\, Q_O V \tag{9.50}$$

where V is the reactor volume. Typical units for Q_O are gmol m^{-3} s^{-1}. $\Delta \hat{H}_{rxn}$ in Eq. (9.50) is negative because the reaction is exothermic. Substituting this equation into Eq. (9.49) gives:

$$\hat{Q} = (460 \text{ kJ gmol}^{-1})\, Q_O V \tag{9.51}$$

If q_O is the *specific oxygen uptake rate*, or the rate of oxygen consumption per cell, $Q_O = q_O x$, where x is the cell concentration. Typical units for q_O are gmol g^{-1} s^{-1}. Therefore:

$$\hat{Q} = (460 \text{ kJ gmol}^{-1})\, q_O x V \tag{9.52}$$

Substituting this into Eq. (9.19) gives:

$$(460 \text{ kJ gmol}^{-1})\, q_O x V = UA\,\Delta T \tag{9.53}$$

The fastest rate of heat transfer occurs when the temperature difference between the fermenter contents and the cooling water is at its maximum. This occurs when the cooling water is at its inlet temperature, that is, $\Delta T = (T_F - T_{ci})$ where T_F is the fermentation temperature and T_{ci} is the water inlet temperature. Therefore, assuming that the cooling water remains at temperature T_{ci}, we can derive from Eq. (9.53) an equation for the hypothetical maximum possible cell concentration supported by the heat transfer system:

$$x_{max} = \frac{UA(T_F - T_{ci})}{(460 \text{ kJ gmol}^{-1})\, q_O V} \tag{9.54}$$

It is undesirable for the biomass concentration in fermenters to be limited by the heat transfer capacity of the cooling system. Ideally, the extent of growth should be limited by other factors, such as the amount of substrate provided. Therefore, if the maximum cell concentration estimated using Eq. (9.54) is lower than that desired from the process, the heat transfer facilities must be improved. For example, the area A could be increased by installing a longer cooling coil, or the overall heat transfer coefficient could be improved by increasing the level of turbulence. Equation (9.54) was derived for fermenters in which shaft work could be ignored; if the stirrer adds significantly to the total heat load, x_{max} will be smaller than that estimated using Eq. (9.54).

9.7 HYDRODYNAMIC CONSIDERATIONS WITH COOLING COILS

When cooling coils are used for temperature control in bioreactors, there is an interaction between heat transfer processes and the hydrodynamic conditions in the vessel. Heat transfer coefficients are sensitive to the levels of turbulence generated and the distribution of fluid velocities in the tank. On the other hand, the presence of massive internal structures such as cooling coils and their position in the fermenter have a significant effect on hydrodynamics.

Helical cooling coils are usually positioned inside fermenters close to the walls of the vessel. This allows the largest possible pipe lengths to be fitted into the tank. Proximity to the fermenter wall also minimises the stresses imparted on the coil during vigorous agitation and allows secure mounting of the coil bank to either the vessel walls or the baffles. Nevertheless, cooling coils are not mounted flush against the sides of fermenters: a gap between the coil and the tank wall is necessary to allow fluid flow around the entire heat transfer surface. This arrangement has its disadvantages, however, as fluid velocities in the gap can be reduced significantly compared with the bulk flow. This effect is particularly severe with viscous broths: as fluid near the coil cools and becomes even more viscous, flow can deteriorate substantially so that stagnant zones develop between the coil and the vessel wall. If fluid circulation is very slow in this region, the cells trapped there may become starved of oxygen. For coils mounted with a clearance of one-tenth the tank diameter between the coil bank and the fermenter wall, the gap between these structures accounts for about 20% of the vessel volume. Therefore, a significant proportion of the vessel contents may be subjected to nonoptimal culture conditions due to poor local mixing. When processes are scaled up from small vessels with external water jackets (refer to Figure 9.1(a)) to large vessels containing cooling coils, the changes in fluid circulation patterns and reduction in mixing effectiveness can be substantial. To avoid this problem, small-scale prototype equipment should use the same heat transfer geometry as larger-scale vessels.

The effectiveness of heat transfer in fermenters is influenced by the type of impeller used and the flow pattern generated. Because cooling coils are located near the vessel walls, the hydrodynamic conditions in this region play a key role in heat transfer. As illustrated in Figure 8.7, radial-flow impellers such as Rushton turbines generate high-speed jets of turbulent fluid that are directed radially outward towards the walls of the vessel. Therefore, the discharge stream from these impellers impinges directly on the cooling coils at high velocity. Because the rate of heat transfer depends strongly on the velocity of fluid sweeping the surface of the coil, this is very effective for heat transfer. In contrast, as shown in Figure 8.8, the discharge jet from axial-flow impellers is directed primarily downward towards the vessel floor. Accordingly, with axial-flow impellers, the velocity of fluid flow near the walls and cooling coil is considerably lower than in the impeller discharge stream. The potential for laminar rather than turbulent flow near the coils, and the possibility of stagnant zones developing at the perimeter of the tank, are greater with axial-flow impellers than with radial-flow impellers.

Correlations, such as Eqs. (9.36) and (9.37), provide an estimate of the average heat transfer coefficient in fermenters. However, the magnitude of heat transfer coefficients depends on the hydrodynamics of fluid flow near the heat transfer surface. As hydrodynamic conditions vary considerably in stirred vessels, there is a distribution of local coefficient values at different locations. As shown in Figure 9.18(a) for a vessel stirred with a single Rushton turbine, the heat transfer coefficient at the tank wall varies with height in the vessel and is maximum adjacent to the impeller. Heat transfer is most rapid in the

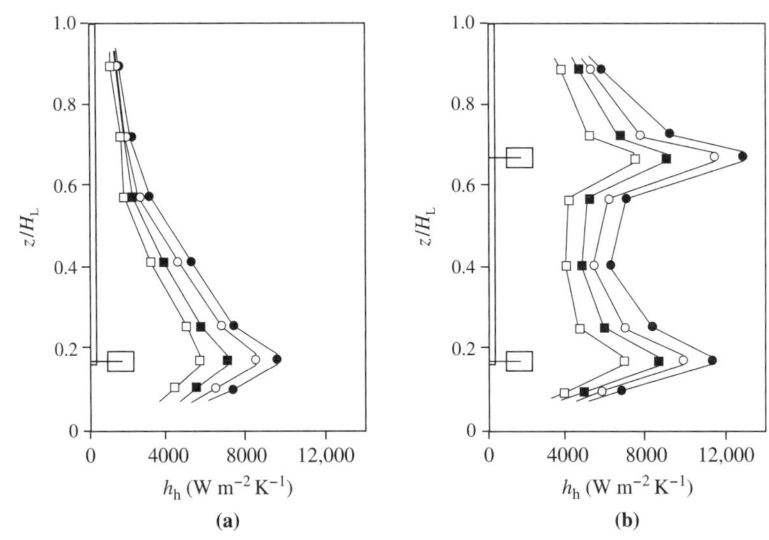

FIGURE 9.18 Profiles of the local heat transfer coefficient h_h at the wall of an ungassed stirred tank at different heights and stirrer speeds for: (a) a single Rushton turbine, and (b) dual Rushton turbines. The stirrer speeds were: (□) 300 rpm; (■) 400 rpm; (○) 500 rpm; and (●) 600 rpm. z is the height from the vessel floor; H_L is the liquid height. The single impeller is located at height $z/H_L = 0.17$; the second impeller in (b) is located at $z/HL = 0.67$. *Reprinted from J. Karcz, Studies of local heat transfer in a gas–liquid system agitated by double disc turbines in a slender vessel,* Chem. Eng. J. *72, 217–227. © 1999, with permission from Elsevier Science.*

region where the impeller discharge stream hits the wall; the heat transfer coefficient also increases with stirrer speed. The distribution of heat transfer coefficients in a vessel with dual Rushton turbines is shown in Figure 9.18(b). In this case, regions of high local values develop near each impeller, indicating elevated rates of heat transfer at two separate planes in the tank.

SUMMARY OF CHAPTER 9

After studying Chapter 9, you should:

- Be able to describe the equipment used for heat exchange in industrial bioprocesses
- Understand the mechanisms of *conduction* and *convection* in heat transfer
- Know *Fourier's law* of conduction in terms of the *thermal conductivity* of materials
- Understand the importance of *thermal boundary layers* in convective heat transfer
- Know the *heat transfer design equation* and the meaning of the *overall heat transfer coefficient, U*
- Understand how the overall heat transfer coefficient can be expressed in terms of the individual resistances to heat transfer
- Know how *individual heat transfer coefficients* are estimated
- Know how to incorporate *fouling factors* into heat transfer analysis
- Be able to carry out simple calculations for design of heat transfer systems
- Understand the interrelationships between heat transfer, hydrodynamics, and culture performance

PROBLEMS

9.1. Rate of conduction

(a) A furnace wall is constructed of firebrick 15 cm thick. The temperature inside the wall is 700°C; the temperature outside is 80°C. If the thermal conductivity of the brick under these conditions is 0.3 W m^{-1} K^{-1}, what is the rate of heat loss through 1.5 m^2 of wall surface?

(b) The 1.5-m^2 area in part (a) is insulated with 4-cm-thick asbestos with thermal conductivity 0.1 W m^{-1} K^{-1}. What is the rate of heat loss now?

9.2. Overall heat transfer coefficient

Heat is transferred from one fluid to a second fluid across a metal wall. The film coefficients are 1.2 and 1.7 kW m^{-2} K^{-1}. The metal is 6 mm thick and has a thermal conductivity of 19 W m^{-1} K^{-1}. On one side of the wall there is a scale deposit with a fouling factor estimated at 830 W m^{-2} K^{-1}. What is the overall heat transfer coefficient?

9.3. Cocurrent versus countercurrent flow

Vegetable oil used in the manufacture of margarine needs to be cooled from 105°C to 55°C in a double-pipe heat exchanger. The inlet temperature of the cooling water is 20°C and the outlet temperature is 43°C. All else being equal, what heat transfer area is required if the heat exchanger is operated with cocurrent flow compared with that required for countercurrent flow?

9.4. Kitchen hot water

The hot water heater in your house is in a closet near the laundry. The kitchen is some distance away from the laundry on the opposite side of the house. When you turn on the hot water tap in the kitchen, it takes about 20 s for the hot water to arrive from the heater. In that time you are able to collect 2.5 litres of cold water from the hot water tap. The diameter of the copper pipe delivering water from the heater to the kitchen is 10 mm.

(a) What is the length of the hot water pipe to the kitchen?

(b) Water leaves the heater at 75°C. If the ambient temperature is 12°C and the overall heat transfer coefficient for heat loss from the hot water pipe is 90 W m^{-2} K^{-1}, what is the temperature of the hot water arriving in your kitchen? Use C_p water = 4.2 kJ kg^{-1} $°C^{-1}$. (Iterative solution may be required.)

9.5. Heat losses from a steam pipe on a windy day

Steam used to sterilise fermentation equipment is piped 20 m from a factory boiler room through a 6-cm-diameter tube. The surface temperature of the steam pipe is roughly constant at 118°C. Most of the length of the tube lies unprotected and exposed to the outside weather. The air temperature is 5°C and the wind speed is 7.5 m s^{-1}. The properties of air at 330 K are:

$$C_p = 1007 \text{ J kg}^{-1} \text{ °C}^{-1}$$
$$k_{fb} = 0.0283 \text{ W m}^{-1} \text{ °C}^{-1}$$
$$\rho = 1.076 \text{ kg m}^{-3}$$
$$\mu_b = 1.99 \times 10^{-5} \text{ kg m}^{-1} \text{ s}^{-1}$$

What is the rate of heat loss from the pipe?

9.6. Double-pipe heat exchanger

In an amino acid fermentation factory, molasses solution must be heated from 12°C to 30°C in a double-pipe heat exchanger at a rate of at least 19 kg min^{-1}. The heat is supplied by water at 63°C; an outlet water temperature of 33°C is desired. The overall heat transfer coefficient is 12 kW m^{-2} $°C^{-1}$ and the exchanger is operated with countercurrent flow. The specific heat capacities of water and molasses solution are 4.2 kJ kg^{-1} $°C^{-1}$ and 3.7 kJ kg^{-1} $°C^{-1}$, respectively.

(a) Each stackable unit of a commercially available double-pipe assembly contains 0.02 m^2 of heat transfer surface. How many units need to be purchased and connected together for this process?

(b) For the number of units determined in (a), what flow rate of molasses can be treated?

(c) For the number of units determined in (a), what flow rate of water is required?

9.7. Water heater on vacation

You are about to leave on a 4-day all-expenses-paid trip to Tahiti to attend a bioprocess engineering conference. It is midwinter in your home town of Hobart, Tasmania, and you are trying to decide whether to turn off your storage hot water heater while you are away. The average midwinter temperature in July is 5°C, your hot water heater has a capacity of 110 litres, and the temperature of the hot water is 68°C. You estimate that the overall heat transfer coefficient for loss of heat from the hot water tank to the atmosphere is 2 W m^{-2} K^{-1}. The surface area of the tank is 1.4 m^2. Which is more economical: turning off the heater to save the cost of maintaining the temperature at 68°C while you are away, or leaving the heater on to save the cost of heating up the entire contents of the tank from 5°C after you return home? Use C_p water = 4.18 kJ kg^{-1} °C^{-1}.

9.8. Fouling and pipe wall resistances

A copper pipe of diameter 3.5 cm and wall thickness 4 mm is used in an antibiotic factory to convey hot water a distance of 60 m at a flow rate of 20 litres s^{-1}. The inlet temperature of the water is 90°C. Due to heat losses to the atmosphere, the outlet water temperature is 84°C. The ambient temperature is 25°C. The inside of the pipe is covered with a layer of fouling with a fouling factor of 7500 W m^{-2} °C^{-1}.

(a) What is the rate of heat loss from the water?

(b) What proportion of the total resistance to heat transfer is provided by the fouling layer?

(c) If the copper pipe were replaced by a clean stainless steel pipe of the same thickness, what proportion of the total resistance to heat transfer would be provided by the pipe wall?

9.9. Effect of cooling-coil length on coolant requirements

A fermenter is maintained at 35°C by water circulating at a rate of 0.5 kg s^{-1} in a cooling coil inside the vessel. The inlet and outlet temperatures of the water are 8°C and 15°C, respectively.

The length of the cooling coil is increased by 50%. In order to maintain the same fermentation temperature, the rate of heat removal must be kept the same. Determine the new cooling-water flow rate and outlet temperature by carrying out the following calculations. The heat capacity of the cooling water can be taken as 4.18 kJ kg^{-1} °C^{-1}.

(a) From a steady-state energy balance on the cooling water, calculate the rate of cooling with the original coil.

(b) Determine the mean temperature difference ΔT with the original coil.

(c) Evaluate UA for the original coil.

(d) If the length of the coil is increased by 50%, the area available for heat transfer, A', is also increased by 50% so that $A' = 1.5A$. The value of the overall heat transfer coefficient is not expected to change very much. For the new coil, what is the value of UA'?

(e) Estimate the new cooling-water outlet temperature. (Iterative solution may be required.)

(f) By how much are the cooling water requirements reduced after the new coil is installed?

9.10. Fermenter cooling coil

A fermentation factory is set up in Belize to produce an anti-UV compound from a newly discovered bacterial strain. The anti-UV compound will be used in the Bahamas for manufacture of an improved sunscreen lotion. A cheap second-hand bioreactor is purchased from a cash-strapped pharmaceutical company in the Cayman Islands, but the reactor must be equipped with a new cooling coil. The fermentation is to be carried out at 35°C. Cooling water is available from a nearby mountain stream at an average temperature of 15°C; however local environmental laws in Belize prohibit the dumping of water back into the stream if it is above 25°C. Based on the heat of reaction for the bacterial culture, the cooling requirements are 15.5 kW. The overall heat transfer coefficient is estimated to be 340 W m^{-2} K^{-1}. If a company in Yucatan can supply stainless steel pipe with diameter 4 cm for fabricating the cooling coil, what length is required?

9.11. Effect of fouling on heat transfer resistance

In current service, 20 kg s^{-1} of cooling water at 12°C must be circulated through a thin-walled coil inside a fermenter to maintain the temperature at 37°C. The coil is 150 m long with pipe diameter 12 cm; the exit water temperature is 28°C. After the inner and outer surfaces of the coil are cleaned it is found that only 13 kg s^{-1} of cooling water is required to control the fermentation temperature.

(a) Calculate the overall heat transfer coefficient before cleaning.

(b) What is the outlet water temperature after cleaning?

(c) What fraction of the total resistance to heat transfer before cleaning was due to fouling deposits?

9.12. Preheating of nutrient medium

Nutrient medium is to be heated from 10°C to 28°C in a single-pass countercurrent shell-and-tube heat exchanger before being pumped into a fed-batch fermenter. Medium passes through the tubes of the exchanger; the shell-side fluid is water that enters with flow rate 3×10^4 kg h^{-1} and temperature 60°C. Preheated medium is required at a rate of 50 m^3 h^{-1}. The density, viscosity, and heat capacity of the medium are the same as water; the thermal conductivity of the medium is 0.54 W m^{-1} °C^{-1}.

It is proposed to use 30 steel tubes with inner diameter 5 cm; the tubes will be arranged in line. The pipe walls are 5 mm thick; the thermal conductivity of the metal is 50 W m^{-1} °C^{-1}. The maximum linear shell-side fluid velocity is about 0.15 m s^{-1}. Estimate the tube length required by carrying out the following calculations.

(a) What is the rate of heat transfer?

(b) Calculate the individual heat transfer coefficients for the tube- and shell-side fluids.

(c) Calculate the overall heat transfer coefficient.

(d) Calculate the log-mean temperature difference.

(e) Determine the heat transfer area.

(f) What tube length is required?

9.13. Cooling coil design

A fermenter of diameter 4 m and height 8 m is being constructed for production of riboflavin using *Eremothecium ashbyii*. The rate of metabolic heat generation is expected to be 520 kW, with an additional power input by stirring of 1 metric horsepower per 1000 litres of broth. The optimum fermentation temperature is 29°C. The fermenter will be cooled using a stainless steel cooling coil with inner pipe diameter 4.2 cm and wall thickness 7 mm. Cooling water at 14°C is available at a rate of 50 m^3 h^{-1}. The fermentation broth is stirred using an impeller of diameter 1.9 m at a speed of 50 rpm. The broth has the following properties: $C_p = 4.0$ kJ kg^{-1} °C^{-1}, $k_{fb} = 0.62$ W m^{-1} °C^{-1}, $\rho = 10^3$ kg m^{-3}, and $\mu_b = 0.05$ Pa s. The specific heat capacity of the cooling water is 4.2 kJ kg^{-1} °C^{-1}, the density and viscosity are 10^3 kg m^{-3} and 10^{-3} kg m^{-1} s^{-1}, respectively, and the thermal conductivity is 0.66 W m^{-1} °C^{-1}. The thermal conductivity of the stainless steel pipe is 17 W m^{-1} °C^{-1}. A fouling factor of 2.5 kW m^{-2} °C^{-1} is expected. Design an appropriate cooling coil for this fermentation by answering the following questions.

(a) What is the exit cooling water temperature?

(b) What is the value of the fermenter-side heat transfer coefficient? Assume that the viscosity at the wall of the coil is 15% higher than the bulk fluid viscosity.
(c) What is the value of the tube-side heat transfer coefficient?
(d) What is the overall heat transfer coefficient?
(e) What fraction of the overall heat transfer resistance is due to the tube-side liquid film?
(f) What fraction of the overall heat transfer resistance is due to the stainless steel pipe?
(g) What fraction of the overall heat transfer resistance is due to fouling?
(h) What is the temperature-difference driving force for this system?
(i) What heat transfer area is required?
(j) What length of cooling-water pipe is required?
(k) The cooling coil must be wound into a helix before fitting into the fermenter. It is planned to mount the coil by attaching it to the baffles. As the baffle width is one-tenth the tank diameter, the diameter of the helix will be around 3.2 m. How many coils will be formed?
(l) The coils must be spaced at least 10 cm apart to allow cleaning and to reduce the dampening effect of the coil on fluid velocity in the vessel. Will the cooling coil fit into the fermenter?

9.14. Suitability of an existing cooling coil

An enzyme manufacturer in the same industrial park as your antibiotic factory has a reconditioned 20-m^3 fermenter for sale. You are in the market for a cheap 20-m^3 fermenter; however the vessel on offer is fitted with a 45-m helical cooling coil with pipe diameter 7.5 cm. You propose to use the fermenter for your newest production organism, which is known to have a maximum oxygen demand of 90 mol m^{-3} h^{-1} at its optimum culture temperature of 28°C. You consider that the 3-m-diameter vessel should be stirred with a 1-m-diameter Rushton turbine operated at an average speed of 50 rpm. The fermentation fluid can be assumed to have the properties of water. If 20 m^3 h^{-1} cooling water is available at 12°C, should you make an offer for the second-hand fermenter and cooling coil?

9.15. Heat transfer and cooling water in fermenter design

A 100-m^3 fermenter of diameter 5 m is stirred using a 1.7-m Rushton turbine operated at 80 rpm. The culture fluid has the following properties:

$$C_p = 4.2 \text{ kJ kg}^{-1} \text{ °C}^{-1}$$
$$k_{fb} = 0.6 \text{ W m}^{-1} \text{ °C}^{-1}$$
$$\rho = 10^3 \text{ kg m}^{-3}$$
$$\mu_b = 10^{-3} \text{ N s m}^{-2}$$

Assume that the viscosity at the wall is equal to the bulk fluid viscosity.

Heat is generated by the fermentation at a rate of 2500 kW. This heat is removed to cooling water flowing in a helical stainless steel coil inside the vessel. The inner diameter of the coil pipe is 12 cm and the wall thickness is 6 mm. The thermal conductivity of the stainless steel is 20 W m^{-1} °C^{-1}. There are no fouling layers present and the tube-side heat transfer coefficient can be neglected. The fermentation temperature is 30°C. Cooling water enters the coil at 10°C at a flow rate of 1.5×10^5 kg h^{-1}.

(a) Calculate the fermenter-side heat transfer coefficient.
(b) Calculate the overall heat transfer coefficient, U.
(c) What proportion of the total resistance to heat transfer is due to the pipe wall?
(d) From the equations in Section 8.5, what is the contribution of shaft work to the cooling requirements for this fermenter?
(e) Calculate the outlet cooling-water temperature.
(f) Estimate the length of cooling coil needed.
(g) If the cooling water flow rate could be increased by 50%, what effect would this have on the length of cooling coil required?

9.16. Test for heat transfer limitation

An 8-m^3 stirred fermenter is used to culture *Gibberella fujikuroi* for production of gibberellic acid. The liquid medium contains 12 g l^{-1} glucose. Under optimal conditions, 1.0 g dry weight of cells is produced for every 2.2 g of glucose consumed. The culture-specific oxygen demand is 7.5 mmol per g dry weight per h. To achieve the maximum yield of gibberellic acid, the fermentation temperature must be held constant at 32°C. The fermenter is equipped with a helical cooling coil of length 55 m and pipe diameter 5 cm. Under usual fermenter operating conditions, the overall heat transfer coefficient is 250 W m^{-2} °C^{-1}. If the water pumped through the coil has an inlet temperature of 15°C, does the heat transfer system support complete consumption of the substrate?

9.17. Optimum stirrer speed for removal of heat from viscous broth

The viscosity of a fermentation broth containing exopolysaccharide is about 10,000 cP. The broth is stirred in an aerated 10-m^3 fermenter of diameter 2.3 m using a single 0.78-m-diameter Rushton turbine. Other properties of the broth are as follows:

$$C_p = 2 \text{ kJ kg}^{-1}\,°\text{C}^{-1}$$
$$k_{fb} = 2 \text{ W m}^{-1}\,°\text{C}^{-1}$$
$$\rho = 10^3 \text{ kg m}^{-3}$$

The fermenter is equipped with an internal cooling coil providing a heat transfer area of 14 m^2; the average temperature difference for heat transfer is 20°C. Neglect any variation of viscosity at the wall of the coil. Assume that the power dissipated in aerated broth is 40% lower than in ungassed liquid.

(a) Using logarithmic coordinates, plot $\hat{Q}$ for several stirrer speeds between 0.5 and 10 s^{-1}.
(b) From the equations presented in Section 8.5, calculate $\hat{W}_s$, the power dissipated by the stirrer, as a function of stirrer speed. Plot these values on the same graph as $\hat{Q}$.
(c) If evaporation, heat losses, and other factors have a negligible effect on the heat load, the difference between $\hat{Q}$ and $\hat{W}_s$ is equal to the rate of removal of metabolic heat from the fermenter. Plot the rate of metabolic heat removal as a function of stirrer speed.
(d) At what stirrer speed is the removal of metabolic heat most rapid?
(e) The specific rate of oxygen consumption is 6 mmol g^{-1} h^{-1}. If the fermenter is operated at the stirrer speed identified in (d), what is the maximum cell concentration?
(f) How do you interpret the intersection of the curves for $\hat{Q}$ and $\hat{W}_s$ at high stirrer speed in terms of the capacity of the system to handle exothermic reactions?

References

[1] J.M. Coulson, J.F. Richardson, J.R. Backhurst, J.H. Harker, Coulson and Richardson's Chemical Engineering, vol. 1, sixth ed., Butterworth-Heinemann, 1999 (Chapter 9).
[2] E.N. Sieder, G.E. Tate, Heat transfer and pressure drop of liquids in tubes, Ind. Eng. Chem. 28 (1936) 1429–1435.
[3] G.F. Hewitt, G.L. Shires, T.R. Bott, Process Heat Transfer, CRC Press; 1994.
[4] J.P. Holman, Heat Transfer, eighth ed., McGraw-Hill, 1997.
[5] T.H. Chilton, T.B. Drew, R.H. Jebens, Heat transfer coefficients in agitated vessels, Ind. Eng. Chem 36 (1944) 510–516.
[6] W.L. McCabe, J.C. Smith, P. Harriott, Unit Operations of Chemical Engineering, sixth ed., Section III, McGraw-Hill, 2001.
[7] G.J. Xu, Y.M. Li, Z.Z. Hou, L.F. Feng, K. Wang, Gas–liquid dispersion and mixing characteristics and heat transfer in a stirred vessel, Can. J. Chem. Eng. 75 (1997) 299–306.

Suggestions for Further Reading

Heat Transfer Principles and Applications

See also references 1 through 4.
Çengel, Y. A. (2003). *Heat Transfer: A Practical Approach* (2nd ed.). McGraw-Hill.
Perry's Chemical Engineers' Handbook (2008) (8th ed., Section 5). McGraw-Hill.
Thomas, L. C. (1992). *Heat Transfer*. Prentice Hall.

Heat Transfer Equipment

Perry's Chemical Engineers' Handbook (2008) (8th ed., Section 11). McGraw-Hill.

Fouling

Bott, T. R. (1995). *Fouling of Heat Exchangers*. Elsevier.

Heat Transfer in Bioprocessing

Atkinson, B., & Mavituna, F. (1991). *Biochemical Engineering and Biotechnology Handbook* (2nd ed., Chapter 14). Macmillan.
Brain, T. J. S., & Man, K. L. (1989). Heat transfer in stirred tank bioreactors. *Chem. Eng. Prog.*, *85*(7), 76–80.
Kelly, W. J., & Humphrey, A. E. (1998). Computational fluid dynamics model for predicting flow of viscous fluids in a large fermentor with hydrofoil flow impellers and internal cooling coils. *Biotechnol. Prog.*, *14*, 248–258.
Swartz, J. R. (1985). Heat management in fermentation processes. In M. Moo-Young (Ed.), *Comprehensive Biotechnology* (vol. 2, pp. 299–303). Pergamon Press.

CHAPTER

10

Mass Transfer

Mass transfer occurs in mixtures containing concentration variations. For example, when dye is dropped into a pail of water, mass transfer processes are responsible for the movement of dye molecules through the water until equilibrium is established and the concentration is uniform. Mass is transferred from one location to another under the influence of a concentration difference or concentration gradient in the system. There are many situations in bioprocessing where the concentrations of compounds are not uniform; we rely on mechanisms of mass transfer to transport material from regions of high concentration to regions where the concentration is initially low.

An important example of mass transfer in bioprocessing is the supply of oxygen in fermenters for aerobic culture. The concentration of oxygen at the surface of air bubbles is high compared with that in the bulk of the liquid; this concentration gradient promotes oxygen transfer from the bubbles into the medium. Another example is the extraction of penicillin from fermentation liquor using organic solvents such as butyl acetate. When solvent is added to the broth, the relatively high concentration of penicillin in the aqueous phase and low concentration in the organic phase causes mass transfer of penicillin into the solvent. Solvent extraction is an effective downstream processing technique because it selectively removes the desired product from the fermentation fluid.

Mass transfer plays a vital role in many reaction systems. As the distance between the reactants and the site of reaction becomes greater, the rate of mass transfer is more likely to influence or control the conversion rate. Taking again the example of oxygen in aerobic cultures, if mass transfer of oxygen from the bubbles is slow, the rate of cell metabolism will become dependent on the rate of oxygen supply from the gas phase. Because oxygen is a critical component of aerobic fermentations and is so sparingly soluble in aqueous solutions, much of our interest in mass transfer lies with the transfer of oxygen across gas–liquid interfaces. However, liquid–solid mass transfer can also be important in systems containing clumps, pellets, flocs, or films of cells or enzymes. In these cases, nutrients in the liquid phase must be transported into the solid before they can be utilised in reaction. Unless mass transfer is rapid, the supply of nutrients will limit the rate of biological conversion.

In solids and quiescent fluids, mass transfer occurs as a result of molecular diffusion. However, most mass transfer systems contain moving fluid so that mass transfer by molecular motion is supplemented by convective transfer. There is an enormous variety of circumstances in which convective mass transfer takes place. In this chapter, we will consider the theory of mass transfer with applications relevant to bioprocessing.

10.1 MOLECULAR DIFFUSION

Molecular diffusion is the movement of component molecules in a mixture under the influence of a concentration difference in the system. Diffusion of molecules occurs in the direction required to destroy the concentration gradient, that is, from regions of high concentration to regions of low concentration. If the gradient is maintained by constantly supplying material to the region of high concentration and removing it from the region of low concentration, diffusion will be continuous. This situation is often exploited in mass transfer operations and reaction systems.

10.1.1 Diffusion Theory

In this text, we confine our discussion of diffusion to *binary mixtures*—that is, mixtures or solutions containing only two components. Consider a closed system containing molecular components A and B. Initially, the concentration of A in the system is not uniform; as indicated in Figure 10.1, concentration C_A varies from C_{A1} to C_{A2} as a function of distance

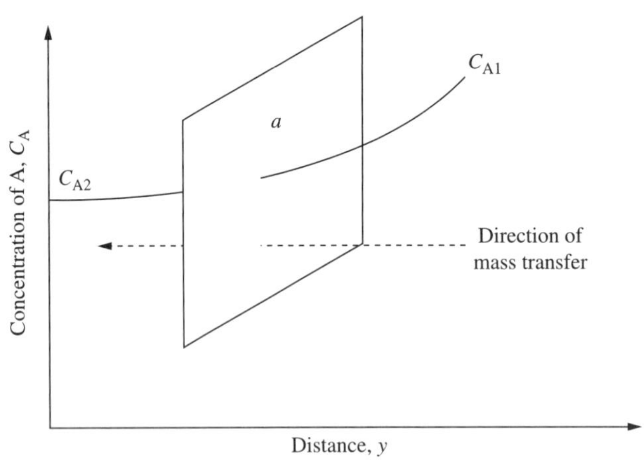

FIGURE 10.1 Concentration gradient of component A inducing mass transfer across area *a*.

y. In response to this concentration gradient, molecules of A will diffuse away from the region of high concentration until eventually the whole system acquires uniform composition. If there is no large-scale fluid motion in the system (e.g., due to stirring), mixing occurs solely by molecular movement.

Assume that mass transfer of A occurs across area a perpendicular to the direction of diffusion. In single-phase systems, the rate of mass transfer due to molecular diffusion is given by *Fick's law of diffusion*, which states that the mass flux is proportional to the concentration gradient:

$$J_A = \frac{N_A}{a} = -\mathscr{D}_{AB}\frac{dC_A}{dy} \tag{10.1}$$

In Eq. (10.1), J_A is the *mass flux* of component A, N_A is the *rate of mass transfer* of component A, a is the area across which mass transfer occurs, $\mathscr{D}_{AB}$ is the *binary diffusion coefficient* or *diffusivity* of component A in a mixture of A and B, C_A is the concentration of component A, and y is distance. dC_A/dy is the *concentration gradient*, or change in concentration of A with distance.

As indicated in Eq. (10.1), mass flux is defined as the rate of mass transfer per unit area perpendicular to the direction of movement; typical units for J_A are gmol s^{-1} m^{-2}. Corresponding units for N_A are gmol s^{-1}, for C_A gmol m^{-3}, and for $\mathscr{D}_{AB}$ m^2 s^{-1}. Mass rather than mole units may be used for J_A, N_A, and C_A; Eq. (10.1) holds in either case. Equation (10.1) indicates that the rate of diffusion can be enhanced by increasing the area available for mass transfer, the concentration gradient in the system, or the magnitude of the diffusion coefficient. The negative sign in Eq. (10.1) indicates that the direction of mass transfer is always from high concentration to low concentration, opposite to the direction of the concentration gradient. In other words, if the slope of C_A versus y is positive as in Figure 10.1, the direction of mass transfer is in the negative y-direction, and vice versa.

The diffusion coefficient $\mathscr{D}_{AB}$ is a property of materials. Values can be found in handbooks. $\mathscr{D}_{AB}$ reflects the ease with which diffusion takes place. Its value depends on both components of the mixture; for example, the diffusivity of carbon dioxide in water will be different from the diffusivity of carbon dioxide in another solvent such as ethanol. The value of $\mathscr{D}_{AB}$ is also dependent on temperature. The diffusivity of gases varies with pressure; for liquids there is an approximate linear dependence on concentration. Diffusivities are several orders of magnitude smaller for diffusion in liquids than in gases. For example, $\mathscr{D}_{AB}$ for oxygen in air at 25°C and 1 atm is 2.1×10^{-5} m^2 s^{-1}, whereas $\mathscr{D}_{AB}$ for oxygen in water at 25°C and 1 atm is 2.5×10^{-9} m^2 s^{-1}. For dilute concentrations of glucose in water at 25°C, $\mathscr{D}_{AB}$ is 6.9×10^{-10} m^2 s^{-1}.

When diffusivity values are not available for the materials, temperature, or pressure of interest, $\mathscr{D}_{AB}$ can be estimated using equations. Relationships for calculating diffusivities are available from other references [1–3]. The theory of diffusion in liquids is not as well advanced as diffusion in gases; there are also fewer experimental data available for liquid systems.

10.1.2 Analogy between Mass, Heat, and Momentum Transfer

There is a close similarity between the processes of mass, heat, and momentum transfer occurring as a result of molecular motion. This is suggested by the form of the equations for mass, heat, and momentum fluxes:

$$J_A = -\mathscr{D}_{AB}\frac{dC_A}{dy} \tag{10.2}$$

$$\hat{q} = -k\frac{dT}{dy} \tag{9.2}$$

and

$$\tau = -\mu\frac{dv}{dy} \tag{7.6}$$

The three processes represented above are quite different at the molecular level, but the basic equations have the same form. In each case, flux in the y-direction is directly proportional to the driving force (either dC_A/dy, dT/dy, or dv/dy), with the proportionality constant ($\mathscr{D}_{AB}$, k, or μ) a physical property of the material. The negative signs in Eqs. (10.2), (9.2), and (7.6) indicate that transfer of mass, heat, or momentum is always in the direction opposite to that of increasing concentration, temperature, or velocity. The similarity in the form of the three rate equations makes it possible in some situations to apply an analysis of one process to either of the other two.

The analogy of Eqs. (10.2), (9.2), and (7.6) is valid for transport of mass, heat, and momentum resulting from motion or vibration of molecules. Extension of the analogy to turbulent flow is generally valid for heat and mass transfer; however the analogy with momentum transfer presents a number of difficulties. Several analogy theories have been proposed in the chemical engineering literature to describe simultaneous transport phenomena in turbulent systems. Details are presented elsewhere [3, 4].

10.2 ROLE OF DIFFUSION IN BIOPROCESSING

Fluid mixing is carried out in most industrial processes where mass transfer takes place. Bulk fluid motion results in more rapid large-scale mixing than does molecular diffusion: why then is diffusive transport still important? These are some areas of bioprocessing where diffusion plays a major role.

- *Scale of mixing*. As discussed in Section 8.9, turbulence in fluids produces bulk mixing on a scale equal to the smallest eddy size. Within the smallest eddies, flow is largely streamline so that further mixing must occur by diffusion of the fluid components. Mixing on a molecular scale therefore relies on diffusion as the final step in the mixing process.
- *Solid-phase reaction*. In biological systems, reactions are sometimes mediated by catalysts in solid form (e.g., clumps, flocs, and films of cells) and by immobilised enzyme and cell particles. When cells or enzymes are clumped together into a solid particle,

substrates must be transported into the solid before reaction can take place. Mass transfer within solid particles is usually unassisted by bulk fluid convection; therefore, the only mechanism for intraparticle mass transfer is molecular diffusion. As the reaction proceeds, diffusion is also responsible for the removal of product molecules away from the site of reaction. As discussed more fully in Chapter 13, when reaction is coupled with diffusion, the overall reaction rate can be reduced significantly if diffusion is slow.

- *Mass transfer across phase boundaries*. Mass transfer between phases occurs often in bioprocessing. Oxygen transfer from gas bubbles to fermentation broth, penicillin recovery from aqueous to organic liquid, and glucose uptake from liquid medium into mould pellets are typical examples. When different phases come into contact, the fluid velocity near the phase interface is decreased significantly and diffusion becomes crucial for mass transfer. This is discussed further in the next section.

10.3 FILM THEORY

The *two-film theory* is a useful model for mass transfer between phases. Mass transfer of solute from one phase to another involves transport from the bulk of one phase to the *phase boundary* or *interface*, then movement from the interface into the bulk of the second phase. The film theory is based on the idea that a fluid film or *mass transfer boundary layer* forms wherever there is contact between two phases.

Let us consider mass transfer of component A across the phase boundary represented in Figure 10.2. Assume that the phases are two immiscible liquids such as water and chloroform, and that A is initially at higher concentration in the aqueous phase than in the

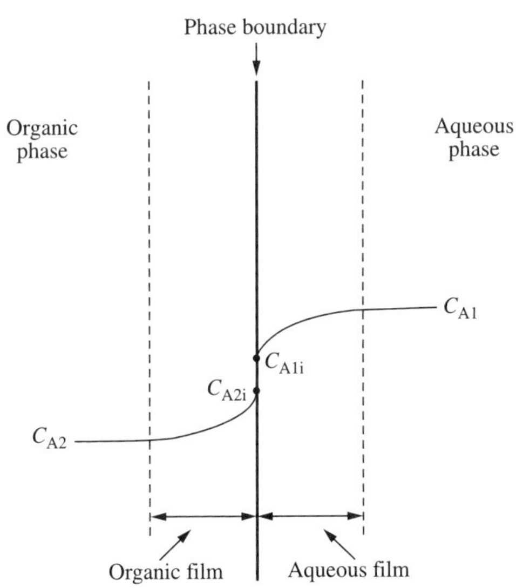

FIGURE 10.2 Film resistance to mass transfer between two immiscible liquids.

organic phase. Each phase is well mixed and in turbulent flow. The concentration of A in the bulk aqueous phase is C_{A1}; the concentration of A in the bulk organic phase is C_{A2}.

According to the film theory, turbulence in each fluid dies out at the phase boundary. A thin film of relatively stagnant fluid exists on either side of the interface; mass transfer through this film is effected solely by molecular diffusion. The concentration of A changes near the interface as indicated in Figure 10.2; C_{A1i} is the interfacial concentration of A in the aqueous phase; C_{A2i} is the interfacial concentration of A in the organic phase. Most of the resistance to mass transfer resides in the liquid films rather than in the bulk liquid. For practical purposes it is generally assumed that there is negligible resistance to transport at the interface itself; this is equivalent to assuming that the phases are in equilibrium at the plane of contact. The difference between C_{A1i} and C_{A2i} at the interface accounts for the possibility that, at equilibrium, A may be more soluble in one phase than in the other. For example, if A were acetic acid in contact at the interface with both water and chloroform, the equilibrium concentration in water would be greater than in chloroform by a factor of between 5 and 10. C_{A1i} would then be significantly higher than C_{A2i}.

Even though the bulk liquids in Figure 10.2 may be well mixed, diffusion of component A is crucial for mass transfer because the local fluid velocities approach zero at the interface. The film theory as described above is applied extensively in analysis of mass transfer, although it is a greatly simplified representation. There are other models of mass transfer in fluids that lead to more realistic mathematical outcomes than the film theory [1, 4]. Nevertheless, irrespective of how mass transfer is visualised, diffusion is always an important mechanism of mass transfer close to the interface between fluids.

10.4 CONVECTIVE MASS TRANSFER

The term *convective mass transfer* refers to mass transfer occurring in the presence of bulk fluid motion. Molecular diffusion will occur whenever there is a concentration gradient; however if the bulk fluid is also moving, the overall rate of mass transfer will be higher due to the contribution of convective currents. Analysis of mass transfer is most important in multiphase systems where interfacial boundary layers provide significant mass transfer resistance. Let us develop an expression for the rate of mass transfer that is applicable to mass transfer boundary layers.

The rate of mass transfer is directly proportional to the area available for transfer and the driving force for the transfer process. This can be expressed as:

$$\text{Transfer rate} \propto \text{transfer area} \times \text{driving force} \tag{10.3}$$

The proportionality coefficient in this equation is called the *mass transfer coefficient*, so that:

$$\text{Transfer rate} = \text{mass transfer coefficient} \times \text{transfer area} \times \text{driving force} \tag{10.4}$$

For each fluid at a phase boundary, the driving force for mass transfer of component A through the boundary layer can be expressed in terms of the concentration difference of

A across the fluid film. Therefore, the rate of mass transfer from the bulk fluid through the boundary layer to the interface is:

$$N_A = ka\,\Delta C_A = ka\,(C_{Ab} - C_{Ai}) \tag{10.5}$$

where N_A is the rate of mass transfer of component A, k is the mass transfer coefficient, a is the area available for mass transfer, C_{Ab} is the bulk concentration of component A away from the phase boundary, and C_{Ai} is the concentration of A at the interface. Equation (10.5) is usually used to represent the *volumetric rate of mass transfer*, so units of N_A are, for example, gmol $m^{-3}\,s^{-1}$.

Consistent with this representation, a is the interfacial area per unit volume with dimensions L^{-1} and units of, for example, $m^2\,m^{-3}$ or m^{-1}. The dimensions of the mass transfer coefficient k are LT^{-1}; the SI units for k are m s^{-1}. Equation (10.5) indicates that the rate of convective mass transfer can be enhanced by increasing the area available for mass transfer, the concentration difference between the bulk fluid and the interface, and the magnitude of the mass transfer coefficient. By analogy with Eq. (9.11) for heat transfer, Eq. (10.5) can also be written in the form:

$$N_A = \frac{\Delta C_A}{R_m} \tag{10.6}$$

where R_m is the resistance to mass transfer:

$$R_m = \frac{1}{k\,a} \tag{10.7}$$

Mass transfer coupled with fluid flow is a more complicated process than diffusive mass transfer. The value of the mass transfer coefficient k reflects the contribution to mass transfer from all the processes in the system that influence the boundary layer. Like the heat transfer coefficient in Chapter 9, k depends on the combined effects of the flow velocity, the geometry of the mass transfer system, and fluid properties such as viscosity and diffusivity. Because the hydrodynamics of most practical systems are not easily characterised, k cannot be calculated reliably from first principles. Instead, it is measured experimentally or estimated using correlations available from the literature. In general, reducing the thickness of the boundary layer or increasing the diffusion coefficient in the film will enhance the value of k, thus improving the rate of mass transfer.

In bioprocessing, three mass transfer situations that involve multiple phases are *liquid–solid mass transfer, liquid–liquid mass transfer* between immiscible solutions, and *gas–liquid mass transfer*. Use of Eq. (10.5) to determine the rate of mass transfer in these systems is discussed in the following sections.

10.4.1 Liquid–Solid Mass Transfer

Mass transfer between a moving liquid and a solid is important in biological processing in a variety of applications. Transport of substrates to solid-phase cell or enzyme catalysts has already been mentioned. Adsorption of molecules onto surfaces, such as in chromatography, requires transport from the liquid phase to a solid; liquid–solid mass transfer is

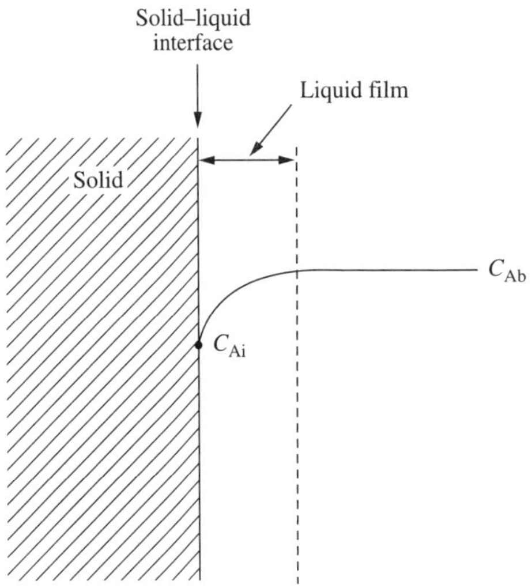

FIGURE 10.3 Concentration gradient for liquid–solid mass transfer.

also important in crystallisation as molecules move from the liquid to the face of the growing crystal. Conversely, the process of dissolving a solid in liquid requires liquid–solid mass transfer directed away from the solid surface.

Let us assume that component A is required for reaction at the surface of a solid. The situation at the interface between flowing liquid containing A and the solid is illustrated in Figure 10.3. Near the interface, the fluid velocity is reduced and a boundary layer develops. As A is consumed by reaction at the surface, the local concentration of A decreases and a concentration gradient is established through the film. The concentration difference between the bulk liquid and the phase interface drives mass transfer of A from the liquid to the solid, allowing the reaction to continue. If the solid is nonporous, A does not penetrate further than the surface. The concentration of A at the phase boundary is C_{Ai}; the concentration of A in the bulk liquid outside the film is C_{Ab}. If a is the liquid–solid interfacial area per unit volume, the volumetric rate of mass transfer can be determined from Eq. (10.5) as:

$$N_A = k_L\, a\,(C_{Ab} - C_{Ai}) \tag{10.8}$$

where k_L is the *liquid-phase mass transfer coefficient*.

Application of Eq. (10.8) requires knowledge of the mass transfer coefficient, the interfacial area between the phases, the bulk concentration of A, and the concentration of A at the interface. The value of the mass transfer coefficient is either measured or calculated using published correlations. In most cases, the area a can be determined from the shape and size of the solid. Bulk concentrations are generally easy to measure; however, estimation of the interfacial concentration C_{Ai} is much more difficult, as measuring compositions at phase boundaries is not straightforward experimentally. To overcome this problem, we must consider the processes in the system that occur in conjunction with mass transfer. In the example of Figure 10.3, transport of A is linked to reaction at the surface of the solid, so C_{Ai} will depend on the rate of consumption of A at the interface. In practice, we can calculate the rate of mass transfer of A in this situation only if we have information about the rate of reaction at the solid surface. Simultaneous reaction and mass transfer occur in many bioprocesses as outlined in more detail in Chapter 13.

10.4.2 Liquid–Liquid Mass Transfer

Liquid–liquid mass transfer between immiscible solvents is most often encountered in the product recovery stages of bioprocessing. Organic solvents are used to isolate antibiotics, steroids, and alkaloids from fermentation broths; two-phase aqueous systems are useful for protein purification. Liquid–liquid mass transfer is also important when hydrocarbons are used as substrates in fermentation, for example, to produce microbial biomass for single-cell protein. Water-immiscible organic solvents are of increasing interest for enzyme and whole-cell biocatalysis: two-phase reaction systems can be used to overcome problems with poor substrate solubility and product toxicity, and can shift chemical equilibria for enhanced yields and selectivity in metabolic reactions.

The situation at the interface between two immiscible liquids is shown in Figure 10.2. Component A is present at bulk concentration C_{A1} in one phase; this concentration falls to C_{A1i} at the interface. In the other liquid, the concentration of A falls from C_{A2i} at the interface to C_{A2} in the bulk. The rate of mass transfer N_A in each liquid phase is expressed using Eq. (10.5):

$$N_{A1} = k_{L1}\, a\, (C_{A1} - C_{A1i}) \tag{10.9}$$

and

$$N_{A2} = k_{L2}\, a\, (C_{A2i} - C_{A2}) \tag{10.10}$$

where k_L is the liquid-phase mass transfer coefficient, and subscripts 1 and 2 refer to the two liquid phases. Application of Eqs. (10.9) and (10.10) is difficult because interfacial concentrations are not measured easily, as mentioned in Section 10.4.1. However, in this case, C_{A1i} and C_{A2i} can be eliminated by considering the physical situation at the interface and manipulating the equations.

First, let us recognise that at steady state, because there can be no accumulation of A at the interface or anywhere else in the system, any A transported through liquid 1 must also be transported through liquid 2. This means that N_{A1} in Eq. (10.9) must be equal to N_{A2} in Eq. (10.10) so that $N_{A1} = N_{A2} = N_A$. We can then rearrange Eqs. (10.9) and (10.10):

$$\frac{N_A}{k_{L1}\, a} = C_{A1} - C_{A1i} \tag{10.11}$$

and

$$\frac{N_A}{k_{L2}\, a} = C_{A2i} - C_{A2} \tag{10.12}$$

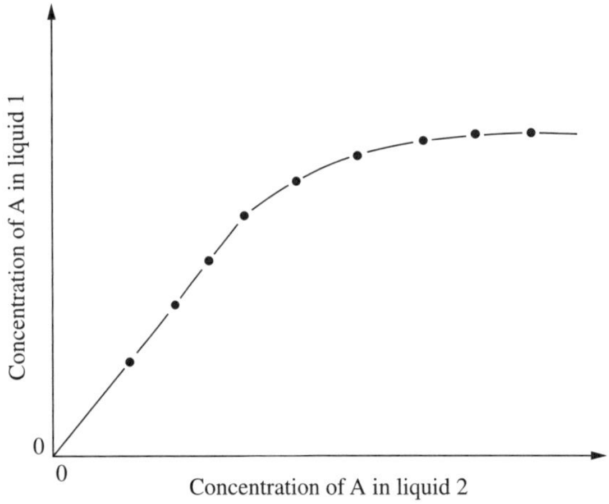

FIGURE 10.4 Equilibrium curve for solute A in two immiscible solvents 1 and 2.

Normally, it can be assumed that there is negligible resistance to mass transfer at the actual interface—that is, within distances corresponding to molecular free paths on either side of the phase boundary. This is equivalent to assuming that the phases are in equilibrium *at the interface*; therefore, C_{A1i} and C_{A2i} are equilibrium concentrations.

The assumption of phase-boundary equilibrium has been subjected to many tests. As a result, it is known that there are special circumstances, such as when there is adsorption of material at the interface, for which the assumption is invalid. However, in ordinary situations, the evidence is that equilibrium does exist at the interface between phases. Note that we are not proposing to relate bulk concentrations C_{A1} and C_{A2} using equilibrium relationships, only C_{A1i} and C_{A2i}. If the bulk liquids were in equilibrium, no net mass transfer would take place.

A typical equilibrium curve relating the concentrations of solute A in two immiscible liquid phases is shown in Figure 10.4. The points making up the curve are obtained readily from experiments; alternatively, equilibrium data can be found in handbooks. The equilibrium distribution of one solute between two phases can be described using the *distribution law*. At equilibrium, the ratio of the solute concentrations in the two phases is equal to the *distribution coefficient* or *partition coefficient*, *m*. As shown in Figure 10.4, when the concentration of A is low, the equilibrium curve is approximately a straight line so that *m* is constant. The distribution law is accurate only if both solvents are immiscible and there is no chemical reaction.

If C_{A1i} and C_{A2i} are equilibrium concentrations, they can be related using the distribution coefficient *m*:

$$m = \frac{C_{A1i}}{C_{A2i}} \tag{10.13}$$

such that

$$C_{A1i} = m\, C_{A2i} \tag{10.14}$$

and

$$C_{A2i} = \frac{C_{A1i}}{m} \tag{10.15}$$

Equations (10.14) and (10.15) can now be used to eliminate the interfacial concentrations from Eqs. (10.11) and (10.12). First, we make a direct substitution:

$$\frac{N_A}{k_{L1}\,a} = C_{A1} - m\,C_{A2i} \tag{10.16}$$

and

$$\frac{N_A}{k_{L2}\,a} = \frac{C_{A1i}}{m} - C_{A2} \tag{10.17}$$

If we now multiply Eq. (10.12) by m:

$$\frac{m\,N_A}{k_{L2}\,a} = m\,C_{A2i} - m\,C_{A2} \tag{10.18}$$

and divide Eq. (10.11) by m:

$$\frac{N_A}{m\,k_{L1}\,a} = \frac{C_{A1}}{m} - \frac{C_{A1i}}{m} \tag{10.19}$$

and add Eq. (10.16) to Eq. (10.18), and Eq. (10.17) to Eq. (10.19), we eliminate the interfacial concentration terms completely:

$$N_A\left(\frac{1}{k_{L1}\,a} + \frac{m}{k_{L2}\,a}\right) = C_{A1} - m\,C_{A2} \tag{10.20}$$

$$N_A\left(\frac{1}{m\,k_{L1}\,a} + \frac{1}{k_{L2}\,a}\right) = \frac{C_{A1}}{m} - C_{A2} \tag{10.21}$$

Equations (10.20) and (10.21) combine the mass transfer resistances in the two liquid films, and relate the rate of mass transfer N_A to the bulk fluid concentrations C_{A1} and C_{A2}. The bracketed expressions for the combined mass transfer coefficients are used to define the *overall liquid-phase mass transfer coefficient*, K_L. Depending on the terms used to represent the concentration difference, we can define two overall mass transfer coefficients:

$$\frac{1}{K_{L1}\,a} = \frac{1}{k_{L1}\,a} + \frac{m}{k_{L2}\,a} \tag{10.22}$$

and

$$\frac{1}{K_{L2}\,a} = \frac{1}{m\,k_{L1}\,a} + \frac{1}{k_{L2}\,a} \tag{10.23}$$

where K_{L1} is the overall mass transfer coefficient based on the bulk concentration in liquid 1, and K_{L2} is the overall mass transfer coefficient based on the bulk concentration in liquid 2.

We can now summarise the results to obtain two equations for the mass transfer rate at the interfacial boundary in liquid–liquid systems:

$$N_A = K_{L1}\,a\,(C_{A1} - m\,C_{A2}) \tag{10.24}$$

and

$$N_A = K_{L2}\,a\left(\frac{C_{A1}}{m} - C_{A2}\right) \tag{10.25}$$

where K_{L1} and K_{L2} are given by Eqs. (10.22) and (10.23). Use of either of these two equations requires knowledge of the concentrations of A in the bulk fluids, the partition coefficient m, the interfacial area a between the two liquid phases, and the value of either K_{L1} or K_{L2}. C_{A1} and C_{A2} are generally easy to measure. m can also be measured or is found in handbooks of physical properties. The overall mass transfer coefficients can be measured experimentally or are estimated from published correlations for k_{L1} and k_{L2} in the literature. The only remaining parameter is the interfacial area, a. In many applications of liquid–liquid mass transfer, it may be difficult to know how much interfacial area is available between the phases. For example, liquid–liquid extraction is often carried out in stirred tanks where an impeller is used to disperse and mix droplets of one phase through the other. The interfacial area in these circumstances will depend on the size, shape, and number of the droplets, which depend in turn on the intensity of agitation and properties of the fluid. Because these factors also affect the value of k_L, correlations for mass transfer coefficients in liquid–liquid systems are often given in terms of $k_L a$ as a combined parameter. For convenience, the combined term $k_L a$ is then referred to as the mass transfer coefficient.

Equations (10.24) and (10.25) indicate that the rate of mass transfer between two liquid phases is not dependent simply on the concentration difference: the equilibrium relationship is also an important factor. According to Eq. (10.24), the driving force for transfer of A from liquid 1 to liquid 2 is the difference between the bulk concentration C_{A1} and *the concentration of A in liquid 1 that would be in equilibrium with bulk concentration C_{A2} in liquid 2.* Similarly, the driving force for mass transfer according to Eq. (10.25) is the difference between C_{A2} and the concentration of A in liquid 2 that would be in equilibrium with C_{A1} in liquid 1.

10.4.3 Gas–Liquid Mass Transfer

Gas–liquid mass transfer is of great importance in bioprocessing because of the requirement for oxygen in aerobic cell cultures. Transfer of a solute such as oxygen from gas to liquid is analysed in a similar way to that presented previously for liquid–solid and liquid–liquid mass transfer.

Figure 10.5 shows the situation at an interface between gas and liquid phases containing component A. Let us assume that A is transferred from the gas phase into the liquid. The concentration of A in the liquid is C_{AL} in the bulk and C_{ALi} at the interface. In the gas, the concentration is C_{AG} in the bulk and C_{AGi} at the interface.

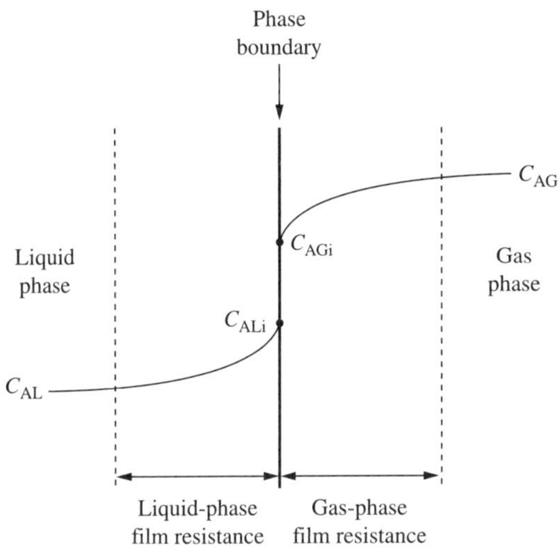

FIGURE 10.5 Concentration gradients for gas–liquid mass transfer.

From Eq. (10.5), the rate of mass transfer of A through the gas boundary layer is:

$$N_{AG} = k_G a (C_{AG} - C_{AGi}) \tag{10.26}$$

and the rate of mass transfer of A through the liquid boundary layer is:

$$N_{AL} = k_L a (C_{ALi} - C_{AL}) \tag{10.27}$$

where k_G is the gas-phase mass transfer coefficient and k_L is the liquid-phase mass transfer coefficient. To eliminate C_{AGi} and C_{ALi}, we must manipulate the equations as discussed in Section 10.4.2.

If we assume that equilibrium exists at the interface, C_{AGi} and C_{ALi} can be related. For dilute concentrations of most gases and for a wide range of concentration for some gases, the equilibrium concentration in the gas phase is a linear function of liquid concentration. Therefore, we can write:

$$C_{AGi} = m\, C_{ALi} \tag{10.28}$$

or, alternatively:

$$C_{ALi} = \frac{C_{AGi}}{m} \tag{10.29}$$

where m is the distribution factor. These equilibrium relationships can be incorporated into Eqs. (10.26) and (10.27) at steady state using procedures that parallel those applied already for liquid–liquid mass transfer. The results are also similar to Eqs. (10.20) and (10.21):

$$N_A \left(\frac{1}{k_G a} + \frac{m}{k_L a} \right) = C_{AG} - m\, C_{AL} \tag{10.30}$$

$$N_A\left(\frac{1}{m\,k_G\,a}+\frac{1}{k_L\,a}\right)=\frac{C_{AG}}{m}-C_{AL} \tag{10.31}$$

The combined mass transfer coefficients in Eqs. (10.30) and (10.31) can be used to define overall mass transfer coefficients. The *overall gas-phase mass transfer coefficient* K_G is defined by the equation:

$$\frac{1}{K_G\,a}=\frac{1}{k_G\,a}+\frac{m}{k_L\,a} \tag{10.32}$$

and the *overall liquid-phase mass transfer coefficient* K_L is defined as:

$$\frac{1}{K_L\,a}=\frac{1}{m\,k_G\,a}+\frac{1}{k_L\,a} \tag{10.33}$$

The rate of mass transfer in gas–liquid systems can therefore be expressed using either of two equations:

$$N_A=K_G\,a\,(C_{AG}-m\,C_{AL}) \tag{10.34}$$

or

$$N_A=K_L\,a\left(\frac{C_{AG}}{m}-C_{AL}\right) \tag{10.35}$$

Equations (10.34) and (10.35) are usually expressed using equilibrium concentrations. mC_{AL} is equal to C^*_{AG}, the gas-phase concentration of A in equilibrium with C_{AL}, and C_{AG}/m is equal to C^*_{AL}, the liquid-phase concentration of A in equilibrium with C_{AG}. Equations (10.34) and (10.35) become:

$$N_A=K_G\,a\,(C_{AG}-C^*_{AG}) \tag{10.36}$$

and

$$N_A=K_L\,a\,(C^*_{AL}-C_{AL}) \tag{10.37}$$

Equations (10.36) and (10.37) can be simplified for systems in which most of the resistance to mass transfer lies in either the gas-phase interfacial film or the liquid-phase interfacial film. When solute A is very soluble in the liquid, for example in transfer of ammonia to water, the liquid-side resistance is small compared with that posed by the gas interfacial film. From Eq. (10.7), if the liquid-side resistance is small, $k_L\,a$ must be relatively large. From Eq. (10.32), $K_G a$ is then approximately equal to $k_G a$. Using this result in Eq. (10.36) gives:

$$N_A=k_G\,a\,(C_{AG}-C^*_{AG}) \tag{10.38}$$

Conversely, if A is poorly soluble in the liquid (e.g., oxygen in aqueous solution), the liquid-phase mass transfer resistance dominates and $k_G a$ is much larger than $k_L a$. From

Eq. (10.33), this means that $K_L a$ is approximately equal to $k_L a$ and Eq. (10.37) can be simplified to:

$$N_A = k_L a(C_{AL}^* - C_{AL}) \tag{10.39}$$

Because gas–liquid oxygen transfer plays a crucial role in many bioprocesses, we will make further use of Eq. (10.39) in subsequent sections of this chapter. In practical applications, obtaining the values of C_{AL} and C_{AL}^* for use in Eq. (10.39) is reasonably straightforward. However, as described in relation to liquid–liquid mass transfer in Section 10.4.2, it is generally difficult to estimate the interfacial area a. When gas is sparged through a liquid, the interfacial area will depend on the size and number of bubbles present, which in turn depend on many other factors such as medium composition, stirrer speed, and gas flow rate. Because k_L is also affected by these parameters, k_L and a are usually combined together and the combined term $k_L a$ referred to as the mass transfer coefficient.

10.5 OXYGEN UPTAKE IN CELL CULTURES

Cells in aerobic culture take up oxygen from the liquid phase. The rate of oxygen transfer from gas to liquid is therefore of prime importance, especially in dense cell cultures where the demand for dissolved oxygen is high. An expression for the rate of oxygen transfer from gas to liquid is given by Eq. (10.39): N_A is the rate of oxygen transfer per unit volume of fluid (gmol m^{-3} s^{-1}), k_L is the liquid-phase mass transfer coefficient (m s^{-1}), a is the gas–liquid interfacial area per unit volume of fluid (m^2 m^{-3}), C_{AL} is the oxygen concentration in the broth (gmol m^{-3}), and C_{AL}^* is the oxygen concentration in the broth in equilibrium with the gas phase (gmol m^{-3}). The equilibrium concentration C_{AL}^* is also known as the *saturation concentration* or *solubility* of oxygen in the broth. C_{AL}^* represents the maximum possible oxygen concentration that can occur in the liquid under the prevailing gas-phase composition, temperature, and pressure. The difference $(C_{AL}^* - C_{AL})$ between the maximum possible and actual oxygen concentrations in the liquid represents the *concentration-difference driving force* for mass transfer.

The solubility of oxygen in aqueous solutions at ambient temperature and pressure is less than 10 ppm. This amount of oxygen is quickly consumed in aerobic cultures and must be replaced constantly by sparging. An actively respiring cell population can consume the entire oxygen content of a culture broth within a few seconds; therefore, the maximum amount of oxygen that can be dissolved in the medium must be transferred from the gas phase 10 to 15 times per minute. This is not an easy task because the low solubility of oxygen guarantees that the concentration difference $(C_{AL}^* - C_{AL})$ is always very small. Design of fermenters for aerobic culture must take these factors into account and provide optimum mass transfer conditions.

10.5.1 Factors Affecting Cellular Oxygen Demand

The rate at which oxygen is consumed by cells in fermenters determines the rate at which oxygen must be transferred from the gas phase to the liquid phase. Many factors influence oxygen demand: the most important of these are cell species, culture growth phase, and the nature of the carbon source provided in the medium. In batch culture, the rate of oxygen uptake varies with time. The reasons for this are twofold. First, the concentration of cells increases during the course of batch culture and the total rate of oxygen uptake is proportional to the number of cells present. In addition, the rate of oxygen consumption per cell, known as the *specific oxygen uptake rate*, also varies, often reaching a maximum during the early stages of cell growth. If Q_O is the oxygen uptake rate per volume of broth and q_O is the specific oxygen uptake rate:

$$Q_O = q_O x \tag{10.40}$$

where x is cell concentration. Typical units for q_O are gmol g^{-1} s^{-1}, and for Q_O, gmol l^{-1} s^{-1}. Typical profiles of Q_O, q_O, and x during batch culture of microbial, plant, and animal cells are shown in Figure 10.6.

The inherent demand of an organism for oxygen (q_O) depends primarily on the biochemical nature of the cell and its nutritional environment. However, when the level of dissolved oxygen in the medium falls below a certain point, the specific rate of oxygen uptake is also dependent on the oxygen concentration in the liquid, C_{AL}. The dependence of q_O on C_{AL} is shown in Figure 10.7. If C_{AL} is above the *critical oxygen concentration* C_{crit}, q_O is a constant maximum and independent of C_{AL}. If C_{AL} is below C_{crit}, q_O is approximately linearly dependent on oxygen concentration.

To eliminate oxygen limitations and allow cell metabolism to function at its fastest, the dissolved oxygen concentration at every point in the fermenter must be above C_{crit}. The exact value of C_{crit} depends on the organism, but usually falls between 5% and 10% of air saturation under average operating conditions. For cells with relatively high C_{crit} levels, the task of transferring sufficient oxygen to maintain $C_{AL} > C_{crit}$ is more challenging than for cultures with low C_{crit} values.

The choice of substrate used in the fermentation medium can also affect the oxygen demand. Because glucose is generally consumed more rapidly than other sugars, rates of oxygen uptake are often higher when glucose is used. For example, maximum oxygen consumption rates of 5.5, 6.1, and 12 mmol l^{-1} h^{-1} have been observed for *Penicillium* mould growing on lactose, sucrose, and glucose, respectively [5]. As discussed in Section 4.6, oxygen requirements for cell growth and product formation also depend on the degree of reduction of the substrate. From electron balance considerations, the specific oxygen demand is greater when carbon substrates with a high degree of reduction are used. Therefore, specific oxygen uptake rates tend to be higher in cultures growing on alcohol or alkane hydrocarbons compared with carbohydrates.

Typical maximum q_O and Q_O values observed during batch culture of various organisms are listed in Table 10.1. Although specific oxygen uptake rates depend on the medium and culture conditions, plant and animal cells generally have significantly lower oxygen requirements than microbial cells.

10.5.2 Oxygen Transfer from Gas Bubble to Cell

In aerobic cultures, oxygen molecules must overcome a series of transport resistances before being utilised by the cells. Eight mass transfer steps involved in transport of oxygen

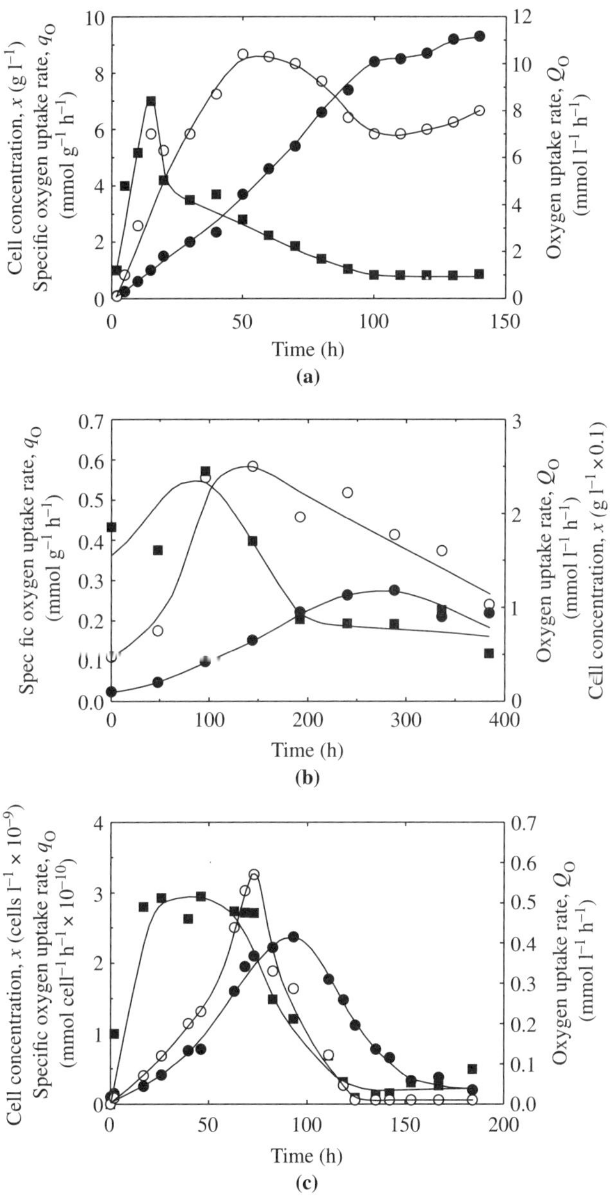

FIGURE 10.6 Variations in the oxygen uptake rate Q_O (○), the specific oxygen uptake rate q_O (■), and the cell concentration x (●) during batch culture of microbial, plant, and animal cells.
(a) Streptomyces aureofaciens. *Data from A.L. Jensen, J.S. Schultz, and P. Shu, 1966, Scale-up of antibiotic fermentations by control of oxygen utilization.* Biotechnol. Bioeng. *8, 525–537.*
(b) *Catharanthus roseus. Data from J.B. Snape, N.H. Thomas, and J.A. Callow, 1989, How suspension cultures of* Catharanthus roseus *respond to oxygen limitation: small-scale tests with applications to large-scale cultures.* Biotechnol. Bioeng. *34, 1058–1062.*
(c) Mouse–mouse hybridoma cells. *Data from S. Tatiraju, M. Soroush, and R. Mutharasan, 1999, Multi-rate nonlinear state and parameter estimation in a bioreactor.* Biotechnol. Bioeng. *63, 22–32.*

from the interior of gas bubbles to the site of intracellular reaction are represented diagrammatically in Figure 10.8. They are:

1. Transfer from the interior of the bubble to the gas–liquid interface
2. Movement across the gas–liquid interface
3. Diffusion through the relatively stagnant liquid film surrounding the bubble
4. Transport through the bulk liquid

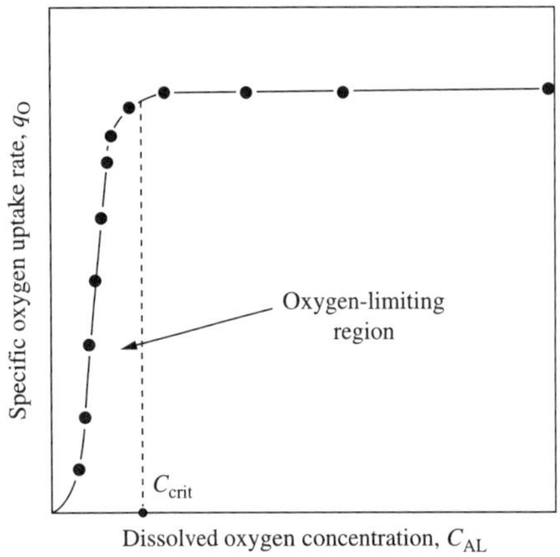

FIGURE 10.7 Relationship between the specific rate of oxygen consumption by cells and dissolved oxygen concentration.

TABLE 10.1 Typical Oxygen Uptake Rates in Different Cell Cultures

Type of cell culture	Carbon source	Maximum specific oxygen uptake rate, q_O (mmol g^{-1} dry weight h^{-1})	Maximum specific oxygen uptake rate, q_O (mmol $cell^{-1}$ h^{-1})	Maximum volumetric oxygen uptake rate, Q_O (mmol l^{-1} h^{-1})	Reference
MICROBIAL					
Aerobacter aerogenes	Peptone	–	3.2×10^{-11}	7.4	[6]
Aspergillus niger	Glucose	1.6	–	8.8	[7]
Bacillus subtilis	Peptone	–	1.5×10^{-10}	–	[8]
Beneckea natriegens	*n*-Propanol	12	–	6.0	[9]
Escherichia coli	Peptone	–	3.2×10^{-11}	5.0	[6]
Penicillium chrysogenum	Lactose	1.2	–	30	[10]
Saccharomyces cerevisiae	Ethanol	10	–	40	[11]
Streptomyces aureofaciens	Corn starch	7.0	–	10	[12]
Streptomyces coelicolor	Glucose	7.4	–	5.5	[13]
Streptomyces griseus	Meat extract	4.1	–	16	[14]
Xanthomonas campestris	Glucose	4.5	–	11	[15]
PLANT					
Catharanthus roseus	Sucrose	0.45	–	2.7	[16]
Nicotiana tabacum	Sucrose	0.90	–	1.0	[17]
ANIMAL					
Chinese hamster ovary (CHO)	Glucose/glutamine	–	2.9×10^{-10}	0.60	[18]
Hybridoma	Glucose/glutamine	–	2.9×10^{-10}	0.57	[19]

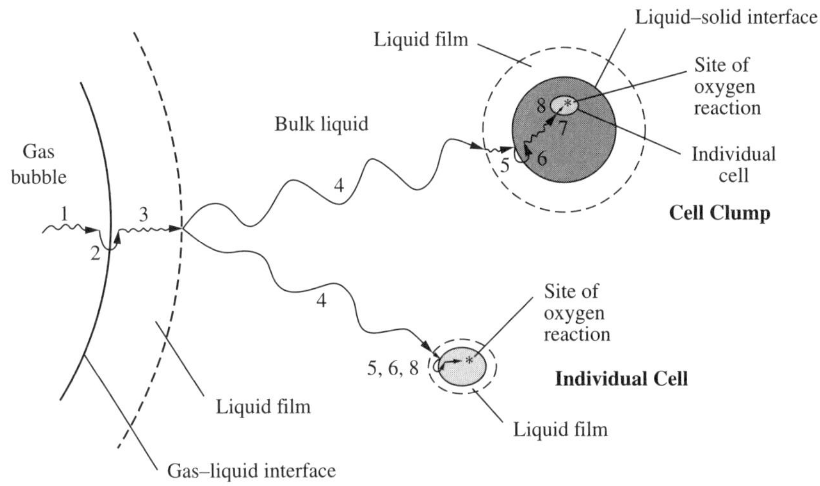

FIGURE 10.8 Steps for transfer of oxygen from gas bubble to cell.

5. Diffusion through the relatively stagnant liquid film surrounding the cells
6. Movement across the liquid–cell interface
7. If the cells are in a floc, clump, or solid particle, diffusion through the solid to the individual cell
8. Transport through the cytoplasm to the site of reaction

Note that resistance due to the gas boundary layer on the inside of the bubble has been neglected; because of the low solubility of oxygen in aqueous solutions, we can assume that the liquid-film resistance dominates gas–liquid mass transfer (see Section 10.4.3). If the cells are individually suspended in liquid rather than in a clump, step (7) disappears.

The relative magnitudes of the various mass transfer resistances depend on the composition and rheological properties of the liquid, the mixing intensity, the size of the bubbles, the size of any cell clumps, interfacial adsorption characteristics, and other factors. However, for most bioreactors the following analysis is valid.

1. Transfer through the bulk gas phase in the bubble is relatively fast.
2. The gas–liquid interface itself contributes negligible resistance.
3. The liquid film around the bubbles is a major resistance to oxygen transfer.
4. In a well-mixed fermenter, concentration gradients in the bulk liquid are minimised and mass transfer resistance in this region is small. However, rapid mixing can be difficult to achieve in viscous fermentation broths; if this is the case, oxygen transfer resistance in the bulk liquid may be important.
5. Because single cells are much smaller than gas bubbles, the liquid film surrounding each cell is much thinner than that around the bubbles. The effect of the cell liquid film on mass transfer can generally be neglected. On the other hand, if the cells form large clumps, the liquid-film resistance around the clumps can be significant.
6. Resistance at the liquid–cell interface is negligible.

7. When the cells are in clumps, intraparticle resistance is likely to be significant as oxygen has to diffuse through the solid pellet to reach the cells in the interior. The magnitude of this resistance depends on the size and properties of the cell clumps.
8. Intracellular oxygen transfer resistance is negligible because of the small distances involved.

When cells are dispersed in the liquid and the bulk fermentation broth is well mixed, *the major resistance to oxygen transfer is the liquid film surrounding the gas bubbles.* Transport through this film becomes the rate-limiting step in the complete process, and controls the overall mass transfer rate. Consequently, the rate of oxygen transfer from the bubbles all the way to the cells is dominated by the rate of step (3). The mass transfer rate for this step is represented by Eq. (10.39).

At steady state there can be no accumulation of oxygen at any location in the fermenter; therefore, the rate of oxygen transfer from the bubbles must be equal to the rate of oxygen consumption by the cells. If we make N_A in Eq. (10.39) equal to Q_O in Eq. (10.40) we obtain the following equation:

$$k_L a\,(C^*_{AL} - C_{AL}) = q_O\, x \tag{10.41}$$

The mass transfer coefficient $k_L a$ is used to characterise the oxygen transfer capability of fermenters. If $k_L a$ for a particular system is small, the ability of the reactor to deliver oxygen to the cells is limited. We can predict the response of the system to changes in mass transfer conditions using Eq. (10.41). For example, if the rate of cell metabolism remains unchanged but $k_L a$ is increased (e.g., by raising the stirrer speed to reduce the thickness of the boundary layer around the bubbles), the dissolved oxygen concentration C_{AL} must rise in order for the left side of Eq. (10.41) to remain equal to the right side. Similarly, if the rate of oxygen consumption by the cells accelerates while $k_L a$ is unaffected, C_{AL} must decrease.

We can use Eq. (10.41) to deduce some important relationships for fermenters. First, let us estimate the maximum cell concentration that can be supported by the fermenter's oxygen transfer system. For a given set of operating conditions, the maximum rate of oxygen transfer occurs when the concentration-difference driving force $(C^*_{AL} - C_{AL})$ is highest—that is, when the concentration of dissolved oxygen C_{AL} is zero. Therefore from Eq. (10.41), the maximum cell concentration that can be supported by oxygen transfer in the fermenter is:

$$x_{max} = \frac{k_L a\ C^*_{AL}}{q_O} \tag{10.42}$$

It is generally undesirable for cell density to be limited by the rate of mass transfer. Therefore, if x_{max} estimated using Eq. (10.42) is lower than the cell concentration required in the fermentation process, $k_L a$ must be improved. Note that the cell concentration in Eq. (10.42) is a theoretical maximum corresponding to operation of the system at its maximum oxygen transfer rate. Cell concentrations approaching x_{max} will be achieved only if all other culture conditions are favourable and if sufficient time and substrates are provided.

Comparison of x_{max} values evaluated using Eqs. (9.54) and (10.42) can be used to gauge the relative effectiveness of heat and mass transfer in aerobic fermentation. For example, if

x_{max} from Eq. (10.42) is small while x_{max} calculated from heat transfer considerations is large, we would know that mass transfer is more likely to limit biomass growth than heat transfer. If both x_{max} values are greater than that desired for the process, heat and mass transfer can be considered adequate.

Eq. (10.42) is a useful hypothetical relationship; however, as indicated in Figure 10.7, operation of culture systems at a dissolved oxygen concentration of zero is not advisable because the specific oxygen uptake rate depends on oxygen concentration. Accordingly, another important parameter is the minimum k_La required to maintain $C_{AL} > C_{crit}$ in the fermenter. This can be determined from Eq. (10.41) as:

$$(k_La)_{crit} = \frac{q_O\, x}{(C^*_{AL} - C_{crit})} \tag{10.43}$$

EXAMPLE 10.1 CELL CONCENTRATION IN AEROBIC CULTURE

A strain of *Azotobacter vinelandii* is cultured in a 15-m^3 stirred fermenter for alginate production. Under current operating conditions, k_La is 0.17 s^{-1}. The solubility of oxygen in the broth is approximately 8×10^{-3} kg m^{-3}.

(a) The specific rate of oxygen uptake is 12.5 mmol g^{-1} h^{-1}. What is the maximum cell concentration supported by oxygen transfer in the fermenter?

(b) The bacteria suffer growth inhibition after copper sulphate is accidentally added to the fermentation broth just after the start of the culture. This causes a reduction in the oxygen uptake rate to 3 mmol g^{-1} h^{-1}. What maximum cell concentration can now be supported by oxygen transfer in the fermenter?

Solution

(a) From Eq. (10.42):

$$x_{max} = \frac{0.17\ s^{-1}\,(8 \times 10^{-3}\ kg\ m^{-3})}{\frac{12.5\ mmol}{g\ h} \cdot \left|\frac{1\ h}{3600\ s}\right| \cdot \left|\frac{1\ gmol}{1000\ mmol}\right| \cdot \left|\frac{32\ g}{1\ gmol}\right| \cdot \left|\frac{1\ kg}{1000\ g}\right|}$$

$$x_{max} = 1.2 \times 10^4\ g\ m^{-3} = 12\ g\ l^{-1}$$

The maximum cell concentration supported by oxygen transfer in the fermenter is 12 g l^{-1}.

(b) Assume that addition of copper sulphate does not affect C^*_{AL} or k_La.

$$x_{max} = \frac{0.17\ s^{-1}\,(8 \times 10^{-3}\ kg\ m^{-3})}{\frac{3\ mmol}{g\ h} \cdot \left|\frac{1\ h}{3600\ s}\right| \cdot \left|\frac{1\ gmol}{1000\ mmol}\right| \cdot \left|\frac{32\ g}{1\ gmol}\right| \cdot \left|\frac{1\ kg}{1000\ g}\right|}$$

$$x_{max} = 5.0 \times 10^4\ g\ m^{-3} = 50\ g\ l^{-1}$$

The maximum cell concentration supported by oxygen transfer in the fermenter after addition of copper sulphate is 50 g l^{-1}.

To assess the oxygen transfer capability of a particular fermenter and for application of Eqs. (10.42) and (10.43), it is important that the actual $k_L a$ developed in the fermenter vessel be known. Methods for measuring $k_L a$ in bioprocesses are outlined in Section 10.10. Application of Eqs. (10.42) and (10.43) also requires knowledge of the oxygen solubility C^*_{AL} and the specific oxygen uptake rate q_O. Evaluation of these parameters is described in Sections 10.8 and 10.11.

10.6 FACTORS AFFECTING OXYGEN TRANSFER IN FERMENTERS

The rate of oxygen transfer in fermentation broths is influenced by several physical and chemical factors that change the value of k_L, the value of a, or the driving force for mass transfer $(C^*_{AL} - C_{AL})$. As a general rule of thumb, k_L in fermentation liquids is about 3 to 4×10^{-4} m s^{-1} for bubbles greater than 2 to 3 mm diameter; this can be reduced to 1×10^{-4} m s^{-1} for smaller bubbles depending on bubble rigidity. Once the bubbles are above 2 to 3 mm in size, k_L is relatively constant and insensitive to conditions. If substantial improvement in mass transfer rates is required, it is usually more productive to focus on increasing the interfacial area a. Operating values of the combined coefficient $k_L a$ span a wide range in bioreactors over about three orders of magnitude; this is due mainly to the large variation in a. In production-scale fermenters, the value of $k_L a$ is typically in the range 0.02 s^{-1} to 0.25 s^{-1}.

In this section, several aspects of fermenter design and operation are discussed in terms of their effect on oxygen mass transfer.

10.6.1 Bubbles

The efficiency of gas–liquid mass transfer depends to a large extent on the characteristics of the bubbles dispersed in the liquid medium. Bubble behaviour exerts a strong influence on the value of $k_L a$: some properties of bubbles affect mainly the magnitude of k_L, whereas others change the interfacial area a.

Large-scale aerobic cultures are carried out most commonly in stirred fermenters. In these vessels, oxygen is supplied to the medium by sparging swarms of air bubbles underneath the impeller. The action of the impeller then creates a dispersion of gas throughout the vessel. In small laboratory-scale fermenters, all of the liquid is close to the impeller; therefore, bubbles in these systems are subjected frequently to severe distortions as they interact with turbulent liquid currents in the vessel. In contrast, bubbles in most industrial stirred tanks spend a large proportion of their time floating relatively free and unimpeded through the liquid after initial dispersion at the impeller. Liquid in large fermenters away from the impeller does not possess sufficient energy for continuous break-up of bubbles. This is a consequence of scale; most laboratory fermenters operate with stirrer power between 10 and 20 kW m^{-3}, whereas large agitated vessels operate at 0.5 to 5 kW m^{-3}. The result is that virtually all large commercial-size stirred tank reactors operate mostly in the free-bubble-rise regime [20].

The most important property of air bubbles in fermenters is their size. For a given volume of gas, more interfacial area a is provided if the gas is dispersed into many small bubbles rather than a few large ones; therefore a major goal in bioreactor design is a high level of gas dispersion. However, there are other important benefits associated with small bubbles. Small bubbles have correspondingly slow bubble-rise velocities; consequently they stay in the liquid longer, allowing more time for oxygen to dissolve. Small bubbles therefore create high *gas hold-up*, which is defined as the fraction of the working volume of the reactor occupied by entrained gas:

$$\varepsilon = \frac{V_G}{V_T} = \frac{V_G}{V_L + V_G} \tag{10.44}$$

where ε is the gas hold-up, V_T is the total fluid volume (gas + liquid), V_G is the volume of gas bubbles in the reactor, and V_L is the volume of liquid. Because the total interfacial area for oxygen transfer depends on the total volume of gas in the system as well as on the average bubble size, high mass transfer rates are achieved at high gas hold-ups.

Gas hold-up values are very difficult to predict and may be anything from very low (0.01) up to a maximum in commercial-scale stirred fermenters of about 0.2. Under normal operating conditions, a significant fraction of the oxygen in fermentation vessels is contained in the gas hold-up. For example, if the culture is sparged with air and the broth saturated with dissolved oxygen, for a gas hold-up of only 0.03, about half the total oxygen in the system is in the gas phase.

While it is desirable to have small bubbles, there are practical limits. Bubbles $\ll 1$ mm diameter can become a nuisance in bioreactors. The oxygen concentration in these bubbles equilibrates with that in the medium within seconds, so that the gas hold-up no longer reflects the capacity of the system for mass transfer [21]. Problems with very small bubbles are exacerbated in viscous non-Newtonian broths: tiny bubbles remain lodged in these fluids for long periods of time because their rise velocity is reduced. As a rule of thumb, relatively large bubbles must be employed in viscous cultures.

Bubble size also affects the value of k_L. In most fermentation broths, if the bubbles have diameters less than 2 to 3 mm, surface tension effects dominate the behaviour of the bubble surface. As a result, the bubbles behave as rigid spheres with immobile surfaces and no internal gas circulation. A rigid bubble surface gives lower k_L values; k_L decreases with decreasing bubble diameter below about 2 to 3 mm. In contrast, depending on the liquid properties, bubbles in fermentation media with sizes greater than about 3 mm develop internal circulation and relatively mobile surfaces. Bubbles with mobile surfaces are able to wobble and move in spirals during free rise; this behaviour has a marked beneficial effect on k_L and the rate of mass transfer.

To summarise the influence of bubble size on oxygen transfer, small bubbles are generally beneficial because they provide higher gas hold-ups and greater interfacial surface area compared with large bubbles. However, k_L for bubbles less than about 2 to 3 mm in diameter is reduced due to surface effects. Very small bubbles $\ll 1$ mm should be avoided, especially in viscous broths.

10.6.2 Sparging, Stirring, and Medium Properties

In this section we consider the physical processes in fermenters and system properties that affect bubble size and the value of k_La.

Bubble Formation

In fermenters, air bubbles are formed at the sparger. Several types of sparger are in common use. *Porous spargers* of sintered metal, glass, or ceramic are applied mainly in small-scale systems; gas throughput is limited because the sparger poses a high resistance to flow. Cells growing through the fine holes and blocking the sparger can also be a problem. *Orifice spargers*, also known as perforated pipes, are constructed by making small holes in piping that is then fashioned into a ring or cross and placed in the reactor. The individual holes on orifice spargers should be large enough to minimise blockages. *Point* or *nozzle spargers* are used in many agitated fermenters from laboratory to production scale. These spargers consist of a single open pipe or partially closed pipe providing a point-source stream of air bubbles. Advantages compared with other sparger designs include low resistance to gas flow and small risk of blockage.

Bubbles leaving the sparger usually fall within a relatively narrow size range depending on the sparger type. This size range is a significant parameter in the design of air-driven fermenters such as bubble and airlift columns (Section 14.2 in Chapter 14) because there is no other mechanism for bubble dispersion in these reactors. However in stirred vessels, design of the sparger and the mechanics of bubble formation are of secondary importance compared with the effects of the impeller. As a result of continual bubble break-up and dispersion by the impeller and coalescence from bubble collisions, the bubble sizes in stirred reactors often bear little relation to those formed at the sparger.

Gas Dispersion

The two-phase flow patterns set up in stirred vessels with gassing have been described in Chapter 8 (Sections 8.3.4 and 8.4). The effectiveness of bubble break-up and dispersion depends on the relative rates of stirring and gas flow; the balance between these operating parameters determines whether *impeller flooding*, *impeller loading*, or *complete gas dispersion* occurs. Flooding should be avoided because an impeller surrounded by gas no longer contacts the liquid properly, resulting in poor mixing and gas dispersion. At stirrer speeds above that required to prevent flooding, gas is increasingly recirculated around the tank.

Gas dispersion in stirred vessels takes place mainly in the immediate vicinity of the impeller. The formation and function of *ventilated cavities* behind the impeller blades are discussed in Section 8.4. Gas from the sparger together with a large fraction of the recirculating gas in the system is entrained in these cavities. As the impeller blades rotate at high speed, small gas bubbles are thrown out from the back of the cavities into the bulk liquid. Because bubbles formed at the sparger are immediately drawn into the impeller zone, dispersion of gas in stirred vessels is largely independent of sparger design. As long as the sparger is located under the stirrer, it has been shown that sparger type does not affect mass transfer significantly.

Bubble Coalescence

Coalescence of small bubbles into bigger bubbles is generally undesirable for oxygen transfer because the total interfacial area and gas hold-up are reduced. The frequency of bubble coalescence depends mainly on the liquid properties. In *coalescing liquids* such as pure water, a large fraction of bubble collisions results in the formation of bigger bubbles.

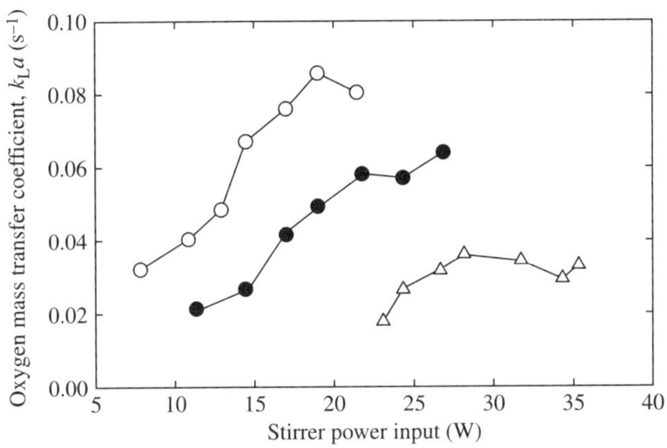

FIGURE 10.9 Effect of solution composition on k_La in a stirred tank at constant gas flow rate: (●) water (coalescing); (○) 5% Na_2SO_4 in water (noncoalescing); and (△) 0.7% w/w carboxymethyl cellulose in water (viscous, pseudoplastic).
Data from S.J. Arjunwadkar, K. Sarvanan, P.R. Kulkarni, and A.B. Pandit, 1998, Gas–liquid mass transfer in dual impeller bioreactor. Biochem. Eng. J. *1, 99–106.*

In contrast, in *noncoalescing liquids*, colliding bubbles tend to bounce off each other due to surface tension effects and do not coalesce readily. In noncoalescing liquids in stirred vessels, the bubbles sizes remain close to those produced at the back of the ventilated cavities behind the impeller blades.

The coalescence properties of liquids depend on the liquid composition. Compared with pure water, the presence of salts and ions suppresses coalescence; therefore, simple fermentation media are usually noncoalescing to some extent depending on composition. This is an advantage for oxygen transfer. The addition of ions to water in sparged vessels markedly reduces the average bubble size and increases the gas hold-up, so much so that the interfacial area *a* in water containing salts may be up to 10 times greater than that obtained without salts [22].

Experimental results for the effect of solution composition on k_La are shown in Figure 10.9. The presence of solutes has a significant impact on the rate of oxygen transfer. The results for water and 5% Na_2SO_4 salt in water illustrate the effect of liquid coalescence properties: k_La is lower in water than in noncoalescing salt solution. Because the composition and therefore the coalescence properties of fermentation broths vary with time during cell cultures, k_La can also be expected to vary accordingly.

Viscosity

The rheology of fluids has a significant effect on bubble size, gas hold-up, and k_La. With increasing viscosity, the thickness of the fluid boundary layers surrounding the bubbles increases, thus increasing the resistance to oxygen transfer. High viscosity also dampens turbulence, changes the size and structure of the ventilated cavities at the impeller blades, and reduces the effectiveness of gas dispersion. Except in low-viscosity fluids where these effects are negligible, k_La decreases with increasing liquid viscosity. As an example, as shown in Figure 10.9, k_La in pseudoplastic carboxymethyl cellulose solution is significantly lower than in water or salt solution at the same fermenter power input. In non-Newtonian fluids, sharp reductions in k_La have been found to occur at apparent viscosities above 10 to 2000 mPa s [23].

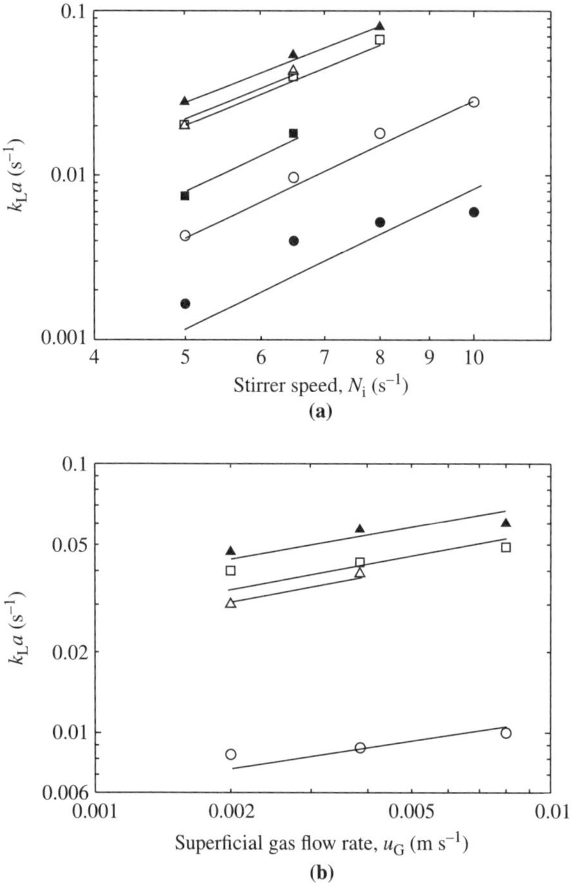

FIGURE 10.10 Dependence of k_La on operating conditions in a stirred tank. The data are plotted using logarithmic coordinates. (a) Effect of stirrer speed N_i at constant gas velocity. (b) Effect of gas flow rate u_G at constant stirrer speed. The symbols represent Newtonian fluids with viscosities: (△) 0.91 mPa s; (▲) 1.3 mPa s; (□) 2.1 mPa s; (■) 5.1 mPa s; (○) 13.3 mPa s; and (●) 70.2 mPa s.

From H. Yagi and F. Yoshida, 1975, Gas absorption by Newtonian and non-Newtonian fluids in sparged agitated vessels. Ind. Eng. Chem. Process Des. Dev. *14, 488–493.*

Stirrer Speed and Gas Flow Rate

Under normal fermenter operating conditions, increasing the stirrer speed and gas flow rate improves the value of k_La. Typical data for k_La in Newtonian fluids of varying viscosity are shown in Figure 10.10. The strong dependence of k_La on stirrer speed is evident from Figure 10.10(a): in this system, doubling the stirrer speed N_i resulted in an average 4.6-fold increase in k_La. Increasing the gas flow rate is a less effective strategy for improving k_La: for example, in the system represented in Figure 10.10(b), doubling the gas flow rate increased k_La by only about 20%. Moreover, in most systems there is limited practical scope for increasing k_La by increasing the gas flow: as discussed in Sections 8.3.4 and 8.4, impeller flooding occurs at high gas flow rates unless the impeller is able to disperse all the gas impinging on it. At very high gassing rates, the liquid contents can be blown out of the fermenter. In viscous non-Newtonian fluids, the dependence of k_La on stirrer speed is generally weaker than in low-viscosity systems while the effect of gas velocity is similar.

10.6.3 Antifoam Agents

Most cell cultures produce a variety of foam-producing and foam-stabilising agents, such as proteins, polysaccharides, and fatty acids. Foam build-up in fermenters is very common, particularly in aerobic systems. Foaming causes a range of reactor operating problems; foam control is therefore an important consideration in fermentation design. Excessive foam overflowing from the top of the fermenter provides an entry route for contaminating organisms and causes blockage of outlet gas lines. Liquid and cells trapped in the foam represent an effective loss of bioreactor volume, as conditions in the foam may not be favourable for metabolic activity. To make matters worse, fragile cells can be damaged by collapsing foam. To accommodate foam layers as well as the increase in fluid volume in aerated vessels due to gas hold-up, a space of 20 to 30% of the tank volume must be left between the top of the liquid and the vessel headplate when setting up fermenters. Foaming is exacerbated by high gas flow rates and high stirrer speeds.

Addition of special antifoam compounds to the medium is the most common method of reducing foam build-up in fermenters. However, antifoam agents affect the surface chemistry of bubbles and their tendency to coalesce, and have a significant effect on k_La. Most antifoam agents are strong surface-tension-lowering substances. Decrease in surface tension reduces the average bubble diameter, thus producing higher values of a. However, this is countered by a reduction in the mobility of the gas–liquid interface, which lowers the value of k_L. With most silicon-based antifoams, the decrease in k_L is generally larger than the increase in a so that, overall, k_La is reduced [24, 25]. Typical data for surface tension and k_La as a function of antifoam concentration are shown in Figure 10.11. In this experiment, a reduction in k_La of almost 50% occurred after addition of only a small amount of antifoam. In some cases, the decrease in the rate of oxygen transfer is dramatic, by up to a factor of 5 to 10.

To maintain the noncoalescing character of the medium and high k_La values, mechanical rather than chemical methods of disrupting foam are preferred because the properties of the liquid are not changed. Mechanical foam breakers, such as high-speed discs rotating at the top of the vessel and centrifugal foam destroyers, are suitable when foam development is moderate. However, some of these devices need large quantities of power to operate in commercial-scale vessels; in addition, their limited foam-destroying capacity is a problem with highly foaming cultures. In many cases, use of chemical antifoam agents is unavoidable.

10.6.4 Temperature

The temperature of aerobic fermentations affects both the solubility of oxygen C^*_{AL} and the mass transfer coefficient k_L. Increasing the temperature causes C^*_{AL} to drop, so that the driving force for mass transfer ($C^*_{AL} - C_{AL}$) is reduced. At the same time, the diffusivity of oxygen in the liquid film surrounding the bubbles is increased, resulting in an increase in k_L. The net effect of temperature on oxygen transfer depends on the range of temperature considered. For temperatures between 10°C and 40°C, which includes the operating range for most fermentations, an increase in temperature is more likely to increase the rate of

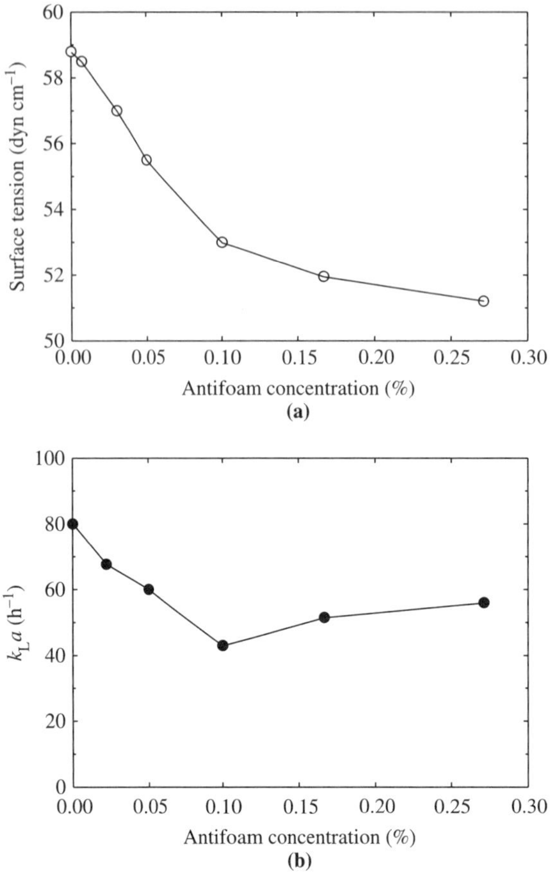

FIGURE 10.11 Effect of antifoam concentration on: (a) surface tension, and (b) $k_L a$, in a *Penicillium chrysogenum* fermentation broth.
Data from F.H. Deindoerfer and E.L. Gaden, 1955, Effects of liquid physical properties on oxygen transfer in penicillin fermentation. Appl. Microbiol. *3, 253–257; with unit conversion from R.K. Finn, 1954, Agitation–aeration in the laboratory and in industry.* Bact. Rev. *18, 254–274.*

oxygen transfer. Above 40°C, the solubility of oxygen drops significantly, adversely affecting the driving force and rate of mass transfer.

10.6.5 Gas Pressure and Oxygen Partial Pressure

The pressure and oxygen partial pressure of the gas used to aerate fermenters affect the value of C^*_{AL}. The equilibrium relationship between these parameters for dilute liquid solutions is given by *Henry's law*:

$$p_{AG} = p_T \, y_{AG} = H C^*_{AL} \tag{10.45}$$

where p_{AG} is the *partial pressure* of component A in the gas phase, p_T is the total gas pressure, y_{AG} is the mole fraction of A in the gas phase, and C^*_{AL} is the solubility of component A in the liquid. H is *Henry's constant*, which is a function of temperature. From Eq. (10.45), if the total gas pressure p_T or the concentration of oxygen in the gas y_{AG} is increased at constant temperature, C^*_{AL} and therefore the mass transfer driving force $(C^*_{AL} - C_{AL})$ also increase.

In some cell cultures, oxygen-enriched air or pure oxygen may be used instead of air to improve oxygen transfer. The effect on oxygen solubility can be determined using Henry's law. At a fixed temperature so that H remains constant, if p_T and y_{AG} are varied from condition 1 to condition 2, from Eq. (10.45):

$$\frac{C^*_{AL2}}{C^*_{AL1}} = \frac{p_{AG2}}{p_{AG1}} = \frac{p_{T2}\, y_{AG2}}{p_{T1}\, y_{AG1}} \tag{10.46}$$

According to the *International Critical Tables* [26], the mole fraction y_{AG1} of oxygen in air is 0.2099. If pure oxygen is used instead of air, y_{AG2} is 1. From Eq. (10.46), if the gases are applied at the same total pressure:

$$\frac{C^*_{AL2}}{C^*_{AL1}} = \frac{1}{0.2099} = 4.8 \tag{10.47}$$

Therefore, sparging pure oxygen instead of air at the same total pressure and temperature increases the solubility of oxygen by a factor of 4.8. Alternatively, the solubility can be increased by sparging compressed air at higher total pressure p_T. Both these strategies increase the operating cost of the fermenter; it is also possible in some cases that the culture will suffer inhibitory effects from exposure to very high oxygen concentrations.

10.6.6 Presence of Cells and Macromolecules

Oxygen transfer is influenced by the presence of cells in fermentation broths. The effect depends on the morphology of the organism and the cell concentration. Cells with complex morphology, such as branched hyphae, generally lead to lower oxygen transfer rates by interfering with bubble break-up and promoting coalescence. Cells, proteins, and other molecules that adsorb at gas–liquid interfaces also cause *interfacial blanketing*, which reduces the effective contact area between gas and liquid. Macromolecules and very small particles accumulating at the bubble surface reduce the mobility of the interface, thus lowering k_L, but may also decrease coalescence, thereby increasing a. The quantitative effect of interfacial blanketing is highly system-specific.

Because cell, substrate, and product concentrations, and therefore the viscosity and coalescence properties of the fluid, change during batch fermentations, the value of $k_L a$ also varies. An example of change in $k_L a$ as a result of these combined factors is shown in Figure 10.12.

10.7 MEASURING DISSOLVED OXYGEN CONCENTRATION

The concentration of dissolved oxygen C_{AL} in fermenters is normally measured using a *dissolved oxygen electrode*. There are two types in common use: *galvanic electrodes* and *polarographic electrodes*. Details of the construction and operating principles of oxygen probes can be found in the literature (e.g., [27, 28]). In both designs, a membrane permeable to oxygen separates the fermentation fluid from the electrode. As illustrated in Figure 10.13, oxygen

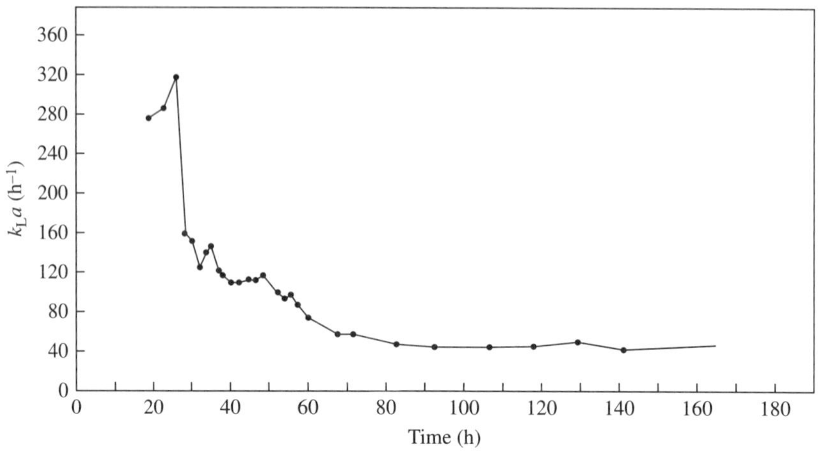

FIGURE 10.12 Variation in k_La during a batch 300-1 streptomycete fermentation.
From C.M. Tuffile and F. Pinho, 1970, Determination of oxygen transfer coefficients in viscous streptomycete fermentations. Biotechnol. Bioeng. *12, 849–871.*

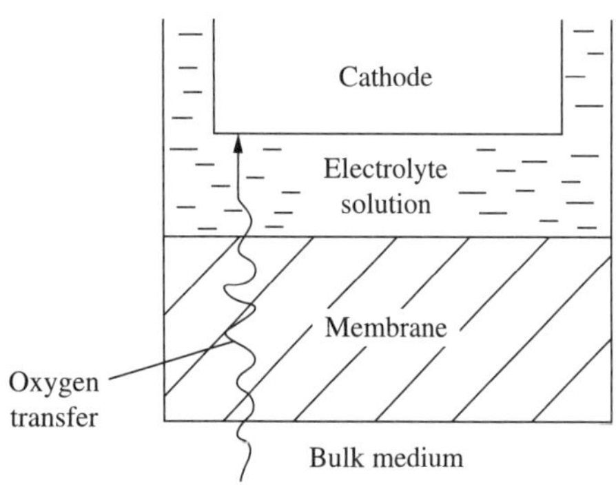

FIGURE 10.13 Diffusion of oxygen from the bulk liquid to the cathode of an oxygen electrode.

diffuses through the membrane to the cathode, where it reacts to produce a current between the anode and cathode proportional to the oxygen partial pressure in the fermentation broth. An electrolyte solution in the electrode supplies ions that take part in the reaction and must be replenished at regular intervals. Steam-sterilisable or autoclavable probes are available commercially and are inserted directly into fermentation vessels for online monitoring of dissolved oxygen. The probe should be located to avoid the direct impingement of bubbles on the electrode membrane as this distorts the probe signal. Repeated calibration of dissolved oxygen probes is necessary; fouling by cells attaching to the membrane surface, electronic noise due to air bubbles passing close to the membrane, and signal drift are the main operating problems.

As indicated in Figure 10.13, the supply of oxygen from the bulk medium to the cathode in oxygen probes is a mass transfer process. Because there is no bulk fluid motion in the membrane or electrolyte solution, operation of the probe relies on diffusion of oxygen across these thicknesses. This takes time, so the response of an electrode to sudden changes in dissolved oxygen level is subject to delay. This does not affect many applications of oxygen probes in fermenters, as changes in dissolved oxygen tension during cell culture are normally relatively slow. However, as described in more detail in Section 10.10.2 (Electrode Response Time and Liquid Boundary Layers subsection), electrode dynamics can have a significant influence on the measurement of oxygen uptake rates and k_La in culture broths. Microprobes for dissolved oxygen measurement are also available. The smaller cathode size and lower oxygen consumption by these instruments means that their response is much quicker than the more robust devices used in fermentation vessels.

Both galvanic and polarographic electrodes measure the *oxygen tension* or partial pressure of dissolved oxygen in the fermentation broth, not the dissolved oxygen concentration. To convert oxygen tension to dissolved oxygen concentration, it is necessary to know the solubility of oxygen in the liquid at the temperature and pressure of measurement. For example, if the dissolved oxygen tension measured using an oxygen electrode is 60% air saturation and the solubility of oxygen in the solution under air is 8.0×10^{-3} kg m^{-3}, the dissolved oxygen concentration is $(0.6 \times 8.0 \times 10^{-3})$ kg m^{-3} $= 4.8 \times 10^{-3}$ kg m^{-3}.

10.8 ESTIMATING OXYGEN SOLUBILITY

The concentration difference $(C^*_{AL} - C_{AL})$ is the driving force for oxygen mass transfer. Because this difference is usually very small, it is important that the solubility C^*_{AL} be known accurately. Otherwise, small errors in C^*_{AL} will result in large errors in $(C^*_{AL} - C_{AL})$.

Air is used to provide oxygen in most industrial fermentations. Values for the solubility of oxygen in water at various temperatures and 1 atm air pressure are listed in Table 10.2. However, fermentations are not carried out using pure water, and the gas composition and pressure can be varied. Because the presence of dissolved material in the liquid and the oxygen partial pressure in the gas phase affect oxygen solubility, the values in Table 10.2 may not be directly applicable to bioprocessing systems.

10.8.1 Effect of Oxygen Partial Pressure

As indicated in Henry's law, Eq. (10.45), oxygen solubility is directly proportional to the total gas pressure and the mole fraction of oxygen in the gas phase. The solubility of oxygen in water as a function of these variables can be determined using Eq. (10.45) and the values for Henry's constant listed in Table 10.2.

10.8.2 Effect of Temperature

The variation of oxygen solubility with temperature is shown in Table 10.2 for water in the range 0 to 40°C. Oxygen solubility falls with increasing temperature. The solubility of

TABLE 10.2 Solubility of Oxygen in Water under 1 atm Air Pressure

Temperature (°C)	Oxygen solubility under 1 atm air pressure (kg m^{-3})	Henry's constant (m^3 atm gmol^{-1})
0	1.48×10^{-2}	0.454
10	1.15×10^{-2}	0.582
15	1.04×10^{-2}	0.646
20	9.45×10^{-3}	0.710
25	8.69×10^{-3}	0.774
26	8.55×10^{-3}	0.787
27	8.42×10^{-3}	0.797
28	8.29×10^{-3}	0.810
29	8.17×10^{-3}	0.822
30	8.05×10^{-3}	0.835
35	7.52×10^{-3}	0.893
40	7.07×10^{-3}	0.950

Calculated from data in International Critical Tables, *1928, vol. III, McGraw-Hill, New York, p. 257.*

oxygen from air in pure water between 0°C and 36°C has been correlated using the following equation [29]:

$$C^*_{AL} = 14.161 - 0.3943\,T + 0.007714\,T^2 - 0.0000646\,T^3 \tag{10.48}$$

where C^*_{AL} is oxygen solubility in units of mg l^{-1}, and T is temperature in °C.

10.8.3 Effect of Solutes

The presence of solutes such as salts, acids, and sugars affects the solubility of oxygen in water as indicated in Tables 10.3 and 10.4. These data show that the solubility of oxygen is reduced by the addition of ions and sugars that are normally required in fermentation media. Quicker et al. [30] have developed an empirical correlation to correct values of oxygen solubility in water for the effects of cations, anions, and sugars:

$$\log_{10}\left(\frac{C^*_{AL0}}{C^*_{AL}}\right) = 0.5\sum_i H_i z_i^2 C_{iL} + \sum_j K_j C_{jL} \tag{10.49}$$

where

C^*_{AL0} = oxygen solubility at zero solute concentration (mol m^{-3})
C^*_{AL} = oxygen solubility in the presence of solutes (mol m^{-3})
H_i = constant for ionic component i (m^3 mol^{-1})
z_i = charge (valence) of ionic component i

TABLE 10.3 Solubility of Oxygen in Aqueous Solutions at 25°C under 1 atm Oxygen Pressure

	Oxygen solubility at 25°C under 1 atm oxygen pressure (kg m^{-3})		
Concentration (M)	**HCl**	**½ H_2SO_4**	**NaCl**
0	4.14×10^{-2}	4.14×10^{-2}	4.14×10^{-2}
0.5	3.87×10^{-2}	3.77×10^{-2}	3.43×10^{-2}
1.0	3.75×10^{-2}	3.60×10^{-2}	2.91×10^{-2}
2.0	3.50×10^{-2}	3.28×10^{-2}	2.07×10^{-2}

Calculated from data in International Critical Tables, *1928, vol. III, McGraw-Hill, New York, p. 271.*

TABLE 10.4 Solubility of Oxygen in Aqueous Solutions of Sugars under 1 atm Oxygen Pressure

Sugar	Concentration (gmol per kg H_2O)	Temperature (°C)	Oxygen solubility under 1 atm oxygen pressure (kg m^{-3})
Glucose	0	20	4.50×10^{-2}
	0.7	20	3.81×10^{-2}
	1.5	20	3.18×10^{-2}
	3.0	20	2.54×10^{-2}
Sucrose	0	15	4.95×10^{-2}
	0.4	15	4.25×10^{-2}
	0.9	15	3.47×10^{-2}
	1.2	15	3.08×10^{-2}

Calculated from data in International Critical Tables, *1928, vol. III, McGraw-Hill, New York, p. 272.*

C_{iL} = concentration of ionic component i in the liquid (mol m^{-3})
K_j = constant for nonionic component j (m^3 mol^{-1})
C_{jL} = concentration of nonionic component j in the liquid (mol m^{-3})

Values of H_i and K_j for use in Eq. (10.49) are listed in Table 10.5. In a typical fermentation medium, the oxygen solubility is between 5% and 25% lower than in water as a result of solute effects.

10.9 MASS TRANSFER CORRELATIONS FOR OXYGEN TRANSFER

In general, there are two approaches to evaluating mass transfer coefficients: calculation using empirical correlations, and experimental measurement. In both cases, for gas–liquid mass transfer, separate determination of k_L and a is laborious and sometimes impossible. It is convenient, therefore, to evaluate k_La as a combined term. In this section we consider

TABLE 10.5 Values of H_i and K_j in Eq. (10.49) at 25°C

Cation	$H_i \times 10^3$ (m^3 mol^{-1})	Anion	$H_i \times 10^3$ (m^3 mol^{-1})	Sugar	$K_j \times 10^3$ (m^3 mol^{-1})
H^+	−0.774	OH^-	0.941	Glucose	0.119
K^+	−0.596	Cl^-	0.844	Lactose	0.197
Na^+	−0.550	CO_3^{2-}	0.485	Sucrose	0.149*
NH_4^+	−0.720	SO_4^{2-}	0.453		
Mg^{2+}	−0.314	NO_3^-	0.802		
Ca^{2+}	−0.303	HCO_3^-	1.058		
Mn^{2+}	−0.311	$H_2PO_4^-$	1.037		
		HPO_4^{2-}	0.485		
		PO_4^{3-}	0.320		

Approximately valid for sucrose concentrations up to about 200 g l^{-1}.

From A. Schumpe, I. Adler, and W.-D. Deckwer, 1978, Solubility of oxygen in electrolyte solutions. Biotechnol. Bioeng. *20, 145–150; and G. Quicker, A. Schumpe, B. König, and W.-D. Deckwer, 1981, Comparison of measured and calculated oxygen solubilities in fermentation media.* Biotechnol. Bioeng. *23, 635–650.*

calculation of $k_L a$ using published correlations. Experimental methods for measuring $k_L a$ are described later in Section 10.10.

The value of $k_L a$ in fermenters depends on the fluid properties and the prevailing hydrodynamic conditions. Relationships between $k_L a$ and parameters such as liquid density, viscosity, oxygen diffusivity, bubble diameter, and fluid velocity have been investigated extensively. The results of these studies in the form of empirical correlations between mass transfer coefficients and important operating variables are available in the literature. Theoretically, these correlations allow prediction of mass transfer coefficients based on information gathered from a large number of previous experiments. In practice, however, the accuracy of published correlations for $k_L a$ applied to biological systems is generally poor.

The main reason is that oxygen transfer is strongly affected by the additives present in fermentation media. Because fermentation liquids contain varying levels of substrates, products, salts, surface-active agents, and cells, the surface chemistry of bubbles, and therefore the mass transfer situation, become very complex. Most available correlations for oxygen transfer coefficients were determined using air in pure water; however, the presence of additives in water affects the value of $k_L a$ significantly and it is very difficult to make corrections for different liquid compositions. The effective mass transfer area a is also subject to interfacial blanketing of the bubble–liquid surface by cells and other components of the broth. Prediction of $k_L a$ under these conditions is problematic.

When mass transfer coefficients are required for large-scale equipment, another factor related to hydrodynamic conditions limits the applicability of published correlations. Most studies of oxygen transfer have been carried out in laboratory-scale stirred reactors, which are characterised by high turbulence throughout most of the vessel. In contrast, intense turbulence in large-scale fermenters is limited to the impeller region; flow in the remainder of the tank is much slower. As a result, because of the different hydrodynamic regimes present in small- and large-scale vessels, mass transfer correlations developed in the laboratory tend to overestimate the oxygen transfer capacity of commercial-scale systems.

Although published mass transfer correlations cannot be applied directly to fermentation systems, there is a general consensus in the literature about the form of the equations and the relationship between k_La and reactor operating conditions. The most successful correlations are dimensional equations of the form [22]:

$$k_La = A\left(\frac{P_T}{V_L}\right)^{\alpha} u_G^{\beta} \tag{10.50}$$

In Eq. (10.50), k_La is the oxygen transfer coefficient, P_T is the total power dissipated, and V_L is the liquid volume. u_G is the *superficial gas velocity*, which is defined as the volumetric gas flow rate divided by the cross-sectional area of the fermenter. All of the hydrodynamic effects of flow and turbulence on bubble dispersion and the mass transfer boundary layer are represented by the power term, P_T. The total power dissipated is calculated as the sum of the power input by stirring under gassed conditions (Section 8.5.3) and, if it makes a significant contribution, the power associated with isothermal expansion of the sparged gas (Section 8.6).

Note that correlations of the type represented by Eq. (10.50) are independent of the sparger or stirrer design: the power input determines k_La independent of stirrer type. A, α, and β are constants. The values of α and β are largely insensitive to broth properties and usually fall within the range 0.2 to 1.0. In contrast, the dimensional parameter A varies significantly with liquid composition and is sensitive to the coalescing properties and cell content of culture broths. Because both exponents α and β are typically <1 in value, increasing k_La by raising either the air flow rate or power input becomes progressively less efficient and more costly as the inputs increase.

For viscous and non-Newtonian fluids, a modified form of Eq. (10.50) can be used to incorporate explicitly the effect of viscosity on the mass transfer coefficient:

$$k_La = B\left(\frac{P_T}{V_L}\right)^{\alpha} u_G^{\beta}\, \mu_a^{-\delta} \tag{10.51}$$

where B is a modified constant reflecting the properties of the liquid other than viscosity, μ_a is the apparent viscosity (Section 7.5), and δ is a constant typically in the range 0.5 to 1.3.

10.10 MEASUREMENT OF k_La

Because of the difficulties associated with using correlations to predict k_La in bioreactors (Section 10.9), oxygen transfer coefficients are routinely determined experimentally. This is not without its own problems however, as discussed below. Whichever method is used to measure k_La, the measurement conditions should match those applied in the fermenter. Techniques for measuring k_La have been reviewed in the literature [22, 31, 32].

10.10.1 Oxygen Balance Method

The steady-state oxygen balance method is the most reliable procedure for estimating k_La, and allows determination from a single-point measurement. An important advantage is that the method can be applied to fermenters during normal operation. It is strongly dependent, however, on accurate measurement of the inlet and outlet gas composition, flow rate, pressure, and temperature. Large errors as high as ±100% can be introduced if the measurement techniques are inadequate. Considerations for the design and operation of laboratory equipment to ensure accurate results are described by Brooks et al. [33].

To determine k_La, the oxygen contents of the gas streams flowing to and from the fermenter are measured. From a mass balance at steady state:

$$N_A = \frac{1}{V_L}\left[(F_g\,C_{AG})_i - (F_g\,C_{AG})_o\right] \tag{10.52}$$

where N_A is the volumetric rate of oxygen transfer, V_L is the volume of liquid in the fermenter, F_g is the volumetric gas flow rate, C_{AG} is the gas-phase concentration of oxygen, and subscripts i and o refer to the inlet and outlet gas streams, respectively. The first bracketed term on the right side represents the rate at which oxygen enters the fermenter in the inlet gas stream; the second term is the rate at which oxygen leaves. The difference between them is the rate at which oxygen is transferred out of the gas into the liquid. Because gas concentrations are generally measured as partial pressures, the ideal gas law Eq. (2.35) can be incorporated into Eq. (10.52) to obtain an alternative expression:

$$N_A = \frac{1}{R\,V_L}\left[\left(\frac{F_g\,p_{AG}}{T}\right)_i - \left(\frac{F_g\,p_{AG}}{T}\right)_o\right] \tag{10.53}$$

where R is the ideal gas constant (Appendix B), p_{AG} is the oxygen partial pressure in the gas, and T is the absolute temperature. Because there is often not a great difference between the amounts of oxygen in the gas streams entering and leaving fermenters, especially in small-scale systems, application of Eq. (10.53) involves subtracting two numbers of similar magnitude. To minimise the associated error, p_{AG} is usually measured using mass spectrometry or similar high-sensitivity technique. The temperature and flow rate of the gases must also be measured carefully to determine an accurate value of N_A. Once N_A is known, if C_{AL} in the fermentation broth is measured using a dissolved oxygen electrode and C^*_{AL} is evaluated as described in Section 10.8, k_La can then be determined from Eq. (10.39). The value of k_La will vary depending on the stirrer speed and air flow rate used during the measurement and the properties of the culture broth.

There are several assumptions inherent in the equations used in the oxygen balance method:

1. The liquid phase is well mixed.
2. The gas phase is well mixed.
3. The pressure is constant throughout the vessel.

These assumptions relate to application of Eq. (10.39) to determine k_La from the measured results for N_A. Assumption (1) allows us to use a single C_{AL} value to represent the concentration of dissolved oxygen in the fermentation broth. It applies reasonably well in small vessels where there is a relatively high stirrer power input per unit volume. However, the liquid phase in large-scale fermenters may not be well mixed, particularly if the culture broth is viscous. Assumptions (2) and (3) are required for evaluation of the oxygen solubility, C^*_{AL}. Because solubility depends on the gas-phase composition and pressure, we assume that these properties are the same throughout the tank. Assumption (2) is valid when the bubble gas composition is uniform throughout the vessel and the same as that in the outlet gas stream. This occurs most readily in small, intensely agitated vessels with high levels of gas recirculation; accordingly, assumption (2) applies generally in laboratory-scale fermenters. In contrast, the higher bubble residence times and lower gas recirculation levels in large vessels mean that the gas phase is less likely to be well mixed. Assumption (3) is also generally valid in small-scale vessels. However, in large, tall fermenters, there may be a considerable difference in hydrostatic pressure between the top and bottom of the tank due to the liquid weight. Issues affecting the analysis of oxygen transfer in large fermenters are considered in more detail in Section 10.12.

EXAMPLE 10.2 STEADY-STATE k_La MEASUREMENT

A 20-litre stirred fermenter containing *Bacillus thuringiensis* is used to produce a microbial insecticide. The oxygen balance method is applied to determine k_La. The fermenter operating pressure is 150 kPa and the culture temperature is 30°C. The oxygen tension in the broth is measured as 82% using a probe calibrated to 100% in situ using water and air at 30°C and 150 kPa. The solubility of oxygen in the culture fluid is the same as in water. Air is sparged into the vessel; the inlet gas flow rate measured outside the fermenter at 1 atm pressure and 22°C is $0.23\ \mathrm{l\ s^{-1}}$. The exit gas from the fermenter contains 20.1% oxygen and has a flow rate of $8.9\ \mathrm{l\ min^{-1}}$.

(a) Calculate the volumetric rate of oxygen uptake by the culture.

(b) What is the value of k_La?

Solution

(a) From Table A.5 in Appendix A, the fermenter operating pressure is:

$$150 \times 10^3\ \mathrm{Pa} \cdot \left| \frac{1\ \mathrm{atm}}{1.013 \times 10^5\ \mathrm{Pa}} \right| = 1.48\ \mathrm{atm}$$

The oxygen partial pressure in the inlet air at 1 atm is 0.2099 atm (Section 10.6.5). From Appendix B, $R = 0.082057\ \mathrm{l\ atm\ K^{-1}\ gmol^{-1}}$. Using Eq. (10.53):

$$N_A = \frac{1}{0.082057\ \mathrm{l\ atm\ K^{-1}\ gmol^{-1}}\ (20\ \mathrm{l})} \left[\left(\frac{0.23\ \mathrm{l\ s^{-1}}\ (0.2099\ \mathrm{atm})}{(22 + 273.15)\ \mathrm{K}} \right) - \left(\frac{8.9\ \mathrm{l\ min^{-1}} \cdot \left| \frac{1\ \mathrm{min}}{60\ \mathrm{s}} \right| \cdot (0.201)\ (1.48\ \mathrm{atm})}{(30 + 273.15)\ \mathrm{K}} \right) \right]$$

$$N_A = \frac{1}{0.082057\ \text{l atm K}^{-1}\ \text{gmol}^{-1}\ (20\ \text{l})}\left[(1.636\times10^{-4}) - (1.456\times10^{-4})\right]\ \text{l atm K}^{-1}\ \text{s}^{-1}$$

$$N_A = 1.1\times10^{-5}\ \text{gmol l}^{-1}\ \text{s}^{-1}$$

Because, at steady state, the rate of oxygen transfer is equal to the rate of oxygen uptake by the cells, the volumetric rate of oxygen uptake by the culture is 1.1×10^{-5} gmol l^{-1} s^{-1}.

(b) Assume that the gas phase is well mixed so that the oxygen concentration in the bubbles contacting the liquid is the same as in the outlet gas, that is, 20.1%. As the difference in oxygen concentration between the inlet and outlet gas streams is small, we can also consider the composition of the gas phase to be constant throughout the fermenter. From Table 10.2, the solubility of oxygen in water at 30°C and 1 atm air pressure is 8.05×10^{-3} kg m^{-3} = 8.05×10^{-3} g l^{-1}. Using Eq. (10.46) to determine the solubility at the fermenter operating pressure of 1.48 atm and gas-phase oxygen mole fraction of 0.201:

$$C^*_{AL2} = \frac{p_{T2}\, y_{AG2}}{p_{T1}\, y_{AG1}} C^*_{AL1} = \frac{(1.48\ \text{atm})\ 0.201}{(1\ \text{atm})\ 0.2099}\ 8.05\times10^{-3}\ \text{g l}^{-1} = 0.0114\ \text{g l}^{-1}$$

C_{AL} in the fermenter is 82% of the oxygen solubility at 30°C and 1.48 atm air pressure. From Eq. (10.45), solubility is proportional to total pressure; therefore:

$$C_{AL} = 0.82\ \frac{1.48\ \text{atm}}{1\ \text{atm}}\ 8.05\times10^{-3}\ \text{g l}^{-1} = 9.77\times10^{-3}\ \text{g l}^{-1}$$

Applying these results in Eq. (10.39):

$$k_L a = \frac{1.1\times10^{-5}\ \text{gmol l}^{-1}\ \text{s}^{-1} \cdot \left|\dfrac{32\ \text{g}}{1\ \text{gmol}}\right|}{0.0114\ \text{g l}^{-1} - 9.77\times10^{-3}\ \text{g l}^{-1}}$$

$$k_L a = 0.22\ \text{s}^{-1}$$

The value of $k_L a$ is 0.22 s^{-1}.

Even when high-sensitivity gas measuring equipment is used, the oxygen balance method is not readily applicable to cultures with low cell growth and oxygen uptake rates, for example, plant and animal cell cultures and aerobic waste treatment systems. When oxygen uptake is slow, the difference in oxygen content between the inlet and outlet gas streams can become diminishingly small, resulting in unacceptable levels of error. In these circumstances, other methods for measuring $k_L a$ must be considered.

10.10.2 Dynamic Method

In this method for estimating $k_L a$, changes in dissolved oxygen tension are measured using an oxygen electrode after a step change in aeration conditions in the fermenter. The results are interpreted using unsteady-state mass balance equations to obtain the value of

k_La. The main advantage of the dynamic method over the steady-state technique is the comparatively low cost of the analytical equipment needed. The measurement is also independent of the oxygen solubility and can be carried out even if C^*_{AL} is unknown. In practice, the dynamic method is best suited for measuring k_La in relatively small vessels. It is used commonly with laboratory-scale fermenters.

Although the dynamic method is simple and easy to perform experimentally, it can give very inaccurate results unless several aspects of the measurement system are examined and characterised. These include the response time of the dissolved oxygen electrode, the effect of liquid boundary layers at the probe surface, and gas-phase dynamics in the vessel. These features of the measurement system should be assessed in conjunction with the dynamic method.

Simple Dynamic Method

The simplest version of the dynamic method will be described here. This method gives reasonable results for k_La only if the following assumptions are valid:

1. The liquid phase is well mixed.
2. The response time of the dissolved oxygen electrode is much smaller than $1/k_La$ (see Section 10.10.2, Electrode Response Time and Liquid Boundary Layers subsection).
3. The measurement is performed at sufficiently high stirrer speed to eliminate liquid boundary layers at the surface of the oxygen probe (see Section 10.10.2, Electrode Response Time and Liquid Boundary Layers subsection).
4. Gas-phase dynamics can be ignored (see Section 10.10.2, Gas-Phase Dynamics subsection).

Assumption (1) is required for C_{AL} measurements at a fixed position in the reactor to represent the dissolved oxygen concentration throughout the vessel, so that results for k_La are not dependent on electrode location. This is relatively easy to achieve in small, vigorously agitated fermenters but may not occur in large vessels containing viscous broths. Assumptions (2) and (3) refer to the results of test experiments that should be carried out in conjunction with the simple dynamic method. These assumptions are more likely to hold using fast oxygen electrodes in vigorously agitated, low-viscosity fluids when the value of k_La is relatively low. Assumption (4) refers to the properties of the gas phase during the measurement period. Gas hold-up and gas mixing have a significant influence in most applications of the dynamic method; however, we neglect these effects here for simplicity. Gas-phase dynamics are discussed later in more detail (Section 10.10.2, Gas-Phase Dynamics subsection). Assumption (4) is likely to be valid only in small vessels containing low-viscosity fluids with low gas hold-up, low levels of gas recirculation, and relatively high gas throughput. If any of the above four assumptions do not hold, the simple dynamic method will not provide an accurate indication of k_La and alternative procedures should be considered.

To measure k_La using the simple dynamic method, the fermenter containing culture broth is stirred and sparged at fixed rates so that the dissolved oxygen concentration C_{AL} is constant. At time t_0, the broth is deoxygenated, either by sparging nitrogen into the vessel or, as indicated in Figure 10.14, by stopping the air flow and allowing the culture to

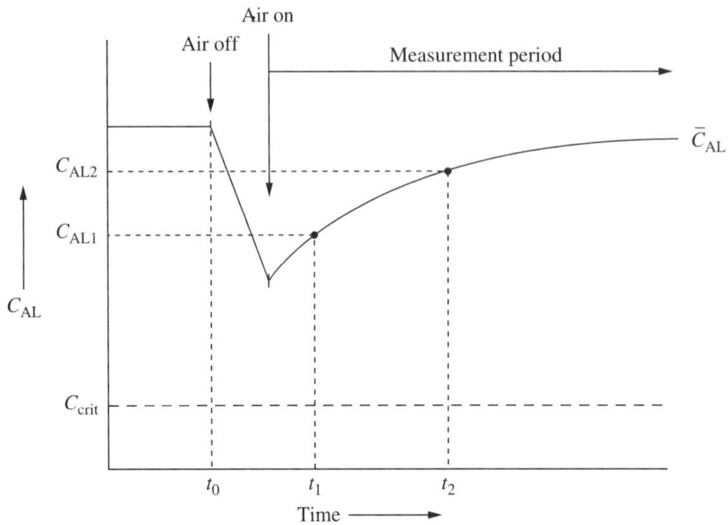

FIGURE 10.14 Variation of dissolved oxygen concentration for the dynamic measurement of k_La.

consume the available oxygen in solution. Air is then pumped into the broth at a constant flow rate and the increase in C_{AL} is measured using a dissolved oxygen probe as a function of time. It is important that the oxygen concentration remains above the critical level C_{crit} so that the rate of oxygen uptake by the cells remains independent of the dissolved oxygen tension. Assuming that reoxygenation of the broth is fast relative to cell growth, the dissolved oxygen level will soon reach a steady-state value $\bar{C}_{AL}$, which reflects a balance between oxygen supply and oxygen consumption in the system. C_{AL1} and C_{AL2} are two oxygen concentrations measured during reoxygenation at times t_1 and t_2, respectively. We can develop an equation for k_La in terms of these experimental data.

During the reoxygenation step, the system is not at steady state. The rate of change in dissolved oxygen concentration is equal to the rate of oxygen transfer from the gas to the liquid, minus the rate of oxygen consumption by the cells:

$$\frac{dC_{AL}}{dt} = k_La\,(C^*_{AL} - C_{AL}) - q_O\,x \tag{10.54}$$

where q_Ox is the rate of oxygen consumption. We can determine an expression for q_Ox by considering the final steady-state dissolved oxygen concentration, $\bar{C}_{AL}$. When $C_{AL} = \bar{C}_{AL}$, $dC_A/dt = 0$ because there is no change in C_{AL} with time. Therefore, from Eq. (10.54):

$$q_O\,x = k_La\,(C^*_{AL} - \bar{C}_{AL}) \tag{10.55}$$

Substituting this result into Eq. (10.54) and cancelling the $k_La\ C^*_{AL}$ terms gives:

$$\frac{dC_{AL}}{dt} = k_La\,(\bar{C}_{AL} - C_{AL}) \tag{10.56}$$

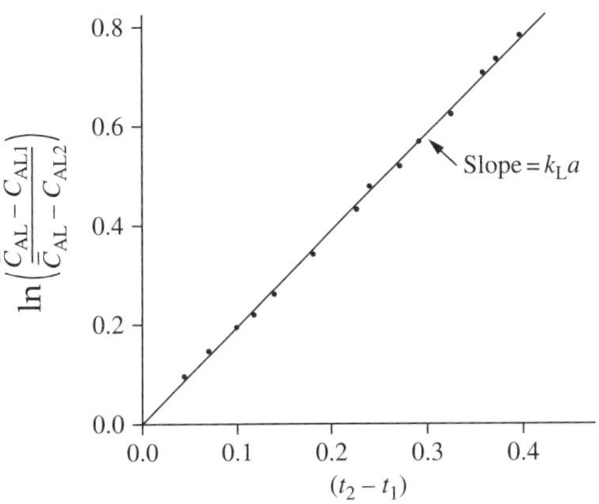

FIGURE 10.15 Evaluating k_La from data measured using the dynamic method.

Assuming k_La is constant with time, we can integrate Eq. (10.56) between t_1 and t_2 using the integration rules described in Appendix E. The resulting equation for k_La is:

$$k_La = \frac{\ln\left(\frac{\overline{C}_{AL} - C_{AL1}}{\overline{C}_{AL} - C_{AL2}}\right)}{t_2 - t_1} \tag{10.57}$$

Using Eq. (10.57), k_La can be estimated using two points from Figure 10.14 or, more accurately, from several values of (C_{AL1}, t_1) and (C_{AL2}, t_2). When $\ln\left(\frac{\overline{C}_{AL} - C_{AL1}}{\overline{C}_{AL} - C_{AL2}}\right)$ is plotted against $(t_2 - t_1)$ as shown in Figure 10.15, the slope is k_La. The value obtained for k_La reflects the operating stirrer speed, the air flow rate during the reoxygenation step, and the properties of the culture broth. Equation (10.57) can be applied to actively respiring cultures or to systems without oxygen uptake. In the latter case, $\overline{C}_{AL} = C^*_{AL}$ and nitrogen sparging is required for the deoxygenation step of the procedure.

Before we accept the value of k_La determined using this technique, the validity of assumptions (1) through (4) during the measurement must be checked. Assumption (1) can be verified by repeating the k_La measurement using different probe locations. The remaining three assumptions are addressed in the following sections.

Electrode Response Time and Liquid Boundary Layers

The dynamic method relies on the measurement of changes in dissolved oxygen tension after a step change in fermenter aeration conditions. Problems can arise with this approach if the oxygen electrode is slow to respond to the increase in liquid-phase oxygen levels as the culture broth is reoxygenated. As described in Section 10.7, because several mass transfer steps are involved in the operation of oxygen probes, it may be impossible to achieve a fast electrode response to changes in dissolved oxygen tension. If the electrode response is slower than the actual increase in oxygen concentration, the measured values of C_{AL} will reflect the response characteristics of the probe rather than the change in oxygen

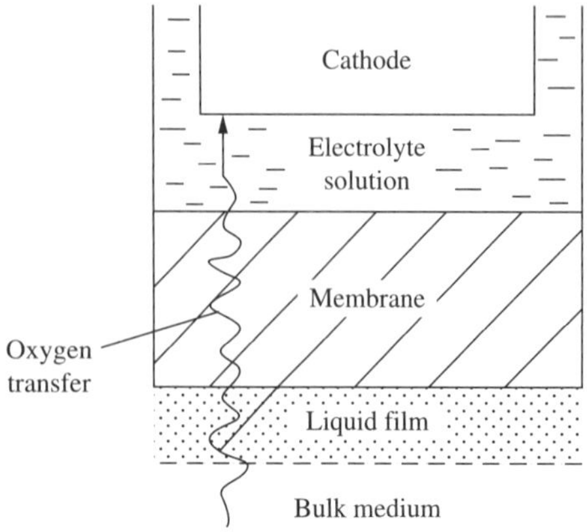

FIGURE 10.16 Development of a liquid film at the surface of an oxygen probe.

concentration in the fermenter. Therefore, the *electrode response time* should always be measured as part of the dynamic method for determining $k_L a$.

As well as the mass transfer resistances represented in Figure 10.13, development of a liquid boundary layer at the membrane–liquid interface, as illustrated in Figure 10.16, may also affect the response time of the electrode. Whether or not a boundary layer is present depends on the flow conditions, liquid properties, and rate of oxygen consumption at the probe cathode. Boundary layer development can be negligible for probes with small cathodes applied in well-mixed, low-viscosity liquids such as water. In some cases, for example, when using microprobes, boundary layers do not develop even in unagitated liquids. However, in most applications, particularly if the liquid is a viscous fermentation broth, liquid films at the probe surface are of concern. When a liquid film is present, the response of the probe is slower than in its absence and the dissolved oxygen readings will vary with stirrer speed. Both these outcomes are undesirable and reduce the accuracy of the dynamic method.

The electrode response time and liquid boundary layer effects can be measured in test experiments. These experiments should be carried out under conditions as close as possible to those applied for $k_L a$ measurement. The same gas flow rate and culture broth are used. The fermenter is prepared by sparging with air to give a constant dissolved oxygen tension, $\overline{C}_{AL}$. The oxygen electrode is equilibrated in a separate, vigorously agitated, nitrogen-sparged vessel providing a 0% oxygen environment. The probe is then transferred quickly from the nitrogen-sparged vessel to the fermenter: this procedure exposes the probe to a step change in dissolved oxygen tension from 0% to $\overline{C}_{AL}$. The response of the probe is recorded. The procedure is repeated using a range of stirrer speeds in the fermenter.

Typical results from the test experiments in low-viscosity fluid are shown in Figure 10.17. The electrode is transferred to the fermenter at time zero. After being at 0% oxygen, the probe takes some time to record a steady new signal corresponding to $\overline{C}_{AL}$. The response of the probe becomes faster as the stirrer speed N_i is increased, reflecting a progressive reduction in the thickness of the liquid boundary layer at the probe surface. At sufficiently high stirrer speed, no further change in electrode response is observed with additional increase in agitation rate, indicating that the boundary layer has been

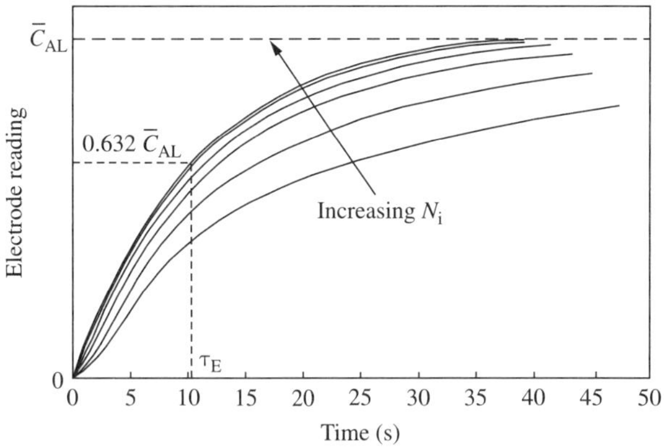

FIGURE 10.17 Typical electrode response curves after a step change in dissolved oxygen tension at different stirrer speeds, N_i. In this example, the electrode response time is just over 10 s.

eliminated. Under these conditions, the response curve represents the dynamics of the electrode itself, independent of external liquid velocity.

The response of dissolved oxygen electrodes is usually assumed to follow first-order kinetics. Accordingly, the *electrode response time* τ_E is defined as the time taken for the probe to indicate 63.2% of the total step change in dissolved oxygen level. As shown in Figure 10.17, the response time can be obtained from the response curves measured at high stirrer speeds in the absence of liquid films. For commercially available steam-sterilisable electrodes, response times are usually in the range of 10 to 100 s. However, faster nonautoclavable electrodes with response times of 2 to 3 s are also available, as are microelectrodes that respond even more rapidly. Some oxygen electrodes have responses that deviate substantially from first-order (e.g., the electrode may have a tailing response so that it slows excessively as the new dissolved oxygen level is approached), or there may be a significant difference between the response times for upward and downward step changes. Such electrodes are not suitable for dynamic k_La measurements.

In viscous fermentation broths, it may be impossible to eliminate liquid boundary layers in the test experiments, even at high stirrer speeds. This makes estimation of the electrode response time difficult using fermentation fluid. Instead, τ_E may be evaluated using water, with the assumption that the electrode response does not depend on the measurement fluid. Experiments in water at different stirrer speeds are still required to ensure that τ_E is determined in the absence of boundary layer effects. Although the value of τ_E may be obtained in water, because the development of liquid films depends strongly on the properties of the fluid, measurements in water do not provide information about the elimination of boundary layers in the actual fermentation broth.

The results from the test experiments are used to check the validity of two of the assumptions involved in the simple dynamic method. Assumption (2) (Section 10.10.2, Simple Dynamic Method subsection) is valid if the electrode response time is small compared with the rate of oxygen transfer; this is checked by comparing the value of τ_E with $1/k_La$. The error in k_La has been estimated to be <6% for $\tau_E \leq 1/k_La$ and <3% for $\tau_E \leq 0.2/k_La$, so that commercial electrodes with response times between 2 and 3 s can be used to measure k_La values up to about 0.1 s^{-1} [22]. For assumption (3) to be valid, only stirrer speeds above that eliminating liquid boundary layers at the probe surface in culture broth can be used to determine k_La. If either assumption (2) or (3) does not hold, the simple dynamic method cannot be used to measure k_La. Factors involved in assumption (4) are outlined in the following section.

Gas-Phase Dynamics

The term *gas-phase dynamics* refers to changes with time in the properties of a gas dispersion, including the number and size of the bubbles and the gas composition. Gas-phase dynamics can have a substantial influence on the results of k_La measurement. Because of the complexity and uncertainty associated with gas flow patterns, it is difficult to accurately account for these effects except in the simplest situations. Problems associated with gas-phase behaviour can make the dynamic method an impractical technique for determining k_La. As the average residence time of bubbles in the liquid increases, so does the influence of gas dynamics on measured k_La values.

In the dynamic method, a change in aeration conditions is used as the basis for evaluating k_La. However, although the inlet gas flow rate and composition may be altered quickly, this does not necessarily result in an immediate change in the gas hold-up and composition of the bubbles dispersed in the liquid. Depending on system variables such as the extent of gas recirculation, coalescence properties of the liquid, and fluid viscosity, some time is required for a new gas hold-up and gas-phase composition to be established. Therefore, because the driving force for oxygen transfer depends on the gas-phase oxygen concentration, and as k_La varies with the volume of gas hold-up through its dependence on the interfacial area a, the oxygen transfer conditions and k_La itself are likely to be changing during the measurement period. Until a new steady state is established within the dispersed gas phase, the measured values of C_{AL} represent not only the kinetics of oxygen transfer but also the gas-phase dynamics in the fermenter.

Let us consider the two methods commonly used to deoxygenate the culture broth for dynamic k_La measurement. Both these procedures affect the state of the gas dispersion at the start of the measurement period.

- *Nitrogen sparging.* In this version of the simple dynamic method, nitrogen is sparged into the broth at t_0 to achieve an initial reduction in dissolved oxygen tension. Depending on the duration of nitrogen sparging, at the end of the deoxygenation step we can assume at least that the gas hold-up contains more nitrogen and less oxygen than in air. In the presence of an active cell culture, the bubbles will also contain carbon dioxide and, depending on the humidity of the inlet gas, water vapour. If, at the commencement of the measurement period, the gas supply is suddenly switched to air at the same flow rate, the gas hold-up volume, and therefore the gas–liquid interfacial area a, will remain roughly the same, but the composition of gas in the bubbles will start to change. As a result of gas mixing and dispersion at the impeller, bubble coalescence, and gas recirculation, the incoming air mixes with the preexisting nitrogen-rich hold-up until all the excess nitrogen from the deoxygenation step is flushed out of the system. Until this process is complete, the measured C_{AL} values will be influenced by the changing gas-phase composition.
- *De-gassing.* An alternative procedure for deoxygenation of active broth is to switch off the normal air supply to the fermenter at t_0 (Figure 10.14), thus allowing oxygen

consumption by the cells to reduce the dissolved oxygen tension. Depending on the extent of gas recirculation and the time required for bubbles to escape the liquid, the gas hold-up volume will be reduced during the deoxygenation step. When aeration is recommenced, the gas hold-up and gas–liquid interfacial area a must be reestablished before k_La becomes constant. Until this occurs, the measured C_{AL} values will be affected by changes in the gas hold-up.

Transient gas-phase conditions are created using both strategies commonly applied for deoxygenation during dynamic k_La measurement. Even if the fermenter is relatively small, gas-phase transitions may continue to occur for a substantial proportion of the measurement period, thus compromising the accuracy of measured k_La values. If nitrogen sparging is used, the high-nitrogen hold-up is more readily flushed out of the liquid if the hold-up volume is low, the gas flow rate is high, and the liquid has a low viscosity so that bubbles disengage quickly and are not trapped in the fluid. Bubble recirculation from the bulk liquid back to the impeller should also be minimal to allow nitrogen-rich bubbles to escape quickly from the vessel. As gas recirculation occurs to a greater extent in intensely agitated tanks, the effect of gas-phase dynamics after nitrogen sparging is more significant in systems operated with high stirrer power per unit volume. If the de-gassing procedure is used instead, a new steady-state gas hold-up is established more quickly if the hold-up volume is low.

As nitrogen sparging and switching off the air supply leave the gas phase in significantly different states after culture deoxygenation, a comparison of k_La values determined using both deoxygenation methods under otherwise identical conditions may show whether gas dynamics are important. If the k_La values are similar, this suggests that changes in gas composition and hold-up occur rapidly relative to the rate of gas–liquid oxygen transfer, giving us more confidence in the measurement technique. However, the limitations imposed by gas-phase dynamics are difficult to overcome and it is possible that the simple dynamic method will be unable to provide reliable k_La results in particular applications.

EXAMPLE 10.3 ESTIMATING k_La USING THE SIMPLE DYNAMIC METHOD

A stirred fermenter is used to culture haematopoietic cells isolated from umbilical cord blood. The liquid volume is 15 litres. The simple dynamic method is used to determine k_La. The air flow is shut off for a few minutes and the dissolved oxygen level drops; the air supply is then reconnected at a flow rate of 0.25 l s^{-1}. The following results are obtained at a stirrer speed of 50 rpm.

Time (s)	5	20
Oxygen tension (% air saturation)	50	66

When steady state is established, the dissolved oxygen tension is 78% air saturation. In separate test experiments, the electrode response to a step change in oxygen tension did not vary with stirrer speed above 40 rpm. The probe response time under these conditions was 2.8 s. When the

k_La measurement was repeated using nitrogen sparging to deoxygenate the culture, the results for oxygen tension as a function of time were similar to those listed. Estimate k_La.

Solution

$\bar{C}_{AL} = 78\%$ air saturation. Let us define $t_1 = 5$ s, $C_{AL1} = 50\%$, $t_2 = 20$ s, and $C_{AL2} = 66\%$. From Eq. (10.57):

$$k_La = \frac{\ln\left(\frac{78-50}{78-66}\right)}{(20-5)\,\mathrm{s}} = 0.056\,\mathrm{s}^{-1}$$

Before we can be confident about this value for k_La, we must consider the electrode response time, the presence of liquid films at the surface of the probe, and the influence of gas-phase dynamics. The results from the test experiments indicate that there are no liquid film effects at 50 rpm. For $\tau_E = 2.8$ s and $1/k_La = 17.9$ s, $\tau_E = 0.16/k_La$. From Section 10.10.2 (Electrode Response Time and Liquid Boundary Layers subsection), as $\tau_E < 0.2/k_La$, τ_E is small enough that the error associated with the electrode response can be neglected. Because the measured results for oxygen tension were similar using two different deoxygenation methods, the effect of gas-phase dynamics can also be neglected. Therefore, k_La is 0.056 s^{-1}.

Modified Dynamic Methods

When the electrode response is slow, if liquid boundary layers at the probe surface cannot be eliminated, or if the effects of gas-phase dynamics are significant, the simple dynamic method (Section 10.10.2, Simple Dynamic Method subsection) is not suitable for estimating k_La. Sterilisable dissolved oxygen electrodes used in fermenters typically have relatively long response times, and measuring k_La under nonsterile conditions with a fast, nonautoclavable electrode is not always practical. It is also sometimes impossible to eliminate liquid boundary layers during k_La measurement, particularly in viscous fluids; moreover, stirrer speeds may be restricted in some cultures to avoid cell damage. In general, unless there is evidence to the contrary, gas-phase dynamics are expected to influence dynamic k_La measurements and often cannot be neglected. Large errors greater than 100% have been found to occur when the simple dynamic method is applied without correction for these effects.

Several modified procedures have been developed to account for the experimental factors that affect the results of the simple dynamic method. Equations describing the mass transfer processes responsible for electrode lag and liquid film resistance can be included in the models used to evaluate k_La [34]. Alternatively, after applying moment analysis of the response curves, the data for dissolved oxygen tension can be normalised based on empirical observation of any electrode, liquid film, or gas mixing effects [35]. Accounting for gas-phase dynamics and imperfections in gas mixing is the most challenging correction required. Gas-phase oxygen concentrations estimated as a function of time and location have been incorporated into the model equations [36]; the experimental methods can also be modified to alleviate the uncertainty about gas-phase properties [37].

A useful variation of the dynamic methods discussed so far is the *dynamic pressure method*. In this technique, a step change in aeration conditions is achieved, not by altering the composition of the inlet gas stream, but by imposing a step change in fermenter pressure. Temporarily closing and then constricting the gas outlet during sparging is an effective method for inducing the required pressure change, including in large-scale vessels. A relatively small pressure increase of about 20% is sufficient. The k_La value is evaluated from measurements of the dissolved oxygen tension as the system moves towards steady state at the new pressure. The advantage of this approach is that the oxygen concentration is changed in all the bubbles in the vessel simultaneously, thus avoiding the problems of transient gas-phase composition or loss of gas hold-up that occur with the simple dynamic method. However, the expansion or shrinkage of bubbles immediately after a sudden pressure change can cause difficulties if the bubble sizes are slow to restabilise; the oxygen probe must also be unaffected by pressure variation. Further details of this method are available [38, 39]. The dynamic pressure method applied in conjunction with equations to correct for electrode and liquid film effects is widely considered to be the most reliable approach for dynamic k_La measurement.

10.10.3 Sulphite Oxidation

This method is based on oxidation of sodium sulphite to sulphate in the presence of a catalyst such as Cu^{2+} or Co^{2+}. Although the sulphite method has been used extensively, the results appear to depend on operating conditions in an unknown way and usually give higher k_La values than other techniques. Accordingly, its application is discouraged [22]. Because salt solutions are used, the average bubble size is affected by changes in the coalescence properties of the liquid and the results obtained have limited applicability to real fermentation broths.

10.11 MEASUREMENT OF THE SPECIFIC OXYGEN UPTAKE RATE, q_O

Application of Eqs. (10.42) and (10.43) for analysis of the mass transfer performance of fermenters requires knowledge of the specific oxygen uptake rate, q_O. This parameter reflects the intrinsic requirement of an organism for oxygen to support growth and product synthesis. It can be measured using several experimental techniques.

The oxygen balance method for measuring k_La (Section 10.10.1) also allows us to evaluate q_O. At steady state, the volumetric rate of oxygen uptake by the cells is equal to the volumetric rate of gas–liquid oxygen transfer, N_A (Section 10.5.2). Once N_A is determined using Eq. (10.53), q_O is found by dividing N_A by the cell concentration x. For example, for N_A with units of gmol l^{-1} s^{-1} and x with units of g l^{-1}, q_O is obtained with units of gmol g^{-1} s^{-1}. As discussed in Section 10.10.1, the accuracy of this technique depends strongly on accurate measurement of the inlet and outlet gas composition, flow rate, pressure, and temperature, and on the validity of the assumptions used to derive the mass transfer equations.

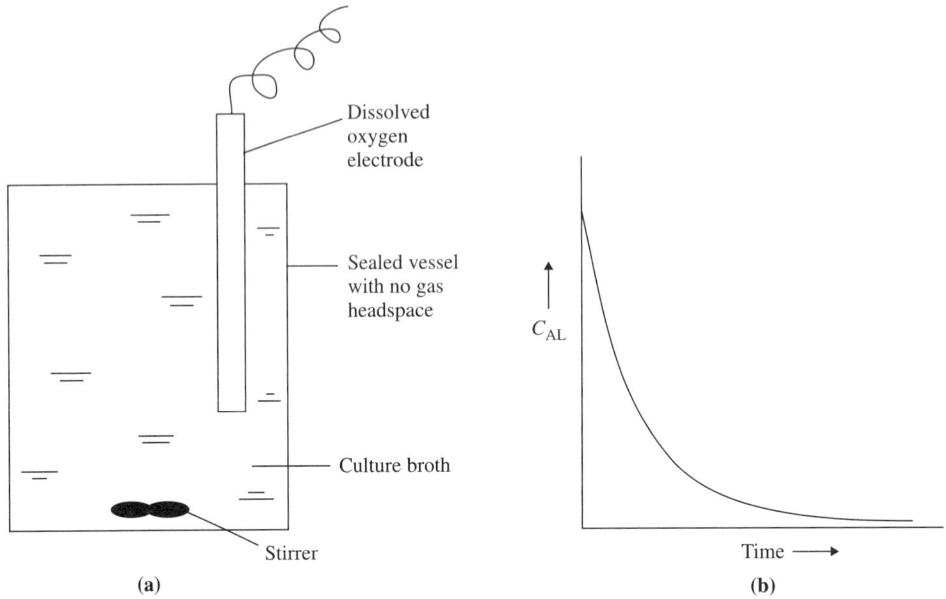

FIGURE 10.18 Dynamic method for measurement of q_O. (a) Sealed experimental chamber equipped with an oxygen electrode and stirrer and containing culture broth. (b) Measured data for dissolved oxygen concentration C_{AL} versus time.

Dynamic methods can also be used to measure q_O. A sample of culture broth containing a known cell concentration x is placed in a small chamber equipped with an oxygen electrode and stirrer as shown in Figure 10.18(a). The broth is sparged with air: at the commencement of the measurement period, the air flow is stopped and the vessel sealed to make it airtight. The decline in dissolved oxygen concentration C_{AL} due to oxygen uptake by the cells is recorded using the oxygen electrode as shown in Figure 10.18(b). The initial slope of the curve of C_{AL} versus time gives the volumetric rate of oxygen uptake by the cells, Q_O. Dividing Q_O by the cell concentration x gives the specific rate of oxygen uptake, q_O. All gas bubbles must be removed from the liquid before the measurements are started; the sealed chamber must also be airtight without any gas headspace so that additional oxygen cannot enter the liquid during data collection.

Factors similar to those outlined in Section 10.10.2 for dynamic measurement of k_La also affect the accuracy of q_O obtained using this technique. The electrode response must be relatively fast and liquid boundary layers at the probe surface must be eliminated by operating the stirrer at a sufficiently high speed. Fortunately, unlike for k_La measurement, the concentration of cells in the chamber can be adjusted to make it easier to comply with these requirements. If necessary, a relatively dilute cell suspension can be used to reduce the speed of oxygen uptake so that a relatively slow electrode response does not affect the results and any liquid boundary layers are more readily removed. The size of the vessel is also typically very small to minimise the effects of gas-phase dynamics and mixing characteristics.

10.12 PRACTICAL ASPECTS OF OXYGEN TRANSFER IN LARGE FERMENTERS

Special difficulties are associated with measuring oxygen transfer rates and k_La in large fermenters. These problems arise mainly because significant gradients of liquid- and gas-phase composition and other properties develop with increasing scale.

10.12.1 Liquid Mixing

In our discussion of oxygen transfer so far, we have assumed that the liquid phase is perfectly mixed and that k_La is constant throughout the entire reactor. This requires that turbulence and rates of turbulence kinetic energy dissipation are uniformly distributed. These conditions occur reasonably well in laboratory-scale stirred reactors, which are characterised by high turbulence throughout most of the vessel. In contrast, as discussed in Section 8.11, perfect mixing is difficult to achieve in commercial-scale reactors and, as illustrated in Figures 8.16 and 8.23, turbulence is far from uniformly distributed. Most of the oxygen transfer in industrial-scale fermenters takes place in the region near the impeller. In much of the rest of the vessel, the bubbles are in free rise and the liquid velocity is significantly reduced, especially if the viscosity is high. The composition of the liquid phase, which has a strong influence on bubble interfacial properties and k_La, can also vary significantly within large vessels.

Imperfect mixing in large-scale fermenters has a number of consequences for the quantification of oxygen transfer. First, different values of k_La and a range of oxygen uptake rates could be determined depending on where in the tank the dissolved oxygen concentration C_{AL} is measured. Accordingly, k_La values from a particular measurement location may not be representative of overall oxygen transfer conditions. Moreover, it may be possible to increase oxygen transfer significantly by improving bulk mixing conditions. Although relationships such as Eq. (10.50) predict that k_La will be the same as long as the power and gas flow rate remain constant, if different degrees of mixing are achieved for the same power input, a change in measured k_La values could be observed as conditions in the vessel become more homogeneous.

Equations (10.50) and (10.51) apply to well-mixed systems. For production-scale bioreactors, a better representation of oxygen transfer may be achieved using a two-compartment model of the vessel, with different correlations applied to the mixed zone close to the impeller and the bubble zone away from the impeller [40].

10.12.2 Gas Mixing

In analysis of oxygen transfer, it is often assumed that the gas phase is well mixed—that is, the gas composition is uniform and equal to that in the outlet gas stream. In large fermenters, these conditions may not be met due to substantial depletion of oxygen in the gas phase during passage of the bubbles from the bottom to the top of the tank. This creates an axial gradient of gas-phase concentration down the height of the vessel. Differences in composition between the inlet and outlet gas streams are greater in large tanks than at small scales. Modified mass transfer models that include the effects of plug flow or plug flow with axial dispersion in the gas phase have been used to better represent the gas mixing conditions in large-scale fermenters [36, 41].

10.12.3 Pressure Effects

Even when there is rapid mixing in large-scale fermenters, variations in gas-phase pressure occur due to static pressure changes down the height of the vessel. The pressure at the bottom of tall vessels is higher than at the top due to the weight of the liquid. The static pressure difference p_s is given by the equation:

$$p_s = \rho g H_L \tag{10.58}$$

where ρ is the liquid density, g is gravitational acceleration, and H_L is the liquid height. As the solubility of oxygen is sensitive to gas-phase pressure and oxygen partial pressure, significant variation in these conditions between the top and bottom of the vessel affects the value of C^*_{AL} used in mass transfer calculations. Allowance can be made for this in models of the mass transfer process [42]; alternatively, an average concentration-difference driving force $(C^*_{AL} - C_{AL})$ across the system can be determined. A suitable average is the *logarithmic-mean concentration difference*, $(C^*_{AL} - C_{AL})_{lm}$:

$$(C^*_{AL} - C_{AL})_{lm} = \frac{(C^*_{AL} - C_{AL})_o - (C^*_{AL} - C_{AL})_i}{\ln\left[\frac{(C^*_{AL} - C_{AL})_o}{(C^*_{AL} - C_{AL})_i}\right]} \tag{10.59}$$

In Eq. (10.59), subscripts i and o represent conditions at the inlet and outlet ends of the vessel, respectively.

10.12.4 Interaction between Oxygen Transfer and Heat Transfer

As discussed in Chapter 9, heat transfer is a critical function in bioreactors. In large-scale vessels, the fermentation broth must be cooled to remove the heat generated by metabolism and thus prevent the culture temperature rising to deleterious levels. Because the rate of metabolic heat generation in aerobic cultures is directly proportional to the rate of oxygen consumption by the cells (Section 5.9.2), oxygen transfer and heat transfer are closely related. Rapid oxygen uptake can create major heat removal problems. For highly aerobic cultures in large fermenters that deliver high oxygen transfer rates, heat transfer can become the limiting factor affecting the maximum feasible rate of reaction.

The heat transfer situation is usually most severe towards the end of the culture cycle when the volumetric rate of oxygen uptake, Q_O, is greatest. In some cases, it may be sensible to slow down the rate of oxygen consumption by the culture to avoid the necessity of installing expensive heat transfer equipment and processes. Therefore, if strategies such as increasing the stirrer speed, gas flow rate, pressure, and oxygen partial pressure (Section 10.6) are undertaken to improve $k_L a$ and the oxygen transfer driving force, the consequent extra heat burden must be borne in mind. Heat and oxygen transfer are linked and should be considered together, especially in large-scale operations.

10.13 ALTERNATIVE METHODS FOR OXYGENATION WITHOUT SPARGING

In small-scale bioreactors or when shear-sensitive organisms such as animal cells are being cultured, alternative methods for providing oxygen are sometimes used. As outlined in Section 8.16.1 (Cell Damage from Bursting Bubbles subsection), the large forces generated by bubbles bursting at the surface of sparged cell cultures can cause very high rates of animal cell damage. For that reason, aeration by other means may be required or preferred.

An alternative to gas sparging is *surface aeration*. Using this approach, gas containing oxygen is flushed through the headspace of the reactor above the liquid; oxygen is then transferred to the liquid through the upper surface of the culture broth. Surface aeration contributes to oxygen transfer even when the liquid is aerated by sparging; however, in vigorously agitated systems, its contribution is small compared with the high oxygenation rates achieved using entrained bubbles. The rate of oxygen transfer during surface aeration can be described using the equations in Section 10.4.3, with k_La in Eq. (10.39) representing the conditions in the liquid boundary layer at the liquid–headspace interface. The value of k_La and the rate of surface aeration increase with stirrer speed. The height of the impeller above the vessel floor may also be important because it affects the fluid velocity at the liquid surface.

Even though animal cell cultures have relatively low oxygen requirements as indicated earlier in Table 10.1, surface aeration can be inadequate for supporting growth to high cell densities. For suspension culture of baby hamster kidney cells, k_La values for surface aeration have been reported in the range 2.8×10^{-5}–$1.1 \times 10^{-3}\ s^{-1}$, whereas k_La for aeration of the same cultures by sparging is 2.8×10^{-4}–$6.9 \times 10^{-3}\ s^{-1}$ [43]. Because mild agitation and reduced air flow rates are used with animal cells in sparged systems to avoid hydrodynamic and bubble damage, these k_La values with sparging are significantly lower than the range of 0.02 to 0.25 s^{-1} typically found in production-scale fermenters used for microbial culture. Relative to the liquid volume, the area available for oxygen transfer by surface aeration declines significantly as the size of the fermenter increases. As a result, surface aeration is unsuitable for large-scale operations or must be supplemented by other techniques.

Membrane tubing aeration is another bubble-free option for oxygenation of cultures. Aeration is achieved by gas exchange through silicone or microporous polypropylene or Teflon tubing immersed in the culture broth. Gas flowing in the tube diffuses through the tube walls and into the medium under a concentration-difference driving force; air or oxygen-enriched air may be used in the tubing. For bubble-free aeration, the gas pressure inside the tubing must remain below the bubble point to avoid bubbles forming on the outside of the tube walls. In membrane aeration, the main resistances to oxygen transfer are the tube wall itself and the liquid film surrounding the outside of the tubing. To prevent the cells from settling and to promote mixing, the tubing assembly may be kept in motion as an effective stirrer; this also enhances oxygen transfer through the liquid boundary layer at the tube surface. Membrane aeration is used mainly for small-scale animal cell culture.

10.14 OXYGEN TRANSFER IN SHAKE FLASKS

Shake flasks, *conical flasks*, or *Erlenmeyer flasks* are employed commonly in laboratories for microbial and plant cell culture. Their application for animal cell culture is less frequent. Shake flasks with capacity from 25 ml to 5 litres are used for a wide range of experimental purposes, including screening and assessing cell lines, testing the effects of culture conditions, and bioprocess development.

A typical shake flask is shown in Figure 10.19. Shake flasks have flat bottoms and sloping sides and can be made of glass, plastic, or metal. The flask opening is of variable width and is fitted with a porous cap or plug closure. Gases are exchanged through the closure without the introduction of contaminating organisms. Flasks containing culture broth are placed on shaking tables or in incubator–shakers; the shaking movement is responsible for mixing and mass transfer in the flask. Shaking can be performed using either an *orbital* motion, which is also called *rotary* or *gyratory shaking*, or a linear *reciprocating* (back-and-forth) motion. Orbital shakers are becoming more common than reciprocating machines.

10.14.1 Oxygen Transfer through the Flask Closure

The first mass transfer resistance encountered in the delivery of oxygen to shake-flask cultures is the flask closure. A variety of porous materials is used to stopper shake flasks, including cotton, cotton wool bound with cheesecloth, and silicone sponge. Aluminium foil may also be used to cover the stopper and upper neck of the flask. The rate of oxygen transfer through the closure depends on the diffusion coefficient for oxygen in the material, the width of the neck opening, and the stopper depth. The diffusion coefficient varies with the porosity or bulk density of the stopper material. Wetting the stopper increases significantly the resistance to gas transfer as well as the risk of culture contamination, and can be a problem when flasks are shaken vigorously.

The flask closure affects the composition of gas in the headspace of the flask. In addition to impeding oxygen transfer into the culture, the closure prevents rapid escape of carbon dioxide and other gases generated by the cells. Volatile substances such as ethanol may also build up in the gas phase.

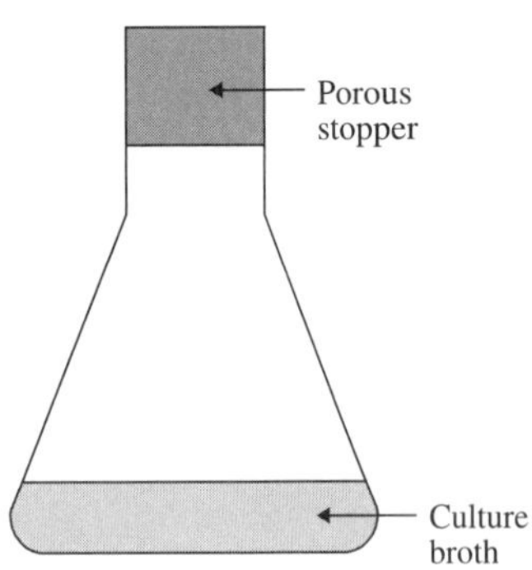

FIGURE 10.19 Typical shake flask for cell culture.

Although several types of flask stopper have been investigated for their oxygen transfer characteristics, it is unclear which type of closure gives the best results. In some cases, for example with large flasks or if the liquid is shaken very vigorously, the flask closure becomes the dominant resistance for oxygen transfer to the cells.

10.14.2 Oxygen Transfer within the Flask

The culture broth in shake flasks is supplied with oxygen by surface aeration from the gas atmosphere in the flask. Within the flask, the main resistance to oxygen transfer is the liquid-phase boundary layer at the gas–liquid interface. The rate of oxygen transfer from the gas phase is represented by Eq. (10.39) and depends on the area available for oxygen transfer, the liquid velocity, the viscosity, diffusion coefficient, and surface mobility at the phase boundary, and the concentration-difference driving force ($C^*_{AL} - C_{AL}$). Although oxygen transfer in shake flasks does not depend on the coalescing properties of the liquid because of the absence of bubbles, electrolytes and other components in the broth may affect the oxygen solubility and surface properties at the interface.

The area available for oxygen transfer in shake flasks is not simply equal to the surface area of the resting fluid as represented in Figure 10.19. When a flask is shaken on an orbital shaker, the liquid is distributed within the flask as shown in Figure 10.20: liquid is thrown up onto the walls due to the centrifugal forces associated with flask rotation. As the liquid swirls around the flask, a thin film is deposited on the flask wall and is replaced with each rotation. Because this film is available for oxygen transfer into the liquid, the mass transfer area a at any given time includes the surface area of liquid film on the flask wall. The overall rate of oxygen transfer depends on the rate of generation of fresh liquid surface, or the frequency with which the liquid film is replenished.

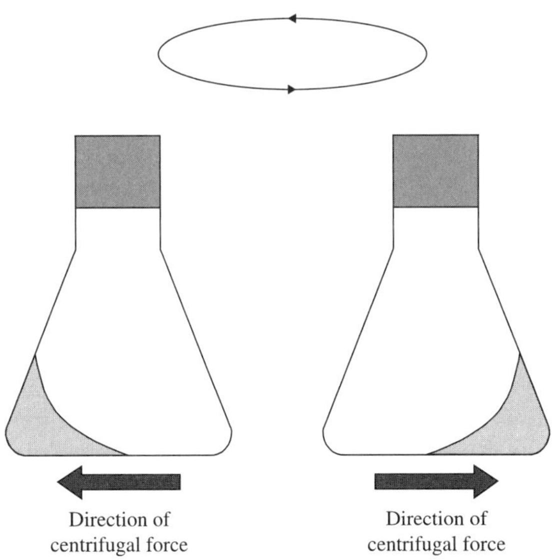

FIGURE 10.20 Distribution and movement of liquid in a shake flask on an orbital shaker.

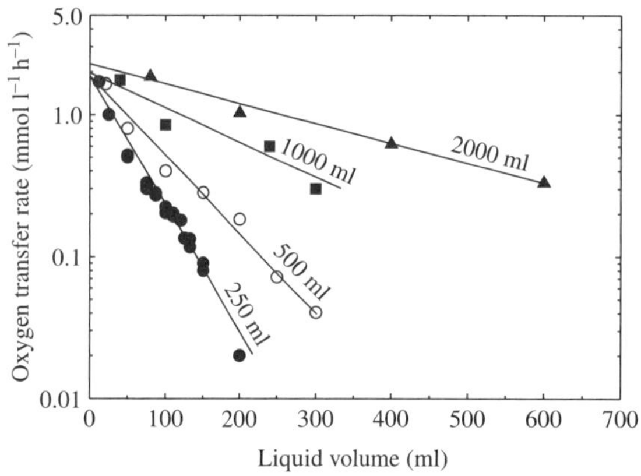

FIGURE 10.21 Effect of flask size and liquid volume on the rate of oxygen transfer. Shake flasks of nominal size—(●) 250 ml, (○) 500 ml, (■) 1000 ml, and (▲) 2000 ml—were tested on a reciprocating shaker with amplitude 3 in. operated at 96 rpm. The data are plotted using semi-logarithmic coordinates.
Data from M.A. Auro, H.M. Hodge, and N.G. Roth, 1957, Oxygen absorption rates in shaken flasks. Ind. Eng. Chem. *49, 1237–1238.*

Oxygen transfer in shake flasks is sensitive to operating conditions that affect the development of liquid films at the flask wall. Factors influencing the value of *a* include:

- Flask shape
- Flask size
- Surface properties of the flask walls (e.g., hydrophilic or hydrophobic)
- Shaking speed
- Flask displacement during shaking
- Liquid volume
- Liquid properties (e.g., viscosity)

The area of liquid film per unit volume of fluid is enhanced using large flasks with small liquid volumes on shakers with large displacement distance per rotation operated at high speed. As indicated in Figure 10.21, the rate of oxygen transfer is strongly dependent on flask size and liquid volume. The importance of limiting the liquid volume in shake flasks is evident in Figure 10.21; for example, using >100 ml of medium in a 250-ml flask reduces the mass transfer coefficient substantially compared with 50 ml. It is also important that the flask material support the development of a liquid film on the walls; hydrophobic materials such as plastic are therefore not recommended when oxygen transfer is critical. The difference between oxygen transfer rates in hydrophilic and hydrophobic flasks is illustrated in Figure 10.22. When the culture's oxygen requirements are high, gas–liquid mass transfer can be enhanced in shake flasks by the use of baffles, creases, or dimples in the flask walls. These indentations break up the swirling motion of the liquid, increase the level of liquid splashing onto the flask walls, and thus improve aeration. A disadvantage of using baffles is the increased risk of wetting the flask closure.

Because surface-to-volume ratios decrease with increasing liquid volume, shake-flask culture and surface aeration are practical only at relatively small scales. For some cultures, the maximum rate of oxygen transfer in shake flasks is not sufficient to meet the cellular oxygen demand. In these cases, shake flasks offer insufficient oxygen supply and the culture performance is limited by oxygen transfer.

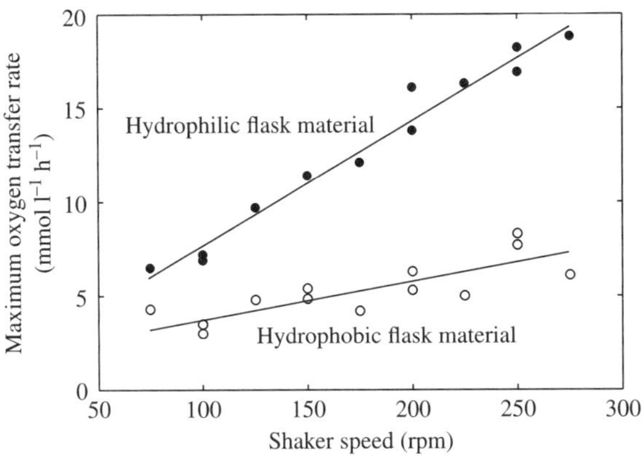

FIGURE 10.22 Effect of flask material properties on the rate of oxygen transfer in 250-ml shake flasks on an orbital shaker.
Data from U. Maier and J. Büchs, 2001, Characterisation of the gas–liquid mass transfer in shaking bioreactors. Biochem. Eng. J. *7, 99–106.*

SUMMARY OF CHAPTER 10

At the end of Chapter 10 you should:

- Be able to describe the *two-film theory* of mass transfer between phases
- Know *Fick's law* in terms of the *binary diffusion coefficient*, $\mathscr{D}_{AB}$
- Be able to describe in simple terms the mathematical analogy between mass, heat, and momentum transfer
- Know the equation for the rate of gas–liquid oxygen transfer in terms of the *mass transfer coefficient* k_La and the *concentration-difference driving force*
- Understand the importance of the *critical oxygen concentration*
- Be able to identify the steps that are most likely to present major resistances to oxygen transfer from bubbles to cells
- Understand how oxygen transfer and k_La can limit the biomass density in fermenters
- Understand the mechanisms of *gas dispersion* and *coalescence* in stirred fermenters and the importance of bubble size in determining *gas hold-up* and k_La
- Know how k_La depends on bioreactor operating conditions such as the stirrer speed, power input, gas flow rate, and liquid properties such as viscosity
- Know how temperature, total pressure, oxygen partial pressure, and the presence of dissolved and suspended material affect the rate of oxygen transfer and the solubility of oxygen in fermentation broths
- Be able to apply the *oxygen balance method* and the *simple dynamic method* for experimental determination of k_La, with understanding of their advantages and limitations
- Know how the specific oxygen uptake rate q_O can be measured in cell cultures
- Understand the particular problems affecting assessment of oxygen transfer in large fermenters
- Be familiar with techniques for culture aeration that do not involve gas sparging
- Understand the mechanisms of oxygen transfer in shake flasks

PROBLEMS

10.1 Rate-controlling processes in fermentation

Serratia marcescens bacteria are used for the production of threonine. The maximum specific oxygen uptake rate of *S. marcescens* in batch culture is 5 mmol O_2 g^{-1} h^{-1}. It is planned to operate the fermenter to achieve a maximum cell density of 40 g l^{-1}. At the fermentation temperature and pressure, the solubility of oxygen in the culture liquid is 8×10^{-3} kg m^{-3}. At a particular stirrer speed, $k_L a$ is 0.15 s^{-1}. Under these conditions, will the rate of cell metabolism be limited by mass transfer or depend solely on metabolic kinetics?

10.2 Test for oxygen limitation

An 8-m^3 stirred fermenter is used to culture *Agrobacterium* sp. ATCC 31750 for production of curdlan. The liquid medium contains 80 g l^{-1} sucrose. Under optimal conditions, 1.0 g dry weight of cells is produced for every 4.2 g of sucrose consumed. The fermenter is sparged with air at 1.5 atm pressure, and the specific oxygen demand is 7.5 mmol per g dry weight per h. To achieve the maximum yield of curdlan, the fermentation temperature is held constant at 32°C. The solubility of oxygen in the fermentation broth is 15% lower than in water due to solute effects. If the maximum $k_L a$ that can be achieved is 0.10 s^{-1}, does the fermenter's mass transfer capacity support complete consumption of substrate?

10.3 $k_L a$ required to maintain critical oxygen concentration

A genetically engineered strain of yeast is cultured in a bioreactor at 30°C for production of heterologous protein. The oxygen requirement is 80 mmol l^{-1} h^{-1}; the critical oxygen concentration is 0.004 mM. The solubility of oxygen in the fermentation broth is estimated to be 10% lower than in water due to solute effects.

(a) What is the minimum mass transfer coefficient necessary to sustain this culture with dissolved oxygen levels above critical if the reactor is sparged with air at approximately 1 atm pressure?

(b) What mass transfer coefficient is required if pure oxygen is used instead of air?

10.4 Oxygen transfer with different impellers

A 10-m^3 stirred fermenter with liquid height 2.3 m is used to culture *Trichoderma reesei* for production of cellulase. The density of the culture fluid is 1000 kg m^{-3}. An equation for the oxygen transfer coefficient as a function of operating variables has been developed for *T. reesei* broth:

$$k_L a = 2.5 \times 10^{-3} \left(\frac{P_T}{V_L}\right)^{0.7} u_G^{0.3}$$

where $k_L a$ has units of s^{-1}, P_T is the total power input in W, V_L is the liquid volume in m^3, and u_G is the superficial gas velocity in m s^{-1}. The fermenter is sparged using a gas flow rate of 0.6 vvm (vvm means volume of gas per volume of liquid per minute). The vessel is stirred with a single impeller but two alternative impeller designs, a Rushton turbine and a curved-blade disc turbine, are available. Both impellers are sized and operated so that their ungassed power draw is 9 kW.

(a) If the power loss with gassing is 50% for the Rushton and 5% for the curved-blade turbine, compare the $k_L a$ values achieved using each impeller.

(b) What is the percentage contribution to P_T from gassing with the two different impellers?

(c) If the cell concentration is limited to 15 g l^{-1} using the Rushton turbine because of mass transfer effects, estimate the maximum possible cell concentration with the curved-blade disc turbine.

It is decided to install the Rushton turbine, but to compensate for the effect on k_La of its loss of power with gassing by increasing the gas flow rate.

(d) Estimate the gas flow rate required to obtain the same maximum cell concentration using the Rushton turbine as that achieved with the curved-blade disc turbine. Express your answer in vvm. What are your assumptions? (Iterative solution may be required.)

10.5 Foam control and oxygen transfer

Foaming is controlled routinely in fermenters using a foam sensor and pump for automatic addition of antifoam agent. As shown in Figure 10P5.1, the foam sensor is located at the top of the vessel above the liquid surface. When a head of foam builds up so that foam contacts the lower tip of the sensor, an electrical signal is sent to the pump to add antifoam. The antifoam agent destroys the foam, the foam height is therefore reduced, contact with the foam sensor is broken, and the pump supplying the antifoam agent is switched off. Further build-up of foam reactivates the control process.

If the position of the foam sensor is fixed, when the gap between the liquid surface and sensor is reduced by raising the liquid height, a smaller foam build-up is tolerated before antifoam agent is added. Therefore, antifoam addition will be triggered more often if the working volume of the vessel is increased. Although a greater fermenter working volume means that more cells and/or product are formed, addition of excessive antifoam

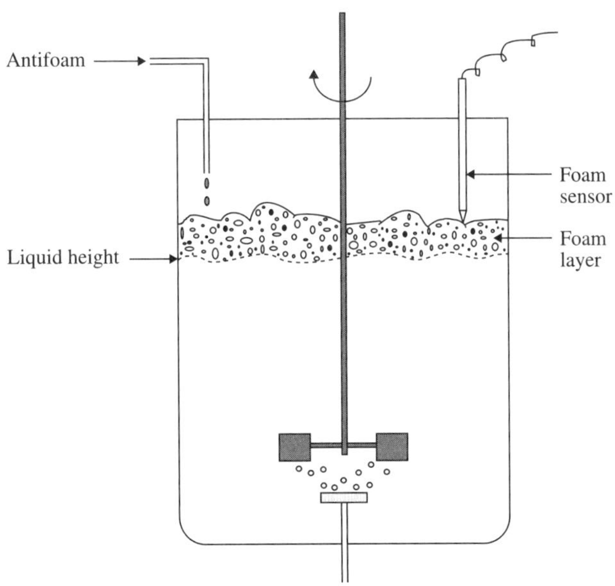

FIGURE 10P5.1 Stirred fermenter with automatic foam control system.

agent could reduce k_La significantly, thereby increasing the likelihood of mass transfer limitations.

A stirred fermenter of diameter 1.5 m is used to culture *Bacillus licheniformis* for production of serine alkaline protease. The fermenter is operated five times with automatic antifoam addition using five different liquid heights. The position of the foam sensor is the same in each fermentation. The volume of antifoam added and the k_La at the end of the culture period are recorded.

Liquid height (m)	Antifoam added (l)	k_La (s^{-1})
1.10	0.16	0.016
1.29	0.28	0.013
1.37	1.2	0.012
1.52	1.8	0.012
1.64	2.4	0.0094

Under ideal conditions, the maximum specific oxygen uptake rate for *B. licheniformis* is 2.6 mmol g^{-1} h^{-1}. When glucose is used as the carbon source at an initial concentration of 20 g l^{-1}, a maximum of 0.32 g of cells are produced for each g of glucose consumed, and 0.055 g of protease is produced per g of biomass formed. The solubility of oxygen in the broth is estimated as 7.8 g m^{-3}.

(a) Using the k_La values associated with each level of antifoam addition, estimate the maximum cell concentrations supported by oxygen transfer as a function of liquid height. Assume that antifoam exerts a much stronger influence on k_La than on other properties of the system such as oxygen solubility and specific oxygen uptake rate.

(b) Calculate the maximum mass of cells and maximum mass of protease that can be produced based on the oxygen transfer capacity of the fermenter as a function of liquid height.

(c) Is protease production limited by oxygen transfer at any of the liquid heights tested?

(d) What operating liquid height would you recommend for this fermentation process? Explain your answer.

10.6 Improving the rate of oxygen transfer

Rifamycin is produced in a 17-m^3 stirred fermenter using a mycelial culture, *Nocardia mediterranei*. The fermenter is sparged with air under slight pressure so the solubility of oxygen in the broth is 10.7 g m^{-3}. Data obtained during operation of the fermenter are shown in Figure 10P6.1. After about 147 hours of culture, vegetable oil is added to the broth to disperse a thick build-up of foam. This has a severe effect on the oxygen transfer coefficient and reduces the dissolved oxygen tension.

(a) Calculate the steady-state oxygen transfer rate before and after addition of the vegetable oil.

(b) The relationship between k_La and the fermenter operating conditions is:

$$k_La \propto \left(\frac{P_T}{V_L}\right)^{0.5} u_G{}^{0.3}$$

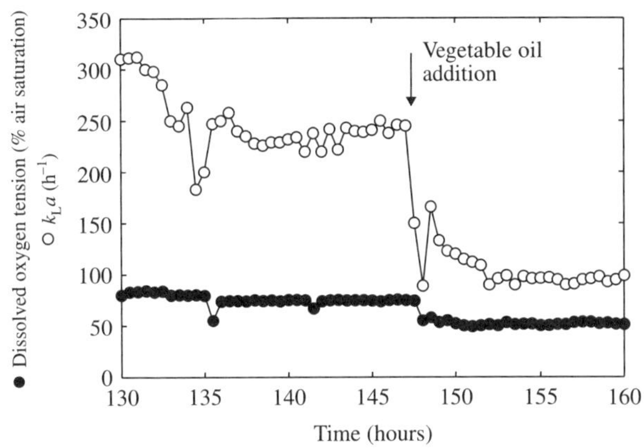

FIGURE 10P6.1 Online time-course data from a stirred fermenter used for rifamycin production.

Because increasing the gas flow rate would aggravate the problems with foaming, it is decided to restore the value of k_La by increasing the power input by stirring. If the power contribution from gas sparging is negligible, by how much does the stirrer power need to be increased to overcome the effects of the vegetable oil on k_La?

(c) To save the cost of increasing the power input, instead of (b), it is decided to improve oxygen transfer by sparging the fermenter with oxygen-enriched air. The total gas flow rate and pressure are unchanged. To restore the rate of oxygen transfer after vegetable oil addition to that before oil was added, what volume percentage of oxygen is required in the sparge gas if the desired dissolved oxygen concentration in the broth is 6.2×10^{-3} kg m^{-3}?

10.7 Oxygen transfer for different cell types

The specific oxygen demands and critical oxygen concentrations for typical microbial, plant, and animal cell cultures are listed below.

Cell culture	q_O	C_{crit} **(mmol l^{-1})**
Escherichia coli	8.5 mmol (g dry weight)$^{-1}$ h^{-1}	0.0082
Vitis vinifera (grape)	0.60 mmol (g dry weight)$^{-1}$ h^{-1}	0.055
Chinese hamster ovary (CHO)	3.0×10^{-10} mmol cell^{-1} h^{-1}	0.020

(a) Estimate the k_La required to achieve cell concentrations of 25 g dry weight l^{-1} for *E. coli* and *V. vinifera* and 3.0×10^9 cells l^{-1} for CHO cells, while maintaining the dissolved oxygen concentration above critical. The oxygen solubility in the media used for the cultures is 7.2×10^{-3} kg m^{-3}.

(b) The relationship between k_La and the power input to a 1-m^3 stirred bioreactor is:

$$k_La \propto \left(\frac{P_T}{V_L}\right)^{0.5}$$

Compare the bioreactor power requirements for culture of the three different cell types under the conditions described in (a).

10.8 Single-point k_La determination using the oxygen balance method

A 200-litre stirred fermenter contains a batch culture of *Bacillus subtilis* bacteria at 28°C. Air at 20°C is pumped into the vessel at a rate of 1 vvm (vvm means volume of gas per volume of liquid per minute). The average pressure in the fermenter is 1 atm. The volumetric flow rate of off-gas from the fermenter is measured as $189 \ l \ min^{-1}$. The exit gas stream is analysed for oxygen and is found to contain 20.1% O_2. The dissolved oxygen concentration in the broth is measured using an oxygen electrode as 52% air saturation. The solubility of oxygen in the fermentation broth at 28°C and 1 atm air pressure is $7.8 \times 10^{-3} \ kg \ m^{-3}$.

(a) Calculate the oxygen transfer rate.

(b) Determine the value of k_La for the system.

(c) The oxygen analyser used to measure the exit gas composition was incorrectly calibrated. If the oxygen content has been overestimated by 10%, what error is associated with the result for k_La?

10.9 Steady-state k_La measurement

Escherichia coli bacteria are cultured at 35°C and 1 atm pressure in a 500-litre fermenter using the following medium:

Component	Concentration ($g \ l^{-1}$)
glucose	20
sucrose	8.5
$CaCO_3$	1.3
$(NH_4)_2SO_4$	1.3
Na_2HPO_4	0.09
KH_2PO_4	0.12

Air at 25°C and 1 atm is sparged into the vessel at a rate of $0.4 \ m^3 \ min^{-1}$. The dissolved oxygen tension measured using a polarographic electrode calibrated in situ in sterile culture medium is 45% air saturation. The gas flow rate leaving the fermenter is measured using a rotary gas meter as $6.3 \ l \ s^{-1}$. The oxygen concentration in the off-gas is 19.7%.

(a) Estimate the solubility of oxygen in the fermentation broth. What are your assumptions?

(b) What is the oxygen transfer rate?

(c) Determine the value of k_La.

(d) Estimate the maximum cell concentration that can be supported by oxygen transfer in this fermenter if the specific oxygen demand of the *E. coli* strain is $5.4 \ mmol \ g^{-1} \ h^{-1}$.

(e) If the biomass yield from the combined sugar substrates is $0.5 \ g \ g^{-1}$, is growth in the culture limited by oxygen transfer or substrate availability?

10.10 Oxygen transfer in a pressure vessel

A fermenter of diameter 3.6 m and liquid height 6.1 m is used for production of ustilagic acid by *Ustilago zeae*. The pressure at the top of the fermenter is 1.4 atma. The vessel is stirred using dual Rushton turbines and the fermentation temperature is 29°C. The dissolved oxygen tension is measured using two electrodes: one electrode is located near the top of the tank, the other is located near the bottom. Both electrodes are calibrated in situ in sterile culture medium. The dissolved oxygen reading at the top of the fermenter is 50% air saturation; the reading at the bottom is 65% air saturation. The fermenter is sparged with air at 20°C at a flow rate of 30 m^3 min^{-1} measured at atmospheric pressure. Off-gas leaving the vessel at a rate of 20.5 m^3 min^{-1} contains 17.2% oxygen. The solubility of oxygen in the fermentation broth is not significantly different from that in water. The density of the culture broth is 1000 kg m^{-3}.

(a) What is the oxygen transfer rate?

(b) Estimate the pressure at the bottom of the tank.

(c) The gas phase in large fermenters is often assumed to exhibit plug flow. Under these conditions, no gas mixing occurs so that the gas composition at the bottom of the tank is equal to that in the inlet gas stream, while the gas composition at the top of the tank is equal to that in the outlet gas stream. For the gas phase in plug flow, estimate the oxygen solubility at the top and bottom of the tank.

(d) What is the value of k_La?

(e) If the cell concentration is 16 g l^{-1}, what is the specific oxygen demand?

(f) Industrial fermentation vessels are rated for operation at elevated pressures so they can withstand steam sterilisation. Accordingly, the fermenter used for ustilagic acid production can be operated safely at a maximum pressure of 2.7 atma. Assuming that respiration by *U. zeae* and the value of k_La are relatively insensitive to pressure, what maximum cell concentration can be supported by oxygen transfer in the fermenter after the pressure is raised?

10.11 Dynamic k_La measurement

The simple dynamic method is used to measure k_La in a fermenter operated at 30°C and 1 atm pressure. Data for the dissolved oxygen concentration as a function of time during the reoxygenation step are as follows.

Time (s)	C_{AL} (% air saturation)
10	43.5
15	53.5
20	60.0
30	67.5
40	70.5
50	72.0
70	73.0
100	73.5
130	73.5

(a) Calculate the value of k_La.

(b) What additional experiments are required to check the reliability of this k_La result?

10.12 k_La measurement using the dynamic pressure method

The dynamic pressure method is applied for measurement of k_La in a 3000-l stirred fermenter containing a suspension culture of *Micrococcus glutamicus*. The stirrer is operated at 60 rpm and the gas flow rate is fixed at 800 l min^{-1}. The following dissolved oxygen concentrations are measured using a polarographic dissolved oxygen electrode after a step increase in fermenter pressure.

Time (s)	C_{AL} (% air saturation)
6	50.0
10	56.1
25	63.0
40	64.7

The steady-state dissolved oxygen tension at the end of the dynamic response is 66% air saturation.

(a) Estimate the value of k_La.

(b) An error is made determining the steady-state oxygen level, which is taken as 70% instead of 66% air saturation. What effect does this 6% error in $\overline{C}_{AL}$ have on the result for k_La?

(c) At the end of the k_La experiment, the electrode response time is measured by observing the output after a step change in dissolved oxygen tension from 0 to 100% air saturation. The stirrer speeds tested are 40, 50, and 60 rpm. Figure 10P12.1 at the bottom of page shows a chart recording of the results at 60 rpm; the results at 50 rpm are not significantly different. From this information, how much confidence do you have in the k_La measurements? Explain your answer.

10.13 Surface versus bubble aeration

Hematopoietic cells used in cancer treatment are cultured at 37°C in an 8.5-cm diameter bioreactor with working volume 500 ml. The culture fluid is mixed using a stirrer speed of 30 rpm. The reactor is operated at ambient pressure.

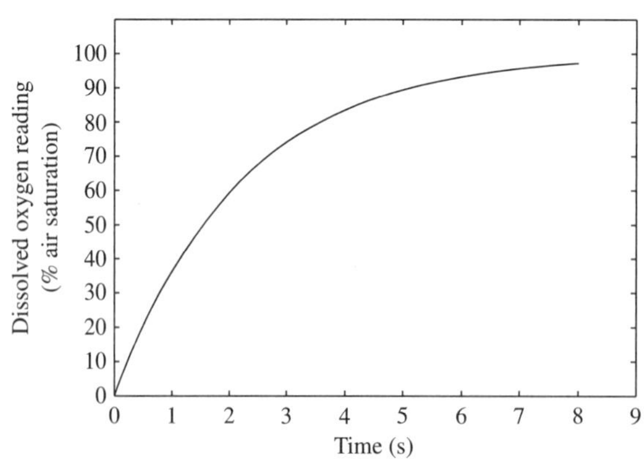

FIGURE 10P12.1 Chart recording of the electrode response at 60 rpm to a step change in dissolved oxygen tension.

(a) The dissolved oxygen tension is controlled at 50% air saturation using surface aeration only. A 50:20:30 mixture of air, oxygen, and nitrogen is passed at a fixed flow rate through the headspace. The specific oxygen uptake rate for hematopoietic cells is 7.7×10^{-12} g cell^{-1} h^{-1} and the cell concentration is 1.1×10^9 cells l^{-1}. Estimate the value of k_La for surface aeration.

(b) Instead of surface aeration, the bioreactor is sparged gently with air. When the dissolved oxygen tension is maintained at the critical level of 8% air saturation, the cell concentration is 3.9×10^9 cells l^{-1}. What is the k_La for bubble aeration?

(c) Surface aeration is preferred for this shear-sensitive culture, but the surface k_La needs improvement. For the gassing conditions applied in (a), estimate the vessel diameter required for surface aeration to achieve the same k_La obtained with sparging. What are your assumptions?

10.14 Shake-flask aeration

A mixed culture of heterotrophic microorganisms isolated from the Roman baths at Bath is prepared for bioleaching of manganese ore. One hundred ml of molasses medium is used in 300-ml flasks with 4-cm-long silicone sponge stoppers. The width of the flask opening is 3.2 cm. The cultures are incubated at 30°C on an orbital shaker operated at 80 rpm.

(a) With the flask closure removed, k_La for gas–liquid mass transfer is estimated using the dynamic method. During the reoxygenation step, the dissolved oxygen tension measured using a small, rapid-response electrode is 65% air saturation after 5 s and 75% after 30 s. The steady-state oxygen tension is 90% air saturation. What is the resistance to oxygen transfer in the flask? What are your assumptions?

(b) An expression for the mass transfer coefficient K_c for the flask closure is:

$$K_c = \frac{\mathscr{D}_e A_c}{L_c V_G}$$

where $\mathscr{D}_e$ is the effective diffusion coefficient of oxygen in the closure material, A_c is the cross-sectional area of the closure, L_c is the closure length, and V_G is the volume of gas in the flask. If $\mathscr{D}_e$ for silicone sponge is 20.8 cm^2 s^{-1}, what resistance to oxygen transfer is provided by the flask closure?

(c) What proportion of the total resistance to oxygen transfer does the flask closure represent?

(d) It is decided to improve the rate of gas–liquid oxygen transfer so that the resistance to oxygen transfer within the flask is approximately equal to that of the flask closure. A study of the dependence of k_La on shake-flask operating parameters yields the relationship:

$$k_La \propto N^{1.2} \left(\frac{V_F}{V_L}\right)^{0.85}$$

where N is the shaker speed in rpm, V_F is the flask size in ml, and V_L is the liquid volume in ml. If the shaker speed can be increased to a maximum of 150 rpm:

(i) What size flask is needed if the culture volume remains at 100 ml?

(ii) If 300-ml flasks are the only shake flasks available, what culture volume should be used?

References

[1] R.E. Treybal, Mass-Transfer Operations, third ed., McGraw-Hill, 1980.
[2] Perry's Chemical Engineers' Handbook, eighth ed., McGraw-Hill, 2008.
[3] T.K. Sherwood, R.L. Pigford, C.R. Wilke, Mass Transfer, McGraw-Hill, 1975.
[4] J.M. Coulson, J.F. Richardson, J.R. Backhurst, J.H. Harker, Coulson and Richardson's Chemical Engineering, vol. 1, sixth ed., Butterworth-Heinemann, 1999 (Chapters 10 and 12).
[5] M.J. Johnson, Metabolism of penicillin-producing molds, Ann. N.Y. Acad. Sci. 48 (1946) 57–66.
[6] C.E. Clifton, A comparison of the metabolic activities of *Aerobacter aerogenes, Eberthella typhi* and *Escherichia coli*, J. Bacteriol. 33 (1937) 145–162.
[7] G.C. Paul, M.A. Priede, C.R. Thomas, Relationship between morphology and citric acid production in submerged *Aspergillus niger* fermentations, Biochem. Eng. J. 3 (1999) 121–129.
[8] O. Rahn, G.L. Richardson, Oxygen demand and oxygen supply, J. Bacteriol. 41 (1941) 225–249.
[9] P. Gikas, A.G. Livingston, Use of specific ATP concentration and specific oxygen uptake rate to determine parameters of a structured model of biomass growth, Enzyme Microb. Technol. 22 (1998) 500–510.
[10] E.B. Chain, G. Gualandi, G. Morisi, Aeration studies. IV. Aeration conditions in 3000-liter submerged fermentations with various microorganisms, Biotechnol. Bioeng. 8 (1966) 595–619.
[11] A.H.E. Bijkerk, R.J. Hall, A mechanistic model of the aerobic growth of *Saccharomyces cerevisiae*, Biotechnol. Bioeng. 19 (1977) 267–296.
[12] A.L. Jensen, J.S. Schultz, P. Shu, Scale-up of antibiotic fermentations by control of oxygen utilization, Biotechnol. Bioeng. 8 (1966) 525–537.
[13] K. Ozergin-Ulgen, F. Mavituna, Oxygen transfer and uptake in *Streptomyces coelicolor* A3(2) culture in a batch bioreactor, J. Chem. Technol. Biotechnol. 73 (1998) 243–250.
[14] W.H. Bartholomew, E.O. Karow, M.R. Sfat, R.H. Wilhelm, Oxygen transfer and agitation in submerged fermentations, Ind. Eng. Chem. 42 (1950) 1801–1809.
[15] A. Pinches, L.J. Pallent, Rate and yield relationships in the production of xanthan gum by batch fermentations using complex and chemically defined growth media, Biotechnol. Bioeng. 28 (1986) 1484–1496.
[16] P.A. Bond, M.W. Fowler, A.H. Scragg, Growth of *Catharanthus roseus* cell suspensions in bioreactors: on-line analysis of oxygen and carbon dioxide levels in inlet and outlet gas streams, Biotechnol. Lett. 10 (1988) 713–718.
[17] A. Kato, S. Nagai, Energetics of tobacco cells, *Nicotiana tabacum* L., growing on sucrose medium, Eur. J. Appl. Microbiol. Biotechnol. 7 (1979) 219–225.
[18] P. Ducommun, P.-A. Ruffieux, M.-P. Furter, I. Marison, U. von Stockar, A new method for on-line measurement of the volumetric oxygen uptake rate in membrane aerated animal cell cultures, J. Biotechnol. 78 (2000) 139–147.
[19] S. Tatiraju, M. Soroush, R. Mutharasan, Multi-rate nonlinear state and parameter estimation in a bioreactor, Biotechnol. Bioeng. 63 (1999) 22–32.
[20] S.P.S. Andrew, Gas–liquid mass transfer in microbiological reactors, Trans. IChemE 60 (1982) 3–13.
[21] J.J. Heijnen, K. van't Riet, A.J. Wolthuis, Influence of very small bubbles on the dynamic k_LA measurement in viscous gas–liquid systems, Biotechnol. Bioeng. 22 (1980) 1945–1956.
[22] K. van't Riet, Review of measuring methods and results in nonviscous gas–liquid mass transfer in stirred vessels, Ind. Eng. Chem. Process Des. Dev. 18 (1979) 357–364.
[23] A. Ogut, R.T. Hatch, Oxygen transfer into Newtonian and non-Newtonian fluids in mechanically agitated vessels, Can. J. Chem. Eng. 66 (1988) 79–85.
[24] Y. Kawase, M. Moo-Young, The effect of antifoam agents on mass transfer in bioreactors, Bioprocess Eng. 5 (1990) 169–173.
[25] A. Prins, K. van't Riet, Proteins and surface effects in fermentation: foam, antifoam and mass transfer, Trends Biotechnol. 5 (1987) 296–301.
[26] International Critical Tables (1926–1930) McGraw-Hill; first electronic ed., International Critical Tables of Numerical Data, Physics, Chemistry and Technology (2003), Knovel.
[27] Y.H. Lee, G.T. Tsao, Dissolved oxygen electrodes, Adv. Biochem. Eng. 13 (1979) 35–86.
[28] V. Linek, J. Sinkule, V. Vacek, Dissolved oxygen probes, in: M. Moo-Young (Ed.), Comprehensive Biotechnology, vol. 4, Pergamon Press, 1985, pp. 363–394.

[29] G.A. Truesdale, A.L. Downing, G.F. Lowden, The solubility of oxygen in pure water and sea-water, J. Appl. Chem. 5 (1955) 53–62.

[30] G. Quicker, A. Schumpe, B. König, W.-D. Deckwer, Comparison of measured and calculated oxygen solubilities in fermentation media, Biotechnol. Bioeng. 23 (1981) 635–650.

[31] P.R. Gogate, A.B. Pandit, Survey of measurement techniques for gas–liquid mass transfer coefficient in bioreactors, Biochem. Eng. J. 4 (1999) 7–15.

[32] M. Sobotka, A. Prokop, I.J. Dunn, A. Einsele, Review of methods for the measurement of oxygen transfer in microbial systems, Ann. Rep. Ferm. Proc. 5 (1982) 127–210.

[33] J.D. Brooks, D.G. Maclennan, J.P. Barford, R.J. Hall, Design of laboratory continuous-culture equipment for accurate gaseous metabolism measurements, Biotechnol. Bioeng. 24 (1982) 847–856.

[34] V. Linek, V. Vacek, Oxygen electrode response lag induced by liquid film resistance against oxygen transfer, Biotechnol. Bioeng. 18 (1976) 1537–1555.

[35] N.D.P. Dang, D.A. Karrer, I.J. Dunn, Oxygen transfer coefficients by dynamic model moment analysis, Biotechnol. Bioeng. 19 (1977) 853–865.

[36] M. Nocentini, Mass transfer in gas–liquid, multiple-impeller stirred vessels, Trans. IChemE 68 (1990) 287–294.

[37] V. Linek, V. Vacek, P. Beneš, A critical review and experimental verification of the correct use of the dynamic method for the determination of oxygen transfer in aerated agitated vessels to water, electrolyte solutions and viscous liquids, Chem. Eng. J. 34 (1987) 11–34.

[38] V. Linek, P. Beneš, V. Vacek, Dynamic pressure method for k_La measurement in large-scale bioreactors, Biotechnol. Bioeng. 33 (1989) 1406–1412.

[39] V. Linek, T. Moucha, M. Doušová, J. Sinkule, Measurement of k_La by dynamic pressure method in pilot-plant fermentor, Biotechnol. Bioeng. 43 (1994) 477–482.

[40] N.M.G. Oosterhuis, N.W.F. Kossen, Oxygen transfer in a production scale bioreactor, Chem. Eng. Res. Des. 61 (1983) 308–312.

[41] S. Shioya, I.J. Dunn, Model comparisons for dynamic k_La measurements with incompletely mixed phases, Chem. Eng. Commun. 3 (1979) 41–52.

[42] K. Petera, P. Ditl, Effect of pressure profile on evaluation of volumetric mass transfer coefficient in k_La bioreactors, Biochem. Eng. J. 5 (2000) 23–27.

[43] R.E. Spier, B. Griffiths, An examination of the data and concepts germane to the oxygenation of cultured animal cells, Develop. Biol. Standard. 55 (1984) 81–92.

Suggestions for Further Reading

Mass Transfer Theory

See also references 1, 3, and 4.

McCabe, W. L., Smith, J. C., & Harriott, P. (2005). *Unit Operations of Chemical Engineering* (7th ed., Chapter 17). McGraw-Hill.

Oxygen Transfer in Fermenters

van't Riet, K. (1983). Mass transfer in fermentation. *Trends Biotechnol.*, *1*, 113–119.

Measurement of k_La *and* q_O

See also references 31–42.

Linek, V., Sinkule, J., & Beneš, P. (1991). Critical assessment of gassing-in methods for measuring k_La in fermentors. *Biotechnol. Bioeng.*, *38*, 323–330.

Pouliot, K., Thibault, J., Garnier, A., & Acuña Leiva, G. (2000). K_La evaluation during the course of fermentation using data reconciliation techniques. *Bioprocess Eng.*, *23*, 565–573.

Ruffieux, P.-A., von Stockar, U., & Marison, I. W. (1998). Measurement of volumetric (OUR) and determination of specific (qO_2) oxygen uptake rates in animal cell cultures. *J. Biotechnol.*, *63*, 85–95.

Tobajas, M., & García-Calvo, E. (2000). Comparison of experimental methods for determination of the volumetric mass transfer coefficient in fermentation processes. *Heat Mass Transfer*, *36*, 201–207.

Oxygen Transfer in Shake Flasks

Maier, U., & Büchs, J. (2001). Characterisation of the gas–liquid mass transfer in shaking bioreactors. *Biochem. Eng. J.*, *7*, 99–106.

Mrotzek, C., Anderlei, T., Henzler, H.-J., & Büchs, J. (2001). Mass transfer resistance of sterile plugs in shaking bioreactors. *Biochem. Eng. J.*, *7*, 107–112.

Schultz, J. S. (1964). Cotton closure as an aeration barrier in shaken flask fermentations. *Appl. Microbiol.*, *12*, 305–310.

van Suijdam, J. C., Kossen, N.W.F., & Joha, A. C. (1978). Model for oxygen transfer in a shake flask. *Biotechnol. Bioeng.*, *20*, 1695–1709.

CHAPTER

11

Unit Operations

Bioprocesses treat raw materials and generate useful products. Individual operations or steps in the process that alter the properties of materials are called unit operations. Although the specific objectives of bioprocesses vary from factory to factory, each processing scheme can be viewed as a series of component operations that appear again and again in different systems. For example, most bioprocesses involve one or more of the following unit operations: adsorption, centrifugation, chromatography, crystallisation, dialysis, distillation, drying, evaporation, filtration, flocculation, flotation, homogenisation, humidification, microfiltration, milling, precipitation, sedimentation, solvent extraction, and ultrafiltration.

As an illustration, the sequence of unit operations used for the manufacture of particular enzymes is shown in the flow sheet of Figure 11.1. Although the same operations are involved in other processes, the order in which they are carried out, the conditions used, and the materials handled account for the differences in final results. However, the engineering principles for design of unit operations are independent of specific industries or applications.

In a typical fermentation process, raw materials are altered most significantly by the reactions occurring in the fermenter. However physical changes before and after fermentation are also important to prepare the substrates for reaction and to extract and purify the desired product from the culture broth. The term 'unit operation' usually refers to processes that cause physical modifications to materials, such as a change of phase or component concentration. Chemical or biochemical transformations are the subject of reaction engineering, which is considered in detail in Chapters 12 through 14.

11.1 OVERVIEW OF DOWNSTREAM PROCESSING

In fermentation broths, the desired product is present within a complex mixture of many components. Any treatment of the culture broth after fermentation to concentrate and purify the product is known as *downstream processing*. In most cases, downstream processing requires only physical modification of the broth material rather than further

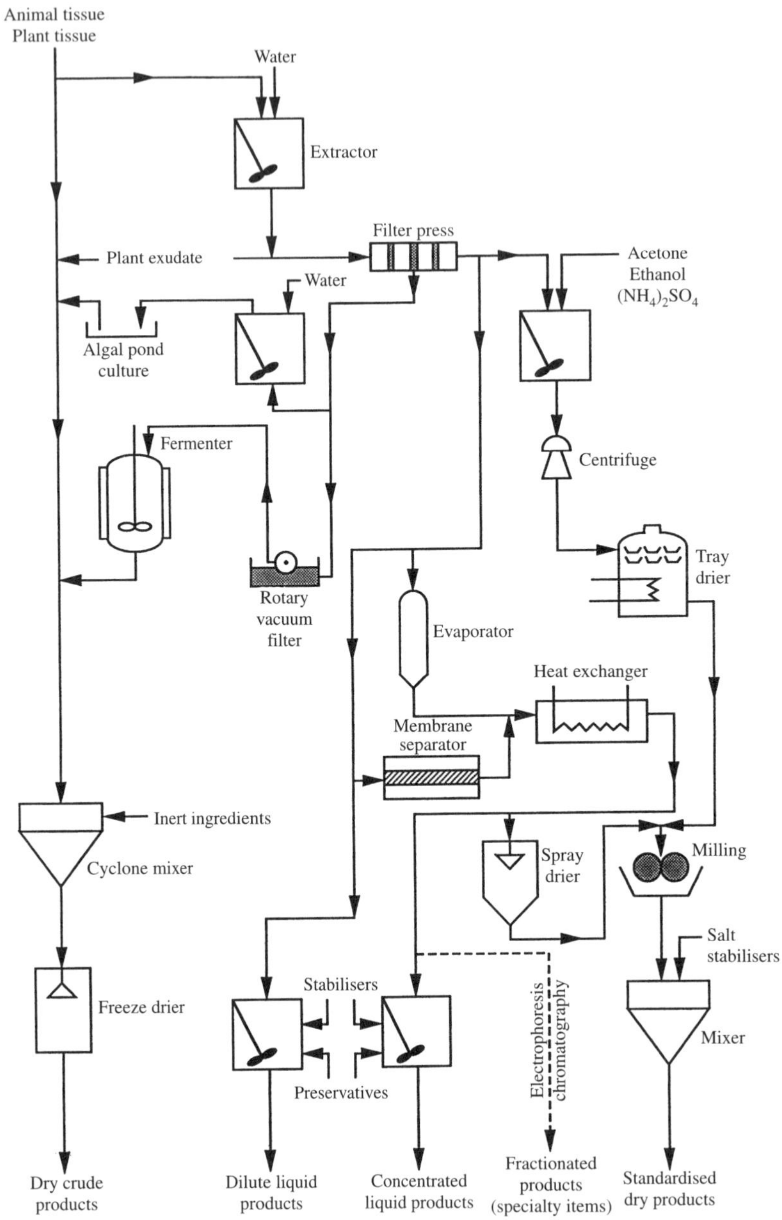

FIGURE 11.1 Typical unit operations used in the manufacture of enzymes. *From B. Atkinson, and F. Mavituna, 1991,* Biochemical Engineering and Biotechnology Handbook, *2nd ed., Macmillan, Basingstoke; and W.T. Faith, C.E. Neubeck, and E.T. Reese, 1971, Production and applications of enzymes.* Adv. Biochem. Eng. *1, 77–111.*

chemical or biochemical transformation. Nevertheless, there are several reasons why downstream processing is often technically very challenging.

- *Fermentation products are formed in dilute solution.* Water is the main component of cell culture media and, therefore, of harvested fermentation broth. Many other components are also present in the broth mixture, providing a wide range of contaminating

substances that must be removed to isolate the desired product. In general, purification from dilute solutions involves more recovery steps and higher costs than when the product is available in a concentrated form with fewer impurities.
- *Biological products are labile*. Products produced in fermentations are sensitive to temperature and can be degraded by exposure to solvents, strong acids and bases, and high salt concentrations. This restricts the range of downstream processing operations that can be applied for product recovery.
- *Harvested fermentation broths are susceptible to contamination*. Typically, once the broth is removed from the controlled environment of the fermenter, aseptic conditions are no longer maintained and the material is subject to degradation by the activity of contaminating organisms. Unless downstream processing occurs rapidly and without delay, product quality can deteriorate significantly.

Although each recovery scheme will be different, the sequence of steps in downstream processing can be generalised depending on whether the biomass itself is the desired product (e.g., bakers' yeast), whether the product is contained within the cells (e.g., enzymes and recombinant proteins), or whether the product accumulates outside the cells in the fermentation liquor (e.g., ethanol, antibiotics, and monoclonal antibodies). General schemes for these three types of downstream processing operation are represented in Figure 11.2 and involve the following major steps.

1. *Cell removal*. A common first step in product recovery is the removal of cells from the fermentation liquor. If the cells are the product, little or no further downstream processing is required. If the product is contained within the biomass, harvesting the cells from the large volume of fermentation liquid removes many of the impurities present and concentrates the product substantially. Removal of the cells can also assist the recovery of products from the liquid phase. Filtration, microfiltration, and centrifugation are typical unit operations for cell removal.
2. *Cell disruption and cell debris removal*. These downstream processing steps are required when the product is located inside the cells. Unit operations such as high-pressure homogenisation are used to break open the cells and release their contents for subsequent purification. The cell debris generated during cell disruption is separated from the product by filtration, microfiltration, or centrifugation.
3. *Primary isolation*. A wide variety of techniques is available for primary isolation of fermentation products from cell homogenate or cell-free broth. The methods used depend on the physical and chemical properties of the product and surrounding material. The aim of primary isolation is to remove components with properties that are substantially different from those of the product. Typically, processes for primary isolation are relatively nonselective; however significant increases in product quality and concentration can be accomplished. Unit operations such as solvent extraction, aqueous two-phase liquid extraction, adsorption, precipitation, and ultrafiltration are used for primary isolation.

   There are special challenges associated with the design and operation of primary isolation processes in large-scale production systems. When the product is extracellular, large volumes of culture liquid must be treated at this stage. As intermediate storage of this liquid is impractical and disposal expensive, the processes and equipment used for

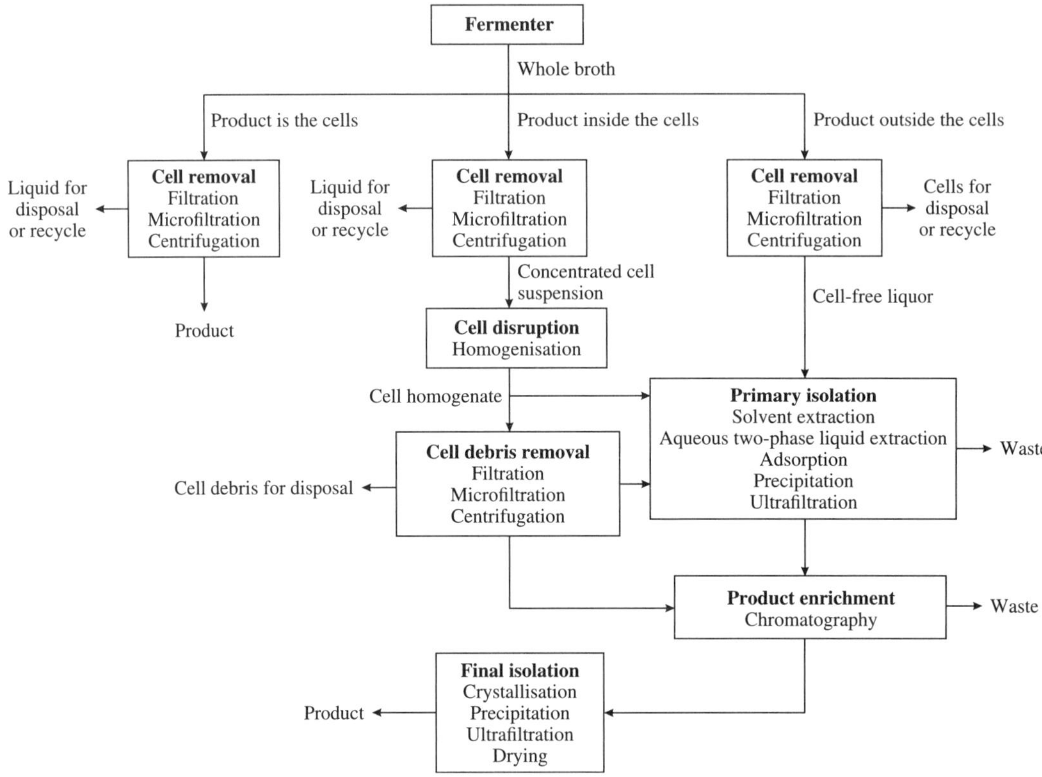

FIGURE 11.2 Generalised downstream processing schemes for cells as product, products located inside the cells, and products located outside the cells in the fermentation liquor.

primary isolation must be robust and reliable to minimise broth spoilage and product deterioration in the event of equipment breakdown or process malfunction. It is essential that the operations used for primary isolation be able to treat the fermentation liquor at the rate it is generated. The large liquid volumes involved also mean that low-energy, low-cost processes are required. A desirable feature of primary isolation processes is that a significant reduction in liquid volume is achieved. This reduces the equipment size and operating costs associated with subsequent recovery steps.

4. *Product enrichment*. Processes for product enrichment are highly selective and are designed to separate the product from impurities with properties close to those of the product. Chromatography is a typical unit operation used at this stage of product resolution.
5. *Final isolation*. The form of the product and final purity required vary considerably depending on the product application. Ultrafiltration for liquid products, and crystallisation followed by centrifugation or filtration and drying for solid products, are typical operations used for final processing of high-quality materials such as pharmaceuticals.

TABLE 11.1 Typical Profile of Product Quality during Downstream Processing

Step	Typical Unit Operation	Product Concentration (g l^{-1})	Product Quality (%)
Harvest broth	–	0.1–5	0.1–1
Cell removal	Filtration	1–5	0.2–2
Primary isolation	Extraction	5–50	1–10
Product enrichment	Chromatography	50–200	50–80
Final isolation	Crystallisation	50–200	90–100

Adapted from P.A. Belter, E.L. Cussler, and W.-S. Hu, 1988, Bioseparations: Downstream Processing For Biotechnology, *John Wiley, New York.*

A typical profile of product concentration and quality through the various stages of downstream processing is given in Table 11.1.

The performance of downstream processing operations can be characterised quantitatively using two parameters, the *concentration factor* δ and the *separation factor* α. These parameters are defined as:

$$\delta = \frac{\text{concentration of product after treatment}}{\text{concentration of product before treatment}} \tag{11.1}$$

and

$$\alpha = \frac{\left(\dfrac{\text{concentration of product}}{\text{concentration of contaminant}}\right)_{\text{after treatment}}}{\left(\dfrac{\text{concentration of product}}{\text{concentration of contaminant}}\right)_{\text{before treatment}}} \tag{11.2}$$

A concentration factor of >1 indicates that the product is enriched during the treatment process. The separation factor differs from the concentration factor by representing the change in product concentration relative to that of some key contaminating compound. Individual downstream processing operations may achieve high concentration factors even though separation of the desired product from a particular contaminant remains relatively poor. On the other hand, highly selective recovery methods give high values of α, but this may be accompanied by only a modest increase in product concentration.

Downstream processing can account for a substantial proportion of the total production cost of a fermentation product. For example, the ratio of fermentation cost to product recovery cost is approximately 60:40 for antibiotics such as penicillin. For newer antibiotics this ratio is reversed; product recovery is more costly than fermentation. Many modern products of biotechnology such as recombinant proteins and monoclonal antibodies require expensive downstream processing that can account for 80 to 90% of the total process cost. Starting product levels before recovery have a strong influence on downstream costs; purification is more expensive when the initial concentration of product in the biomass or fermentation broth is low. As illustrated in Figure 11.3 for several products of bioprocessing, the higher the starting concentration, the cheaper is the final material.

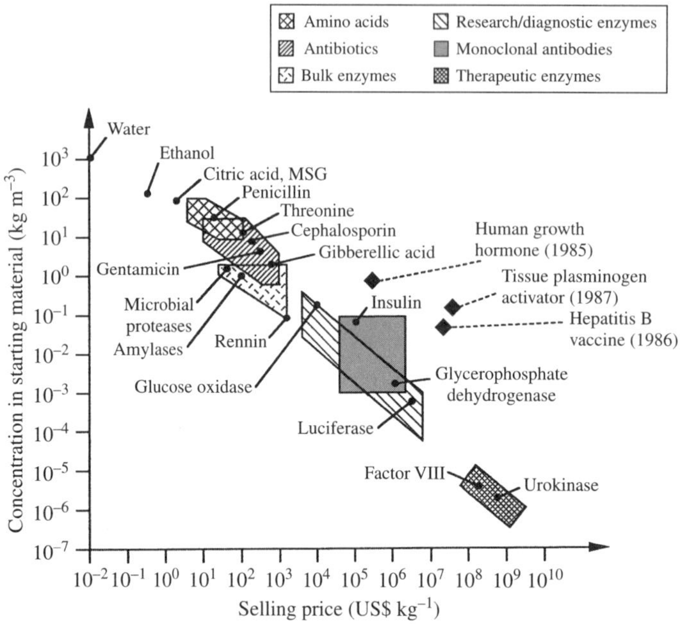

FIGURE 11.3 Relationship between selling price and concentration before downstream recovery for several products of bioprocessing. *From J.L. Dwyer, 1984, Scaling up bioproduct separation with high performance liquid chromatography.* Bio/Technology *2, 957–964; and J. van Brunt, 1988,* How big is big enough? Bio/Technology *6, 479–485.*

Because each downstream processing step involves some loss of product, total losses can be substantial for multistep procedures. For example, if 80% of the product is retained at each purification step, after a six-step process only $(0.8)^6$ or about 26% of the initial product remains. If the starting concentration is very low, more recovery stages are required with higher attendant losses and costs. This situation can be improved either by enhancing the synthesis of product during fermentation or by developing better downstream processing techniques that minimise product loss.

There is an extensive literature on downstream processing, much of it dealing with recent advances. To cover thoroughly all unit operations used in bioprocessing is beyond the scope of this book. Rather than attempt such treatment, this chapter considers the engineering principles of a selection of unit operations commonly applied for recovery of fermentation products. Information about other unit operations can be found in the references listed at the end of the chapter.

11.2 OVERVIEW OF CELL REMOVAL OPERATIONS

One of the first steps in downstream processing is the removal of cells from the culture liquid. This is the case if the cells themselves are the product, or if the product is an intra- or extracellular metabolite. Although whole broth processing without cell removal is possible, it is not commonly pursued.

The major process options for cell removal are *filtration*, *microfiltration*, and *centrifugation*. Broadly, separations using filtration and microfiltration are based on particle size, whereas centrifugation relies on particle density. However, other factors such as the shape, compressibility, and surface charge of the cells and the density and viscosity of the broth also exert a considerable influence on the performance of these operations. Cell removal from microbial fermentation broths can be technically very challenging because of the small size, low density, and gelatinous nature of many industrial microorganisms. In contrast, due mainly to the low numbers of cells generated, cell removal after mammalian cell culture is comparatively straightforward.

Prior to filtration or centrifugation, it may be necessary to pretreat or *precondition* the fermentation broth to improve the efficiency of cell separation. Heating is used to denature proteins and enhance the filterability of mycelial broths, such as in penicillin production. Pretreatments that reduce the viscosity of the broth are beneficial for both filtration and centrifugation. Alternatively, electrolytes or polymeric flocculants may be added to promote aggregation of cells and colloids into larger, denser particles that are easier to filter or centrifuge. Filter aids, which are solid particles used to increase the speed of filtration, are often applied to fermentation broths prior to filtering (Section 11.3.1). Changing the duration or conditions of the fermentation can sometimes assist subsequent cell removal by modifying the composition and viscosity of the medium and properties of the cells.

The general features of unit operations used for cell removal are described here.

- *Filtration*. In conventional filtration, cell solids are retained on a filter cloth to form a porous cake while liquid filtrate passes through the cloth. The process generates a relatively dry, friable cake of packed cells; however, the liquid filtrate usually contains a small proportion of solids that escape through the filter cloth. This is reflected in the filtrate clarity, which generally decreases if filter aids are used to improve the rate of filtration. Large-scale filtration is difficult to perform under sterile conditions. Filtration is not a practical option for cell removal if filter aid is required to achieve an acceptable filtration rate and the product is intracellular or the cells themselves are the fermentation product. This is because contamination of the filtered cells with foreign particles is inappropriate unless the cells are waste by-products; the presence of filter aid can also cause equipment problems if the cells must be disrupted mechanically to release intracellular material. Microfiltration and centrifugation are better options for cell removal under these circumstances.
- *Microfiltration*. Microfiltration uses microporous membranes and cross-flow filtration methods to recover the cells as a fluid concentrate. The maximum cell concentrations generated range from about 10% w/v for gelatinous solids up to 60 to 70% w/v for more rigid particles. It is generally not necessary to precondition broths treated using microfiltration techniques. Because filter aids are also not required, microfiltration is an attractive option for harvesting cells containing intracellular products or cells as product. Because cell recovery using microfiltration is typically close to 100%, the liquid filtrate or permeate produced by microfiltration is of much greater clarity than that generated by conventional filtration using filter aids. This is an advantage when the product is extracellular, as the filtrate contains minimal contaminating components. A potential disadvantage is that unacceptably high amounts of extracellular product may be entrained in the cell concentrate stream. Microfiltration can be carried out under sterile conditions. In many cases, microfiltration is less expensive than filtration or centrifugation, although exact cost comparisons depend on the particular process.

- *Centrifugation*. This can be an effective strategy for cell recovery when the cells are too small and difficult to filter using conventional filtration. In principle, centrifugation is suitable for recovery of cells as product and for preprocessing of both intracellular and extracellular components. Centrifugation of fermentation broths produces a thick, concentrated cell sludge or cream that contains more extracellular liquid than is produced in conventional filtration. The liquid supernatant may be either cloudy or clear depending on the operating conditions. The use of flocculants and other broth-conditioning agents improves the performance of centrifuges; however, the presence of these additives in either the cell concentrate or liquid discharge may be undesirable for particular products. Steam-sterilisable centrifuges are available for separations that must be carried out under aseptic conditions. Aerosol generation associated with the operation of high-speed centrifuges can create health and safety problems depending on the organism and products being separated. The equipment costs for centrifugation tend to be greater than for filtration and microfiltration.
- *Other operations*. Although filtration, microfiltration, and centrifugation account for the vast majority of cell separations in bioprocessing, other methods are also available. *Gravity settling* or *sedimentation* is suitable for cells that form aggregates or flocs of sufficient size and density to settle quickly under gravity. Polyvalent agents and polymers may be added to increase particle coagulation and improve the rate of sedimentation. Gravity settling is used typically in large-scale waste treatment processes for liquid clarification, and in beer brewing processes with flocculent strains of yeast. Another cell recovery process is *foam flotation*, which relies on the selective adsorption or attachment of cells to gas bubbles rising through liquid. Surfactants may be used to increase the number of cells associated with the bubbles. The cells are recovered by skimming the foam layers that collect on top of the liquid. The effectiveness of foam flotation has been reported for a range of different microorganisms [1].

Filtration and centrifugation operations are considered in greater detail in Sections 11.3 and 11.4. As one of a group of similar membrane processes, microfiltration is discussed further in Section 11.10. Decisions about which type of cell removal operation to use for a particular application are usually made after a considerable amount of experimental testing.

11.3 FILTRATION

In conventional filtration, solid particles are separated from a fluid–solid mixture by forcing the fluid through a *filter medium* or *filter cloth* that retains the particles. Solids are deposited on the filter and, as the deposit or *filter cake* increases in depth, pose a resistance to further filtration. Filtration can be performed using either vacuum or positive-pressure equipment. The pressure difference exerted across the filter to separate fluid from the solids is called the filtration *pressure drop*.

The ease of filtration depends on the properties of the solid and fluid. Filtration of crystalline, incompressible solids in low-viscosity liquids is relatively straightforward. In contrast, fermentation broths can be difficult to filter because of the small size and gelatinous nature of the cells and the viscous non-Newtonian behaviour of the broth. Most microbial filter cakes are *compressible*, that is, the porosity of the cake declines as the pressure drop across the filter increases. This can be a major problem causing reduced filtration rates and product loss. Filtration of fermentation broths is carried out typically under nonaseptic conditions; therefore, the process must be efficient and reliable to avoid undue contamination and the degradation of labile products.

11.3.1 Filter Aids

Filter aids such as diatomaceous earth have found widespread use in the fermentation industry to improve the efficiency of filtration. Diatomaceous earth, also known as kieselguhr, is the fused skeletal remains of diatoms. Packed beds of granulated kieselguhr have very high porosity; as little as 15% of the total volume of packed kieselguhr is solid, the rest is empty space. Such high porosity facilitates liquid flow around the particles and improves the rate of filtration. Kieselguhr with average particle sizes of 10 to 25 μm are used with fermentation broths.

Filter aids are applied in two ways. As shown in Figure 11.4, filter aid can be used as a precoat on the filter medium to prevent blockage or 'blinding' of the filter by solids that

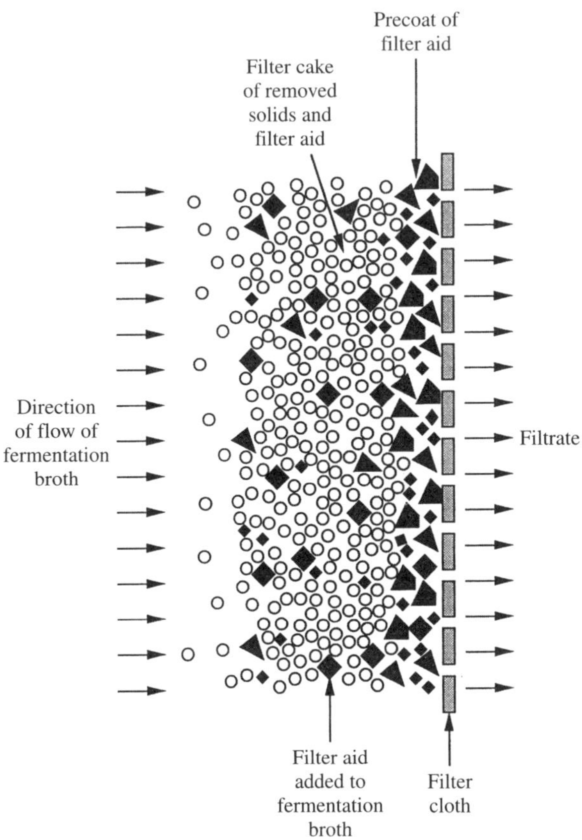

FIGURE 11.4 Use of filter aid in filtration of fermentation broth.

would otherwise wedge themselves into the pores of the cloth. Filter aid can also be added to the fermentation broth, typically at concentrations of 1 to 5% w/w, to increase the porosity of the cake as it forms. Filter aids add to the cost of filtration; the minimum quantity needed to achieve the desired result must be established experimentally. Kieselguhr absorbs liquid; therefore, if the fermentation product is in the liquid phase, some will be lost. Another disadvantage is that filtrate clarity is reduced when the rate of filtration is increased using filter aid. Disposal of waste cell material is more difficult if it contains kieselguhr; for example, biomass cannot be used as animal feed unless the filter aid is removed. Use of filter aids is not appropriate if the cells themselves are the product of the fermentation or when the cells require further processing after filtration for recovery of intracellular material.

11.3.2 Filtration Equipment

Plate filters are suitable for filtration of small fermentation batches; this type of filter gradually accumulates biomass and must be periodically opened and cleared of filter cake. Larger processes require continuous filters. *Rotary drum vacuum filters*, such as that shown in Figure 11.5, are the most widely used filtration devices in the fermentation

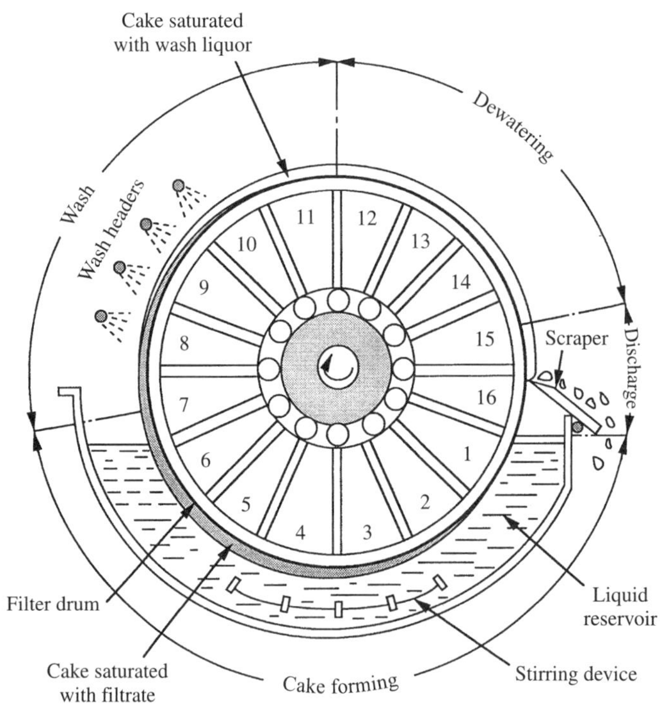

FIGURE 11.5 Continuous rotary drum vacuum filter.
From G.G. Brown, A.S. Foust, D.L. Katz, R. Schneidewind, R.R. White, W.P. Wood, G.M. Brown, L.E. Brownell, J.J. Martin, G.B. Williams, J.T. Banchero, and J.L. York, 1950, Unit Operations, *John Wiley, New York.*

industry. A horizontal drum 0.5 to 3 m in diameter is covered with filter cloth and rotated slowly at 0.1 to 2 rpm. The cloth is partially immersed in an agitated reservoir containing material to be filtered. As a section of drum enters the liquid, a vacuum is applied from the interior of the drum. A cake forms on the face of the cloth while liquid is drawn through internal pipes to a collection tank. As the drum rotates out of the reservoir, the surface of the filter is sprayed with wash liquid, which is drawn through the cloth and collected in a separate holding tank. After washing, the cake is dewatered by continued application of the vacuum. The vacuum is turned off as the drum reaches the discharge zone where the cake is removed by means of a scraper, knife, or strings. Air pressure may be applied at this stage to help dislodge the filter cake from the cloth. After the cake is removed, the drum reenters the reservoir for another filtration cycle.

11.3.3 Filtration Theory

Filtration theory is used to estimate the rate of filtration. For a given pressure drop across the filter, the rate of filtration is greatest just as filtering begins. This is because the resistance to filtration is at a minimum when there are no deposited solids. The orientation of particles in the initial cake deposit is very important and can influence significantly the structure and permeability of the whole filter bed. Excessive pressure drops and high initial rates of filtration cause plugging of the filter cloth and a very high subsequent resistance to flow. Flow resistance due to the filter cloth can be considered constant if particles do not penetrate the material; however, the resistance due to the cake increases with cake thickness.

The rate of filtration is usually measured as the rate at which liquid filtrate is collected. The filtration rate depends on the area of the filter cloth, the viscosity of the fluid, the pressure difference across the filter, and the resistance to filtration offered by the cloth and deposited solids. At any instant during filtration, the rate of filtration is given by the equation:

$$\frac{1}{A}\frac{\mathrm{d}V_\mathrm{f}}{\mathrm{d}t} = \frac{\Delta p}{\mu_\mathrm{f}\left[\alpha\left(\frac{M_\mathrm{c}}{A}\right) + R_\mathrm{m}\right]} \tag{11.3}$$

where A is the filter area, V_f is the volume of filtrate, t is the filtration time, Δp is the pressure drop across the filter, μ_f is the filtrate viscosity, M_c is the total mass of solids in the cake, α is the average *specific cake resistance*, and R_m is the *filter medium resistance*. R_m includes the effect of the filter cloth and any particles wedged in it during the initial stages of filtration. α has dimensions LM^{-1}; R_m has dimensions L^{-1}. $\mathrm{d}V_\mathrm{f}/\mathrm{d}t$ is the filtrate flow rate or *volumetric rate of filtration*. The capital cost of the filter depends on A; the bigger the area required to achieve a given filtration rate, the larger are the equipment and related investment. α is a measure of the resistance of the filter cake to flow; its value depends on the shape and size of the particles, the size of the interstitial spaces between them, and the mechanical stability of the cake. Resistance due to the filter medium is often negligible compared with the cake resistance, which is represented by the term $\alpha(M_\mathrm{c}/A)$.

If the filter cake is incompressible, the specific cake resistance α does not vary with the pressure drop across the filter. However, cakes from fermentation broths are seldom incompressible; as cell cakes compress with increasing Δp, filtration rates decline. For a compressible cake, α can be related to Δp empirically as follows:

$$\alpha = \alpha'(\Delta p)^s \tag{11.4}$$

where s is the cake *compressibility* and α' is a constant dependent largely on the size and morphology of the particles in the cake. The value of s is zero for rigid incompressible solids; for highly compressible material s is close to unity. α is also related to the average properties of the particles in the cake as follows:

$$\alpha = \frac{K_v a^2 (1 - \varepsilon)}{\varepsilon^3 \rho_p} \tag{11.5}$$

where K_v is a factor dependent on the shape of the particles, a is the particle specific surface area:

$$a = \frac{\text{surface area of a single particle}}{\text{volume of a single particle}} \tag{11.6}$$

ε is the *porosity* of the cake:

$$\varepsilon = \frac{\text{total volume of the cake} - \text{volume of solids in the cake}}{\text{total volume of the cake}} \tag{11.7}$$

and ρ_p is the density of the particles. For compressible cakes, both ε and K_v depend on the filtration pressure drop.

It is useful to consider methods for improving the rate of filtration. A number of strategies can be deduced from the relationship between the variables in Eq. (11.3).

- *Increase the filter area A*. When all other parameters remain constant, the rate of filtration is improved if A is increased. However, this requires installation of larger filtration equipment and greater capital cost.
- *Increase the filtration pressure drop* Δp. The problem with this approach for compressible cakes is that α increases with Δp as indicated in Eq. (11.4), and higher α results in lower filtration rates. In practice, pressure drops are usually kept below about 0.5 atm to minimise cake resistance. Improving the filtration rate by increasing the pressure drop can only be achieved by reducing s, the compressibility of the cake. Addition of filter aid to the broth may reduce s to some extent.
- *Reduce the cake mass* M_c. This is achieved in continuous rotary filtration by reducing the thickness of the cake deposited per revolution of the drum, and ensuring that the scraper leaves minimal cake residue on the filter cloth.
- *Reduce the liquid viscosity* μ_f. Material to be filtered is sometimes diluted if the starting viscosity is very high.
- *Reduce the specific cake resistance* α. From Eq. (11.5), possible methods of reducing α for compressible cakes are as follows:
  - *Increase the porosity* ε. Cake porosity usually decreases as cells are filtered. Application of filter aid reduces this effect.

- *Reduce the shape factor of the particles K_v.* In the case of mycelial broths, it may be possible to change the morphology of the cells by manipulating the fermentation conditions.
- *Reduce the specific surface area of the particles a.* Increasing the average size of the particles and minimising variations in particle size reduce the value of *a*. Changes in fermentation conditions and broth pretreatment are used to achieve these effects.

Integration of Eq. (11.3) allows us to calculate the time required to filter a given volume of material. Before carrying out the integration, let us substitute an expression for the mass of solids in the cake as a function of the filtrate volume:

$$M_c = c\,V_f \tag{11.8}$$

In Eq. (11.8), c is the mass of solids deposited per volume of filtrate; this term is related to the concentration of solids in the material to be filtered. Substituting Eq. (11.8) into Eq. (11.3), the expression for the rate of filtration becomes:

$$\frac{1}{A}\frac{dV_f}{dt} = \frac{\Delta p}{\mu_f\left[\alpha\left(\frac{cV_f}{A}\right) + R_m\right]} \tag{11.9}$$

Equation (11.9) can be interpreted according to the *general rate principle*, which equates the rate of a process to the ratio of the driving force and the resistance. As Δp is the driving force for filtration, the resistances offered by the filter cake and filter medium are represented by the terms summed together in the denominator of Eq. (11.9). The relative contributions of these two resistances can be estimated using the equations:

$$\text{Proportion of the total resistance due to the filter cake} = \frac{\alpha\left(\frac{cV_f}{A}\right)}{\alpha\left(\frac{cV_f}{A}\right) + R_m} \tag{11.10}$$

$$\text{Proportion of the total resistance due to the filter medium} = \frac{R_m}{\alpha\left(\frac{cV_f}{A}\right) + R_m} \tag{11.11}$$

A filter can be operated in two different ways. If the pressure drop across the filter is kept constant, the filtration rate will become progressively smaller as resistance due to the cake increases. Alternatively, in constant-rate filtration, the flow rate is maintained by gradually increasing the pressure drop. Filtrations are most commonly carried out at constant pressure. When this is the case, Eq. (11.9) can be integrated directly because V_f and t are the only variables: for a given filtration device and material to be filtered, each of the remaining parameters is constant.

It is convenient for integration to write Eq. (11.9) in its reciprocal form:

$$A\frac{dt}{dV_f} = \mu_f \alpha c \frac{V_f}{A\,\Delta p} + \frac{\mu_f R_m}{\Delta p} \tag{11.12}$$

Separating variables and placing constant terms outside of the integrals:

$$A\int dt = \frac{\mu_f \alpha c}{A\,\Delta p}\int V_f\,dV_f + \frac{\mu_f R_m}{\Delta p}\int dV_f \tag{11.13}$$

At the beginning of filtration $t = 0$ and $V_f = 0$; this is the initial condition for integration. Carrying out the integration gives:

$$At = \frac{\mu_f \alpha c}{2A\,\Delta p}V_f^2 + \frac{\mu_f R_m}{\Delta p}V_f \tag{11.14}$$

Thus, for constant-pressure filtration, Eq. (11.14) can be used to calculate either the filtrate volume V_f or the filtration time t, provided all the constants are known. However, α and R_m for a particular filtration must be evaluated beforehand.

For experimental determination of α and R_m, Eq. (11.14) is rearranged by dividing both sides of the equation by AV_f:

$$\frac{t}{V_f} = \frac{\mu_f \alpha c}{2A^2\Delta p}V_f + \frac{\mu_f R_m}{A\,\Delta p} \tag{11.15}$$

Equation (11.15) can be written more simply as:

$$\frac{t}{V_f} = K_1 V_f + K_2 \tag{11.16}$$

where:

$$K_1 = \frac{\mu_f \alpha c}{2A^2\Delta p} \tag{11.17}$$

and

$$K_2 = \frac{\mu_f R_m}{A\,\Delta p} \tag{11.18}$$

K_1 and K_2 are constant during constant-pressure filtration. Therefore, from Eq. (11.16), a straight line is generated when t/V_f is plotted against V_f. The slope K_1 depends on the filtration pressure drop and the properties of the cake; the intercept K_2 also depends on the pressure drop but is independent of cake properties. α is calculated from the slope; R_m is determined from the intercept. Equation (11.16) is valid for compressible and incompressible cakes; however K_1 for compressible cakes becomes a more complex function of Δp than is directly apparent from Eq. (11.17) because of the dependence of α on pressure.

Equation (11.15) is the basic filtration equation for industrial-scale equipment such as the rotary drum vacuum filter shown in Figure 11.5. The rate of filtration during cake formation and washing determines the size of the filter required for the process. Simple modifications to Eq. (11.15) give equations that are applicable specifically to rotary drum filters [2].

EXAMPLE 11.1 FILTRATION OF MYCELIAL BROTH

A 30-ml sample of broth from a penicillin fermentation is filtered in the laboratory on a 3-cm^2 filter at a pressure drop of 5 psi. The filtration time is 4.5 min. Previous studies have shown that *Penicillium chrysogenum* filter cake is compressible with $s = 0.5$. If 500 litres of broth from a pilot-scale fermenter must be filtered in 1 hour, what size filter is required if the pressure drop is:

(a) 10 psi?
(b) 5 psi?

The resistance due to the filter medium is negligible.

Solution

The properties of the filtrate and mycelial cake can be determined from the results of the laboratory experiment. If R_m is zero in Eq. (11.15):

$$\frac{t}{V_f} = \frac{\mu_f \alpha c}{2A^2 \Delta p} V_f$$

Substituting the expression for α for a compressible cake from Eq. (11.4):

$$\frac{t}{V_f} = \frac{\mu_f \alpha' (\Delta p)^{s-1} c}{2A^2} V_f \tag{1}$$

Rearranging gives:

$$\mu_f \alpha' c = \frac{2A^2 t}{(\Delta p)^{s-1} V_f^2}$$

Substituting values:

$$\mu_f \alpha' c = \frac{2\,(3\,\text{cm}^2)^2\,(4.5\,\text{min})}{(5\,\text{psi})^{0.5-1}\,(30\,\text{cm}^3)^2} = 0.201\,\text{cm}^{-2}\,\text{psi}^{0.5}\,\text{min}$$

This value for $\mu_f \alpha' c$ is used to evaluate the area required for pilot-scale filtration. From (1):

$$A^2 = \frac{\mu_f \alpha' c\,(\Delta p)^{s-1}}{2t} V_f^2$$

Therefore:

$$A = \left(\frac{\mu_f \alpha' c\,(\Delta p)^{s-1}}{2t}\right)^{1/2} V_f$$

(a) Substituting values with $\Delta p = 10$ psi:

$$A = \left(\frac{0.201\,\text{cm}^{-2}\,\text{psi}^{0.5}\,\text{min}\,(10\,\text{psi})^{0.5-1}}{2\,(1\,\text{h}) \cdot \left|\frac{60\,\text{min}}{1\,\text{h}}\right|}\right)^{1/2} (500\,\text{l}) \cdot \left|\frac{1000\,\text{cm}^3}{1\,\text{l}}\right|$$

$$A = 1.15 \times 10^4\,\text{cm}^2 = 1.15\,\text{m}^2$$

(b) When $\Delta p = 5$ psi:

$$A = \left(\frac{0.201\,\text{cm}^{-2}\,\text{psi}^{0.5}\,\text{min}\,(5\,\text{psi})^{0.5-1}}{2\,(1\,\text{h}) \cdot \left| \frac{60\,\text{min}}{1\,\text{h}} \right|} \right)^{1/2} (500\,\text{l}) \cdot \left| \frac{1000\,\text{cm}^3}{1\,\text{l}} \right|$$

$$A = 1.37 \times 10^4\,\text{cm}^2 = 1.37\,\text{m}^2$$

Halving the pressure drop increases the area required by only 20% because at 5 psi the cake is less compressed and more porous than at 10 psi.

Example 11.1 underlines the importance of laboratory testing in the design of filtration systems. Experiments are required to evaluate properties of the cake such as compressibility and specific resistance; these parameters cannot be calculated reliably from theory. Experimental observations are also necessary to evaluate a wide range of other important filtration characteristics that do not appear in the equations. These include filtrate clarity, ease of washing, dryness of the final cake, ease of cake removal, and the effects of filter aids and broth pretreatment. It is essential that the laboratory tests be conducted with the same materials as the large-scale process. Variables such as temperature, age of the broth, and the presence of contaminants and cell debris have significant effects on filtration characteristics.

In general, fungal mycelia are filtered relatively easily because the porosity of mycelial filter cakes is sufficiently high. Yeast and bacteria are much more difficult to handle because of their small size. Alternative microfiltration methods that eliminate the filter cake are being adopted increasingly. As described further in Section 11.10, microfiltration is achieved by developing large cross-flow fluid velocities across the filter surface while the velocity normal to the surface is relatively small. Build-up of filter cake and problems with high cake resistance are therefore avoided. Microfiltration is applied routinely in downstream processing for separation of fine particles such as yeast, bacteria, and debris from disrupted cells.

11.4 CENTRIFUGATION

Centrifugation is used to separate materials of different density by application of a force greater than gravity. In downstream processing, centrifugation is used to remove cells from fermentation broths, to eliminate cell debris, to collect precipitates and crystals, and to separate phases after liquid extraction. Centrifugation may also be applied in other areas of bioprocessing, such as to clarify molasses used in fermentation media and in the production of wort for brewing.

Steam-sterilisable centrifuges are applied when either the separated cells or fermentation liquid is recycled back to the fermenter, or when product contamination must be prevented. Industrial centrifuges generate large amounts of heat due to friction; it is therefore necessary to have good ventilation and cooling. Aerosols created by fast-spinning centrifuges have been known to cause infections and allergic reactions in factory workers so that isolation cabinets are required for certain applications.

Centrifugation is most effective when the density difference between the particles and liquid is great, the particles are large, and the liquid viscosity is low. It is also assisted by a large centrifuge radius and high rotational speed. However, in centrifugation of biological solids such as cells, the particles are very small, the viscosity of the medium can be relatively high, and the particle density is very similar to that of the suspending fluid. These disadvantages are overcome easily in the laboratory with small centrifuges operated at high speed. However, problems arise in industrial centrifugation when large quantities of material must be treated.

Centrifuge capacity cannot be increased by simply increasing the size of the equipment without limit; mechanical stresses in centrifuges increase in proportion to (radius)^2 so that safe operating speeds are substantially lower in large equipment. The need for continuous throughput of material in industrial applications also restricts practical operating speeds. To overcome these difficulties, a range of centrifuges has been developed for bioprocessing. The types of centrifuge commonly used in industrial operations are described in the next section.

11.4.1 Centrifuge Equipment

Centrifuge equipment is classified according to internal structure. The *tubular bowl centrifuge* has the simplest configuration and is employed widely in the food and pharmaceutical industries. Feed enters under pressure through a nozzle at the bottom, is accelerated to rotor speed, and moves upward through the cylindrical bowl. As the bowl rotates, particles travelling upward are spun out and collide with the walls of the bowl as illustrated schematically in Figure 11.6. Solids are removed from the liquid if they move with sufficient velocity to reach the wall of the bowl within the residence time of the liquid in the machine. As the feed rate is increased, the liquid layer moving up the wall of the

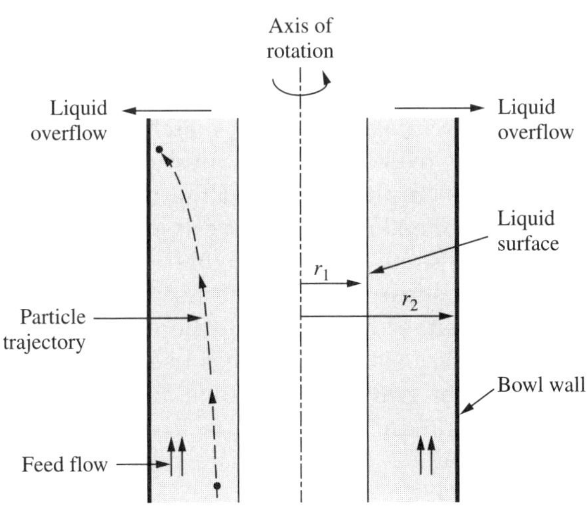

FIGURE 11.6 Separation of solids in a tubular bowl centrifuge.

centrifuge becomes thicker; this reduces the performance of the centrifuge by increasing the distance a particle must travel to reach the wall. Liquid from the feed spills over a weir at the top of the bowl; solids that have collided with the walls are collected separately. When the thickness of sediment collecting in the bowl reaches the position of the liquid-overflow weir, separation efficiency declines rapidly. This limits the capacity of the centrifuge. Tubular centrifuges are applied mainly for difficult separations requiring high centrifugal forces. Solids in tubular centrifuges are accelerated by forces between 13,000 and 16,000 times the force of gravity.

A type of narrow tubular bowl centrifuge is the *ultracentrifuge*. This device is used for recovery of fine precipitates from high-density solutions, for breaking down emulsions, and for separation of colloidal particles such as ribosomes and mitochondria. It produces centrifugal forces 10^5 to 10^6 times the force of gravity. The bowl is usually air-driven and operated at low pressure or in an atmosphere of nitrogen to reduce the generation of frictional heat. The main commercial application of ultracentrifuges has been in the production of vaccines to separate viral particles from cell debris. Typically, ultracentrifuges are operated in batch mode, so their processing capacity is restricted by the need to empty the bowl manually. Continuous ultracentrifuges are available commercially; however safe operating speeds for these machines are not as high as for batch equipment.

An alternative to the tubular centrifuge is the *disc stack bowl centrifuge*. Disc stack centrifuges are common in bioprocessing. There are many types of disc centrifuge; the principal difference between them is the method used to discharge the accumulated solids. In simple disc centrifuges, solids must be removed periodically by hand. Continuous or intermittent discharge of solids is possible in a variety of disc centrifuges without reducing the bowl speed. Some centrifuges are equipped with peripheral nozzles for continuous solids removal; others have valves for intermittent discharge. Another method is to concentrate the solids in the periphery of the bowl and then discharge them at the top of the centrifuge using a paring device; the equipment configuration for this mode of operation is shown in Figure 11.7. A disadvantage of centrifuges with automatic discharge of solids is that the solids must remain sufficiently wet to flow through the machine. Extra nozzles may be provided for cleaning the bowl should blockage of the system occur.

Disc stack centrifuges contain conical sheets of metal called discs that are stacked one on top of the other with clearances as small as 0.3 mm. The discs rotate with the bowl and their function is to split the liquid into thin layers. As shown in Figure 11.8, feed is released near the bottom of the centrifuge and travels upward through matching holes in the discs. Between the discs, heavy components of the feed are thrown outward under the influence of centrifugal forces while lighter liquid is displaced toward the centre of the bowl. As they are flung out, the solids strike the undersides of the overlying discs and slide down to the bottom edge of the bowl. At the same time, the lighter liquid flows inward over the upper surfaces of the discs to be discharged from the top of the bowl. Heavier liquid containing solids can be discharged either at the top of the centrifuge or through nozzles around the periphery of the bowl. Disc stack centrifuges used in bioprocessing typically develop forces of 5000 to 15,000 times gravity. As a guide, the minimum solid–liquid density difference for successful separation in a disc stack centrifuge is approximately 0.01 to 0.03 kg m^{-3}. In practical operations at appropriate flow rates, the minimum particle diameter separated is about 0.5 μm [3].

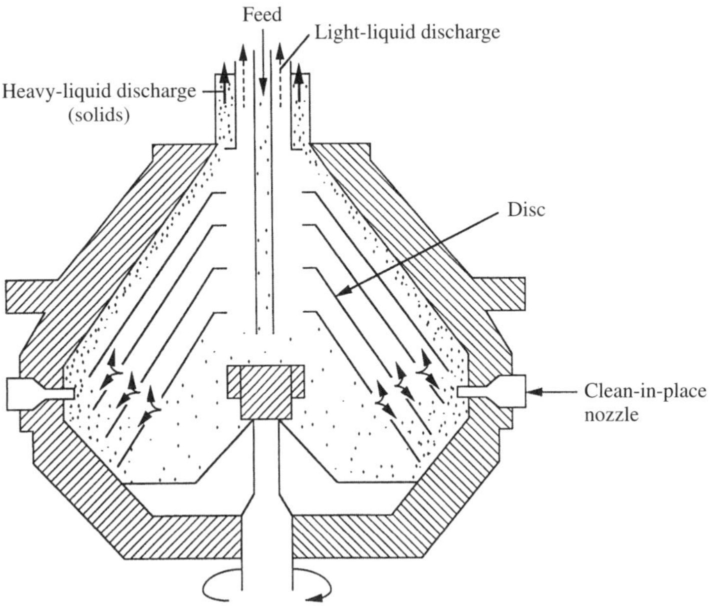

FIGURE 11.7 Disc stack bowl centrifuge with continuous discharge of solids.

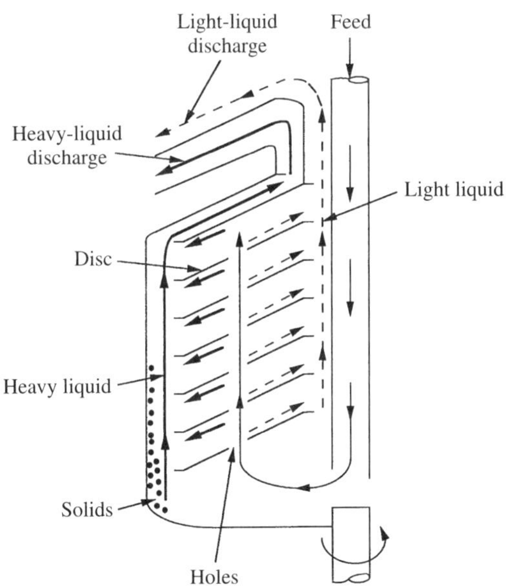

FIGURE 11.8 Mechanism of solids separation in a disc stack bowl centrifuge.
From C.J. Geankoplis, 1983, Transport Processes and Unit Operations, *2nd ed., Allyn and Bacon, Boston.*

The performance characteristics of tubular bowl and disc stack centrifuges used for industrial separations are summarised in Table 11.2. In general, as the size of the centrifuge increases, the practical operating speed is reduced and the maximum centrifugal force decreases.

TABLE 11.2 Performance Characteristics of Tubular Bowl and Disc Stack Centrifuges

Centrifuge Type	Bowl Diameter (cm)	Speed (rpm)	Maximum Centrifugal Force (× gravity)	Liquid Throughput ($l\ min^{-1}$)	Solids Throughput ($kg\ min^{-1}$)	Motor Size (kW)
Tubular bowl	10	15,000	13,200	0.4–40		1.5
	13	15,000	15,900	0.8–80		2.2
Disc stack	18	12,000	14,300	0.4–40		0.25
	33	7500	10,400	20–200		4.5
	61	4000	5500	80–800		5.6
Disc stack with nozzle discharge	25	10,000	14,200	40–150	1.5–15	15
	41	6250	8900	100–550	7–70	30
	69	4200	6750	150–1500	15–190	95
	76	3300	4600	150–1500	15–190	95

Data from Perry's Chemical Engineers' Handbook, *1997, 7th ed., McGraw-Hill, New York.*

11.4.2 Centrifugation Theory

The particle velocity achieved in a particular centrifuge compared with the settling velocity under gravity characterises the effectiveness of centrifugation. The terminal velocity during gravity settling of a small spherical particle in dilute suspension is given by Stoke's law:

$$u_g = \frac{\rho_p - \rho_L}{18\,\mu} D_p^2\, g \tag{11.19}$$

where u_g is the sedimentation velocity under gravity, ρ_p is the density of the particle, ρ_L is the density of the liquid, μ is the viscosity of the liquid, D_p is the particle diameter, and g is gravitational acceleration. In a centrifuge, the corresponding terminal velocity is:

$$u_c = \frac{\rho_p - \rho_L}{18\,\mu} D_p^2\, \omega^2 r \tag{11.20}$$

where u_c is the particle velocity in the centrifuge, ω is the angular velocity of the bowl in units of rad s^{-1}, and r is the radius of the centrifuge drum. The ratio of the velocity in the centrifuge to the velocity under gravity is called the *centrifuge effect* or *g-number*, and is usually denoted Z. Therefore:

$$Z = \frac{\omega^2 r}{g} \tag{11.21}$$

The force developed in a centrifuge is Z times the force of gravity, and is often expressed as so many g-forces. Industrial centrifuges have Z factors up to about 16,000; for small laboratory centrifuges, Z may be up to 500,000 [4].

Sedimentation occurs in a centrifuge as particles moving away from the centre of rotation collide with the walls of the centrifuge bowl. Increasing the velocity of motion will improve the rate of sedimentation. From Eq. (11.20), the particle velocity in a given centrifuge can be increased by:

- Increasing the centrifuge speed, ω
- Increasing the particle diameter, D_p
- Increasing the density difference between the particle and liquid, $\rho_p - \rho_L$
- Decreasing the viscosity of the suspending fluid, μ

Whether the particles reach the walls of the bowl also depends on the time of exposure to the centrifugal force. In batch centrifuges such as those used in the laboratory, centrifuge time is increased by running the equipment longer. In continuous-flow devices such as disc stack centrifuges equipped for continuous solids discharge, the residence time is increased by decreasing the feed flow rate.

The performance of centrifuges of different size can be compared using a parameter called the *sigma factor* Σ. Physically, Σ represents the cross-sectional area of a gravity settler with the same sedimentation characteristics as the centrifuge. For continuous centrifuges, Σ is related to the feed rate of material as follows:

$$\Sigma = \frac{Q}{2u_g} \tag{11.22}$$

where Q is the volumetric feed rate and u_g is the terminal velocity of the particles in a gravitational field as given by Eq. (11.19). If two centrifuges perform with equal effectiveness:

$$\frac{Q_1}{\Sigma_1} = \frac{Q_2}{\Sigma_2} \tag{11.23}$$

where subscripts 1 and 2 denote the two centrifuges. Equation (11.23) can be used to scale up centrifuge equipment. Equations for evaluating Σ depend on the centrifuge design. For a disc stack bowl centrifuge [5]:

$$\Sigma = \frac{2\pi\omega^2(N-1)}{3g\tan\theta}(r_2^3 - r_1^3) \tag{11.24}$$

where ω is the angular velocity in rad s^{-1}, N is the number of discs in the stack, r_2 is the outer radius of the discs, r_1 is the inner radius of the discs, g is gravitational acceleration, and θ is the half-cone angle of the discs. For a tubular bowl centrifuge, the following equation is accurate to within 4% [6]:

$$\Sigma = \frac{\pi\omega^2 b}{2g}(3r_2^2 + r_1^2) \tag{11.25}$$

where b is the length of the bowl, r_1 is the radius of the liquid surface, and r_2 is the radius of the inner wall of the bowl (Figure 11.6). Because r_1 and r_2 in a tubular bowl centrifuge are about equal, Eq. (11.25) can be approximated as:

$$\Sigma = \frac{2\pi\omega^2 b r^2}{g} \tag{11.26}$$

where r is an average radius roughly equal to either r_1 or r_2.

The equations for Σ are based on ideal operating conditions. Because different types of centrifuge deviate to varying degrees from ideal operation, Eq. (11.23) cannot generally be used to compare different centrifuge configurations. The performance of any centrifuge can deviate from that predicted theoretically due to factors such as the particle shape and size distribution, particle aggregation, nonuniform flow distribution in the centrifuge, and interaction between the particles during sedimentation. Experimental tests must be performed to account for these factors.

EXAMPLE 11.2 CELL RECOVERY IN A DISC STACK CENTRIFUGE

A continuous disc stack centrifuge is operated at 5000 rpm for separation of bakers' yeast. At a feed rate of 60 l min^{-1}, 50% of the cells are recovered. For operation at constant centrifuge speed, solids recovery is inversely proportional to the flow rate.

(a) What flow rate is required to achieve 90% cell recovery if the centrifuge speed is maintained at 5000 rpm?

(b) What operating speed is required to achieve 90% recovery at a feed rate of 60 l min^{-1}?

Solution

(a) If solids recovery is inversely proportional to feed rate, the flow rate required is:

$$\frac{50\%}{90\%}(60\ \text{l min}^{-1}) = 33.3\ \text{l min}^{-1}$$

(b) Equation (11.23) relates the operating characteristics of centrifuges achieving the same separation. From (a), 90% recovery is achieved at $Q_1 = 33.3$ l min^{-1} and $\omega_1 = 5000$ rpm. For $Q_2 = 60$ l min^{-1}, from Eq. (11.23):

$$\frac{Q_1}{Q_2} = \frac{\Sigma_1}{\Sigma_2} = \frac{33.3\ \text{l min}^{-1}}{60\ \text{l min}^{-1}} = 0.56$$

Because the same centrifuge is used and all the geometric parameters are the same, from Eq. (11.24):

$$\frac{\Sigma_1}{\Sigma_2} = \frac{\omega_1^2}{\omega_2^2} = 0.56$$

The ratio of ω_1^2 to ω_2^2 is 0.56, irrespective of the units used to express angular velocity. Therefore, using units of rpm:

$$\omega_2^2 = \frac{\omega_1^2}{0.56} = \frac{(5000\ \text{rpm})^2}{0.56} = 4.46 \times 10^7\ \text{rpm}^2$$

Taking the square root, $\omega_2 = 6680$ rpm.

11.5 CELL DISRUPTION

Downstream processing of fermentation broths usually begins with removal of the cells by filtration or centrifugation. The next step depends on the location of the desired product. For substances such as ethanol, citric acid, and antibiotics that are excreted from cells, the product is recovered from the cell-free broth using unit operations such as those described later in this chapter. In these cases, the biomass separated from the liquid is discarded or sold as a by-product. For products such as enzymes and recombinant proteins that remain in the biomass, the harvested cells must be broken open to release the desired material. A variety of methods is available to disrupt cells. Mechanical options include grinding with abrasives, high-speed agitation, high-pressure pumping, and ultrasound. Nonmechanical methods such as osmotic shock, freezing and thawing, enzymic digestion of the cell walls, and treatment with solvents and detergents can also be applied.

A widely used technique for cell disruption is high-pressure homogenisation. The forces generated in this treatment are sufficient to completely disrupt many types of cell. A common apparatus for homogenisation of cells is the Gaulin homogeniser. As indicated in Figure 11.9, this high-pressure pump incorporates an adjustable valve with restricted orifice through which cells are forced at pressures up to 550 atm. The homogeniser is of general applicability for disruption of all types of cell; however the homogenising valve can become blocked when used with highly filamentous organisms. Greater disruption is achieved by maintaining a small gap around the valve so that the cells strike the impact ring with high velocity. The exact mechanisms responsible for cell disruption in the Gaulin homogeniser remain a subject of debate. Cavitation, fluid shear, impact, and pressure shock may all play a role in the breakage process.

Because the desired degree of cell disruption and product release generally is not achieved during a single pass through the homogeniser, multiple passes are required. Experimental results for the amount of protein released from yeast cells as a function of

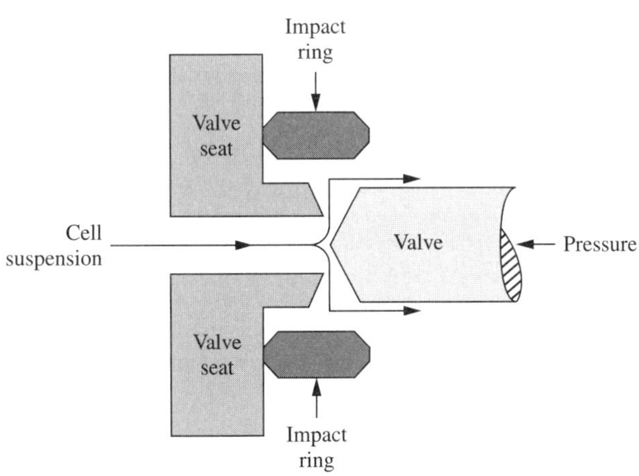

FIGURE 11.9 Cell disruption in a high-pressure homogeniser.

the number of times the suspension is passed through the equipment are shown in Figure 11.10. The extent of cell disruption in a Gaulin homogeniser is related to operating conditions by the equation [7]:

$$\ln\left(\frac{R_{max}}{R_{max} - R}\right) = k N p^{\alpha} \tag{11.27}$$

In Eq. (11.27), R_{max} is the maximum amount of protein available for release, R is the amount of protein released after N passes through the homogeniser, k is a temperature-dependent rate constant, and p is the operating pressure. Both k and α vary with cell type. The exponent α is a measure of the resistance of the cells to disruption; values of α range between 0.9 and 2.9 for bacteria and yeast [8]. However, because the exponent for a particular organism depends on the strength of the cell wall, α varies to some extent with growth conditions and phase of growth [9]. For practical purposes, α is considered independent of operating pressure, although some variation has been observed over very wide pressure ranges [10, 11].

Homogeniser efficiency has been found to decrease slightly at high cell concentrations [12]; otherwise Eq. (11.27) applies irrespective of the concentration of cells in the homogeniser feed. The dependence of protein release on pressure in Eq. (11.27) suggests that high-pressure operation is beneficial. In theory, complete disruption can be achieved in a single pass through the homogeniser if the pressure is sufficiently high; in practice, however, this rarely occurs and multiple passes are required. Reduction in the number of passes used is desirable because multiple passes produce fine cell debris, which can cause problems in subsequent clarification steps.

Protein release depends markedly on temperature; protein recovery increases at elevated temperatures up to 50°C. However, cooling to between 0 and 4°C is recommended during operation of homogenisers to minimise protein denaturation. Oxidation of certain

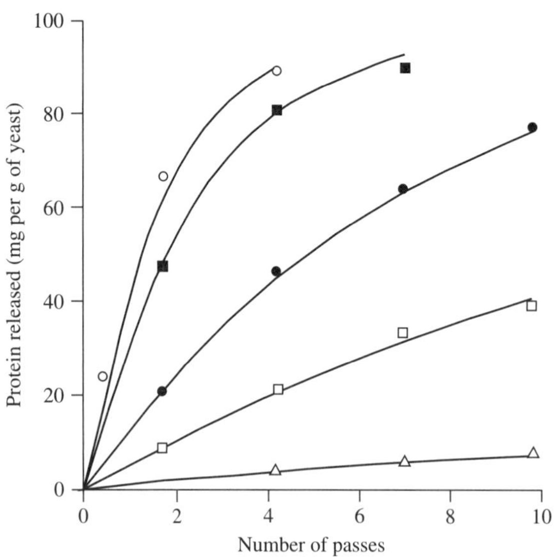

FIGURE 11.10 Protein release from yeast cells at 30°C as a function of operating pressure and the equivalent number of passes in recycle operation through a Gaulin homogeniser: (△) 100 $kg_f\ cm^{-2}$, (□) 200 $kg_f\ cm^{-2}$, (●) 270 $kg_f\ cm^{-2}$, (■) 400 $kg_f\ cm^{-2}$, (○) 460 $kg_f\ cm^{-2}$.
Adapted from P.J. Hetherington, M. Follows, P. Dunnill, and M.D. Lilly, 1971, Release of protein from baker's yeast (Saccharomyces cerevisiae) *by disruption in an industrial homogeniser.* Trans. IChemE *49, 142–148.*

intracellular products is known to occur during normal operation of the Gaulin homogeniser. To minimise this damage, the cells may be suspended prior to disruption in a buffer containing redox and other protective agents. Procedures for scale-up of homogenisers are not well developed. Accordingly, methods that work well in the laboratory may give variable results when used at a larger scale.

11.6 THE IDEAL STAGE CONCEPT

So far, we have considered only the initial steps of downstream processing: cell isolation and disruption. An important group of unit operations used for primary isolation of fermentation products relies on mass transfer to achieve separation of components between phases. Equipment for these separations may consist of a series of stages in which the phases make intimate contact so that material can be transferred between them. Even if the equipment itself is not constructed in discrete stages, mass transfer can be considered to occur in stages. The effectiveness of each stage in accomplishing mass transfer depends on many factors, including the equipment design, the physical properties of the phases, and equilibrium relationships.

Consider the simple mixer–settler device of Figure 11.11 for extraction of component A from one liquid phase to another. The mass transfer principles of liquid–liquid extraction have already been described in Chapter 10; unit operations for extraction are also discussed in Section 11.7. Here, we will use the mixer–settler to explain the general concept of an ideal stage. Two immiscible liquids enter the mixing vessel with volumetric flow rates L_{i1} and L_{i2}; these streams contain A at concentrations C_{Ai1} and C_{Ai2}, respectively. The two liquid phases are mixed together vigorously and A is transferred from one phase to the other. The mixture then passes to a settler where phase separation is allowed to occur under the influence of gravity. The flow rate of light liquid out of the settler is L_{o1}; the flow rate of heavy liquid is L_{o2}. The concentrations of A in these streams are C_{Ao1} and C_{Ao2}, respectively.

Operation of the mixer–settler relies on the liquids entering not being in equilibrium as far as the concentration of A is concerned, that is, C_{Ai1} and C_{Ai2} are not equilibrium concentrations. This means that a driving force exists in the mixer for change of concentration as A is transferred from one liquid to the other. Depending on the ability of the mixer to

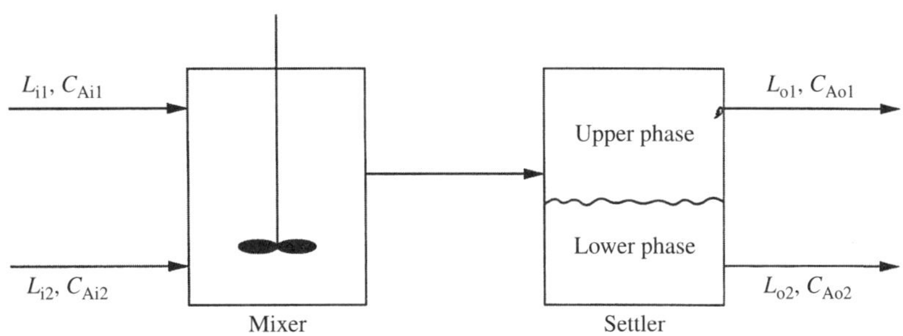

FIGURE 11.11 An ideal stage.

promote mass transfer between the liquids and the effectiveness of the settler in allowing phase separation, when the liquids leave the system they will have been brought closer to equilibrium. If the mixer–settler were an *ideal stage*, the two streams leaving it would be in equilibrium. Under these conditions, no further mass transfer of A would result from additional contact between the phases; C_{Ao1} and C_{Ao2} would be equilibrium concentrations and the device would be operating at maximum efficiency. In reality, stages are not ideal; the change in concentration achieved in a real stage is always less than in an ideal stage so that extra stages are necessary to achieve the desired separation. The relative performance of an actual stage compared with that of an ideal stage is expressed as the *stage efficiency*.

The concept of ideal stages is used in design calculations for several unit operations. The important elements in analysis of these operations are material balances, energy balances if applicable, and the equilibrium relationships between phases. Application of these principles is illustrated in the following sections.

11.7 AQUEOUS TWO-PHASE LIQUID EXTRACTION

Liquid extraction is used to isolate many pharmaceutical products from animal and plant sources. In liquid extraction of fermentation products, components dissolved in liquid are recovered by transfer into an appropriate solvent. Extraction of penicillin from aqueous broth using solvents such as butyl acetate, amyl acetate, or methyl isobutyl ketone, and isolation of erythromycin using pentyl or amyl acetate are examples. Solvent extraction techniques are also applied for recovery of steroids, purification of vitamin B_{12} from microbial sources, and isolation of alkaloids such as morphine and codeine from raw plant material.

The simplest equipment for liquid extraction is the separating funnel used for laboratory-scale product recovery. Liquids forming two distinct phases are shaken together in the separating funnel; solute in dilute solution in one solvent transfers to the other solvent to form a more concentrated solution. The two phases are then allowed to separate and the heavy phase is withdrawn from the bottom of the funnel. The phase containing the solute in concentrated form is processed further to purify the product. Whatever apparatus is used for extraction, it is important that contact between the liquid phases is maximised by vigorous mixing and turbulence to facilitate solute transfer.

Extraction with organic solvents is a major separation technique in bioprocessing, particularly for recovery of antibiotics. However, organic solvents are unsuitable for isolation of proteins and other sensitive products. Techniques have been developed for aqueous two-phase extraction of these molecules. Aqueous solvents that form two distinct phases provide favourable conditions for separation of proteins, cell fragments, and organelles with protection of their biological activity. Two-phase aqueous systems are produced when particular polymers or a polymer and salt are dissolved together in water above certain concentrations. The liquid partitions into two phases, each containing 85 to 99% water.

Some components used to form aqueous two-phase systems are listed in Table 11.3. When added to these mixtures, biomolecules and cell fragments partition between the phases: by selecting appropriate conditions, cell fragments can be confined to one phase

TABLE 11.3 Examples of Aqueous Two-Phase Systems

Component 1	Component 2
Polyethylene glycol	Dextran
	Polyvinyl alcohol
	Polyvinylpyrrolidone
	Ficoll
	Potassium phosphate
	Ammonium sulphate
	Magnesium sulphate
	Sodium sulphate
Polypropylene glycol	Polyvinyl alcohol
	Polyvinylpyrrolidone
	Dextran
	Methoxypolyethylene glycol
	Potassium phosphate
Ficoll	Dextran
Methylcellulose	Dextran
	Hydroxypropyldextran
	Polyvinylpyrrolidone

From M.R. Kula, 1985, Liquid–liquid extraction of biopolymers. In: M. Moo-Young, Ed., Comprehensive Biotechnology, *vol. 2, pp. 451–471, Pergamon Press, Oxford.*

while the desired protein partitions into the other phase. Aqueous two-phase separations are of special interest for extraction of enzymes and recombinant proteins from cell debris produced by cell disruption. After partitioning, the product is removed from the extracting phase using other unit operations such as precipitation or crystallisation.

The extent of differential partitioning between phases depends on the equilibrium relationship for the system. The *partition coefficient* K is defined as:

$$K = \frac{C_{Au}}{C_{Al}} \tag{11.28}$$

where C_{Au} is the equilibrium concentration of component A in the upper phase and C_{Al} is the equilibrium concentration of A in the lower phase. If $K > 1$, component A favours the upper phase; if $K < 1$, A is concentrated in the lower phase. In many aqueous systems, K is constant over a wide range of concentrations.

Partitioning is influenced by the size, electric charge, and hydrophobicity of the particles or solute molecules; biospecific affinity for one of the polymers may also play a role in some systems. The surface free energy of the phase components and the ionic composition of the liquids are of paramount importance in determining separation; K is related to both these parameters. Partitioning is also affected by other sometimes interdependent factors so that it is impossible to predict partition coefficients from molecular properties. For single-stage extraction of enzymes, partition coefficients ≥3 are normally required [3].

Even when the partition coefficient is low, good *product recovery* or *yield* can be achieved by using a large volume of the phase preferred by the solute. The yield of A in the upper phase, Y_{Au}, is defined as:

$$Y_{Au} = \frac{V_u C_{Au}}{V_0 C_{A0}} = \frac{V_u C_{Au}}{V_u C_{Au} + V_l C_{Al}} \tag{11.29}$$

where V_u is the volume of the upper phase, V_l is the volume of the lower phase, V_0 is the original volume of solution containing the product, and C_{A0} is the original product concentration in that solution. The yield of A in the lower phase, Y_{Al}, is defined as:

$$Y_{Al} = \frac{V_l C_{Al}}{V_0 C_{A0}} = \frac{V_l C_{Al}}{V_u C_{Au} + V_l C_{Al}} \tag{11.30}$$

The maximum possible yield for an ideal extraction stage can be evaluated using Eqs. (11.29) and (11.30) and the equilibrium partition coefficient, K. Dividing both the numerator and denominator of Eq. (11.29) by C_{Au} and recognising that, at equilibrium, C_{Al}/C_{Au} is equal to $1/K$:

$$Y_{Au} = \frac{V_u}{V_u + \dfrac{V_l}{K}} \tag{11.31}$$

Similarly, dividing the numerator and denominator of Eq. (11.30) by C_{Al} gives:

$$Y_{Al} = \frac{V_l}{V_u K + V_l} \tag{11.32}$$

EXAMPLE 11.3 ENZYME RECOVERY USING AQUEOUS EXTRACTION

Aqueous two-phase extraction is used to recover α-amylase from solution. A polyethylene glycol–dextran mixture is added and the solution separates into two phases. The partition coefficient is 4.2. Calculate the maximum possible enzyme recovery when:

(a) The volume ratio of the upper to lower phase is 5.0
(b) The volume ratio of the upper to lower phase is 0.5

Solution

As the partition coefficient is greater than 1, amylase prefers the upper phase. The yield at equilibrium is therefore calculated for the upper phase. Dividing both the numerator and denominator of Eq. (11.31) by V_l gives:

$$Y_{Au} = \frac{\dfrac{V_u}{V_l}}{\dfrac{V_u}{V_l} + \dfrac{1}{K}}$$

(a) $\frac{V_u}{V_l} = 5.0$. Therefore:

$$Y_{Au} = \frac{5.0}{5.0 + \frac{1}{4.2}} = 0.95$$

The maximum enzyme recovery is 95%.

(b) $\frac{V_u}{V_l} = 0.5$. Therefore:

$$Y_{Au} = \frac{0.5}{0.5 + \frac{1}{4.2}} = 0.68$$

The maximum enzyme recovery is 68%. Increasing the relative volume of the extracting phase enhances recovery.

Another parameter used to characterise two-phase partitioning is the *concentration factor* δ_c, which is defined as the ratio of the product concentration in the preferred phase to the initial product concentration:

$$\delta_c = \frac{C_{Al}}{C_{A0}} \text{ (when product partitions to the lower phase)} \tag{11.33}$$

$$\delta_c = \frac{C_{Au}}{C_{A0}} \text{ (when product partitions to the upper phase)} \tag{11.34}$$

Aqueous extraction in polyethylene glycol–salt mixtures is an effective technique for separating proteins from cell debris. In this system, debris partitions to the lower phase while most target proteins are recovered from the upper phase. Extractions can be carried out in single-stage operations such as that depicted in Figure 11.11 using a polymer mixture that provides a suitable partition coefficient. Equilibrium is approached in extraction operations but rarely reached. For industrial application, it is important to consider the time taken for mass transfer and the ease of mechanical separation of the phases. The rate of approach to equilibrium depends on the surface area available for exchange between the phases; this is maximised by rapid mixing.

Separation of the phases is sometimes a problem because of the low interfacial tension between aqueous phases; however, very rapid large-scale extractions can be achieved by combining mixed vessels with centrifugal separators. In many cases, recovery and concentration of product with yields exceeding 90% can be achieved using a single extraction step [3]. When single-stage extraction does not give sufficient recovery, repeated extractions can be carried out in a chain or cascade of contacting and separation units.

11.8 PRECIPITATION

Precipitation is a well-established and widely used method for recovering proteins from cell homogenates and culture broths. Precipitation is also applied in downstream processing of other fermentation products such as citric acid and antibiotics, and may be used to remove unwanted contaminants such as pigments, calcium, proteins, and nucleic acids from process streams. Typically, addition of *precipitants* such as salts, solvents, and polymers, and changes in the pH, ionic strength, or temperature of the solution, are used to reduce the solubility of the product, causing it to precipitate in the form of insoluble particles. The precipitated solids are then recovered by filtration, microfiltration, or centrifugation.

Precipitation is applied commonly during the early stages of downstream processing because large reductions in liquid volume can be achieved, thereby minimising the size and cost of the equipment required for subsequent processing steps. However, although the degree of product concentration effected by precipitation is substantial due to the removal of large amounts of water, only partial purification from complex mixtures such as cell extracts is achieved because many other substances in addition to the product are present in the precipitate. The separation factors (Section 11.1) obtained using precipitation are therefore relatively modest compared with those achieved using more selective operations such as chromatography.

11.8.1 Protein Structure and Surface Chemistry

Because precipitation is used commonly for large-scale recovery of proteins, we will focus our attention on protein applications. Proteins treated using precipitation include natural products such as enzymes and food proteins from microbial, plant, and animal sources, as well as recombinant proteins produced using genetically engineered organisms. The tendency of a protein to precipitate depends on the properties of the solvent, such as pH, ionic strength, and dielectric constant, and the size, charge, and hydrophobicity of the protein. In particular, the structure and surface chemistry of proteins play a critical role in precipitation. Proteins remain in solution only if it is energetically more favourable for them to be surrounded by solvent than to be aggregated with other protein molecules into a solid phase.

Both attractive and repulsive forces exist between neighbouring protein molecules in solution. The balance of these forces, that is, whether attraction overcomes repulsion or vice versa, determines whether or not the protein will precipitate. Some of the sources of attraction and repulsion between protein molecules are illustrated in Figure 11.12. In aqueous solution, water-soluble globular proteins are folded so that most of their polar, hydrophilic amino acid side chains are presented on the protein surface, while most of the hydrophobic residues, such as those associated with phenylalanine, tryptophan, leucine, valine, and methionine, are shielded within. However, as indicated schematically in Figure 11.12(a), a typical protein molecule will have some hydrophobic zones on its surface. Globular proteins are also characterised by large numbers of nonuniformly distributed surface charges. When two protein molecules are brought together, they are attracted to each other due to interactions between opposite surface charges and any exposed hydrophobic regions.

To understand the repulsive forces between protein molecules, let us consider the interactions between soluble proteins and their solvent. In most cases, proteins are surrounded

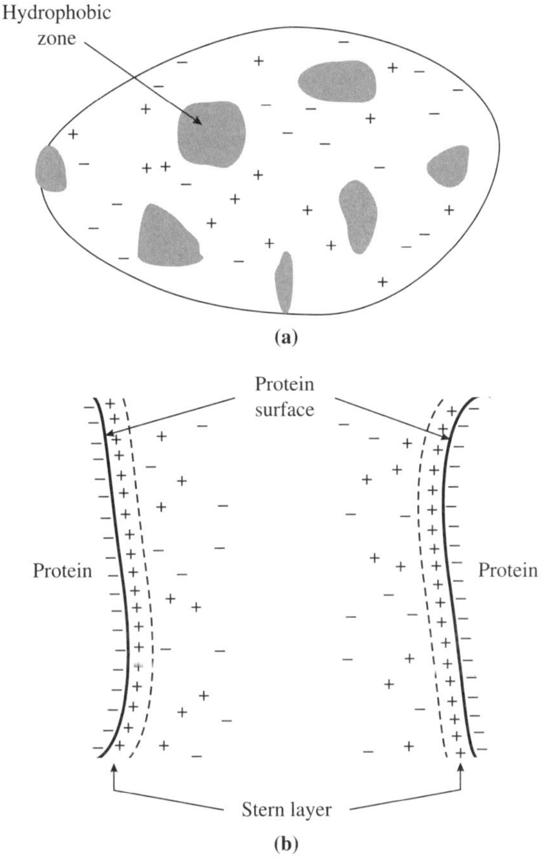

FIGURE 11.12 Surface chemistry of globular proteins. (a) Hydrophobic zones and nonuniform charge distribution on the surface of a protein. (b) Formation of the Stern layer of counter-ions and the electrical double layer around two neighbouring soluble proteins.

by charged layers provided by the solvent. The *isoelectric point* of a protein is the pH at which the protein carries zero net charge. Under these conditions, the positive and negative charges on the protein surface are balanced. Although proteins can have isoelectric points anywhere between about 1 and 12, the isoelectric points of many proteins range between 4 and 6. Therefore, when suspended in electrolyte at neutral or near-neutral pH, the surfaces of these proteins have a net negative charge. As a consequence, the protein will attract positive ions from the solution to form a close layer of counter-ions over the molecule surface. This is called the *Stern layer*, and is illustrated in Figure 11.12(b). Further out, a more diffuse and mobile layer of counter-ions also develops to form an *electrical double layer* surrounding the protein. When two similarly charged protein molecules are brought together, although they may be attracted by ionic and hydrophobic interactions with each other's surfaces, they are also repulsed by the interaction of their electrical double layers.

Another source of repulsion between proteins in aqueous solution is the hydration layers that form around protein molecules. Water attracted to the large exposed hydrophilic regions of globular proteins becomes organised at the protein surface to form a hydration shell consisting of both tightly bound and loosely bound water molecules. This shell of water represents an additional barrier to protein–protein aggregation.

The thickness and structure of the Stern and diffuse ion layers and the stability of the hydration shell determine the strength of the repulsive forces between protein molecules that keep the protein in solution. However, these properties can be modified in the presence of salts, acids/bases, organic solvents, and polymers. Additives such as these are often used to reduce protein solubility and induce precipitation in downstream processing, as described in the following sections.

11.8.2 Precipitation Methods

In this section, we consider the principal techniques used to reduce the solubility of materials in precipitation operations, thereby inducing precipitation. For protein recovery, precipitation is aimed at separating the protein from solution without inducing major or irreversible changes to the protein structure or biological function.

Salting-Out

High salt concentrations promote the aggregation and precipitation of proteins. This phenomenon is considered to occur as a result of disruption of the hydration barriers between protein molecules, as salt causes water surrounding the protein to move into the bulk solution. The hydrophobic zones of the protein surface thus become exposed, providing sites of attraction between neighbouring protein molecules. The success of salting-out therefore depends on the hydrophobicity of the protein: proteins with few exposed hydrophobic regions tend to remain in solution even at high salt concentrations. Additional mechanisms for increasing the affinity of neighbouring protein molecules, such as direct binding between the salt and protein molecules, may also be involved in salting-out.

At high ionic strengths, the *solubility* of a pure protein depends on the ionic strength of the solution according to the Cohn equation [13]:

$$\ln S = \beta - K\,I \tag{11.35}$$

where S is the solubility of the protein at ionic strength I, and β and K are constants. The solubility of a protein in a particular solvent is the maximum concentration of dissolved protein that can be achieved under given conditions of temperature and pressure. The *ionic strength* of a solution is defined as:

$$I = \frac{1}{2}\Sigma z_i^2 C_i \tag{11.36}$$

where z_i is the charge or valence of ionic component i, C_i is the molar concentration of component i in the solution, and the symbol Σ means the sum for all ionic components. Typical units for C_i and I are mol l^{-1} or mol m^{-3}. If the ionic strength of the solution is altered by changing the concentration of only one salt, Eqs. (11.35) and (11.36) can be combined and simplified to give the empirical equation:

$$\ln S = \beta_S - K_S\,C_S \tag{11.37}$$

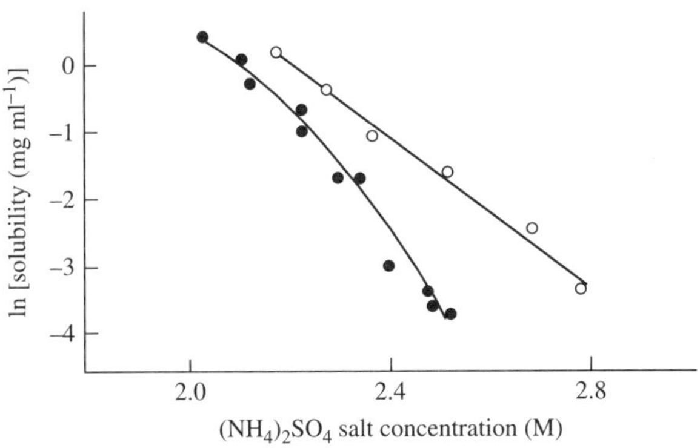

FIGURE 11.13 Solubility of rabbit muscle aldolase in ammonium sulphate solutions at pH 7.0: (○) pure aldolase, and (●) aldolase in a mixture with pyruvate kinase, glyceraldehyde phosphate dehydrogenase, and lactate dehydrogenase. *Adapted from R.K. Scopes, 1994,* Protein Purification: Principles and Practice, *3rd ed., Springer, New York.*

where β_S and K_S are constants and C_S is the concentration of the salt used to adjust the ionic strength of the solution. For a pure protein, the *salting-out constant* K_S depends on the properties of the protein and the salt, but is relatively insensitive to pH and temperature. In contrast, the constant β_S varies strongly with pH, temperature, and the nature of the protein but is essentially independent of the salt. In general, for proteins with similar chemical properties, high-molecular-weight proteins require less salt for precipitation than small proteins.

Equations (11.35) and (11.36) apply only at high ionic strength. Therefore, according to the relationship represented in Eq. (11.37), a plot of ln S versus C_S at high salt concentrations should give a straight line with negative slope. Experimental data showing the change in protein solubility with salt concentration are shown in Figure 11.13. Although Eq. (11.37) holds for many pure proteins, it does not necessarily apply to protein mixtures. This is because coprecipitation can occur with the other proteins present, depending on their properties. The effect in a protein mixture is that the solubility of a particular protein will be reduced relative to that predicted using Eq. (11.37), as illustrated in Figure 11.13.

An implication of Eqs. (11.35) and (11.37) is that, at constant temperature, pH, and ionic strength, protein solubility is also constant. Although this is true for many proteins, for others such as bovine serum albumin and α-chymotrypsin, solubility depends on the initial concentration of protein present. Cases for which the Cohn equation does not fully represent protein solubility relationships are discussed further in the literature [14].

Salts with high values of K_S are more effective for salting-out than salts with low K_S. The Hofmeister series gives the relative effectiveness of different anions for protein precipitation:

$$\text{citrate} > \text{phosphate} > \text{sulphate} > \text{acetate} > \text{chloride} > \text{nitrate} > \text{thiocyanate} \qquad (11.38)$$

Because polyvalent cations such as calcium and magnesium depress the value of K_S, ammonium, potassium, and sodium salts are used widely. Protein precipitation at an industrial scale requires large amounts of salt; therefore, it is important that the salt used be inexpensive. The salt must also be highly soluble in aqueous solutions to achieve the concentrations required, and should not have a large heat of solution (Section 5.4.3) to avoid temperature increases that might denature the product or affect its solubility. For these reasons, ammonium sulphate $(NH_4)_2SO_4$ is the most frequently applied precipitant for protein salting-out. The solubility of $(NH_4)_2SO_4$ in water at temperatures of 0 to 30°C is approximately 530 g l^{-1} or 4 M. $(NH_4)_2SO_4$ does have some disadvantages, however: it is corrosive, releases ammonia at high pH, and can cause toxic reactions in animals if residues are allowed to remain in proteins destined for therapeutic use.

When salts are dissolved into solutions at the high concentrations necessary for protein salting-out, the solution volume increases. A formula for estimating the mass G_S of $(NH_4)_2SO_4$ that must be added to 1 litre of solution at 20°C to change the salt concentration from molar concentration M_1 to molar concentration M_2 is [15]:

$$G_S = \frac{533\,(M_2 - M_1)}{4.05 - 0.3\,M_2} \tag{11.39}$$

where G_S is in grams. According to Eq. (11.39), even though the molecular weight of $(NH_4)_2SO_4$ is 132.1, approximately 142 g of salt must be added to 1 litre of water to make a 1-M solution of $(NH_4)_2SO_4$. The reason is that the volume of the solution increases to 1.08 litres due to addition of the salt. The increase in solution volume per kg of $(NH_4)_2SO_4$ added is approximately 0.54 l. As a consequence, protein concentrations based on the total solution volume are effectively reduced as salt is added.

Precipitates formed by salting-out contain large quantities of salt as well as protein; therefore, the precipitate must be *desalted* to remove the residual salt. This is usually achieved using dialysis, diafiltration, or gel chromatography. Some loss of protein activity can be expected during salting-out. In general, the maximum concentration of salt used is determined by balancing the desire to purify the protein with the need to avoid protein denaturation.

High salt concentrations only reduce the solubility of proteins; they do not cause precipitation of all the protein present. Sequential precipitation operations using increasing salt concentrations may be required to recover the desired quantity of protein from solution.

EXAMPLE 11.4 ANTIBODY RECOVERY BY SALTING-OUT

The solubility of pure IgA monoclonal antibody in 1.4 M $(NH_4)_2SO_4$ is measured as 770 μg ml^{-1}; the solubility in 3.0 M $(NH_4)_2SO_4$ is found to be 29 μg ml^{-1}. A solution containing 180 g of pure antibody is treated for protein recovery using 2.0 M $(NH_4)_2SO_4$ in a total volume of 1000 l. What percentage of the antibody is precipitated?

Solution

The solubility of antibody in 2.0 M $(NH_4)_2SO_4$ can be determined using Eq. (11.37) after the values of β_S and K_S are evaluated from the experimental data provided. Equation (11.37) has the same form as Eq. (3.17) with β_S equivalent to ln B and $-K_S$ equivalent to A. The values of β_S and K_S can therefore be determined using the equations outlined in Section 3.4.2. From Eq. (3.18):

$$-K_S = \frac{\ln (S_2/S_1)}{C_{S2} - C_{S1}}$$

Substituting parameter values gives:

$$-K_S = \frac{\ln (29/770)}{(3.0 - 1.4)\,\text{M}} = -2.05\,\text{M}^{-1}$$

Therefore, $K_S = 2.05\ \text{M}^{-1}$. From Eq. (3.19):

$$\beta_S = \ln S_1 - (-K_S)\, C_{S1}$$

or

$$\beta_S = \ln 770 + (2.05\ \text{M}^{-1})\, 1.4\ \text{M} = 9.52$$

Therefore, when μg ml^{-1} is used as the units for S, $\beta_S = 9.52$. Applying these parameter values in Eq. (11.37) with $C_S = 2.0$ M:

$$\ln S = 9.52 - (2.05\ \text{M}^{-1})\, 2.0\ \text{M} = 5.42$$

From Eqs. (E.1) and (E.2) in Appendix E:

$$S = e^{5.42} = 226\ \mu\text{g ml}^{-1}$$

Therefore, the solubility of antibody in 2.0 M $(NH_4)_2SO_4$ is 226 μg ml^{-1}. The total concentration of antibody present in the treatment solution is:

$$\frac{180\ \text{g}}{1000\ \text{l}} \cdot \left|\frac{10^6\ \mu\text{g}}{1\ \text{g}}\right| \cdot \left|\frac{1\ \text{l}}{1000\ \text{ml}}\right| = 180\ \mu\text{g ml}^{-1}$$

As the solubility of antibody in 2.0 M $(NH_4)_2SO_4$ is greater than the concentration of antibody to be treated, precipitation will not occur. A higher $(NH_4)_2SO_4$ concentration is needed to recover this protein.

Isoelectric Precipitation

A protein at its isoelectric point has zero net charge. Under these conditions, the electrical double layer shown in Figure 11.12(b) is weakened, thereby reducing the electrostatic repulsion between neighbouring protein molecules. In a solution of relatively low ionic strength, this reduction in repulsive forces may be sufficient to allow protein precipitation. As illustrated in Figure 11.14, most globular proteins have minimum solubility at or near their isoelectric point. In many cases, protein precipitation can be induced by lowering the pH to between 4.0 and 6.0 provided the protein is stable and does not denature under these conditions. An advantage of isoelectric precipitation compared with salting-out is that subsequent desalting of the precipitate is not required. Isoelectric precipitation may be used in conjunction with other precipitation techniques using salt, organic solvents, or polymer to achieve higher levels of product recovery.

Precipitation with Organic Solvents

Addition of a water-soluble, weakly polar solvent such as ethanol or acetone to aqueous protein solutions will generally produce a protein precipitate. As indicated in Table 11.4, organic solvents have lower *dielectric constant* or *permittivity* than water; this means that they store less electrostatic energy than water. Dielectric constants for aqueous mixtures of organic solvents are listed in the *International Critical Tables* [16]. The attraction between polar groups is stronger in solvents of low dielectric constant and weaker in solvents of

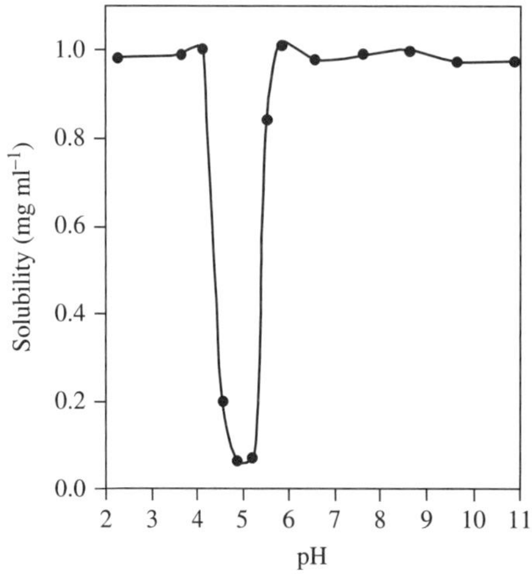

FIGURE 11.14 Solubility of soybean vicilin (7S) protein as a function of pH. The protein isoelectric point is close to pH 5.0. *Adapted from Y.J. Yuan, O.D. Velev, K. Chen, B.E. Campbell, E.W. Kaler, and A.M. Lenhoff, 2002, Effect of pH and Ca^{2+}-induced associations of soybean proteins.* J. Agric. Food Chem. *50, 4953–4958.*

TABLE 11.4 Dielectric Constants of Water and Organic Solvents at 20°C

Solvent	Relative Dielectric Constant*
Acetone	21.01
Diethyl ether	4.2666
Dimethylformamide	38.25
Dioxane	2.2189
Ethanol	25.3
Methanol	33.0
2-Propanol	20.18
Water	80.1

**Ratio of the dielectric constant to the dielectric constant of a vacuum. From* Handbook of Chemistry and Physics, *2001, 82nd ed., CRC Press, Boca Raton, FL.*

high dielectric constant. Therefore, in the presence of organic solvent, oppositely charged groups of proteins experience greater attractive forces than in water, resulting in protein aggregation.

An additional effect is that organic solvents reduce the level of hydration of globular proteins by displacing water molecules into the bulk solution. This exposes interactive sites on the protein surface, thus reducing the solubility. All other properties being equal, the larger the protein the lower is the concentration of organic solvent required for precipitation. Addition of organic solvents may be combined with salting-out and pH adjustment to improve the yield of precipitated protein.

A major concern with organic solvent precipitation is denaturation of the protein. To avoid protein structural damage, precipitation must be carried out at reduced temperatures, usually less than 10°C. Fortunately, the extent of precipitation increases at lower temperatures. Typically, addition of organic solvent to aqueous solutions causes the temperature to rise due to the heat of mixing (Section 5.4.3); therefore, slow rates of addition and efficient cooling are required. In large-scale operations, the flammability of organic solvents can present safety problems. On the other hand, high volatility also assists the subsequent removal and recovery of solvent from the precipitate.

Precipitation with Polymers

Increasingly, nonionic, water-soluble polymers such as polyethylene glycol are being applied for protein precipitation. The effect of these polymers on the surface chemistry of proteins is thought to be similar to that of organic solvents. As polyethylene glycol is a dense, viscous liquid, practical concentrations for protein precipitation are limited to a maximum of about 20% w/v. Proteins of high molecular weight are precipitated at lower polyethylene glycol concentrations than proteins of low molecular weight. An advantage compared with salting-out or organic solvent precipitation is that many globular proteins are stabilised by interaction with polyethylene glycol. Residual polymer incorporated into the precipitate can be removed using unit operations such as adsorption, aqueous two-phase liquid extraction, or ultrafiltration.

11.8.3 Selective Precipitation

A characteristic of most precipitation operations is that they are relatively nonselective. In other words, the precipitate contains many other proteins and contaminating components that must be removed in further processing. However, methods are being developed to improve the specificity of precipitation. The terms *selective*, *fractional*, and *differential precipitation* refer to precipitation techniques aimed at separating different proteins from each other. Under ideal conditions, selective precipitation can generate high-purity homogeneous protein products from crude biological mixtures with yields of greater than 90%.

Affinity precipitation offers a high degree of specificity by making use of binding molecules called *ligands* capable of selective and reversible binding with particular proteins. The ligand may be attached to an insoluble matrix or solid to facilitate separation of the protein from solution. Alternatively, the ligand may be borne on a water-soluble polymer, surfactant, or other material that is readily precipitated by addition of salt or change in pH. Potential ligands include bifunctional molecules such as *p*-aminobenzamidine, lectins, and antibodies. If the protein is an enzyme, substrates, inhibitors, and co-enzymes may also be employed in selective precipitation schemes. After the precipitate is recovered from solution, the protein is eluted from the ligand under suitable conditions and the ligand recycled for further use. A major disadvantage of affinity precipitation is the high cost and biodegradability of ligand-based precipitants. On the other hand, the activity of proteins recovered after precipitation is generally high.

Protein-binding dyes can also be used for differential protein precipitation. Although considerably less selective than affinity ligands, these precipitants bind with some degree of specificity to certain classes of protein. For example, yellow acridine dye has been found to precipitate all plasma proteins except IgG immunoglobulins, which remain in solution without denaturation. Other dyes interact with binding sites on enzymes to cause protein crosslinking and precipitation. Alternatively, low concentrations of heavy metal ions are known to selectively precipitate proteins. Interactions between Fe^{3+}, Cd^{2+}, Zn^{2+}, or Cu^{2+} and carboxyl, imidazole, or sulphydryl residues can be used to insolubilise specific proteins and fractionate protein mixtures. An advantage of dye and metal-ion precipitants compared with biological ligands is their lower cost and improved stability in crude cell lysates.

Another type of selective precipitation uses temperature, pH, or solvent treatment to denature and precipitate unwanted proteins from solution. Denatured proteins lose tertiary structure due to unravelling of their polypeptide chains; this increases the tendency of the protein to precipitate. Great care must be exercised with this approach, as the solution also contains the desired protein that must be protected from denaturation. Selective denaturation works only if the desired protein is robust and can withstand relatively harsh conditions. Experiments are required to identify treatments that produce the maximum loss or precipitation of unwanted proteins, while maintaining an acceptable percentage of the desired protein in solution in a biologically active form.

11.8.4 Practical Aspects of Precipitation Operations

As indicated in Sections 11.8.2 and 11.8.3, precipitation depends on the nature and concentration of the protein, precipitant, and other components present, and on the temperature, pH, and ionic strength of the solution. In practice, several additional factors have a significant influence on the effectiveness of the process and the quality of precipitate formed.

Precipitates are harvested using unit operations such as filtration, microfiltration, and centrifugation. Precipitate properties such as particle size and size distribution, density, and mechanical strength determine the ease of precipitate recovery. As outlined in Sections 11.3.3 and 11.4.2, solid–liquid separations using filtration and centrifugation are more effective if the particle size is increased; therefore, the formation of large precipitate particles is desirable. Precipitates with a high proportion of particles less than 1 μm in size are generally considered unsuitable for subsequent processing. Several common precipitants such as ammonium sulphate generate protein precipitates with densities not much different from that of the suspending fluid, adding to the challenges associated with particle sedimentation in industrial centrifuges. Many protein precipitates are also colloidal or gelatinous in nature, which creates extra equipment and handling difficulties.

Most precipitation operations are carried out in batch in stirred tanks. Precipitants such as salt or organic solvent are added to the protein solution and mixed in by stirring. The formation of precipitates involves a series of steps, from initial *nucleation* to particle *growth* and *aging*. Mixing within the vessel plays a critical role at all stages of the precipitation process.

- *Nucleation*. This is the primary step in precipitation, resulting in the formation of submicron-sized particles. As nucleation is often very rapid and can occur within a few seconds, good mixing is essential to ensure that the precipitant being added to the solution is distributed to all regions of the vessel at uniform concentration. Because the solubilities of the product and contaminating proteins depend on the precipitant concentration, if good mixing is not achieved, the precipitate formed may contain much less product than expected due to coprecipitation of undesirable components in local regions of the vessel. In some processes, because of the influence of mixing on precipitate formation, different results are produced depending on the rate at which precipitant is added to the protein solution. Providing good mixing is of particular concern for proteins with rapid nucleation kinetics. The risk of protein denaturation is also greater if mixing is poor, as high local concentrations of salt or organic solvent may persist near the precipitant feed point.

- *Growth*. Growth of the precipitate depends initially on diffusion processes, as solute molecules in the vicinity of the nuclei diffuse to the particles. Aggregates of size 0.1 to 1 μm are formed in this way. Subsequent growth may be promoted by mixing, which causes the particles to collide with each other and agglomerate. A balance must be struck, however, as collisions also cause particle erosion and break-up. Precipitates subjected to vigorous mixing and high levels of fluid shear during growth tend to be smaller than those produced at lower shear rates. This effect is illustrated in Figure 11.15.
- *Ageing or ripening*. During this stage of precipitation, the precipitate particles continue to grow in size but also become stronger and more dense. Adjustments in chemical composition may occur as some components in the precipitate resolubilise. Less intense mixing is required during the ageing process to avoid particle disruption. Accordingly, the final stages of precipitation may be carried out without stirring or with very low levels of agitation. As shown in Figure 11.16, ageing precipitates approach a maximum stable size that is dependent on the mixing conditions. Fully aged particles have improved mechanical strength, which provides resistance to breakage during subsequent handling and processing.

After the precipitate is harvested, the protein must be separated from the precipitant and any contaminants incorporated into the particles. In most cases, for economical

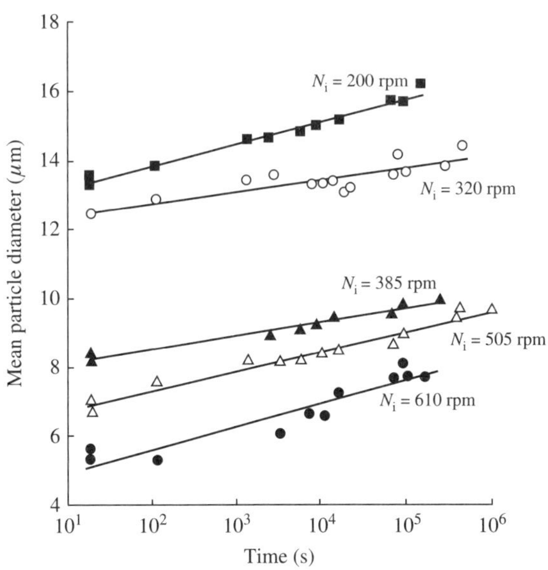

FIGURE 11.15 The effect of stirrer speed N_i on growth of casein precipitates. The precipitates were formed in a stirred vessel by salting-out with ammonium sulphate, then aged without stirring. *Adapted from M. Hoare, 1982, Protein precipitation and precipitate ageing. Part I: Salting-out and ageing of casein precipitates.* Trans. IChemE *60, 79–87.*

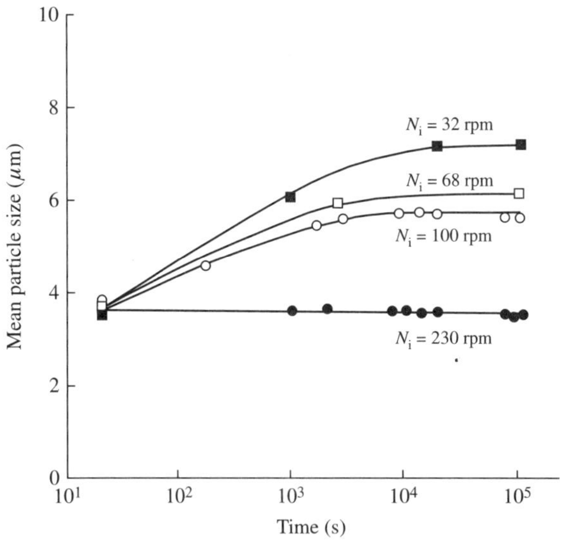

FIGURE 11.16 The effect of stirrer speed N_i on the size of casein precipitates during ageing. *Adapted from M. Hoare, 1982, Protein precipitation and precipitate ageing. Part I: Salting-out and ageing of casein precipitates.* Trans. IChemE *60, 79–87.*

reasons, the bulk of the precipitant is recovered and reused. The solid precipitated particles are washed and then redissolved in a small volume of buffer to form a concentrated protein solution. Additional downstream processing using operations such as chromatography and ultrafiltration may be carried out to remove any residual precipitant and further purify the protein product.

11.9 ADSORPTION

Adsorption is a surface phenomenon whereby components of a gas or liquid are concentrated on the surface of solid particles or at fluid interfaces. Adsorption is the result of electrostatic, van der Waals, reactive, or other binding forces between individual atoms, ions, or molecules. Four types of adsorption can be distinguished: exchange, physical, chemical, and nonspecific.

Adsorption is used to isolate products from dilute fermentation liquors. Several different adsorption operations are used in bioprocessing, particularly for medical and pharmaceutical products. Ion-exchange adsorption is established practice for recovery of amino acids, proteins, antibiotics, and vitamins. Adsorption on activated charcoal is a method of long standing for purification of citric acid; adsorption of organic chemicals on charcoal or porous polymer adsorbents is common in waste water treatment. Adsorption is gaining increasing application primarily because of its suitability for protein isolation.

In adsorption operations, the substance being concentrated on the surface is called the *adsorbate*; the material to which the adsorbate binds is the *adsorbent*. Ideal adsorbent materials have a high surface area per unit volume; this can be achieved if the solid contains a network of fine internal pores that provide an extremely large internal surface area. Carbons and synthetic resins based on styrene, divinylbenzene, or acrylamide polymers are used commonly for adsorption of biological molecules. Commercially available adsorbents are porous granular or gel resins with void volumes of 30 to 50% and pore diameters generally less than 0.01 mm. As an example, the total surface area in particles of activated carbon ranges from 450 to 1800 $m^2\ g^{-1}$. Not all of this area is necessarily available for adsorption; adsorbate molecules only have access to surfaces in pores of appropriate diameter.

11.9.1 Adsorption Operations

A typical adsorption operation consists of the following stages: a *contacting* or adsorption step which loads solute on to the adsorptive resin, a *washing* step to remove residual unadsorbed material, *desorption* or *elution* of adsorbate with a suitable solvent, *washing* to remove residual eluant, and *regeneration* of the adsorption resin to its original condition. Because adsorbate is bound to the resin by physical or ionic forces, the conditions used for desorption must overcome these forces. Desorption is normally accomplished by feeding a stream of different ionic strength or pH; elution with organic solvent or reaction of the sorbed material may be necessary in some applications. Eluant containing stripped solute in concentrated form is processed to recover the adsorbate; operations for final purification include spray drying, precipitation, and crystallisation. After elution, the adsorbent undergoes regenerative treatment to remove any impurities and regain adsorptive capacity. Despite regeneration, performance of the resin will decrease with use as complete removal of adsorbed material is impossible. Accordingly, after a few regenerations the adsorbent is replaced.

11.9.2 Equilibrium Relationships for Adsorption

Equilibrium relationships determine the extent to which material can be adsorbed on a particular surface. When an adsorbate and adsorbent are at equilibrium, there is a defined distribution of solute between the solid and fluid phases and no further net adsorption occurs. Adsorption equilibrium data are available as *adsorption isotherms*. For adsorbate A, an isotherm gives the concentration of A in the adsorbed phase versus the concentration in the unadsorbed phase at a given temperature. Adsorption isotherms are useful for selecting the most appropriate adsorbent; they also play a crucial role in predicting the performance of adsorption systems.

Several types of equilibrium isotherm have been developed to describe adsorption relationships. However, no single model is universally applicable; all involve assumptions that may or may not be valid in particular cases. One of the simplest adsorption isotherms that accurately describes certain practical systems is the *Langmuir isotherm* shown in Figure 11.17(a). The Langmuir isotherm can be expressed as follows:

$$C_{AS}^{*} = \frac{C_{ASm}\, K_A C_A^{*}}{1 + K_A C_A^{*}} \tag{11.40}$$

In Eq. (11.40), C_{AS}^{*} is the equilibrium concentration or loading of A on the adsorbent in units of, for example, kg of solute per kg of solid or kg of solute per m^3 of solid. C_{ASm} is the maximum loading of adsorbate corresponding to complete monolayer coverage of all available adsorption sites, C_A^{*} is the equilibrium concentration of solute in the fluid phase in units of, for example, kg m^{-3}, and K_A is a constant. Because different units are used for fluid- and solid-phase concentrations, K_A usually has units such as $m^3\ kg^{-1}$ of solid.

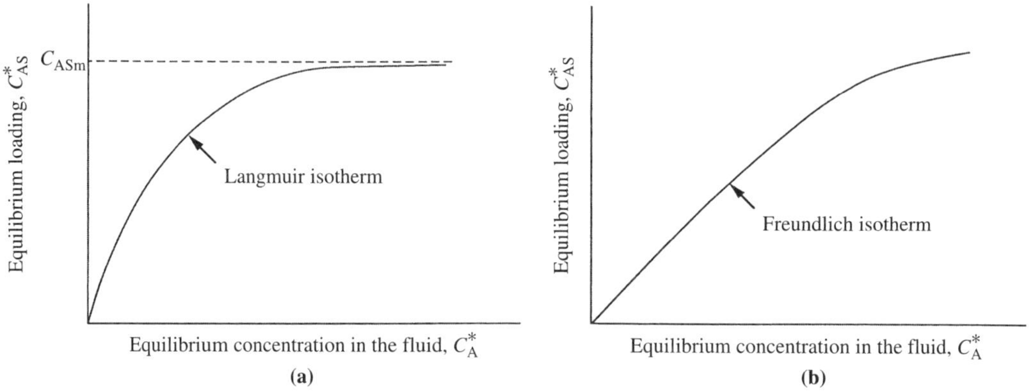

FIGURE 11.17 Adsorption isotherms.

Theoretically, Langmuir adsorption is applicable to systems where:

- Adsorbed molecules form no more than a monolayer on the surface
- Each site for adsorption is equivalent in terms of adsorption energy
- There are no interactions between adjacent adsorbed molecules

In many experimental systems at least one of these conditions is not met. For example, many commercial adsorbents possess highly irregular surfaces so that adsorption is favoured at particular points or 'strong sites' on the surface. Accordingly, each site is not equivalent. In addition, interactions between adsorbed molecules exist for almost all real adsorption systems. Recognition of these and other factors has led to the application of other adsorption isotherms.

Of particular interest because of its widespread use in liquid–solid systems is the *Freundlich isotherm*, described by the relationship:

$$C^*_{AS} = K_F \, C_A^{*1/n} \tag{11.41}$$

K_F and n are constants characteristic of the particular adsorption system; the dimensions of K_F depend on the dimensions of C^*_{AS} and C^*_A and the value of n. If adsorption is favourable, $n > 1$; if adsorption is unfavourable, $n < 1$. The form of the Freundlich isotherm is shown in Figure 11.17(b). Equation (11.41) applies to adsorption of a wide variety of antibiotics, hormones, and steroids.

There are many other forms of adsorption isotherm giving different curves for C^*_{AS} versus C^*_A [17]. Because the exact mechanisms of adsorption are not well understood, adsorption equilibrium data must be determined experimentally.

EXAMPLE 11.5 ANTIBODY RECOVERY BY ADSORPTION

Cell-free fermentation liquor contains 8×10^{-5} mol l^{-1} immunoglobulin G. It is proposed to recover at least 90% of this antibody by adsorption on synthetic, nonpolar resin. Experimental equilibrium data are correlated as follows:

$$C^*_{AS} = 5.5 \times 10^{-5} \, C_A^{*0.35}$$

where C^*_{AS} is the moles of solute adsorbed per cm^3 of adsorbent and C^*_A is the liquid-phase solute concentration in mol l^{-1}. What minimum quantity of resin is required to treat 2 m^3 of fermentation liquor in a single-stage mixed tank?

Solution

The minimum quantity of resin is required when equilibrium occurs. If 90% of the antibiotic is adsorbed, the residual concentration in the liquid is:

$$\frac{(100-90)\%}{100\%}(8\times10^{-5}\ \text{mol l}^{-1}) = 8\times10^{-6}\ \text{mol l}^{-1}$$

Substituting this value for C^*_A in the isotherm expression gives the equilibrium loading of immunoglobulin:

$$C^*_{AS} = 5.5\times10^{-5}(8\times10^{-6})^{0.35} = 9.05\times10^{-7}\ \text{mol cm}^{-3}$$

The amount of adsorbed antibody is:

$$\frac{90\%}{100\%}(8\times10^{-5}\ \text{mol l}^{-1})\,(2\ \text{m}^3)\cdot\left|\frac{1000\ \text{l}}{1\ \text{m}^3}\right| = 0.144\ \text{mol}$$

Therefore, the mass of adsorbent needed is:

$$\frac{0.144\ \text{mol}}{9.05\times10^{-7}\ \text{mol cm}^{-3}} = 1.59\times10^5\ \text{cm}^3$$

The minimum quantity of resin required is 1.6×10^5 cm^3, or 0.16 m^3.

11.9.3 Performance Characteristics of Fixed-Bed Adsorbers

Various types of equipment have been developed for adsorption operations, including fixed beds, moving beds, fluidised beds, and stirred tank contactors. Of these, fixed-bed adsorbers are most commonly applied; the adsorption area available per unit volume is greater in fixed beds than in most other configurations. A fixed-bed adsorber is a vertical column or tube packed with adsorbent particles. Commercial adsorption operations are performed as unsteady-state processes; liquid containing solute is passed through the bed and the loading or amount of product retained in the column increases with time.

Operation of a downflow fixed-bed adsorber is illustrated in Figure 11.18. Liquid solution containing adsorbate at concentration C_{Ai} is fed at the top of a packed column that is initially free of adsorbate. At first, adsorbent resin at the top of the column takes up solute rapidly; solution passing through the column becomes depleted of solute and leaves the system with effluent concentration close to zero. As the flow of solution continues, the region of the bed where most adsorption occurs, the *adsorption zone*, moves down the column as the top resin becomes saturated with solute in equilibrium with the entering liquid concentration C_{Ai}. Movement of the adsorption zone usually occurs at a speed much lower than the velocity of fluid through the bed, and is called the *adsorption wave*. Eventually the lower edge of the adsorption zone reaches the bottom of the bed, the resin is almost completely saturated, and the concentration of solute in the effluent starts to rise

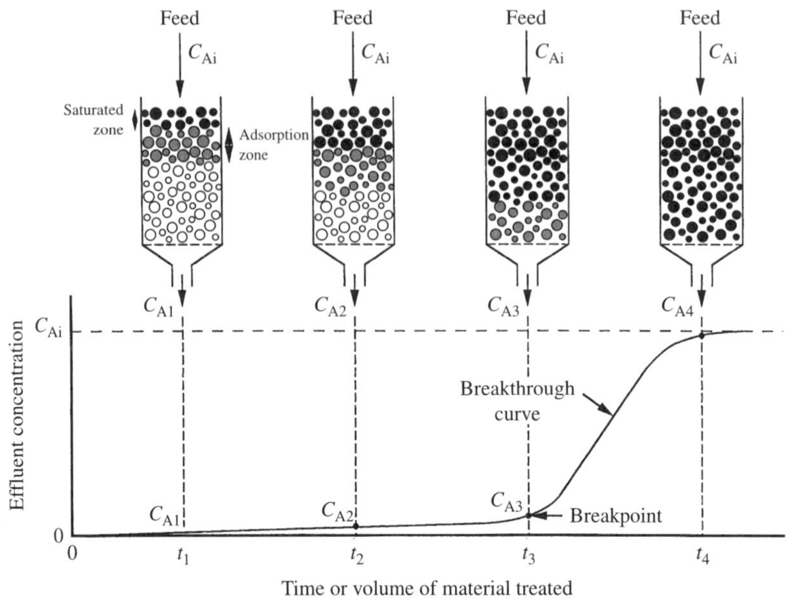

FIGURE 11.18 Movement of the adsorption zone and development of the breakthrough curve for a fixed-bed adsorber.

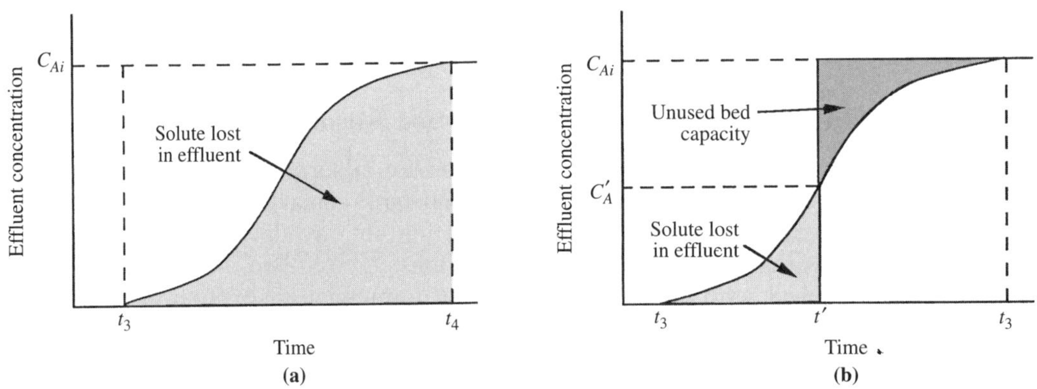

FIGURE 11.19 Relationship between the breakthrough curve, loss of solute in the effluent, and unused column capacity.

appreciably; this is called the *breakpoint*. As the adsorption zone passes through the bottom of the bed, the resin can no longer adsorb solute and the effluent concentration rises to the inlet value, C_{Ai}. At this time the bed is completely saturated with adsorbate. The curve in Figure 11.18 showing the effluent concentration as a function of time or volume of material processed is known as the *breakthrough curve*.

The shape of the breakthrough curve greatly influences the design and operation of fixed-bed adsorbers. Figure 11.19 shows the portion of the breakthrough curve between times t_3 and t_4 when solute appears in the column effluent. The amount of solute lost in the effluent is given by the area under the breakthrough curve. As indicated in Figure 11.19(a), if adsorption continues until the entire bed is saturated and the effluent concentration equals C_{Ai}, a considerable amount of solute is wasted. To avoid this, adsorption operations are usually stopped before the bed is completely saturated. As shown in Figure 11.19(b), if adsorption is halted at time t' when the effluent concentration is C'_A, only a small amount of solute is wasted compared with the process of Figure 11.19(a). The disadvantage is that some portion of the bed capacity is unused, as represented by the shaded area above the breakthrough curve. Because of the importance of the breakthrough curve in determining schedules of operation, much effort has been given to its prediction and analysis of the factors affecting it. This is discussed further in the next section.

11.9.4 Engineering Analysis of Fixed-Bed Adsorbers

For design of fixed-bed adsorbers, the quantity of resin and the time required for adsorption of a given quantity of solute must be determined. Design procedures involve predicting the shape of the breakthrough curve and the time of appearance of the breakpoint. The form of the breakthrough curve is influenced by factors such as the feed rate, the concentration of solute in the feed, the nature of the adsorption equilibrium, and the rate of adsorption.

The performance of commercial adsorbers is controlled by the rate of adsorption. This in turn depends on mass transfer processes within and outside of the adsorbent particles. The usual approach to adsorber design is to conduct extensive pilot studies to examine the effects of major system variables. However, the duration and cost of these experimental studies can be minimised by prior mathematical analysis of the process. Because fixed-bed adsorption is an unsteady-state process and the equations for adsorption isotherms are generally nonlinear, the calculations involved in engineering analysis are relatively complex compared with many other unit operations. Nonhomogeneous packing in adsorption beds and the difficulty of obtaining reproducible results in apparently identical beds add to these problems. It is beyond the scope of this text to consider design procedures in any depth, as considerable effort and research are required to establish predictive models for adsorption systems. However, a simplified engineering analysis is presented here.

Let us consider the processes that cause changes in the liquid-phase concentration of adsorbate in a fixed-bed adsorber. The aim of this analysis is to derive an equation for the effluent concentration as a function of time, that is, the breakthrough curve. The technique used is the mass balance. Consider the column packed with adsorbent resin shown in Figure 11.20. Liquid containing solute A is fed at the top of the column and flows down the bed. The total length of the bed is L. At distance z from the top is a section of column around which we can perform an unsteady-state mass balance. We will assume that this section is very thin so that z is approximately the same anywhere in the section. The system boundary is indicated in Figure 11.20 by dashed lines; four streams representing flow of material cross the boundary. The general mass balance equation given in Chapter 4 can be applied to solute A.

$$\left\{\begin{array}{c}\text{mass in}\\\text{through}\\\text{system}\\\text{boundaries}\end{array}\right\} - \left\{\begin{array}{c}\text{mass out}\\\text{through}\\\text{system}\\\text{boundaries}\end{array}\right\} + \left\{\begin{array}{c}\text{mass}\\\text{generated}\\\text{within}\\\text{system}\end{array}\right\} - \left\{\begin{array}{c}\text{mass}\\\text{consumed}\\\text{within}\\\text{system}\end{array}\right\} = \left\{\begin{array}{c}\text{mass}\\\text{accumulated}\\\text{within}\\\text{system}\end{array}\right\} \tag{4.1}$$

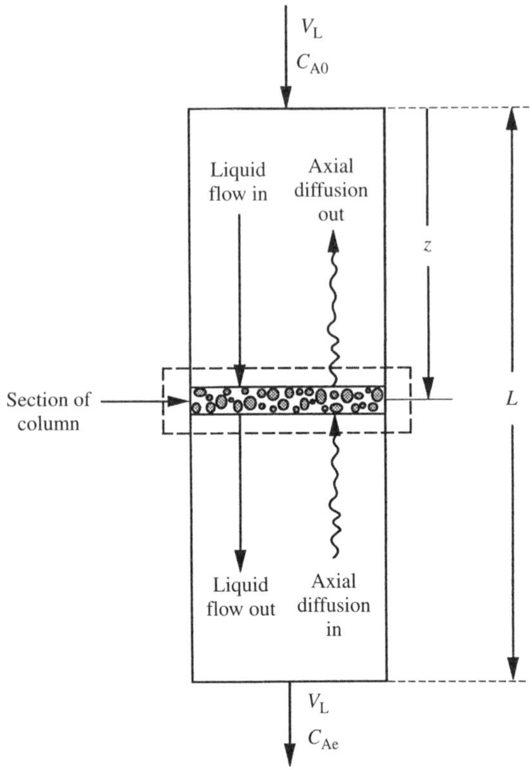

FIGURE 11.20 Fixed-bed adsorber for mass balance analysis.

Let us consider each term of Eq. (4.1) to see how it applies to solute A in the designated section of the column. First, we assume that no chemical reaction takes place so A can be neither generated nor consumed; therefore, the third and fourth terms of Eq. (4.1) are zero. On the left side, this leaves only the input and output terms. What are the mechanisms for input of component A to the section? A is brought into the section largely as a result of liquid flow down the column; this is indicated in Figure 11.20 by the solid arrow entering the section. Other input mechanisms are related to local mixing and diffusion processes within the interstices or gaps between the resin particles. For example, some A may enter the section from the region just below it by countercurrent diffusion against the direction of flow; this is indicated in Figure 11.20 by the wiggly arrow. Let us now consider the movement of A out of the system. The mechanisms for removal of A are the same as for entry: A is carried out of the section by liquid flow and by axial transfer along the length of the tube against the direction of flow. These processes are also indicated in Figure 11.20. The remaining term in Eq. (4.1) is the accumulation of A. A will accumulate within the section due to adsorption on the interior and exterior surfaces of the adsorbent particles. A may also accumulate in liquid trapped within the interstitial spaces or gaps between the resin particles.

When appropriate mathematical expressions for the rates of flow, axial dispersion, and accumulation are substituted into Eq. (4.1), the following equation is obtained:

$$\underset{\text{axial dispersion}}{\mathscr{D}_{AZ}\frac{\partial^2 C_A}{\partial z^2}} + \underset{\text{flow}}{u\frac{-\partial C_A}{\partial z}} = \underset{\text{accumulation in the interstices}}{\frac{\partial C_A}{\partial t}} + \underset{\text{accumulation by adsorption}}{\left(\frac{1-\varepsilon}{\varepsilon}\right)\frac{\partial C_{AS}}{\partial t}} \tag{11.42}$$

In Eq. (11.42), C_A is the concentration of A in the liquid, z is distance from the top of the bed, t is time, and C_{AS} is the average concentration of A in the solid phase. $\mathscr{D}_{AZ}$ is the *effective axial dispersion coefficient* for A in the column. In most packed beds $\mathscr{D}_{AZ}$ is substantially greater than the molecular diffusion coefficient; the value of $\mathscr{D}_{AZ}$ incorporates the effects of axial mixing in the column as the solid particles interrupt smooth liquid flow. ε is the bed *void fraction*, which is defined as:

$$\varepsilon = \frac{V_T - V_s}{V_T} \tag{11.43}$$

where V_T is the total volume of the column and V_s is the volume of the resin particles. u is the interstitial liquid velocity, defined as:

$$u = \frac{F_L}{\varepsilon A_c} \tag{11.44}$$

where F_L is the volumetric liquid flow rate and A_c is the cross-sectional area of the column. In Eq. (11.42), $\partial/\partial t$ denotes the partial differential with respect to time, $\partial/\partial z$ denotes the partial differential with respect to distance, and $\partial^2/\partial z^2$ denotes the second partial differential with respect to distance. Although Eq. (11.42) looks relatively complicated, it is useful to recognise the physical meaning of its components. As indicated below each term of the equation, the rates of axial dispersion, flow, and accumulation in the liquid and solid phases are represented. Accumulation is equal to the sum of the net rates of axial diffusion and flow into the system.

There are four variables in Eq. (11.42): the concentration of A in the liquid, C_A; the concentration of A in the solid, C_{AS}; the distance from the top of the column, z; and time, t. The other parameters can be considered constant. C_A and C_{AS} vary with time of operation and depth in the column. Theoretically, with the aid of further information about the system, Eq. (11.42) can be solved to provide an equation for the breakthrough curve, that is, C_A (at $z = L$) versus t. However, solution of Eq. (11.42) is difficult.

To assist the analysis, simplifying assumptions are often made. For example, it is normally assumed that dilute solutions are being processed; this results in nearly isothermal operation and eliminates the need for an accompanying energy balance for the system. In many cases the axial diffusion term can be neglected; axial dispersion is significant only at low flow rates. If the interstitial fluid content of the bed is small compared with the total volume of feed, accumulation of A between the particles can also be neglected. With these simplifications, the first and third terms of Eq. (11.42) are eliminated and the design equation is reduced to:

$$u\frac{-\partial C_A}{\partial z} = \left(\frac{1-\varepsilon}{\varepsilon}\right)\frac{\partial C_{AS}}{\partial t} \tag{11.45}$$

To progress further with the analysis, information about $\partial C_{AS}/\partial t$ is required. This term represents the rate of change of solid-phase adsorbate concentration and depends on the overall rate at which adsorption takes place. The rate of adsorption depends on two factors: the rate at which solute is transferred from the liquid to the solid by mass transfer mechanisms, and the rate of the actual adsorption or attachment process. The mass transfer pathway for adsorbate is analogous to that described in Section 10.5.2 for oxygen transfer. There are up to five steps that can pose significant resistance to adsorption as indicated in Figure 11.21. They are:

1. Transfer from the bulk liquid to the liquid boundary layer surrounding the particle
2. Diffusion through the relatively stagnant liquid film surrounding the particle
3. Transfer through the liquid in the pores of the particle to the internal surfaces
4. The actual adsorption process
5. Surface diffusion along the internal pore surfaces (i.e., migration of adsorbate molecules within the surface without prior desorption)

Normally, relatively little adsorption occurs on the outer surface of the particle compared to within the vast network of internal pores; accordingly, external adsorption is not shown in Figure 11.21. Bulk transfer of solute is usually rapid because of mixing and convective flow of liquid passing over the solid; the effect of step (1) on the overall adsorption

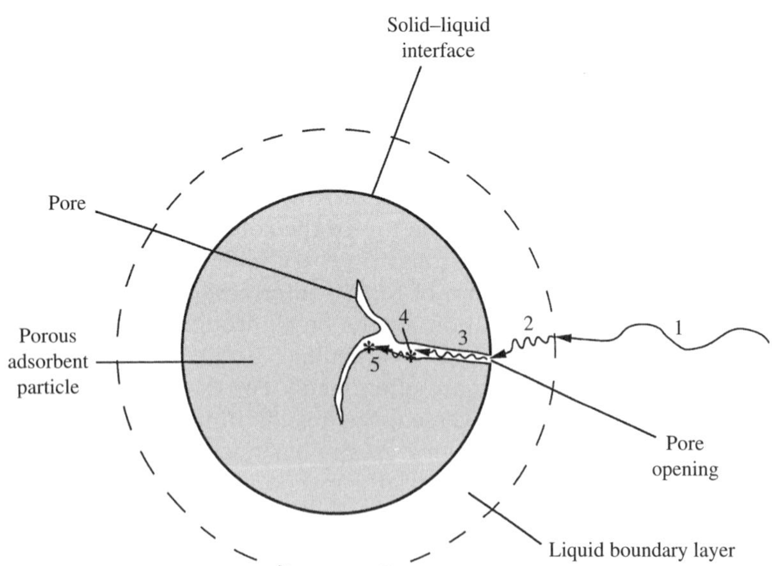

FIGURE 11.21 Graphic of steps involved in adsorption of solute from liquid to porous adsorbent particle.

rate can therefore be neglected. The adsorption step itself is sometimes very slow and can become the rate-limiting process; however, in most cases, adsorption occurs relatively quickly so that step (4) is not rate-controlling. Step (2) represents the major external resistance to mass transfer, while steps (3) and (5) represent the major internal resistances. Any or all of these steps can control the overall rate of adsorption depending on the situation. Rate-controlling steps are usually identified experimentally using a small column with packing identical to the industrial-scale system; mass transfer coefficients can then be measured under appropriate flow conditions. Unfortunately, however, it is possible that the rate-controlling step changes as the process is scaled up, making rational design difficult.

Greatest simplification of Eq. (11.45) is obtained when the overall rate of mass transfer from the liquid to the internal surfaces is represented by a single equation. For example, by analogy with Eq. (10.37), we can write:

$$\frac{\partial C_{AS}}{\partial t} = \frac{K_L a}{1-\varepsilon}(C_A - C_A^*) \tag{11.46}$$

where K_L is the overall mass transfer coefficient, a is the surface area of the solid per unit volume, C_A is the liquid-phase concentration of A, and C_A^* is the liquid-phase concentration of A in equilibrium with C_{AS}. The value of K_L will depend on the properties of the liquid and the flow conditions; C_A^* can be related to C_{AS} by the equilibrium isotherm. In the end, after the differential equations are simplified as much as possible and then integrated with appropriate boundary conditions (usually with a computer), the result is a relationship between the effluent concentration (C_A at $z = L$) and time. The height of the column required to achieve a certain recovery of solute can also be evaluated.

As mentioned already in this section, there are many uncertainties associated with the design and scale-up of adsorption systems. The approach described here will give only an approximate indication of the necessary design and operating parameters. Other methods involving various simplifying assumptions can be employed [2]. The preceding analysis highlights the important role played by mass transfer in practical adsorption operations. *Equilibrium is seldom achieved in commercial adsorption systems; performance is controlled by the overall rate of adsorption.* Therefore, the parameters determining the economics of adsorption are mostly those affecting mass transfer. Improvement of adsorber operation is achieved by enhancing rates of diffusion and reducing the resistances to mass transfer.

11.10 MEMBRANE FILTRATION

Membrane filtration is a versatile unit operation that can be used at various stages of downstream processing to separate, concentrate, and purify products. Applications include removal of cells and cell debris, concentration of proteins, removal of viruses, recovery of precipitates, solvent clarification, buffer exchange, and desalting. As in conventional filtration (Section 11.3), membrane filtration relies on pressure-driven fluid flow to separate components in a mixture. A pressure drop exerted across a porous polymer membrane causes small particles or molecules to permeate through the membrane while components of larger size are retained. In contrast with conventional filtration, separations are possible at the molecular scale between components that remain mostly in solution. Membrane separations are generally not highly selective, as different substances with similar properties may be co-retained or allowed to pass with the desired product. Membrane filtration is particularly suited, however, for product concentration applications.

Membrane filtration offers several advantages compared with other unit operations used to concentrate bioproducts. Process energy requirements are low, separations can be accomplished under aseptic conditions, and exposure to harsh chemicals is not required. Either batch or continuous processing may be used and scale-up is relatively straightforward. The principal disadvantages relate to the robustness and reliability of the membranes used. For example, membranes are susceptible to fouling, have limited resistance to cleaning chemicals, solvents, and wide pH ranges, and can be damaged by fluctuations in operating pressure.

Membrane filtration operations are classified according to the size of the particles or molecules retained by the membrane, the operating pressures used, and the nature of the material filtered. These features are summarised in Figure 11.22.

- *Microfiltration*. This is used to remove particulate matter such as cells and cell debris of size 0.2 to 10 μm from fermentation broths and cell homogenates. Typical microfiltration membranes have nominal pore diameters of 0.05 to 5 μm. The application range for microfiltration overlaps somewhat with that of conventional filtration (Section 11.3) and centrifugation (Section 11.4).
- *Ultrafiltration*. Membranes for ultrafiltration have pores with nominal diameter between 0.001 μm (10 Å) and 0.1 μm (1000 Å). Ultrafiltration is used for removing colloids, large molecules such as proteins and polysaccharides, viruses, and inclusion bodies from smaller solutes and water. Ultrafiltration membranes are capable of retaining

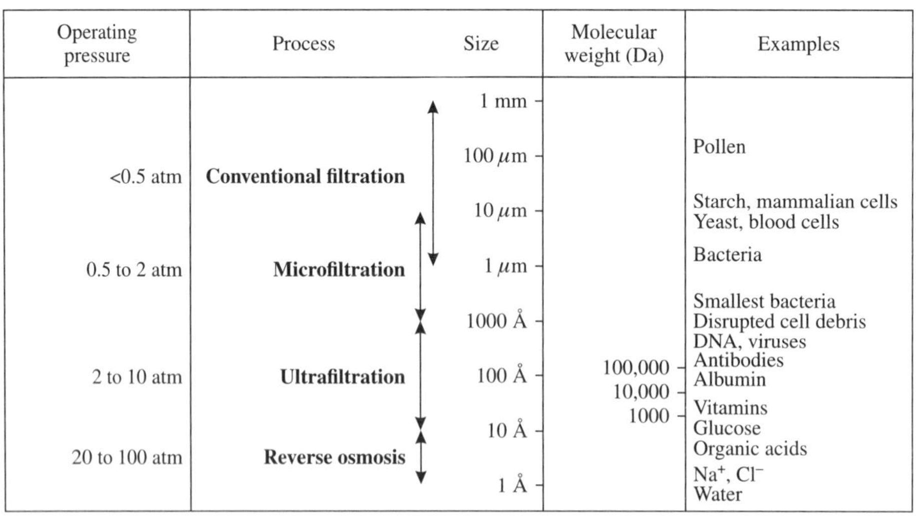

FIGURE 11.22 Summary of size ranges and applications of filtration processes.

macrosolutes with molecular weights of 10^3 to 10^6 daltons. To some extent, the applications of ultrafiltration overlap with those of ultracentrifugation (Section 11.4.1). The boundary between microfiltration and ultrafiltration is not precise, as various criteria such as pore size range, membrane structure, and type of material treated are used to define the two processes. For example, ultrafiltration membranes capable of retaining proteins of molecular weight around 100,000 daltons are used commonly in cell recovery operations.

- *Reverse osmosis*. Membranes for reverse osmosis or *hyperfiltration* have nominal pore diameters of 1 to 10 Å. Reverse osmosis is used to separate water from dissolved salts and other microsolutes. Reverse osmosis is not as frequently applied as microfiltration and ultrafiltration for downstream processing of biological products. Accordingly, we will not consider reverse osmosis operations in detail.

Other specialised membrane filtration methods have also been developed. For example, *high-performance tangential-flow filtration* is an enhanced ultrafiltration process capable of accomplishing protein–protein fractionations using charged membranes with much higher selectivity than normal ultrafiltration. A modified form of reverse osmosis is *nanofiltration*, which uses charged membranes to separate ions and small solutes such as sugars based on combined charge and size effects. *Membrane affinity filtration* and *membrane affinity chromatography* use membranes activated with coupling molecules called *ligands* capable of selective and reversible binding with specific proteins. Affinity techniques allow both concentration and purification of protein mixtures.

In addition to the applications already mentioned, several other membrane-based unit operations are used in bioprocessing, including dialysis, electrodialysis, and osmosis. However, these processes are not driven by a pressure drop across the membrane and rely instead on other variables such as differences in solute concentration or electromotive force. Here, we focus our attention on membrane filtration operations driven by pressure.

Membrane filtration is normally accomplished using *cross-flow* or *tangential-flow* filtration, whereby the flow of feed material is directed parallel to the membrane. This mode of operation is illustrated in Figure 11.23(a). The material passing through the filter is called the *permeate* or *filtrate*; the retained components are known as the *retentate* or *concentrate*. Cross-flow filtration is different from conventional 'dead-end' filtration represented in Figure 11.23(b) and described in Section 11.3. The principal benefit of cross-flow operation is that the material prevented from passing through the membrane is swept away from the membrane surface. Whereas conventional filtration is characterised by a build-up of retained solids or cake on the filter cloth, cross-flow filtration provides a mechanism for cleaning the membrane surface and thus reducing the resistance to further filtration.

The performance of tangential-flow processes is strongly dependent on the rate at which retained molecules are transported away from the membrane into the bulk fluid, thereby minimising the accumulation of material at the membrane. This aspect of membrane operations is considered further in Section 11.10.2. Because the rate of permeate flow through the membrane is usually much lower than the fluid cross-flow rate, membrane filtrations are almost always operated with retentate recirculation. This brings the material to be filtered in contact with the membrane a number of times, thus providing the opportunity for separation of a substantial permeate volume.

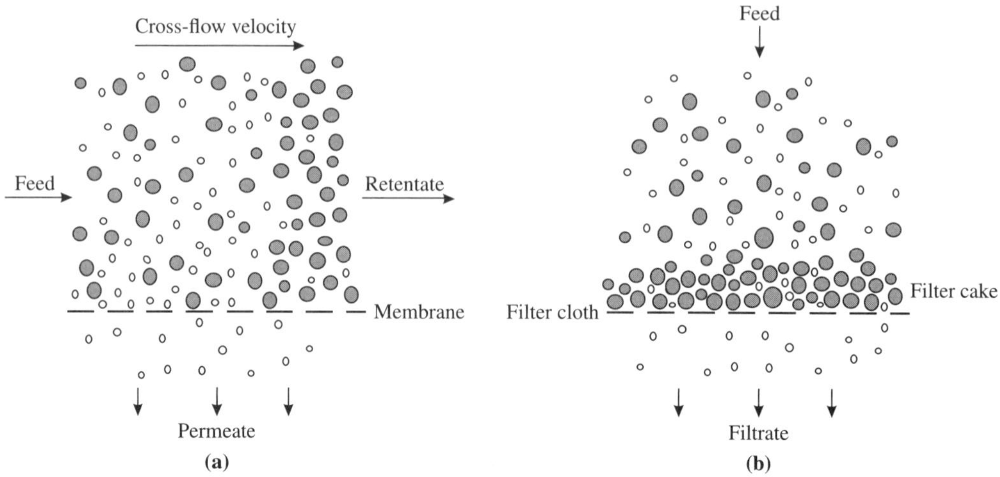

FIGURE 11.23 Operating schemes for: (a) cross-flow membrane filtration and (b) conventional 'dead-end' filtration.

11.10.1 Membrane Properties

Membranes for filtration processes are manufactured from a wide variety of materials, including polysulfone, polyethersulfone, polyamide, polyacrylonitrile, polyvinyl chloride, regenerated cellulose, cellulose acetate, and ceramics. The ideal membrane has well-defined particle or solute rejection characteristics, is resistant to variations in temperature, pH, and operating pressure, allows high rates of filtration, has high mechanical strength, and is easy and economical to produce. There are two main categories of membrane structure: *symmetric* or *homogeneous* and *asymmetric* or *anisotropic*.

- *Symmetric membranes*. The pores in symmetric membranes have close to a uniform diameter throughout the depth of the membrane. This allows the entire membrane to act as a selective barrier to the passage of particles or molecules. Flow through symmetric membranes can proceed in either direction. As well as retaining material on the surface of the filter, symmetric membranes also tend to retain components of approximately the same size as the pores within the membrane itself. In this sense, the membrane acts as a *depth filter* as well as a *screen* or *surface filter*. As it is difficult to remove particles trapped within the membrane, filtration using symmetric membranes can become progressively less efficient as the membrane becomes irreversibly blocked.
- *Asymmetric membranes*. These are comprised of an ultra-thin ($0.1-1$ μm) skin layer with very small pores supported by a thick macroporous support. The diameter of the pores therefore changes considerably through the depth of the membrane. During filtration, flow through asymmetric membranes proceeds in one direction only from the skin layer to the macroporous layer. The molecular exclusion characteristics of asymmetric membranes are determined by the size of the pores in the skin layer; however, once the solutes for passage through the membrane are selected at the upper surface, the macrovoids in the lower depths of the membrane provide high permeability and allow

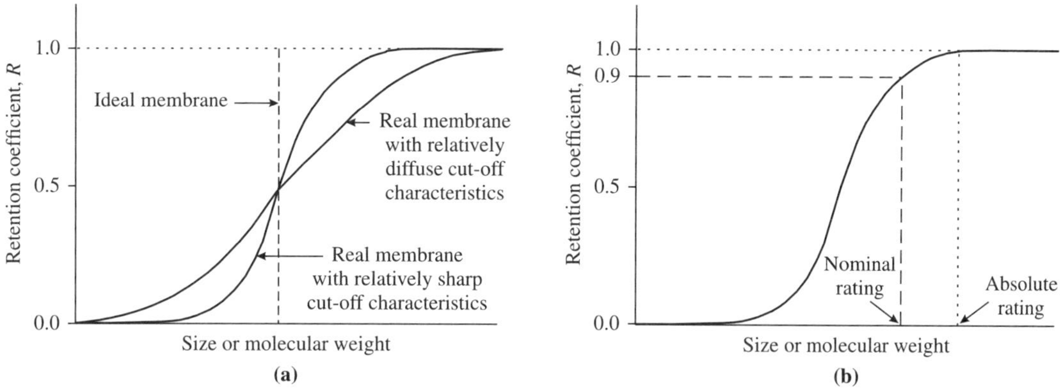

FIGURE 11.24 (a) Ideal and real membrane retention characteristic and (b) absolute and nominal ratings for a real membrane.

rapid filtration rates. Asymmetric membranes operate as screen filters by retaining material at the surface and not within the membrane itself. Accordingly, asymmetric membranes rarely block in the same way as do symmetric membranes. Cleaning is also relatively straightforward, as only the surface and not the entire filter volume requires treatment to remove residual material.

A key parameter of membrane performance is its *retention characteristics*. An ideal membrane would be capable of rejecting completely all particles or solutes above a specified size or molecular weight, while passing completely all species below that size. However, because the pores in real membranes are not all exactly the same diameter, the cut-off is imperfect. As a result, some solutes may be present after filtration in both the permeate and retentate. The selectivity of the membrane with respect to a given solute is quantified using the *rejection* or *retention coefficient*, R:

$$R = \frac{C_R - C_P}{C_R} = 1 - \frac{C_P}{C_R} \tag{11.47}$$

where C_R is the concentration of solute in the retentate and C_P is the concentration of solute in the permeate at any point during the filtration process. If the solute is rejected completely by the membrane and retained in the retentate, $C_P = 0$ and $R = 1$. Conversely, if a solute is not rejected at all by the membrane and passes through freely, its concentration on both sides of the membrane will be the same. Under these conditions, $C_R = C_P$ and $R = 0$.

Typical membrane retention characteristics are illustrated in Figure 11.24(a). Whereas an ideal membrane can accomplish sharply defined separations, real membranes have more diffuse cut-off characteristics depending on whether the membrane pore size distribution is broad or narrow. The sharp cut-off properties shown for an ideal membrane do not occur in practice. Membranes are *rated* by their manufacturers to give an indication of the minimum particle size or solute molecular weight retained [18, 19]. An *absolute rating* means that the membrane can be expected to retain all particles above its rated pore size. For example, an absolute membrane rating of 0.45 μm implies that particles with size greater than 0.45 μm will not pass through under standardised operating conditions. The retention characteristics of a membrane with an absolute rating are indicated in Figure 11.24(b).

Alternatively, retention properties can be specified using a *nominal rating*. This is used mainly with ultrafiltration membranes and is expressed as a *molecular weight cut-off*. As shown in Figure 11.24(b), the molecular weight cut-off usually refers to the molecular weight of a test solute that is 90% retained by the membrane. Therefore, if a membrane with a molecular weight cut-off of 10,000 is used to filter a protein or other solute of molecular weight 10,000, roughly 90% of the solute can be expected to be found in the retentate and 10% in the permeate. Use of the 90% retention criterion is not standard, however, as different membrane manufacturers use criteria ranging from 50 to 95% retention to establish the molecular weight cut-off. In any case, the molecular weight cut-off serves only as an approximate indication of the membrane retention characteristics. Other factors such as the shape, deformability, charge, hydrophobicity, and concentration of the solute, and the pH, ionic strength, and composition of the solvent, also affect membrane separations [18, 19]. For example, linear, flexible molecules tend to pass through membranes to a greater extent than molecules of the same molecular weight but with more complex globular or branched structure. The retention properties of a particular solute and membrane are best determined experimentally.

Membranes are characterised by other important properties in addition to pore size. The rate of filtration per unit area of membrane depends on a range of pore and membrane characteristics, such as the *pore size distribution* (range of pores sizes present), *pore density* (number of pores per unit area of membrane), *surface porosity* (fraction of the membrane surface occupied by pores), and *membrane porosity* (fraction of the membrane volume occupied by pores). Typical ultrafiltration membranes have pore densities of 10^7 to 10^{10} cm^{-2}. The surface porosity of microfiltration membranes is usually within the range 25 to 40%; in contrast, ultrafiltration membranes have considerably lower surface porosities of less than 10%. The larger and more numerous the pores per unit area of membrane, the greater is the rate of permeate flow.

11.10.2 Membrane Filtration Theory

An important objective in the design of membrane filtration systems is estimation of the membrane area required to achieve a certain rate of filtration. The membrane area dictates the size and number of filtration units needed to achieve the desired process throughput. Alternatively, the time taken to filter a given volume of material using equipment that is already installed may be of most interest. In either case, a key parameter in analysis of membrane filtration is the *permeate flux*, J, which is defined as the permeate flow rate per unit area of membrane:

$$J = \frac{F_{\mathrm{P}}}{A} \tag{11.48}$$

where F_P is the volumetric flow rate of permeate and A is the membrane area. Typical units for permeate flux are $l\,m^{-2}\,h^{-1}$, $m^3\,m^{-2}\,s^{-1}$, or, more conveniently, $m\,s^{-1}$. Several theories have been developed to relate J to operating conditions such as the pressure exerted across the membrane, the cross-flow fluid velocity, and the properties of the material being filtered. Two models describing membrane filtration are outlined after we consider some basic features of membrane filter operation.

Transmembrane Pressure

The principal driving force for membrane filtration is the pressure difference Δp exerted across the membrane. For most applications, Δp is equal to the difference between the pressure on the feed side of the membrane, p_F, and the pressure on the permeate side, p_P:

$$\Delta p = p_F - p_P \tag{11.49}$$

However, in some situations, the osmotic pressure of the solutions on either side of the membrane also affects the net driving force for filtration:

$$\Delta p = (p_F - p_P) - (\pi_F - \pi_P) \tag{11.50}$$

where π_F is the osmotic pressure of the feed solution and π_P is the osmotic pressure of the permeate. The osmotic pressure π of any species in dilute solution can be approximated using the van't Hoff equation:

$$\pi = iCRT \tag{11.51}$$

where C is the molar concentration of the solute, R is the ideal gas constant (Appendix B), T is the absolute temperature, and i is the number of ions formed as the solute molecule dissociates (e.g., $i = 2$ for NaCl and $i = 3$ for $CaCl_2$). The influence of osmotic pressure can be significant when the concentration of low-molecular-weight solutes such as salts is different on each side of the membrane, for example in reverse osmosis applications. However, in micro- and ultrafiltration, the effect of osmotic pressure is usually negligible even if the concentrations of larger molecules such as proteins and sugars are different in the feed and permeate. For these applications, Δp is given by Eq. (11.49).

Although the permeate pressure p_P is generally constant during filtration, because flow of the feed material across the membrane surface results in a pressure drop, p_F may vary significantly between the filter inlet and outlet. This is illustrated in Figure 11.25. The value of p_F for application of Eq. (11.49) or (11.50) is usually estimated by averaging the inlet and outlet pressures:

$$p_F = \frac{p_i + p_o}{2} \tag{11.52}$$

where p_i is the membrane feed inlet pressure and p_o is the outlet pressure.

Concentration Polarisation

When a clean fluid such as water is driven through a microporous membrane, the permeate flux depends on the number, diameter, and length of the pores, the pressure

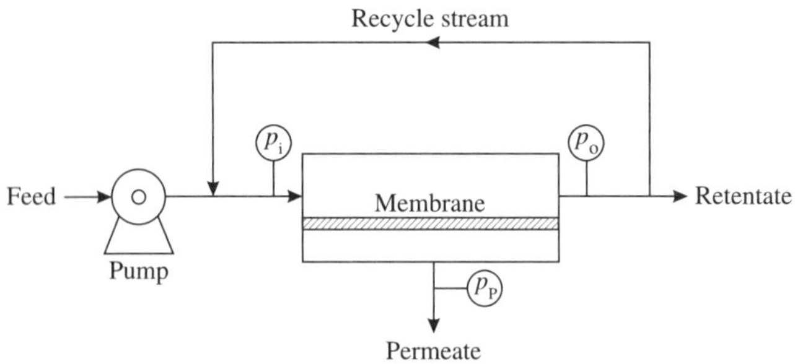

FIGURE 11.25 Flow sheet for membrane filtration, showing feed inlet and outlet pressures and permeate pressure.

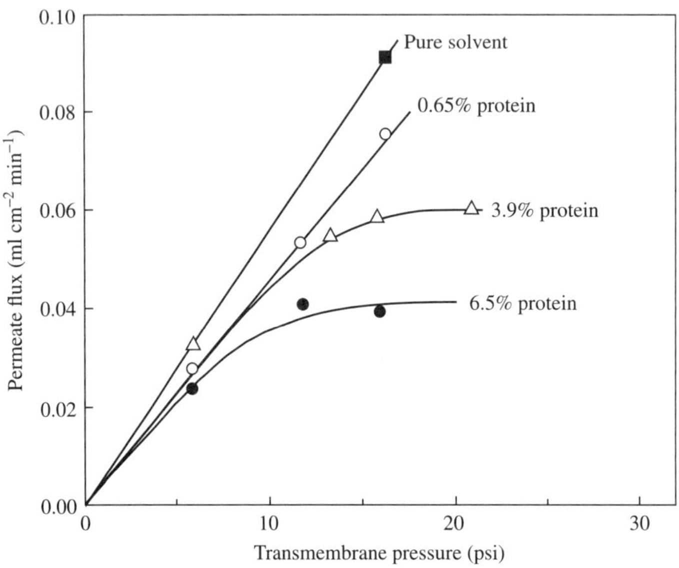

FIGURE 11.26 Relationship between permeate flux and transmembrane pressure for pure solvent (0.9% saline) and solutions of serum albumin protein.
Adapted from A.S. Michaels, 1968, Ultrafiltration. In: E.S. Perry, Ed., Progress in Separation and Purification, *vol. 1, pp. 297–334, John Wiley, New York.*

difference exerted across the membrane, and the viscosity of the fluid. The relationship between these variables is given by the Hagen–Poiseuille equation, which represents flow of liquid through pores of cylindrical geometry:

$$J = \frac{\Delta p \, d_p^2 \, \varepsilon}{32 \, x \, \mu_p} \tag{11.53}$$

In Eq. (11.53), J is the permeate flux, Δp is the transmembrane pressure, d_p is the diameter of the pores, ε is the surface porosity of the membrane, x is the pore length, and μ_p is the viscosity of the permeating fluid. ε is a dimensionless variable representing the fraction of the membrane surface occupied by pores.

According to Eq. (11.53), the permeate flux is directly proportional to the transmembrane pressure difference. Therefore, a plot of J versus Δp should give a straight line passing through the origin. As shown in Figure 11.26, this is generally the case for pure solvents such as water and for dilute solutions of microsolutes that pass readily through the membrane without retention. However, the relationship of Eq. (11.53) does not hold for fluids containing components such as cells or proteins that are retained by the membrane. As indicated by the data in Figure 11.26 for protein solutions, the permeate flux may be approximately linearly dependent on transmembrane pressure at low values of Δp but soon becomes independent of pressure. Deviation from the Hagen–Poiseuille equation increases as the concentration of protein in the feed material increases.

The nonlinear, asymptotic pressure–flux relationships shown in Figure 11.26 are due to the effects of *concentration polarisation*, which refers to the accumulation of retained particles or solutes near the surface of membranes. As liquid is drawn through the pores under the influence of the pressure gradient, solutes are transported toward the membrane surface by convective flow. However, as particles and high-molecular-weight substances are rejected by the membrane, these components remain near the membrane surface to form a thin cake or gel layer. Therefore, even though fluid cross-flow is a key feature of membrane filtration and prevents the formation of a thick filter cake, build-up of retained components near the membrane surface cannot be avoided completely. This layer of material is known as the *concentration polarisation layer* and is formed during normal membrane filtration operations.

Concentration polarisation has a major impact on the performance of membrane processes. It poses a substantial resistance to filtration and thus reduces the membrane flux. For solutes that are not completely rejected by the membrane, concentration polarisation can also alter the membrane rejection characteristics. For example, high solute concentrations at the membrane surface provide a significant driving force for passage through the membrane, thus allowing greater solute leakage into the permeate and yielding lower membrane retention coefficients than if concentration polarisation did not occur. As a result, membranes that retain certain proteins under conditions of low concentration polarisation may become almost completely nonretentive if the level of concentration polarisation increases substantially. High solute concentrations at the membrane can also result in a local increase in the feed osmotic pressure, π_F. As indicated in Eq. (11.50), this has the effect of lowering the overall transmembrane pressure, which may then lead to a reduction in permeate flux.

Resistances-in-Series Model

According to the *general rate principle*, the rate of any process is equal to the ratio of the driving force and the resistance. Therefore, as Δp is the driving force for membrane filtration, the permeate flux can be expressed using the equation:

$$J = \frac{\Delta p}{R_T} \tag{11.54}$$

where R_T is the total resistance to permeation through the membrane. As in conventional filtration (Section 11.3.3), the total resistance is equal to the sum of the individual resistances represented by the membrane itself and the material accumulated in the concentration polarisation layer. Therefore, Eq. (11.54) can be written as:

$$J = \frac{\Delta p}{R_M + R_C} \tag{11.55}$$

where R_M is the intrinsic membrane resistance and R_C is the resistance due to concentration polarisation. The value of R_M is determined experimentally by measuring J as a function of Δp in the absence of a concentration polarisation layer, for example, using pure water as the feed material:

$$R_M = \frac{\Delta p}{J_{water}} \tag{11.56}$$

The magnitude of R_C depends on several factors affecting the thickness and permeability of the concentration polarisation layer, such as the transmembrane pressure, cross-flow velocity, fluid viscosity, temperature, and solute concentration. Typical experimental data for R_C and R_M are shown in Figure 11.27. R_C increases with pressure but decreases with increasing cross-flow fluid velocity. In contrast, R_M is essentially independent of these two parameters. As shown in Figure 11.27(a), at low pressures the concentration polarisation resistance is small relative to the membrane resistance; however R_C soon dominates R_M as Δp is increased. R_C also tends to dominate when low cross-flow fluid velocities provide inadequate removal of accumulated solutes from the membrane surface. Accordingly, the value of R_C can be much greater than R_M at low fluid flow rates, as illustrated in Figure 11.27(b).

The dependence of R_C on the operating pressure drop can be incorporated into Eq. (11.55):

$$J = \frac{\Delta p}{R_M + R'_C \Delta p} \tag{11.57}$$

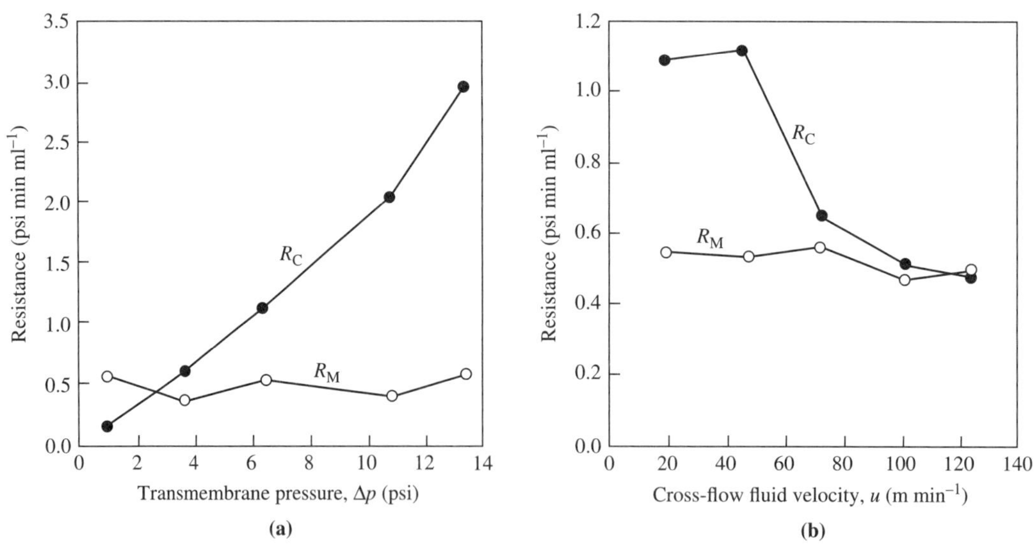

FIGURE 11.27 Individual values of the membrane resistance R_M and the concentration polarisation resistance R_C as a function of: (a) transmembrane pressure, and (b) cross-flow fluid velocity. The data were measured during microfiltration of *Escherichia coli* cells on a ceramic membrane filter.
Adapted from S.-L. Li, K.-S. Chou, J.-Y. Lin, H.-W. Yen, and I.-M. Chu, 1996, Study on the microfiltration of Escherichia coli-*containing fermentation broth by a ceramic membrane filter.* J. Membrane Sci. *110, 203–210.*

where R'_C is equal to R_C divided by Δp. At low pressures, $R'_C\Delta p$ can be expected to be much smaller than R_M, leaving J in Eq. (11.57) roughly equal to $\Delta p/R_M$ or directly proportional to Δp. This region of operation is represented in Figure 11.26 by the straight-line relationship between permeate flux and pressure at low pressure values. Conversely, at high pressures $R'_C\Delta p$ becomes much larger than R_M, which thus can be neglected in Eq. (11.57). Under these circumstances, the Δp terms in the numerator and denominator cancel and J is independent of pressure. This situation is illustrated in Figure 11.26 by the flattening-out of the flux curves at high operating pressures.

Mass Transfer Model

As an alternative to the resistances-in-series model, the effect of concentration polarisation on membrane filtration can be analysed in terms of mass transfer theory and hydrodynamic boundary layer effects. This conceptual approach draws on the *film theory* of mass transfer described in Section 10.3.

As illustrated in Figure 11.28, the accumulation of solute near the membrane gives rise to a solute concentration gradient in the vicinity of the membrane surface. The highest solute concentration occurs at the membrane wall and is denoted C_w. If the fluid away from the membrane is well mixed, the solute concentration is constant throughout the bulk fluid and equal to C_B. The thickness of the concentration boundary layer, δ, is defined by the distance over which the solute concentration gradient exists. At any instant in time, the concentration gradient depends on the balance between the rates of solute transport to and from the membrane surface. At steady state:

$$JC - JC_P - \mathscr{D}\frac{dC}{dy} = 0 \tag{11.58}$$

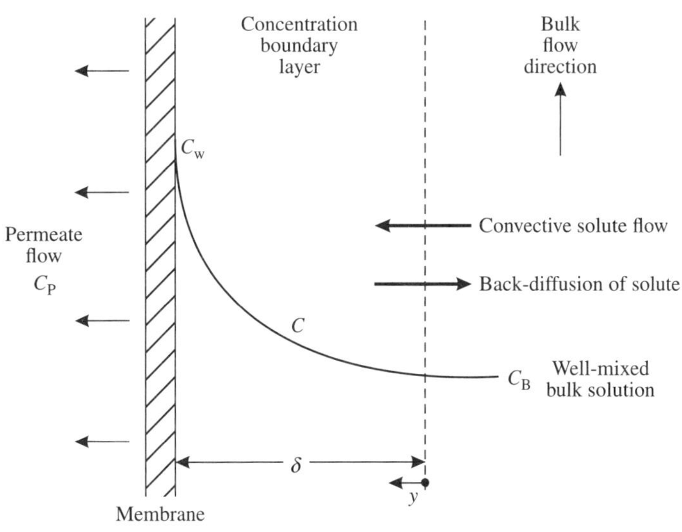

FIGURE 11.28 Concentration polarisation of rejected particles or solute and the development of a concentration boundary layer at the membrane surface.

where J is the permeate flux, C is the solute concentration in the boundary layer, C_P is the solute concentration in the permeate, $\mathscr{D}$ is the solute diffusion coefficient, and dC/dy is the concentration gradient or change in solute concentration C with distance y.

The first term in Eq. (11.58) represents the rate of solute transport to the membrane surface by pressure-driven convective flow, the second term is the rate of solute removal in the permeate, and the third term is the rate of solute back-diffusion from the region of high concentration in the boundary layer to the bulk fluid. The rate of diffusion is expressed using Fick's law as outlined in Section 10.1.1. Assuming that the diffusivity $\mathscr{D}$ is independent of solute concentration, Eq. (11.58) can be solved by integrating over the thickness of the boundary layer. From Figure 11.28, at $y=0$, $C=C_B$ and at $y=\delta$, $C=C_w$. Therefore:

$$J = \frac{\mathscr{D}}{\delta} \ln\left(\frac{C_w - C_P}{C_B - C_P}\right) \tag{11.59}$$

The ratio $\mathscr{D}/\delta$ has the same dimensions as the permeate flux J and corresponds to the mass transfer coefficient, k. Accordingly, Eq. (11.59) can be written as:

$$J = k \ln\left(\frac{C_w - C_P}{C_B - C_P}\right) \tag{11.60}$$

Like other mass transfer coefficients, the value of k depends on the flow geometry, the flow velocity, and the properties of the fluid. The dimensions of k are LT^{-1}; therefore, typical units are $m\ s^{-1}$. If the membrane has a high rejection coefficient for the solute so that C_P is negligible, Eq. (11.60) reduces to:

$$J = k \ln\left(\frac{C_w}{C_B}\right) \tag{11.61}$$

In general, the preceding equations apply at a fixed operating pressure. If the pressure is changed, a new steady-state concentration gradient is established, affecting the value of C_w. Raising the pressure increases C_w: from Eqs. (11.59) through (11.61), this improves the membrane flux. However, there is a limit to which J can be enhanced using this approach. If the pressure continues to rise, C_w reaches a limiting value C_G at which gelation or precipitation of solute occurs at the membrane surface. This condition is known as *gel polarisation*. The *gelation concentration* C_G depends on the physiochemical properties of the solute, such as its size, shape, and degree of solvation, and can be estimated experimentally for cells and macromolecular solutes. Typical C_G values are 45 to 80% w/w on a wet weight basis for microorganisms, 10 to 45% w/w for proteins, and 5 to 20% w/w for polysaccharides [18].

If solute gelation occurs at the membrane, C_w cannot be increased by raising the operating pressure. As a result, the permeate flux becomes independent of pressure. Increasing the pressure increases the thickness of the gel layer, but there is no beneficial effect on flux. This phenomenon is represented in Figure 11.26 by the flattening-out of the flux curves at high pressures. If the other operating variables remain constant, the permeate flux is limited to that given by Eq. (11.61) with $C_w = C_G$:

$$J = k \ln\left(\frac{C_G}{C_B}\right) \tag{11.62}$$

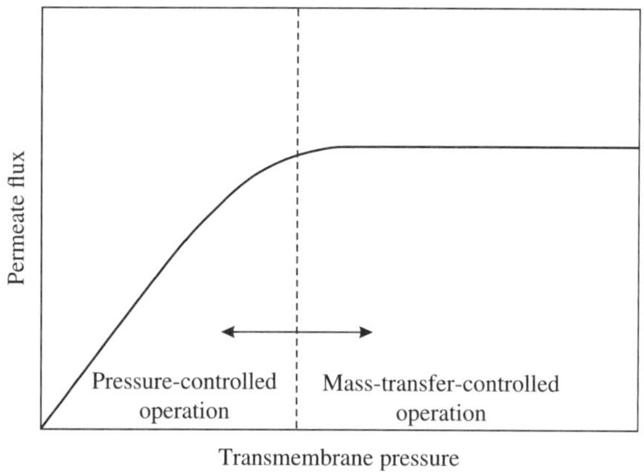

FIGURE 11.29 Generalised operating regions for membrane filtration.

The relationship between permeate flux and operating pressure is summarised in Figure 11.29 for solutions containing solute retained by the membrane. At low pressures, the flux is dependent on the operating pressure. At higher pressures, increasing the pressure has no effect on the permeate flux. Equation (11.62) can be used to estimate the permeate flux in the pressure-independent region of membrane operation. From Eq. (11.62), for a given feed material (i.e., for fixed C_B and C_G), the only way to improve J is to increase the value of the mass transfer coefficient k. The permeate flux in the pressure-independent region is therefore considered to be *mass-transfer controlled*.

As k represents the mass transfer coefficient for back-transfer of solute, at high pressures, the rate of membrane filtration is controlled by the rate at which retained material is transported *from* the membrane *back* into the bulk fluid. Accordingly, operating strategies that assist back-transport of concentrated solutes, such as using faster fluid flow rates, stirring the feed material, or increasing the temperature, generally are successful in improving filtration. These methods for enhancing membrane operations are discussed later in more detail (Section 11.10.2, Improving the Rate of Membrane Filtration subsection). In contrast, increasing the transmembrane pressure, which provides a higher driving force for filtration but does not address the concomitant need for greater back-diffusion, is ultimately ineffectual for increasing the rate of filtration. Rather, increasing the pressure drop results in a greater build-up of retained material at the membrane. For similar reasons, according to the mass transfer model, increasing the porosity of the membrane to reduce the resistance to permeate flow is also unlikely to improve filtration rates. Unless back-diffusion is enhanced, the end result is only a thicker concentration polarisation layer.

Practical application of Eq. (11.62) requires estimation of the mass transfer coefficient, k. Empirical correlations are available to allow calculation of k for membrane filtration systems. Mass transfer correlations for flow in narrow channels are expressed using the following dimensionless groups:

$$Sh = \text{Sherwood number} = \frac{k\, d_h}{\mathscr{D}} \tag{11.63}$$

$$Re = \text{Reynolds number} = \frac{d_h u \rho}{\mu} \tag{11.64}$$

$$Sc = \text{Schmidt number} = \frac{\mu}{\rho \mathscr{D}} \tag{11.65}$$

The parameters in Eqs. (11.63) through (11.65) are as follows: k is the mass transfer coefficient, $\mathscr{D}$ is the solute diffusivity, u is the linear flow velocity of fluid across the membrane, ρ is the density of the fluid, μ is the fluid viscosity, and d_h is the hydraulic diameter of the flow channel. If filtration takes place in tubular membranes with circular cross-section, d_h is equal to the tube diameter. Alternatively, if the flow channel has a rectangular cross-section, d_h is approximately equal to $2b$, where b is the channel height. The mass transfer coefficient is obtained from correlations between the dimensionless groups in the form:

$$Sh = M\, Re^{\alpha} Sc^{\beta} \left(\frac{d_h}{L}\right)^{\omega} \tag{11.66}$$

where M, α, β, and ω are correlation parameters and L is the length of the flow channel. The values of M, α, β, and ω depend on the flow conditions, but have also been found in ultrafiltration studies to vary with other operating conditions. Representative parameter values for laminar and turbulent flow in open channels and for flow in channels containing turbulence-inducing mesh screens or spacers are listed in Table 11.5. Mass transfer correlations for other membrane filtration geometries such as stirred vessels are also available [20].

Reasonable agreement has been found between experimentally measured filtration rates during ultrafiltration of macromolecular solutions and those estimated using Eq. (11.66). However, the agreement between experiment and theory is not so close for microfiltration of colloids and particles such as cells. In these systems, the experimental permeate flux is often 10 to 100 times greater than that predicted using mass transfer coefficients calculated from conventional correlation equations. The reasons for this are presently not entirely clear.

The mass transfer model has several additional shortcomings that limit its utility and reliability. For example, although the gelation concentration C_G is assumed to be independent of operating conditions, in practice C_G varies with the membrane type, bulk solute concentration, and flow velocity. Furthermore, permeate fluxes in the pressure-independent region have been found to differ by several-fold when membranes of different porosity are

TABLE 11.5 Parameter Values in Eq. (11.66)

Flow Conditions	M	α	β	ω	Range of Validity
Open channel					
Laminar	1.86	0.43	0.33	0.33	$Re < 2100$
Turbulent	0.023	0.89	0.3	0	$Re > 2100$
Mesh/spacer-filled channel	0.0096	0.5	0.6	0	–

Data from J.A. Howell, 1993, Design of membrane systems. In: J.A. Howell, V. Sanchez, and R.W. Field, Eds., Membranes in Bioprocessing: Theory and Applications, *pp. 141–202, Chapman and Hall, London.*

used. These findings are inconsistent with simple mass transfer models and have led to the development of alternative theoretical approaches. Further details of other mathematical models for membrane filtration in both the pressure-controlled and mass-transfer-controlled operating regions can be found elsewhere [18, 19, 21].

EXAMPLE 11.6 ULTRAFILTRATION OF POLYSACCHARIDE

Clarified fermentation broth contains a polysaccharide product with a gelation concentration of 25 g l^{-1}. The fluid density is 1020 kg m^{-3}, the viscosity is 1.8 cP, and the polysaccharide diffusivity is 5.6×10^{-11} m^2 s^{-1}. The product is harvested from the broth using ultrafiltration at a fluid velocity of 0.34 m s^{-1} in open membrane tubes of diameter 2.4 cm and length 2.0 m. Estimate the permeate flux if the filter is operated under gel polarisation conditions and the polysaccharide concentration in the feed is 12 g l^{-1}.

Solution

From Eq. (11.65):

$$Sc = \frac{1.8\text{ cP} \cdot \left|\frac{10^{-3}\text{ kg m}^{-1}\text{ s}^{-1}}{1\text{ cP}}\right|}{1020\text{ kg m}^{-3}\,(5.6 \times 10^{-11}\text{ m}^2\text{ s}^{-1})} = 3.15 \times 10^4$$

From Eq. (11.64):

$$Re = \frac{2.4 \times 10^{-2}\text{ m}\,(0.34\text{ m s}^{-1})\,(1020\text{ kg m}^{-3})}{1.8\text{ cP} \cdot \left|\frac{10^{-3}\text{ kg m}^{-1}\text{ s}^{-1}}{1\text{ cP}}\right|} = 4624$$

The minimum Re for turbulent flow in pipes is about 4000 (Section 7.2.3). As Re in the membrane tubes is >4000, the parameter values in Table 11.5 for turbulent flow in open channels can be used to evaluate the mass transfer coefficient. From Eq. (11.66):

$$Sh = 0.023\,(4624)^{0.89}\,(3.15 \times 10^4)^{0.3} = 940$$

Rearranging Eq. (11.63) and solving for k:

$$k = \frac{Sh\,\mathscr{D}}{d_h} = \frac{940\,(5.6 \times 10^{-11}\text{ m}^2\text{ s}^{-1})}{2.4 \times 10^{-2}\text{ m}} = 2.19 \times 10^{-6}\text{ m s}^{-1}$$

The permeate flux at a polysaccharide concentration of 12 g l^{-1} is obtained using Eq. (11.62):

$$J = (2.19 \times 10^{-6}\text{ m s}^{-1})\,\ln\left(\frac{25\text{ g l}^{-1}}{12\text{ g l}^{-1}}\right) = 1.6 \times 10^{-6}\text{ m s}^{-1}$$

The permeate flux is 1.6×10^{-6} m s^{-1} or 1.6×10^{-6} m^3 m^{-2} s^{-1}.

Improving the Rate of Membrane Filtration

According to the mass transfer model of membrane filtration, significant improvements in permeate flux do not necessarily occur if the pressure drop is increased or membranes

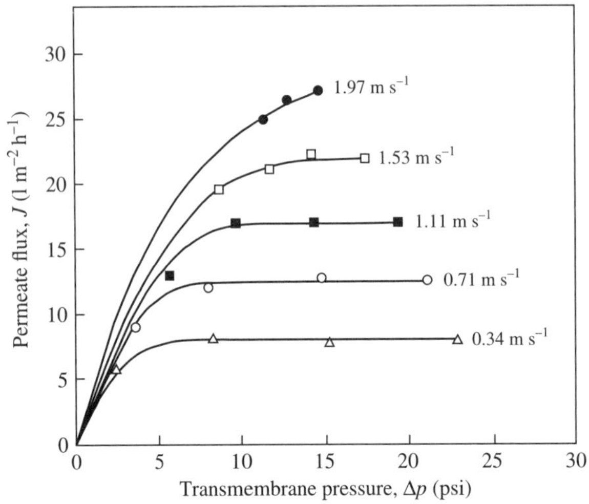

FIGURE 11.30 Effect of increasing cross-flow velocity on permeate flux. The data were measured during ultrafiltration of concentrated skim milk at 60°C.
Data from M. Cheryan, 1998, Ultrafiltration and Microfiltration Handbook, *Technomic, Lancaster, PA.*

with higher porosity are used. Under gel polarisation conditions, alternative strategies that improve the mass transfer of solutes away from the membrane are required.

- *Flow velocity.* Mass transfer is enhanced as the linear flow velocity across the membrane is increased. As indicated by Eqs. (11.63) and (11.66), the value of the mass transfer coefficient k is affected by flow velocity through the dependence of Sh on Re. This dependence is significantly stronger for turbulent than for laminar flow. From the values of α in Table 11.5, k varies with flow velocity raised to the power of 0.89 under turbulent conditions compared with 0.43 for laminar flow. Permeate flux is therefore enhanced considerably by the development of turbulence, which can be achieved by rapid pumping or stirring of the fluid and/or rotating or vibrating the membrane. Turbulent flow helps remove accumulated material from the membrane surface, reduces the thickness of the concentration boundary layer, and thus lowers the resistance to permeate flux. Experimental data illustrating the beneficial effects of increasing the cross-flow velocity are shown in Figure 11.30. Higher linear flow velocities across the membrane can be achieved by raising the volumetric flow rate or by reducing the height or diameter of the flow channels. As the increase in energy costs for pumping generally is lower for the latter option, membrane filtration equipment is constructed typically with very narrow flow channels.
- *Solute concentration.* According to Eq. (11.62), under gel-polarisation conditions J increases exponentially as C_B decreases. Experimental data demonstrating this strong relationship between solute concentration and permeate flux are plotted using semi-logarithmic coordinates in Figure 11.31. Because the concentration of solute affects several other properties of the fluid, such as viscosity, density, and solute diffusivity, the influence of C_B on flux can be more complex than that illustrated in Figure 11.31 due to associated changes in the mass transfer coefficient. Overall, however, reducing the concentration of solute in the feed can improve the rate of membrane filtration substantially.

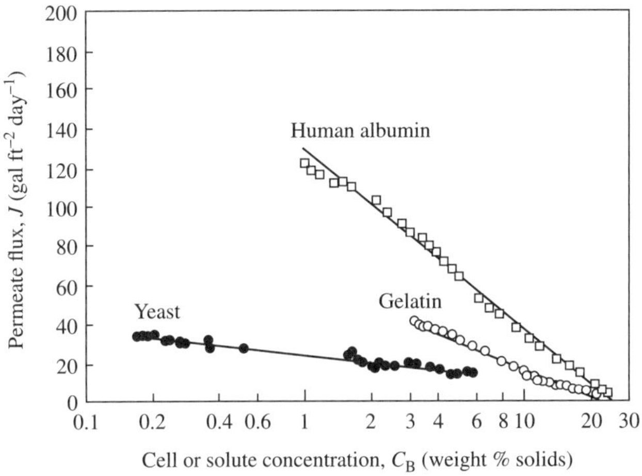

FIGURE 11.31 Variation of permeate flux with cell or solute concentration. Permeate flux decreases exponentially as the solute concentration increases.
Data from M.C. Porter, 1997, Membrane filtration. In: P.A. Schweitzer, Ed., Handbook of Separation Techniques for Chemical Engineers, *3rd ed., pp. 2-3–2-101, McGraw-Hill, New York.*

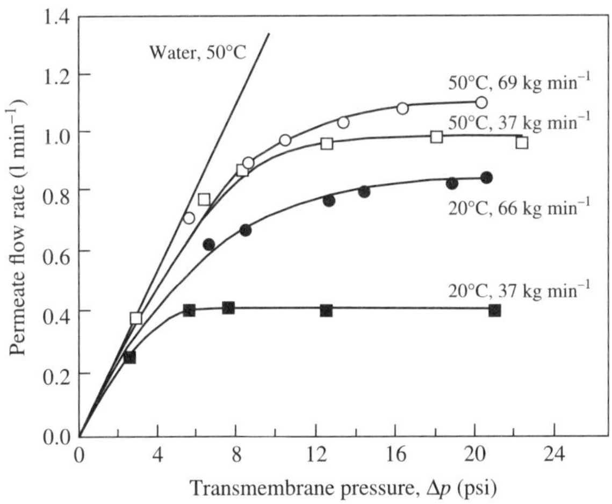

FIGURE 11.32 Effect of temperature on permeate flow rate during ultrafiltration of aqueous soybean extract. The symbols represent different temperatures and flow rates of soybean extract.
Data from M. Cheryan, 1998, Ultrafiltration and Microfiltration Handbook, *Technomic, Lancaster, PA.*

- *Temperature.* In general, higher temperatures result in higher membrane flux. This applies in both the pressure-controlled and mass-transfer-controlled regions of operation, and is due mainly to the reduction in fluid viscosity with increasing temperature. In the pressure-independent region, high temperatures also increase the diffusivity of the solute, thus increasing the mass transfer coefficient k. The effect of temperature on permeate flux can be significant, as illustrated in Figure 11.32. Membrane filtration systems are usually operated at the highest temperature that can be used without degrading the material being filtered or exceeding the functional limits of the membrane material.
- *Length of the flow channel.* From Eq. (11.66) and Table 11.5, when flow across the membrane is laminar, the mass transfer coefficient is inversely related to the flow channel length L. Therefore, higher permeate flux can be achieved under laminar flow conditions using short rather than long feed flow channels.

11.10.3 Fouling

When micro- or ultrafiltration systems are operated over a period of time, there is a progressive decline in the permeate flux. The filtration data in Figure 11.33 illustrate this effect. The cause is *membrane fouling*. Fouling is a major limiting factor in application of membrane technology. The consequence of membrane fouling and reduced permeate flux is higher equipment costs, as larger or greater numbers of membrane filtration units are needed to achieve a given process throughput. Fouling can also substantially alter the retention characteristics of the membrane due to obstruction of the membrane pores.

Fouling is a different phenomenon from concentration and gel polarisation, although all these processes reduce the permeate flux. Fouling is irreversible under the normal range of operating conditions and cannot be alleviated by reducing the transmembrane pressure, lowering the feed solute concentration, increasing the fluid flow rate, or otherwise improving mass transfer conditions at the membrane surface. Increasing the permeate flux through a fouled membrane is usually only achieved by increasing the applied pressure drop.

Membrane fouling is considered to be a two-step process. The first step occurs within the first few minutes of contact with the feed solution and involves the adsorption of foulants onto the membrane and pore surfaces. This process is influenced by the surface charge, chemical properties, and physical morphology of the membrane. The second step, which is characterised by a relatively slow decline in permeation rate, has been ascribed to plugging of the membrane pores and/or formation of a consolidated or solidified gelatinous layer at the membrane surface. The second stage of fouling is very unpredictable, being dependent on a wide range of parameters including the membrane composition, nature of the retained material, temperature, operating pressure, and solvent properties. However the results can be severe, with flux reductions of up to 90% occurring within a few days.

The equations derived in Section 11.10.2 (Mass Transfer Model subsection), for estimating the permeate flux do not take into account the declining rate of permeate flow due to

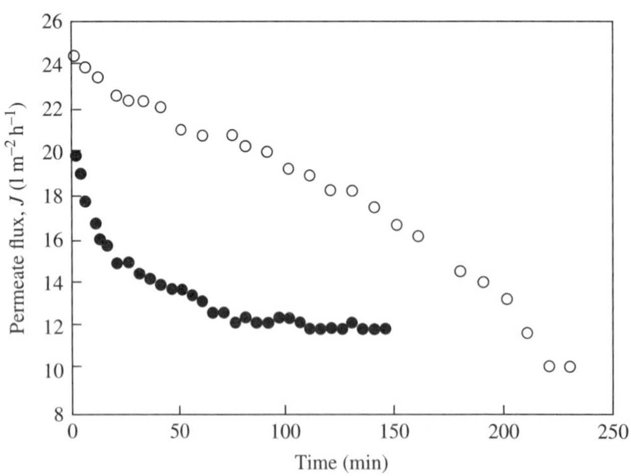

FIGURE 11.33 Reduction in permeate flux with operating time as a result of membrane fouling. The data were measured during ultrafiltration of (○) carrot juice, and (●) orange juice.
Data from A. Cassano, E. Drioli, G. Galaverna, R. Marchelli, G. Di Silvestro, and P. Cagnasso, 2003, Clarification and concentration of citrus and carrot juices by integrated membrane processes. J. Food Eng. *57, 153–163.*

fouling. However, fouling can be incorporated into the resistances-in-series model of Section 11.10.2 (Resistances-in-Series Model subsection), by adding an extra resistance term to Eq. (11.55):

$$J = \frac{\Delta p}{R_M + R_C + R_F} \tag{11.67}$$

where R_F is the resistance due to fouling. Because R_F reflects the physicochemical interactions between foulants and the membrane, its magnitude is relatively unaffected by operating parameters such as the transmembrane pressure and flow velocity. Table 11.6 lists values of R_M, R_C, and R_F reported in the literature for various membrane filtration applications. In most cases, the membrane resistance R_M is small relative to both the concentration polarisation resistance R_C and the fouling resistance R_F. For those applications where the fouling resistance dominates, the rate of filtration can be increased most effectively by preventing fouling or improving techniques for membrane cleaning.

Pretreatment of the feed can be used to reduce fouling. Heat treatment and adjusting the pH or ionic strength of the solution may help prevent foulant adsorption to the membrane. Prefiltration or centrifugation to remove oils and fats is also known to alleviate fouling. Many chemical antifoam agents applied in fermentations to suppress foaming (Section 10.6.3) have a strong tendency to foul hydrophobic microporous membranes, even at low concentrations. If a culture broth is to be treated for product recovery using micro- or ultrafiltration, particular attention is needed to select an appropriate antifoam agent and reduce the concentration used to minimise membrane fouling.

To avoid the worst effects, cleaning and scouring procedures are applied routinely to membranes to remove adherent material. Ideally, cleaning is carried out in place to avoid having to dismantle the membrane unit. Back-flushing, pulsing, shocking, and washing the membrane using high liquid velocities and pressure drops are conventional cleaning techniques. The permeate flow may be reversed periodically to dislodge accumulated

TABLE 11.6 Resistances to Membrane Filtration According to Eq. (11.67) for Various Applications

Feed Material	Membrane Material	R_M (kPa l^{-1} m^2 h)	R_C (kPa l^{-1} m^2 h)	R_F (kPa l^{-1} m^2 h)
Blood, bovine	Polyvinyl alcohol	0.06	0.16	0.015
Ethanol broth	Ceramic	0.11	2.0	2.3
Ovalbumin	Polyacrylonitrile	1.2	7.9	9.4
Ovalbumin	Polysulfone	0.51	9.2	0.91
Passionfruit juice	Polysulfone	0.27	1.9	1.4
Polyvinylpyrrolidone 360				
0.1%	Polyethersulfone	0.14	2.2	0.53
2.0%	Polyethersulfone	0.14	7.9	6.7
Wheat starch effluent	Polysulfone	0.68	2.7	4.9

From M. Cheryan, 1998, Ultrafiltration and Microfiltration Handbook, *Technomic, Lancaster, PA.*

deposits; typical back-wash frequencies are 1 to 10 times per minute. Back-flushing can be performed also using pressurised air instead of liquid. Maintenance of high membrane fluxes or significant restoration of membrane performance can be achieved using these methods. However, some fouling deposits are very difficult to remove and require treatment with harsh chemicals or abrasives that shorten the lifespan of the membrane considerably. Overall, the need for membrane cleaning reduces the operating time available for filtration and adds a further cost to the process.

11.10.4 Membrane Filtration Equipment

Membrane filtration is carried out using modular equipment, which is available in a variety of configurations. The modules must physically support the membrane, allow a high surface area of membrane to be packed within a small volume, facilitate mass transfer between the membrane and feed stream, and minimise the risk of particle plugging or clogging.

As outlined in Section 11.10.2 (Mass Transfer Model and Improving the Rate of Membrane Filtration subsections), filtration is most effective when fluid flow across the membrane is turbulent. To achieve high fluid velocities and maximise the contact area between the fluid and membrane, the feed stream is divided into a large number of narrow channels that are encased by membrane. Adjacent membrane surfaces may be separated by only a few millimetres, which restricts the size of the solid particles that can be processed without channel blockage. In cross-section, the channels are either open to allow free flow of the feed, or contain turbulence-promoting devices such as mesh screens and spacers to improve mass transfer at the membrane surface. Open channels are less susceptible to clogging and operate with smaller pressure drop between the feed inlet and outlet; however, the increase in membrane flux resulting from enhanced turbulence in the feed stream can outweigh the disadvantages of using channel screens.

Examples of commonly used membrane modules are shown in Figure 11.34.

Flat Sheet

Flat sheet membranes are housed in *plate-and-frame* or *cassette* devices. The separation between individual membranes for flow of the feed material is typically 0.5 to 2.0 mm. As illustrated in Figure 11.34(a), the membranes are supported on the permeate side by plates and rigid porous spacers, and several membrane-and-plate units are sandwiched together to form a module or cartridge. Depending on the separation between the membrane sheets, membrane surface areas per unit volume in cassette units are in the range 300 to 500 $m^2\ m^{-3}$. A disadvantage of plate-and-frame filters is that the membranes cannot be cleaned by back-flushing, as each membrane is supported on one side only.

Spiral

As shown in Figure 11.34(b), spiral-wound cartridges are constructed from pairs of flat membranes that are separated and supported by mesh screens. The membranes in each pair are arranged with their active surfaces facing away from each other, so that the spacer screen defines the permeate flow path. Several pairs of flat membranes are stacked

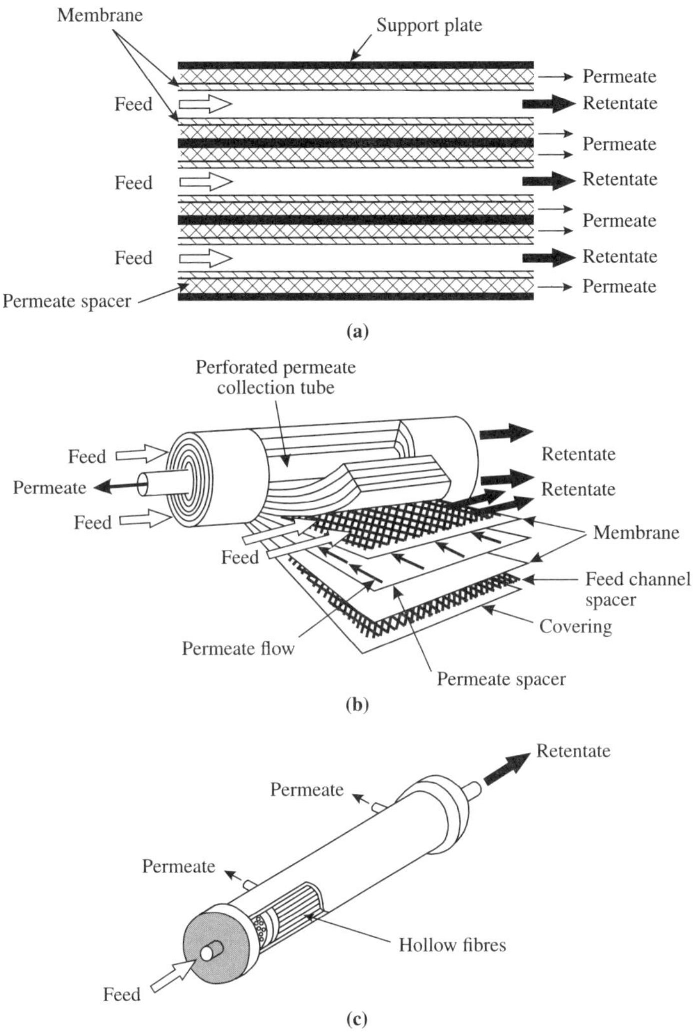

FIGURE 11.34 Common configurations of membrane filtration modules: (a) flat sheet membrane filtration in a plate-and-frame filter module; (b) internal structure of a spiral-wound membrane filter module; and (c) hollow fibre membrane filter module.

together, using more mesh-like spacers to separate the pairs and define the feed channel. The entire membrane assembly is then wound around a perforated central tube to form a spiral and fitted into a hollow tubular steel or plastic casing. Feed enters at one end of the casing and flows across the mesh spacers and membrane surfaces down the length of the spiral. Permeate is forced through the individual membranes and flows toward a central permeate collection tube. High membrane packing densities can be achieved in spiral filtration modules, giving relatively high membrane surface areas per unit volume of 600 to 1000 $m^2\ m^{-3}$.

Spiral units usually operate in the turbulent flow regime and provide effective mass transfer with relatively low energy consumption. They are susceptible to particle clogging, however, as narrow 0.25- to 0.50-mm flow paths are used between adjacent membranes. Prefiltration of the feed to remove particulate matter is often carried out to prevent blockage of the flow channels. Spiral-wound modules also can be difficult to clean after clogging and difficult to sterilise for aseptic operation.

Hollow Fibre

Hollow fibre modules typically contain 50 to 5000 self-supporting, narrow-bore membrane tubes, as shown schematically in Figure 11.34(c). The diameter of the tubes is between 0.2 and 2 mm and the wall or membrane thickness is 0.1 to 0.4 mm. Feed enters at one end of the module and retentate leaves at the other end. A pair of permeate exit ports, one at each end of the module, allows the unit to be operated using either cocurrent or countercurrent retentate and permeate flows.

Of all the available module geometries, hollow fibre systems offer the highest membrane surface areas per unit volume of around 1200 $m^2 m^{-3}$. This is due to the narrow bore of the individual tubes and the tight fibre packing within the cartridge. However, hollow fibres are also very susceptible to blockage by suspended particles. Accordingly, the feed is usually prefiltered to remove particles of size one-tenth the channel diameter or greater. Because the membranes are well supported, they are relatively easy to clean by applying back-pressure to reverse the direction of permeate flow.

Tubular

Tubular modules are similar to hollow fibre modules except that the tube diameters are considerably larger at 3 to 25 mm. This has the advantage of reducing particulate clogging. Cleaning is also easier as sponges or cleaning rods can be passed through the bore of the tubes. However, the membrane packing density in tubular modules is low, giving surface areas per unit volume of less than 100 $m^2 m^{-3}$.

Special Modules

Because the performance of membrane filters depends on the rate of mass transfer at the membrane surface, special module geometries and operating procedures have been developed to enhance mass transfer effects. Modules in which the membrane spins or vibrates, or in which fluid turbulence and flow instabilities are induced by other design features, offer the possibility of significantly improved permeate flux. Further details about specialty or advanced membrane modules are available in the literature [18, 19, 22].

11.10.5 Membrane Filtration Operations

There are three major types of membrane filtration application: *concentration*, *fractionation*, and *diafiltration*.

Concentration

A typical flow sheet for concentration operations is shown in Figure 11.35. The aim is to recover particles such as cells or macromolecules such as proteins as a concentrate in a relatively small volume of fluid. The desired product is contained in the filtration retentate.

Assuming that the densities of the feed, retentate and permeate are the same, the mass balance equation for total mass in membrane filtration operations is:

$$V_0 = V_R + V_P \tag{11.68}$$

where V_0 is the initial volume of feed material, V_R is the volume of retentate, and V_P is the volume of permeate. The mass balance equation for any dissolved solute is:

$$V_0C_0 = V_RC_R + V_PC_P \tag{11.69}$$

where C_0 is the solute concentration in the feed, C_R is the solute concentration in the retentate, and C_P is the solute concentration in the permeate.

Several parameters are used to describe the performance of concentration filtrations. The *volume concentration ratio*, *VCR*, is defined as:

$$VCR = \frac{V_0}{V_R} \tag{11.70}$$

The fractional *yield* Y_R of a feed component that is recovered in the retentate stream is:

$$Y_R = \frac{C_R V_R}{C_0 V_0} \tag{11.71}$$

At any point in time, the concentration of a particular component in the retentate depends on the volume concentration ratio *VCR* and the membrane retention coefficient *R* for that component [23]:

$$C_R = C_0 \, (VCR)^R \tag{11.72}$$

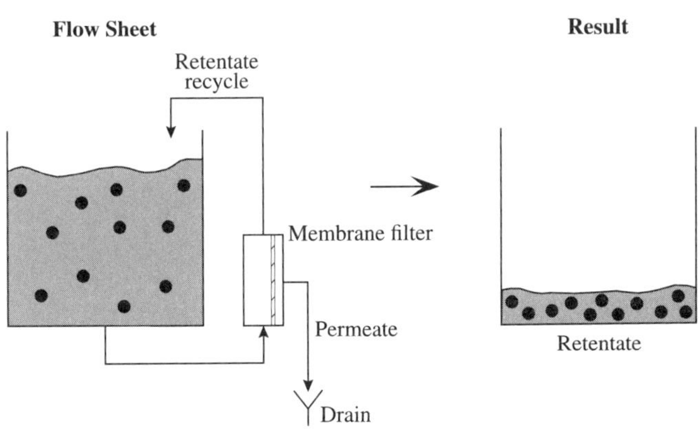

FIGURE 11.35 Membrane filtration used for concentration of particles or macromolecules in the retentate.

where R is defined in Eq. (11.47). Combining Eqs. (11.70), (11.71), and (11.72) gives an alternative expression for Y_R:

$$Y_R = (VCR)^{R-1} \tag{11.73}$$

For solutes that are rejected completely by the membrane, $R = 1$. Combining Eqs. (11.70) and (11.72) for these solutes gives:

$$C_R = \frac{C_0 V_0}{V_R} \quad \text{for } R = 1 \tag{11.74}$$

Conversely, if a solute is not rejected by the membrane and passes through freely to the permeate, $R = 0$. Therefore, from Eq. (11.72):

$$C_R = C_0 \quad \text{for } R = 0 \tag{11.75}$$

EXAMPLE 11.7 PROTEIN RECOVERY FROM CELL HOMOGENATE

Cells cultured in a bioreactor are homogenised and the cell debris removed by centrifugation. A recombinant protein present in the clarified homogenate is concentrated using ultrafiltration. For every 100 l of homogenate, 5 l of protein concentrate is required. The concentrations of recombinant protein, polysaccharide, and glucose in the clarified homogenate are 1 mg l^{-1}, 25 mg l^{-1}, and 2 g l^{-1}, respectively. The ultrafiltration membrane chosen for this separation gives retention coefficients of 1 for the recombinant protein, 0.6 for the polysaccharide, and 0 for glucose. Fouling has a negligible effect on solute rejection in this system.

(a) What are the concentrations of recombinant protein, polysaccharide, and glucose in the retentate?

(b) What fraction of the polysaccharide present in the homogenate is retained in the protein concentrate?

Solution

(a) From Eq. (11.70):

$$VCR = \frac{100\ \text{l}}{5\ \text{l}} = 20$$

Using Eq. (11.72):

$$C_R = 1\ \text{mg l}^{-1}\ (20)^1 = 20\ \text{mg l}^{-1} \quad \text{for recombinant protein}$$

$$C_R = 25\ \text{mg l}^{-1}\ (20)^{0.6} = 151\ \text{mg l}^{-1} \quad \text{for polysaccharide}$$

$$C_R = 2\ \text{g l}^{-1}\ (20)^0 = 2\ \text{g l}^{-1} \quad \text{for glucose}$$

(b) The fraction of polysaccharide retained in the concentrate is calculated using Eq. (11.73):

$$Y_R = (20)^{0.6-1} = 0.30$$

Fractionation

In this type of operation, the permeate contains the product. Fractionation is used to recover low-molecular-weight products by separating them from cells or molecules of higher molecular weight. The flow sheet for fractionation is shown in Figure 11.36.

The mass balance relationships of Eqs. (11.68) and (11.69) apply to fractionation operations. The fractional *yield* Y_P of a feed component recovered in the permeate stream can be determined as:

$$Y_P = \frac{C_P V_P}{C_0 V_0} \tag{11.76}$$

Diafiltration

The aim of diafiltration is to replace the solvent or buffer used to suspend cells or macromolecules. The main applications are desalting of proteins and buffer exchange, for example, between chromatography steps. As shown in Figure 11.37, new solvent or buffer

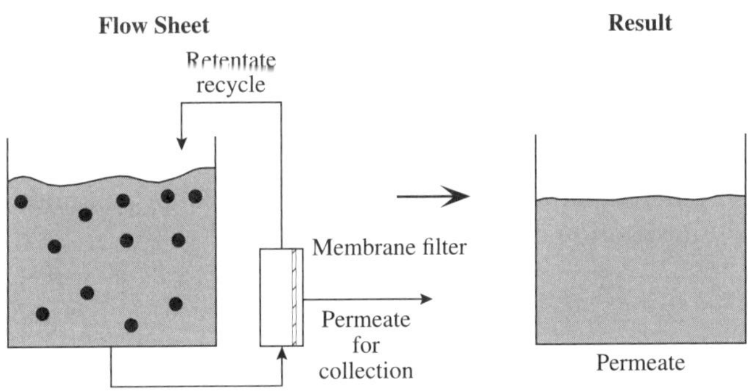

FIGURE 11.36 Membrane filtration used for fractionation of low-molecular-weight solutes in the permeate.

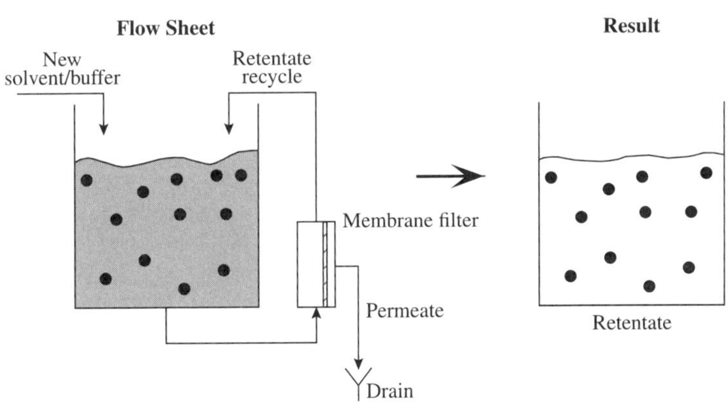

FIGURE 11.37 Membrane filtration used for diafiltration to replace the suspending solvent with new solvent or buffer.

is added to the feed material during membrane filtration. As a result, as well as separating particles or high-molecular-weight components from smaller solutes such as salts, the original particle- or macrosolute-suspending solution is replaced by the new solvent. The undesired microsolutes are discarded in the permeate.

In constant-volume diafiltration, the volume of added solvent is equal to the volume of permeate removed, making the final volume of solution the same as the starting volume, V_0. The change in solute concentration during continuous, constant-volume diafiltration can be determined from an unsteady-state mass balance on solute around the system shown in Figure 11.37 using the mathematical procedures described in Chapter 6:

$$V_0 \frac{dC_R}{dt} = -F_P\, C_P \tag{11.77}$$

where V_0 is the constant volume of the solution, C_R is the solute concentration in the retained solution, F_P is the volumetric flow rate of permeate, and C_P is the solute concentration in the permeate. An equation relating C_P to C_R and the solute rejection coefficient R is obtained from Eq. (11.47):

$$C_P = C_R\,(1 - R) \tag{11.78}$$

Therefore, Eq. (11.77) becomes:

$$V_0 \frac{dC_R}{dt} = -F_P\, C_R\,(1 - R) \tag{11.79}$$

When F_P and R are constant, the differential equation contains only two variables, C_R and t. Therefore, Eq. (11.79) can be solved by integration after separating the variables:

$$\int \frac{dC_R}{C_R} = \int \frac{-F_P\,(1 - R)}{V_0}\, dt \tag{11.80}$$

Using integration rules (E.27) and (E.24) from Appendix E:

$$\ln C_R = \frac{-F_P\,(1 - R)}{V_0}\, t + K \tag{11.81}$$

where K is the integration constant. Applying the initial condition:

$$\text{at } t = 0 \quad C_R = C_0 \tag{11.82}$$

to Eq. (11.81) gives:

$$\ln C_0 = K \tag{11.83}$$

After substituting this value for K into Eq. (11.81), the solution is:

$$\ln \frac{C_R}{C_0} = \frac{-F_P\,(1 - R)}{V_0}\, t \tag{11.84}$$

or

$$C_R = C_0 \, e^{\frac{-F_P(1-R)\,t}{V_0}} \tag{11.85}$$

In continuous, constant-volume diafiltration, the permeate flow rate F_P is equal to the flow rate of added solvent, F_D. Therefore, Eq. (11.85) can be written:

$$C_R = C_0 \, e^{\frac{-F_D(1-R)\,t}{V_0}} \tag{11.86}$$

Furthermore, the volume of diafiltration solvent added, V_D, is equal to $F_D t$. Therefore, Eq. (11.86) becomes:

$$C_R = C_0 \, e^{\frac{-V_D(1-R)}{V_0}} \tag{11.87}$$

Diafiltration is usually carried out using membranes with retention coefficients of zero for undesired microsolutes such as salt. In this case, Eq. (11.87) becomes:

$$C_R = C_0 \, e^{\frac{-V_D}{V_0}} \quad \text{for } R = 0 \tag{11.88}$$

To minimise the volume of diafiltration solvent required and the volume of permeate that must be disposed of, cell suspensions and macromolecule solutions are often concentrated into small volumes prior to diafiltration. The drawback associated with this strategy is a significantly lower diafiltration rate. As described in Section 11.10.2 (Improving the Rate of Membrane Filtration subsection), the rate of membrane filtration reduces considerably as the concentration of retained solute is increased.

EXAMPLE 11.8 DIAFILTRATION OF PROTEIN SOLUTION

A protein solution contains 1.95 M KCl. Constant-volume diafiltration using a membrane with retention coefficient $R = 0$ for KCl and $R = 1$ for protein is used to desalt 2000 l of the solution. The filter is operated so that the permeate flux is $20 \text{ l m}^{-2} \text{ h}^{-1}$. The total membrane area is 65 m^2. Water is added until the KCl concentration is reduced to 0.01 M.

(a) What time is required for this process?
(b) What volume of waste permeate is generated?

Solution

(a) From Eq. (11.48), the permeate flow rate is:

$$F_P = JA = 20 \text{ l m}^{-2} \text{ h}^{-1} \, (65 \text{ m}^2) = 1300 \text{ l h}^{-1}$$

For constant-volume diafiltration, the flow rate of added solvent F_D must be equal to the permeate flow rate. Therefore, from Eq. (11.86) with $R = 0$ for KCl:

$$0.01 \text{ M} = 1.95 \text{ M} \, e^{\frac{-(1300 \text{ l h}^{-1})\,t}{2000 \text{ l}}}$$

$$5.13 \times 10^{-3} = e^{-0.65t}$$

where t has units of h. Solving for t:

$$\ln(5.13 \times 10^{-3}) = -0.65t$$

$$t = 8.1 \text{ h}$$

The diafiltration process takes 8.1 h to complete.

(b) The volume of waste permeate generated $= F_P t = 1300 \text{ l h}^{-1} (8.1 \text{ h}) = 10{,}530$ litres. This is a large amount of waste relative to the volume of protein solution treated.

11.10.6 Process Configurations

Membrane filtration can be carried out using various process configurations and modes of operation, including single-pass, batch, fed-batch, continuous, and multistage. Flow sheets for these operating systems are shown in Figure 11.38.

Single-Pass

Single-pass operations do not involve recirculation of the retentate. A flow sheet for single-pass membrane filtration is shown in Figure 11.38(a). Because the permeate flux in membrane filtration is usually very low compared with the flow rate across the membrane surface, single-pass processes achieve only small values of the volume concentration ratio, *VCR*. Therefore, the utility of single-pass operations is limited to situations in which the rate of throughput is low and a large membrane surface area is available.

Batch

Batch operations, such as that represented in Figure 11.38(b), are commonly used in laboratory- and pilot-scale filtrations and for industrial separations where the permeate contains the desired product. Recirculation of the retentate is used to achieve the desired volume concentration ratio; however, because all of the retentate is recycled, components retained by the membrane have a high average residence time in the system. Batch operations are therefore unsuitable for retention of products that are susceptible to damage from repeated pumping.

During batch processing, the concentration of retained solute on the feed side of the membrane increases with time as permeate is removed. This causes the flux to decline as the filtration proceeds: as indicated in Eqs. (11.59) through (11.62), permeate flux J decreases with increasing solute concentration C_B. However, because the highest retained solute concentration is not reached until the very end of the process, a significant proportion of the permeate can be removed when the solute concentration is still relatively low. The average permeate flux achieved in batch filtration is therefore higher than for other operating modes. As a consequence, the membrane area required to process a given quantity of material is relatively low.

The changes in retained volume and solute concentration with time during batch filtration can be determined using the unsteady-state mass balance procedures outlined in Chapter 6. Assuming that the density of the retained solution is constant and equal to that

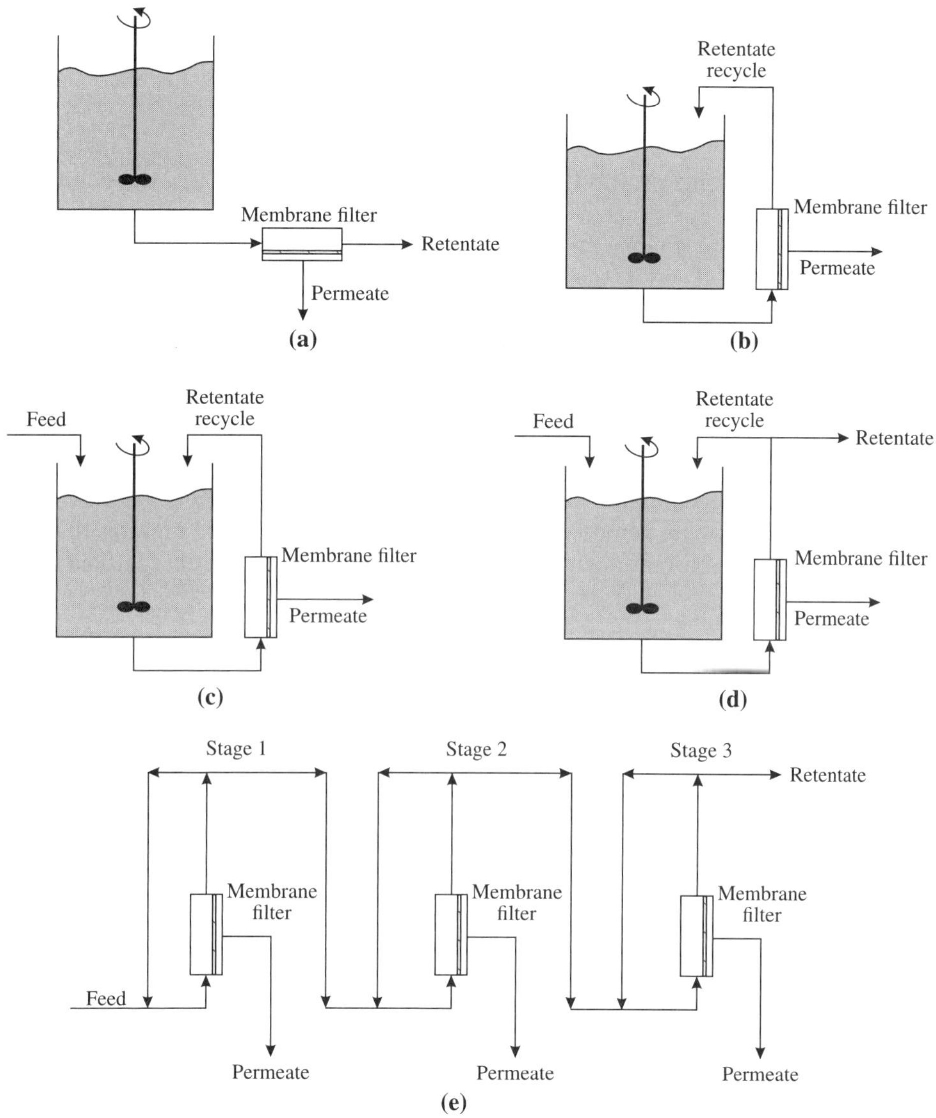

FIGURE 11.38 Modes of operation for membrane filtration processes: (a) single-pass; (b) batch; (c) fed-batch; (d) continuous; (e) multistage.

of the permeate, an equation representing the total mass balance around the system in Figure 11.38(b) is:

$$\frac{dV_R}{dt} = -A\,J \tag{11.89}$$

where V_R is the total retained solution volume, t is time, A is the membrane area, and J is the permeate flux. A similar unsteady-state mass balance for the retained solute is:

$$\frac{d(V_R C_B)}{dt} = -A\,J\,C_P \tag{11.90}$$

where C_B is the solute concentration in the retained solution and C_P is the solute concentration in the permeate.

Let us assume that the change in retained solute concentration per pass of the membrane is small compared with the solute concentration in the retentate. Actually, at any moment in time, the concentration of solute in the retentate leaving the filter will be greater than that in the fluid entering because permeate is being removed through the membrane. However, because this effect is relatively small during each passage through the membrane module, in practice the system as a whole can be considered *well mixed*. Under these conditions, at any point in time the bulk solute concentration C_B is approximately equal to the retentate concentration C_R leaving the membrane. This assumption is generally reasonable because the recirculation flow rate is usually much faster than the rate of permeate removal through the membrane. After making the substitution $C_R = C_B$, C_P can be expressed in terms of the membrane rejection coefficient R using Eq. (11.47):

$$C_P = C_B(1 - R) \tag{11.91}$$

Therefore, Eq. (11.90) becomes:

$$\frac{d(V_R C_B)}{dt} = -A\,J\,C_B(1 - R) \tag{11.92}$$

Both V_R and C_B vary with time during batch filtration. The left side of Eq. (11.92) can be expanded using the product rule for derivatives (Appendix E):

$$V_R\frac{dC_B}{dt} + C_B\frac{dV_R}{dt} = -A\,J\,C_B(1 - R) \tag{11.93}$$

Substituting for dV_R/dt from Eq. (11.89) gives:

$$V_R\frac{dC_B}{dt} - A\,J\,C_B = -A\,J\,C_B(1 - R) \tag{11.94}$$

After cancelling the terms common to both sides of the equation:

$$V_R\frac{dC_B}{dt} = A\,J\,C_B R \tag{11.95}$$

Equations (11.89) and (11.95) are two first-order differential equations that must be solved simultaneously to obtain values of V_R and C_B as a function of time. Let us consider the case in which solute is retained completely by the membrane, that is, $R = 1$. Let us assume also that the relationship between J and the solute concentration C_B can be expressed using Eq. (11.62) for filtration under gel polarisation conditions; this assumption

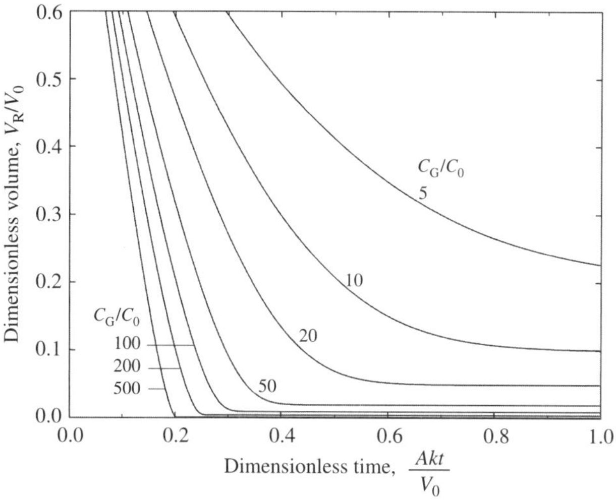

FIGURE 11.39 Change in retained volume during batch membrane filtration as a function of the dimensionless gelation concentration and dimensionless time. The curves were obtained by numerical integration of the unsteady-state mass balance equations describing batch filtration for solute with retention coefficient $R = 1$ in the absence of membrane fouling.

ignores the effect of membrane fouling on filtration rate. Equations (11.89) and (11.95) can be solved using numerical integration techniques with the initial conditions:

$$\text{at } t = 0 \quad V_R = V_0 \tag{11.96}$$

$$\text{at } t = 0 \quad C_B = C_0 \tag{11.97}$$

where V_0 is the initial volume of solution to be filtered and C_0 is the initial solute concentration. Solutions to the differential equations are shown in Figure 11.39 in terms of dimensionless variables for retentate volume, solute concentration, and time. The particular solution for a given process also depends on the value of the gelation constant C_G from Eq. (11.62), which is represented in the figure as a dimensionless variable. The results in Figure 11.39 can be used to evaluate the area required for batch filtration or the time required to filter a given volume of material, as illustrated in Example 11.9.

Fed-batch

Fed-batch processing is shown in Figure 11.38(c). Extra feed material is added to the system during fed-batch operations. This increases both the duration of the process and the number of times retained solutes are passed through the recycle pump, thus increasing the risk of damage if the product is sensitive to pumping. Fed-batch systems are commonly used in industry when the permeate is the desired product. Fed-batch filtration is an unsteady-state process: retained solutes accumulate in the system with time. Unsteady-state mass balance methods similar to those applied for batch operations can be used to derive equations for the retentate solute concentration.

Continuous

Continuous processing is the most commonly applied operating mode for large-scale membrane filtrations. As indicated in Figure 11.38(d), a proportion of the retentate is harvested continuously from the process while the remainder is recycled. To keep the total volume constant and the process at steady state, the flow rate of feed material into the process is maintained equal to the sum of the flow rates of permeate and harvested retentate. Assuming that the densities of the feed, retentate, and permeate are the same, the steady-state balance equations for total mass and mass of solute are:

$$F_0 = F_R + F_P \tag{11.98}$$

and

$$F_0 C_0 = F_R C_R + F_P C_P \tag{11.99}$$

where F_0 is the volumetric flow rate of feed material into the system, F_R is the volumetric flow rate of harvested retentate, F_P is the volumetric flow rate of permeate, C_0 is the solute concentration in the feed, C_R is the solute concentration in the harvested retentate, and C_P is the solute concentration in the permeate. The volume concentration ratio VCR during continuous operations is usually expressed in terms of flow rates rather than volumes. By analogy with Eq. (11.70):

$$VCR = \frac{F_0}{F_R} = \frac{F_R + F_P}{F_R} \tag{11.100}$$

The permeate flow rate F_P is related to the flux J by Eq. (11.48). As in our analysis of batch operations (Section 11.10.6, Batch subsection), if we assume that the change in solute concentration per pass of the membrane is relatively small, the overall filtration system can be considered well mixed and the bulk solute concentration C_B contacting the membrane is approximately equal to the retentate concentration C_R. Under these conditions and in the absence of fouling, J under gel polarisation conditions is related to C_R using Eq. (11.62) with $C_B = C_R$. Application of these relationships to estimate the membrane area required for continuous filtration is illustrated in Example 11.9.

At all times during continuous filtration, the membrane is operated at the retained solute concentration required for the final retentate. This is necessary as retentate is harvested continuously during the process. Continuous filters therefore operate under conditions equivalent to those achieved at the end of batch operations when the permeate flux is lowest. As a result, the flux achieved in continuous operations is not as high as that achieved in batch processing and the membrane area required is greater.

Multistage

Multistage or cascade operations are used for large-scale continuous processing. Figure 11.38(e) shows a closed multistage system with three filtration units. The units are operated in series with respect to retentate flow and in parallel with respect to permeate. Other arrangements, for example, with additional feeding at each stage or using the permeate from later stages to dilute the feed of earlier stages, are also possible. Cascade systems are often used when multiple membrane filtrations are required: for example, microfiltration to remove cells from fermentation broth, followed by ultrafiltration to harvest and concentrate an extracellular protein.

EXAMPLE 11.9 MEMBRANE SURFACE AREA REQUIREMENTS

A protein is present in 8 m^3 of fermentation liquor at a concentration of 0.6 g l^{-1}. Ultrafiltration under gel polarisation conditions is used to concentrate the protein by a factor of five. The gelation concentration is estimated as 25 g l^{-1}. The filtration is carried out using a spiral-wound module with a mass transfer coefficient of 3.4×10^{-6} m s^{-1}. The protein is rejected completely by the membrane and the effects of fouling are assumed to be negligible.

(a) What membrane area is required for batch ultrafiltration if the process must be completed within 1 hour?

(b) What membrane area is required if the filtration is carried out as a continuous operation over 1 hour with a feed flow rate of 8 m^3 h^{-1}?

Solution

(a) Figure 11.39 is used to estimate the membrane area required for batch filtration. The dimensionless gelation concentration for this system is:

$$\frac{C_G}{C_0} = \frac{25 \text{ g l}^{-1}}{0.6 \text{ g l}^{-1}} = 42$$

For well-mixed operation, $C_B = C_R$. Therefore, from Eq. (11.74) for $R = 1$:

$$\frac{V_R}{V_0} = \frac{C_0}{C_B} = \frac{0.6 \text{ g l}^{-1}}{5 \times 0.6 \text{ g l}^{-1}} = 0.2$$

Interpolating Figure 11.39 for these parameter values:

$$\frac{A\,k\,t}{V_0} \simeq 0.27$$

Therefore:

$$A = \frac{0.27\,V_0}{k\,t} = \frac{0.27\,(8 \text{ m}^3)}{(3.4 \times 10^{-6} \text{ m s}^{-1})\,1 \text{ h} \cdot \left|\frac{3600 \text{ s}}{1 \text{ h}}\right|} = 176 \text{ m}^2$$

The membrane area required for batch filtration is 176 m^2.

(b) From Eq. (11.99), if the protein is fully retained so that $C_P = 0$:

$$F_R = \frac{F_0 C_0}{C_R} = \frac{8 \text{ m}^3 \text{ h}^{-1}\,(0.6 \text{ g l}^{-1})}{5 \times 0.6 \text{ g l}^{-1}} = 1.6 \text{ m}^3 \text{ h}^{-1}$$

Using Eq. (11.98):

$$F_P = F_0 - F_R = (8 - 1.6) \text{ m}^3 \text{ h}^{-1} = 6.4 \text{ m}^3 \text{ h}^{-1}$$

For well-mixed operation, $C_B = C_R = 5 \times C_0$. The steady-state value for the permeate flux is calculated using Eq. (11.62):

$$J = (3.4 \times 10^{-6}\ \mathrm{m\ s^{-1}})\ \ln\left(\frac{25\ \mathrm{g\ l^{-1}}}{5 \times 0.6\ \mathrm{g\ l^{-1}}}\right) = 7.2 \times 10^{-6}\ \mathrm{m\ s^{-1}}$$

The membrane area is found by applying Eq. (11.48):

$$A = \frac{F_P}{J} = \frac{6.4\ \mathrm{m^3\ h^{-1}} \cdot \left|\frac{1\ \mathrm{h}}{3600\ \mathrm{s}}\right|}{7.2 \times 10^{-6}\ \mathrm{m\ s^{-1}}} = 247\ \mathrm{m^2}$$

The membrane area required for continuous filtration is 247 m^2. This is significantly greater than the area needed to accomplish the same process using batch processing.

11.10.7 Membrane Filtration Scale-Up

It is impossible to predict all aspects of membrane performance from theoretical analysis. Accordingly, extensive experimental testing is required for scale-up of membrane filters. A major unpredictable element is the effect of fouling on long-term membrane function. To assess the rate of fouling and the response of the filter to cleaning regimes, membrane performance is tested in the laboratory over many operating cycles including a series of cleaning steps.

Because membrane facilities are built on a modular basis, the operating conditions for industrial-scale processing can be determined using the feed material and a single membrane module. The effect of basic operating parameters such as transmembrane pressure, fluid velocity, solute concentration, and temperature on solute retention and permeate flux is determined in the laboratory or pilot plant. The membrane material, pore size, module geometry, and, ideally, the channel height and length, are kept the same in the large- and small-scale systems. The pilot module should be capable of handling the feed volume per unit membrane area required in the large-scale process. Provided that the transmembrane pressure and fluid flow regime are similar in the pilot- and large-scale units, design data measured in the laboratory can be expected to apply reasonably well to industrial operations.

Hollow fibre and tubular systems are scaled up by installing more membrane fibres of the same length rather than by increasing the size of the individual tubes. As pumping costs can be substantial in large-scale operations, the maximum allowable pressure drop between the module inlet and outlet may be the limiting factor in the scale-up process. The capacity of industrial filtration systems is enhanced by increasing the number of membrane modules in service rather than by developing larger individual filtration units.

11.11 CHROMATOGRAPHY

Chromatography is a separation procedure for resolving mixtures into individual components. Many of the principles described in Section 11.9 for adsorption apply also to

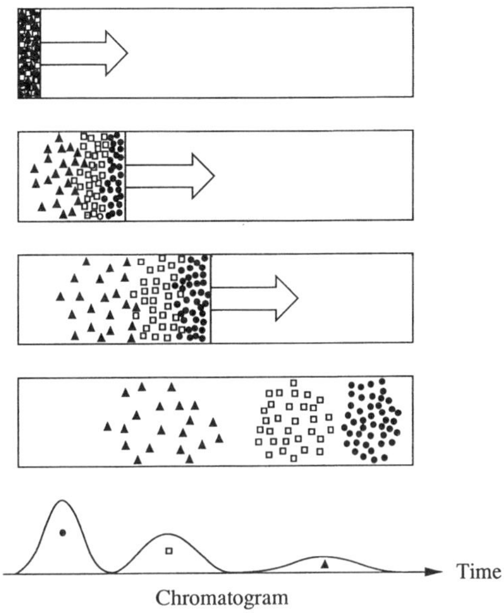

FIGURE 11.40 Chromatographic separation of components in a mixture. Three different solutes are shown schematically as circles, squares, and triangles.
From P.A. Belter, E.L. Cussler, and W.-S. Hu, 1988, Bioseparations: Downstream Processing for Biotechnology, *John Wiley, New York.*

chromatography. The basis of chromatography is *differential migration*, that is, the selective retardation of solute molecules during passage through a bed of resin particles. A schematic representation of chromatography is given in Figure 11.40; this diagram shows the separation of three solutes from a mixture injected into a column. As solvent flows through the column, the solutes travel at different speeds depending on their relative affinities for the resin particles. As a result, they will be separated and appear for collection at the end of the column at different times. The pattern of solute peaks emerging from a chromatography column is called a *chromatogram*. The fluid carrying solutes through the column or used for elution is known as the *mobile phase*; the material that stays inside the column and effects the separation is called the *stationary phase*.

In *gas chromatography* (GC), the mobile phase is a gas. Gas chromatography is used widely as an analytical tool for separating relatively volatile components such as alcohols, ketones, aldehydes, and many other organic and inorganic compounds. However, of greater relevance to bioprocessing is *liquid chromatography*, which can take a variety of forms. Liquid chromatography finds application both as a laboratory method for sample analysis and as a preparative technique for large-scale purification of biomolecules. There have been rapid developments in the technology of liquid chromatography aimed at isolation of recombinant products from genetically engineered organisms. As a high-resolution technique, chromatography is suitable for recovery of high-purity therapeutics and pharmaceuticals.

Chromatographic methods available for purification of proteins, peptides, amino acids, nucleic acids, alkaloids, vitamins, steroids, and many other biological materials include *adsorption chromatography*, *partition chromatography*, *ion-exchange chromatography*, *gel chromatography*, and *affinity chromatography*. These methods differ in the principal mechanism by which molecules are retarded in the chromatography column.

- *Adsorption chromatography*. Biological molecules have varying tendencies to adsorb on polar adsorbents such as silica gel, alumina, diatomaceous earth, and charcoal. Performance of the adsorbent relies strongly on the chemical composition of the surface, which is determined by the types and concentrations of exposed atoms or groups. The order of elution of sample components depends primarily on molecule polarity. Because the mobile phase is in competition with solute for the adsorption sites, solvent properties are also important. Polarity scales for solvents are available to aid mobile-phase selection for adsorption chromatography [24].
- *Partition chromatography*. Partition chromatography relies on the unequal distribution of solute between two immiscible solvents. This is achieved by fixing one solvent (the stationary phase) to a support and passing the other solvent containing solute over it. The solvents make intimate contact, allowing multiple extractions of solute to occur. Several methods are available to chemically bond the stationary solvent to supports such as silica [25]. When the stationary phase is more polar than the mobile phase, the technique is called *normal-phase chromatography*. When nonpolar compounds are being separated, it is usual to use a stationary phase that is less polar than the mobile phase; this is called *reverse-phase chromatography*. A common stationary phase for reverse-phase chromatography is hydrocarbon with 8 or 18 carbons bonded to silica gel: these materials are called C_8 and C_{18} packings, respectively. The solvent systems most frequently used are water–acetonitrile and water–methanol; aqueous buffers are also employed to suppress ionisation of sample components. Elution is generally in order of increasing solute hydrophobicity.
- *Ion-exchange chromatography*. The basis of separation in this procedure is electrostatic attraction between the solute and dense clusters of charged groups on the column packing. Ion-exchange chromatography can give high resolution of macromolecules and is used commercially for fractionation of antibiotics and proteins. Column packings for separation of low-molecular-weight compounds include silica, glass, and polystyrene; carboxymethyl and diethylaminoethyl groups attached to cellulose, agarose, or dextran provide suitable resins for protein chromatography. Solutes are eluted by changing the pH or ionic strength of the liquid phase; salt gradients are the most common way of eluting proteins from ion exchangers. Practical aspects of protein ion-exchange chromatography are described in greater detail elsewhere [15].
- *Gel chromatography*. This technique is also known as *molecular-sieve chromatography, exclusion chromatography, gel filtration*, and *gel-permeation chromatography*. Molecules in solution are separated using a column packed with gel particles of defined porosity. The gels most often applied are cross-linked dextran, agarose, and polyacrylamide. The speed with which components travel through the column depends on their effective molecular size. Large molecules are completely excluded from the gel matrix and move rapidly through the column to appear first in the chromatogram. Small molecules are able to penetrate the pores of the packing, traverse the column very slowly, and appear last in the chromatogram. Molecules of intermediate size enter the pores but spend less time there than the small solutes. Gel filtration can be used to separate proteins and lipophilic compounds. Large-scale gel-filtration columns are operated with upward-flow elution.

- *Affinity chromatography*. This separation technique exploits the binding specificity of biomolecules. Enzymes, hormones, receptors, antibodies, antigens, binding proteins, lectins, nucleic acids, vitamins, whole cells, and other components capable of specific and reversible binding are amenable to highly selective affinity purification. The column packing is prepared by linking a binding molecule or ligand to an insoluble support; when a sample is passed through the column, only solutes with appreciable affinity for the ligand are retained. The ligand must be attached to the support in such a way that its binding properties are not seriously affected; molecules called *spacer arms* are often used to set the ligand away from the support and make it more accessible to the solute. Many ready-made support–ligand preparations are available commercially and are suitable for a wide range of proteins. The conditions for elution depend on the specific binding complex formed: elution usually involves a change in pH, ionic strength, or buffer composition. Enzyme proteins can be desorbed using a compound with higher affinity for the enzyme than the ligand, for example, a substrate or substrate analogue. Affinity chromatography using antibody ligands is called *immunoaffinity chromatography*.

In this section we will consider the principles of liquid chromatography for separation of biological molecules such as proteins and amino acids. The choice of stationary phase will depend to a large extent on the type of chromatography employed; however certain basic requirements must be met. For high capacity, the solid support must be porous with high internal surface area; it must also be insoluble and chemically stable during operation and cleaning. Ideally, the particles should exhibit high mechanical strength and show little or no nonspecific binding. The low rigidity of many porous gels was initially a problem in industrial-scale chromatography; the weight of the packing material in large columns and the pressures developed during flow tended to compress the packing and impede operation. Many macroporous gels and composite materials of improved rigidity are now available for industrial use. Nevertheless, pressure may still be a limiting factor affecting column operation in some applications.

Two methods for carrying out chromatographic separations are high-performance liquid chromatography (HPLC) and fast protein liquid chromatography (FPLC). In principle, any of the types of chromatography described earlier can be executed using HPLC and FPLC techniques. Specialised equipment for HPLC and FPLC allows automated injection of samples, rapid flow of material through the column, collection of the separated fractions, and data analysis. Chromatographic separations traditionally performed under atmospheric pressure in vertical columns with manual sample feed and gravity elution are carried out faster and with better resolution using densely packed columns and high flow rates in HPLC and FPLC systems.

The differences between HPLC and FPLC lie in the flow rates and pressures used, the size of the packing material, and the resolution accomplished. In general, HPLC instruments are designed for small-scale, high-resolution analytical applications; FPLC is tailored for large-scale, preparative purifications. To achieve the high resolutions characteristic of HPLC, stationary-phase particles 2 to 5 μm in diameter are commonly used. Because the particles are so small, HPLC systems are operated under high pressure (5–10 MPa) to achieve flow rates of 1 to 5 ml min^{-1}. FPLC instruments are not able to develop such high pressures and are therefore operated at 1 to 2 MPa with column packings of larger size. Resolution is poorer using FPLC compared with HPLC; accordingly, it is common practice to collect only the central peak of the solute pulse emerging from the end of the column and to recycle or discard the leading and trailing edges. FPLC equipment is particularly suited to protein separations; many gels used for gel chromatography and affinity chromatography are compressible and cannot withstand the high pressures exerted in HPLC.

Chromatography is essentially a batch operation; however industrial chromatography systems can be monitored and controlled for easy automation. Cleaning the column in place is difficult. Depending on the nature of the impurities contained in the samples, relatively harsh treatments using concentrated salt or dilute alkali solutions are required; these may affect the swelling of the gel beads and, therefore, liquid flow in the column. Regeneration in place is necessary as the repacking of large columns is laborious and time consuming. Repeated use of chromatographic columns is essential because of their high cost.

11.11.1 Differential Migration

Differential migration provides the basis for chromatographic separations and is represented diagrammatically in Figure 11.41. A solution contains two solutes A and B that have different equilibrium affinities for a particular stationary phase. For the sake of brevity, let us say that the solutes are adsorbed on the stationary phase, although they may be adsorbed, bound, or entrapped depending on the type of chromatography employed. Assume that A is adsorbed more strongly than B. If a small quantity of solution is passed through the column so that only a limited depth of packing is saturated, both solutes will be retained in the bed as shown in Figure 11.41(a).

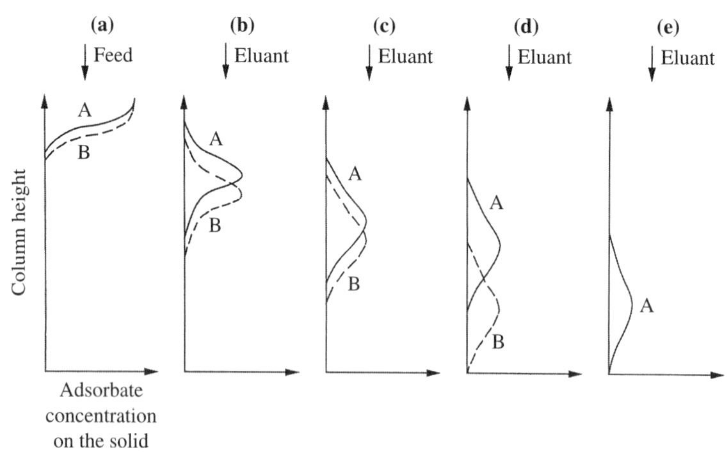

FIGURE 11.41 Differential migration of two solutes A and B.

A suitable eluant is now passed through the bed. As shown in Figures 11.41(b–e), both solutes will be alternately adsorbed and desorbed at lower positions in the column as the flow of eluant continues. Because solute B is more easily desorbed than A, it moves forward more rapidly. The difference in migration velocity between the two solutes is related to the difference in equilibrium distributions between the stationary and mobile phases. In Figure 11.41(e), solute B has been separated from A and washed out of the system.

Several parameters are used to characterise differential migration. An important variable is the volume V_e of eluting solvent required to carry the solute through the column until it emerges at its maximum concentration. Each component separated by chromatography has a different elution volume. Another parameter commonly used to characterise elution is the *capacity factor*, *k*:

$$k = \frac{V_e - V_o}{V_o} \tag{11.101}$$

where V_o is the void volume in the column, that is, the volume of liquid in the column outside of the particles. For two solutes, the ratio of their capacity factors k_1 and k_2 is called the *selectivity* or *relative retention*, δ:

$$\delta = \frac{k_2}{k_1} \tag{11.102}$$

Equations (11.101) and (11.102) are normally applied to adsorption, partition, ion-exchange, and affinity chromatography.

In gel chromatography where separation is a function of the effective molecular size, the elution volume is easily related to certain physical properties of the gel column. The total volume of a gel column is:

$$V_T = V_o + V_i + V_s \tag{11.103}$$

where V_T is the total volume, V_o is the void volume outside of the particles, V_i is the internal volume of liquid within the pores of the particles, and V_s is the volume of the gel itself. The outer volume V_o can be determined by measuring the elution volume of a substance that is completely excluded from the stationary phase: a solute that does not penetrate the gel can be washed from the column using a volume of liquid equal to V_o. V_o is usually about one-third V_T. Solutes that are only partly excluded from the stationary phase elute with a volume described by the following equation:

$$V_e = V_o + K_p V_i \tag{11.104}$$

where K_p is the *gel partition coefficient*, defined as the fraction of the internal volume available to the solute. For large molecules that do not penetrate the solid, $K_p = 0$. From Eq. (11.104):

$$K_p = \frac{V_e - V_o}{V_i} \tag{11.105}$$

K_p is a convenient parameter for comparing the separation results obtained with different gel-chromatography columns; it is independent of the column size and packing density.

However, experimental determination of K_p depends on knowledge of V_i, which is difficult to measure accurately. V_i is usually calculated using the equation:

$$V_i = a\,W_r \tag{11.106}$$

where a is the mass of dry gel and W_r is the *water regain value*, which is defined as the volume of water taken up per mass of dry gel. The value for W_r is generally specified by the gel manufacturer. If, as is often the case, the gel is supplied already wet and swollen, the value of a is unknown and V_i is determined using the following equation:

$$V_i = \frac{W_r \rho_g}{(1 + W_r \rho_w)}(V_T - V_o) \tag{11.107}$$

where ρ_g is the density of wet gel and ρ_w is the density of water.

EXAMPLE 11.10 HORMONE SEPARATION USING GEL CHROMATOGRAPHY

A pilot-scale gel-chromatography column packed with Sephacryl resin is used to separate two hormones, A and B. The column is 5 cm in diameter and 0.3 m high and the void volume is 1.9×10^{-4} m^3. The water regain value of the gel is 3×10^{-3} m^3 kg^{-1} dry Sephacryl; the density of wet gel is 1.25×10^3 kg m^{-3}. The partition coefficient for hormone A is 0.38; the partition coefficient for hormone B is 0.15. If the eluant flow rate is 0.7 l h^{-1}, what is the retention time for each hormone?

Solution

The total column volume is:

$$V_T = \pi r^2 h = \pi\,(2.5 \times 10^{-2}\ \text{m})^2\,(0.3\ \text{m}) = 5.89 \times 10^{-4}\ \text{m}^3$$

$V_o = 1.9 \times 10^{-4}$ m^3 and $\rho_w = 1000$ kg m^{-3}. From Eq. (11.107):

$$V_i = \frac{3 \times 10^{-3}\ \text{m}^3\ \text{kg}^{-1}\,(1.25 \times 10^3\ \text{kg m}^{-3})}{1 + (3 \times 10^{-3}\ \text{m}^3\ \text{kg}^{-1})\,(1000\ \text{kg m}^{-3})}\,(5.89 \times 10^{-4}\ \text{m}^3 - 1.9 \times 10^{-4}\ \text{m}^3)$$

$$V_i = 3.74 \times 10^{-4}\ \text{m}^3$$

$K_{pA} = 0.38$ and $K_{pB} = 0.15$. Therefore, from Eq. (11.104):

$$V_{eA} = 1.9 \times 10^{-4}\ \text{m}^3 + 0.38\,(3.74 \times 10^{-4}\ \text{m}^3) = 3.32 \times 10^{-4}\ \text{m}^3$$
$$V_{eB} = 1.9 \times 10^{-4}\ \text{m}^3 + 0.15\,(3.74 \times 10^{-4}\ \text{m}^3) = 2.46 \times 10^{-4}\ \text{m}^3$$

The times associated with these elution volumes are:

$$t_A = \frac{3.32 \times 10^{-4}\ \text{m}^3}{0.7\ \text{l h}^{-1} \cdot \left|\frac{1\ \text{m}^3}{1000\ \text{l}}\right| \cdot \left|\frac{1\ \text{h}}{60\ \text{min}}\right|} = 28\ \text{min}$$

$$t_B = \frac{2.46 \times 10^{-4}\ \text{m}^3}{0.7\ \text{l h}^{-1} \cdot \left|\frac{1\ \text{m}^3}{1000\ \text{l}}\right| \cdot \left|\frac{1\ \text{h}}{60\ \text{min}}\right|} = 21\ \text{min}$$

11.11.2 Zone Spreading

The effectiveness of chromatography depends not only on differential migration but on whether the elution bands for individual solutes remain compact and without overlap. Ideally, each solute should pass out of the column at a different instant in time. In practice, elution bands spread out somewhat so that each solute takes a finite period of time to pass across the end of the column. Zone spreading is not so important when migration rates vary widely because there is little chance that the solute peaks will overlap. However, if the molecules to be separated have similar structure, migration rates will also be similar and zone spreading must be carefully controlled.

As illustrated in Figure 11.40, typical chromatogram elution bands have a peak of high concentration at or about the centre of the pulse but are of finite width as the concentration trails off to zero before and after the peak. Spreading of the solute peak is caused by several factors represented schematically in Figure 11.42.

- *Axial diffusion.* As solute is carried through the column, molecular diffusion of solute will occur from regions of high concentration to regions of low concentration. Diffusion in the axial direction (i.e., up and down the length of the tube) is indicated in Figure 11.42(a) by broken lines with arrows. Axial diffusion broadens the solute peak by transporting material upstream and downstream away from the region of greatest concentration.
- *Eddy diffusion.* In columns packed with solid particles, the flow paths of liquid through the bed can be highly variable. As indicated in Figure 11.42(a), some liquid will flow almost directly through the bed while other liquid will take longer and more tortuous paths through the gaps or interstices between the particles. Accordingly, some solute molecules carried in the fluid will move slower than the average rate of progress through the column while others will take shorter paths and move ahead of the

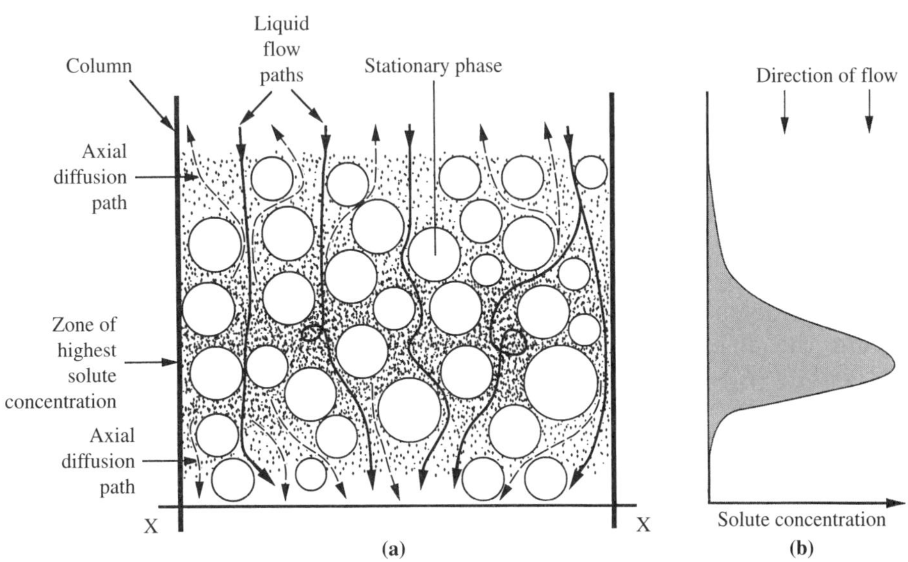

FIGURE 11.42 Zone spreading in a chromatography column.

average. The result is spreading of the solute band. Differential motion of material due to erratic local variations in flow velocity is known as *eddy diffusion*.

- *Local nonequilibrium effects*. In most columns, lack of equilibrium is the most important factor affecting zone spreading, although perhaps the most difficult to understand. Consider the situation at position X indicated in Figure 11.42(a). A solute pulse is passing through the column; as shown in Figure 11.42(b), the concentration within this pulse increases from the front edge to a maximum near the centre and then decreases to zero. As the solute pulse moves down the column, an initial gradual increase in solute concentration will be experienced at X. In response to this increase in mobile-phase solute concentration, solute will bind to the stationary phase and the stationary-phase concentration will start to increase toward an appropriate equilibrium value. Equilibrium is not established immediately however; it takes time for the solute to undergo the mass transfer steps from liquid to solid as outlined in Section 11.9.4. Indeed, before equilibrium can be established, the mobile-phase concentration increases again as the centre of the solute pulse moves closer to X. Because the concentration in the mobile phase is continuously increasing, equilibrium at X always remains out of reach and the stationary-phase concentration lags behind equilibrium values. As a consequence, a higher concentration of solute remains in the liquid than if equilibrium were established, and the front edge of the solute pulse effectively moves ahead faster than the remainder of the pulse. As the peak of the solute pulse passes X, the mobile-phase concentration starts to decrease with time. In response, the solid phase must divest itself of solute to reach equilibrium with the lower liquid-phase concentrations. However, again, because of delays due to mass transfer, equilibrium cannot be established with the continuously changing liquid concentration. As the solute pulse passes and the liquid concentration at X falls to zero, the solid phase still contains solute molecules that continue to be released into the liquid. Consequently, the rear of the solute pulse is effectively stretched out until the stationary phase reaches equilibrium with the liquid.

In general, conditions that improve mass transfer will increase the rate at which equilibrium is approached between the phases, thus minimising zone spreading. For example, increasing the particle surface area per unit volume facilitates mass transfer and reduces nonequilibrium effects; surface area is usually increased by using smaller particles. On the other hand, increasing the liquid flow rate will exacerbate nonequilibrium effects as the rate of adsorption fails even more to keep up with concentration changes in the mobile phase. Viscous solutions give rise to considerable zone broadening as a result of slower mass transfer rates; zone spreading is also more pronounced if the solute molecules are large. Changes in temperature can affect zone spreading in several ways. Because the viscosity is reduced at elevated temperatures, heating the column often decreases zone spreading. However, rates of axial diffusion increase at higher temperatures so that the overall effect depends on the system and temperature range tested.

11.11.3 Theoretical Plates in Chromatography

The concept of theoretical plates is often used to analyse zone broadening in chromatography. The idea is essentially the same as that described in Section 11.6 for an ideal equilibrium stage. The chromatography column is considered to be made up of a number of segments or plates of height H; the magnitude of H is of the same order as the diameter of the resin particles. Within each segment, equilibrium is supposed to exist.

As in adsorption operations, equilibrium is not often achieved in chromatography so that the theoretical plate concept does not accurately reflect conditions in the column. Nevertheless, the idea of theoretical plates is applied extensively, mainly because it provides a parameter, the plate height H, that can be used to characterise zone spreading. Use of the plate height, which is also known as the *height equivalent to a theoretical plate* (*HETP*), is acceptable practice in chromatography design even though it is based on a poor model of column operation. HETP is a measure of zone broadening; in general, the lower the HETP value, the narrower is the solute peak.

HETP depends on various processes that occur during elution of a chromatography sample. A popular and simple expression for HETP takes the form:

$$H = \frac{A}{u} + B\,u + C \tag{11.108}$$

where H is the plate height, u is the linear liquid velocity, and A, B, and C are experimentally determined kinetic constants. A, B, and C include the effects of liquid–solid mass transfer, forward and backward axial dispersion, and nonideal distribution of liquid around the packing. As outlined in Section 11.9.4, overall rates of solute adsorption and desorption in chromatography depend mainly on mass transfer processes. The values of A, B, and C can be reduced by improving mass transfer between the liquid and solid phases, resulting in a decrease in HETP and better column performance. Equation (11.108) and other HETP models are discussed further in other references [26, 27].

HETP for a particular component is related to the elution volume and width of the solute peak as it appears on the chromatogram. If, as shown in Figure 11.43(a), the pulse has the standard symmetrical form of a normal distribution around a mean value $\bar{x}$, the *number of theoretical plates* can be calculated as follows:

$$N = 16\left(\frac{V_e}{w}\right)^2 \tag{11.109}$$

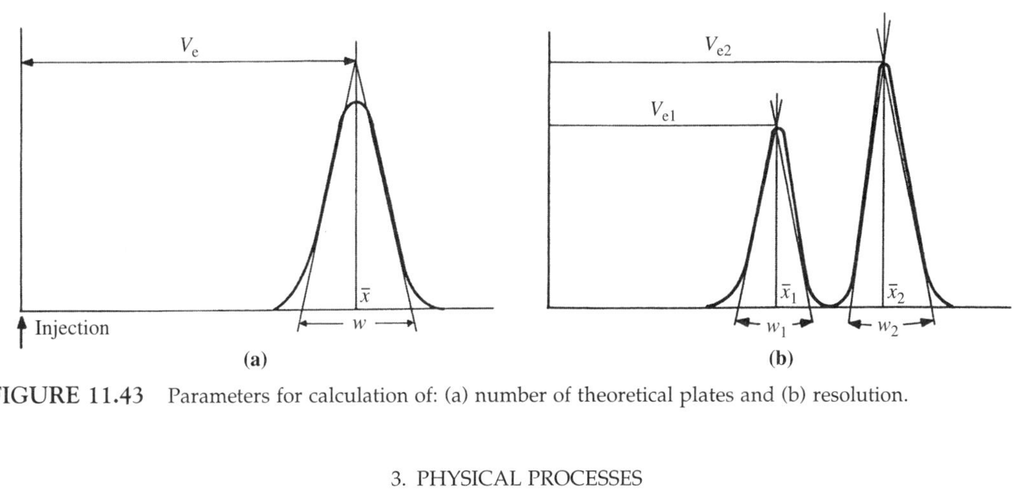

FIGURE 11.43 Parameters for calculation of: (a) number of theoretical plates and (b) resolution.

where N is the number of theoretical plates, V_e is the distance on the chromatogram corresponding to the elution volume of the solute, and w is the baseline width of the peak between lines drawn as tangents to the inflection points of the curve. Equation (11.109) applies if the sample is introduced into the column as a narrow pulse. The number of theoretical plates is related to HETP as follows:

$$N = \frac{L}{H} \tag{11.110}$$

where L is the length of the column. For a given column, the greater the number of theoretical plates, the greater is the number of ideal equilibrium stages in the system and the more efficient is the separation. Values of H and N vary for a particular column depending on the component being separated.

11.11.4 Resolution

Resolution is a measure of zone overlap in chromatography and an indicator of column effectiveness. For separation of two components, resolution is given by the following equation:

$$R_N = \frac{2(V_{e2} - V_{e1})}{(w_1 + w_2)} \tag{11.111}$$

where R_N is the resolution, V_{e1} and V_{e2} are distances on the chromatogram corresponding to the elution volumes for components 1 and 2, and w_1 and w_2 are the baseline widths of the chromatogram peaks as shown in Figure 11.43(b). Column resolution is a dimensionless quantity; the greater the value of R_N, the more separated are the two solute peaks. An R_N value of 1.5 corresponds to a baseline resolution of 99.8% or almost complete separation; when $R_N = 1.0$, the two peaks overlap by about 2% of the total peak area.

Column resolution can be expressed in terms of HETP. Assuming w_1 and w_2 are approximately equal, Eq. (11.111) becomes:

$$R_N = \frac{(V_{e2} - V_{e1})}{w_2} \tag{11.112}$$

Substituting for w_2 from Eq. (11.109):

$$R_N = \frac{1}{4}\sqrt{N}\left(\frac{V_{e2} - V_{e1}}{V_{e2}}\right) \tag{11.113}$$

The term $\frac{V_{e2} - V_{e1}}{V_{e2}}$ can be expressed in terms of k_2 and δ from Eqs. (11.101) and (11.102), so that Eq. (11.113) becomes:

$$R_N = \frac{1}{4}\sqrt{N}\left(\frac{\delta - 1}{\delta}\right)\left(\frac{k_2}{k_2 + 1}\right) \tag{11.114}$$

Using the expression for N from Eq. (11.110), the equation for column resolution is:

$$R_N = \frac{1}{4}\sqrt{\frac{L}{H}}\left(\frac{\delta - 1}{\delta}\right)\left(\frac{k_2}{k_2 + 1}\right) \tag{11.115}$$

As is apparent from Eq. (11.115), peak resolution increases as a function of $\sqrt{L}$, where L is the column length. Resolution also increases with decreasing HETP; therefore, any enhancement of mass transfer conditions that reduces H will also improve resolution. Derivation of Eq. (11.115) involves Eq. (11.109), which applies to chromatography systems where a relatively small quantity of sample is injected rapidly. Resolution is sensitive to increases in sample size; as the amount of sample increases, the resolution declines. In laboratory analytical work, it is common to use extremely small sample volumes, of the order of microlitres. However, depending on the type of chromatography used for production-scale purification, sample volumes of 5 to 20% of the column volume and higher are used. Because resolution under these conditions is relatively poor, when the solute peak is collected for recovery of product, the central portion of the peak is retained while the leading and trailing edges are recycled back to the feed.

11.11.5 Chromatography Scale-Up

The aim in scale-up of chromatography is to retain the resolution and solute recovery achieved using a small-scale column while increasing the throughput of material. Strategies for scale-up must take into account the dominance of mass transfer effects in chromatography separations.

The easiest approach to scale-up is to simply increase the flow rate through the column; however, this gives unsatisfactory results. Raising the liquid velocity increases zone spreading and produces high pressures in the column that compress the stationary phase and cause pumping difficulties. The pressure drop problem can be alleviated by increasing the particle size, but this hinders mass transfer and so decreases resolution. Increasing the column length can help regain any resolution lost due to increase of the flow rate or particle diameter; however increasing L also has a strong effect in raising the pressure drop through the column.

The solution for scale-up is to keep the same column length, linear flow velocity, and particle size as in the small column, but increase the column diameter. The larger capacity of the column is therefore due solely to its greater cross-sectional area. The volumetric flow rate is increased in proportion to the column volume, so that:

$$\frac{Q_1}{V_1} = \frac{Q_2}{V_2} \tag{11.116}$$

In Eq. (11.116), Q_1 and V_1 are the volumetric flow rate and column volume giving adequate performance at the smaller scale, and Q_2 and V_2 are the volumetric flow rate and column volume for the large-scale process. If the length of the two columns is kept the same, Eq. (11.116) can be written as:

$$\frac{Q_2}{Q_1} = \left(\frac{D_2}{D_1}\right)^2 \tag{11.117}$$

If scale-up is carried out using this approach, all the important parameters affecting the packing matrix, liquid flow, mass transfer, and equilibrium conditions are kept constant. Similar column performance can therefore be expected. Because the liquid distribution in large-diameter packed columns tends to be poor, care must be taken to ensure that liquid is fed evenly over the entire column cross-section after scale-up.

In practice, variations in column properties and efficiency do occur with scale-up. As an example, compressible solids such as those used in gel chromatography get better support from the column wall in small-diameter columns than in large-diameter columns; as a result, lower linear flow rates must be used in large-scale systems. An advantage of using gels of high mechanical strength in laboratory systems is that they allow more direct scale-up to commercial operations. The elasticity and compressibility of gels used for fractionation of high-molecular-weight proteins preclude the use of long columns in large-scale processes; bed heights in these systems are normally restricted to 0.6 to 1.0 m.

11.12 CRYSTALLISATION

Crystallisation is used for *polishing* or final purification of a wide range of fermentation products. Many pharmaceuticals and fine biochemicals, including antibiotics, amino acids, β-carotene, organic acids, and vitamins, are marketed in the form of crystals. As an example, Figure 11.44 shows crystals of the amino acid glycine. Methods for bulk crystallisation of proteins from fermentation broths are also being developed. As well as for polishing, crystallisation can be used within downstream processing schemes for recovery and concentration of components from the liquid phase.

The formation of solids from solution is a feature common to both precipitation (Section 11.8) and crystallisation. However, an important difference is that, in contrast to precipitates, crystals often achieve very high ($\geq$99.5%) levels of purity. In many cases, the degree of purification achieved using crystallisation is so substantial that other more expensive unit operations, such as chromatography, are not required. From the original solution, a two-phase slurry or *magma* is formed containing *mother liquor* and solid crystals of varying size. The crystals are harvested from the magma using unit operations such as filtration and

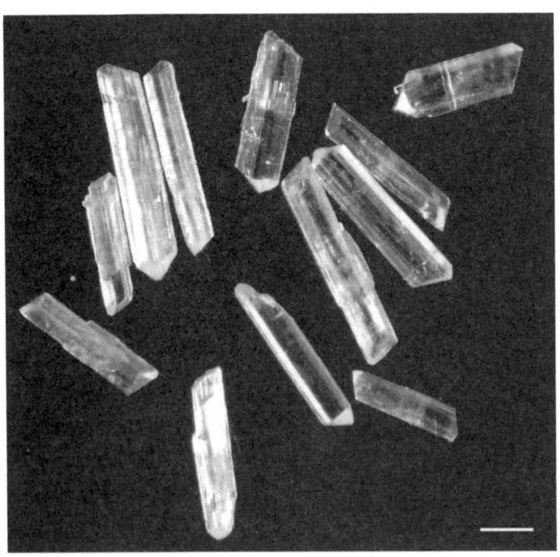

FIGURE 11.44 Glycine crystals. The scale bar represents 2 mm.
Photograph by H. Zhao.

centrifugation then washed and dried prior to packaging. Crystallisation is generally carried out at low temperatures, thus minimising thermal damage to heat-labile products.

Mass or bulk crystallisation of large quantities of crystals is a complex process. Although small numbers of crystals can be produced relatively easily in the laboratory, upscaling to industrial-size equipment is a challenge. Crystallisation is a two-phase, multicomponent, surface-dependent process that involves simultaneous heat and mass transfer and is sensitive to a range of conditions such as fluid hydrodynamics, particle mechanics, vessel geometry, and mode of operation. Many aspects of large-scale crystallisation are closer to art than science. In addition, most of our engineering understanding of crystallisation has been derived from studies of inorganic salts rather than biological molecules. Although many of the principles of crystallisation are the same irrespective of the type of product, there are some features of biological systems that require special attention.

11.12.1 Crystal Properties

Crystals are characterised in terms of their *shape*, *size*, *size distribution*, and *purity*. All of these properties affect the end-use quality of the product; several also influence the ease of crystal recovery and handling.

Shape and Size

Crystals are composed of molecules, atoms, or ions arranged in an orderly, repetitive, three-dimensional array or *space lattice*. The distances between constituent units within crystals of a given substance are fixed and characteristic of that material. Under ideal growth conditions, this gives rise to reproducible and well-defined crystal shapes.

Crystals forming freely from solution appear as polyhedrons bounded by planar faces. For a given material, although the angles between the crystal faces remain constant, the space lattice allows for variation in the relative sizes of the faces. Accordingly, variation in *crystal habit* may be observed between individual crystals of a particular substance. As an example, different growth habits for hexagonal crystals are illustrated in Figure 11.45. Crystal habit is constrained by internal geometry but is subject to external factors such as the solvent used, impurities present, and conditions such as temperature and pH that affect the rate of crystal growth. For some fermentation products such as amino acids and antibiotics, it may be necessary to produce crystals with a specific habit. As solution

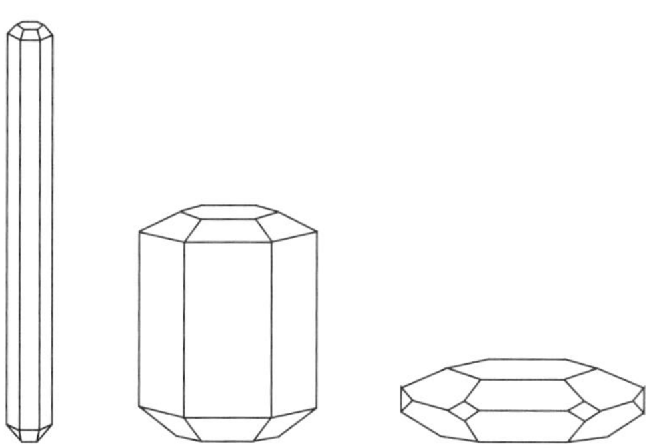

FIGURE 11.45 Three different crystal habits for hexagonal crystals.

TABLE 11.7 Properties of Crystal Products Affected by Crystal Size, Shape, and Size Distribution

End-use and Handling Property	Processing Property
Bulk density	Filtration rate
Solids flow characteristics	Centrifugation rate
Residual moisture content	Entrainment of liquid after dewatering
Caking properties in storage	
Rate of dissolution	
Dustiness	
Appearance	
Fluidisation properties	
Pneumatic handling properties	

impurities have the ability to bind to and thus selectively alter the growth of one or more crystal faces, they can strongly influence crystal habit even when present at only trace concentrations. Impurities with this effect are known as *habit modifiers*.

The shape and size of individual crystals affect important physical properties of crystalline products. Some of these are listed in Table 11.7. The ease with which crystals are recovered from magma also depends on their shape and size. For example, crystals with chunky, compact habits are easier to filter, centrifuge, wash, and dry than long, needle-like, or twinned crystals that are easily broken, or broad, flat crystals that are difficult to separate from each other.

Under ideal conditions, a crystal will maintain constant habit and geometric proportions during its growth. Such crystals are called *invariant*. The size of an invariant crystal can be expressed using only a single dimension known as the *characteristic length, L*. The value of L is often measured using sieve analysis as described further in the next section, where L is related to the sieve aperture size. The properties of individual crystals, such as surface area, volume, and mass, are defined in terms of the characteristic length and appropriate *shape factors*:

$$A_C = k_a L^2 \tag{11.118}$$

$$V_C = k_v L^3 \tag{11.119}$$

$$M_C = \rho V_C = \rho k_v L^3 \tag{11.120}$$

where A_C is the surface area of the crystal, k_a is the area shape factor, L is the characteristic length, V_C is the volume of the crystal, k_v is the volume shape factor, M_C is the mass of the crystal, and ρ is the crystal density. Methods for measuring crystal shape factors are described in the literature [28].

The shape and size of crystals can be affected detrimentally by *agglomeration* or the bonding together of individual crystals into larger particles. Another crystal property that particularly affects biological molecules is *polymorphism*, which is the occurrence of multiple crystal forms. The ability of biological materials to undergo polymorphic crystallisation reflects the diversity of forces, including hydrogen bonding, van der Waals forces, and hydrophobic/hydrophilic interactions, that affect the molecular conformation and packing properties of substances such as proteins, amino acids, fatty acids, and lipids. For some bioproducts, complex polymorphic transitions occur during crystallisation, as thermodynamically unstable crystals produced in the early stages of the process transform into more stable forms either in the solid state or after redissolution. Crystallisation conditions must be used to control polymorphic systems so that stable crystals of the desired shape and size are produced.

Size Distribution

Crystallisation processes generate particles with a range of sizes. Crystal size distribution affects several important product and processing properties (Table 11.7). For example, difficulties with filtration are minimised, and the tendency of crystals to bind together or *cake* into large solid lumps during storage is reduced, if the crystals are of relatively uniform size. Therefore, an important objective in crystallisation operations is to produce crystals with as narrow a size range as possible.

Several measurement techniques are available for measuring crystal size distribution. The simplest and most widely used is *sieving* or *screening*. Sieving is applied mostly for sizing dry particles; however, wet sieving using liquid to assist particle movement through the screens may be used to separate small particles 10 to 45 μm in size. Other methods for measuring particle size include Coulter counter, light scattering, and sedimentation techniques [28–30]. Each method gives particle size information based on particular properties of the particle; for example, sieving discriminates between particles depending on their linear dimensions whereas Coulter counters measure particle volume.

In industrial processes, the crystal size distribution is determined typically by sieving and measuring the weight of the crystals retained on the screens. Standard screens are available to separate particles with sizes between about 37 μm and 108 mm. The screens are made of woven wire with carefully calibrated spaces between the wires. Common sieve series are the *U.S. sieve series* and the *Tyler standard screen series*. The characteristics of these screens are listed in Appendix F.

For analysis of crystal size distribution, a set of standard screens is arranged in order in a stack with the smallest-aperture screen at the bottom and the largest at the top. A pan is placed under the bottom sieve to catch the finest particles. A sample of crystals is placed on the top screen and the entire stack is shaken. Crystals of varying size move down through the stack until they reach a screen with small enough aperture to prevent further passage. The particles retained on each screen and in the pan are then collected and weighed. Sieving must be continued until the sizing process is complete, that is, until all the particles have been retained by a screen through which they cannot pass. As illustrated in Figure 11.46, particles with a variety of shapes and sizes are able to pass through mesh screens to give the same size measurement in sieve analysis. Accordingly, sieving provides only an approximate indication of crystal size based on a single linear dimension of the particle.

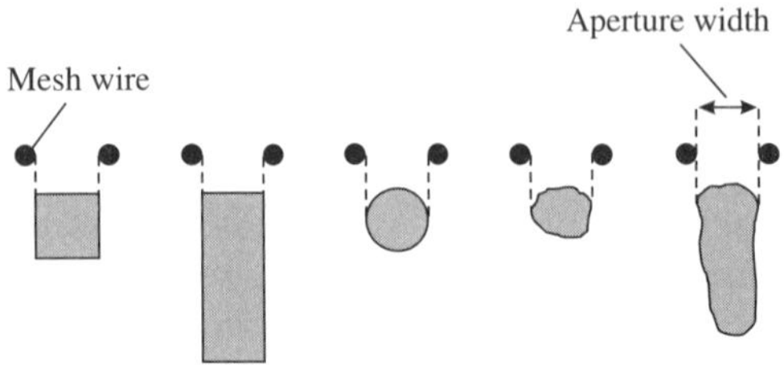

FIGURE 11.46 Different particle shapes and sizes giving the same size measurement by sieving.

The *mass fraction* of the crystals retained at a particular screen is calculated as:

$$\Delta w = \frac{\Delta M}{M_T} \tag{11.121}$$

where Δw is the mass fraction, ΔM is the mass of particles retained on the screen, and M_T is the total sample mass. For *differential analysis* of the crystal size distribution, individual mass fractions are reported as a function of particle size. Particles on a particular screen can have any size between the aperture of the screen retaining the particle and the aperture of the screen above it. Because the difference in aperture size between adjacent screens may not be the same throughout the stack, the crystal size distribution is determined by dividing the mass fraction retained at each screen by the increment in aperture size between that screen and the one immediately above it. This type of analysis gives information about the *mass density distribution*:

$$m = \frac{\Delta w}{\Delta L} \tag{11.122}$$

where m is the mass density and ΔL is the increment in aperture size from above. Mass density has dimensions L^{-1} so typical units are mm^{-1} or μm^{-1}. Because mass fractions and mass densities refer to a range of particle sizes, differential size distributions may be represented using histogram plots. More typically, however, the results are plotted as line graphs against the *average particle size* L_{av} retained on each screen. The particle size corresponding to the maximum value of the mass density is called the *dominant size*, L_D. Typical results from screening are analysed in Example 11.11.

EXAMPLE 11.11 CRYSTAL SIEVING ANALYSIS

Crystals of sebacic acid are fractionated using a series of sieves. The results are shown in the table.

Tyler Screen	Mass Retained (g)
24	0
28	0.05
35	1.05
48	14.3

65	31.8
100	21.6
150	13.8
200	4.4
Pan	4.2

(a) Plot the mass density distribution using a histogram plot.
(b) Plot the mass density distribution using a line plot.
(c) What is the dominant crystal size?

Solution

Aperture sizes for the different Tyler screens used are obtained from Table F.2 in Appendix F. The total sample mass of 91.2 g is obtained by adding the masses retained at each screen and in the pan. The mass fraction of the sample retained on each screen is calculated using Eq. (11.121). For example, the mass fraction retained on Tyler screen 100 is:

$$\Delta w = \frac{21.6\text{ g}}{91.2\text{ g}} = 0.237$$

This result means that crystals with sizes between 147 and 208 μm account for 23.7% of the total mass of the sample. The mass density m is calculated by dividing the mass fraction Δw at each screen by the increment in aperture size from above, ΔL. For example, the mass density for Tyler screen 100 is:

$$m = \frac{0.237}{(208 - 147)\ \mu\text{m}} = 3.9 \times 10^{-3}\ \mu\text{m}^{-1}$$

The average particle size retained on each screen is calculated by taking the average of the aperture sizes of that screen and the one above it. For example, the average size of the crystals retained on Tyler screen 28 is:

$$L_{av} = \frac{(701 + 589)\ \mu\text{m}}{2} = 645\ \mu\text{m}$$

The mass density and average size of the crystals collected in the pan are calculated as if the aperture size of the pan is zero. The complete results are listed in the following table.

Tyler Mesh	Aperture Size (μm)	Increment in Aperture Size From Above, ΔL (μm)	Mass Retained, ΔM (g)	Mass Fraction Retained, Δw	Mass Density, m (μm^{-1})	Average Particle Size Retained, L_{av} (μm)
24	701	–	0	0	0	–
28	589	112	0.05	5.5×10^{-4}	4.9×10^{-6}	645
35	417	172	1.05	0.012	7.0×10^{-5}	503
48	295	122	14.3	0.157	1.3×10^{-3}	356
65	208	87	31.8	0.349	4.0×10^{-3}	252

(Continued)

100	147	61	21.6	0.237	3.9×10^{-3}	178
150	104	43	13.8	0.151	3.5×10^{-3}	126
200	74	30	4.4	0.048	1.6×10^{-3}	89
Pan	–	74	4.2	0.046	6.2×10^{-4}	37
Total	–	–	91.2	1.00	–	–

(a) The mass density is plotted versus particle size as a histogram plot in Figure 11.47.
(b) The mass density is plotted versus the average particle size as a line plot in Figure 11.48.
(c) From Figure 11.48, the dominant crystal size L_D corresponding to the maximum value of m is about 240 μm.

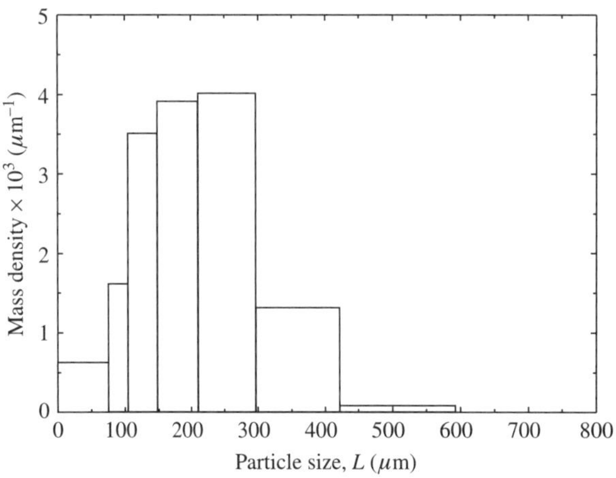

FIGURE 11.47 Histogram plot of the mass density distribution.

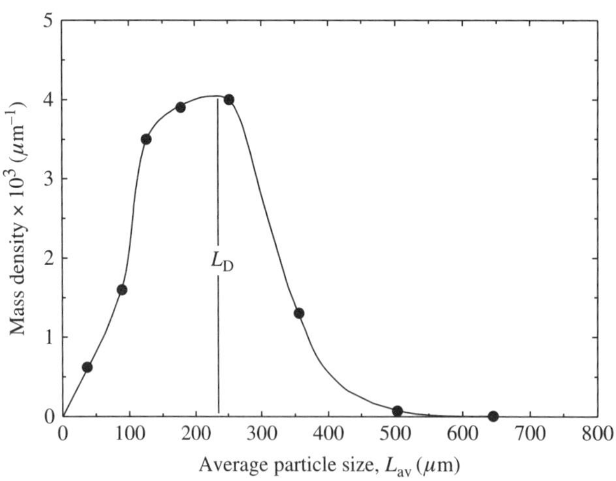

FIGURE 11.48 Line plot of the mass density distribution

The crystal size distribution in Example 11.11 was determined on a mass basis. This follows naturally from sieving analysis, which gives the masses of particles retained on the screens. However, particle size distributions are also commonly represented in terms of the *number* of particles rather than their mass. The conversion between mass and number of particles uses the relationship of Eq. (11.120) and the volume shape factor k_v. If ΔM is the mass of crystals retained on a particular screen, the number of particles ΔN on that screen can be estimated as:

$$\Delta N = \frac{\Delta M}{\rho k_v L_{\text{av}}^3} \tag{11.123}$$

where ρ is the crystal density, k_v is the particle volume shape factor, and L_{av} is the average size of the particles retained on the screen. By analogy with the mass density in Eq. (11.122), the *crystal number density* or *crystal population density* n is evaluated as:

$$n = \frac{\Delta N}{\Delta L} = \frac{\Delta M / \Delta L}{\rho k_v L_{\text{av}}^3} \tag{11.124}$$

where ΔL is the size increment corresponding to ΔN. Using Eq. (11.121), an alternative expression for n is:

$$n = \frac{(M_{\text{T}}\, \Delta w)/\Delta L}{\rho k_v L_{\text{av}}^3} \tag{11.125}$$

The dimensions of n defined by Eqs. (11.124) and (11.125) are L^{-1}; typical units are mm^{-1} or μm^{-1}. However, for analysis of crystal production in liquid systems, the total mass of crystals M_{T} may be expressed on a volumetric basis as a crystal mass concentration or *slurry density* with dimensions ML^{-3}. In this case, the dimensions of n determined using Eq. (11.125) become L^{-4}, so that units for n include $\text{mm}^{-1}\,\text{l}^{-1}$, $\mu\text{m}^{-1}\,\text{l}^{-1}$, and m^{-4}. Evaluation of the crystal population density is illustrated in Example 11.12.

EXAMPLE 11.12 CRYSTAL POPULATION DENSITY

Salicylic acid crystals are harvested from magma with a slurry density of 120 g l^{-1}. The crystals are washed, dried, and size-fractionated by sieving. The mass fraction of crystals with sizes between 210 and 250 μm is 0.14. The density of the crystals is 1.44 g cm^{-3} and the volume shape factor is 0.55. What is the population density of crystals of size 230 μm?

Solution

The average size of crystals in the 210 to 250 μm sample is:

$$L_{\text{av}} = \frac{(210 + 250)\ \mu\text{m}}{2} = 230\ \mu\text{m}$$

Therefore, the crystal population density can be evaluated using Eq. (11.125):

$$n = \frac{\dfrac{(120 \text{ g l}^{-1})\,(0.14)}{(250-210)\ \mu\text{m}}}{(1.44 \text{ g cm}^{-3}) \cdot \left|\dfrac{1 \text{ cm}}{10^4\ \mu\text{m}}\right|^3 \cdot (0.55)\,(230)^3\ \mu\text{m}^3} = 4.4 \times 10^4\ \mu\text{m}^{-1}\,\text{l}^{-1}$$

or:

$$n = 4.4 \times 10^4\ \mu\text{m}^{-1}\,\text{l}^{-1} \cdot \left|\frac{1000 \text{ l}}{1 \text{ m}^3}\right| \cdot \left|\frac{10^6\ \mu\text{m}}{1 \text{ m}}\right| = 4.4 \times 10^{13} \text{ m}^{-4}$$

The particle size distribution developed during crystallisation reflects the balance between formation of new crystals and growth of existing crystals. If tiny new crystals are constantly generated, a large number of crystals will be produced, the average particle size will be small, and the size distribution is likely to be broad. On the other hand, if few new particles are produced but the existing particles are allowed to grow in size, the number of crystals will be relatively small, the average particle size will be large, and the size distribution is likely to be narrow. Control over the relative rates of crystal formation and growth is therefore necessary to produce crystals with specified size and size distribution characteristics.

Purity

If crystals are allowed to grow slowly under constant conditions, their purity can reach very high values of 99.5 to 99.8%. However, in industrial practice, varying levels of impurities are carried over from the mother liquor into the crystalline product. The mechanisms by which impurities infiltrate crystalline materials include:

- Adsorption or deposition of impurities on crystal surfaces
- Entrapment of mother liquor between crystals forming agglomerates
- Inclusions of small quantities of mother liquor within crystals
- Substitution of impurity molecules within the crystal lattice

Impurities present on the external surfaces are removed by washing the crystals with fresh solvent during recovery by filtration or centrifugation. If agglomeration is responsible for low purity, it may be possible to reduce agglomerate formation by manipulating the crystallisation conditions. Liquid inclusions within the crystals are minimised by lowering the rate of crystal growth. The substitution of impurities within crystals is a more difficult problem that arises when the impurities present have properties similar to those of the crystal-forming solute. In this case, the impurity must be removed by dissolving the crystals in a new solvent that differentially affects the solubility of the solute and impurity. If the impurity is more soluble in the new solvent than the desired product, *recrystallisation* will yield a fresh population of crystals with improved purity. Recrystallisation is sometimes carried out several times before crystals of the desired purity are obtained.

11.12.2 Crystal Formation and Growth

The critical phenomena in crystallisation processes are the formation of new crystals and crystal growth. These processes determine the crystal size and size distribution. The driving force for both is the level of solute supersaturation in the solution.

Supersaturation

The *solubility* of a substance in a particular solvent is the maximum concentration of the solute that can be dissolved under given conditions of temperature and pressure while maintaining thermodynamic stability. A solution containing this maximum concentration is called a *saturated solution*. In many cases, it is relatively easy to generate solutions containing more dissolved solute than the saturation concentration; however, such *supersaturated solutions* are thermodynamically unstable. Some supersaturated solutions are *metastable*, meaning that the solution is capable of resisting small disturbances but becomes unstable when subjected to large perturbations.

Because solubility data in the literature usually refer to pure solutes and solvents, and as impurities have a significant affect on solution properties, solubility curves for practical systems must be determined experimentally. Typical results for dissolved solute concentration as a function of solution temperature are shown in Figure 11.49. The *solubility curve* represents the condition of thermodynamic equilibrium between the solid and liquid phases. Below the solubility curve, the solution is *undersaturated* and crystallisation is impossible. The broken line above the solubility curve is the *supersolubility curve*, which indicates the temperatures and solute concentrations at which spontaneous crystal formation is likely to occur. This curve is determined experimentally by slowly reducing the temperature or increasing the solute concentration in a crystal-free solution until new crystals are formed. The supersolubility curve is not as well defined as the solubility curve: its position may change with conditions such as the intensity of agitation. The zone between the solubility and supersolubility

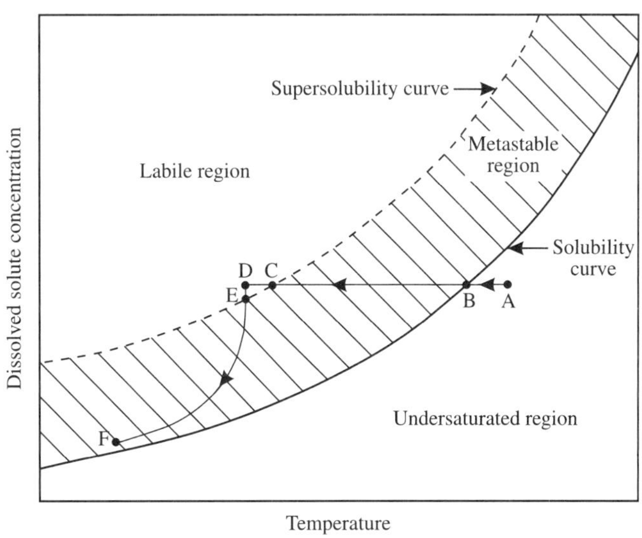

FIGURE 11.49 Solubility diagram showing curves for solubility and supersolubility as a function of temperature. A hypothetical pathway for batch crystallisation without the use of seed crystals is indicated between points A and F.

curves is known as the *metastable region* of supersaturation. In this region, although spontaneous crystal formation is unlikely, any crystals already present will grow in size. As crystallisation operations are carried out mainly within the metastable region, the width of this zone has a considerable influence on the ease with which crystallisation can be controlled. Above the supersolubility curve is the *unstable* or *labile region* of supersaturation. In this zone, spontaneous crystal generation occurs.

Let us consider how variation in temperature affects the solution conditions in Figure 11.49. As an undersaturated solution represented by point A is cooled, the system enters the metastable zone at point B and becomes saturated and then supersaturated. However, crystal formation does not occur until the temperature reaches point C, which represents the *metastable limit*. Generation of crystals in the labile region around point D may reduce the dissolved solute concentration sufficiently so that the solution returns to the metastable zone at point E. In this region, crystallisation occurs by growth of the new crystals that were formed in the labile zone. If the temperature is reduced further, the dissolved solute concentration becomes progressively lower as more and more solute leaves the solution and is incorporated into the crystals. In this example, the crystallisation process is stopped at point F.

Methods other than cooling may be used to induce supersaturation. From point A in Figure 11.49, supersaturation is also achieved by increasing the dissolved solute concentration at constant temperature. Typically, this is accomplished by evaporating the solvent. In industrial practice, a combination of cooling and evaporation is often employed. Supersaturation may also be induced by adding substances that reduce the solubility of the solute, or using chemical reaction to form a less soluble product. In crystallisation, *salting-out* refers to the use of nonelectrolytic additives such as organic solvents as well as electrolytes such as salts to lower the solubility of solute. Addition of water-miscible solvents is often used to induce crystallisation of organic compounds. For example, acetone, iso-propanol, methanol, ethanol, and ethanol–amine solutions are effective for crystallisation of several antibiotics and amino acids from aqueous solution.

The level of supersaturation existing in a solution can be quantified in several different ways. The most common expressions for supersaturation are the *supersaturation driving force* ΔC, the *supersaturation ratio* S, and the *relative supersaturation* σ. These parameters are defined as:

$$\Delta C = C - C^* \tag{11.126}$$

$$S = \frac{C}{C^*} \tag{11.127}$$

and

$$\sigma = \frac{C - C^*}{C^*} \tag{11.128}$$

where C is the solute concentration and C^* is the solubility. Combining Eqs. (11.126), (11.127), and (11.128) yields the relationship:

$$\sigma = \frac{\Delta C}{C^*} = S - 1 \tag{11.129}$$

Methods for measuring solubility and supersaturation in liquid solutions are described in the literature [31].

Nucleation

The formation of a new crystal from a liquid or amorphous phase is called *nucleation*. Different nucleation mechanisms have been observed, as summarised in Figure 11.50.

Primary nucleation refers to nucleation processes that occur in a previously crystal-free solution. As described in relation to Figure 11.49 primary nucleation is achieved by moving a solution into the labile region of supersaturation. Solute molecules gather together under these conditions to form clusters, which become ordered and adopt the structural geometry of the crystalline form. Primary nucleation is further classified as *homogeneous*, meaning that crystal formation occurs spontaneously in a perfectly clean solution free of impurities, or *heterogeneous*, where crystal formation is induced by the presence of inert foreign particles such as dust. Homogeneous nucleation is not relevant in industrial crystallisation because perfectly clean solutions are virtually impossible to attain. Therefore, any primary nucleation that occurs can be assumed to be heterogeneous. Primary nucleation has been studied extensively and many of its features are well understood.

Secondary nucleation is the generation of new crystals by crystals already present in the suspension. Secondary nucleation is thought to occur as a result of several factors, including the dislodgment of extremely small crystals from the surface of larger crystals, contact due to collisions between crystals or between crystals and the vessel walls and impeller, and crystal attrition or breakage. *Contact* or *collision nucleation*, which is also known as *collision breeding*, is considered the most important mechanism of secondary nucleation for materials of moderate-to-high solubility. Collision nucleation occurs more frequently when the concentration of suspended crystals and the intensity of agitation are high.

Generally, secondary nucleation occurs at much lower supersaturation levels than primary nucleation. To minimise the range of crystal sizes produced, excessive generation of new crystals is undesirable in most crystallisation operations. Therefore, ideally, supersaturation levels are minimised and controlled, the labile region of supersaturation is avoided, and operation of the crystalliser is confined to within the metastable region. Under these conditions, supersaturation is constrained to levels considerably lower than those required for primary nucleation so that secondary nucleation dominates. This situation is typical for industrial crystallisation of relatively soluble materials. For biological and organic substances with low solubility in aqueous solution, high levels of supersaturation supporting primary nucleation occur more readily and may be difficult to control.

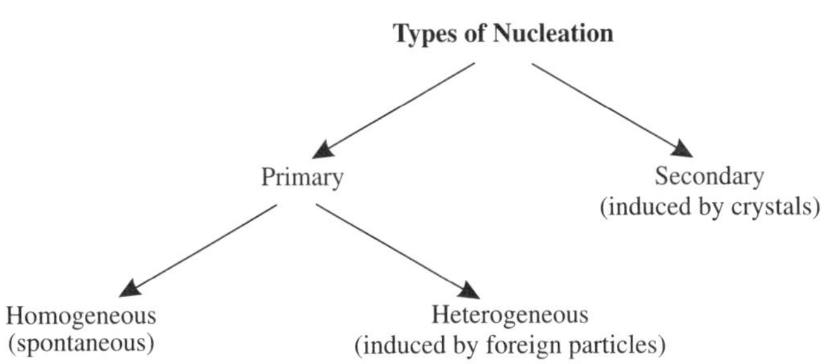

FIGURE 11.50 Graphic of the types of crystal nucleation.

Accordingly, both primary and secondary nucleation can be important in crystallisation of fermentation products.

The rate at which new crystals are formed in a supersaturated solution can be represented by the empirical equation:

$$B = K_N(C - C^*)^b \tag{11.130}$$

where B is the *birth rate* of new crystals, K_N is the *nucleation rate constant*, C is the concentration of dissolved solute, C^* is the solubility of solute at the solution temperature, and b is the *order of nucleation*. K_N and b depend on the physical properties of the system, including the temperature, impurities, and operating conditions such as agitation intensity. In systems where secondary nucleation plays a key role, K_N also depends on the concentration of crystals in the suspension. B represents the number of nuclei formed per unit time and has dimensions T^{-1}; B can also be expressed on a volumetric basis with dimensions $L^{-3}T^{-1}$. The driving force for nucleation is the degree of supersaturation, $C - C^*$. Practical methods for measuring crystal nucleation rates are described in the literature [28].

Nucleation conditions and kinetics are more complex for biological molecules that exhibit polymorphism such as amino acids and fatty acids. Different polymorphic crystal forms have different solubilities and nucleation rates, and one or more forms may nucleate preferentially depending on the crystallisation conditions. In general, metastable polymorphs have faster nucleation rates than stable forms. As a consequence, metastable polymorphs tend to form quickly but later redissolve in favour of the more stable form. Although the degree of supersaturation remains a controlling parameter in nucleation of polymorphic systems, additional factors such as the rate of solution cooling and choice of solvent also influence the nucleation rate and final crystal morphology.

Crystal Growth

Crystal growth refers to the increase in crystal size due to deposition of solute molecules in layers on the crystal surfaces. A schematic illustration of growth at two opposing crystal faces is shown in Figure 11.51. Mathematically, the rate of crystal growth can be defined in various ways. If L is the characteristic length of the crystal, the rate of solute deposition at the crystal faces perpendicular to L is expressed as:

$$G = \frac{dL}{dt} \tag{11.131}$$

where G is the *linear growth rate* of an individual crystal and t is time. The dimensions of G are LT^{-1}; typical units are m s^{-1} or μm h^{-1}. Table 11.8 lists some examples of linear growth rates reported for crystals of biological materials. Methods for measuring the rate of crystal growth in single- and multicrystal systems are described elsewhere [28].

Crystal growth rate can also be expressed in terms of the *mass deposition rate* or *mass growth rate*, R_G. Differentiating Eq. (11.120) with respect to time gives an expression for the rate of increase in crystal mass M_C:

$$R_G = \frac{dM_C}{dt} = 3\rho k_v L^2 \frac{dL}{dt} \tag{11.132}$$

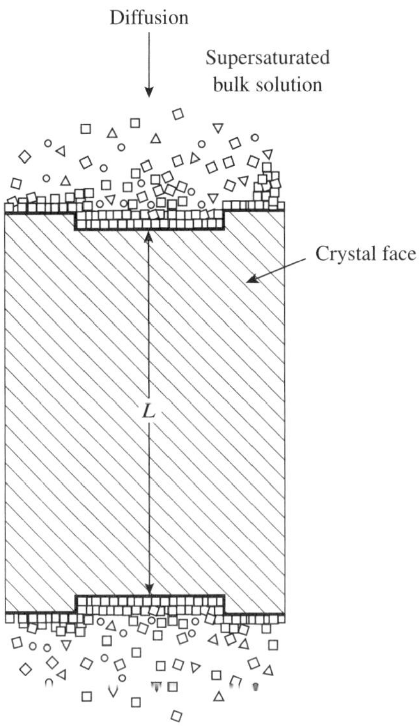

FIGURE 11.51 Growth of a crystal by mass transfer and layer-by-layer integration of solute molecules at opposing crystal faces.

where ρ is the crystal density and k_v is the volume shape factor. The dimensions of R_G are MT^{-1}; typical units are $kg\ s^{-1}$. Substituting Eq. (11.131) into Eq. (11.132) gives an expression relating the mass growth rate R_G to the linear growth rate G:

$$R_G = 3\rho k_v L^2 G \tag{11.133}$$

Also, from Eq. (11.118):

$$L^2 = \frac{A_C}{k_a} \tag{11.134}$$

where k_a is the area shape factor and A_C is the surface area of the crystal. Substituting Eq. (11.134) into Eq. (11.133) gives an alternative expression:

$$R_G = \frac{3\rho k_v A_C G}{k_a} \tag{11.135}$$

For geometrically similar crystals of the same material grown from birth in the same solution, it has been shown that when the growth rate is defined by Eq. (11.131), the rate of growth of all crystals is the same irrespective of their size [39]. This result is known as the *delta L* or *ΔL law*. The ΔL law is a reasonable generalisation for many industrial crystallisation processes and is often assumed in crystallisation modelling. However, many

TABLE 11.8 Linear Crystal Growth Rates

Substance	Temperature (°C)	Supersaturation[a]	Growth Rate (m s^{-1})	Reference
L-Asparagine monohydrate: (101) face	36	1.19	2.4×10^{-8}	[32]
Canavalin	20	1.5	2.6×10^{-9}	[33]
S-Carboxymethyl-L-cysteine	26	2.0	8.3×10^{-9}	[34]
Citric acid monohydrate	25	1.05	3.0×10^{-8}	[31]
	30	1.01	1.0×10^{-8}	[31]
	30	1.05	4.0×10^{-8}	[31]
Glycine	24	1.1	1.7×10^{-8}	[35]
L-Histidine, B form	20	1.13	4.2×10^{-8}	[31]
Lipase	28	2.0	5.1×10^{-9}	[36]
Lysozyme: (110) face	22	10	7.4×10^{-8}	[37]
Ovalbumin	30	3.0	2.8×10^{-10}	[38]
Sucrose	30	1.13	[b] 1.1×10^{-8}	[31]
	30	1.27	[b] 2.1×10^{-8}	[31]
	70	1.09	9.5×10^{-8}	[31]
	70	1.15	1.5×10^{-7}	[31]

[a] *Supersaturation expressed as* C/C*, *where* C *is the concentration of dissolved solute and* C* *is the equilibrium solubility.*
[b] *Growth rate probably size-dependent.*

crystals are known to violate the ΔL law and exhibit *size-dependent growth*, usually with large crystals growing more rapidly than small crystals. Alternatively, some crystal systems exhibit *growth dispersion*, which means that crystals of the same size grow at different rates. In general, the ΔL law does not apply if surface defects or dislocations affect the growth rates of particular crystal faces, if the crystals are very large, or if movement of the crystals in the suspension is very rapid.

A useful model of crystal growth is the *diffusion–integration theory*, which considers growth to be a two-step process. According to this model, mass transfer of solute to the crystal surface is followed by a surface reaction resulting in the molecular integration of solute into the crystal lattice. The combined kinetics of these two processes is represented by the empirical equation:

$$R_G = K_{GM} A_C (C - C^*)^g \tag{11.136}$$

where R_G is the mass growth rate defined by Eq. (11.132), K_{GM} is the *overall growth coefficient*, A_C is the crystal surface area, C is the concentration of dissolved solute, C^* is the solubility of solute at the solution temperature, and g is the *order of growth*. The value of K_{GM} depends on system properties such as temperature, agitation conditions, physical properties of the fluid, and the presence of impurities. An alternative expression for the rate of crystal growth is obtained by substituting for R_G in Eq. (11.136) from Eq. (11.135):

$$\frac{3\rho\, k_v A_C G}{k_a} = K_{GM} A_C (C - C^*)^g \tag{11.137}$$

Rearranging gives:

$$G = \frac{K_{GM}\, k_a}{3\rho k_v}\,(C - C^*)^g \tag{11.138}$$

or

$$G = K_{GL}(C - C^*)^g \tag{11.139}$$

where the *overall growth coefficient* K_{GL} is equal to the combined terms:

$$K_{GL} = \frac{K_{GM}\, k_a}{3\rho k_v} \tag{11.140}$$

Like Eq. (11.136), Eq. (11.139) is a simple, commonly applied empirical equation representing the rate of crystal growth. Several alternative crystal growth models have been developed to account for features such as size-dependent growth rate and growth dispersion, and to describe the separate rates of solute diffusion and integration [30].

To produce a narrow distribution of crystal sizes, nucleation should be minimised while a relatively rapid rate of crystal growth is maintained. However, as indicated in Eqs. (11.130), (11.136), and (11.139), the driving force for both nucleation and growth is the degree of supersaturation, $C - C^*$. Accordingly, it can be difficult to control these two processes differentially. As indicated in Figure 11.52, a range of complex interactions and feedback relationships between crystal parameters affect the outcome of crystallisation

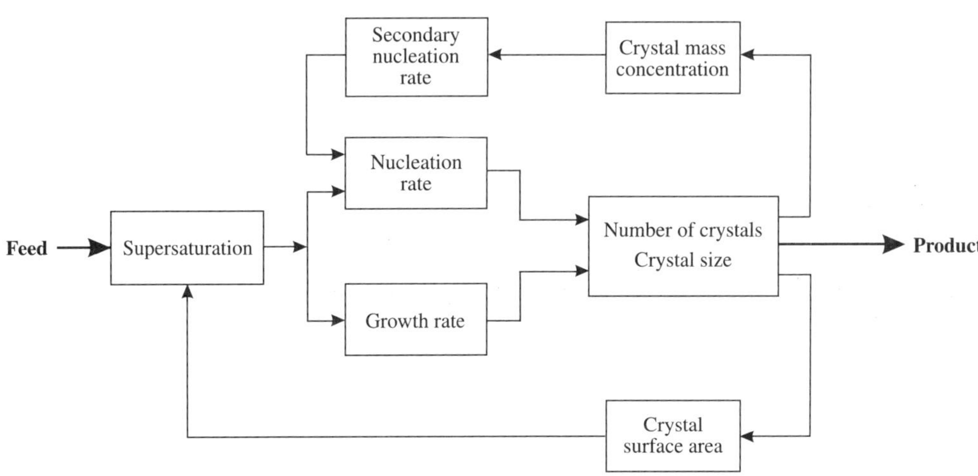

FIGURE 11.52 Interactions between crystallisation parameters affecting the outcome of crystallisation operations.

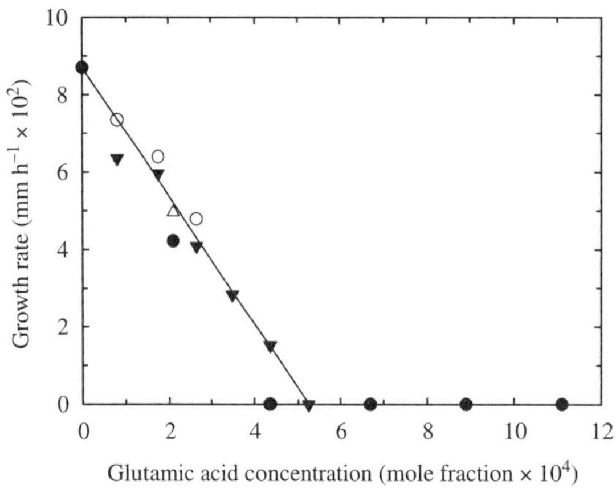

FIGURE 11.53 Effect of L-glutamic acid on the growth rate of the (101) face of L-asparagine monohydrate crystals. The symbols represent different individual crystals.
Data from S.N. Black, R.J. Davey, and M. Halcrow, 1986, The kinetics of crystal growth in the presence of tailor-made additives. J. Crystal Growth *79, 765–774.*

processes. The balance between nucleation and growth determines the number, size, and size distribution of the crystals. These parameters are related to the mass concentration of crystals in the magma, which plays a role in collision breeding or secondary nucleation that impacts the overall nucleation rate. A further feedback loop involves the number and size of crystals, the resulting crystal surface area, and the supersaturation level. For example, large crystals have less surface area per unit mass than small crystals. Therefore, because the mass growth rate depends on the crystal area as well as supersaturation as indicated in Eq. (11.136), higher supersaturation levels are required to achieve the same mass production rate of large compared with small crystals. More complex interactions not shown in Figure 11.52 include the effects of fluid hydrodynamics and crystal breakage and agglomeration.

To provide better control over the crystal size distribution, small *seed crystals* are often added to crystallisers. This overcomes the need to induce spontaneous nucleation by operating in the labile region (refer to Figure 11.49). Seed crystals with a narrow size distribution are added to the system as soon as the solution becomes supersaturated.

The crystal growth rate for biological materials is affected by interactions between the solute and impurities. For example, residual fermentation sugars, soluble proteinaceous material, and coexisting amino acids can have a significant effect on the purification of amino acids by crystallisation. In some cases, crystallisation is enhanced by the presence of coexisting substances; in other cases, crystal growth is slower and smaller crystals with inferior properties are produced [40]. The data in Figure 11.53 show how the presence of a coexisting amino acid can reduce to zero the growth rate of a particular face of L-asparagine crystals, illustrating the major impact of an impurity on crystal growth, habit, and morphology. Interactions between the solute and solvent and the competitive crystallisation behaviour of different polymorphic forms also influence crystal growth rates. The crystal growth behaviour of biological substances is often significantly more complex than for inorganic solutes.

11.12.3 Crystallisation Equipment

A wide variety of equipment and processing systems is used for industrial crystallisation. The essential requirements are a means for generating supersaturation, and a vessel in which the crystals can be held for a sufficient time to allow growth to the desired size. Crystallisation equipment is classified according to the method of generating supersaturation in the solution.

Cooling crystallisers use temperature reduction to induce supersaturation. To control the cooling rate, the vessel may be equipped with a jacket or cooling coil; alternatively, the solution may be pumped through an external heat exchanger. Agitation improves both the crystal size distribution and purity by reducing agglomeration and the entrapment of mother liquor within agglomerates. A particular problem in cooling crystallisers is *encrustation*, which is the build-up of hard crystalline deposits on the internal surfaces of the equipment. Encrustation is also known as *incrustation*, *salting*, *fouling*, or *scaling*. Because the degree of supersaturation is greatest at the low-temperature surfaces of the cooling equipment, crystals tend to form most readily at those locations. Eventually, as the surfaces become covered with thick layers of deposited crystal, the cooling equipment ceases to function effectively. To reduce encrustation and excessive supersaturation in cooling crystallisers, the temperature difference between the cooling fluid and crystal solution should be minimal, and the heat transfer surfaces should be as flat and smooth as possible. Alternative approaches to cooling that avoid the use of solid surfaces, such as directly adding coolant to the solution, may also be applied.

Evaporative crystallisers are used when supersaturation is more readily achieved by concentrating the solution rather than by lowering the temperature. Magma circulates around steam pipes that are within the crystalliser or through external steam-heated tubes. *Evaporative–cooling* or *vacuum crystallisers* use vacuum cooling to bring the solution to a lower temperature and simultaneously evaporate the solvent. As the pressure inside the vessel is reduced, the solution begins to boil and cool. Crystallisation that relies on salting-out or reaction to initiate supersaturation is generally carried out in tank crystallisers, preferably with forced circulation of the magma. Salting-out is commonly combined with cooling, evaporative, or vacuum methods for better control over supersaturation levels.

Irrespective of the type of crystalliser employed or the method of inducing supersaturation, the quality of the crystals produced can be improved substantially if the equipment allows crystal size fractionation and/or destruction of excess crystal nuclei. In this way, the size distribution can be controlled to within relatively narrow limits. *Classifying crystallisers* help achieve this by partially classifying the crystals within the vessel on the basis of size. By means of solution recirculation and particle fluidisation within the vessel, small crystals are retained in the crystalliser for further growth while crystals of a specified minimum size are allowed to leave in the product stream. Alternatively, excess fine crystals may be separated from larger crystals within the crystalliser and removed from the top of the vessel.

Biological products are treated using all types of crystalliser equipment. Cooling crystallisers are applied for crystallisation of fatty acids, lipids, marine oils, amino acids, and some organic acids. Evaporative or vacuum crystallisation is used for recovery of citric and other organic acids and monosodium glutamate. Small quantities of fine biochemicals and specialty products may be crystallised using simple static or stirred tank crystallisers with surface cooling or evaporation.

Further information about the different equipment configurations used for industrial crystallisation is available in the literature [31, 41].

11.12.4 Crystallisation Operations

Crystallisation can be carried out using either batch or continuous operations. Continuous crystallisation is most commonly applied for treatment of large volumes of material; batch operations are more suitable for producing small-to-medium quantities of crystals.

Continuous Crystallisation

Compared with batch crystallisation, continuous operations offer better control of process parameters affecting crystal quality and process efficiency. In particular, the crystal size distribution can be manipulated by changing operating conditions that influence crystal nucleation and growth, such as the solution residence time in the crystalliser and the crystal slurry density.

Four flow sheets for continuously operated crystallisers are shown in Figure 11.54. In the simplest configuration of Figure 11.54(a), product crystals and mother liquor leave the vessel at the same rate as feed solution enters. This type of crystalliser operation is known as *mixed suspension–mixed product removal*, or *MSMPR*. As no distinction is made in the outflow between the solid and liquid phases, the crystals and mother liquor remain in the crystalliser for approximately the same period of time.

As shown in Figure 11.54(b), MSMPR operation can be modified by allowing mother liquor to leave separately from the crystal product stream at a flow rate that is independent of the rate of product withdrawal. The advantages of this type of operation are that excess fine crystals from unwanted nucleation can be removed with the liquor overflow to reduce the spread of product crystal sizes, and crystals can be retained in the vessel for longer than the liquor residence time. If the liquor overflow is recycled back to the crystalliser as shown in Figure 11.54(c), it may be used to agitate or circulate the crystal suspension and thus improve crystal quality. Recirculation into the bottom of an *elutriation leg* extending from the base of the crystalliser fluidises the crystals and allows hydraulic

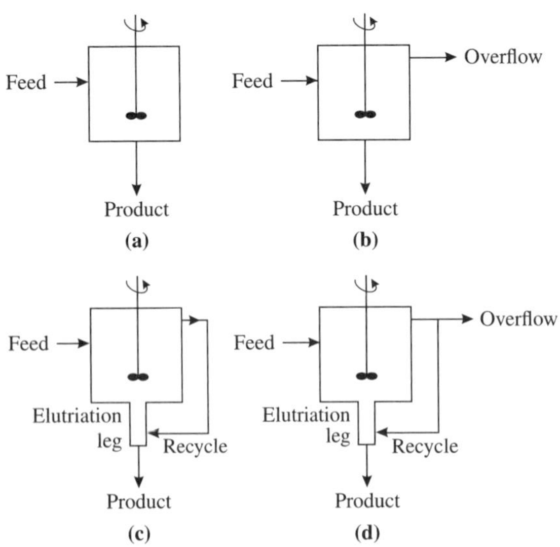

FIGURE 11.54 Flow sheets for the four types of continuous crystalliser operation: (a) mixed suspension–mixed product removal (MSMPR); (b) MSMPR with liquor overflow; (c) mixed suspension with liquor recycle and classified product removal; (d) mixed suspension with liquor overflow, partial liquor recycle, and classified product removal.
From J.W. Mullin, 1979, Crystallization. In: Kirk–Othmer Encyclopedia of Chemical Technology, *3rd ed., vol. 7, pp. 243–285, John Wiley, New York.*

classification or sorting of the particles based on size. In this way, small crystals making their way down to the product stream outlet are carried back by the recycled fluid into the main part of the vessel for further growth. If the overflow liquor contains very small crystals, the suspension can be heated prior to reentering the crystalliser to redissolve the solids. If only a proportion of the overflow liquor is recirculated back to the crystallisation vessel, the crystalliser flow sheet is as shown in Figure 11.54(d).

Ideally, continuous crystallisers should be operated at steady state. In practice, however, instabilities occur and continuous systems often display unsteady-state behaviour. Instabilities occur more frequently at high slurry densities and when classified product removal is employed. All continuous crystallisers must be shut down periodically to remove fouling deposits and encrustation in the heat exchangers. As continuous crystallisers are essentially self-seeded with crystals that develop during the process, it is possible after prolonged operation that undesirable crystal forms begin to dominate the resident population. This problem contributes to the need for regular equipment shut-down and cleaning.

Crystal population balance techniques can be applied to crystallisers to determine the relationship between operating conditions and the crystal size distribution produced. Let us consider a balance on the total number of crystals of size between L and $L + \Delta L$ in a crystalliser. This balance can be written as:

$$\left\{\begin{array}{c}\text{number of}\\ \text{crystals}\\ \text{growing}\\ \text{into the}\\ \text{size range}\end{array}\right\} - \left\{\begin{array}{c}\text{number of}\\ \text{crystals}\\ \text{growing}\\ \text{out of the}\\ \text{size range}\end{array}\right\} + \left\{\begin{array}{c}\text{number of}\\ \text{crystals}\\ \text{within the}\\ \text{size range}\\ \text{flowing in}\end{array}\right\} - \left\{\begin{array}{c}\text{number of}\\ \text{crystals}\\ \text{within the}\\ \text{size range}\\ \text{flowing out}\end{array}\right\} = \left\{\begin{array}{c}\text{number of}\\ \text{crystals}\\ \text{accumulated}\\ \text{in the}\\ \text{size range}\end{array}\right\} \quad (11.141)$$

Eq. (11.141) accounts for several mechanisms by which crystals enter and leave the population of particles in the crystalliser with sizes between L and $L + \Delta L$.

First, smaller crystals can grow to reach sizes within this range, as represented by the first term in the equation. Crystals already in the size range of interest can grow bigger and thus leave the population, as represented by the second term. Third, crystals of size between L and $L + \Delta L$ may enter the crystalliser in the feed, as indicated by the third term. Finally, crystals of size between L and $L + \Delta L$ may flow out of the system in the product stream, as indicated by the fourth term of the equation. The balance of these terms gives the accumulation term on the right side of the equation, which can be either positive or negative.

Let us apply this general population balance equation to the simplest form of continuous crystalliser operation, mixed suspension–mixed product removal or MSMPR, shown in Figure 11.54(a). MSMPR crystallisers are often used in the laboratory for crystallisation studies as well as for large-scale industrial crystal production. Population balance techniques allow us to derive equations relating the crystal size distribution expressed as the crystal population density n (Section 11.12.1, Size Distribution subsection) to the size and growth rate of the crystals and the flow rate of material through the crystalliser. The analysis involves the following assumptions:

1. MSMPR operation is maintained without product classification, fines removal, or recycle.
2. The feed does not contain crystals.

3. The system operates at steady state.
4. The system is well mixed.
5. All crystals have the same shape and can be characterised by a single linear dimension, L.
6. No crystal attrition occurs.
7. The crystal growth rate is independent of size (i.e., the ΔL law (Section 11.12.2, Crystal Growth subsection) applies).

Assumptions (2) and (3) allow us to simplify Eq. (11.141). If there are no crystals in the feed stream, the third term of the equation is zero. In addition, at steady state, the accumulation term becomes zero. Therefore, Eq. (11.141) reduces to:

$$\left\{\begin{array}{c}\text{number of crystals}\\\text{growing into the}\\\text{size range}\end{array}\right\}-\left\{\begin{array}{c}\text{number of crystals}\\\text{growing out of the}\\\text{size range}\end{array}\right\}-\left\{\begin{array}{c}\text{number of crystals}\\\text{within the size range}\\\text{flowing out}\end{array}\right\}=0 \tag{11.142}$$

We need to express each of the terms in Eq. (11.142) using crystal properties and system operating parameters. If ΔN is the number of crystals per unit volume growing into the size range over time interval Δt, the first term of Eq. (11.142) can be written as:

$$\left\{\begin{array}{c}\text{number of crystals}\\\text{growing into the}\\\text{size range}\end{array}\right\}=V\Delta N=V\frac{\Delta N}{\Delta L}\cdot\frac{\Delta L}{\Delta t}\cdot\Delta t \tag{11.143}$$

where V is the volume of the system. For small ΔL and Δt, $\Delta L/\Delta t$ is equal to the crystal growth rate G defined in Eq. (11.131). Making this substitution and using the definition of the population density n from Eq. (11.124), Eq. (11.143) becomes:

$$\left\{\begin{array}{c}\text{number of crystals}\\\text{growing into the}\\\text{size range}\end{array}\right\}=(V\,n\,G\,\Delta t)\big|_L \tag{11.144}$$

where $(V\,n\,G\,\Delta t)\big|_L$ means $V\,n\,G\,\Delta t$ evaluated for crystals of size L, that is, for crystals growing into the size range between L and $L+\Delta L$. Note that n in Eq. (11.144) is the crystal population density expressed on a per unit volume basis.

A similar expression can be obtained for the second term of Eq. (11.142):

$$\left\{\begin{array}{c}\text{number of crystals}\\\text{growing out of the}\\\text{size range}\end{array}\right\}=(V\,n\,G\,\Delta t)\big|_{L+\Delta L} \tag{11.145}$$

where $(V\,n\,G\,\Delta t)|_{L+\Delta L}$ means $V\,n\,G\,\Delta t$ evaluated for crystals of size $L+\Delta L$, that is, for crystals growing out of the size range between L and $L+\Delta L$. The number of crystals of the desired size flowing out in the product stream during time interval Δt is:

$$\left\{\begin{array}{c}\text{number of crystals}\\\text{within the size range}\\\text{flowing out}\end{array}\right\}=F\Delta N\Delta t=F\frac{\Delta N}{\Delta L}\cdot\Delta L\Delta t=F\,\overline{n}\,\Delta L\,\Delta t \tag{11.146}$$

where F is the volumetric flow rate of the product stream and $\overline{n}$ is the average population density for crystals in the size range between L and $L + \Delta L$. Substituting Eqs. (11.144), (11.145), and (11.146) into Eq. (11.142) gives:

$$(V\,n\,G\,\Delta t)\big|_L - (V\,n\,G\,\Delta t)\big|_{L+\Delta L} - F\,\overline{n}\,\Delta L\,\Delta t = 0 \tag{11.147}$$

According to assumption (7), the growth rate G is independent of crystal size. As V and Δt are also independent of size, Eq. (11.147) can be written as:

$$V\,G\,\Delta t\,(n\big|_L - n\big|_{L+\Delta L}) - F\,\overline{n}\,\Delta L\,\Delta t = 0 \tag{11.148}$$

Cancelling the Δt terms, dividing by ΔL, and rearranging gives:

$$\frac{(n\big|_L - n\big|_{L+\Delta L})}{\Delta L} - \frac{F}{V\,G}\overline{n} = 0 \tag{11.149}$$

Eq. (11.149) is valid for crystals with sizes between L and $L + \Delta L$. To develop an equation that applies to any crystal size, we must shrink ΔL to zero. As ΔL approaches zero, the average value $\overline{n}$ over the size range becomes a point value n. Using the rules of calculus and the definition of the derivative (Appendix E), taking the limit of Eq. (11.149) as $\Delta L \to 0$ gives:

$$-\frac{\mathrm{d}n}{\mathrm{d}L} - \frac{F}{V\,G}n = 0 \tag{11.150}$$

or

$$\frac{\mathrm{d}n}{\mathrm{d}L} + \frac{F}{V\,G}n = 0 \tag{11.151}$$

Equation (11.151) is a first-order differential equation representing the crystal population balance for an MSMPR crystalliser. Solving the equation by integration gives:

$$n = n_0\,e^{-FL/VG} \tag{11.152}$$

where n_0 is the crystal population density per unit volume for particles with size $L = 0$, that is, newly nucleated crystals. An expression for n_0 in terms of the rates of nucleation and growth is:

$$n_0 = \frac{B}{G} \tag{11.153}$$

where B is the nucleation rate.

Equation (11.152) relates the crystal size L to the volumetric crystal population density n produced in an MSMPR crystalliser. As n and L can be measured experimentally, for example, using sieving analysis as illustrated in Example 11.12 (with L_{av} representing L), Eqs. (11.152) and (11.153) allow us to estimate the crystal growth rate G and nucleation rate B from experimental data and crystalliser operating parameters. The form of Eq. (11.152) indicates that a plot of n versus L on semi-logarithmic coordinates should give a straight line with slope equal to $-F/VG$ and intercept at $L = 0$ equal to n_0. Once n_0 is known, the nucleation rate B can be obtained using Eq. (11.153). These calculations are illustrated in Example 11.13.

Further mathematical analysis of the equations for the MSMPR crystal size distribution yields several additional useful relationships. Some of these are summarised in Table 11.9. The *cumulative undersize fraction* in Table 11.9 is the fraction of the crystal population with size less than L. Another property that can be derived from the population balance is the *dominant size* L_D (see Section 11.12.1, Size Distribution subsection), which is the crystal size corresponding to the maximum value of the mass density m:

$$L_D = \frac{3VG}{F} \tag{11.154}$$

TABLE 11.9 Properties of MSMPR Crystallisers Derived from Population Balance Analysis

Property	Total per Unit Volume in an MSMPR Crystalliser	Cumulative Undersize Fraction in an MSMPR Crystalliser*
Number of crystals	$\frac{VG\,n_0}{F}$	$1 - e^{-\chi}$
Mass of crystals	$6\,k_v\,\rho\,n_0\left(\frac{VG}{F}\right)^4$	$1 - e^{-\chi}\left(1 + \chi + \frac{\chi^2}{2} + \frac{\chi^3}{6}\right)$

**The dimensionless quantity* $\chi = \frac{FL}{VG}$.

EXAMPLE 11.13 EVALUATING CRYSTAL PROPERTIES FROM EXPERIMENTAL DATA

An MSMPR crystalliser is used for production of organic acid crystals. The crystalliser volume is 120 l and the product stream flow rate is 35 l h^{-1}. The density of the crystals is 1.56 g cm^{-3} and the volume shape factor is 0.7. The following results are obtained from sieve analysis of a 1-litre sample of slurry.

Tyler Screen	Mass Retained (g)
20	0
28	5.9
35	11.1
48	14.4
100	10.6
Pan	2.3

(a) Determine the crystal growth rate.
(b) Determine the nucleation rate.
(c) What is the rate of crystal production?

Solution

(a) The crystal growth rate is determined according to Eq. (11.152) by plotting n versus L on semi-logarithmic coordinates. The parameters n and L are evaluated from the sieve data using the methods described in Examples 11.11 and 11.12.

Tyler Mesh	Aperture Size (μm)	Increment in Aperture Size From Above, ΔL (μm)	Mass Retained, ΔM (g)	Average Particle Size Retained, L_{av} (μm)	Population Density, n (m^{-4})
20	833	–	0	–	0
28	589	244	5.9	711	6.16×10^{10}
35	417	172	11.1	503	4.64×10^{11}
48	295	122	14.4	356	2.40×10^{12}
100	147	148	10.6	221	6.08×10^{12}
Pan	–	147	2.3	73.5	3.61×10^{13}

Values for n are calculated using Eq. (11.124). As an example, the value of n for crystals on the Tyler 28 screen is:

$$n = \frac{\dfrac{5.9\text{ g}}{244\ \mu\text{m} \cdot \left|\dfrac{1\text{ m}}{10^6\ \mu\text{m}}\right|}}{1.56\text{ g cm}^{-3} \cdot \left|\dfrac{100\text{ cm}}{1\text{ m}}\right|^3 \cdot (0.7)\left(711\ \mu\text{m} \cdot \left|\dfrac{1\text{ m}}{10^6\ \mu\text{m}}\right|\right)^3} = 6.16 \times 10^7\text{ m}^{-1}$$

This is the crystal population density obtained from analysis of crystals in a 1-litre sample of slurry. Expressing n on a volumetric basis:

$$n = \frac{6.16 \times 10^7\text{ m}^{-1}}{1\,\text{l} \cdot \left|\dfrac{1\text{ m}^3}{1000\,\text{l}}\right|} = 6.16 \times 10^{10}\text{ m}^{-4}$$

Values of n for the other screens are evaluated similarly. A semi-log plot of n versus L_{av} is shown in Figure 11.55. The slope of the line in Figure 11.55 is $-9.85 \times 10^{-3}\ \mu\text{m}^{-1}$ and the intercept is $6.76 \times 10^{13}\text{ m}^{-4}$. Therefore, from Eq. (11.152):

$$\frac{-F}{V\,G} = -9.85 \times 10^{-3}\ \mu\text{m}^{-1}$$

and

$$n_0 = 6.76 \times 10^{13}\text{ m}^{-4}$$

Evaluating the growth rate G from the slope:

$$G = \frac{-F}{V(-9.85 \times 10^{-3}\ \mu\text{m}^{-1})} = \frac{35\,\text{l h}^{-1}}{120\,\text{l}(9.85 \times 10^{-3}\ \mu\text{m}^{-1})} = 29.6\ \mu\text{m h}^{-1}$$

The crystal growth rate is $30\ \mu\text{m h}^{-1}$.

(b) The nucleation rate is determined from the values of n_0 and G using Eq. (11.153):

$$B = n_0\,G = (6.76 \times 10^{13}\text{ m}^{-4})\ 29.6\ \mu\text{m h}^{-1} \cdot \left|\frac{1\text{ m}}{10^6\ \mu\text{m}}\right| = 2.00 \times 10^9\text{ m}^{-3}\text{ h}^{-1}$$

The crystal nucleation rate is $2.0 \times 10^9\text{ m}^{-3}\text{ h}^{-1}$.

(c) In 1 litre of slurry, the total mass of crystals recovered was 44.3 g. The mass concentration of crystals in the magma is therefore 44.3 g l^{-1} or 44.3 kg m^{-3}. An additional check can be performed using the expression for the total mass of crystals per unit volume given in Table 11.9:

$$\text{Total mass of crystals per unit volume} = 6\,k_v\,\rho\,n_0\left(\frac{VG}{F}\right)^4 = 6\,k_v\,\rho\,n_0\left(\frac{-1}{\text{slope}}\right)^4$$

$$= 6\,(0.7)\left(1.56\ \text{g cm}^{-3}\cdot\left|\frac{100\ \text{cm}}{1\ \text{m}}\right|^3\right)(6.76\times10^{13}\ \text{m}^{-4})\left(\frac{-1}{-9.85\times10^{-3}\ \mu\text{m}^{-1}\cdot\left|\frac{10^6\ \mu\text{m}}{1\ \text{m}}\right|}\right)^4$$

$$= 4.71\times10^4\ \text{g m}^{-3} = 47.1\ \text{kg m}^{-3}$$

This is close to 44.3 kg m^{-3}, considering that the data were smoothed to determine the slope and intercept. If the crystalliser is well mixed, the concentration of crystals in the product stream is the same as that in the vessel. The rate at which crystals are produced is equal to the mass concentration of crystals multiplied by the product stream flow rate, F.

$$\text{Crystal production rate} = 44.3\ \text{kg m}^{-3}\,(35\ \text{l h}^{-1})\cdot\left|\frac{1\ \text{m}^3}{1000\ \text{l}}\right| = 1.55\ \text{kg h}^{-1}$$

The rate of crystal production is 1.6 kg h^{-1}.

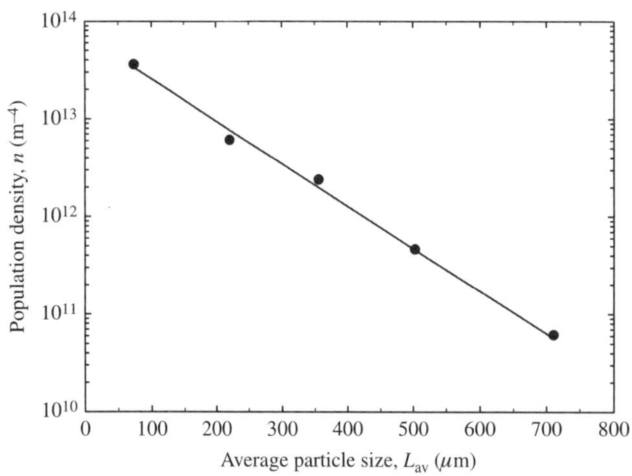

FIGURE 11.55 Population density plot on semi-logarithmic coordinates.

According to Eq. (11.152), a plot of n versus L on semi-logarithmic coordinates should give a straight line for MSMPR operations. Sometimes, however, a straight line is not found. One explanation for this is that the system may not obey the ΔL law—that is, size-dependent growth or growth dispersion may have occurred (Section 11.12.2, Crystal

Growth subsection). Imperfect mixing in the crystalliser and withdrawal of a product stream that is not representative of the bulk crystal slurry are also deviations from ideal operating conditions. Achieving perfect mixing in a solids suspension so that the withdrawal stream contains the complete range of particle sizes present in the vessel can be challenging, particularly at large scales.

Although the crystal size distribution in an MSMPR crystalliser is dependent on operating conditions such as the flow rate and slurry density, the effect of changing these parameters is often relatively small. More control over crystal size is obtained by altering the equipment design to include features such as classified product removal, fines destruction, and liquor recycle as discussed in relation to Figure 11.54. Population balance techniques are applied to these modified process configurations in the literature [29].

Batch Crystallisation

Batch crystallisers offer simplicity and flexibility of operation and require considerably less capital investment than continuous systems. In bioprocessing, batch crystallisation is used extensively for purification and final recovery of fermentation products such as antibiotics and amino acids. Unlike continuous crystallisers that are effectively self-seeded because of the continuing presence of crystals in the vessel, batch crystallisers are usually seeded deliberately for production of each batch of product.

Batch crystallisation is inherently an unsteady-state process. Although population balance techniques similar to those used in the previous section for MSMPR crystallisers can be applied to batch operations [29], solution of the resulting partial differential equations is not straightforward for practical purposes as the properties of the system are time-dependent. Explicit results for the crystal size distribution in batch systems are therefore not readily obtained. However, other useful information and operating guidelines can be determined. Equations for the optimum cooling program and optimum rate of solvent evaporation to achieve constant supersaturation conditions in batch crystallisers are available in the literature [29, 31].

11.13 DRYING

Drying is often the last step in downstream processing for recovery of solid products from fermentation. The aim is to remove relatively small amounts of residual water or solvent from materials such as crystals, precipitates, and cell biomass, thus rendering the product suitable for packaging and storage. Drying may be necessary to minimise chemical or physical degradation of solids during storage, for example, due to oxidation or aggregation. Drying is a relative term: material containing 0 to 20% water by weight may be considered dry depending on the product and the specifications for acceptable product quality. Drying is an energy-intensive unit operation; accordingly, drier effectiveness and energy efficiency are of concern for economical processing.

In this section, we will consider drying to be the removal of water from solid material into air; however, the same general principles apply for other liquids and gases. Drying is achieved by vaporising liquid water or ice contained within a solid and then removing the vapour. In many drying operations, a stream of hot air supplies the heat needed for vaporisation and the means for transporting the water vapour away from the solid. Particular attention is required when drying heat-sensitive biological materials to ensure that thermal degradation does not occur. Vacuum drying and freeze-drying are used to dry fermentation products such as proteins, vitamins, vaccines, steroids, and cells at temperatures below 0°C to protect their biological properties and activity. Freeze-drying is considered in more detail in Section 11.13.5.

11.13.1 Water Content of Air

Moist or *humid* air is a mixture of dry air and water vapour. The *humidity* of air, also known as the *absolute humidity* or *humidity ratio*, is a dimensionless parameter defined as:

$$\text{Humidity} = \frac{M_w}{M_a} \tag{11.155}$$

where M_w is the mass of water vapour carried by mass M_a of dry air. Humidity is measured using instruments called *hygrometers*.

The total pressure of humid air is equal to the sum of the partial pressures of its constituents, including water vapour. The partial pressure of water vapour in air, p_w, depends on the molar concentration of water in the gas phase:

$$p_w = y_w p_T \tag{11.156}$$

where y_w is the mole fraction of water vapour in the air–vapour mixture and p_T is the total pressure. Air is said to be *saturated* with water vapour at a particular temperature and pressure if its humidity is the maximum it can be under those conditions. Addition of further water vapour to saturated air results in the condensation of liquid water in the form of droplets or a mist. Under saturation conditions, the partial pressure of water vapour in air is equal to the *saturation vapour pressure* p_{sw} of pure water at that temperature. Values for the saturation vapour pressure of water as a function of temperature are listed in Table D.1 in Appendix D.

If a mixture of water vapour in air is cooled, the temperature at which the mixture becomes saturated is called the *dew point* or *saturation temperature*. The dew point is the temperature at which pure water exerts a vapour pressure equal to the partial pressure of water vapour in the mixture. The *relative humidity* of air is defined as the ratio of the partial pressure of water vapour in the air to the saturation vapour pressure of pure water at the same temperature, expressed as a percentage:

$$\text{Relative humidity} = \frac{p_w}{p_{sw}} \times 100\% \tag{11.157}$$

The thermodynamic properties of air and its associated water content, including the relationships between temperature, humidity, enthalpy, and specific volume, are represented in *psychrometric charts*. Examples of psychrometric charts for air and water vapour mixtures can be found in handbooks (e.g., [42]).

11.13.2 Water Content of Solids

At a given temperature, the moisture content of a wet solid depends on the humidity of the atmosphere surrounding it. When a solid is brought into contact with a relatively large amount of air, the conditions of the air remain essentially constant even if the water content of the solid changes. After a sufficient period, equilibrium between the air and wet solid is reached. Results at a fixed temperature for the *equilibrium moisture content* of the solid as a function of air relative humidity are presented as an *equilibrium moisture content isotherm* or *equilibrium water sorption isotherm*. Measured equilibrium moisture content isotherms for yeast cells, insulin, lysozyme crystals, and benzyl penicillin are shown in Figure 11.56(a). The equilibrium moisture content of solids is usually expressed on a dry weight basis using units of, for example, g of water per 100 g of dry solid.

Data for equilibrium water sorption isotherms are obtained by exposing solid material to air of varying relative humidity and measuring the equilibrium water content of the solid at

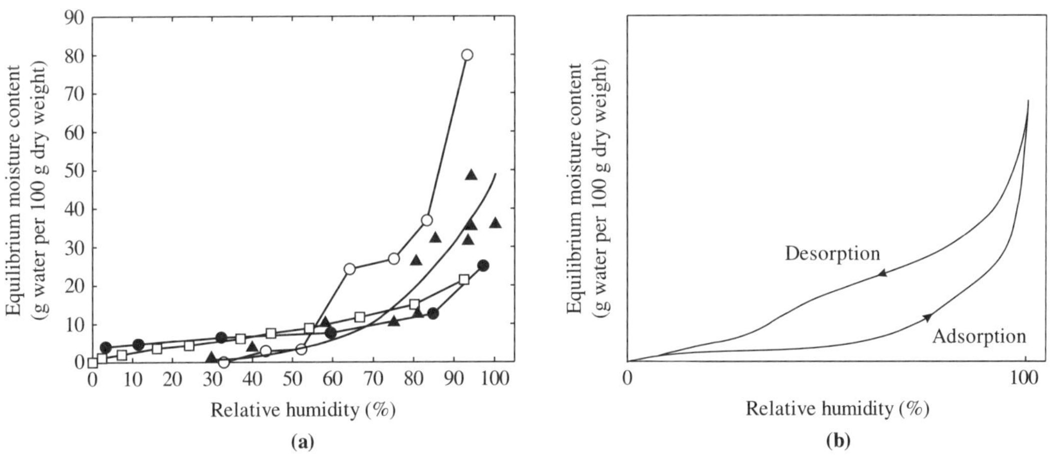

FIGURE 11.56 Equilibrium moisture content isotherms.
(a) Experimentally measured isotherms (●). *Saccharomyces cerevisiae* yeast during hydration at 30°C. *Data from S. Koga, A. Echigo, and K. Nunomura, 1966, Physical properties of cell water in partially dried Saccharomyces cerevisiae.* Biophys. J. *6, 665–674.*
(□) Crystalline insulin at 25°C. *Data from M.J. Pikal and D.R. Rigsbee, 1997, The stability of insulin in crystalline and amorphous solids: observation of greater stability for the amorphous form.* Pharm. Res. *14, 1379–1387.*
(▲) Lysozyme crystals at 22°C. *Data from E.T. White, W.H. Tan, J.M. Ang, S. Tait, and J.D. Litster, 2007, The density of a protein crystal.* Powder Technol. *179, 55–58.*
(○) Sodium benzyl penicillin at 25°C. *Data from N.A. Visalakshi, T.T. Mariappan, H. Bhutani, and S. Singh, 2005, Behavior of moisture gain and equilibrium moisture contents (EMC) of various drug substances and correlation with compendial information on hygroscopicity and loss on drying.* Pharm. Dev. Technol. *10, 489–497.*
(b) Isotherms for moisture adsorption (hydration) and desorption (dehydration) showing water sorption hysteresis.

constant temperature. Exposure to air of progressively increasing humidity gives an *adsorption* or *hydration isotherm* as the solid gains water to equilibrate with the air. Conversely, exposure to air of progressively decreasing humidity gives a *desorption* or *dehydration isotherm* as the solid dries to achieve equilibrium. Some materials display hysteresis, in that the adsorption and desorption isotherms do not coincide, as shown in Figure 11.56(b). This occurs when wetting or drying leads to irreversible changes in the structure of the solid. For drying applications, the desorption isotherm is of greater importance than the adsorption isotherm. Sorption isotherms are also dependent on external pressure; however, for practical purposes, this effect is usually neglected. Equilibrium moisture content isotherms cannot be predicted from theoretical consideration of the material of interest: they must be determined by experiment. Methods for measuring sorption isotherms are described elsewhere [42].

Equilibrium water sorption isotherms are important in drying operations because they indicate how dry a solid can become if it is brought into contact with air of a given relative humidity. After drying, the moisture content of a solid cannot be less than the equilibrium moisture content on the isotherm corresponding to the relative humidity of the air entering the drier. For example, from Figure 11.56(a), using air at 30°C with relative humidity 60%, the yeast cells tested cannot be dried to water contents below the equilibrium value of about 7.5 g per 100 g dry cells.

EXAMPLE 11.14 MOISTURE CONTENT AFTER DRYING

A filter cake produced by vacuum filtration of bakers' yeast (*Saccharomyces cerevisiae*) contains 38 g of water per 100 g of dry cells. The cells are dried before packaging. Atmospheric air at 10°C and 90% relative humidity is heated to 30°C at constant pressure for use in a tray drier. Assuming that equilibrium is reached during the drying process and that the equilibrium moisture content isotherm for yeast shown in Figure 11.56(a) applies, what is the moisture content of the cells after drying?

Solution

From Table D.1 in Appendix D, the saturation vapour pressure p_{sw} of water at 10°C is 1.227 kPa. Therefore, from Eq. (11.157), the partial pressure of water vapour in air at 10°C and 90% relative humidity is:

$$p_w = \frac{90\%}{100\%}(1.227 \text{ kPa}) = 1.104 \text{ kPa}$$

From Eq. (11.156), this partial pressure is proportional to the mole fraction of water vapour in the air. When the air is heated to 30°C, because the composition of the air–vapour mixture and the total pressure remain constant, p_w in the air at 30°C is also 1.104 kPa. From Table D.1 in Appendix D, the saturation vapour pressure p_{sw} of water at 30°C is 4.24 kPa. Therefore, using Eq. (11.157) for the heated air:

$$\text{Relative humidity} = \frac{1.104 \text{ kPa}}{4.24 \text{ kPa}} \times 100\% = 26.0\%$$

From Figure 11.56(a), the equilibrium moisture content of yeast exposed to air at 30°C and 26% relative humidity is around 6 g of water per 100 g dry weight. Note that the initial water content of the wet yeast does not figure in this calculation.

The moisture content of the cells after drying is 6 g of water per 100 g dry cells.

The distribution and binding strength of water within wet solids vary depending on the properties of the material. Many crystalline solids do not contain water within their lattice structures, which generally are packed too tightly to admit hydration. Therefore, most of the moisture removed during drying of crystals is from the surface of the particles. However, some crystalline solids form crystal hydrates as water molecules are incorporated either on a stoichiometric or nonstoichiometric basis within the solid. Water of crystallisation is generally relatively immobile, being held within the solid by very strong water–solid interactions. Nevertheless, it can be removed by drying if sufficient heat and time are provided. Compared with crystals of inorganic compounds, protein crystals contain unusually large amounts of water within the crystal lattice: water contents of around 50% are not uncommon. In general, however, the amount of water incorporated within the bulk structure of solids increases as the degree of crystallinity is reduced.

In Figure 11.56(a), benzyl penicillin displays the characteristics of a moderately *hygroscopic* solid, as it has an enhanced ability to attract and hold water molecules from the surrounding atmosphere, resulting in a marked increase in equilibrium moisture content at relative humidity levels above about 50%. As amorphous and partially amorphous solids allow penetration and dissolution of water within their matrix, these materials typically contain significant amounts of internal moisture as well as surface water. The strength of water–solid interactions in amorphous solids depends on whether the solid is polar or nonpolar, and the plasticising effect of water entering the solid structure on the mobility of both the moisture and solid components. Porosity is also an important factor determining moisture distribution and water mobility in solids. The interconnected pores and channels of porous solids fill with water after wetting and capillary flow mechanisms contribute to the process of water removal in drying. In contrast, molecular diffusion may be the only mechanism available for water removal from nonporous solids that lack internal flow channels.

11.13.3 Drying Kinetics and Mechanisms

Drying is a complex process involving simultaneous heat and mass transfer. Because the physical properties of solids may be changed during drying, predicting the rate of drying from theoretical principles is often impossible. Drying occurs by vaporising water using heat. As the heat is usually provided by a hot gas stream, convective heat transfer is required to heat the outer surface of the solid while conductive heat transfer allows penetration of heat within the material. Mass transfer is also important, as water within the solid must be transported to the surface as either liquid or vapour before being removed into the gaseous environment.

The rate of drying N is defined as the rate at which the mass of water associated with a wet solid reduces with time:

$$N = -M_s \frac{dX}{dt} \tag{11.158}$$

where M_s is the mass of completely dry solid, X is the moisture content of the solid expressed on a dry mass basis (e.g., kg kg^{-1} of dry solid), and t is time. The dimensions of N are MT^{-1}; typical units are kg h^{-1}. Because dX/dt is negative during drying, the minus sign in Eq. (11.158) is required to make N a positive quantity. The rate of drying can also be expressed on a unit area basis as a flux n_a with units of, for example, kg m^{-2} h^{-1}:

$$n_a = -\frac{M_s}{A}\frac{dX}{dt} \tag{11.159}$$

where A is the area available for evaporation. Alternatively, the specific drying rate per unit mass of dry solid, n_m, is:

$$n_m = -\frac{dX}{dt} \tag{11.160}$$

Typical units for n_m are kg kg^{-1} h^{-1}.

Drying Curve

The kinetics of drying are assessed by plotting experimentally determined values for the rate of drying against the moisture content of the solid X. The result is a *drying rate curve* as illustrated in Figure 11.57. The shape of the drying rate curve depends on the material being dried, its size and thickness, and the drying conditions. Drying rate curves are measured using *constant drying conditions*, that is, constant air temperature, humidity, flow rate, and flow direction. The moisture content and other properties of the solid change under constant drying conditions, which refer only to the gas phase.

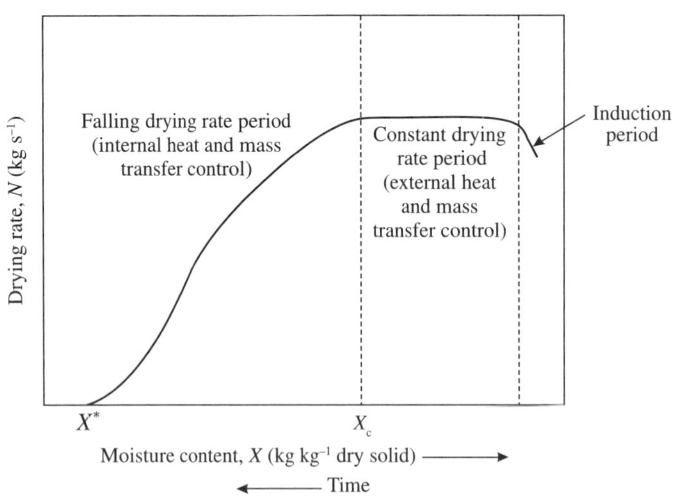

FIGURE 11.57 Drying rate curve for constant drying conditions.

As drying commences, the water content of the solid is high. After an initial warming-up or induction period, the first water to evaporate comes from the surfaces of the wet solid that are in direct contact with the air stream moving across them. If water is supplied by mass transfer from within the solid to the surface at a sufficiently rapid rate, the surface remains saturated with water and a *constant drying rate period* ensues. This period is represented in Figure 11.57 by the flat section of the curve. During the constant drying rate period, the surfaces of the solid remain wet so that free water is always available for evaporation, and heat and mass transfer take place at the surface. Accordingly, the resistances to heat and mass transfer are located within the external gas boundary layer surrounding the material.

With continued drying, as the water content of the solid decreases, a *critical moisture content* X_c is reached. The rate of drying begins to decline at this point as the process enters the *falling drying rate period*. The reduction in drying rate below X_c reflects a change in the heat and/or mass transfer conditions in the system. At first during the falling rate period, the drying surface becomes partially unsaturated and there is no longer a continuous or nearly continuous liquid film at the surface of the solid. This situation extends gradually until the entire surface becomes dry. As a solid layer of dried material builds up at and then below the surface, heat must be transferred by conduction to the remaining water further inside the solid; in effect, the evaporating surface recedes into the material as drying proceeds. Because the dried solid near the surface is generally a poor conductor of heat, the rate of heat transfer declines progressively.

As the surfaces become dry, water must be transported from deeper within the solid to the surface before it can be removed into the gas phase. As heat and mass transfer become controlled completely by internal resistances, the drying rate continues to decline and the process is characterised as *diffusion* or *hindered drying*. The moisture content of the solid reduces further until it approaches the equilibrium moisture content X^*. As described in Section 11.13.2, this is the lowest moisture content that can be achieved at the temperature and relative humidity used for the drying process.

The curve in Figure 11.57 is somewhat idealised: significant deviations can occur for particular materials and constant drying conditions are often not applied in practical drying operations. The critical moisture content X_c is a property of the material being dried but also varies with material thickness and drying rate: it reflects the magnitude of the heat and mass transfer resistances in the solid. For example, X_c can be reduced by decreasing the thickness of the solid, as this reduces internal transfer resistance. Many materials exhibit a constant drying rate period; however, for some solids, internal heat and/or mass transfer always determines the rate of drying so a constant drying rate is never achieved. Constant drying rates also do not occur if the starting moisture content of the solid is less than X_c.

If structural or chemical modifications occur within the solid as drying takes place, the drying rate curve may be significantly more complex than that shown in Figure 11.57, with many inflections or abrupt changes as the nature of the material is altered. During drying, the solid may shrink, expand, harden, become more or less porous, or change its crystallinity; as a result, its properties such as thermal conductivity and moisture diffusivity can change with drying time. In nonporous colloidal solids, shrinkage of the rapidly drying outer layers of material and the accompanying decline in moisture diffusivity result in *case hardening*, as the solid develops a skin or crust at the surface that is virtually impenetrable to water so that further drying is prevented.

Mechanisms of Moisture Transport in Solids

Several mechanisms of moisture transport operate during drying to deliver water from within the solid to the surface. These include:

- Molecular diffusion of liquid water
- Capillary flow of liquid water within porous solids
- Molecular diffusion of vapour evaporated within the solid
- Convective transport of vapour evaporated within the solid

In porous solids, moisture is usually transported more effectively by capillary forces than by diffusion, depending on the pore size. Capillary flow relies on the pressure differences that occur within solids as a result of surface tension effects in very small pores. If the rate of water vaporisation within the solid exceeds the rate of vapour transport to the surroundings, mass transfer is affected by the resulting build-up of pressure inside the material. Pressure gradients can also drive mass transfer if shrinkage of the solid occurs during drying.

Drying Kinetics during Constant Rate Drying

The rate equation for heat transfer during drying is analogous to that derived in Chapter 9:

$$\hat{Q} = U A_h (T_a - T) \tag{11.161}$$

where $\hat{Q}$ is the rate of heat transfer, U is the overall heat transfer coefficient, A_h is the area available for heat transfer, T_a is the air temperature, and T is the temperature of the solid surface that is drying. During the constant drying rate period, convective heat transfer is the principal transport mechanism and the gas film boundary layer external to the solid provides the main heat transfer resistance. Under these conditions, U in Eq. (11.161) can be replaced by h_s, the individual heat transfer coefficient for the fluid boundary layer at the solid surface:

$$\hat{Q} = h_s A_h (T_a - T) \tag{11.162}$$

If the heat transferred during drying is used solely to evaporate water, the rate of vapour production can be related to $\hat{Q}$ using the *latent heat of vaporisation* (Section 5.4.2):

$$\text{Rate of production of water vapour} = \frac{h_s A_h (T_a - T)}{\Delta h_v} \tag{11.163}$$

where Δh_v is the latent heat of vaporisation at temperature T.

An equation similar to those derived in Chapter 10 can be used to represent the rate of mass transfer of water being evaporated from the solid surface during constant rate drying:

$$N_A = k_G a \, (Y - Y_a) \tag{11.164}$$

where N_A is the rate of mass transfer, k_G is the mass transfer coefficient for the gas boundary layer with dimensions $ML^{-2}T^{-1}$, a is the area available for mass transfer, Y is the humidity of air in equilibrium with water at the surface temperature of the solid, and Y_a is the humidity of the air.

At steady state, the rate of production of water vapour due to input of latent heat is equal to the rate at which the vapour is removed from the solid by mass transfer. Both these rates are also equal to the constant drying rate N_c. Combining Eqs. (11.163) and (11.164) gives:

$$N_c = \frac{h_s A_h (T_a - T)}{\Delta h_v} = k_G a\,(X - X_a) \tag{11.165}$$

To apply Eq. (11.165), the heat and/or mass transfer coefficient must be determined for the particular drying equipment and operating conditions used. Calculation of the drying rate is usually based on the heat transfer component of Eq. (11.165), as application of the mass transfer equation is less straightforward. Because the distribution of moisture in the solid changes during drying, there is considerable uncertainty about the driving force for mass transfer at any given time.

Drying Time

The time required to achieve a desired state of dryness can be found by integrating the expressions for drying rate with respect to time. Under constant drying conditions and *during the constant drying rate period*, from Eq. (11.158):

$$N_c = -M_s \frac{dX}{dt} \tag{11.166}$$

As N_c and M_s are constant during constant rate drying, the only variables in Eq. (11.166) are X and t. Separating variables and integrating gives:

$$\int_0^{t_1} dt = \frac{-M_s}{N_c} \int_{X_0}^{X_1} dX \tag{11.167}$$

or

$$\Delta t = \frac{M_s}{N_c}(X_0 - X_1) \tag{11.168}$$

Equation (11.168) is used to estimate Δt, the time required to dry solids from an initial moisture content of X_0 to a final moisture content of X_1 when the drying rate is constant. From the definition of drying rate in Eq. (11.158), X_0 and X_1 are moisture contents expressed on a dry mass basis using units of, for example, kg kg^{-1} of dry solid.

During the falling drying rate period, the drying rate N is no longer constant. Equations for drying time during this period can be developed depending on the relationship between N and X and the properties of the solid. Kinetic models for predicting the drying rate curve, including during the falling rate period when internal heat and mass transfer mechanisms are limiting, are described elsewhere [43].

EXAMPLE 11.15 DRYING TIME DURING CONSTANT RATE DRYING

Precipitated enzyme is filtered and the filter solids washed and dried before packaging. Washed filter cake containing 10 kg of dry solids and 15% water measured on a wet basis is dried in a tray drier under constant drying conditions. The critical moisture content is 6%, dry basis. The area available for drying is 1.2 m^2. The air temperature in the drier is 35°C. At the air humidity used, the surface temperature of the wet solids is 28°C. The heat transfer coefficient is 25 J m^{-2} s^{-1} °C^{-1}. What drying time is required to reduce the moisture content to 8%, wet basis?

Solution

The initial and final moisture contents expressed on a wet basis must be converted to a dry basis:

$$15\%\ \text{wet basis} = \frac{15\ \text{g water}}{100\ \text{g wet solid}} = \frac{15\ \text{g water}}{15\ \text{g water} + 85\ \text{g dry solid}}$$

$$X_0 = \frac{15\ \text{g water}}{85\ \text{g dry solid}} = 0.176$$

Similarly:

$$8\%\ \text{wet basis} = \frac{8\ \text{g water}}{8\ \text{g water} + 92\ \text{g dry solid}}$$

$$X_1 = \frac{8\ \text{g water}}{92\ \text{g dry solid}} = 0.087$$

As X_1 is greater than the critical moisture content $X_c = 0.06$, the entire drying operation takes place with constant drying rate. Equation (11.165) is used to determine the value of N_c. From Table D.1 in Appendix D, the latent heat of vaporisation Δh_v for water at 28°C, the temperature of the surface of the solids where evaporation takes place, is 2435.4 kJ kg^{-1}. Therefore:

$$N_c = \frac{25\ \text{J m}^{-2}\ \text{s}^{-1}\ {}^\circ\text{C}^{-1}\ (1.2\ \text{m}^2)\ (35-28)^\circ\text{C}}{2435.4 \times 10^3\ \text{J kg}^{-1}} = 8.62 \times 10^{-5}\ \text{kg s}^{-1}$$

Applying Eq. (11.168) to calculate the drying time:

$$\Delta t = \frac{10\ \text{kg}}{8.62 \times 10^{-5}\ \text{kg s}^{-1}} \cdot \left|\frac{1\ \text{h}}{3600\ \text{s}}\right| \cdot (0.176 - 0.087) = 2.87\ \text{h}$$

The time required for drying is 2.9 h.

11.13.4 Drying Equipment

A diverse range of equipment is used industrially for drying operations [42], including tray, screen-conveyor, screw-conveyor, rotary drum, tunnel, bin, tower, spray, fluidised bed, and flash driers. Some driers have a direct mode of heating, whereby air entering the drier is brought into contact with the wet solid. Other types of equipment apply indirect heating to the drying material through a metal wall or tray. Some drier installations use a combination of direct and indirect heating.

Most driers operate at or close to atmospheric pressure. However, tray and enclosed rotary driers may be operated under vacuum, generally with indirect heating. The advantage of using vacuum drying is that evaporation of water occurs at lower temperatures when the pressure is reduced; for example, the boiling point of water at 6 kPa or 0.06 atm is only about 36°C (Table D.2, Appendix D). This makes vacuum operation suitable for processing heat-sensitive fermentation products. Rates of drying are also enhanced under vacuum compared with atmospheric pressure. The water vapour produced during vacuum drying is usually condensed during operation of the drier to maintain the vacuum. As an alternative to vacuum drying, flash or spray drying may be suitable for heat-labile solids because drying in these systems occurs very rapidly, usually within 0.5 to 6 seconds, so that thermal damage from prolonged exposure to heat is avoided. The creation of dust during drying is a potential concern for processing of fermentation products and may influence the choice of drier equipment. For materials with biological activity, exposure to the large amounts of fine particle dust generated by, for example, some rotary driers, is undesirable.

Large-scale driers cannot be designed or sized using theoretical analysis alone. The drying properties of numerous batches of material must be assessed experimentally. Scale-up of drying requires appropriate laboratory- and pilot-scale testing to characterise the material being dried and the transport processes that occur. To improve energy use and cost-effectiveness, the operating efficiency of large-scale drying equipment can be improved using measures such as preheating the inlet air with hot exhaust air, recycling some of the exhaust air, and reducing air leakage.

11.13.5 Freeze-Drying

Freeze-drying, also known as *lyophilisation* or *cryodesiccation*, is used to dry unstable or heat-sensitive products at low temperature, thus protecting the material from heat damage and chemical decomposition. The wet solids are placed in vials that are partially stoppered so that water vapour can escape. The material is frozen and then exposed to low pressure, which causes the frozen water within the solid to *sublimate* directly to vapour without passing through a liquid phase. Sublimation of ice crystals leaves networks of cavities within the solid, so that the dried material has a porous, friable structure with high internal surface area. Because drying takes place at temperatures below 0°C, damage to biological molecules is minimised and any volatile substances are retained.

Freeze-drying is used commonly in the pharmaceutical industry; it is also used for drying some foods and for downstream processing of proteins, vaccines, and vitamins. However, the energy required for freeze-drying is substantially greater than for other drying methods, and the time required for drying is generally longer. The drying time for freeze-drying is roughly proportional to the material thickness raised to the power 1.5 to 2.0 [44, 45].

Freeze-drying comprises three steps:

1. Freezing
2. Sublimation or primary drying
3. Desorption or secondary drying

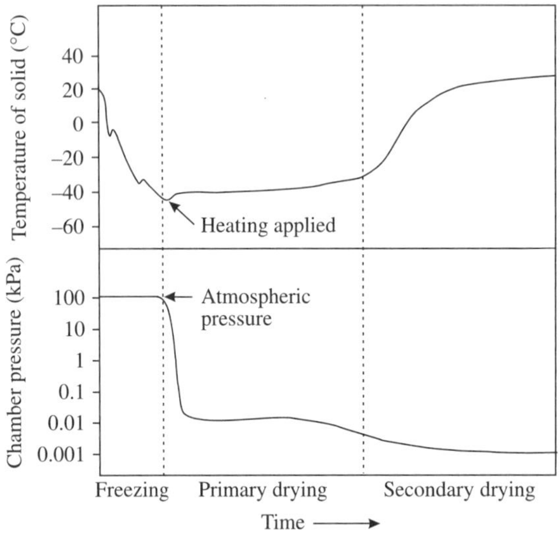

FIGURE 11.58 Variation of solids temperature and chamber pressure during a complete freeze-drying cycle.

The changes in operating variables during an entire freeze-drying cycle are illustrated in Figure 11.58. Freezing takes place at roughly atmospheric pressure as the temperature is progressively reduced. When the temperature is low enough to ensure that liquid water does not form when the pressure is reduced, primary drying is initiated by dropping the chamber pressure. Under these conditions, ice is sublimated to water vapour, which is transported away from the solid material and condensed. After the ice crystals are removed, the temperature of the material being dried increases and a period of secondary drying begins. Unfrozen water is removed during the secondary drying process at low pressure. The duration of a complete freeze-drying cycle is typically 24 to 48 hours.

Freezing

The first stage of freeze-drying is cooling of the wet solid so that the material solidifies completely. Typically, temperatures of −40°C to −80°C are used and the freezing step takes about 2 hours. Freezing is crucial because the microstructure formed determines to a large extent the quality of the final freeze-dried product. Information about the freezing behaviour of the solid is required, including whether the material forms a crystalline or amorphous matrix, and the maximum temperature that can be used while ensuring that water in the system sublimates during primary drying. To consider these points further, we need to understand the phenomena associated with freezing of water and solutions containing water and solutes.

Pure water can exist in three phases: solid or ice, liquid, and gas or vapour. The *phase diagram* for water showing the relationship between these phases and the prevailing temperature and vapour pressure is shown in Figure 11.59. At the *triple point* indicated as TP in Figure 11.59, ice, liquid water, and water vapour coexist in equilibrium. The triple point for water occurs at a temperature of 0.0098°C and a water vapour pressure of 0.61 kPa. Along the lines shown in Figure 11.59, two phases exist together at equilibrium. Liquid and solid coexist at the combinations of temperature and pressure represented by line

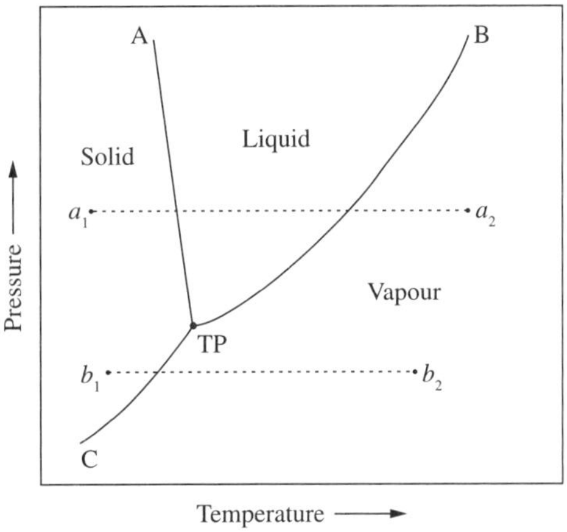

FIGURE 11.59 Phase diagram for water.

A–TP, liquid and vapour coexist at the conditions represented by line B–TP, and solid and vapour coexist at the conditions represented by line C–TP. If the temperature is raised at constant pressure, the condition of the system can be followed by moving horizontally across the phase diagram.

As an example, starting with ice at point a_1 on the diagram, adding heat increases the temperature, melts the ice as the phase boundary with liquid water is encountered, and then moves the system into the liquid phase. Further temperature increase causes the liquid water to evaporate as it passes the liquid–gas phase boundary to finish at point a_2 as water vapour. If the same process is repeated at a lower pressure below the triple point, starting with ice at point b_1, adding heat and raising the temperature causes a direct phase change from solid to vapour as the phase boundary C–TP is encountered and the ice sublimates to vapour without passing through a liquid phase. If the water vapour is heated further, the process is completed at b_2.

For freeze-drying of pure water, it is essential to cool the system to below the triple point so that sublimation rather than melting occurs when the pressure is reduced for primary drying. However, water in wet solids is not often present in pure form: it exists instead as a solution containing dissolved solutes. In an *ideal two-component solution*, both the water and solute may crystallise during freezing. A typical freezing curve for such a system is shown in Figure 11.60. Here, cooling starts with liquid solution at point *a* and continues to point *b*. As point *b* is below the equilibrium freezing point of the solution, the solution at *b* is *supercooled*. Supercooling induces the nucleation of ice crystals. The phase change from liquid water to solid ice results in an increase in temperature as the *latent heat of fusion* Δh_f (Section 5.4.2) is released. As cooling continues to point *c*, the ice crystals grow in size, there is less and less liquid water present, and the solute concentration in the remaining liquid increases progressively as more water is crystallised. The liquid solution approaches saturation, meaning that no further increase in solute concentration can occur. At point *c*, the solute begins to crystallise. A *eutectic point* T_e may be reached, which means that the water and solute solidify as if they were a single pure compound and the material

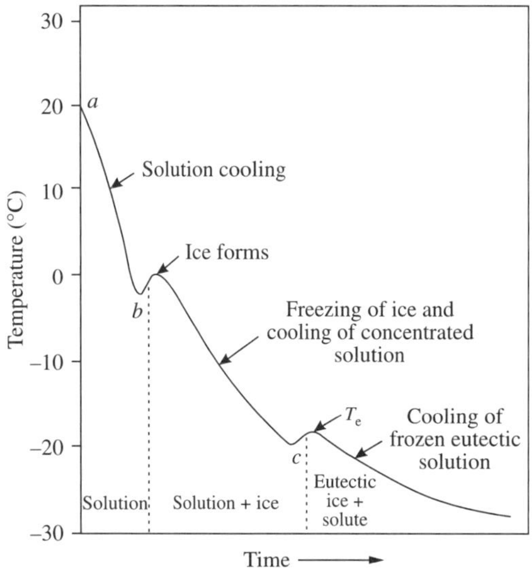

FIGURE 11.60 Freezing curve for an ideal two-component solution forming a eutectic solution.

becomes wholly crystalline. After eutectic freezing is complete at T_e, no liquid remains in the system. Further reduction in temperature cools the frozen mixture.

The eutectic temperature is important for freeze-drying as it is the maximum temperature that can be allowed to occur during primary drying. If the temperature exceeds T_e, melting occurs and liquid is formed, and drying takes place from the liquid phase rather than from the solid phase. As a consequence, the open porous structure characteristic of freeze-dried solids is not produced.

Not all aqueous solutions exhibit eutectic behaviour, as some solutes do not crystallise during freezing. In amorphous systems, after ice begins to form and the solution becomes more and more concentrated, the solution viscosity increases so that the material becomes syrupy and then rubbery. If the temperature is reduced further, a *glass transition* occurs at temperature T_g as the solution changes from a viscous liquid to a glass-like material. The glass transition temperature represents the maximum temperature suitable for primary drying of noncrystalline systems because, above that temperature, liquid is present. If the temperature during primary drying exceeds either the eutectic or glass transition point, the dry solid may collapse. This occurs when the material remaining after sublimation of ice is not sufficiently rigid to support its own weight. Primary drying at temperatures below T_e or T_g is necessary to ensure that the residual dried material is not able to flow and thereby destroy the microstructure of the crystalline or amorphous product.

Primary Drying

During primary drying, the pressure is reduced to below the triple point of water and heat is supplied to the material to provide the *latent heat of sublimation* (Section 5.4.2). This heat is required to vaporise the frozen water. Typically, primary drying takes around 10 hours. For crystalline eutectic systems, primary drying normally removes about 95% of the water contained in the solid.

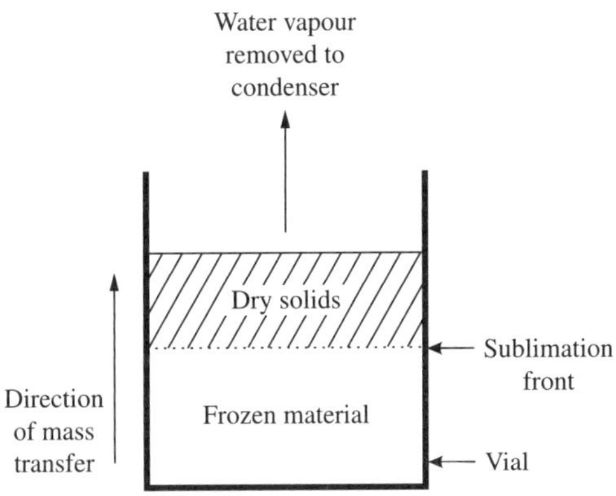

FIGURE 11.61 Mass transfer and the sublimation front during primary drying.

The situation within the solid during primary drying is shown in Figure 11.61. Vaporised water leaves the solid and is removed to a condenser; this allows the low vapour pressure in the chamber to be maintained. As water is removed from the top of the solid, a layer of dried material is formed. With further drying, the thickness of this layer increases while the thickness of the remaining frozen material containing water decreases. As a result, the sublimation front where vaporisation takes place moves down from the top of the solid through the depth of the drying material.

Heat transfer is often the rate-limiting step in freeze-drying. Heat for sublimation is provided either from below through the frozen material or from above through the low-pressure atmosphere and layer of dried solids, or from both directions. To avoid melting, the temperature in all parts of the solid must be maintained at less than the eutectic or glass transition temperature. Consequently, the thermal gradient through the material that provides the driving force for heat transfer at the sublimation front remains low to avoid overheating the solid at the bottom of the vial. In addition, the partial vacuum in the freeze-drying chamber has a substantial insulating effect in terms of heat transfer through the atmosphere from any heat sources above the vial. The low density of gas in the chamber means that convective heat transfer is minimal.

Mass transfer is required in primary drying to transport water vapour out of the solid. The primary resistance to mass transfer is the increasing thickness of dried solid that forms above the sublimation front. It is important, therefore, to minimise the depth of solid applied for freeze-drying. When mass transfer is the rate-limiting step, the drying rate decreases with time as the thickness of the dried layer increases. Mass transfer rates are affected by the conditions used in the freezing step of the freeze-drying process. The formation of large ice crystals by slow freezing creates a dry solid of greater porosity after sublimation compared with that created after the formation of small ice crystals by rapid freezing. Mass transfer of water vapour through the solid by either convective flow or molecular diffusion is greater when the porosity is high.

Secondary Drying

When all the ice crystals have been removed by sublimation, if heat continues to be provided after primary drying, the temperature of the solid rises as heat is no longer needed for the phase change from ice to vapour. Secondary drying thus commences to remove any residual water from the solid. When solutes crystallise during freezing, virtually all of the water present in the wet solid is transformed to eutectic ice and is removed during primary drying. Secondary drying is therefore more important when solutes form an amorphous material during freezing, as frozen amorphous solids can contain significant amounts of water that are not removed by sublimation. Secondary drying of these materials relies on the removal of water by molecular diffusion through the glassy frozen matrix and can be a relatively slow process taking 10 to 12 hours. The temperatures used during secondary drying are higher and the pressures lower than for primary drying. After secondary drying, the residual water content of most dried biological materials is reduced to 1 to 4% by weight.

SUMMARY OF CHAPTER 11

At the end of Chapter 11 you should:

- Know what a *unit operation* is
- Be able to describe the five major steps involved in *downstream processing* schemes for fermentation products
- Understand the factors affecting the choice of *cell removal operations* for treating fermentation broths
- Understand the theory and practice of conventional *filtration*
- Understand the principles of *centrifugation*, including scale-up considerations
- Be familiar with methods used for *cell disruption*
- Understand the concept of an *ideal stage*
- Be able to analyse *aqueous two-phase extractions* in terms of the *equilibrium partition coefficient*, *product yield*, and *concentration factor*
- Understand the mechanisms of large-scale *precipitation* for recovery of proteins from solution
- Understand the principles of *adsorption* operations and the design of *fixed-bed adsorbers*
- Know the different types of *membrane filtration* used in downstream processing, understand the importance of *concentration polarisation* and *gel polarisation*, and know the implications of *mass transfer control* and *fouling* for membrane filter performance
- Be able to analyse processes for *concentration*, *fractionation*, and *diafiltration* using membrane filtration and batch and continuous modes of operation
- Know the different types of *chromatography* used to separate biomolecules
- Be familiar with concepts such as *differential migration*, *zone spreading*, and *resolution* in chromatography, know what operating conditions enhance chromatography performance, and be able to describe the *scale-up procedures* for chromatography columns
- Understand the principles of *crystallisation*, including characterisation of the *size* and *size distribution* of crystals, the mechanisms of *crystal nucleation* and *crystal growth*, and operation of *continuous mixed suspension–mixed product removal (MSMPR)* crystallisers
- Know the role of *drying* operations in downstream processing, understand the relationship between *humidity* and the *equilibrium moisture content* of solids, and be able to describe the factors affecting *drying rate*
- Know the basic principles of *freeze-drying*

PROBLEMS

11.1 Overall product recovery

Supercoiled plasmid DNA for gene therapy applications is produced using high-yielding recombinant bacteria. After the cell culture step, the DNA is recovered using a series of six unit operations. The processes used and the percentage product recovery for each step are as follows: microfiltration 89%, cell disruption 74%, precipitation 71%, ion-exchange chromatography 82%, gel chromatography 68%. What fraction of the plasmid DNA produced by the cells is available at the end of downstream processing?

11.2 Product yield from transgenic goats' milk

Goats are genetically engineered to produce a therapeutic protein, anti-thrombin III, in their milk. Twenty thousand litres of milk containing 10 g l^{-1} anti-thrombin III are treated each week to recover and purify the product. Fat is removed from the milk by skimming and centrifugation with a product yield of 75%. Caseins are then removed by salt precipitation and filtration with a product yield of 60% for these techniques. The clear whey fraction is treated further using gel chromatography and affinity chromatography; 85% yield is achieved in each of the chromatography steps.

(a) What is the overall product yield?

(b) How much purified anti-thrombin III is produced per week?

(c) It is decided to improve the salt precipitation and filtration step to minimise product loss. If the target anti-thrombin III production rate is 80 kg $week^{-1}$ from 20,000 l of milk, by how much must the product yield for this stage of the process be increased above the current 60%?

11.3 Laboratory algal filtration

A suspension of red *Porphyridium cruentum* microalgal cells is filtered in the laboratory using a 6-cm-diameter Büchner funnel and a vacuum pressure of 5 psi. The suspension contains 7.5 g of cells per litre of filtrate. The viscosity of the filtrate is 10^{-3} Pa s. The following data are measured.

Time (min)	1	2	3	4	5
Filtrate volume (ml)	124	191	245	289	330

(a) What proportion of the total resistance to filtration is due to the cell cake after 3 minutes of filtration?

(b) It is proposed to use the same Büchner funnel to filter a *P. cruentum* suspension with a higher cell concentration. What maximum cell density (mass of cells per litre of filtrate) can be handled if 200 ml of filtrate must be collected within 5 min?

11.4 Filtration of plant cells

Suspended plant cells expressing a therapeutic Fab′ antibody fragment are filtered using a laboratory filter of area 78 cm^2. A pressure drop of 9 psi is applied. The filtrate has a viscosity of 5 cP and the suspension deposits 25 g of cells per litre of filtrate. The results from the filtration experiments are correlated according to the equation:

$$\frac{t}{V_f} = K_1 V_f + K_2$$

where t is the filtration time, V_f is the filtrate volume, $K_1 = 8.3 \times 10^{-6}$ min cm^{-6}, and $K_2 = 4.2 \times 10^{-3}$ min cm^{-3}.

(a) Determine the specific cake resistance and filter medium resistance.

(b) The same plant cell suspension is processed using a pressure drop of 9 psi on a filter of area 4 m^2. Estimate the time required to collect 6000 litres of filtrate.

11.5 Bacterial filtration

A suspension of *Bacillus subtilis* cells is filtered under constant pressure for recovery of protease. A pilot-scale filter is used to measure the filtration properties. The filter area is 0.25 m^2, the pressure drop is 360 mmHg, and the filtrate viscosity is 4.0 cP. The cell suspension deposits 22 g of cake per litre of filtrate. The following data are measured.

Time (min)	2	3	6	10	15	20
Filtrate volume (l)	9.8	12.1	18.0	23.8	29.9	37.5

(a) Determine the specific cake resistance and filter medium resistance.

(b) Based on the data obtained in the pilot-scale study, what size filter is required to process 4000 l of cell suspension in 30 min at a pressure drop of 360 mmHg?

11.6 Filtration of mycelial suspensions

Pelleted and filamentous forms of *Streptomyces griseus* are filtered separately using a small laboratory filter of area 1.8 cm^2. The mass of wet solids per volume of filtrate is 0.25 g ml^{-1} for the pelleted cells and 0.1 g ml^{-1} for the filamentous culture. The viscosity of the filtrate is 1.4 cP. Five filtration experiments at different pressures are carried out with each suspension. The following results are obtained.

	Pressure Drop (mmHg)				
	100	**250**	**350**	**550**	**750**
Filtrate Volume (ml) for Pelleted Suspension	**Time (s)**				
10	22	12	9	7	5
15	52	26	20	14	12
20	90	49	36	28	22
25	144	75	60	43	34
30	200	110	88	63	51
35	285	149	119	84	70
40	368	193	154	110	90
45	452	240	195	140	113
50	–	301	238	175	141

Filtrate Volume (ml) for Filamentous Suspension	Pressure Drop (mmHg)				
	100	250	350	550	750
	Time (s)				
10	36	22	17	13	11
15	82	47	40	31	25
20	144	85	71	53	46
25	226	132	111	85	70
30	327	194	157	121	100
35	447	262	215	166	139
40	–	341	282	222	180
45	–	434	353	277	229
50	–	–	442	338	283

(a) Evaluate the specific cake resistance as a function of pressure for each culture.
(b) Determine the compressibility for each culture.
(c) A filter press with area 15 m^2 is used to process 20 m^3 of filamentous *S. griseus* culture. If the filtration must be completed in one hour, what pressure drop is required?

11.7 Rotary drum vacuum filtration

Continuous rotary vacuum filtration can be analysed by considering each revolution of the drum as a stationary batch filtration. Per revolution, each cm^2 of filter cloth is used to form cake only for the period of time it spends submerged in the liquid reservoir. A rotary drum vacuum filter with drum diameter 1.5 m and filter width 1.2 m is used to filter starch from an aqueous slurry. The pressure drop is kept constant at 4.5 psi; the filter operates with 30% of the filter cloth submerged. Resistance due to the filter medium is negligible. Laboratory tests with a 5-cm^2 filter have shown that 500 ml of slurry can be filtered in 23.5 min at a pressure drop of 12 psi; the starch cake was also found to be compressible with $s = 0.57$. Use the following steps to determine the drum speed required to produce 20 m^3 of filtered liquid per hour.

(a) Evaluate $\mu_f \alpha' c$ from the laboratory test data.
(b) If N is the drum speed in revolutions per hour, what is the cycle time?
(c) From (b), for what period of time per revolution is each cm^2 of filter cloth used for cake formation?
(d) What volume of filtrate must be filtered per revolution to achieve the desired rate of 20 m^3 per hour?
(e) Apply Eq. (11.15) to a single revolution of the drum to evaluate N.
(f) The liquid level is raised so that the fraction of submerged filter area increases from 30 to 50%. What drum speed is required under these conditions?

11.8 Centrifugation of yeast

Yeast cells must be separated from a fermentation broth. Assume that the cells are spherical with diameter 5 μm and density 1.06 g cm^{-3}. The viscosity of the culture broth is

1.36×10^{-3} N s m^{-2}. At the temperature of separation, the density of the suspending fluid is 0.997 g cm^{-3}. Five hundred litres of broth must be treated every hour.

(a) Specify Σ for a suitably sized disc stack centrifuge.

(b) The small size and low density of microbial cells are disadvantages in centrifugation. If instead of yeast, quartz particles of diameter 0.1 mm and specific gravity 2.0 are separated from the culture liquid, by how much is Σ reduced?

11.9 Centrifugation of food particles

Small food particles with diameter 10^{-2} mm and density 1.03 g cm^{-3} are suspended in liquid of density 1.00 g cm^{-3}. The viscosity of the liquid is 1.25 mPa s. A tubular bowl centrifuge of length 70 cm and radius 11.5 cm is used to separate the particles. If the centrifuge is operated at 10,000 rpm, estimate the feed flow rate at which the food particles are just removed from the suspension.

11.10 Scale-up of disc stack centrifuge

A pilot-scale disc stack centrifuge is tested for recovery of bacteria. The centrifuge contains 25 discs with inner and outer diameters of 2 cm and 10 cm, respectively. The half-cone angle is 35°. When operated at a speed of 3000 rpm with a feed rate of 3.5 litres min^{-1}, 70% of the cells are recovered. If a bigger centrifuge is to be used for industrial treatment of 80 litres min^{-1}, what operating speed is required to achieve the same sedimentation performance if the larger centrifuge contains 55 discs with outer diameter 15 cm, inner diameter 4.7 cm, and half-cone angle 45°?

11.11 Centrifugation of yeast and cell debris

A tubular bowl centrifuge is used to concentrate a suspension of genetically engineered yeast containing a new recombinant protein. At a speed of 12,000 rpm, the centrifuge treats 3 l of broth per min with satisfactory results. It is proposed to use the same centrifuge to separate cell debris from the homogenate produced by mechanical disruption of the yeast. If the average size of the debris is one-third that of the yeast and the viscosity of the homogenate is five times greater than the cell suspension, what flow rate can be handled if the centrifuge is operated at the same speed?

11.12 Centrifuge throughput

Escherichia coli bacteria are harvested from a fermentation broth using a disc stack centrifuge with 72 discs operated at 7300 rpm. The outer diameter of the discs is 15 cm, the inner diameter is 7.5 cm, and the half-cone angle is 36°. The broth viscosity is 1.25 cP, the liquid density is 1025 kg m^{-3}, and the smallest cell size is 1.1 μm. The density of *E. coli* cells is determined by laboratory measurement to be 1.06 g cm^{-3}. Assume that the centrifuge operates ideally.

(a) Estimate the maximum throughput of broth in litres h^{-1} that can be handled by this system while still achieving the desired separation.

(b) The measurement of *E. coli* cell density has a small error of $\pm 2\%$ associated with it. If the particle density is actually 2% lower than the value provided by the laboratory, what effect does this have on the estimated maximum throughput? What conclusions about predicting centrifuge performance can you draw from this result?

11.13 Cell disruption

Micrococcus bacteria are disrupted at 5°C in a Gaulin homogeniser operated at pressures between 200 and 550 kg$_f$ cm^{-2}. The data obtained for protein release as a function of number of passes through the homogeniser are listed in the table.

Number of Passes	Pressure Drop ($kg_f\ cm^{-2}$) 200	300	400	500	550
	% Protein Release				
1	5.0	13.5	23.3	36.0	42.0
2	9.5	23.5	40.0	58.5	66.0
3	14.0	33.5	52.5	75.0	83.7
4	18.0	43.0	66.6	82.5	88.5
5	22.0	47.5	73.0	88.5	94.5
6	26.0	55.0	79.5	91.3	–

(a) How many passes are required to achieve 80% protein release at an operating pressure of 460 $kg_f\ cm^{-2}$?

(b) Estimate the pressure required to deliver 70% protein recovery in only two passes.

11.14 Disruption of cells cultured under different conditions

Candida utilis cells are cultivated in different fermenters using either repeated batch or continuous culture. An experiment is conducted using a single pass in a Gaulin homogeniser to disrupt cells from the two culture systems. The following results are obtained.

Type of Culture	Percentage of Protein Released: Pressure = 57 MPa	Pressure = 89 MPa
Repeated batch	66	84
Continuous	30	55

(a) How many passes are required to achieve 90% protein release from the batch-culture cells at an operating pressure of 70 MPa?

(b) What percentage protein is released from the continuous-culture cells using the number of passes determined in (a) and the same operating pressure?

(c) What operating pressure is required to release 90% of the protein contained in the cells from continuous culture in five passes?

11.15 Enzyme purification using two-phase aqueous partitioning

Leucine dehydrogenase is recovered from 150 litres of *Bacillus cereus* homogenate using an aqueous two-phase polyethylene glycol–salt system. The homogenate initially contains 3.2 units of enzyme ml^{-1}. A polyethylene glycol–salt mixture is added and two phases form. The enzyme partition coefficient is 3.5.

(a) What volume ratio of upper and lower phases must be chosen to achieve 80% recovery of enzyme in a single extraction step?

(b) If the volume of the lower phase is 100 litres, what is the concentration factor for 80% recovery?

11.16 Recovery of viral particles

Cells of the fall armyworm *Spodoptera frugiperda* are cultured in a fermenter to produce viral particles for insecticide production. Virus is released into the culture broth after lysis of the host cells. The initial culture volume is 5 litres. An aqueous two-phase polymer solution of volume 2 litres is added to this liquid; the volume of the bottom phase is 1 litre. The virus partition coefficient is 10^{-2}.

(a) What is the yield of virus at equilibrium?

(b) Write a mass balance for the viral particles in terms of the concentrations and volumes of the phases, equating the amounts of virus present before and after addition of polymer solution.

(c) Derive an equation for the concentration factor in terms of the liquid volumes and partition coefficient only.

(d) Calculate the concentration factor for the viral extraction.

11.17 Enzyme salting-out

The solubility of cellulase enzyme in a buffered extract of *Trichoderma viride* cells is determined in the laboratory as a function of ammonium sulphate concentration. At $(NH_4)_2SO_4$ concentrations of 1.8, 2.2, and 2.5 M, the solubilities are 44, 0.58, and 0.07 mg l^{-1}, respectively. *T. viride* cells produced by fermentation are homogenised to generate 800 litres of buffered extract containing 25 mg l^{-1} of soluble cellulase. Cellulase is precipitated by adding $(NH_4)_2SO_4$ to the extract to give a concentration of 2.0 M and a final solution volume of 930 l.

(a) Estimate the mass of enzyme recovered.

(b) What is the yield or percentage recovery of cellulase in this precipitation process?

(c) The residual cellulase in solution is treated for further precipitation. Assuming negligible volume change for the second increment in salt content, estimate the concentration of $(NH_4)_2SO_4$ required to recover 90% of the residual cellulase.

11.18 Precipitation of monoclonal antibody

Human IgM monoclonal antibody is produced by hybridoma cells in a stirred fermenter. Antibody secreted into the culture medium reaches a concentration of 120 μg ml^{-1} of cell-free broth. The product is concentrated by precipitation immediately after removal of the cells. The yield of antibody recovered in the precipitate is determined as a percentage of the initial mass of antibody present in the broth. The precipitate purity or percentage of the precipitate weight that is IgM antibody is also measured. Various precipitation methods are tested and the following results are obtained.

Precipitant	Concentration of Precipitant (% w/v)	pH	Antibody Yield (%)	Precipitate Purity (% w/w)
Ammonium sulphate	18	7.2	0.0	–
	24	7.2	49	90
	29	7.2	75	51
	31	7.2	83	18
Polyethylene glycol	8	5.5	75	95
	12	5.5	96	60
Ethanol	25	7.2	0.0	–

(a) From the data provided, what are the advantages and disadvantages of using 12% w/v polyethylene glycol for this precipitation process compared with 8% w/v?

(b) Using % w/v as the unit for salt concentration and $\mu g\ g^{-1}$ of cell-free broth as the unit for protein solubility, derive an empirical equation for IgM solubility as a function of ammonium sulphate concentration at pH 7.2. The density of cell-free broth can be taken as 1000 kg m^{-3}.

(c) Predict the antibody yield using 27% w/v $(NH_4)_2SO_4$.

(d) What mass of antibody will remain in solution if 27% w/v $(NH_4)_2SO_4$ is used to treat 100 l of cell-free broth?

(e) The residual solution after salting-out with 27% w/v $(NH_4)_2SO_4$ is treated for further antibody recovery.

- **(i)** If a total yield of 94% is required after both precipitation steps, estimate the concentration of $(NH_4)_2SO_4$ that must be applied in the second salting-out stage.
- **(ii)** What mass of antibody do you expect to recover in the second precipitation step?
- **(iii)** When the second-stage precipitation is carried out using the $(NH_4)_2SO_4$ concentration determined in (i), it is found that the amount of antibody contained in the precipitate is somewhat more than expected. What possible reasons can you give for this?

11.19 Enzyme ultrafiltration

A solution containing 0.1% w/v polygalacturonase enzyme must be concentrated to 3% w/v using a continuous ultrafiltration system. The flow rate of the feed solution is 2500 l h^{-1}. Two types of membrane module are being considered for this separation: a hollow fibre unit operated to give a mass transfer coefficient of 10 l $m^{-2}\ h^{-1}$ and a plate-and-frame device with a mass transfer coefficient of 6.5 l $m^{-2}\ h^{-1}$. The enzyme is completely rejected by the membrane and the filtration is carried out at steady state under gel polarisation conditions with negligible fouling. The gelation concentration for polygalacturonase can be taken as 45% w/v.

(a) What is the permeate flow rate?

(b) Estimate the permeate flux delivered by the two types of membrane device.

(c) What membrane areas are required for the hollow fibre and plate-and-frame systems?

11.20 Antibody concentration with membrane fouling

A humanised monoclonal antibody is produced using hybridoma cell culture. A solution containing 150 mg l^{-1} of antibody is concentrated tenfold using steady-state continuous ultrafiltration. The pressure at the inlet to the membrane module is 450 kPa; the pressure at the outlet is 200 kPa. The permeate flow is open to the atmosphere. The membrane resistance determined using water is 0.24 kPa $m^2\ h\ l^{-1}$. The relationship between antibody concentration and the resistance due to concentration polarisation is determined in laboratory experiments as:

$$R_C = 4.5 + 4.33 \times 10^{-3}\, C_B$$

where C_B is the antibody concentration in units of mg l^{-1} and R_C has units of kPa $m^2\ h\ l^{-1}$.

(a) Relative to the water flux, by how much is the permeate flux reduced by concentration polarisation?

(b) Soon after ultrafiltration commences, the membrane becomes fouled. The fouling resistance adds about 5.3 kPa $m^2\ h\ l^{-1}$ to the total resistance to filtration. By how much does this reduce the permeate flux relative to that achieved without fouling?

11.21 Clarification of wine

A boutique winery in the Coonawarra region of South Australia uses microfiltration to clarify its premium vintage shiraz before aging. The unfiltered wine contains 7.5 g l^{-1} yeast cells. For every cubic metre of unfiltered wine, 0.05 m^3 of concentrated yeast retentate is produced. The permeate contains a negligible number of cells. After microfiltration, the shiraz is aged in oak and then bottled.

(a) What is the yeast concentration in the retentate?

(b) If a carton of bottled wine contains one dozen 750-ml bottles and the winery produces 1500 cartons of shiraz per year, how many litres of concentrated yeast slurry must be disposed of per annum?

11.22 Fractionation of cell homogenate

Recombinant insulin is produced using *E. coli* fermentation. The harvested cells are homogenised using a bead mill to release the protein inclusion bodies. The cell homogenate is then treated with CNBr to cleave the cell proteins into small fragments, leaving the proinsulin as a large, uncleaved molecule. The CNBr-treated homogenate contains several components, including 5% w/w cell debris, 8% w/w proinsulin, 5% w/w cleaved proteins, and 0.5% w/w sugar. The mixture is treated using membrane filtration to remove the smaller molecules and fragments. The membrane chosen for this application gives retention coefficients of 0.95 for the cell debris, 0.90 for proinsulin, 0.25 for cleaved protein, and 0 for sugar. The effect of fouling on solute retention is negligible. The cell debris is concentrated fivefold by this process.

(a) What is the composition of the retentate?

(b) What is the composition of the permeate?

11.23 Ultrafiltration pressure

Penicillinase from *Streptomyces aureus* is concentrated by continuous ultrafiltration using three separate plate-and-frame membrane modules. The membrane area in each module is 58 m^2, the channel height is 3 mm, and the channel length is 50 cm. The steady-state concentration of penicillinase in the retentate is 8 g l^{-1}. The viscosity of the solution is 1.0 cP and the density is 1000 kg m^{-3}. The effective diffusion coefficient for penicillinase is 3.3×10^{-7} cm^2 s^{-1} and its gelation concentration is approximately 300 g l^{-1}. Ultrafiltration is carried out at steady state using a transmembrane pressure of 0.7 bar and a linear cross-membrane velocity of 65 cm s^{-1}. The entire filtration process generates 8.7 m^3 h^{-1} of permeate containing a negligible amount of penicillinase. Can the rate of filtration in this system be improved by increasing the operating pressure drop?

11.24 Washing thawed red blood cells

Blood donated by volunteers is frozen to prolong its shelf life. Glycerol is added to the blood as a cryoprotectant prior to freezing. After the blood is thawed and before its transfusion into patients, the glycerol and any haemoglobin released from older, fragile blood cells during the thawing process must be removed. Constant-volume diafiltration is applied using physiological saline to wash the cells free of these contaminants. The membrane pore size is 0.2 μm, which allows both glycerol and haemoglobin to pass freely into the permeate while the blood cells are retained completely. Before diafiltration, the blood glycerol concentration is 1.7 M and the released haemoglobin concentration is

$5.5\ g\ l^{-1}$. The washed blood must contain no more than 0.1 M glycerol. Diafiltration is carried out on 500-ml batches of blood using hollow fibre membranes.

(a) What volume of saline is required per batch of blood?

(b) What is the concentration of released haemoglobin in the washed blood?

(c) If the diafiltration process is halted after only 500 ml of saline is added, what concentrations of glycerol and released haemoglobin are present in the blood?

11.25 Diafiltration of recombinant protein

A glycosylated recombinant protein is produced using mink lung epithelial cells. In the first step of protein recovery from the culture liquor, the protein is precipitated by addition of $(NH_4)_2SO_4$ and the precipitate harvested by centrifugation. When redissolved in buffer, the precipitate yields 150 litres of solution containing $18\ mg\ l^{-1}$ of recombinant protein and $50\ g\ l^{-1}$ of $(NH_4)_2SO_4$. Constant-volume diafiltration is used to desalt the protein and replace the suspending buffer with a tris(hydroxymethyl)aminomethane (Tris) solution prior to additional protein purification by chromatography. Diafiltration is carried out using a $2.5\text{-}m^2$ membrane that fully retains the protein while allowing $(NH_4)_2SO_4$ and Tris to pass through freely. The membrane filter is operated under gel polarisation conditions with a mass transfer coefficient of $1.4 \times 10^{-5}\ m\ s^{-1}$. The gelation concentration of the recombinant protein is $120\ mg\ l^{-1}$. The $(NH_4)_2SO_4$ concentration in the final protein solution must be no greater than $0.1\ g\ l^{-1}$. Fouling effects can be neglected.

(a) What volume of Tris solution is needed?

(b) Determine the time required for diafiltration.

(c) To reduce the volume of Tris consumed in the diafiltration process, a new protocol is developed whereby the protein precipitate is redissolved in only 30 l of buffer rather than 150 l. With this modification, the initial protein solution is five times more concentrated and contains $90\ mg\ l^{-1}$ of recombinant protein and $250\ g\ l^{-1}$ of $(NH_4)_2SO_4$. The same membrane system is used to accomplish the diafiltration step under gel polarisation conditions.

(i) By how much is the required volume of Tris solution reduced relative to that estimated in (a)?

(ii) How does the new protocol affect the rate of diafiltration?

(iii) What is the duration of the diafiltration process using the new protocol?

11.26 Time for batch ultrafiltration

A solution of recombinant interleukin-3 produced using mammalian cell culture is concentrated by batch ultrafiltration. The solution viscosity is 1.2 mPa s and the density is $1.0\ g\ cm^{-3}$. A battery of spiral-wound membrane modules containing a total membrane area of $250\ m^2$ is available for the process. The feed channel height in the modules is 2.5 mm and mesh spacers are installed to improve mass transfer. The cross-flow liquid velocity is $45\ cm\ s^{-1}$. The membrane molecular weight cut-off is chosen so that the permeate contains negligible interleukin. Interleukin is concentrated by a factor of 20 under gel polarisation conditions from an initial concentration of 5 units ml^{-1}. The diffusivity of interleukin-3 in aqueous solution is estimated as $11.5 \times 10^{-7}\ cm^2\ s^{-1}$ and the gelation concentration is 2000 units ml^{-1}. Fouling effects are negligible. What time is required to treat $15\ m^3$ of dilute interleukin solution?

11.27 Combined ultrafiltration and diafiltration

Recombinant somatotropin is overproduced in *E. coli* in the form of intracellular protein inclusion bodies. After cell disruption and removal of other insoluble material, the homogenate is treated with a detergent to solubilise the somatotropin protein. The result is 5000 l of solution containing 5% w/v somatotropin and 1.2% w/v detergent. Ultrafiltration is used in a two-step process to concentrate the somatotropin and wash out the detergent by diafiltration. The membrane has a 10,000-dalton molecular weight cut-off to achieve full retention of the somatotropin. The rejection coefficient for the detergent is zero. Hollow fibre modules are operated under gel polarisation conditions to give a mass transfer coefficient of 2.2×10^{-6} m s^{-1}. The gelation concentration for somatotropin is about 50% w/w. The effects of fouling can be neglected.

(a) The somatotropin is concentrated by reducing the volume of the protein solution to 1000 l using batch ultrafiltration. The detergent is then removed by constant-volume diafiltration with buffer.

- **(i)** What membrane area is required if the batch concentration step must be completed within 2 hours?
- **(ii)** What volume of diafiltration buffer is required if the residual detergent concentration must be no greater than 0.002% w/v?

(b) Instead of performing diafiltration after the protein concentration step as in (a), diafiltration is carried out first. What volume of buffer is required in this case to reduce the detergent concentration to 0.002% w/v?

11.28 Scale-up of virus ultrafiltration

Ultrafiltration is used to concentrate a suspension of baboon endogenous virus under mass-transfer-controlled conditions. A spiral-wound membrane cartridge is used in pilot experiments. At a cross-flow fluid velocity of 0.45 m s^{-1}, the permeate flux is 27 l m^{-2} h^{-1}. For fluid velocities within the range 0.40 to 3.5 m s^{-1}, the flux is proportional to the flow rate raised to the power 0.66. The membrane viral retention coefficient is 1. In large-scale operations, the same membrane modules are operated in continuous mode with a cross-flow velocity of 2.2 m s^{-1}. Fouling effects are negligible. If 100 m^3 of feed solution must be concentrated by a factor of 2.5 within 1 h, estimate the total membrane area required.

11.29 Scale-up of enzyme chromatography

A laboratory column of internal diameter 4.5 cm and length 25 cm is used to purify leucine dehydrogenase by anion exchange chromatography. During elution, 0.6 g of enzyme is collected in 35 min. The process is scaled up for production of 15 g h^{-1} of purified leucine dehydrogenase from the same feed solution. If the linear fluid velocity is kept constant, what is the size of the large chromatography column?

11.30 Gel chromatography scale-up

Gel chromatography is used for commercial-scale purification of a proteinaceous diphtheria toxoid from *Corynebacterium diphtheriae* supernatant. In the laboratory, a small column of 1.5 cm inner diameter and height 0.4 m is packed with 10 g dry Sephadex gel; the void volume is measured as 23 ml. A sample containing the toxoid and impurities is injected into the column. At a liquid flow rate of 14 ml min^{-1}, the elution volume for the toxoid is 29 ml and the elution volume for the principal impurity is 45 ml. A column of height 0.6 m and diameter 0.5 m is available for large-scale gel chromatography. The same type of packing is used; the void fraction and the ratio of pore volume to total bed volume remain the same as in the laboratory-scale column. The liquid flow rate in the large column is scaled up in proportion to the column cross-sectional area; the flow patterns in both columns can be assumed identical. The water regain value for the packing is given by the manufacturer as 0.0035 $m^3\ kg^{-1}$ dry gel.

(a) Which is the larger molecule, the diphtheria toxoid or the principal impurity?
(b) Determine the partition coefficients for the toxoid and impurity.
(c) Estimate the elution volumes in the commercial-scale column.
(d) What is the volumetric flow rate in the large column?
(e) Estimate the retention time of toxoid in the large column.

11.31 Protein separation using affinity chromatography

Human insulin A and B from recombinant *Escherichia coli* are separated using pilot-scale affinity chromatography. Laboratory studies show that the capacity factors for the proteins are 0.85 for the A-chain and 1.05 for the B-chain. The dependence of HETP on liquid velocity satisfies the following type of equation:

$$H = \frac{A}{u} + Bu + C$$

where u is the linear liquid velocity and A, B, and C are constants. Values of A, B, and C for the insulin system are $2 \times 10^{-9}\ m^2\ s^{-1}$, 1.5 s, and 5.7×10^{-5} m, respectively. Two columns with inner diameter 25 cm are available for the process; one is 1.0 m high, the other is 0.7 m high.

(a) Plot the relationship between H and u from $u = 0.1 \times 10^{-4}\ m\ s^{-1}$ to $u = 2 \times 10^{-4}\ m\ s^{-1}$.
(b) What is the minimum HETP? At what liquid velocity is the minimum HETP obtained?
(c) If the larger column is used with a liquid flow rate of 0.31 litres min^{-1}, will the two insulin chains be separated completely?
(d) If the smaller column is used, what is the maximum liquid flow rate that will give complete separation?

11.32 Crystal size distribution from screen analysis

Cephalosporin C salt crystals are subjected to screen analysis for measurement of the crystal size distribution. The following results are obtained.

U.S. Sieve	Mass Retained (g)
No. 30	0
No. 40	0.5
No. 50	3.05
No. 60	8.3
No. 70	10.8
No. 100	15.6
No. 140	16.8
No. 200	9.0
No. 270	6.4
Pan	4.2

(a) Plot the mass density distribution as a function of average particle size.
(b) What is the dominant crystal size?

11.33 Crystal growth and nucleation rates

Crystals of citric acid monohydrate leaving an MSMPR crystalliser are washed, dried, and subjected to screen analysis. The following results are obtained from a 1-litre sample of crystal slurry.

Tyler Screen	Mass Retained (g)
14 mesh	0
16 mesh	0.77
20 mesh	4.6
28 mesh	8.5
32 mesh	13.2
35 mesh	18.1
42 mesh	23.8
Pan	0

The density of the citric acid crystals is 1.542 g cm^{-3} and the volume shape factor is 0.775. The crystalliser has a volume of 10 m^3 and the operating flow rate is 40 l min^{-1}.
(a) Determine the crystal growth rate.
(b) What is the nucleation rate?
(c) What is the operating magma density in the crystalliser?

11.34 Bench-scale continuous crystalliser

A bench-scale mixed suspension–mixed product removal (MSMPR) crystalliser is set up in the laboratory to study the crystallisation of insulin. The volume of the crystalliser is 2.5 l and the operating flow rate is 0.4 l h^{-1}. The insulin crystals produced have a density of 1.25 g cm^{-3} and a volume shape factor of 0.60. The steady-state crystal slurry density is 280 g l^{-1}. A sample of the crystals is analysed by sieving using Tyler screens. The following results are obtained.

Tyler Mesh	Mass Percent
60	0
65	1.9
80	7.1
100	23.5
150	40.4
170	13.6
200	13.5
Pan	0

(a) What is the growth rate of the crystals?
(b) What is the crystal nucleation rate?

11.35 Continuous crystalliser specification

A mixed suspension–mixed product removal (MSMPR) crystalliser is used to produce itaconic acid crystals from *Aspergillus terreus* fermentation broth. The crystal density is 1690 kg m^{-3} and the volume shape factor is 0.5. The crystal production rate is 350 kg h^{-1}. Bench-scale studies have shown that a nucleation rate of 4.5×10^7 m^{-3} s^{-1} and a crystal linear growth rate of 2.1×10^{-9} m s^{-1} can be expected. The crystalliser is operated with a solution feed rate of 1.8 m^3 h^{-1}.

(a) What is the operating slurry density in the crystalliser?
(b) Determine the crystalliser volume.
(c) What is the dominant crystal size?

11.36 Drying of benzyl penicillin

Sodium benzyl penicillin must be dried to a moisture content of 20% measured on a wet weight basis. Air at 25°C is used for the drying operation at 1 atm pressure in a fluidised bed drier.

(a) What relative humidity of air is required?
(b) What is the mole fraction of water in the air used for drying?
(c) What is the humidity of the air?

Use the isotherm data for sodium benzyl penicillin in Figure 11.56(a) and assume that equilibrium is reached.

11.37 Drying rate for filter cake

A cake of filtered solids is dried using a screen tray drier. The dimensions of the cake are diameter 30 cm and thickness 5 mm. Air at 35°C is passed over both the top and bottom of the cake on the screen. The critical moisture content of the cake is 5% measured on a wet basis and the density of completely dried material is 1380 kg m^{-3}. At the operating air flow rate, the heat transfer coefficient for gas-phase heat transfer at the surface of the solids is 45 W m^{-2} °C^{-1}. At the air humidity used, the temperature of the wet cake surface is 26°C.

(a) What is the rate of drying?
(b) What time is required to dry the cake from a moisture content of 15% wet basis to a moisture content of 5% wet basis?

11.38 Drying time for protein crystals

Insulin crystals are dried in air in a batch drier under constant drying conditions. The moisture content associated with 20 kg dry weight of crystals is reduced from 18 g per 100 g dry solids to 8 g per 100 g dry solids in 4.5 hours. The area available for drying is 0.5 m^2. The critical moisture content for this system is 5 g per 100 g dry solids. If the air temperature is 28°C and the surface temperature of the wet crystals is 20°C, what is the heat transfer coefficient for this drying operation?

11.39 Drying time for solid precipitate

Precipitated protein is centrifuged to a water content of 35% measured on a dry mass basis. It is then dried in a batch drier for 6 hours under constant drying conditions to a water content of 15%, dry basis. The critical moisture content for the solids is 8%, dry basis. How long would it take to dry a sample of the same size from a water content of 35% to 10% under the same conditions?

11.40 Drying of amino acid crystals

A batch of wet asparagine crystals containing 32 kg of dry solids and 25% moisture measured on a wet basis is dried in a tunnel drier for 3 hours using air at 28°C. The critical moisture content of the crystals is 10% dry basis. The drying area available is 12 m^2. The heat transfer coefficient for gas-phase heat transfer to the surface of the crystals is 32 W m^{-2} °C^{-1}. At the air humidity used, the temperature at the wet crystal surface is 24°C and the crystal equilibrium moisture content is 7.5%, dry basis.

(a) What is the drying rate?

(b) What is the moisture content of the dried crystals?

References

[1] K. Schügerl, Recovery of proteins and microorganisms from cultivation media by foam flotation, Adv. Biochem. Eng./Biotechnol. 68 (2000) 191–233.

[2] P.A. Belter, E.L. Cussler, W.-S. Hu, Bioseparations: Downstream Processing for Biotechnology, John Wiley, 1988.

[3] M.-R. Kula, Recovery operations, in: H.-J. Rehm, G. Reed (Eds.), Biotechnology, vol. 2, VCH, 1985, pp. 725–760.

[4] H.-W. Hsu, Separations by Centrifugal Phenomena, John Wiley, 1981.

[5] C.M. Ambler, The evaluation of centrifuge performance, Chem. Eng. Prog. 48 (1952) 150–158.

[6] W.W.-F. Leung, Centrifugation, in: P.A. Schweitzer (Ed.), Handbook of Separation Techniques for Chemical Engineers, third ed., McGraw-Hill, 1997, pp. 4-63–4-96.

[7] P.J. Hetherington, M. Follows, P. Dunnill, M.D. Lilly, Release of protein from baker's yeast (*Saccharomyces cerevisiae*) by disruption in an industrial homogeniser, Trans. IChE 49 (1971) 142–148.

[8] E. Keshavarz-Moore, Cell disruption: a practical approach, in: M.S. Verrall (Ed.), Downstream Processing of Natural Products, John Wiley, 1996, pp. 41–52.

[9] C.R. Engler, C.W. Robinson, Effects of organism type and growth conditions on cell disruption by impingement, Biotechnol. Lett. 3 (1981) 83–88.

[10] P. Dunnill, M.D. Lilly, Protein extraction and recovery from microbial cells, in: S.R. Tannenbaum, D.I.C. Wang (Eds.), Single-Cell Protein II, MIT Press, 1975, pp. 179–207.

[11] C.R. Engler, C.W. Robinson, Disruption of *Candida utilis* cells in high pressure flow devices, Biotechnol. Bioeng. 23 (1981) 765–780.

[12] A.R. Kleinig, C.J. Mansell, Q.D. Nguyen, A. Badalyan, A.P.J. Middelberg, Influence of broth dilution on the disruption of *Escherichia coli*, Biotechnol. Technique 9 (1995) 759–762.

[13] E.J. Cohn, J.D. Ferry, Interactions of proteins with ions and dipolar ions, in: E.J. Cohn, J.T. Edsall (Eds.), Proteins, Amino Acids and Peptides, Reinhold, 1943.

[14] Y.-C. Shih, J.M. Prausnitz, H.W. Blanch, Some characteristics of protein precipitation by salts, Biotechnol. Bioeng. 40 (1992) 1155–1164.

[15] R.K. Scopes, Protein Purification: Principles and Practice, third ed., Springer-Verlag, 1994.

[16] International Critical Tables (1926–1930), McGraw-Hill; 1st electronic ed., International Critical Tables of Numerical Data, Physics, Chemistry and Technology, Knovel, 2003.

[17] J.F. Richardson, J.H. Harker, Coulson and Richardson's Chemical Engineering, vol. 2, fifth ed., Butterworth-Heinemann, 2002 (Chapters 17 and 18).

[18] M. Cheryan, Ultrafiltration and Microfiltration Handbook, Technomic, 1998.

[19] L.J. Zeman, A.L. Zydney, Microfiltration and Ultrafiltration, Marcel Dekker, 1996.

[20] W.F. Blatt, A. Dravid, A.S. Michaels, L. Nelsen, Solute polarization and cake formation in membrane ultrafiltration: causes, consequences, and control techniques, in: J.E. Finn (Ed.), Membrane Science and Technology, Plenum, 1970, pp. 47–97.

[21] M.S. Le, J.A. Howell, Ultrafiltration, in: M. Moo-Young (Ed.), Comprehensive Biotechnology, vol. 2, Pergamon Press, 1985, pp. 383–409.
[22] T. Alex, H. Haughney, New membrane-based technologies for the pharmaceutical industry, in: T.H. Meltzer, M.W. Jornitz (Eds.), Filtration in the Biopharmaceutical Industry, Marcel Dekker, 1998, pp. 745–782.
[23] M. Cheryan, Ultrafiltration Handbook, Technomic, 1986.
[24] L.R. Snyder, Classification of the solvent properties of common liquids, J. Chromatog. 92 (1974) 223–230.
[25] E.L. Johnson, R. Stevenson, Basic Liquid Chromatography, Varian Associates, 1978.
[26] J.C. Giddings, Dynamics of Chromatography, Part I, Marcel Dekker, 1965.
[27] E. Heftmann (Ed.), Chromatography, second ed., Reinhold, 1967.
[28] J. Garside, A. Mersmann, J. Nyvlt, Measurement of Crystal Growth and Nucleation Rates, second ed., Institution of Chemical Engineers, 2002.
[29] A.D. Randolph, M.A. Larson, Theory of Particulate Processes, second ed., Academic Press, 1988.
[30] S.J. Jančić, P.A.M. Grootscholten, Industrial Crystallization, Delft University Press, 1984.
[31] J.W. Mullin, Crystallization, fourth ed., Butterworth-Heinemann, 2001.
[32] S.N. Black, R.J. Davey, M. Halcrow, The kinetics of crystal growth in the presence of tailor-made additives, J. Crystal Growth 79 (1986) 765–774.
[33] R.C. DeMattei, R.S. Feigelson, Growth rate study of canavalin single crystals, J. Crystal Growth 97 (1989) 333–336.
[34] K. Toyokura, K. Mizukawa, M. Kurotani, Crystal growth of L-SCMC seeds in a DL-SCMC solution of pH 0.5, in: A.S. Myerson, D.A. Green, P. Meenan (Eds.), Crystal Growth of Organic Materials, American Chemical Society, 1996, pp. 72–77.
[35] D.J. Kirwan, I.B. Feins, A.J. Mahajan, Crystal growth kinetics of complex organic compounds, in: A.S. Myerson, D.A. Green, P. Meenan (Eds.), Crystal Growth of Organic Materials, American Chemical Society, 1996, pp. 116–121.
[36] C. Jacobsen, J. Garside, M. Hoare, Nucleation and growth of microbial lipase crystals from clarified concentrated fermentation broths, Biotechnol. Bioeng. 57 (1998) 666–675.
[37] E. Forsythe, M.L. Pusey, The effects of temperature and NaCl concentration on tetragonal lysozyme face growth rates, J. Crystal Growth 139 (1994) 89–94.
[38] R.A. Judge, M.R. Johns, E.T. White, Protein purification by bulk crystallization: the recovery of ovalbumin, Biotechnol. Bioeng. 48 (1995) 316–323.
[39] W.L. McCabe, Crystal growth in aqueous solutions, parts I and II, Ind. Eng. Chem. 21 (1929) 30–33,112–119.
[40] K. Yamada, S. Kinoshita, T. Tsunoda, K. Aida (Eds.), Microbial Production of Amino Acids, Kodansha, 1972.
[41] Perry's Chemical Engineers' Handbook, eighth ed., McGraw-Hill, 2008, pp. 18-39–18-59.
[42] Perry's Chemical Engineers' Handbook, Section 12, eighth ed., McGraw-Hill, 2008.
[43] C.J. Geankoplis, Transport Processes and Separation Process Principles, fourth ed., Prentice Hall, 2003 (Chapter 9).
[44] W.L. McCabe, J.C. Smith, P. Harriott, Unit Operations of Chemical Engineering, seventh ed., McGraw-Hill, 2005 (Chapter 24).
[45] J.W. Snowman, Lyophilization, in: M.S. Verrall (Ed.), Downstream Processing of Natural Products, John Wiley, 1996, pp. 275–299.

Suggestions for Further Reading

There is an extensive literature on downstream processing of biomolecules. The following is a small selection.

Downstream Processing

See also references [2] and [3].

Asenjo, J. A. (Ed.), (1990). *Separation Processes in Biotechnology*. Marcel Dekker.
Atkinson, B., & Mavituna, F. (1991). *Biochemical Engineering and Biotechnology Handbook* (2nd ed., Chapters 16 and 17). Macmillan.
Desai, M. A. (Ed.), (2010). *Downstream Processing of Proteins: Methods and Protocols*. Humana Press.

Goldberg, E. (Ed.), (1997). *Handbook of Downstream Processing*. Chapman and Hall.

Harrison, R. G., Todd, P., Rudge, S. R., & Petrides, D. P. (2003). *Bioseparations Science and Engineering*. Oxford University Press.

Ladisch, M. R. (2001). *Bioseparations Engineering: Principles, Practice, and Economics*. John Wiley.

Verrall, M. S. (Ed.), (1996). *Downstream Processing of Natural Products: A Practical Handbook*. John Wiley.

Filtration

See also reference [2].

McCabe, W. L., Smith, J. C., & Harriott, P. (2005). *Unit Operations of Chemical Engineering* (7th ed., pp. 1006–1033). McGraw-Hill.

Oolman, T., & Liu, T. -C. (1991). Filtration properties of mycelial microbial broths. *Biotechnol. Prog.*, *7*, 534–539.

Richardson, J. F., & Harker, J. H. (2002). *Coulson and Richardson's Chemical Engineering*, vol. 2 (5th ed., Chapter 7). Butterworth-Heinemann.

Wakeman, R. J., & Tarleton, E. S. (1999). *Filtration*. Elsevier Science, Oxford.

Centrifugation

See also references [2] and [4].

Axelsson, H. A. C. (1985). Centrifugation. In M. Moo-Young (Ed.), *Comprehensive Biotechnology*, vol. 2 (pp. 325–346). Pergamon Press.

Leung, W. W.-F. (2007). *Centrifugal Separations in Biotechnology*. Academic Press.

Richardson, J. F., & Harker, J. H. (2002). *Coulson and Richardson's Chemical Engineering*, vol. 2 (5th ed., Chapter 9). Butterworth-Heinemann.

Cell Disruption

See also references [7] through [12].

Chisti, Y., & Moo-Young, M. (1986). Disruption of microbial cells for intracellular products. *Enzyme Microb. Technol.*, *8*, 194–204.

Engler, C. R. (1985). Disruption of microbial cells. In M. Moo-Young (Ed.), *Comprehensive Biotechnology*, vol. 2 (pp. 305–324). Pergamon Press.

Kula, M.-R., & Schütte, H. (1987). Purification of proteins and the disruption of microbial cells. *Biotechnol. Prog.*, *3*, 31–42.

Lander, R., Manger, W., Scouloudi, M., Ku, A., Davis, C., & Lee, A. (2000). Gaulin homogenization: a mechanistic study. *Biotechnol. Prog.*, *16*, 80–85.

Middelberg, A. P. J. (1995). Process-scale disruption of microorganisms. *Biotechnol. Adv.*, *13*, 491–551.

Aqueous Two-Phase Liquid Extraction

Albertsson, P.-Å. (1971). *Partition of Cell Particles and Macromolecules* (2nd ed.). John Wiley.

Diamond, A. D., & Hsu, J. T. (1992). Aqueous two-phase systems for biomolecule separation. *Adv. Biochem. Eng./Biotechnol.*, *47*, 89–135.

Kula, M.-R. (1985). Liquid–liquid extraction of biopolymers. In M. Moo-Young (Ed.), *Comprehensive Biotechnology* vol. 2 (pp. 451–471). Pergamon Press.

Precipitation

See also references [14] and [15].

Bell, D. J., Hoare, M., & Dunnill, P. (1983). The formation of protein precipitates and their centrifugal recovery. *Adv. Biochem. Eng./Biotechnol.*, *26*, 1–71.

Ladisch, M. R. (2001). *Bioseparations Engineering: Principles, Practice, and Economics* (Chapter 4). John Wiley.

Niederauer, M. Q., & Glatz, C. E. (1992). Selective precipitation. *Adv. Biochem. Eng./Biotechnol.*, *47*, 159–188.

Rothstein, F. (1994). Differential precipitation of proteins. In R. G. Harrison (Ed.), *Protein Purification Process Engineering* (pp. 115–208). Marcel Dekker.

Adsorption

See also references [2] and [15].

Arnold, F. H., Blanch, H. W., & Wilke, C. R. (1985). Analysis of affinity separations. I. Predicting the performance of affinity adsorbers. *Chem. Eng. J.*, *30*, B9–B23.

Hines, A. L., & Maddox, R. N. (1985). *Mass Transfer: Fundamentals and Applications*. Prentice Hall.

Slejko, F. L. (Ed.), (1985). *Adsorption Technology*. Marcel Dekker.

Membrane Filtration

See also references [18] through [23].

Cardew, P. T., & Le, M. S. (1998). *Membrane Processes: A Technology Guide*. Royal Society of Chemistry.

Dosmar, M., & Brose, D. (1998). Crossflow ultrafiltration. In T. H. Meltzer, & M. W. Jornitz (Eds.), *Filtration in the Biopharmaceutical Industry* (pp. 493–532). Marcel Dekker.

Porter, M. C. (1997). Membrane filtration. In P. A. Schweitzer (Ed.), *Handbook of Separation Techniques for Chemical Engineers* (3rd ed., pp. 2-3–2-101). McGraw-Hill.

van Reis, R., & Zydney, A. L. (1999). Protein ultrafiltration. In M. C. Flickinger, & S. W. Drew (Eds.), *Encyclopedia of Bioprocess Technology: Fermentation, Biocatalysis, and Bioseparation* (pp. 2197–2214). John Wiley.

Chromatography

See also references [26] and [27].

Chisti, Y., & Moo-Young, M. (1990). Large scale protein separations: engineering aspects of chromatography. *Biotechnol. Adv.*, *8*, 699–708.

Delaney, R. A. M. (1980). Industrial gel filtration of proteins. In R. A. Grant (Ed.), *Applied Protein Chemistry* (pp. 233–280). Applied Science.

Janson, J. -C., & Hedman, P. (1982). Large-scale chromatography of proteins. *Adv. Biochem. Eng.*, *25*, 43–99.

Ladisch, M. R. (2001). *Bioseparations Engineering: Principles, Practice, and Economics* (Chapters 5–9). John Wiley.

Robinson, P. J., Wheatley, M. A., Janson, J.-C., Dunnill, P., & Lilly, M. D. (1974). Pilot scale affinity chromatography: purification of β-galactosidase. *Biotechnol. Bioeng.*, *16*, 1103–1112.

Verrall, M. S. (Ed.), (1996). *Downstream Processing of Natural Products: A Practical Handbook* (Chapters 12–15). John Wiley.

Wheelwright, S. M. (1991). *Protein Purification: Design and Scale-Up of Downstream Processing*. John Wiley.

Crystallisation

See also references [28] through [31].

Ladisch, M. R. (2001). *Bioseparations Engineering: Principles, Practice, and Economics* (Chapter 4). John Wiley.

Richardson, J. F., & Harker, J. H. (2002). *Coulson and Richardson's Chemical Engineering*, vol. 2 (5th ed., Chapter 15). Butterworth-Heinemann.

Drying

See also references [42] through [45].

Gatlin, L. A., & Nail, S. L. (1994). Freeze drying: a practical overview. In R. G. Harrison (Ed.), *Protein Purification Process Engineering* (pp. 317–367). Marcel Dekker.

Harrison, R. G., Todd, P., Rudge, S. R., & Petrides, D. P. (2003). *Bioseparations Science and Engineering* (Chapter 10). Oxford University Press.

PART 4

REACTIONS AND REACTORS

CHAPTER

12

Homogeneous Reactions

The heart of a typical bioprocess is the reactor or fermenter. Flanked by unit operations that carry out physical changes for medium preparation and recovery of products, the reactor is where the major chemical and biochemical transformations occur. In many bioprocesses, characteristics of the reaction determine to a large extent the economic feasibility of the project.

Of most interest in biological systems are *catalytic* reactions. By definition, a catalyst is a substance that affects the rate of reaction without altering the reaction equilibrium or undergoing permanent change itself. Enzymes, enzyme complexes, cell organelles, and whole cells perform catalytic roles; the latter may be viable or nonviable, growing or nongrowing. Biocatalysts can be of microbial, plant, or animal origin. Cell growth is an *autocatalytic reaction*: this means that the catalyst is a product of the reaction. Properties such as the reaction rate and yield of product from substrate characterise the performance of catalytic reactions. Knowledge of these parameters is crucial for the design and operation of reactors.

In engineering analysis of catalytic reactions, a distinction is made between *homogeneous* and *heterogeneous* reactions. A reaction is homogeneous if the temperature and all concentrations in the system are uniform. Most enzyme reactions and fermentations carried out in mixed vessels fall into this category. In contrast, heterogeneous reactions take place in the presence of concentration or temperature gradients. Analysis of heterogeneous reactions requires application of mass transfer principles in conjunction with reaction theory. Heterogeneous reactions are the subject of Chapter 13.

Here we consider the basic aspects of reaction theory that allow us to quantify the extent and speed of homogeneous reactions, and to identify the important factors affecting reaction rate.

12.1 BASIC REACTION THEORY

Reaction theory has two fundamental parts: *reaction thermodynamics* and *reaction kinetics*. Reaction thermodynamics is concerned with *how far* the reaction can proceed: no matter how fast a reaction is, it cannot continue beyond the point of chemical equilibrium. On the other hand, reaction kinetics is concerned with the *rate* at which reactions occur.

12.1.1 Reaction Thermodynamics

Consider a reversible reaction represented by the following equation:

$$A + bB \rightleftharpoons yY + zZ \tag{12.1}$$

A, B, Y, and Z are chemical species; b, y, and z are stoichiometric coefficients. If the components are left in a closed system, the reaction proceeds until *thermodynamic equilibrium* is reached. At equilibrium, there is no net driving force for further change: the reaction has reached the limit of its capacity for chemical transformation in the absence of perturbation. The equilibrium concentrations of reactants and products are related by the *equilibrium constant*, K_{eq}. For the reaction of Eq. (12.1):

$$K_{eq} = \frac{[Y]_e^{\,y}\ [Z]_e^{\,z}}{[A]_e\ [B]_e^{\,b}} \tag{12.2}$$

where $[\ ]_e$ denotes the molar concentration (gmol l^{-1}) of A, B, Y, or Z at equilibrium. In aqueous systems, if water, H^+ ions, or solid substances are involved in the reaction, these components are not included when applying Eq. (12.2).

The value of K_{eq} is greater when equilibrium favours the products of reaction and smaller when equilibrium favours the reactants. K_{eq} varies with temperature as follows:

$$\ln K_{eq} = \frac{-\Delta G^{\circ}_{rxn}}{RT} \tag{12.3}$$

where ΔG°_{rxn} is the *change in standard free energy* per mole of A reacted, R is the ideal gas constant, and T is absolute temperature. Values of R are listed in Appendix B. The superscript $^{\circ}$ in ΔG°_{rxn} indicates standard conditions. Usually, the standard condition for a substance is its most stable form at 1 atm pressure and 25°C; however, for biochemical reactions occurring in solution, other standard conditions may be used [1]. ΔG°_{rxn} is equal to the difference between the standard free energies of formation of the products and reactants:

$$\Delta G^{\circ}_{rxn} = y\,G^{\circ}_{Y} + z\,G^{\circ}_{Z} - G^{\circ}_{A} - b\,G^{\circ}_{B} \tag{12.4}$$

where G° is the *standard free energy of formation*. Values of G° are available in handbooks such as those listed in Section 2.6.

Free energy G is related to enthalpy H, entropy S, and absolute temperature T as follows:

$$\Delta G = \Delta H - T\Delta S \tag{12.5}$$

Therefore, from Eq. (12.3):

$$\ln K_{eq} = \frac{-\Delta H^{\circ}_{rxn}}{RT} + \frac{\Delta S^{\circ}_{rxn}}{R} \tag{12.6}$$

Thus, for exothermic reactions with negative ΔH°_{rxn} (Section 5.8), K_{eq} decreases with increasing temperature. For endothermic reactions and positive ΔH°_{rxn}, K_{eq} increases with increasing temperature.

EXAMPLE 12.1 EFFECT OF TEMPERATURE ON GLUCOSE ISOMERISATION

Glucose isomerase is used extensively in the United States for production of high-fructose syrup. The reaction is:

$$\text{glucose} \rightleftharpoons \text{fructose}$$

ΔH°_{rxn} for this reaction is 5.73 kJ gmol^{-1}; ΔS°_{rxn} is 0.0176 kJ gmol^{-1} K^{-1}.

(a) Calculate the equilibrium constants at 50°C and 75°C.

(b) A company aims to develop a sweeter mixture of sugars, that is, one with a higher concentration of fructose. Considering equilibrium only, would it be more desirable to operate the reaction at 50°C or 75°C?

Solution

(a) Convert the temperatures from degrees Celsius to Kelvin using the formula of Eq. (2.27):

$$T = 50°\text{C} = 323.15\ \text{K}$$

$$T = 75°\text{C} = 348.15\ \text{K}$$

From Appendix B, $R = 8.3144\ \text{J gmol}^{-1}\ \text{K}^{-1} = 8.3144 \times 10^{-3}\ \text{kJ gmol}^{-1}\ \text{K}^{-1}$. Using Eq. (12.6):

$$\ln K_{eq}\,(50°\text{C}) = \frac{-5.73\ \text{kJ gmol}^{-1}}{(8.3144 \times 10^{-3}\ \text{kJ gmol}^{-1}\ \text{K}^{-1})\ 323.15\ \text{K}} + \frac{0.0176\ \text{kJ gmol}^{-1}\ \text{K}^{-1}}{8.3144 \times 10^{-3}\ \text{kJ gmol}^{-1}\ \text{K}^{-1}}$$

$$K_{eq}\,(50°\text{C}) = 0.98$$

Similarly for $T = 75°\text{C}$:

$$\ln K_{eq}\,(75°\text{C}) = \frac{-5.73\ \text{kJ gmol}^{-1}}{(8.3144 \times 10^{-3}\ \text{kJ gmol}^{-1}\ \text{K}^{-1})\ 348.15\ \text{K}} + \frac{0.0176\ \text{kJ gmol}^{-1}\ \text{K}^{-1}}{8.3144 \times 10^{-3}\ \text{kJ gmol}^{-1}\ \text{K}^{-1}}$$

$$K_{eq}\,(75°\text{C}) = 1.15$$

(b) As K_{eq} increases, the fraction of fructose in the equilibrium mixture increases. Therefore, from an equilibrium point of view, it is more desirable to operate the reactor at 75°C. However, other factors such as enzyme deactivation at high temperatures should also be considered.

A limited number of commercially important enzyme conversions, such as glucose isomerisation and starch hydrolysis, are treated as reversible reactions. In these systems, the reaction mixture at equilibrium contains significant amounts of reactants as well as products. However, for many reactions, ΔG°_{rxn} is negative and large in magnitude. As a result, K_{eq} is also very large, the reaction strongly favours the products rather than the reactants, and the reaction is regarded as *irreversible*.

Most industrial enzyme and cell conversions fall into this category. For example, the equilibrium constant for sucrose hydrolysis by invertase is about 10^4; for fermentation of glucose to ethanol and carbon dioxide, K_{eq} is about 10^{30}. The equilibrium ratio of products to reactants is so overwhelmingly large for these reactions that they are considered to proceed to completion (i.e., the reaction stops only when the concentration of one of the reactants falls to zero). Equilibrium thermodynamics therefore has only limited application in industrial enzyme and cell reactions. Moreover, the thermodynamic principles outlined in this section apply only to closed systems; true thermodynamic equilibrium does not exist in living cells that exchange matter with their surroundings. Metabolic processes exist in cells in a dynamic state: the products formed are continually removed or broken down so that the reactions are driven forward. Most reactions in large-scale biological processes proceed to completion within a finite period of time.

If we know that complete conversion will eventually take place, the most useful reaction parameter is the rate at which the transformation proceeds. In addition, because reaction networks in cells are complex and involve many branched and alternative pathways, another important characteristic is the proportion of the reactant consumed that is channelled into synthesis of the desired products. These properties of reactions are discussed in the remainder of this chapter.

12.1.2 Reaction Yield

The extent to which reactants are converted to products is expressed as the reaction *yield*. Generally speaking, yield is the amount of product formed or accumulated per amount of reactant provided or consumed. Unfortunately, there is no strict definition of yield; several different yield parameters are applicable in different situations. The terms used to express yield in this text do not necessarily have universal acceptance and are defined here for our convenience. Be prepared for other books to use different definitions.

Consider the simple enzyme reaction:

$$\text{L-histidine} \longrightarrow \text{urocanic acid} + NH_3 \tag{12.7}$$

catalysed by histidase. According to the reaction stoichiometry, 1 gmol of urocanic acid is produced for each gmol of L-histidine consumed; the yield of urocanic acid from histidine is therefore 1 gmol gmol^{-1}. However, let us assume that the histidase used in this reaction is contaminated with another enzyme, histidine decarboxylase. Histidine decarboxylase catalyses the following reaction:

$$\text{L-histidine} \longrightarrow \text{histamine} + CO_2 \tag{12.8}$$

If both enzymes are active, some L-histidine will react with histidase according to Eq. (12.7), while some will be decarboxylated according to Eq. (12.8). After addition of the enzymes to the substrate, analysis of the reaction mixture shows that 1 gmol of urocanic acid and 1 gmol of histamine are produced for every 2 gmol of histidine consumed. The *observed* or *apparent yield* of urocanic acid from L-histidine is 1 gmol/2 gmol = 0.5 gmol gmol^{-1}. The observed yield of 0.5 gmol gmol^{-1} is different from the *stoichiometric, true*, or *theoretical yield* of 1 gmol gmol^{-1} calculated from the reaction stoichiometry of Eq. (12.7) because the reactant was channelled into two separate reaction pathways. An analogous situation arises if product rather that reactant is consumed in other reactions; in this case, the observed yield of product is lower than the theoretical yield. *When reactants or products are involved in additional reactions, the observed yield may be different from the theoretical yield.*

This analysis leads to two useful definitions of yield for reaction systems:

$$\begin{pmatrix}\text{true, stoichiometric, or}\\ \text{theoretical yield}\end{pmatrix} = \frac{\begin{pmatrix}\text{mass or moles of}\\ \text{product formed}\end{pmatrix}}{\begin{pmatrix}\text{mass or moles of reactant used}\\ \text{to form that particular product}\end{pmatrix}} \tag{12.9}$$

and

$$(\text{observed or apparent yield}) = \frac{(\text{mass or moles of product present})}{(\text{mass or moles of reactant consumed})} \tag{12.10}$$

There is a third type of yield applicable in certain situations. For reactions with incomplete conversion of reactant, it may be of interest to specify the amount of product formed per amount of reactant *provided to the reaction* rather than actually consumed. For example, consider the isomerisation reaction catalysed by glucose isomerase:

$$\text{glucose} \rightleftarrows \text{fructose} \tag{12.11}$$

The reaction is carried out in a closed reactor with pure enzyme. At equilibrium the sugar mixture contains 55 mol% glucose and 45 mol% fructose. The *theoretical yield* of fructose from glucose is 1 gmol gmol^{-1} because, from stoichiometry, formation of 1 gmol of fructose requires 1 gmol of glucose. The *observed yield* would also be 1 gmol gmol^{-1} if the reaction occurs in isolation. However, if the reaction is started with glucose only present, the equilibrium yield of fructose per gmol of glucose added to the reactor is 0.45 gmol gmol^{-1}, because a residual 0.55 gmol of glucose remains unreacted in the reaction mixture. This type of yield for incomplete reactions may be denoted *gross yield*:

$$\text{gross yield} = \frac{(\text{mass or moles of product present})}{(\text{mass or moles of reactant provided to the reaction})} \tag{12.12}$$

EXAMPLE 12.2 INCOMPLETE ENZYME REACTION

An enzyme catalyses the reaction:

$$A \rightleftarrows B$$

At equilibrium, the reaction mixture contains 6.3 g l^{-1} of A and 3.7 g l^{-1} of B.

(a) What is the equilibrium constant?
(b) If the reaction starts with A only, what is the yield of B from A at equilibrium?

Solution

(a) From stoichiometry, the molecular weights of A and B must be equal. Therefore the ratio of molar concentrations of A and B is equal to the ratio of mass concentrations. From Eq. (12.2):

$$K_{eq} = \frac{[B]_e}{[A]_e} = \frac{3.7\ \text{g l}^{-1}}{6.3\ \text{g l}^{-1}} = 0.59$$

(b) Total mass is conserved in chemical reactions. Using a basis of 1 litre, as the total mass at equilibrium is (6.3 g + 3.7 g) = 10 g, if the reactions starts with A only, 10 g of A must have been provided. As 6.3 g of A remain at equilibrium, 3.7 g of A must have been consumed during in reaction.
From stoichiometry, the true yield of B from A is 1 gmol gmol^{-1}.
The observed yield is 3.7 g of B/3.7 g of A = 1 g g^{-1} or, in this case, 1 gmol gmol^{-1}.
The gross yield is 3.7 g of B/10 g of A = 0.37 g g^{-1} or 0.37 gmol gmol^{-1}.

12.1.3 Reaction Rate

Consider the general irreversible reaction:

$$a\,\text{A} + b\,\text{B} \longrightarrow y\,\text{Y} + z\,\text{Z} \tag{12.13}$$

The rate of this reaction can be represented by the rate of conversion of compound A. Let us use the symbol R_A to denote the *rate of reaction with respect to A*. The units of R_A are, for example, kg s^{-1}.

How do we measure reaction rates? For a general reaction system that is open to the surroundings, the rate of reaction can be related to the rate of change of mass in the system using the unsteady-state mass balance equation derived in Chapter 6:

$$\frac{dM}{dt} = \hat{M}_i - \hat{M}_o + R_G - R_C \tag{6.5}$$

In Eq. (6.5), M is mass, t is time, $\hat{M}_i$ is the mass flow rate into the system, $\hat{M}_o$ is the mass flow rate out of the system, R_G is the mass rate of generation by reaction, and R_C is the mass rate of consumption by reaction. Let us apply Eq. (6.5) to compound A, assuming that the reaction of Eq. (12.13) is the only reaction taking place that involves A. The rate of consumption R_C is equal to R_A, and $R_G = 0$. The mass balance equation becomes:

$$\frac{dM_A}{dt} = \hat{M}_{Ai} - \hat{M}_{Ao} - R_A \tag{12.14}$$

Therefore, the rate of reaction R_A can be determined if we measure the rate of change in the mass of A in the system, dM_A/dt, and the rates of flow of A into and out of the system, $\hat{M}_{Ai}$ and $\hat{M}_{Ao}$. *In a closed system* where $\hat{M}_{Ai} = \hat{M}_{Ao} = 0$, Eq. (12.14) becomes:

$$R_A = \frac{-dM_A}{dt} \tag{12.15}$$

and the reaction rate can be measured simply by monitoring the rate of change in the mass of A in the system. Most measurements of reaction rate are carried out in closed systems so that the data can be analysed according to Eq. (12.15). dM_A/dt is negative when A is consumed by reaction; therefore the minus sign in Eq. (12.15) is necessary to make R_A a positive quantity. Rate of reaction is sometimes called *reaction velocity*. Reaction velocity can also be measured in terms of components B, Y, or Z. In a closed system:

$$R_B = \frac{-dM_B}{dt} \quad R_Y = \frac{dM_Y}{dt} \quad R_Z = \frac{dM_Z}{dt} \tag{12.16}$$

where M_B, M_Y, and M_Z are masses of B, Y, and Z, respectively. When reporting reaction rate, the reactant being monitored should be specified. Because R_Y and R_Z are based on product accumulation, these reaction rates are called *production rates* or *productivity*.

Reaction rates can be expressed using different measurement bases. In bioprocess engineering, there are three distinct ways of expressing reaction rate that can be applied in different situations.

1. *Total rate*. Total reaction rate is defined in Eqs. (12.15) and (12.16) and is expressed as either mass or moles per unit time. Total rate is useful for specifying the output of a particular reactor or manufacturing plant. Production rates for factories are often expressed as total rates; for example: "The production rate is 100,000 tonnes per year". If additional reactors are built so that the reaction volume in the plant is increased, then clearly the total reaction rate would increase. Similarly, if the amount of cells or enzyme used in each reactor were also increased, then the total production rate would be improved even further.
2. *Volumetric rate*. Because the total mass of reactant converted in a reaction mixture depends on the size of the system, it is often convenient to specify the reaction rate as a rate per unit volume. Units of volumetric rate are, for example, kg m^{-3} s^{-1}. The rate of reaction expressed on a volumetric basis is used to account for differences in volume between reaction systems. Therefore, if the reaction mixture in a closed system has volume V:

$$r_A = \frac{R_A}{V} = \frac{-1}{V}\frac{dM_A}{dt} \tag{12.17}$$

where r_A is the volumetric rate of reaction with respect to A. When V is constant, Eq. (12.17) can be written:

$$r_A = \frac{-dC_A}{dt} \tag{12.18}$$

where C_A is the concentration of A in units of, for example, kg m^{-3}. Volumetric rates are particularly useful for comparing the performance of reactors of different size. A common objective in optimising reaction processes is to maximise the volumetric productivity so that the desired total production rate can be achieved using reactors of minimum size and therefore minimum cost.
3. *Specific rate*. Biological reactions involve enzyme and cell catalysts. Because the total rate of conversion depends on the amount of catalyst present, it is sometimes useful to

specify reaction rate as the rate per quantity of enzyme or cells involved in the reaction. In a closed system, the specific reaction rate is:

$$r_A = -\left(\frac{1}{E} \text{ or } \frac{1}{X}\right)\frac{dM_A}{dt} \tag{12.19}$$

where r_A is the specific rate of reaction with respect to A, E is the quantity of enzyme, X is the quantity of cells, and dM_A/dt is the rate of change of the mass of A in the system. As the quantity of cells is usually expressed as mass, the units of specific rate for a cell-catalysed reaction would be, for example, kg (kg cells)$^{-1}$ s^{-1} or simply s^{-1}. On the other hand, the mass of a particular enzyme added to a reaction is rarely known; most commercial enzyme preparations contain several components in unknown and variable proportions depending on the batch obtained from the manufacturer. To overcome these difficulties, enzyme quantity is often expressed as *units of activity* measured under specified conditions. One unit of enzyme is usually taken to be the amount that catalyses the conversion of 1 μmole of substrate per minute at the optimal temperature, pH, and substrate concentration. Therefore, if E in Eq. (12.19) is expressed as units of enzyme activity, the specific rate of reaction could be reported as, for example, kg (unit enzyme)$^{-1}$ s^{-1}. In a closed system where the volume of the reaction mixture remains constant, an alternative expression for the specific reaction rate is:

$$r_A = -\left(\frac{1}{e} \text{ or } \frac{1}{x}\right)\frac{dC_A}{dt} \tag{12.20}$$

where e is enzyme concentration and x is cell concentration.

Volumetric and total rates are not a direct reflection of catalyst performance; this is represented by the specific rate. Specific rates are employed when comparing different enzymes or cells. Under most circumstances, the specific rate is not dependent on the size of the system or the amount of catalyst present. Some care is necessary when interpreting results for reaction rate. For example, if two fermentations are carried out with different cell lines and the volumetric rate of reaction is greater in the first fermentation than in the second, you should not jump to the conclusion that the cell line in the first experiment is 'better', or capable of greater metabolic activity. It could be that the faster volumetric rate is due to the first fermenter being operated at a higher cell density than the second, leading to measurement of a more rapid rate per unit volume. Different strains of organism should be compared in terms of their specific reaction rates, not in terms of the volumetric or total rate.

Total, volumetric, and specific productivities are interrelated concepts in process design. For example, high total productivity could be achieved with a catalyst of low specific activity if the reactor is loaded with a high catalyst concentration. If this is not possible, the volumetric productivity will be relatively low and a larger reactor is required to achieve the desired total productivity. In this book, the symbol R_A will be used to denote total reaction rate with respect to component A; r_A represents either volumetric or specific rate.

12.1.4 Reaction Kinetics

As reactions proceed in closed systems, the concentrations of the reactants decrease. In general, the rate of reaction depends on reactant concentration; therefore, the rate of conversion usually decreases during reactions. Reaction rates also vary with temperature; most reactions speed up considerably as the temperature rises. *Reaction kinetics* refers to the relationship between the rate of reaction and the conditions that affect it, such as reactant concentration and temperature. These relationships are conveniently described using *kinetic expressions* or *kinetic equations.*

Consider again the general irreversible reaction of Eq. (12.13). The volumetric rate of this reaction can be expressed as a function of reactant concentration using the following mathematical form:

$$r_A = k\, C_A^{\alpha}\, C_B^{\beta} \tag{12.21}$$

where k is the *rate constant* or *rate coefficient* for the reaction, C_A is the concentration of reactant A, and C_B is the concentration of reactant B. By definition, the rate constant is independent of the concentrations of the reacting species but is dependent on other variables that influence reaction rate, such as temperature. When the kinetic equation has the form of Eq. (12.21), the reaction is said to be of *order* α with respect to component A and order β with respect to B. The order of the overall reaction is $(\alpha + \beta)$. It is not usually possible to predict the order of reactions from stoichiometry. The mechanism of single reactions and the functional form of the kinetic expression must be determined by experiment. The dimensions and units of k depend on the order of the reaction.

12.1.5 Effect of Temperature on Reaction Rate

Temperature has a significant kinetic effect on reactions. Variation of the rate constant k with temperature is described by the *Arrhenius equation*:

$$k = A\, e^{-E/RT} \tag{12.22}$$

where k is the rate constant, A is the *Arrhenius constant* or *frequency factor*, E is the *activation energy* for the reaction, R is the ideal gas constant, and T is absolute temperature. Values of R are listed in Appendix B. For many reactions, the value of E is positive and large so that, as T increases, k increases very rapidly. Taking the natural logarithm of both sides of Eq. (12.22):

$$\ln k = \ln A - \frac{E}{RT} \tag{12.23}$$

Therefore, a plot of $\ln k$ versus $1/T$ gives a straight line with slope $-E/R$.

12.2 CALCULATION OF REACTION RATES FROM EXPERIMENTAL DATA

As outlined in Section 12.1.3, the volumetric rate of reaction in a closed system can be found by measuring the rate of change in the mass of reactant present, provided the reactant is involved in only one reaction. Most kinetic studies of biological reactions are carried out in closed systems with a constant volume of reaction mixture; therefore, Eq. (12.18) can be used to evaluate the volumetric reaction rate. The concentration of a

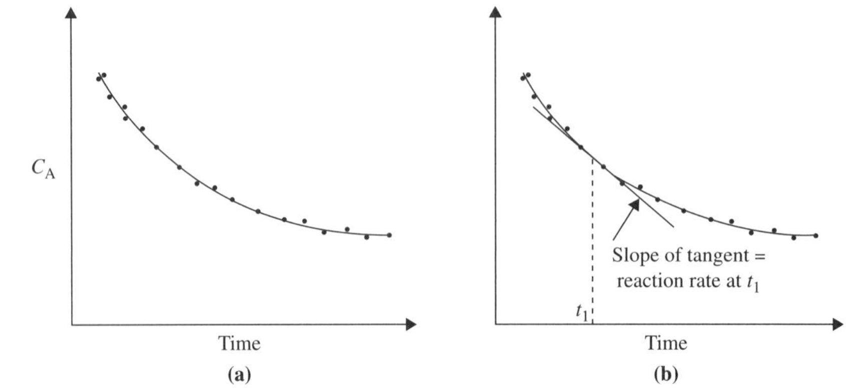

FIGURE 12.1 (a) Change in reactant concentration with time during reaction. (b) Graphical differentiation of concentration data by drawing a tangent.

particular reactant or product is measured as a function of time. For a reactant such as A in Eq. (12.13), the results will be similar to those shown in Figure 12.1(a): the concentration will decrease with time. The volumetric rate of reaction is equal to dC_A/dt, which can be evaluated as the slope of a smooth curve drawn through the data. The slope of the curve in Figure 12.1(a) changes with time; the reaction rate is greater at the beginning of the experiment than at the end.

One obvious way to determine reaction rate is to draw tangents to the curve of Figure 12.1(a) at various times and evaluate the slopes of the tangents; this is shown in Figure 12.1(b). If you have ever attempted this you will know that, although correct in principle, it can be an extremely difficult exercise. Drawing tangents to curves is a highly subjective procedure prone to great inaccuracy, even with special drawing tools designed for the purpose. The results depend strongly on the way the data are smoothed and the appearance of the curve at the points chosen. More reliable techniques are available for *graphical differentiation* of rate data. Graphical differentiation is valid only if the data can be presumed to differentiate smoothly.

12.2.1 Average Rate–Equal Area Method

This technique for determining rates is based on the *average rate–equal area construction* and will be demonstrated using data for oxygen uptake by cells in suspension culture. Results from the measurement of oxygen concentration in a closed system as a function of time are listed in the first two columns of Table 12.1. The average rate–equal area method involves the following steps.

- Tabulate values of ΔC_A and Δt for each time interval as shown in Table 12.1. ΔC_A values are negative because C_A decreases over each interval.
- Calculate the average oxygen uptake rate, $\Delta C_A/\Delta t$, for each time interval.
- Plot $\Delta C_A/\Delta t$ on linear graph paper. Over each time interval, draw a horizontal line to represent $\Delta C_A/\Delta t$ for that interval; this is shown in Figure 12.2.

TABLE 12.1 Graphical Differentiation Using the Average Rate–Equal Area Construction

Time (t, min)	Oxygen Concentration (C_A, ppm)	ΔC_A	Δt	$\Delta C_A/\Delta t$	dC_A/dt
0.0	8.00				−0.59
		−0.45	1.0	−0.45	
1.0	7.55				−0.38
		−0.33	1.0	−0.33	
2.0	7.22				−0.29
		−0.26	1.0	−0.26	
3.0	6.96				−0.23
		−0.20	1.0	−0.20	
4.0	6.76				−0.18
		−0.15	1.0	−0.15	
5.0	6.61				−0.14
		−0.12	1.0	−0.12	
6.0	6.49				−0.11
		−0.16	2.0	−0.08	
8.0	6.33				−0.06
		−0.08	2.0	−0.04	
10.0	6.25				−0.02

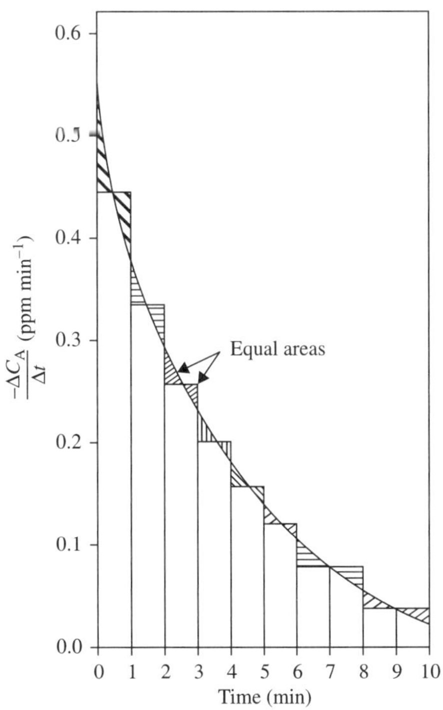

FIGURE 12.2 Graphical differentiation using the average rate–equal area construction.

- Draw a smooth curve to cut the horizontal lines in such a manner that the shaded areas above and below the curve are equal for each time interval. The curve thus developed gives values of dC_A/dt for all points in time. Results for dC_A/dt at the times of sampling can be read from the curve and are tabulated in Table 12.1.

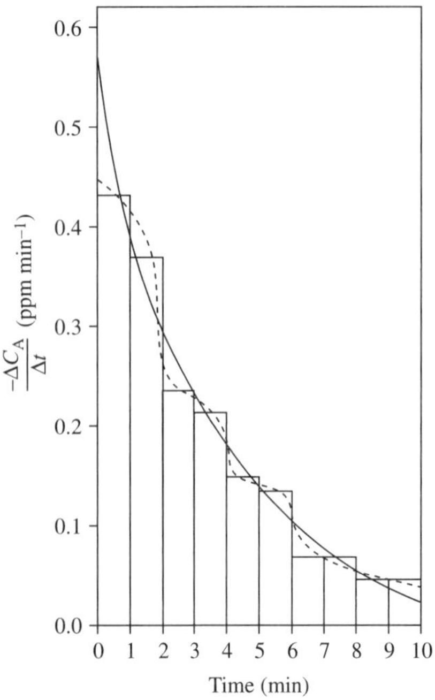

FIGURE 12.3 Average rate—equal area method for data with experimental error.

A disadvantage of the average rate—equal area method is that it is not easily applied if the data show scatter. If the concentration measurements were not very accurate, the horizontal lines representing $\Delta C_A/\Delta t$ might be located as shown in Figure 12.3. A curve equalising areas above and below each $\Delta C_A/\Delta t$ line would then show complex behaviour as indicated by the dashed line in Figure 12.3. Experience suggests that this is not a realistic representation of reaction rate. Because of the inaccuracies in measured data, we need several concentration measurements to define a change in rate. The data of Figure 12.3 are better represented using a smooth curve to equalise as far as possible the areas above and below adjacent groups of horizontal lines. For data showing even greater scatter, it may be necessary to average consecutive pairs of $\Delta C_A/\Delta t$ values to simplify the graphical analysis.

A second graphical differentiation technique for evaluating dC_A/dt is described in the following section.

12.2.2 Midpoint Slope Method

In this method, the raw data are smoothed and values tabulated at intervals. The midpoint slope method is illustrated using the same data as that analysed in Section 12.2.1.

- Plot the raw concentration data as a function of time and draw a smooth curve through the points. This is shown in Figure 12.4.
- Mark off the smoothed curve at time intervals of ε. ε should be chosen so that the number of intervals is less than the number of datum points measured; the less accurate the data, the fewer should be the intervals. In this example, ε is taken as 1.0 min until

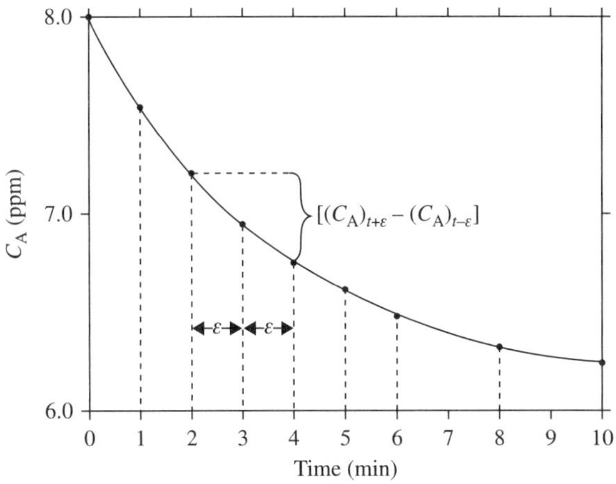

FIGURE 12.4 Graphical differentiation using the midpoint slope method.

TABLE 12.2 Graphical Differentiation Using the Midpoint Slope Method

Time (t, min)	Oxygen Concentration (C_A, ppm)	ε	$[(C_A)_{t+\varepsilon} - (C_A)_{t-\varepsilon}]$	dC_A/dt
0.0	8.00	1.0	–	–
1.0	7.55	1.0	−0.78	−0.39
2.0	7.22	1.0	−0.59	−0.30
3.0	6.96	1.0	−0.46	−0.23
4.0	6.76	1.0	−0.35	−0.18
5.0	6.61	1.0	−0.27	−0.14
6.0	6.49	1.0	−0.22	−0.11
8.0	6.33	2.0	−0.24	−0.06
10.0	6.25	2.0	–	–

$t = 6$ min; thereafter $\varepsilon = 2.0$ min. The intervals are marked in Figure 12.4 as dashed lines. Values of ε are entered in Table 12.2.

- In the midpoint slope method, rates are calculated midway between two adjacent intervals of size ε. Therefore, the first rate determination is made for $t = 1$ min. Calculate the differences $[(C_A)_{t+\varepsilon} - (C_A)_{t-\varepsilon}]$ from Figure 12.4, where $(C_A)_{t+\varepsilon}$ denotes the concentration of A at time $t + \varepsilon$, and $(C_A)_{t-\varepsilon}$ denotes the concentration at time $t - \varepsilon$. A difference calculation is illustrated in Figure 12.4 for $t = 3$ min. Note that the concentrations are not taken from the list of original data but are read from the smoothed curve through the points. When $t = 6$ min, $\varepsilon = 1.0$; concentrations for the difference calculation are read from the curve at $t - \varepsilon = 5$ min and $t + \varepsilon = 7$ min. For the last rate determination at $t = 8$ min, $\varepsilon = 2$ and the concentrations are read from the curve at $t - \varepsilon = 6$ min and $t + \varepsilon = 10$ min.

- Determine the slope or rate using the central-difference formula:

$$\frac{dC_A}{dt} = \frac{[(C_A)_{t+\varepsilon} - (C_A)_{t-\varepsilon}]}{2\varepsilon} \tag{12.24}$$

The results are listed in Table 12.2.

The values of dC_A/dt calculated using the two differentiation methods (Tables 12.1 and 12.2) compare favourably. Application of both methods using the same data allows checking of the results.

12.3 GENERAL REACTION KINETICS FOR BIOLOGICAL SYSTEMS

The kinetics of many biological reactions are either zero-order, first-order, or a combination of these called Michaelis–Menten kinetics. Kinetic expressions for biological systems are examined in this section.

12.3.1 Zero-Order Kinetics

If a reaction obeys zero-order kinetics, the reaction rate is independent of reactant concentration. The kinetic expression is:

$$r_A = k_0 \tag{12.25}$$

where r_A is the volumetric rate of reaction with respect to component A and k_0 is the *zero-order rate constant*. k_0 as defined in Eq. (12.25) is a volumetric rate constant with units of, for example, mol m^{-3} s^{-1}. Because the volumetric rate of a catalytic reaction depends on the amount of catalyst present, when Eq. (12.25) is used to represent the rate of an enzyme or cell reaction, the value of k_0 includes the effect of catalyst concentration as well as the specific rate of reaction. We could write:

$$k_0 = k_0' \, e \quad \text{or} \quad k_0 = k_0'' \, x \tag{12.26}$$

where k_0' is the specific zero-order rate constant for enzyme reaction and e is the concentration of enzyme. Correspondingly, k_0'' is the specific zero-order rate constant for cell reaction and x is the cell concentration.

Let us assume that we have measured concentration data for reactant A as a function of time and wish to determine the appropriate kinetic constant for the reaction. The rate of reaction can be evaluated as the rate of change of C_A using the methods for graphical differentiation described in Section 12.2. Once r_A is found, if the reaction is zero-order, r_A will be constant and equal to k_0 at all times during the reaction, thus allowing k_0 to be determined. However, as an alternative approach, because the kinetic expression for zero-order reaction is relatively simple, rather than differentiate the concentration data it is easier to integrate the rate equation to obtain an equation for C_A as a function of time. The experimental data can then be checked against the integrated equation. Combining

Eqs. (12.18) and (12.25), the rate equation for a zero-order reaction in a closed, constant-volume system is:

$$\frac{-dC_A}{dt} = k_0 \tag{12.27}$$

Separating variables and integrating with initial condition $C_A = C_{A0}$ at $t = 0$ gives:

$$\int dC_A = \int -k_0\,dt \tag{12.28}$$

or

$$C_A - C_{A0} = -k_0 t \tag{12.29}$$

Rearranging gives an equation for C_A:

$$C_A = C_{A0} - k_0 t \tag{12.30}$$

Therefore, when the reaction is zero-order, a plot of C_A versus time gives a straight line with slope $-k_0$. Application of Eq. (12.30) is illustrated in Example 12.3.

EXAMPLE 12.3 KINETICS OF OXYGEN UPTAKE

Serratia marcescens is cultured in minimal medium in a small stirred fermenter. Oxygen consumption is measured at a cell concentration of 22.7 g l^{-1} dry weight.

Time (min)	Oxygen Concentration (mmol l^{-1})
0	0.25
2	0.23
5	0.21
8	0.20
10	0.18
12	0.16
15	0.15

(a) Determine the rate constant for oxygen uptake.
(b) If the cell concentration is reduced to 12 g l^{-1}, what is the value of the rate constant?

Solution

(a) As outlined in Section 10.5.1, microbial oxygen consumption is a zero-order reaction over a wide range of oxygen concentrations above C_{crit}. To test if the measured data are consistent with the zero-order kinetic model of Eq. (12.30), oxygen concentration is plotted as a function of time as shown in Figure 12.5.

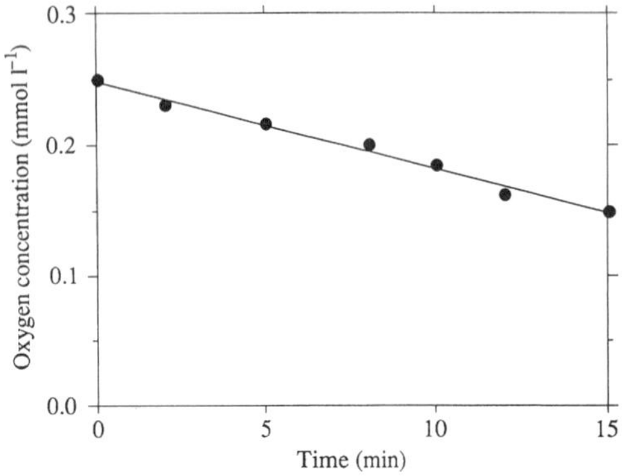

FIGURE 12.5 Kinetic analysis of oxygen uptake.

A zero-order model using a straight line to connect the points fits the data well. The slope is -6.7×10^{-3} mmol l^{-1} min^{-1}; therefore, $k_0 = 6.7 \times 10^{-3}$ mmol l^{-1} min^{-1}.

(b) For the same cells cultured under the same conditions, from Eq. (12.26), k_0 is directly proportional to the number of cells present. Therefore, at a cell concentration of 12 g l^{-1}:

$$k_0 = \frac{12 \text{ g l}^{-1}}{22.7 \text{ g l}^{-1}} (6.7 \times 10^{-3} \text{ mmol l}^{-1} \text{ min}^{-1})$$

$$k_0 = 3.5 \times 10^{-3} \text{ mmol l}^{-1} \text{ min}^{-1}$$

12.3.2 First-Order Kinetics

If a reaction obeys first-order kinetics, the relationship between reaction rate and reactant concentration is as follows:

$$r_A = k_1 C_A \tag{12.31}$$

where r_A is the volumetric rate of reaction with respect to reactant A, k_1 is the *first-order rate constant*, and C_A is the concentration of A. k_1 has dimensions T^{-1} and units of, for example, s^{-1}. Like the zero-order rate constant in Section 12.3.1, the value of k_1 depends on the catalyst concentration.

If a first-order reaction takes place in a closed, constant-volume system, from Eqs. (12.18) and (12.31):

$$\frac{dC_A}{dt} = -k_1 C_A \tag{12.32}$$

Separating variables and integrating with initial condition $C_A = C_{A0}$ at $t = 0$ gives:

$$\int \frac{dC_A}{C_A} = \int -k_1 dt \tag{12.33}$$

or

$$\ln C_A = \ln C_{A0} - k_1 t \tag{12.34}$$

Rearranging gives an equation for C_A as a function of time:

$$C_A = C_{A0}\, e^{-k_1 t} \tag{12.35}$$

According to Eq. (12.34), if a reaction follows first-order kinetics, a plot of C_A versus time on semi-logarithmic coordinates gives a straight line with slope $-k_1$ and intercept C_{A0}. Analysis of first-order kinetics is illustrated in Example 12.4.

EXAMPLE 12.4 KINETICS OF CRUDE OIL DEGRADATION

Soil contaminated with crude oil is treated using a mixed population of indigenous bacteria and fungi. The concentration of total petroleum hydrocarbons in a soil sample is measured as a function of time over a 6-week period.

Time (days)	Total Petroleum Hydrocarbon Concentration (mg kg^{-1})
0	1375
7	802
14	695
21	588
28	417
35	356
42	275

(a) Determine the rate constant for petroleum degradation.
(b) Estimate the contaminant concentration after 16 days.

Solution

(a) Test whether petroleum degradation can be modelled as a first-order reaction. A semi-log plot of total petroleum hydrocarbon concentration versus time is shown in Figure 12.6.

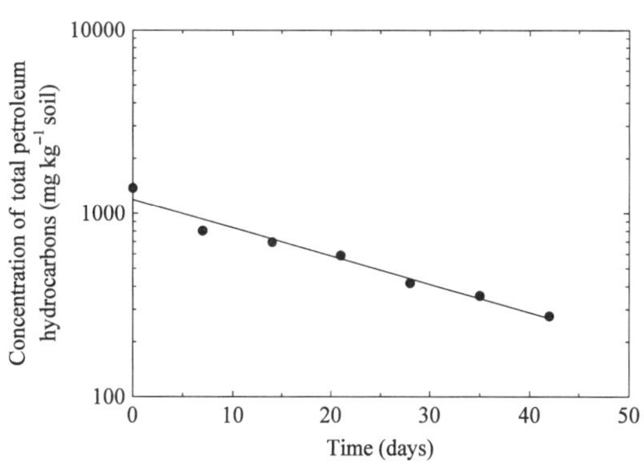

FIGURE 12.6 Graphic of the kinetic analysis of petroleum degradation.

The first-order model gives a straight line that fits the data well. The slope and intercept are evaluated as described in Section 3.4.2. The slope is equal to $-k_1$ and the intercept is C_{A0}. Therefore, $k_1 = 0.0355\ \text{day}^{-1}$ and $C_{A0} = 1192\ \text{mg kg}^{-1}$.

(b) Applying Eq. (12.35), the kinetic equation is:

$$C_A = 1192\, e^{-0.0355t}$$

where C_A has units of mg kg^{-1} and t has units of days. Therefore, after 16 days, $C_A = 675\ \text{mg kg}^{-1}$.

12.3.3 Michaelis–Menten Kinetics

The kinetics of most enzyme reactions are reasonably well represented by the *Michaelis–Menten equation*:

$$r_A = \frac{v_{max}\, C_A}{K_m + C_A} \tag{12.36}$$

where r_A is the volumetric rate of reaction with respect to reactant A, C_A is the concentration of A, v_{max} is the *maximum rate of reaction*, and K_m is the *Michaelis constant*. v_{max} has the same dimensions as r_A; K_m has the same dimensions as C_A. Typical units for v_{max} are mol m^{-3} s^{-1}; typical units for K_m are mol m^{-3}. As defined in Eq. (12.36), v_{max} is a volumetric rate that is proportional to the amount of active enzyme present. The Michaelis constant K_m is equal to the reactant concentration at which $r_A = v_{max}/2$. K_m is independent of enzyme concentration but varies from one enzyme to another and with different substrates for the same enzyme. Values of K_m for some enzyme–substrate systems are listed in Table 12.3. K_m and other enzyme properties depend on the source of the enzyme.

If we adopt conventional symbols for biological reactions and call reactant A the *substrate*, Eq. (12.36) can be rewritten in the familiar form:

$$v = \frac{v_{max}\, s}{K_m + s} \tag{12.37}$$

where v is the volumetric rate of reaction and s is the substrate concentration. The biochemical basis of the Michaelis–Menten equation will not be covered here; discussion of enzyme reaction models and the assumptions involved in derivation of Eq. (12.37) can be found elsewhere [2, 3]. Suffice it to say here that the simplest reaction sequence that accounts for the kinetic properties of many enzymes is:

$$\text{E} + \text{S} \underset{k_{-1}}{\overset{k_1}{\rightleftharpoons}} \text{ES} \xrightarrow{k_2} \text{E} + \text{P} \tag{12.38}$$

where E is enzyme, S is substrate, and P is product. ES is the *enzyme–substrate complex*. As expected in catalytic reactions, enzyme E is recovered at the end of the reaction. Binding of substrate to the enzyme in the first step is considered reversible with forward reaction constant k_1 and reverse reaction constant k_{-1}. Decomposition of the enzyme–substrate complex to give the product is an irreversible reaction with rate constant k_2; k_2 is known as the *turnover*

TABLE 12.3 Michaelis Constants for Some Enzyme–Substrate Systems

Enzyme	Source	Substrate	K_m (mM)
Alcohol dehydrogenase	*Saccharomyces cerevisiae*	Ethanol	13.0
α-Amylase	*Bacillus stearothermophilus*	Starch	1.0
	Porcine pancreas	Starch	0.4
β-Amylase	Sweet potato	Amylose	0.07
Aspartase	*Bacillus cadaveris*	L-Aspartate	30.0
β-Galactosidase	*Escherichia coli*	Lactose	3.85
Glucose oxidase	*Aspergillus niger*	D-Glucose	33.0
	Penicillium notatum	D-Glucose	9.6
Histidase	*Pseudomonas fluorescens*	L-Histidine	8.9
Invertase	*Saccharomyces cerevisiae*	Sucrose	9.1
	Neurospora crassa	Sucrose	6.1
Lactate dehydrogenase	*Bacillus subtilis*	Lactate	30.0
Penicillinase	*Bacillus licheniformis*	Benzylpenicillin	0.049
Urease	Jack bean	Urea	10.5

From B. Atkinson and F. Mavituna, 1991, Biochemical Engineering and Biotechnology Handbook, *2nd ed., Macmillan, Basingstoke.*

number as it defines the number of substrate molecules converted to product per unit time by an enzyme saturated with substrate. The turnover number is sometimes referred to as the *catalytic constant* k_{cat}. The dimensions of k_{cat} are T^{-1} and the units are, for example, s^{-1}. Analysis of the reaction sequence yields the relationship:

$$v_{max} = k_{cat}\, e_a \tag{12.39}$$

where e_a is the concentration of active enzyme and v_{max} and e_a are expressed in Eq. (12.39) using molar units. Values of k_{cat} range widely for different enzymes from about 50 min^{-1} to 10^7 min^{-1}.

The definition of K_m as the substrate concentration at which $v = v_{max}/2$ is equivalent to saying that K_m is the substrate concentration at which half of the enzyme's active sites are saturated with substrate. K_m is therefore considered a relative measure of the *substrate binding affinity* or the stability of the enzyme–substrate complex: lower K_m values imply higher enzyme affinity for the substrate. The *catalytic efficiency* of an enzyme is defined as the ratio k_{cat}/K_m with units of, for example, mol l^{-1} s^{-1}. Catalytic efficiency is often used to compare the utilisation of different substrates by a particular enzyme and is a measure of the *substrate specificity* or relative suitability of a substrate for reaction with the enzyme. Substrates with higher catalytic efficiency are more favourable.

An essential feature of Michaelis–Menten kinetics is that the catalyst becomes saturated at high substrate concentrations. Figure 12.7 shows the form of Eq. (12.37); the reaction

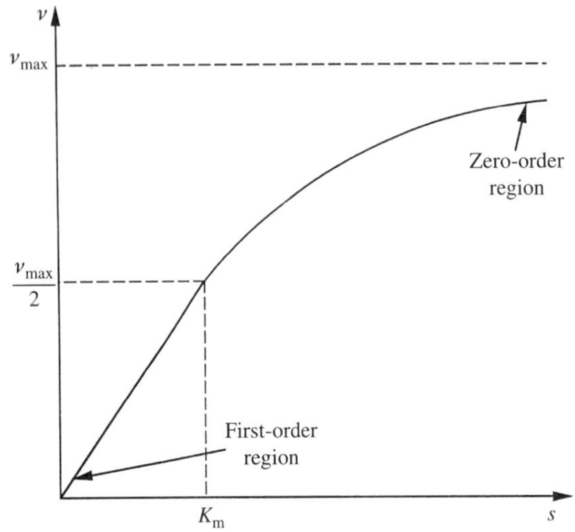

FIGURE 12.7 Michaelis–Menten plot.

rate v does not increase indefinitely with substrate concentration but approaches a limit, v_{max}. When $v = v_{max}$, all the enzyme is bound to substrate in the form of the enzyme–substrate complex. At high substrate concentrations $s \gg K_m$, K_m in the denominator of Eq. (12.37) is negligibly small compared with s so we can write:

$$v \approx \frac{v_{max} s}{s} \tag{12.40}$$

or

$$v \approx v_{max} \tag{12.41}$$

Therefore, at high substrate concentrations, the reaction rate approaches a constant value independent of substrate concentration; in this concentration range, the reaction is essentially *zero order* with respect to substrate. On the other hand, at low substrate concentrations $s \ll K_m$, the value of s in the denominator of Eq. (12.37) is negligible compared with K_m, and Eq. (12.37) can be simplified to:

$$v \approx \frac{v_{max}}{K_m} s \tag{12.42}$$

The ratio of constants v_{max}/K_m is, in effect, a first-order rate coefficient for the reaction. Therefore, at low substrate concentrations there is an approximate linear dependence of reaction rate on s; in this concentration range, Michaelis–Menten reactions are essentially *first order* with respect to substrate.

The rate of enzyme reactions depends on the amount of enzyme present as indicated by Eq. (12.39). However, enzymes are not always available in pure form so that e_a may be unknown. In this case, the amount of enzyme can be expressed as *units of activity*; the specific activity of an enzyme–protein mixture could be reported, for example, as units of activity per mg of protein. The *international unit of enzyme activity*, which is abbreviated IU or U, is the amount of enzyme required to convert 1 μmole of substrate into products per minute under standard conditions. Alternatively, the SI unit for enzyme activity is the katal, which is defined as the amount of enzyme required to convert 1 mole of substrate per second. The abbreviation for katal is kat. Enzyme concentration can therefore be expressed using units of, for example, U ml^{-1} or kat l^{-1}.

The Michaelis–Menten equation is a satisfactory description of the kinetics of many industrial enzymes, although there are exceptions such as glucose isomerase and amyloglucosidase. Procedures for checking whether a particular reaction follows Michaelis–Menten kinetics and for evaluating v_{max} and K_m from experimental data are described in Section 12.4. More complex kinetic equations must be applied if there are multiple substrates [2–4]. Modified kinetic expressions for enzymes subject to inhibition and other forms of regulation are described in Section 12.5.

12.3.4 Effect of Conditions on Enzyme Reaction Rate

As well as substrate concentration, other conditions such as temperature and pH also influence the rate of enzyme reaction. For an enzyme with a single rate-controlling step, the effect of temperature is described reasonably well using the Arrhenius expression of Eq. (12.22) with v_{max} substituted for k. An example showing the relationship between temperature and the rate of sucrose inversion by yeast invertase is given in Figure 12.8. Activation energies for enzyme reactions are of the order 40 to 80 kJ mol^{-1} [5]; as a rough guide, this means that a 10°C rise in temperature between 20°C and 30°C will increase the rate of reaction by a factor of 2 to 3.

Although an Arrhenius-type relationship between temperature and rate of reaction is observed for enzymes, the temperature range over which Eq. (12.22) is applicable is quite limited. Many proteins start to denature at 45 to 50°C; if the temperature is raised higher than

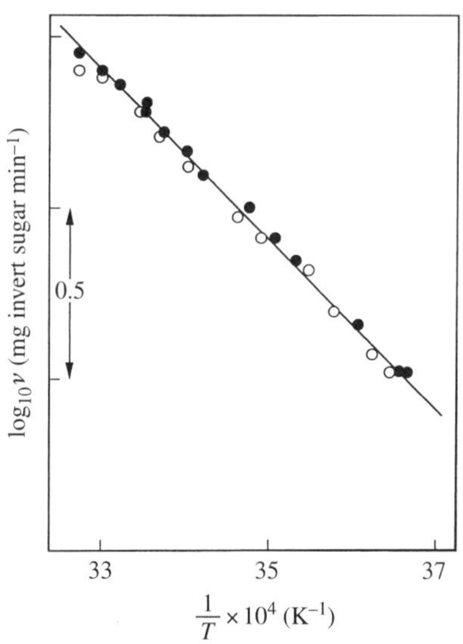

FIGURE 12.8 Arrhenius plot for inversion of sucrose by yeast invertase.
From I.W. Sizer, 1943, Effects of temperature on enzyme kinetics. Adv. Enzymol. *3, 35–62.*

this, thermal deactivation occurs and the reaction velocity drops quickly. Figure 12.9 illustrates how the Arrhenius relationship breaks down at high temperatures. In this experiment, the Arrhenius equation was obeyed between temperatures of about 0°C ($T = 273.15$ K; $1/T = 3.66 \times 10^{-3}$ K^{-1}) and about 53°C ($T = 326.15$ K; $1/T = 3.07 \times 10^{-3}$ K^{-1}). However, with further increase in temperature, the reaction rate declined rapidly due to thermal deactivation. Enzyme stability and the rate of enzyme deactivation are important factors affecting overall catalytic performance in enzyme reactors. This topic is discussed further in Section 12.5.

pH has a pronounced effect on enzyme kinetics, as illustrated in Figure 12.10. Typically, the reaction rate is maximum at some optimal pH and declines sharply if the pH is shifted either side of the optimum value. Ionic strength and water activity can also have considerable influence on the rate of enzyme reaction.

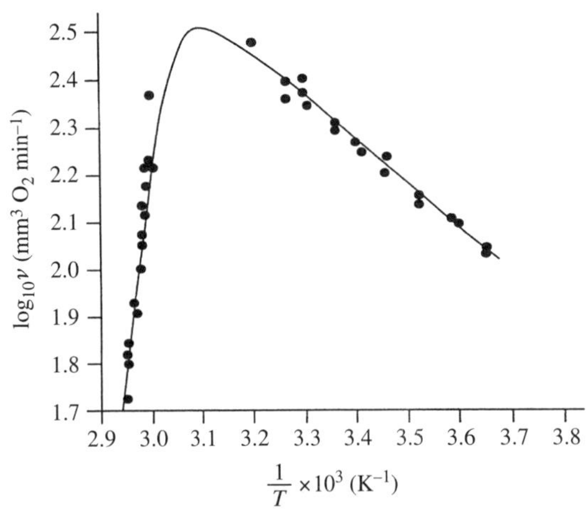

FIGURE 12.9 Arrhenius plot for catalase. The enzyme breaks down at high temperatures.
From I.W. Sizer, 1944, Temperature activation and inactivation of the crystalline catalase–hydrogen peroxide system. J. Biol. Chem. *154, 461–473.*

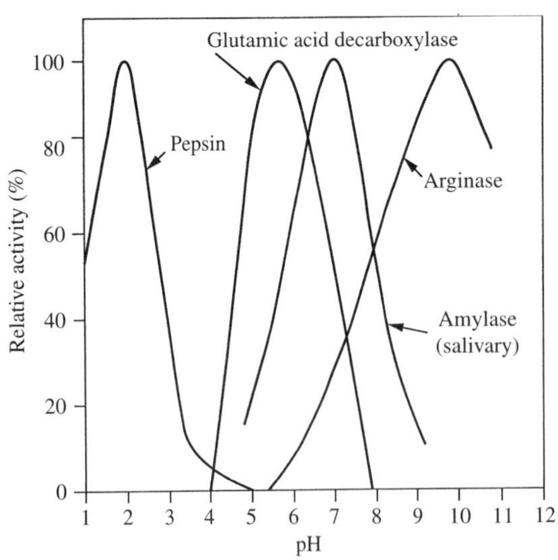

FIGURE 12.10 Effect of pH on enzyme activity.
From J.S. Fruton and S. Simmonds, 1958, General Biochemistry, *2nd ed., John Wiley, New York.*

12.4 DETERMINING ENZYME KINETIC CONSTANTS FROM BATCH DATA

To specify fully the kinetics of Michaelis–Menten reactions (see Section 12.3.3), two rate constants, v_{max} and K_m, must be evaluated. Estimating kinetic parameters for Michaelis–Menten reactions is not as straightforward as for zero- and first-order reactions. Several graphical methods are available; unfortunately some do not give accurate results.

The first step in kinetic analysis of enzyme reactions is to obtain data for the rate of reaction v as a function of substrate concentration s. Rates of reaction can be determined from batch concentration data as described in Section 12.2. Typically, only *initial rate data* are used. This means that several batch experiments are carried out with different initial substrate concentrations; from each set of data the reaction rate is evaluated at time zero. The initial rates and corresponding initial substrate concentrations are used as (v, s) pairs that can then be plotted in various ways to determine v_{max} and K_m. Initial rate data are preferred for analysis of enzyme reactions because experimental conditions, such as the enzyme and substrate concentrations, are known most accurately at the beginning of reactions.

12.4.1 Michaelis–Menten Plot

This simple procedure involves plotting (v, s) values directly as shown in Figure 12.7. v_{max} and K_m can be estimated roughly from this graph; v_{max} is the rate as $s \to \infty$ and K_m is the value of s at $v = v_{max}/2$. The accuracy of this method is usually poor because of the difficulty of extrapolating to v_{max}.

12.4.2 Lineweaver–Burk Plot

This method uses a linearisation procedure to give a straight-line plot from which v_{max} and K_m can be determined. Inverting Eq. (12.37) gives:

$$\frac{1}{v} = \frac{K_m}{v_{max}\, s} + \frac{1}{v_{max}} \tag{12.43}$$

Therefore, a plot of $1/v$ versus $1/s$ should give a straight line with slope K_m/v_{max} and intercept $1/v_{max}$. This double-reciprocal plot is known as the *Lineweaver–Burk plot* and is found frequently in the literature on enzyme kinetics. However, the linearisation process used in this method distorts the experimental error in v (Section 3.3.4) so that these errors are amplified at low substrate concentrations. As a consequence, the Lineweaver–Burk plot often gives inaccurate results and is therefore not recommended [3].

12.4.3 Eadie–Hofstee Plot

If Eq. (12.43) is multiplied by $v\left(\frac{v_{max}}{K_m}\right)$ and then rearranged, another linearised form of the Michaelis–Menten equation is obtained:

$$\frac{v}{s} = \frac{v_{max}}{K_m} - \frac{v}{K_m} \tag{12.44}$$

According to Eq. (12.44), a plot of v/s versus v gives a straight line with slope $-1/K_m$ and intercept v_{max}/K_m. This is called the *Eadie–Hofstee plot*. As with the Lineweaver–Burk plot, the Eadie–Hofstee linearisation distorts errors in the data so that the method has reduced accuracy.

12.4.4 Langmuir Plot

Multiplying Eq. (12.43) by s produces the linearised form of the Michaelis–Menten equation according to Langmuir:

$$\frac{s}{v} = \frac{K_m}{v_{max}} + \frac{s}{v_{max}} \tag{12.45}$$

Therefore, a *Langmuir plot* of s/v versus s should give a straight line with slope $1/v_{max}$ and intercept K_m/v_{max}. Linearisation of data for the Langmuir plot minimises distortions in experimental error. Accordingly, its use for evaluation of v_{max} and K_m is recommended [6]. The Langmuir plot is also known as the *Hanes–Woolf plot*.

12.4.5 Direct Linear Plot

A different method for plotting enzyme kinetic data has been proposed by Eisenthal and Cornish-Bowden [7]. For each observation, the reaction rate v is plotted on the vertical axis against s on the negative horizontal axis. This is shown in Figure 12.11 for four pairs of (v, s) data. A straight line is then drawn to join corresponding ($-s$, v) points. In the absence of experimental error, lines for each ($-s$, v) pair intersect at a unique point,

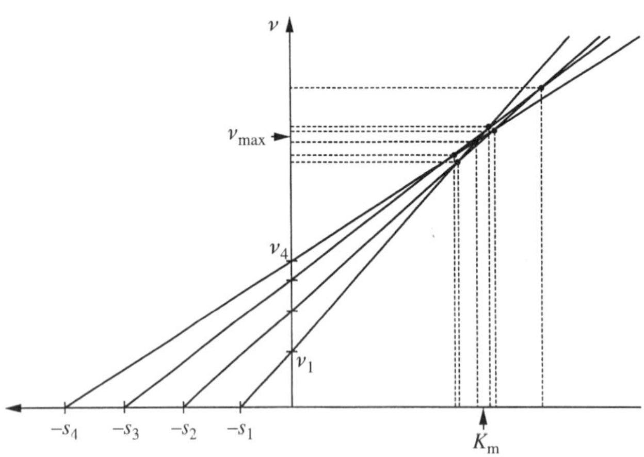

FIGURE 12.11 Direct linear plot for determination of enzyme kinetic parameters. *From R. Eisenthal and A. Cornish-Bowden, 1974, The direct linear plot: a new graphical procedure for estimating enzyme kinetic parameters.* Biochem. J. *139, 715–720.*

(K_m, v_{max}). When real data containing errors are plotted, a family of intersection points is obtained as shown in Figure 12.11. Each intersection gives one estimate of v_{max} and K_m; the median or middle v_{max} and K_m values are taken as the kinetic parameters for the reaction. This method is relatively insensitive to individual erroneous readings that may be far from the correct values. However, a disadvantage of the procedure is that deviations from Michaelis–Menten behaviour are not easily detected. It is recommended, therefore, for enzymes that are known to obey Michaelis–Menten kinetics.

12.5 REGULATION OF ENZYME ACTIVITY

In cells, enzyme activity is subject to diverse control mechanisms that regulate the throughput of material in metabolic pathways. One way to modify the rate of a particular enzyme reaction is to change the total amount of enzyme present by inducing or repressing the transcription of DNA and/or translation of RNA required for enzyme synthesis. This allows relatively slow or long-term control of enzyme activity; however, it does not provide the rapid responses often observed in metabolic systems. Various additional mechanisms are available for controlling the activity of enzymes that are already present without the need for significant changes in enzyme concentration. Metabolites and other effector molecules interact with enzymes to change properties such as the availability of active sites for reaction, the affinity of the substrate for enzyme binding, and the ability of the enzyme to form and release product. These interactions allow the activity of enzymes to be altered relatively quickly in situ.

Any molecule that reduces the rate of an enzyme reaction is called an *inhibitor*. Reaction products, substrate analogues, synthetic molecules, metabolic intermediates, and heavy metals are examples of enzyme inhibitors; substrates can also act as inhibitors. Binding between the inhibitor and enzyme may be *reversible* or *irreversible*.

Enzyme inhibition and regulation are of greater relevance for the understanding of cellular metabolism than for the operation of enzyme reactors in bioprocessing. When enzymes are used for product manufacture, unless the substrate or product of the reaction has a regulatory function affecting enzyme activity, inhibitors and other effector molecules are generally not available because they are not added to the reaction mixture. In contrast, enzyme reactions take place within cells in the presence of a wide range of potential ligands. Several types of enzyme inhibition and activation have been identified, as outlined in the following sections. The many forms of enzyme regulation are treated in more detail elsewhere [2].

12.5.1 Reversible Inhibition

There are several mechanisms of reversible enzyme inhibition. The effects of this type of regulation on enzyme reaction rate are described using modified versions of the Michaelis–Menten equation.

Competitive Inhibition

Substances that cause competitive enzyme inhibition have some degree of molecular similarity with the substrate. These properties allow the inhibitor to compete with substrate for binding to the active reaction sites on the enzyme, thus preventing some of the enzyme from forming the enzyme–substrate complex. Binding of inhibitor to enzyme is noncovalent and reversible so that an equilibrium is established between the free and bound forms of the inhibitor. Competitive inhibitors bind to free enzyme E but not to the enzyme–substrate complex ES. Interaction between the enzyme and inhibitor takes place at the same time as conversion of substrate S to product P according to the usual reaction sequence of Eq. (12.38):

$$\begin{array}{l} \mathrm{E + S \rightleftarrows ES \longrightarrow E + P} \\ \mathrm{E + I \rightleftarrows EI} \end{array} \tag{12.46}$$

where I is the inhibitor and EI is the *enzyme–inhibitor complex*. The equation for the rate of enzyme reaction with competitive inhibition is:

$$v = \frac{v_{\max}\, s}{K_{\mathrm{m}}\left(1 + \dfrac{i}{K_{\mathrm{i}}}\right) + s} \tag{12.47}$$

where i is the inhibitor concentration. K_{i} is the *inhibitor coefficient* or *dissociation constant* for the enzyme–inhibitor binding reaction:

$$K_{\mathrm{i}} = \frac{[\mathrm{E}]_{\mathrm{e}}\,[\mathrm{I}]_{\mathrm{e}}}{[\mathrm{EI}]_{\mathrm{e}}} \tag{12.48}$$

where $[\;]_{\mathrm{e}}$ denotes molar concentration at equilibrium. Comparison of Eq. (12.47) with Eq. (12.37) for Michaelis–Menten kinetics without inhibition shows that competitive enzyme inhibition does not affect $v_{\max}$ but changes the apparent value of K_{m}:

$$K_{\mathrm{m,app}} = K_{\mathrm{m}}\left(1 + \frac{i}{K_{\mathrm{i}}}\right) \tag{12.49}$$

where $K_{\mathrm{m,app}}$ is the apparent value of K_{m}. The effect of competitive inhibition on the rate of enzyme reaction can be overcome by increasing the concentration of substrate.

Noncompetitive Inhibition

Noncompetitive inhibitors have an affinity for binding both the free enzyme and the ES complex, so that the inhibitor and substrate may bind simultaneously with the enzyme to form an inactive ternary complex EIS. The inhibitor binding site on the enzyme is located away from the active reaction site. Binding of the inhibitor is noncovalent and reversible and does not affect the affinity of the substrate for enzyme binding. The reactions involved in noncompetitive inhibition are:

$$\begin{array}{l} \mathrm{E + S \rightleftarrows ES \longrightarrow E + P} \\ \mathrm{E + I \rightleftarrows EI} \\ \mathrm{EI + S \rightleftarrows EIS} \\ \mathrm{ES + I \rightleftarrows EIS} \end{array} \tag{12.50}$$

Noncompetitive inhibition reduces the effective maximum rate of enzyme reaction while K_m remains unchanged. The equation for the rate of enzyme reaction with noncompetitive inhibition is:

$$v = \frac{v_{max}}{\left(1 + \frac{i}{K_i}\right)} \frac{s}{K_m + s} \tag{12.51}$$

where i is the inhibitor concentration and K_i is the inhibitor coefficient:

$$K_i = \frac{[E]_e\,[I]_e}{[EI]_e} = \frac{[ES]_e\,[I]_e}{[EIS]_e} \tag{12.52}$$

From Eq. (12.51), the apparent maximum reaction velocity $v_{max,app}$ is:

$$v_{max,app} = \frac{v_{max}}{\left(1 + \frac{i}{K_i}\right)} \tag{12.53}$$

The decline in v_{max} reflects the reduced concentration of ES in the reaction mixture as some ES is converted to the inactive complex, EIS. Because inhibitor binding does not interfere with the formation of ES, noncompetitive inhibition cannot be overcome by increasing the substrate concentration.

Uncompetitive Inhibition

Uncompetitive inhibitors do not bind to free enzyme but affect enzyme reactions by binding to the enzyme–substrate complex at locations away from the active site. Inhibitor binding to ES is noncovalent and reversible and produces an inactive ternary complex EIS:

$$\begin{aligned} &E + S \rightleftarrows ES \longrightarrow E + P \\ &ES + I \rightleftarrows EIS \end{aligned} \tag{12.54}$$

Uncompetitive inhibition decreases both v_{max} and K_m. The equation for the rate of enzyme reaction with uncompetitive inhibition is:

$$v = \frac{v_{max}}{\left(1 + \frac{i}{K_i}\right)} \frac{s}{\left[\frac{K_m}{\left(1 + \frac{i}{K_i}\right)} + s\right]} \tag{12.55}$$

where:

$$K_i = \frac{[ES]_e\,[I]_e}{[EIS]_e} \tag{12.56}$$

From Eq. (12.55):

$$v_{\text{max,app}} = \frac{v_{\text{max}}}{\left(1 + \frac{i}{K_i}\right)} \tag{12.57}$$

and

$$K_{\text{m,app}} = \frac{K_m}{\left(1 + \frac{i}{K_i}\right)} \tag{12.58}$$

Because the inhibitor does not interfere with formation of the enzyme–substrate complex, uncompetitive inhibition cannot be overcome by increasing the substrate concentration.

Substrate inhibition is a special case of uncompetitive inhibition. High substrate concentrations can inhibit enzyme activity if more than one substrate molecule binds to an active site that is meant to accommodate only one. For example, different regions of multiple molecules of substrate may bind to different moieties within a single active site on the enzyme. If the resulting enzyme–substrate complex is unreactive, the rate of the enzyme reaction is reduced. The reaction sequence for substrate inhibition is:

$$\begin{aligned} &E + S \rightleftarrows ES \longrightarrow E + P \\ &ES + S \rightleftarrows ESS \end{aligned} \tag{12.59}$$

where ESS is an inactive substrate complex. The equation for the rate of enzyme reaction with substrate inhibition is:

$$v = \frac{v_{\text{max}}}{\left(1 + \frac{K_1}{s} + \frac{s}{K_2}\right)} \tag{12.60}$$

where K_1 is the dissociation constant for the reaction forming ES and K_2 is the dissociation constant for the reaction forming ESS:

$$K_1 = \frac{[E]_e\,[S]_e}{[ES]_e} \tag{12.61}$$

$$K_2 = \frac{[ES]_e\,[S]_e}{[ESS]_e} \tag{12.62}$$

Figure 12.12 shows the effect of substrate inhibition on the rate of enzyme reaction. The rate increases with substrate concentration at relatively low substrate levels, but passes through a maximum at substrate concentration s_{max} as inhibition causes a progressive decline in reaction rate with further increase in s. The value of s_{max} is related to K_1 and K_2 by the equation:

$$s_{\text{max}} = \sqrt{K_1\,K_2} \tag{12.63}$$

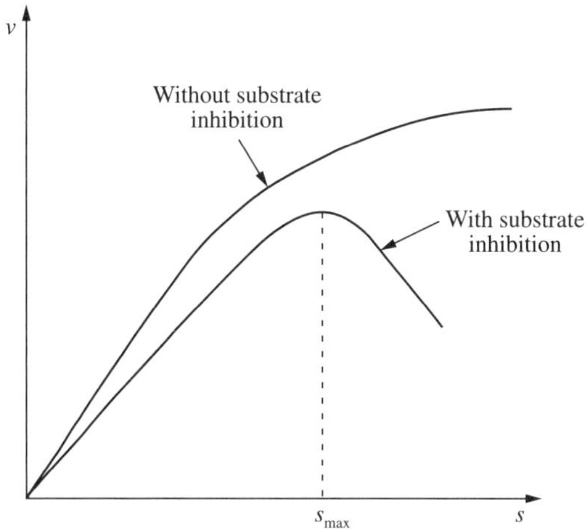

FIGURE 12.12 Relationship between enzyme reaction rate and substrate concentration with and without substrate inhibition.

Partial Inhibition

In the models for enzyme inhibition considered so far, binding between the inhibitor and enzyme was considered to block completely the formation of product P. However, it is possible that turnover may still occur after inhibitor binding even though the rate of product formation is reduced significantly; for example:

$$EIS \rightleftharpoons EI + P \tag{12.64}$$

Partial inhibition does not occur often but is characterised by the maintenance of enzyme activity even at high inhibitor concentrations.

12.5.2 Irreversible Inhibition

Irreversible inhibition occurs when an inhibitor binds with an enzyme but does not dissociate from it under reaction conditions or within the reaction time-frame. In some cases, the inhibitor binds covalently so that product formation is prevented permanently. Michaelis–Menten kinetics do not apply to irreversible inhibition. The rate of inhibitor binding to the enzyme determines the effectiveness of the inhibition. Accordingly, the extent to which the enzyme reaction rate is slowed depends on the enzyme and inhibitor concentrations but is independent of substrate concentration.

12.5.3 Allosteric Regulation

Allosteric regulation of enzymes is crucial for the control of cellular metabolism. Allosteric regulation occurs when an activator or inhibitor molecule binds at a specific regulatory site on the enzyme and induces conformational or electrostatic changes that either enhance or reduce enzyme activity. Not all enzymes possess sites for allosteric binding; those that do are called *allosteric enzymes*. Allosteric enzymes typically comprise multiple protein subunits. Ligands that bind to allosteric enzymes and affect binding at a different site on the enzyme are known as *effectors*. *Homotropic* regulation occurs when a substrate also acts as an effector and influences the binding of further substrate molecules. *Heterotropic* regulation occurs when the effector and substrate are different entities.

When binding of an effector at one site on the enzyme alters the binding affinity at another site, the binding sites are said to be acting *cooperatively*. Positive allosteric modulation or *allosteric activation* occurs when binding of the effector enhances the enzyme's affinity for other ligands. Conversely, negative allosteric modulation or *allosteric inhibition* occurs when the affinity for other ligands is reduced by effector binding.

Allosteric enzymes do not follow Michaelis–Menten kinetics. When enzyme binding sites display cooperativity so that effector binding influences substrate affinity, the *Hill equation* is the simplest model representing the relationship between enzyme activity and substrate concentration:

$$v = \frac{v_{\max}\, s^n}{K_\text{h} + s^n} \tag{12.65}$$

where K_h is the *Hill constant* and n is the *Hill coefficient*. The Hill constant is related to the enzyme–substrate dissociation constant and reflects the affinity of the enzyme for a particular substrate. The units of K_h are those of (concentration)^n. From Eq. (12.65), it follows that:

$$K_\text{h} = s^n \quad \text{when } v = \frac{v_{\max}}{2} \tag{12.66}$$

The Hill coefficient n is an index of the cooperativity between binding sites on an oligomeric enzyme. When $n = 1$, there is no cooperativity and Eq. (12.65) reduces to the Michaelis–Menten relationship of Eq. (12.37). For $n > 1$, the cooperativity is positive and the benefit from effector binding increases as the value of n increases. The maximum possible value of n is the number of substrate-binding sites on the enzyme. If cooperativity is positive but of low strength, n will be greater than unity but less than the number of substrate-binding sites and may be a noninteger number. For $n < 1$, the cooperativity is negative and effector binding reduces substrate affinity and enzyme activity.

Graphical representation of Eq. (12.65) is shown in Figure 12.13. Values of $n > 1$ give the curve a sigmoidal shape; the greater the value of $n > 1$, the more pronounced is the sigmoidicity of the curve and the deviation from Michaelis–Menten kinetics ($n = 1$). The advantage of a sigmoidal saturation curve is that, in the intermediate range of substrate concentrations where the sigmoidal curve is steeper than for Michaelis–Menten kinetics, the sensitivity of the enzyme to substrate concentration is enhanced. In this region, a small change in substrate concentration produces a larger change in reaction velocity than for the same enzyme with no cooperativity between binding sites. As a result, greater changes in enzyme activity can occur over a narrower range of substrate concentration. Allosteric enzymes that exhibit negative cooperativity give reaction rate curves similar to that shown in Figure 12.13 for $n = 0.5$. For these enzymes, in the intermediate range of substrate concentration, enzyme activity is relatively insensitive to small changes in substrate concentration.

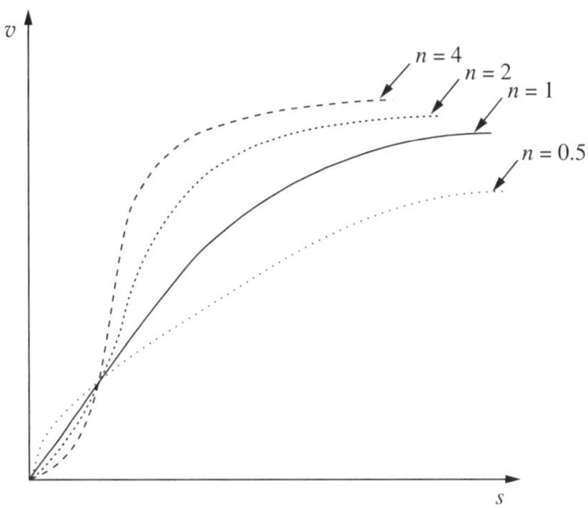

FIGURE 12.13 Reaction rate curves for allosteric enzymes exhibiting no ($n = 1$), positive ($n > 1$), and negative ($n < 1$) cooperativity.

A linearised form of the Hill equation is obtained by rearranging Eq. (12.65) and taking logarithms:

$$\ln\left(\frac{v}{v_{max} - v}\right) = n \ln s - \ln K_h \tag{12.67}$$

Therefore, the values of n and K_h for a particular allosteric enzyme can be determined from the slope and intercept of the straight line generated when $v/(v_{max} - v)$ is plotted versus s on log–log coordinates. Analysis of experimental data using Eq. (12.67) sometimes shows deviations from linearity at low substrate concentrations; however, measured results typically conform reasonably well to Eq. (12.67) at reaction velocities between $0.1v_{max}$ and $0.9v_{max}$.

Several other models for allosteric enzyme kinetics have been developed. However, a major problem with the application of more complex mechanistic models of cooperativity and allosterism is that the kinetic parameters required become difficult to estimate and check for accuracy.

Allosteric enzymes are vitally important in metabolic control. In particular, they allow feedback inhibition of the enzyme cascades responsible for catabolism and biosynthesis. Typically, a metabolite produced at the end of a cascade will function as a heterotropic inhibitor of an enzyme that is active earlier in the pathway. In this way, build-up of excess product through overactivity of the pathway can be moderated as the product regulates substrate binding affinity and enzyme activity near the beginning of the cascade, thus reducing the rate of product synthesis.

12.6 KINETICS OF ENZYME DEACTIVATION

Enzymes are protein molecules of complex configuration that can be destabilised by relatively weak forces. During the course of enzyme-catalysed reactions, enzyme deactivation occurs at a rate that is dependent on the structure of the enzyme and the reaction conditions. Environmental factors affecting enzyme stability include temperature, pH, ionic strength, mechanical forces, and the presence of denaturants such as solvents, detergents, and heavy metals. Because the amount of active enzyme can decline considerably during reaction, in many applications, the kinetics of enzyme deactivation are just as important as the kinetics of the reaction itself.

In the simplest model of enzyme deactivation, active enzyme E_a undergoes irreversible transformation to an inactive form E_i:

$$E_a \longrightarrow E_i \tag{12.68}$$

The rate of deactivation is generally considered to be first order in active enzyme concentration:

$$r_d = k_d\, e_a \tag{12.69}$$

where r_d is the volumetric rate of deactivation, e_a is the active enzyme concentration, and k_d is the *deactivation rate constant*. In a closed system where enzyme deactivation is the only process affecting the concentration of active enzyme:

$$r_d = \frac{-de_a}{dt} = k_d\, e_a \tag{12.70}$$

Integration of Eq. (12.70) gives an expression for active enzyme concentration as a function of time:

$$e_a = e_{a0}\, e^{-k_d t} \tag{12.71}$$

where e_{a0} is the concentration of active enzyme at time zero. According to Eq. (12.71), the concentration of active enzyme decreases exponentially with time; the greatest rate of enzyme deactivation occurs when e_a is high.

As indicated in Eq. (12.39), the value of v_{max} for enzyme reaction depends on the concentration of active enzyme present. Therefore, as e_a declines due to deactivation, v_{max} is also reduced. We can estimate the variation of v_{max} with time by substituting into Eq. (12.39) the expression for e_a from Eq. (12.71):

$$v_{max} = k_{cat}\, e_{a0}\, e^{-k_d t} = v_{max0}\, e^{-k_d t} \tag{12.72}$$

where v_{max0} is the initial value of v_{max} before deactivation occurs.

The stability of enzymes is reported frequently in terms of *half-life*. Half-life is the time required for half the enzyme activity to be lost as a result of deactivation; after one half-life, the active enzyme concentration equals $e_{a0}/2$. Substituting $e_a = e_{a0}/2$ into Eq. (12.71), taking logarithms, and rearranging yields the following expression:

$$t_h = \frac{\ln 2}{k_d} \tag{12.73}$$

where t_h is the enzyme half-life.

The rate at which enzymes deactivate depends strongly on temperature. This dependency is generally well described using the Arrhenius equation (Section 12.1.5):

$$k_d = A\, e^{-E_d/RT} \tag{12.74}$$

where A is the Arrhenius constant or frequency factor, E_d is the *activation energy for enzyme deactivation*, R is the ideal gas constant, and T is absolute temperature. Values of R are listed in Appendix B. According to Eq. (12.74), as T increases, the rate of enzyme deactivation increases exponentially. Values of E_d are high, of the order 170 to 400 kJ gmol^{-1} for many enzymes [5]. Accordingly, a temperature rise of 10°C between 30°C and 40°C will increase the rate of enzyme deactivation by a factor of between 10 and 150. The stimulatory effect of increasing temperature on the rate of enzyme reaction has already been described in Section 12.3.4. However, as shown here, raising the temperature also reduces the amount of active enzyme present. It is clear that temperature has a critical effect on enzyme kinetics.

EXAMPLE 12.5 ENZYME HALF-LIFE

Amyloglucosidase from *Endomycopsis bispora* is immobilised in very small polyacrylamide gel beads. The activities of immobilised and soluble enzyme are compared at 80°C. Initial rate data are measured at a fixed substrate concentration with the following results.

	Enzyme Activity (μmol ml^{-1} min^{-1})	
Time (min)	**Soluble Enzyme**	**Immobilised Enzyme**
0	0.86	0.45
3	0.79	0.44
6	0.70	0.43
9	0.65	0.43
15	0.58	0.41
20	0.46	0.40
25	0.41	0.39
30	–	0.38
40	–	0.37

What is the half-life for each form of enzyme?

Solution

From Eq. (12.37), at any fixed substrate concentration, the rate of enzyme reaction v is directly proportional to v_{max}. Therefore, k_d can be determined from Eq. (12.72) using enzyme activity v instead of v_{max}. Making this change and expressing Eq. (12.72) using logarithms gives:

$$\ln v = \ln v_0 - k_d\, t$$

where v_0 is the initial enzyme activity before deactivation. So, if deactivation follows a first-order model, a semi-log plot of reaction rate versus time should give a straight line with slope $-k_d$. The experimental data are plotted in Figure 12.14.

From the slopes, k_d for soluble enzyme is 0.0296 min^{-1} and k_d for immobilised enzyme is 0.0051 min^{-1}. Applying Eq. (12.73) for half-life:

$$t_h\,(\text{soluble}) = \frac{\ln 2}{0.0296\ \text{min}^{-1}} = 23\ \text{min}$$

$$t_h\,(\text{immobilised}) = \frac{\ln 2}{0.0051\ \text{min}^{-1}} = 136\ \text{min}$$

Immobilisation enhances the stability of the enzyme significantly.

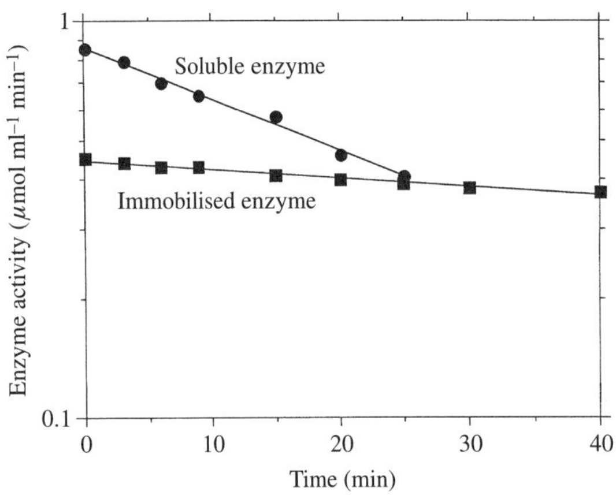

FIGURE 12.14 Kinetic analysis of enzyme deactivation.

12.7 YIELDS IN CELL CULTURE

The basic concept of reaction yield was introduced in Section 12.1.2 for simple one-step reactions. When we consider processes such as cell growth, we are in effect lumping together many individual enzyme and chemical conversions. Despite this complexity, yield principles can be applied to cell metabolism to relate the flow of substrate in metabolic pathways to the formation of biomass and other products. Yields that are frequently reported and of particular significance are expressed using *yield coefficients* or *yield factors*. Several yield coefficients, such as the yield of biomass from substrate, the yield of biomass from oxygen, and the yield of product from substrate, are in common use. Yield coefficients allow us to quantify the nutrient requirements and production characteristics of organisms.

Some metabolic yield coefficients—the biomass yield Y_{XS}, the product yield Y_{PS}, and the respiratory quotient RQ—were introduced in Section 4.6 in Chapter 4. The definition of yield coefficients can be generalised as follows:

$$Y_{JK} = \frac{-\Delta J}{\Delta K} \tag{12.75}$$

TABLE 12.4 Some Metabolic Yield Coefficients

Symbol	Definition
Y_{XS}	Mass or moles of biomass produced per unit mass or mole of substrate consumed; moles of biomass can be calculated from the 'molecular formula' for biomass (see Section 4.6.1)
Y_{PS}	Mass or moles of product formed per unit mass or mole of substrate consumed
Y_{PX}	Mass or moles of product formed per unit mass or mole of biomass formed
Y_{XO}	Mass or moles of biomass formed per unit mass or mole of oxygen consumed
Y_{CS}	Mass or moles of carbon dioxide formed per unit mass or mole of substrate consumed
RQ	Moles of carbon dioxide formed per mole of oxygen consumed; this yield is called the *respiratory quotient*
Y_{ATP}	Mass or moles of biomass formed per mole of ATP formed
Y_{kcal}	Mass or moles of biomass formed per kilocalorie of heat evolved during fermentation

where Y_{JK} is the yield factor, J and K are substances involved in metabolism, ΔJ is the mass or moles of J produced, and ΔK is the mass or moles of K consumed. The negative sign is required in Eq. (12.75) because ΔK for a consumed substance is negative in value and yield is calculated as a positive quantity. A list of frequently used yield coefficients is given in Table 12.4. Note that in some cases, such as Y_{PX}, both substances represented by the yield coefficient are products of metabolism. Although the term 'yield' usually refers to the amount of product formed per amount of reactant, yields can also be used to relate other quantities. Some yield coefficients are based on parameters such as the amount of ATP formed or heat evolved during metabolism.

12.7.1 Overall and Instantaneous Yields

A problem with application of Eq. (12.75) is that values of ΔJ and ΔK depend on the time period over which they are measured. In batch culture, ΔJ and ΔK can be calculated as the difference between initial and final states; this gives an *overall yield* representing an average value over the entire culture period. Alternatively, ΔJ and ΔK can be determined between two other points in time; this calculation might produce a different result for Y_{JK}. Yields can vary during culture and it is sometimes necessary to evaluate the *instantaneous yield* at a particular point in time. If r_J and r_K are the volumetric rates of production and consumption of J and K, respectively, in a closed, constant-volume reactor, the instantaneous yield can be calculated as follows:

$$Y_{JK} = \lim_{\Delta K \to 0} \frac{-\Delta J}{\Delta K} = \frac{-dJ}{dK} = \frac{\frac{-dJ}{dt}}{\frac{dK}{dt}} = \frac{r_J}{r_K} \tag{12.76}$$

For example, Y_{XS} at a particular instant in time is defined as:

$$Y_{XS} = \frac{r_X}{r_S} = \frac{\text{growth rate}}{\text{substrate consumption rate}} \tag{12.77}$$

When yields for fermentation are reported, the time or time period to which they refer should also be stated.

12.7.2 Theoretical and Observed Yields

As described in Section 12.1.2, it is necessary to distinguish between theoretical and observed yields. This is particularly important for cell metabolism because there are always many reactions occurring at the same time; theoretical and observed yields are therefore very likely to differ. Consider the example of biomass yield from substrate, Y_{XS}. If the total mass of substrate consumed is S_T, some proportion of S_T equal to S_G will be used for growth while the remainder, S_R, is channelled into other products and metabolic activities not related to growth. Therefore, the observed biomass yield based on total substrate consumption is:

$$Y'_{XS} = \frac{-\Delta X}{\Delta S_T} = \frac{-\Delta X}{\Delta S_G + \Delta S_R} \tag{12.78}$$

where ΔX is the amount of biomass produced and Y'_{XS} is the *observed biomass yield from substrate*. Values of observed biomass yields for several organisms and substrates are listed in Table 12.5. In comparison, the *true* or *theoretical biomass yield* from substrate is:

$$Y_{XS} = \frac{-\Delta X}{\Delta S_G} \tag{12.79}$$

TABLE 12.5 Observed Biomass Yields for Several Microorganisms and Substrates

Microorganism	**Substrate**	**Observed Biomass Yield Y'_{XS} (g g^{-1})**
Aerobacter cloacae	Glucose	0.44
Penicillium chrysogenum	Glucose	0.43
Candida utilis	Glucose	0.51
	Acetic acid	0.36
	Ethanol	0.68
Candida intermedia	*n*-Alkanes (C_{16}–C_{22})	0.81
Pseudomonas sp.	Methanol	0.41
Methylococcus sp.	Methane	1.01

From S.J. Pirt, 1975, Principles of Microbe and Cell Cultivation, *Blackwell Scientific, Oxford.*

as ΔS_G is the mass of substrate actually directed into biomass production. Theoretical yields are sometimes referred to as *maximum possible yields* because they represent the yield in the absence of competing reactions. However, because of the complexity of metabolism, ΔS_G for cell growth is often unknown and the observed biomass yield may be the only biomass yield available.

EXAMPLE 12.6 YIELDS IN ACETIC ACID PRODUCTION

The reaction equation for aerobic production of acetic acid from ethanol is:

$$\underset{\text{(ethanol)}}{C_2H_5OH} + O_2 \longrightarrow \underset{\text{(acetic acid)}}{CH_3CO_2H} + H_2O$$

Acetobacter aceti bacteria are added to vigorously aerated medium containing $10\ \mathrm{g\,l^{-1}}$ ethanol. After some time, the ethanol concentration is $2\ \mathrm{g\,l^{-1}}$ and $7.5\ \mathrm{g\,l^{-1}}$ of acetic acid is produced. How does the observed yield of acetic acid from ethanol compare with the theoretical yield?

Solution

Using a basis of 1 litre, the observed yield over the entire culture period is obtained from application of Eq. (12.10):

$$Y'_{PS} = \frac{7.5\ \mathrm{g}}{(10-2)\ \mathrm{g}} = 0.94\ \mathrm{g\,g^{-1}}$$

The theoretical yield is based on the mass of ethanol actually used for synthesis of acetic acid. From the stoichiometric equation:

$$Y_{PS} = \frac{1\ \text{gmol acetic acid}}{1\ \text{gmol ethanol}} = \frac{60\ \mathrm{g}}{46\ \mathrm{g}} = 1.30\ \mathrm{g\,g^{-1}}$$

The observed yield is 72% of the theoretical yield.

12.8 CELL GROWTH KINETICS

The kinetics of cell growth are expressed using equations similar to those presented in Section 12.3. From a mathematical point of view, there is little difference between the kinetic equations for enzymes and cells; after all, cell metabolism depends on the integrated action of a multitude of enzymes.

12.8.1 Batch Growth

When cells are grown in batch culture, several phases of cell growth are observed. A typical growth curve showing the changes in cell concentration with time is illustrated in Figure 12.15. The different phases of growth are more readily distinguished when the logarithm of viable cell concentration is plotted against time; alternatively, a semi-log plot can be used. The rate of cell growth varies depending on the growth phase. During the lag phase immediately after inoculation of the culture, the rate of growth is essentially zero.

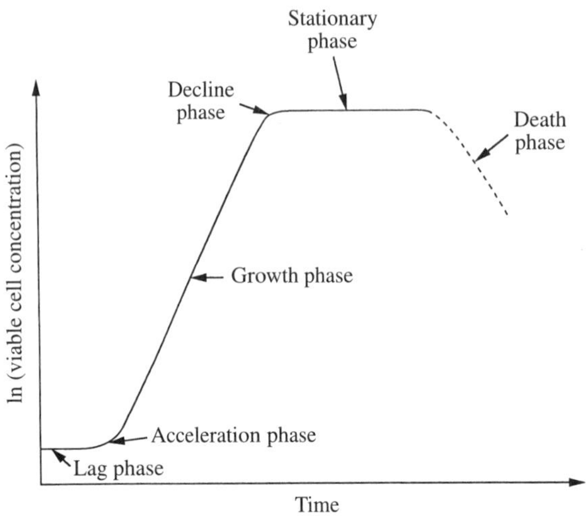

FIGURE 12.15 Typical batch growth curve.

TABLE 12.6 Summary of Batch Cell Growth

Phase	Description	Specific Growth Rate
Lag	Cells adapt to the new environment; no or very little growth	$\mu \approx 0$
Acceleration	Growth starts	$\mu < \mu_{max}$
Growth	Growth achieves its maximum rate	$\mu \approx \mu_{max}$
Decline	Growth slows due to nutrient exhaustion or build-up of inhibitory products	$\mu < \mu_{max}$
Stationary	Growth ceases	$\mu = 0$
Death	Cells lose viability and lyse	$\mu < 0$

Cells use the lag phase to adapt to their new environment: new enzymes or structural components may be synthesised but the concentration of cells does not increase. Following the lag period, growth starts in the acceleration phase and continues through the growth and decline phases. If growth is exponential, the growth phase appears as a straight line on a semi-log plot. At the end of the growth phase, as nutrients in the culture medium become depleted or inhibitory products accumulate, growth slows down and the cells enter the decline phase. After this transition period, the stationary phase is reached during which no further growth occurs. Some cultures exhibit a death phase as the cells lose viability or are destroyed by lysis. Table 12.6 provides a summary of growth and metabolic activity during the phases of batch culture.

During the growth and decline phases, the rate of cell growth is described by the equation:

$$r_X = \mu x \tag{12.80}$$

where r_X is the volumetric rate of biomass production with units of, for example, kg m^{-3} h^{-1}, x is the viable cell concentration with units of, for example, kg m^{-3}, and μ is the *specific growth rate*. The specific growth rate has dimensions T^{-1} and units of, for example, h^{-1}. Equation (12.80) has the same form as Eq. (12.31); cell growth is therefore considered a *first-order autocatalytic reaction*.

In a closed system where growth is the only process affecting cell concentration, the rate of growth r_X is equal to the rate of change of cell concentration:

$$r_X = \frac{dx}{dt} \tag{12.81}$$

Combining Eqs. (12.80) and (12.81) gives:

$$\frac{dx}{dt} = \mu x \tag{12.82}$$

If μ is constant we can integrate Eq. (12.82) directly. Separating variables and integrating with initial condition $x = x_0$ at $t = 0$ gives:

$$\int \frac{dx}{x} = \int \mu \, dt \tag{12.83}$$

or

$$\ln x = \ln x_0 + \mu t \tag{12.84}$$

where x_0 is the viable cell concentration at time zero. Rearranging gives an equation for cell concentration x as a function of time during batch growth:

$$x = x_0 \, e^{\mu t} \tag{12.85}$$

Equation (12.85) represents *exponential growth*. According to Eq. (12.84), a plot of ln x versus time gives a straight line with slope μ. Because the relationship of Eq. (12.84) is strictly valid only if μ is unchanging, a plot of ln x versus t is often used to assess whether the specific growth rate is constant. As illustrated in Figure 12.15, μ is usually constant during the growth phase. It is always advisable to prepare a semi-log plot of cell concentration before identifying phases of growth. As illustrated in Figure 3.7, if cell concentrations are plotted on linear coordinates, growth often appears slow at the beginning of the culture. We might be tempted to conclude there was a lag phase of 1 to 2 hours during the culture represented in Figure 3.7(a). However, when the same data are plotted using logarithms as shown in Figure 3.7(b), it is clear that the culture did not experience a lag phase. Exponential growth always appears much slower at the beginning of the culture because the number of cells present is small.

Cell growth rates are often expressed in terms of the *doubling time* t_d. An expression for doubling time can be derived from Eq. (12.85). Starting with a cell concentration of x_0, the concentration at $t = t_d$ is $2x_0$. Substituting these values into Eq. (12.85):

$$2x_0 = x_0 \, e^{\mu t_d} \tag{12.86}$$

Cancelling x_0 gives:

$$2 = e^{\mu t_d} \tag{12.87}$$

Taking the natural logarithm of both sides:

$$\ln 2 = \mu t_d \tag{12.88}$$

or

$$t_d = \frac{\ln 2}{\mu} \tag{12.89}$$

Because the relationships used to derive Eq. (12.89) require the assumption of constant μ, doubling time is a valid representation of the growth rate only when μ is constant.

12.8.2 Balanced Growth

In an environment favourable for growth, cells regulate their metabolism and adjust the rates of various internal reactions so that a condition of *balanced growth* occurs. During balanced growth, the composition of the biomass remains constant. Balanced growth means that the cell is able to modulate the effects of external perturbations and keep the biomass composition steady despite changes in environmental conditions.

For the biomass composition to remain constant during growth, the specific rate of production of each component in the culture must be equal to the cell specific growth rate μ:

$$r_Z = \mu z \tag{12.90}$$

where Z is a cellular constituent such as protein, RNA, polysaccharide, and so on, r_Z is the volumetric rate of production of Z, and z is the concentration of Z in the reactor volume. Therefore, during balanced growth, the doubling time for each cell component must be equal to t_d for growth. Balanced growth cannot be achieved if environmental changes affect the rate of growth. In most cultures, balanced growth occurs at the same time as exponential growth.

12.8.3 Effect of Substrate Concentration

During the growth and decline phases of batch culture, the specific growth rate of the cells depends on the concentration of nutrients in the medium. Often, a single substrate exerts a dominant influence on the rate of growth; this component is known as the *growth-rate-limiting substrate* or, more simply, the *growth-limiting substrate*. The growth-limiting substrate is often the carbon or nitrogen source, although in some cases it is oxygen or another oxidant such as nitrate. During balanced growth, the specific growth rate is related to the concentration of the growth-limiting substrate by a homologue of the Michaelis–Menten expression, the *Monod equation*:

$$\mu = \frac{\mu_{max} s}{K_S + s} \tag{12.91}$$

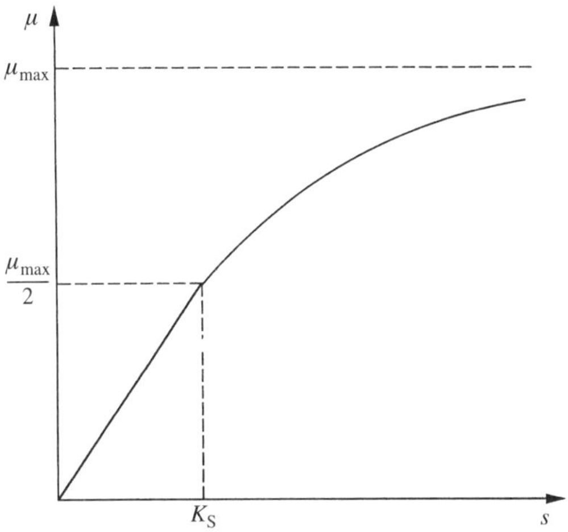

FIGURE 12.16 The relationship between the specific growth rate and the concentration of growth-limiting substrate in cell culture.

In Eq. (12.91), s is the concentration of growth-limiting substrate, μ_{max} is the *maximum specific growth rate*, and K_S is the *substrate constant*. μ_{max} has dimensions T^{-1}; K_S has the same dimensions as substrate concentration. The form of Eq. (12.91) is shown in Figure 12.16. μ_{max} and K_S are intrinsic parameters of the cell–substrate system; values of K_S for several organisms are listed in Table 12.7.

If μ is dependent on substrate concentration as indicated in Eq. (12.91), how can μ remain constant during cell growth as explained in Section 12.8.1 if substrate is being consumed and s is decreasing continuously throughout the growth period? How can Eq. (12.91) be consistent with the observation of exponential growth with constant μ that is typical of so many cell cultures? The answer lies with the relative magnitudes of K_S and s in Eq. (12.91). Typical values of K_S are very small, of the order of mg per litre for carbohydrate substrates and μg per litre for other compounds such as amino acids. The concentration of the growth-limiting substrate in culture media is normally much greater than K_S. As a result, $K_S \ll s$ in Eq. (12.91) and can be neglected, the s terms in the numerator and denominator cancel, and the specific growth rate is effectively independent of substrate concentration until s reaches very low values. Therefore, $\mu \approx \mu_{max}$ and growth follows exponential kinetics as long as s remains greater than about $10K_S$, which is usually for most of the culture period. This explains why μ remains constant and equal to μ_{max} until the medium is virtually exhausted of substrate. When s finally falls below $10K_S$, the transition from growth to stationary phase can be very abrupt as the very small quantity of remaining substrate is consumed rapidly by the large number of cells present.

The Monod equation is by far the most frequently used expression relating growth rate to substrate concentration. However, it is valid only for balanced growth and should not be applied when growth conditions are changing rapidly. There are also other restrictions; for example, the Monod equation has been found to have limited applicability at extremely low substrate levels. When growth is inhibited by high substrate or product concentrations, extra terms can be added to the Monod equation to account for these effects. Several other kinetic expressions have been developed for cell growth; these provide better correlations for experimental data in certain situations [8–11].

TABLE 12.7 K_S Values for Several Organisms

Microorganism (genus)	Limiting Substrate	K_S (mg l^{-1})
Saccharomyces	Glucose	25
Escherichia	Glucose	4.0
	Lactose	20
	Phosphate	1.6
Aspergillus	Glucose	5.0
Candida	Glycerol	4.5
	Oxygen	0.042–0.45
Pseudomonas	Methanol	0.7
	Methane	0.4
Klebsiella	Carbon dioxide	0.4
	Magnesium	0.56
	Potassium	0.39
	Sulphate	2.7
Hansenula	Methanol	120.0
	Ribose	3.0
Cryptococcus	Thiamine	1.4×10^{-7}

From S.J. Pirt, 1975, Principles of Microbe and Cell Cultivation, *Blackwell Scientific; and D.I.C. Wang, C.L. Cooney, A.L. Demain, P. Dunnill, A.E. Humphrey, and M.D. Lilly, 1979,* Fermentation and Enzyme Technology, *John Wiley, New York.*

12.9 GROWTH KINETICS WITH PLASMID INSTABILITY

A potential problem in culture of recombinant organisms is plasmid loss or inactivation. Plasmid instability occurs in individual cells which, by reproducing, can generate a large plasmid-free population in the reactor and reduce the overall rate of synthesis of plasmid-encoded products. Plasmid instability occurs as a result of DNA mutation or defective plasmid segregation. For segregational stability, the total number of plasmids present in the culture must double once per generation and the plasmid copies must be distributed equally between mother and daughter cells.

A simple model has been developed for batch culture to describe changes in the fraction of plasmid-bearing cells as a function of time [12]. The important parameters in this model are the probability of plasmid loss per generation of cells, and the difference in growth rate between plasmid-bearing and plasmid-free cells. Exponential growth of the host cells is assumed. If x^+ is the concentration of plasmid-carrying cells and x^- is the concentration of plasmid-free cells, the rates at which the two cell populations grow are:

$$r_{X^+} = (1 - p)\, \mu^+ x^+ \tag{12.92}$$

and

$$r_{X^-} = p\,\mu^+ x^+ + \mu^- x^- \tag{12.93}$$

where r_{X^+} is the rate of growth of the plasmid-bearing population, r_{X^-} is the rate of growth of the plasmid-free population, p is the probability of plasmid loss per cell division ($p \leq 1$), μ^+ is the specific growth rate of plasmid-carrying cells, and μ^- is the specific growth rate of plasmid-free cells.

The model assumes that all plasmid-containing cells are identical in growth rate and probability of plasmid loss; this is the same as assuming that all plasmid-containing cells have the same copy number. By comparing Eq. (12.92) with Eq. (12.80) we can see that the rate of growth of the plasmid-bearing population is reduced by ($p\,\mu^+ x^+$). This is because some of the progeny of plasmid-bearing cells do not contain plasmid and do not join the plasmid-bearing population. On the other hand, growth of the plasmid-free population has two contributions as indicated in Eq. (12.93). Existing plasmid-free cells grow with specific growth rate μ^-; in addition, this population is supplemented by the generation of plasmid-free cells due to defective plasmid segregation by plasmid-carrying cells.

At any time, the fraction of cells in the culture with plasmid is:

$$F = \frac{x^+}{x^+ + x^-} \tag{12.94}$$

In batch culture where rates of growth can be determined by monitoring cell concentration, $r_{X^+} = dx^+/dt$ and $r_{X^-} = dx^-/dt$. Therefore, Eqs. (12.92) and (12.93) can be integrated simultaneously with initial conditions $x^+ = x_0^+$ and $x^- = x_0^-$ at $t = 0$. After n generations of plasmid-containing cells:

$$F = \frac{1 - \alpha - p}{1 - \alpha - 2^{n(\alpha + p - 1)}p} \tag{12.95}$$

where

$$\alpha = \frac{\mu^-}{\mu^+} \tag{12.96}$$

and

$$n = \frac{\mu^+ t}{\ln 2} \tag{12.97}$$

The value of F depends on α, the ratio of the specific growth rates of plasmid-free and plasmid-carrying cells. In the absence of selection pressure, the presence of plasmid usually reduces the growth rate of organisms due to the additional metabolic requirements imposed by the plasmid DNA. Therefore α is usually >1.

In general, the difference between μ^+ and μ^- becomes more pronounced as the size of the plasmid or copy number increases. Some values of α from the literature are listed in Table 12.8; typically $1.0 < \alpha < 2.0$. Under selection pressure α may equal zero; if the plasmid encodes biosynthetic enzymes for essential nutrients, the loss of plasmid may result

TABLE 12.8 Relative Growth Rates of Plasmid-Free and Plasmid-Carrying Cells

Organism	Plasmid	$\alpha = \frac{\mu^-}{\mu^+}$
Escherichia coli C600	F′ lac	0.99–1.10
E. coli K12 EC1055	R1*drd*-19	1.03–1.12
E. coli K12 IR713	TP120 (various)	1.50–2.31
E. coli JC7623	Col E1	1.29
	Col E1 derivative TnA insertion (various)	1.15–1.54
	Col E1 deletion mutant (various)	1.06–1.65
Pseudomonas aeruginosa PA01	TOL	2.00

Source: From T. Imanaka and S. Aiba, 1981, A perspective on the application of genetic engineering: stability of recombinant plasmid. Ann. N.Y. Acad. Sci. *369, 1–14.*

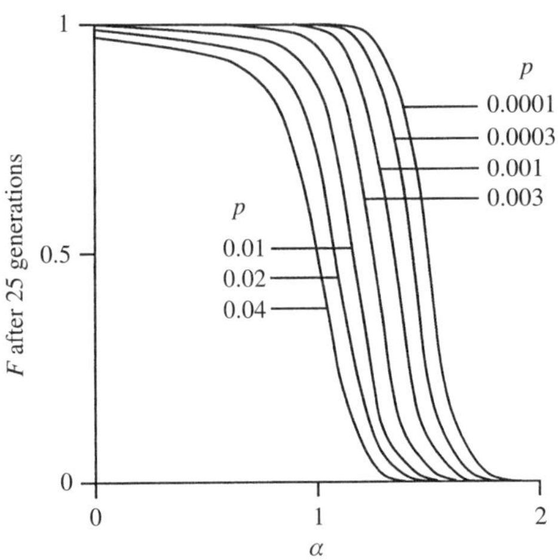

FIGURE 12.17 Fraction of plasmid-carrying cells in batch culture after 25 generations.
From T. Imanaka and S. Aiba, 1981, A perspective on the application of genetic engineering: stability of recombinant plasmid. Ann. N.Y. Acad. Sci. *369, 1–14.*

in $\mu^- = 0$. When this is the case, F remains close to 1 as the plasmid-free population cannot reproduce. F also depends on p, the probability of plasmid loss per generation, which can be as high as 0.1 if segregation occurs. When mutation or random insertions or deletions are the only cause of plasmid instability, p is usually much lower at about 10^{-6}. Plasmid fragmentation within a host cell can occur with higher frequency if the cloning vector is inherently unstable.

Batch culture of microorganisms usually requires 25 cell generations or more. Results for F after 25 generations have been calculated from Eq. (12.95) and are shown in Figure 12.17 as a function of p and α. F deteriorates substantially as α increases from 1.0 to 2.0. Cultures with $p < 0.01$ and $\alpha < 1$ are relatively stable, with F after 25 generations remaining close to 1. Further application of Eq. (12.95) is illustrated in Example 12.7.

EXAMPLE 12.7 PLASMID INSTABILITY IN BATCH CULTURE

A plasmid-containing strain of *E. coli* is used to produce recombinant protein in a 250-litre fermenter. The probability of plasmid loss per generation is 0.005. The specific growth rate of plasmid-free cells is 1.4 h^{-1}; the specific growth rate of plasmid-bearing cells is 1.2 h^{-1}. Estimate the fraction of plasmid-bearing cells after 18 h of growth if the inoculum contains only cells with plasmid.

Solution

The number of generations of plasmid-carrying cells after 18 h is calculated from Eq. (12.97):

$$n = \frac{(1.2\ \text{h}^{-1})\ 18\ \text{h}}{\ln 2} = 31$$

Substituting this into Eq. (12.95) with $p = 0.005$ and $\alpha = 1.4\ \text{h}^{-1}/1.2\ \text{h}^{-1} = 1.17$:

$$F = \frac{1 - 1.17 - 0.005}{1 - 1.17 - 2^{31(1.17+0.005-1)}\ 0.005} = 0.45$$

Therefore, after 18 h only 45% of the cells contain plasmid.

Alternative models for growth with plasmid instability have been developed [13]; some include equations for substrate utilisation and product formation [14]. A weakness in the simple model presented here is the assumption that all plasmid-containing cells are the same. In reality, there are differences in copy number and therefore specific growth rate between plasmid-carrying cells; the probability of plasmid loss also varies from cell to cell. More complex models that recognise the segregated nature of plasmid populations are available [15, 16].

12.10 PRODUCTION KINETICS IN CELL CULTURE

In this section we consider the kinetics of production of low-molecular-weight compounds, such as ethanol, amino acids, antibiotics, and vitamins, which are excreted from cells in culture. As indicated in Table 12.9, fermentation products can be classified according to the relationship between product synthesis and energy generation in the cell [10, 17]. Compounds in the first category are formed directly as end- or by-products of energy metabolism; these materials are synthesised using pathways that produce ATP. The second class of product is partly linked to energy generation but requires additional energy for synthesis. Formation of other products such as antibiotics involves reactions far removed from energy metabolism.

Irrespective of the class of product, the rate of product formation in cell culture can be expressed as a function of biomass concentration:

$$r_P = q_P x \qquad (12.98)$$

TABLE 12.9 Classification of Low-Molecular-Weight Fermentation Products

Class of Metabolite	Examples
Products directly associated with generation of energy in the cell	Ethanol, acetic acid, gluconic acid, acetone, butanol, lactic acid, other products of anaerobic fermentation
Products indirectly associated with energy generation	Amino acids and their products, citric acid, nucleotides
Products for which there is no clear direct or indirect coupling to energy generation	Penicillin, streptomycin, vitamins

where r_P is the volumetric rate of product formation with units of, for example, kg m^{-3} s^{-1}, x is biomass concentration, and q_P is the *specific rate of product formation* with dimensions T^{-1}. q_P can be evaluated at any time during fermentation as the ratio of production rate and biomass concentration; q_P is not necessarily constant during batch culture. Depending on whether the product is linked to energy metabolism or not, we can develop equations for q_P as a function of growth rate and other metabolic parameters.

12.10.1 Product Formation Directly Coupled with Energy Metabolism

For products formed using pathways that generate ATP, the rate of production is related to the cellular energy demand. Growth is usually the major energy-requiring function of cells; therefore, if production is coupled to energy metabolism, product will be formed whenever there is growth. However, ATP is also required for other activities called *maintenance*. Examples of maintenance functions include cell motility, turnover of cellular components, and adjustment of membrane potential and internal pH. Maintenance activities are carried out by living cells even in the absence of growth. Products synthesised in energy pathways will be produced whenever maintenance functions are carried out because ATP is required. Kinetic expressions for the rate of product formation must account for growth-associated and maintenance-associated production, as in the following equation:

$$r_P = Y_{PX} r_X + m_P x \tag{12.99}$$

In Eq. (12.99), r_X is the volumetric rate of biomass formation, Y_{PX} is the theoretical or true yield of product from biomass, m_P is the *specific rate of product formation due to maintenance*, and x is biomass concentration. m_P has dimensions T^{-1} and typical units of kg product (kg biomass)$^{-1}$ s^{-1}. Equation (12.99) states that the rate of product formation depends partly on the rate of growth but partly also on cell concentration. Applying Eq. (12.80) to Eq. (12.99):

$$r_P = (Y_{PX}\mu + m_P)x \tag{12.100}$$

Comparison of Eqs. (12.98) and (12.100) shows that, for products coupled to energy metabolism, q_P is equal to a combination of growth-associated and nongrowth-associated terms:

$$q_P = Y_{PX}\mu + m_P \tag{12.101}$$

12.10.2 Product Formation Indirectly Coupled with Energy Metabolism

When product is synthesised partly in metabolic pathways used for energy generation and partly in other pathways requiring energy, the relationship between product formation and growth can be complicated. We will not attempt to develop equations for q_P for this type of product. A generalised treatment of product formation indirectly coupled to energy metabolism is given by Roels and Kossen [10].

12.10.3 Product Formation Not Coupled with Energy Metabolism

Production not involving energy metabolism is difficult to relate to growth because growth and product synthesis are somewhat dissociated. However, in some cases, the rate of formation of nongrowth-associated product is directly proportional to the biomass concentration, so that the production rate defined in Eq. (12.98) can be applied with constant q_P. Sometimes q_P is a complex function of the growth rate and must be expressed using empirical equations derived from experiment. An example is penicillin synthesis; equations for the rate of penicillin production as a function of biomass concentration and specific growth rate have been derived by Heijnen et al. [18].

12.11 KINETICS OF SUBSTRATE UPTAKE IN CELL CULTURE

Cells consume substrate from the external environment and channel it into different metabolic pathways. Some substrate may be directed into growth and product synthesis; another fraction is used to generate energy for maintenance activities. Substrate requirements for maintenance vary considerably depending on the organism and culture conditions; a complete account of substrate uptake should include a maintenance component. The specific rate of substrate uptake for maintenance activities is known as the *maintenance coefficient*, m_S. The dimensions of m_S are T^{-1}; typical units are kg substrate (kg biomass)$^{-1}$ s^{-1}. Some examples of maintenance coefficients for various microorganisms are listed in Table 12.10. The ionic strength of the culture medium exerts a considerable influence on the value of m_S; significant amounts of energy are needed to maintain concentration gradients across cell membranes. The physiological significance of m_S has been the subject of much debate; there are indications that m_S for a particular organism may not be constant at all possible growth rates.

The rate of substrate uptake during cell culture can be expressed as a function of biomass concentration using an equation analogous to Eq. (12.98):

$$r_S = q_S x \tag{12.102}$$

where r_S is the volumetric rate of substrate consumption with units of, for example, kg m^{-3} s^{-1}, q_S is the *specific rate of substrate uptake*, and x is the biomass concentration. Like q_P, q_S has dimensions T^{-1}. In this section, we will develop equations for q_S as a function of growth rate and other relevant metabolic parameters.

TABLE 12.10 Maintenance Coefficients for Several Microorganisms with Glucose as Energy Source

Microorganism	Growth Conditions	m_S (kg substrate (kg cells)$^{-1}$ h^{-1})
Saccharomyces cerevisiae	Anaerobic	0.036
	Anaerobic, 1.0 M NaCl	0.360
Azotobacter vinelandii	Nitrogen fixing, 0.2 atm dissolved oxygen tension	1.5
	Nitrogen fixing, 0.02 atm dissolved oxygen tension	0.15
Klebsiella aerogenes	Anaerobic, tryptophan-limited, 2 g l^{-1} NH_4Cl	2.88
	Anaerobic, tryptophan-limited, 4 g l^{-1} NH_4Cl	3.69
Lactobacillus casei		0.135
Aerobacter cloacae	Aerobic, glucose-limited	0.094
Penicillium chrysogenum	Aerobic	0.022

From S.J. Pirt, 1975, Principles of Microbe and Cell Cultivation, *Blackwell Scientific, Oxford.*

12.11.1 Substrate Uptake in the Absence of Extracellular Product Formation

In some cultures there is little or no extracellular product formation; for example, biomass itself is the product of fermentation in the manufacture of bakers' yeast and single-cell protein. In the absence of extracellular product synthesis, we assume that all substrate entering the cell is used for growth and maintenance functions. The rates of these cell activities are related as follows:

$$r_S = \frac{r_X}{Y_{XS}} + m_S x \qquad (12.103)$$

In Eq. (12.103), r_X is the volumetric rate of biomass production, Y_{XS} is the true yield of biomass from substrate, m_S is the maintenance coefficient, and x is the biomass concentration. Equation (12.103) states that the rate of substrate uptake depends partly on the rate of growth but also varies with cell concentration directly. When r_X is expressed using Eq. (12.80), Eq. (12.103) becomes:

$$r_S = \left(\frac{\mu}{Y_{XS}} + m_S\right) x \qquad (12.104)$$

If we now express μ as a function of substrate concentration using Eq. (12.91), Eq. (12.104) becomes:

$$r_S = \left[\frac{\mu_{max} s}{Y_{XS}(K_S + s)} + m_S\right] x \qquad (12.105)$$

When s is zero, Eq. (12.105) predicts that substrate consumption will proceed at a rate equal to $m_S x$. Substrate uptake in the absence of substrate is impossible; this feature of Eq. (12.105) is therefore unrealistic. The problem arises because of implicit assumptions we have made about the nature of maintenance activities. It can be shown, however, that

Eq. (12.105) is a realistic description of substrate uptake as long as there is external substrate available. When the substrate is exhausted, maintenance energy is generally supplied by endogenous metabolism.

12.11.2 Substrate Uptake with Extracellular Product Formation

Patterns of substrate flow in cells synthesising products depend on whether product formation is coupled to energy metabolism. When products are formed in energy-generating pathways, for example, in anaerobic culture, product synthesis is an unavoidable consequence of cell growth and maintenance. As illustrated in Figure 12.18(a), there is no separate flow of substrate into the cell for product synthesis; product is formed from the substrate taken up to support growth and maintenance. Substrate consumed for maintenance does not contribute to growth; it therefore constitutes a separate substrate flow into the cell. In contrast, when production is not linked or only partly linked to energy metabolism, all or some of the substrate required for product synthesis is in addition to and separate from that needed for growth and maintenance. Flow of substrate in this case is illustrated in Figure 12.18(b).

When products are directly linked to energy generation, equations for the rate of substrate consumption do not include a separate term for production: the substrate requirements for product formation are already accounted for in the terms representing

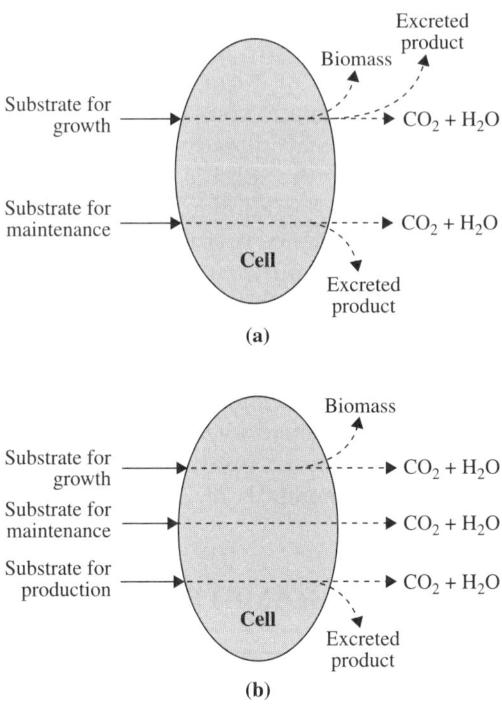

FIGURE 12.18 Substrate uptake with product formation: (a) product formation directly coupled to energy metabolism; (b) product formation not directly coupled to energy metabolism.

growth- and maintenance-associated substrate uptake. Accordingly, the equations presented in Section 12.11.1 for substrate consumption in the absence of extracellular product formation apply; the rate of substrate uptake is related to growth and maintenance requirements by Eqs. (12.104) and (12.105).

In cultures where product synthesis is only indirectly coupled to energy metabolism, the rate of substrate consumption is a function of three factors: the growth rate, the rate of product formation, and the rate of substrate uptake for maintenance. These different cell functions can be related using yield and maintenance coefficients:

$$r_S = \frac{r_X}{Y_{XS}} + \frac{r_P}{Y_{PS}} + m_S x \tag{12.106}$$

where r_S is the volumetric rate of substrate consumption, r_X is the volumetric rate of biomass production, r_P is the volumetric rate of product formation, Y_{XS} is the true yield of biomass from substrate, Y_{PS} is the true yield of product from substrate, m_S is the maintenance coefficient, and x is the biomass concentration. If we express r_X and r_P using Eqs. (12.80) and (12.98), respectively:

$$r_S = \left(\frac{\mu}{Y_{XS}} + \frac{q_P}{Y_{PS}} + m_S\right)x \tag{12.107}$$

12.12 EFFECT OF CULTURE CONDITIONS ON CELL KINETICS

Temperature has a marked effect on metabolic rate. Temperature influences reaction rates directly according to the Arrhenius equation (Section 12.1.5); it can also change the configuration of cell constituents, especially proteins and membrane components. In general, the effect of temperature on growth is similar to that already described in Section 12.3.4 for enzymes. An approximate twofold increase in the specific growth rate of cells occurs for every 10°C rise in temperature, until structural breakdown of cell proteins and lipids starts to occur. Like other rate constants, the maintenance coefficient m_S has an Arrhenius-type temperature dependence [19]; this can have a significant kinetic effect on cultures where turnover of macromolecules is an important contribution to maintenance requirements. In contrast, temperature has only a minor effect on the biomass yield coefficient, Y_{XS} [19]. Other cellular responses to temperature are described elsewhere [1, 20, 21].

The growth rate of cells depends on medium pH in much the same way as enzyme activity (Section 12.3.4). Maximum growth rate is usually maintained over 1 to 2 pH units but declines with further variation. pH also affects the profile of product synthesis in anaerobic culture and can change maintenance energy requirements [1, 20, 21].

12.13 DETERMINING CELL KINETIC PARAMETERS FROM BATCH DATA

To apply the equations presented in Sections 12.8 through 12.11 to real fermentations, we must know the kinetic and yield parameters for the system and have information about the rates of growth, substrate uptake, and product formation. Batch culture is the most frequently applied method for investigating kinetic behaviour, but it is not always the best. Methods for determining reaction parameters from batch data are described in the following sections.

12.13.1 Rates of Growth, Product Formation, and Substrate Uptake

Determining growth rates in cell culture requires measurement of cell concentration. Many different experimental procedures are applied for biomass estimation [20, 22]. Direct measurement can be made of cell number, dry or wet cell mass, packed cell volume, or culture turbidity; alternatively, indirect estimates are obtained from measurements of product formation, heat evolution, or cell composition. Cell viability is usually evaluated using plating or staining techniques. Each method for biomass estimation will give somewhat different results. For example, the rate of growth determined using cell dry weight may differ from that based on cell number because dry weight in the culture can increase without a corresponding increase in the number of cells.

Irrespective of how cell concentration is measured, the techniques described in Section 12.2 for graphical differentiation of concentration data are suitable for determining volumetric growth rates in batch culture. The results will depend to some extent on how the data are smoothed. For the reasons described in Section 12.8.3, there is usually a relatively abrupt change in growth rate between the growth and stationary phases; this feature requires extra care for accurate differentiation of batch growth curves. As discussed in Section 3.3.1, an advantage of hand-smoothing is that it provides us with an opportunity to judge the significance of individual datum points. Once the volumetric growth rate r_X is known, the specific growth rate μ can be obtained by dividing r_X by the cell concentration.

During the exponential growth phase of batch culture, an alternative method can be applied to calculate μ. Assuming that growth is represented by the first-order model of Eq. (12.80), the integrated relationship of Eq. (12.84) or (12.85) allows us to obtain μ directly. During the exponential growth phase when μ is essentially constant, a plot of ln x versus time gives a straight line with slope μ. This is illustrated in Example 12.8.

EXAMPLE 12.8 HYBRIDOMA DOUBLING TIME

A mouse–mouse hybridoma cell line is used to produce monoclonal antibody. Growth in batch culture is monitored with the following results.

Time (days)	Cell Concentration (cells ml^{-1} × 10^{-6})
0.0	0.45
0.2	0.52
0.5	0.65
1.0	0.81
1.5	1.22

(Continued)

2.0	1.77
2.5	2.13
3.0	3.55
3.5	4.02
4.0	3.77
4.5	2.20

(a) Determine the specific growth rate during the growth phase.
(b) What is the culture doubling time?

Solution

(a) The data are plotted as a semi-log graph in Figure 12.19.
No lag phase is evident. As Eq. (12.84) applies only when μ is constant, that is, during exponential growth, we must determine which datum points belong to the exponential growth phase. In Figure 12.19, the final three points appear to belong to the decline and death phases of the culture. Fitting a straight line to the remaining data up to and including $t = 3.0$ days gives a slope of 0.67 day^{-1}. Therefore, $\mu = 0.67$ day^{-1}.

(b) From Eq. (12.89):

$$t_d = \frac{\ln 2}{0.67 \text{ day}^{-1}} = 1.0 \text{ day}$$

This doubling time applies only during the exponential growth phase of the culture.

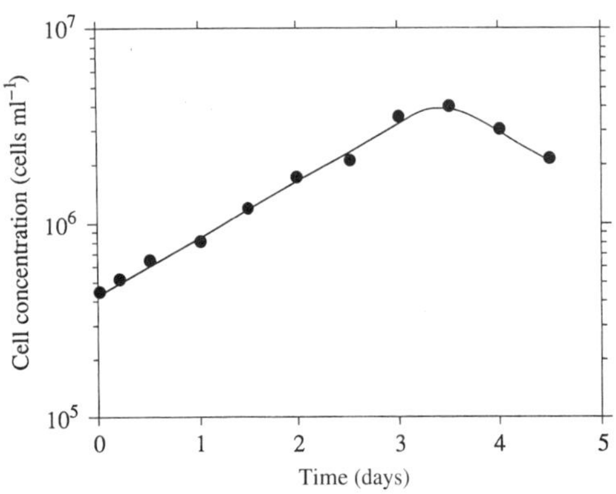

FIGURE 12.19 Calculation of the specific growth rate for hybridoma cells.

The volumetric rates of product formation and substrate uptake, r_P and r_S, can be evaluated by graphical differentiation of product and substrate concentration data, respectively. The specific product formation rate q_P and specific substrate uptake rate q_S are obtained by dividing the respective volumetric rates by the cell concentration.

12.13.2 μ_{max} and K_S

The Monod equation for the specific growth rate, Eq. (12.91), is analogous mathematically to the Michaelis–Menten expression for enzyme kinetics. In principle, therefore, the techniques described in Section 12.4 for determining v_{max} and K_m for enzyme reaction can be applied for evaluation of μ_{max} and K_S. However, because values of K_S in cell culture are usually very small, accurate determination of this parameter from batch data is difficult. Better estimation of K_S can be made using continuous culture of cells as discussed in Chapter 14. On the other hand, measurement of μ_{max} from batch data is relatively straightforward. As described in Section 12.8.3, if all nutrients are present in excess, the specific growth rate during exponential growth is equal to the maximum specific growth rate. Therefore, the specific growth rate calculated using the procedure of Example 12.8 can be considered equal to μ_{max}.

12.14 EFFECT OF MAINTENANCE ON YIELDS

True yields such as Y_{XS}, Y_{PX}, and Y_{PS} are often difficult to evaluate. True yields are essentially stoichiometric coefficients (Section 4.6); however, the stoichiometry of biomass production and product formation is known only for relatively simple fermentations. If the metabolic pathways are complex, stoichiometric calculations become too complicated. However, theoretical yields can be related to observed yields such as Y'_{XS}, Y'_{PX}, and Y'_{PS}, which are more easily determined.

12.14.1 Observed Yields

Expressions for observed yield coefficients are obtained by applying Eq. (12.76):

$$Y'_{XS} = \frac{-dX}{dS} = \frac{r_X}{r_S} \tag{12.108}$$

$$Y'_{PX} = \frac{dP}{dX} = \frac{r_P}{r_X} \tag{12.109}$$

and

$$Y'_{PS} = \frac{-dP}{dS} = \frac{r_P}{r_S} \tag{12.110}$$

where X, S, and P are masses of cells, substrate, and product, respectively, and r_X, r_S, and r_P are observed rates evaluated from experimental data. Therefore, observed yield coefficients can be determined by plotting ΔX, ΔS, or ΔP against each other and evaluating the slope as illustrated in Figure 12.20. Alternatively, observed yield coefficients at a particular instant in time can be calculated as the ratio of rates evaluated at that instant. Observed yields are not necessarily constant throughout batch culture; in some cases they exhibit significant dependence on growth rate and environmental parameters such as substrate concentration. Nevertheless, for many cultures, the observed biomass yield Y'_{XS} is approximately constant during batch growth. Because of the errors in experimental data, considerable uncertainty is usually associated with measured yield coefficients.

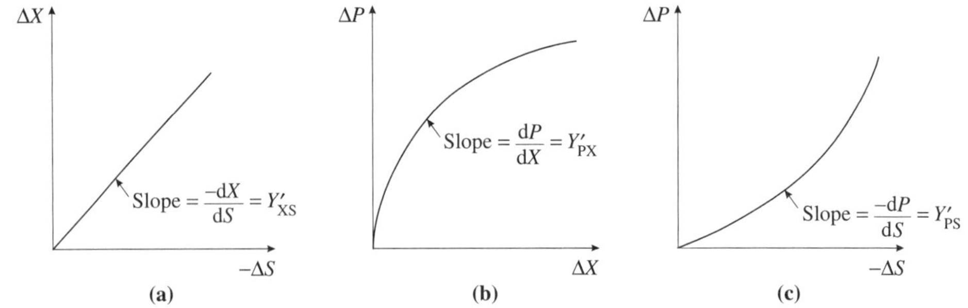

FIGURE 12.20 Evaluation of observed yields in batch culture from cell, substrate, and product concentrations.

12.14.2 Biomass Yield from Substrate

Equations for the true biomass yield can be determined for systems without extracellular product formation or when product synthesis is directly coupled to energy metabolism. Substituting expressions for r_X and r_S from Eqs. (12.80) and (12.104) into Eq. (12.108) gives:

$$Y'_{XS} = \frac{\mu}{\left(\dfrac{\mu}{Y_{XS}} + m_S\right)} \tag{12.111}$$

Inverting Eq. (12.111) produces the expression:

$$\frac{1}{Y'_{XS}} = \frac{1}{Y_{XS}} + \frac{m_S}{\mu} \tag{12.112}$$

Therefore, if Y_{XS} and m_S are relatively constant, a plot of $1/Y'_{XS}$ versus $1/\mu$ gives a straight line with slope m_S and intercept $1/Y_{XS}$. Equation (12.112) is not generally applied to batch growth data; under typical batch conditions, μ does not vary from μ_{max} for much of the culture period so it is difficult to plot Y'_{XS} as a function of specific growth rate. We will revisit Eq. (12.112) when we consider continuous cell culture in Chapter 14. As a rule of thumb, the true biomass yield from glucose under aerobic conditions is around 0.5 g g^{-1}.

In processes such as the production of bakers' yeast and single-cell protein where the required product is biomass, it is desirable to maximise the actual or observed yield of cells from substrate. The true yield Y_{XS} is limited by stoichiometric considerations. However, from Eq. (12.111), Y'_{XS} can be improved by decreasing the maintenance coefficient. This is achieved by lowering the temperature of fermentation, using a medium of lower ionic strength, or applying a different organism or strain with lower maintenance energy requirements. Assuming that these changes do not also alter the growth rate, they can be employed to improve the biomass yield.

When a culture produces compounds that are not directly coupled with energy metabolism, Eqs. (12.111) and (12.112) do not apply because a different expression for r_S must be used in Eq. (12.108). Determination of true yields and maintenance coefficients is more difficult in this case because of the number of terms involved.

12.14.3 Product Yield from Biomass

The observed yield of product from biomass Y'_{PX} is defined in Eq. (12.109). When product synthesis is directly coupled to energy metabolism, r_P is given by Eq. (12.100). Substituting this and Eq. (12.80) into Eq. (12.109) gives:

$$Y'_{PX} = Y_{PX} + \frac{m_P}{\mu} \tag{12.113}$$

The extent of deviation of Y'_{PX} from Y_{PX} depends on the relative magnitudes of m_P and μ. To increase the observed yield of product for a particular process, m_P should be increased and μ decreased. Equation (12.113) does not apply to products not directly coupled to energy metabolism; we do not have a general expression for r_P in terms of the true yield coefficient for this class of product.

12.14.4 Product Yield from Substrate

The observed product yield from substrate Y'_{PS} is defined in Eq. (12.110). For products coupled to energy generation, expressions for r_P and r_S are available from Eqs. (12.100) and (12.104). Therefore:

$$Y'_{PS} = \frac{Y_{PX}\mu + m_P}{\left(\frac{\mu}{Y_{XS}} + m_S\right)} \tag{12.114}$$

In many anaerobic fermentations such as ethanol production, the yield of product from substrate is a critical factor affecting process economics. At high Y'_{PS}, more ethanol is produced per mass of carbohydrate consumed and the overall cost of production is reduced. Growth rate has a strong effect on Y'_{PS} for ethanol. Because Y'_{PS} is low when $\mu = \mu_{max}$, it is desirable to reduce the specific growth rate of the cells. Low growth rates can be achieved by depriving the cells of some essential nutrient, for example, a nitrogen source, or by immobilising the cells to prevent growth. As outlined in Chapter 14, continuous culture provides more opportunity for manipulating rates of growth than batch culture. Increasing the rate of maintenance activity relative to growth will also enhance the product yield. This can be done by using a medium of high ionic strength, raising the temperature, or selecting a mutant or different organism with high maintenance requirements.

The effect of growth rate and maintenance on Y'_{PS} is difficult to determine for products not directly coupled with energy metabolism unless information is available about the effect of these parameters on q_P.

12.15 KINETICS OF CELL DEATH

The kinetics of cell death are an important consideration in the design of sterilisation processes and for analysis of cell cultures where substantial loss of viability is expected. In a lethal environment, cells in a population do not die all at once; deactivation of the culture occurs over a finite period of time depending on the initial number of viable cells and the severity of the conditions imposed. Loss of cell viability can be described mathematically in much the same way as enzyme deactivation (Section 12.6). Cell death is considered to be a first-order process:

$$r_d = k_d N \tag{12.115}$$

where r_d is the rate of cell death, N is the number of viable cells, and k_d is the *specific death constant*. Alternatively, the rate of cell death can be expressed in terms of cell concentration rather than cell number:

$$r_d = k_d x \tag{12.116}$$

where k_d is the specific death constant based on cell concentration and x is the concentration of viable cells.

In a closed system where cell death is the only process affecting the concentration of viable cells, the rate of cell death is equal to the rate of decrease in the number of cells. Therefore, using Eq. (12.115):

$$r_d = \frac{-dN}{dt} = k_d N \tag{12.117}$$

If k_d is constant, we can integrate Eq. (12.117) to derive an expression for N as a function of time:

$$N = N_0 e^{-k_d t} \tag{12.118}$$

where N_0 is the number of viable cells at time zero. Taking natural logarithms of both sides of Eq. (12.118) gives:

$$\ln N = \ln N_0 - k_d t \tag{12.119}$$

According to Eq. (12.119), if first-order death kinetics apply, a plot of ln N versus t gives a straight line with slope $-k_d$. Experimental measurements have confirmed the relationship of Eq. (12.119) for many organisms in a vegetative state; as an example, results for the thermal death of *Escherichia coli* at various temperatures are shown in Figure 12.21. However, first-order death kinetics do not always hold, particularly for bacterial spores immediately after exposure to heat.

Like other kinetic constants, the value of the specific death constant k_d depends on temperature. This effect can be described using the Arrhenius relationship of Eq. (12.74). Typical E_d values for the thermal destruction of microorganisms are high, of the order 250 to 290 kJ gmol^{-1} [23]. Therefore, small increases in temperature have a significant effect on k_d and the rate of cell death.

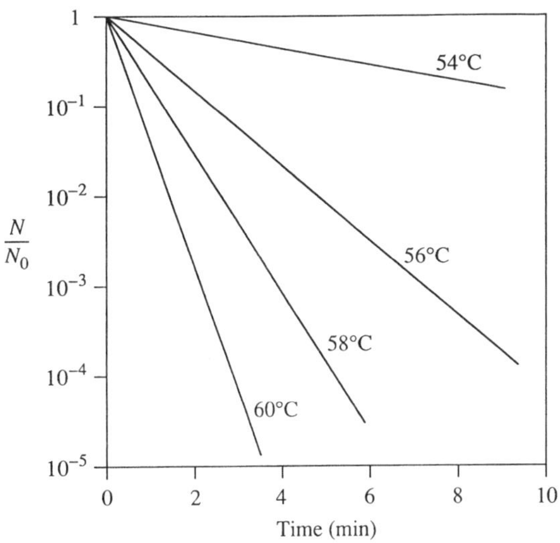

FIGURE 12.21 Relationship between temperature and the rate of thermal death for vegetative *Escherichia coli* cells.
From S. Aiba, A.E. Humphrey, and N.F. Millis, 1965, Biochemical Engineering, *Academic Press, New York.*

EXAMPLE 12.9 THERMAL DEATH KINETICS

The number of viable spores of a new strain of *Bacillus subtilis* is measured as a function of time at various temperatures.

	Number of Spores at:			
Time (min)	$T = 85°C$	$T = 90°C$	$T = 110°C$	$T = 120°C$
0.0	2.40×10^9	2.40×10^9	2.40×10^9	2.40×10^9
0.5	2.39×10^9	2.38×10^9	1.08×10^9	2.05×10^7
1.0	2.37×10^9	2.30×10^9	4.80×10^8	1.75×10^5
1.5	–	2.29×10^9	2.20×10^8	1.30×10^3
2.0	2.33×10^9	2.21×10^9	9.85×10^7	–
3.0	2.32×10^9	2.17×10^9	2.01×10^7	–
4.0	2.28×10^9	2.12×10^9	4.41×10^6	–
6.0	2.20×10^9	1.95×10^9	1.62×10^5	–
8.0	2.19×10^9	1.87×10^9	6.88×10^3	–
9.0	2.16×10^9	1.79×10^9	–	–

(a) Determine the activation energy for thermal death of *B. subtilis* spores.
(b) What is the specific death constant at 100°C?
(c) Estimate the time required to kill 99% of spores in a sample at 100°C.

Solution

(a) A semi-log plot of the number of viable spores versus time is shown in Figure 12.22 for each of the four temperatures. From Eq. (12.119), the slopes of the lines in Figure 12.22 are equal to $-k_d$ at the various temperatures. Fitting straight lines to the data gives the following results:

$$k_d\ (85°C) = 0.012\ \text{min}^{-1}$$
$$k_d\ (90°C) = 0.032\ \text{min}^{-1}$$
$$k_d\ (110°C) = 1.60\ \text{min}^{-1}$$
$$k_d\ (120°C) = 9.61\ \text{min}^{-1}$$

The relationship between k_d and absolute temperature is given by Eq. (12.74). Therefore, a semi-log plot of k_d versus $1/T$ should yield a straight line with slope $= -E_d/R$ where T is absolute temperature. Temperature is converted from degrees Celsius to Kelvin using the formula of Eq. (2.27); the results for k_d are plotted against $1/T$ in units of K^{-1} in Figure 12.23. The slope is −27,030 K. From Appendix B, $R = 8.3144\ \text{J K}^{-1}\ \text{gmol}^{-1}$. Therefore:

$$E_d = 27{,}030\ \text{K}\ (8.3144\ \text{J K}^{-1}\ \text{gmol}^{-1}) = 2.25 \times 10^5\ \text{J gmol}^{-1}$$
$$E_d = 225\ \text{kJ gmol}^{-1}$$

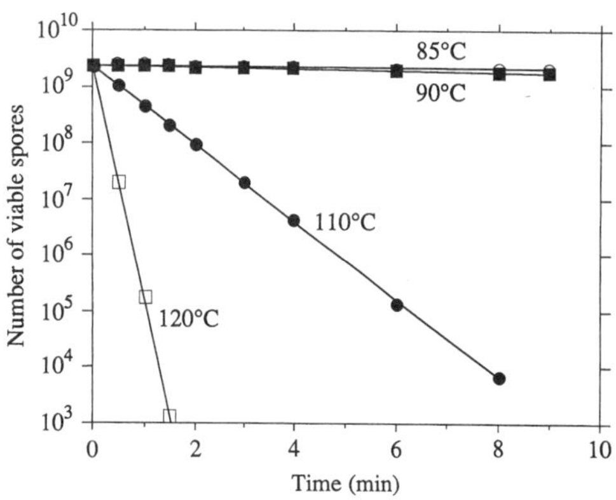

FIGURE 12.22 The thermal death of *Bacillus subtilis* spores.

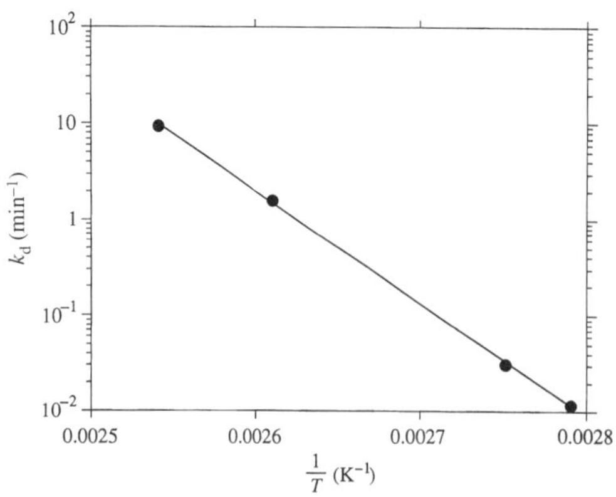

FIGURE 12.23 Calculation of kinetic parameters for thermal death of spores.

(b) The equation for the line in Figure 12.23 is:

$$k_d = 6.52 \times 10^{30}\, e^{-27{,}030/T}$$

where k_d has units of min^{-1} and T has units of K. Therefore, at $T = 100°\text{C} = 373.15$ K, $k_d = 0.23\ \text{min}^{-1}$.

(c) From Eq. (12.119):

$$t = \frac{-(\ln N - \ln N_0)}{k_d}$$

or

$$t = \frac{-\ln\left(\dfrac{N}{N_0}\right)}{k_d}$$

For N equal to 1% of N_0, $N/N_0 = 0.01$. At 100°C, $k_d = 0.23\ \text{min}^{-1}$ and the time required to kill 99% of the spores is:

$$t = \frac{-\ln(0.01)}{0.23\ \text{min}^{-1}} = 20\ \text{min}$$

When contaminating organisms in culture media are being killed by heat sterilisation, nutrients in the medium may also be destroyed. The sensitivity of nutrient molecules to temperature is described by the Arrhenius equation of Eq. (12.74). Values of the activation energy E_d for thermal destruction of vitamins and amino acids are 84 to 92 kJ gmol^{-1}; for proteins, E_d is about 165 kJ gmol^{-1} [23]. Because these values are somewhat lower than typical E_d values for the thermal death of microorganisms, raising the temperature has a greater effect on cell death than nutrient destruction. This means that sterilisation at higher temperatures for shorter periods of time has the advantage of killing cells with limited destruction of medium components.

12.16 METABOLIC ENGINEERING

Metabolic engineering is concerned with the behaviour and properties of reaction networks, including their structure, stoichiometry, kinetics, and control. In bioprocessing, metabolic engineering is used to redesign metabolic pathways so that the performance of cell cultures is improved. An understanding of the operation of reaction networks allows us to identify which reactions in metabolism are the best to target for genetic manipulation to enhance cellular properties. The improvements that can be achieved using metabolic engineering include:

- Increased yield and productivity of compounds synthesised by cells
- Synthesis of new products not previously produced by a particular organism
- Greater resistance to bioprocessing conditions, for example, tolerance of hypoxia
- Uptake and metabolism of substrates not previously assimilated
- Enhanced biological degradation of pollutants

Several other well-established strategies may be used to improve culture performance in bioprocesses. These include making operational changes to preculture and bioreactor conditions to optimise cell function, applying random mutagenesis and strain improvement programs to modify cellular phenotype, and using recombinant DNA technology to express new genes in cells, delete genes, amplify endogenous gene expression, and modulate enzyme activity. However, a feature of metabolic engineering that distinguishes it from these alternative approaches is that a detailed analysis of the metabolic pathways responsible for the cell property of interest is carried out before any changes are made to the organism or its environment. Alteration of single conversion steps has been shown to be generally ineffective for achieving significant changes in metabolism. Accordingly, the emphasis in metabolic engineering is on the functioning of integrated metabolic networks rather than single or a few enzyme reactions. As such, metabolic engineering is a step towards a systems biology approach to organism development.

As an example, consider the reaction network shown in Figure 12.24. From substrate A, two major products, J and L, are formed. Commercially, L is much more valuable than J, but J and L are normally produced in about equal amounts. This is indicated by the approximately equal thickness of the arrows leading to them in Figure 12.24(a). Without a systems analysis of the reaction network, it is decided to reduce the production of J by knocking out the gene coding for the enzyme responsible for converting A to B. This is considered an appropriate strategy as it prevents the wasteful channelling of A into the pathway leading to J. The outcome of this genetic manipulation is shown in Figure 12.24(b). Eliminating the reaction A→B also prevents the reaction B→G. As indicated by the altered thickness of several of the reaction arrows in Figure 12.24(b), the cell has compensated for the loss of this direct pathway to J by adjusting the throughput of other reactions in the network. The overall result is that production of L is virtually unchanged. Such an outcome is often observed after single-step modifications to metabolic pathways. Better understanding of the interconnectedness and influence of reactions across the entire network is required to make more informed decisions about genetic engineering strategies. Typically, modification of several reactions steps is required for effective adjustment of the selectivity and yield of metabolic pathways.

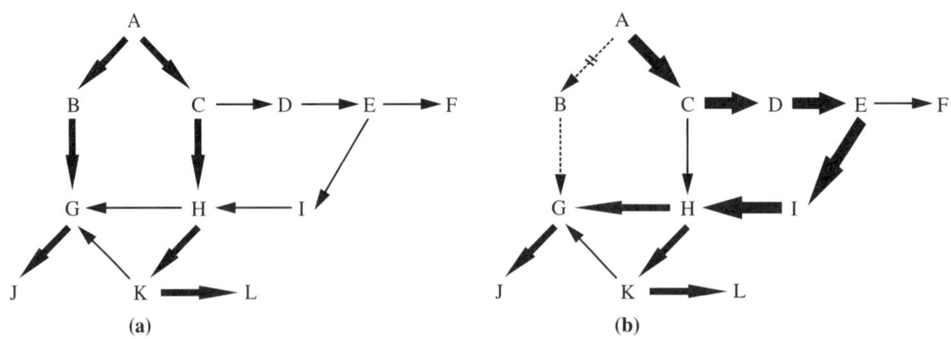

FIGURE 12.24 Reaction network for production of products J and L from substrate A: (a) before genetic modification of the cells, and (b) after genetic modification to prevent the conversion of A to B.

Implementation of metabolic engineering requires skills in engineering and biology. After pathways are subjected to engineering analysis to identify the reactions, enzymes, and gene targets for directed modification, molecular biology and recombinant DNA techniques are applied to remove or amplify the appropriate macromolecular components. In keeping with the topic and scope of this book, here we focus mainly on those aspects of metabolic engineering requiring the application of engineering methods and analysis.

12.16.1 Metabolic Flux Analysis

You are no doubt familiar with metabolic pathway diagrams such as that shown in Figure 12.25. This diagram presents several reaction networks that operate during anaerobic

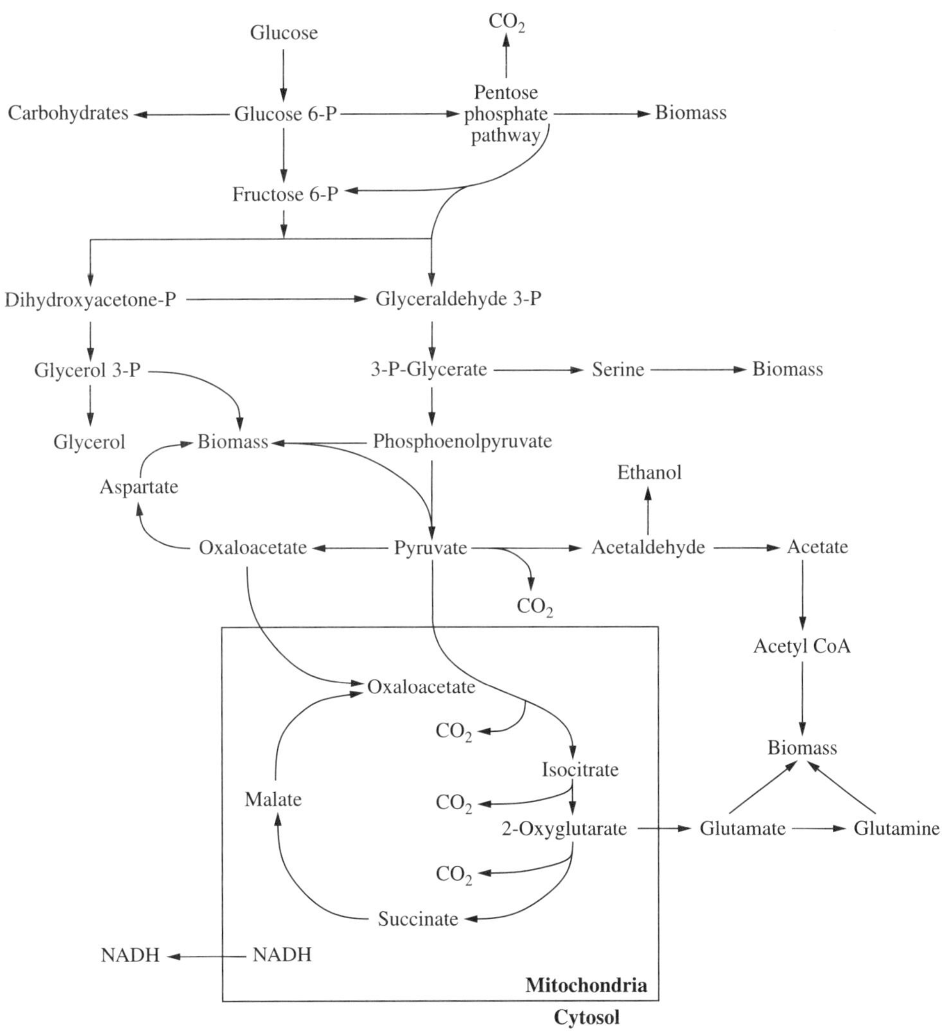

FIGURE 12.25 Metabolic pathways for anaerobic growth of *Saccharomyces cerevisiae*.

growth of *Saccharomyces cerevisiae*. Metabolic maps contain much useful information about the intermediates formed during metabolism and their reactive relationships with each other. However, they do not contain information about whether particular sections of the pathway are more or less active under actual culture conditions, or the relative contributions of specific reaction steps to the overall outcome of metabolism. For example, although the reactions required for synthesis of glycerol from glucose are represented in Figure 12.25, it is unclear if glycerol is a major or minor product of metabolism. Other information about operation of the network, such as the extent to which acetate synthesis detracts from ethanol production, and the amount of material directed into the pentose phosphate pathway relative to that processed by glycolysis, is not available without further investigation and analysis.

Figure 12.26 shows the results of a *metabolic flux analysis* for *Saccharomyces cerevisiae* cultured under anaerobic conditions. The term *metabolic flux* denotes the rate at which material is processed through a metabolic pathway. The flux in each branch of the reaction network is shown in Figure 12.26 for a particular set of culture conditions. The numbers at each conversion step indicate the flux of carbon in moles through that step relative to a flux of 100 for glucose uptake into the cell. A negative value for a particular conversion means that the net reaction operates in the reverse direction to that shown. The quantitative information included in Figure 12.26 reflects the relative engagement or participation of all of the major pathways in the network. The results show, for example, that glycerol is a relatively minor product of metabolism, as the carbon flux for glycerol synthesis is only 9.43/52.76 or 18% of the flux for ethanol production. Other details of interest, for example that acetate synthesis detracts negligibly from the carbon flow into ethanol, and that the pentose phosphate pathway draws only 8% of the carbon channelled into glycolysis, are also indicated.

The greatest benefits of metabolic flux analysis are obtained when several flux maps are compared after cultures are subjected to environmental or genetic perturbation. For instance, a comparison of fluxes under culture conditions that promote low and high yeast growth rates can be used to show how metabolites are redirected from one region of the pathway to another to accommodate the need for faster biomass production. Flux analysis is also used to evaluate cellular responses at the molecular level to amplification or deletion of genes for specific enzymes in pathways. Principal branch points in reaction networks and the effects of enzyme and pathway regulation can be identified from flux analyses applied before and after genetic manipulation.

Overall Approach

How are metabolic fluxes such as those shown in Figure 12.26 evaluated? Because the rates of reactions occurring internally within the cell are quantified in flux analysis, it is reasonable to expect that experimental measurement of the intracellular metabolites and enzyme reactions would be necessary. However, a major strength of flux analysis is that it does not require kinetic information about individual reactions or the measurement of intermediate concentrations to reconstruct network properties. In any case, the complexity of metabolic networks, the difficulty of measuring metabolite levels and enzyme kinetic parameters in vivo, and our incomplete understanding of the reaction sequences and enzymes involved in some areas of metabolism, make this approach either impossible or

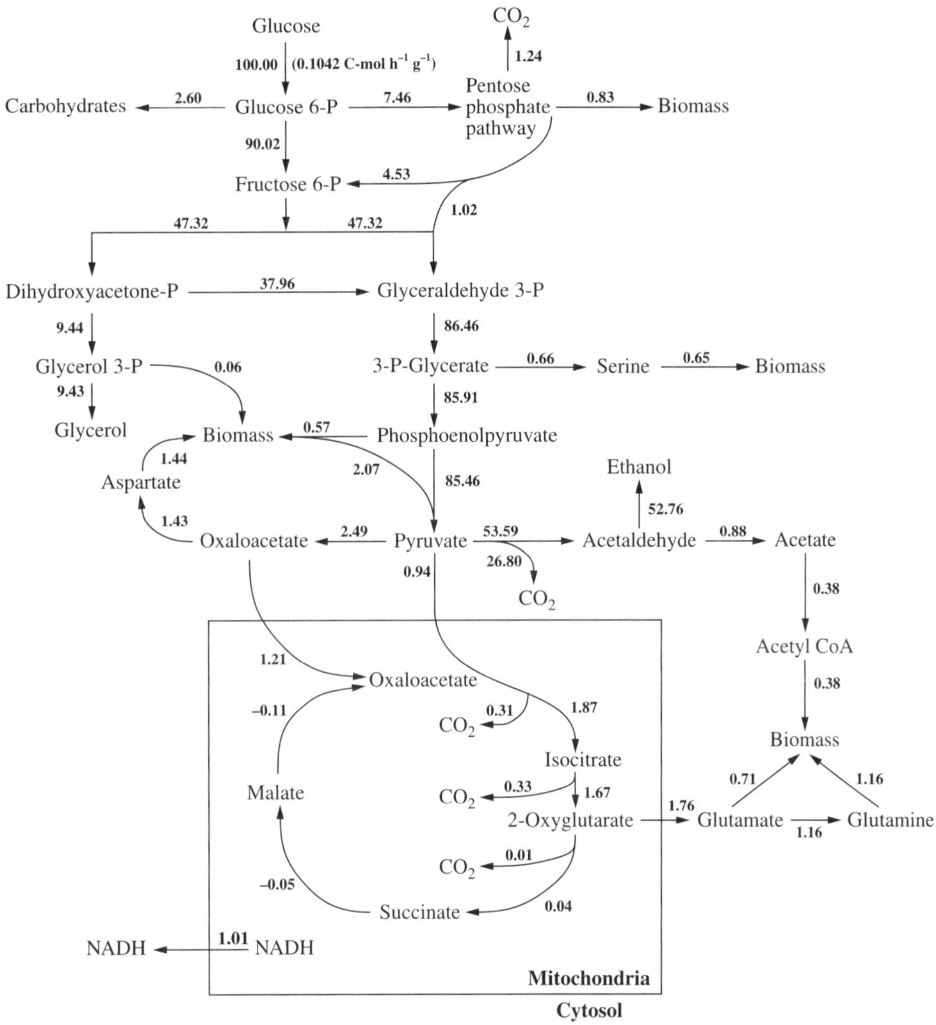

FIGURE 12.26 Metabolic pathways for anaerobic growth of *Saccharomyces cerevisiae* showing the fluxes of carbon in individual branches of the pathway relative to a flux of 100 for glucose uptake into the cell. The specific rate of glucose uptake was 0.1042 C-mol h^{-1} per g of biomass.
From T.L. Nissen, U. Schulze, J. Nielsen, and J. Villadsen, 1997, Flux distributions in anaerobic, glucose-limited continuous cultures of Saccharomyces cerevisiae. Microbiology *143, 203–218. Copyright 1997. Reproduced with permission of the Society for General Microbiology.*

very onerous in practice. Rather, in flux analysis, a limited number of experimental measurements, which could include substrate consumption, product formation, and biomass composition, is used in conjunction with a simplified stoichiometric analysis of the pathways to determine the internal reaction rates. In essence, highly complex, nonlinear, and intricately regulated reaction systems are reduced to a set of linear algebraic equations that can be solved relatively easily.

The methods employed in metabolic flux analysis draw on those introduced in Chapter 4 to evaluate the stoichiometry of cell growth and product synthesis. Here, we extend the macroscopic description of cell metabolism developed in Section 4.6 to include reactions that occur wholly within the cell. As intracellular reactions link the consumption of substrate to the production of biomass and extracellular products, determining their fluxes provides an understanding of how the overall outcome of metabolism is achieved. In Sections 12.10 and 12.11, we employed the notion that the rates of different processes, such as substrate consumption and product formation, could be related using appropriate yield coefficients. As molar yield coefficients and stoichiometric coefficients are, conceptually, the same thing (Sections 4.6.3 and 4.6.4), it follows that individual reaction rates in a network can be related using stoichiometric coefficients. We will make further use of these ideas in the following sections.

The key steps in metabolic flux analysis are the construction of a model reaction network for the metabolic pathways of interest, and the application of stoichiometric and mass-balance constraints and experimental data to evaluate intracellular fluxes. Several fundamental principles underpin flux analysis.

- *Mass and carbon balances.* Generally, flux analysis refers to the flux of carbon within metabolic pathways. Mass balances on intracellular metabolites show how carbon provided in the substrate is distributed between the different branches and reactions of the pathway. As the network model developed for flux analysis must satisfy the law of conservation of mass, mass is conserved across the entire system and at each reaction step.
- *Chemical energy balance.* Metabolic reactions generate and consume chemical energy through the interconversion of ATP, ADP, AMP, and similar compounds. Metabolic flux analysis can be carried out using mass balancing alone; however, when energy carriers are included in the model, the amount of energy generated in one section of the metabolic network must equal the amount consumed in other reactions.
- *Redox balance.* Redox carriers such as NADH, NADPH, and $FADH_2$ are involved in many cellular reactions. When redox carriers are included in flux calculations, redox generated in one part of the reaction network must be consumed in the remainder of the network.
- *Reaction thermodynamics.* Reactions in the flux model must be thermodynamically feasible. For example, if the results of flux analysis show a reverse flux through an irreversible reaction step, important thermodynamic principles have been violated. Such a result indicates an error in the analysis.

Network Definition and Simplification

Thousands of individual reactions are involved in cellular metabolism. Defining the network of reactions to be included in metabolic flux analysis is therefore a major undertaking. Substantial prior knowledge about the structure of the pathways, the components involved, and the links between metabolic subnetworks is required. Because flux analysis relies on carbon balancing, the fate of all C atoms must be defined within the metabolic model. If there are gaps in the reaction network involving unknown sources or sinks of

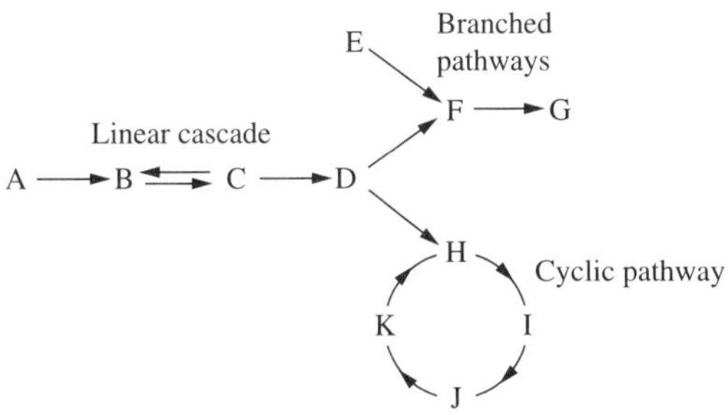

FIGURE 12.27 Types of reaction sequence in metabolic pathways.

carbon, the results obtained from flux analysis will be inconsistent with the behaviour of real cell cultures and thus of little value.

Most biochemical pathways contain ambiguities that need to be resolved when developing a network model. For example, several alternative pathways may be available for assimilation of some substrates and synthesis of particular intermediates; the contributions of different isoenzymes and the effects of reaction compartmentation may also require consideration. After the metabolic pathways of interest are identified, the next step is to simplify the reaction network so that a manageable set of equations is obtained. To achieve this, some intermediates are removed from consideration and complex subnetworks may be lumped together to streamline the analysis.

Three types of reaction sequence are commonly found in metabolic pathways. As shown in Figure 12.27, metabolic reactions occur in *linear* cascades, *branched* or *split* pathways, and *cyclic* pathways. In linear sequences, the product of one reaction serves as the reactant for the following reaction and no alternative or additional reactions take place. Branched pathways can be either *diverging* or *converging*. In a diverging pathway, a metabolite functions as the reactant in at least two different conversions to form multiple products. For example, metabolite D in Figure 12.27 is a diverging pathway branch point, producing metabolites F and H that then proceed along different reaction routes. In a converging pathway, a metabolite is produced in at least two different reactions, as shown for metabolite F. Some metabolites act simultaneously as both diverging and converging branch points. In cyclic pathways, several metabolites serve as both reactants and products and are regenerated during each passage of the cycle.

Reactions in linear cascades are not represented individually in metabolic flux analysis; instead, sequential reactions are lumped together into a single reaction step. This form of reaction grouping reduces significantly the number of reactions and intermediates in the metabolic model without affecting the overall results. Consider the simple metabolic pathway shown in Figure 12.28(a). Substrate S enters a cell and is converted to product P in a linear, unbranched sequence involving two intermediates, I_1 and I_2, and three reaction steps. S is provided in the medium outside of the cell; P is excreted from the cell into the medium. I_1 and I_2 are internal intermediates that remain within the cell. No other products are formed and the stoichiometry for each reaction step is given in Figure 12.28(a). The rate of substrate uptake is r_S and the rate of product excretion is r_P. We will define the *forward rates* of reaction for the three internal reaction steps as r_1, r_2, and r_3. According to this definition, r_1 is the rate of conversion of S, r_2 is the rate of conversion of I_1, and r_3 is the rate of conversion of I_2.

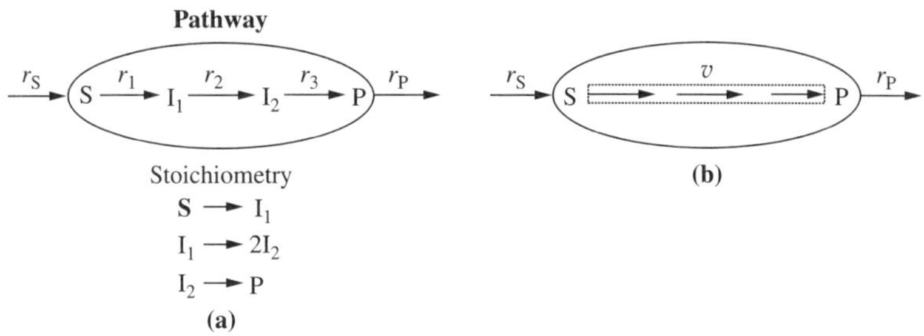

FIGURE 12.28 (a) Linear, unbranched reaction pathway with stoichiometry for the individual reaction steps. (b) Grouping of reactions in a linear, unbranched pathway.

A central tenet of metabolic flux analysis is that *metabolic pathways can be considered to operate under steady-state conditions.* This is a key simplifying assumption that allows us to calculate the fluxes in metabolic networks. When the system is at steady state, there can be no accumulation or depletion of intermediates and the concentrations of all metabolites remain constant. Furthermore, the total flux leading to a particular metabolite must be equal to the total flux leading away from that metabolite. Because, at steady-state, the reaction rates on either side of intermediates are balanced, metabolic flux analysis is sometimes called *flux balance analysis*.

The assumption of steady state is generally accepted as valid because the rate of turnover of most metabolic intermediates is high relative to the amounts of those metabolites present in the cells at any given time. Therefore, because metabolism occurs on a relatively fast time-scale, any transients after perturbation are very rapid relative to the rate of cell growth or metabolic regulation. The steady-state assumption, which is sometimes called the *pseudo-steady-state assumption*, means that any experimental culture data used in metabolic flux analysis should be measured under steady-state conditions. For suspended microbial cell cultures, this involves the operation of continuous *chemostat* bioreactors; chemostat operations are described in detail in Chapter 14. During chemostat culture, the concentrations of cells and all substrates and products in the medium do not vary with time. In some cases, batch or fed-batch culture data are used in flux analysis, as the condition of *balanced growth* during batch culture (Section 12.8.2) and the *quasi-steady-state* condition achieved in fed-batch operations (Section 14.5.3) may be considered sufficient approximations of steady state. Different experimental devices (e.g., perfusion bioreactors) are required for other cell systems such as plants and animal tissues to provide approximate steady-state conditions.

If the metabolic pathway in Figure 12.28(a) operates at steady state, there can be no accumulation or depletion of any of the intermediates in the pathway. If r_S and r_P are expressed in mass terms and r_1, r_2, and r_3 represent mass rates of reaction in units of, for example, g s^{-1}, we can say immediately that:

$$r_S = r_1 = r_2 = r_3 = r_P = v \tag{12.120}$$

because, at steady state, the mass rate of reaction of any intermediate in the sequence must be equal to the mass rate of reaction of all the other intermediates to avoid metabolite accumulation or depletion. Substrate uptake and product excretion must also occur at this rate to maintain constant concentrations within the cell. Accordingly, v in Eq. (12.120) can be used to represent the flux of mass through the entire pathway from S to P. Similarly, Eq. (12.120) also applies if r_S, r_P, r_1, r_2, and r_3 represent the rates at which carbon is taken up, excreted, and transferred between intermediates. As an atom, carbon remains unchanged in chemical processes and must flow through the sequence at a constant rate under steady-state conditions. In this case, v in Eq. (12.120) represents the flux of C through the entire pathway.

Our results so far are consistent with Table 4.1 in Chapter 4, which indicates that steady-state balance equations can be applied to total mass and atoms during reactions. However, the number of moles of chemical compounds is generally not conserved in reaction. Therefore, if r_S, r_P, r_1, r_2, and r_3 are molar rates in units of, for example, mol s^{-1}, information about the stoichiometry of the reactions is required to assess the relative molar fluxes through the pathway. From the individual reaction stoichiometries provided in Figure 12.28(a), the overall stoichiometry for conversion of 1 mole of S to P is $S \rightarrow 2P$ based on the stoichiometric sequence:

$$S \longrightarrow I_1 \longrightarrow 2I_2 \longrightarrow 2P \tag{12.121}$$

As molar rates of forward reaction, r_1, r_2, and r_3 are defined as follows:

$$\begin{aligned} r_1 &= \text{moles of S converted per unit time} \\ r_2 &= \text{moles of } I_1 \text{ converted per unit time} \\ r_3 &= \text{moles of } I_2 \text{ converted per unit time} \end{aligned} \tag{12.122}$$

Let us perform mass balances on each of the components of the reaction sequence to find the relationships between these reaction rates. We will use the general unsteady-state mass balance equation derived in Chapter 6:

$$\frac{dM}{dt} = \hat{M}_i - \hat{M}_o + R_G - R_C \tag{6.5}$$

where M is mass, t is time, $\hat{M}_i$ is the mass flow rate entering the system, $\hat{M}_o$ is the mass flow rate leaving the system, R_G is the rate of generation by chemical reaction, and R_C is the rate of consumption by chemical reaction. The system in this case is the cell. Applying Eq. (6.5) to each component of the pathway in Figure 12.28(a):

$$\begin{aligned} \frac{dM_S}{dt} &= (MW)_S(r_S) - 0 + 0 - (MW)_S(r_1) \\ \frac{dM_{I1}}{dt} &= 0 - 0 + (MW)_{I1}(r_1) - (MW)_{I1}(r_2) \\ \frac{dM_{I2}}{dt} &= 0 - 0 + (MW)_{I2}(2r_2) - (MW)_{I2}(r_3) \\ \frac{dM_P}{dt} &= 0 - (MW)_P(r_P) + (MW)_P(r_3) - 0 \end{aligned} \tag{12.123}$$

In Eq. (12.123), MW denotes molecular weight and subscripts S, I1, I2, and P denote substrate, I_1, I_2, and product, respectively. Because the equations are mass balances, the molar reaction rates r are converted to mass rates using the molecular weights of the compounds to which they refer. The stoichiometric coefficients for each component in the reaction sequence have also been applied. For example, in the mass balance for I_1, the mass rate of generation of I_1 by chemical reaction is shown as $(MW)_{I1}(r_1)$, where r_1 is used to represent the molar rate of production of I_1. However, as indicated in Eq. (12.122), r_1 is defined in terms of the moles of S reacted. We can use r_1 in the mass balance for I_1 because we know from the reaction stoichiometry that conversion of 1 mole of substrate produces 1 mole of I_1. The molar rate of S conversion is therefore equal to the molar rate of I_1 production. The stoichiometric coefficient 2 appears in the mass balance equation for I_2. Because, from Eq. (12.122), r_2 is the number of moles of I_1 reacted per unit time, from the stoichiometry of conversion of I_1 to I_2, the number of moles of I_2 formed per unit time must be $2r_2$. Therefore, $2r_2$ appears in the balance for I_2 as the molar rate of generation of I_2 by reaction. The molar rate at which I_2 is converted to P has been defined as r_3 in Eq. (12.122). Accordingly, a stoichiometric coefficient is not required for the consumption term in the balance for I_2.

At steady state, the derivatives with respect to time in Eq. (12.123) are zero as there can be no change in mass of any component. Applying this result allows us to cancel the molecular weight terms and simplify the equations to give:

$$\begin{aligned} r_S &= r_1 \\ r_1 &= r_2 \\ 2r_2 &= r_3 \\ r_3 &= r_P \end{aligned} \tag{12.124}$$

or

$$r_S = r_1 = r_2 = 0.5r_3 = 0.5r_P = v \tag{12.125}$$

Eq. (12.125) applies when the reaction rates are expressed as molar rates.

Our analysis of the reaction sequence of Figure 12.28(a) has demonstrated that the fluxes of mass, carbon atoms, and moles through a linear, unbranched metabolic pathway can be represented using a single rate v, as illustrated in Figure 12.28(b). Consistent with our definition of reaction rate, v is a forward rate, that is, v represents the rate of conversion of S. Whereas r denoted the reaction velocity for each individual reaction, v may represent the overall rate of grouped reactions, or the reaction velocity across several reaction steps. We will continue to use r_S and r_P for the rates of substrate uptake and product excretion, respectively, consistent with the previous sections of this chapter. These rates are different from the internal flux v in that they are generally readily measurable.

When v is expressed in terms of mass or C atoms, from Eq. (12.120), v is equal to each of the individual reaction rates at each step of the pathway, irrespective of stoichiometry. However, metabolic flux analysis is generally carried out using fluxes expressed as molar rates. When molar rates are used, the relationship between the individual reaction rates and v for grouped reactions depends on the reaction stoichiometry, as indicated by the example in Eq. (12.125). Use of v in units of mol s^{-1} to represent the overall flux of moles from S to P does not imply that the molar rate of conversion of all the intermediates in the grouping is also v mol s^{-1}. Using v to denote the molar rate of the grouped reactions does mean, however, that the overall rate of conversion of S to P can be represented as v moles of S per second, provided that the correct stoichiometry for the conversion, in this case $S \rightarrow 2P$, is used in conjunction with that rate.

Grouping of reactions allows us to omit intermediates along linear sections of pathways so that only intermediates at branch points need be considered. For example, depending on the purpose and scope of the flux analysis, the reaction sequence in glycolysis:

$$\text{glyceraldehyde 3-phosphate} \rightarrow \text{1,3-diphosphoglycerate} \rightarrow \text{3-phosphoglycerate}$$
$$\rightarrow \text{2-phosphoglycerate} \rightarrow \text{phosphoenolpyruvate} \rightarrow \text{pyruvate}$$

might be represented as:

$$\text{glyceraldehyde 3-phosphate} \rightarrow \text{3-phosphoglycerate} \rightarrow \text{phosphoenolpyruvate} \rightarrow \text{pyruvate}$$

because the intermediates 1,3-diphosphoglycerate and 2-phosphoglycerate are not branch points leading to other metabolic pathways. These intermediates can therefore be grouped into lumped reactions, as shown earlier in the flux map of Figure 12.26. Other metabolic reactions may also be rationalised. For example, depending on the aim of the analysis, fatty acid synthesis can be lumped into a single reaction if the fractions of different fatty acids in the cell are known. The individual steps in biosynthetic pathways for organic acids such as fumarate and succinate may also be grouped for convenience, provided that incorporation of nitrogen and incorporation or evolution of CO_2 are accounted for in the lumped reaction. For flux analyses with energy and redox balances, the involvement of ATP and/or NADH must also be included in any grouped representation of reactions.

Once the reactions to be included in the network are determined, a complete list of the metabolites involved and their chemical reaction equations is required. The stoichiometry of the reactions, either actual or lumped, must be accurate and the reaction equations must be balanced. All C atoms must be accounted for within the simplified metabolic model. As mentioned above, metabolic flux analysis is generally carried out using fluxes expressed in moles. However, because we are interested primarily in the distribution of carbon within pathways, flux is sometimes reported in terms of moles of C rather than moles of chemical compounds. The reaction equations and stoichiometry may be modified to reflect this, using chemical formulae for C-containing compounds based on 1 atom of C, and writing each reaction for conversion of 1 reactant mole of C. As an example, the reaction equation for conversion of glucose to ethanol is usually written using a basis of 1 mole of glucose:

$$C_6H_{12}O_6 \longrightarrow 2\,C_2H_6O + 2\,CO_2 \tag{12.126}$$

where $C_6H_{12}O_6$ is glucose and C_2H_6O is ethanol. This equation rewritten using a basis of 1 mole of carbon is:

$$CH_2O \longrightarrow 0.67\,CH_3O_{0.5} + 0.33\,CO_2 \tag{12.127}$$

where CH_2O represents glucose and $CH_3O_{0.5}$ represents ethanol. In Eq. (12.127), the formulae for all compounds are based on 1 C atom or 1 *C-mol*, where 1 C-mol is 6.02×10^{23} carbon atoms. This affects the stoichiometric coefficients required to achieve elemental balance in the reaction equation. The reaction of Eq. (12.127) starts with one-sixth of the mass of glucose used in Eq. (12.126); otherwise however, the two reactions are the same. The advantage of writing reaction equations in C-mol is that information about the amount of carbon transferred from a reactant to a particular product is contained in the stoichiometric coefficients alone. From Eq. (12.127), we can see immediately that two-thirds of the C in glucose is transferred to ethanol. If we were working with Eq. (12.126) instead, this information is available only after we multiply the stoichiometric coefficient for ethanol (2) by the number of C atoms in one molecule of ethanol (2) and divide the result by the number of C atoms in one molecule of glucose (6). Representing reactions in C-mol is helpful for assessing the flux of carbon through metabolic pathways. A disadvantage is that C-mol formulae for reactants and products may not be recognised as easily as their full molecular formulae.

Fluxes in metabolic engineering are normally expressed as specific rates (Section 12.1.3). For example, a flux may be reported in units of mol (g dry weight of cells)$^{-1}$ h^{-1} or, alternatively, as C-mol (g biomass)$^{-1}$ h^{-1} or C-mol (C-mol biomass)$^{-1}$ h^{-1}. One C-mol of biomass is represented by the chemical 'formula' for dry cells normalised to one atom of carbon, $CH_{\alpha}O_{\beta}N_{\delta}$, as described in Section 4.6.1 of Chapter 4. For ash-free cellular biomass of average C, H, O, and N composition, 1 C-mol of biomass weighs about 24.6 g.

Formulating the Equations Using Matrices

Consider now the branched metabolic pathway of Figure 12.29, where the pathway from S splits at intermediate I to form two products, P_1 and P_2, which are excreted at rates r_{P1} and r_{P2}, respectively. The pathways from S to I, I to P_1, and I to P_2 may each comprise multiple reaction steps, but these have been grouped as discussed in the previous section so that the fluxes are represented by forward rates v_1, v_2, and v_3 through the three branches. The stoichiometry of the grouped reactions for each branch is shown in Figure 12.29. We will carry out the analysis using rates expressed in units of mol g^{-1} h^{-1}. To keep it simple, the pathway does not involve energy or redox conversions.

We will use this metabolic pathway to illustrate the mathematical techniques employed for metabolic flux analysis. The aim is to calculate the internal fluxes v_1, v_2, and v_3. Although the solution to this problem is trivial, it is a useful exercise to follow the same

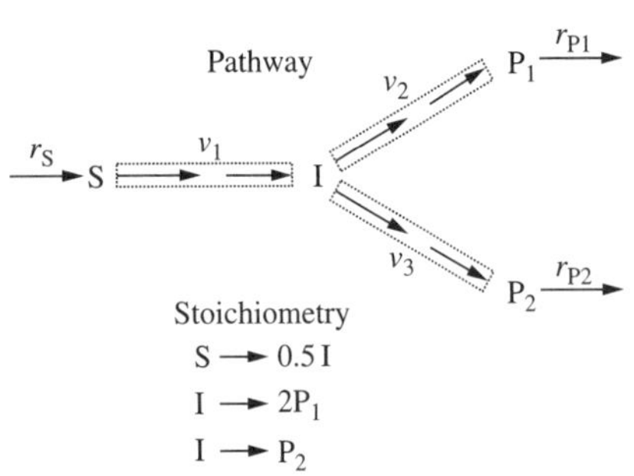

FIGURE 12.29 Diverging branched pathway with stoichiometry for the grouped reactions in each branch.

methods applied to larger, more complex systems. Using Eq. (6.5) and a basis of 1 g of cellular biomass, the unsteady-state mass balance equations for S, I, P_1, and P_2 are:

$$
\begin{aligned}
\frac{dM_S}{dt} &= (MW)_S(r_S) - 0 + 0 - (MW)_S(v_1) \\
\frac{dM_I}{dt} &= 0 - 0 + (MW)_{I1}(0.5v_1) - (MW)_{I1}(v_2) - (MW)_{I1}(v_3) \\
\frac{dM_{P1}}{dt} &= 0 - (MW)_{P1}(r_{P1}) + (MW)_{P1}(2v_2) - 0 \\
\frac{dM_{P2}}{dt} &= 0 - (MW)_{P2}(r_{P2}) + (MW)_{P2}(v_3) - 0
\end{aligned}
\tag{12.128}
$$

As explained in relation to Eq. (12.123), the stoichiometric coefficients 0.5 and 2 are required in the mass balance equations for I and P_1 to represent the rates of generation of these components, consistent with the definition of the rates as forward reaction rates. At steady state, Eq. (12.128) becomes:

$$
\begin{aligned}
r_S - v_1 &= 0 \\
0.5v_1 - v_2 - v_3 &= 0 \\
2v_2 - r_{P1} &= 0 \\
v_3 - r_{P2} &= 0
\end{aligned}
\tag{12.129}
$$

These linear algebraic equations can be arranged in matrix form. Let us define a column vector **v** containing the six rates as elements:

$$
\mathbf{v} = \begin{bmatrix} r_S \\ v_1 \\ v_2 \\ v_3 \\ r_{P1} \\ r_{P2} \end{bmatrix}
\tag{12.130}
$$

and a matrix **S** containing the coefficients for the rates in Eq. (12.129):

$$
\mathbf{S} = \begin{bmatrix} 1 & -1 & 0 & 0 & 0 & 0 \\ 0 & 0.5 & -1 & -1 & 0 & 0 \\ 0 & 0 & 2 & 0 & -1 & 0 \\ 0 & 0 & 0 & 1 & 0 & -1 \end{bmatrix}
\tag{12.131}
$$

Each row in **S** represents a mass balance equation so, in this example, there are four rows. Each column in **S** contains coefficients for each of the six rates that define vector **v**; therefore, in this case, there are six columns. The first column in **S** contains the coefficients for r_S, the second for v_1, the third for v_2, and so on for the remainder of **v**. Typically, **S** for large metabolic networks is a relatively sparse matrix with many zero elements as most metabolic reactions involve only a few components. Note that **S** can be defined in different ways and the arrangement of elements in the matrix of Eq. (12.131) is sometimes referred to in the literature as $\mathbf{S}^T$, or the transpose (Appendix E) of the stoichiometric matrix.

Using matrix notation, Eq. (12.129) can be written succinctly as:

$$\mathbf{Sv} = 0 \tag{12.132}$$

Multiplication of **S** and **v** is carried out using the rules for matrix operations described in Section E.4 of Appendix E. The rules for multiplying a matrix and column vector given in Eq. (E.46) can be used to verify that Eqs. (12.129) and (12.132) are equivalent.

Underdetermined and Overdetermined Systems

There are six rates or fluxes in Eq. (12.129) but only four mass balance equations. For any metabolic network, if there are M linear algebraic equations, one for each of M metabolites, and R reaction fluxes, the degree of freedom F in the system is:

$$F = R - M \tag{12.133}$$

Typically, the number of reactions R in a metabolic pathway is greater than the number of metabolites M. Such a system where $F > 0$ is *underdetermined*, as there is not enough information available to solve for all the rates in the mass balance equations. To allow solution of Eq. (12.132), some of the rate elements in **v** must be measured, or additional constraints must be introduced into the equations, before the remaining rates can be determined. If F rates are measured and therefore known independently, the system becomes *determined* and a single unique solution can be found for the remaining fluxes. If more than F rates are measured, the system is *overdetermined*. An overdetermined system is not easily achieved in metabolic flux analysis but is desirable because the additional information can be used for checking purposes. This is discussed further in Section 12.16.1 (Validation subsection).

For our problem represented by Eqs. (12.129) and (12.132), let us bring the system to a determined state so that a unique solution can be found for the fluxes. According to Eq. (12.133), $F = 6 - 4 = 2$, so we need to measure two of the rates in **v**. As it is much easier to observe external rather than internal fluxes, r_S, r_{P1}, and r_{P2} are obvious candidates for experimental measurement. We only need two of these, however, so let us choose to measure r_{P1} and r_{P2}. These rates can be grouped into a new vector for measured rates $\mathbf{v_m}$, while the remaining unknown rates are collected into another vector $\mathbf{v_c}$ for flux calculation. We can write Eq. (12.132) as:

$$\mathbf{Sv} = \mathbf{S_m v_m} + \mathbf{S_c v_c} = 0 \tag{12.134}$$

where

$$\mathbf{v_m} = \begin{bmatrix} r_{P1} \\ r_{P2} \end{bmatrix} \tag{12.135}$$

and

$$\mathbf{v_c} = \begin{bmatrix} r_S \\ v_1 \\ v_2 \\ v_3 \end{bmatrix} \tag{12.136}$$

From the coefficients in Eq. (12.129) for r_{P1} and r_{P2}:

$$\mathbf{S_m} = \begin{bmatrix} 0 & 0 \\ 0 & 0 \\ -1 & 0 \\ 0 & -1 \end{bmatrix} \tag{12.137}$$

and

$$\mathbf{S_c} = \begin{bmatrix} 1 & -1 & 0 & 0 \\ 0 & 0.5 & -1 & -1 \\ 0 & 0 & 2 & 0 \\ 0 & 0 & 0 & 1 \end{bmatrix} \tag{12.138}$$

Both matrices $\mathbf{S_m}$ and $\mathbf{S_c}$ contain four rows, one for each mass balance equation. The elements in the two columns of $\mathbf{S_m}$ are the coefficients in the mass balance equations for the measured rates, r_{P1} and r_{P2}, respectively. The four columns of $\mathbf{S_c}$ contain the coefficients for the four rates in vector $\mathbf{v_c}$, that is, r_S, v_1, v_2, and v_3, in that order.

Solution of this problem to find the internal fluxes v amounts to solving for the unknown elements of vector $\mathbf{v_c}$. Once we have measured r_{P1} and r_{P2}, the elements of $\mathbf{v_m}$, $\mathbf{S_m}$, and $\mathbf{S_c}$ are known. We can write Eq. (12.134) as:

$$\mathbf{S_c v_c} = -\mathbf{S_m v_m} \tag{12.139}$$

or, using the rules for matrix operations described in Appendix E:

$$\mathbf{v_c} = -\mathbf{S_c}^{-1}\mathbf{S_m v_m} \tag{12.140}$$

The solution for $\mathbf{v_c}$ requires calculation of the inverse of matrix $\mathbf{S_c}$. As described in Section E.4 of Appendix E, the inverse of a matrix can be determined only if the matrix is square and of full rank. In our example, $\mathbf{S_c}$ is a square 4×4 matrix: this is a consequence of the system being determined so that the number of unknown rates is equal to the number of equations available to solve for them. To be of full rank, the matrix must contain no row or column that is a linear combination of other rows or columns. The determinant of a matrix of full rank is nonzero (Section E.4); therefore, calculating the determinant allows us to check whether $\mathbf{S_c}$ can be inverted.

Finding the determinants of matrices becomes increasingly complex as the size of the matrix increases. However, commonly available spreadsheet programs perform the calculation; specialised computer programs for metabolic flux analysis may also be used. For

$\mathbf{S_c}$ defined in Eq. (12.138), applying appropriate software gives det $(\mathbf{S_c}) = 1.0$. As this determinant $\neq 0$, we know that $\mathbf{S_c}$ is invertible. Further application of matrix solution software gives the inverse of $\mathbf{S_c}$ as:

$$\mathbf{S_c}^{-1} = \begin{bmatrix} 1 & 2 & 1 & 2 \\ 0 & 2 & 1 & 2 \\ 0 & 0 & 0.5 & 0 \\ 0 & 0 & 0 & 1 \end{bmatrix} \tag{12.141}$$

This result can be verified by checking that the product $\mathbf{S_c}\,\mathbf{S_c}^{-1}$ is equal to the identity matrix, **I** (Section E.4, Appendix E).

If matrix $\mathbf{S_c}$ is found to be singular (i.e., det $(\mathbf{S_c}) = 0$) so that $\mathbf{S_c}$ cannot be inverted, the solution cannot proceed without modifying the equations. When $\mathbf{S_c}$ is singular, it is likely that one or more of the mass balance equations used to formulate the problem is a linear combination of one or more of the other equations. In this situation, although the system may appear to be determined, it is in fact underdetermined. When large numbers of metabolic reactions are being analysed, difficulties with linearly dependent elements in the equations occur with some frequency. To give organisms flexibility under different environmental conditions, cellular metabolism contains many complementary and redundant pathways that serve the same function as others if they operate at the same time. Particular areas of metabolism, such as ammonia assimilation, anaplerotic pathways, isoenzyme reactions, and transhydrogenase reactions, lend themselves to representation using linearly dependent equations. Intracellular assays may be required to resolve the singularity created by related pathways, by providing additional information about their relative activities. Alternatively, complementary pathways might be lumped together into a single reaction. Balances on *conserved moieties*, such as ATP, ADP, and AMP, and cofactor pairs such as NAD and NADH, are linearly dependent because the total concentration of conserved moieties remains constant even though there is interchange between different members of the group. Therefore, balances on only one compound in conserved groups are included in flux analysis to avoid matrix singularity. Identifying and eliminating dependent equations by changing the metabolic network or introducing other information into the solution procedure are often required to render $\mathbf{S_c}$ invertible. Sometimes, even if the full matrix of coefficients **S** is nonsingular, the choice of rates for experimental measurement renders $\mathbf{S_c}$ singular, indicating that one or more of the measured rates is redundant. When this occurs, a different combination of measurements is required to allow solution of the flux balance.

Solution of the Equations

Let us assume that the measured results for r_{P1} and r_{P2} are $r_{P1} = 0.05$ mol g^{-1} h^{-1} and $r_{P2} = 0.08$ mol g^{-1} h^{-1}. Therefore, we can write Eq. (12.135) as:

$$\mathbf{v_m} = \begin{bmatrix} 0.05 \\ 0.08 \end{bmatrix} \text{mol g}^{-1}\,\text{h}^{-1} \tag{12.142}$$

Applying Eqs. (12.137), (12.141), and (12.142) in Eq. (12.140) gives the following equation for $\mathbf{v}_c$:

$$\mathbf{v}_c = -\begin{bmatrix} 1 & 2 & 1 & 2 \\ 0 & 2 & 1 & 2 \\ 0 & 0 & 0.5 & 0 \\ 0 & 0 & 0 & 1 \end{bmatrix} \begin{bmatrix} 0 & 0 \\ 0 & 0 \\ -1 & 0 \\ 0 & -1 \end{bmatrix} \begin{bmatrix} 0.05 \\ 0.08 \end{bmatrix} \text{mol g}^{-1}\text{ h}^{-1} \tag{12.143}$$

Using the associative multiplication rule Eq. (E.39) from Appendix E:

$$\mathbf{v}_c = -\begin{bmatrix} 1 & 2 & 1 & 2 \\ 0 & 2 & 1 & 2 \\ 0 & 0 & 0.5 & 0 \\ 0 & 0 & 0 & 1 \end{bmatrix} \begin{bmatrix} 0 \\ 0 \\ -0.05 \\ -0.08 \end{bmatrix} \text{mol g}^{-1}\text{ h}^{-1} \tag{12.144}$$

Further multiplication gives:

$$\mathbf{v}_c = -\begin{bmatrix} -0.21 \\ -0.21 \\ -0.025 \\ -0.08 \end{bmatrix} \text{mol g}^{-1}\text{ h}^{-1} \tag{12.145}$$

or

$$\mathbf{v}_c = \begin{bmatrix} 0.21 \\ 0.21 \\ 0.025 \\ 0.08 \end{bmatrix} \text{mol g}^{-1}\text{ h}^{-1} \tag{12.146}$$

Equation (12.146) provides the values for the unknown fluxes. From the definition of $\mathbf{v}_c$ in Eq. (12.136), the results are $r_S = 0.21$ mol g^{-1} h^{-1}, $v_1 = 0.21$ mol g^{-1} h^{-1}, $v_2 = 0.025$ mol g^{-1} h^{-1}, and $v_3 = 0.08$ mol g^{-1} h^{-1}. You can see easily that these results, together with the measured values for r_{P1} and r_{P2}, satisfy the mass balances of Eq. (12.129).

Product yields in cell culture are controlled by the *flux split ratio* or *flux partitioning ratio* Φ at critical branch points in the metabolic network. From our analysis of the pathway in Figure 12.29, we can calculate the flux split ratio between the two products P_1 and P_2 at branch point I:

$$\Phi_{P1} = \frac{\text{flux to P}_1}{\text{flux to P}_1 + \text{flux to P}_2} = \frac{v_2}{v_2 + v_3} = \frac{0.025}{0.025 + 0.08} = 0.24 \tag{12.147}$$

and

$$\Phi_{P2} = \frac{\text{flux to P}_2}{\text{flux to P}_1 + \text{flux to P}_2} = \frac{v_3}{v_2 + v_3} = \frac{0.08}{0.025 + 0.08} = 0.76 \tag{12.148}$$

The problem we have just solved was very simple. Once the measured results for r_{P1} and r_{P2} were known, the algebraic equations of Eq. (12.129) could have been solved easily to arrive at the solution of Eq. (12.146) without the need to formulate the problem using matrices. However, the procedure described here illustrates the mathematical approach required when realistic metabolic networks involving 20 to 50 fluxes are investigated. Genome-scale stoichiometric models involving more than 2000 reactions have been developed for organisms such as *Escherichia coli* [24]. In such cases, simultaneous solution of the mass balance equations is difficult without the use of matrix methods. Flux analysis of a real metabolic pathway is illustrated in Example 12.10.

Energy and Redox Balances and Biomass Production

Metabolic flux analysis can be carried out using mass balance equations alone, as illustrated in the above example. However, adding balance equations for cofactors such as energy and redox carriers provides extra constraints on the system, reduces the degree of freedom, and thus may be required for solution. It also ensures that sufficient energy and reducing equivalents are available for the reactions to proceed and can be useful for linking fluxes in one subnetwork of metabolism to another. However, if energy and redox carriers are included, all reactions involving these cofactors must be represented to avoid substantial errors in flux estimation. This can be problematic, as the involvement of cofactors in every area of cellular metabolism may not be known. The need to account completely for all instances of ATP generation and consumption and all redox transactions is a common source of error in flux analysis.

As discussed in Section 12.16.1 (Underdetermined and Overdetermined Systems subsection) balances for only one member of conserved cofactor groups are used to avoid generating linearly dependent equations and matrix singularity. Furthermore, for simplicity, all energy carriers are usually represented as ATP and all redox carriers as NADH, even though other energy carriers such as GTP and other reducing agents such as NADPH and $FADH_2$ are involved in metabolism. To account for cellular maintenance requirements and the operation of futile cycles, a reaction may be included in the analysis to dissipate excess ATP.

Biomass is often included in flux analysis as a product of metabolism. Cell growth provides a sink for ATP and precursor compounds from catabolic pathways, and for building-block metabolites such as amino acids, nucleotides, and lipids. A stoichiometric equation for production of biomass represented as $CH_{\alpha}O_{\beta}N_{\delta}$ is determined from the composition of the cells (Section 4.6.1) or from information about the metabolic requirements for synthesis of various biomass constituents. For example, growth can be simulated by creating a lumped reaction equation that draws metabolites such as polysaccharides, proteins, DNA, RNA, and lipids from their respective biosynthetic pathways at ratios that reflect the composition of the cells. The rate of the growth reaction may then be scaled so that the flux is equal to the exponential growth rate of the organism. Modelling biomass synthesis can be the most challenging aspect of network definition for metabolic flux analysis.

EXAMPLE 12.10 FLUX ANALYSIS FOR MIXED ACID FERMENTATION

The metabolic pathways for anaerobic mixed acid fermentation have been studied extensively. Under certain culture conditions, lactic acid bacteria catabolise glucose to produce lactic acid, formic acid, acetic acid, and ethanol. The pathway for this fermentation is shown in

Figure 12.30. For cells, the main purpose of the pathway is to generate ATP for growth. There is no or very little exchange of intermediates to anabolic metabolism; the pathway is also self-sufficient in reducing equivalents.

A chemostat is used to obtain experimental results for culture of *Lactococcus lactis* subsp. *lactis* under steady-state conditions. The rate of glucose uptake is measured as 0.01 mol h^{-1} per g dry weight of cells. The yield of lactic acid from glucose is found to be 0.49 mol mol^{-1} and the yield of formic acid from glucose is 0.41 mol mol^{-1}.

(a) Draw a simplified pathway diagram suitable for metabolic flux analysis.
(b) Perform a flux balance on the simplified metabolic network. Label the pathway with the flux results in units of C-mol g^{-1} h^{-1}.
(c) What is the flux split ratio at pyruvate for lactate production?
(d) What is the flux split ratio at acetyl co-enzyme A (acetyl CoA) for acetate and ethanol production?
(e) Estimate the observed yield of ethanol from glucose in mol mol^{-1}.
(f) What is the rate of ATP generation for cell growth and maintenance?

Solution

(a) In this fermentation, intermediates from glycolysis are not used for biosynthesis and growth. Therefore, the pathway from glucose to pyruvate can be considered a linear sequence of reactions without branch points. The individual glycolytic reactions are grouped into a single reaction step while preserving the net cofactor requirements of glycolysis. The major branch points of the pathway are at pyruvate and acetyl CoA. The steps from acetyl CoA to acetate, and from acetyl CoA to ethanol, are also linear sequences and can be lumped into two grouped reactions. The resulting simplified pathway is shown in Figure 12.31.
(b) The stoichiometric equations for the reactions in the simplified pathway can be found from biochemistry texts and are listed below. The equations are written on a mole basis.

$$\text{glucose} + 2\,\text{ADP} + 2\,\text{NAD} \longrightarrow 2\,\text{pyruvate} + 2\,\text{ATP} + 2\,\text{NADH} + 2\,H_2O$$
$$\text{pyruvate} + \text{NADH} \longrightarrow \text{lactate} + \text{NAD}$$
$$\text{pyruvate} + \text{CoA} + \text{NAD} \longrightarrow \text{acetyl CoA} + CO_2 + \text{NADH}$$
$$\text{pyruvate} + \text{CoA} \longrightarrow \text{formate} + \text{acetyl CoA}$$
$$\text{acetyl CoA} + \text{ADP} \longrightarrow \text{acetate} + \text{CoA} + \text{ATP}$$
$$\text{acetyl CoA} + 2\,\text{NADH} \longrightarrow \text{ethanol} + \text{CoA} + 2\,\text{NAD}$$
$$\text{ATP} \longrightarrow \text{ADP}$$

The overall stoichiometry and cofactor requirements for glycolytic conversion of glucose to pyruvate are well known. The stoichiometric coefficients in the remaining reactions are unity, except that 2 NADH and 2 NAD are involved in the lumped reaction for conversion of acetyl CoA to ethanol. The forward rates of the internal reactions are labelled v_1 to v_7 in Figure 12.31. The rates of substrate uptake and product excretion are labelled r_G, r_L, r_C, r_F, r_A, and r_E for glucose, lactate, CO_2, formate, acetate, and ethanol, respectively. Because the pathway generates a net positive amount of ATP, a reaction ATP→ADP representing ATP requirements for growth and maintenance is included so that the mass balance for ATP may be closed.

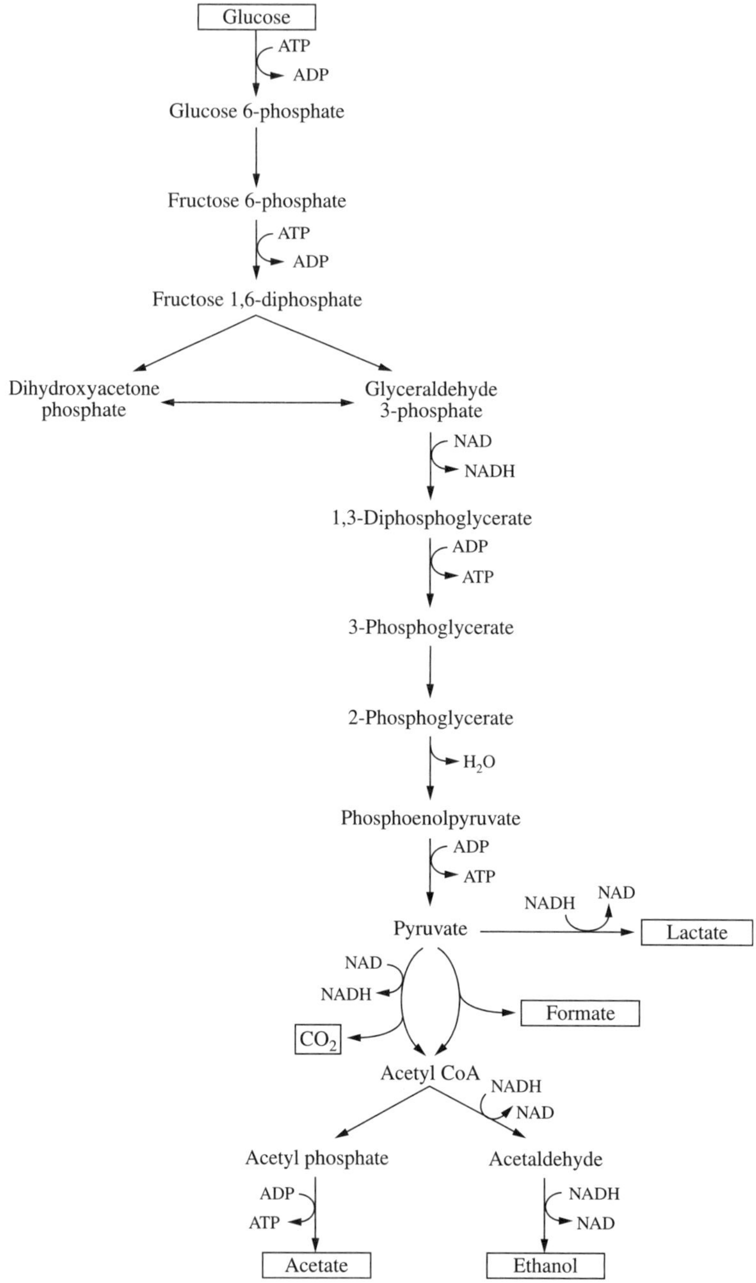

FIGURE 12.30 Metabolic pathway for mixed acid fermentation.

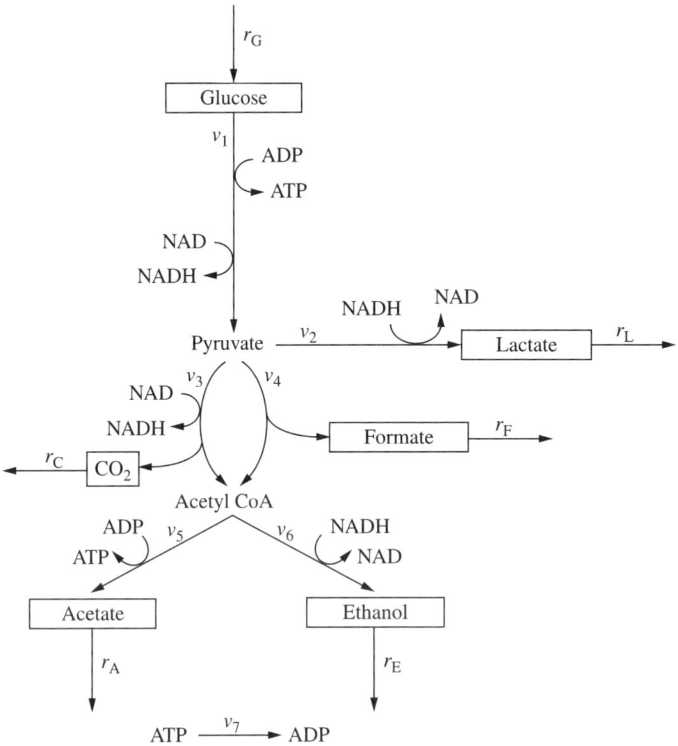

FIGURE 12.31 Graphic of simplified metabolic pathway for mixed acid fermentation.

Steady-state mass balances are performed for glucose and the excreted products, the internal intermediates at branch points, and the redox and energy carriers:

$$\begin{aligned}
&\text{glucose:}\ \ r_G - v_1 = 0\\
&\text{pyruvate:}\ \ 2v_1 - v_2 - v_3 - v_4 = 0\\
&\text{lactate:}\ \ v_2 - r_L = 0\\
&CO_2\text{:}\ \ v_3 - r_C = 0\\
&\text{formate:}\ \ v_4 - r_F = 0\\
&\text{acetyl CoA:}\ \ v_3 + v_4 - v_5 - v_6 = 0\\
&\text{acetate:}\ \ v_5 - r_A = 0\\
&\text{ethanol:}\ \ v_6 - r_E = 0\\
&\text{NADH:}\ \ 2v_1 - v_2 + v_3 - 2v_6 = 0\\
&\text{ATP:}\ \ 2v_1 + v_5 - v_7 = 0
\end{aligned}$$

As outlined in Section 12.16.1 (Underdetermined and Overdetermined Systems subsection) balances are performed for only one component involved in redox transfer and one component involved in energy transfer; balances for NAD and ADP are therefore not included. In total, there are 7 internal (v) and 6 external (r) fluxes, making a total of 13 unknown rates. There are 10 metabolites/cofactors giving 10 mass balance equations. From Eq. (12.133), the degree of freedom F for the system is:

$$F = 13 - 10 = 3$$

As three experimental measurements are available from chemostat cultures, the system is determined and can be solved to find a unique solution. The measured value for r_G is 0.01 mol g^{-1} h^{-1}. From the yield coefficients for lactate and formate and the definition of Y_{PS} in Table 12.4:

$$r_L = Y_{PS} r_G = 0.49 \text{ mol mol}^{-1}(0.01 \text{ mol g}^{-1} \text{ h}^{-1}) = 0.0049 \text{ mol g}^{-1} \text{ h}^{-1}$$
$$r_F = Y_{PS} r_G = 0.041 \text{ mol mol}^{-1}(0.01 \text{ mol g}^{-1} \text{ h}^{-1}) = 0.0041 \text{ mol g}^{-1} \text{ h}^{-1}$$

The vector of measured fluxes $\mathbf{v_m}$ and the corresponding matrix $\mathbf{S_m}$ are:

$$\mathbf{v_m} = \begin{bmatrix} r_G \\ r_L \\ r_F \end{bmatrix} = \begin{bmatrix} 0.01 \\ 0.0049 \\ 0.0041 \end{bmatrix} \text{ mol g}^{-1} \text{ h}^{-1} \quad \mathbf{S_m} = \begin{bmatrix} 1 & 0 & 0 \\ 0 & 0 & 0 \\ 0 & -1 & 0 \\ 0 & 0 & 0 \\ 0 & 0 & -1 \\ 0 & 0 & 0 \\ 0 & 0 & 0 \\ 0 & 0 & 0 \\ 0 & 0 & 0 \\ 0 & 0 & 0 \end{bmatrix}$$

The elements of $\mathbf{S_m}$ are the coefficients in the mass balance equations for r_G, r_L, and r_F. Each row in $\mathbf{S_m}$ represents a mass balance equation; the first column in $\mathbf{S_m}$ contains the coefficients for r_G, the second for r_L, and the third for r_F. The remaining 10 rates define the vector of calculated fluxes, $\mathbf{v_c}$. The corresponding matrix of coefficients $\mathbf{S_c}$ is obtained from the mass balance equations:

$$\mathbf{v_c} = \begin{bmatrix} v_1 \\ v_2 \\ v_3 \\ v_4 \\ v_5 \\ v_6 \\ v_7 \\ r_C \\ r_A \\ r_E \end{bmatrix} \quad \mathbf{S_c} = \begin{bmatrix} -1 & 0 & 0 & 0 & 0 & 0 & 0 & 0 & 0 & 0 \\ 2 & -1 & -1 & -1 & 0 & 0 & 0 & 0 & 0 & 0 \\ 0 & 1 & 0 & 0 & 0 & 0 & 0 & 0 & 0 & 0 \\ 0 & 0 & 1 & 0 & 0 & 0 & 0 & -1 & 0 & 0 \\ 0 & 0 & 0 & 1 & 0 & 0 & 0 & 0 & 0 & 0 \\ 0 & 0 & 1 & 1 & -1 & -1 & 0 & 0 & 0 & 0 \\ 0 & 0 & 0 & 0 & 1 & 0 & 0 & 0 & -1 & 0 \\ 0 & 0 & 0 & 0 & 0 & 1 & 0 & 0 & 0 & -1 \\ 2 & -1 & 1 & 0 & 0 & -2 & 0 & 0 & 0 & 0 \\ 2 & 0 & 0 & 0 & 1 & 0 & -1 & 0 & 0 & 0 \end{bmatrix}$$

$\mathbf{S_c}$ is nonsingular with det $(\mathbf{S_c}) = -2$. Matrix calculation software is used to determine the inverse of $\mathbf{S_c}$:

$$\mathbf{S_c}^{-1} = \begin{bmatrix} -1 & 0 & 0 & 0 & 0 & 0 & 0 & 0 & 0 & 0 \\ 0 & 0 & 1 & 0 & 0 & 0 & 0 & 0 & 0 & 0 \\ -2 & -1 & -1 & 0 & -1 & 0 & 0 & 0 & 0 & 0 \\ 0 & 0 & 0 & 0 & 1 & 0 & 0 & 0 & 0 & 0 \\ 0 & -0.5 & 0 & 0 & 0.5 & -1 & 0 & 0 & 0.5 & 0 \\ -2 & -0.5 & -1 & 0 & -0.5 & 0 & 0 & 0 & -0.5 & 0 \\ -2 & -0.5 & 0 & 0 & 0.5 & -1 & 0 & 0 & 0.5 & -1 \\ -2 & -1 & -1 & -1 & -1 & 0 & 0 & 0 & 0 & 0 \\ 0 & -0.5 & 0 & 0 & 0.5 & -1 & -1 & 0 & 0.5 & 0 \\ -2 & -0.5 & -1 & 0 & -0.5 & 0 & 0 & -1 & -0.5 & 0 \end{bmatrix}$$

Applying Eq. (12.140):

$$\mathbf{v_c} = -\begin{bmatrix} -1 & 0 & 0 & 0 & 0 & 0 & 0 & 0 & 0 & 0 \\ 0 & 0 & 1 & 0 & 0 & 0 & 0 & 0 & 0 & 0 \\ -2 & -1 & -1 & 0 & -1 & 0 & 0 & 0 & 0 & 0 \\ 0 & 0 & 0 & 0 & 1 & 0 & 0 & 0 & 0 & 0 \\ 0 & -0.5 & 0 & 0 & 0.5 & -1 & 0 & 0 & 0.5 & 0 \\ -2 & -0.5 & -1 & 0 & -0.5 & 0 & 0 & 0 & -0.5 & 0 \\ -2 & -0.5 & 0 & 0 & 0.5 & -1 & 0 & 0 & 0.5 & -1 \\ -2 & -1 & -1 & -1 & -1 & 0 & 0 & 0 & 0 & 0 \\ 0 & -0.5 & 0 & 0 & 0.5 & -1 & -1 & 0 & 0.5 & 0 \\ -2 & -0.5 & -1 & 0 & -0.5 & 0 & 0 & -1 & -0.5 & 0 \end{bmatrix} \begin{bmatrix} 1 & 0 & 0 \\ 0 & 0 & 0 \\ 0 & -1 & 0 \\ 0 & 0 & 0 \\ 0 & 0 & -1 \\ 0 & 0 & 0 \\ 0 & 0 & 0 \\ 0 & 0 & 0 \\ 0 & 0 & 0 \\ 0 & 0 & 0 \end{bmatrix} \begin{bmatrix} 0.01 \\ 0.0049 \\ 0.0041 \end{bmatrix} \text{mol g}^{-1}\,\text{h}^{-1}$$

Multiplying $\mathbf{S_m}$ with $\mathbf{v_m}$ first:

$$\mathbf{v_c} = -\begin{bmatrix} -1 & 0 & 0 & 0 & 0 & 0 & 0 & 0 & 0 & 0 \\ 0 & 0 & 1 & 0 & 0 & 0 & 0 & 0 & 0 & 0 \\ -2 & -1 & -1 & 0 & -1 & 0 & 0 & 0 & 0 & 0 \\ 0 & 0 & 0 & 0 & 1 & 0 & 0 & 0 & 0 & 0 \\ 0 & -0.5 & 0 & 0 & 0.5 & -1 & 0 & 0 & 0.5 & 0 \\ -2 & -0.5 & -1 & 0 & -0.5 & 0 & 0 & 0 & -0.5 & 0 \\ -2 & -0.5 & 0 & 0 & 0.5 & -1 & 0 & 0 & 0.5 & -1 \\ -2 & -1 & -1 & -1 & -1 & 0 & 0 & 0 & 0 & 0 \\ 0 & -0.5 & 0 & 0 & 0.5 & -1 & -1 & 0 & 0.5 & 0 \\ -2 & -0.5 & -1 & 0 & -0.5 & 0 & 0 & -1 & -0.5 & 0 \end{bmatrix} \begin{bmatrix} 0.01 \\ 0 \\ -0.0049 \\ 0 \\ -0.0041 \\ 0 \\ 0 \\ 0 \\ 0 \\ 0 \end{bmatrix} \text{mol g}^{-1}\,\text{h}^{-1}$$

Completing the matrix multiplication gives:

$$\mathbf{v}_c = \begin{bmatrix} 0.01 \\ 0.0049 \\ 0.0110 \\ 0.0041 \\ 0.0021 \\ 0.0131 \\ 0.0221 \\ 0.0110 \\ 0.0021 \\ 0.0131 \end{bmatrix} \text{mol g}^{-1}\,\text{h}^{-1} = \begin{bmatrix} v_1 \\ v_2 \\ v_3 \\ v_4 \\ v_5 \\ v_6 \\ v_7 \\ r_C \\ r_A \\ r_E \end{bmatrix}$$

All the internal and external fluxes are now known in units of mol $g^{-1}\,h^{-1}$. To convert to C-mol $g^{-1}\,h^{-1}$, each result except v_7 is multiplied by the number of C atoms in the molecule of the compound to which the flux refers:

glucose = 6
pyruvate = 3
lactate = 3
CO_2 = 1
formate = 1
acetyl CoA = 2
acetate = 2
ethanol = 2

The number of C atoms in acetyl CoA is taken as two, as glucose carbon is distributed only to the acetyl group with CoA acting as a conserved cofactor. The results are:

$$\begin{bmatrix} v_1 \\ v_2 \\ v_3 \\ v_4 \\ v_5 \\ v_6 \\ r_G \\ r_L \\ r_C \\ r_F \\ r_A \\ r_E \end{bmatrix} = \begin{bmatrix} 0.06 \\ 0.015 \\ 0.033 \\ 0.012 \\ 0.004 \\ 0.026 \\ 0.06 \\ 0.015 \\ 0.011 \\ 0.004 \\ 0.004 \\ 0.026 \end{bmatrix} \text{C-mol g}^{-1}\,\text{h}^{-1}$$

Because ATP is not a component of the carbon balance, the rate of ATP conversion to ADP, v_7, is expressed in units of mol $g^{-1}\,h^{-1}$ rather than C-mol $g^{-1}\,h^{-1}$. The results are shown in Figure 12.32.

(c) The flux split ratio at pyruvate for lactate production is:

$$\Phi_L = \frac{\text{flux to lactate}}{\text{flux to lactate} + \text{flux to acetyl CoA}} = \frac{0.015}{0.015 + 0.033 + 0.012} = 0.25$$

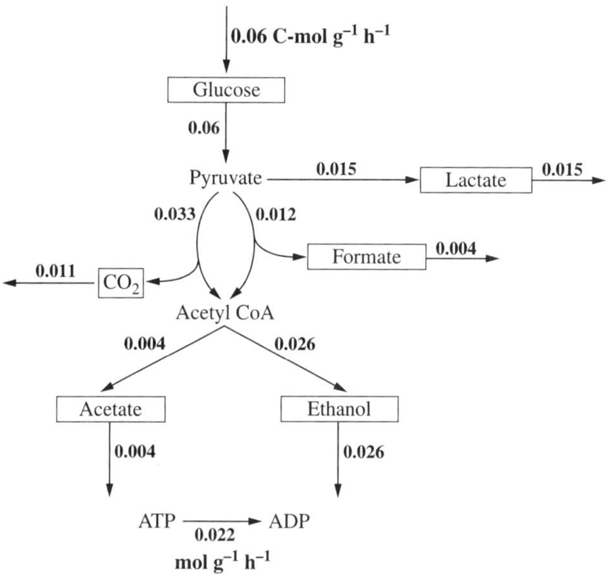

FIGURE 12.32 Simplified metabolic pathway for mixed acid fermentation showing carbon fluxes in C-mol g^{-1} h^{-1} and the ATP flux in mol g^{-1} h^{-1}.

(d) The flux split ratio at acetyl CoA is:

$$\Phi_A = \frac{\text{flux to acetate}}{\text{flux to acetate + flux to ethanol}} = \frac{0.004}{0.004 + 0.026} = 0.13 \text{ for acetate production}$$

$$\Phi_E = \frac{\text{flux to ethanol}}{\text{flux to acetate + flux to ethanol}} = \frac{0.026}{0.004 + 0.026} = 0.87 \text{ for ethanol production}$$

(e) From Eq. (12.110), the observed yield of ethanol from glucose is:

$$Y'_{PS} = \frac{r_E}{r_G} = \frac{0.0131 \text{ mol g}^{-1}\text{ h}^{-1}}{0.01 \text{ mol g}^{-1}\text{ h}^{-1}} = 1.3 \text{ mol mol}^{-1}$$

(f) The rate of ATP generation in the pathway is equal to the rate of ATP consumption for growth $= v_7 = 0.022$ mol g^{-1} h^{-1}.

Example 12.10 shows how a complete metabolic flux analysis can be performed on a stoichiometric network that is exactly determined. However, it is rare and generally undesirable for flux analyses to be carried out using mass balance equations and only enough measured data to give a unique solution. There are two major reasons for this.

1. For most metabolic networks, the number of metabolites M is substantially less than the number of reactions R so that, from Eq. (12.133), the system is underdetermined. The number of fluxes that can be found by measuring external substrate uptake and product excretion rates is limited and often insufficient to bring an extended metabolic network to a determined state. Other approaches and methods for analysing the system are required.

2. The results from flux analysis should be checked against independent experimental data additional to those used as constraints in the analysis. In other words, an overdetermined system where there are more measurements than degrees of freedom is required to allow validation of the network model.

Both these issues are addressed in the following sections.

Additional Constraints and Objective Functions

Most network models defined for metabolic flux analysis are underdetermined. In these circumstances, a unique solution to the flux distribution cannot be found; rather, there exists an infinite number of solutions that satisfy all of the balance equations available. Additional constraints are needed to reduce the size of the solution space and progress with the analysis. These constraints restrict the values of the fluxes to within prescribed limits so that realistic solutions reflecting the actual biological properties of the system can be found.

When metabolic fluxes are calculated, a positive value of v indicates that the reaction operates in the forward direction whereas a negative value indicates that the reaction operates in the reverse direction. For reversible reactions, both positive and negative values may be acceptable, as both forward and reverse reactions are possible thermodynamically. In this sense, v for reversible reactions is unrestricted:

$$-\infty < v < \infty \tag{12.149}$$

In contrast, many metabolic reactions are known to be irreversible, so that *thermodynamic constraints* can be imposed regarding the direction of the reaction and therefore the value of v. For example:

$$0 < v < \infty \tag{12.150}$$

may be applied as a constraint for an irreversible reaction that can proceed only in the forward direction, whereas:

$$-\infty < v < 0 \tag{12.151}$$

is valid for an irreversible reaction that can proceed only in the reverse direction. Such constraints ensure that the reactions and cyclic pathways included in the model are thermodynamically feasible. Additional *capacity constraints* reflecting restrictions on enzyme or cell function are imposed if, for example, a maximum limit on v is known:

$$0 < v < v_{\text{max}} \tag{12.152}$$

Other *regulatory constraints* relating to gene expression may also be applied. For instance, the fluxes for certain metabolic reactions may be set to zero based on transcriptional data if the gene coding for the enzyme responsible is found to be inactive.

Even after such constraints are included in the model, metabolic systems often remain underdetermined. However, a certain type of solution may be found by identifying an *objective function* and solving the equations using computational methods known as *linear programming* or *linear optimisation*. Examples of objective functions include cell growth, substrate uptake, synthesis of a particular metabolite, generation or utilisation of ATP, and generation or utilisation of NADH. In linear programming, numerical algorithms are applied to find the set of feasible steady-state fluxes that maximises or minimises the objective function chosen. A common approach is to solve for the fluxes that maximise the rate of cell growth.

Validation

Results from metabolic flux analysis that are not validated using experimental data are not considered reliable. Therefore, independent experimental information in addition to that used to solve for the fluxes is needed to check the model and calculations. However, as discussed in the previous section, most systems used for flux analysis are already underdetermined even after all feasible external rates have been quantified. To fulfil the requirement for additional data, the intracellular properties of metabolic pathways must be measured. This has led to the development of experimental and computational methods for the resolution of internal fluxes using carbon isotope labelling and analytical techniques such as mass spectrometry (MS) and nuclear magnetic resonance (NMR) spectroscopy.

When cells are supplied with substrate carrying a positional label such as ^{13}C, the label is passed to metabolites derived from the substrate. The resulting pattern of molecular labelling can be used to determine the relative activities of different metabolic pathways in the cell. For example, glucose taken up by *E. coli* is utilised in three metabolic pathways that operate simultaneously: the Embden–Meyerhof pathway, the Entner–Doudoroff pathway, and the pentose phosphate pathway. If the first carbon position of glucose is labelled uniformly with ^{13}C, pyruvate molecules produced in the three pathways will be labelled in different ways as illustrated in Figure 12.33. For labelled glucose entering the Embden–Meyerhof pathway, half of the resulting pyruvate is labelled with ^{13}C in the third carbon position while the remaining half is unlabelled. If labelled glucose is metabolised in the Entner–Doudoroff pathway, half of the pyruvate contains ^{13}C in the first

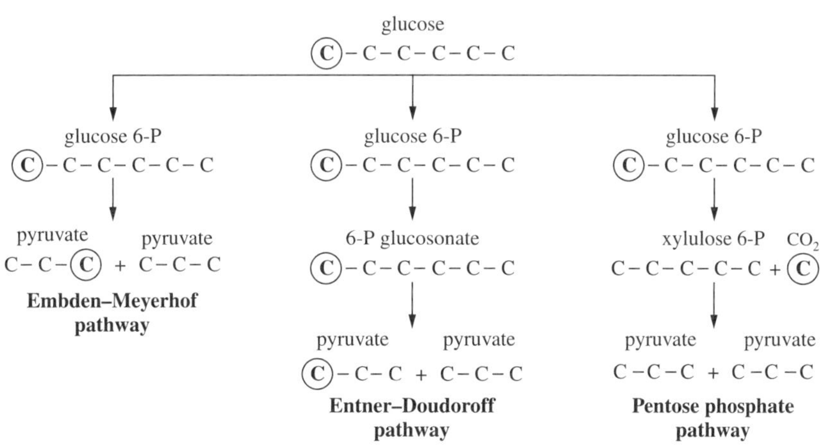

FIGURE 12.33 Fate of ^{13}C in labelled glucose metabolised to pyruvate. The ^{13}C label is indicated by a circle.

position while the other half is unlabelled. Labelled glucose metabolised in the pentose phosphate pathway produces unlabelled pyruvate as the ^{13}C label is passed to CO_2.

In principle, therefore, the relative fluxes through the three pathways can be determined by measuring the fractions of pyruvate molecules with ^{13}C labels in the first and third positions, for example, using MS or NMR, and applying appropriate mass balance equations for the carbon atoms. Because the carbon backbone of molecules in central carbon metabolism is preserved in the amino acids, similar analysis is also possible using measurements of amino acid labelling. An advantage of this approach is that amino acids are present in much larger quantities in cells than metabolic intermediates such as pyruvate. Alternative labelling choices provide information about the relative participation of other pathways. The internal fluxes determined in this way can be used to check the reliability of calculated results from metabolic flux analysis as well as provide additional insights into pathway operation.

When extra information about pathway function is obtained from intracellular measurements, the system may reach the desirable state of being overdetermined. If this occurs, there is sufficient redundant data available to perform computational checks on all the information used in the model. This improves estimation of the nonmeasured fluxes and minimises the effects of experimental error in the measured fluxes. If experimental error can be assumed to be equally distributed among the constraints imposed, a *least-squares solution* for the fluxes may be derived. In these calculations, the redundant measurements are used to adjust the measured fluxes depending on statistical assessment of their standard deviation (Section 3.1.5) or prevailing level of measurement noise.

Several other computational and statistical techniques are applied routinely to test and validate the outcome of metabolic flux analysis and improve the reliability of the results. The magnitude of the experimental error associated with measured reaction rates can have a significant effect on the accuracy of calculated fluxes. The *condition number* for the system, which is calculated from the matrix of stoichiometric coefficients, is a measure of the sensitivity of the results to small changes in the parameters used in the analysis. A system is considered ill-conditioned if the condition number is too high; this indicates that the model requires restructuring to avoid such heightened sensitivity. It is important to use accurate data and, because the influence of experimental error varies for different measured rates, to choose those rates for measurement with errors that have a relatively low impact on the flux results.

Limitations of Metabolic Flux Analysis

Certain features of metabolic pathways, such as reaction cycles and diverging pathways that converge at a later point in the network, present difficulties in flux analysis. The functioning of reversible reactions cannot be determined because the analysis provides information only about the net fluxes between metabolites. The size of metabolite pools is also not evaluated. Some of these limitations can be overcome by monitoring intracellular reactions using experimental techniques such as those described in the previous section.

Because it is based on stoichiometric information and the functioning of metabolism at steady state, flux analysis can only describe the properties of cells under the fixed conditions represented by the mass balance equations and measured rates. Although it can reveal important branch points and reactions that might be suitable targets for genetic manipulation to improve metabolic outcomes, flux analysis does not predict the effect of altering cell properties. After particular genes are deleted or amplified following flux analysis, the calculations and experimental measurements must be repeated to determine the redistribution of fluxes in the modified organism. Moreover, flux analysis provides no information about the kinetic response of metabolism to external perturbation and cannot be used to predict how metabolic processes are affected by enzyme regulation. Some of these issues are addressed using *dynamic flux balance models*, which combine analysis of stoichiometric equations for intracellular metabolism with unsteady-state mass balances on key extracellular substrates and products. More ambitious, however, is *metabolic control analysis*, which makes the connection between metabolic fluxes and enzyme activity and regulation. Some elements of metabolic control analysis are described in the following section.

12.16.2 Metabolic Control Analysis

So far in our treatment of metabolic engineering, we have seen how the fluxes of carbon through metabolic pathways can be estimated, and how this information reveals the relative contributions of particular reactions in the metabolic network. Metabolic flux analysis gives a snap-shot of pathway function under specific culture conditions, but falls short of predicting the response of metabolism to genetic, chemical, or physical change that may affect the cells. Consequently, we have not yet found a solution to the problem presented in Figure 12.24, where genetic modification to eliminate a particular metabolic reaction was ineffective for reducing the synthesis of an unwanted product.

One of the many possible explanations for the response shown in Figure 12.24(b) is that the increase in concentration of metabolite A, or the decrease in concentration of metabolite B, following knockout of A→B stimulated the activity of several enzymes, including those responsible for the reactions A→C and H→G. As a result, the impediment to production of G and therefore J was overcome. To understand this further, the relationships between metabolic fluxes, enzyme activities, and metabolite and effector concentrations must be determined. Metabolic control analysis offers a range of concepts and mathematical theorems for evaluating the control functions of metabolic pathways. Here we consider some of its basic features; further information is available in the literature [25, 26].

Metabolic control analysis makes use of a set of sensitivity parameters called *control coefficients* that quantify the fractional change in a specified property of a metabolic pathway in response to a small fractional change in the activity of an enzyme. The general formula for any control coefficient ξ is:

$$\xi = \lim_{\Delta y \to 0} \frac{\dfrac{\Delta x}{x}}{\dfrac{\Delta y}{y}} \tag{12.153}$$

where $\Delta x/x$ is the fractional change in system property x in response to a fractional change $\Delta y/y$ in parameter y, evaluated in the limit as Δy approaches zero. The limit condition means that control coefficients describe responses to very small changes in system parameters. Control coefficients are dimensionless variables that are independent of the units used to express x and y. Eq. (12.153) can be written as:

$$\xi = \frac{\frac{dx}{x}}{\frac{dy}{y}} = \frac{y\,dx}{x\,dy} = \frac{d\ln x}{d\ln y} \tag{12.154}$$

Several specific control coefficients have been defined. *Flux control coefficients* quantify the effect of changes in the activities of individual enzymes on the metabolic flux through a pathway or branch of a pathway. Flux control coefficients can take either positive or negative values, as an increase in the relative activity of a specific enzyme can either enhance or diminish a particular flux. The greater the absolute value of the flux control coefficient for a particular enzyme–flux combination, the greater is the influence exerted by that enzyme on that flux. In a similar way, *concentration control coefficients* describe the sensitivity of metabolite concentrations to changes in enzyme activity. The concentration of a particular metabolite in a pathway may increase or decrease as the activity of an enzyme changes, and the magnitude of the concentration change will vary for different enzyme–metabolite combinations.

Elasticity coefficients are different from control coefficients in that they are properties of individual enzymes rather than of the metabolic system as a whole. Elasticity coefficients describe the effect of a change in the concentration of a metabolite or effector on the rate of an enzyme reaction when the concentrations of all remaining metabolites and effectors are held constant. Elasticity coefficients are positive when enzyme activity is stimulated by an increase in the concentration of a substrate or enzyme activator, and negative when enzyme activity is reduced by an increased concentration of an inhibitor or reaction product. Although kinetic expressions describing the effect of substrate and inhibitor concentrations on enzyme reaction rates have been developed as outlined in Sections 12.3.3 and 12.5, these equations are considered to have limited applicability to enzymes in vivo. The presence of multiple substrates, relatively high product concentrations, and a variety of effector molecules in the intracellular environment makes kinetic models determined in vitro from initial rate data potentially unsuitable for metabolic analysis. Evaluation of the elasticities of metabolic enzymes under the conditions prevailing during in vivo operation of metabolic pathways is the preferred approach. Several methods have been developed to measure control and elasticity coefficients [27].

The interactions that occur between fluxes, enzymes, and metabolites and effector molecules are illustrated in Figure 12.34. Enzyme activity has a direct influence on the flux of pathways; this effect is described by the flux control coefficients. Enzyme activity also affects the concentrations of metabolites in the pathway, as represented by the concentration control coefficients. However, metabolites and other effectors exert an indirect influence on flux through their effect on enzyme activity, as measured by the elasticity

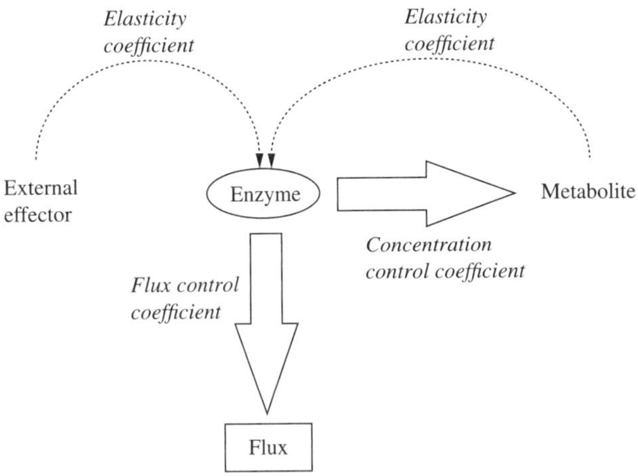

FIGURE 12.34 Interactions between metabolic flux, enzyme activity, and metabolites and external effectors.

coefficients. Some metabolites are not only subjected to biochemical conversion in metabolism but are also information carriers with roles in enzyme regulation.

Control and elasticity coefficients are subject to certain constraints and relationships that can be expressed mathematically. These provide information about how fluxes are affected by enzyme regulation. An implication of the *flux control summation theorem* that has been confirmed experimentally is that flux control is shared or distributed between many enzymes in metabolic pathways. Individual enzymes generally do not have large flux control coefficients; instead, near-zero and very small coefficients are common. As a result, the presence of a single enzyme playing a dominant rate-limiting or 'bottleneck' role for pathway control is relatively rare. This explains why manipulation of only one or two genes in cells often fails to influence metabolic flux, as illustrated in our hypothetical example of Figure 12.24.

To achieve significant change, coordinated modification of several enzymes is likely to be required, where the sum of the flux control coefficients is sufficiently high. An implication of the *flux control connectivity theorem*, which relates the elasticity coefficients to the flux control coefficients, is that enzymes with large elasticity coefficients, such as allosteric enzymes (Section 12.5.3), do not necessarily play an important role in the control of metabolic flux. The value of the flux control coefficient must also be taken into account as a measure of the enzyme's influence on the pathway. There is a tendency for enzymes with relatively large elasticity coefficients to have relatively small flux control coefficients, and vice versa, so that enzymes with small elasticity coefficients have a considerable influence on flux [27, 28]. This result from metabolic control analysis highlights the distinction between regulatory effects and flux control, in that enzymes subject to substantial regulation are not necessarily the best targets for genetic engineering strategies aimed at altering pathway function.

SUMMARY OF CHAPTER 12

At the end of Chapter 12 you should:

- Understand the difference between *reversible* and *irreversible reactions*, and the limitations of equilibrium thermodynamics in representing industrial cell culture and enzyme reactions
- Know the meaning of *total rate*, *volumetric rate*, and *specific rate* when describing the rate of reactions
- Be able to calculate reaction rates from batch concentration data using *graphical differentiation*
- Be familiar with kinetic relationships for *zero-order*, *first-order*, and *Michaelis–Menten* reactions
- Be able to quantify the effect of temperature on biological reaction rates
- Know how to determine the enzyme kinetic parameters v_{max} and K_m from batch concentration data
- Understand the effects of *inhibition* and *allosteric regulation* on enzyme kinetics
- Know how the kinetics of *enzyme deactivation* affect the rate of enzyme reactions
- Be able to calculate *yield coefficients* for cell cultures
- Understand the mathematical relationships that describe *cell growth kinetics*
- Be able to evaluate the specific rates of growth, product formation, and substrate uptake in batch cell cultures
- Know the effect of *maintenance activities* on growth, product synthesis, and substrate utilisation in cells
- Understand the kinetics of *cell death* and the thermal destruction of cells
- Know the goals, assumptions, and solution techniques for *metabolic flux analysis*
- Understand the importance of experimental data in flux balancing for *underdetermined* and *overdetermined* systems
- Know the basic concepts of *metabolic control analysis* for evaluating the sensitivity of fluxes to enzyme activities and metabolite concentrations

PROBLEMS

12.1 Reaction equilibrium

Calculate equilibrium constants for the following reactions under standard conditions:

(a) glutamine + $H_2O \longrightarrow$ glutamate + NH_4^+

$\Delta G^{\circ}_{rxn} = -14.1 \text{ kJ mol}^{-1}$

(b) malate $\longrightarrow$ fumarate + H_2O

$\Delta G^{\circ}_{rxn} = 3.2 \text{ kJ mol}^{-1}$

Can either of these reactions be considered irreversible?

12.2 Equilibrium yield

The following reaction catalysed by phosphoglucomutase occurs during breakdown of glycogen:

$$\text{glucose 1-phosphate} \rightleftarrows \text{glucose 6-phosphate}$$

A reaction is started by adding phosphoglucomutase to 0.04 gmol of glucose 1-phosphate in 1 litre of solution at 25°C. The reaction proceeds to equilibrium, at which the concentration of glucose 1-phosphate is 0.002 M and the concentration of glucose 6-phosphate is 0.038 M.

(a) Calculate the equilibrium constant.
(b) What is the theoretical yield?
(c) What is the yield based on the amount of reactant supplied?

12.3 Reaction rate

(a) The volume of a fermenter is doubled while keeping the cell concentration and other fermentation conditions the same.
  (i) How is the volumetric productivity affected?
  (ii) How is the specific productivity affected?
  (iii) How is the total productivity affected?
(b) If, instead of (a), the cell concentration were doubled, what effect would this have on the volumetric, specific, and total productivities?
(c) A fermenter produces 100 kg of lysine per day.
  (i) If the volumetric productivity is 0.8 g l^{-1} h^{-1}, what is the volume of the fermenter?
  (ii) The cell concentration is 20 g l^{-1} dry weight. Calculate the specific productivity.

12.4 Enzyme kinetics

Lactase, also known as β-galactosidase, catalyses the hydrolysis of lactose to produce glucose and galactose from milk and whey. Experiments are carried out to determine the kinetic parameters for the enzyme. The initial rate data are as follows.

Lactose Concentration (mol $l^{-1} \times 10^2$)	Initial Reaction Velocity (mol l^{-1} $min^{-1} \times 10^3$)
2.50	1.94
2.27	1.91
1.84	1.85
1.35	1.80
1.25	1.78
0.730	1.46
0.460	1.17
0.204	0.779

Evaluate v_{max} and K_m.

12.5 Enzyme kinetics after site-specific mutagenesis

Cyclophosphamide (CPA) is an anticancer prodrug that requires activation in the liver by cytochrome P450 2B enzymes for production of cytotoxic metabolites. Site-specific mutagenesis is used to alter the amino acid sequence of P450 2B1 in an attempt to improve the kinetics of CPA activation. The rate of reaction of CPA is studied using rat P450 2B1 and a site-specific variant of P450 2B1 produced using *Escherichia coli*. The results are as follows.

Initial CPA Concentration (mM)	Initial Reaction Velocity (mol min^{-1} mol^{-1} P450) Rat P450 2B1	Variant P450 2B1
0.3	5.82	17.5
0.5	9.03	24.5
0.8	12.7	24.0
1.5	17.1	23.9
3	20.2	27.3
5	27.8	33.1
7	31.5	27.7

(a) Determine the kinetic constants for the rat and site-specific variant enzymes.

(b) When CPA is administered to cancer patients, the peak plasma concentration of the drug is relatively low at around 100 to 200 μM. Are the kinetic properties of the variant P450 2B1 better than those of rat P450 2B1 for CPA activation in this situation? Explain your answer.

(c) Did manipulation of the enzyme using site-specific mutagenesis improve the catalytic efficiency for CPA activation?

12.6 Kinetic properties of pheromone-degrading enzyme

An enzyme that degrades pheromone is isolated from the sensory hairs of the silk moth, *Antheraea polyphemus*. The kinetics of the reaction are studied at pH 7.2 using a fixed enzyme concentration and temperatures of 10°C to 40°C. The following table lists results for reaction velocity as a function of substrate concentration.

Pheromone Concentration (μmol l^{-1})	Initial Reaction Velocity (μmol l^{-1} s^{-1}) $T = 10°C$	$T = 20°C$	$T = 30°C$	$T = 40°C$
0.5	3.0×10^{-6}	5.5×10^{-6}	4.2×10^{-6}	7.7×10^{-6}
1.0	5.1×10^{-6}	9.2×10^{-6}	9.5×10^{-6}	1.2×10^{-5}
1.0	4.2×10^{-6}	9.7×10^{-6}	8.9×10^{-6}	1.6×10^{-5}
1.5	6.1×10^{-6}	1.1×10^{-5}	1.3×10^{-5}	1.8×10^{-5}
2.2	7.1×10^{-6}	1.6×10^{-5}	9.8×10^{-6}	2.1×10^{-5}
5.5	1.1×10^{-5}	1.5×10^{-5}	1.9×10^{-5}	3.8×10^{-5}
5.5	9.8×10^{-6}	1.9×10^{-5}	2.6×10^{-5}	3.0×10^{-5}
11	1.2×10^{-5}	2.2×10^{-5}	2.9×10^{-5}	3.9×10^{-5}
11	9.5×10^{-6}	2.1×10^{-5}	2.5×10^{-5}	3.6×10^{-5}

(a) Determine v_{max} and K_m at the four reaction temperatures.

(b) Determine the activation energy for this enzyme reaction.

12.7 Enzyme substrate specificity

Xylose isomerase extracted from the thermophile *Thermoanaerobacterium thermosulfurigenes* is capable of using glucose and xylose as substrates. Applying each substrate separately, the rates of reaction are measured at 60°C using an enzyme concentration of 0.06 mg ml^{-1}.

Glucose Concentration (mM)	Initial Reaction Velocity (μmol ml^{-1} min^{-1})
80	0.151
100	0.194
300	0.385
500	0.355
750	0.389
1000	0.433
1400	0.445

Xylose Concentration (mM)	Initial Reaction Velocity (μmol ml^{-1} min^{-1})
20	0.320
50	0.521
100	0.699
300	1.11
500	1.19
700	1.33
900	1.13

(a) What is the relative catalytic efficiency of the enzyme using xylose as substrate compared with glucose?

(b) What conclusions can you draw about the substrate specificity of this enzyme?

12.8 Effect of temperature on the hydrolysis of starch

α-Amylase from malt is used to hydrolyse starch. The dependence of the initial reaction rate on temperature is determined experimentally. Results measured at fixed starch and enzyme concentrations are listed in the following table.

Temperature (°C)	Rate of Glucose Production (mmol m^{-3} s^{-1})
20	0.31
30	0.66
40	1.20
60	6.33

(a) Determine the activation energy for this reaction.

(b) α-Amylase is used to break down starch in baby food. It is proposed to conduct the reaction at a relatively high temperature so that the viscosity is reduced. What is the reaction rate at 55°C compared with 25°C?

(c) Thermal deactivation of this enzyme is described by the equation:

$$k_d = 2.25 \times 10^{27}\, e^{-41{,}630/RT}$$

where k_d is the deactivation rate constant in h^{-1}, R is the ideal gas constant in cal $gmol^{-1}$ K^{-1}, and T is temperature in K. What is the half-life of the enzyme at 55°C compared with 25°C? Which of these two operating temperatures is more practical for processing baby food? Explain your answer.

12.9 Optimum temperature for enzymatic hydrolysis of cellulose

The production of glucose from lignocellulosic materials is a major hurdle in the production of biofuels from renewable resources. β-Glucosidase from *Aspergillus niger* is investigated as a potential component in an enzyme cocktail for cellulose hydrolysis at elevated temperatures. The activity of 5.1 mg l^{-1} of β-glucosidase is measured at temperatures from 30°C to 70°C using cellobiose as substrate. The results are as follows.

Temperature (°C)	v_{max} (mmol l^{-1} min^{-1})
30	456
45	1250
50	1590
60	2900
70	567

(a) Estimate the optimum temperature for cellobiose hydrolysis.
(b) Evaluate the Arrhenius parameters for thermal activation of β-glucosidase.
(c) What is the maximum rate of cellobiose hydrolysis at 55°C?

12.10 Enzyme reaction and deactivation

Lipase is being investigated as an additive to laundry detergent for removal of stains from fabric. The general reaction is:

$$\text{fats} \longrightarrow \text{fatty acids} + \text{glycerol}$$

The Michaelis constant for pancreatic lipase is 5 mM. At 60°C, lipase is subject to deactivation with a half-life of 8 min. Fat hydrolysis is carried out in a well-mixed batch reactor that simulates a top-loading washing machine. The initial fat concentration is 45 gmol m^{-3}. At the beginning of the reaction, the rate of hydrolysis is 0.07 mmol l^{-1} s^{-1}. How long does it take for the enzyme to hydrolyse 80% of the fat present?

12.11 Effect of amino acid composition and metal binding on enzyme stability

Genomic analysis has revealed that enzymes from hyperthermophilic microorganisms contain lower amounts of the thermo-labile amino acids, glutamine (Gln) and asparagine (Asn), than enzymes from mesophiles. It is thought that this adaptation may contribute to the thermostability of enzymes in high-temperature environments above 50°C. A variant form of xylose isomerase containing a low (Gln + Asn) content is developed using site-directed mutagenesis. Experiments are conducted to measure the activity of native and variant xylose isomerase at 68°C using an enzyme concentration of 0.5 mg ml^{-1}. Because binding with metal ions also has a potential stabilising effect on this enzyme, the activity of native xylose isomerase is measured in the absence and presence of

0.015 mM Mn^{2+}. The reactions are performed with excess substrate and give the following results.

Native Enzyme		Variant Enzyme		Native Enzyme + Mn^{2+}	
Time (min)	Ratio of Activity to Initial Activity	Time (min)	Ratio of Activity to Initial Activity	Time (min)	Ratio of Activity to Initial Activity
0	1	0	1	0	1
10	0.924	5	0.976	10	0.994
30	0.795	20	0.831	20	0.834
45	0.622	50	0.608	50	0.712
60	0.513	140	0.236	180	0.275
120	0.305	200	0.149	240	0.197
150	0.225	–	–	–	–

(a) Does deactivation of this enzyme at 68°C follow first-order kinetics under each of the conditions tested?

(b) Compare the half life of the variant enzyme with those of the native enzyme with and without Mn^{2+}.

(c) What conclusions can you draw about the relative effectiveness of the strategies used to enhance the stability of this enzyme?

12.12 Growth parameters for recombinant *E. coli*

Escherichia coli is used for production of recombinant porcine growth hormone. The bacteria are grown aerobically in batch culture with glucose as the growth-limiting substrate. Cell and substrate concentrations are measured as a function of culture time with the following results.

Time (h)	Cell Concentration, x (kg m^{-3})	Substrate Concentration, s (kg m^{-3})
0.0	0.20	25.0
0.33	0.21	24.8
0.5	0.22	24.8
0.75	0.32	24.6
1.0	0.47	24.3
1.5	1.00	23.3
2.0	2.10	20.7
2.5	4.42	15.7

(Continued)

2.8	6.9	10.2
3.0	9.4	5.2
3.1	10.9	1.65
3.2	11.6	0.2
3.5	11.7	0.0
3.7	11.6	0.0

(a) Plot μ as a function of time.
(b) What is the value of μ_{max}?
(c) What is the observed biomass yield from substrate? Is Y'_{XS} constant during the culture?

12.13 Growth parameters for hairy roots

Hairy roots are produced by genetic transformation of plants using *Agrobacterium rhizogenes*. The following biomass and sugar concentrations are obtained during batch culture of *Atropa belladonna* hairy roots in a bubble column fermenter.

Time (days)	Biomass Concentration ($g\ l^{-1}$ dry weight)	Sugar Concentration ($g\ l^{-1}$)
0	0.64	30.0
5	1.95	27.4
10	4.21	23.6
15	5.54	21.0
20	6.98	18.4
25	9.50	14.8
30	10.3	13.3
35	12.0	9.7
40	12.7	8.0
45	13.1	6.8
50	13.5	5.7
55	13.7	5.1

(a) Plot μ as a function of culture time. When does the maximum specific growth rate occur?
(b) Plot the specific rate of sugar uptake as a function of time.
(c) What is the observed biomass yield from substrate? Is Y'_{XS} constant during the culture?

12.14 Kinetics of diatom growth and silicate uptake

Growth and nutrient uptake in batch cultures of the freshwater diatom, *Cyclotella meneghiniana*, are studied under silicate-limiting conditions. Unbuffered freshwater medium containing 25 μM silicate is inoculated with cells. Samples are taken over a period of 4 days for measurement of cell and silicate concentrations. The results are as follows.

Time (days)	Cell Concentration (cells l^{-1})	Silicate Concentration (μM)
0	4.41×10^5	8.00
0.5	5.53×10^5	7.97
1.0	1.31×10^6	7.72
1.5	3.00×10^6	7.59
2.0	4.82×10^6	6.96
2.5	1.12×10^7	5.33
3.0	1.67×10^7	4.63
4.0	2.57×10^7	1.99

(a) Does this culture exhibit exponential growth?
(b) What is the value of μ_{max}?
(c) Is there a lag phase?
(d) What is the observed biomass yield from substrate?
(e) Is the observed biomass yield from substrate constant during the culture?

12.15 Algal batch growth kinetics

Chlorella sp. algae are used to produce lipids for the manufacture of biodiesel. The kinetics of algal growth are determined in batch culture under phosphate-limiting conditions at an illumination intensity of approximately 60 μmol m^{-2} s^{-1} and a photoperiod of 14 h light:10 h dark. The following results are obtained from five different cultures at temperatures from 19°C to 28.5°C, including data from a replicate culture at 25°C.

$T = 19°C$		$T = 20°C$		$T = 25°C$		$T = 25°C$		$T = 28.5°C$	
Time (days)	Cell Concentration (cells l^{-1})	Time (days)	Cell Concentration (cells l^{-1})	Time (days)	Cell Concentration (cells l^{-1})	Time (days)	Cell Concentration (cells l^{-1})	Time (days)	Cell Concentration (cells l^{-1})
0	1.30×10^5	0	1.30×10^5	0	1.35×10^5	0	1.35×10^5	0	1.30×10^5
0.2	1.87×10^5	0.5	3.99×10^5	0.5	2.58×10^5	0.5	2.95×10^5	1.0	1.87×10^5
0.5	2.44×10^5	1.0	6.97×10^5	1.0	9.49×10^5	1.0	1.02×10^6	1.5	2.44×10^5
1.0	6.12×10^5	1.5	1.14×10^6	1.5	2.91×10^6	1.5	1.34×10^6	2.6	6.12×10^5
1.4	1.34×10^6	2.1	4.88×10^6	2.1	7.66×10^6	2.1	9.03×10^6	–	–
1.8	1.72×10^6	2.1	3.98×10^6	2.6	1.77×10^7	2.6	6.26×10^7	–	–
2.2	3.92×10^6	3.5	4.12×10^7	3.8	2.03×10^8	3.8	2.96×10^8	–	–
2.5	7.01×10^6	–	–	–	–	–	–	–	–
3.0	1.75×10^7	–	–	–	–	–	–	–	–
3.8	4.70×10^7	–	–	–	–	–	–	–	–

(a) Do the cultures exhibit exponential growth at each of the temperatures tested?

(b) Evaluate the maximum specific growth rates and doubling times for each of the five algal cultures.

(c) Use the data to determine an expression for μ_{max} as a function of temperature during thermal activation of growth.

(d) Estimate the maximum specific growth rate at 22°C.

(e) Evaluate the amount of algal cells produced if a 1.6-m^3 bioreactor is inoculated at a cell density of 2×10^5 cells l^{-1} and the culture grows exponentially at 22°C for a period of 2 days and 6 hours.

12.16 Kinetics of batch cell culture with nisin production

Lactococcus lactis subsp. *lactis* produces nisin, a biological food preservative with bactericidal properties. Medium containing 10 g l^{-1} sucrose is inoculated with bacteria and the culture is monitored for 24 h. The pH is controlled at 6.80 using 10 M NaOH. The following results are recorded.

Time (h)	Cell Concentration (g l^{-1} dry weight)	Sugar Concentration (g l^{-1})	Nisin Concentration (IU ml^{-1})
0.0	0.02	9.87	0
1.0	0.03	9.77	278
2.0	0.03	9.38	357
3.0	0.042	9.45	564
4.0	0.21	9.02	662
4.5	0.33	8.55	695
5.0	0.35	9.12	1213
5.5	0.38	8.09	1341
6.0	1.13	7.58	1546
6.5	1.95	6.24	1574
7.0	3.66	1.30	1693
7.5	4.09	1.14	1678
8.0	4.23	0.05	1793
8.5	4.07	0.02	1733
9.0	3.85	0.02	1567
9.5	2.66	0.015	1430
10.0	2.42	0.018	1390
11.0	2.03	0	–
12.0	1.54	0	1220
14.0	1.89	0	995
16.0	1.45	0	–
24.0	1.67	0	617

(a) Plot all of the data on a single graph, using multiple vertical axes to show the cell, sugar, and nisin concentrations as a function of time.

(i) What is the relationship between nisin production and growth?

(ii) *L. lactis* produces lactic acid as a product of energy metabolism. Release of lactic acid into the medium can lower the pH significantly and cause premature cessation of growth. Do the measured data show any evidence of this? Explain your answer.

(iii) Is 24 hours an appropriate batch culture duration for nisin production? Explain your answer.

(iv) What is the observed overall yield of biomass from substrate? Is this a meaningful parameter for characterising the extent of sugar conversion to biomass in this culture? Explain your answer.

(b) Plot the data for cell concentration as a function of time on semi-logarithmic coordinates.

(i) Does the culture undergo a lag phase? If so, what is the duration of the lag phase?

(ii) Does exponential growth occur during this culture? If so, over what time period?

(iii) Develop an equation for cell concentration as a function of time during the growth phase.

(iv) What is the maximum specific growth rate?

(v) What is the culture doubling time? When does this doubling time apply?

(vi) Does the culture reach stationary phase? Explain your answer.

(vii) Over what period does the culture undergo a decline phase?

(viii) Determine the specific cell death constant for the culture.

(ix) At the end of the growth phase, what is the observed yield of biomass from substrate based on the initial concentrations of biomass and substrate? Compare this with the answer to (a) (iv) above and comment on any difference.

(x) At the end of the growth phase, what is the observed yield of nisin from substrate? Use a conversion factor of 40×10^6 IU per g of nisin.

(xi) At the end of the growth phase, what is the observed yield of nisin from biomass in units of mg g^{-1} dry weight?

(c) Analyse the data for sugar concentration to determine the rate of substrate consumption as a function of culture time. Plot the results.

(i) What is the maximum rate of substrate consumption?

(ii) When does the maximum rate of substrate consumption occur?

(d) Plot the specific rate of substrate consumption as a function of time.

(i) What is the maximum specific rate of substrate consumption? When does it occur? How much confidence do you have in your answer?

(ii) What is the specific rate of substrate consumption during the growth phase?

(e) Analyse the data for nisin concentration to determine the rate of nisin production as a function of culture time. Plot the results.

(i) Estimate the maximum nisin productivity. Express your answer in units of mg l^{-1} h^{-1}.

(ii) When does the maximum nisin productivity occur?

(f) Plot the specific rate of nisin production as a function of time.

(i) What is the maximum specific rate of nisin production? Express your answer in units of mg h^{-1} g^{-1} dry weight.

(ii) When does this maximum occur?

(g) Plot the rate of biomass production versus the rate of substrate consumption for the culture.

   (i) During growth when the rate of biomass production is positive rather than negative, can the observed biomass yield from substrate be considered constant?

   (ii) How does the value for Y'_{XS} determined graphically for the entire growth phase compare with the results in (a) and (b) above? Explain any differences.

(h) Plot the rate of nisin production versus the rate of substrate consumption, and the rate of nisin production versus the rate of biomass production, using two separate graphs. During the culture period when the rate of nisin production is positive rather than negative, can the observed yield coefficients Y'_{PS} and Y'_{PX} be considered constant during batch culture?

12.17 Yeast culture and astaxanthin production

The yeast *Phaffia rhodozyma* produces the carotenoid pigment astaxanthin in the absence of light. The cells are grown in batch culture using medium containing 40 g l^{-1} glucose. Cell, substrate, and product concentrations are measured as a function of culture time. The results are as follows.

Time (h)	Cell Concentration ($g\ l^{-1}$ dry weight)	Sugar Concentration ($g\ l^{-1}$)	Astaxanthin Concentration ($mg\ l^{-1}$)
0	0.01	40	0
5	0.019	39.96	0.19
10	0.029	39.90	0.32
12	0.029	39.91	0.45
20	0.074	39.65	0.60
24	0.124	39.14	0.72
30	0.157	38.95	0.81
35	0.356	38.22	0.87
45	0.906	35.24	1.01
50	2.28	32.40	1.29
60	3.67	17.05	1.35
70	7.17	0	1.33
78	6.59	0	1.48

(a) Does the culture exhibit a lag phase?

(b) Does growth of *Phaffia rhodozyma* follow first-order kinetics?

(c) Evaluate μ_{max} and the culture doubling time. Over what period are these parameters valid representations of growth?

(d) From a plot of astaxanthin concentration versus time, estimate the initial rate of astaxanthin synthesis. Express your answer in units of mg l^{-1} h^{-1}.

(e) Using the result from (d), estimate the initial specific rate of astaxanthin production. What confidence do you have in this result?

(f) Plot Δx versus Δs to determine the observed biomass yield from glucose. Use the initial cell and substrate concentrations to calculate Δx and Δs. During the growth phase, how does the observed biomass yield from glucose vary with culture time?

(g) What is the overall product yield from glucose?

(h) What is the overall product yield from biomass?

12.18 Ethanol fermentation by yeast and bacteria

Ethanol is produced by anaerobic fermentation of glucose by *Saccharomyces cerevisiae*. For the particular strain of *S. cerevisiae* employed, the maintenance coefficient is 0.18 kg kg^{-1} h^{-1}, Y_{XS} is 0.11 kg kg^{-1}, Y_{PX} is 3.9 kg kg^{-1}, and μ_{max} is 0.4 h^{-1}. It is decided to investigate the possibility of using *Zymomonas mobilis* bacteria instead of yeast for making ethanol. *Z. mobilis* is known to produce ethanol under anaerobic conditions using a different metabolic pathway to that employed by yeast. Typical values of Y_{XS} are lower than for yeast at about 0.06 kg kg^{-1}; on the other hand, the maintenance coefficient is higher at 2.2 kg kg^{-1} h^{-1}. Y_{PX} for *Z. mobilis* is 7.7 kg kg^{-1}; μ_{max} is 0.3 h^{-1}.

(a) From stoichiometry, what is the maximum theoretical yield of ethanol from glucose?

(b) Y'_{PS} is maximum and equal to the theoretical yield when there is no growth and all substrate entering the cell is used for maintenance activities. If ethanol is the sole extracellular product of energy-yielding metabolism, calculate m_P for each organism.

(c) *S. cerevisiae* and *Z. mobilis* are cultured in batch fermenters. Predict the observed product yield from substrate for the two cultures.

(d) What is the efficiency of ethanol production by the two organisms? Efficiency is defined as the observed product yield from substrate divided by the maximum or theoretical product yield.

(e) How does the specific rate of ethanol production by *Z. mobilis* compare with that by *S. cerevisiae*?

(f) Using Eq. (12.101), compare the proportions of growth-associated and nongrowth-associated ethanol production by *Z. mobilis* and *S. cerevisiae*. For which organism is nongrowth-associated production more substantial?

(g) To achieve the same volumetric ethanol productivity from the two cultures, what relative concentrations of yeast and bacteria are required?

(h) In the absence of growth, the efficiency of ethanol production is the same in both cultures. Under these conditions, if yeast and bacteria are employed at the same concentration, what size fermenter is required for the yeast culture compared with that required for the bacterial culture to achieve the same total productivity?

(i) Predict the observed biomass yield from substrate for the two organisms. For which organism is biomass disposal less of a problem?

(j) Make a recommendation about which organism is better suited for industrial ethanol production. Explain your answer.

12.19 Plasmid loss during culture maintenance

A stock culture of plasmid-containing *Streptococcus cremoris* cells is maintained with regular subculturing for a period of 28 days. After this time, the fraction of plasmid-carrying cells is found to be 0.66. The specific growth rate of plasmid-free cells at the storage temperature is 0.033 h^{-1}; the specific growth rate of plasmid-containing cells is 0.025 h^{-1}. If all the cells initially contained plasmid, estimate the frequency of plasmid loss per generation.

12.20 Medium sterilisation

A steam steriliser is used to sterilise liquid medium for fermentation. The initial concentration of contaminating organisms is 10^8 per litre. For design purposes, the final acceptable level of contamination is usually taken to be 10^{-3} cells; this corresponds to a risk that one batch in a thousand will remain contaminated even after sterilisation is complete. For how long should 1 m^3 of medium be treated if the sterilisation temperature is:

(a) 80°C?

(b) 121°C?

(c) 140°C?

To be safe, assume that the contaminants present are spores of *Bacillus stearothermophilus*, one of the most heat-resistant microorganisms known. For these spores, the activation energy for thermal death is 283 kJ $gmol^{-1}$ and the Arrhenius constant is $10^{36.2}$ s^{-1} [29].

12.21 Effect of medium osmolarity on growth and death of hybridoma cells

Medium osmolarity is an important variable in the design of serum-free media for mammalian cell culture. A murine hybridoma cell line synthesising IgG_1 antibody is grown in culture media adjusted to three different osmolarities by addition of NaCl or sucrose. Growth and viability of the cells in batch culture are monitored over a period of 300 hours. Results for viable cell concentration as a function of culture time are as follows.

	Viable Cell Concentration (cells ml^{-1})		
Time (h)	Osmolarity = 290 mOsm l^{-1}	Osmolarity = 380 mOsm l^{-1}	Osmolarity = 435 mOsm l^{-1}
0	4.2×10^4	4.2×10^4	4.2×10^4
25	9.8×10^4	6.3×10^4	4.8×10^4
50	2.2×10^5	1.2×10^5	7.0×10^4
75	6.5×10^5	2.2×10^5	9.7×10^4
100	1.1×10^6	4.1×10^5	1.6×10^5
125	1.0×10^6	5.1×10^5	2.5×10^5
150	9.7×10^5	7.9×10^5	4.1×10^5
175	7.8×10^5	7.7×10^5	5.9×10^5
200	7.0×10^5	6.0×10^5	5.4×10^5
225	7.0×10^5	5.5×10^5	4.4×10^5
250	6.5×10^5	3.5×10^5	2.6×10^5
275	4.7×10^5	1.7×10^5	1.2×10^5
300	4.2×10^5	1.3×10^5	8.8×10^4

(a) Does medium osmolarity affect whether the hybridoma cells undergo a lag phase?

(b) How does medium osmolarity affect the maximum specific growth rate of the cells?

(c) How does medium osmolarity affect the maximum cell concentration achieved and the time taken to reach maximum cell density?

(d) What effect does medium osmolarity have on the specific death rate of the cells during the culture decline phase?

References

[1] B. Atkinson, F. Mavituna, Biochemical Engineering and Biotechnology Handbook, second ed., Macmillan, 1991.
[2] R.L. Stein, Kinetics of Enzyme Action, John Wiley, 2011.
[3] A. Cornish-Bowden, C.W. Wharton, Enzyme Kinetics, IRL Press, 1988.
[4] M. Dixon, E.C. Webb, Enzymes, second ed., Longmans, 1964.
[5] I.W. Sizer, Effects of temperature on enzyme kinetics, Adv. Enzymol 3 (1943) 35–62.
[6] A. Moser, Rate equations for enzyme kinetics, in: H.-J. Rehm, G. Reed (Eds.), Biotechnology, vol. 2, VCH, 1985, pp. 199–226.
[7] R. Eisenthal, A. Cornish-Bowden, The direct linear plot: a new graphical procedure for estimating enzyme kinetic parameters, Biochem. J. 139 (1974) 715–720.
[8] A. Moser, Kinetics of batch fermentations, in: H.-J. Rehm, G. Reed (Eds.), Biotechnology, vol. 2, VCH, 1985, pp. 243–283.
[9] J.E. Bailey, D.F. Ollis, Biochemical Engineering Fundamentals, second ed., McGraw-Hill, 1986. (Chapter 7).
[10] J.A. Roels, N.W.F. Kossen, On the modelling of microbial metabolism, Prog. Ind. Microbiol 14 (1978) 95–203.
[11] M.L. Shuler, F. Kargi, Bioprocess Engineering: Basic Concepts, second ed., Prentice Hall, 2002. (Chapter 6).
[12] T. Imanaka, S. Aiba, A perspective on the application of genetic engineering: stability of recombinant plasmid, Ann. N.Y. Acad. Sci 369 (1981) 1–14.
[13] N.S. Cooper, M.E. Brown, C.A. Caulcott, A mathematical model for analysing plasmid stability in microorganisms, J. Gen. Microbiol. 133 (1987) 1871–1880.
[14] D.F. Ollis, H.-T. Chang, Batch fermentation kinetics with (unstable) recombinant cultures, Biotechnol. Bioeng. 24 (1982) 2583–2586.
[15] J.E. Bailey, M. Hjortso, S.B. Lee, F. Srienc, Kinetics of product formation and plasmid segregation in recombinant microbial populations, Ann. N.Y. Acad. Sci 413 (1983) 71–87.
[16] K.D. Wittrup, J.E. Bailey, A segregated model of recombinant multicopy plasmid propagation, Biotechnol. Bioeng. 31 (1988) 304–310.
[17] A.H. Stouthamer, H.W. van Verseveld, Stoichiometry of microbial growth, in: M. Moo-Young (Ed.), Comprehensive Biotechnology, vol. 1, Pergamon, 1985, pp. 215–238.
[18] J.J. Heijnen, J.A. Roels, A.H. Stouthamer, Application of balancing methods in modeling the penicillin fermentation, Biotechnol. Bioeng. 21 (1979) 2175–2201.
[19] J.J. Heijnen, J.A. Roels, A macroscopic model describing yield and maintenance relationships in aerobic fermentation processes, Biotechnol. Bioeng 23 (1981) 739–763.
[20] S.J. Pirt, Principles of Microbe and Cell Cultivation, Blackwell Scientific, 1975.
[21] R.G. Forage, D.E.F. Harrison, D.E. Pitt, Effect of environment on microbial activity, in: M. Moo-Young (Ed.), Comprehensive Biotechnology, vol. 1, Pergamon, 1985, pp. 251–280.
[22] D.I.C. Wang, C.L. Cooney, A.L. Demain, P. Dunnill, A.E. Humphrey, M.D. Lilly, Fermentation and Enzyme Technology, John Wiley, 1979.
[23] C.L. Cooney, Media sterilization, in: M. Moo-Young (Ed.), Comprehensive Biotechnology, vol. 2, Pergamon, 1985, pp. 287–298.
[24] A.M. Feist, C.S. Henry, J.L. Reed, M. Krummenacker, A.R. Joyce, P.D. Karp, L.J. Broadbelt, V. Hatzimanikatis, B.Ø. Palsson, A genome-scale metabolic reconstruction for *Escherichia coli* K-12 MG1655 that accounts for 1260 ORFs and thermodynamic information, Mol. Syst. Biol. 3 (2007) 121.
[25] G.N. Stephanopoulos, A.A. Aristidou, J. Nielsen, Metabolic Engineering: Principles and Methodologies, Academic Press, 1998.
[26] D. Fell, Understanding the Control of Metabolism, Portland Press, 1997.
[27] D.A. Fell, Metabolic control analysis: a survey of its theoretical and experimental development, Biochem. J. 286 (1992) 313–330.
[28] J.R. Small, D.A. Fell, Metabolic control analysis: sensitivity of control coefficients to elasticities, Eur. J. Biochem. 191 (1990) 413–420.
[29] F.H. Deindoerfer, A.E. Humphrey, Analytical method for calculating heat sterilization times, Appl. Microbiol. 7 (1959) 256–264.

Suggestions for Further Reading

Reaction Thermodynamics

Lehninger, A. L. (1965). *Bioenergetics*. W.A. Benjamin.

General Reaction Kinetics

Froment, G. F., Bischoff, K. B., & De Wilde, J. (2010). *Chemical Reactor Analysis and Design* (3rd ed., Chapter 1). John Wiley.

Holland, C. D., & Anthony, R. G. (1989). *Fundamentals of Chemical Reaction Engineering* (2nd ed., Chapter 1). Prentice Hall.

Levenspiel, O. (1999). *Chemical Reaction Engineering* (3rd ed., Chapters 2 and 3). John Wiley.

Graphical Differentiation

Churchill, S. W. (1974). *The Interpretation and Use of Rate Data: The Rate Concept*. McGraw-Hill.

Hougen, O. A., Watson, K. M., & Ragatz, R. A. (1962). *Chemical Process Principles, Part I* (2nd ed., Chapter 1). John Wiley.

Enzyme Kinetics and Deactivation

See also references [2–6].

Bisswanger, H. (2008). *Enzyme Kinetics: Principles and Methods* (2nd ed.). Wiley-VCH.

Copeland, R. A. (2000). *Enzymes* (2nd ed.). Wiley-VCH.

Lencki, R. W., Arul, J., & Neufeld, R. J. (1992). Effect of subunit dissociation, denaturation, aggregation, coagulation, and decomposition on enzyme inactivation kinetics. Parts I and II. *Biotechnol. Bioeng.*, *40*, 1421–1434.

Cell Kinetics and Yield

See also references [1], [8–11], and [17–22].

Roels, J.A. (1983). *Energetics and Kinetics in Biotechnology*. Elsevier Biomedical.

Stouthamer, A. H. (1979). Energy production, growth, and product formation by microorganisms. In O. K. Sebek & A. I. Laskin (Eds.), *Genetics of Industrial Microorganisms*. American Society for Microbiology.

van't Riet, K., & Tramper, J. (1991). *Basic Bioreactor Design*. Marcel Dekker.

Growth Kinetics with Plasmid Instability

See also references [12–16].

Hjortso, M. A., & Bailey, J. E. (1984). Plasmic stability in budding yeast populations: steady-state growth with selection pressure. *Biotechnol. Bioeng.*, *26*, 528–536.

Sardonini, C. A., & DiBiasio, D. (1987). A model for growth of *Saccharomyces cerevisiae* containing a recombinant plasmid in selective media. *Biotechnol. Bioeng.*, *29*, 469–475.

Srienc, F., Campbell, J. L., & Bailey, J. E. (1986). Analysis of unstable recombinant *Saccharomyces cerevisiae* population growth in selective medium. *Biotechnol. Bioeng.*, *18*, 996–1006.

Cell Death Kinetics

See also references [23] and [29].

Aiba, S., Humphrey, A. E., & Millis, N. F. (1965). *Biochemical Engineering*. Academic Press.

Richards, J. W. (1968). *Introduction to Industrial Sterilization*. Academic Press.

Metabolic Engineering

See also references [24–28].

Kim, H. U., Kim, T. Y., & Lee, S. Y. (2008). Metabolic flux analysis and metabolic engineering of microorganisms. *Mol. BioSyst.*, *4*, 113–120.
Sauer, U. (2006). Metabolic networks in motion: ^{13}C-based flux analysis. *Mol. Syst. Biol.* doi:10.1038/msb4100109.
Shimizu, H. (2002). Metabolic engineering: integrating methodologies of molecular breeding and bioprocess systems engineering. *J. Biosci. Bioeng.*, *94*, 563–573.
Smolke, C. D. (Ed.), (2010). *Metabolic Pathway Engineering Handbook: Fundamentals.* CRC Press.

CHAPTER

13

Heterogeneous Reactions

In Chapter 12, we considered the effect of variables such as substrate concentration and temperature on the rate of enzyme and cell reactions. The reaction systems analysed were assumed to be homogeneous: any local variations in concentration or the rate of conversion were not examined. Yet, in many bioprocesses, concentrations of substrates and products differ from point to point in the reaction mixture. Concentration gradients arise in single-phase systems when mixing is poor; if different phases are present, local variations in composition can be considerable. As described in Chapter 10, concentration gradients can be expected to occur within boundary layers around gas bubbles and solids. More severe gradients are found inside solid biocatalysts such as cell flocs, pellets, and biofilms, and in immobilised cell and enzyme particles.

Reactions occurring in the presence of significant concentration or temperature gradients are called *heterogeneous reactions*. Because biological reactions are not typically associated with large temperature gradients, we confine our attention in this chapter to concentration effects. When heterogeneous reactions occur in solid catalysts, not all reactive molecules are available for immediate conversion. Reaction takes place only after the reactants are transported to the site of reaction. Thus, mass transfer processes can have a considerable influence on the overall conversion rate.

Because, in general, reaction rates depend on the concentration of the reactants present, when there are spatial variations in concentration, the kinetic analysis becomes more complex than when the concentrations are uniform. The principles of homogeneous reaction and the equations outlined in Chapter 12 remain valid for heterogeneous systems; however, the concentrations used in the equations must be those actually prevailing at the site of reaction. For solid biocatalysts, the concentration of substrate at each point inside the solid must be known for evaluation of local rates of conversion. In most cases these concentrations cannot be measured; fortunately, however, they can be estimated using diffusion–reaction theory.

In this chapter, methods are presented for analysing reactions affected by mass transfer. The mathematics required is more advanced than that applied elsewhere in this book; however, attention can be directed to the results of the analysis rather than to the mathematical derivations. The practical outcome of this chapter is simple criteria for assessing mass transfer limitations in heterogeneous reaction systems. These criteria can be used directly in experimental design and data analysis.

13.1 HETEROGENEOUS REACTIONS IN BIOPROCESSING

Reactions involving solid-phase catalysts are important in bioprocessing. Macroscopic flocs, clumps, and pellets are produced naturally by certain bacteria and fungi; mycelial pellets are common in antibiotic fermentations. Some cells grow as biofilms on reactor walls; others form slimes such as in waste treatment processes. Plant cell suspensions invariably contain aggregates; microorganisms in soil crumbs play a crucial role in environmental bioremediation. In tissue engineering, animal cells are cultured in three-dimensional scaffolds for surgical transplantation and organ repair. More traditionally, many food fermentations involve microorganisms attached to solid particles. In all of these systems, the rate of reaction depends on the rate of reactant mass transfer outside and within the solid catalyst.

If cells or enzymes do not spontaneously form clumps or attach to solid surfaces, they can be induced to do so using *immobilisation* techniques. Many procedures are available for artificial immobilisation of cells and enzymes; the results of two commonly used methods are illustrated in Figure 13.1. As shown in Figure 13.1(a), cells and enzymes can be immobilised by entrapment within gels such as alginate, agarose, and carrageenan. Cells or enzymes are mixed with liquified gel before it is hardened or cross-linked and formed into small particles. The gel polymer must be porous and relatively soft to allow diffusion of reactants and products to and from the interior of the particle. As shown in Figure 13.1(b), an alternative to gel immobilisation is entrapment within porous solids such as ceramics, porous glass, and resin beads. Enzymes or cells migrate into the pores of these particles and attach to the internal surfaces; substrate must diffuse through the pores for reaction to occur. In both immobilisation systems, the sites of reaction are distributed throughout the particle. Thus, a catalyst particle of higher activity can be formed by increasing the loading of cells or enzyme per volume of matrix.

Immobilised biocatalysts have many advantages in large-scale processing. One of the most important is continuous operation using the same catalytic material. For enzymes, an additional advantage is that immobilisation often enhances stability and increases the enzyme half-life (Section 12.6). Further discussion of immobilisation methods and the rationale behind cell and enzyme immobilisation can be found in many articles and books; a selection of references is given at the end of this chapter.

In Chapter 12, enzymes and cells were considered as biological catalysts. For heterogeneous reactions involving a solid phase, the term *catalyst* is also used to refer to the entire catalytically active body, such as a particle or biofilm. Engineering analysis of heterogeneous reactions applies equally well to naturally occurring solid catalysts and artificially immobilised cells and enzymes.

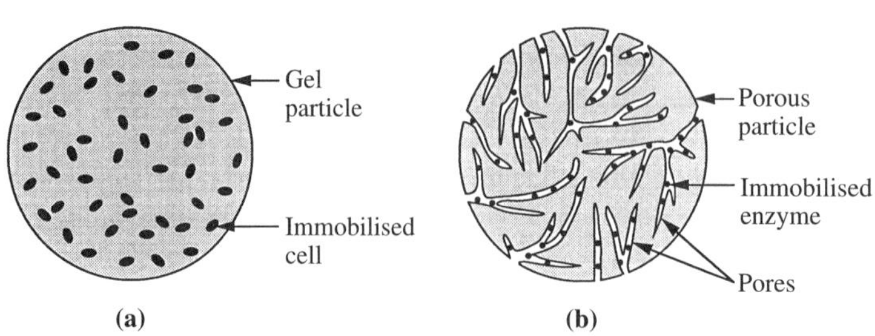

FIGURE 13.1 Immobilised biocatalysts: (a) cells entrapped in soft gel; (b) enzymes attached to the internal surfaces of a porous solid.

13.2 CONCENTRATION GRADIENTS AND REACTION RATES IN SOLID CATALYSTS

Consider a spherical catalyst of radius R immersed in well-mixed liquid containing substrate A. In the bulk liquid away from the particle, the substrate concentration is uniform and equal to C_{Ab}. If the particle were inactive, after some time the concentration of substrate inside the solid would reach a constant value in equilibrium with C_{Ab}. However, when substrate is consumed by reaction, its concentration C_A decreases within the particle as shown in Figure 13.2. If immobilised cells or enzymes are distributed uniformly within the catalyst, the concentration profile is symmetrical with a minimum at the centre. Mass transfer of substrate to reaction sites in the particle is driven by the concentration difference between the bulk solution and particle interior.

In the bulk liquid, substrate is carried rapidly by convective currents. However, as substrate molecules approach the solid they must be transported from the bulk liquid across the relatively stagnant boundary layer to the solid surface; this process is called *external mass transfer*. A concentration gradient develops across the boundary layer from C_{Ab} in the bulk liquid to C_{As} at the solid–liquid interface. If the particle were nonporous and all the enzyme or cells were confined to its outer surface, external mass transfer would be the only transport process required. More often, reaction takes place inside the particle so that *internal mass transfer* must also occur through the solid.

Although the form of the concentration gradient shown in Figure 13.2 is typical, other variations are possible. If mass transfer is much slower than reaction, all the substrate entering the particle may be consumed before reaching the centre. In this case, the concentration falls to zero within the solid as illustrated in Figure 13.3(a). Cells or enzyme near the centre are starved of substrate and the core of the particle becomes inactive. In the examples of Figures 13.3(b) and 13.3(c), the *partition coefficient* for the substrate is not equal to unity. This means that, at equilibrium and in the absence of reaction, the concentration

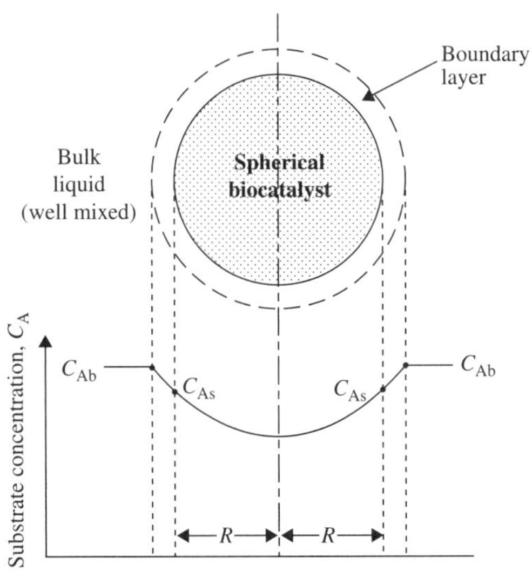

FIGURE 13.2 Typical substrate concentration profile for a spherical biocatalyst.

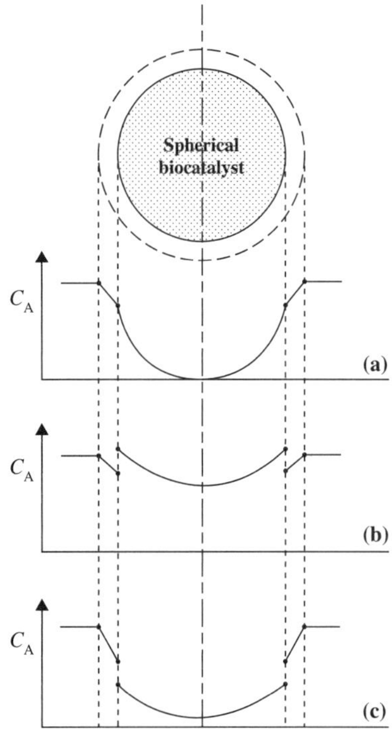

FIGURE 13.3 Variations in substrate concentration profile in spherical biocatalysts.

of substrate in the solid is higher or lower than in the liquid. In Figure 13.3(b), the discontinuity of concentration at the solid–liquid interface shows that substrate distributes preferentially to the solid phase. Conversely, Figure 13.3(c) shows the concentrations when substrate is attracted more to the liquid than to the solid. The effect of mass transfer on intraparticle concentration can be magnified or diminished by substrate partitioning. Partitioning is important when the substrate and solid are charged or if strong hydrophobic interactions cause repulsion or attraction. Because most materials used for cell and enzyme immobilisation are very porous and contain a high percentage of water, partition effects can often be neglected. In our treatment of heterogeneous reaction, we will assume that partitioning is not significant.

13.2.1 True and Observed Reaction Rates

Because concentrations vary in solid catalysts, local rates of reaction also vary depending on the position within the particle. Even for zero-order reactions, the reaction rate changes with position if substrate is exhausted. Each cell or enzyme molecule within the solid responds to the substrate concentration at its location with a rate of reaction determined by its kinetic properties. This local rate of reaction is known as the *true rate* or *intrinsic rate*. Like any reaction rate, intrinsic rates can be expressed as total, volumetric, or specific rates as described in Section 12.1.3. The relationship between true reaction rate and local substrate concentration follows the principles outlined in Chapter 12 for homogeneous reactions.

True local reaction rates are difficult to measure in solid catalysts without altering the reaction conditions. It is relatively easy, however, to measure the overall reaction rate for the entire catalyst. In a closed system, the rate of disappearance of substrate from the bulk liquid must equal the overall rate of conversion by reaction; in heterogeneous systems, this is called the *observed rate*. It is important to remember that the observed rate is not usually equal to the true activity of any cell or enzyme in the particle. Because substrate levels are reduced inside solid catalysts compared with those in the external medium, we expect the observed rate to be lower than the rate that would occur if the entire particle were exposed to the bulk liquid. The relationship between observed reaction rate and bulk substrate concentration is not as simple as in homogeneous reactions. Equations for the observed rate of heterogeneous reactions also involve mass transfer parameters.

True reaction rates depend on the kinetic parameters of the cells or enzyme. For example, the rate of reaction of an immobilised enzyme obeying Michaelis–Menten kinetics (Section 12.3.3) depends on the values of v_{max} and K_m for the enzyme in its immobilised state. These parameters are sometimes called *true kinetic parameters* or *intrinsic kinetic parameters*. Because kinetic parameters can be altered during immobilisation, for example, due to cell or enzyme damage, conformational change, and steric hindrance, values measured before immobilisation may not apply. True kinetic parameters for immobilised biocatalysts can be difficult to determine because any measured reaction rates incorporate mass transfer effects. The problem of evaluating true kinetic parameters is discussed further in Section 13.9.

13.2.2 Interaction between Mass Transfer and Reaction

Rates of reaction and substrate mass transfer are not independent in heterogeneous systems. The rate of mass transfer depends on the concentration gradient established in the system; this in turn depends on the rate of substrate depletion by reaction. On the other hand, the rate of reaction depends on the availability of substrate, which depends on the rate of mass transfer.

One of the objectives in analysing heterogeneous reactions is to determine the relative influences of mass transfer and reaction on observed reaction rates. One can conceive, for example, that if a reaction proceeds slowly even in the presence of adequate substrate, it is likely that mass transfer will be rapid enough to meet the demands of the reaction. In this case, the observed rate would reflect more directly the reaction process rather than mass transfer. Conversely, if the reaction tends to be very rapid, it is likely that mass transfer will be too slow to supply substrate at the rate required. The observed rate would then reflect strongly the rate of mass transfer. As will be shown in the remainder of this chapter, there are mathematical criteria for assessing the extent to which mass transfer influences the observed reaction rate. Reactions that are significantly affected are called *mass transfer limited* or *diffusion limited* reactions. It is also possible to distinguish the relative influence of internal and external mass transfer on the observed rate of reaction. Improving mass transfer and eliminating mass transfer restrictions are desired objectives in heterogeneous catalysis. Once the effect and location of major mass transfer resistances are identified, we can then devise strategies for their alleviation.

13.3 INTERNAL MASS TRANSFER AND REACTION

Let us now concentrate on the processes of mass transfer and reaction occurring within a solid biocatalyst; external mass transfer will be examined later in the chapter. The equations and procedures used in this analysis depend on the geometry of the system and the reaction kinetics. First, let us consider the case of cells or enzymes immobilised in spherical particles.

13.3.1 Steady-State Shell Mass Balance

Mathematical analysis of heterogeneous reactions involves a technique called the *shell mass balance*. In this section, we will perform a shell mass balance on a spherical catalyst particle of radius R. Imagine a thin spherical shell of thickness Δr located at radius r from the centre, as shown in Figure 13.4. It may be helpful to think of this shell as the thin wall of a ping-pong ball encased inside and concentric with a larger cricket ball of radius R. Substrate diffusing into the sphere must cross the shell to reach the centre.

A mass balance of substrate is performed around the shell by considering the processes of mass transfer and reaction occurring at radius r. The system considered for the mass balance is the shell only; the remainder of the sphere is ignored for the moment. Substrate diffuses into the shell at radius $(r + \Delta r)$ and leaves at radius r; within the shell, immobilised cells or enzyme consume substrate by reaction. The flow of mass through the shell is analysed using the general mass balance equation derived in Chapter 4:

$$\left\{\begin{array}{c}\text{mass in}\\\text{through}\\\text{system}\\\text{boundaries}\end{array}\right\} - \left\{\begin{array}{c}\text{mass out}\\\text{through}\\\text{system}\\\text{boundaries}\end{array}\right\} + \left\{\begin{array}{c}\text{mass}\\\text{generated}\\\text{within}\\\text{system}\end{array}\right\} - \left\{\begin{array}{c}\text{mass}\\\text{consumed}\\\text{within}\\\text{system}\end{array}\right\} = \left\{\begin{array}{c}\text{mass}\\\text{accumulated}\\\text{within}\\\text{system}\end{array}\right\} \quad (4.1)$$

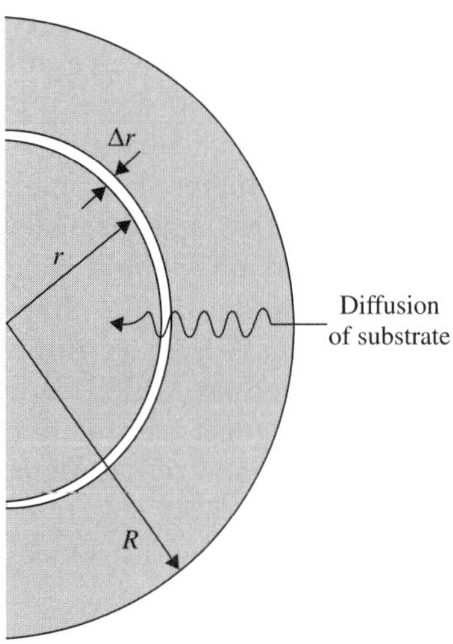

FIGURE 13.4 Shell mass balance on a spherical particle.

Before application of Eq. (4.1), certain assumptions must be made so that each term in the equation can be expressed mathematically [1].

1. *The particle is isothermal.* The kinetic parameters for enzyme and cell reactions vary considerably with temperature. If temperature in the particle is nonuniform, different values of the kinetic parameters must be applied. However, as temperature gradients generated by immobilised cells and enzymes are generally negligible, assuming constant temperature throughout the particle is reasonable and greatly simplifies the mathematical analysis.
2. *Mass transfer occurs by diffusion only.* We will assume that the particle is impermeable to flow, so that convection within the pores is negligible. This assumption is valid for many solid-phase biocatalysts. However, some anomalies have been reported [2, 3]. Depending on pore size, pressure gradients can induce convection of liquid through the particle and this enhances significantly the supply of nutrients. When convective transport occurs, the analysis of mass transfer and reaction presented in this chapter must be modified [4–6].
3. *Diffusion can be described using Fick's law with constant effective diffusivity.* We will assume that diffusive transport through the particle is governed by Fick's law (Section 10.1.1). Interaction of the substrate with other concentration gradients, and phenomena affecting the transport of charged species, are ignored. Fick's law will be applied using the *effective diffusivity* of substrate in the solid, $\mathscr{D}_{Ae}$. The value of $\mathscr{D}_{Ae}$ is a complex function of the molecular diffusion characteristics of the substrate, the tortuousness of the diffusion path within the solid, and the fraction of the particle volume available for diffusion. We will assume that $\mathscr{D}_{Ae}$ is constant and independent of substrate concentration in the particle; this means that $\mathscr{D}_{Ae}$ does not change with position.
4. *The particle is homogeneous.* Immobilised enzymes or cells are assumed to be distributed uniformly within the particle. Properties of the immobilisation matrix are also considered to be uniform.
5. *The substrate partition coefficient is unity.* This assumption is valid for most substrates and particles, and ensures that there is no discontinuity of concentration at the solid–liquid interface.
6. *The particle is at steady state.* This assumption is usually valid if there is no change in activity of the catalyst due to, for example, enzyme deactivation, cell growth, or cell differentiation. It is not valid when the system exhibits rapid transients such as when cells quickly consume and store substrates for subsequent metabolism.
7. *Substrate concentration varies with a single spatial variable.* For the sphere of Figure 13.4, we will assume that the substrate concentration varies only in the radial direction, and that substrate diffuses radially through the particle from the external surface toward the centre.

Equation (4.1) will be applied for our shell mass balance according to these assumptions. Substrate is transported into and out of the shell by diffusion; therefore, the first and second terms are expressed using Fick's law with constant effective diffusivity. The third term is zero as no substrate is generated. Substrate is consumed by reaction inside the shell at a rate equal to the volumetric rate of reaction r_A multiplied by the volume of the shell. According to assumption (6), the system is at steady state. Therefore, its mass and composition must be unchanging, substrate cannot accumulate in the shell, and the right side of Eq. (4.1) is zero. As outlined below, after substituting the appropriate expressions and applying calculus to reduce the dimension of the shell to an infinitesimal thickness, the result of the shell mass balance is a second-order differential equation for substrate concentration as a function of radius in the particle.

For a shell mass balance on substrate A, the terms of Eq. (4.1) are as follows:

Rate of input by diffusion: $\left(\mathscr{D}_{Ae}\dfrac{dC_A}{dr}4\pi r^2\right)\Bigg|_{r+\Delta r}$

Rate of output by diffusion: $\left(\mathscr{D}_{Ae}\dfrac{dC_A}{dr}4\pi r^2\right)\Bigg|_{r}$

Rate of generation: 0

Rate of consumption by reaction: $r_A\ 4\pi r^2\Delta r$

Rate of accumulation at steady state: 0

$\mathscr{D}_{Ae}$ is the effective diffusivity of substrate A in the particle, C_A is the concentration of A in the particle, r is the distance measured radially from the centre, Δr is the thickness of the shell, and r_A is the rate of reaction *per unit volume of particle*. Each term in the mass balance equation has dimensions MT^{-1} or NT^{-1} and units of, for example, $kg\,h^{-1}$ or $gmol\,s^{-1}$. The first two terms are derived from Fick's law of Eq. (10.1); the area of the spherical shell available for diffusion is $4\pi r^2$. The term

$$\left(\mathscr{D}_{Ae}\frac{dC_A}{dr}4\pi r^2\right)\Bigg|_{r+\Delta r}$$

means $\left(\mathscr{D}_{Ae}\dfrac{dC_A}{dr}4\pi r^2\right)$ evaluated at radius $(r+\Delta r)$, and

$$\left(\mathscr{D}_{Ae}\frac{dC_A}{dr}4\pi r^2\right)\Bigg|_{r}$$

means $\left(\mathscr{D}_{Ae}\dfrac{dC_A}{dr}4\pi r^2\right)$ evaluated at r. The volume of the shell is $4\pi r^2\Delta r$.

After substituting these terms into Eq. (4.1), we obtain the following steady-state mass balance equation:

$$\left(\mathscr{D}_{Ae}\frac{dC_A}{dr}4\pi r^2\right)\Bigg|_{r+\Delta r}-\left(\mathscr{D}_{Ae}\frac{dC_A}{dr}4\pi r^2\right)\Bigg|_{r}-r_A 4\pi r^2\Delta r=0 \tag{13.1}$$

Dividing each term by $4\pi\Delta r$ gives:

$$\frac{\left(\mathscr{D}_{Ae}\dfrac{dC_A}{dr}r^2\right)\Bigg|_{r+\Delta r}-\left(\mathscr{D}_{Ae}\dfrac{dC_A}{dr}r^2\right)\Bigg|_{r}}{\Delta r}-r_A r^2=0 \tag{13.2}$$

Eq. (13.2) can be written in the form:

$$\frac{\Delta\left(\mathscr{D}_{Ae}\dfrac{dC_A}{dr}r^2\right)}{\Delta r}-r_A r^2=0 \tag{13.3}$$

where $\Delta\left(\mathscr{D}_{Ae}\frac{dC_A}{dr}r^2\right)$ means the change in $\left(\mathscr{D}_{Ae}\frac{dC_A}{dr}r^2\right)$ across Δr.

Equation (13.3) is valid for a spherical shell of thickness Δr. To develop an equation that applies to any *point* in the sphere, we must shrink Δr to zero. As Δr appears only in the first term of Eq. (13.3), taking the limit of Eq. (13.3) as $\Delta r \rightarrow 0$ gives:

$$\lim_{\Delta r \rightarrow 0} \frac{\Delta\left(\mathscr{D}_{Ae}\frac{dC_A}{dr}r^2\right)}{\Delta r} - r_A r^2 = 0 \tag{13.4}$$

Invoking the definition of the derivative from Section E.2 of Appendix E, Eq. (13.4) is identical to the second-order differential equation:

$$\frac{d}{dr}\left(\mathscr{D}_{Ae}\frac{dC_A}{dr}r^2\right) - r_A r^2 = 0 \tag{13.5}$$

According to assumption (3), $\mathscr{D}_{Ae}$ is independent of r and can be moved outside of the differential:

$$\mathscr{D}_{Ae}\frac{d}{dr}\left(\frac{dC_A}{dr}r^2\right) - r_A r^2 = 0 \tag{13.6}$$

In Eq. (13.6) we have a differential equation representing diffusion and reaction in a spherical biocatalyst. That Eq. (13.6) is a second-order differential equation becomes clear if the first term is written in its expanded form:

$$\mathscr{D}_{Ae}\left(\frac{d^2C_A}{dr^2}r^2 + 2r\frac{dC_A}{dr}\right) - r_A r^2 = 0 \tag{13.7}$$

In principle, Eq. (13.7) can be solved by integration to yield an expression for the concentration profile in the particle, C_A as a function of r. However, we cannot integrate Eq. (13.7) as it stands because the reaction rate r_A is in most cases a function of C_A. Let us consider solutions of Eq. (13.7) with r_A for first-order, zero-order, and Michaelis–Menten kinetics (Section 12.3).

13.3.2 Concentration Profile: First-Order Kinetics and Spherical Geometry

For first-order kinetics, Eq. (13.7) becomes:

$$\mathscr{D}_{Ae}\left(\frac{d^2C_A}{dr^2}r^2 + 2r\frac{dC_A}{dr}\right) - k_1 C_A r^2 = 0 \tag{13.8}$$

where k_1 is the intrinsic first-order rate constant with dimensions T^{-1}. For biocatalytic reactions, k_1 depends on the amount of cells or enzyme in the particle. According to assumptions (1), (3), and (4) in Section 13.3.1, k_1 and $\mathscr{D}_{Ae}$ for a given particle can be considered constant. Accordingly, as the only variables in Eq. (13.8) are C_A and r, the equation is ready for integration. Because Eq. (13.8) is a second-order differential equation we need two boundary conditions to solve it. These are:

$$C_A = C_{As} \quad \text{at } r = R \tag{13.9}$$

$$\frac{dC_A}{dr} = 0 \quad \text{at } r = 0 \tag{13.10}$$

where C_{As} is the concentration of substrate at the outer surface of the particle. For the present, we will assume that C_{As} is known or can be measured. Equation (13.10) is called the *symmetry condition*. As indicated in Figures 13.2 and 13.3, for a particle with uniform properties, the substrate concentration profile is symmetrical about the centre of the sphere, with a minimum at the centre so that $dC_A/dr = 0$ at $r = 0$. Integration of Eq. (13.8) with boundary conditions Eqs. (13.9) and (13.10) gives the following expression for substrate concentration as a function of radius [7]:

$$C_A = C_{As}\frac{R}{r}\frac{\sinh\left(r\sqrt{k_1/\mathscr{D}_{Ae}}\right)}{\sinh\left(R\sqrt{k_1/\mathscr{D}_{Ae}}\right)} \tag{13.11}$$

In Eq. (13.11), sinh is the abbreviation for *hyperbolic sine*; sinh x is defined as:

$$\sinh x = \frac{e^x - e^{-x}}{2} \tag{13.12}$$

Equation (13.11) may appear complex but contains simple exponential terms relating C_A and r, with $\mathscr{D}_{Ae}$ representing the rate of mass transfer and k_1 representing the rate of reaction.

EXAMPLE 13.1 CONCENTRATION PROFILE FOR IMMOBILISED ENZYME

Enzyme is immobilised in agarose beads of diameter 8 mm. The concentration of enzyme in the beads is 0.018 kg of protein per m^3 of gel. Ten beads are immersed in a well-mixed solution containing substrate at a concentration of 3.2×10^{-3} kg m^{-3}. The effective diffusivity of substrate in the agarose gel is 2.1×10^{-9} m^2 s^{-1}. The kinetics of the enzyme reaction can be approximated as first-order with specific rate constant 3.11×10^5 s^{-1} per kg of protein. Mass transfer effects outside the particles are negligible. Plot the steady-state substrate concentration profile inside the beads as a function of particle radius.

Solution

$R = 4 \times 10^{-3}$ m; $\mathscr{D}_{Ae} = 2.1 \times 10^{-9}$ m^2 s^{-1}. In the absence of external mass transfer effects, $C_{As} = 3.2 \times 10^{-3}$ kg m^{-3}. To determine the substrate concentration profile, we consider mass transfer and reaction in a single bead.

$$\text{Volume per bead} = \frac{4}{3}\pi R^3 = \frac{4}{3}\pi\,(4 \times 10^{-3}\text{ m})^3 = 2.68 \times 10^{-7}\text{ m}^3$$

$$\text{Amount of enzyme per bead} = 2.68 \times 10^{-7}\text{ m}^3\,(0.018\text{ kg m}^{-3}) = 4.82 \times 10^{-9}\text{ kg}$$

Therefore:

$$k_1 = 3.11 \times 10^5 \text{ s}^{-1} \text{ kg}^{-1}(4.82 \times 10^{-9} \text{ kg}) = 0.0015 \text{ s}^{-1}$$

Intraparticle substrate concentrations are calculated as a function of radius using Eq. (13.11). Terms in Eq. (13.11) include:

$$R\sqrt{\frac{k_1}{\mathcal{D}_{Ae}}} = (4 \times 10^{-3} \text{ m})\sqrt{\frac{0.0015 \text{ s}^{-1}}{2.1 \times 10^{-9} \text{ m}^2 \text{ s}^{-1}}} = 3.381$$

and

$$\sinh\left(R\sqrt{k_1/\mathcal{D}_{Ae}}\right) = \frac{e^{3.381} - e^{-3.381}}{2} = 14.68$$

Results for C_A as a function of r are as follows.

r (m)	C_A (kg m^{-3})
0.005×10^{-3}	7.37×10^{-4}
0.5×10^{-3}	7.59×10^{-4}
1.0×10^{-3}	8.28×10^{-4}
2.0×10^{-3}	1.14×10^{-3}
2.5×10^{-3}	1.42×10^{-3}
3.0×10^{-3}	1.82×10^{-3}
3.5×10^{-3}	2.39×10^{-3}
4.0×10^{-3}	3.20×10^{-3}

The results are plotted in Figure 13.5. The substrate concentration is reduced inside the particle to reach a minimum of 7.4×10^{-4} kg m^{-3} at the centre.

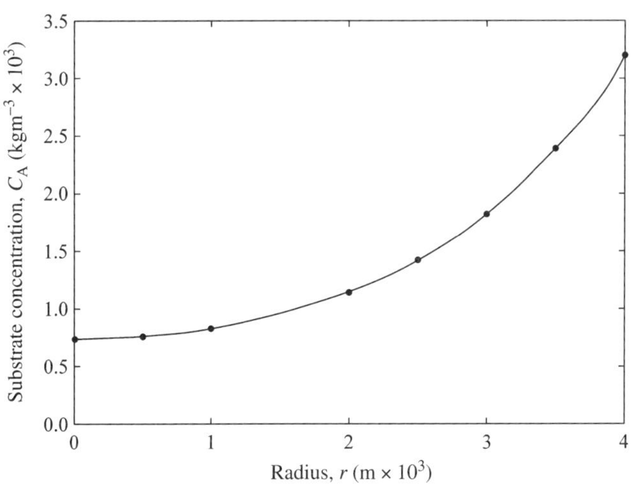

FIGURE 13.5 Substrate concentration profile in an immobilised enzyme bead.

13.3.3 Concentration Profile: Zero-Order Kinetics and Spherical Geometry

For zero-order kinetics, Eq. (13.7) becomes:

$$\mathscr{D}_{Ae}\left(\frac{d^2C_A}{dr^2}r^2 + 2r\frac{dC_A}{dr}\right) - k_0 r^2 = 0 \tag{13.13}$$

where k_0 is the intrinsic zero-order rate constant with units of, for example, gmol s^{-1} m^{-3} of particle. Like k_1 for first-order reactions, k_0 varies with cell or enzyme density.

Zero-order reactions are unique in that, provided substrate is present, the reaction rate is independent of substrate concentration. In solving Eq. (13.13), we must account for the possibility that substrate becomes depleted within the particle. As illustrated in Figure 13.6, if we assume that this occurs at some radius R_0, the rate of reaction for $0 < r \leq R_0$ is zero. Everywhere else inside the particle (i.e., $r > R_0$) the volumetric reaction rate is constant and equal to k_0 irrespective of substrate concentration. For this situation, the boundary conditions are:

$$C_A = C_{As} \quad \text{at } r = R \tag{13.9}$$

$$\frac{dC_A}{dr} = 0 \quad \text{at } r = R_0 \tag{13.14}$$

Solution of Eq. (13.13) with these boundary conditions gives the following expression for C_A as a function of r [7]:

$$C_A = C_{As} + \frac{k_0 R^2}{6\mathscr{D}_{Ae}}\left(\frac{r^2}{R^2} - 1 + \frac{2R_0^3}{rR^2} - \frac{2R_0^3}{R^3}\right) \tag{13.15}$$

Eq. (13.15) is difficult to apply in practice because R_0 is generally not known. However, the equation can be simplified if C_A remains > 0 everywhere so that R_0 no longer exists. Substituting $R_0 = 0$ into Eq. (13.15) gives:

$$C_A = C_{As} + \frac{k_0}{6\mathscr{D}_{Ae}}(r^2 - R^2) \tag{13.16}$$

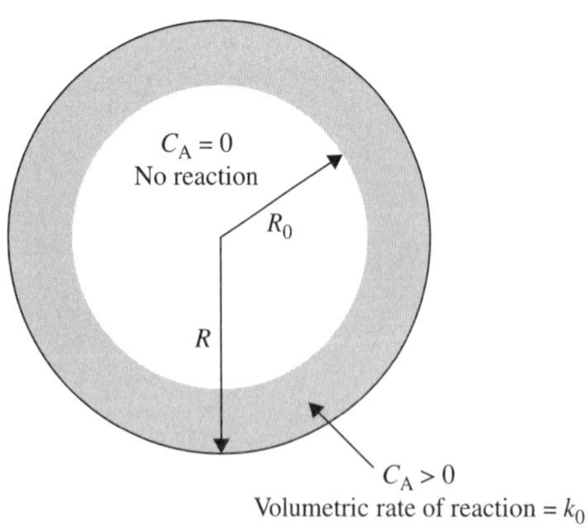

FIGURE 13.6 Concentration and reaction zones in a spherical particle with zero-order reaction. Substrate is depleted at radius R_0.

In bioprocessing applications, it is important that the core of catalyst particles does not become starved of substrate. The likelihood of this happening increases with the size of the particle. For zero-order reactions, we can calculate the maximum particle radius for which C_A remains > 0. In such a particle, substrate is depleted just at the centre point. Therefore, calculating R from Eq. (13.16) with $C_A = r = 0$:

$$R_{max} = \sqrt{\frac{6\mathscr{D}_{Ae}C_{As}}{k_0}} \tag{13.17}$$

where R_{max} is the maximum particle radius for $C_A > 0$.

EXAMPLE 13.2 MAXIMUM PARTICLE SIZE FOR ZERO-ORDER REACTION

Nonviable yeast cells are immobilised in alginate beads. The beads are stirred in glucose medium under anaerobic conditions. The effective diffusivity of glucose in the beads depends on cell density according to the relationship:

$$\mathscr{D}_{Ae} = 6.33 - 7.17y_C$$

where $\mathscr{D}_{Ae}$ is the effective diffusivity $\times 10^{10}$ m^2 s^{-1} and y_C is the weight fraction of yeast in the gel. The rate of glucose consumption can be assumed to be zero-order; the reaction rate constant at a yeast density in the beads of 15 wt% is 0.5 g min^{-1} l^{-1}. For the catalyst to be utilised effectively, the concentration of glucose inside the particles should remain above zero.

(a) Plot the maximum allowable particle size as a function of the bulk glucose concentration between 5 g l^{-1} and 60 g l^{-1}.

(b) For medium containing 30 g l^{-1} glucose, plot R_{max} as a function of cell loading between 10 and 45 wt%.

Solution

(a) Using the equation provided, at $y_C = 0.15$, $\mathscr{D}_{Ae} = 5.25 \times 10^{-10}$ m^2 s^{-1}. Converting k_0 to units of kg, m, s:

$$k_0 = 0.5\ \text{g min}^{-1}\ \text{l}^{-1} \cdot \left|\frac{1\ \text{kg}}{1000\ \text{g}}\right| \cdot \left|\frac{1\ \text{min}}{60\ \text{s}}\right| \cdot \left|\frac{1000\ \text{l}}{1\ \text{m}^3}\right|$$

$$k_0 = 8.33 \times 10^{-3}\ \text{kg s}^{-1}\ \text{m}^{-3}$$

Assume that C_{As} is equal to the bulk glucose concentration. R_{max} is calculated from Eq. (13.17).

C_{As} (kg m^{-3})	R_{max} (m)
5	1.38×10^{-3}
15	2.38×10^{-3}
25	3.07×10^{-3}
45	4.13×10^{-3}
60	4.76×10^{-3}

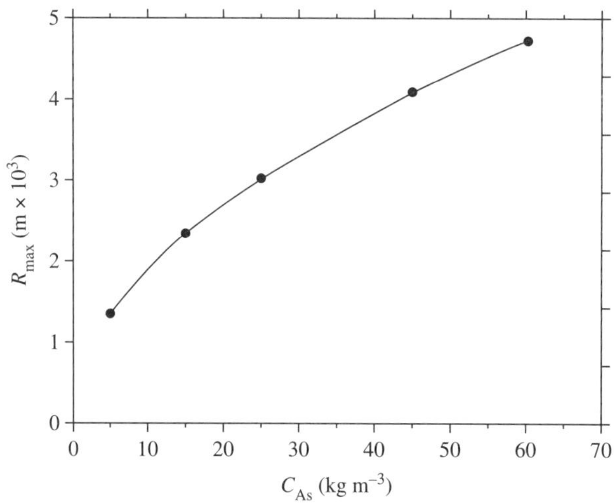

FIGURE 13.7 Maximum particle radius as a function of external substrate concentration.

The results are plotted in Figure 13.7. At low external glucose concentrations, the particles are restricted to small radii. As C_{As} increases, the driving force for diffusion increases so that larger particles may be used.

(b) $C_{As} = 30$ g l^{-1} = 30 kg m^{-3}. As y_C varies, the values of $\mathscr{D}_{Ae}$ and k_0 are affected. Changes in $\mathscr{D}_{Ae}$ can be calculated from the equation provided. We assume that k_0 is directly proportional to the cell density as described in Eq. (12.26), that is, there is no steric hindrance or interaction between the cells as y_C increases. R_{max} is calculated as a function of y_C using Eq. (13.17) and the corresponding values of $\mathscr{D}_{Ae}$ and k_0.

y_C	$\mathscr{D}_{Ae}$ (m^2 s^{-1})	k_0 (kg m^{-3} s^{-1})	R_{max} (m)
0.1	5.61×10^{-10}	5.56×10^{-3}	4.26×10^{-3}
0.2	4.90×10^{-10}	1.11×10^{-2}	2.82×10^{-3}
0.3	4.18×10^{-10}	1.67×10^{-2}	2.12×10^{-3}
0.4	3.46×10^{-10}	2.22×10^{-2}	1.67×10^{-3}
0.45	3.10×10^{-10}	2.50×10^{-2}	1.49×10^{-3}

The results are plotted in Figure 13.8. As y_C increases, $\mathscr{D}_{Ae}$ declines and k_0 increases. Reducing $\mathscr{D}_{Ae}$ lowers the rate of diffusion into the particles; raising k_0 increases the demand for substrate. Therefore, increasing the cell density exacerbates the limiting effect of mass transfer on the reaction rate. To ensure adequate supply of substrate under these conditions, the particle size must be reduced.

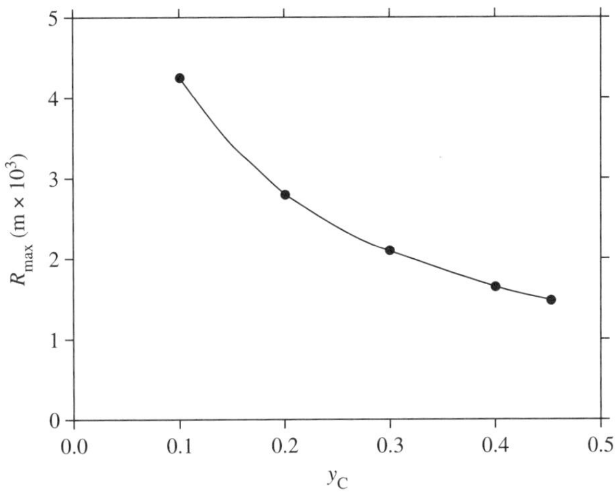

FIGURE 13.8 Maximum particle radius as a function of cell density.

13.3.4 Concentration Profile: Michaelis–Menten Kinetics and Spherical Geometry

If reaction in the particle follows Michaelis–Menten kinetics, r_A takes the form of Eq. (12.36) and Eq. (13.7) becomes:

$$\mathscr{D}_{Ae}\left(\frac{d^2C_A}{dr^2}r^2 + 2r\frac{dC_A}{dr}\right) - \frac{v_{max}C_A}{K_m + C_A}r^2 = 0 \tag{13.18}$$

where v_{max} and K_m are the intrinsic kinetic parameters for the reaction. In Eq. (13.18), v_{max} has units of, for example, kg s^{-1} m^{-3} of particle; the value of v_{max} depends on the cell or enzyme density.

Owing to the nonlinearity of the Michaelis–Menten expression, simple analytical integration of Eq. (13.18) is not possible. However, results for C_A as a function of r can be obtained using numerical methods, usually by computer. Because Michaelis–Menten kinetics lie between zero- and first-order kinetics (Section 12.3.3), the explicit solutions found in Sections 13.3.2 and 13.3.3 for first- and zero-order reactions can be used to estimate the upper and lower limits of the concentration profile for Michaelis–Menten reactions.

Concentration profiles calculated from the equations presented in this section have been verified experimentally in several studies. Using special microelectrodes with tip diameters of the order of 1 μm, it is possible to measure the concentrations of oxygen and ions inside soft solids and cell slimes. As an example, the oxygen concentrations measured in agarose beads containing immobilised enzyme are shown in Figure 13.9. The experimental data are very close to the calculated concentration profiles. Similar results have been found in other systems [8–10].

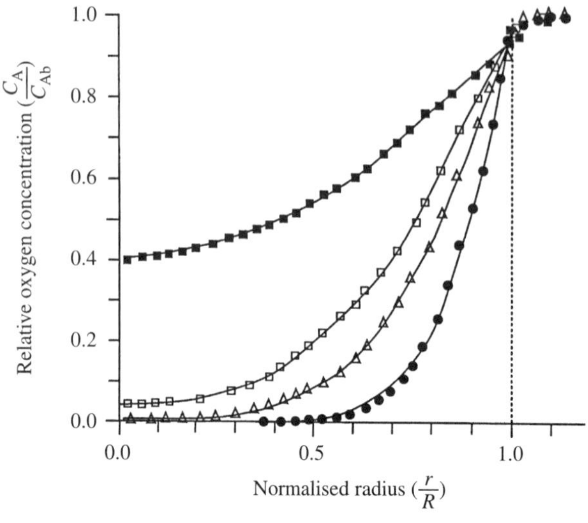

FIGURE 13.9 Measured and calculated oxygen concentrations in a spherical agarose bead containing immobilised enzyme. Particle diameter = 4 mm; C_{Ab} = 0.2 mol m^{-3}. The enzyme loadings are: 0.0025 kg m^{-3} of gel (■); 0.005 kg m^{-3} of gel (□); 0.0125 kg m^{-3} of gel (△); and 0.025 kg m^{-3} of gel (●). Measured concentrations are shown using symbols; calculated profiles are shown as lines.

From C.M. Hooijmans, S.G.M. Geraats, and K.Ch. A.M. Luyben, 1990, Use of an oxygen microsensor for the determination of intrinsic kinetic parameters of an immobilised oxygen reducing enzyme. Biotechnol. Bioeng. *35, 1078–1087.*

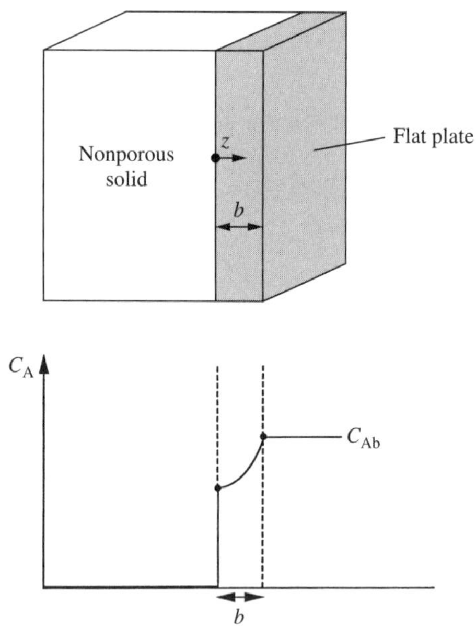

FIGURE 13.10 Substrate concentration profile in an infinite flat plate without boundary-layer effects.

13.3.5 Concentration Profiles for Other Geometries

Our attention so far has been focussed on spherical catalysts. However, equations similar to Eq. (13.7) can be obtained from shell mass balances on other geometries. Of all other shapes, the one of most relevance to bioprocessing is the flat plate. A typical substrate concentration profile for this geometry without external boundary layer effects is illustrated in Figure 13.10. Equations for flat-plate geometry are used to analyse reactions in cell films attached to inert solids; in this case, the biofilm constitutes the flat plate. Even if the surface supporting the biofilm is curved rather than flat, if the biofilm thickness b is very small compared with the radius of curvature, equations for flat-plate geometry are applicable.

To simplify the mathematical treatment and to keep the problem one-dimensional (as required by assumption (7) of Section 13.3.1), the flat plate is assumed to have infinite length. In practice, this assumption is reasonable if its length is much greater than its thickness. If not, it must be assumed that the ends of the plate are sealed to eliminate axial concentration gradients. The boundary conditions for solution of the differential equation for diffusion and reaction in a flat plate are analogous to Eqs. (13.9) and (13.10):

$$C_A = C_{As} \quad \text{at } z = b \tag{13.19}$$

$$\frac{dC_A}{dz} = 0 \quad \text{at } z = 0 \tag{13.20}$$

where C_{As} is the concentration of A at the solid–liquid interface, z is distance measured from the inner surface of the plate (Figure 13.10), and b is the plate thickness. Equations for the steady-state concentration profiles for first- and zero-order kinetics and spherical and flat-plate geometries are summarised in Table 13.1.

Another catalyst shape of some interest in bioprocessing is the hollow cylinder, which is used in analysis of hollow-fibre membrane reactors. However, because of its relatively limited application, we will not consider this geometry further.

TABLE 13.1 Steady-State Concentration Profiles

First-order reaction: $r_A = k_1 C_A$	
Sphere[a]	$C_A = C_{As}\dfrac{R}{r}\dfrac{\sinh\left(r\sqrt{k_1/\mathscr{D}_{Ae}}\right)}{\sinh\left(R\sqrt{k_1/\mathscr{D}_{Ae}}\right)}$
Flat plate[b]	$C_A = C_{As}\dfrac{\cosh\left(z\sqrt{k_1/\mathscr{D}_{Ae}}\right)}{\cosh\left(b\sqrt{k_1/\mathscr{D}_{Ae}}\right)}$
Zero-order reaction: $r_A = k_0$	
Sphere[c]	$C_A = C_{As} + \dfrac{k_0}{6\mathscr{D}_{Ae}}(r^2 - R^2)$
Flat plate[c]	$C_A = C_{As} + \dfrac{k_0}{2\mathscr{D}_{Ae}}(z^2 - b^2)$

[a] *Sinh is the abbreviation for hyperbolic sine. Sinh x is defined as:*

$$\sinh x = \frac{e^x - e^{-x}}{2}$$

[b] *Cosh is the abbreviation for hyperbolic cosine. Cosh x is defined as:*

$$\cosh x = \frac{e^x + e^{-x}}{2}$$

[c] *For $C_A > 0$ everywhere within the catalyst.*

13.3.6 Prediction of the Observed Reaction Rate

Equations for intracatalyst substrate concentrations such as those in Table 13.1 allow us to predict the overall or observed rate of reaction in the catalyst. Let us consider the situation for spherical particles and first-order, zero-order, and Michaelis–Menten kinetics. Analogous equations can be derived for other geometries.

1. *First-order kinetics.* The rate of reaction at any point in a spherical catalyst depends on the first-order kinetic constant k_1 and the concentration of substrate at that point. The overall rate for the entire particle is equal to the sum of all the rates at every location in the solid. This sum is equivalent mathematically to integrating the expression k_1C_A over the entire particle volume, taking into account the variation of C_A with radius expressed in Eq. (13.11). The result is an equation for the observed reaction rate $r_{A,obs}$ in a single particle:

$$r_{A,obs} = 4\pi R\mathscr{D}_{Ae}C_{As}\left[R\sqrt{k_1/\mathscr{D}_{Ae}}\coth\left(R\sqrt{k_1/\mathscr{D}_{Ae}}\right) - 1\right] \tag{13.21}$$

where coth is the abbreviation for *hyperbolic cotangent* defined as:

$$\coth x = \frac{e^x + e^{-x}}{e^x - e^{-x}} \tag{13.22}$$

Note that, because the observed reaction rate was found by integrating the local reaction rate over the volume of a single particle, $r_{A,obs}$ in Eq. (13.21) is the observed reaction rate *per particle* with units of, for example, kg s^{-1}.

2. *Zero-order kinetics.* As long as substrate is present, zero-order reactions occur at a fixed rate that is independent of substrate concentration. Therefore, if $C_A > 0$ everywhere in the particle, the observed rate of reaction is equal to the zero-order rate constant k_0 multiplied by the particle volume:

$$r_{A,obs} = \frac{4}{3}\pi R^3 k_0 \tag{13.23}$$

However, if C_A falls to zero at some radius R_0 in the particle, the inner volume $(4/3)\pi R_0^3$ is inactive. In this case, the rate of reaction is equal to k_0 multiplied by the active particle volume:

$$r_{A,obs} = \left(\frac{4}{3}\pi R^3 - \frac{4}{3}\pi R_0^3\right)k_0 = \frac{4}{3}\pi\,(R^3 - R_0^3)k_0 \tag{13.24}$$

$r_{A,obs}$ calculated using Eqs. (13.23) and (13.24) is the observed reaction rate per particle with units of, for example, kg s^{-1}.

3. *Michaelis–Menten kinetics.* Observed rates for Michaelis–Menten reactions cannot be expressed explicitly because we do not have an equation for C_A as a function of radius. $r_{A,obs}$ can be evaluated, however, using numerical methods.

13.4 THE THIELE MODULUS AND EFFECTIVENESS FACTOR

Charts based on the equations of the previous section allow us to determine $r_{A,obs}$ relative to r^*_{As}, the reaction rate that would occur if all cells or enzyme were exposed to the external substrate concentration. Differences between $r_{A,obs}$ and r^*_{As} show immediately the extent to which reaction is affected by internal mass transfer. The theoretical basis for comparing $r_{A,obs}$ and r^*_{As} is described in the following sections.

13.4.1 First-Order Kinetics

If a catalyst particle is unaffected by mass transfer, the concentration of substrate inside the particle is constant and equal to the surface concentration, C_{As}. Thus, the rate of reaction per particle for first-order kinetics without internal mass transfer effects is $k_1 C_{As}$ multiplied by the particle volume:

$$r^*_{As} = \frac{4}{3}\pi R^3 k_1 C_{As} \tag{13.25}$$

The extent to which $r_{A,obs}$ is different from r^*_{As} is expressed using the *internal effectiveness factor* η_i:

$$\eta_i = \frac{r_{A,obs}}{r^*_{As}} = \frac{(\text{observed rate})}{\left(\begin{array}{c}\text{rate that would occur if } C_A = C_{As}\\ \text{everywhere in the particle}\end{array}\right)} \tag{13.26}$$

In the absence of mass transfer limitations, $r_{A,obs} = r^*_{As}$ and $\eta_i = 1$. When mass transfer effects reduce $r_{A,obs}$, then $\eta_i < 1$. For calculation of η_i, the same units should be used for $r_{A,obs}$ and r^*_{As}, for example, kg s^{-1} per particle, gmol s^{-1} m^{-3}, and so on. We can substitute expressions for $r_{A,obs}$ and r^*_{As} from Eqs. (13.21) and (13.25) into Eq. (13.26) to derive an expression for η_{i1}, the internal effectiveness factor for first-order reaction:

$$\eta_{i1} = \frac{3\mathscr{D}_{Ae}}{R^2\, k_1}\left[R\sqrt{k_1/\mathscr{D}_{Ae}}\,\coth\left(R\sqrt{k_1/\mathscr{D}_{Ae}}\right) - 1\right] \tag{13.27}$$

Thus, the internal effectiveness factor for first-order reaction depends on R, k_1, and $\mathscr{D}_{Ae}$. These parameters are usually grouped together to form a dimensionless variable called the *Thiele modulus*. There are several definitions of the Thiele modulus in the literature. As it was formulated originally [11], application of the modulus was cumbersome because a separate definition was required for different reaction kinetics and catalyst geometries. Generalised moduli that apply to any catalyst shape and reaction kinetics have since been proposed [12–14].

The *generalised Thiele modulus* ϕ is defined as:

$$\phi = \frac{V_p}{S_x}\,\frac{r_A|_{C_{As}}}{\sqrt{2}}\left(\int_{C_{A,eq}}^{C_{As}} \mathscr{D}_{Ae}\, r_A \mathrm{d}C_A\right)^{-1/2} \tag{13.28}$$

where V_p is the catalyst volume, S_x is the catalyst external surface area, C_{As} is the substrate concentration at the surface of the catalyst, r_A is the reaction rate, $r_A|_{C_{As}}$ is the reaction rate when $C_A = C_{As}$, $\mathscr{D}_{Ae}$ is the effective diffusivity of substrate, and $C_{A,eq}$ is the equilibrium substrate concentration. As explained in Section 12.1.1, fermentations and many enzyme reactions are irreversible so that $C_{A,eq}$ is zero for most biological applications. From geometry, $V_p/S_x = R/3$ for spheres and b for flat plates. Expressions determined from Eq. (13.28) for first-order, zero-order, and Michaelis–Menten kinetics are listed in

TABLE 13.2 Generalised Thiele Moduli

First-order reaction: $r_A = k_1 C_A$

$$\phi_1 = \frac{V_p}{S_x}\sqrt{\frac{k_1}{\mathscr{D}_{Ae}}}$$

Sphere $$\phi_1 = \frac{R}{3}\sqrt{\frac{k_1}{\mathscr{D}_{Ae}}}$$

Flat plate $$\phi_1 = b\sqrt{\frac{k_1}{\mathscr{D}_{Ae}}}$$

Zero-order reaction: $r_A = k_0$

$$\phi_0 = \frac{1}{\sqrt{2}}\frac{V_p}{S_x}\sqrt{\frac{k_0}{\mathscr{D}_{Ae}C_{As}}}$$

Sphere $$\phi_0 = \frac{R}{3\sqrt{2}}\sqrt{\frac{k_0}{\mathscr{D}_{Ae}C_{As}}}$$

Flat plate $$\phi_0 = \frac{b}{\sqrt{2}}\sqrt{\frac{k_0}{\mathscr{D}_{Ae}C_{As}}}$$

Michaelis–Menten reaction: $r_A = \frac{v_{max}C_A}{K_m + C_A}$

$$\phi_m = \frac{1}{\sqrt{2}}\frac{V_p}{S_x}\sqrt{\frac{v_{max}}{\mathscr{D}_{Ae}C_{As}}}\left(\frac{1}{1+\beta}\right)\left[1+\beta\ln\left(\frac{\beta}{1+\beta}\right)\right]^{-1/2}$$

$$\beta = \frac{K_m}{C_{As}}$$

Sphere $$\phi_m = \frac{R}{3\sqrt{2}}\sqrt{\frac{v_{max}}{\mathscr{D}_{Ae}C_{As}}}\left(\frac{1}{1+\beta}\right)\left[1+\beta\ln\left(\frac{\beta}{1+\beta}\right)\right]^{-1/2}$$

Flat plate $$\phi_m = \frac{b}{\sqrt{2}}\sqrt{\frac{v_{max}}{\mathscr{D}_{Ae}C_{As}}}\left(\frac{1}{1+\beta}\right)\left[1+\beta\ln\left(\frac{\beta}{1+\beta}\right)\right]^{-1/2}$$

Table 13.2 as ϕ_1, ϕ_0, and ϕ_m, respectively. ϕ represents the important parameters affecting mass transfer and reaction in heterogeneous systems: the catalyst size (R or b), the effective diffusivity ($\mathscr{D}_{Ae}$), the surface substrate concentration (C_{As}), and the intrinsic rate parameters (k_0, k_1, or v_{max} and K_m). Only the Thiele modulus for first-order reaction does not depend on the substrate concentration.

When the parameters R, k_1, and $\mathscr{D}_{Ae}$ in Eq. (13.27) are grouped together as ϕ_1, the result is:

$$\eta_{i1} = \frac{1}{3\phi_1^2}(3\phi_1 \coth 3\phi_1 - 1) \tag{13.29}$$

where coth is defined in Eq. (13.22). Equation (13.29) applies to spherical geometry and first-order reaction; the analogous equation for flat plates is listed in Table 13.3. Plots of η_{i1} versus ϕ_1 for sphere, cylinder, and flat-plate catalysts are shown in Figure 13.11. The curves coincide exactly for $\phi_1 \to 0$ and $\phi_1 \to \infty$, and fall within 10 to 15% for the remainder of the range. Figure 13.11 can be used to evaluate η_{i1} for any catalyst shape provided that ϕ_1 is calculated using Eq. (13.28). Because of the errors involved in estimating the parameters defining ϕ_1, it has been suggested that effectiveness factor curves such as those in Figure 13.11 be viewed as diffuse bands rather than precise functions [15].

TABLE 13.3 Effectiveness Factors (ϕ for each geometry and kinetic order is defined in Table 13.2)

First-order reaction: $r_A = k_1 C_A$		
Sphere[a]	$\eta_{i1} = \frac{1}{3\phi_1^2}(3\phi_1 \coth 3\phi_1 - 1)$	
Flat plate[b]	$\eta_{i1} = \frac{\tanh \phi_1}{\phi_1}$	
Zero-order reaction: $r_A = k_0$		
Sphere[c]	$\eta_{i0} = 1$	for $0 < \phi_0 \leq 0.577$
	$\eta_{i0} = 1 - \left[\frac{1}{2} + \cos\left(\frac{\Psi + 4\pi}{3}\right)\right]^3$ where $\Psi = \cos^{-1}\left(\frac{2}{3\phi_0^2} - 1\right)$	for $\phi_0 > 0.577$
Flat plate	$\eta_{i0} = 1$	for $0 < \phi_0 \leq 1$
	$\eta_{i0} = \frac{1}{\phi_0}$	for $\phi_0 > 1$

[a] *Coth is the abbreviation for hyperbolic cotangent. Coth x is defined as:*

$$\coth x = \frac{e^x + e^{-x}}{e^x - e^{-x}}$$

[b] *Tanh is the abbreviation for hyperbolic tangent. Tanh x is defined as:*

$$\tanh x = \frac{e^x - e^{-x}}{e^x + e^{-x}}$$

[c] *Cos is the abbreviation for cosine. The notation* $\cos^{-1}$ x *(or arccos x) denotes any angle the cosine of which is x. Angles used to determine cos and* $\cos^{-1}$ *are in radians.*

If the first-order Thiele modulus is known, we can use Figure 13.11 to find the internal effectiveness factor. Equations (13.25) and (13.26) can then be applied to predict the observed reaction rate for the catalyst. At low values of $\phi_1 < 0.3$, $\eta_{i1} \approx 1$ and the rate of reaction is not adversely affected by internal mass transfer. However as ϕ_1 increases above 0.3, η_{i1} falls as mass transfer limitations come into play. Therefore, the value of the Thiele modulus indicates immediately whether the rate of reaction is diminished due to diffusional effects, or whether the catalyst is performing at its maximum rate at the prevailing surface substrate concentration. For strong diffusional limitations at $\phi_1 > 10$, η_{i1} for all geometries can be estimated as:

$$\eta_{i1} \approx \frac{1}{\phi_1} \tag{13.30}$$

13.4.2 Zero-Order Kinetics

If substrate is present throughout the catalyst, evaluation of the internal effectiveness factor η_{i0} for zero-order reactions is straightforward. Under these conditions, the reaction rate is independent of substrate concentration and the reaction proceeds at the same rate

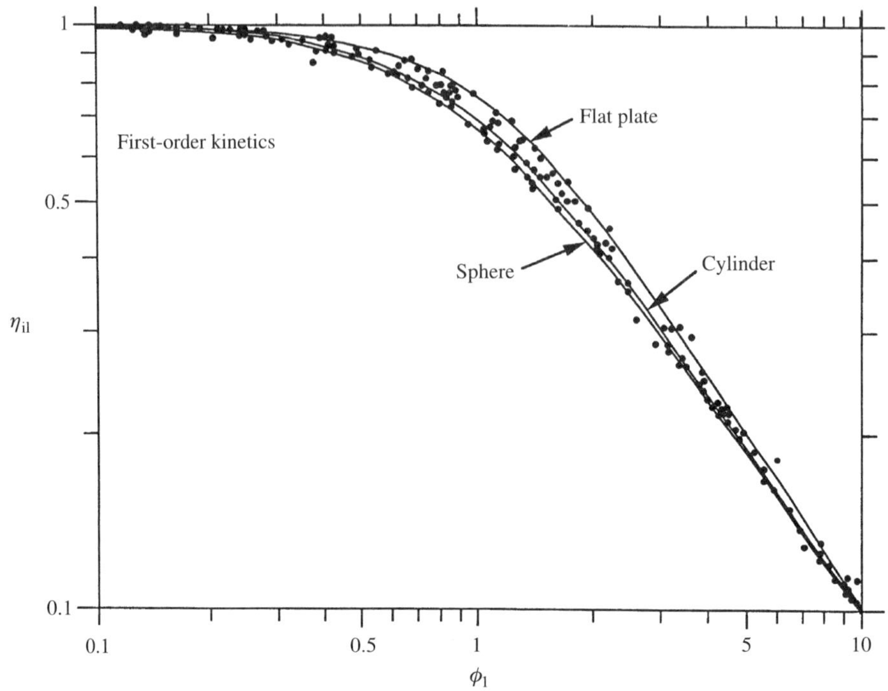

FIGURE 13.11 Internal effectiveness factor η_{i1} as a function of the generalised Thiele modulus ϕ_1 for first-order kinetics and spherical, cylindrical, and flat-plate geometries. The dots represent calculations on finite or hollow cylinders and parallelepipeds.
From R. Aris, 1975, Mathematical Theory of Diffusion and Reaction in Permeable Catalysts, vol. 1, *Oxford University Press, London.*

that would occur if $C_A = C_{As}$ throughout in the particle. Therefore, from Eq. (13.26), $\eta_{i0} = 1$ and:

$$r_{A,obs} = r_{As}^* = \frac{4}{3}\pi R^3 k_0 \tag{13.31}$$

If C_A falls to zero within the particle, the effectiveness factor must be evaluated differently. In this case, $r_{A,obs}$ is given by Eq. (13.24) and the internal effectiveness factor is:

$$\eta_{i0} = \frac{\frac{4}{3}\pi\,(R^3 - R_0^3)\,k_0}{\frac{4}{3}\pi R^3 k_0} = 1 - \left(\frac{R_0}{R}\right)^3 \tag{13.32}$$

According to the previous analysis, to evaluate η_{i0} for zero-order kinetics, first we must know whether or not substrate is depleted in the catalyst and, if it is, the value of R_0. Usually this information is not available because we cannot easily measure intraparticle concentrations. Fortunately, further mathematical analysis [7] overcomes this problem by representing the system in terms of measurable properties such as R, $\mathscr{D}_{Ae}$, C_{As}, and k_0 rather than R_0. These parameters define the Thiele modulus for zero-order reaction, ϕ_0.

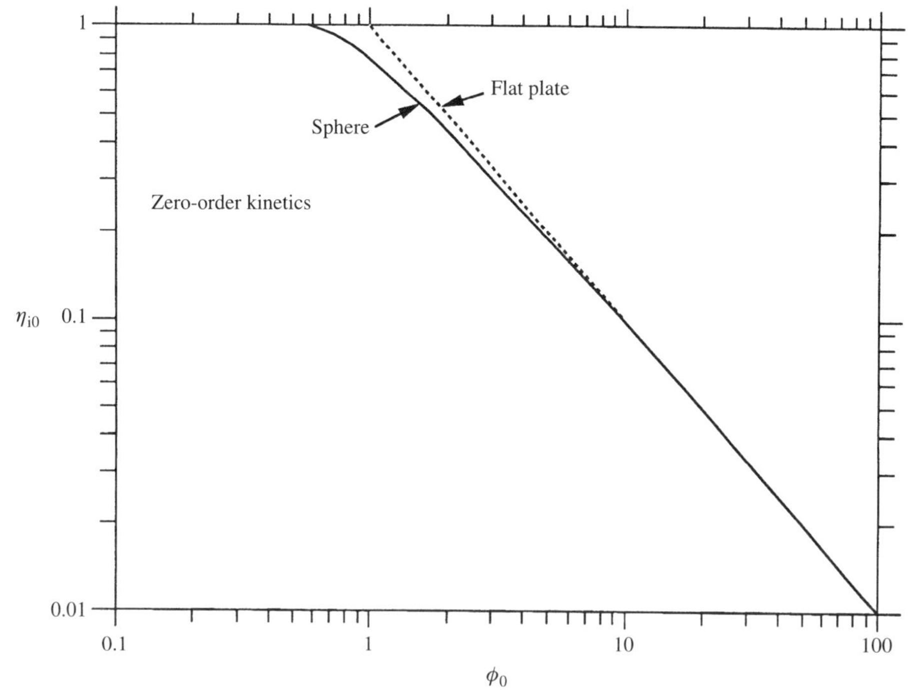

FIGURE 13.12 Internal effectiveness factor η_{i0} as a function of the generalised Thiele modulus ϕ_0 for zero-order kinetics and spherical and flat-plate geometries.

The results are summarised in Table 13.3 and Figure 13.12. C_A remains >0 and $\eta_{i0}=1$ for $0<\phi_0\leq 0.577$. For $\phi_0>0.577$, η_{i0} declines as more and more of the particle becomes inactive. Effectiveness factors for flat-plate systems are also shown in Table 13.3 and Figure 13.12. In flat biofilms, $\phi_0=1$ represents the threshold condition for substrate depletion. As shown in Figure 13.12, the η_{i0} curves for spherical and flat-plate geometries coincide exactly at small and large values of ϕ_0.

13.4.3 Michaelis–Menten Kinetics

For a spherical catalyst, the rate of Michaelis–Menten reaction in the absence of internal mass transfer effects is:

$$r^*_{As} = \frac{4}{3}\pi R^3\left(\frac{v_{max}C_{As}}{K_m + C_{As}}\right) \tag{13.33}$$

Our analysis cannot proceed further, however, because we do not have an equation for $r_{A,obs}$. Accordingly, we cannot develop an analytical expression for the effectiveness factor for Michaelis–Menten kinetics, η_{im}, as a function of ϕ_m. Diffusion–reaction equations for Michaelis–Menten kinetics are generally solved by numerical computation. As an

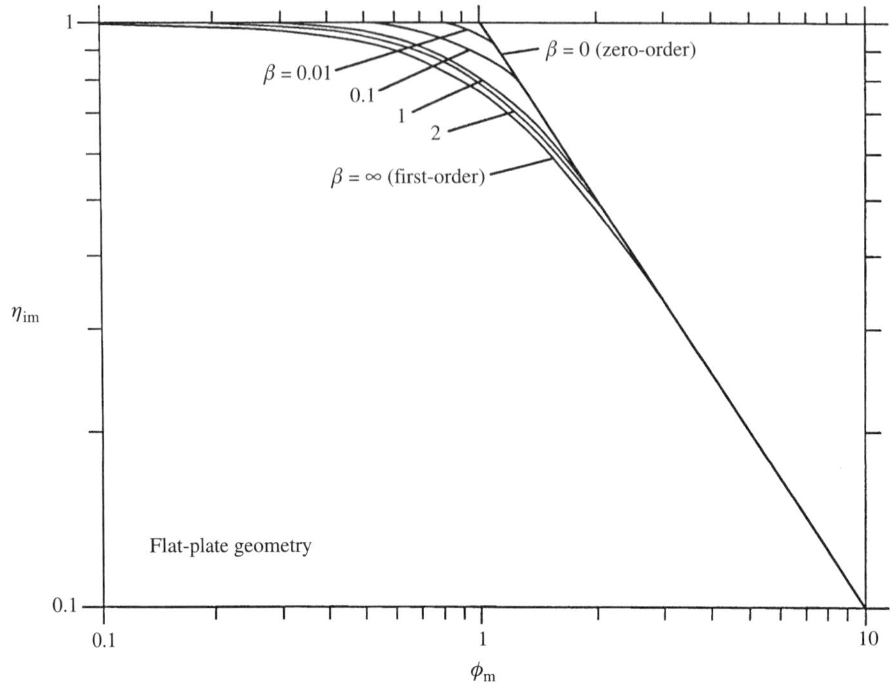

FIGURE 13.13 Internal effectiveness factor η_{im} as a function of the generalised Thiele modulus ϕ_m and parameter β for Michaelis–Menten kinetics and flat-plate geometry. $\beta = K_m/C_{As}$.
From R. Aris, 1975, Mathematical Theory of Diffusion and Reaction in Permeable Catalysts, vol. 1, *Oxford University Press, London.*

example, the results for flat-plate geometry are shown in Figure 13.13 as a function of the parameter β, which is equal to the ratio K_m/C_{As}.

We can obtain approximate values for η_{im} by considering the zero- and first-order asymptotes of the Michaelis–Menten equation. As indicated in Figure 13.13, the curves for η_{im} fall between the lines for zero- and first-order reactions depending on the value of β. As $\beta \to \infty$, Michaelis–Menten kinetics can be approximated as first-order and the internal effectiveness factor can be evaluated from Figure 13.11 with $k_1 = v_{max}/K_m$. When $\beta \to 0$, zero-order kinetics and Figure 13.12 apply with $k_0 = v_{max}$. Effectiveness factors for β between zero and infinity must be evaluated using numerical methods.

Use of the generalised Thiele modulus for Michaelis–Menten reactions (Table 13.2) eliminates almost all variation in the internal effectiveness factor with changing β, except in the vicinity of $\phi_m = 1$. As in Figures 13.11 and 13.12, the generalised modulus also brings together effectiveness-factor curves for all shapes of catalyst at the two asymptotes $\phi_m \to 0$ and $\phi_m \to \infty$. Therefore, Figure 13.13 is valid for spherical catalysts if ϕ_m is much less or much greater than 1. It should be noted however that variation between geometries in the intermediate region around $\phi_m = 1$ can be significant [16].

If the values of ϕ_m and β are such that Michaelis–Menten kinetics cannot be approximated as either zero- or first-order, η_{im} may be estimated using an equation proposed by Moo-Young and Kobayashi [17]:

$$\eta_{im} = \frac{\eta_{i0} + \beta\eta_{i1}}{1+\beta} \tag{13.34}$$

where $\beta = K_m/C_{As}$. η_{i0} is the zero-order internal effectiveness factor obtained using values of ϕ_0 evaluated with $k_0 = v_{max}$, and η_{i1} is the first-order effectiveness factor obtained using ϕ_1 calculated with $k_1 = v_{max}/K_m$. For flat-plate geometry, the largest deviation of Eq. (13.34) from exact values of η_{im} is 0.089; this occurs at $\phi_m = 1$ and $\beta = 0.2$. For spherical geometry, the greatest deviations occur around $\phi_m = 1.7$ and $\beta = 0.3$; the maximum error in this region is 0.09. Further details can be found in the original paper [17].

EXAMPLE 13.3 REACTION RATES FOR FREE AND IMMOBILISED ENZYME

Invertase is immobilised in ion-exchange resin of average diameter 1 mm. The amount of enzyme in the beads is measured by protein assay as 0.05 kg m^{-3} of resin. A small column reactor is packed with 20 cm^3 of beads. Seventy-five ml of a 16-mM sucrose solution is pumped rapidly through the bed. In another reactor, an identical quantity of free enzyme is mixed into the same volume of sucrose solution. The kinetic parameters for the immobilised enzyme can be assumed to be equal to those for the free enzyme: K_m is 8.8 mM and the turnover number is 2.4×10^{-3} gmol of sucrose (g enzyme)$^{-1}$ s^{-1}. The effective diffusivity of sucrose in the ion-exchange resin is 2×10^{-6} cm^2 s^{-1}.

(a) What is the rate of reaction using free enzyme?
(b) What is the rate of reaction using immobilised enzyme?

Solution

The invertase reaction is:

$$\underset{\text{sucrose}}{C_{12}H_{22}O_{11}} + H_2O \longrightarrow \underset{\text{glucose}}{C_6H_{12}O_6} + \underset{\text{fructose}}{C_6H_{12}O_6}$$

Convert the data provided to units of gmol, m, s:

$$K_m = \frac{8.8 \times 10^{-3}\ \text{gmol}}{\text{litre}} \cdot \left|\frac{1000\ \text{litres}}{1\ \text{m}^3}\right| = 8.8\ \text{gmol m}^{-3}$$

$$\mathscr{D}_{Ae} = 2 \times 10^{-6}\ \text{cm}^2\ \text{s}^{-1} \cdot \left|\frac{1\ \text{m}}{100\ \text{cm}}\right|^2 = 2 \times 10^{-10}\ \text{m}^2\ \text{s}^{-1}$$

$$R = \frac{1\ \text{mm}}{2} \cdot \left|\frac{1\ \text{m}}{10^3\ \text{mm}}\right| = 5 \times 10^{-4}\ \text{m}$$

If flow through the reactor is rapid, we can assume that C_{As} is equal to the bulk sucrose concentration C_{Ab}:

$$C_{As} = C_{Ab} = 16\ \text{mM} = \frac{16 \times 10^{-3}\ \text{gmol}}{\text{litre}} \cdot \left|\frac{1000\ \text{litres}}{1\ \text{m}^3}\right| = 16\ \text{gmol m}^{-3}$$

Also:

$$\text{Mass of enzyme} = \frac{0.05\ \text{kg}}{\text{m}^3}\ (20\ \text{cm}^3) \cdot \left|\frac{1\ \text{m}}{100\ \text{cm}}\right|^3 = 10^{-6}\ \text{kg}$$

(a) In the free enzyme reactor:

$$\text{Enzyme concentration} = \frac{10^{-6}\ \text{kg}}{75\ \text{cm}^3} \cdot \left|\frac{100\ \text{cm}}{1\ \text{m}}\right|^3 = 1.33 \times 10^{-2}\ \text{kg m}^{-3}$$

From Eq. (12.39), v_{max} is obtained by multiplying the turnover number by the concentration of active enzyme. Assuming that all of the enzyme present is active:

$$v_{max} = \frac{2.4 \times 10^{-3}\ \text{gmol}}{\text{g s}} (1.33 \times 10^{-2}\ \text{kg m}^{-3}) \cdot \left|\frac{1000\ \text{g}}{1\ \text{kg}}\right| = 3.19 \times 10^{-2}\ \text{gmol s}^{-1}\ \text{m}^{-3}$$

The volumetric rate of reaction is given by the Michaelis–Menten equation. As the free enzyme reaction takes place at uniform sucrose concentration C_{Ab}:

$$v = \frac{v_{max} C_{Ab}}{K_m + C_{Ab}} = \frac{3.19 \times 10^{-2}\ \text{gmol s}^{-1}\ \text{m}^{-3}\ (16\ \text{gmol m}^{-3})}{(8.8\ \text{gmol m}^{-3} + 16\ \text{gmol m}^{-3})} = 2.06 \times 10^{-2}\ \text{gmol s}^{-1}\ \text{m}^{-3}$$

Multiplying by the liquid volume gives the total rate of reaction:

$$v = 2.06 \times 10^{-2}\ \text{gmol s}^{-1}\ \text{m}^{-3}\ (75\ \text{cm}^3) \cdot \left|\frac{1\ \text{m}}{100\ \text{cm}}\right|^3 = 1.55 \times 10^{-6}\ \text{gmol s}^{-1}$$

The total rate of reaction using free enzyme is 1.55×10^{-6} gmol s^{-1}.

(b) In the equations for heterogeneous reaction, v_{max} is expressed on a catalyst volume basis. Therefore:

$$v_{max} = \frac{2.4 \times 10^{-3}\ \text{gmol}}{\text{g s}}\ (0.05\ \text{kg m}^{-3}) \cdot \left|\frac{1000\ \text{g}}{1\ \text{kg}}\right| = 0.12\ \text{gmol s}^{-1}\ \text{m}^{-3}\ \text{of resin}$$

To determine the effect of mass transfer we must calculate η_{im}. The method used depends on the values of β and ϕ_m:

$$\beta = \frac{K_m}{C_{As}} = \frac{8.8\ \text{gmol m}^{-3}}{16\ \text{gmol m}^{-3}} = 0.55$$

From Table 13.2 for Michaelis–Menten kinetics and spherical geometry:

$$\phi_m = \frac{R}{3\sqrt{2}}\sqrt{\frac{v_{max}}{\mathscr{D}_{Ae}C_{As}}}\left(\frac{1}{1+\beta}\right)\left[1+\beta\ln\left(\frac{\beta}{1+\beta}\right)\right]^{-1/2}$$

$$\phi_m = \frac{5\times10^{-4}\ \text{m}}{3\sqrt{2}}\sqrt{\frac{0.12\ \text{gmol s}^{-1}\ \text{m}^{-3}}{(2\times10^{-10}\ \text{m}^2\ \text{s}^{-1})(16\ \text{gmol m}^{-3})}}\left(\frac{1}{1+0.55}\right)\left[1+0.55\ln\left(\frac{0.55}{1+0.55}\right)\right]^{-1/2}$$

$$\phi_m = 0.71$$

Because both β and ϕ_m have intermediate values, Figure 13.13 cannot be applied for spherical geometry. Instead, we can use Eq. (13.34). From Table 13.2 for zero-order kinetics and spherical geometry:

$$\phi_0 = \frac{R}{3\sqrt{2}}\sqrt{\frac{k_0}{\mathscr{D}_{Ae}C_{As}}}$$

Using $k_0 = v_{max}$:

$$\phi_0 = \frac{R}{3\sqrt{2}}\sqrt{\frac{v_{max}}{\mathscr{D}_{Ae}C_{As}}}$$

$$\phi_0 = \frac{5\times10^{-4}\ \text{m}}{3\sqrt{2}}\sqrt{\frac{0.12\ \text{gmol s}^{-1}\ \text{m}^{-3}}{(2\times10^{-10}\ \text{m}^2\ \text{s}^{-1})(16\ \text{gmol m}^{-3})}}$$

$$\phi_0 = 0.72$$

From Figure 13.12 or Table 13.3, at this value of ϕ_0, $\eta_{i0} = 0.93$. Similarly, for first-order kinetics and spherical geometry with $k_1 = v_{max}/K_m$:

$$\phi_1 = \frac{R}{3}\sqrt{\frac{v_{max}}{K_m\mathscr{D}_{Ae}}}$$

$$\phi_1 = \frac{5\times10^{-4}\ \text{m}}{3}\sqrt{\frac{0.12\ \text{gmol m}^{-3}\ \text{s}^{-1}}{(8.8\ \text{gmol m}^{-3})(2\times10^{-10}\ \text{m}^2\ \text{s}^{-1})}}$$

$$\phi_1 = 1.4$$

From Figure 13.11 or Table 13.3, $\eta_{i1} = 0.54$. Substituting these results into Eq. (13.34) gives:

$$\eta_{im} = \frac{0.93 + 0.55\,(0.54)}{1+0.55} = 0.79$$

The total rate of reaction for immobilised enzyme without diffusional limitations is evaluated using the Michaelis–Menten expression with substrate concentration C_{As}:

$$r^*_{As} = \frac{v_{max}C_{As}}{K_m + C_{As}} = \frac{0.12\ \text{gmol s}^{-1}\ \text{m}^{-3}\ (20\ \text{cm}^3)\cdot\left|\frac{1\ \text{m}}{100\ \text{cm}}\right|^3(16\ \text{gmol m}^{-3})}{(8.8\ \text{gmol m}^{-3} + 16\ \text{gmol m}^{-3})}$$

$$= 1.55\times10^{-6}\ \text{gmol s}^{-1}$$

Not surprisingly as the kinetic parameters were assumed to be equal, this is the same rate as that found using the same quantity of free enzyme. The effectiveness factor indicates that the rate of reaction for the immobilised enzyme is 79% of that occurring in the absence of mass transfer effects. Applying Eq. (13.26):

$$r_{A,obs} = \eta_{im} r^*_{As} = 0.79\,(1.55 \times 10^{-6}\ \text{gmol s}^{-1}) = 1.22 \times 10^{-6}\ \text{gmol s}^{-1}$$

The total rate of reaction using immobilised enzyme is 1.22×10^{-6} gmol s^{-1}.

13.4.4 The Observable Thiele Modulus

Diffusion–reaction theory as presented in the previous sections allows us to quantify the effect of mass transfer on the rate of enzyme and cell reactions. However, a shortcoming of the methods outlined so far is that they are useful only if we know the true kinetic parameters for the reaction: k_0, k_1, or v_{max} and K_m. In many cases these values are not known and, as discussed in Section 13.9, can be difficult to evaluate. One way to circumvent this problem is to apply the *observable Thiele modulus* Φ, sometimes called *Weisz's modulus* [18], which is defined as:

$$\Phi = \left(\frac{V_p}{S_x}\right)^2 \frac{r_{A,obs}}{\mathscr{D}_{Ae} C_{As}} \tag{13.35}$$

where V_p is the catalyst volume, S_x is the catalyst external surface area, $r_{A,obs}$ is the observed reaction rate *per unit volume of catalyst*, $\mathscr{D}_{Ae}$ is the effective diffusivity of substrate, and C_{As} is the substrate concentration at the catalyst surface. Expressions for Φ for spheres and flat plates are listed in Table 13.4. Evaluation of the observable Thiele modulus does not rely on prior knowledge of the kinetic parameters; Φ is defined in terms of the measured or observed reaction rate, $r_{A,obs}$.

For the observable Thiele modulus to be useful, we need to relate Φ to the internal effectiveness factor η_i. Some mathematical consideration of the equations already presented in this chapter yields the following relationships for first-order, zero-order, and Michaelis–Menten kinetics:

$$\text{First order:}\quad \Phi = \phi_1^2 \eta_{i1} \tag{13.36}$$

$$\text{Zero order:}\quad \Phi = 2\phi_0^2 \eta_{i0} \tag{13.37}$$

TABLE 13.4 Observable Thiele Moduli

Sphere	$\Phi = \left(\frac{R}{3}\right)^2 \frac{r_{A,obs}}{\mathscr{D}_{Ae} C_{As}}$
Flat plate	$\Phi = b^2 \frac{r_{A,obs}}{\mathscr{D}_{Ae} C_{As}}$

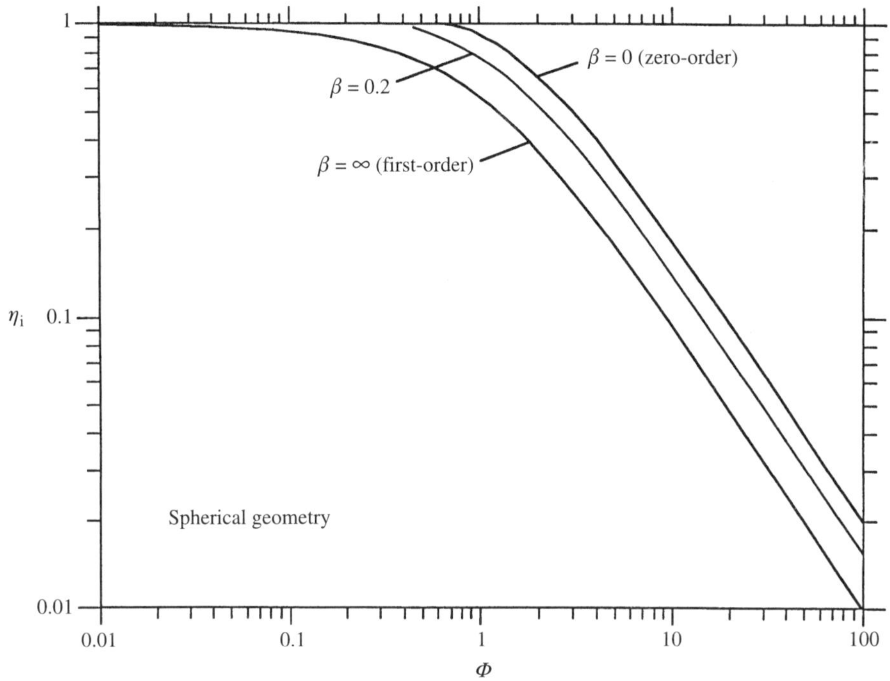

FIGURE 13.14 Internal effectiveness factor η_i as a function of the observable Thiele modulus Φ for spherical geometry and first-order, zero-order, and Michaelis–Menten kinetics. $\beta = K_m/C_{As}$.
From W.H. Pitcher, 1975, Design and operation of immobilized enzyme reactors. In: R.A. Messing (Ed.), Immobilized Enzymes for Industrial Reactors, *pp. 151–199, Academic Press, New York.*

$$\text{Michaelis–Menten:} \quad \Phi = 2\phi_m^2 \eta_{im}(1+\beta)\left[1 + \beta \ln\left(\frac{\beta}{1+\beta}\right)\right] \tag{13.38}$$

Equations (13.36) through (13.38) apply to all catalyst geometries and allow us to develop plots of Φ versus η_i using the relationships between ϕ and η_i developed in the previous sections and represented in Figures 13.11 through 13.13. The results for spherical catalysts and first-order, zero-order, and Michaelis–Menten kinetics are given in Figure 13.14; similar results for flat-plate geometry are shown in Figure 13.15. All curves for β between zero and infinity are bracketed by the first- and zero-order lines. For each value of β, curves for all geometries coincide in the asymptotic regions $\Phi \to 0$ and $\Phi \to \infty$. At intermediate values of Φ, the variation between effectiveness factors for different geometries can be significant. For $\Phi > 10$:

$$\text{First-order kinetics:} \quad \eta_{i1} \approx \frac{1}{\Phi} \tag{13.39}$$

$$\text{Zero-order kinetics:} \quad \eta_{i0} \approx \frac{2}{\Phi} \tag{13.40}$$

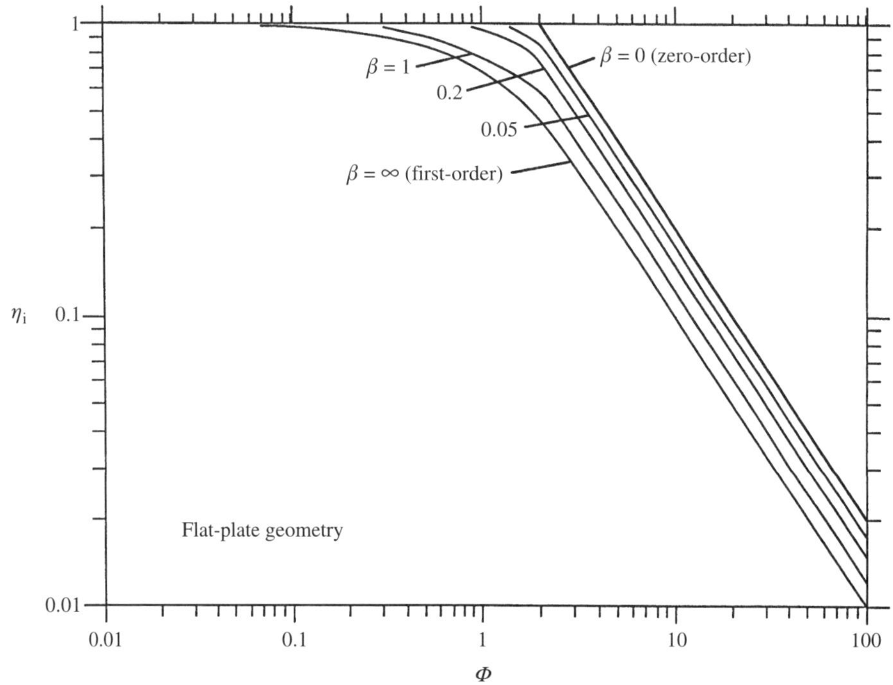

FIGURE 13.15 Internal effectiveness factor η_i as a function of the observable Thiele modulus Φ for flat-plate geometry and first-order, zero-order, and Michaelis–Menten kinetics. $\beta = K_m / C_{As}$.
From W.H. Pitcher, 1975, Design and operation of immobilized enzyme reactors. In: R.A. Messing (Ed.), Immobilized Enzymes for Industrial Reactors, *pp. 151–199, Academic Press, New York.*

Although Φ is an observable modulus and is independent of kinetic parameters, use of Figures 13.14 and 13.15 for Michaelis–Menten reactions requires knowledge of K_m for evaluation of β. This makes application of the observable Thiele modulus difficult for Michaelis–Menten kinetics. However, we know that effectiveness factors for Michaelis–Menten reactions lie between the first- and zero-order curves of Figures 13.14 and 13.15; therefore, we can always estimate the upper and lower bounds of η_{im}.

13.4.5 Weisz's Criteria

The following general observations can be made from Figures 13.14 and 13.15.

1. If $\Phi < 0.3$, $\eta_i \approx 1$ and internal mass transfer limitations are insignificant.
2. If $\Phi > 3$, η_i is substantially < 1 and internal mass transfer limitations are significant.

These statements are known as *Weisz's criteria*. They are valid for all geometries and reaction kinetics. For Φ in the intermediate range $0.3 \leq \Phi \leq 3$, closer analysis is required to determine the influence of mass transfer on reaction rate.

EXAMPLE 13.4 INTERNAL OXYGEN TRANSFER TO IMMOBILISED CELLS

Baby hamster kidney cells are immobilised in alginate beads. The average particle diameter is 5 mm. The bulk oxygen concentration in the medium is 8×10^{-3} kg m^{-3}, the rate of oxygen consumption by the cells is 8.4×10^{-5} kg s^{-1} m^{-3} of catalyst, and the effective diffusivity of oxygen in the beads is 1.88×10^{-9} m^2 s^{-1}. Assume that the oxygen concentration at the surface of the catalyst is equal to the bulk concentration, and that oxygen uptake follows zero-order kinetics.

(a) Are internal mass transfer effects significant?
(b) What reaction rate would be observed if diffusional resistances were eliminated?

Solution

(a) To assess the effect of internal mass transfer on the reaction rate, we need to calculate the observable Thiele modulus. From Table 13.4 for spherical geometry:

$$\Phi = \left(\frac{R}{3}\right)^2 \frac{r_{A,obs}}{\mathscr{D}_{Ae} C_{As}}$$

For:

$$R = \frac{5 \times 10^{-3}\ \text{m}}{2} = 2.5 \times 10^{-3}\ \text{m}$$

and $C_{As} = C_{Ab}$:

$$\Phi = \left(\frac{2.5 \times 10^{-3}\ \text{m}}{3}\right)^2 \frac{8.4 \times 10^{-5}\ \text{kg s}^{-1}\ \text{m}^{-3}}{(1.88 \times 10^{-9}\ \text{m}^2\ \text{s}^{-1})\ (8 \times 10^{-3}\ \text{kg m}^{-3})} = 3.9$$

From Weisz's criteria, internal mass transfer effects are significant.

(b) For spherical catalysts and zero-order reaction, from Figure 13.14, at $\Phi = 3.9$, $\eta_{i0} = 0.40$. From Eq. (13.26), the reaction rate without diffusional restrictions is:

$$r^*_{As} = \frac{r_{A,obs}}{\eta_i} = \frac{8.4 \times 10^{-5}\ \text{kg s}^{-1}\ \text{m}^{-3}}{0.40} = 2.1 \times 10^{-4}\ \text{kg s}^{-1}\ \text{m}^{-3}\ \text{of catalyst}$$

13.4.6 Minimum Intracatalyst Substrate Concentration

It is sometimes of interest to know the minimum substrate concentration, $C_{A,min}$, inside solid catalysts. We can use this information to check, for example, that the concentration does not fall below some critical value for cell metabolism. $C_{A,min}$ is estimated easily for zero-order reactions. If Φ is such that $\eta_{i0} < 1$, we know immediately that $C_{A,min}$ is zero because $\eta_{i0} = 1$ if $C_A > 0$ throughout the particle. For $\eta_{i0} = 1$, simple manipulation of the equations already presented in this chapter allows us to estimate $C_{A,min}$ for zero-order reactions. The results are summarised in Table 13.5.

TABLE 13.5 Minimum Intracatalyst Substrate Concentration for Zero-Order Kinetics (Φ for each geometry is defined in Table 13.4)

Sphere	
$C_{A,min} = 0$	for $\Phi \geq 0.667$
$C_{A,min} = C_{As}\left(1 - \frac{3}{2}\Phi\right)$	for $\Phi < 0.667$
Flat plate	
$C_{A,min} = 0$	for $\Phi \geq 2$
$C_{A,min} = C_{As}\left(1 - \frac{1}{2}\Phi\right)$	for $\Phi < 2$

13.5 EXTERNAL MASS TRANSFER

Many of the equations in Sections 13.3 and 13.4 contain the term C_{As}, the concentration of substrate A at the external surface of the catalyst. This term made its way into the analysis in the boundary conditions used for solution of the shell mass balance. So far, we have assumed that C_{As} is a known quantity. However, because surface concentrations are very difficult to measure accurately, we must find ways to estimate C_{As} using theoretical principles.

During reaction, a reduction in external substrate concentration may occur as a liquid boundary layer develops at the surface of the solid catalyst (Figure 13.2). In the absence of a boundary layer, the substrate concentration at the solid surface C_{As} is equal to the substrate concentration in the bulk liquid C_{Ab}, which is easily measured. When the boundary layer is present, C_{As} takes some value less than C_{Ab}. The rate of mass transfer across the boundary layer is given by the following equation derived from Eq. (10.8) in Chapter 10:

$$N_A = k_S\, a\, (C_{Ab} - C_{As}) \tag{13.41}$$

where N_A is the rate of mass transfer, k_S is the *liquid-phase mass transfer coefficient* with dimensions LT^{-1}, and a is the external surface area of the catalyst. If N_A is expressed per volume of catalyst with units of, for example, kgmol s^{-1} m^{-3}, to be consistent, a must also be expressed on a catalyst-volume basis with units of, for example, $m^2\,m^{-3}$ or m^{-1}. Therefore, using the previous notation of this chapter, a in Eq. (13.41) is equal to S_x/V_p for the catalyst. At steady state, the rate of substrate transfer across the boundary layer must be equal to the rate of substrate consumption by the catalyst, $r_{A,obs}$. Therefore:

$$r_{A,obs} = k_S \frac{S_x}{V_p} (C_{Ab} - C_{As}) \tag{13.42}$$

where $r_{A,obs}$ is the observed reaction rate per volume of catalyst. Rearranging gives:

$$\frac{C_{As}}{C_{Ab}} = 1 - \frac{V_p\, r_{A,obs}}{S_x\, k_S\, C_{Ab}} \tag{13.43}$$

Equation (13.43) can be used to evaluate C_{As} before applying the equations in the previous sections to calculate internal substrate concentrations and effectiveness factors. The magnitude of external mass transfer effects can be gauged from Eq. (13.43). $C_{As}/C_{Ab} \approx 1$ indicates no or negligible external mass transfer limitations, as the substrate concentration at the surface is approximately equal to that in the bulk. On the other hand, $C_{As}/C_{Ab} \ll 1$ indicates a very steep concentration gradient in the boundary layer and severe external mass transfer effects. We can define from Eq. (13.43) an *observable modulus for external mass transfer*, Ω:

$$\Omega = \frac{V_P\, r_{A,obs}}{S_x\, k_S\, C_{Ab}} \tag{13.44}$$

Expressions for Ω for spherical and flat-plate geometries are listed in Table 13.6. The following criteria are used to assess the extent of external mass transfer effects.

1. If $\Omega \ll 1$, $C_{As} \approx C_{Ab}$ and external mass transfer effects are insignificant.
2. Otherwise, $C_{As} < C_{Ab}$ and external mass transfer effects are significant.

For reactions affected by both internal and external mass transfer restrictions, we can define a *total effectiveness factor* η_T:

$$\eta_T = \frac{r_{A,obs}}{r^*_{Ab}} = \frac{(\text{observed rate})}{\left(\begin{array}{c}\text{rate that would occur if } C_A = C_{Ab}\\ \text{everywhere in the particle}\end{array}\right)} \tag{13.45}$$

η_T can be related to the internal effectiveness factor η_i by rewriting Equation (13.45) as:

$$\eta_T = \left(\frac{r_{A,obs}}{r^*_{As}}\right)\left(\frac{r^*_{As}}{r^*_{Ab}}\right) = \eta_i\, \eta_e \tag{13.46}$$

where η_e is the *external effectiveness factor* and η_i is defined in Eq. (13.26). Therefore, η_e has the following meaning:

$$\eta_e = \frac{r^*_{As}}{r^*_{Ab}} = \frac{\left(\begin{array}{c}\text{rate that would occur if } C_A = C_{As}\\ \text{everywhere in the particle}\end{array}\right)}{\left(\begin{array}{c}\text{rate that would occur if } C_A = C_{Ab}\\ \text{everywhere in the particle}\end{array}\right)} \tag{13.47}$$

Expressions for η_e for first-order, zero-order, and Michaelis–Menten kinetics are listed in Table 13.7. For zero-order reactions, $\eta_{e0} = 1$ as long as $C_{As} > 0$ and $C_{Ab} > 0$. However, $\eta_{e0} = 1$ does not imply that an external boundary layer does not exist. Because r^*_{As} and r^*_{Ab} are independent of C_A for zero-order kinetics, $\eta_{e0} = 1$ even when there is a reduction in

TABLE 13.6 Observable Moduli for External Mass Transfer

Sphere	$\Omega = \frac{R\, r_{A,obs}}{3\, k_S\, C_{Ab}}$
Flat plate	$\Omega = b\frac{r_{A,obs}}{k_S\, C_{Ab}}$

TABLE 13.7 External Effectiveness Factors

First-order reaction: $r_A = k_1 C_A$	$\eta_{e1} = \dfrac{C_{As}}{C_{Ab}}$
Zero-order reaction: $r_A = k_0$	$\eta_{e0} = 1$
Michaelis–Menten reaction: $r_A = \dfrac{v_{max} C_A}{K_m + C_A}$	$\eta_{em} = \dfrac{C_{As}(K_m + C_{Ab})}{C_{Ab}(K_m + C_{As})}$

concentration across the boundary layer. Furthermore, $\eta_{e0} = 1$ does not imply that eliminating the external boundary layer could not improve the observed reaction rate. Removing the boundary layer would increase the value of C_{As}, thus establishing a greater driving force for internal mass transfer and reducing the likelihood of C_A falling to zero inside the particle.

EXAMPLE 13.5 EFFECT OF MASS TRANSFER ON BACTERIAL DENITRIFICATION

Denitrifying bacteria are immobilised in gel beads and used in a stirred reactor for removal of nitrate from groundwater. At a nitrate concentration of 3 g m^{-3}, the conversion rate is 0.011 g s^{-1} m^{-3} of catalyst. The effective diffusivity of nitrate in the gel is 1.5×10^{-9} m^2 s^{-1}, the beads are 6 mm in diameter, and the liquid–solid mass transfer coefficient is 10^{-5} m s^{-1}. K_m for the immobilised bacteria is approximately 25 g m^{-3}.

(a) Does external mass transfer influence the reaction rate?
(b) Are internal mass transfer effects significant?
(c) By how much would the reaction rate be improved if both the internal and external mass transfer resistances were eliminated?

Solution

(a)

$$R = \frac{6 \times 10^{-3}\ \text{m}}{2} = 3 \times 10^{-3}\ \text{m}$$

The effect of external mass transfer is found by calculating Ω. From Table 13.6 for spherical geometry:

$$\Omega = \frac{R\, r_{A,obs}}{3\, k_S C_{Ab}} = \frac{3 \times 10^{-3}\ \text{m}}{3} \frac{0.011\ \text{g s}^{-1}\ \text{m}^{-3}}{(10^{-5}\ \text{m s}^{-1})\,(3\ \text{g m}^{-3})} = 0.37$$

As this value of Ω is relatively large, external mass transfer effects are significant.

(b) From Eq. (13.43):

$$\frac{C_{As}}{C_{Ab}} = 1 - \Omega = 0.63$$

$$C_{As} = 0.63\, C_{Ab} = 0.63\,(3\ \text{g m}^{-3}) = 1.9\ \text{g m}^{-3}$$

The observable Thiele modulus is calculated with $C_{As} = 1.9\ \text{g m}^{-3}$ using the equation for spheres from Table 13.4:

$$\Phi = \left(\frac{R}{3}\right)^2 \frac{r_{A,obs}}{\mathscr{D}_{Ae} C_{As}} = \left(\frac{3 \times 10^{-3}\ \text{m}}{3}\right)^2 \frac{0.011\ \text{g s}^{-1}\ \text{m}^{-3}}{(1.5 \times 10^{-9}\ \text{m}^2\ \text{s}^{-1})\ (1.9\ \text{g m}^{-3})} = 3.9$$

From Weisz's criteria (Section 13.4.5), as $\Phi > 3$, internal mass transfer effects are significant.

(c) The reaction rate in the absence of mass transfer effects is r^*_{Ab}, which is related to $r_{A,obs}$ by Eq. (13.45). Therefore, we can calculate r^*_{Ab} if we know η_T. Because the nitrate concentration is much smaller than K_m, we can assume that the reaction operates in the first-order regime. From Table 13.7 for first-order kinetics:

$$\eta_{e1} = \frac{C_{As}}{C_{Ab}} = 0.63$$

From Figure 13.14 for spherical geometry, at $\Phi = 3.9$, $\eta_{i1} = 0.21$. Therefore, using Eq. (13.46):

$$\eta_{T1} = \eta_{i1}\ \eta_{e1} = 0.21\ (0.63) = 0.13$$

From Eq. (13.45):

$$r^*_{Ab} = \frac{r_{A,obs}}{\eta_{T1}} = \frac{0.011\ \text{g s}^{-1}\ \text{m}^{-3}}{0.13} = 0.085\ \text{g s}^{-1}\ \text{m}^{-3}$$

If the internal and external mass transfer resistances were eliminated, the reaction rate would be increased by a factor of 0.085/0.011, or about 7.7.

13.6 LIQUID–SOLID MASS TRANSFER CORRELATIONS

The mass transfer coefficient k_S must be known before we can account for external mass transfer effects. k_S depends on liquid properties such as viscosity, density, and diffusivity as well as the reactor hydrodynamics. It is difficult to determine k_S accurately, especially for particles that are neutrally buoyant. However, values can be estimated using correlations from the literature; these are usually accurate under the conditions specified to within 10 to 20%. Selected correlations are presented below. Further details can be found in chemical engineering texts [19, 20].

Correlations for k_S are expressed using the following dimensionless groups:

$$Re_p = \text{(particle) Reynolds number} = \frac{D_p\ u_{pL}\ \rho_L}{\mu_L} \tag{13.48}$$

$$Sc = \text{Schmidt number} = \frac{\mu_L}{\rho_L\ \mathscr{D}_{AL}} \tag{13.49}$$

$$Sh = \text{Sherwood number} = \frac{k_S D_p}{\mathscr{D}_{AL}} \tag{13.50}$$

and

$$Gr = \text{Grashof number} = \frac{g\, D_p^3\, \rho_L\, (\rho_p - \rho_L)}{\mu_L^2} \tag{13.51}$$

In Eqs. (13.48) through (13.51), D_p is the particle diameter, u_{pL} is the linear velocity of the particle relative to the bulk liquid, ρ_L is liquid density, μ_L is liquid viscosity, $\mathscr{D}_{AL}$ is the molecular diffusivity of component A in the liquid, k_S is the liquid–solid mass transfer coefficient, g is gravitational acceleration, and ρ_p is the particle density. The Sherwood number contains the mass transfer coefficient and represents the ratio of overall mass transfer rate to diffusive mass transfer rate through the boundary layer. The Schmidt number represents the ratio of momentum diffusivity to mass diffusivity and is evaluated from the physical properties of the system. At constant temperature, pressure, and composition, *Sc* is constant for Newtonian fluids. The Grashof number represents the ratio of gravitational forces to viscous forces and is important when the particles are neutrally buoyant. The form of the correlation used to evaluate *Sh* and therefore k_S depends on the configuration of the mass transfer system, the flow conditions, and other factors.

13.6.1 Free-Moving Spherical Particles

The equations presented here apply to solid particles suspended in stirred vessels. The rate of mass transfer depends on the velocity of the solid relative to the liquid; this is known as the *slip velocity*. The slip velocity is difficult to measure in suspensions and must therefore be estimated before calculating k_S. The following equations allow evaluation of the particle Reynolds number Re_p, which incorporates the slip velocity u_{pL} [7]:

$$\text{For } Gr < 36 \qquad Re_p = \frac{Gr}{18} \tag{13.52}$$

$$\text{For } 36 < Gr < 8 \times 10^4 \qquad Re_p = 0.153\, Gr^{0.71} \tag{13.53}$$

$$\text{For } 8 \times 10^4 < Gr < 3 \times 10^9 \qquad Re_p = 1.74\, Gr^{0.5} \tag{13.54}$$

Once Re_p is known, *Sh* is determined using equations such as [7, 19–22]:

$$\text{For } Re_p Sc < 10^4 \qquad Sh = \sqrt{4 + 1.21\, (Re_p Sc)^{0.67}} \tag{13.55}$$

$$\text{For } Re_p < 10^3 \qquad Sh = 2 + 0.6\, Re_p^{0.5} Sc^{0.33} \tag{13.56}$$

13.6.2 Spherical Particles in a Packed Bed

k_S in a packed bed depends on the properties of the liquid and the velocity of flow around the particles. For the range $10 < Re_p < 10^4$, the Sherwood number in packed beds has been correlated by the equation [23]:

$$Sh = 0.95\, Re_p^{0.5} Sc^{0.33} \tag{13.57}$$

13.7 EXPERIMENTAL ASPECTS

Applying diffusion–reaction theory to real biocatalysts requires prior measurement of several parameters. In this section we consider some experimental aspects of the analysis of heterogeneous reactions.

13.7.1 Observed Reaction Rate

Calculation of Weisz's modulus Φ and the external modulus Ω requires knowledge of $r_{A,obs}$, the observed rate of reaction per unit volume of catalyst. This information can be obtained as the rate at which substrate is converted in a batch reaction system using the methods described in Section 12.2.

When substrate levels change relatively slowly during reaction, batch rate data can be obtained by removing samples of the reaction mixture at various times and analysing for substrate concentration. However if substrate is consumed rapidly, or if oxygen uptake rates are required, continuous in situ monitoring is necessary. The equipment shown in Figure 13.16 is configured for measurement of oxygen consumption by immobilised cells or enzymes. Catalyst particles are packed into a column and liquid medium is recirculated through the packed bed. Initially, air or oxygen is sparged into a stirred vessel connected to the column and the dissolved oxygen tension is measured using an oxygen electrode (Section 10.7). The catalyst is allowed to reach steady state at a particular oxygen tension; the gas supply is then switched off, the system is closed, and the initial rate of oxygen uptake is recorded. Observed rates must be measured after the catalyst reaches steady state so that the results do not reflect rapidly changing transient conditions. The system must also be airtight when closed so that oxygen from the atmosphere does not enter the vessel and affect the dissolved oxygen readings.

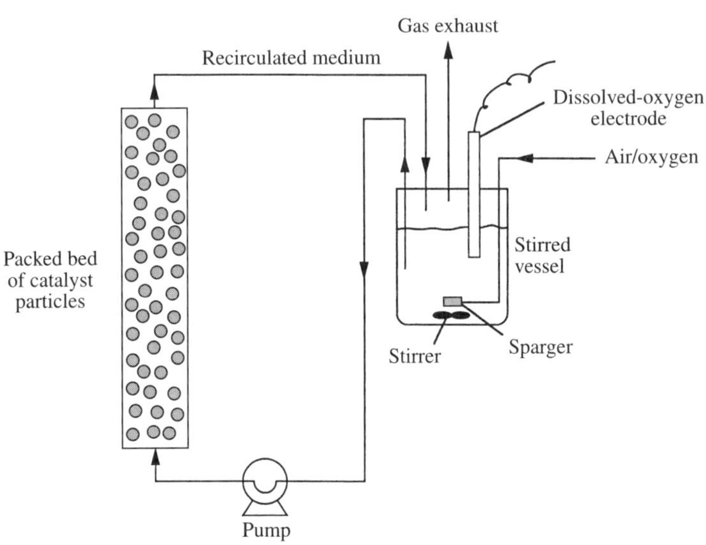

FIGURE 13.16 Batch recirculation reactor for measuring the rate of oxygen uptake by immobilised cells or enzymes.

For analysis of the rate data, the system of Figure 13.16 is assumed to be well mixed so that the oxygen tension measured by the electrode is the same everywhere in the bulk liquid. Actually, oxygen concentrations entering and leaving the packed bed will be different because oxygen is consumed by the catalyst. However, if only 1 to 2% of the oxygen entering the column is consumed during each pass of liquid through the bed, from a practical point of view, the system can be regarded as perfectly mixed. For this to occur, the recirculation flow rate must be sufficiently high [24]. The stirred apparatus shown in Figure 10.18 can also be used for measuring oxygen uptake rates by suspended catalyst particles. However, if minimisation of external mass transfer effects is required during the measurement, it is more difficult to achieve adequate slip velocities using particles in suspension than in the fixed-bed device of Figure 13.16. In oxygen uptake experiments, temperature control is very important as the solubility and mass transfer properties of oxygen are temperature dependent (Sections 10.6.4 and 10.8.2).

Observed reaction rates for catalytic reactions depend on the cell or enzyme loading in the catalyst. Therefore, the results of experimental measurements apply only to the cell or enzyme density tested. Increasing the amount of cells or enzyme in the catalyst will increase $r_{A,obs}$ and the likelihood of mass transfer restrictions.

13.7.2 Effective Diffusivity

The value of the effective diffusivity $\mathscr{D}_{Ae}$ reflects the ease with which compound A is transported within the catalyst matrix and is strongly dependent on the pore structure of the solid. Effective diffusivities are normally lower than corresponding molecular diffusivities in water because porous solids offer more resistance to diffusion. Some experimental $\mathscr{D}_{Ae}$ values for selected biocatalyst systems are listed in Table 13.8.

Techniques for measuring effective diffusivity are described in several papers [28–32, 35]; however, accurate measurement of $\mathscr{D}_{Ae}$ is difficult in most systems. During the measurement, external mass transfer limitations must be overcome using high liquid flow rates around the catalyst. If experimental values of $\mathscr{D}_{Ae}$ are higher than the molecular diffusion coefficient in water, this may indicate the presence of convective fluid currents within the catalyst [3–6]. As cells can pose a significant barrier to diffusion, $\mathscr{D}_{Ae}$ for immobilised cell preparations must be determined with the biomass present.

13.8 MINIMISING MASS TRANSFER EFFECTS

To improve overall rates of reaction in bioprocesses, mass transfer restrictions must be minimised or eliminated. In this section, we consider practical ways of achieving this objective based on the equations presented in Tables 13.4 and 13.6.

TABLE 13.8 Effective Diffusivity Values

Substance	Solid	Temperature	$\mathscr{D}_{Ae}(m^2\ s^{-1})$	$\frac{\mathscr{D}_{Ae}}{\mathscr{D}_{Aw}}$ [a]	Reference
Oxygen	Agar (2% w/v) containing *Candida lipolytica* cells	30°C	1.94×10^{-9}	0.70	[25]
	Microbial aggregates from domestic waste treatment plant	20°C	1.37×10^{-9}	0.62	[26]
	Trickling-filter slime	25°C	0.82×10^{-9}	–	[27]
Glucose	Microbial aggregates from domestic waste treatment plant	20°C	0.25×10^{-9}	0.37	[26]
	Glass fibre discs containing *Saccharomyces uvarum* cells	30°C	0.30×10^{-9}	0.43	[3]
	Calcium alginate (3 wt%)	30°C	0.62×10^{-9}	0.87	[28]
	Calcium alginate (2.4–2.8 wt%) containing 50 wt% bakers' yeast	30°C	0.26×10^{-9}	0.37	[28]
Sucrose	Calcium alginate (2% w/v)	25°C	0.48×10^{-9}	0.86	[29]
	Calcium alginate (2% w/v) containing 12.5% (v/v) *Catharanthus roseus* cells	25°C	0.14×10^{-9}	0.25	[29]
L-Tryptophan	Calcium alginate (2%)	30°C	0.67×10^{-9}	1.0	[30]
	κ-Carrageenan (4%)	30°C	0.58×10^{-9}	0.88	[31]
Lactate	Agar (1%) containing 1% Ehrlich ascites tumour cells	37°C	1.4×10^{-9}	0.97	[32]
	Agar (1%) containing 6% Ehrlich ascites tumour cells	37°C	0.7×10^{-9}	0.48	[32]
Ethanol	κ-Carrageenan (4%)	30°C	1.01×10^{-9}	0.92	[31]
	Calcium alginate (1.4–3.8 wt%)	30°C	1.25×10^{-9}	0.92	[28]
	Calcium alginate (2.4–2.8 wt%) containing 50 wt% bakers' yeast	30°C	0.45×10^{-9}	0.33	[28]
Nitrate	Compressed film of nitrifying organisms	–	1.4×10^{-9}	0.90	[26]
Ammonia	Compressed film of nitrifying organisms	–	1.3×10^{-9}	0.80	[26]
Bovine serum albumin	Agarose (5.5% v/v)	20°C	0.023×10^{-9}	0.38	[33]
	Agarose (7.2% v/v)	20°C	0.016×10^{-9}	0.28	[33]
Lysozyme	Cation exchange resin, pH 7	–	0.05×10^{-9}	0.45	[34]

[a] *$\mathscr{D}_{Aw}$ is the molecular diffusivity in water at the temperature of measurement.*

13.8.1 Minimising Internal Mass Transfer Effects

Internal mass transfer effects are eliminated when the internal effectiveness factor η_i is equal to 1. η_i approaches unity as the observable Thiele modulus Φ decreases. From Table 13.4, Φ is decreased by:

1. Reducing the observed reaction rate $r_{A,obs}$
2. Reducing the size of the catalyst
3. Increasing the effective diffusivity $\mathscr{D}_{Ae}$
4. Increasing the surface substrate concentration C_{As}

All of these changes impact directly on the effectiveness of substrate mass transfer within the particle.

Paradoxically, reducing the reaction rate $r_{A,obs}$ improves the effectiveness of mass transfer aimed at increasing the reaction rate. When the catalyst is very active with a high demand for substrate, mass transfer is likely to be slow relative to reaction, so that steep concentration gradients are produced. However, limiting the reaction rate by operating at suboptimum conditions or using an organism or enzyme with low intrinsic activity does not achieve the overall goal of higher conversion rates. Because $r_{A,obs}$ is the reaction rate per unit volume of catalyst, another way of reducing $r_{A,obs}$ is to reduce the cell or enzyme loading in the solid. This reduces the demand for substrate per particle so that mass transfer has a better chance of supplying substrate at a sufficient rate. Therefore, if the same mass of cells or enzyme is distributed between more particles, the rate of conversion will increase. The trade-off, however, is that using more particles may mean that a larger reactor is required.

Because Φ is proportional to the catalyst size squared (R^2 for spheres or b^2 for flat plates), reducing the catalyst size has a more dramatic effect on Φ than changes in any other variable. It is therefore a good way to improve the reaction rate. In principle, mass transfer limitations can be overcome completely if the particle size is decreased sufficiently. However, it is often extremely difficult in practice to reduce particle dimensions to this extent [36]. Even if mass transfer effects are eliminated, operation of large-scale reactors with tiny, highly compressible gel particles raises new problems. Some degree of internal mass transfer restriction must usually be tolerated.

13.8.2 Minimising External Mass Transfer Effects

External mass transfer effects decrease as the observable modulus Ω is reduced. From the equations in Table 13.6, this is achieved by:

1. Reducing the observed reaction rate $r_{A,obs}$
2. Reducing the size of the catalyst
3. Increasing the mass transfer coefficient k_S
4. Increasing the bulk substrate concentration C_{Ab}

Decreasing the catalyst size and increasing the mass transfer coefficient reduce the thickness of the boundary layer and facilitate external mass transfer. k_S is increased most readily by raising the liquid velocity outside of the catalyst. However, as k_S is a function of other system properties such as the liquid viscosity, density, and substrate diffusivity,

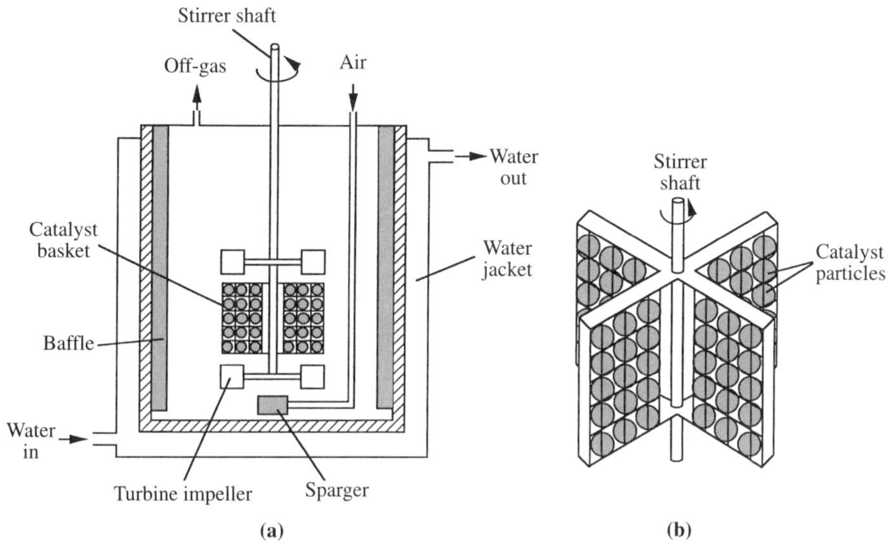

FIGURE 13.17 Spinning basket reactor for minimising external mass transfer effects: (a) reactor configuration; (b) detail of the spinning basket.
From D.G. Tajbl, J.B. Simons, and J.J. Carberry, 1966, Heterogeneous catalysis in a continuous stirred tank reactor. Ind. Eng. Chem. Fund. 5, 171–175

changes in these variables can also reduce boundary layer effects to some extent. External mass transfer is more rapid at high bulk substrate concentrations; the higher the concentration, the greater is the driving force for mass transfer across the boundary layer. By reducing the demand for substrate, decreasing $r_{A,obs}$ as described in the previous section also reduces external mass transfer limitations.

In large-scale reactors, problems with external mass transfer may be unavoidable if sufficiently high liquid velocities cannot be achieved. However, when evaluating biocatalyst kinetics in the laboratory, it is advisable to eliminate fluid boundary layers to simplify analysis of the data. Several laboratory reactor configurations allow almost complete elimination of interparticle and interphase concentration gradients [37]. Recycle reactors such as that shown in Figure 13.16 have been employed extensively for study of immobilised cell and enzyme reactions; operation with high liquid velocity through the bed reduces boundary layer effects. Another suitable configuration is the *spinning basket reactor* [38, 39] illustrated in Figure 13.17. By rotating the baskets at high speed, high relative velocities are achieved between the particles and the surrounding fluid. The slip velocities obtained in this apparatus are significantly greater than those found using freely suspended particles. Another laboratory design aimed at increasing the slip velocity is the stirred vessel shown in Figure 13.18. In this system, catalyst particles are held relatively stationary in a wire mesh cage while liquid is agitated at high speed.

The elimination of external mass transfer effects can be verified by calculating the observable parameter Ω as described in Section 13.5. However, if the mass transfer coefficient is not known accurately, an experimental test may be used instead [24]. Consider again the apparatus of Figure 13.16. If boundary layers around the particles affect the rate

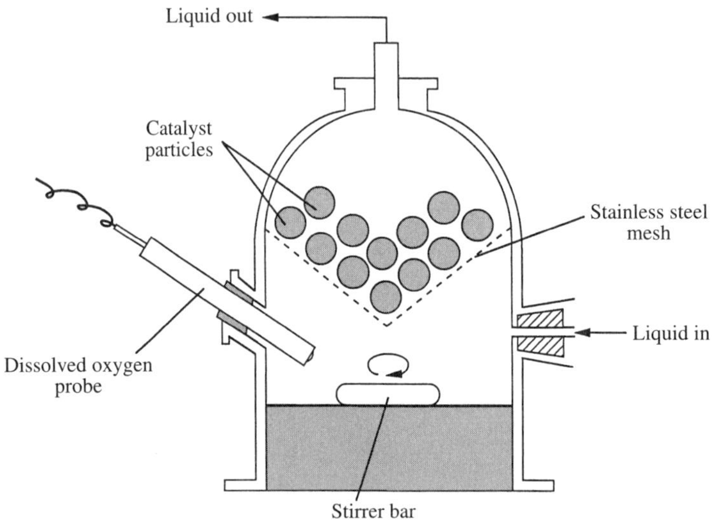

FIGURE 13.18 Stirred laboratory reactor for minimising external mass transfer effects.
From K. Sato and K. Toda, 1983, Oxygen uptake rate of immobilized growing Candida lipolytica. J. Ferment. Technol. *61, 239–245.*

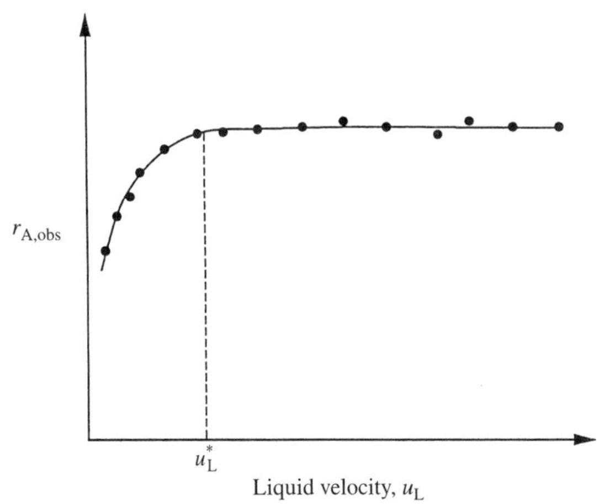

FIGURE 13.19 Relationship between observed reaction rate and external liquid velocity for reduction of external mass transfer effects.

of reaction, increasing the liquid velocity through the bed will improve conversion rates by reducing the boundary layer thickness and bringing C_{As} closer to C_{Ab}. At sufficiently high liquid velocity, external mass transfer effects may be removed. When this occurs, further increases in pump speed do not change the overall reaction rate, as illustrated in Figure 13.19. Therefore, if we can identify a liquid velocity u_L^* at which the reaction rate becomes independent of liquid velocity, operation at $u_L > u_L^*$ will ensure that $\eta_e = 1$ and external mass transfer effects are eliminated. For stirred reactors, a similar relationship holds between $r_{A,obs}$ and agitation speed.

13.9 EVALUATING TRUE KINETIC PARAMETERS

The intrinsic kinetics of zero-order, first-order, and Michaelis–Menten reactions are represented by the parameters k_0, k_1, and v_{max} and K_m. In general, it cannot be assumed that the values of these parameters will be the same before and after cell or enzyme immobilisation: significant changes can be wrought during the immobilisation process. As an example, Figure 13.20 shows Lineweaver–Burk plots (Section 12.4.2) for free and immobilised β-galactosidase enzyme. According to Eq. (12.43), the slopes and intercepts of the lines in Figure 13.20 indicate the values of K_m/v_{max} and $1/v_{max}$, respectively. Compared with the results shown for free enzyme, the steeper slopes and higher intercepts obtained for the immobilised enzyme indicate that immobilisation reduces v_{max}; this is a commonly observed result. The value of K_m can also be affected [9].

As described in Sections 12.3 and 12.4, kinetic parameters for homogeneous reactions can be determined directly from experimental rate data. However, evaluating the true kinetic parameters of immobilised cells and enzymes is somewhat more difficult. The observed rate of reaction is not the true rate at all points in the catalyst; mass transfer processes effectively 'mask' the true kinetic behaviour. Accordingly, v_{max} and K_m for immobilised catalysts cannot be estimated using the classical plots described in Section 12.4. Under the influence of mass transfer, these plots no longer give straight lines over the entire range of substrate concentration [40, 41].

As illustrated in Figure 13.20 for β-galactosidase, Lineweaver–Burk plots for immobilised enzymes are nonlinear. Often in such plots, however, the deviation from linearity is obscured by the scatter in real experimental data and the distortion of errors due to the Lineweaver–Burk linearisation (Sections 12.4.2 and 3.3.4). At low substrate concentrations (i.e., large values of $1/s$), the Lineweaver–Burk plot appears linear as the reaction exhibits

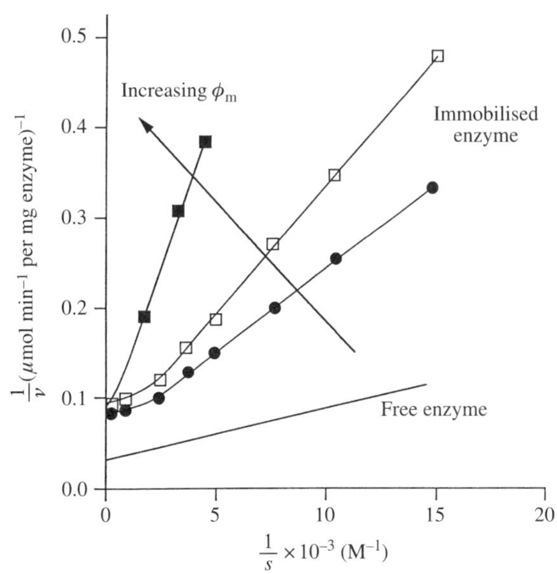

FIGURE 13.20 Lineweaver–Burk plots for free and immobilised β-galactosidase. Enzyme concentrations within the gel are: 0.10 mg ml^{-1} (●); 0.17 mg ml^{-1} (□); and 0.50 mg ml^{-1} (■). *Data from P.S. Bunting and K.J. Laidler, 1972, Kinetic studies on solid-supported β-galactosidase.* Biochemistry *11, 4477–4483.*

approximate first-order kinetics. By inference then, the apparent linearity of Lineweaver–Burk or similar plots cannot be considered sufficient evidence of the absence of mass transfer restrictions. Even if we mistakenly interpret one of the immobilised enzyme curves in Figure 13.20 as a straight line and evaluate the apparent values of v_{max} and K_m, we can check whether or not we have found the true kinetic parameters by changing the particle radius and substrate concentration over a wide range of values. True enzyme kinetic parameters do not vary with these conditions that affect mass transfer into the catalyst. Therefore, if the slope and/or intercept changes with the Thiele modulus, as indicated in Figure 13.20, it is soon evident that the system is subject to diffusional limitations. Studies have shown that the effects of diffusion are more pronounced in Eadie–Hofstee plots than in Lineweaver–Burk or Langmuir plots; however, all three for immobilised enzymes can be approximated by straight lines over certain intervals.

Although the Lineweaver–Burk plots for immobilised β-galactosidase are nonlinear in Figure 13.20, we should not conclude that the immobilised enzyme fails to obey Michaelis–Menten kinetics. The kinetic form of reactions is generally maintained after immobilisation of cells and enzymes [9]. The nonlinearity of the Lineweaver–Burk plots is due instead to the effect of mass transfer on the measured reaction rate.

Several methods have been proposed for determining v_{max} and K_m in heterogeneous catalysts [40–43]. The most straightforward approach is experimental: it involves reducing the particle size and catalyst loading and increasing the external liquid velocity to eliminate all mass transfer resistances. The measured rate data can then be analysed for kinetic parameters as if the reaction were homogeneous. However, because it is usually very difficult or impossible to completely remove intraparticle mass transfer effects, procedures involving a series of experiments coupled with theoretical analysis have also been proposed. In these methods, rate data are collected at high and low substrate concentrations using different particle sizes. At high substrate levels, it is assumed that the reaction is zero-order with $\eta_i = 1$; at low substrate concentrations, first-order kinetics are assumed. These assumptions simplify the analysis but may not always be valid.

When adequate computing facilities are available, true values of v_{max} and K_m can be extracted from diffusion-limited data using iterative calculations based on numerical integration and nonlinear regression [7, 9]. Many iteration loops may be required before convergence to the final parameter values.

13.10 GENERAL COMMENTS ON HETEROGENEOUS REACTIONS IN BIOPROCESSING

Before concluding this chapter, some general observations and rules of thumb for heterogeneous reactions are outlined.

- *Importance of oxygen mass transfer limitations* In aerobic reactions, mass transfer of oxygen is more likely to limit the rate of reaction than mass transfer of most other substrates. The reason is the poor solubility of oxygen in aqueous solutions. Whereas the sugar concentration in a typical fermentation broth is around 20 to 50 kg m^{-3}, the dissolved oxygen concentration under 1 atm of air pressure is limited to about

8×10^{-3} kg m^{-3}. Therefore, because C_{As} is so low for oxygen, the observable Thiele modulus Φ (refer to Table 13.4) can be several orders of magnitude greater than for other substrates. In anaerobic systems, the substrate most likely to be affected by mass transfer limitations is more difficult to identify.

- *Relationship between the internal effectiveness factor and the substrate concentration gradient* Depending on the reaction kinetics, the severity of intraparticle concentration gradients can be inferred from the value of the internal effectiveness factor. For first-order kinetics, because the reaction rate is proportional to the substrate concentration, $\eta_{i1} = 1$ implies that concentration gradients do not exist in the catalyst. Conversely, if $\eta_{i1} < 1$, we can conclude that concentration gradients are present. In contrast, for zero-order reactions, $\eta_{i0} = 1$ does not imply that gradients are absent because, as long as $C_A > 0$, the reaction rate is unaffected by any reduction in substrate concentration. Concentration gradients can be so steep that C_A is reduced to almost zero within the catalyst, but η_{i0} will remain equal to 1. On the other hand, $\eta_{i0} < 1$ implies that the concentration gradient is very severe and that some fraction of the particle volume is starved of substrate.
- *Relative importance of internal and external mass transfer limitations* For porous catalysts, it has been demonstrated with realistic values of mass transfer and diffusion parameters that external mass transfer limitations do not exist unless internal limitations are also present [44]. Concentration differences between the bulk liquid and external catalyst surface are never observed without larger internal gradients developing within the particle. On the other hand, if internal limitations are known to be present, external limitations may or may not be important depending on conditions. Significant external mass transfer effects can occur when reaction does not take place inside the catalyst, for example, if cells or enzymes are attached only to the exterior surface.
- *Operation of catalytic reactors* Certain solid-phase properties are desirable for operation of immobilised cell and enzyme reactors. For example, in packed bed reactors, large, rigid, and uniformly shaped particles promote well-distributed and stable liquid flow. Solids in packed columns should also have sufficient mechanical strength to withstand their own weight. These requirements are in direct conflict with those needed for rapid intraparticle mass transfer, as diffusion is facilitated in particles that are small, soft, and porous. Because blockages and large pressure drops through the bed must be avoided, mass transfer rates are usually compromised. In stirred reactors, soft, porous gels are readily destroyed at the agitation speeds needed to eliminate external boundary layer effects.
- *Product effects* Products formed by reaction inside catalysts must diffuse out under the influence of a concentration gradient between the interior of the catalyst and the bulk liquid. The concentration profile for product is the reverse of that for substrate: the concentration is highest at the centre of the catalyst and lowest in the bulk liquid. If activity of the cells or enzyme is affected by product inhibition, high intraparticle product concentrations may inhibit progress of the reaction. Immobilised enzymes that produce or consume H^+ ions are often affected in this way. Because enzyme reactions are very sensitive to pH (Section 12.3.4), small local variations in intraparticle pH due to slow diffusion of ions can have a significant influence on the reaction rate [45].

SUMMARY OF CHAPTER 13

At the end of Chapter 13 you should:

- Know what heterogeneous reactions are and where they occur in bioprocessing
- Understand the difference between *observed* and *true reaction rates*
- Know how concentration gradients arise in solid-phase catalysts
- Understand the concept of the *effectiveness factor*
- Be able to apply the *generalised Thiele modulus* and *observable Thiele modulus* to determine the effect of internal mass transfer on reaction rate
- Be able to quantify external mass transfer effects from measured data
- Know how to minimise internal and external mass transfer restrictions
- Understand that it is generally difficult to determine true kinetic parameters for heterogeneous reactions

PROBLEMS

13.1 Diffusion and reaction in a waste treatment lagoon

Industrial waste water is often treated in large shallow lagoons. Consider such a lagoon as shown in Figure 13P1.1 covering land of area A. Microorganisms form a sludge layer of thickness L at the bottom of the lagoon; this sludge remains essentially undisturbed by movement of the liquid. At steady state, waste water is fed into the lagoon so that the bulk concentration of digestible substrate in the liquid remains constant at s_b. Cells consume substrate diffusing into the sludge layer, thereby establishing a concentration gradient across thickness L. The microorganisms can be assumed to be distributed uniformly in the sludge. As indicated in Figure 13P1.1, distance from the bottom of the lagoon is measured by coordinate z.

(a) Set up a shell mass balance on substrate by considering a thin slice of sludge of thickness Δz perpendicular to the direction of diffusion. The rate of microbial reaction per unit volume of sludge is:

$$r_S = k_1 s$$

where s is the concentration of substrate in the sludge layer (gmol cm^{-3}) and k_1 is the first-order rate constant (s^{-1}). The effective diffusivity of substrate in the sludge is $\mathscr{D}_{Se}$. Obtain a differential equation relating s and z.

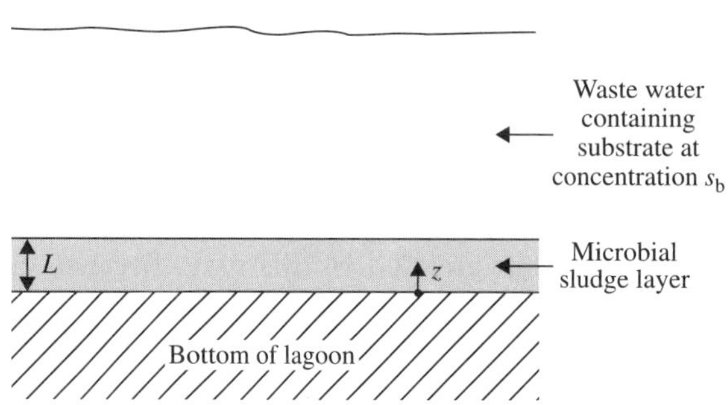

FIGURE 13P1.1 Lagoon for waste water treatment.

Hint: Area A is constant for flat-plate geometry and can be cancelled from all terms of the mass balance equation.

(b) External mass transfer effects at the liquid–sludge interface are negligible. What are the boundary conditions for this problem?

(c) The differential equation obtained in (a) is solved by making the substitution:

$$s = Ne^{pz}$$

where N and p are constants.

(i) Substitute this expression for s into the differential equation derived in (a) to obtain an equation for p. Remember that $\sqrt{p^2} = p$ or $-p$.

(ii) Because there are two possible values of p, let:

$$s = Ne^{pz} + Me^{-pz}$$

Apply the boundary condition at $z = 0$ to this expression to obtain a relationship between N and M.

(iii) Use the boundary condition at $z = L$ to find N and M explicitly. Obtain an expression for s as a function of z.

(iv) Use the definition of cosh x:

$$\cosh x = \frac{e^x + e^{-x}}{2}$$

to prove that:

$$\frac{s}{s_b} = \frac{\cosh\left(z\sqrt{k_1/\mathscr{D}_{Se}}\right)}{\cosh\left(L\sqrt{k_1/\mathscr{D}_{Se}}\right)}$$

(d) At steady state, the rate of substrate consumption must be equal to the rate at which substrate enters the sludge. As substrate enters the sludge by diffusion, the overall rate of reaction can be evaluated using Fick's law:

$$r_{A,obs} = \mathscr{D}_{Se}\, A \left.\frac{ds}{dz}\right|_{z=L}$$

where $\left.\frac{ds}{dz}\right|_{z=L}$ means ds/dz evaluated at $z = L$. Use the equation for s from (c) to derive an equation for $r_{A,obs}$.

Hint: The derivative of $\cosh(ax) = a \sinh(ax)$ where a is a constant and:

$$\sinh x = \frac{e^x - e^{-x}}{2}$$

(e) Show from the result of (d) that the internal effectiveness factor is given by the expression:

$$\eta_{i1} = \frac{\tanh \phi_1}{\phi_1}$$

where

$$\phi_1 = L\sqrt{\frac{k_1}{\mathcal{D}_{Se}}}$$

and

$$\tanh x = \frac{\sinh x}{\cosh x}$$

(f) Plot the substrate concentration profiles through a sludge layer of thickness 2 cm for the following sets of conditions:

	Condition		
	1	**2**	**3**
k_1 (s^{-1})	4.7×10^{-8}	2.0×10^{-7}	1.5×10^{-4}
$\mathcal{D}_{Se}$ ($cm^2\ s^{-1}$)	7.5×10^{-7}	2.0×10^{-7}	6.0×10^{-6}

Take s_b to be 10^{-5} gmol cm^{-3}. Label the profiles with corresponding values of ϕ_1 and η_{i1}. Comment on the general relationship between ϕ_1, the shape of the concentration profile, and the value of η_{i1}.

13.2 Oxygen concentration profile in an immobilised enzyme catalyst

L-Lactate 2-monooxygenase from *Mycobacterium smegmatis* is immobilised in spherical agarose beads. The enzyme catalyses the reaction:

$$\underset{\text{lactic acid}}{C_3H_6O_3} + O_2 \longrightarrow \underset{\text{acetic acid}}{C_2H_4O_2} + CO_2 + H_2O$$

Agarose beads 4 mm in diameter are immersed in a well-mixed solution containing 0.5 mM oxygen. A high lactic acid concentration is provided so that oxygen is the rate-limiting substrate. The effective diffusivity of oxygen in agarose is 2.1×10^{-9} $m^2\ s^{-1}$. K_m for the immobilised enzyme is 0.015 mM and v_{max} is 0.12 mol s^{-1} per kg of enzyme. The beads contain 0.012 kg of enzyme m^{-3} of gel. External mass transfer effects are negligible.

(a) Plot the oxygen concentration profile inside the beads.

(b) What fraction of the catalyst volume is active?

(c) Determine the largest bead size that allows the maximum overall rate of conversion.

13.3 Effect of oxygen transfer on recombinant cells

Recombinant *E. coli* cells contain a plasmid derived from pBR322 incorporating genes for the enzymes β-lactamase and catechol 2,3-dioxygenase from *Pseudomonas putida*. To produce the desired enzymes, the organism requires aerobic conditions. The cells are immobilised in spherical beads of carrageenan gel. The effective diffusivity of oxygen is 1.4×10^{-9} $m^2\ s^{-1}$. Oxygen uptake is zero-order with intrinsic rate constant 10^{-3} mol s^{-1} m^{-3} of particle. The concentration of oxygen at the surface of the catalyst is 8×10^{-3} kg m^{-3}. Cell growth is negligible.

(a) What is the maximum particle diameter for aerobic conditions throughout the catalyst particles?

(b) For particles half the diameter calculated in (a), what is the minimum oxygen concentration in the beads?

(c) The density of cells in the gel is reduced by a factor of five. If the specific activity of the cells is independent of cell loading, what is the maximum particle size for aerobic conditions?

13.4 Ammonia oxidation by immobilised cells

Thiosphaera pantotropha is being investigated for aerobic oxidation of ammonia to nitrite for waste water treatment. The organism is immobilised in spherical agarose particles of diameter 3 mm. The effective diffusivity of oxygen in the particles is 1.9×10^{-9} m^2 s^{-1}. The immobilised cells are placed in a flow chamber for measurement of oxygen uptake rate. Using published correlations, the liquid–solid mass transfer coefficient for oxygen is calculated as 6×10^{-5} m s^{-1}. When the bulk oxygen concentration is 6×10^{-3} kg m^{-3}, the observed rate of oxygen consumption is 2.2×10^{-5} kg s^{-1} m^{-3} of catalyst.

(a) What effect does external mass transfer have on the respiration rate?

(b) What is the effectiveness factor?

(c) For optimal activity of *T. pantotropha*, oxygen levels must be kept above the critical level, 1.2×10^{-3} kg m^{-3}. Is this condition satisfied?

13.5 Microcarrier culture and external mass transfer

Mammalian cells form a monolayer on the surface of microcarrier beads of diameter 120 μm and density 1.2×10^3 kg m^{-3}. The culture is maintained in spinner flasks in serum-free medium of viscosity 10^{-3} N s m^{-2} and density 10^3 kg m^{-3}. The diffusivity of oxygen in the medium is 2.3×10^{-9} m^2 s^{-1}. The observed rate of oxygen uptake is 0.015 mol s^{-1} m^{-3} at a bulk oxygen concentration of 0.2 mol m^{-3}. What effect does external mass transfer have on the reaction rate?

13.6 Immobilised enzyme reaction kinetics

Invertase catalyses the reaction:

$$\underset{\text{sucrose}}{C_{12}H_{22}O_{11}} + H_2O \longrightarrow \underset{\text{glucose}}{C_6H_{12}O_6} + \underset{\text{fructose}}{C_6H_{12}O_6}$$

Invertase from *Aspergillus oryzae* is immobilised in porous resin particles of diameter 1.6 mm at a density of 0.1 μmol of enzyme g^{-1}. The effective diffusivity of sucrose in the resin is 1.3×10^{-11} m^2 s^{-1}. The resin is placed in a spinning basket reactor operated so that external mass transfer effects are eliminated. At a sucrose concentration of 0.85 kg m^{-3}, the observed rate of conversion is 1.25×10^{-3} kg s^{-1} m^{-3} of resin. K_m for the immobilised enzyme is 3.5 kg m^{-3}.

(a) Calculate the effectiveness factor.

(b) Determine the true first-order reaction constant for immobilised invertase.

(c) Assume that the specific enzyme activity is not affected by steric hindrance or conformational changes as the enzyme loading increases. This means that k_1 should be directly proportional to the enzyme concentration in the resin. Plot the changes in effectiveness factor and observed reaction rate as a function of enzyme loading from 0.01 μmol g^{-1} to 2.0 μmol g^{-1}. Comment on the relative benefit of increasing the concentration of enzyme in the resin.

13.7 Mass transfer effects in plant cell culture

Suspended *Catharanthus roseus* cells form spherical aggregates approximately 1.5 mm in diameter. Oxygen uptake is measured using the apparatus of Figure 13.16; medium is recirculated with a superficial liquid velocity of 0.83 cm s^{-1}. At a bulk concentration of 8 mg l^{-1}, oxygen is consumed at a rate of 0.28 mg per g wet weight of cells per hour. Assume that the density and viscosity of the medium are similar to water and the specific gravity of wet cells is 1. The effective diffusivity of oxygen in the aggregates is 9×10^{-6} cm^2 s^{-1}, or half that in the medium. Oxygen uptake follows zero-order kinetics.

(a) Does external mass transfer affect the oxygen uptake rate?

(b) To what extent does internal mass transfer affect oxygen uptake?

(c) Roughly, what would you expect the profile of oxygen concentration to be within the aggregates?

13.8 Respiration in mycelial pellets

Aspergillus niger cells are observed to form self-immobilised aggregates of average diameter 5 mm. The effective diffusivity of oxygen in the aggregates is 1.75×10^{-9} m^2 s^{-1}. In a fixed-bed reactor, the oxygen consumption rate at a bulk oxygen concentration of 8×10^{-3} kg m^{-3} is 8.7×10^{-5} kg s^{-1} m^{-3} of biomass. The liquid–solid mass transfer coefficient is 3.8×10^{-5} m s^{-1}.

(a) Is oxygen uptake affected by external mass transfer?

(b) What is the external effectiveness factor?

(c) What reaction rate would be observed if both internal and external mass transfer resistances were eliminated?

(d) If only external mass transfer effects were removed, what would be the reaction rate?

13.9 Effect of mass transfer on glyphosate removal

Cultured soybean (*Glycine max*) cells are immobilised in spherical agarose beads of average radius 3.5 mm. In phytoremediation experiments, the beads are placed in a column reactor and medium containing 60 mg l^{-1} glyphosate is recirculated through the bed using a peristaltic pump. Consumption of glyphosate is considered to be a first-order reaction. The rate of glyphosate uptake is measured as a function of liquid recirculation rate.

Liquid Flow Rate (cm^3 s^{-1})	Glyphosate Uptake Rate per Unit Volume of Catalyst (kg s^{-1} m^{-3})
10	8.3×10^{-6}
15	1.9×10^{-5}
20	3.6×10^{-5}
25	6.1×10^{-5}
30	8.3×10^{-5}
35	8.4×10^{-5}
45	8.4×10^{-5}
55	8.4×10^{-5}

(a) What is the minimum liquid flow rate required to eliminate external mass transfer effects? Explain your answer.

(b) The immobilised cell reactor is operated using a recirculation flow rate of 40 $cm^3 s^{-1}$. If the effective diffusivity of glyphosate in the beads is 0.9×10^{-9} $m^2 s^{-1}$, estimate the maximum volumetric rate of glyphosate uptake for this system in the absence of mass transfer limitations.

13.10 Mass transfer effects in tissue-engineered cartilage

A layer of cartilage 3 mm thick is produced using human chondrocytes seeded at high density into a polymer mesh scaffold. The scaffold is attached to a nonporous polymer base simulating the mechanical properties of bone. Oxygen is provided to the cells by sparging nutrient medium with air in a separate vessel and bathing the surface of the cartilage tissue with the medium in a perfusion bioreactor. The concentration of dissolved oxygen in the medium is maintained at 7×10^{-3} kg m^{-3} and the effective diffusivity of oxygen in cartilage is 1.0×10^{-9} $m^2 s^{-1}$. The rate of oxygen consumption by the chondrocytes is measured as 6.6×10^{-6} kg s^{-1} m^{-3} of cartilage. The mass transfer coefficient for transport of oxygen through the liquid boundary layer at the cartilage surface is 2×10^{-5} m s^{-1}. Oxygen uptake follows zero-order kinetics.

(a) Is the cartilage culture affected by oxygen mass transfer limitations?

(b) Estimate the oxygen uptake rate in the absence of internal and external mass transfer effects.

(c) What is the lowest oxygen concentration within the cartilage tissue?

(d) The flow rate of medium over the tissue is increased so that any boundary layers at the surface of the cartilage are eliminated. The thickness of the cartilage layer is also reduced to 2 mm while keeping the chondrocyte density constant. Estimate the culture oxygen demand under these conditions. (Iterative solution may be required.)

(e) When external boundary layers are removed, what maximum thickness of cartilage can be used while avoiding oxygen depletion in the tissue?

13.11 Oxygen uptake by immobilised bacteria

Bacteria are immobilised in spherical agarose beads of diameter 3 mm. The effective diffusivity of oxygen in the beads is 1.9×10^{-9} $m^2 s^{-1}$. The immobilised cells are placed in a stirred bioreactor and the rate of oxygen uptake is measured as 2.2×10^{-5} kg s^{-1} m^{-3} of catalyst at a bulk oxygen concentration of 6×10^{-3} kg m^{-3}. At the operating stirrer speed in the bioreactor, the liquid–solid mass transfer coefficient is estimated as 6×10^{-5} m s^{-1}. Oxygen uptake can be considered a zero-order reaction.

(a) What effect does external mass transfer have on the rate of oxygen uptake?

(b) What are the values of the internal, external, and total effectiveness factors?

(c) The oxygen concentration within the particles must be kept above 4×10^{-4} kg m^{-3} for optimal culture performance. Is this condition satisfied?

13.12 Three-dimensional culture of mesenchymal stem cells

Human mesenchymal stem cells isolated from adipose tissue are immobilised in spherical alginate beads of diameter 3.5 mm. The beads are placed in a column reactor similar to that shown in Figure 13.16. The pump is set to deliver a medium recirculation flow rate of $0.033\ l\ s^{-1}$. At this liquid flow rate, the mass transfer coefficient for external oxygen transfer is calculated using literature correlations for spherical particles in a packed bed as $3.6 \times 10^{-5}\ m\ s^{-1}$. The effective diffusivity of oxygen in the alginate beads containing cells is $9 \times 10^{-6}\ cm^2\ s^{-1}$, and the bulk dissolved oxygen concentration is $8 \times 10^{-3}\ kg\ m^{-3}$. The rate of oxygen consumption per unit volume of catalyst is measured using the dissolved oxygen electrode as $7.6 \times 10^{-5}\ kg\ s^{-1}\ m^{-3}$. Consumption of oxygen by the cells can be considered a zero-order reaction.

(a) Is the rate of oxygen uptake affected by external mass transfer effects?

(b) Is the rate of oxygen uptake affected by internal mass transfer effects?

(c) Are cells at the centre of the beads supplied with oxygen?

(d) What is the maximum oxygen consumption rate that could be achieved using this immobilised cell system if the dissolved oxygen concentration in the bulk liquid, $8 \times 10^{-3}\ kg\ m^{-3}$, were present at all points within all of the beads?

(e) In the absence of an external boundary layer, what is the maximum bead diameter that supports a finite (>0) concentration of oxygen at all points in the catalyst?

13.13 Diffusion and reaction of glucose and oxygen

Microbial cells are immobilised in spherical gel beads and cultured in a batch reactor under aerobic conditions. The medium used contains $20\ g\ l^{-1}$ of glucose. The effective diffusivity of glucose in the beads is $0.42 \times 10^{-9}\ m^2\ s^{-1}$; the effective diffusivity of oxygen is $1.8 \times 10^{-9}\ m^2\ s^{-1}$. Oxygen uptake follows zero-order kinetics; as the culture progresses and the concentration of glucose in the medium is reduced to low levels, glucose uptake can be considered a first-order reaction. The gas–liquid mass transfer capacity of the reactor is such that the dissolved oxygen concentration in the medium is maintained at the saturation value of $8 \times 10^{-3}\ kg\ m^{-3}$ throughout the culture period. According to the stoichiometry of growth for this culture, for every mole of glucose consumed, 2.7 moles of oxygen are required. Oxygen uptake is strongly limited by internal mass transfer effects; however, vigorous mixing in the reactor is effective in eliminating external boundary layers. Estimate the medium glucose concentration that must be reached before mass transfer of glucose exerts a limiting effect on the reaction rate similar to that prevailing for oxygen. What are the implications of your answer for operation of immobilised cell reactors?

13.14 Uptake of growth factor by neural stem cell spheroids

Neural stem cells cultured in vitro form compact spheroids of diameter 500 μm and density $1.15\ g\ cm^{-3}$. The spheroids are suspended and gently agitated in nutrient medium of viscosity 1 cP and density $1\ g\ cm^{-3}$. The stem cells consume complex growth factor proteins that are provided in the medium at a concentration of $10\ ng\ ml^{-1}$. The diffusivity of growth factor in aqueous solution is $3.3 \times 10^{-6}\ cm^2\ s^{-1}$; the effective diffusivity of growth factor in the spheroids is $5 \times 10^{-11}\ m^2\ s^{-1}$. When 10 mg of spheroids are cultured in 1 ml of nutrient medium, the medium is exhausted of growth factor after 15 hours. Cell growth during this period is negligible. Utilisation of growth factor by the cells can be considered to follow first-order kinetics.

(a) Does external mass transfer affect the rate of growth factor uptake?

(b) Does internal mass transfer affect the rate of growth factor uptake?

(c) By what factor would the rate of growth factor consumption be increased in the absence of all mass transfer resistances?

13.15 Maximum reaction rate for immobilised enzyme

Penicillin-G amidase is immobilised in commercial macroporous carrier beads of diameter 100 μm. The immobilised enzyme is used for large-scale penicillin hydrolysis in a well-mixed enzyme reactor that eliminates external boundary layer effects. Penicillin-G substrate is provided as a 2.68-mM solution. The molecular diffusivity of penicillin-G in water is 4.0×10^{-6} cm^2 s^{-1}; the effective diffusivity in the carrier beads is estimated to be 45% of that value. The observed rate of penicillin-G conversion is 125 U cm^{-3} of catalyst, where 1 U is defined as 1 μmol min^{-1}.

(a) By what factor could the rate of penicillin-G conversion be increased if internal mass transfer effects were eliminated?

(b) The Michaelis constant K_m for freely suspended penicillin-G amidase is 13 μM. In the absence of information about the kinetic parameters after immobilisation, this value is assumed to apply to the immobilised enzyme. Estimate the maximum rate of conversion of penicillin-G that could be achieved in the absence of mass transfer effects.

References

[1] S.F. Karel, S.B. Libicki, C.R. Robertson, The immobilization of whole cells: engineering principles, Chem. Eng. Sci. 40 (1985) 1321–1354.

[2] R. Wittler, H. Baumgartl, D.W. Lübbers, K. Schügerl, Investigations of oxygen transfer into *Penicillium chrysogenum* pellets by microprobe measurements, Biotechnol. Bioeng. 28 (1986) 1024–1036.

[3] V. Bringi, B.E. Dale, Experimental and theoretical evidence for convective nutrient transport in an immobilized cell support, Biotechnol. Prog. 6 (1990) 205–209.

[4] A. Nir, L.M. Pismen, Simultaneous intraparticle forced convection, diffusion and reaction in a porous catalyst, Chem. Eng. Sci. 32 (1977) 35–41.

[5] A.E. Rodrigues, J.M. Orfao, A. Zoulalian, Intraparticle convection, diffusion and zero order reaction in porous catalysts, Chem. Eng. Commun. 27 (1984) 327–337.

[6] G. Stephanopoulos, K. Tsiveriotis, The effect of intraparticle convection on nutrient transport in porous biological pellets, Chem. Eng. Sci. 44 (1989) 2031–2039.

[7] K. van't Riet, J. Tramper, Basic Bioreactor Design, Marcel Dekker, 1991.

[8] C.M. Hooijmans, S.G.M. Geraats, E.W.J. van Neil, L.A. Robertson, J.J. Heijnen, K.Ch.A.M. Luyben, Determination of growth and coupled nitrification/denitrification by immobilized *Thiosphaera pantotropha* using measurement and modeling of oxygen profiles, Biotechnol. Bioeng. 36 (1990) 931–939.

[9] C.M. Hooijmans, S.G.M. Geraats, K.Ch.A.M. Luyben, Use of an oxygen microsensor for the determination of intrinsic kinetic parameters of an immobilized oxygen reducing enzyme, Biotechnol. Bioeng. 35 (1990) 1078–1087.

[10] D. de Beer, J.C. van den Heuvel, Gradients in immobilized biological systems, Anal. Chim. Acta 213 (1988) 259–265.

[11] E.W. Thiele, Relation between catalytic activity and size of particle, Ind. Eng. Chem. 31 (1939) 916–920.

[12] R. Aris, A normalization for the Thiele modulus, Ind. Eng. Chem. Fund. 4 (1965) 227–229.

[13] K.B. Bischoff, Effectiveness factors for general reaction rate forms, AIChE J. 11 (1965) 351–355.

[14] G.F. Froment, K.B. Bischoff, J. De Wilde, Chemical Reactor Analysis and Design, third ed., John Wiley, 2010 (Chapter 3).

[15] R. Aris, Introduction to the Analysis of Chemical Reactors, Prentice Hall, 1965.

[16] R. Aris, Mathematical Theory of Diffusion and Reaction in Permeable Catalysts, vol. 1, Oxford University Press, 1975.

[17] M. Moo-Young, T. Kobayashi, Effectiveness factors for immobilized-enzyme reactions, Can. J. Chem. Eng. 50 (1972) 162–167.

[18] P.B. Weisz, Diffusion and chemical transformation: an interdisciplinary excursion, Science 179 (1973) 433–440.
[19] T.K. Sherwood, R.L. Pigford, C.R. Wilke, Mass Transfer, McGraw-Hill, 1975 (Chapter 6).
[20] W.L. McCabe, J.C. Smith, P. Harriott, Unit Operations of Chemical Engineering, seventh ed., McGraw-Hill, 2005 (Section IV).
[21] P.L.T. Brian, H.B. Hales, Effects of transpiration and changing diameter on heat and mass transfer to spheres, AIChE J. 15 (1969) 419–425.
[22] W.E. Ranz, W.R. Marshall, Evaporation from drops. Parts I and II, Chem. Eng. Prog. 48 (1952) 141–146, 173–180.
[23] M. Moo-Young, H.W. Blanch, Design of biochemical reactors: mass transfer criteria for simple and complex systems, Adv. Biochem. Eng. 19 (1981) 1–69.
[24] J.R. Ford, A.H. Lambert, W. Cohen, R.P. Chambers, Recirculation reactor system for kinetic studies of immobilized enzymes, Biotechnol. Bioeng. Symp. 3 (1972) 267–284.
[25] K. Sato, K. Toda, Oxygen uptake rate of immobilized growing *Candida lipolytica*, J. Ferment. Technol. 61 (1983) 239–245.
[26] J.V. Matson, W.G. Characklis, Diffusion into microbial aggregates, Water Res. 10 (1976) 877–885.
[27] Y.S. Chen, H.R. Bungay, Microelectrode studies of oxygen transfer in trickling filter slimes, Biotechnol. Bioeng. 23 (1981) 781–792.
[28] A. Axelsson, B. Persson, Determination of effective diffusion coefficients in calcium alginate gel plates with varying yeast cell content, Appl. Biochem. Biotechnol. 18 (1988) 231–250.
[29] H.T. Pu, R.Y.K. Yang, Diffusion of sucrose and yohimbine in calcium alginate gel beads with or without entrapped plant cells, Biotechnol. Bioeng. 32 (1988) 891–896.
[30] H. Tanaka, M. Matsumura, I.A. Veliky, Diffusion characteristics of substrates in Ca-alginate gel beads, Biotechnol. Bioeng. 26 (1984) 53–58.
[31] C.D. Scott, C.A. Woodward, J.E. Thompson, Solute diffusion in biocatalyst gel beads containing biocatalysis and other additives, Enzyme Microb. Technol. 11 (1989) 258–263.
[32] T.J. Chresand, B.E. Dale, S.L. Hanson, R.J. Gillies, A stirred bath technique for diffusivity measurements in cell matrices, Biotechnol. Bioeng. 32 (1988) 1029–1036.
[33] E.M. Johnson, D.A. Berk, R.K. Jain, W.M. Deen, Hindered diffusion in agarose gels: test of effective medium model, Biophys. J. 70 (1996) 1017–1026.
[34] S.R. Dziennik, E.B. Belcher, G.A. Barker, M.J. DeBergalis, S.E. Fernandez, A.M. Lenhoff, Nondiffusive mechanisms enhance protein uptake rates in ion exchange particles, Proc. Natl. Acad. Sci. USA 100 (2003) 420–425.
[35] S.H. Omar, Oxygen diffusion through gels employed for immobilization: Parts 1 and 2, Appl. Microbiol. Biotechnol. 40 (1993) 1–6, 173–181.
[36] B.J. Rovito, J.R. Kittrell, Film and pore diffusion studies with immobilized glucose oxidase, Biotechnol. Bioeng. 15 (1973) 143–161.
[37] Y.T. Shah, Gas–Liquid–Solid Reactor Design, McGraw-Hill, 1979.
[38] J.J. Carberry, Designing laboratory catalytic reactors, Ind. Eng. Chem. 56 (1964) 39–46.
[39] D.G. Tajbl, J.B. Simons, J.J. Carberry, Heterogeneous catalysis in a continuous stirred tank reactor, Ind. Eng. Chem. Fund. 5 (1966) 171–175.
[40] B.K. Hamilton, C.R. Gardner, C.K. Colton, Effect of diffusional limitations on Lineweaver–Burk plots for immobilized enzymes, AIChE J. 20 (1974) 503–510.
[41] J.-M. Engasser, C. Horvath, Effect of internal diffusion in heterogeneous enzyme systems: evaluation of true kinetic parameters and substrate diffusivity, J. Theor. Biol. 42 (1973) 137–155.
[42] D.S. Clark, J.E. Bailey, Structure–function relationships in immobilized chymotrypsin catalysis, Biotechnol. Bioeng. 25 (1983) 1027–1047.
[43] G.K. Lee, R.A. Lesch, P.J. Reilly, Estimation of intrinsic kinetic constants for pore diffusion-limited immobilized enzyme reactions, Biotechnol. Bioeng. 23 (1981) 487–497.
[44] E.E. Petersen, Chemical Reaction Analysis, Prentice Hall, 1965.
[45] P.S. Stewart, C.R. Robertson, Product inhibition of immobilized *Escherichia coli* arising from mass transfer limitation, App. Environ. Microbiol. 54 (1988) 2464–2471.

Suggestions for Further Reading

Immobilised Cells and Enzymes

Buchholz, K., Kasche, V., & Bornscheuer, U. T. (2005). *Biocatalysts and Enzyme Technology*. Wiley-VCH.

de Bont, J. A. M., Visser, J., Mattiasson, B., & Tramper, J. (Eds.), (1990). *Physiology of Immobilized Cells,* Elsevier.

Guisan, J. M. (Ed.), (2010). *Immobilization of Enzymes and Cells,* Humana Press.

Katchalski-Katzir, E. (1993). Immobilized enzymes—learning from past successes and failures. *Trends Biotechnol., 11*, 471–478.

Nedović, V., & Willaert, R. (2010). *Fundamentals of Cell Immobilisation Biotechnology*. Kluwer Academic.

Tischer, W., & Kasche, V. (1999). Immobilized enzymes: crystals or carriers? *Trends Biotechnol., 17*, 326–335.

Webb, C., & Dervakos, G. A. (1996). *Studies in Viable Cell Immobilization*. Academic Press.

Engineering Analysis of Mass Transfer and Reaction

See also references [1], [7], [14], and [40–43].

Engasser, J.-M., & Horvath, C. (1976). Diffusion and kinetics with immobilized enzymes. *Appl. Biochem. Biotechnol., 1*, 127–220.

Satterfield, C. N. (1970). *Mass Transfer in Heterogeneous Catalysis*. MIT Press.

C H A P T E R

14

Reactor Engineering

The reactor is the heart of any fermentation or enzyme conversion process. Designing bioreactors is a complex task, relying on scientific and engineering principles and many rules of thumb. Decisions made in reactor design have a significant impact on overall process performance, yet there are no simple or standard design procedures available for specifying all aspects of the vessel and its operation. Knowledge of reaction kinetics is essential for understanding how biological reactors work. Other areas of bioprocess engineering, such as mass and energy balances, mixing, mass transfer, and heat transfer, are also required.

The performance of bioreactors depends on several key aspects of their design and operation.

- *Reactor configuration.* For example, should the reactor be a stirred tank or an air-driven vessel without mechanical agitation?
- *Reactor size.* What size reactor is required to achieve the desired rate of production?
- *Mode of operation.* Will the reactor be operated batch-wise or as a continuous flow process? Should substrate be fed intermittently? Should the reactor be operated alone or in series with others?
- *Processing conditions inside the reactor.* What reaction conditions, such as temperature, pH, and dissolved oxygen tension, should be maintained in the vessel? How will these parameters be monitored and controlled to optimise process performance? How will contamination be avoided?

This chapter brings together many of the principles covered in the previous chapters of this book. Here, we consider a range of bioreactor configurations and their applications. Mass balance techniques will be used to predict the outcome of enzyme and cellular reactions for different modes of equipment operation. The design of sterilisation systems is included as a critical element of industrial bioreactor facilities. We will also consider the broader economic and environmental aspects of bioreactor design and operation and the implications for sustainable bioprocessing.

14.1 BIOREACTOR ENGINEERING IN PERSPECTIVE

Before starting to design a bioreactor, some objectives have to be defined. Simple aims like 'Produce 1 g of monoclonal antibody per day', or 'Produce 10,000 tonnes of amino acid per year', provide the starting point. Other objectives are also relevant: in industrial processes, the product should be made at the lowest possible cost to maximise the company's commercial advantage. In some cases, economic objectives are overridden by safety concerns, the requirement for high product purity, regulatory considerations, or the need to minimise environmental impact. The final reactor design will be a reflection of all these process requirements and, in most cases, represents a compromise solution to conflicting demands.

In this section, we will consider the various contributions to bioprocessing costs for different types of product, and examine the importance of reactor engineering in improving overall process performance. As shown in Figure 14.1, the value of products made by bioprocessing covers a wide range. Typically, products with the highest value are those from mammalian cell culture, such as therapeutic proteins and monoclonal antibodies. At the opposite end of the scale is treatment of waste, where the overriding objective is minimal financial outlay for the desired level of purity. To reduce the cost of any bioprocess, it is first necessary to identify which aspects of it are cost-determining. The breakdown of production costs varies from process to process; however, a general scheme is shown in Figure 14.2. The following components are important:

1. Research and development
2. The fermentation or reaction step
3. Downstream processing, including waste treatment or disposal
4. Administration and marketing

In most bioprocesses, the cost of administration and marketing (4) is relatively small. Products for which the cost of reaction (2) dominates include biomass such as bakers'

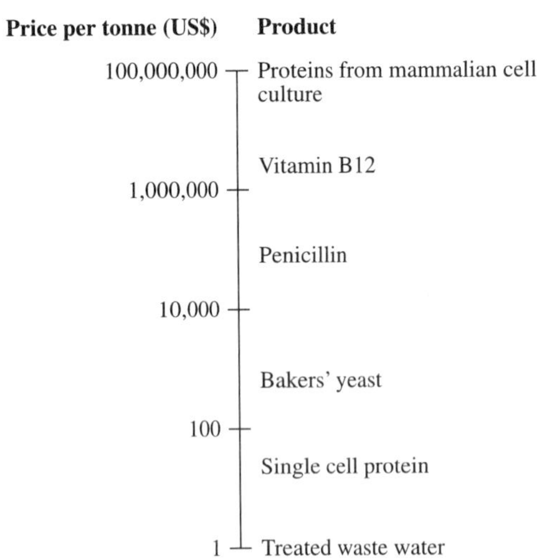

FIGURE 14.1 Range of value of fermentation products. *From P.N. Royce, 1993, A discussion of recent developments in fermentation monitoring and control from a practical perspective.* Crit. Rev. Biotechnol. *13*, 117–149.

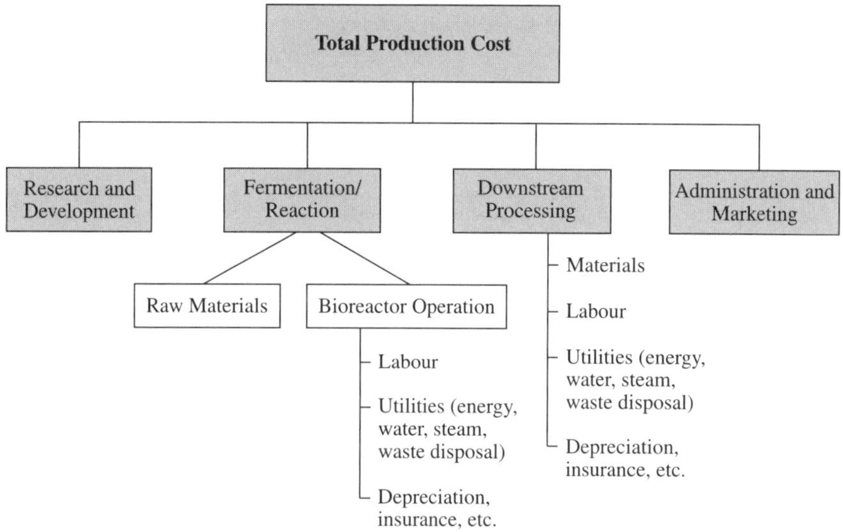

FIGURE 14.2 Contributions to total production cost in bioprocessing.

yeast and single cell protein, catabolic metabolites such as ethanol and lactic acid, and bioconversion products such as high-fructose corn syrup and 6-aminopenicillanic acid. Intracellular products such as proteins have high downstream processing costs (3) compared with reaction; other examples in this category are antibiotics, vitamins, and amino acids. For new, high-value biotechnology products such as recombinant proteins and antibodies, actual processing costs (2 and 3) are only a small part of the total because of the enormous investment required for research and development (1), including regulatory approval. Getting the product into the marketplace quickly is the most important cost-saving measure in these cases; any savings made by improving the efficiency of the reactor are generally trivial in comparison. Nevertheless, for the majority of fermentation products outside of this new, high-value category, bioprocessing costs make a significant contribution to the final price.

If the reaction step (2) dominates the cost structure, as indicated in Figure 14.2 this may be because of the high cost of the raw materials required or the high cost of bioreactor operation. The relative contribution of these two factors depends on the process and, to some extent, the geographical location of the production facility, as labour and energy costs vary from place to place. In general, however, for production of high-value, low-yield products such as antibiotics, vitamins, enzymes, and pigments, the raw materials or fermentation media typically represent 60 to 90% of the reaction cost [1]. For low-value, high-yield metabolites such as ethanol, citric acid, biomass, and lactic acid, raw material costs range from 40% of the cost of reaction for citric acid to about 70% for ethanol produced from molasses [1, 2].

As indicated in Figure 14.3, identifying the cost structure of bioprocesses assists in defining the objectives for bioreactor design. Even if bioreactor operation itself is not cost-determining, aspects of reactor design may still be important. If the cost of research

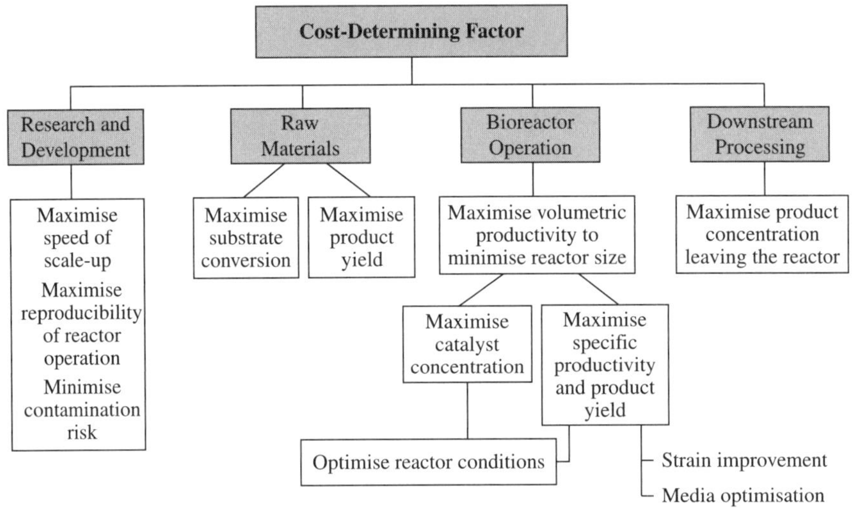

FIGURE 14.3 Strategies for bioreactor design after identification of the cost-determining factors in the process.

and development dominates, design of the reactor is directed toward achieving rapid scale-up; this is more important than maximising conversion or minimising operating costs. The implications for bioreactor design in this situation are discussed further in Section 14.7.4. For new biotechnology products intended for therapeutic use, regulatory guidelines require that the entire production scheme be validated and process control guaranteed for consistent quality and safety; the reproducibility of reactor operation is therefore critical. When the cost of raw materials is significant, maximising the conversion of substrate and yield of product in the reactor has high priority. If downstream processing is expensive, the reactor is designed and operated to maximise the product concentration leaving the vessel; this cuts the cost of recovering the product from dilute solutions. When bioreactor operation dominates the cost structure, the reactor should be as small as possible as this reduces both the capital cost of the equipment and the cost of day-to-day operations.

If analysis of the process indicates that a small bioreactor is required to minimise the most significant costs, to achieve the desired total production rate using a small vessel, the volumetric productivity of the reactor must be sufficiently high (see Section 12.1.3). Volumetric productivity depends on the concentration of the catalyst and its specific rate of production. Therefore, as indicated in Figure 14.3, to achieve high volumetric rates, the reactor must allow maximum catalyst activity at the highest practical catalyst concentration. For tightly packed cells or cell organelles, the physical limit on concentration is of the order of 200 kg of dry weight m^{-3}; for enzymes in solution, the maximum concentration depends on the solubility of enzyme in the reaction mixture. The extent to which these limiting concentrations can be approached depends on how well the reactor is performing. For example, if mixing or mass transfer is inadequate, oxygen or nutrient starvation will reduce the maximum cell density that can be achieved. Alternatively, if shear levels in the reactor are too high, cells will be disrupted and enzymes inactivated so that the effective concentration of catalyst is reduced.

Maximum specific productivity is obtained when the catalyst is capable of high rates of production and conditions in the reactor allow the best possible catalytic function. For simple metabolites such as ethanol, butanol, and acetic acid that are linked to energy production in the cell, the maximum theoretical yield is limited by the thermodynamic and stoichiometric principles outlined in Section 4.6. Accordingly, there is little scope for increasing the production titres of these materials; reduced production costs and commercial advantage rely mostly on improvements in reactor operation that allow the system to achieve close to the maximum theoretical yield. In contrast, it is not unusual for strain improvement and medium optimisation programs to improve the yields of antibiotics and enzymes by over 100-fold, particularly during the early stages of process development. Therefore, for these products, identification of high-producing strains and optimal environmental conditions is initially more rewarding than improving the bioreactor design and operation.

14.2 BIOREACTOR CONFIGURATIONS

The cylindrical tank, either stirred or unstirred, is the most common reactor in bioprocessing. Yet, a vast array of fermenter configurations is in use in different bioprocess industries. Novel bioreactors are constantly being developed for special applications and new forms of biocatalyst. Much of the challenge in reactor design lies in the provision of adequate mixing and aeration for the large number of fermentations that require oxygen. In contrast, reactors for anaerobic culture are often very simple in construction without sparging or agitation. In the following discussion of bioreactor configurations, we will assume that aerobic operation is required.

14.2.1 Stirred Tank

A conventional stirred, aerated bioreactor is shown schematically in Figure 14.4. Mixing and bubble dispersion are achieved by mechanical agitation; this requires a relatively high input of energy per unit volume. Baffles are used in stirred reactors to reduce vortexing. A wide variety of impeller sizes and shapes is available to produce different flow patterns inside the vessel; in tall fermenters, installation of multiple impellers improves mixing. The mixing and mass transfer functions of stirred reactors are described in detail in Chapters 8 and 10.

Typically, only 70 to 80% of the volume of stirred reactors is filled with liquid. This allows adequate headspace for disengagement of liquid droplets from the exhaust gas and to accommodate any foam that may develop. If foaming is a problem, a supplementary impeller called a *foam breaker* may be installed as shown in Figure 14.4. Alternatively, the upper impeller of multiple-impeller stirring systems can be selected and positioned to maximise foam disruption [3]. Chemical antifoam agents are also used; however, because

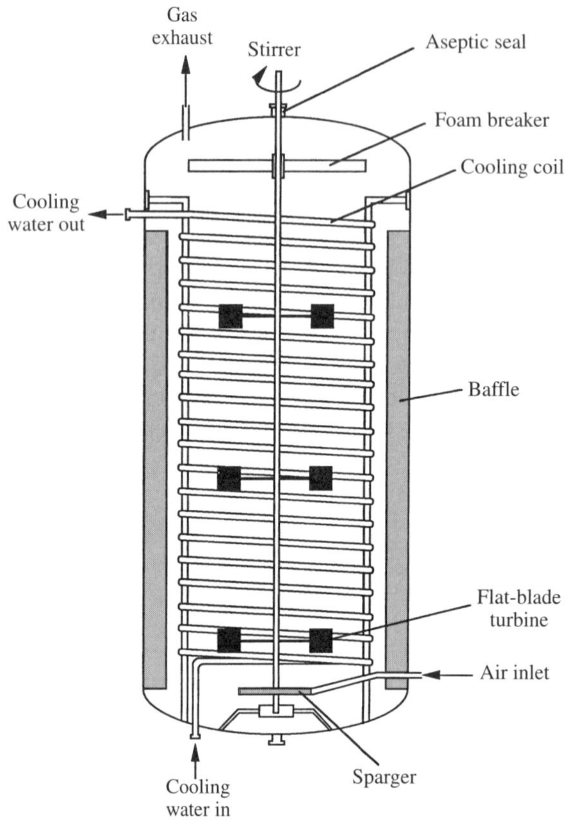

FIGURE 14.4 Typical stirred tank fermenter for aerobic culture.

antifoams reduce the rate of oxygen transfer (Section 10.6.3), mechanical foam dispersal is generally preferred if it can be made sufficiently effective.

The aspect ratio of stirred vessels (i.e., the ratio of height to diameter) can be varied over a wide range. The least expensive shape to build has an aspect ratio of about 1; this shape has the smallest surface area and therefore requires the least material to construct for a given volume. However, when aeration is required, the aspect ratio is usually increased. This provides for longer contact times between the rising bubbles and the liquid and produces a greater hydrostatic pressure at the bottom of the vessel for increased oxygen solubility.

As shown in Figure 14.4, temperature control and heat transfer in stirred vessels can be accomplished using internal cooling coils. Alternative cooling equipment for bioreactors is illustrated in Figure 9.1. The relative advantages and disadvantages of different heat exchange systems for bioreactors are discussed in Section 9.1.1.

Stirred fermenters are used for free and immobilised enzyme reactions, and for culture of suspended and immobilised cells. Care is required with particulate catalysts that may be damaged or destroyed by the impeller at high speeds. As discussed in Section 8.16, high levels of shear can also damage sensitive cells, particularly in plant and animal cell cultures.

14.2.2 Bubble Column

Alternatives to the stirred reactor include vessels with no mechanical agitation. In bubble column reactors, aeration and mixing are achieved by gas sparging: this requires considerably less energy than mechanical stirring. Bubble columns are applied industrially for production of bakers' yeast, beer, and vinegar, and for treatment of waste water.

Bubble columns are structurally very simple. As shown in Figure 14.5, they are generally cylindrical vessels with height greater than twice the diameter. Other than a sparger for entry of compressed air, bubble columns typically have no internal structures. A height-to-diameter ratio of about 3:1 is common in bakers' yeast production; for other applications, towers with height-to-diameter ratios of 6:1 have been used. Perforated horizontal plates are sometimes installed in tall bubble columns to break up and redistribute coalesced bubbles. The advantages of bubble columns include low capital cost, lack of moving parts, and satisfactory heat and mass transfer performance. As in stirred vessels, foaming can be a problem requiring mechanical dispersal or addition of antifoam to the medium.

Bubble column hydrodynamics and mass transfer characteristics depend entirely on the behaviour of the bubbles released from the sparger. Different flow regimes occur depending on the gas flow rate, sparger design, column diameter, and medium properties such as viscosity. *Homogeneous flow* occurs only at low gas flow rates and when bubbles leaving the sparger are evenly distributed across the column cross-section. In homogeneous flow, all bubbles rise with the same upward velocity and there is little or no backmixing of the gas phase. Liquid mixing in this flow regime is limited, as it arises solely from entrainment in

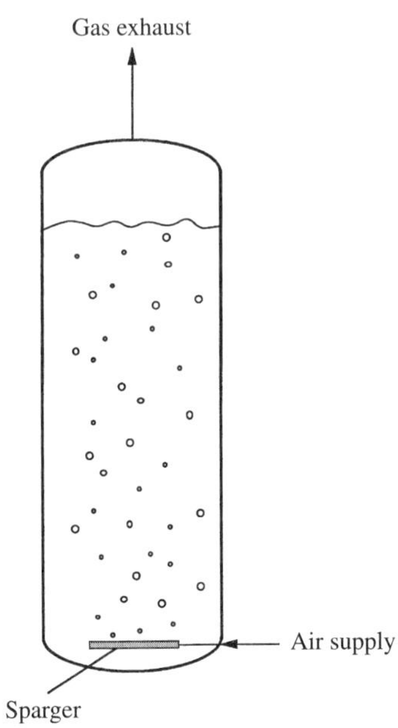

FIGURE 14.5 Bubble column bioreactor.

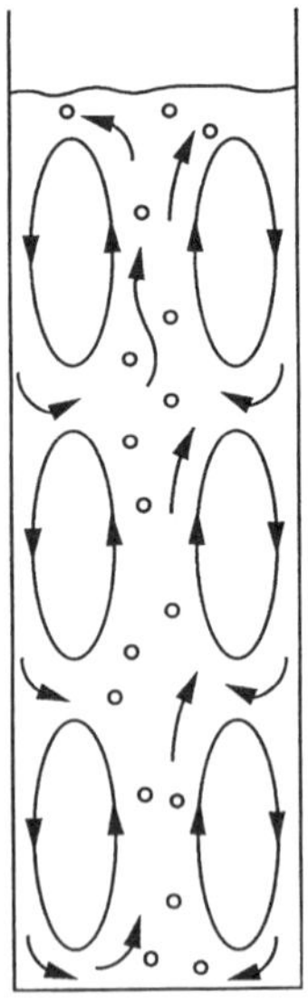

FIGURE 14.6 Heterogeneous flow in a bubble column.

the wakes of the bubbles. In contrast, under normal operating conditions at higher gas velocities, large chaotic circulatory flow cells develop and *heterogeneous flow* occurs as illustrated in Figure 14.6. In this regime, bubbles and liquid tend to rise up the centre of the column while a corresponding downflow of liquid occurs near the walls. This liquid circulation entrains bubbles so that some backmixing of gas occurs.

The liquid mixing time in bubble columns depends on the flow regime. For heterogeneous flow, the following equation has been proposed [4] for the upward liquid velocity at the centre of the column for $0.1 < D < 7.5$ m and $0 < u_G < 0.4$ m s^{-1}:

$$u_L = 0.9\,(g\,D\,u_G)^{0.33} \tag{14.1}$$

where u_L is the linear liquid velocity, g is gravitational acceleration, D is the column diameter, and u_G is the gas superficial velocity. u_G is equal to the volumetric gas flow rate at atmospheric pressure divided by the reactor cross-sectional area. From this equation for u_L, an expression for the mixing time t_m (Section 8.10) in bubble columns can be obtained [5]:

$$t_m = 11\frac{H}{D}(g\,u_G\,D^{-2})^{-0.33} \tag{14.2}$$

where H is the height of the bubble column.

As discussed in Section 10.6.1, the value of the gas–liquid mass transfer coefficient in bioreactors depends largely on the diameter of the bubbles and the gas hold-up achieved. In bubble columns containing nonviscous liquids, these variables depend solely on the gas flow rate. However, as exact bubble sizes and liquid circulation patterns are impossible to predict in bubble columns, accurate estimation of the mass transfer coefficient is difficult. The following correlation has been proposed for nonviscous media in heterogeneous flow [4, 5]:

$$k_{\mathrm{L}}a \approx 0.32\, u_{\mathrm{G}}^{0.7} \tag{14.3}$$

where $k_{\mathrm{L}}a$ is the combined volumetric mass transfer coefficient and u_{G} is the gas superficial velocity. Equation (14.3) is valid for bubbles with mean diameter about 6 mm, $0.08\ \mathrm{m} < D < 11.6\ \mathrm{m}$, $0.3\ \mathrm{m} < H < 21\ \mathrm{m}$, and $0 < u_{\mathrm{G}} < 0.3\ \mathrm{m\ s^{-1}}$. If smaller bubbles are produced at the sparger and the medium is noncoalescing, $k_{\mathrm{L}}a$ will be larger than the value calculated using Eq. (14.3), especially at low values of u_{G} less than about $10^{-2}\ \mathrm{m\ s^{-1}}$ [4].

14.2.3 Airlift Reactor

As in bubble columns, mixing in airlift reactors is accomplished without mechanical agitation. Several types of airlift reactor are in use. Their distinguishing feature compared with bubble columns is that the patterns of liquid flow are more defined, owing to the physical separation of upflowing and downflowing streams. As shown in Figure 14.7, gas is sparged into only part of the vessel cross-section called the *riser*. The resulting gas hold-up and decreased fluid density cause liquid in the riser to move upward. As gas bubbles disengage from the liquid at the top of the vessel, heavier bubble-free liquid is left to recirculate through the *downcomer*. Thus, liquid circulation in airlift reactors results from the density difference between the riser and downcomer.

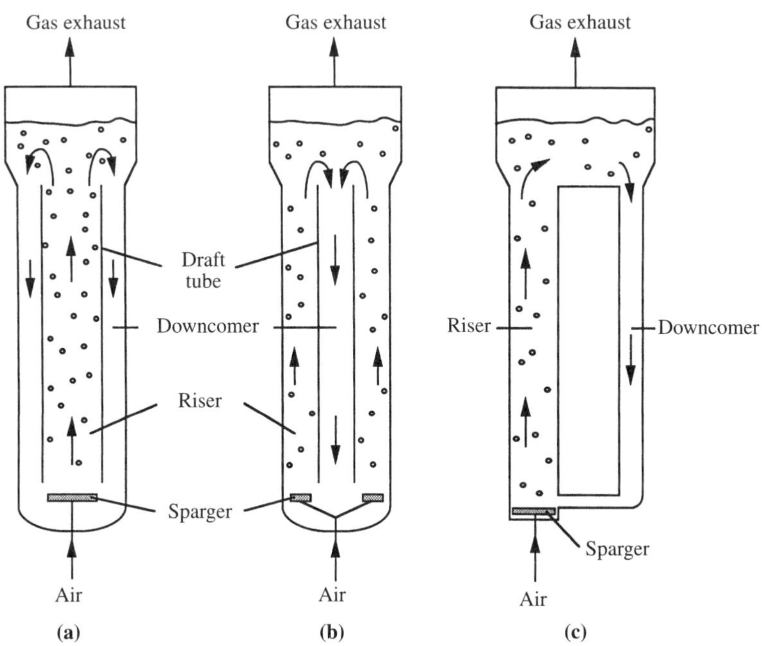

FIGURE 14.7 Airlift reactor configurations: (a) and (b) internal loop vessels; (c) external loop airlift.

Figure 14.7 illustrates the most common airlift configurations. In the *internal loop vessels* of Figures 14.7(a) and 14.7(b), the riser and downcomer are separated by an internal baffle or *draft tube*; air may be sparged into either the draft tube or the annulus. In the *external loop* or *outer loop* airlift of Figure 14.7(c), separate vertical tubes are connected by short horizontal sections at the top and bottom. Because the riser and downcomer are further apart in external loop vessels, gas disengagement is more effective than in internal loop devices. Fewer bubbles are carried into the downcomer, the density difference between fluids in the riser and downcomer is greater, and circulation of liquid in the vessel is faster. Accordingly, mixing is usually better in external loop than in internal loop reactors.

Airlift reactors generally provide better mixing than bubble columns, except at low liquid velocities when circulatory flow similar to that shown in Figure 14.6 develops. Compared with bubble columns, the airlift configuration confers a degree of stability to the liquid flow patterns produced; therefore, higher gas flow rates can be used without incurring operating problems such as slug flow or spray formation. Several empirical correlations have been developed for calculation of the liquid velocity, circulation time, and mixing time in airlift reactors; however there is considerable discrepancy between the results [6]. Equations derived from hydrodynamic models are also available [6, 7]; these are usually relatively complex and, because liquid velocity and gas hold-up are not independent, require iterative numerical solution.

Gas hold-up and gas–liquid mass transfer rates in internal loop airlifts are similar to those in bubble columns [6]. However, in external loop devices, near-complete bubble disengagement at the top of the vessel decreases the gas hold-up [8, 9] so that mass transfer rates at identical gas velocities are lower than in bubble columns [6]. Therefore, by comparison with Eq. (14.3) for bubble columns, for external loop airlifts:

$$k_L a < 0.32\, u_G^{0.7} \tag{14.4}$$

Several other empirical mass transfer correlations have been developed for Newtonian and non-Newtonian fluids in airlift reactors [6].

The performance of airlift devices is influenced significantly by the details of vessel construction [6, 10, 11]. For example, in internal loop airlifts, changing the distance between the lower edge of the draft tube and the base of the reactor alters the pressure drop in this region and affects the liquid velocity and gas hold-up. The depth of draft tube submersion from the top of the liquid also influences mixing and mass transfer characteristics.

Airlift reactors have been applied for production of single cell protein from methanol and gas oil, in municipal and industrial waste treatment, and for plant and animal cell culture. Large airlift reactors with capacities ranging up to thousands of cubic metres have been constructed. Tall internal loop airlifts built underground are known as *deep shaft reactors*; the very high hydrostatic pressure at the bottom of these vessels improves gas–liquid mass transfer considerably. The height of airlift reactors is typically about 10 times the diameter; for deep shaft systems, the height-to-diameter ratio may be increased up to 100.

14.2.4 Stirred and Air-Driven Reactors: Comparison of Operating Characteristics

For low-viscosity fluids, adequate mixing and mass transfer can be achieved in stirred tanks, bubble columns, and airlift vessels. When a large fermenter (50–500 m^3) is required for low-viscosity culture, a bubble column is an attractive choice because it is simple and cheap to install and operate. Mechanically agitated reactors are impractical at volumes greater than about 500 m^3 as the power required to achieve adequate mixing becomes extremely high (Section 8.11).

If the culture has high viscosity, sufficient mixing and mass transfer cannot be provided using air-driven reactors. Stirred vessels are more suitable for viscous liquids because greater power can be input by mechanical agitation. Nevertheless, mass transfer rates decline rapidly in stirred vessels at viscosities over 50 to 100 cP [5].

Heat transfer can be an important consideration in the choice between air-driven and stirred reactors. Mechanical agitation generates much more heat than sparging of compressed gas. When the heat of reaction is high, such as for production of single cell protein from methanol, the removal of frictional stirrer heat can be a problem so that air-driven reactors may be preferred.

Stirred tank and air-driven vessels account for the vast majority of bioreactor configurations used for aerobic culture. However, other reactor configurations may be used in particular processes.

14.2.5 Packed Bed

Packed bed reactors are used with immobilised or particulate biocatalysts. The reactor consists of a tube, usually vertical, packed with catalyst particles. Medium can be fed at either the top or bottom of the column and forms a continuous liquid phase between the particles. Damage due to particle attrition is minimal in packed beds compared with agitation of solid catalysts in stirred vessels. Packed bed reactors have been applied commercially with immobilised cells and enzymes for production of aspartate and fumarate, conversion of penicillin to 6-aminopenicillanic acid, and resolution of amino acid isomers.

Mass transfer between the liquid medium and solid catalyst is facilitated by high liquid flow rates through the bed. To achieve these conditions, packed beds are often operated with liquid recycle as shown in Figure 14.8. The catalyst is prevented from leaving the column by screens at the liquid outlet. The particles used should be relatively incompressible and able to withstand their own weight in the column without deforming and occluding the liquid flow. The recirculating medium must also be clean and free of debris to avoid clogging the bed. Aeration is generally accomplished in a separate vessel; if air is sparged directly into the bed, bubble coalescence produces gas pockets between the particles and flow *channelling* or maldistribution occurs. Packed beds are unsuitable for processes that produce large quantities of carbon dioxide or other gases that can become trapped in the packing.

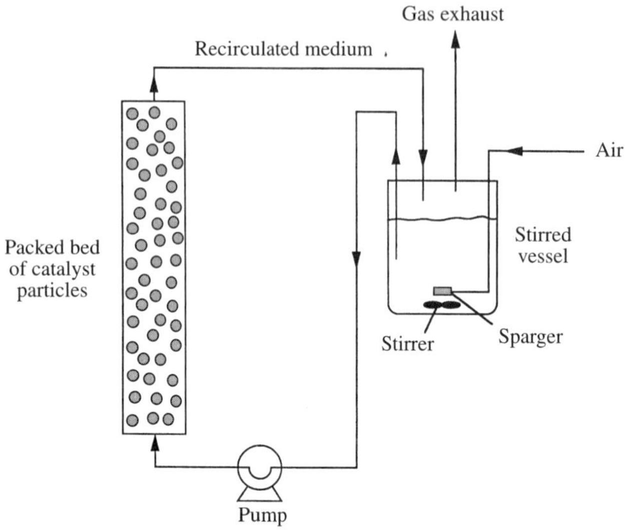

FIGURE 14.8 Packed bed reactor with medium recycle.

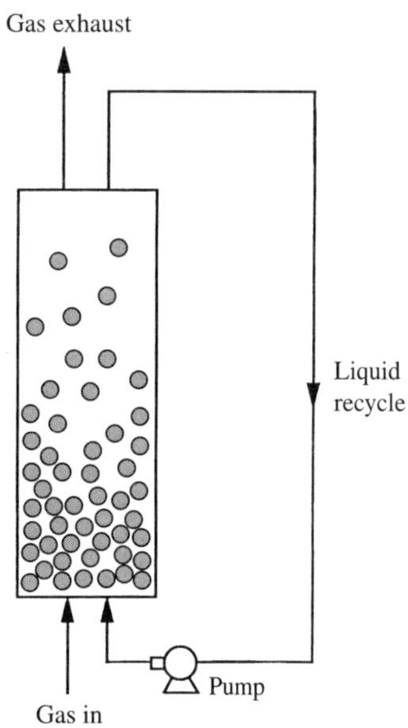

FIGURE 14.9 Fluidised bed reactor.

14.2.6 Fluidised Bed

When packed beds are operated in upflow mode with catalyst beads of appropriate size and density, the bed expands at high liquid flow rates due to upward motion of the particles. This is the basis for operation of fluidised bed reactors as illustrated in Figure 14.9. Because particles in fluidised beds are in constant motion, clogging of the bed and flow channelling are avoided so that air can be introduced directly into the column. Fluidised

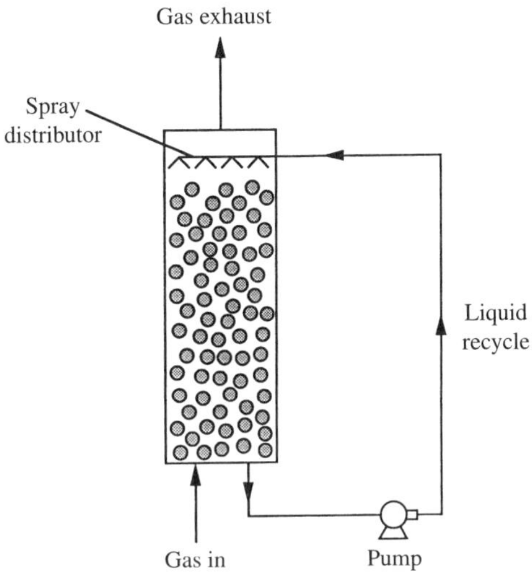

FIGURE 14.10 Trickle bed reactor.

bed reactors are used in waste treatment with sand or similar material supporting mixed microbial populations. They are also used with flocculating organisms in brewing and for production of vinegar.

14.2.7 Trickle Bed

The trickle bed reactor is another variation of the packed bed. As illustrated in Figure 14.10, liquid is sprayed onto the top of the stationary packing and trickles down through the bed in small rivulets. Air may be introduced at the base; because the liquid phase is not continuous throughout the column, air and other gases move with relative ease around the packing. Trickle bed reactors are used widely for aerobic waste water treatment.

14.3 PRACTICAL CONSIDERATIONS FOR BIOREACTOR CONSTRUCTION

Industrial bioreactors for sterile operation are usually designed as steel pressure vessels capable of withstanding full vacuum and up to about 3 atm of positive pressure at 150 to 180°C. A hole is provided at the top of large vessels to allow workers entry into the tank for cleaning and maintenance; on smaller vessels, the top is removable. Flat headplates are commonly used with laboratory-scale fermenters; for larger vessels, a domed construction is less expensive. Large fermenters are equipped with a lighted vertical sight-glass for inspecting the contents of the reactor. Nozzles for medium, antifoam, and acid and alkali addition, air exhaust pipes, a pressure gauge, and a rupture disc for emergency pressure release are normally located on the headplate. Side ports for pH, temperature, and dissolved oxygen sensors are a minimum requirement; a steam-sterilisable sample outlet should also be provided. The vessel must be fully draining via a harvest nozzle located at the lowest point of the reactor. If the vessel is mechanically agitated, either a top- or bottom-entering stirrer is installed (Section 8.2.4).

14.3.1 Aseptic Operation

Most fermentations outside of the food and beverage industry are carried out using pure cultures and aseptic conditions. Keeping the bioreactor free of unwanted organisms is especially important for slow-growing cultures that can be quickly overrun by contamination. Fermenters must be capable of operating aseptically for a number of days, sometimes months.

Typically, 3 to 5% of fermentations in an industrial plant are lost due to failure of sterilisation procedures. However, the frequency and causes of contamination vary considerably from process to process. For example, the nature of the product in antibiotic fermentations affords some protection from contamination; fewer than 2% of production-scale antibiotic fermentations are lost through contamination by microorganisms or phage [12]. Higher contamination rates occur in processes using complex, nutrient-rich media and relatively slow-growing cells such as mammalian cells; a contamination rate of 17% has been reported for industrial-scale production of β-interferon from human fibroblasts cultured in 50-litre bioreactors [13].

Industrial fermenters are designed for in situ steam sterilisation under pressure. The vessel should have a minimum number of internal structures, ports, nozzles, connections, and other attachments to ensure that steam reaches all parts of the equipment. For effective sterilisation, all air in the vessel and pipe connections must be displaced by steam. The reactor should be free of crevices and stagnant areas where liquid or solids can accumulate; for this reason, polished welded joints are used in preference to other coupling methods. Small cracks or gaps in joints and fine fissures in welds are a haven for microbial contaminants and are avoided in fermenter construction whenever possible. After sterilisation, all nutrient medium and air entering the fermenter must be sterile. As soon as the flow of steam into the fermenter is stopped, sterile air is introduced to maintain a slight positive pressure in the vessel and discourage entry of airborne contaminants. Filters preventing the passage of microorganisms are fitted to exhaust gas lines; this serves to contain the culture inside the fermenter and insures against contamination should there be a drop in operating pressure.

The flow of liquids to and from the fermenter is controlled using valves. Because valves are a potential entry point for contaminants, their construction must be suitable for aseptic operation. Common designs such as simple gate and globe valves have a tendency to leak around the valve stem and accumulate broth solids in the closing mechanism. Although used in the fermentation industry, they are unsuitable if a high level of sterility is required. Pinch and diaphragm valves such as those shown in Figures 14.11 and 14.12 are recommended for fermenter construction. These designs make use of flexible sleeves or diaphragms so that the closing mechanism is isolated from the contents of the pipe and there are no dead spaces in the valve structure. Rubber or neoprene capable of withstanding repeated sterilisation cycles is used to fashion the valve closure; the main drawback is

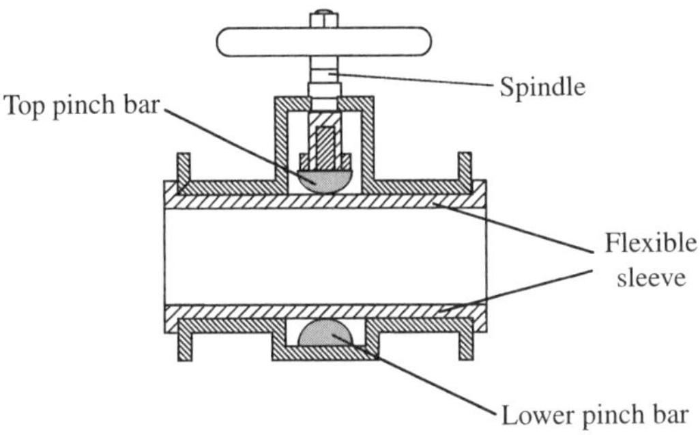

FIGURE 14.11 Pinch valve.

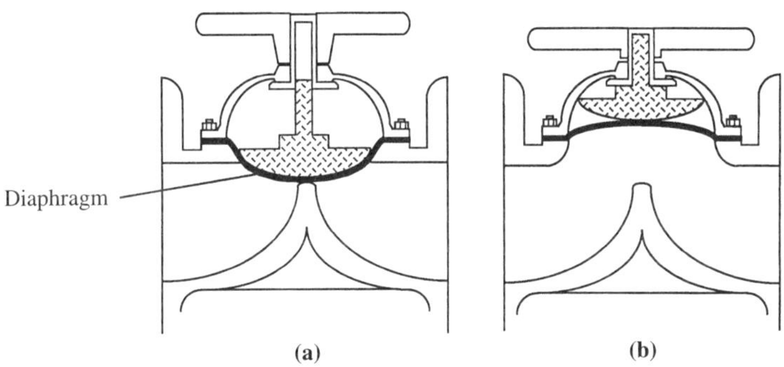

FIGURE 14.12 Weir-type diaphragm valve in (a) closed and (b) open positions.

that these components must be checked regularly for wear to avoid valve failure. To minimise costs, ball and plug valves are also used in fermenter construction.

With stirred reactors, another potential entry point for contamination is where the stirrer shaft enters the vessel. The gap between the rotating stirrer shaft and the fermenter body must be sealed; if the fermenter is operated for long periods, wear at the seal opens the way for airborne contaminants. Several types of stirrer seal have been developed to prevent contamination. On large fermenters, mechanical seals are commonly used [14]. In these devices, one part of the assembly is stationary while the other rotates on the shaft; the precision-machined surfaces of the two components are pressed together by springs or expanding bellows and cooled and lubricated with water. Mechanical seals with running surfaces of silicon carbide paired with tungsten carbide are often specified for fermenter application. Stirrer seals are especially critical if the reactor is designed with a bottom-entering stirrer; in this case, double mechanical seals may be installed to prevent fluid leakage. On smaller vessels, magnetic drives can be used to couple the stirrer shaft with the motor; with these devices, the stirrer shaft does not pierce the fermenter body. A magnet in a housing on the outside of the fermenter is driven by the stirrer motor; inside, another magnet is attached to the end of the stirrer shaft and held in place by bearings. Sufficient power can be transmitted using magnetic drives to agitate vessels up to at least 800 litres in size [15]. However, the suitability of magnetic drives for viscous broths, especially when high oxygen transfer rates are required, is limited.

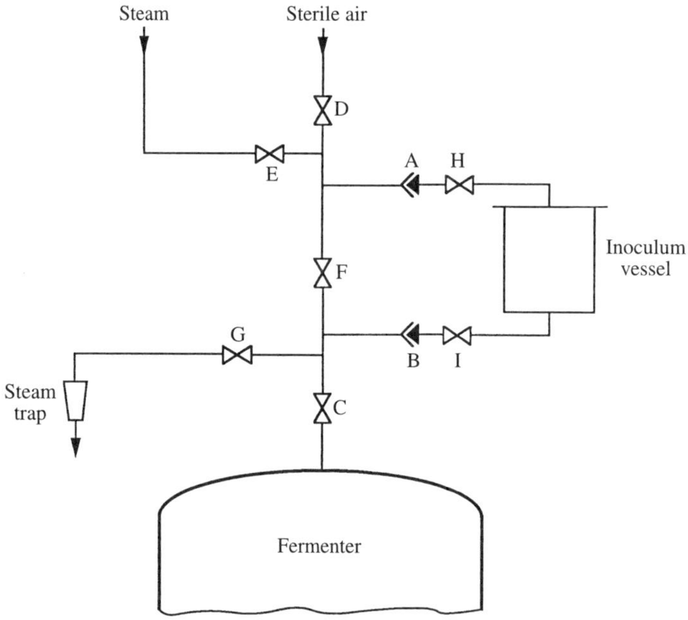

FIGURE 14.13 Pipe and valve connections for aseptic transfer of inoculum to a large-scale fermenter. *From A. Parker, 1958, Sterilization of equipment, air and media. In: R. Steel (Ed.),* Biochemical Engineering, *pp. 97–121, Heywood, London.*

14.3.2 Fermenter Inoculation and Sampling

Consideration must be given in the design of fermenters to aseptic inoculation and sample removal. Inocula for large-scale fermentations are transferred from smaller reactors: to prevent contamination during this operation, both vessels are maintained under positive air pressure. The simplest aseptic transfer method is to pressurise the inoculum vessel using sterile air: culture is then effectively blown into the larger fermenter. An example of the pipe and valve connections required for this type of transfer is shown in Figure 14.13.

The fermenter and its piping and the inoculum tank and its piping, including valves H and I, are sterilised separately before the cells are added to the inoculum tank. With valves H and I closed, the small vessel is joined to the fermenter at connections A and B. Because these connectors were open prior to being joined, they must be sterilised before the inoculum tank is opened. With valves D, H, I, and C closed and A and B slightly open, steam flows through E, F, and G and bleeds slowly from A and B. After about 20 minutes of steam sterilisation, valves E and G and connectors A and B are closed and the path from the inoculum tank to the fermenter is now sterile. Valves D and C are opened for flow of sterile air into the fermenter to cool the line under positive pressure. Valve F is then closed, valves H and I are opened, and sterile air is used to force the contents of the inoculum tank into the fermenter. The line between the vessels is emptied of most residual liquid by blowing through with sterile air. Valves D, C, H, and I are then closed to isolate both the fermenter and the empty inoculum tank that can now be disconnected at A and B.

Sampling ports are fitted to fermenters to allow removal of broth for analysis. An arrangement for sampling that preserves aseptic operation is shown in Figure 14.14. Initially, valves A and D are closed; valves B and C are open to maintain a steam barrier

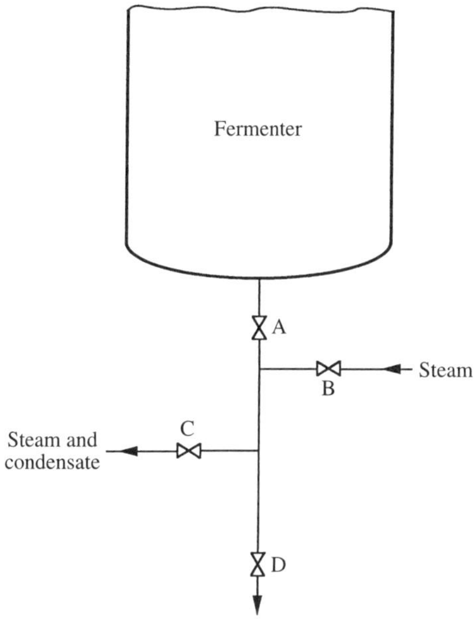

FIGURE 14.14 Pipe and valve connections for a simple fermenter sampling port.

between the reactor and the outside environment. Valve C is then closed, valve B partially closed, and valve D partially opened to allow steam and condensate to bleed from the sampling port D. For sampling, A is opened briefly to cool the pipe and carry away any condensate that would dilute the sample; this broth is discarded. Valve B is then closed and a sample collected through D. When sampling is complete, valve A is closed and B opened for resterilisation of the sample line; this prevents any contaminants that entered while D was open from travelling up to the fermenter. Valve D is then closed and valve C reopened.

14.3.3 Materials of Construction

Fermenters are constructed from materials that can withstand repeated steam sterilisation and cleaning cycles. Materials contacting the fermentation medium and broth should also be nonreactive and nonabsorptive. Glass is used to construct fermenters of capacity up to about 30 litres. The advantages of glass are that it is smooth, nontoxic, corrosion-proof, and transparent for easy inspection of the vessel contents. Because entry ports are required for medium, inoculum, air, and instruments such as pH and temperature sensors, glass fermenters are usually equipped with stainless steel headplates containing many screw fittings.

Most pilot- and large-scale fermenters are made of corrosion-resistant stainless steel, although mild steel with stainless steel cladding has also been used. Cheaper grades of stainless steel may be used for the jacket and other surfaces isolated from the broth. Copper and copper-containing materials must be avoided in all parts of the fermenter contacting the culture because of the toxic effect on cells. Interior steel surfaces are polished to a bright 'mirror' finish to facilitate cleaning and sterilisation of the reactor; welds on the interior of the vessel are ground flush before polishing. Electropolishing is preferred over mechanical polishing, which leaves tiny ridges and grooves in the metal that accumulate dirt and microorganisms.

14.3.4 Impeller, Baffle, and Sparger Design

The impeller and baffles determine the effectiveness of mixing and oxygen transfer in stirred bioreactors. Their design features are described in Sections 8.2 through 8.4. In air-driven bioreactors without mechanical stirring, the sparger plays a direct role in achieving mixing and oxygen transfer. Several sparger configurations used in bioreactors are described in Sections 8.2.3 and 10.6.2 (Bubble Formation subsection). Other types of sparger have also been developed. In *two-phase ejector–injectors*, gas and liquid are pumped concurrently through a nozzle to produce tiny bubbles; in *combined sparger–agitator* systems for smaller stirred fermenters, a hollow stirrer shaft is used to deliver air beneath the impeller. Irrespective of the sparger design used, provision should be made for in-place cleaning of the interior of the pipe as cell growth through the pores or holes of the sparger can lead to blockage.

14.3.5 Evaporation Control

Aerobic cultures are sparged continuously with air. As most components of air are inert and leave directly through the exhaust gas line, when air entering the fermenter is dry, water is stripped continually from the medium and leaves as vapour in the off-gas. Over a period of days, evaporative water losses can be significant. This problem is more pronounced in air-driven reactors because the gas flow rates required for good mixing and mass transfer are higher than those typically used in stirred reactors.

To combat evaporation problems, the air sparged into fermenters may be prehumidified by bubbling through columns of water outside the fermenter: humid air entering the fermenter has less capacity for evaporation than dry air. Fermenters are also equipped with water-cooled condensers that return to the broth any vapours carried by the exit gas. Evaporation can be a particular problem when the fermentation products or substrates are more volatile than water. For example, *Acetobacter* species are used to produce acetic acid from ethanol in a highly aerobic process requiring large quantities of air. It has been reported for stirred tank reactors operated at air flow rates between 0.5 and 1.0 vvm (vvm means volume of gas per volume of liquid per minute) that from a starting alcohol concentration of 5% (v/v), 30 to 50% of the substrate may be lost within 48 hours due to evaporation [16].

14.4 MONITORING AND CONTROL OF BIOREACTORS

The environment inside bioreactors should allow optimal catalytic activity. Parameters such as temperature, pH, dissolved oxygen concentration, medium flow rate, stirrer speed, and sparging rate have a significant effect on the outcome of fermentation and enzyme reactions. To provide the desired environment, system properties must be monitored and control action taken to rectify any deviations from the desired values. Fermentation monitoring and control is an active area of research aimed at improving the performance of bioprocesses and achieving reliable fermenter operation.

Various levels of process control exist in the fermentation industry. Manual control is the simplest, requiring a human operator to manipulate devices such as pumps, motors, and valves. Automatic feedback control is also used to maintain parameters at specified values. With the increasing application of computers, there is scope for implementing advanced control and optimisation strategies based on fermentation models.

14.4.1 Fermentation Monitoring

Any attempt to understand or control the state of a fermentation depends on knowledge of the critical variables that affect the process. These parameters can be grouped into three categories: physical, chemical, and biological. Examples of process variables in each group are given in Table 14.1. Many of the physical measurements listed are well established in the fermentation industry; others are currently being developed or are the focus of research into new instrumentation.

Despite the importance of fermentation monitoring, industrially reliable instruments and sensors capable of rapid, accurate, and direct measurements are not available for many process variables. For effective control of fermentations based on measured data, the time taken to complete the measurement should be consistent with the rate of change of the variable being monitored. For example, in a typical fermentation, the time scale for changes in pH and dissolved oxygen tension is several minutes, while the time scale for changes in culture fluorescence is less than 1 second. For other variables, such as biomass

TABLE 14.1 Parameters Measured or Controlled in Bioreactors

Physical	Chemical	Biological
Temperature	pH	Biomass concentration
Pressure	Dissolved O_2 tension	Enzyme concentration
Reactor weight	Dissolved CO_2 concentration	Biomass composition (DNA, RNA, protein, lipid, carbohydrate, ATP/ADP/AMP, NAD/NADH levels, etc.)
Liquid level	Redox potential	
Foam level	Exit gas composition	
Agitator speed	Conductivity	Viability
Power consumption	Broth composition (substrate, product, ion concentrations, etc.)	Morphology
Gas flow rate		
Medium flow rate		
Culture viscosity		
Gas hold-up		

concentration, an hour or more may pass before measurable changes occur. Ideally, measurements should be made in situ and online (i.e., in or near the reactor during operation) so that the result is available for timely control action.

Many important variables, such as biomass concentration and broth composition, are generally not measured online because of the lack of appropriate instruments. Instead, samples are removed from the reactor and taken to the laboratory for offline analysis. Because fermentation conditions can change during the time required for laboratory analysis, control actions based on laboratory measurements are not as effective as when online results are used. For offline measurement of variables such as biomass carbohydrate, protein, phosphate and lipid concentrations, enzyme activity, and broth rheology, samples are usually taken manually every 4 to 8 h and the results are available 2 to 24 h later.

Examples of measurements that can be made online in industrial fermentations are temperature, pressure, pH, dissolved oxygen tension, flow rate, stirrer speed, power consumption, foam level, broth weight, and gas composition. The instruments for taking these measurements are relatively commonplace and detailed descriptions can be found elsewhere [17–19]. The availability of an online measurement or its use in the laboratory does not necessarily mean it is applied in commercial-scale processing. Measurement devices used in industry must meet stringent performance criteria that reflect the cost of installation and the economic, safety, and environmental consequences of instrument failure during fermentation. These criteria may include, for example, accuracy to within 1 to 2% of full scale, reliable operation for at least 80% of the time, low maintenance needs, steam sterilisability, simple and fast calibration, and maximum drift of less than 1 to 2% of full scale [1].

Sterilisable pH probes have been proven reliable if properly grounded and are used widely in industrial fermentations; their low failure rate is assisted by replacement every four or five fermentations. However, a variety of problems are associated in the industrial environment with electrodes for dissolved carbon dioxide, redox potential, and specific ions; these instruments are therefore confined mostly to laboratory applications. Drift and interference from air bubbles are well-known limitations with dissolved oxygen probes in industrial fermenters; deposits and microbial fouling of the membrane also reduce measurement reliability. Even if electrodes are steam-sterilisable, the life of the probe can be reduced significantly by repeated sterilisation cycles.

The development of new online measurements for chemical and biological variables is a challenging area of bioprocess engineering research. One of the areas targeted is *biosensors* [20, 21]. There are many different biosensor designs; however, in general, biosensors incorporate a biological sensing element in close proximity or integrated with a signal transducer. As illustrated schematically in Figure 14.15, the binding specificity of particular biological molecules immobilised on the probe is used to 'sense' or recognise target species in the fermentation broth. Ultimately, an electrical signal is generated after interaction between the sensing molecules and analyte is detected by the transducer.

Biosensors with enzyme-, antibody-, and cell-based sensing elements have been tested for a wide range of substrates and products including glucose, sucrose, ethanol, acetic acid, ammonia, penicillin, urea, riboflavin, cholesterol, and amino acids. However, although several biosensors are available commercially, they are not generally used for routine monitoring of large-scale fermentations because of intractable problems with robustness, sterilisation, limited life span, and long-term stability. At this stage, because of the sensitivity of

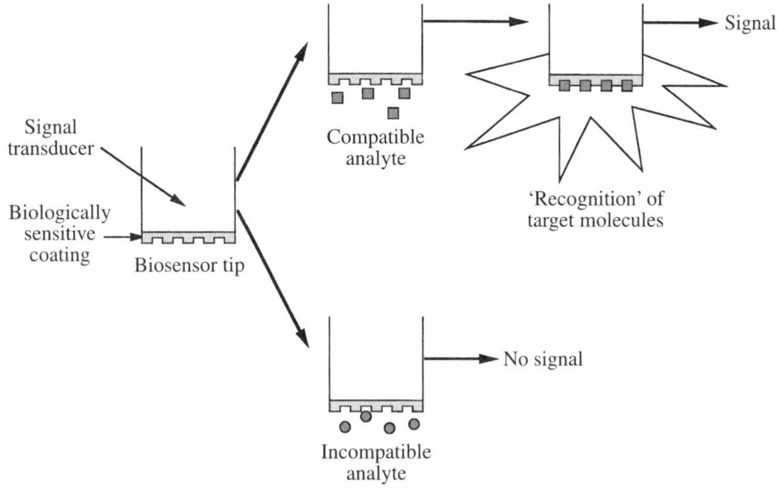

FIGURE 14.15 Operating principle for biosensors.
From E.A.H. Hall, 1991, Biosensors, *Prentice Hall, Englewood Cliffs, NJ.*

their biological components, biosensors appear to be more suitable for medical diagnostics and environmental analyses than in situ bioprocess applications. Nevertheless, biosensors may be used for rapid analysis of broth samples removed from fermenters.

Another approach to online chemical and biological measurements involves the use of automatic sampling devices linked to analytical equipment for high-performance liquid chromatography (HPLC), image analysis, nuclear magnetic resonance (NMR) spectroscopy, flow cytometry, or fluorometry. For these techniques to work, methods must be available for online aseptic sampling and analysis without blockage or interference from gas bubbles and cell solids. Development in this area has centred around *flow injection analysis,* a sample-handling technique for removing cell-free medium from the fermenter and delivering a pulse of analyte to ex situ measurement devices. Several procedures based on culture turbidity, light scattering, fluorescence, calorimetry, piezoelectric membranes, radio frequency dielectric permittivity, and acoustics have been developed; however, they have not proved sufficiently accurate and reliable under most industrial conditions. Air bubbles and background particles readily interfere with optical readings, while the relationship between biomass density and other culture properties such as fluorescence is affected by pH, dissolved oxygen tension, and substrate levels [19]. Other difficulties have also been encountered in the development of automatic sampling devices: the most important of these is the high risk of contamination and blockage. Further details about online measurement techniques can be found elsewhere [22, 23].

In a large fermentation factory, thousands of different variables may be monitored at any given time. The traditional device for recording online process information is the chart recorder; however, with increasing application of computers in the fermentation industry, digital data logging is widespread. The enormous quantity of information now obtained from continuous monitoring of bioprocesses means that management and effective utilisation of these large amounts of data are important challenges in the industry.

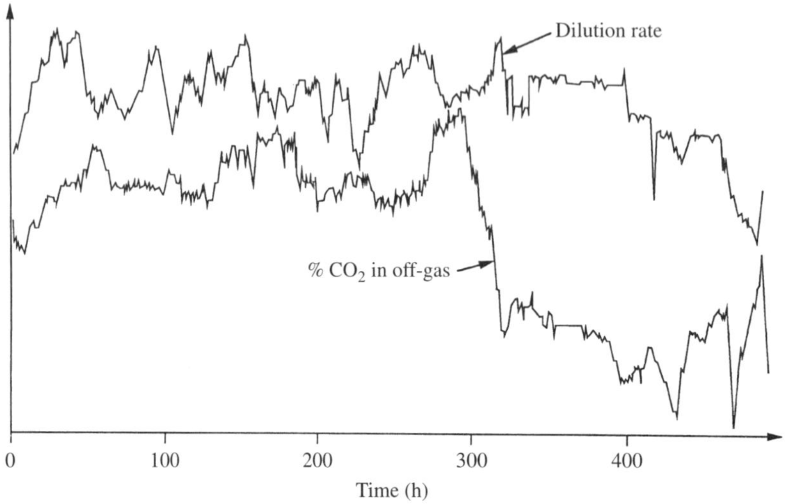

FIGURE 14.16 Online measurements of dilution rate and off-gas carbon dioxide during an industrial mycelial fermentation. *From G.A. Montague, A.J. Morris, and A.C. Ward, 1989, Fermentation monitoring and control: a perspective.* Biotechnol. Genet. Eng. Rev. *7, 147–188.*

14.4.2 Measurement Analysis

Any attempt to analyse or apply the results of fermentation monitoring must consider the errors and spurious or transient effects incorporated into the data. Noise and variability are particular problems with certain fermentation measurements; for example, probes used for pH and dissolved oxygen measurements are exposed to rapid fluctuations and heterogeneities in the broth so that noise can seriously affect the accuracy of point readings. Figure 14.16 shows typical results from online measurement of dilution rate and carbon dioxide evolution in a production-scale fermenter. In many cases, *signal conditioning* or *smoothing* must be carried out to reduce the noise in data before they can be applied for process control or modelling. Most data acquisition and logging systems contain facilities for signal conditioning. Unwanted pseudo-random noise can be filtered out using analogue filter circuits or by averaging values over successive measurements. Alternatively, unfiltered signals can be digitised and filtering algorithms applied using computer software. As measurement drift cannot be corrected using electronic circuitry, instruments must be periodically recalibrated during long fermentations to avoid loss of accuracy.

Raw data are sometimes used to calculate *derived variables* that characterise the performance of the fermenter. The most common derived variables are the oxygen uptake rate, the rate of carbon dioxide evolution, the respiratory quotient RQ (Section 4.6.1), and the oxygen transfer coefficient k_La (Chapter 10). As the oxygen uptake rate is usually determined during cell culture from the difference between two quantities of similar magnitude (i.e., the gas inlet and outlet oxygen levels, Section 10.10.1), noise in this variable can be significant [24], thus affecting the quality of other dependent variables such as RQ and k_La.

14.4.3 Fault Analysis

Faults in reactor operation affect 15 to 20% of fermentations [25]. Fermentation measurements can be used to detect faults; for example, signals from a flow sensor could be used to

TABLE 14.2 Reliability of Fermentation Equipment

Equipment	Reliability	Mean Time Between Failures (weeks)
Temperature probe	–	150–200
pH probe	98%	9–48
Dissolved oxygen probe	50–80%	9–20
Mass spectrometer	–	10–50
Paramagnetic O_2 analyser	–	24
Infrared CO_2 analyser	–	52

From S.W. Carleysmith, 1987, Monitoring of bioprocessing, In: J.R. Leigh (Ed.), Modelling and Control of Fermentation Processes, *pp. 97–117, Peter Peregrinus, London; and P.N. Royce, 1993, A discussion of recent developments in fermentation monitoring and control from a practical perspective,* Crit. Rev. Biotechnol. *13, 117–149.*

detect blockages in a pipe and trigger an alarm in the factory control room. Normally however, the sensors themselves are the most likely components to fail. The reliability and frequency of failure of some fermentation instruments are listed in Table 14.2. The failure of a sensor might be detected as an unexpected change or rate of change in its signal, or a change in the noise characteristics. Several approaches can be used to reduce the impact of faults on large-scale fermentations, including comparison of current measurements with historical values, cross-checking between independent measurements, using multiple and back-up sensors, and building hardware redundancy into computer systems and power supplies.

14.4.4 Process Modelling

In modern approaches to fermentation control, a reasonably accurate mathematical model of the reaction and reactor environment is required. Using process models, we can progress beyond environmental control of bioreactors into the realm of direct biological control. Development of fermentation models is aided by information from measurements taken during process operation.

Models are mathematical relationships between variables. Traditionally, models are based on a combination of 'theoretical' relationships that provide the structure of the model, and experimental observations that provide the numerical values of coefficients. For biological processes, specifying the model structure can be difficult because of the complexity of cellular processes and the large number of environmental factors that affect cell culture. Usually, bioprocess models are much-simplified approximate representations deduced from observation rather than from theoretical laws of science. As an example, a frequently used mathematical model for batch fermentation consists of the Monod equation for growth and an expression for the rate of substrate consumption as a function of biomass concentration:

$$\frac{dx}{dt} = \mu\, x = \frac{\mu_{max}\, s\, x}{K_S + s} \tag{14.5}$$

$$\frac{-\mathrm{d}s}{\mathrm{d}t} = \frac{\mu x}{Y_{XS}} + m_S x \tag{14.6}$$

This model represents a combination of Eqs. (12.80), (12.91), and (12.103). The form of the equations was determined from experimental observation of a large number of different culture systems. In principle, once values of the parameters μ_{max}, K_S, Y_{XS}, and m_S have been determined, we can use the model to estimate the cell concentration x and substrate concentration s as a function of time. A common problem with fermentation models is that the model parameters can be difficult to measure, or they tend to change with time during the reaction process.

We have already encountered several model equations in this book; examples include the stoichiometric equations of Chapter 4 relating the masses of substrates, biomass, and products during reaction, and the kinetic, yield, and maintenance relationships of Chapter 12. Other models were derived in Chapter 13 for heterogeneous reactions. The dependence of culture parameters such as cell concentration on physical conditions such as heat and mass transfer rates in bioreactors is represented in simple form by the equations of Sections 9.6.2 and 10.5.2. Process models vary in form but have the unifying feature that they predict outputs from a set of inputs. When models are used for fermentation control, they are usually based at least in part on mass and energy balance equations for the system.

Development of a comprehensive model covering all key aspects of a particular bioprocess and capable of predicting the effects of a wide range of culture variables is a demanding exercise. Accurate models applicable to a range of process conditions are rare. As many aspects of fermentation are poorly understood, it is difficult to devise mathematical models covering these areas. For example, the response of cells to spatial variations in dissolved oxygen and substrate levels in fermenters, or the effect of impeller design on microbial growth and productivity, is not generally incorporated into models because the subject has been inadequately studied. Evidence that all important fermentation variables have not yet been identified is the significant batch-to-batch variation that occurs in the fermentation industry.

14.4.5 State Estimation

As described in Section 14.4.1, it is generally not possible to measure online all the key variables or states of a fermentation process. Often, considerable delays are involved in offline measurement of important variables such as biomass, substrate, and product concentrations. Such delays make effective control of the reactor difficult if control action is dependent on the value of these parameters but must be undertaken more quickly than offline analysis allows. One approach to this problem is to use available online measurements in conjunction with mathematical models of the process to estimate unknown variables. The computer programs and numerical procedures developed to achieve this are called *software sensors*, *estimators*, or *observers*. The *Kalman filter* is a well-known type of observer applicable to linear process equations; nonlinear systems can be treated using the *extended Kalman filter* [26]. The success of Kalman filters and other observers depends largely on the accuracy and robustness of the process model used.

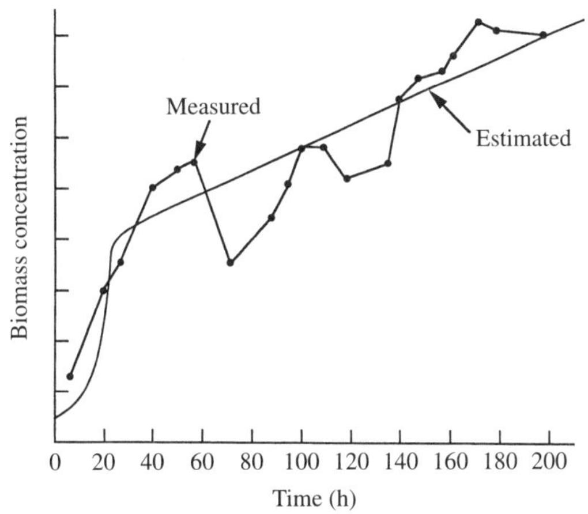

FIGURE 14.17 Measured and estimated biomass concentrations during the course of a large-scale fed-batch penicillin fermentation.
From G.A. Montague, A.J. Morris, and J.R. Bush, 1988, Considerations in control scheme development for fermentation process control. IEEE Contr. Sys. Mag. *8, April, 44–48.*

Several techniques for state estimation are available. As a simple example, online measurements of carbon dioxide in fermenter off-gas can be applied with an appropriate process model to estimate the biomass concentration during penicillin fermentation [27]. As shown in Figure 14.17, the results were satisfactory for 100-m^3 fed-batch fermentations over a period of more than 8 days. Direct state estimation can achieve reasonable results as long as the process model remains valid and the estimator is able to reduce measurement noise. However, if major fluctuations occur in operating conditions, or if cell properties and model parameters change with time, the model may become inadequate and the accuracy of the estimation will decline. *Adaptive estimators* are used to adjust faulty model parameters as the fermentation proceeds, to alleviate problems caused by error in the equations. Offline measurements can also be incorporated into the procedure as they become available to improve the accuracy and reliability of parameter estimation. Another technique involves generic software sensors that use generally structured models rather than model equations specific to the process of interest. Process characteristics are then 'learnt' or incorporated into the model structure as information becomes available from on- and offline measurements. The primary advantage of generic sensors is that the software development time is significantly reduced [28].

Mass-balancing techniques are another approach to state estimation. As described in Chapter 4, biomass concentration can be estimated from stoichiometry and other process measurements using elemental balances. This method is suitable for fermentation of defined media, but is difficult to apply with complex medium ingredients such as molasses, casein hydrolysate, soybean meal, and corn steep liquor that have undefined elemental composition. Another disadvantage is that the accuracy of biomass estimation often depends on the measurement of substrate or product concentration that must be done offline.

14.4.6 Feedback Control

Let us assume that we wish to maintain the pH in a bioreactor at a constant value against a variety of disturbances, for example, from metabolic activity. One of the simplest

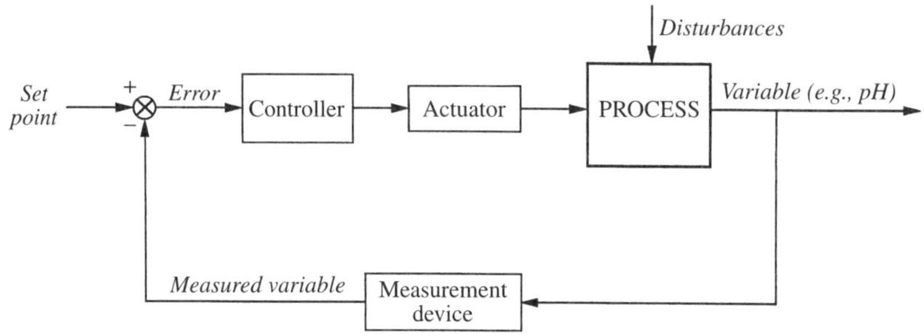

FIGURE 14.18 Components of a feedback control loop.

control schemes is a conventional *feedback control loop*, the basic elements of which are shown in Figure 14.18. A measurement device senses the value of the pH and sends the signal to a controller. At the controller, the measured value is compared with the desired value known as the *set point*. The difference between the measured and desired values is the *error*, which is used by the controller to determine what action must be taken to correct the process. The controller may be a person who monitors the process measurements and decides what to do; more often, the controller is an automatic electronic, pneumatic, or computer device. The controller produces a signal that is transmitted to the actuator, which executes the control action.

In a typical system for pH control, an electrode serves as the measurement device and a pump connected to a reservoir of acid or alkali serves as the actuator. Simple *on–off control* is generally sufficient for pH; if the measured pH falls below the set point, the controller switches on the pump that adds alkali to the fermenter. When enough alkali is added and the pH returned to the set value, the pump is switched off. Small deviations from the set point are usually tolerated in on–off control to avoid rapid switching and problems due to measurement delay.

On–off control is used when the actuator is an on–off device, such as a single-speed pump. If the actuator is able to function over a continuous range, such as a variable-speed pump for supply of cooling water or a multiposition valve allowing flow of air, it is common to use *proportional-integral-derivative* or *PID control*. With PID control, the control action is determined in proportion to the error, the integral of the error, and the derivative of the error with respect to time. The relative weightings given to these functions determine the response of the controller and the overall 'strength' of the control action. PID control is used to determine, for example, whether the pump speed for flow of water in a fermenter cooling coil should be increased by a small or large amount when a certain increase in reactor temperature is detected above the set point. Proper adjustment of PID controllers usually provides excellent regulation of measured variables. However, poorly tuned controllers can destabilise a process and cause continuous or accentuated fluctuations in culture conditions. Considerations for adjusting PID controllers are covered in texts on process control, for example [29].

Close fermentation control requires the simultaneous monitoring and adjustment of many parameters. Instead of individual controllers for each function, it is becoming commonplace to use a single computer or microprocessor for several feedback control loops. The computer logs measurements from a range of sensors in a time sequence and generates electronic signals that may be used directly or indirectly to drive various actuators. The application of computers requires digitisation of signals from the sensor; after digital-to-analogue conversion of the output, the computer or microprocessor can provide the same PID functions as a conventional analogue controller.

14.4.7 Indirect Metabolic Control

Maintaining particular values of temperature or pH in a fermenter can be considered an indirect approach to bioprocess control. The real objective is to optimise the performance of the catalyst and maximise production of the desired product. Certain derived variables such as the oxygen uptake rate and respiratory quotient reflect at least to some extent the biological state of the culture. It can be advantageous to base control actions on the deviation of these metabolic variables from their desired values rather than on environmental conditions.

Indirect metabolic control is often used in fed-batch culture of bakers' yeast. Due to the Crabtree effect, yeast metabolism can switch from respiratory to fermentative mode depending on the prevailing glucose and dissolved oxygen concentrations. The maximum biomass yield from substrate is achieved at relatively low glucose concentrations in the presence of adequate oxygen; fermentative metabolism occurs if the glucose concentration rises above a certain level even though oxygen may be present. Fermentative metabolism should be avoided for biomass production because the yield of cells is reduced as ethanol and carbon dioxide are formed as end-products. The respiratory quotient RQ is a convenient indicator of the metabolic state of the cells in this process: RQ values above about 1 indicate ethanol formation. In industrial bakers' yeast production, the feed rate of glucose to the culture is controlled to maintain RQ values within the desired range.

14.4.8 Programmed Control

Because of the inherent time-varying character of batch and fed-batch fermentations, maintaining a constant environment or constant values of metabolic variables is not always the optimal control strategy. Depending on the process, changes in variables such as pH and temperature at critical times can improve the production rate and yield. Varying the rate of feed is important in fed-batch bakers' yeast fermentations to minimise the Crabtree effect and maximise biomass production; feed rate is also manipulated in *E. coli* fermentations to reduce by-product synthesis. In secondary metabolite fermentations, the specific growth rate should be high at the start of the culture but, at high cell densities, different conditions are required to slow growth and stimulate product formation. Similar strategies are needed to optimise protein synthesis from recombinant organisms. The expression of recombinant product is usually avoided at the start of the culture to prevent any adverse effects on cell growth. However, later in the batch, an inducer may be added to switch on protein synthesis.

For many bioprocesses, a particular time sequence of pH, temperature, dissolved oxygen tension, feed rate, and other variables is required to develop the culture in a way that maximises productivity. A control strategy that can accommodate wide-ranging changes in fermentation variables is *programmed control*, also known as *batch fermenter scheduling*. In programmed control, the control policy consists of a schedule of control functions to be implemented at various times during the process. This type of control requires a detailed understanding of the requirements of the process at various stages and a reasonably complete and accurate mathematical model of the system.

14.4.9 Application of Artificial Intelligence in Bioprocess Control

Several different approaches to bioprocess control have been described in the preceding sections. However, fermentation systems are by nature multivariable with nonlinear and time-varying properties, so that conventional control strategies may not be totally satisfactory. In the industrial environment, additional problems are caused by unexpected disturbances that affect process operation. Most bioprocesses cannot be described exactly by a mathematical model; it is also difficult to identify beforehand the optimum values of metabolic or environmental parameters and controller response functions. The flexibility of computer software is a significant advantage in complex situations; researchers continue to make progress in developing optimal, adaptive, and self-tuning algorithms for bioprocess control. However, although these techniques offer much for improving process performance, the usual mathematical approach may not be the best way to solve bioprocess control problems.

A relatively recent development in control engineering is the application of artificial intelligence techniques, especially *knowledge-based expert systems*. The term 'expert system' usually refers to computer software that processes linguistically formulated 'knowledge' about a particular subject; this knowledge is represented by simple rules. The most important step in building a useful expert system is the extraction of heuristic or subjective rules of thumb from experimental data and human experts. Use can be made of the ever-increasing quantity of measured data from industrial fermentations to synthesise the wide range of rules required for expert systems. The knowledge available is encoded using a computer as IF/THEN rules; for example, IF the cell density is high, THEN dilution with sterile water is recommended, or IF the dissolved oxygen concentration increases quickly AND the rate of carbon dioxide evolution decreases quickly AND the sugar concentration drops AND the pH increases, THEN broth harvest is advised. Information about the progress of highly productive fermentations is stored in the knowledge base.

As measured data become available from fermentations in progress, pattern recognition techniques are applied to assess the results and, in conjunction with the knowledge base and its rules, handle a variety of operating problems or disturbances. The expert system can also learn new information about process behaviour by upgrading its knowledge base. To maximise the potential of expert systems for intelligent supervisory control of fermentation processes, large and representative databases of microbiological information and engineering knowledge must be established for use in rule formulation and interpretation of process phenomena. Expert systems can also be applied for fault diagnosis, estimating unmeasurable fermentation properties, reconciling contradictory data, and computer-aided modelling of metabolism [30–33].

Another area of artificial intelligence with applications in fermentation control is the theory of *neural networks*. Neural networks are particularly suited for extracting useful information from complex and uncertain data such as fermentation measurements, and for formulating generalisations from previous experience. They offer the ability to learn complex, nonlinear relationships between variables and may therefore be useful in the development of fermentation models and for estimating unknown fermentation parameters. Neural network technology is based on an analogy with the brain in that information is stored in the form of connected computational elements or weights (synapses) between artificial neurons.

The most commonly used neural structure is the feed-forward network in which neurons are arranged in layers; incoming signals at the input layer are fed forward through the network connections to the output layer. The topology of the network provides it with powerful data-processing capabilities. To solve a problem, a network structure is chosen and examples of the knowledge to be acquired are shown to the network. It adjusts the synaptic strength of its neural connections so that, in effect, the knowledge is integrated within the structure. When the real data set is presented to the system, the network is able to predict outcomes based on the learning set. For example, information about feed flow rate and substrate and biomass concentrations as a function of time can be used to develop a neural network for analysis of transient behaviour in continuous fermenters that is then able to predict future changes in substrate and cell concentrations [34–36].

14.5 IDEAL REACTOR OPERATION

So far in this chapter we have considered the configuration of bioreactors and aspects of their construction and control. Another important factor affecting reactor performance is mode of operation. There are three principal modes of bioreactor operation: batch, fed-batch, and continuous. The choice of operating mode has a significant effect on substrate conversion, product concentration, susceptibility to contamination, and process reliability.

Characteristics such as the final biomass, substrate, and product concentrations and the time required for conversion can be determined for different reactor operating schemes using mass balances. For a general reaction system, we can relate the rates of change of component masses in the system to the rate of reaction using Eq. (6.5) from Chapter 6:

$$\frac{dM}{dt} = \hat{M}_i - \hat{M}_o + R_G - R_C \tag{6.5}$$

where M is the mass of component A in the vessel, t is time, $\hat{M}_i$ is the mass flow rate of A entering the reactor, $\hat{M}_o$ is the mass flow rate of A leaving, R_G is the mass rate of generation of A by reaction, and R_C is the mass rate of consumption of A by reaction. In this section, we consider the application of Eq. (6.5) to batch, fed-batch, and continuous reactors for enzyme conversion and cell culture.

14.5.1 Batch Operation of a Mixed Reactor

Batch processes operate in closed systems; substrate is added at the beginning of the process and products removed only at the end. Aerobic reactions are not batch operations in the strictest sense; the low solubility of oxygen in aqueous media means that it must be supplied continuously while carbon dioxide and other off-gases are removed. However, bioreactors with neither input nor output of liquid or solid material are classified as batch. If there are no leaks or evaporation from the vessel, the liquid volume in batch reactors can be considered constant.

Most commercial bioreactors are mixed vessels operated in batch. The classic mixed reactor is the stirred tank; however mixed reactors can also be of bubble column, airlift, or other configuration as long as the concentrations of substrate, product, and catalyst inside the vessel are uniform. The cost of running a batch reactor depends on the time taken to achieve the desired product concentration or level of substrate conversion; operating costs are reduced if the reaction is completed quickly. It is therefore useful to be able to predict the time required for batch reactions.

Enzyme Reaction

Let us apply Eq. (6.5) to the limiting substrate in a batch enzyme reactor such as that shown in Figure 14.19. $\hat{M}_i = \hat{M}_o = 0$ because there is no substrate flow into or out of the vessel. The mass of substrate in the reactor, M, is equal to the substrate concentration s multiplied by the liquid volume V. As substrate is not generated in the reaction, $R_G = 0$. The rate of substrate consumption R_C is equal to the volumetric rate of enzyme reaction v multiplied by V; for Michaelis–Menten kinetics, v is given by Eq. (12.37). Therefore, from Eq. (6.5), the mass balance is:

$$\frac{d(sV)}{dt} = \frac{-v_{max}s}{K_m + s}V \tag{14.7}$$

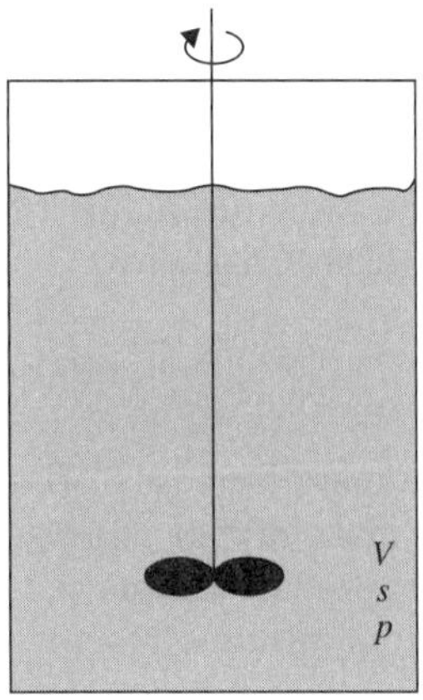

FIGURE 14.19 Flow sheet for a stirred batch enzyme reactor.

where v_{max} is the maximum rate of enzyme reaction and K_m is the Michaelis constant (Section 12.3.3). Because V is constant in batch reactors, we can take V outside of the differential and cancel it from both sides of the equation to give:

$$\frac{ds}{dt} = \frac{-v_{max}\, s}{K_m + s} \tag{14.8}$$

Integration of this differential equation provides an expression for the batch reaction time. Assuming that v_{max} and K_m are constant during the reaction, separating variables gives:

$$-\int dt = \int \frac{K_m + s}{v_{max}\, s}\, ds \tag{14.9}$$

Integrating with initial condition $s = s_0$ at $t = 0$ gives:

$$t_b = \frac{K_m}{v_{max}} \ln\frac{s_0}{s_f} + \frac{s_0 - s_f}{v_{max}} \tag{14.10}$$

where t_b is the batch reaction time required to reduce the substrate concentration from s_0 to s_f. The batch reaction time required to produce a certain concentration of product can be determined from Eq. (14.10) and stoichiometric relationships.

EXAMPLE 14.1 TIME COURSE FOR BATCH ENZYME CONVERSION

An enzyme is used to produce a compound used in the manufacture of sunscreen lotion. v_{max} for the enzyme is 2.5 mmol $m^{-3}\,s^{-1}$; K_m is 8.9 mM. The initial concentration of substrate is 12 mM. Plot the time required for batch reaction as a function of substrate conversion.

Solution

$s_0 = 12$ mM. Converting the units of v_{max} to mM h^{-1}:

$$v_{max} = 2.5\ \text{mmol m}^{-3}\,\text{s}^{-1} \cdot \left|\frac{3600\ \text{s}}{1\ \text{h}}\right| \cdot \left|\frac{1\ \text{m}^3}{1000\ \text{l}}\right| = 9\ \text{mmol l}^{-1}\,\text{h}^{-1} = 9\ \text{mM h}^{-1}$$

The results from application of Eq. (14.10) are tabulated on the next page and plotted in Figure 14.20.

Substrate Conversion (%)	s_f (mM)	t_b (h)
0	12.0	0.00
10	10.8	0.24
20	9.6	0.49
40	7.2	1.04
50	6.0	1.35

(*Continued*)

60	4.8	1.71
80	2.4	2.66
90	1.2	3.48
95	0.60	4.23
99	0.12	5.87

At high substrate conversions, the time required to achieve an incremental increase in conversion is greater than at low conversions. Accordingly, the benefit gained from conversions above 80 to 90% must be weighed against the significantly greater reaction time, and therefore reactor operating costs, involved.

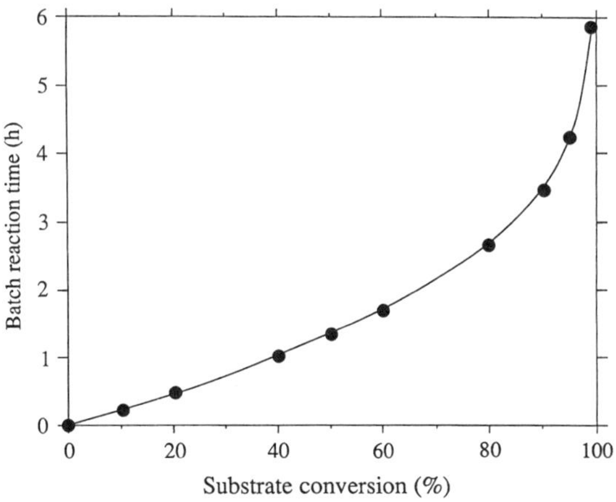

FIGURE 14.20 Batch reaction time as a function of substrate conversion for enzyme reaction in a mixed vessel.

As discussed in Section 12.6, enzymes are subject to deactivation. Accordingly, the concentration of active enzyme in the reactor, and therefore the value of v_{max}, may change during reaction. When deactivation is significant, the variation of v_{max} with time can be expressed using Eq. (12.72) so that Eq. (14.8) becomes:

$$\frac{ds}{dt} = \frac{-v_{max0}\, e^{-k_d t} s}{K_m + s} \tag{14.11}$$

where v_{max0} is the value of v_{max} before deactivation occurs and k_d is the first-order deactivation rate constant. Separating variables gives:

$$-\int e^{-k_d t}\, dt = \int \frac{K_m + s}{v_{max0}\, s}\, ds \tag{14.12}$$

Integrating Eq. (14.12) with initial condition $s = s_0$ at $t = 0$ gives:

$$t_b = \frac{-1}{k_d} \ln\left[1 - k_d\left(\frac{K_m}{v_{max0}} \ln\frac{s_0}{s_f} + \frac{s_0 - s_f}{v_{max0}}\right)\right] \tag{14.13}$$

where t_b is the batch reaction time and s_f is the final substrate concentration.

EXAMPLE 14.2 BATCH REACTION TIME WITH ENZYME DEACTIVATION

The enzyme of Example 14.1 deactivates with half-life 4.4 h. Compare with Figure 14.20 the batch reaction time required to achieve 90% substrate conversion.

Solution

$s_0 = 12$ mM; $v_{max0} = 9$ mM h^{-1}; $K_m = 8.9$ mM; $s_f = (0.1\ s_0) = 1.2$ mM. The deactivation rate constant is calculated from the half-life t_h using Eq. (12.73):

$$k_d = \frac{\ln 2}{t_h} = \frac{\ln 2}{4.4\ \text{h}} = 0.158\ \text{h}^{-1}$$

Substituting values into Eq. (14.13) gives:

$$t_b = \frac{-1}{0.158\ \text{h}^{-1}} \ln\left[1 - 0.158\ \text{h}^{-1}\left(\frac{8.9\ \text{mM}}{9\ \text{mM h}^{-1}} \ln\frac{12\ \text{mM}}{1.2\ \text{mM}} + \frac{(12-1.2)\ \text{mM}}{9\ \text{mM h}^{-1}}\right)\right] = 5.0\ \text{h}$$

The time required to achieve 90% conversion is 5.0 h. Therefore, enzyme deactivation increases the batch reaction time from 3.5 h (Example 14.1) to 5.0 h.

For reactions with immobilised enzyme, Eq. (14.8) must be modified to account for the effect of mass transfer limitations on the rate of enzyme reaction:

$$\frac{ds}{dt} = -\eta_T \frac{v_{max} s}{K_m + s} \tag{14.14}$$

where η_T is the total effectiveness factor incorporating internal and external mass transfer restrictions (Section 13.5), s is the bulk substrate concentration, and v_{max} and K_m are the intrinsic kinetic parameters. In principal, integration of Eq. (14.14) allows evaluation of t_b; however, because η_T is a function of s, integration is not straightforward.

Cell Culture

A similar analysis can be applied to fermentation processes to evaluate batch culture times. Let us perform a mass balance on cells in a well-mixed batch fermenter using Eq. (6.5). A typical flow sheet for this system is shown in Figure 14.21. $\hat{M}_i = \hat{M}_o = 0$ because cells do not flow into or out of the vessel. The mass of cells in the reactor, M, is equal to the cell concentration x multiplied by the liquid volume V. The mass rate of cell growth R_G is equal to $r_X V$ where r_X is the volumetric rate of growth. From Eq. (12.80), $r_X = \mu x$ where μ is the specific growth rate. If cell death takes place in the reactor alongside growth, $R_C = r_d V$ where r_d is the volumetric rate of cell death. From Eq. (12.116), r_d can be expressed using the first-order equation $r_d = k_d x$, where k_d is the specific death constant. Therefore, Eq. (6.5) for cells in a batch reactor is:

$$\frac{d(xV)}{dt} = \mu x V - k_d x V \tag{14.15}$$

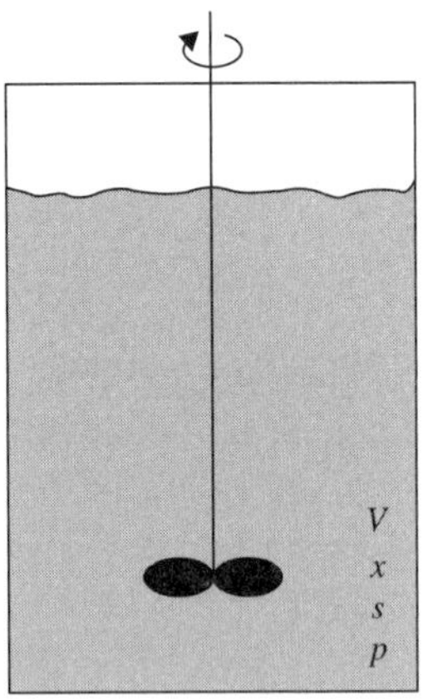

FIGURE 14.21 Flow sheet for a stirred batch fermenter.

For V constant, Eq. (14.15) becomes:

$$\frac{dx}{dt} = (\mu - k_d)\,x \tag{14.16}$$

Because μ in batch culture remains approximately constant and equal to μ_{max} for most of the growth period (Section 12.8.3), if k_d likewise remains constant, we can integrate Eq. (14.16) directly to find the relationship between batch culture time and cell concentration. Using the initial condition $x = x_0$ at $t = 0$:

$$x = x_0\, e^{(\mu_{max} - k_d)t} \tag{14.17}$$

If x_f is the final biomass concentration after batch culture time t_b, rearrangement of Eq. (14.17) gives:

$$t_b = \frac{1}{\mu_{max} - k_d} \ln \frac{x_f}{x_0} \tag{14.18}$$

If the rate of cell death is negligible compared with growth, $k_d \ll \mu_{max}$ and Eqs. (14.17) and (14.18) reduce to:

$$x = x_0\, e^{\mu_{max} t} \tag{14.19}$$

and

$$t_b = \frac{1}{\mu_{max}} \ln \frac{x_f}{x_0} \tag{14.20}$$

Equations (14.18) and (14.20) allow us to calculate the batch culture time required to achieve cell density x_f starting from cell density x_0.

The time required for batch culture can also be related to substrate conversion and product concentration using expressions for the rates of substrate uptake and product formation derived in Chapter 12. First, let us apply Eq. (6.5) to the growth-limiting substrate in a batch fermentation. $\hat{M}_i = \hat{M}_o = 0$ because substrate does not flow into or out of the reactor;

the mass of substrate in the reactor, M, is equal to sV where s is substrate concentration and V is liquid volume. Substrate is not generated; therefore $R_G = 0$. R_C is equal to $r_S V$ where r_S is the volumetric rate of substrate uptake. As discussed in Section 12.11, the expression for r_S depends on whether extracellular product is formed by the culture and the relationship between product synthesis and energy generation in the cell. If product is formed but is not directly coupled with energy metabolism, r_S is given by Eq. (12.107) and:

$$R_C = r_S V = \left(\frac{\mu}{Y_{XS}} + \frac{q_P}{Y_{PS}} + m_S \right) xV \tag{14.21}$$

where μ is the specific growth rate, Y_{XS} is the true biomass yield from substrate, q_P is the specific rate of product formation not directly linked with energy metabolism, Y_{PS} is the true product yield from substrate, and m_S is the maintenance coefficient. Therefore, from Eq. (6.5), the mass balance equation for substrate is:

$$\frac{d(sV)}{dt} = -\left(\frac{\mu}{Y_{XS}} + \frac{q_P}{Y_{PS}} + m_S \right) xV \tag{14.22}$$

For μ equal to μ_{max} and V constant, we can write Eq. (14.22) as:

$$\frac{ds}{dt} = -\left(\frac{\mu_{max}}{Y_{XS}} + \frac{q_P}{Y_{PS}} + m_S \right) x \tag{14.23}$$

Because x is a function of time, we must substitute an expression for x into Eq. (14.23) before it can be integrated. When $\mu = \mu_{max}$ and assuming that cell death is negligible, x is given by Eq. (14.19). Therefore, Eq. (14.23) becomes:

$$\frac{ds}{dt} = -\left(\frac{\mu_{max}}{Y_{XS}} + \frac{q_P}{Y_{PS}} + m_S \right) x_0 \, e^{\mu_{max} t} \tag{14.24}$$

If all of the bracketed terms are constant during culture, Eq. (14.24) can be integrated directly with initial condition $s = s_0$ at $t = 0$ to obtain the following equation:

$$t_b = \frac{1}{\mu_{max}} \ln \left[1 + \frac{s_0 - s_f}{\left(\frac{1}{Y_{XS}} + \frac{q_P}{\mu_{max} Y_{PS}} + \frac{m_S}{\mu_{max}} \right) x_0} \right] \tag{14.25}$$

where t_b is the batch culture time and s_f is the final substrate concentration. Evaluating q_P for products indirectly coupled or not related at all to energy metabolism requires further analysis (Sections 12.10.2 and 12.10.3). However, if no product is formed or if production is directly linked with energy metabolism, the expression for the rate of substrate consumption r_S does not contain a separate term for product synthesis (Sections 12.11.1 and 12.11.2) and Eq. (14.25) can be simplified to:

$$t_b = \frac{1}{\mu_{max}} \ln \left[1 + \frac{s_0 - s_f}{\left(\frac{1}{Y_{XS}} + \frac{m_S}{\mu_{max}} \right) x_0} \right] \tag{14.26}$$

If, in addition, maintenance requirements can be neglected:

$$t_b = \frac{1}{\mu_{max}} \ln \left[1 + \frac{Y_{XS}}{x_0}(s_0 - s_f) \right] \tag{14.27}$$

To obtain an expression for batch culture time as a function of product concentration, we must apply Eq. (6.5) to the product. Again, $\hat{M}_i = \hat{M}_o = 0$. The mass of product in the reactor, M, is equal to pV where p is product concentration and V is liquid volume. Assuming that product is not consumed, $R_C = 0$. R_G is equal to $r_P V$ where r_P is the volumetric rate of product formation. According to Eq. (12.98), for all types of product $r_P = q_P x$ where q_P is the specific rate of product formation. Therefore:

$$R_G = r_P V = q_P x V \tag{14.28}$$

From Eq. (6.5), the mass balance equation for product is:

$$\frac{d(pV)}{dt} = q_P x V \tag{14.29}$$

If cell death is negligible and μ is equal to μ_{max}, x is given by Eq. (14.19). Therefore, for V constant, we can write Eq. (14.29) as:

$$\frac{dp}{dt} = q_P \, x_0 \, e^{\mu_{max} t} \tag{14.30}$$

If q_P is constant, Eq. (14.30) can be integrated directly with initial condition $p = p_0$ at $t = 0$ to obtain the following equation for batch culture time as a function of the final product concentration p_f:

$$t_b = \frac{1}{\mu_{max}} \ln \left[1 + \frac{\mu_{max}}{x_0 \, q_P}(p_f - p_0) \right] \tag{14.31}$$

To summarise the equations for batch culture time, the time required to achieve a certain biomass density can be evaluated using Eq. (14.18) or (14.20). If cell death is negligible, the time needed for a particular level of substrate conversion is calculated using Eq. (14.25), (14.26), or (14.27) depending on the type of product formed and the importance of maintenance metabolism. For negligible cell death and constant q_P, the batch time required to achieve a particular product concentration can be found from Eq. (14.31). Application of these equations is illustrated in Example 14.3.

EXAMPLE 14.3 BATCH CULTURE TIME

Zymomonas mobilis is used to convert glucose to ethanol in a batch fermenter under anaerobic conditions. The yield of biomass from substrate is 0.06 g g^{-1}; Y_{PX} is 7.7 g g^{-1}. The maintenance coefficient is 2.2 g g^{-1} h^{-1}; the specific rate of product formation due to maintenance is 1.1 h^{-1}. The maximum specific growth rate of *Z. mobilis* is approximately 0.3 h^{-1}. Five grams of bacteria are inoculated into 50 litres of medium containing 12 g l^{-1} glucose. Determine the batch culture times required to:

(a) Produce 10 g of biomass
(b) Achieve 90% substrate conversion
(c) Produce 100 g of ethanol

Solution

$Y_{XS} = 0.06\ \text{g g}^{-1}$; $Y_{PX} = 7.7\ \text{g g}^{-1}$; $\mu_{max} = 0.3\ \text{h}^{-1}$; $m_S = 2.2\ \text{g g}^{-1}\ \text{h}^{-1}$; $m_P = 1.1\ \text{h}^{-1}$; $s_0 = 12\ \text{g l}^{-1}$.

$$x_0 = \frac{5\ \text{g}}{50\ \text{l}} = 0.1\ \text{g l}^{-1}$$

(a) If 10 g of biomass are produced, the final amount of biomass present is (10 + 5) g = 15 g. Therefore:

$$x_f = \frac{15\ \text{g}}{50\ \text{l}} = 0.3\ \text{g l}^{-1}$$

From Eq. (14.20):

$$t_b = \frac{1}{0.3\ \text{h}^{-1}} \ln \frac{0.3\ \text{g l}^{-1}}{0.1\ \text{g l}^{-1}} = 3.7\ \text{h}$$

The batch culture time required to produce 10 g of biomass is 3.7 h.

(b) If 90% of the substrate is converted, $s_f = (0.1\ s_0) = 1.2\ \text{g l}^{-1}$. Ethanol synthesis is directly coupled to energy metabolism in the cell; therefore, Eq. (14.26) applies:

$$t_b = \frac{1}{0.3\ \text{h}^{-1}} \ln \left[1 + \frac{(12 - 1.2)\ \text{g l}^{-1}}{\left(\frac{1}{0.06\ \text{g g}^{-1}} + \frac{2.2\ \text{g g}^{-1}\ \text{h}^{-1}}{0.3\ \text{h}^{-1}} \right) 0.1\ \text{g l}^{-1}} \right] = 5.7\ \text{h}$$

The batch culture time required to achieve 90% substrate conversion is 5.7 h.

(c) q_P is calculated using Eq. (12.101). For batch culture with $\mu = \mu_{max}$:

$$q_P = 7.7\ \text{g g}^{-1}\ (0.3\ \text{h}^{-1}) + 1.1\ \text{h}^{-1} = 3.4\ \text{h}^{-1}$$

As no product is present initially, $p_0 = 0$. For production of 100 g ethanol:

$$p_f = \frac{100\ \text{g}}{50\ \text{l}} = 2.0\ \text{g l}^{-1}$$

From Eq. (14.31):

$$t_b = \frac{1}{0.3\ \text{h}^{-1}} \ln \left[1 + \frac{0.3\ \text{h}^{-1}}{(0.1\ \text{g l}^{-1})(3.4\ \text{h}^{-1})} (2\ \text{g l}^{-1}) \right] = 3.4\ \text{h}$$

The batch culture time required to produce 100 g of ethanol is 3.4 h.

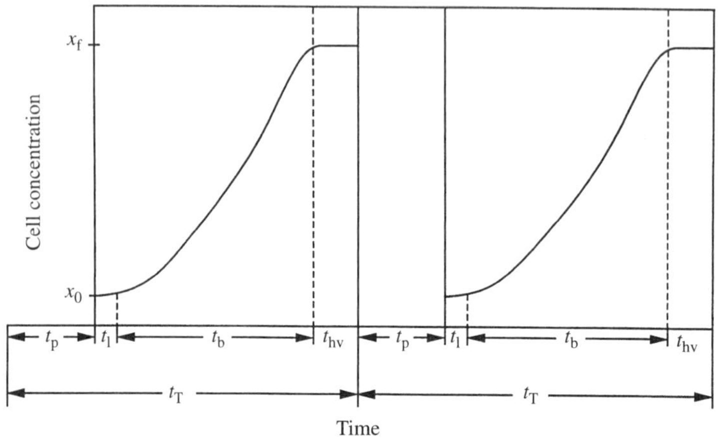

FIGURE 14.22 Preparation, lag, reaction, and harvest times in operation of a batch fermenter.

14.5.2 Total Time for Batch Reaction Cycle

In the above analysis, t_b represented the time required for batch enzyme or cell conversion. In practice, batch operations involve lengthy unproductive periods in addition to t_b. Following the enzyme reaction or fermentation, time t_{hv} is taken to harvest the contents of the reactor and time t_p is needed to clean, sterilise, or otherwise prepare the reactor for the next batch. For cell culture, a lag time of duration t_l often occurs after inoculation of the cells (Section 12.8.1), during which no growth or product formation occurs. These time periods are illustrated for fermentation processes in Figure 14.22. The total *downtime* t_{dn} associated with batch reactor operation is:

$$t_{dn} = t_{hv} + t_p + t_l \tag{14.32}$$

and the total batch reaction time t_T is:

$$t_T = t_b + t_{dn} \tag{14.33}$$

14.5.3 Fed-Batch Operation of a Mixed Reactor

In fed-batch operations, intermittent or continuous feeding of nutrients is used to supplement the reactor contents and provide control over the substrate concentration. By starting with a relatively dilute solution of substrate and adding more nutrients as the conversion proceeds, high growth rates are avoided. This is important, for example, in cultures where the oxygen demand during fast growth is too high for the mass transfer capabilities of the reactor, or when high substrate concentrations are inhibitory or switch on undesirable metabolic pathways. Fed-batch culture is used extensively in the production of bakers' yeast to overcome catabolite repression and control oxygen demand; it is also used routinely for penicillin production. Space must be allowed in fed-batch reactors for the addition of fresh medium; in some cases, a portion of the broth may be removed before injection of additional material. The flow rate and timing of the feed are often determined by monitoring parameters such as the dissolved oxygen tension or exhaust gas composition. As enzyme reactions are rarely carried out as fed-batch operations, we will consider fed-batch reactors for cell culture only.

The flow sheet for a well-mixed fed-batch fermenter is shown in Figure 14.23. The volumetric flow rate of entering feed is F; the concentrations of biomass, growth-limiting substrate, and product in this stream are x_i, s_i, and p_i, respectively. We will assume that F is constant. Owing to input of the feed, the liquid volume V is not constant. Equations for fed-batch culture are derived by carrying out unsteady-state mass balances.

The unsteady-state mass balance equation for total mass in a flow reactor was derived in Chapter 6:

$$\frac{d(\rho V)}{dt} = F_i \rho_i - F_o \rho_o \tag{6.6}$$

where ρ is the density of the reactor contents, V is the liquid volume in the reactor, F_i and F_o are the input and output mass flow rates, and ρ_i and ρ_o are the densities of the input and output streams, respectively. For the fed-batch reactor of Figure 14.23, $F_o = 0$ and $F_i = F$. With dilute solutions such as those used in bioprocessing, we can assume that ρ is constant and that $\rho_i = \rho$; density can then be taken outside of the differential and cancelled through the equation. Therefore, Eq. (6.6) for fed-batch fermenters is:

$$\frac{dV}{dt} = F \tag{14.34}$$

A similar mass balance can be performed for cells based on Eq. (6.5). In fed-batch operations, $\hat{M}_o = 0$. $\hat{M}_i$ is equal to the feed flow rate F multiplied by the cell concentration x_i in the feed. As in Section 14.5.1 (Cell Culture subsection), the mass of cells in the reactor, M, is equal to xV where x is the cell concentration and V is the liquid volume; the rate of biomass generation R_G is equal to $\mu x V$ where μ is the specific growth rate; and the rate of cell death R_C is equal to $k_d x V$ where k_d is the specific death constant. Applying these terms in Eq. (6.5) gives:

$$\frac{d(xV)}{dt} = F x_i + \mu x V - k_d x V \tag{14.35}$$

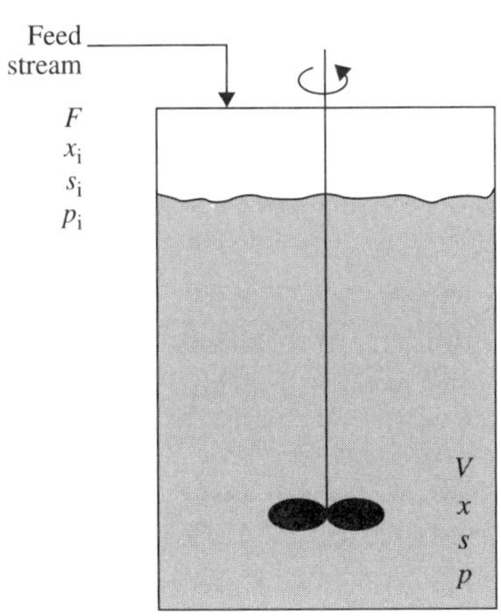

FIGURE 14.23 Flow sheet for a stirred fed-batch fermenter.

Because V in fed-batch culture is not constant, it cannot be taken outside of the differential in Eq. (14.35). We can expand the differential using the product rule of Eq. (E.22) in Appendix E. After grouping terms this gives:

$$x\frac{dV}{dt} + V\frac{dx}{dt} = Fx_i + (\mu - k_d)\,xV \tag{14.36}$$

Applying Eq. (14.34) to Eq. (14.36):

$$xF + V\frac{dx}{dt} = Fx_i + (\mu - k_d)\,xV \tag{14.37}$$

Dividing through by V and rearranging gives:

$$\frac{dx}{dt} = \frac{F}{V}x_i + x\left(\mu - k_d - \frac{F}{V}\right) \tag{14.38}$$

Let us define the *dilution rate* D with dimensions T^{-1}:

$$D = \frac{F}{V} \tag{14.39}$$

In fed-batch systems, V increases with time; therefore, if F is constant, D decreases as the reaction proceeds. Applying Eq. (14.39) to Eq. (14.38):

$$\frac{dx}{dt} = Dx_i + x\,(\mu - k_d - D) \tag{14.40}$$

Equation (14.40) can be simplified for most applications. Usually the feed material is sterile so that $x_i = 0$. If, in addition, the rate of cell death is negligible compared with growth so that $k_d \ll \mu$, Eq. (14.40) becomes:

$$\frac{dx}{dt} = x\,(\mu - D) \tag{14.41}$$

Let us now apply Eq. (6.5) to the limiting substrate in our fed-batch reactor. In this case, $\hat{M}_o$ and R_G are zero and the mass flow rate of substrate entering the reactor, $\hat{M}_i$, is equal to Fs_i. The mass of substrate in the reactor, M, is equal to the substrate concentration s multiplied by the volume V. For fermentations producing product not directly coupled with energy metabolism, R_C is given by Eq. (14.21). Substituting these terms into Eq. (6.5) gives:

$$\frac{d(sV)}{dt} = Fs_i - \left(\frac{\mu}{Y_{XS}} + \frac{q_P}{Y_{PS}} + m_S\right)xV \tag{14.42}$$

where μ is the specific growth rate, Y_{XS} is the true biomass yield from substrate, q_P is the specific rate of product formation, Y_{PS} is the true product yield from substrate, and m_S is the maintenance coefficient. Expanding the differential and applying Eqs. (14.34) and (14.39) gives:

$$\frac{ds}{dt} = D\,(s_i - s) - \left(\frac{\mu}{Y_{XS}} + \frac{q_P}{Y_{PS}} + m_S\right)x \tag{14.43}$$

Equations (14.41) and (14.43) are differential equations for the rates of change of the cell and substrate concentrations in fed-batch reactors. Because D is a function of time, integration of these equations is more complicated than for batch reactors. However, we can derive analytical expressions for fed-batch culture if we simplify Eqs. (14.41) and (14.43). Here, we examine the situation where the reactor is operated first in batch until a high cell density is achieved and the substrate is virtually exhausted. When this condition is reached, fed-batch operation is started with medium flow rate F. As a result, the cell concentration x is maintained high and approximately constant so that $\mathrm{d}x/\mathrm{d}t \approx 0$. From Eq. (14.41), if $\mathrm{d}x/\mathrm{d}t \approx 0$, $\mu \approx D$. Therefore, substituting $\mu \approx D$ into the Monod expression of Eq. (12.91):

$$D \approx \frac{\mu_{\max} s}{K_S + s} \tag{14.44}$$

Rearrangement of Eq. (14.44) gives an expression for the substrate concentration as a function of dilution rate:

$$s \approx \frac{D K_S}{\mu_{\max} - D} \tag{14.45}$$

Let us assume that the culture does not produce product or, if there is product formation, that it is directly linked with energy generation. If maintenance requirements can also be neglected, Eq. (14.43) can be simplified to:

$$\frac{\mathrm{d}s}{\mathrm{d}t} = D\,(s_i - s) - \frac{\mu x}{Y_{XS}} \tag{14.46}$$

When the cell density in the reactor is high, virtually all substrate entering the vessel is consumed immediately; therefore, $s \ll s_i$ and $\mathrm{d}s/\mathrm{d}t \approx 0$. Applying these relationships with $\mu \approx D$ to Eq. (14.46), we obtain:

$$x \approx Y_{XS}\, s_i \tag{14.47}$$

For product synthesis directly coupled with energy metabolism, Eq. (14.47) allows us to derive an approximate expression for the product concentration in fed-batch reactors. Assuming that the feed does not contain product:

$$p \approx Y_{PS}\, s_i \tag{14.48}$$

Even though the cell concentration remains virtually unchanged with $\mathrm{d}x/\mathrm{d}t \approx 0$, because the liquid volume increases with time in fed-batch reactors, the total mass of cells also increases. Consider the rate of increase of total biomass in the reactor, $\mathrm{d}X/\mathrm{d}t$, where X is equal to xV. Using the results of Eqs. (14.34) and (14.47) with $\mathrm{d}x/\mathrm{d}t \approx 0$:

$$\frac{\mathrm{d}X}{\mathrm{d}t} = \frac{\mathrm{d}(xV)}{\mathrm{d}t} = x\frac{\mathrm{d}V}{\mathrm{d}t} + V\frac{\mathrm{d}x}{\mathrm{d}t} = Y_{XS}\, s_i\, F \tag{14.49}$$

Equation (14.49) can now be integrated with initial condition $X = X_0$ at the start of liquid flow to give:

$$X = X_0 + (Y_{XS}\, s_i\, F)\, t_{fb} \tag{14.50}$$

where t_{fb} is the fed-batch time after commencement of feeding. Equation (14.50) indicates that, for Y_{XS}, s_i, and F constant, the total biomass in fed-batch fermenters increases as a linear function of time.

Under conditions of high biomass density and almost complete depletion of substrate, a *quasi-steady-state* condition prevails in fed-batch reactors where $dx/dt \approx 0$, $ds/dt \approx 0$, and $dp/dt \approx 0$. At quasi-steady state, Eqs. (14.47) and (14.45) can be used to calculate the biomass and substrate concentrations in reactors where cell death and maintenance requirements are negligible and product is either absent or directly coupled with energy metabolism. Equation (14.48) allows calculation of the product concentration for metabolites directly coupled with energy metabolism. At quasi-steady state, the specific growth rate μ and the dilution rate F/V are approximately equal; therefore as V increases, the growth rate decreases. When fed-batch operation is used for production of biomass such as bakers' yeast, it is useful to be able to predict the total mass of cells in the reactor as a function of time. An expression for total biomass is given by Eq. (14.50). Note that under quasi-steady-state conditions, x, s, and p are almost constant, but μ, V, D, and X are changing. Further details of fed-batch operation are given by Pirt [37].

14.5.4 Continuous Operation of a Mixed Reactor

Bioreactors are operated continuously in a few bioprocess industries such as brewing, production of bakers' yeast, and waste treatment. Enzyme conversions can also be carried out using continuous systems. The flow sheet for a continuous mixed reactor is shown in Figure 14.24. If the vessel is well mixed, the product stream has the same composition as the liquid in the reactor (Section 6.4). Therefore, when continuous reactors are used with freely suspended cells or enzymes, the catalyst is continuously withdrawn from the vessel in the product stream. For enzymes this is a serious shortcoming as catalyst is not produced by the reaction; however, in cell culture, growth supplies additional cells to replace those removed. Continuous reactors are used with free enzymes only if the enzyme is inexpensive and can be added continuously to maintain the catalyst concentration. With

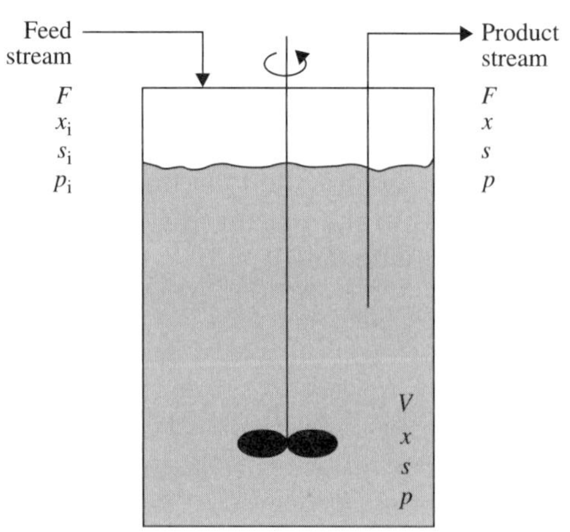

FIGURE 14.24 Flow sheet for a continuous stirred tank fermenter.

more costly enzymes, continuous operation may be feasible if the enzyme is immobilised and retained inside the vessel. Well-mixed continuous reactors are often referred to using the abbreviation CSTR, meaning *continuous stirred tank reactor*. The term CSTF is also sometimes used to denote *continuous stirred tank fermenter*.

Different steady-state operating strategies are available for continuous fermenters. In a *chemostat* the liquid volume is kept constant by setting the inlet and outlet flow rates equal. The dilution rate is therefore constant and steady state is achieved as concentrations in the chemostat adjust themselves to the feed rate. In a *turbidostat*, the liquid volume is kept constant by setting the outlet flow rate equal to the inlet flow rate; however, the inlet flow rate is adjusted to keep the biomass concentration constant. Thus, in a turbidostat, the dilution rate adjusts to the steady-state value required to achieve the desired biomass concentration. Turbidostats require more complex monitoring and control systems than chemostats and are not used at large scales. Accordingly, we will concentrate here on chemostat operation.

Characteristic operating parameters for continuous reactors are the dilution rate D defined in Eq. (14.39) and the average *residence time* τ. These parameters are related as follows:

$$\tau = \frac{1}{D} = \frac{V}{F} \tag{14.51}$$

In continuous reactor operation, the amount of material that can be processed over a given period of time is represented by the flow rate F. Therefore, for a given throughput, the reactor size V and associated capital and operating costs are minimised when τ is made as small as possible. Continuous reactor theory allows us to determine relationships between τ (or D) and the steady-state cell, substrate, and product concentrations in the reactor. This theory is based on steady-state mass balances derived using Eq. (6.5).

Enzyme Reaction

Let us apply Eq. (6.5) to the limiting substrate in a continuous enzyme reactor operated at steady state. The mass flow rate of substrate entering the reactor, $\hat{M}_i$, is equal to Fs_i; $\hat{M}_o$, the mass flow rate of substrate leaving, is Fs. As substrate is not generated in the reactor, $R_G = 0$. The rate of substrate consumption R_C is equal to the volumetric rate of enzyme reaction v multiplied by V; for Michaelis–Menten kinetics, v is given by Eq. (12.37). The left side of Eq. (6.5) is zero when the system is operated at steady state. Therefore, the steady-state substrate mass balance equation for continuous enzyme reaction is:

$$Fs_i - Fs - \frac{v_{max}s}{K_m + s}V = 0 \tag{14.52}$$

where v_{max} is the maximum rate of reaction and K_m is the Michaelis constant. For reactions with free enzyme, we assume that the enzyme lost in the product stream is replaced continuously so that v_{max} remains constant and steady state is achieved. Dividing through by V and applying the definition of the dilution rate from Eq. (14.39) gives:

$$D\,(s_i - s) = \frac{v_{max}s}{K_m + s} \tag{14.53}$$

If v_{max}, K_m, and s_i are known, Eq. (14.53) can be used to calculate the dilution rate required to achieve a particular level of substrate conversion. The steady-state product concentration can then be evaluated from stoichiometry.

For reactions with immobilised enzymes, Eq. (14.53) must be modified to account for mass transfer effects on the rate of enzyme reaction:

$$D\,(s_i - s) = \frac{\eta_T v_{max} s}{K_m + s} \tag{14.54}$$

where η_T is the total effectiveness factor (Section 13.5), s is the bulk substrate concentration, and v_{max} and K_m are the intrinsic kinetic parameters. η_T can be calculated for constant s using the theory for heterogeneous reactions outlined in Chapter 13.

EXAMPLE 14.4 IMMOBILISED ENZYME REACTION IN A CSTR

Mushroom tyrosinase is immobilised in spherical beads of diameter 2 mm for conversion of tyrosine to dihydroxyphenylalanine (DOPA) in a continuous, well-mixed bubble column. The Michaelis constant for the immobilised enzyme is 2 gmol m^{-3}. A solution containing 15 gmol m^{-3} tyrosine is fed into the reactor; the desired level of substrate conversion is 99%. The reactor is loaded with beads at a density of 0.25 m^3 m^{-3}; all enzyme is retained within the vessel. The intrinsic v_{max} for the immobilised enzyme is 1.5×10^{-2} gmol s^{-1} per m^3 of catalyst. The effective diffusivity of tyrosine in the beads is 7×10^{-10} m^2 s^{-1}. External mass transfer effects are negligible. Immobilisation stabilises the enzyme so that deactivation is minimal over the operating period. Determine the reactor volume needed to treat 18 m^3 of tyrosine solution per day.

Solution

$K_m = 2$ gmol m^{-3}; $v_{max} = 1.5 \times 10^{-2}$ gmol s^{-1} m^{-3}; $R = 10^{-3}$ m; $\mathcal{D}_{Ae} = 7 \times 10^{-10}$ m^2 s^{-1}; $s_i =$ 15 gmol m^{-3}. Converting the feed flow rate to units of m^3 s^{-1}:

$$F = 18\ \text{m}^3\ \text{day}^{-1} \cdot \left|\frac{1\ \text{day}}{24\ \text{h}}\right| \cdot \left|\frac{1\ \text{h}}{3600\ \text{s}}\right| = 2.08 \times 10^{-4}\ \text{m}^3\ \text{s}^{-1}$$

For 99% conversion of tyrosine, the outlet and therefore the internal substrate concentration $s =$ (0.01 s_i) = 0.15 gmol m^{-3}. As $s \ll K_m$, we can assume that the reaction follows first-order kinetics (Section 12.3.3) with $k_1 = v_{max}/K_m = 7.5 \times 10^{-3}$ s^{-1}. The first-order Thiele modulus for spherical catalysts is calculated from Table 13.2:

$$\phi_1 = \frac{R}{3}\sqrt{\frac{k_1}{\mathcal{D}_{Ae}}} = \frac{10^{-3}\ \text{m}}{3}\sqrt{\frac{7.5 \times 10^{-3}\ \text{s}^{-1}}{7 \times 10^{-10}\ \text{m}^2\ \text{s}^{-1}}} = 1.09$$

From Table 13.3 or Figure 13.11, $\eta_{i1} = 0.64$. As there is negligible external mass transfer resistance, $\eta_{e1} = 1$ and, from Eq. (13.46), $\eta_T = 0.64$. Substituting values into Eq. (14.54) gives:

$$D\,(15 - 0.15)\ \text{gmol m}^{-3} = \frac{0.64\,(1.5 \times 10^{-2}\ \text{gmol s}^{-1}\ \text{m}^{-3})\,(0.15\ \text{gmol m}^{-3})}{2\ \text{gmol m}^{-3} + 0.15\ \text{gmol m}^{-3}}$$

$$D = 4.51 \times 10^{-5}\ \text{s}^{-1}$$

From Eq. (14.51):

$$V = \frac{F}{D} = \frac{2.08 \times 10^{-4}\ \mathrm{m^3\ s^{-1}}}{4.51 \times 10^{-5}\ \mathrm{s^{-1}}} = 4.6\ \mathrm{m^3}$$

The reactor volume needed to treat 18 $\mathrm{m^3}$ of tyrosine solution per day is 4.6 $\mathrm{m^3}$.

Cell Culture

Let us consider the reactor of Figure 14.24 operated as a continuous fermenter and apply Eq. (6.5) for steady-state mass balances on biomass, substrate, and product.

For biomass, $\hat{M}_i$ in Eq. (6.5) is the mass flow rate of cells entering the reactor: $\hat{M}_i = Fx_i$. $\hat{M}_o$ is the mass flow rate of cells leaving: $\hat{M}_o = Fx$. The other terms in Eq. (6.5) are the same as in Section 14.5.1 (Cell Culture subsection): $R_G = \mu x V$ where μ is the specific growth rate and V is the liquid volume, and $R_C = k_d x V$ where k_d is the specific death constant. At steady state, the left side of Eq. (6.5) is zero. Therefore, the steady-state mass balance equation for biomass is:

$$Fx_i - Fx + \mu x V - k_d x V = 0 \qquad (14.55)$$

Usually, the feed stream in continuous culture is sterile so that $x_i = 0$. If, in addition, the rate of cell death is negligible compared with growth, $k_d \ll \mu$ and Eq. (14.55) becomes:

$$\mu x V = Fx \qquad (14.56)$$

Cancelling x from both sides, dividing by V, and applying the definition of dilution rate from Eq. (14.39) gives:

$$\mu = D \qquad (14.57)$$

Applying Eq. (14.57) to the Monod expression of Eq. (12.91) gives an equation for the steady-state concentration of limiting substrate in the reactor:

$$s = \frac{D K_S}{\mu_{max} - D} \qquad (14.58)$$

Let us now apply Eq. (6.5) at steady state to the limiting substrate. In this case, $\hat{M}_i = Fs_i$, $\hat{M}_o = Fs$, and $R_G = 0$. If product is formed but is not directly coupled with energy metabolism, R_C is given by Eq. (14.21). Therefore, the steady-state mass balance equation for substrate is:

$$Fs_i - Fs - \left(\frac{\mu}{Y_{XS}} + \frac{q_P}{Y_{PS}} + m_S \right) xV = 0 \qquad (14.59)$$

where μ is the specific growth rate, Y_{XS} is the true biomass yield from substrate, q_P is the specific rate of product formation not directly linked with energy metabolism, Y_{PS} is the true product yield from substrate, and m_S is the maintenance coefficient. In Eq. (14.59), we can divide through by V, substitute the definition of dilution rate from Eq. (14.39), and replace μ with D according to Eq. (14.57). Rearrangement then gives the following expression for the steady-state cell concentration x:

$$x = \frac{D(s_i - s)}{\frac{D}{Y_{XS}} + \frac{q_P}{Y_{PS}} + m_S} \tag{14.60}$$

Equation (14.60) can be simplified if there is no product synthesis or if production is directly linked with energy metabolism:

$$x = \frac{D(s_i - s)}{\frac{D}{Y_{XS}} + m_S} \tag{14.61}$$

If, in addition, maintenance effects can be ignored, Eq. (14.61) becomes:

$$x = (s_i - s)Y_{XS} \tag{14.62}$$

Substituting for s from Eq. (14.58), we obtain an expression for the steady-state cell concentration in a CSTR:

$$x = \left(s_i - \frac{DK_S}{\mu_{max} - D}\right)Y_{XS} \tag{14.63}$$

Equation (14.63) is valid at steady state in the absence of maintenance requirements and when product synthesis is either absent or directly linked with energy metabolism.

We can also apply Eq. (6.5) for a steady-state mass balance on fermentation product. In this case, $\hat{M}_i = Fp_i$ and $\hat{M}_o = Fp$. R_G is given by Eq. (14.28) and $R_C = 0$. Therefore, Eq. (6.5) becomes:

$$Fp_i - Fp + q_P xV = 0 \tag{14.64}$$

where q_P is the specific rate of formation for all classes of product. Dividing through by V, substituting the definition of the dilution rate from Eq. (14.39), and rearranging gives an expression for the steady-state product concentration as a function of biomass concentration x:

$$p = p_i + \frac{q_P x}{D} \tag{14.65}$$

x in Eq. (14.65) can be evaluated from Eq. (14.60), (14.61), or (14.62). Evaluation of q_P depends on the type of product formed (Section 12.10).

Equation (14.58) is an explicit expression for the steady-state substrate concentration in a chemostat. The steady-state biomass concentration x can be evaluated from Eq. (14.60), (14.61), or (14.62); the choice of expression for biomass depends on the relative significance of maintenance metabolism and the type of product, if any, produced. If q_P is known, the product concentration in the reactor can be evaluated from Eq. (14.65). In the simplest case when products are either absent or directly linked to energy generation and maintenance effects can be neglected, the chemostat is represented by Eqs. (14.58) and (14.63). The form of these equations is shown in Figure 14.25. At low feed rates

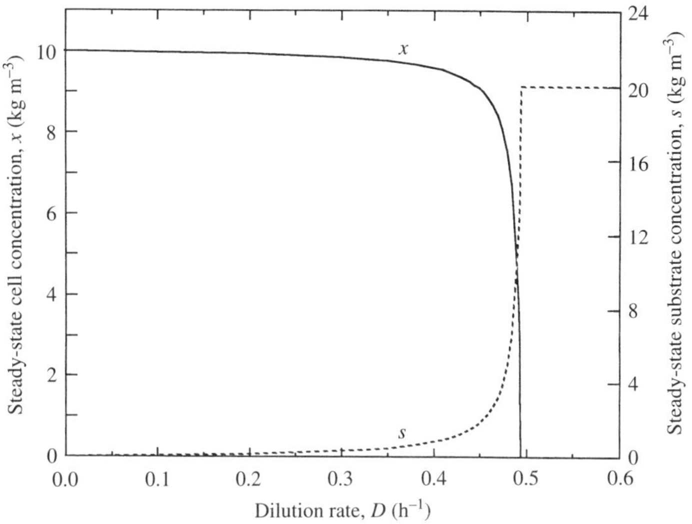

FIGURE 14.25 Steady-state cell and substrate concentrations as a function of dilution rate in a chemostat. The curves were calculated using the following parameter values: $\mu_{max} = 0.5\ h^{-1}$, $K_S = 0.2\ kg\ m^{-3}$, $Y_{XS} = 0.5\ kg\ kg^{-1}$, and $s_i = 20\ kg\ m^{-3}$.

(i.e., $D \to 0$) nearly all the substrate is consumed at steady state so that, from Eq. (14.62), $x \approx s_i Y_{XS}$. As D increases, s increases slowly at first and then more rapidly as D approaches μ_{max}; correspondingly, x decreases so that $x \to 0$ as $D \to \mu_{max}$. The condition at high dilution rate whereby x reduces to zero is known as *washout*; washout of cells occurs when the rate of cell removal in the reactor outlet stream is greater than the rate of generation by growth. For systems with negligible maintenance requirements and either zero or energy-associated product formation, the critical dilution rate D_{crit} at which the steady-state biomass concentration just becomes zero can be estimated by substituting $x = 0$ into Eq. (14.63) and solving for D:

$$D_{crit} = \frac{\mu_{max} s_i}{K_S + s_i} \tag{14.66}$$

For most cell cultures $K_S \ll s_i$; therefore $D_{crit} \approx \mu_{max}$. To avoid washout of cells from the chemostat, the operating dilution rate must always be less than D_{crit}. Near washout, the system is very sensitive to small changes in D that cause relatively large shifts in x and s.

The rate of biomass production in a CSTR is equal to the rate at which cells leave the reactor, Fx. The volumetric productivity Q_X is therefore equal to Fx divided by V. Applying the definition of the dilution rate from Eq. (14.39):

$$Q_X = \frac{Fx}{V} = Dx \tag{14.67}$$

where Q_X is the volumetric rate of biomass production. Similarly the volumetric rate of product formation Q_P is:

$$Q_P = \frac{Fp}{V} = Dp \tag{14.68}$$

When maintenance requirements are negligible and product formation is either absent or energy-associated, we can substitute into Eq. (14.67) the expression for x from Eq. (14.63):

$$Q_X = D\left(s_i - \frac{D K_S}{\mu_{max} - D}\right) Y_{XS} \tag{14.69}$$

This relationship between Q_X and D is shown in Figure 14.26. The rate of biomass production reaches a maximum at the optimum dilution rate for biomass productivity D_{opt}; therefore, at D_{opt} the slope $dQ_X/dD = 0$. Differentiating Eq. (14.69) with respect to D and equating to zero provides an expression for D_{opt}:

$$D_{opt} = \mu_{max}\left(1 - \sqrt{\frac{K_S}{K_S + s_i}}\right) \tag{14.70}$$

Operation of a chemostat at D_{opt} gives the maximum rate of biomass production from the reactor. However, because D_{opt} is usually very close to D_{crit}, it may not be practical to operate at D_{opt}. Small variations of dilution rate in this region can cause large fluctuations in x and s and, unless the dilution rate is controlled very precisely, washout may occur.

Excellent agreement between chemostat theory and experimental results has been found for many culture systems. When deviations occur, they are due primarily to imperfect operation of the reactor. For example, if the vessel is not well mixed, some liquid will have higher residence time in the reactor than the rest and the concentrations will not be uniform. Under these conditions, the equations derived in this section do not hold. Similarly, if cells adhere to glass or metal surfaces in the reactor and produce wall growth, biomass will be retained in the vessel and washout will not occur even at high dilution rates. Other deviations occur if inadequate time is allowed for the system to reach steady state.

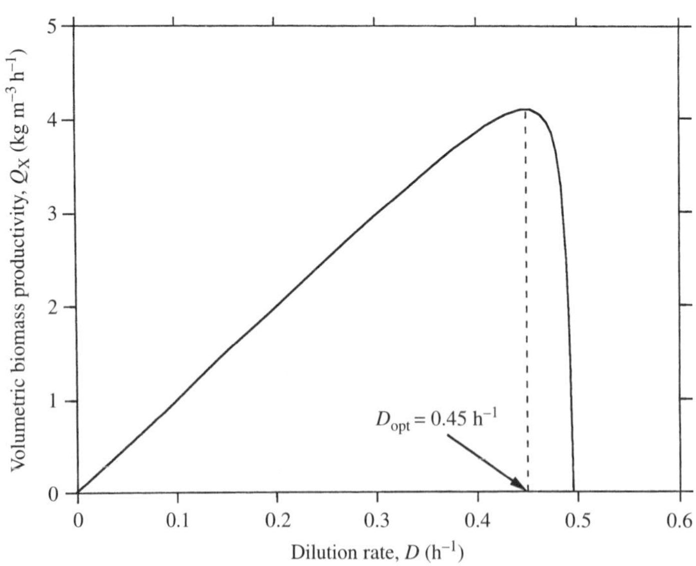

FIGURE 14.26 Steady-state volumetric biomass productivity as a function of dilution rate in a chemostat. The curve was calculated using the following parameter values: $\mu_{max} = 0.5$ h^{-1}, $K_S = 0.2$ kg m^{-3}, $Y_{XS} = 0.5$ kg kg^{-1}, and $s_i = 20$ kg m^{-3}.

EXAMPLE 14.5 STEADY-STATE CONCENTRATIONS IN A CHEMOSTAT

The *Zymomonas mobilis* cells of Example 14.3 are used for chemostat culture in a 60-m^3 fermenter. The feed contains 12 $g\,l^{-1}$ of glucose; K_S for the organism is 0.2 $g\,l^{-1}$.

(a) What flow rate is required for a steady-state substrate concentration of 1.5 $g\,l^{-1}$?
(b) At the flow rate of (a), what is the cell density?
(c) At the flow rate of (a), what concentration of ethanol is produced?

Solution

$Y_{XS} = 0.06\,g\,g^{-1}$; $Y_{PX} = 7.7\,g\,g^{-1}$; $\mu_{max} = 0.3\,h^{-1}$; $K_S = 0.2\,g\,l^{-1}$; $m_S = 2.2\,g\,g^{-1}\,h^{-1}$; $s_i = 12\,g\,l^{-1}$; $V = 60\,m^3$. From Example 14.3, $q_P = 3.4\,h^{-1}$. From the general definition of yield in Section 12.7.1, we can deduce that $Y_{PS} = Y_{PX}\,Y_{XS} = 0.46\,g\,g^{-1}$.

(a) $s = 1.5\,g\,l^{-1}$. Combining the Monod expression of Eq. (12.91) with Eq. (14.57) gives an expression for D:

$$D = \frac{\mu_{max}s}{K_S + s} = \frac{(0.3\,h^{-1})(1.5\,g\,l^{-1})}{0.2\,g\,l^{-1} + 1.5\,g\,l^{-1}} = 0.26\,h^{-1}$$

From the definition of dilution rate in Eq. (14.39):

$$F = D\,V = (0.26\,h^{-1})(60\,m^3) = 15.6\,m^3\,h^{-1}$$

The flow rate giving a steady-state substrate concentration of 1.5 $g\,l^{-1}$ is 15.6 $m^3\,h^{-1}$.

(b) When product synthesis is coupled with energy metabolism, as is the case for ethanol, x is evaluated using Eq. (14.61). Therefore:

$$x = \frac{(0.26\,h^{-1})(12 - 1.5)\,g\,l^{-1}}{\left(\frac{0.26\,h^{-1}}{0.06\,g\,g^{-1}} + 2.2\,g\,g^{-1}\,h^{-1}\right)} = 0.42\,g\,l^{-1}$$

The steady-state cell density at a flow rate of 15.6 $m^3\,h^{-1}$ is 0.42 $g\,l^{-1}$.

(c) Assuming that ethanol is not present in the feed, $p_i = 0$. The steady-state product concentration is given by Eq. (14.65):

$$p = \frac{(3.4\,h^{-1})(0.42\,g\,l^{-1})}{0.26\,h^{-1}} = 5.5\,g\,l^{-1}$$

The steady-state ethanol concentration at a flow rate of 15.6 $m^3\,h^{-1}$ is 5.5 $g\,l^{-1}$.

EXAMPLE 14.6 SUBSTRATE CONVERSION AND BIOMASS PRODUCTIVITY IN A CHEMOSTAT

A 5-m^3 fermenter is operated continuously using a feed substrate concentration of 20 $kg\,m^{-3}$. The microorganism cultivated in the reactor has the following characteristics: $\mu_{max} = 0.45\,h^{-1}$; $K_S = 0.8\,kg\,m^{-3}$; $Y_{XS} = 0.55\,kg\,kg^{-1}$.

(a) What feed flow rate is required to achieve 90% substrate conversion?
(b) How does the biomass productivity at 90% substrate conversion compare with the maximum possible?

Solution

(a) For 90% substrate conversion, $s = (0.1\ s_i) = 2\ \text{kg m}^{-3}$. Combining the Monod expression of Eq. (12.91) with Eq. (14.57) gives an expression for D:

$$D = \frac{\mu_{max} s}{K_S + s} = \frac{(0.45\ \text{h}^{-1})(2\ \text{kg m}^{-3})}{0.8\ \text{kg m}^{-3} + 2\ \text{kg m}^{-3}} = 0.32\ \text{h}^{-1}$$

From Eq. (14.39):

$$F = D\,V = (0.32\ \text{h}^{-1})(5\ \text{m}^3) = 1.6\ \text{m}^3\ \text{h}^{-1}$$

The feed flow rate required for 90% substrate conversion is $1.6\ \text{m}^3\ \text{h}^{-1}$.

(b) Assuming that maintenance requirements and product formation are negligible, from Eq. (14.69):

$$Q_X = 0.32\ \text{h}^{-1}\left(20\ \text{kg m}^{-3} - \frac{(0.32\ \text{h}^{-1})(0.8\ \text{kg m}^{-3})}{(0.45\ \text{h}^{-1} - 0.32\ \text{h}^{-1})}\right)0.55\ \text{kg kg}^{-1} = 3.17\ \text{kg m}^{-3}\ \text{h}^{-1}$$

The maximum biomass productivity occurs at D_{opt}, which can be evaluated using Eq. (14.70):

$$D_{opt} = 0.45\ \text{h}^{-1}\left(1 - \sqrt{\frac{0.8\ \text{kg m}^{-3}}{0.8\ \text{kg m}^{-3} + 20\ \text{kg m}^{-3}}}\right) = 0.36\ \text{h}^{-1}$$

The maximum biomass productivity is determined from Eq. (14.69) with $D = D_{opt}$:

$$Q_{X,max} = 0.36\ \text{h}^{-1}\left(20\ \text{kg m}^{-3} - \frac{(0.36\ \text{h}^{-1})(0.8\ \text{kg m}^{-3})}{(0.45\ \text{h}^{-1} - 0.36\ \text{h}^{-1})}\right)0.55\ \text{kg kg}^{-1} = 3.33\ \text{kg m}^{-3}\ \text{h}^{-1}$$

Therefore, the biomass productivity at 90% substrate conversion is $3.17/3.33 \times 100\% = 95\%$ of the theoretical maximum.

14.5.5 Chemostat with Immobilised Cells

Consider the continuous stirred tank immobilised cell fermenter shown in Figure 14.27. Spherical particles containing cells are kept suspended and well mixed by the stirrer. The concentration of immobilised cells per unit volume of liquid in the reactor is x_{im}. Let us assume that x_{im} is constant; this is achieved if all particles are retained in the vessel and any cells produced by immobilised cell growth are released into the medium. The concentration of suspended cells is x_s. We will assume that the intrinsic specific growth rates of suspended and immobilised cells are the same and equal to μ. Suspended cells are removed from the reactor in the product stream; immobilised cells are retained inside the vessel. For simplicity, let us assume that cell death and maintenance requirements are

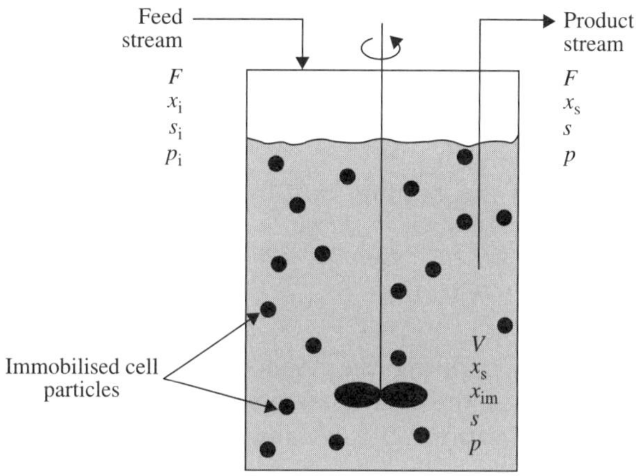

FIGURE 14.27 Flow sheet for a continuous stirred tank fermenter with immobilised cells.

negligible, the reactor feed is sterile, and any product synthesis is directly coupled with energy metabolism.

The system shown in Figure 14.27 reaches steady state. The relationships between the operating variables and the concentrations inside the reactor can be determined using mass balances. Let us consider a mass balance on suspended cells. At steady state, the mass balance equation is similar to Eq. (14.55) except that x_i is zero for sterile feed and cell death is assumed to be negligible. In addition, as suspended cells are produced by growth of both the suspended and immobilised cell populations, the equation for the immobilised cell fermenter must contain two cell generation terms instead of one:

$$-Fx_s + \mu x_s V + \mu x_{im} V = 0 \tag{14.71}$$

If diffusional limitations affect the growth rate of the immobilised cells, μx_{im} must be replaced by $\eta_T \mu x_{im}$ where η_T is the total effectiveness factor defined in Section 13.5. Dividing through by V and applying the definition of the dilution rate from Eq. (14.39) gives:

$$Dx_s = \mu\,(x_s + \eta_T x_{im}) \tag{14.72}$$

or

$$D = \mu\left(1 + \frac{\eta_T x_{im}}{x_s}\right) \tag{14.73}$$

For $x_{im} = 0$, Eq. (14.73) reduces to Eq. (14.57) for a chemostat containing suspended cells only: $\mu = D$.

The steady-state mass balance equation for the limiting substrate can be derived from Eq. (6.5) with $\hat{M}_i = Fs_i$, $\hat{M}_o = Fs$, and $R_G = 0$. Both cell populations consume substrate: in the absence of product- and maintenance-associated substrate requirements, the rate of substrate consumption R_C can be related directly to the growth rates of the immobilised and suspended cells using the biomass yield coefficient Y_{XS}. If we assume that the value of Y_{XS} is the same for all cells, by analogy with Eq. (14.59), the mass balance equation for substrate is:

$$Fs_i - Fs - \frac{\mu x_s}{Y_{XS}}V - \frac{\eta_T \mu x_{im}}{Y_{XS}}V = 0 \tag{14.74}$$

Dividing through by V, applying the definition of the dilution rate from Eq. (14.39), and rearranging gives:

$$D(s_i - s) = \frac{\mu}{Y_{XS}}(x_s + \eta_T x_{im}) \tag{14.75}$$

By manipulating Eqs. (14.73) and (14.75) and substituting the Monod expression for μ from Eq. (12.91), the following relationship between the steady-state substrate concentration s, dilution rate D, and immobilised cell concentration x_{im} is obtained:

$$\frac{\mu_{max}s}{K_S + s} = \frac{D(s_i - s)Y_{XS}}{(s_i - s)Y_{XS} + \eta_T x_{im}} \tag{14.76}$$

The form of Eq. (14.76) is shown in Figure 14.28 as a graph of the percentage substrate conversion versus dilution rate. For a chemostat with suspended cells only (i.e., $x_{im} = 0$), at steady state $D = \mu$ and the maximum operating dilution rate D_{crit} is limited by the maximum specific growth rate of the cells. From Eq. (14.73), for any $x_{im} > 0$, D at steady state in the immobilised cell reactor is greater than μ. Accordingly, the dilution rate is no longer limited by the maximum specific growth rate of the cells and, as shown in Figure 14.28, immobilised cell chemostats can be operated at D considerably greater than D_{crit} without washout. At a given dilution rate, the presence of immobilised cells also improves substrate conversion and reduces the amount of substrate lost in the product stream.

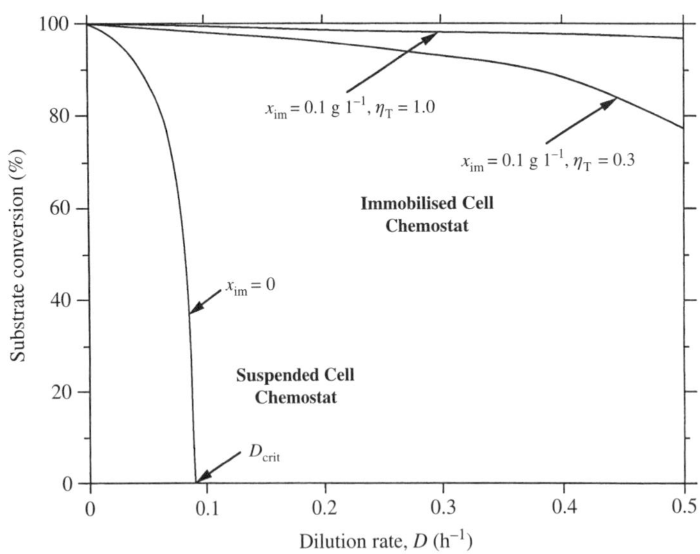

FIGURE 14.28 Steady-state substrate conversion in a chemostat as a function of dilution rate with and without immobilised cells. The curves were calculated using the following parameter values: $\mu_{max} = 0.1\ h^{-1}$, $K_S = 10^{-3}\ g\ l^{-1}$, $Y_{XS} = 0.5\ g\ g^{-1}$, and $s_i = 8 \times 10^{-3}\ g\ l^{-1}$.

However, reaction rates with immobilised cells can be reduced significantly by the effects of mass transfer in and around the particles. As illustrated in Figure 14.28, at the same concentration of immobilised cells, the substrate conversion at $\eta_T = 1$ is greater than at lower values of η_T when mass transfer limitations are significant.

14.5.6 Chemostat Cascade

The joining together of two or more CSTRs in series produces a multistage process in which conditions such as pH, temperature, and medium composition can be varied in each reactor. This is advantageous if the reactor conditions required for growth are different from those required for product synthesis, for example, in the production of recombinant proteins and many metabolites not directly linked with energy metabolism. One way of operating a two-stage chemostat cascade is shown in Figure 14.29. In this process, the product stream from the first reactor feeds directly into the second reactor. Substrate leaving the first reactor at concentration s_1 is converted in the second tank so that $s_2 < s_1$ and $p_2 > p_1$. In some applications, the second CSTR is supplemented with fresh medium containing nutrients, inducers, or inhibitors for optimal product formation.

Design equations for cell and enzyme CSTR cascades can be derived as a simple extension of the theory developed in Section 14.5.4; the same mass balance principles are applied and steady state is assumed at each reactor stage. Details can be found in other references [5, 37, 38]. In fermenter cascades, cells entering the second and subsequent vessels may go through periods of unbalanced growth as they adapt to the new environmental conditions in each reactor. Therefore, the use of simple unstructured metabolic models such as those outlined in this text does not always give accurate results. Nevertheless, it can be shown that the total reactor residence time required to achieve a given degree of substrate conversion is significantly lower using two CSTRs in series than if only one CSTR were used. In other words, the total reactor volume required is reduced using two smaller tanks in series compared with a single large tank. Usually however, only two to four reactors in series are justified as the benefits associated with adding successive stages diminish significantly [39].

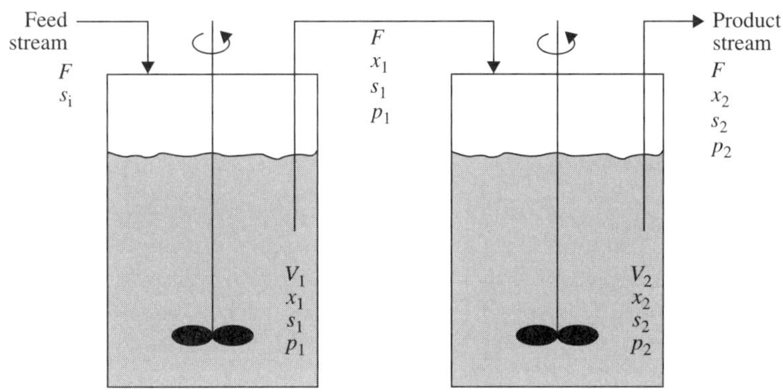

FIGURE 14.29 Flow sheet for a cascade of two continuous stirred tank fermenters.

14.5.7 Chemostat with Cell Recycle

The cell concentration in a single chemostat can be increased by recycling the biomass in the product stream back to the reactor. With more catalyst present in the vessel, higher rates of substrate utilisation and product formation can be achieved. The critical dilution rate for washout is also increased, thus allowing greater operating flexibility.

There are several ways by which cells can be recycled in fermentation processes. *External biomass feedback* is illustrated in Figure 14.30. In this scheme, a cell separator such as a centrifuge or settling tank is used to concentrate the biomass leaving the reactor. A portion of the concentrate is recycled back to the CSTR with flow rate F_r and cell concentration x_r. Such systems can be operated under steady-state conditions and are used extensively in biological waste treatment. Another way of achieving biomass feedback is *perfusion culture* or *internal biomass feedback*. This operating scheme is used often for mammalian cell culture and is illustrated in Figure 14.31. Depletion of nutrients and

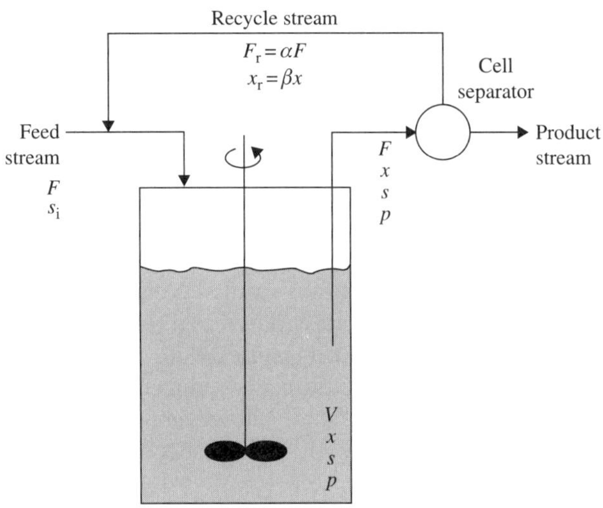

FIGURE 14.30 Flow sheet for a continuous stirred tank fermenter with external biomass feedback.

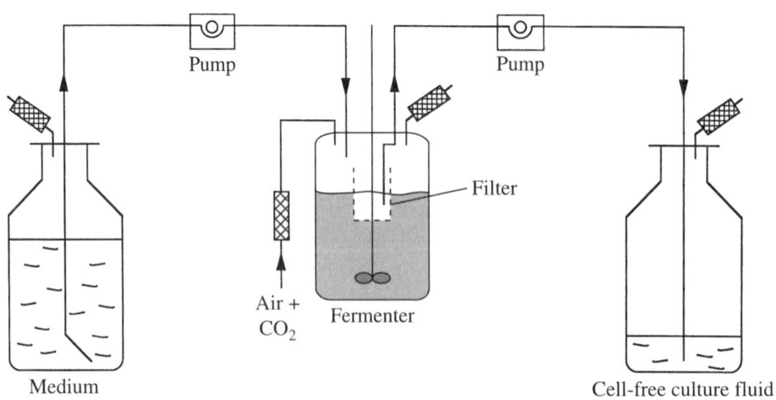

FIGURE 14.31 Flow sheet for a continuous perfusion reactor with internal biomass feedback.

accumulation of inhibitory products limit batch cell densities for many animal cell lines to about 10^6 cells ml^{-1}. The cell density and therefore the volumetric productivity of these cultures can be increased by retaining the biomass in the reactor while fresh medium is added and spent broth removed. As indicated in Figure 14.31, cells in a perfusion reactor are physically retained in the vessel by a mechanical device such as a filter. Liquid throughput is thus achieved without continuous removal or dilution of the cells so that cell concentrations in excess of 10^7 cells ml^{-1} can be obtained. However, a common problem associated with perfusion systems is blocking or blinding of the filter.

With growing cells, if all the biomass is recycled or retained in the reactor, the cell concentration will increase with time and steady state will not be achieved. Therefore, for steady-state operation, some proportion of the biomass must be removed from the system. Chemostat reactors with cell recycle are analysed using the same mass balance techniques applied in Section 14.5.4 (Cell Culture subsection) [37, 38]. Typical results for biomass concentration and biomass productivity with and without cell recycle are shown in Figure 14.32. With cell recycle, because the recycled cells are an additional source of biomass in the reactor, washout occurs at dilution rates greater than the maximum specific growth rate. If α is the recycle ratio:

$$\alpha = \frac{F_r}{F} \tag{14.77}$$

and β is the biomass concentration factor:

$$\beta = \frac{x_r}{x} \tag{14.78}$$

the critical dilution rate with cell recycle is increased by a factor of $1/(1 + \alpha - \alpha\beta)$ relative to that in a simple chemostat without cell recycle. Figure 14.32 also shows that, at the same dilution rate, the biomass productivity is also greater in recycle systems by the same factor.

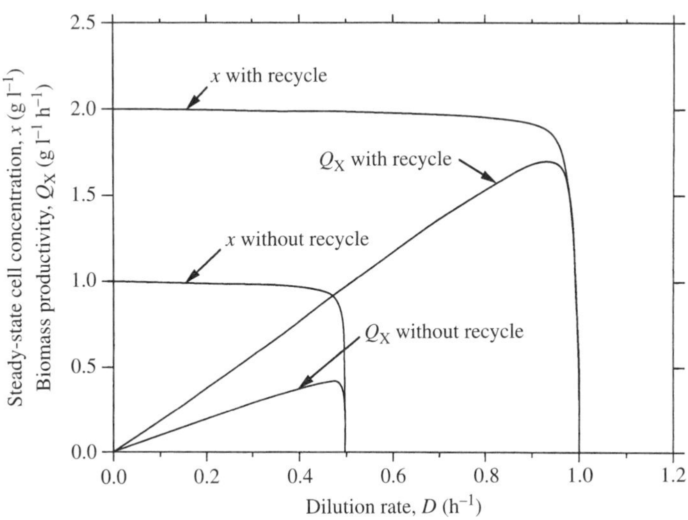

FIGURE 14.32 Steady-state biomass concentration and volumetric biomass productivity for a chemostat with and without cell recycle. The curves were calculated using the following parameter values: $\mu_{max} = 0.5\ h^{-1}$, $K_S = 0.01\ g\ l^{-1}$, $Y_{XS} = 0.5\ g\ g^{-1}$, $s_i = 2\ g\ l^{-1}$, $\alpha = 0.5$, and $\beta = 2.0$.

14.5.8 Continuous Operation of a Plug Flow Reactor

Plug flow operation is an alternative to mixed operation for continuous reactors. No mixing occurs in an ideal plug flow reactor; liquid entering the reactor passes through as a discrete 'plug' and does not interact with neighbouring fluid elements. This is achieved at high flow rates that minimise backmixing and variations in liquid velocity. Plug flow occurs most readily in column or tubular reactors such as that shown in Figure 14.33. Plug flow reactors can be operated in upflow or downflow mode or, in some cases, horizontally. Plug flow tubular reactors are known by the abbreviation PFTR.

Liquid in a PFTR flows at constant velocity; therefore all parts of the liquid have identical residence time in the reactor. As reaction in the vessel proceeds, concentration gradients of substrate and product develop in the direction of flow. At the feed end of the PFTR, the substrate concentration will be high and the product concentration low because the reaction mixture has just entered the vessel. At the outlet end of the tube, the substrate concentration will be low and the product level high.

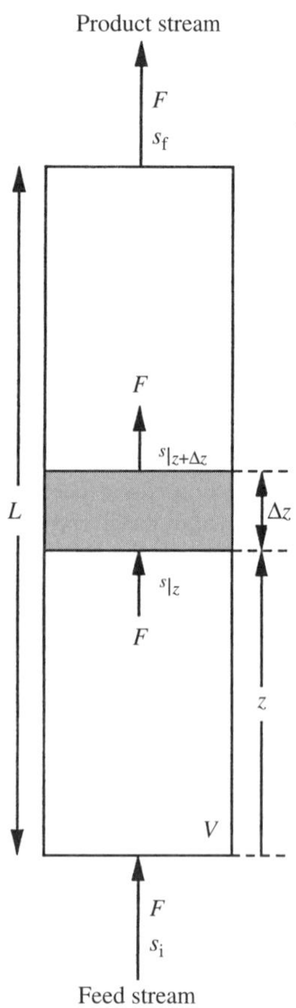

FIGURE 14.33 Flow sheet for a continuous plug flow tubular reactor.

Consider operation of the PFTR shown in Figure 14.33. The volumetric liquid flow rate through the vessel is F and the feed stream contains substrate at concentration s_i. At the reactor outlet, the substrate concentration is s_f. This exit concentration can be related to the inlet conditions and reactor residence time using mass balance techniques. Let us first consider plug flow operation for enzyme reaction.

Enzyme Reaction

To develop equations for a plug flow enzyme reactor, we must consider a small section of the reactor of length Δz as indicated in Figure 14.33. This section is located at distance z from the feed point. Let us perform a steady-state mass balance on substrate around the section using Eq. (6.5).

In Eq. (6.5), $\hat{M}_i$ is the mass flow rate of substrate entering the system; therefore $\hat{M}_i = Fs|_z$ where F is the volumetric flow rate through the reactor and $s|_z$ is the substrate concentration at z. Similarly, $\hat{M}_o$, the mass flow rate of substrate leaving the section, is $Fs|_{z+\Delta z}$. Substrate is not generated in the reaction, so $R_G = 0$. The rate of substrate consumption R_C is equal to the volumetric rate of enzyme reaction v multiplied by the volume of the section; for Michaelis–Menten kinetics, v is given by Eq. (12.37). The section volume is equal to $A\Delta z$ where A is the cross-sectional area of the reactor. At steady state, the left side of Eq. (6.5) is zero. Therefore, the mass balance equation for substrate is:

$$Fs\big|_z - Fs\big|_{z+\Delta z} - \frac{v_{\max} s}{K_m + s} A\,\Delta z = 0 \tag{14.79}$$

where $v_{\max}$ is the maximum rate of enzyme reaction and K_m is the Michaelis constant. Dividing through by $A\Delta z$ and rearranging gives:

$$\frac{F\left(s\big|_{z+\Delta z} - s\big|_z\right)}{A\,\Delta z} = \frac{-v_{\max} s}{K_m + s} \tag{14.80}$$

The volumetric flow rate F divided by the reactor cross-sectional area A is equal to the superficial velocity through the column, u. Therefore:

$$\frac{u\left(s\big|_{z+\Delta z} - s\big|_z\right)}{\Delta z} = \frac{-v_{\max} s}{K_m + s} \tag{14.81}$$

For F and A constant, u is also constant. Equation (14.81) applies to any section in the reactor of thickness Δz. To obtain an equation that is valid at any point in the reactor, we must take the limit as $\Delta z \to 0$:

$$u\left(\lim_{\Delta z \to 0} \frac{s\big|_{z+\Delta z} - s\big|_z}{\Delta z}\right) = \frac{-v_{\max} s}{K_m + s} \tag{14.82}$$

Applying the definition of the differential from Eq. (E.13) in Appendix E:

$$u\frac{ds}{dz} = \frac{-v_{\max} s}{K_m + s} \tag{14.83}$$

Equation (14.83) is a differential equation for the substrate concentration gradient through the length of a plug flow reactor. Assuming that u and the kinetic parameters are constant, Eq. (14.83) is ready for integration. Separating variables and integrating with boundary condition $s = s_i$ at $z = 0$ gives an expression for the reactor length L required to achieve an outlet substrate concentration of s_f:

$$L = u\left[\frac{K_m}{v_{max}}\ln\frac{s_i}{s_f} + \frac{s_i - s_f}{v_{max}}\right] \tag{14.84}$$

The residence time τ for continuous reactors is defined in Eq. (14.51). If we divide V and F in Eq. (14.51) by A, we can express the residence time for plug flow reactors in terms of parameters L and u:

$$\tau = \frac{V}{F} = \frac{\left(\frac{V}{A}\right)}{\left(\frac{F}{A}\right)} = \frac{L}{u} \tag{14.85}$$

Therefore, Eq. (14.84) can be written as:

$$\tau = \frac{K_m}{v_{max}}\ln\frac{s_i}{s_f} + \frac{s_i - s_f}{v_{max}} \tag{14.86}$$

Equations (14.84) and (14.86) allow us to calculate the reactor length and residence time required to achieve conversion of substrate from concentration s_i to s_f at flow rate u. Note that the form of Eq. (14.86) is identical to that of Eq. (14.10) for a batch enzyme reactor. As in batch reactors where the concentrations of substrate and product vary continuously during the reaction period, the concentrations in plug flow reactors also change continuously as material moves from the inlet to the outlet of the vessel. Thus, plug flow operation can be seen as a way of simulating batch culture in a continuous flow system.

Plug flow operation is generally impractical for enzyme conversions unless the enzyme is immobilised and retained inside the vessel. For immobilised enzyme reactions affected by diffusion, Eq. (14.83) must be modified to account for mass transfer effects:

$$u\frac{ds}{dz} = -\eta_T\frac{v_{max}s}{K_m + s} \tag{14.87}$$

where η_T is the total effectiveness factor representing internal and external mass transfer limitations (Section 13.5), s is the bulk substrate concentration, and v_{max} and K_m are intrinsic enzyme kinetic parameters. Because η_T is a function of s, Eq. (14.87) cannot be integrated directly.

Plug flow operation with immobilised enzymes is most likely to be approached in packed bed reactors such as that shown in Figure 14.8. However, the presence of packing in the column can cause substantial backmixing and axial dispersion of the liquid and thus interfere with ideal plug flow. Nevertheless, application of the equations developed in this section can give satisfactory results for the design of fixed bed immobilised enzyme reactors.

EXAMPLE 14.7 PLUG FLOW REACTOR FOR IMMOBILISED ENZYME

Immobilised lactase is used to hydrolyse lactose in dairy waste to glucose and galactose. The enzyme is immobilised in resin particles and packed into a 0.5-m^3 column. The total effectiveness factor for the system is close to unity, K_m for the immobilised enzyme is 1.32 kg m^{-3}, and v_{max} is 45 kg m^{-3} h^{-1}. The lactose concentration in the feed stream is 9.5 kg m^{-3}. A substrate conversion of 98% is required. The column is operated under plug flow conditions for a total of 310 days per year.

(a) At what flow rate should the reactor be operated?
(b) How many tonnes of glucose are produced per year?

Solution

(a) For 98% substrate conversion, $s_f = (0.02\ s_i) = 0.19$ kg m^{-3}. Substituting parameter values into Eq. (14.86) gives:

$$\tau = \frac{1.32 \text{ kg m}^{-3}}{45 \text{ kg m}^{-3}\text{ h}^{-1}} \ln\left(\frac{9.5 \text{ kg m}^{-3}}{0.19 \text{ kg m}^{-3}}\right) + \frac{(9.5 - 0.19) \text{ kg m}^{-3}}{45 \text{ kg m}^{-3}\text{ h}^{-1}} = 0.32 \text{ h}$$

From Eq. (14.51):

$$F = \frac{V}{\tau} = \frac{0.5 \text{ m}^3}{0.32 \text{ h}} = 1.56 \text{ m}^3\text{ h}^{-1}$$

The flow rate required is 1.6 m^3 h^{-1}.

(b) The rate of lactose conversion is equal to the difference between the inlet and outlet mass flow rates of lactose:

$$F(s_i - s_f) = 1.56 \text{ m}^3\text{ h}^{-1}(9.5 - 0.19) \text{ kg m}^{-3} = 14.5 \text{ kg h}^{-1}$$

Converting this to an annual rate based on 310 days of operation per year and a molecular weight for lactose of 342:

$$\text{Rate of lactose conversion} = 14.5 \text{ kg h}^{-1} \cdot \left|\frac{24 \text{ h}}{1 \text{ day}}\right| \cdot \left|\frac{310 \text{ days}}{1 \text{ year}}\right| \cdot \left|\frac{1 \text{ kgmol}}{342 \text{ kg}}\right| = 315 \text{ kgmol year}^{-1}$$

The enzyme reaction for lactase is:

$$\text{lactose} + H_2O \longrightarrow \text{glucose} + \text{galactose}$$

Therefore, from reaction stoichiometry, 315 kgmol of glucose are produced per year. The molecular weight of glucose is 180; therefore:

$$\text{Rate of glucose production} = 315 \text{ kgmol year}^{-1} \cdot \left|\frac{180 \text{ kg}}{1 \text{ kgmol}}\right| \cdot \left|\frac{1 \text{ tonne}}{1000 \text{ kg}}\right| = 56.7 \text{ tonne year}^{-1}$$

The amount of glucose produced per year is 57 tonnes.

Cell Culture

Analysis of plug flow reactors for cell culture follows the same procedure as for enzyme reaction. If the cell specific growth rate is constant and equal to μ_{max} throughout the reactor and cell death can be neglected, the equations for the reactor residence time are analogous to those derived in Section 14.5.1 (Cell Culture subsection) for batch fermentation:

$$\tau = \frac{1}{\mu_{max}} \ln \frac{x_f}{x_i} \tag{14.88}$$

where τ is the reactor residence time defined in Eq. (14.51), x_i is the biomass concentration at the inlet, and x_f is the biomass concentration at the outlet. The form of Eq. (14.88) is identical to that of Eq. (14.20) for batch culture time.

Plug flow operation is not suitable for cultivation of suspended cells unless the biomass is recycled or there is continuous inoculation of the vessel. Plug flow operation with cell recycle is used in large-scale waste water treatment; however these applications are limited. Plug flow reactors may be suitable for immobilised cell reactions when the catalyst is packed into a fixed bed as shown in Figure 14.8. Even so, operating problems such as those mentioned in Section 14.2.5 mean that PFTRs are rarely employed for industrial fermentations.

14.5.9 Comparison between Major Modes of Reactor Operation

The relative performance of batch, CSTR, and PFTR reactors can be considered from a theoretical point of view in terms of the substrate conversions and product concentrations achieved using vessels of the same size. Because the total reactor volume is not fully utilised at all times during fed-batch operation, it is difficult to include this mode of operation in a general comparison.

As indicated in Section 14.5.8, the kinetic characteristics of PFTRs are the same as batch reactors: the residence time required for conversion in a plug flow reactor is therefore the same as in a mixed vessel operated in batch. It can also be shown theoretically that as the number of stages in a CSTR cascade increases, the conversion characteristics of the entire system approach those of an ideal plug flow or mixed batch reactor. This is shown diagrammatically in Figure 14.34. The smooth dashed curve represents the progressive decrease in substrate concentration with time spent in a PFTR or batch reactor; the concentration is reduced from s_i at the inlet to s_f at the outlet. In a single well-mixed CSTR operated with the same inlet and outlet concentrations, because conditions in the vessel are uniform, there is a step change in substrate concentration as soon as the feed enters the reactor. In a cascade of CSTRs, the concentration is uniform in each reactor but there is a stepwise drop in concentration between each stage. As illustrated in Figure 14.34, the greater the number of units in a CSTR cascade, the closer the concentration profile approaches plug flow or batch behaviour.

The benefits associated with particular reactor designs or modes of operation depend on the reaction kinetics. For zero-order reactions, there is no difference between single batch, CSTR, and PFTR reactors in terms of the overall conversion rate. However, for most reactions including first-order and Michaelis–Menten conversions, the rate of reaction

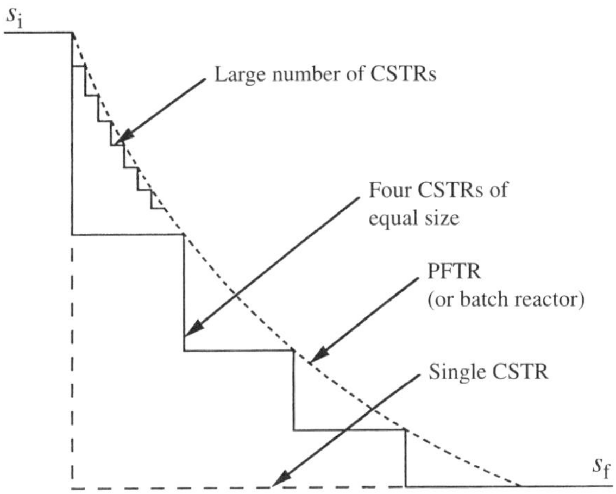

FIGURE 14.34 Concentration changes in PFTR, single CSTR, and multiple CSTR vessels.

decreases as the concentration of substrate decreases. The reaction rate is therefore high at the start of batch reaction or at the entrance to a plug flow reactor because the substrate concentration there is at its greatest. Subsequently, the reaction velocity falls gradually as substrate is consumed. In contrast, substrate entering a CSTR is immediately diluted to the final or outlet steady-state concentration so that the rate of reaction is comparatively low for the entire reactor.

Accordingly, for first-order and Michaelis–Menten reactions, CSTRs achieve lower substrate conversions and lower product concentrations than batch reactors or PFTRs of the same volume. In practice, batch processing is usually preferred to PFTR systems for enzyme reactions because of the operating problems mentioned in Section 14.2.5. However, as discussed in Section 14.5.2, the total time for batch operation depends on the duration of the downtime between batches as well as on the actual conversion time. Because the length of downtime varies considerably from system to system, we cannot account for it here in a general way. The downtime between batches should be minimised as much as possible to maintain high overall production rates.

The comparison between reactors yields a different result if the reaction is autocatalytic, such as in cell cultures. At the beginning of batch culture, the rate of substrate conversion is low because relatively few cells are present: as catalyst is produced during fermentation, the volumetric reaction rate increases as the conversion proceeds. However, in CSTR operations, substrate entering the vessel is exposed immediately to a relatively high biomass concentration so that the rate of conversion is also high. Rates of conversion in chemostats operated close to the optimum dilution rate for biomass productivity (Section 14.5.4, Cell Culture subsection) are often 10 to 20 times greater than in PFTR or batch reactors. This rate advantage may disappear if the steady-state substrate concentration in the CSTR is so small that, despite the higher biomass levels present, the conversion rate is lower than the average in a batch or PFTR device. Nevertheless, for most fermentations, CSTRs offer significant theoretical advantages over other modes of reactor operation.

Despite the productivity benefits associated with CSTRs, an overwhelming majority of commercial fermentations are conducted in batch. The reasons lie with the practical advantages associated with batch culture. The risk of contamination is lower in batch systems than in continuous flow reactors; equipment and control failures during long-term continuous operation are also potential problems. Continuous fermentation is feasible only when the cells are genetically stable; if developed strains revert to more rapidly growing mutants, the culture can become dominated over time by the revertant cells. Because freshly produced inocula are used in batch fermentations, closer control over the genetic characteristics of the culture is achieved. Continuous culture is not suitable for the production of metabolites normally formed near stationary phase when the culture growth rate is low; as mentioned above, the productivity in a batch reactor is likely to be greater than in a CSTR under these conditions. Production can be much more flexible using batch processing; for example, different products each with small market volumes can be made in different batches. In contrast, continuous fermentations must be operated for lengthy periods to reap the full benefits of their high productivity.

14.5.10 Evaluation of Kinetic and Yield Parameters in Chemostat Culture

In a steady-state chemostat with sterile feed and negligible cell death, as indicated in Eq. (14.57), the specific growth rate μ is equal to the dilution rate D. This relationship is useful for determining the kinetic and yield parameters that characterise cellular reactions. Combining the Monod expression of Eq. (12.91) with Eq. (14.57) gives an equation for D in chemostat cultures:

$$D = \frac{\mu_{max} s}{K_S + s} \tag{14.89}$$

where μ_{max} is the maximum specific growth rate, K_S is the substrate constant, and s is the steady-state substrate concentration in the reactor.

Equation (14.89) is analogous mathematically to the Michaelis–Menten expression for enzyme kinetics, Eq. (12.37). If s is measured at various dilution rates, the same techniques described in Section 12.4 for determining v_{max} and K_m can be applied for evaluation of μ_{max} and K_S. Rearrangement of Eq. (14.89) gives the following linearised equations that can be used for Lineweaver–Burk, Eadie–Hofstee, and Langmuir plots, respectively:

$$\frac{1}{D} = \frac{K_S}{\mu_{max} s} + \frac{1}{\mu_{max}} \tag{14.90}$$

$$\frac{D}{s} = \frac{\mu_{max}}{K_S} - \frac{D}{K_S} \tag{14.91}$$

and

$$\frac{s}{D} = \frac{K_S}{\mu_{max}} + \frac{s}{\mu_{max}} \tag{14.92}$$

For example, according to Eq. (14.90), μ_{max} and K_S can be determined from the slope and intercept of a plot of $1/D$ versus $1/s$. The comments made in Sections 12.4.2 through 12.4.4 about the distortion of experimental error with linearisation of measured data apply also to Eqs. (14.90) through (14.92).

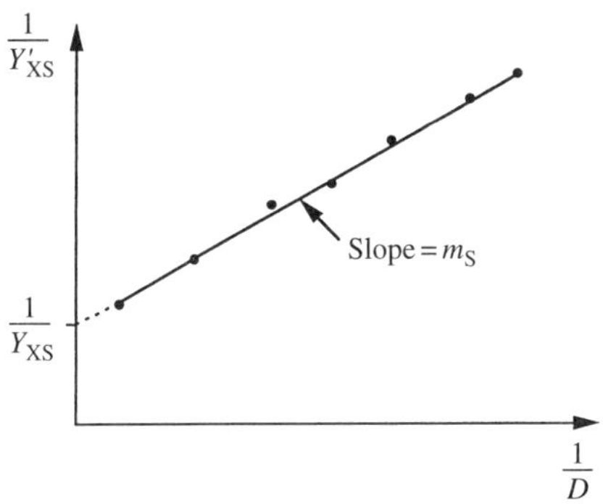

FIGURE 14.35 Graphical determination of the maintenance coefficient m_S and true biomass yield Y_{XS} using data from chemostat culture.

Chemostat operation is also convenient for determining true yields and maintenance coefficients for cell cultures. An expression relating these parameters to the specific growth rate is given by Eq. (12.112). In chemostat culture with $\mu = D$, Eq. (12.112) becomes:

$$\frac{1}{Y'_{XS}} = \frac{1}{Y_{XS}} + \frac{m_S}{D} \tag{14.93}$$

where Y'_{XS} is the observed biomass yield from substrate, Y_{XS} is the true biomass yield from substrate, and m_S is the maintenance coefficient. Therefore, as shown in Figure 14.35, a plot of $1/Y'_{XS}$ versus $1/D$ gives a straight line with slope m_S and intercept $1/Y_{XS}$. In a chemostat with sterile feed, the observed biomass yield from substrate Y'_{XS} is obtained as follows:

$$Y'_{XS} = \frac{x}{s_i - s} \tag{14.94}$$

where x and s are measured steady-state cell and substrate concentrations, respectively, and s_i is the inlet substrate concentration.

14.6 STERILISATION

Commercial fermentations typically require thousands of litres of liquid medium and millions of litres of air. For processes operated with axenic cultures, these raw materials must be provided free of contaminating organisms. Of all the methods available for sterilisation, including chemical treatment, exposure to ultraviolet, gamma, and X-ray radiation, sonication, filtration, and heating, only the last two are used in large-scale operations. Aspects of fermenter design and construction for aseptic operation were described in Sections 14.3.1 and 14.3.2. Here, we consider the design of sterilisation systems for liquids and gases.

14.6.1 Batch Heat Sterilisation of Liquids

Liquid medium is most commonly sterilised in batch in the vessel where it will be used. The liquid is heated to the sterilisation temperature by introducing steam into the

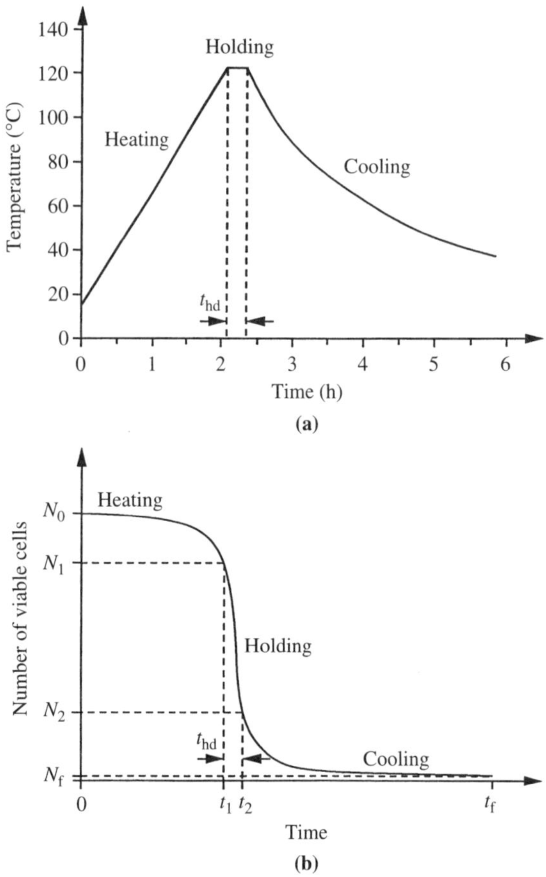

FIGURE 14.36 (a) Variation of temperature with time for batch sterilisation of liquid medium. (b) Reduction in the number of viable cells with time during batch sterilisation.

coils or jacket of the vessel (Figure 9.1); alternatively, steam is bubbled directly into the medium, or the vessel is heated electrically. If direct steam injection is used, allowance must be made for dilution of the medium by condensate, which typically adds 10 to 20% to the liquid volume. The quality of the steam used for direct injection must also be sufficiently high to avoid contaminating the medium with metal ions or organics. A typical temperature–time profile for batch sterilisation is shown in Figure 14.36(a). Depending on the rate of heat transfer from the steam or electrical element, raising the temperature of the medium in large fermenters can take a considerable period of time. Once the holding or sterilisation temperature is reached, the temperature is held constant for time t_{hd}. Cooling water in the coils or jacket of the fermenter is then used to reduce the medium temperature to the required level.

For operation of batch sterilisation systems, we must be able to estimate the holding time required to achieve the desired level of cell destruction. As well as destroying contaminant organisms, heat sterilisation is also capable of destroying nutrients in the medium. To minimise this loss, holding times at the highest sterilisation temperature should be kept as short as possible. Cell death occurs at all times during batch sterilisation, including within the heating-up and cooling-down periods. The holding time t_{hd} can be minimised by taking into account the extent of cell destruction achieved during those periods.

Let us denote the number of contaminants present in the raw medium N_0. As indicated in Figure 14.36(b), during the heating period this number is reduced to N_1. At the end of the holding period, the cell number is N_2. The final cell number after cooling is N_f. Ideally, N_f is zero: at the end of the sterilisation cycle we want no contaminants to be present. However, because absolute sterility would require an infinitely long sterilisation time, it is theoretically impossible to achieve. Normally, the target level of contamination is expressed as a fraction of a cell, which is related to the *probability* of contamination. For example, we could aim for $N_f = 10^{-3}$; this means that we accept the risk that one batch in 1000 will not be sterile at the end of the process. Once N_0 and N_f are known, we can determine the holding time required to reduce the number of cells from N_1 to N_2 by considering the kinetics of cell death.

The rate of heat sterilisation is governed by the equations for thermal death outlined in Section 12.15. In a batch vessel where cell death is the only process affecting the number of viable cells, from Eq. (12.117):

$$\frac{dN}{dt} = -k_d N \tag{14.95}$$

where N is the number of viable cells, t is time, and k_d is the specific death constant. Equation (14.95) applies to each stage of the batch sterilisation cycle: heating, holding, and cooling. However, because k_d is a strong function of temperature, direct integration of Eq. (14.95) is valid only when the temperature is constant, that is, during the holding period. The result is:

$$\ln \frac{N_1}{N_2} = k_d t_{hd} \tag{14.96}$$

or

$$t_{hd} = \frac{\ln \frac{N_1}{N_2}}{k_d} \tag{14.97}$$

where t_{hd} is the holding time, N_1 is the number of viable cells at the start of holding, and N_2 is the number of viable cells at the end of holding. k_d is evaluated as a function of temperature using the Arrhenius equation from Chapter 12:

$$k_d = A\,e^{-E_d/RT} \tag{12.74}$$

where A is the Arrhenius constant or frequency factor, E_d is the activation energy for thermal cell death, R is the ideal gas constant (Appendix B), and T is absolute temperature.

To use Eq. (14.97) we must know N_1 and N_2. These numbers are determined by considering the effect of cell death during the heating and cooling periods when the temperature is not constant. Combining Eqs. (14.95) and (12.74) gives:

$$\frac{dN}{dt} = -A\,e^{-E_d/RT} N \tag{14.98}$$

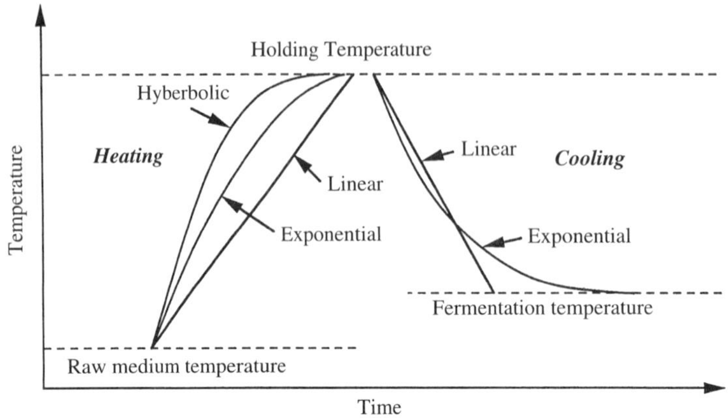

FIGURE 14.37 Graph of generalised temperature–time profiles for the heating and cooling stages of a batch sterilisation cycle. *From F.H. Deindoerfer and A.E. Humphrey, 1959, Analytical method for calculating heat sterilization times.* Appl. Microbiol. *7*, *256–264.*

Integration of Eq. (14.98) gives for the heating period:

$$\ln\frac{N_0}{N_1} = \int_0^{t_1} A e^{-E_d/RT}\, dt \tag{14.99}$$

and for the cooling period:

$$\ln\frac{N_2}{N_f} = \int_{t_2}^{t_f} A e^{-E_d/RT}\, dt \tag{14.100}$$

where t_1 is the time at the end of heating, t_2 is the time at the end of holding, and t_f is the time at the end of cooling. We cannot complete integration of these equations until we know how the temperature varies with time during the heating and cooling periods.

As outlined in Chapter 6, unsteady-state temperatures during heating and cooling can be determined from the heat transfer properties of the system. Depending on how heating and cooling are achieved, the general form of equations for temperature as a function of time is either linear, exponential, or hyperbolic, as shown in Figure 14.37 and Table 14.3. Applying an appropriate expression for T in Eq. (14.99) from Table 14.3 allows the cell number N_1 at the start of the holding period to be determined. Similarly, applying an appropriate expression for T in Eq. (14.100) for cooling allows evaluation of N_2 at the end of the holding period. Use of these results for N_1 and N_2 in Eq. (14.97) completes the holding time calculation.

Normally, cell death at temperatures below about 100°C is minimal. However, when heating and cooling are relatively slow, temperatures can remain elevated and close to the holding temperature for considerable periods of time. As a result, the reduction in cell numbers outside of the holding period is significant. Usually, the holding time is of the order of minutes, whereas heating and cooling of large volumes of liquid can take hours. Further information and sample calculations for batch sterilisation are given by Aiba, Humphrey, and Millis [40] and Richards [41].

The design procedures outlined in this section apply to batch sterilisation of medium where the temperature is uniform throughout the vessel. However, if the liquid contains

TABLE 14.3 General Equations for Temperature as a Function of Time during the Heating and Cooling Periods of Batch Sterilisation

Heat Transfer Method	Temperature–time Profile
HEATING	
Direct sparging with steam	$T = T_0 \left(1 + \frac{\frac{h\hat{M}_s t}{M_m C_p T_0}}{1 + \frac{\hat{M}_s}{M_m} t} \right)$ (hyperbolic)
Electrical heating	$T = T_0 \left(1 + \frac{\hat{Q} t}{M_m C_p T_0} \right)$ (linear)
Heat transfer from isothermal steam	$T = T_S \left[1 + \frac{T_0 - T_S}{T_S} e^{\left(\frac{-UAt}{M_m C_p} \right)} \right]$ (exponential)
COOLING	
Heat transfer to nonisothermal cooling water	$T = T_{ci} \left\{ 1 + \frac{T_0 - T_{ci}}{T_{ci}} e^{\left[\left(\frac{-\hat{M}_w C_{pw} t}{M_m C_p} \right) \left(1 - e^{\left[\frac{-UA}{\hat{M}_w C_{pw}} \right]} \right) \right]} \right\}$ (exponential)

A = surface area for heat transfer; C_p = specific heat capacity of medium; C_{pw} = specific heat capacity of cooling water; h = specific enthalpy difference between the steam and raw medium; M_m = initial mass of medium; $\hat{M}_s$ = mass flow rate of steam; $\hat{M}_w$ = mass flow rate of cooling water; $\hat{Q}$ = rate of heat transfer; T = temperature; T_0 = initial medium temperature; T_{ci} = inlet temperature of cooling water; T_S = steam temperature; t = time; U = overall heat transfer coefficient.

From F.H. Deindoerfer and A.E. Humphrey, 1959, Analytical method for calculating heat sterilization times. Appl. Microbiol. *7, 256–264.*

solid particles (e.g., in the form of cell flocs or pellets), temperature gradients may develop. Because heat transfer within particles is slower than in liquid, the temperature at the centre of the particles will be lower than that in the liquid for some proportion of the sterilisation time. As a result, cell death inside the particles does not occur as rapidly as in the liquid. Longer holding times are required to treat solid-phase substrates and media containing particles.

When heat sterilisation is scaled up to larger volumes, longer treatment times are needed to achieve the same sterilisation results at the same holding temperature. As indicated by the equations in this section, steriliser design is based on the number of contaminating organisms, not their concentration. Because the initial number of organisms in a given raw medium increases in direct proportion to the liquid volume, to obtain the same final N_f in a larger volume, a greater number of cells must be destroyed. Scale-up also affects the temperature–time profiles for heating and cooling. Heat transfer rates depend on the equipment used; however, heating and cooling of larger volumes is likely to take more time. Sustained elevated temperatures during heating and cooling are damaging to vitamins, proteins, and sugars in nutrient solutions and reduce the quality of the medium [42]. Because it is necessary to hold larger volumes of medium for longer periods of time, this problem is exacerbated with scale-up.

14.6.2 Continuous Heat Sterilisation of Liquids

Continuous sterilisation, particularly a high-temperature, short-exposure-time process, can reduce thermal damage to the medium significantly compared with batch sterilisation, while achieving high levels of cell destruction. Other advantages include improved steam economy and more reliable scale-up. The amount of steam needed for continuous sterilisation is 20 to 25% of that used in batch processes; the time required is also significantly reduced because heating and cooling are virtually instantaneous.

Typical equipment configurations for continuous sterilisation are shown in Figure 14.38. In Figure 14.38(a), raw medium entering the system is first preheated by hot, sterile medium in a heat exchanger; this reduces the subsequent steam requirements for heating and cools the sterile medium. Steam is then injected directly into the medium as it flows through a pipe; as a result, the temperature of the medium rises almost instantaneously to the desired sterilisation temperature. The time of exposure to this temperature depends on the length of pipe in the holding section of the steriliser. After sterilisation, the medium is cooled instantly by flash cooling, which is achieved by passing the liquid through an expansion valve into a vacuum chamber. Further cooling takes place in the heat exchanger where residual heat is used to preheat incoming medium. The sterile medium is then ready for use in the fermenter.

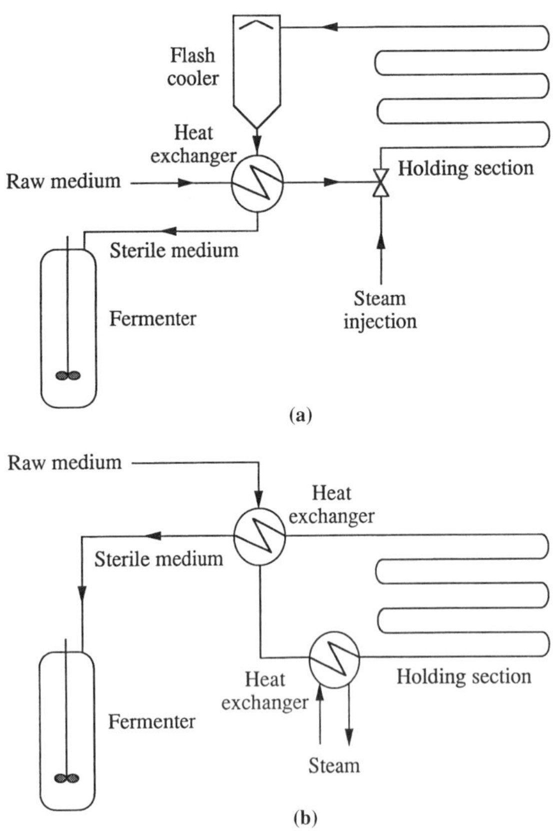

FIGURE 14.38 Continuous sterilising equipment: (a) continuous steam injection with flash cooling; (b) heat transfer using heat exchangers.

Figure 14.38(b) shows an alternative sterilisation scheme based on heat exchange between the medium and steam. Raw medium is preheated using hot, sterile medium in a heat exchanger and is brought to the sterilisation temperature by further heat exchange with steam. The sterilisation temperature is maintained in the holding section; the sterile medium is then cooled by heat exchange with incoming medium before being used in the fermenter. Heat exchange systems are more expensive to construct than injection devices; fouling of the internal heat exchange surfaces (Section 9.4.4) also reduces the efficiency of heat transfer between cleanings. On the other hand, a disadvantage associated with steam injection is dilution of the medium by condensate; foaming from direct steam injection can also cause problems with operation of the flash cooler. As indicated in Figure 14.39, the rates of heating and cooling in continuous sterilisers are much more rapid than in batch systems (Figure 14.36). Accordingly, in the design of continuous sterilisers, contributions to cell death outside of the holding period are generally ignored.

An important variable affecting the performance of continuous sterilisers is the nature of the fluid flow in the system. Ideally, all fluid entering the equipment at a particular instant should spend the same time in the steriliser and exit the system at the same time. Unless this occurs, we cannot fully control the time spent in the steriliser by all fluid elements. No mixing should take place in the tubes: if fluid nearer the entrance of the pipe mixes with fluid ahead of it, there is a risk that contaminants will be transferred to the outlet of the steriliser. The type of flow in pipes where there is neither mixing nor variation in fluid velocity is called *plug flow*, as already described in Section 14.5.8 for plug flow reactors. Plug flow is an ideal flow pattern: in reality, fluid elements in pipes have a range of different velocities. As illustrated in Figure 14.40, flow tends to be faster through the centre of pipes than near the walls. Plug flow is approached in pipes at high Reynolds numbers (Section 7.2.3) above about 2×10^4. However, although operation at high Reynolds number is used to minimise fluid mixing and velocity variation in continuous sterilisers, deviations from ideal plug flow are inevitable.

Deviation from plug flow behaviour is characterised by *axial dispersion* in the system, that is, the degree to which mixing occurs along the length or axis of the pipe. Axial dispersion is a critical factor affecting the design of continuous sterilisers. If axial dispersion is substantial, the performance of the steriliser is reduced. The relative importance of axial

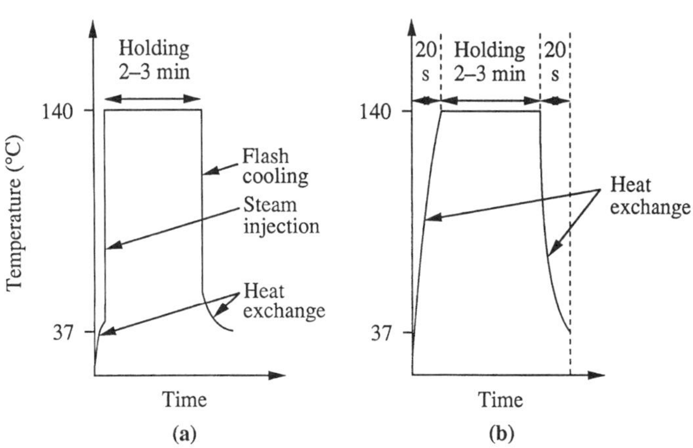

FIGURE 14.39 Variation of temperature with time in the continuous sterilisers of Figure 14.38: (a) continuous steam injection with flash cooling; (b) heat transfer using heat exchangers.

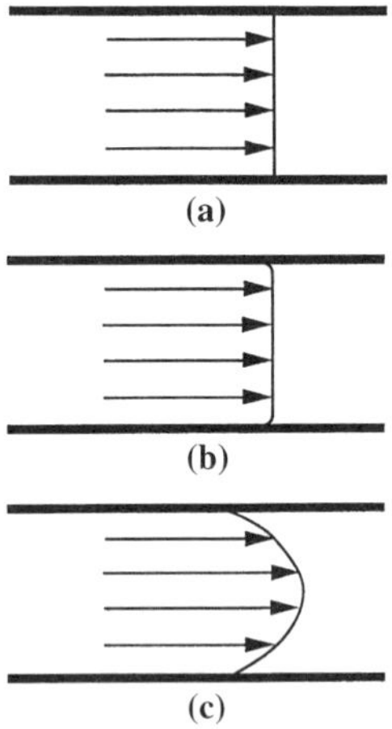

FIGURE 14.40 Velocity distributions for flow in pipes. (a) In plug flow, the fluid velocity is the same across the diameter of the pipe as indicated by the arrows of equal length. (b) In fully-developed turbulent flow, the velocity distribution approaches that of plug flow; however there is some reduction of flow speed at the walls. (c) In laminar flow, the fluid velocity is lowest at the walls of the pipe and highest along the central axis of the tube.

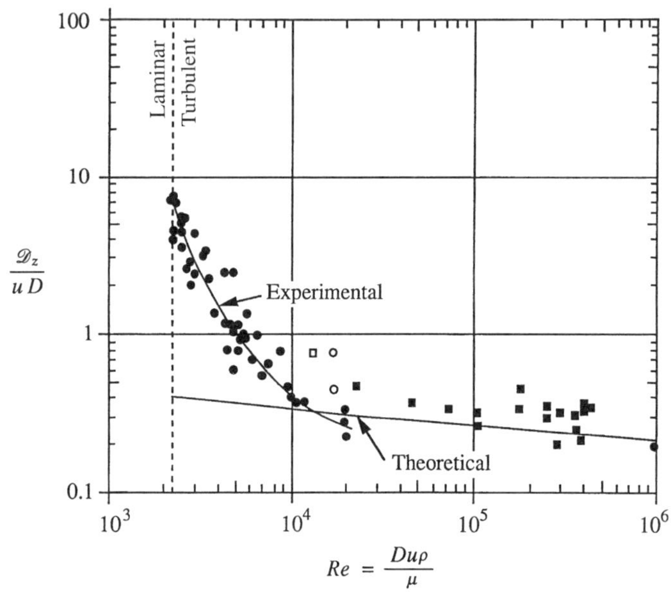

FIGURE 14.41 Correlation for determining the axial dispersion coefficient in turbulent pipe flow. *Re* is the Reynolds number, *D* is the pipe diameter, *u* is the average linear fluid velocity, ρ is the fluid density, μ is the fluid viscosity, and $\mathscr{D}_z$ is the axial dispersion coefficient. Data were measured using single fluids in: (●) straight pipes; (■) pipes with bends; (□) artificially roughened pipe; and (○) curved pipe. *Reprinted (adapted) with permission from O. Levenspiel, Longitudinal mixing of fluids flowing in circular pipes.* Ind. Eng. Chem. *50, 343–346. Copyright 1958, American Chemical Society.*

dispersion and bulk flow in the transfer of material through pipes is represented by a dimensionless variable called the *Peclet number*:

$$Pe = \frac{uL}{\mathscr{D}_z} \tag{14.101}$$

where *Pe* is the Peclet number, *u* is the average linear fluid velocity, *L* is the pipe length, and $\mathscr{D}_z$ is the *axial dispersion coefficient*. For perfect plug flow, $\mathscr{D}_z$ is zero and *Pe* is infinitely large. In practice, Peclet numbers between 3 and 600 are typical. The value of $\mathscr{D}_z$ for a

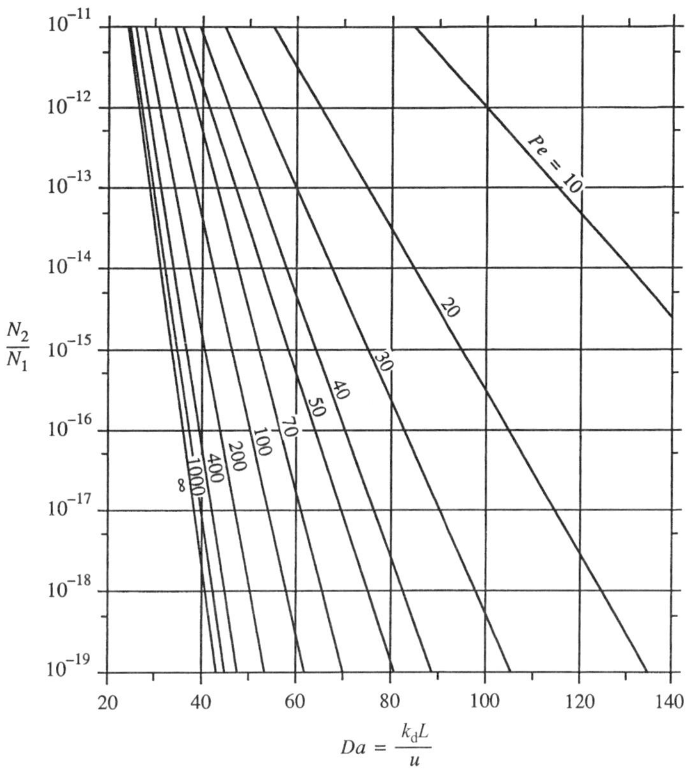

FIGURE 14.42 Thermal destruction of contaminating organisms as a function of the Peclet number *Pe* and Damköhler number *Da*. N_1 is the number of viable cells entering the holding section of the steriliser; N_2 is the number of cells leaving. *From S. Aiba, A.E. Humphrey, and N.F. Millis, 1965,* Biochemical Engineering, *Academic Press, New York.*

particular system depends on the Reynolds number and pipe geometry; a correlation from the engineering literature for evaluating $\mathscr{D}_z$ is shown in Figure 14.41.

Once the Peclet number has been calculated from Eq. (14.101), the extent of cell destruction in the steriliser can be related to the cell specific death constant k_d using Figure 14.42. In this figure, N_1 is the number of viable cells entering the steriliser, N_2 is the number of cells leaving, *Pe* is the Peclet number as defined by Eq. (14.101), and *Da* is another dimensionless number called the *Damköhler number*:

$$Da = \frac{k_d L}{u} \tag{14.102}$$

where k_d is the specific death constant, L is the length of the holding pipe, and u is the average linear liquid velocity. The relationship between k_d and temperature is given by Eq. (12.74). The lower the value of N_2/N_1, the greater is the level of cell destruction. Figure 14.42 shows that, at any given sterilisation temperature defining the value of k_d and therefore *Da*, the performance of the steriliser declines significantly as the Peclet number decreases, reflecting the detrimental effect of axial dispersion on steriliser efficiency. Design calculations for a continuous steriliser are illustrated in Example 14.8.

EXAMPLE 14.8 HOLDING TEMPERATURE IN A CONTINUOUS STERILISER

Liquid medium at a flow rate of 2 $m^3 h^{-1}$ is to be sterilised by heat exchange with steam in a continuous steriliser. The medium contains bacterial spores at a concentration of 5×10^{12} m^{-3}. Values of the activation energy and Arrhenius constant for thermal destruction of these contaminants are 283 kJ $gmol^{-1}$ and 5.7×10^{39} h^{-1}, respectively. A contamination risk of one organism surviving every 60 days of operation is considered acceptable. The steriliser pipe has an inner diameter of 0.1 m and the length of the holding section is 24 m. The density of the medium is 1000 kg m^{-3} and the viscosity is 3.6 kg m^{-1} h^{-1}. What sterilising temperature is required?

Solution

The desired level of cell destruction is evaluated using a basis of 60 days. Ignoring any cell death in the heating and cooling sections, the number of cells entering the holding section over 60 days is:

$$N_1 = 2\ \text{m}^3\ \text{h}^{-1}(5 \times 10^{12}\ \text{m}^{-3}) \cdot \left|\frac{24\ \text{h}}{1\ \text{day}}\right| \cdot (60\ \text{days}) = 1.44 \times 10^{16}$$

N_2, the acceptable number of cells leaving during this period, is 1. Therefore:

$$\frac{N_2}{N_1} = \frac{1}{1.44 \times 10^{16}} = 6.9 \times 10^{-17}$$

The linear velocity u in the steriliser is equal to the volumetric flow rate divided by the cross-sectional area of the pipe:

$$u = \frac{2\ \text{m}^3\ \text{h}^{-1}}{\pi \left(\frac{0.1\ \text{m}}{2}\right)^2} = 254.6\ \text{m h}^{-1}$$

Calculating the Reynolds number for pipe flow using Eq. (7.1):

$$Re = \frac{D u \rho}{\mu} = \frac{(0.1\ \text{m})\ (254.6\ \text{m h}^{-1})\ (1000\ \text{kg m}^{-3})}{3.6\ \text{kg m}^{-1}\ \text{h}^{-1}} = 7.07 \times 10^3$$

For this value of Re, we can determine $\mathscr{D}_z$ from Figure 14.41 using either the experimental or theoretical curve. Let us choose the experimental curve as this gives a larger value of $\mathscr{D}_z$ and a smaller value of Pe; the steriliser design will thus be more conservative. Therefore, for $\mathscr{D}_z/uD = 0.65$:

$$\mathscr{D}_z = 0.65\ (254.6\ \text{m h}^{-1})(0.1\ \text{m}) = 16.5\ \text{m}^2\ \text{h}^{-1}$$

From Eq. (14.101):

$$Pe = \frac{uL}{\mathscr{D}_z} = \frac{(254.6\ \text{m h}^{-1})(24\ \text{m})}{16.5\ \text{m}^2\ \text{h}^{-1}} = 370$$

Using Figure 14.42, we can determine the value of k_d for the desired level of cell destruction. For $N_2/N_1 = 6.9 \times 10^{-17}$ and $Pe = 370$, the corresponding value of Da is about 42. Therefore, from Eq. (14.102):

$$k_d = \frac{u\,Da}{L} = \frac{(254.6\text{ m h}^{-1})(42)}{24\text{ m}} = 445.6\text{ h}^{-1}$$

The sterilisation temperature can be evaluated from the Arrhenius equation after rearranging Eq. (12.74). Dividing both sides by A and taking natural logarithms gives:

$$\ln\frac{k_d}{A} = \frac{-E_d}{RT}$$

Therefore:

$$T = \frac{\left(\frac{-E_d}{R}\right)}{\ln\left(\frac{k_d}{A}\right)}$$

$E_d = 283$ kJ gmol^{-1} $= 283 \times 10^3$ J gmol^{-1}; $A = 5.7 \times 10^{39}$ h^{-1}. From Appendix B, the ideal gas constant R is 8.3144 J K^{-1} gmol^{-1}. Therefore:

$$T = \frac{\left(\frac{-283 \times 10^3\text{ J gmol}^{-1}}{8.3144\text{ J K}^{-1}\text{ gmol}^{-1}}\right)}{\ln\left(\frac{445.6\text{ h}^{-1}}{5.7 \times 10^{39}\text{ h}^{-1}}\right)} = 398.4\text{ K}$$

Using the conversion between K and °C given in Eq. (2.27), $T = 125$°C. Therefore, the sterilisation temperature required is 125°C.

Heating and cooling in continuous sterilisers are so rapid that in design calculations they are considered instantaneous. While reducing nutrient deterioration, this feature of the process can cause problems if there are solids present in the medium. During heating, the temperature at the core of solid particles remains lower than in the medium. Because of the extremely short contact times in continuous sterilisers compared with batch systems, there is a much greater risk that particles will not be heated thoroughly and will therefore not be properly sterilised. It is important that raw medium be clarified as much as possible before it enters a continuous steriliser.

14.6.3 Filter Sterilisation of Liquids

Sometimes, fermentation media or selected medium ingredients are sterilised by filtration rather than by heat. Medium containing heat-labile components such as serum, differentiation factors, enzymes, or other proteins is easily destroyed by heat and must be sterilised by other means. Typically, the membranes used for filter sterilisation of liquids are made of cellulose esters or other polymers and have pore diameters between 0.2 and 0.45 μm. As medium passes through the filter, bacteria and other particles with dimensions greater than the pore size are screened out and collect on the surface of the membrane. The small pore sizes used in liquid filtration mean that the membranes readily become blocked unless the medium is prefiltered to remove any large particles. To achieve high filtration flow rates, large membrane surface areas are required. The membranes themselves must be sterilised by steam or radiation before use; alternatively, disposable filters and filter cartridges are used. Filtration using pore sizes of 0.2 to 0.45 μm may not be as effective or reliable as heat sterilisation because viruses and mycoplasma are able to pass through the membranes. However, the availability of 0.1-μm sterilising-grade membrane filters for mycoplasma removal has seen the widespread adoption of filter sterilisation in many mammalian cell culture processes where complex, proteinaceous media are used routinely.

14.6.4 Sterilisation of Air

The number of microbial cells in air is of the order 10^3 to 10^4 m^{-3} [40]. Filtration is the most common method used to sterilise air in large-scale bioprocesses; heat sterilisation of gases is economically impractical. *Depth filters* consisting of compacted beds or pads of fibrous material such as glass wool have been used widely in the fermentation industry. Distances between the fibres in depth filters are typically 2 to 10 μm, or up to 10 times greater than the dimensions of the bacteria and spores to be removed. Airborne particles penetrate the bed to various depths before their passage through the filter is arrested; the depth of the filter required to produce air of sufficient quality depends on the operating flow rate and the incoming level of contamination. Cells are collected in depth filters by a combination of impaction, interception, electrostatic effects, and, for particles smaller than about 1.0 μm, diffusion to the fibres. Depth filters do not perform well if there are large fluctuations in air flow rate or if the air is wet: liquid condensing in the filter increases the pressure drop, causes channelling of the gas flow, and provides a pathway for organisms collected on the fibres to grow through the bed.

Increasingly, depth filters are being replaced in industrial applications by membrane cartridge filters. These filters use steam-sterilisable or disposable polymeric membranes that act as surface filters trapping contaminants as on a sieve. Membrane filter cartridges typically contain a pleated, hydrophobic filter with small and uniformly sized pores of diameter 0.45 μm or less. The hydrophobic nature of the surface minimises problems with filter wetting while the pleated configuration allows a high filtration area to be packed into a small cartridge volume. Prefilters built into the cartridge or located upstream reduce fouling of the membrane by removing large particles, oil, water droplets, and foam from the incoming gas.

Filters are also used to sterilise off-gases leaving fermenters. In this case, the objective is to prevent release into the atmosphere of any organisms entrained in aerosols in the headspace of the reactor. The concentration of cells in unfiltered fermenter off-gas is several times greater than in air. Containment is particularly important when the organisms used in fermentation processes are potentially harmful to plant personnel or the environment. Companies operating fermentations with pathogenic or recombinant strains are required by regulatory authorities to prevent the cells from escaping into the atmosphere.

14.7 SUSTAINABLE BIOPROCESSING

The environmental impacts and energy costs of industrial activities are of increasing concern to the companies involved, governmental and regulatory agencies, and the general public. Analysis of waste generation and the conservation of resources and energy are now integral to the design procedures employed for bioreactors and other components of bioprocessing schemes. The specific areas addressed to improve the sustainability of bioprocesses include:

- Saving energy
- Saving water
- Minimising the use of materials
- Minimising greenhouse gas emissions
- Minimising the generation of effluent
- Minimising the environmental impact of effluent

Bioprocesses have many characteristics that render them attractive from an environmental point of view. Compared with chemical manufacturing, biological processing uses non-toxic substrates and relatively mild conditions of temperature, pressure, pH, and ionic strength. Biodegradable catalysts (enzymes and cells) are employed, minimal quantities of organic solvents are required, and the conversions carried out exhibit a high level of specificity. Nevertheless, the production of cellular biomass as a waste product and the substantial energy and water requirements for maintaining aseptic operations make a significant contribution to the overall environmental burden associated with bioprocessing.

Much of sustainable design focuses on minimising waste, including energy and materials. Waste reduction and application of 'lean' manufacturing principles enhance efficiency, productivity, and profits, and so have always been incorporated at some level into engineering design and management procedures. However, other aspects of sustainable design are broader in scope, extending to the assessment and reduction of environmental impacts across the entire life cycle of processes and equipment.

14.7.1 Sources of Waste and Pollutants in Bioprocessing

The main components of solid and liquid effluent streams produced in fermentation processes are cell biomass either before or after disruption, aqueous spent broth, and the buffers, solvents, and salt solutions used in downstream processing. Gas emissions include metabolically produced CO_2 and, in some processes, organic vapours such as ethanol and other volatile compounds. Large quantities of waste water and dilute cleaning solutions are generated by the clean-in-place (CIP) systems used to maintain hygienic conditions in all equipment employed in repeat processing operations. Further waste water is produced by steam-in-place (SIP) facilities for sterilisation of medium, bioreactors, piping, and other equipment such as centrifuges. The disposable plastics and polymers used for liquid and gas filtration and in the analytical laboratory for fermentation monitoring and quality control generate additional solid waste. More broadly, gas emissions from transportation of goods and personnel can also be considered as waste products of bioprocessing.

Although effluent streams from bioprocessing generally do not contain high levels of hazardous components, large volumes of aqueous solutions containing organic and inorganic nutrients are potentially toxic in aquatic environments. Minerals and trace elements including Fe, Cu, Mo, and Mn are components of cell culture media and can become concentrated within the cellular biomass generated during fermentation. The discharge from a single 2-m^3 high-density yeast fermentation has been estimated to contain a significant quantity of around 500 g of trace metals [43]. Some compounds used in downstream processing, such as urea for solubilising bacterial protein inclusion bodies and high-salt buffers for chromatographic separations, may present problems for waste disposal. Other chemicals including antifoams, surfactants, and fatty acids in fermenter effluent cause foaming in waste treatment systems. Effluent from standard cleaning operations comprises dilute solutions containing bleach, phosphoric acid, and/or sodium hydroxide.

Energy is consumed in bioprocesses for heating, cooling, evaporation, pumping, agitation, aeration, and centrifugation and is provided typically in the form of electricity and steam. For mammalian cell culture and biopharmaceutical production, substantial amounts of additional energy are consumed in the distillation processes used to generate purified, pyrogen-free water (also called water-for-injection, or WFI) for medium preparation and final rinsing and flushing of equipment. Much of this energy use is essential and cannot be classified as waste, although steps may be taken to improve the efficiency of energy-consuming operations. Away from the direct requirements of the bioprocess, however, there are other areas of energy consumption in fermentation facilities. These include operation of controlled temperature units for cold and frozen storage, and heating, ventilation, and air conditioning (HVAC) systems in clean rooms. Energy consumption in these areas can represent 50 to 75% of the total energy cost in a biopharmaceutical plant. Such facilities function continuously to preserve cell and chemical stocks and maintain cleanliness and positive air pressures even when production is not taking place. If they are over-designed or operate routinely under capacity, there may be scope to implement energy-saving measures in these areas.

14.7.2 Waste Metrics

Several parameters are used to characterise the utilisation of materials and generation of waste in industrial processing. No single parameter is ideal and none is employed universally. In principle, these metrics can be compared across different manufacturing sectors, companies, and products to gauge their relative environmental impacts.

Calculation of various indices requires definition of the system boundaries and knowledge of the mass flows into and out of the process. The *environmental-* or *E-factor* is defined as:

$$\text{E-factor} = \frac{\text{mass of waste}}{\text{mass of product}} \tag{14.103}$$

For calculation of the E-factor, waste is all the material generated by the process except the desired product. A higher E-factor means more waste and a greater negative environmental impact; the ideal E-factor is zero. An alternative index for waste generation is the *mass intensity*:

$$\text{mass intensity (MI)} = \frac{\text{mass of raw materials}}{\text{mass of product}} \tag{14.104}$$

where the ideal MI is 1. As it was originally formulated, the E-factor did not include water [44]; however, water usage is an important issue in many processes and may be represented in the calculation [45].

The E-factor and related indices do not take into account the nature of the waste generated, particularly its pollutant strength and effect on human health. For example, although the E-factor for a particular process may be high, if the waste consists mainly of water and relatively benign chemicals such as inorganic salts, the effect on the environment will be lower than for a process generating relatively small quantities of toxic effluent. The *environmental quotient* has been proposed to quantify both the relative amount and type of waste produced [44]:

$$\text{environmental quotient (EQ)} = \text{E-factor} \times \text{Q} \tag{14.105}$$

where Q reflects the extent to which the waste presents an environmental problem. As an example, Q may be assigned a value of 1 for NaCl and 100 to 1000 for a toxic heavy metal. The magnitude of Q could be both volume- and location-dependent, as the ease of recycling or disposal of materials depends on those factors. An alternative approach to representing the environmental impact of waste is the *effective mass yield*:

$$\text{effective mass yield (EMY)} = \frac{\text{mass of product}}{\text{mass of non-benign raw materials}} \times 100\% \tag{14.106}$$

where substances such as water, dilute salts, dilute ethanol, dilute acetic acid, and so on, are classified as benign. A problem with indices such as those described in Eqs. (14.105) and (14.106) is that the Q value used for a particular waste component is debatable, and there is no agreed consensus about what constitutes an environmentally benign material.

Typical results from application of the E-factor to different chemical manufacturing industries are shown in Table 14.4. In the table, bulk or industrial chemicals include products such as sulphuric acid, acrylonitrile, bulk polymers, and methanol from carbon monoxide. Specialty or fine chemicals include perfumes, pesticides, coatings, inks, and sealants.

TABLE 14.4 E-Factors for Manufacture of Chemicals and Pharmaceuticals

	Manufacturing Sector			
	Bulk/industrial Chemicals	**Specialty/fine Chemicals**	**Pharmaceuticals**	**Biopharmaceuticals**
E-factor	<1–5	5–50	25–100	2500–10,000
Annual production (tonnes)	10^4–10^6	10^2–10^4	10–10^3	10^{-3}–10
Annual waste production (tonnes)	10^4–5×10^6	5×10^2–5×10^5	250–10^5	2.5–10^5
Product price per kg	Low	Medium	High	High

Data from R.A Sheldon, 2007, The E factor: fifteen years on. Green Chem. *9, 1273–1283; B. Junker, 2010, Minimizing the environmental footprint of bioprocesses. Part 1: Introduction and evaluation of solid-waste disposal.* BioProcess Int. *8(8), 62–72; and S.V. Ho, J.M. McLaughlin, B.W. Cue, and P.J. Dunn, 2010, Environmental considerations in biologics manufacturing.* Green Chem. *12, 755–766.*

Pharmaceuticals refers to small-molecule drugs such as steroids, alkaloids, antipsychotics, analgesics, and enzyme inhibitors that are produced by chemical synthesis rather than biocatalysis. The biopharmaceuticals category includes the production of therapeutic proteins using cell culture and the bioprocessing techniques that are the subject of this book. The results for biopharmaceuticals include process water. In general, E-factors become higher as the process production volume decreases and the product price increases.

Although the E-factors for pharmaceutical and biopharmaceutical production are relatively high, when the low product volumes in these sectors are taken into account, the total quantity of waste generated is, on average, lower than for the other types of chemical production. Nevertheless, the E-factor results draw attention to the relatively inefficient use of materials in the pharmaceuticals and biopharmaceuticals industries. This is due primarily to the use of multistep batch processing and, in the case of biopharmaceuticals, consumption of large amounts of water for equipment cleaning and steam sterilisation. Compared with small-molecule drug manufacture, synthesis of therapeutic proteins using fermentation technology has been estimated to require approximately 10 to 100 times more water per kg of product [46]. The large E-factors for pharmaceutical production can also be interpreted as reflecting the higher selling prices and profit margins in the sector and, consequently, a reduced focus on the economic benefits of lean manufacturing practices. There is considerable scope in bioprocessing for intensification of production systems, for example, using fed-batch operations at high cell concentrations (Section 14.5.3), and for development of recovery and recycling programs to minimise resource consumption and waste generation. Therefore, an objective for improved bioprocess sustainability and reduced environmental impact is to identify strategies for reducing the E-factors associated with enzyme reaction and fermentation processes.

14.7.3 Life Cycle Analysis

The aim of life cycle analysis is to identify and evaluate the environmental impacts associated with product manufacture or process operation. Whereas the metrics described in the previous section are usually applied only to the production process itself, life cycle analysis examines the material and energy uses and releases to the environment from a much broader range of activities beyond the boundaries of the manufacturing plant. All of the 'life stages' of a product may be assessed for their environmental impact, from production or extraction of the raw materials and generation of the power required in the production scheme, to maintenance procedures, packaging, transportation, end use of the product, and waste treatment and disposal. This type of analysis is particularly useful for comparing the relative effects on the environment of two or more products or processes. Specialist software packages have been developed to assist the evaluation.

In principle, the all-encompassing framework of life cycle analysis reduces the risk of obtaining incorrect or misleading results for environmental impact when only some of the operations associated with a product or process are considered. Including all stages in the life cycle prevents environmental burdens being transferred from one stage to another. In practice, carrying out a full life cycle assessment of any production scheme is a lengthy and detailed exercise involving potentially thousands of inputs, outputs, and elements of process information. Determining the system boundaries for the analysis and making decisions about the inclusion or exclusion of particular aspects of the process (the cut-off criteria) can influence the results substantially.

Figure 14.43 shows the elements that might be included in a life cycle analysis of a fermentation product. As well as the production process, where and how the raw materials are sourced, the use of transportation, the generation and use of electricity, and the treatment of waste are all examined. In this example, agricultural production of sugar cane and sugar milling and refining are included as life cycle stages necessary for the production of substrate used in the fermentation. Emissions to the atmosphere and the generation of liquid and solid waste are considered for all steps in the life cycle.

Consensus-based international standards developed by the International Organization for Standardization (ISO) are used in life cycle analysis. The principles outlined in ISO 14040 provide the framework for a systematic approach to gathering and assessing the quantitative data needed to gauge environmental impact in a scientific and transparent way. The *functional unit* to be used in the analysis is determined; for a bioprocess this might be, for example, production of 1 kg of monoclonal antibody or 1 tonne of valine. The functional unit serves as a reference for quantifying all of the input and output streams at each of the life stages in the same way as a basis was chosen for evaluating mass and energy balances in Chapters 4 and 5.

An *inventory* of the mass and energy inputs and environmental releases associated with each step in the life cycle is developed. This process is guided by process flow diagrams and uses data from industry sources, published studies, and specialised life cycle databases. Information covering areas of agriculture, energy generation, base and precious metals, transportation, bulk and speciality chemicals, biomaterials, construction materials, packaging, waste treatment, electronics, and information and communications technology is available from international life cycle database providers. Some of the data required will be location-specific, as agricultural, transport, and energy generation practices vary from country to country and region to region. For analysis of a bioprocess, the productivities, yields, and concentrations associated with operation of the bioreactor provide important information for developing the inventory of inputs and outputs for this stage of the process. Any recycling of heat or water within the production facility is taken into account as the inventory focuses on inputs and outputs that cross the system boundaries.

As many industrial processes produce more than one product, each product can be considered responsible for only a fraction of the resources consumed and emissions generated. *Allocation* of environmental impacts between coproducts can be a challenging aspect of life cycle analysis, as the choice of different apportioning methods can lead to different overall results. An example from bioprocessing is if the cells produced as a by-product of fermentation were sold as animal feed. An effective environmental credit would be included in the life cycle analysis to account for coproduction of this useful product.

The life cycle inventory of inputs and outputs must be translated into environmental impacts or consequences. The inventory items are aggregated into a limited set of recognised *environmental impact categories*, which are also known as 'stressor' categories. There is no standard or universally agreed-on set of impact categories; however, commonly used categories include global warming, ozone depletion, acidification, eutrophication, particulate emissions, depletion of nonrenewable resources, human toxicity, ecotoxicity, and

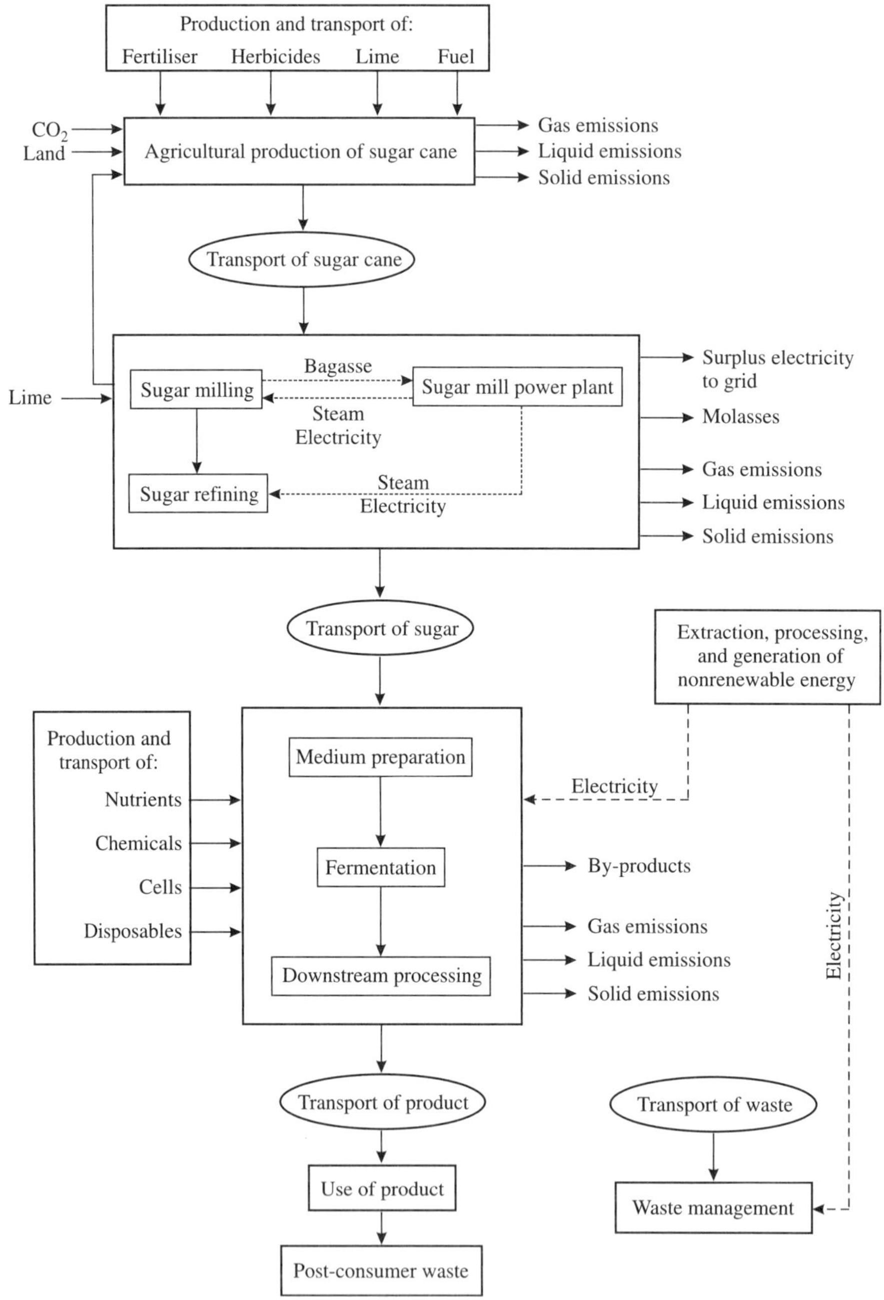

FIGURE 14.43 Schematic diagram of possible life cycle stages for a fermentation product.

land use. As an example, if the fermentation and downstream processes in Figure 14.43 generate CO_2 and waste acidic buffer, the quantities involved would be assigned as contributions to the global warming and acidification impact categories, respectively. Published lists are used to guide the classification of particular materials to individual impact categories; some inventory items may be split between multiple categories. Within each category, some substances will be more damaging to the environment than others. Therefore, *characterisation* or *equivalency factors* are applied to estimate the relative magnitude of the environmental impacts of each of the inventory components. These factors are based on scientific models of the interactions of particular substances with the environment. As an example, for greenhouse gases, the global warming potential of methane is considered to be 56 times greater than that of CO_2, so any methane emissions in the inventory are scored 56 times higher per unit mass than CO_2 emissions. Similarly, acidic substances in liquid effluent might be scored for their acidification potential based on how many H^+ ions they release as well as any technical factors that affect their environmental potency. Databases and life cycle analysis tools provide characterisation factors for the different impact categories. Ongoing improvements in our understanding of cause and effect in environmental damage are reflected in the information available.

Further evaluation may be carried out as part of life cycle analysis to *normalise* the results against benchmark values: this highlights the relative severity of the environmental burden in the different categories. *Weighting* of the categories can also be performed if particular types of environmental impact are of more concern than others. *Aggregation* of all the findings across the impact categories may be used to generate a single index reflecting the overall environmental status of the product or process. As the scientific basis for these three steps is weaker than for the preceding analysis, many life cycle studies are performed only to the characterisation stage of the analysis. Assessing the validity, reliability, and robustness of the results is an important element of life cycle techniques. *Sensitivity analysis* is used to determine whether the results are altered substantially when small changes are made to the input parameters; this identifies which data used in the assessment need to be most accurate. Other analyses may be used to gauge the effect of varying particular data that were estimated or approximated in the calculations, or to determine whether alternative process scenarios and models produce significantly different results.

Typical results from a life cycle analysis are shown in Figure 14.44. In this example, the life cycle environmental impacts of two biofuels, ethanol from sugar beet and methyl ester (biodiesel) from rapeseed, are compared with those of two fossil fuels, petrol (gasoline) and diesel. The environmental effects of the fuels are quantified in terms of their contributions to the depletion of nonrenewable resources, acidification and eutrophication, ecotoxicity, and global warming. In each environmental impact category, the fuel having the largest impact is represented as 100% and the impacts of the other three products are shown relative to that value. The life stages used in the analysis include cultivation and processing of agricultural crops in the case of the biofuels, extraction and refining in the case of the fossil fuels, ethanol fermentation, reaction of rapeseed oil with petrochemical methanol, all transportation steps, final use of the fuels in midsized cars, and application

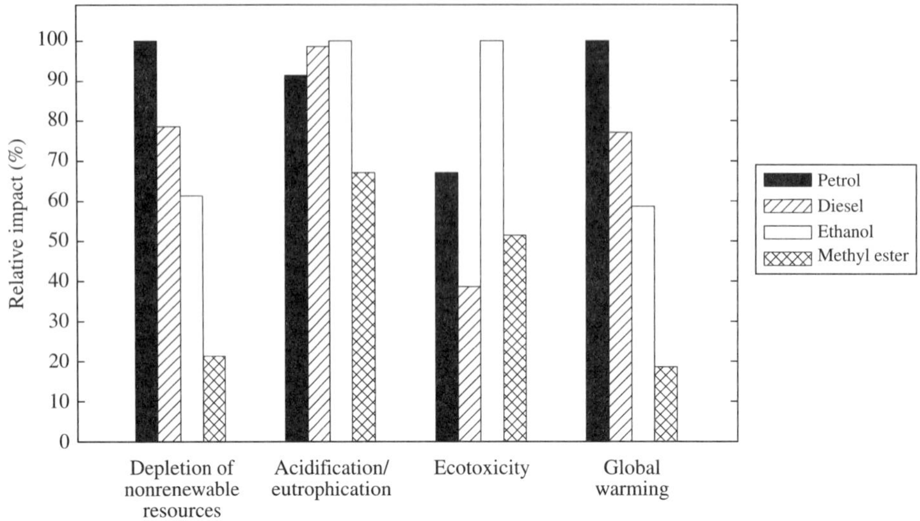

FIGURE 14.44 Results of life cycle analysis showing the relative environmental impacts of two fossil fuels—petrol (gasoline) and diesel—and two biofuels—ethanol from sugar beet and methyl ester from rapeseed.
Data from H. Halleux, S. Lassaux, R. Renzoni, and A. Germain, 2008, Comparative life cycle assessment of two biofuels: ethanol from sugar beet and rapeseed methyl ester. Int. J. LCA *13, 184–190.*

of by-products such as glycerine, rapeseed meal, and residual sugar beet pulp in the chemical industry or as animal feed. Overall, the results indicate that rapeseed methyl ester exerts the least environmental impact in three of the four categories examined, while ethanol offers improvements over the fossil fuels only in terms of nonrenewable resource consumption and global warming. More details of the analysis can be found in the original paper [47]. This example illustrates the type of environmental impact assessment that can be accomplished using life cycle methodology.

14.7.4 Disposable Bioreactors

Earlier in this chapter, practical considerations for the construction of bioreactors were described, such as the use of stainless steel for fabrication of large-scale bioreactors as pressure vessels, and the need to incorporate special design features for in situ steam sterilisation, inoculation, sampling, and cleaning to maintain aseptic conditions (Section 14.3). These requirements apply to the majority of industrial fermentations carried out using equipment and other hardware built to last for many years. However, as outlined in Section 14.1, if a company is launching a new biotechnology product, optimising the bioreactor design and culture performance is often of much lower priority than getting the product to market quickly. Application of ready-made, off-the-shelf disposable bioreactors can be an attractive option in these circumstances. A particular advantage is that none of the cleaning and sterilisation facilities and procedures that account for a substantial proportion of the capital and operating costs of conventional fermentation is required.

Many disposable or *single-use* items are employed routinely in large-scale bioprocessing, including filter membranes and cartridges, depth filter pads, membrane chromatography capsules, and aseptic connectors. However, additional disposable items are available commercially. Plastic process bags with volumes up to 2000 to 3000 litres may replace stainless steel vessels for liquid storage; disposable bioreactor or cell culture bags of capacity up to 500 to 1000 litres are also used in some applications. The size of disposable bioreactors is limited by the strength of the bag material and the narrow range of options available for providing mixing to the culture contents. Bioreactor bags either are placed on a rocking or tilting platform to provide agitation or may be equipped with a disposable internal paddle mixing and sparger system.

The performance of disposable bioreactors in terms of mixing, heat transfer, and oxygen transfer is poor relative to conventional fermenters, as intense agitation and high rates of heat and mass transfer cannot be achieved. Accordingly, their application is limited mainly to animal cell cultures where oxygen demands and heat loads are low and rapid mixing is generally not provided because of the shear sensitivity of the cells (Section 8.16). However, as many new product developments in biotechnology involve animal cell culture, disposable bioreactors are being adopted as an alternative approach to making small-to-medium quantities of recombinant and therapeutic proteins, antibodies, viruses, and vaccines.

There may be substantial cost benefits associated with using disposable bioprocessing systems. This is because the utilities and materials required for sterilisation and cleaning are minimised. If single-use components are applied everywhere throughout the manufacturing plant, a substantial amount of infrastructure normally associated with bioprocessing can be eliminated altogether. Usually, this more than compensates for the costs involved in having to replace disposable items. When sterilisation and cleaning are no longer required, the need for boilers, clean steam generators, steam-in-place and clean-in-place systems, and many other items of utility hardware is removed. Significant savings of energy, water, and cleaning chemicals may follow, as well as savings of time and labour.

An important concern associated with disposable bioprocessing is the environmental cost of single-use plasticware. Up to 90 disposable bags and biocontainers may be required to carry out a single 2000-litre batch culture [48]. Use of disposable equipment creates plastic waste that must be incinerated as biohazardous material or disposed of as landfill after autoclaving or chemical deactivation. The potential for reusing or recycling most disposables is limited by the risk of contamination or biological carry-over and the need to separate the multiple types of plastic present in each item before recycling. Life cycle analysis has been applied to evaluate the environmental impacts associated with using 500-litre stainless steel bioreactors compared with 500-litre disposable bag bioreactors [49]. Despite the environmental burden associated with the generation of plastic waste, when the environmental effects of steel production are taken into account and depending on local conditions for electricity and steam generation and waste disposal, single-use bioreactor systems may have a lower overall impact on the environment compared with stainless steel equipment.

SUMMARY OF CHAPTER 14

Chapter 14 contains a variety of qualitative and quantitative information about the design and operation of bioreactors. After studying this chapter, you should:

- Be able to assess in general terms the influence of reactor engineering on total production costs in bioprocessing
- Be familiar with a range of bioreactor configurations in addition to the standard *stirred tank*, including *bubble column*, *airlift*, *packed bed*, *fluidised bed*, and *trickle bed* designs
- Understand the practical aspects of bioreactor construction, particularly those aimed at maintaining aseptic conditions
- Know the types of measurement required for *fermentation monitoring* and the problems associated with the lack of online analytical methods for important fermentation parameters
- Be familiar with established and alternative approaches to *fermentation control*
- Be able to estimate *batch reaction times* for enzyme and cell reactions
- Be able to predict the performance of *fed-batch reactors* operated under *quasi-steady-state conditions*
- Be able to analyse the operation of *continuous stirred tank reactors* and *continuous plug flow reactors*
- Understand the theoretical advantages and disadvantages of batch and continuous reactors for enzyme and cell reactions
- Know how to use steady-state chemostat data to determine cell kinetic and yield parameters
- Know the design procedures for batch and continuous heat sterilisation of liquid medium
- Be able to describe methods for filter sterilisation of liquid medium and fermentation gases
- Know the principal environmental concerns associated with resource utilisation and waste generation in bioprocesses
- Understand the approaches used in *life cycle analysis* of the environmental impacts of bioprocessing

PROBLEMS

14.1 Economics of batch enzyme conversion

An enzyme is used to convert substrate to a commercial product in a 1600-litre batch reactor. v_{max} for the enzyme is 0.9 g l^{-1} h^{-1}; K_m is 1.5 g l^{-1}. The substrate concentration at the start of the reaction is 3 g l^{-1}. According to the reaction stoichiometry, conversion of 1 g of substrate produces 1.2 g of product. The cost of operating the reactor including labour, maintenance, energy, and other utilities is estimated at $4800 per day. The cost of recovering the product depends on the substrate conversion achieved and the resulting concentration of product in the final reaction mixture. For conversions between 70% and 100%, the cost of downstream processing can be approximated using the equation:

$$C = 155 - 0.33X$$

where C is the cost in \$ per kg of product treated and X is the percentage substrate conversion. Product losses during processing are negligible. The market price for the product is \$750 kg^{-1}. Currently, the enzyme reactor is operated with 75% substrate conversion; however it is proposed to increase this to 90%. Estimate the effect that this will have on the economics of the process.

14.2 Batch production of aspartic acid using cell-bound enzyme

Aspartase enzyme is used industrially for the manufacture of aspartic acid, a component of low-calorie sweetener. Fumaric acid ($C_4H_4O_4$) and ammonia are converted to aspartic acid ($C_4H_7O_4N$) according to the equation:

$$C_4H_4O_4 + NH_3 \rightleftharpoons C_4H_7O_4N$$

Under investigation is a process using aspartase in intact *Bacillus cadaveris* cells. In the substrate range of interest, the conversion can be described using Michaelis–Menten kinetics with $K_m = 4.0 \text{ g l}^{-1}$. The substrate solution contains 15% (w/v) ammonium fumarate; enzyme is added in the form of lyophilised cells and the reaction is stopped when 85% of the substrate is converted. At 32°C, v_{max} for the enzyme is 5.9 $\text{g l}^{-1} \text{h}^{-1}$ and its half-life is 10.5 days. At 37°C, v_{max} increases to 8.5 $\text{g l}^{-1} \text{h}^{-1}$ but the half-life is reduced to 2.3 days.

(a) Which operating temperature would you recommend?

(b) The average downtime between batch reactions is 28 h. At the temperature chosen in (a), calculate the reactor volume required to produce 5000 tonnes of aspartic acid per year.

14.3 Prediction of batch culture time

A strain of *Escherichia coli* has been genetically engineered to produce human protein. A batch culture is started by inoculating 12 g of cells into a 100-litre bubble column fermenter containing 10 g l^{-1} glucose. The culture does not exhibit a lag phase. The maximum specific growth rate of the cells is 0.9 h^{-1}; the biomass yield from glucose is 0.575 g g^{-1}.

(a) Estimate the time required to reach stationary phase.

(b) What will be the final cell density if the fermentation is stopped after only 70% of the substrate is consumed?

14.4 Fed-batch scheduling

Nicotiana tabacum cells are cultured to high density for production of polysaccharide gum. The reactor used is a stirred tank that initially contains 100 litres of medium. The maximum specific growth rate of the culture is 0.18 day^{-1} and the yield of biomass from substrate is 0.5 g g^{-1}. The concentration of growth-limiting substrate in the medium is 3% (w/v). The reactor is inoculated with 1.5 g l^{-1} of cells and operated in batch until the substrate is virtually exhausted; medium flow is then started at a rate of 4 l day^{-1}. Fed-batch operation is carried out for 40 days under quasi-steady-state conditions.

(a) Estimate the batch culture time and the final biomass concentration achieved after the batch culture period.

(b) What is the final mass of cells in the reactor?

(c) The fermenter is available 275 days per year with a downtime between runs of 24 h. How much plant cell biomass is produced annually?

14.5 Fed-batch production of cheese starter culture

Lactobacillus casei is propagated under essentially anaerobic conditions to provide a starter culture for manufacture of Swiss cheese. The culture produces lactic acid as a by-product of energy metabolism. The system has the following characteristics:

$Y_{XS} = 0.23\ \text{kg kg}^{-1}$
$K_S = 0.15\ \text{kg m}^{-3}$
$\mu_{max} = 0.35\ \text{h}^{-1}$
$m_S = 0.135\ \text{kg kg}^{-1}\ \text{h}^{-1}$

A stirred fermenter is operated in fed-batch mode under quasi-steady-state conditions with a feed flow rate of 4 $\text{m}^3\ \text{h}^{-1}$ and feed substrate concentration of 80 kg m^{-3}. After 6 h, the liquid volume is 40 m^3.

(a) What was the initial culture volume?
(b) What is the concentration of substrate at quasi-steady state?
(c) What is the concentration of cells at quasi-steady state?
(d) What mass of cells is produced after 6 h of fed-batch operation?

14.6 Continuous enzyme conversion in a fixed bed reactor

A system is being developed to remove urea from the blood of patients with renal failure. A prototype fixed bed reactor is set up using urease immobilised in 2-mm-diameter gel beads. Buffered urea solution is recycled rapidly through the bed so that the system is well mixed and external mass transfer effects are negligible. The urease reaction is:

$$(NH_2)_2CO + 3H_2O \longrightarrow 2NH_4^+ + HCO_3^- + OH^-$$

K_m for the immobilised urease is 0.54 g l^{-1}. The volume of beads in the reactor is 250 cm^3, the total amount of urease is 10^{-4} g, and the turnover number is 11,000 g NH_4^+ (g enzyme)$^{-1}$ s^{-1}. The effective diffusivity of urea in the gel is $7 \times 10^{-6}\ \text{cm}^2\ \text{s}^{-1}$. The reactor is operated continuously with a liquid volume of 1 litre. The feed stream contains 0.42 g l^{-1} of urea and the desired urea concentration after enzyme treatment is 0.02 g l^{-1}. Ignoring enzyme deactivation, what volume of urea solution can be treated in 30 min?

14.7 Batch and continuous biomass production

Pseudomonas methylotrophus is used to produce single cell protein from methanol in a 1000-m^3 pressure-cycle airlift fermenter. The biomass yield from substrate is 0.41 g g^{-1}, K_S is 0.7 mg l^{-1}, and the maximum specific growth rate is 0.44 h^{-1}. The medium contains 4% (w/v) methanol. A substrate conversion of 98% is desirable. The reactor may be operated in either batch or continuous mode. If operated in batch, an inoculum of 0.01% (w/v) is used and the downtime between batches is 20 h. If continuous operations are used at steady state, a downtime of 25 days is expected per year. Neglecting maintenance requirements, compare the annual biomass production achieved using batch and continuous reactors.

14.8 Bioreactor design for immobilised enzymes

6-Aminopenicillanic acid used to produce semi-synthetic penicillins is prepared by enzymatic hydrolysis of fermentation-derived penicillin-G. Penicillin-G acylase immobilised in alginate is being considered for the process. The immobilised enzyme particles are sufficiently small so that mass transfer does not affect the reaction rate. The starting concentration of penicillin-G is 10% (w/v); because of the high cost of the substrate, 99% conversion is required. Under these conditions, enzymatic conversion of penicillin-G can be considered a first-order reaction. It has not been decided whether a batch, CSTR, or plug flow reactor would be most suitable. The downtime between batch reactions is expected to be 20 h. For the batch and CSTR reactors, the reaction rate constant is $0.8 \times 10^{-4}\ s^{-1}$; in the PFTR, the packing density of enzyme beads can be up to four times greater than in the other reactors. Determine the smallest reactor required to treat 400 tonnes of penicillin-G per year.

14.9 Chemostat culture with protozoa

Tetrahymena thermophila protozoa have a minimum doubling time of 6.5 hours when grown using bacteria as the limiting substrate. The yield of protozoal biomass is 0.33 g per g of bacteria and the substrate constant is $12\ mg\ l^{-1}$. The protozoa are cultured at steady state in a chemostat using a feed stream containing $10\ g\ l^{-1}$ of nonviable bacteria.

(a) What is the maximum dilution rate for operation of the chemostat?

(b) What is the concentration of *T. thermophila* when the operating dilution rate is one-half of the maximum?

(c) What is the concentration of bacteria when the dilution rate is three-quarters of the maximum?

(d) What is the biomass productivity when the dilution rate is one-third of the maximum?

14.10 Two-stage chemostat for secondary metabolite production

A two-stage chemostat system is used for production of secondary metabolite. The volume of each reactor is $0.5\ m^3$; the flow rate of feed is $50\ l\ h^{-1}$. Mycelial growth occurs in the first reactor; the second reactor is used for product synthesis. The concentration of substrate in the feed is $10\ g\ l^{-1}$. Kinetic and yield parameters for the organism are:

$$Y_{XS} = 0.5\ kg\ kg^{-1}$$
$$K_S = 1.0\ kg\ m^{-3}$$
$$\mu_{max} = 0.12\ h^{-1}$$
$$m_S = 0.025\ kg\ kg^{-1}\ h^{-1}$$
$$q_P = 0.16\ kg\ kg^{-1}\ h^{-1}$$
$$Y_{PS} = 0.85\ kg\ kg^{-1}$$

Assume that product synthesis is negligible in the first reactor and growth is negligible in the second reactor.

(a) Determine the cell and substrate concentrations entering the second reactor.

(b) What is the overall substrate conversion?

(c) What is the final concentration of product?

14.11 Growth of algae in a continuous bioreactor with cell recycle

Ammonium-limited growth of the blue–green alga *Oscillatoria agardhii* is studied using a 1-litre chemostat. The temperature of the culture is controlled at 25°C and continuous illumination is applied at an intensity of $37\ \mu E\ m^{-2}\ s^{-1}$. Sterile ammonia solution containing $0.1\ mg\ l^{-1}$ N is fed to the cells. The maximum specific growth rate of the organism is $0.5\ day^{-1}$ and the substrate constant for N is $0.5\ \mu g\ l^{-1}$. The yield of biomass on nitrogen is 18 mg of dry weight per mg of N. The reactor is operated to give a hydraulic residence time of 60 h. No extracellular products are formed.

(a) What is the overall substrate conversion?
(b) What is the biomass productivity?

Intensification of the culture process is investigated using cell recycle. While keeping the feed rate of fresh medium the same as that used without recycle, a recycle stream containing three times the concentration of cells in the reactor outflow is fed back to the vessel at a volumetric flow rate that is one-quarter of that of the fresh feed. The substrate concentration s in a chemostat with cell recycle is given by the equation:

$$s = \frac{D\,K_S\,(1+\alpha-\alpha\beta)}{\mu_{max} - D(1+\alpha-\alpha\beta)}$$

where D is the dilution rate based on the fresh feed flow rate, μ_{max} and K_S are the maximum specific growth rate and substrate constant for the organism, respectively, α is the volumetric recycle ratio, and β is the biomass concentration factor for the recycle stream.

(c) By how much does using cell recycle reduce the size of the bioreactor required to achieve the same level of substrate conversion as that obtained without cell recycle?

14.12 Kinetic analysis of bioremediating bacteria using a chemostat

A strain of *Ancylobacter* bacteria capable of growing on 1,2-dichloroethane is isolated from sediment in the river Rhine. The bacteria are to be used for onsite bioremediation of soil contaminated with chlorinated halogens. Kinetic parameters for the organism are determined using data obtained from chemostat culture. A 1-litre fermenter is used with a feed stream containing 100 μM of 1,2-dichloroethane. Steady-state substrate concentrations are measured as a function of chemostat flow rate.

Flow rate (ml h^{-1})	Substrate Concentration (μM)
10	17.4
15	25.1
20	39.8
25	46.8
30	69.4
35	80.1
50	100

(a) Determine μ_{max} and K_S for this organism.
(b) Determine the maximum practical operating flow rate.

14.13 Kinetic and yield parameters of an auxotrophic mutant

An *Enterobacter aerogenes* auxotroph capable of overproducing threonine has been isolated. The kinetic and yield parameters for this organism are investigated using a 2-litre chemostat fed with medium containing 10 g l^{-1} glucose. Steady-state cell and substrate concentrations are measured at a range of reactor flow rates.

Flow Rate ($l\,h^{-1}$)	Cell Concentration ($g\,l^{-1}$)	Substrate Concentration ($g\,l^{-1}$)
1.0	3.15	0.010
1.4	3.22	0.038
1.6	3.27	0.071
1.7	3.26	0.066
1.8	3.21	0.095
1.9	3.10	0.477

Determine the maximum specific growth rate, the substrate constant, the maintenance coefficient, and the true biomass yield from glucose for this culture.

14.14 Chemostat culture for metabolic engineering

A chemostat of working volume 400 ml is used to obtain steady-state data for metabolic flux analysis. *Lactobacillus rhamnosus* is cultured at pH 6.0 using sterile medium containing 12 $g\,l^{-1}$ glucose and no lactic acid. Lactic acid production is coupled with energy metabolism in this organism. Concentrations of biomass, glucose, and lactic acid are measured at five different operating flow rates.

Flow Rate ($ml\,h^{-1}$)	Steady-state Concentration ($g\,l^{-1}$)		
	Biomass	Glucose	Lactic Acid
48	1.26	0.078	16.4
112	1.33	0.285	13.1
140	1.33	0.466	12.8
164	1.30	0.706	11.4
216	0.94	4.09	8.07

(a) Determine μ_{max} and K_S.
(b) Determine m_S and Y_{XS}.
(c) Determine m_P and Y_{PX}.
(d) Determine Y_{PS}.
(e) Determine the maximum practical operating flow rate for this system.
(f) Determine the operating flow rate for maximum biomass productivity.
(g) At the flow rate determined in (f), what are the steady-state rates of biomass and lactic acid production and substrate consumption?

14.15 Effect of axial dispersion on continuous sterilisation

A 15-m^3 chemostat is operated with dilution rate 0.1 h^{-1}. A continuous steriliser with steam injection and flash cooling delivers sterilised medium to the fermenter. Medium in the holding section of the steriliser is maintained at 130°C. The concentration of contaminants in the raw medium is 10^5 ml^{-1}; an acceptable contamination risk is one organism every 3 months. The Arrhenius constant and activation energy for thermal death are estimated as 7.5×10^{39} h^{-1} and 288.5 kJ $gmol^{-1}$, respectively. The inner diameter of the steriliser pipe is 12 cm. At 130°C, the liquid density is 1000 kg m^{-3} and the viscosity is 4 kg m^{-1} h^{-1}.

(a) Assuming perfect plug flow, determine the length of the holding section.

(b) What length is required if axial dispersion effects are taken into account?

(c) If the steriliser is constructed with the length determined in (a) and operated at 130°C as planned, estimate the frequency of fermenter contamination.

14.16 Contamination frequency after continuous sterilisation

A continuous steriliser is constructed using a 21-metre length of pipe of internal diameter 8 cm. Liquid medium in the pipe is maintained at 128°C using saturated steam. At this temperature, the specific death constant of the contaminating organisms is 340 h^{-1} and the density and viscosity of the medium are 1000 kg m^{-3} and 0.9 cP, respectively. The concentration of contaminants in the raw medium is 6.5×10^5 ml^{-1}. If sterile medium is required on the fermentation floor at a rate of 0.9 m^3 h^{-1}, what is the frequency of contamination in the fermentation factory?

References

[1] P.N. Royce, A discussion of recent developments in fermentation monitoring and control from a practical perspective, Crit. Rev. Biotechnol. 13 (1993) 117–149.

[2] B.L. Maiorella, H.W. Blanch, C.R. Wilke, Economic evaluation of alternative ethanol fermentation processes, Biotechnol. Bioeng. 26 (1984) 1003–1025.

[3] L.A. Boon, F.W.J.M.M. Hoeks, R.G.J.M. van der Lans, W. Bujalski, M.O. Wolff, A.W. Nienow, Comparing a range of impellers for 'stirring as foam disruption', Biochem. Eng. J. 10 (2002) 183–195.

[4] J.J. Heijnen, K. van't Riet, Mass transfer, mixing and heat transfer phenomena in low viscosity bubble column reactors, Chem. Eng. J. 28 (1984) B21–B42.

[5] K. van't Riet, J. Tramper, Basic Bioreactor Design, Marcel Dekker, 1991.

[6] M.Y. Chisti, Airlift Bioreactors, Elsevier Applied Science, 1989.

[7] P. Verlaan, J. Tramper, K. van't Riet, K.Ch.A.M. Luyben, A hydrodynamic model for an airlift-loop bioreactor with external loop, Chem. Eng. J. 33 (1986) B43–B53.

[8] U. Onken, P. Weiland, Hydrodynamics and mass transfer in an airlift loop fermentor, Eur. J. Appl. Microbiol. Biotechnol. 10 (1980) 31–40.

[9] P. Verlaan, J.-C. Vos, K. van't Riet, Hydrodynamics of the flow transition from a bubble column to an airlift-loop reactor, J. Chem. Tech. Biotechnol. 45 (1989) 109–121.

[10] M.H. Siegel, J.C. Merchuk, K. Schügerl, Airlift reactor analysis: interrelationships between riser, downcomer, and gas–liquid separator behavior, including gas recirculation effects, AIChE J. 32 (1986) 1585–1596.

[11] A.B. Russell, C.R. Thomas, M.D. Lilly, The influence of vessel height and top-section size on the hydrodynamic characteristics of airlift fermentors, Biotechnol. Bioeng. 43 (1994) 69–76.

[12] B. Atkinson, F. Mavituna, Biochemical Engineering and Biotechnology Handbook, second ed., Macmillan, 1991.

[13] M. Morandi, A. Valeri, Industrial scale production of β-interferon, Adv. Biochem. Eng./Biotechnol. 37 (1988) 57–72.

[14] B. Cameron, Mechanical seals for bioreactors, Chem. Engr. 442 (November) (1987) 41–42.

[15] Y. Chisti, Assure bioreactor sterility, Chem. Eng. Prog. 88 (September) (1992) 80–85.

[16] T. Akiba, T. Fukimbara, Fermentation of volatile substrate in a tower-type fermenter with a gas entrainment process, J. Ferment. Technol. 51 (1973) 134–141.

[17] P.F. Stanbury, A. Whitaker, S.J. Hall, Principles of Fermentation Technology, second ed., Butterworth-Heinemann, 1995 (Chapter 8).

[18] B. Kristiansen, Instrumentation, in: J. Bu'Lock, B. Kristiansen (Eds.), Basic Biotechnology, Academic Press, 1987, pp. 253–281.
[19] S.W. Carleysmith, Monitoring of bioprocessing, in: J.R. Leigh (Ed.), Modelling and Control of Fermentation Processes, Peter Peregrinus, 1987, pp. 97–117.
[20] J. Cooper, T. Cass (Eds.), Biosensors, second ed., Oxford University Press, 2003.
[21] B.R. Eggins, Chemical Sensors and Biosensors, John Wiley, 2002.
[22] A. Hayward, On-line, *in-situ* measurements within fermenters, in: B. McNeil, L.M. Harvey (Eds.), Practical Fermentation Technology, John Wiley, 2008, pp. 271–288.
[23] D. Dochain (Ed.), Automatic Control of Bioprocesses, John Wiley, 2008.
[24] P.N. Royce, N.F. Thornhill, Analysis of noise and bias in fermentation oxygen uptake rate data, Biotechnol. Bioeng. 40 (1992) 634–637.
[25] M. Meiners, W. Rapmundt, Some practical aspects of computer applications in a fermentor hall, Biotechnol. Bioeng. 25 (1983) 809–844.
[26] R.T.J.M. van der Heijden, C. Hellinga, K.Ch.A.M. Luyben, G. Honderd, State estimators (observers) for the on-line estimation of non-measurable process variables, Trends Biotechnol. 7 (1989) 205–209.
[27] G.A. Montague, A.J. Morris, J.R. Bush, Considerations in control scheme development for fermentation process control, IEEE Contr. Sys. Mag. 8 (April) (1988) 44–48.
[28] G.A. Montague, A.J. Morris, M.T. Tham, Enhancing bioprocess operability with generic software sensors, J. Biotechnol. 25 (1992) 183–201.
[29] D.E. Seborg, T.F. Edgar, D.A. Mellichamp, F.J. Doyle, Process Dynamics and Control, third ed., John Wiley, 2011.
[30] J.-I. Horiuchi, Fuzzy modeling and control of biological processes, J. Biosci. Bioeng. 94 (2002) 574–578.
[31] K.B. Konstantinov, T. Yoshida, Knowledge-based control of fermentation processes, Biotechnol. Bioeng. 39 (1992) 479–486.
[32] J. Glassey, G. Montague, P. Mohan, Issues in the development of an industrial bioprocess advisory system, Trends Biotechnol. 18 (2000) 136–141.
[33] R. Babuška, M.R. Damen, C. Hellinga, H. Maarleveld, Intelligent adaptive control of bioreactors, J. Intellig. Manuf. 14 (2003) 255–265.
[34] Z.K. Nagy, Model based control of a yeast fermentation bioreactor using optimally designed artificial neural networks, Chem. Eng. J. 127 (2007) 95–109.
[35] L.Z. Chen, S.K. Nguang, X.D. Chen, Modelling and Optimization of Biotechnological Processes: Artificial Intelligence Approaches, Springer-Verlag, 2010.
[36] T. Becker, T. Enders, A. Delgado, Dynamic neural networks as a tool for the online optimization of industrial fermentation, Bioprocess Biosyst. Eng. 24 (2002) 347–354.
[37] S.J. Pirt, Principles of Microbe and Cell Cultivation, Blackwell Scientific, 1975.
[38] M.L. Shuler, F. Kargi, Bioprocess Engineering: Basic Concepts, second ed., Prentice Hall, 2002 (Chapter 9).
[39] F. Reusser, Theoretical design of continuous antibiotic fermentation units, Appl. Microbiol. 9 (1961) 361–366.
[40] S. Aiba, A.E. Humphrey, N.F. Millis, Biochemical Engineering, Academic Press, 1965.
[41] J.W. Richards, Introduction to Industrial Sterilisation, Academic Press, 1968.
[42] C.L. Chopra, G.N. Qasi, S.K. Chaturvedi, C.N. Gaind, C.K. Atal, Production of citric acid by submerged fermentation. Effect of medium sterilisation at pilot-plant level, J. Chem. Technol. Biotechnol. 31 (1981) 122–126.
[43] B. Junker, Minimizing the environmental footprint of bioprocesses. Part 1: Introduction and evaluation of solid-waste disposal, BioProcess Int. 8 (8) (2010) 62–72.
[44] R.A Sheldon, The E factor: fifteen years on, Green Chem. 9 (2007) 1273–1283.
[45] M. Lancaster, Green Chemistry: An Introductory Text, Royal Society of Chemistry, 2002.
[46] S.V. Ho, J.M. McLaughlin, B.W. Cue, P.J. Dunn, Environmental considerations in biologics manufacturing, Green Chem. 12 (2010) 755–766.
[47] H. Halleux, S. Lassaux, R. Renzoni, A. Germain, Comparative life cycle assessment of two biofuels: ethanol from sugar beet and rapeseed methyl ester, Int. J. LCA 13 (2008) 184–190.
[48] B. Rawlings, H. Pora, A prescriptive approach to management of solid waste from single-use systems, BioProcess Int. 7 (3) (2009) 40–47.
[49] M. Mauter, Environmental life-cycle assessment of disposable bioreactors, BioProcess Int. 7 (S4) (2009) 18–29.

Suggestions for Further Reading

Reactor Configurations and Operating Characteristics

See also references [5] through [12].

Chisti, M. Y., & Moo-Young, M. (1987). Airlift reactors: characteristics, applications and design considerations. *Chem. Eng. Comm.*, *60*, 195–242.

Cooney, C. L. (1983). Bioreactors: design and operation. *Science*, *219*, 728–733.

Deckwer, W.-D. (1985). Bubble column reactors. In H.-J. Rehm, & G. Reed (Eds.), *Biotechnology* (vol. 2, pp. 445–464). VCH.

Practical Considerations for Reactor Design

See also references [14] and [15].

Chisti, Y. (1992). Build better industrial bioreactors. *Chem. Eng. Prog.*, *88*(January), 55–58.

Fermentation Monitoring and Control

See also references [17] through [36].

Albert, S., & Kinley, R. D. (2001). Multivariate statistical monitoring of batch processes: an industrial case study of fermentation supervision. *Trends Biotechnol.*, *19*, 53–62.

Alford, J. S. (2006). Bioprocess control: advances and challenges. *Comp. Chem. Eng.*, *30*, 1464–1475.

Harms, P., Kostov, Y., & Rao, G. (2002). Bioprocess monitoring. *Curr. Opinion Biotechnol.*, *13*, 124–127.

Lennox, B., Montague, G. A., Hiden, H. G., Kornfeld, G., & Goulding, P. R. (2001). Process monitoring of an industrial fed-batch fermentation. *Biotechnol. Bioeng.*, *74*, 125–135.

Mandenius, C.-F. (2004). Recent developments in the monitoring, modeling and control of biological production systems. *Bioprocess Biosyst. Eng.*, *26*, 347–351.

Richards, J. R. (1987). Principles of control system design. In J. R. Leigh (Ed.), *Modelling and Control of Fermentation Processes* (pp. 189–214). Peter Peregrinus.

Schügerl, K. (1997). *Bioreaction Engineering: Bioprocess Monitoring* (vol. 3). John Wiley.

Shimizu, K. (1993). An overview on the control system design of bioreactors. *Adv. Biochem. Eng./Biotechnol.*, *50*, 65–84.

Sonnleitner, B. (2006). Measurement, monitoring, modelling and control. In C. Ratledge, & B. Kristiansen (Eds.), *Basic Biotechnology* (3rd ed., pp. 251–270). Cambridge University Press.

Sterilisation

See also references [40] and [41].

Bader, F. G. (1986). Sterilization: prevention of contamination. In A. L. Demain, & N. A. Solomon (Eds.), *Manual of Industrial Microbiology and Biotechnology* (pp. 345–362). American Society of Microbiology.

Conway, R. S. (1985). Selection criteria for fermentation air filters. In M. Moo-Young (Ed.), *Comprehensive Biotechnology* (vol. 2, pp. 279–286). Pergamon.

Cooney, C. L. (1985). Media sterilization. In M. Moo-Young (Ed.), *Comprehensive Biotechnology* (vol. 2, pp. 287–298). Pergamon.

Deindoerfer, F. H., & Humphrey, A. E. (1959). Analytical method for calculating heat sterilization times. *Appl. Microbiol*, *7*, 256–264.

Sustainable Bioprocessing

See also references [43] through [49].

Adv. Biochem. Eng./Biotechnol. (2009) vol. 115: Disposable bioreactors.

Dunn, P. J., Wells, A. S., & Williams, M. T. (Eds.), (2010). *Green Chemistry in the Pharmaceutical Industry*. Wiley-VCH.

Heinzle, E., Biwer, A., & Cooney, C. (2006). *Development of Sustainable Bioprocesses*. John Wiley.

Hermann, B. G., Blok, K., & Patel, M. K. (2007). Producing bio-based bulk chemicals using industrial biotechnology saves energy and combats climate change. *Environ. Sci. Technol.*, *41*, 7915–7921.

Lapkin, A., & Constable, D. (Eds.), (2009). *Green Chemistry Metrics*. Wiley-Blackwell.

Scott, C. (2011). Sustainability in bioprocessing. *BioProcess Int.*, *9*(10), 25–36.

APPENDICES

APPENDIX

A

Conversion Factors

Entries in the same row are equivalent. For example, in Table A.1, 1 m = 3.281 ft, 1 mile = 1.609×10^3 m, etc. Exact numerical values are printed in bold type; others are given to four significant figures.

TABLE A.1 Length (L)

Metre (m)	Inch (in.)	Foot (ft)	Mile	Micrometre (μm) (formerly micron)	Angstrom (Å)
1	3.937×10^1	3.281	6.214×10^{-4}	$\mathbf{10^6}$	$\mathbf{10^{10}}$
$\mathbf{2.54 \times 10^{-2}}$	**1**	8.333×10^{-2}	1.578×10^{-5}	$\mathbf{2.54 \times 10^4}$	$\mathbf{2.54 \times 10^8}$
$\mathbf{3.048 \times 10^{-1}}$	$\mathbf{1.2 \times 10^1}$	**1**	1.894×10^{-4}	$\mathbf{3.048 \times 10^5}$	$\mathbf{3.048 \times 10^9}$
1.609×10^3	$\mathbf{6.336 \times 10^4}$	$\mathbf{5.28 \times 10^3}$	**1**	1.609×10^9	1.609×10^{13}
$\mathbf{10^{-6}}$	3.937×10^{-5}	3.281×10^{-6}	6.214×10^{-10}	**1**	$\mathbf{10^4}$
$\mathbf{10^{-10}}$	3.937×10^{-9}	3.281×10^{-10}	6.214×10^{-14}	$\mathbf{10^{-4}}$	**1**

TABLE A.2 Volume (L^3)

Cubic metre (m^3)	Litre (l or L) or cubic decimeter* (dm^3)	Cubic foot (ft^3)	Cubic inch ($in.^3$)	Imperial gallon (UKgal)	U.S. gallon (USgal)
1	$\mathbf{10^3}$	3.531×10^1	6.102×10^4	2.200×10^2	2.642×10^2
$\mathbf{10^{-3}}$	**1**	3.531×10^{-2}	6.102×10^1	2.200×10^{-1}	2.642×10^{-1}
2.832×10^{-2}	2.832×10^1	**1**	$\mathbf{1.728 \times 10^3}$	6.229	7.481
1.639×10^{-5}	1.639×10^{-2}	5.787×10^{-4}	**1**	3.605×10^{-3}	4.329×10^{-3}
4.546×10^{-3}	4.546	1.605×10^{-1}	2.774×10^2	**1**	1.201
3.785×10^{-3}	3.785	1.337×10^{-1}	$\mathbf{2.31 \times 10^2}$	8.327×10^{-1}	**1**

**The litre was defined in 1964 as 1 dm^3 exactly.*

TABLE A.3 Mass (M)

Kilogram (kg)	Gram (g)	Pound (lb)	Ounce (oz)	Tonne (t)	Imperial ton (UK ton)	Atomic mass unit* (u)
1	$\mathbf{10^3}$	2.205	3.527×10^1	$\mathbf{10^{-3}}$	9.842×10^{-4}	6.022×10^{26}
$\mathbf{10^{-3}}$	**1**	2.205×10^{-3}	3.527×10^{-2}	$\mathbf{10^{-6}}$	9.842×10^{-7}	6.022×10^{23}
4.536×10^{-1}	4.536×10^2	**1**	$\mathbf{1.6 \times 10^1}$	4.536×10^{-4}	4.464×10^{-4}	2.732×10^{26}
2.835×10^{-2}	2.835×10^1	$\mathbf{6.25 \times 10^{-2}}$	**1**	2.835×10^{-5}	2.790×10^{-5}	1.707×10^{25}
$\mathbf{10^3}$	$\mathbf{10^6}$	2.205×10^3	3.527×10^4	**1**	9.842×10^{-1}	6.022×10^{29}
1.016×10^3	1.016×10^6	2.240×10^3	3.584×10^4	1.016	**1**	6.119×10^{29}
1.661×10^{-27}	1.661×10^{-24}	3.661×10^{-27}	5.857×10^{-26}	1.661×10^{-30}	1.634×10^{-30}	**1**

**Atomic mass unit (unified); 1 u = 1/12 of the rest mass of a neutral atom of the nuclide ^{12}C in the ground state.*

TABLE A.4 Force (LMT^{-2})

Newton (N, kg m s^{-2})	Kilogram-force (kg_f)	Pound-force (lb_f)	Dyne (dyn, g cm s^{-2})	Poundal (pdl, lb ft s^{-2})
1	1.020×10^{-1}	2.248×10^{-1}	$\mathbf{10^5}$	7.233
9.807	**1**	2.205	9.807×10^5	7.093×10^1
4.448	4.536×10^{-1}	**1**	4.448×10^5	3.217×10^1
$\mathbf{10^{-5}}$	1.020×10^{-6}	2.248×10^{-6}	**1**	7.233×10^{-5}
1.383×10^{-1}	1.410×10^{-2}	3.108×10^{-2}	1.383×10^4	**1**

TABLE A.5 Pressure and Stress ($L^{-1}MT^{-2}$)

Pascal (Pa, N m^{-2}, J m^{-3}, kg m^{-1} s^{-2})	Pound-force per inch^2 (psi, lb_f in.$^{-2}$)	Kilogram-force per metre^2 (kg_f m^{-2})	Standard atmosphere (atm)	Dyne per cm^2 (dyn cm^{-2})	Torr (Torr, mmHg)*	Inches of water** (in. H_2O)	Bar
1	1.450×10^{-4}	1.020×10^{-1}	9.869×10^{-6}	$\mathbf{10^1}$	7.501×10^{-3}	4.015×10^{-3}	$\mathbf{10^{-5}}$
6.895×10^3	**1**	7.031×10^2	6.805×10^{-2}	6.895×10^4	5.171×10^1	2.768×10^1	6.895×10^{-2}
9.807	1.422×10^{-3}	**1**	9.678×10^{-5}	9.807×10^1	7.356×10^{-2}	3.937×10^{-2}	9.807×10^{-5}
1.013×10^5	1.470×10^1	1.033×10^4	**1**	1.013×10^6	$\mathbf{7.6 \times 10^2}$	4.068×10^2	1.013
$\mathbf{10^{-1}}$	1.450×10^{-5}	1.020×10^{-2}	9.869×10^{-7}	**1**	7.501×10^{-4}	4.015×10^{-4}	$\mathbf{10^{-6}}$
1.333×10^2	1.934×10^{-2}	1.360×10^1	1.316×10^{-3}	1.333×10^3	**1**	5.352×10^{-1}	1.333×10^{-3}
2.491×10^2	3.613×10^{-2}	2.540×10^1	2.458×10^{-3}	2.491×10^3	1.868	**1**	2.491×10^{-3}
$\mathbf{10^5}$	1.450×10^1	1.020×10^4	9.869×10^{-1}	$\mathbf{10^6}$	7.501×10^2	4.015×10^2	**1**

**mmHg refers to Hg at 0°C; 1 Torr = 1.00000 mmHg.*
***in. H_2O refers to water at 4°C.*

TABLE A.6 Surface Tension (MT^{-2})

Newton per metre ($N\ m^{-1}$, $kg\ s^{-2}$, $J\ m^{-2}$)	**Dyne per centimetre ($dyn\ cm^{-1}$, $g\ s^{-2}$, $erg\ cm^{-2}$)**	**Kilogram-force per metre ($kg_f\ m^{-1}$)**
1	**10^3**	1.020×10^{-1}
10^{-3}	**1**	1.020×10^{-4}
9.807	9.807×10^3	**1**

TABLE A.7 Energy, Work, and Heat ($L^2 MT^{-2}$)

Joule (J, N m, Pa m^3, W s, kg $m^2\ s^{-2}$)	**Kilocalorie* (kcal)**	**British thermal unit (Btu)**	**Foot pound-force (ft lb_f)**	**Litre atmosphere (l atm)**	**Kilowatt hour (kW h)**	**Erg (dyn cm)**
1	2.388×10^{-4}	9.478×10^{-4}	7.376×10^{-1}	9.869×10^{-3}	2.778×10^{-7}	**10^7**
4.187×10^3	**1**	3.968	3.088×10^3	4.132×10^1	**1.163×10^{-3}**	4.187×10^{10}
1.055×10^3	2.520×10^{-1}	**1**	7.782×10^2	1.041×10^1	2.931×10^{-4}	1.055×10^{10}
1.356	3.238×10^{-4}	1.285 × 10[illegible]	**1**	1.000×10^{-2}	3.766×10^{-7}	1.356×10^7
1.013×10^2	2.420×10^{-2}	9.604×10^{-2}	7.473×10^1	**1**	2.815×10^{-5}	1.013×10^9
3.6×10^6	8.598×10^2	3.412×10^3	2.655×10^6	3.553×10^4	**1**	**3.6×10^{13}**
10^{-7}	2.388×10^{-11}	9.478×10^{-11}	7.376×10^{-8}	9.869×10^{-10}	2.778×10^{-14}	**1**

**International Table kilocalorie ($kcal_{IT}$).*

TABLE A.8 Power (L^2MT^{-3})

Watt (W, J s^{-1}, kg $m^2\ s^{-3}$)	**Kilocalorie per min ($kcal\ min^{-1}$)**	**Foot pound-force per second (ft $lb_f\ s^{-1}$)**	**Horsepower (British) (hp)**	**Metric horsepower**	**British thermal unit per minute ($Btu\ min^{-1}$)**	**Kilogram-force metre per second ($kg_f\ m\ s^{-1}$)**
1	1.433×10^{-2}	7.376×10^{-1}	1.341×10^{-3}	1.360×10^{-3}	5.687×10^{-2}	1.020×10^{-1}
6.978×10^1	**1**	5.147×10^1	9.358×10^{-2}	9.487×10^{-2}	3.968	7.116
1.356	1.943×10^{-2}	**1**	1.818×10^{-3}	1.843×10^{-3}	7.710×10^{-2}	1.383×10^{-1}
7.457×10^2	1.069×10^1	**5.5×10^2**	**1**	1.014	4.241×10^1	7.604×10^1
7.355×10^2	1.054×10^1	5.425×10^2	9.863×10^{-1}	**1**	4.183×10^1	**7.5×10^1**
1.758×10^1	2.520×10^{-1}	1.297×10^1	2.358×10^{-2}	2.391×10^{-2}	**1**	1.793
9.807	1.405×10^{-1}	7.233	1.315×10^{-2}	1.333×10^{-2}	5.577×10^{-1}	**1**

TABLE A.9 Dynamic Viscosity ($L^{-1}MT^{-1}$)

Pascal second (Pa s, N s m^{-2}, kg m^{-1} s^{-1})	Poise (g cm^{-1} s^{-1}, dyn s cm^{-2})	Centipoise (cP)	kg m^{-1} h^{-1}	lb ft^{-1} h^{-1}
1	**10^1**	**10^3**	**3.6×10^3**	2.419×10^3
10^{-1}	**1**	**10^2**	**3.6×10^2**	2.419×10^2
10^{-3}	**10^{-2}**	**1**	**3.6**	2.419
2.778×10^{-4}	2.778×10^{-3}	2.778×10^{-1}	**1**	6.720×10^{-1}
4.134×10^{-4}	4.134×10^{-3}	4.134×10^{-1}	1.488	**1**

TABLE A.10 Plane Angle (1)

Radian (rad)	Revolution (rev)	Degree (°)	Minute (′)	Second (″)
1	1.592×10^{-1}	5.730×10^1	3.438×10^3	2.063×10^5
6.283	**1**	**3.6×10^2**	2.160×10^4	1.296×10^6
1.745×10^{-2}	2.778×10^{-3}	**1**	**6×10^1**	**3.6×10^3**
2.909×10^{-4}	4.630×10^{-5}	1.667×10^{-2}	**1**	**6×10^1**
4.848×10^{-6}	7.716×10^{-7}	2.778×10^{-4}	1.667×10^{-2}	**1**

TABLE A.11 Illuminance (L^{-2} J)

Lux or lumen per metre^2 (lx or lm m^{-2})	Foot-candle (fc, lm ft^{-2})
1	9.290×10^{-2}
1.076×10^1	**1**

APPENDIX

B

Ideal Gas Constant

TABLE B.1 Values of the Ideal Gas Constant, *R*

Energy unit	Temperature unit	Mole unit	*R*
cal	K	gmol	1.9872
J	K	gmol	8.3144
cm^3 atm	K	gmol	82.057
l atm	K	gmol	0.082057
m^3 atm	K	gmol	0.000082057
l mmHg	K	gmol	62.361
l bar	K	gmol	0.083144
$kg_f\ m^{-2}$ l	K	gmol	847.9
$kg_f\ cm^{-2}$ l	K	gmol	0.08479
mmHg ft^3	K	lbmol	998.9
atm ft^3	K	lbmol	1.314
Btu	°R	lbmol	1.9869
psi ft^3	°R	lbmol	10.731
lb_f ft	°R	lbmol	1545
atm ft^3	°R	lbmol	0.7302
in.Hg ft^3	°R	lbmol	21.85
hp h	°R	lbmol	0.0007805
kW h	°R	lbmol	0.0005819
mmHg ft^3	°R	lbmol	555

APPENDIX

C

Physical and Chemical Property Data

TABLE C.1 Atomic Weights and Numbers

Name	Symbol	Relative atomic mass	Atomic number
Actinium	Ac	–	89
Aluminium	Al	26.9815	13
Americium	Am	–	95
Antimony	Sb	121.75	51
Argon	Ar	39.948	18
Arsenic	As	74.9216	33
Astatine	At	–	85
Barium	Ba	137.34	56
Berkelium	Bk	–	97
Beryllium	Be	9.0122	4
Bismuth	Bi	208.98	83
Boron	B	10.811	5
Bromine	Br	79.904	35
Cadmium	Cd	112.40	48
Caesium	Cs	132.905	55
Calcium	Ca	40.08	20
Californium	Cf	–	98
Carbon	C	12.011	6
Cerium	Ce	140.12	58

(Continued)

TABLE C.1 Atomic Weights and Numbers (Continued)

Name	Symbol	Relative atomic mass	Atomic number
Chlorine	Cl	35.453	17
Chromium	Cr	51.996	24
Cobalt	Co	58.9332	27
Copper	Cu	63.546	29
Curium	Cm	–	96
Dysprosium	Dy	162.50	66
Einsteinium	Es	–	99
Erbium	Er	167.26	68
Europium	Eu	151.96	63
Fermium	Fm	–	100
Fluorine	F	18.9984	9
Francium	Fr	–	87
Gadolinium	Gd	157.25	64
Gallium	Ga	69.72	31
Germanium	Ge	72.59	32
Gold	Au	196.967	79
Hafnium	Hf	178.49	72
Helium	He	4.0026	2
Holmium	Ho	164.930	67
Hydrogen	H	1.00797	1
Indium	In	114.82	49
Iodine	I	126.9044	53
Iridium	Ir	192.2	77
Iron	Fe	55.847	26
Krypton	Kr	83.80	36
Lanthanum	La	138.91	57
Lawrencium	Lr	–	103
Lead	Pb	207.19	82
Lithium	Li	6.939	3
Lutetium	Lu	174.97	71

Magnesium	Mg	24.312	12
Manganese	Mn	54.938	25
Mendelevium	Md	–	101
Mercury	Hg	200.59	80
Molybdenum	Mo	95.94	42
Neodymium	Nd	144.24	60
Neon	Ne	20.183	10
Neptunium	Np	–	93
Nickel	Ni	58.71	28
Niobium	Nb	92.906	41
Nitrogen	N	14.0067	7
Nobelium	No	–	102
Osmium	Os	190.2	76
Oxygen	O	15.9994	8
Palladium	Pd	106.4	46
Phosphorus	P	30.9738	15
Platinum	Pt	195.09	78
Plutonium	Pu	–	94
Polonium	Po	–	84
Potassium	K	39.102	19
Praseodymium	Pr	140.907	59
Promethium	Pm	–	61
Protactinium	Pa	–	91
Radium	Ra	–	88
Radon	Rn	–	86
Rhenium	Re	186.2	75
Rhodium	Rh	102.905	45
Rubidium	Rb	85.47	37
Ruthenium	Ru	101.07	44
Samarium	Sm	150.35	62
Scandium	Sc	44.956	21
Selenium	Se	78.96	34
Silicon	Si	28.086	14

(*Continued*)

TABLE C.1 Atomic Weights and Numbers (Continued)

Name	Symbol	Relative atomic mass	Atomic number
Silver	Ag	107.868	47
Sodium	Na	22.9898	11
Strontium	Sr	87.62	38
Sulphur	S	32.064	16
Tantalum	Ta	180.948	73
Technetium	Tc	–	43
Tellurium	Te	127.60	52
Terbium	Tb	158.924	65
Thallium	Tl	204.37	81
Thorium	Th	232.038	90
Thulium	Tm	168.934	69
Tin	Sn	118.69	50
Titanium	Ti	47.90	22
Tungsten	W	183.85	74
Uranium	U	238.03	92
Vanadium	V	50.942	23
Wolfram (Tungsten)	W	183.85	74
Xenon	Xe	131.30	54
Ytterbium	Yb	173.04	70
Yttrium	Y	88.905	39
Zinc	Zn	65.37	30
Zirconium	Zr	91.22	40

Note: Based on the atomic mass of ^{12}C. Values for atomic weights apply to elements as they exist in nature.

TABLE C.2 Degree of Reduction of Biological Materials

Compound	Formula	Degree of reduction γ relative to NH_3	Degree of reduction γ relative to N_2
Acetaldehyde	C_2H_4O	5.00	5.00
Acetic acid	$C_2H_4O_2$	4.00	4.00
Acetone	C_3H_6O	5.33	5.33
Adenine	$C_5H_5N_5$	2.00	5.00
Alanine	$C_3H_7O_2N$	4.00	5.00
Ammonia	NH_3	0	3.00
Arginine	$C_6H_{14}O_2N_4$	3.67	5.67
Asparagine	$C_4H_8O_3N_2$	3.00	4.50
Aspartic acid	$C_4H_7O_4N$	3.00	3.75
n-Butanol	$C_4H_{10}O$	6.00	6.00
Butyraldehyde	C_4H_8O	5.50	5.50
Butyric acid	$C_4H_8O_2$	5.00	5.00
Carbon monoxide	CO	2.00	2.00
Citric acid	$C_6H_8O_7$	3.00	3.00
Cytosine	$C_4H_5ON_3$	2.50	4.75
Ethane	C_2H_6	7.00	7.00
Ethanol	C_2H_6O	6.00	6.00
Ethene	C_2H_4	6.00	6.00
Ethylene glycol	$C_2H_6O_2$	5.00	5.00
Ethyne	C_2H_2	5.00	5.00
Formaldehyde	CH_2O	4.00	4.00
Formic acid	CH_2O_2	2.00	2.00
Fumaric acid	$C_4H_4O_4$	3.00	3.00
Glucitol	$C_6H_{14}O_6$	4.33	4.33
Gluconic acid	$C_6H_{12}O_7$	3.67	3.67
Glucose	$C_6H_{12}O_6$	4.00	4.00
Glutamic acid	$C_5H_9O_4N$	3.60	4.20
Glutamine	$C_5H_{10}O_3N_2$	3.60	4.80
Glycerol	$C_3H_8O_3$	4.67	4.67
Glycine	$C_2H_5O_2N$	3.00	4.50

(*Continued*)

TABLE C.2 Degree of Reduction of Biological Materials (Continued)

Compound	Formula	Degree of reduction γ relative to NH_3	Degree of reduction γ relative to N_2
Graphite	C	4.00	4.00
Guanine	$C_5H_5ON_5$	1.60	4.60
Histidine	$C_6H_9O_2N_3$	3.33	4.83
Hydrogen	H_2	2.00	2.00
Isoleucine	$C_6H_{13}O_2N$	5.00	5.50
Lactic acid	$C_3H_6O_3$	4.00	4.00
Leucine	$C_6H_{13}O_2N$	5.00	5.50
Lysine	$C_6H_{14}O_2N_2$	4.67	5.67
Malic acid	$C_4H_6O_5$	3.00	3.00
Methane	CH_4	8.00	8.00
Methanol	CH_4O	6.00	6.00
Oxalic acid	$C_2H_2O_4$	1.00	1.00
Palmitic acid	$C_{16}H_{32}O_2$	5.75	5.75
Pentane	C_5H_{12}	6.40	6.40
Phenylalanine	$C_9H_{11}O_2N$	4.44	4.78
Proline	$C_5H_9O_2N$	4.40	5.00
Propane	C_3H_8	6.67	6.67
iso-Propanol	C_3H_8O	6.00	6.00
Propionic acid	$C_3H_6O_2$	4.67	4.67
Pyruvic acid	$C_3H_4O_3$	3.33	3.33
Serine	$C_3H_7O_3N$	3.33	4.33
Succinic acid	$C_4H_6O_4$	3.50	3.50
Threonine	$C_4H_9O_3N$	4.00	4.75
Thymine	$C_5H_6O_2N_2$	3.20	4.40
Tryptophan	$C_{11}H_{12}O_2N_2$	4.18	4.73
Tyrosine	$C_9H_{11}O_3N$	4.22	4.56
Uracil	$C_4H_4O_2N_2$	2.50	4.00
Valeric acid	$C_5H_{10}O_2$	5.20	5.20
Valine	$C_5H_{11}O_2N$	4.80	5.40
Biomass	$CH_{1.8}O_{0.5}N_{0.2}$	4.20	4.80

Adapted from J.A. Roels, 1983, Energetics and Kinetics in Biotechnology, *Elsevier, Amsterdam.*

TABLE C.3 Heat Capacities

Compound	State	Temperature (T) unit	a	$b.10^2$	$c.10^5$	$d.10^9$	Temperature range (units of T)
Acetone	l	°C	123.0	18.6			−30–60
Air	g	°C	28.94	0.4147	0.3191	−1.965	0–1500
	g	K	28.09	0.1965	0.4799	−1.965	273–1800
Ammonia	g	°C	35.15	2.954	0.4421	−6.686	0–1200
Ammonium sulphate	c	K	215.9				275–328
Calcium hydroxide	c	K	89.5				276–373
Carbon dioxide	g	°C	36.11	4.233	−2.887	7.464	0–1500
Ethanol	l	°C	103.1				0
	l	°C	158.8				100
	g	°C	61.34	15.72	−8.749	19.83	0–1200
Formaldehyde	g	°C	34.28	4.268	0.000	−8.694	0–1200
n-Hexane	l	°C	216.3				20–100
Hydrogen	g	°C	28.84	0.00765	0.3288	−0.8698	0–1500
Hydrogen chloride	g	°C	29.13	−0.1341	0.9715	−4.335	0–1200
Hydrogen sulphide	g	°C	33.51	1.547	0.3012	−3.292	0–1500
Magnesium chloride	c	K	72.4	1.58			273–991
Methane	g	°C	34.31	5.469	0.3661	−11.00	0–1200
	g	K	19.87	5.021	1.268	−11.00	273–1500
Methanol	l	°C	75.86	16.83			0–65
	g	°C	42.93	8.301	−1.87	−8.03	0–700
Nitric acid	l	°C	110.0				25
Nitrogen	g	°C	29.00	0.2199	0.5723	−2.871	0–1500
Oxygen	g	°C	29.10	1.158	−0.6076	1.311	0–1500
n-Pentane	l	°C	155.4	43.68			0–36
Sulphur							
(rhombic)	c	K	15.2	2.68			273–368
(monoclinic)	c	K	18.3	1.84			368–392
Sulphuric acid	l	°C	139.1	15.59			10–45
Sulphur dioxide	g	°C	38.91	3.904	−3.104	8.606	0–1500
Water	l	°C	75.4				0–100
	g	°C	33.46	0.6880	0.7604	−3.593	0–1500

C_p (J gmol^{-1} °C^{-1}) $= a + bT + cT^2 + dT^3$

Example. For ammonia gas between 0°C and 1200°C:

C_p (J gmol^{-1} °C^{-1}) $= 35.15 + (2.954 \times 10^{-2})T + (0.4421 \times 10^{-5})T^2 - (6.686 \times 10^{-9})T^3$

where T is in °C. Note that some equations require T in K, as indicated.

State: g = gas; l = liquid; c = crystal. Equations for gases apply at 1 atm.

Adapted from R.M. Felder and R.W. Rousseau, 2005, Elementary Principles of Chemical Processes, *3rd ed., John Wiley.*

TABLE C.4 Mean Heat Capacities of Gases

	C_{pm} (J gmol^{-1} °C^{-1})					
T(°C)	Air	O_2	N_2	H_2	CO_2	H_2O
0	29.06	29.24	29.12	28.61	35.96	33.48
18	29.07	29.28	29.12	28.69	36.43	33.51
25	29.07	29.30	29.12	28.72	36.47	33.52
100	29.14	29.53	29.14	28.98	38.17	33.73
200	29.29	29.93	29.23	29.10	40.12	34.10
300	29.51	30.44	29.38	29.15	41.85	34.54

Reference state: $T_{ref} = 0$°C; $P_{ref} = 1$ atm.
Adapted from D.M. Himmelblau, 1974, Basic Principles and Calculations in Chemical Engineering, *3rd ed., Prentice Hall.*

TABLE C.5 Specific Heats of Organic Liquids

Compound	Formula	Temperature (°C)	C_p (cal g^{-1} °C^{-1})
Acetic acid	$C_2H_4O_2$	26–95	0.522
Acetone	C_3H_6O	3–22.6	0.514
		0	0.506
		24.2–49.4	0.538
Acetonitrile	C_2H_3N	21–76	0.541
Benzaldehyde	C_7H_6O	22–172	0.428
Butyl alcohol (*n*-)	$C_4H_{10}O$	2.3	0.526
		19.2	0.563
		21–115	0.687
		30	0.582
Butyric acid (*n*-)	$C_4H_8O_2$	0	0.444
		40	0.501
		20–100	0.515
Carbon tetrachloride	CCl_4	0	0.198
		20	0.201
		30	0.200
Chloroform	$CHCl_3$	0	0.232
		15	0.226
		30	0.234

Cresol	C_7H_8O		
(*o*-)		0–20	0.497
(*m*-)		21–197	0.551
		0–20	0.477
Dichloroacetic acid	$C_2H_2Cl_2O_2$	21–106	0.349
		21–196	0.348
Diethylamine	$C_4H_{11}N$	22.5	0.516
Diethyl malonate	$C_7H_{12}O_4$	20	0.431
Diethyl oxalate	$C_6H_{10}O_4$	20	0.431
Diethyl succinate	$C_8H_{14}O_4$	20	0.450
Dipropyl malonate	$C_9H_{16}O_4$	20	0.431
Dipropyl oxalate (*n*-)	$C_8H_{14}O_4$	20	0.431
Dipropyl succinate	$C_{10}H_{18}O_4$	20	0.450
Ethanol	C_2H_6O	0–98	0.680
Ether	$C_4H_{10}O$	−5	0.525
		0	0.521
		30	0.545
		80	0.687
		120	0.800
		140	0.819
		180	1.037
Ethyl acetate	$C_4H_8O_2$	20	0.457
		20	0.476
Ethylene glycol	$C_2H_6O_2$	−11.1	0.535
		0	0.542
		2.5	0.550
		5.1	0.554
		14.9	0.569
		19.9	0.573
Formic acid	CH_2O_2	0	0.436
		15.5	0.509
		20–100	0.524
Furfural	$C_5H_4O_2$	0	0.367
		20–100	0.416
Glycerol	$C_3H_8O_3$	15–50	0.576
Hexadecane (*n*-)	$C_{16}H_{34}$	0–50	0.496

(*Continued*)

TABLE C.5 Specific Heats of Organic Liquids (Continued)

Compound	Formula	Temperature (°C)	C_p (cal g^{-1} °C^{-1})
Isobutyl acetate	$C_6H_{12}O_2$	20	0.459
Isobutyl alcohol	$C_4H_{10}O$	21–109	0.716
		30	0.603
Isobutyl succinate	$C_{12}H_{22}O_4$	0	0.442
Isobutyric acid	$C_4H_8O_2$	20	0.450
Lauric acid	$C_{12}H_{24}O_2$	40–100	0.572
		57	0.515
Methanol	CH_4O	5–10	0.590
		15–20	0.601
Methyl butyl ketone	$C_6H_{12}O$	21–127	0.553
Methyl ethyl ketone	C_4H_8O	20–78	0.549
Methyl formate	$C_2H_4O_2$	13–29	0.516
Methyl propionate	$C_4H_8O_2$	20	0.459
Palmitic acid	$C_{16}H_{32}O_2$	65–104	0.653
Propionic acid	$C_3H_6O_2$	0	0.444
		20–137	0.560
Propyl acetate (*n*-)	$C_5H_{10}O_2$	20	0.459
Propyl butyrate	$C_7H_{14}O_2$	20	0.459
Propyl formate (*n*-)	$C_4H_8O_2$	20	0.459
Pyridine	C_5H_5N	20	0.405
		21–108	0.431
		0–20	0.395
Quinoline	C_9H_7N	0–20	0.352
Salicylaldehyde	$C_7H_6O_2$	18	0.382
Stearic acid	$C_{18}H_{36}O_2$	75–137	0.550

Adapted from R.H. Perry, D.W. Green, and J.O. Maloney (Eds.), 1984, Chemical Engineers' Handbook, *6th ed., McGraw-Hill.*

TABLE C.6 Specific Heats of Organic Solids

Compound	Formula	Temperature T (°C)	C_p (cal g^{-1} °C^{-1})
Acetic acid	$C_2H_4O_2$	−200 to 25	0.330 + 0.00080T
Acetone	C_3H_6O	−210 to −80	0.540 + 0.0156T
Aniline	C_6H_7N		0.741
Anthraquinone	$C_{14}H_8O_2$	0 to 270	0.258 + 0.00069T
Capric acid	$C_{10}H_{20}O_2$	8	0.695
Chloroacetic acid	$C_2H_3ClO_2$	60	0.363
Crotonic acid	$C_4H_6O_2$	38 to 70	0.520 + 0.00020T
Dextrin	$(C_6H_{10}O_5)_x$	0 to 90	0.291 + 0.00096T
Diphenylamine	$C_{12}H_{11}N$	26	0.337
Erythritol	$C_4H_{10}O_4$	60	0.351
Ethylene glycol	$C_2H_6O_2$	−190 to −40	0.366 + 0.00110T
Formic acid	CH_2O_2	−22	0.387
		0	0.430
D-Glucose	$C_6H_{12}O_6$	0	0.277
		20	0.300
Glutaric acid	$C_5H_8O_4$	20	0.299
Glycerol	$C_3H_8O_3$	0	0.330
Hexadecane	$C_{16}H_{34}$		0.495
Lactose	$C_{12}H_{22}O_{11}$	20	0.287
	$C_{12}H_{22}O_{11}\cdot H_2O$	20	0.299
Lauric acid	$C_{12}H_{24}O_2$	−30 to 40	0.430 + 0.000027T
Levoglucosane	$C_6H_{10}O_5$	40	0.607
Levulose	$C_6H_{12}O_6$	20	0.275
Malonic acid	$C_3H_4O_4$	20	0.275
Maltose	$C_{12}H_{22}O_{11}$	20	0.320
Mannitol	$C_6H_{14}O_6$	0 to 100	0.313 + 0.00025T
Oxalic acid	$C_2H_2O_4$	−200 to 50	0.259 + 0.00076T
	$C_2H_2O_4\cdot 2H_2O$	0	0.338
		50	0.385
		100	0.416
Palmitic acid	$C_{16}H_{32}O_2$	0	0.382
		20	0.430

(Continued)

TABLE C.6 Specific Heats of Organic Solids (Continued)

Compound	Formula	Temperature T (°C)	C_p (cal g^{-1} °C^{-1})
Phenol	C_6H_6O	14 to 26	0.561
Phthalic acid	$C_8H_6O_4$	20	0.232
Picric acid	$C_6H_3N_3O_7$	0	0.240
		50	0.263
		100	0.297
Stearic acid	$C_{18}H_{36}O_2$	15	0.399
Succinic acid	$C_4H_6O_4$	0 to 160	$0.248 + 0.00153T$
Sucrose	$C_{12}H_{22}O_{11}$	20	0.299
Sugar (cane)	$C_{12}H_{22}O_{11}$	22 to 51	0.301
Tartaric acid	$C_4H_6O_6$	36	0.287
	$C_4H_6O_6 \cdot H_2O$	0	0.308
		50	0.366
Urea	CH_4N_2O	20	0.320

Adapted from D.W. Green and R.H. Perry (Eds.), 2008, Perry's Chemical Engineers' Handbook, *8th ed., McGraw-Hill, New York.*

TABLE C.7 Normal Melting Points and Boiling Points, and Standard Heats of Phase Change

Compound	Molecular weight	Melting temperature (°C)	Δh_f at melting point (kJ gmol^{-1})	Normal boiling point (°C)	Δh_v at boiling point (kJ gmol^{-1})
Acetaldehyde	44.05	−123.7		20.2	25.1
Acetic acid	60.05	16.6	12.09	118.2	24.39
Acetone	58.08	−95.0	5.69	56.0	30.2
Ammonia	17.03	−77.8	5.653	−33.43	23.351
Benzaldehyde	106.12	−26.0		179.0	38.40
Carbon dioxide	44.01	−56.6 (at 5.2 atm)	8.33	sublimates at −78°C	
Chloroform	119.39	−63.7		61.0	
Ethanol	46.07	−114.6	5.021	78.5	38.58
Formaldehyde	30.03	−92		−19.3	24.48
Formic acid	46.03	8.30	12.68	100.5	22.25
Glycerol	92.09	18.20	18.30	290.0	
Hydrogen	2.016	−259.19	0.12	−252.76	0.904

Hydrogen chloride	36.47	−114.2	1.99	−85.0	16.1
Hydrogen sulphide	34.08	−85.5	2.38	−60.3	18.67
Methane	16.04	−182.5	0.94	−161.5	8.179
Methanol	32.04	−97.9	3.167	64.7	35.27
Nitric acid	63.02	−41.6	10.47	86	30.30
Nitrogen	28.02	−210.0	0.720	−195.8	5.577
Oxalic acid	90.04			decomposes at 186°C	
Oxygen	32.00	−218.75	0.444	−182.97	6.82
Phenol	94.11	42.5	11.43	181.4	
Phosphoric acid	98.00	42.3	10.54		
Sodium chloride	58.45	808	28.5	1465	170.7
Sodium hydroxide	40.00	319	8.34	1390	
Sulphur					
S_8 rhombic	256.53	113	10.04	444.6	83.7
S_8 monoclinic	256.53	119	14.17	444.6	83.7
Sulphur dioxide	64.07	−75.48	7.402	−10.02	24.91
Sulphuric acid	98.08	10.35	9.87	decomposes at 340°C	
Water	18.016	0.00	6.0095	100.00	40.656

Note: All thermodynamic data are at 1 atm unless otherwise indicated.
Adapted from R.M. Felder and R.W. Rousseau, 2005, Elementary Principles of Chemical Processes*, 3rd ed., John Wiley.*

TABLE C.8 Heats of Combustion

Compound	Formula	Molecular weight	State	Heat of combustion Δh_c° (kJ gmol^{-1})
Acetaldehyde	C_2H_4O	44.053	l	−1166.9
			g	−1192.5
Acetic acid	$C_2H_4O_2$	60.053	l	−874.2
			g	−925.9
Acetone	C_3H_6O	58.080	l	−1789.9
			g	−1820.7
Acetylene	C_2H_2	26.038	g	−1301.1
Adenine	$C_5H_5N_5$	135.128	c	−2778.1
			g	−2886.9
Alanine (D-)	$C_3H_7O_2N$	89.094	c	−1619.7

(*Continued*)

TABLE C.8 Heats of Combustion (Continued)

Compound	Formula	Molecular weight	State	Heat of combustion Δh_c° (kJ gmol^{-1})
Alanine (L-)	$C_3H_7O_2N$	89.094	c	−1576.9
			g	−1715.0
Ammonia	NH_3	17.03	g	−382.6
Ammonium ion	NH_4^+			−383
Arginine (D-)	$C_6H_{14}O_2N_4$	174.203	c	−3738.4
Asparagine (L-)	$C_4H_8O_3N_2$	132.119	c	−1928.0
Aspartic acid (L-)	$C_4H_7O_4N$	133.104	c	−1601.1
Benzaldehyde	C_7H_6O	106.124	l	−3525.1
			g	−3575.4
Butanoic acid	$C_4H_8O_2$	88.106	l	−2183.6
			g	−2241.6
1-Butanol	$C_4H_{10}O$	74.123	l	−2675.9
			g	−2728.2
2-Butanol	$C_4H_{10}O$	74.123	l	−2660.6
			g	−2710.3
Butyric acid	$C_4H_8O_2$	88.106	l	−2183.6
			g	−2241.6
Caffeine	$C_8H_{10}O_2N_4$		s	−4246.5*
Carbon	C	12.011	c	−393.5
Carbon monoxide	CO	28.010	g	−283.0
Citric acid	$C_6H_8O_7$		s	−1962.0
Codeine	$C_{18}H_{21}O_3N{\cdot}H_2O$		s	−9745.7*
Cytosine	$C_4H_5ON_3$	111.103	c	−2067.3
Ethane	C_2H_6	30.070	g	−1560.7
Ethanol	C_2H_6O	46.069	l	−1366.8
			g	−1409.4
Ethylene	C_2H_4	28.054	g	−1411.2
Ethylene glycol	$C_2H_6O_2$	62.068	l	−1189.2
			g	−1257.0
Formaldehyde	CH_2O	30.026	g	−570.7
Formic acid	CH_2O_2	46.026	l	−254.6
			g	−300.7
Fructose (D-)	$C_6H_{12}O_6$		s	−2813.7

Fumaric acid	$C_4H_4O_4$	116.073	c	−1334.0
Galactose (D-)	$C_6H_{12}O_6$		s	−2805.7
Glucose (D-)	$C_6H_{12}O_6$		s	−2805.0
Glutamic acid (L-)	$C_5H_9O_4N$	147.131	c	−2244.1
Glutamine (L-)	$C_5H_{10}O_3N_2$	146.146	c	−2570.3
Glutaric acid	$C_5H_8O_4$	132.116	c	−2150.9
Glycerol	$C_3H_8O_3$	92.095	l	−1655.4
			g	−1741.2
Glycine	$C_2H_5O_2N$	75.067	c	−973.1
Glycogen	$(C_6H_{10}O_5)_x$ per kg		s	−17,530.1*
Guanine	$C_5H_5ON_5$	151.128	c	−2498.2
Hexadecane	$C_{16}H_{34}$	226.446	l	−10,699.2
			g	−10,780.5
Hexadecanoic acid	$C_{16}H_{32}O_2$	256.429	c	−9977.9
			l	−10,031.3
			g	−10,132.3
Histidine (L-)	$C_6H_9O_2N_3$	155.157	c	−3180.6
Hydrogen	H_2	2.016	g	−285.8
Hydrogen sulphide	H_2S	34.08		−562.6
Inositol	$C_6H_{12}O_6$		s	−2772.2*
Isoleucine (L-)	$C_6H_{13}O_2N$	131.175	c	−3581.1
Isoquinoline	C_9H_7N	129.161	l	−4686.5
Lactic acid (D,L-)	$C_3H_6O_3$		l	−1368.3
Lactose	$C_{12}H_{22}O_{11}$		s	−5652.5
Leucine (D-)	$C_6H_{13}O_2N$	131.175	c	−3581.7
Leucine (L-)	$C_6H_{13}O_2N$	131.175	c	−3581.6
Lysine	$C_6H_{14}O_2N_2$	146.189	c	−3683.2
Malic acid (L-)	$C_4H_6O_5$		s	−1328.8
Malonic acid	$C_3H_4O_4$		s	−861.8
Maltose	$C_{12}H_{22}O_{11}$		s	−5649.5
Mannitol (D-)	$C_6H_{14}O_6$		s	−3046.5*
Methane	CH_4	16.043	g	−890.8
Methanol	CH_4O	32.042	l	−726.1
			g	−763.7

(Continued)

TABLE C.8 Heats of Combustion (Continued)

Compound	Formula	Molecular weight	State	Heat of combustion Δh_c° (kJ gmol^{-1})
Morphine	$C_{17}H_{19}O_3N \cdot H_2O$		s	−8986.6*
Nicotine	$C_{10}H_{14}N_2$		l	−5977.8*
Oleic acid	$C_{18}H_{34}O_2$		l	−11,126.5
Oxalic acid	$C_2H_2O_4$	90.036	c	−251.1
Papaverine	$C_{20}H_{21}O_4N$		s	−10,375.8*
Pentane	C_5H_{12}	72.150	l	−3509.0
			g	−3535.6
Phenylalanine (L-)	$C_9H_{11}O_2N$	165.192	c	−4646.8
Phthalic acid	$C_8H_6O_4$	166.133	c	−3223.6
Proline (L-)	$C_5H_9O_2N$	115.132	c	−2741.6
Propane	C_3H_8	44.097	g	−2219.2
1-Propanol	C_3H_8O	60.096	l	−2021.3
			g	−2068.8
2-Propanol	C_3H_8O	60.096	l	−2005.8
			g	−2051.1
Propionic acid	$C_3H_6O_2$	74.079	l	−1527.3
			g	−1584.5
1,2-Propylene glycol	$C_3H_8O_2$	76.095	l	−1838.2
			g	−1902.6
1,3-Propylene glycol	$C_3H_8O_2$	76.095	l	−1859.0
			g	−1931.8
Pyridine	C_5H_5N	79.101	l	−2782.3
			g	−2822.5
Pyrimidine	$C_4H_4N_2$	80.089	l	−2291.6
			g	−2341.6
Salicylic acid	$C_7H_6O_3$	138.123	c	−3022.2
			g	−3117.3
Serine (L-)	$C_3H_7O_3N$	105.094	c	−1448.2
Starch	$(C_6H_{10}O_5)_x$ per kg		s	−17,496.6*
Succinic acid	$C_4H_6O_4$	118.089	c	−1491.0
Sucrose	$C_{12}H_{22}O_{11}$		s	−5644.9
Thebaine	$C_{19}H_{21}O_3N$		s	−10,221.7*
Threonine (L-)	$C_4H_9O_3N$	119.120	c	−2053.1

Thymine	$C_5H_6O_2N_2$	126.115	c	−2362.2
Tryptophan (L-)	$C_{11}H_{12}O_2N_2$	204.229	c	−5628.3
Tyrosine (L-)	$C_9H_{11}O_3N$	181.191	c	−4428.6
Uracil	$C_4H_4O_2N_2$	112.088	c	−1716.3
			g	−1842.8
Urea	CH_4ON_2	60.056	c	−631.6
			g	−719.4
Valine (L-)	$C_5H_{11}O_2N$	117.148	c	−2921.7
			g	−3084.5
Xanthine	$C_5H_4O_2N_4$	152.113	c	−2159.6
Xylose	$C_5H_{10}O_5$		s	−2340.5
Biomass	$CH_{1.8}O_{0.5}N_{0.2}$	24.6	s	−552

Reference conditions: 1 atm and 25°C or 20°C; values marked with an asterisk refer to 20°C.
Products of combustion are taken to be CO_2 (gas), H_2O (liquid), and N_2 (gas); therefore, $\Delta h_c^\circ = 0$ for CO_2 (g), H_2O (l), and N_2 (g).
State: g = gas; l = liquid; c = crystal; s = solid.
Adapted from Handbook of Chemistry and Physics, *1992, 73rd ed., CRC Press, Boca Raton;* Handbook of Chemistry and Physics, *1976, 57th ed., CRC Press, Boca Raton; and R.M. Felder and R.W. Rousseau, 2005,* Elementary Principles of Chemical Processes, *3rd ed., John Wiley, New York.*

APPENDIX

D

Steam Tables

The tables that follow are from R.W. Haywood, *Thermodynamic Tables in SI (Metric) Units*, 1972, 2nd ed., Cambridge University Press, Cambridge.

TABLE D.1 Enthalpy of Saturated Water and Steam (Temperatures from 0.01°C to 100°C)

		Specific enthalpy (kJ kg^{-1})		
Temperature (°C)	**Pressure (kPa)**	**Saturated liquid**	**Evaporation (Δh_v)**	**Saturated vapour**
0.01 (triple point)	0.611	+0.0	2501.6	2501.6
2	0.705	8.4	2496.8	2505.2
4	0.813	16.8	2492.1	2508.9
6	0.935	25.2	2487.4	2512.6
8	1.072	33.6	2482.6	2516.2
10	1.227	42.0	2477.9	2519.9
12	1.401	50.4	2473.2	2523.6
14	1.597	58.8	2468.5	2527.2
16	1.817	67.1	2463.8	2530.9
18	2.062	75.5	2459.0	2534.5
20	2.34	83.9	2454.3	2538.2
22	2.64	83.9	2454.3	2538.2
24	2.98	100.6	2444.9	2545.5
25	3.17	104.8	2442.5	2547.3
26	3.36	108.9	2440.2	2549.1

(*Continued*)

TABLE D.1 Enthalpy of Saturated Water and Steam (Temperatures from 0.01°C to 100°C) (Continued)

		Specific enthalpy (kJ kg^{-1})		
Temperature (°C)	Pressure (kPa)	Saturated liquid	Evaporation (Δh_v)	Saturated vapour
28	3.78	117.3	2435.4	2552.7
30	4.24	125.7	2430.7	2556.4
32	4.75	134.0	2425.9	2560.0
34	5.32	142.4	2421.2	2563.6
36	5.94	150.7	2416.4	2567.2
38	6.62	159.1	2411.7	2570.8
40	7.38	167.5	2406.9	2574.4
42	8.20	175.8	2402.1	2577.9
44	9.10	184.2	2397.3	2581.5
46	10.09	192.5	2392.5	2585.1
48	11.16	200.9	2387.7	2588.6
50	12.34	209.3	2382.9	2592.2
52	13.61	217.6	2378.1	2595.7
54	15.00	226.0	2373.2	2599.2
56	16.51	234.4	2368.4	2602.7
58	18.15	242.7	2363.5	2606.2
60	19.92	251.1	2358.6	2609.7
62	21.84	259.5	2353.7	2613.2
64	23.91	267.8	2348.8	2616.6
66	26.15	276.2	2343.9	2620.1
68	28.56	284.6	2338.9	2623.5
70	31.16	293.0	2334.0	2626.9
72	33.96	301.4	2329.0	2630.3
74	36.96	309.7	2324.0	2633.7
76	40.19	318.1	2318.9	2637.1
78	43.65	326.5	2313.9	2640.4
80	47.36	334.9	2308.8	2643.8
82	51.33	343.3	2303.8	2647.1
84	55.57	351.7	2298.6	2650.4

86	60.11	360.1	2293.5	2653.6
88	64.95	368.5	2288.4	2656.9
90	70.11	376.9	2283.2	2660.1
92	75.61	385.4	2278.0	2663.4
94	81.46	393.8	2272.8	2666.6
96	87.69	402.2	2267.5	2669.7
98	94.30	410.6	2262.2	2672.9
100 (boiling point)	101.325	419.1	2256.9	2676.0

Reference state: Triple point of water: 0.01°C, 0.6112 kPa

TABLE D.2 Enthalpy of Saturated Water and Steam (Pressures from 0.6112 kPa to 22,120 kPa.)

		Specific enthalphy (kJ kg^{-1})		
Pressure (kPa)	Temperature (°C)	Saturated liquid	Evaporation (Δh_v)	Saturated vapour
0.6112 (triple point)	0.01	+0.0	2501.6	2501.6
0.8	3.8	15.8	2492.6	2508.5
1.0	7.0	29.3	2485.0	2514.4
1.4	12.0	50.3	2473.2	2523.5
1.8	15.9	66.5	2464.1	2530.6
2.0	17.5	73.5	2460.2	2533.6
2.4	20.4	85.7	2453.3	2539.0
2.8	23.0	96.2	2447.3	2543.6
3.0	24.1	101.0	2444.6	2545.6
3.5	26.7	111.8	2438.5	2550.4
4.0	29.0	121.4	2433.1	2554.5
4.5	31.0	130.0	2428.2	2558.2
5.0	32.9	137.8	2423.8	2561.6
6	36.2	151.5	2416.0	2567.5
7	39.0	163.4	2409.2	2572.6
8	41.5	173.9	2403.2	2577.1
9	43.8	183.3	2397.9	2581.1
10	45.8	191.8	2392.9	2584.8

(Continued)

TABLE D.2 Enthalpy of Saturated Water and Steam (Pressures from 0.6112 kPa to 22,120 kPa.) (Continued)

		Specific enthalpy (kJ kg^{-1})		
Pressure (kPa)	**Temperature (°C)**	**Saturated liquid**	**Evaporation (Δh_v)**	**Saturated vapour**
12	49.4	206.9	2384.3	2591.2
14	52.6	220.0	2376.7	2596.7
16	55.3	231.6	2370.0	2601.6
18	57.8	242.0	2363.9	2605.9
20	60.1	251.5	2358.4	2609.9
24	64.1	268.2	2348.6	2616.8
28	67.5	282.7	2340.0	2622.7
30	69.1	289.3	2336.1	2625.4
35	72.7	304.3	2327.2	2631.5
40	75.9	317.7	2319.2	2636.9
45	78.7	329.6	2312.0	2641.7
50	81.3	340.6	2305.4	2646.0
55	83.7	350.6	2299.3	2649.9
60	86.0	359.9	2293.6	2653.6
65	88.0	368.6	2288.3	2656.9
70	90.0	376.8	2283.3	2660.1
80	93.5	391.7	2274.1	2665.8
90	96.7	405.2	2265.6	2670.9
100	99.6	417.5	2257.9	2675.4
101.325 (boiling point)	100.0	419.1	2256.9	2676.0
120	104.8	439.4	2244.1	2683.4
140	109.3	458.4	2231.9	2690.3
160	113.3	475.4	2220.9	2696.2
180	116.9	490.7	2210.8	2701.5
200	120.2	504.7	2201.6	2706.3
220	123.3	517.6	2193.0	2710.6
240	126.1	529.6	2184.9	2714.5
260	128.7	540.9	2177.3	2718.2
280	131.2	551.4	2170.1	2721.5

300	133.5	561.4	2163.2	2724.7
320	135.8	570.9	2156.7	2727.6
340	137.9	579.9	2150.4	2730.3
360	139.9	588.5	2144.4	2732.9
380	141.8	596.8	2138.6	2735.3
400	143.6	604.7	2133.0	2737.6
420	145.4	612.3	2127.5	2739.8
440	147.1	619.6	2122.3	2741.9
460	148.7	626.7	2117.2	2743.9
480	150.3	633.5	2112.2	2745.7
500	151.8	640.1	2107.4	2747.5
550	155.5	655.8	2095.9	2751.7
600	158.8	670.4	2085.0	2755.5
650	162.0	684.1	2074.7	2758.9
700	165.0	697.1	2064.9	2762.0
750	167.8	709.3	2055.5	2764.8
800	170.4	720.9	2046.5	2767.5
850	172.9	732.0	2037.9	2769.9
900	175.4	742.6	2029.5	2772.1
950	177.7	752.8	2021.4	2774.2
1000	179.9	762.6	2013.6	2776.2
1100	184.1	781.1	1998.5	2779.7
1200	188.0	798.4	1984.3	2782.7
1300	191.6	814.7	1970.7	2785.4
1400	195.0	830.1	1957.7	2787.8
1500	198.3	844.7	1945.2	2789.9
1600	201.4	858.6	1933.2	2791.7
1700	204.3	871.8	1921.5	2793.4
1800	207.1	884.6	1910.3	2794.8
1900	209.8	896.8	1899.3	2796.1
2000	212.4	908.6	1888.6	2797.2
2200	217.2	931.0	1868.1	2799.1
2400	221.8	951.9	1848.5	2800.4

(Continued)

TABLE D.2 Enthalpy of Saturated Water and Steam (Pressures from 0.6112 kPa to 22,120 kPa.) (Continued)

		Specific enthalpy (kJ kg^{-1})		
Pressure (kPa)	**Temperature (°C)**	**Saturated liquid**	**Evaporation (Δh_v)**	**Saturated vapour**
2600	226.0	971.7	1829.6	2801.4
2800	230.0	990.5	1811.5	2802.0
3000	233.8	1008.4	1793.9	2802.3
3200	237.4	1025.4	1776.9	2802.3
3400	240.9	1041.8	1760.3	2802.1
3600	244.2	1057.6	1744.2	2801.7
3800	247.3	1072.7	1728.4	2801.1
4000	250.3	1087.4	1712.9	2800.3
4200	253.2	1101.6	1697.8	2799.4
4400	256.0	1115.4	1682.9	2798.3
4600	258.8	1128.8	1668.3	2797.1
4800	261.4	1141.8	1653.9	2795.7
5000	263.9	1154.5	1639.7	2794.2
5200	266.4	1166.8	1625.7	2792.6
5400	268.8	1178.9	1611.9	2790.8
5600	271.1	1190.8	1598.2	2789.0
5800	273.3	1202.3	1584.7	2787.0
6000	275.6	1213.7	1571.3	2785.0
6200	277.7	1224.8	1558.0	2782.9
6400	279.8	1235.7	1544.9	2780.6
6600	281.8	1246.5	1531.9	2778.3
6800	283.8	1257.0	1518.9	2775.9
7000	285.8	1267.4	1506.0	2773.5
7200	287.7	1277.6	1493.3	2770.9
7400	289.6	1287.7	1480.5	2768.3
7600	291.4	1297.6	1467.9	2765.5
7800	293.2	1307.4	1455.3	2762.8
8000	295.0	1317.1	1442.8	2759.9
8400	298.4	1336.1	1417.9	2754.0

8800	301.7	1354.6	1393.2	2747.8
9000	303.3	1363.7	1380.9	2744.6
10,000	311.0	1408.0	1319.7	2727.7
11,000	318.0	1450.6	1258.7	2709.3
12,000	324.6	1491.8	1197.4	2689.2
13,000	330.8	1532.0	1135.0	2667.0
14,000	336.6	1571.6	1070.7	2642.4
15,000	342.1	1611.0	1004.0	2615.0
16,000	347.3	1650.5	934.3	2584.9
17,000	352.3	1691.7	859.9	2551.6
18,000	357.0	1734.8	779.1	2513.9
19,000	361.4	1778.7	692.0	2470.6
20,000	365.7	1826.5	591.9	2418.4
21,000	369.8	1886.3	461.3	2347.6
22,000	373.7	2011	185	2196
22,120 (critical point)	374.15	2108	0	2108

Reference state: Triple point of water: 0.01°C, 0.6112 kPa

TABLE D.3 Enthalpy of Superheated Steam

Pressure (kPa)	10	50	100	500	1000	2000	4000	6000	8000	10,000	15,000	20,000	22,120*	30,000	50,000
Saturation temperature (°C)	45.8	81.3	99.6	151.8	179.9	212.4	250.3	275.6	295.0	311.0	342.1	365.7	374.15	–	–
State	Specific enthalpy at saturation ($kJ\ kg^{-1}$)														
Water	191.8	340.6	417.5	640.1	762.6	908.6	1087.4	1213.7	1317.1	1408.0	1611.0	1826.5	2108	–	–
Steam	2584.8	2646.0	2675.4	2747.5	2776.2	2797.2	2800.3	2785.0	2759.9	2727.7	2615.0	2418.4	2108	–	–
Temperature (°C)	Specific enthalpy ($kJ\ kg^{-1}$)														
0	0.0	0.0	0.1	0.5	1.0	2.0	4.0	6.1	8.1	10.1	15.1	20.1	22.2	30.0	49.3
25	104.8	104.8	104.9	105.2	105.7	106.6	108.5	110.3	112.1	114.0	118.6	123.1	125.1	132.2	150.2
50	2593	209.3	209.3	209.7	210.1	211.0	212.7	214.4	216.1	217.8	222.1	226.4	228.2	235.0	251.9
75	2640	313.9	314.0	314.3	314.7	315.5	317.1	318.7	320.3	322.0	326.0	330.0	331.7	338.1	354.2
100	2688	2683	2676	419.4	419.7	420.5	422.0	423.5	425.0	426.5	430.3	434.0	435.7	441.6	456.8
125	2735	2731	2726	525.2	525.5	526.2	527.6	529.0	530.4	531.8	535.3	538.8	540.2	545.8	560.1
150	2783	2780	2776	632.2	632.5	633.1	634.3	635.6	636.8	638.1	641.3	644.5	645.8	650.9	664.1
175	2831	2829	2826	2800	741.1	741.7	742.7	743.8	744.9	746.0	748.7	751.5	752.7	757.2	769.1
200	2880	2878	2875	2855	2827	852.6	853.4	854.2	855.1	855.9	858.1	860.4	861.4	865.2	875.4
225	2928	2927	2925	2909	2886	2834	967.2	967.7	968.2	968.8	970.3	971.8	972.5	975.3	983.4
250	2977	2976	2975	2961	2943	2902	1085.8	1085.8	1085.8	1085.8	1086.2	1086.7	1087.0	1088.4	1093.6
275	3027	3026	3024	3013	2998	2965	2886	1210.8	1210.0	1209.2	1207.7	1206.6	1206.3	1205.6	1206.7
300	3077	3076	3074	3065	3052	3025	2962	2885	2787	1343.4	1338.3	1334.3	1332.8	1328.7	1323.7
325	3127	3126	3125	3116	3106	3083	3031	2970	2899	2811	1486.0	1475.5	1471.8	1461.1	1446.0
350	3177	3177	3176	3168	3159	3139	3095	3046	2990	2926	2695	1647.1	1636.5	1609.9	1576.3
375	3228	3228	3227	3220	3211	3194	3156	3115	3069	3019	2862	2604	2319	1791	1716
400	3280	3279	3278	3272	3264	3249	3216	3180	3142	3100	2979	2820	2733	2162	1878
425	3331	3331	3330	3325	3317	3303	3274	3243	3209	3174	3075	2957	2899	2619	2068
450	3384	3383	3382	3377	3371	3358	3331	3303	3274	3244	3160	3064	3020	2826	2293
475	3436	3436	3435	3430	3424	3412	3388	3363	3337	3310	3237	3157	3120	2969	2522
500	3489	3489	3488	3484	3478	3467	3445	3422	3399	3375	3311	3241	3210	3085	2723
600	3706	3705	3705	3702	3697	3689	3673	3656	3640	3623	3580	3536	3516	3443	3248
700	3929	3929	3928	3926	3923	3916	3904	3892	3879	3867	3835	3804	3790	3740	3610
800	4159	4159	4158	4156	4154	4149	4140	4131	4121	4112	4089	4065	4055	4018	3925

Reference state: Triple point of water: 0.01°C, 0.6112 kPa
Critical isobar.

APPENDIX

E

Mathematical Rules

In this Appendix, some simple rules for logarithms, differentiation, integration, and matrices are presented. Further details of mathematical functions can be found in handbooks, for example [1–5].

E.1 Logarithms

The *natural logarithm* (ln or $\log_e$) is the inverse of the exponential function. Therefore, if:

$$y = \ln x \tag{E.1}$$

then

$$e^y = x \tag{E.2}$$

where the number e is approximately 2.71828. It also follows that:

$$\ln (e^y) = y \tag{E.3}$$

and

$$e^{\ln x} = x \tag{E.4}$$

Natural logarithms are related to *common logarithms*, or logarithms to the base 10 (written as lg, log or $\log_{10}$), as follows:

$$\ln x = \ln 10\,(\log_{10} x) \tag{E.5}$$

Since ln 10 is approximately 2.30259:

$$\ln x = 2.30259 \log_{10} x \tag{E.6}$$

Zero and negative numbers do not have logarithms.

Rules for taking logarithms of products and powers are illustrated below. The logarithm of the product of two numbers is equal to the sum of the logarithms:

$$\ln(ax) = \ln a + \ln x \tag{E.7}$$

When one term of the product involves an exponential function, application of Eqs. (E.7) and (E.3) gives:

$$\ln(be^{ax}) = \ln b + ax \tag{E.8}$$

The logarithm of the quotient of two numbers is equal to the logarithm of the numerator minus the logarithm of the denominator:

$$\ln\left(\frac{a}{x}\right) = \ln a - \ln x \tag{E.9}$$

As an example of this rule, because $\ln 1 = 0$:

$$\ln\left(\frac{1}{x}\right) = -\ln x \tag{E.10}$$

The rule for taking the logarithm of a power function is:

$$\ln(x^b) = b \ln x \tag{E.11}$$

E.2 Differentiation

The derivative of y with respect to x, $\mathrm{d}y/\mathrm{d}x$, is defined as the limit of $\Delta y/\Delta x$ as Δx approaches zero, provided this limit exists:

$$\frac{\mathrm{d}y}{\mathrm{d}x} = \lim_{\Delta x \to 0} \frac{\Delta y}{\Delta x} \tag{E.12}$$

That is:

$$\frac{\mathrm{d}y}{\mathrm{d}x} = \lim_{\Delta x \to 0} \frac{y|_{x+\Delta x} - y|_x}{\Delta x} \tag{E.13}$$

where $y|_x$ means the value of y evaluated at x, and $y|_{x+\Delta x}$ means the value of y evaluated at $x + \Delta x$. The operation of determining the derivative is called differentiation.

There are simple rules for rapid evaluation of derivatives. Derivatives of various functions with respect to x are listed below; in all these equations A is a constant:

$$\frac{\mathrm{d}A}{\mathrm{d}x} = 0 \tag{E.14}$$

$$\frac{\mathrm{d}x}{\mathrm{d}x} = 1 \tag{E.15}$$

$$\frac{\mathrm{d}}{\mathrm{d}x}(e^x) = e^x \tag{E.16}$$

$$\frac{\mathrm{d}}{\mathrm{d}x}(e^{Ax}) = Ae^{Ax} \tag{E.17}$$

and

$$\frac{\mathrm{d}}{\mathrm{d}x}(\ln x) = \frac{1}{x} \tag{E.18}$$

When a function is multiplied by a constant, the constant can be taken out of the differential. For example:

$$\frac{\mathrm{d}}{\mathrm{d}x}(Ax) = A\frac{\mathrm{d}x}{\mathrm{d}x} = A \tag{E.19}$$

and

$$\frac{\mathrm{d}}{\mathrm{d}x}(A \ln x) = A\frac{\mathrm{d}}{\mathrm{d}x}(\ln x) = \frac{A}{x} \tag{E.20}$$

When a function consists of a sum of terms, the derivative of the sum is equal to the sum of the derivatives. Therefore, if $f(x)$ and $g(x)$ are functions of x:

$$\frac{\mathrm{d}}{\mathrm{d}x}\left[f(x) + g(x)\right] = \frac{\mathrm{d}f}{\mathrm{d}x} + \frac{\mathrm{d}g}{\mathrm{d}x} \tag{E.21}$$

To illustrate application of Eq. (E.21), for A and B constants:

$$\frac{\mathrm{d}}{\mathrm{d}x}\left[Ax + e^{Bx}\right] = \frac{\mathrm{d}(Ax)}{\mathrm{d}x} + \frac{\mathrm{d}(e^{Bx})}{\mathrm{d}x} = A + Be^{Bx}$$

When a function consists of terms multiplied together, the *product rule* for derivatives is:

$$\frac{\mathrm{d}}{\mathrm{d}x}\left[f(x) \cdot g(x)\right] = f(x)\frac{\mathrm{d}g}{\mathrm{d}x} + g(x)\frac{\mathrm{d}f}{\mathrm{d}x} \tag{E.22}$$

As an example of the product rule:

$$\frac{\mathrm{d}}{\mathrm{d}x}\left[(Ax) \cdot \ln x\right] = Ax \cdot \frac{\mathrm{d}(\ln x)}{\mathrm{d}x} + \ln x \cdot \frac{\mathrm{d}(Ax)}{\mathrm{d}x} = Ax \cdot \frac{1}{x} + \ln x \cdot (A) = A\,(1 + \ln x)$$

E.3 Integration

The integral of y with respect to x is indicated as $\int y\,\mathrm{d}x$. The function to be integrated (y) is called the *integrand*; the symbol $\int$ is the *integral sign*. Integration is the opposite of differentiation; integration is the process of finding a function from its derivative.

From Eq. (E.14), if the derivative of a constant is zero, the integral of zero must be a constant:

$$\int 0\,\mathrm{d}x = K \tag{E.23}$$

where K is a constant. K is called the *constant of integration*, and appears whenever a function is integrated. For example, the integral of constant A with respect to x is:

$$\int A \, \mathrm{d}x = Ax + K \tag{E.24}$$

We can check that Eq. (E.24) is correct by taking the derivative of the right side and making sure it is equal to the integrand, A. Although the equation:

$$\int A \, \mathrm{d}x = Ax$$

is also correct, addition of K in Eq. (E.24) makes solution of the integral complete. Addition of K accounts for the possibility that the answer we are looking for may have an added constant that disappears when the derivative is taken. Irrespective of the value of K, the derivative of the integral will always be the same because $\mathrm{d}K/\mathrm{d}x = 0$. Extra information is needed to evaluate the magnitude of K; this point is considered further in Chapter 6 where integration is used to solve unsteady-state mass and energy problems.

The integral of $\mathrm{d}y/\mathrm{d}x$ with respect to x is:

$$\int \frac{\mathrm{d}y}{\mathrm{d}x} \mathrm{d}x = y + K \tag{E.25}$$

When a function is multiplied by a constant, the constant can be taken out of the integral. For example, for $f(x)$ a function of x and A a constant:

$$\int A \, f(x) \, \mathrm{d}x = A \int f(x) \, \mathrm{d}x \tag{E.26}$$

Other rules of integration are:

$$\int \frac{\mathrm{d}x}{x} = \int \frac{1}{x} \mathrm{d}x = \ln x + K \tag{E.27}$$

and, for A and B constants:

$$\int \frac{\mathrm{d}x}{A + Bx} = \int \frac{1}{A + Bx} \mathrm{d}x = \frac{1}{B} \ln \, (A + Bx) + K \tag{E.28}$$

The results of Eqs. (E.27) and (E.28) can be confirmed by differentiating the right sides of the equations with respect to x.

E.4 Matrices

A matrix is an array of numbers arranged in m rows and n columns:

$$\mathbf{A} = (a_{ij}) = \begin{bmatrix} a_{11} & a_{12} & \cdots & a_{1n} \\ a_{21} & a_{22} & \cdots & a_{2n} \\ \cdots & \cdots & \cdots & \cdots \\ a_{m1} & a_{m2} & \cdots & a_{mn} \end{bmatrix} \tag{E.29}$$

In Eq. (E.29), **A** is the symbol for the matrix and a_{ij} represents the element in the ith row and jth column, where $i = 1, 2, \ldots. m$ and $j = 1, 2, \ldots. n$. The total number of entries in a matrix is mn. The *dimensions* of a matrix are indicated as $m \times n$. If $n = m$, the matrix is square and of *order n*.

Consider the following 2×2 matrices, $\mathbf{A} = (a_{ij})$ and $\mathbf{B} = (b_{ij})$:

$$\mathbf{A} = \begin{bmatrix} a_{11} & a_{12} \\ a_{21} & a_{22} \end{bmatrix} \tag{E.30}$$

and

$$\mathbf{B} = \begin{bmatrix} b_{11} & b_{12} \\ b_{21} & b_{22} \end{bmatrix} \tag{E.31}$$

A and **B** are *equal* if they are identical, that is, they have the same number of rows, the same number of columns, and all corresponding elements are equal: $a_{ij} = b_{ij}$ for all i and j.

Addition and subtraction of matrices are possible only when they have the same number of rows and the same number of columns. Corresponding elements in the matrices are added or subtracted:

$$\mathbf{A} + \mathbf{B} = \begin{bmatrix} a_{11} & a_{12} \\ a_{21} & a_{22} \end{bmatrix} + \begin{bmatrix} b_{11} & b_{12} \\ b_{21} & b_{22} \end{bmatrix} = \begin{bmatrix} a_{11} + b_{11} & a_{12} + b_{12} \\ a_{21} + b_{21} & a_{22} + b_{22} \end{bmatrix} \tag{E.32}$$

$$\mathbf{A} - \mathbf{B} = \begin{bmatrix} a_{11} & a_{12} \\ a_{21} & a_{22} \end{bmatrix} - \begin{bmatrix} b_{11} & b_{12} \\ b_{21} & b_{22} \end{bmatrix} = \begin{bmatrix} a_{11} - b_{11} & a_{12} - b_{12} \\ a_{21} - b_{21} & a_{22} - b_{22} \end{bmatrix} \tag{E.33}$$

The result of adding or subtracting two $m \times n$ matrices is an $m \times n$ matrix.

Multiplication of a matrix by a constant involves multiplying each element of the matrix by the constant. The dimensions of the matrix are conserved. For example:

$$c\mathbf{A} = c\begin{bmatrix} a_{11} & a_{12} \\ a_{21} & a_{22} \end{bmatrix} = \begin{bmatrix} ca_{11} & ca_{12} \\ ca_{21} & ca_{22} \end{bmatrix} \tag{E.34}$$

where c is a constant.

Multiplication of two matrices **A** and **B** is possible only when the number of columns of **A**, $n_{\mathbf{A}}$, is equal to the number of rows of **B**, $m_{\mathbf{B}}$. The product **AB** has the same number of rows as **A**, $m_{\mathbf{A}}$, and the same number of columns as **B**, $n_{\mathbf{B}}$. The elements of **AB** are obtained by summing the products of elements in a row of **A** by elements in a column of **B**. The *row by column rule* for matrix multiplication is:

$$(ab)_{ij} = \sum_{k=1}^{m_{\mathbf{B}}} a_{ik} b_{kj} \tag{E.35}$$

As an example, consider the following two matrices **A** and **B**:

$$\mathbf{A} = \begin{bmatrix} a_{11} & a_{12} \\ a_{21} & a_{22} \end{bmatrix} \tag{E.36}$$

and

$$\mathbf{B} = \begin{bmatrix} b_{11} & b_{12} & b_{13} \\ b_{21} & b_{22} & b_{23} \end{bmatrix} \tag{E.37}$$

Multiplication gives:

$$\mathbf{AB} = \begin{bmatrix} a_{11} & a_{12} \\ a_{21} & a_{22} \end{bmatrix} \times \begin{bmatrix} b_{11} & b_{12} & b_{13} \\ b_{21} & b_{22} & b_{23} \end{bmatrix} = \begin{bmatrix} a_{11}b_{11} + a_{12}b_{21} & a_{11}b_{12} + a_{12}b_{22} & a_{11}b_{13} + a_{12}b_{23} \\ a_{21}b_{11} + a_{22}b_{21} & a_{21}b_{12} + a_{22}b_{22} & a_{21}b_{13} + a_{22}b_{23} \end{bmatrix} \tag{E.38}$$

In this example, the number of rows in $\mathbf{AB} = 2 = m_\mathbf{A}$ and the number of columns in $\mathbf{AB} = 3 = n_\mathbf{B}$. Matrix multiplication is *not commutative*. This means, in general, that $\mathbf{AB} \neq \mathbf{BA}$. However, matrix multiplication is *associative*:

$$\mathbf{ABC} = (\mathbf{AB})\mathbf{C} = \mathbf{A}(\mathbf{BC}) \tag{E.39}$$

and *distributive*:

$$\mathbf{A}(\mathbf{B} + \mathbf{C}) = \mathbf{AB} + \mathbf{AC} \tag{E.40}$$

$$(\mathbf{A} + \mathbf{B})\mathbf{C} = \mathbf{AC} + \mathbf{BC} \tag{E.41}$$

A *column vector* is a matrix of dimensions $m \times 1$:

$$\mathbf{v} = (v_{i1}) = \begin{bmatrix} v_{11} \\ v_{21} \\ \cdots \\ v_{m1} \end{bmatrix} = \begin{bmatrix} v_1 \\ v_2 \\ \cdots \\ v_m \end{bmatrix} \tag{E.42}$$

A *row vector* has dimensions $1 \times n$:

$$\mathbf{v} = (v_{1j}) = \begin{bmatrix} v_{11} & v_{12} & \cdots & v_{1n} \end{bmatrix} = \begin{bmatrix} v_1 & v_2 & \cdots & v_n \end{bmatrix} \tag{E.43}$$

For multiplication of a matrix by a column vector, the number of rows of the vector must be equal to the number of columns of the matrix. As an example, for:

$$\mathbf{A} = \begin{bmatrix} a_{11} & a_{12} & a_{13} \\ a_{21} & a_{22} & a_{23} \end{bmatrix} \tag{E.44}$$

and

$$\mathbf{v} = \begin{bmatrix} v_1 \\ v_2 \\ v_3 \end{bmatrix} \tag{E.45}$$

then

$$\mathbf{Av} = \begin{bmatrix} a_{11} & a_{12} & a_{13} \\ a_{21} & a_{22} & a_{23} \end{bmatrix} \times \begin{bmatrix} v_1 \\ v_2 \\ v_3 \end{bmatrix} = \begin{bmatrix} a_{11}v_1 + a_{12}v_2 + a_{13}v_3 \\ a_{21}v_1 + a_{22}v_2 + a_{23}v_3 \end{bmatrix} \tag{E.46}$$

The product **Av** is a column vector with the same number of rows as **A**.

The *transpose* of matrix **A** is obtained by interchanging the matrix rows and columns:

$$\mathbf{A}^{\mathrm{T}} = (a_{ij})^{\mathrm{T}} = (a_{ji}) \tag{E.47}$$

where $\mathbf{A}^{\mathrm{T}}$ is the transpose of **A**. Therefore, for **A** given by Eq. (E.44):

$$\mathbf{A}^{\mathrm{T}} = \begin{bmatrix} a_{11} & a_{12} & a_{13} \\ a_{21} & a_{22} & a_{23} \end{bmatrix}^{\mathrm{T}} = \begin{bmatrix} a_{11} & a_{21} \\ a_{12} & a_{22} \\ a_{13} & a_{23} \end{bmatrix} \tag{E.48}$$

The transpose of an $m \times n$ matrix is an $n \times m$ matrix. For two matrices **A** and **B**:

$$(\mathbf{AB})^{\mathrm{T}} = \mathbf{B}^{\mathrm{T}}\mathbf{A}^{\mathrm{T}} \tag{E.49}$$

Determinants are defined only for square matrices. The determinant of an $n \times n$ matrix **A** is denoted det (**A**) or |**A**|. The determinant of a 2×2 matrix is:

$$\det(\mathbf{A}) = \begin{vmatrix} a_{11} & a_{12} \\ a_{21} & a_{22} \end{vmatrix} = a_{11}a_{22} - a_{12}a_{21} \tag{E.50}$$

The determinant of a 3×3 matrix is:

$$\det(\mathbf{A}) = \begin{vmatrix} a_{11} & a_{12} & a_{13} \\ a_{21} & a_{22} & a_{23} \\ a_{31} & a_{32} & a_{33} \end{vmatrix} = a_{31}\begin{vmatrix} a_{12} & a_{13} \\ a_{22} & a_{23} \end{vmatrix} - a_{32}\begin{vmatrix} a_{11} & a_{13} \\ a_{21} & a_{23} \end{vmatrix} + a_{33}\begin{vmatrix} a_{11} & a_{12} \\ a_{21} & a_{22} \end{vmatrix} \tag{E.51}$$

As the numbers of rows and columns increase, calculating the determinant can become complex and lengthy. Software programs are available to analyse large matrix arrays.

A square matrix **A** of order n is said to have an *inverse* or *reciprocal*, $\mathbf{A}^{-1}$, if the product $\mathbf{AA}^{-1}$ is equal to **I**, where **I** is the *identity matrix* of order n. For all n, the diagonal elements of **I** are equal to 1 and the nondiagonal elements are equal to 0:

$$\mathbf{I} = \begin{bmatrix} 1 & 0 & \cdots & 0 \\ 0 & 1 & \cdots & 0 \\ \cdots & \cdots & \cdots & \cdots \\ 0 & 0 & \cdots & 1 \end{bmatrix} \tag{E.52}$$

If the inverse of **A** exists so that $\mathbf{AA}^{-1} = \mathbf{I}$, then $\mathbf{A}^{-1}\mathbf{A} = \mathbf{I}$ also. The inverse of an $n \times n$ square matrix is also an $n \times n$ square matrix. A square matrix **A** has an inverse only if the matrix is *nonsingular*. A matrix is *singular* if its determinant is zero. As examples, the determinants of the following 2×2 matrices are zero:

$$\begin{bmatrix} 0 & 1 \\ 0 & 0 \end{bmatrix} \quad \begin{bmatrix} 1 & 1 \\ 0 & 0 \end{bmatrix} \quad \begin{bmatrix} 0 & 0 \\ 1 & 1 \end{bmatrix} \quad \begin{bmatrix} 1 & 1 \\ 1 & 1 \end{bmatrix}$$

hence, these matrices cannot be inverted. A matrix possessing an inverse is considered to be *invertible*. The inverse of an invertible 2×2 matrix $\mathbf{A}$ is:

$$\mathbf{A}^{-1} = \begin{bmatrix} a_{11} & a_{12} \\ a_{21} & a_{22} \end{bmatrix}^{-1} = \frac{1}{|\mathbf{A}|}\begin{bmatrix} a_{22} & -a_{12} \\ -a_{21} & a_{11} \end{bmatrix} = \frac{1}{(a_{11}a_{22} - a_{12}a_{21})}\begin{bmatrix} a_{22} & -a_{12} \\ -a_{21} & a_{11} \end{bmatrix} \tag{E.53}$$

Software programs are used to calculate the inverse of large matrices. The following relationships involving inverse matrices apply:

$$(\mathbf{AB})^{-1} = \mathbf{B}^{-1}\mathbf{A}^{-1} \tag{E.54}$$

$$(\mathbf{ABC})^{-1} = \mathbf{C}^{-1}\mathbf{B}^{-1}\mathbf{A}^{-1} \tag{E.55}$$

and

$$(\mathbf{A}^{-1})^{\mathrm{T}} = (\mathbf{A}^{\mathrm{T}})^{-1} \tag{E.56}$$

Inverse matrices are used to solve equations involving matrices. For example, for matrices $\mathbf{A}$, $\mathbf{B}$ and $\mathbf{C}$, if:

$$\mathbf{C} = \mathbf{AB} \tag{E.57}$$

then

$$\mathbf{B} = (\mathbf{A}^{-1}\mathbf{A})\mathbf{B} = \mathbf{A}^{-1}(\mathbf{AB}) = \mathbf{A}^{-1}\mathbf{C} \tag{E.58}$$

and

$$\mathbf{A} = \mathbf{A}(\mathbf{BB}^{-1}) = (\mathbf{AB})\mathbf{B}^{-1} = \mathbf{CB}^{-1} \tag{E.59}$$

The *rank* of a matrix is the number of genuinely independent rows or columns in that matrix. Rows or columns are independent if they are not a linear combination of other rows or columns, or equal to another row or column multiplied by a constant. For example, consider the following matrix:

$$\mathbf{A} = \begin{bmatrix} 1 & 4 & 1 \\ 0 & 1 & -1 \end{bmatrix} \tag{E.60}$$

$\mathbf{A}$ has only two independent columns as the third column is equal to 5 multiplied by the first column minus the second column. The rank of an $m \times n$ matrix cannot be greater than m or n. A matrix that has maximum rank is said to have *full rank*; otherwise, the matrix is *rank deficient*. For a square matrix of dimensions $n \times n$, if the matrix has full rank equal to n, it can be shown that the matrix possesses an inverse and that the inverse is unique.

References

[1] D. Zwillinger, CRC Standard Mathematical Tables and Formulae, thirty-second ed., CRC Press, 2012.
[2] A. Jeffrey, H.-H. Dai, Handbook of Mathematical Formulas and Integrals, fourth ed., Academic Press, 2008.
[3] A. Cornish-Bowden, Basic Mathematics for Biochemists, second ed., Oxford University Press, 1999.
[4] J.C. Newby, Mathematics for the Biological Sciences, Clarendon Press, 1992.
[5] J.C. Arya, R.W. Lardner, Mathematics for the Biological Sciences, Prentice Hall, 1979.

APPENDIX

F

U.S. Sieve and Tyler Standard Screen Series

TABLE F.1 U.S. Sieve Series

Sieve designation		Sieve opening		Nominal wire diameter		Approximate Tyler equivalent
Standard	Alternate	mm	in. (approx.)	mm	in. (approx.)	
107.6 mm	4.24 in.	107.6	4.24	6.40	0.2520	
101.6 mm	4 in.	101.6	4.00	6.30	0.2480	
90.5 mm	3½ in.	90.5	3.50	6.08	0.2394	
76.1 mm	3 in.	76.1	3.00	5.80	0.2283	
64.0 mm	2½ in.	64.0	2.50	5.50	0.2165	
53.8 mm	2.12 in.	53.8	2.12	5.15	0.2028	
50.8 mm	2 in.	50.8	2.00	5.05	0.1988	
45.3 mm	1¾ in.	45.3	1.75	4.85	0.1909	
38.1 mm	1½ in.	38.1	1.50	4.59	0.1807	
32.0 mm	1¼ in.	32.0	1.25	4.23	0.1665	
26.9 mm	1.06 in.	26.9	1.06	3.90	0.1535	1.050 in.
25.4 mm	1 in.	25.4	1.00	3.80	0.1496	
22.6 mm	⅞ in.	22.6	0.875	3.50	0.1378	0.883 in.
19.0 mm	¾ in.	19.0	0.750	3.30	0.1299	0.742 in.
16.0 mm	⅝ in.	16.0	0.625	3.00	0.1181	0.624 in.
13.5 mm	0.530 in.	13.5	0.530	2.75	0.1083	0.525 in.

(Continued)

TABLE F.1 U.S. Sieve Series (Continued)

Sieve designation		Sieve opening		Nominal wire diameter		Approximate Tyler equivalent
Standard	Alternate	mm	in. (approx.)	mm	in. (approx.)	
12.7 mm	½ in.	12.7	0.500	2.67	0.1051	
11.2 mm	7/16 in.	11.2	0.438	2.45	0.0965	0.441 in.
9.51 mm	⅜ in.	9.51	0.375	2.27	0.0894	0.371 in.
8.00 mm	5/16 in.	8.00	0.312	2.07	0.0815	2½ mesh
6.73 mm	0.265 in.	6.73	0.265	1.87	0.0736	3 mesh
6.35 mm	¼ in.	6.35	0.250	1.82	0.0717	
5.66 mm	No. 3½	5.66	0.223	1.68	0.0661	3½ mesh
4.76 mm	No. 4	4.76	0.187	1.54	0.0606	4 mesh
4.00 mm	No. 5	4.00	0.157	1.37	0.0539	5 mesh
3.36 mm	No. 6	3.36	0.132	1.23	0.0484	6 mesh
2.83 mm	No. 7	2.83	0.111	1.10	0.0430	7 mesh
2.38 mm	No. 8	2.38	0.0937	1.00	0.0394	8 mesh
2.00 mm	No. 10	2.00	0.0787	0.900	0.0354	9 mesh
1.68 mm	No. 12	1.68	0.0661	0.810	0.0319	10 mesh
1.41 mm	No. 14	1.41	0.0555	0.725	0.0285	12 mesh
1.19 mm	No. 16	1.19	0.0469	0.650	0.0256	14 mesh
1.00 mm	No. 18	1.00	0.0394	0.580	0.0228	16 mesh
841 micron	No. 20	0.841	0.0331	0.510	0.0201	20 mesh
707 micron	No. 25	0.707	0.0278	0.450	0.0177	24 mesh
595 micron	No. 30	0.595	0.0234	0.390	0.0154	28 mesh
500 micron	No. 35	0.500	0.0197	0.340	0.0134	32 mesh
420 micron	No. 40	0.420	0.0165	0.290	0.0114	35 mesh
354 micron	No. 45	0.354	0.0139	0.247	0.0097	42 mesh
297 micron	No. 50	0.297	0.0117	0.215	0.0085	48 mesh
250 micron	No. 60	0.250	0.0098	0.180	0.0071	60 mesh
210 micron	No. 70	0.210	0.0083	0.152	0.0060	65 mesh
177 micron	No. 80	0.177	0.0070	0.131	0.0052	80 mesh
149 micron	No. 100	0.149	0.0059	0.110	0.0043	100 mesh
125 micron	No. 120	0.125	0.0049	0.091	0.0036	115 mesh
105 micron	No. 140	0.105	0.0041	0.076	0.0030	150 mesh

88 micron	No. 170	0.088	0.0035	0.064	0.0025	170 mesh
74 micron	No. 200	0.074	0.0029	0.053	0.0021	200 mesh
63 micron	No. 230	0.063	0.0025	0.044	0.0017	250 mesh
53 micron	No. 270	0.053	0.0021	0.037	0.0015	270 mesh
44 micron	No. 325	0.044	0.0017	0.030	0.0012	325 mesh
37 micron	No. 400	0.037	0.0015	0.025	0.0010	400 mesh

From Perry's Chemical Engineers' Handbook, *1997, 7th ed., pp. 19–20, McGraw-Hill, New York.*

TABLE F.2 Tyler Standard Screen Series

Screen designation	Screen opening		Wire diameter (in.)
	in.	mm	
	1.050	26.67	0.148
	0.883	22.43	0.135
	0.742	10.05	0.135
	0.624	15.85	0.120
	0.525	13.33	0.105
	0.441	11.20	0.105
	0.371	9.423	0.092
2$\frac{1}{2}$ mesh	0.312	7.925	0.088
3 mesh	0.263	6.680	0.070
3$\frac{1}{2}$ mesh	0.221	5.613	0.065
4 mesh	0.185	4.699	0.065
5 mesh	0.156	3.962	0.044
6 mesh	0.131	3.327	0.036
7 mesh	0.110	2.794	0.0328
8 mesh	0.093	2.362	0.032
9 mesh	0.078	1.981	0.033
10 mesh	0.065	1.651	0.035
12 mesh	0.055	1.397	0.028
14 mesh	0.046	1.168	0.025
16 mesh	0.0390	0.991	0.0235
20 mesh	0.0328	0.833	0.0172

(Continued)

TABLE F.2 Tyler Standard Screen Series (Continued)

Screen designation	Screen opening		Wire diameter (in.)
	in.	mm	
24 mesh	0.0276	0.701	0.0141
28 mesh	0.0232	0.589	0.0125
32 mesh	0.0195	0.495	0.0118
35 mesh	0.0164	0.417	0.0122
42 mesh	0.0138	0.351	0.0100
48 mesh	0.0116	0.295	0.0092
60 mesh	0.0097	0.246	0.0070
65 mesh	0.0082	0.208	0.0072
80 mesh	0.0069	0.175	0.0056
100 mesh	0.0058	0.147	0.0042
115 mesh	0.0049	0.124	0.0038
150 mesh	0.0041	0.104	0.0026
170 mesh	0.0035	0.088	0.0024
200 mesh	0.0029	0.074	0.0021
270 mesh	0.0021	0.053	
325 mesh	0.0017	0.044	

From W.L. McCabe, J.C. Smith, and P. Harriott, 2005, Unit Operations of Chemical Engineering, *7th ed., McGraw-Hill, New York.*

Index

Note: Page numbers followed by f indicate figures and t indicate tables

F

G

N

U

V

W

Y

Z